ANIMMA 2023 - Advancements in Nuclear Instrumentation Measurement Methods and their Applications

EPJ Web of Conferences Volume 288 (2023)

Lucca, Italy
12-16 June 2023

Editors:

A. Lyoussi
F. D'Errico
M. Carette
M. Joyce
R. Hodák
I. Jenčič
P. Le Dû
M. Morichi
S. Pospíšil
C. Reynard-Carette
L. Snoj
L. Vermeeren

ISBN: 978-1-7138-8492-7

Printed from e-media with permission by:

Curran Associates, Inc.
57 Morehouse Lane
Red Hook, NY 12571

Some format issues inherent in the e-media version may also appear in this print version.

For additional information, please contact EDP Sciences – Web of Conferences
at the address below.

EDP Sciences – Web of Conferences
17, Avenue du Hoggar
Parc d'Activité de Courtabœuf
BP 112
F-91944 Les Ulis Cedex A
France

Phone: +33 (0) 1 69 18 75 75
Fax: +33 (0) 1 69 28 84 91

contact-edps@webofconferences.org

Additional copies of this publication are available from:

Curran Associates, Inc.
57 Morehouse Lane
Red Hook, NY 12571 USA
Phone: 845-758-0400
Fax: 845-758-2633
Email: curran@proceedings.com
Web: www.proceedings.com

TABLE OF CONTENTS

FUNDAMENTAL PHYSICS

A High-Granularity Timing Detector for the ATLAS Phase-II Upgrade 01001 1
Océane Perrin

Comparison Between GEANT4 and MCNP for Well Logging Applications 01002 7
Geoffrey Varignier, Valentin Fondement, Cédric Carasco, Johann Collot, Bertrand Pérot, Thomas Marchais, Pierre Chuilon, Emmanuel Caroli, Mai-Linh Doan

Detector Development for a Double-Differential Cross Section Experiment with the Emission of Light Charged Particles from High Energy Neutrons 01003 15
M. Dietz, R. Beyer, A. Junghans, R. Nolte, E. Pirovano, Y. Romanets, P. Vaz

The New Readout System for the ALICE Zero Degree Calorimeter in LHC Run 3 01004 19
Paolo De Remigis, Pietro Cortese, Giorgio Cotto, Nora De Marco, Luca Lombardo, Carlo Puggioni, Mario Sitta, Stefan Cristi Zugravel

Trustworthiness of Laser-Induced Breakdown Spectroscopy Predictions Via Simulation-Based Synthetic Data Augmentation and Multitask Learning 01005 22
Riccardo Finotello, Daniel L'Hermite, Celine Quéré, Benjamin Rouge, Mohamed Tamaazousti, Jean-Baptiste Sirven

SPACE SCIENCES AND TECHNOLOGY

Design of the Angle-Resolving Electron Spectrometer Aboard the PRESET Mission 02001 30
Benjamin Dyer, Xingzhi Cheng, Andrei Hanu, Soo Hyun Byun

R&D Status for an Innovative Crystal Calorimeter for the Future Muon Collider 02002................................. 37
C. Cantone, S. Ceravolo, F. Colao, E. Di Meco, E. Diociaiuti, P. Gianotti, A. Liedl, D. Lucchesi, D. Paesani, N. Pastrone

FUSION DIAGNOSTICS AND TECHNOLOGY

Modelling of a SiC Based Detector for the Interpretation of 14.1 MeV Neutrons Measurements 03001 42
Quentin Potiron, Christophe Destouches, Léo Dubus, Michael Houry, Olivier Llido, Abdallah Lyoussi, Laurent Ottaviani, Christelle Reynard-Carette

Characterization of the Fast Neutron Generators for Calibration of Fusion Neutron Diagnostics 03002 47
T. Kormilitsyn, A. Pankratenko, Yu. Kashchuk, S. Obudovsky, R. Rodionov, D. Fridrikhsen, A. Krasilnikov, S. Syromukov, D. Yurkov

Concept of Calorimetry System for ITER 03003 53
Lukasz Tomkow, Eleonore Geulin, Daniel Iglesias, Matthew Clough, George Vayakis

Detection of Fast Neutrons Using 4H-SiC Radiation Detectors 03004 58
Bohumír Zatko, Andrea Šagátová, Eva Kovácová, Andrej Novák, Rudolf Sýkora, Zdenek Kohout, Štepán Polansky

Digitizer Hardware for Magnetic Data Acquisition on COMPASS-U 03005 62
J. Oliveira, A. Torres, A. J. N Batista, J. Sousa, B. Carvalho, A. Havranek, H. Fernandes

Evaluation of the Energy Calibration of Diamond Detectors for Fast Neutrons Applications 03006 68
Giada Gandolfo, Luigi Lepore, Nadia Cherubini, Ermanno Piccinelli, Giuseppe Augusto Marzo

RESEARCH REACTORS AND PARTICLE ACCELERATORS

A Novel Smart Rad-Hard Fast Detection System for Radioactive Ion Beam Tagging and Diagnostics 04001 73
Luis Acosta, Carmen Altana, Giuseppe Cardella, Andrea Castoldi, Michele Costa, Enrico De Filippo, Elena Geraci, Brunilde Gnoffo, Chiara Guazzoni, Cettina Maiolino

Absolute Neutron and Photon Flux Characterization in the CNESTEN's TRIGA Mark II Research Reactor 04002 78
H. Ghninou, A. Gruel, A. Lyoussi, C. Reynard-Carette, C. El Younoussi, B. El Bakkari, Y. Boulaich

Assessment of the Gamma and Neutron Dose Field Around the Closed-Water Activation Loop at the JSI TRIGA Reactor 04003 84
Domen Kotnik, Luka Snoj, Igor Lengar

Characterization of a New Reduced-Height CALORRE Differential Calorimeter for CALOR-I Irradiation in MITR 04004 89
A. Volte, S. Hauptman, D. Carpenter, M. Carette, A. Lyoussi, G. Kohse, C. Reynard-Carette

Defected Fuel Rods Identification in TRIGA Reactors: The Experience at the ENEA Casaccia TRIGA RC-1 Reactor 04005 94
L. Lepore, L. Falconi, V. Fabrizio, A. Roberti, D. Formenton, M. G. Iorio, L. Sperandio

Design and Fabrication of an Axial Neutron Flux Profile Measurement Assembly for the Advanced Test Reactor Critical Facility 04006 98
M. A. Reichenberger, M. R. Holtz, D. Nichols, B. Harris, R. D. Ball, H. Rollins, C. Downey

Design of an Instrumented Fuel Rod for Intra-Pin Neutron Flux Measurements in the CROCUS Experimental Reactor 04007 104
Tom Mager, Jean-Baptiste Valentin, Vincent Lamirand, Thomas Ligonnet, Won Dong Shin, Jean-Marie Fürbringer, Mathieu Hursin, Andreas Pautz

Experimental and Numerical Results for a New Single-Cell Calorimeter Dedicated to Nuclear Heating Rate Measurement 04008 111
J. Rebaud, A. Volte, M. Carette, A. Lyoussi, C. Reynard-Carette

Exploring the ENEA Casaccia and ENEA Frascati Irradiation Capabilities: Status and Perspectives 04009 115
Michele Croia, Nunzio Burgio, Alessandro Ampollini, Maria Denise Astorino, Giulia Bazzano, Barbara Bianchi, Mateo Cesaroni, Luca Falconi, Salvatore Fiore, Luigi Lepore

Flux and Nuclear Heating Rate Calculations for CALORRE Irradiation in MITR 04010 121
S. Hauptman, A. Volte, G. Kohse, M. Carette, A. Lyoussi, C. Reynard-Carette

Fuel Irradiation Devices Test of Feedthroughs Equipped with Optical Fibers in Support of the Development of Innovative Instrumentation 04011 126
S. Gaillot, C. Destouches, G. Cheymol, J. Brinster And Al

The CABRI Fast Neutrons Hodoscope: CABRI Model and Signal-To-Mass Conversion Charts 04012 132
Vincent Chevalier, Jacques Di Salvo

Identified Opportunities in Utilization of the European Research Reactor Fleet as a Part of the TOURR Project 04013 138
Anže Pungercic, Roberta Cirillo, Joanna Walkiewicz, Evžen Novák, Jacek Gajewski, Laszlo Szentmiklósi, Lisanne Van Puyvelde, Jörg Starflinger, Daniel Cano, Georg Pohlner

In-Core Measurement Methodology with Optical Fibers Using Cherenkov Radiation for Transient Induced Power Measurement in the CABRI Experimental Reactor 04014 143
Georges Gay, Adrien Gruel, Romain Boffy, Grégoire De Izarra, Christian Jammes

Investigation of Organic Scintillators for Neutron-Gamma Noise Measurements in a Zero Power Reactor 04015 147
F. B. Darby, O. Pakari, M. Y. Hua, V. Lamirand, S. D. Clarke, A. Pautz, S. A. Pozzi

New Experiment for Measuring the Delayed Neutron Yields of 238U Fission in the Range 1 MeV to 19 MeV 04016 151
D. Belverge, P. Leconte, B. Geslot, G. Kessedjian, P. Mutti, F. Rodiac, E. Pirovano, B. Lutz, X. Ledoux, L. Mathieu, O. Meplan

Optimization of the Magnetic Core of a Linear Variable Differential Transducer 04017 157
Rudi Van Nieuwenhove

POLLEN: A Pile-Oscillator for the BLOOM Experimental Program 04018 163
Thomas Ligonnet, Nicolas Weiss, Vincent Lamirand, Axel Laureau, Andreas Pautz

Qualification of the Industrialized Libera MONACO 3 Digital Acquisition System Dedicated to Fission/Ion Chamber Measurements in Research Reactors 04019 168
Loic Barbot, Gregoire De Izarra, Christophe Destouches, Danilo Bisiach, Manuel Cargnelutti, Sebastjan Zorzut, Matjaž Škabar, Aleš Bardorfer, Aurora Makovac, Jure Trnovec, Peter Paglovec

Radiation Measurements in Reactor Pulse Mode at the JSI TRIGA Reactor – Power Meter Based on Cherenkov Light Intensity Measurements 04020 173
Julijan Peric, Vladimir Radulovic, Loïc Barbot, Grégoire De Izarra

Reactivity Modulation Experiments for Nuclear Data in CROCUS Within a CEA-EPFL Collaboration 04021 180
Benoit Geslot, Vincent Lamirand, Yifeng Jiang, Pierre Leconte, Oskari Pakari, Daniel Godat, Laurent Braun, Pavel Frajtag, Laurent Pantera, Andreas Pautz

Reactor Power Measurement Using N16 and N17 Production Rate in MARIA Reactor 04022 185
Justyna Cybowska

Simultaneous Inter-Calibration of 160 MiMi Neutron Detectors 04023 189
Fanny Vitullo, Vincent Lamirand, Daniel Godat, Andreas Pautz

Specially Developed LR-0 Reactor Graphite Environment for Gen IV Reactor Support and Cross-Section Measurement 04024 197
Tomas Peltan, Eva Vilimova, Tomas Czakoj, Zdenek Matej, Filip Mravec, Frantisek Cvachovec, Jan Simon, Vlastimil Juricek, Michal Kostal

Validation of Monte Carlo Simulations by Experimental Measurements of Neutron-Induced Activation in Cyclotrons 04025 205
Jonathan Collin, Jean-Michel Horodynski, Nicolas Arbor, Massimo Barbagallo, Federico Carminati, Giuliana Galli Carminati, Luca J. Tagliapietra, Abdel-Mjid Nourreddine

Ten Springs of Experiments in CROCUS 04026 211
Vincent Lamirand

NUCLEAR POWER REACTORS AND NUCLEAR FUEL CYCLE

Characterization and Performance of an Acoustic Sensor for Fission Gas Release Characterization Devoted to JHR Environment Measurements 05001 219
F. Baudry, E. Rosenkrantz, P. Combette, D. Fourmentel, C. Destouches, P. V. Bonzon, K. Paumel, J. Y. Ferrandis

Experimental Analysis of Natural Convection in a Differentially Heated Cavity 05002 223
Tommaso Carlesi, Delphine Laboureur, Philippe Planquart, Julio Pacio, Katrien Van Tichelen, Patrice Mégret

Experimental Validation of a CeBr3 Gamma-Ray Logging Probe MCNP Model 05003........ 231
Thomas Marchais, Bertrand Pérot, Pierre-Guy Allinei, Hervé Toubon, Youcef Bensedik, Romain Mieszkalski, Sebastien Hocquet, Dragomir Savov, Matthew Schubert, Bruno Legros

Towards the Enhancement of Nuclear Reactor Safety: Physical Phenomena Involved in the Measurement Accuracy of Boron Concentration Online Monitoring Systems 05004........ 235
Nouhaila Tabti, Adrien Sari, Dominique Tromson

NUCLEAR SAFEGUARDS, HOMELAND SECURITY AND CBRN

A Mobile Radiation Detection System for Security Inspection and Monitoring 06001 240
Luís Marques, Rui Coito, Sofia Fernandes, Tiago Costa, Alberto Vale, Pedro Vaz

A Portable Aerosol Monitor with a Novel Linear Deconvolution Algorithm for Fast and Accurate Detection of Alpha Emitters 06002 245
Luca F. Ferrante Vero, Giacomo P. Manessi, Giuseppe A. Marzo, Gabriele Zorloni

An Extended List-Mode MLEM Algorithm for 3D Compton Image Reconstruction from Multi-View Data 06003 249
Nhan Le, Hichem Snoussi, Zied Hmissi, Alain Iltis, Guillaume Lebonvallet, Ghislain Zeufack

Development of a New Method for the Detection of Illicit Materials Based on the Active Photon Interrogation Method and Photo-Neutron Spectrometry 06004 255
C. Besnard-Vauterin, V. Blideanu, B. Rapp

Development of a Prototype of Monitor for Alpha and Beta Radiation in Water Using New Silicone-Based Contamination-Safe Detectors 06005........ 263
Marcello Lunardon, Paolo Sartori, Pierino De Felice, Luca Stevanato

Field Tests of the MICADO Monitoring Detectors in Real Radwaste Storages 06006........ 269
Paolo Finocchiaro, Luigi Cosentino, Fabio Longhitano, Gaetano Elio Poma, Simone Amaducci, Bernard Dumercq, Christopher Helbert, Lionel Tondut

Hot-Spots Finding with Modular Gamma-Ray System for Sort and Segregate Activities 06007........ 274
Gaetano Elio Poma, Luigi Cosentino, Fabio Longhitano, Paolo Finocchiaro

Neutron Coincidence Counters for Disarmament Verification: Experiments and Modeling 06008 280
Alessandro Borella, Yurdunaz Çelik, Riccardo Rossa

On the Use of Free Code Tools to Simulate the Propagation of Radiation Following Dirty Bomb Explosions in Sensible Contexts 06009 288
Riccardo Quaranta, Gian Marco Ludovici, Guglielmo Manenti, Pasqualino Gaudio, Andrea Malizia

Pulse Shape Discrimination of CLYC, CLLBC and EJ-276 with SiPM Readout 06010 292
Bjoern Seitz, David Bennett, Frank Thomson

Simulation Design of Online Beam Monitoring for the MULTISCAN 3D Laser-Plasma X-Ray Source 06011 297
Eric Garrido, David Tisseur, Cedric Carasco, Bertrand Perot

Thermal Neutron Detection Based on Resistive Gaseous Devices 06012 302
G. Bencivenni, I. Balossino, G. Cibinetto, R. De Oliveira, R. Farinelli, G. Felici, I. Garzia, M. Gatta, M. Giovannetti, S. Gramigna

Flying DEMon, a Newborn Startup for Environmental Monitoring 06013 306
C. Altomare, L. Di Venere, F. Giordano, S. Loporchio, A. Mastroserio, D. Serini

DECOMMISSIONING, DISMANTLING AND REMOTE HANDLING

3D Imaging of a Nuclear Reactor Using Muography Measurements with Micromegas Detectors 07001 311
Baptiste Lefevre, Héctor Gomez, Sébastien Procureur, David Attié, Laurent Gallego, Philippe Gonzales, Marion Lehuraux, Bertrand Lesage, Irakli Mandjavidze, Philippe Mas, Daniel Pomarede

A Comparative Study on the Influence of the Type of Acid Used for 55Fe and 63Ni Determination in Steel Samples Through Liquid Scintillation Counting 07002 318
Maria Letizia Cozzella, Giada Gandolfo, Giovanna La Notte, Giuseppe Augusto Marzo

Assessing Performance of Nucleco Passive/Active System (PANWAS) Through a Long-Term Verification Campaign 07003 323
Filippo Gagliardi, Alessandro Ciotoli, Andrea De Angelis, Edoardo Gorello, Luigi Meschini, Marco Pagliuca, Egidio Mauro

CLEANDEM, a Cyber physicaL Equipment for unmAnned Nuclear DEcommissioning Measurements 07004 327
Maugan Michel, Guillaume Amoyal, Vincent Schoepff

Gamma-Ray Spectroscopy for the Characterization of Uranium Contamination in Nuclear Decommissioning 07005 333
Francisco Salvador, Thomas Marchais, Bertrand Pérot, Pierre-Guy Allinei, Frédéric Morales, Olivier Gueton, Julien Venara, Marie Cuozzo, Frédéric Mayet

Development of a Specialised Calibration and Infra-Red Emissivity Correction to Enable the Temperature Monitoring of Intermediate Level Nuclear Waste Using Thermal Imaging 07006 339
W. Bond, M. Hayes, R. Simpson, S. Korniliou, G. Machin

Multi-Source Contamination Mapping on the Ground: A Novel Approach 07007 345
Dor Raz, Ohad Fitoussi, Nadav Ben-David, Eran Vax, Alon Osovizky, Guy Zaidner

Nanoreinforced Concrete Composites for Possible Nuclear Waste Confinement a Preliminary Study 07008 351
Maria Rita Mancini, Daniele Mirabile Gattia, Fabio Girardi, Donatella Ferri, Giuseppe A. Marzo, Domenica Masci

Neutron Imaging with Micromegas Detectors with Optical Readout 07009 356
A. Cools, S. Aune, F. M. Brunbauer, T. Benoit, A. Corsi, E. Ferrer-Ribas, F. J. Iguaz, C. Mazoin, E. Oliveri, T. Papaevangelou

Neutron-Gamma Survey System for Decommissioning and Dismantling Activities 07010 360
Jessica C. Delgado, Felix Pino, Erica Fanchini, Alessandro Iovane, Daniela Fabris, Sandra Moretto

New 3D Coded Aperture Collimator for X/Gamma-Ray Wide-Field Imaging 07011 365
Yaroslav Susaiev, Vincent Schoepff, Olivier Limousin

Pure Beta Emitters in Water Samples from Neutron Irradiation Facilities: Activity Determination Via Vacuum Distillation and Liquid Scintillation Counting 07012 371
D. Masci, D. Ferri, G. Gandolfo, L. Lepore, G. M. Contessa, A. Pietropaolo, M. Pillon, G. A. Marzo

Seeing Radioactivity: Gamma-Ray Imaging Technique Applied to TRIGA RC-1 Research Reactor in ENEA Casaccia 07013 375
Filippo Gagliardi, Alessandro Ciotoli, Edoardo Gorello, Luigi Meschini, Marco Pagliuca, Egidio Mauro, Nadia Cherubini, Luigi Lepore, Giada Gandolfo, Giuseppe A. Marzo

System-Level Testing and Qualification of Environmental Sensors for Nuclear Decommissioning 07014 382
A. Urena-Acuna, J. Favrichon, E. Cantrel, P. A. Robin, A. Balier, T. Maraine, F. Saigné, J. Boch

The MICADO Integrated Gamma Station for Radioactive Waste Packages Radiological Characterization 07015 386
L. Lepore, G. Gandolfo, G. A. Marzo, N. Cherubini, D. Casagrande, E. Palumbo, N. Ratto, E. Fanchini, A. Pepperosa, F. Giordano, M. Morichi

ENVIRONMENTAL AND MEDICAL SCIENCES

An Electronic Dosimeter for Diagnostic X-Ray Beams Based on a Differential Amplifier Circuit with MOSFETs 09001 391
Luiz C. Gonçalves Filho, Luiz A. P. Santos

An Optocoupler-Based Method for Dosimetry in Low Energy X-Ray Beams 09002 395
Jonatas V. Silva, Luiz C. Gonçalves Filho, Luiz A. P. Santos

Prompt Gamma-Ray Spectroscopy in Conjunction with the Monte Carlo Library Least Squares Approach: Applications to Range Verification in Proton Therapy 09003 399
Kyrre Skjerdal, Toni Kögler, William Lionheart, Kristian Smeland Ytre-Hauge, Ilker Meric

Radiological Characterization of High-Energy Proton Tantalum Targets from CERN-ISOLDE Facility 09004 405
Nabil Menaa, Aurore Boscher, Richard Catherall, Gerald Dumont, Matteo Magistris, Sven De Man, Renaud Mouret, Paolo Pisano, Sebastien Othe, Simon Stegemann

Systematic Radioecological Experiments in the United Arab Emirates (UAE) for Nuclear Safety and Emergency-Preparedness 09005 412
Prasoon Raj, Nemeer Padiyath, Francois Foulon, Maryam Almakrani

An Innovative Detector for X-Ray Beams Made of Polyethylene Terephthalate Built in a 3D Printer 09006 417
Daniel F. Gomes, Edvane B. Silva, Luiz A. P. Santos

Radiation Monitoring with Plant-Based Biotas and an Automated Micronucleus Scoring Approach 09007 421
F. D'Errico, F. Barco, T. Butini, M. G. Cascone, R. Ciolini, M. Quattrocchi, E. Rosellini, T. Hulber, J. A. Torres Novaes, M. N. Xavier, S. De Souza Lalic

CURRENT TRENDS IN DEVELOPMENT RADIATION DETECTORS

Model-Based Deep Learning Algorithm for Pulse Shape Discrimination in High Event Rates 10001 426
Itai Morad, Max Ghelman, Dimitry Ginzburg, Alon Osovizky, Nir Shlezinger

A Model of Spatial Resolution Uncertainty for Compton Camera Imaging 10002 433
Yanting Ma, Joshua Rapp, Petros Boufounos, Hassan Mansour

A New Technique for Quick Identification of Defective Region Inside γ-Ray Detector 10003 441
Biswajit Das, R. Palit, A. Kundu, P. Dey, V. Malik, S. K. Jadav, B. S. Naidu, A. T. Vazhappilly

Active Collimator for the BGO Anti-Compton Shield of the HPGe Clover Detector 10004 446
Biswajit Das, N. R. Khan Chowdhury, S. Saha, Md. R. S. Laskar, F. S. Babra, V. Malik, A. Kundu, P. Dey, S. K. Jadav, B. S. Naidu

Analog Pulse Shape Discrimination Circuit for High Event Rates and Fast Scintillators with a Dynamic Deadtime 10005 451
Itai Morad, Max Ghelman, Alon Osovizky, Amir Ellenbogen, Rami Seif, Eran Vax, Amir Broide, Ron Harn

Characterization and Use of Stilbene Scintillator for Neutron Metrology and Spectrometry from 100 keV to 22 MeV 10006 459
Diane Quevauvillers, Laurent Ottaviani, Michael Petit, Christelle Reynard-Carette

Characterization of 3D Micro-Structured TIMEPIX Detectors for Neutron Imaging 10007 463
Matteo Polo, Roberto Mendicino, Alberto Quaranta, Gian-Franco Dalla Betta

Performance Comparison of Different SiPM Arrays Coupled to Neutron-γ Scintillators 10008 470
Jessica Delgado, Felix Pino, Sandra Moretto, Daniele Corti, Alessandro Griggio, Lucio Pancheri, Daniela Fabris

Comparison of CLYC Vs. CsI in a Hand-Held Detector 10009 477
Monika Risse, Hermann Friedrich, Jeannette Glabian, Theo Köble, Olaf Schumann, Thorsten Teuteberg

Design of a Reliable Numerical Model of a Plastic Scintillation Detector for Beta Spectrometry 10010 482
L. Fleres, J. Venara, M. Cuozzo, C. Mahe, S. Fargier, E. El Haber, F. Carrel

Evaluation of Neutron-Induced Activation Gamma-Rays Creation and Transport in a LaBr3 Inorganic Scintillator with PHITS 10011 488
C. Besnard-Vauterin, R. De Stefano, G. Amoyal, V. Schoepff, J.-C. Angélique

Fast Timing Detectors for the Muon System of a Muon Collider Experiment 10012 494
C. Aimè, S. Calzaferri, D. Fiorina, C. Riccardi, P. Salvini, I. Vai, P. Vitulo

Influence of Base Material Thickness on Spectrometry of Semiconductor Detectors Based on Semi-Insulating GaAs 10013 498
Andrea Šagátová, Nikola Kurucová, Sona Kotorová, Eva Kovácová, Bohumír Zatko

Inorganic Scintillator Surface Enhancements with 2-D Photonic Crystals to Improve Light Collection 10014 502
Stuti Surani, Faruk Logoglu, Patrick Albert, Douglas Wolfe, Marek Flaska

Investigating the Thickness-Dependent Scintillator-PMT Interface Reflection Coefficients with GAGG:Ce3+ Crystals Using the Dual-PMT Setup 10015 509
Faruk Logoglu, Stuti Surani, Marek Flaska

Methodology for an Experimental Drift Velocity Determination Using a µTPC. Example in a 700mbar 95%He+5%CO2 Mixture 10016 514
Thibaut Vinchon, Michaël Petit

Particles Discrimination and Tracking Through a Matrix of Detection Coupled with a Voltage-Controlled Oscillator 10017 519
Karine Coulié, Wenceslas Rahajandraibe, Laurent Ottaviani

Radioactive Direction of Arrival Estimation Using Neural Networks Approach 10018 525
Yossi Salomon, Eran Vax, Alon Osowizky, Yakir Knafo, Nadav Ben David, Dan Vilenchick

Resource Constrained Electronics and Signal Processing for UAV Radiation Sensors 10019 531
Andrea Chierici, Riccardo Ciolini, Andrea Malizia, Francesco D'Errico

Semiconductor-Based Diodes for Tritium Detection 10020 539
Paul Eyméoud, Stéphane Biondo, Vanessa Vervisch, Nadia Grillet, Laurent Ottaviani, Wilfried Vervisch

The High Efficiency fast-Response GAmma (HERGA) Detector Based on SiPM Readout 10021 543
C. Altomare, D. Cerasole, L. Di Venere, E. Fanchini, F. Gargano, F. Giordano, F. Loparco, S. Loporchio, A. Mastroserio, M. Morichi

Optoelectronics for Neutron Detectors Based on Superheated Emulsions 10022 547
Andrea Chierici, Francesco D'Errico

Methodical Optimization of Mutual-Shielding Directional Gamma Detector 10023 554
Vax Eran, Ben David Nadav, Salomon Yossi, Knafo Yakir, Dadon Shay, Stern Adrian, Behar Ehud, Osovizky Alon

EDUCATION, TRAINING AND OUTREACH

Education, Training and Research at Khalifa University in Support of the United Arab Emirates Nuclear Power Program 11001 561
Francois Foulon, Yacine Addad, Saeed Al Ameri, Prasoon Raj, Ahmed Al Kaabi, Akram Al Fantazi

Author Index

Dear Colleagues and Friends, dear ANIMMA attendees,

Nowadays nuclear measurement techniques and associated instrumentation are still in progress, and, within high-energy physics experiments, still bring a key contribution to fundamental research and to our knowledge of the Universe and its fundamental laws, the "physics of the two infinites". They also enable us to control nuclear reactors, to monitor the dosimetry, to contribute in public health in the field of diagnosis and therapy, to ensure some controls and characterization related to transport safety, radioactive wastes and dismantling operations, to check the compliance of components and industrial processes, to monitor pollution, to cope with nuclear proliferation… Not less than that, so to say!

I think we are all well aware of the fact that, over the past few decades, we have made tremendous progress in the development of measuring devices, data processing and analysis tools in many areas of interest such as computing (data mining, Artificial Intelligence, Algorithmic, Machine Learning and Deep Learning), robotics, space, health technologies, energy, environment, and transport. These advances unavoidably affect the requested performances of predictive modelling tools as well as experimental capabilities linked to our fields of activities. In this respect, ANIMMA scientific community has both the legitimacy and the leadership to face these challenges, and continuously develop a relevant international network.

As such, ANIMMA conference and the associated network is a seedbed of new scientific international collaborations dealing with experimental sciences and technologies.

The ANIMMA Conference has always strived to create a special meeting place shared by all those working in nuclear instrumentation and measurement and their applications, as we strongly believe that cross-border exchanges between scientists, engineers and industrialists can only lead to the most developed ideas, the best solutions and the most efficient collaborations and partnerships. Therefore, ANIMMA continues to maintain a high level of scientific and technical quality by presenting not only the latest advances but also the state of the art in each field through the participation of international specialists and experts.

ANIMMA-2023 was held exclusively in-person from June 12th – 16th at Real Collegio in Lucca, Italy (www.animma.com). The conference was locally organized by both CAEN-Sys and University of Pisa in strong partnership with CEA (France), SCK.CEN (Belgium), Aix-Marseille-University (France), IEAP Institute (Czech Republic) and Josef Stefan Institute (Slovenia).

ANIMMA-2023 is the eight edition of a series of conferences with a scientific program dealing with:

- 3 full days dedicated to the brand new ANIMMA International Summer School (AISS) from June 10th to 12th
- A full-day dedicated to 4 international workshops on Monday June 12th
- 3 prestigious keynote lectures on Tuesday June 13th during the opening ceremony
- 3 relevant plenary talks on Wednesday 14th , Thursday 15th and Friday 16th at the start of each day followed by parallel oral sessions
- Intensive oral sessions to promote posters on Tuesday 13th and Wednesday 14th afternoon followed by plenary poster session
- Permanent poster and industrial exhibition from Tuesday 13th to Friday 16th

More than 270 scientific presentations were given during the conference.

ANIMMA-2023 welcomed 403 attendees from 31 countries including 80 graduate students. We awarded 2 best student papers and 3 best posters.

The attendees deserve our warmest thanks and consideration for making this eighth edition a great success in an extraordinary location.

All the committee members are looking forward to welcoming everyone again at ANIMMA2025 in Valencia, Spain.

Prof. Dr. Abdallah LYOUSSI

ANIMMA General Chairman and Program Committee Chair

edp sciences

EPJ Web of Conferences

Statement of Peer review

In submitting conference proceedings to EPJ Web of Conferences, I certify to the Publisher that I adhere to the **Policy on Publishing Integrity** of the journal in order to safeguard good scientific practice in publishing.

1. All articles have been subjected to peer review administered by the proceedings editors.
2. Reviews have been conducted by expert referees, who have been requested to provide unbiased and constructive comments aimed, whenever possible, at improving the work.
3. Proceedings editors have taken all reasonable steps to ensure the quality of the materials they publish and their decision to accept or reject a paper for publication has been based only on the merits of the work and the relevance to the journal.

Title, date and place of the conference

ANIMMA : Advancements in Nuclear Instrumentation Measurement Methods and their Applications

LUCCA, Italy, June 12-16, 2023

Proceedings editor(s):

Abdallah Lyoussi, Francesco D'Errico, Michel Carette, Malcolm Joyce, Rastislav Hodak, Igor Jencic, Patrick Ledu, Massimo Morichi, Stanislav Pospisil, Christelle Reynard-Carette, Luka Snoj, Ludo Vermeeren

Date and editor's signature

October, 16th, 2023

https://doi.org/10.1051/epjconf/202328801001

A High-Granularity Timing Detector for the ATLAS Phase-II upgrade

Océane Perrin*
on behalf of the ATLAS HGTD group
Laboratoire de Physique de Clermont (LPC)
Institut National de Physique nucléaire et de Physique des particules (IN2p3)
(*) oceane.perrin@cern.ch

Abstract— **The High Luminosity Large Hadron Collider (HL-LHC) will reach an integrated luminosity up to 4 000 fb^{-1} from 2029 to 2039. The number of collisions per bunch crossing will significantly increase, rising a challenge in terms of pileup mitigation that will have a severe impact on the ATLAS detector performance. Therefore, the High-Granularity Timing Dectector (HGTD) will be installed in front of the Liquid Argon Calorimeter (LAr) covering the forward region with a pseudo-rapity from 2.4 to 4.0. HGTD will provide a time measurement of the time for Minimum Ionizing Particles (MIP) with a 30 ps per track resolution and will be coupled to the futur tracking detector (ITk) to assign each particle to a vertex. HGTD will be composed of 3.6 million readout channels, including a Low Gain Avalanche Diode (LGAD) based technology sensors and a front-end readout chip (ASIC) based on the 130 nm CMOS technology. These latest will be combined and assembled to produce a module, that will be installed and glued on a support unit. This article presents a general overview of the HGTD project, its status and some highlights of sensitive components.**

Keywords — **HL-LHC, ATLAS, HGTD, timing detector**

I. Introduction

The High-Luminosity Large Hadron Collider (HL-LHC) *[1]* is expected to start operation in 2029. It will reach an instantaneous luminosity of 7.5 x 10^{34} cm^{-2}s^{-1} and will collect in 10 year of operation an integrated luminosity of 4 000 fb^{-1}. To reach this luminosity, an average of 200 collisions per bunch crossing will be necessary, leading to a significant pileup effects.

The ATLAS detector is one of the main experiments of the LHC. As one of the two general-purpose experiments, ATLAS experiment investigates in a wide range of physics, from the Higgs Boson to the dark matter. In order to face the high pileup environenment, the ATLAS detector will be upgraded. The Data AQuisition and the trigger system will be upgraded with a higher granularity and faster electronics. Two new detectors will be installed, a fully silicon based traker detector (ITk) *[2]* and a timing detector HGTD.

II. Motivation

The association of the tracks to primary vertices is one of the key elements for the pileup mitigation. Tracks and vertices are associated if they are geometrically compatible in the longitudinal axis z. The separation of the vertices is ensured while the following equation is satisfyed:

$$\frac{|Z_0 - Z_{vtx}|}{\sigma_{Z_0}} < s$$

with s a significance cut that is typically 2.5 or 3, σ_{z0} the per track resolution of the longitudinal impact parameter Z_0. The spatial resolution of the current ATLAS detector satisfies the previous equation for a luminosity of 140 fb^{-1} under LHC pileup conditions.

ITk will provide tracking performances similar to those in Run II LHC as shown in Fig. 1, for a region with a pseudo-rapidity up to 2.4. In the forward region defined with a pseudo-rapidity from 2.4 to 4.0, the spatial resolution is deteriorating rapidely.

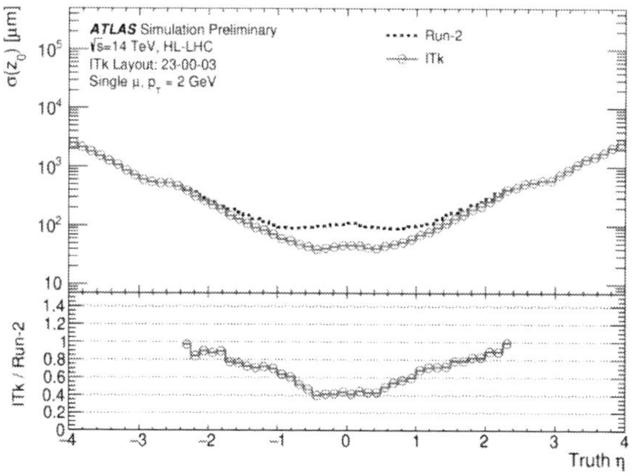

Figure 1: The resolution of the longitudinal track impact parameter Z_0, as function of the pseudo-rapidity [3]

In order to improve the reconstruction of physics objects in the forward region, an additional HGTD detector will be installed and will satisfy the following equation:

$$\frac{t_{track}-t}{\sigma_t}<s$$

with s a significance cut (set at 2 or 3), σ_t the sum in quadrature of the vertex time t and t_{track} the track-time.

III. HGTD DETECTOR

A. Overview and requierements

The HGTD will be installed in the forward region at ±3.5 m from the interaction point, in a gap region between the end-cap calorimeters and the barrel, as shown in Fig. 2. The detector will be installed in a constrained space, its total disk radius goes from 11 cm to 110 cm, with an active radius from 12 cm to 64 cm and its thickness should not exceed 125 mm including 50 mm of moderator to protect the detector from back-scattered neutrons. The active region is designed with a three-ring stucture, the overlapping of each layer has been optimized to meet the time resolution requierements in the strict constrained space. In order to insure the optimal timing resolution and minimize the effect of the pileup, each active layer is rotated by 20 degrees.

Figure 2: Position of HGTD in the ATLAS Detector [3]

The detector has been designed to provide a time resolution of 30 ps per track (50 ps per hit) at its inital state. After a radiation dose of 2.5 x 10^{15} n_{eq} cm^{-2} and 2 MGy, corresponding to the estimated lifetime of the detector, the time resolution should remain below 35 ps per track (70 ps per hit).
Table I summarizes the main parameters of the HGTD.

TABLE I
MAIN PARAMETERS OF HGTD

Pseudo-rapidity coverage	$2.4 < \lvert\eta\rvert < 4.0$
Thickness	75 mm + 50 mm moderator
Position of active layer in z	± 3.5 m
Radial total area	110 mm < r < 1000 mm
Radial active area	120 mm < r < 640 mm
Time resolution per track	35 ps (start) and 70 ps (end)
Radiation tolerance	2.5 x 10^{15} n_{eq} cm^{-2}, 2 MGy
Pixel size	1.3 mm x 1.3 mm
Active sensor size	39 mm × 19.5 mm (30 × 15 pixels)
Minimum charge	4 fC
Active thickness	50 µm
Min. hit efficiency	95 %

B. Detector modules

The active layer of the HGTD detector will be made of 8032 modules. These fundamental units of the HGTD detector are an assembly of two components: two LGAD sensors bump-bonded to two ASICs and a flexible printed circuit board. A flex cable ensures the communication, the power distribution and the data-transfer. The modules, shown in Fig. 3, will be installed and glued on a support unit. The main constrains of the mechanical designed arises from the thickness constrains (7.5 cm) and the optimized geometry in oder to get at least two hits per tracks.

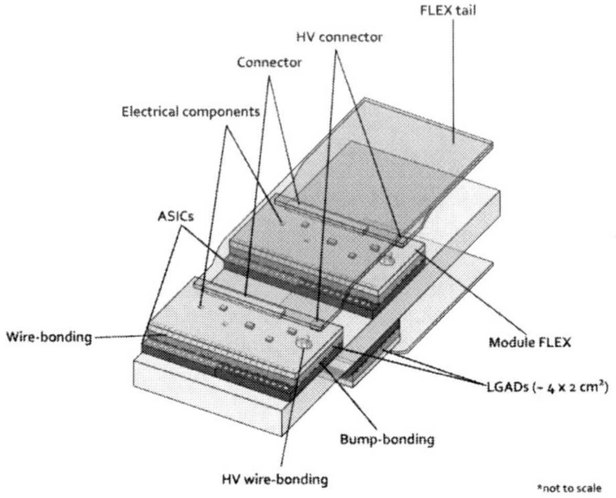

Figure 3: View of an HGTD module [3]

C. Low Gain Avalanche Diode sensor

The HGTD sensor are based of Low Gain Avalanche Diodes (LGAD). Originally designed by the Centro National de Microelectronica Barcelona (CNM) *[4]*, the LGAD technology is a n-on-p planar silicon detector, containing an

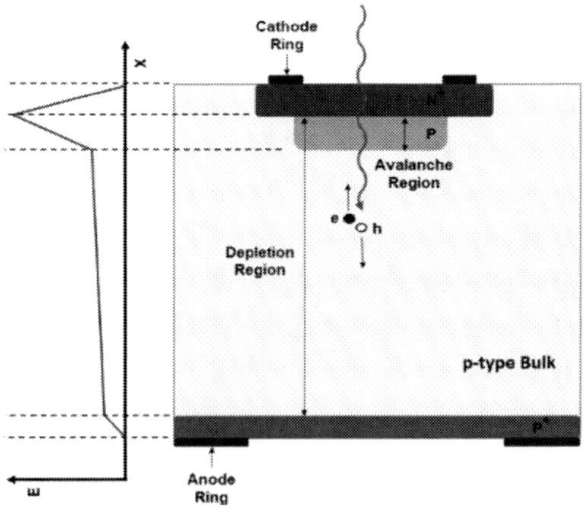

Figure 4: Working principle of the LGAD sensors [3]

extra highly-doped p-layer below the n-p junction in order to provide a charge amplification, referred as the gain of the LGAD.

When a charge particle hits the sensor, an initial current is generated by the drift of the electrons and holes in the silicon. The charge amplification is due to the drift of the electron in the avalanche region, that induces the creation of hole/electron pairs, as shown in Fig. 4. The resulting gain is higher than standard sensors and remains the key ingredient to get the requiered time resolution.

For seven years, LGAD sensors from various vendors have been tested, from CNM (Spain), FBK (Italy), HPK (Japan), IME (China) with different active thicknesses, pad sizes and doping levels. Each LGAD has been exposed to neutrons, protons and X-Rays up to the maximum irradiation level 2.5 x 10^{15} n_{eq} cm^{-2} to simulate their end of life state.

D. Front-end readout chip: ASIC

The read-out of the LGAD will be insured by the front-end electronics ASIC chip designed with 130 nm CMOS technology. The ASIC will have 225 channels arranged in a 15 x 15 matrix. Each channel is made of a preamplifier, a discriminator, and two Time to Digital Converters (TDC) in order to extract and digitize the Time of Arrival (TOA) and the Time Over Threshold (TOT) of each LGAD signal. Then the data is stored in a local buffer until the reception of the trigger. Another function of the ASIC is to provide a bunch per bunch precise measurement of the instantaneous luminosity. The luminosity processing system contained in the digital part of the ASIC, will count the amount of hits and will deliver this data at 40 MHz. Fig. 5 illustrates the full front-end chain.

Four versions of the ASIC has been tested so far. ALTIROC0 and ALTIROC1 were designed to test the analog part of the electronic chain, with limited number of channel. ALTIROC2 is the first full version with 225 channels. ALTIROC3, the newest version is under test. Currently, the pre-production version of ALTIROC is under design.

The main challenge in the design of the ASIC is to reach the smallest contribution to the time resolution in order to match the performances of the LGADs. The main contributions to the time resolution of a hit in the HGTD detector are:

$$\sigma^2_{Total} = \sigma^2_L + \sigma^2_{TW} + \sigma^2_{clock} + \sigma^2_{TDC} + \sigma^2_{jitter}$$

$$\text{with } \sigma_{TW} = \left[\frac{Vth/S}{t_{rise}}\right]_{RMS} \propto \left[\frac{N}{dV/dt}\right]_{RMS}$$

$$\text{and } \sigma_{jitter} = \frac{N}{dV/dt} \approx \frac{t_{rise}}{S/N}$$

where S refers to the signal (proportional the the gain), V_{th} the threshold voltage, t_{rise} the rise time and N the noise. σ^2_L is the landau fluctuation in the energy deposits of particles that cross the sensors, it is related to the thickness of the sensor.

σ^2_{TW} is the variation due to the amplitude of the signal, it expected to be negligible after the Time Walk correction. σ^2_{jitter} corresponds to the variation due to the noise of the signal, it is requiered to be about 25 ps for an injected charge of 4 fC. Finally σ^2_{TDC} and σ^2_{clock} are the contribution from the TDC and the clock, they are expected to be respectively negligible and lower than 15 ps after calibration.

IV. Characterization of the LGAD sensors

A. Performance tests

The characterization of LGADs has been done in laboratories with a ^{90}Sr source or in test beams at DESY or CERN. The setup remains similar for both places. The device, sometimes composed of multiples sensors, is placed in a cooling box. To perform the time characterization of the

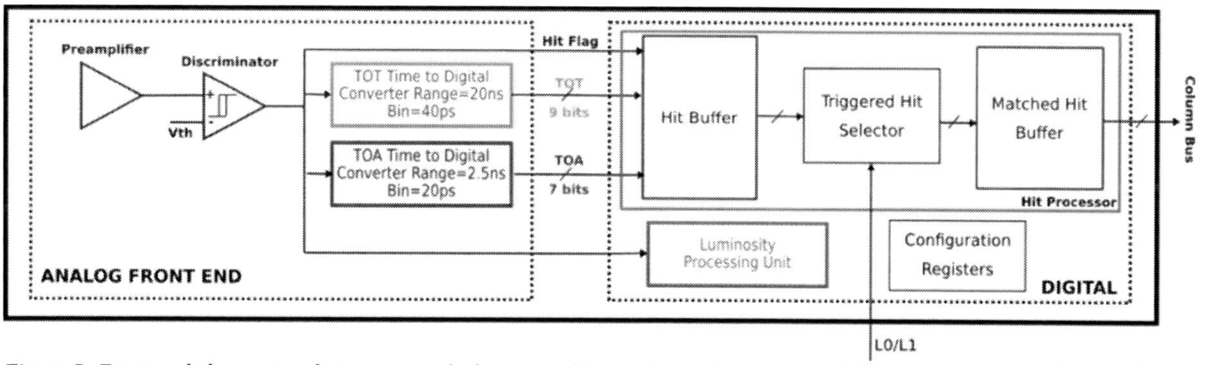

Figure 5: Front-end electronics chain composed of a preamplifier, a discriminator, two TDC to extract respectivelly the TOA and the TOT and the local memory system. This latest is composed of 3 blocks: the hit buffer, the Triggered hit selector and the matched hit buffer. The pink block corresponds to the block that will process the instantaneous luminosity measurement. One ASIC is composed of a matrix 15 x 15 pixels, so 225 channels [3].

sensors, a time reference with at least similar time resolution is provided to the data acquisition system, as a Silicon Photomultiplier (SiPM). Finally, for position-dependent measurement, a beam telescope is used [5]. Fig. 6 illustrates the time resolution as function of the bias voltage for three differents LGAD, irradiated at the maximum fluence. This measurement has been performed at DESY, with a 5 GeV electron beam and at CERN SPS with a 120 GeV pion beam between 2021 and 2022. The dash line represents the maximum requirement of 70 ps. The measurements are performed at temperatures between -43 °C and -24 °C, for three differents type of LGADs: FBK, USTC and IHEP. The study highlights that under these conditions, LGADs are able to achieve the expected collected charge of 4 fC and an efficiency of at least 95 %, uniformly over the sensor surface, as shown in Fig. 7.

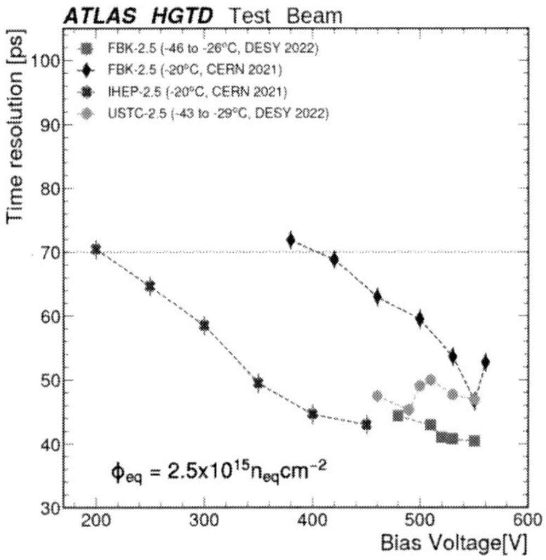

Figure 6: Time resolution as function of the bias voltage for the different LGAD sensors irradiated to the maximal fluence: FBK, IHEP and USTC at DESY with 5 GeV electron beam . The dash line represents the maximum time resolution limit 70 ps [6].

B. Single Event Burnout study

One of the major observations was the Single Event Burnout caracterization. This phenomenon results from an instantaneous alteration in the device response after high energy particle interaction. It leads to a permanent failure of the device. In September 2020, during a perfromance test at DESY, permanent burnouts have been observed on irradiated LGADs. To understand this phenomenon, various test beam campains have been organized to test in total 64 sensors at DESY (3 GeV electrons) and CERN (120 GeV pions). The main objective was to identify robust technologies and identify a safe limit to prevent any failure.

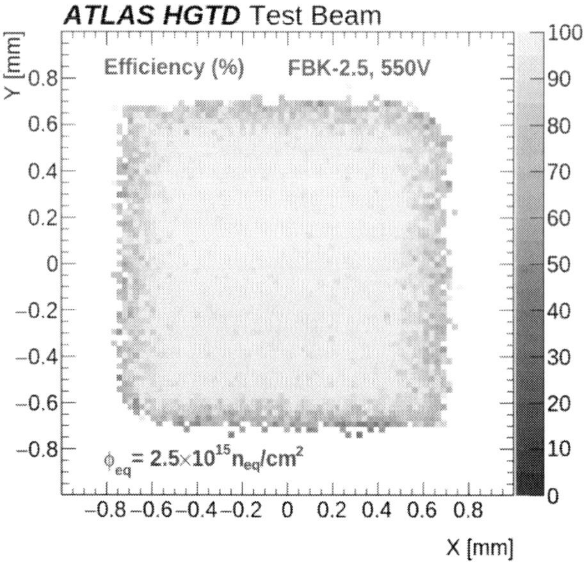

Figure 7: Efficiency as function of the hit position in the sensor plane. Figure made with an FBK sensor, at DESY with 5 GeV electron beam [6].

The setup was designed for these measurements and includes a printed circuit board (PCB) with two different sensors. The state of the sensors was monitored using the current.

It has been shown that the risk of burnout is governed by the electric field, whatever the design of the LGAD is. A critical upper limit of 11 V µm^{-1} was determined, as shown in the Fig. 8. Carbon enriched sensors can be operated with a low electric field which solves the burnout risk.

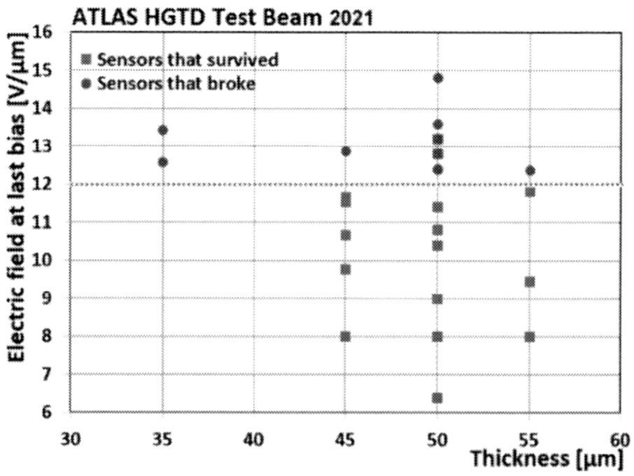

Figure 8: : The electric field in the sensor as function of the thickness. The dash red line indicates the limite 12 V/µm. [7]

 https://doi.org/10.1051/epjconf/202328801001

V. Characterization of the ASIC

A. Performance tests

The characterization of the ASIC alone is performed in laboratory. These tests aim at verifying the main requirement of the read-out chip, including the charge dynamic range, the working temperature and the jitter.
Tests of the ASIC are performed with the following setup. A Field-Programmable Gate Array (FPGA), ensures the voltage supply, the clock generation, the communication with the operator and the data transfer. The tests can be performed either with the ASIC alone using the internal charge pulser or with a LGAD bump bonded with a ^{90}Sr source or a beam. Fig. 9 shows the evolution of the jitter with respect to the charge. Red dots and purple dots represent respectively the results for the assembly ALTIROC and LGAD bump-bonded with Trans-impedance amplifier (TZ) or Voltage Pre-Amplifier (VPA). Green and blue dots are the results for ALTIROC only, using an internal charge pulser. Results satisfy the expected 20 ps of jitter for an injected charge of 10 fC. In addition, ALTIROC manages to detect a minimum injected charge of 4 fC.

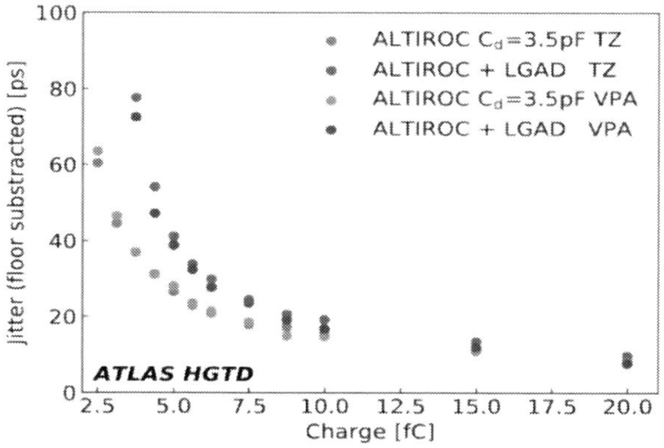

Figure 9: Noise over the TOA measurement. Red and purple results correspond to the ALTIROC+LGAD system respectivelly using only TZ or only VPA amplification. Green and blue dots correpond to ALTIROC results, using an injected charge system, able to reproduce LGAD signal [8]

B. ALTIROC+LGAD at test beam

Since 2019, the ALTIROC+LGAD system is tested in test beams at CERN using the Super Proton Synchrotron or the Proton Synchrotron. Three versions of the ASIC have been tested so far: ALTIROC0, ALTIROC1 and ALTIROC2.
The setup of test beam campains is very similar to the test bench setup, apart from the reconstruction system based on a telescope system to ensure the tracking of particles. The trigger is sent from a reference plan (FEI4) to the telescope for the tracking data and to the ASIC for the timing data throught a FPGA.

Due to the dependancy of the TOA with respect to the amplitude of the injected signal, a time walk correction based on a fit of the TOA with repect to the TOT measurement is used, as shown in Fig. 10 [9]. After this time-walk correction, the estimated resolution of the ASIC+LGAD system reachs $\sim$ 45 ps with a jitter contribution of $\sim$ 40 ps.

The efficiency is defined as the ratio of the reconstructed tracks associated to a signal above the threshold to all reconstructed tracks that traverse the active area.

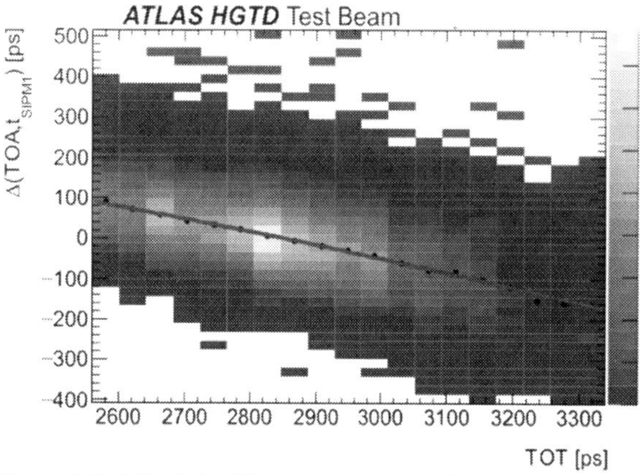

Figure 10: A fit of the TOA with respect to the TOT. This method is used to determine the time-walk correction.

The efficiency map shown in Fig. 11, has been obtained with ALTIROC2 plus a bump-bonded LGAD, performed with 75 GeV charge pions at CERN SPS beam line in Autumn 2022. The ALTIROC was configured with only transimpedance amplifiers, working at room temperature with a charge threshold set at 4.8 fC.

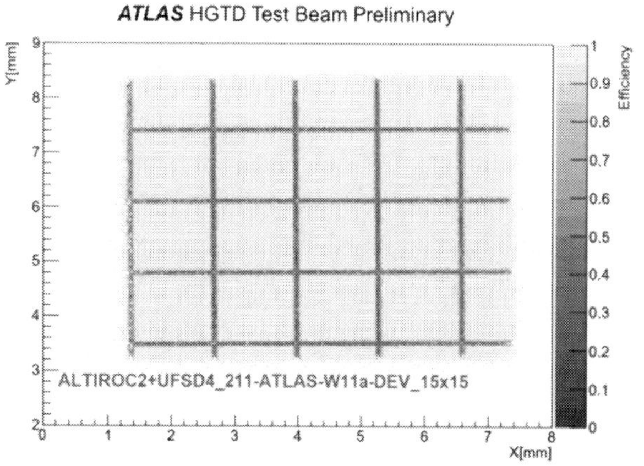

Figure 11: Efficiency map obtained with ALTIROC +LGAD system, at room temperature with 75 GeV charge pions at CERN SPS during automn 2022 test beam campaing.

An efficiency of about 100 % was achieved for each pixel in the non inter-pad region. The inter-pad region is defined as the region between two pixels where the efficiency is lower than 50 %. Fig. 12 shows that this interpad distance measures $\sim 65\,\mu$m.

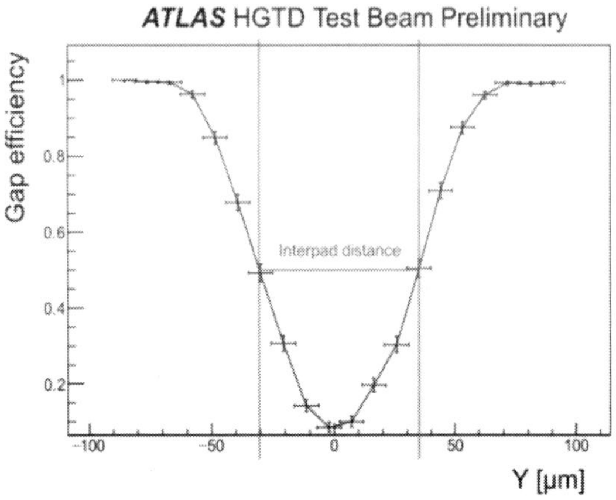

Figure 12: Efficiency versus y position obtained with ALTIROC +LGAD system, at room temperature with 75 GeV charge pions at CERN SPS during automn 2022 test beam campaing . The red line represents the inter-pad distance defined as the 50 % efficiency.

VI. Conclusions

HGTD detector will play a key role to mitigate the effect of the pileup, during the HL-LHC era in the forward region. This detector will improve the reconstruction of physics objects to a similar level of the one expected in the central region. This detector will provide a time resolution of 30 ps per minimum-ionizing particle. It will be made with a sensor based on LGAD technology and the ALTIROC readout electronics. The respective characterization of LGADs and the ASIC satify the HL-LHC requierements. The overall design and tests are in progress, with an installation foreseen in 2026-2028.

References

1 High-Luminosity Large Hadron Collider (HL-LHC): Technical Design Report, CERN-2020-010, I. Zurbano Fernandez, M. Zobov, A. Zlobin, F. Zimmermann, M. Zerlauth et al. https://cds.cern.ch/record/2749422/files/127-117-PB.pdf

2 ATLAS Collaboration, "Technical Design Report for the ATLAS Inner Tracker Pixel Detector," CERN, Geneva, Tech. Rep., Sep 2017. https://cds.cern.ch/record/2285585

3 ATLAS Collaboration. Technical Design Report: A High-Granularity Timing Detector for the ATLAS Phase- II Upgrade. June 2020. https://cds.cern.ch/record/2719855/files/ATLAS-TDR-031.pdf

4 G. Pellegrini et al., Technology developments and first measurements of Low Gain Avalanche Detectors (LGAD) for high energy physics applications, Nucl. Instrum. Meth. A 765 (2014) 12

5 H. Jansen, S. Spannagel, J. Behr, A. Bulgheroni, G. Claus, E. Corrin, D. Cussans, J. Dreyling-Eschweiler, D. Eckstein, T. Eichhorn et al., "Performance of the eudet-type beam telescopes," EPJ Techniques and Instrumentation, vol. 3, pp. 1–20, 2016. https://link.springer.com/content/pdf/10.1140/epjti/s40485-016-0033-2.pdf

6 S. Ali et al. "Performance in beam tests of carbon-enriched irradiated Low Gain Avalanche Detectors for the ATLAS High Granularity Timing Detector". In: Journal of Instrumentation 18.05 (May 2023), P05005. doi: 10.1088/1748-0221/18/05/p05005. https://doi.org/10.1088%2F1748-0221%2F18%2F05%2Fp05005.

7 L.A. Beresford et al. "Destructive breakdown studies of irradiated LGADs at beam tests for the ATLAS HGTD". In: Journal of Instrumentation 18.07 (July 2023), P07030. doi:10.1088/1748-0221/18/07/P07030. Url: https://dx.doi.org/10.1088/1748-0221/18/07/P07030

8 C. Agapopoulou et al. Performance of a front-end prototype ASIC for the ATLAS High Granularity Timing Detector. 2023. arXiv: 2306.08949 [physics.ins-det] https://doi.org/10.48550/arXiv.2306.08949

9 C. Agapopoulou et al., Performance of a Front End prototype ASIC for picosecond precision time measurements with LGAD sensors, JINST 15 (2020) P07007 [2002.06089]

Comparison between GEANT4 and MCNP for well logging applications

Geoffrey Varignier[1,*], Valentin Fondement[2], Cédric Carasco[2], Johann Collot[3], Bertrand Pérot[2], Thomas Marchais[2], Pierre Chuilon[1], Emmanuel Caroli[1], Mai-Linh Doan[4]
[1]TotalEnergies, France
[2]CEA, France
[3] LPSC-IN2P3, France
[4]Univ. Grenoble Alpes, Univ. Savoie Mont Blanc, CNRS, IRD, Univ. Gustave Eiffel, ISTerre, France
(*) geoffrey.varignier@totalenergies.com

Abstract— **MCNP and GEANT4 are two reference Monte Carlo nuclear simulators, MCNP being the standard in the Oil & Gas nuclear logging industry. While performing a simulation benchmark of these two software for the purpose of "Cased Hole" wellbore evaluation, discrepancies between MCNP and GEANT4 were observed: computational experiments were performed first in a theoretical and simplified environment using spherical models, then in a more realistic "Open Hole" wellbore context with simplified logging tools. Results of this comparison show an excellent overall agreement for gamma-gamma physics and an acceptable agreement for neutron-neutron physics. However, the agreement for neutron-gamma physics is satisfactory only for certain lithologies and energy windows, but not acceptable for other operating conditions. These results need to be put in perspective with the current use of nuclear simulation in the logging industry. Indeed, wellbore evaluations rely on charts simulated with Monte Carlo codes in various contexts. In the case of radially heterogeneous environments such as "Cased Hole" wellbores, nuclear simulations are mandatory to precisely determine the radial sensitivity of logging tools via the so-called sensitivity functions. The feasibility of wellbore inversion relies on the physical validity of such sensitivity functions obtained from nuclear simulations. This MCNP vs. GEANT4 benchmark was conducted with the perspective to secure the physical fundamentals used for building the sensitivity functions of logging tools.**

Keywords— **Well logging, Monte-Carlo modeling, MCNP, GEANT4, Gamma-Gamma, Neutron-Neutron, Neutron-Gamma, Sensitivity function, Nuclear logging probes.**

I. INTRODUCTION

Nuclear logging tools have been used in the oil and gas industry for over the last 70 years, mainly to characterize formation reservoirs and to monitor the hydrocarbon fraction in the close vicinity of the wellbore during production. Many logging probes based on different nuclear physics are frequently used for data acquisition. Once calibrated, the data acquired along the wellbore are interpreted into some parameters of interest, such as porosity, density, water saturation, etc.

To understand the sensitivity of nuclear measurements to the target properties and convert them to quantitative ones, the oil industry started using recurrently Monte Carlo nuclear simulations [11]. It enables the 3D modeling of particle interactions with matter through a stochastic approach. Trajectories of millions of particles (neutrons or gamma) are simulated simultaneously. Their interactions with the logging tool itself, borehole fluid, casing, cement and formation are calculated using the principles of nuclear physics. Particles heading to the detectors are counted and their spectra analyzed similarly to what occurs in any logging tool.

Oilfield Services companies and Oil&Gas companies usually use MCNP (Monte Carlo N-Particle) computer code to model the logging tools response to various wellbore contexts. This software is considered standard in the Oil&Gas industry [1]. MCNP is a Monte Carlo nuclear simulation code written in Fortran 90 and C, developed by the Los Alamos National Laboratory (USA). Being developed by a US institution, MCNP is not available all over the world. An alternative to MCNP is GEANT4 (GEometry ANd Tracking) [10], which is another Monte Carlo nuclear simulation toolkit written in C++ and developed by the CERN agency. Although originally developed for high-energy physics, it has been extended to applications in low-energy physics.

The goal of nuclear simulations is to characterize the relationship between the nuclear measurements and the wellbore components: reservoir properties, as well as completion, production fluids, steel casings, and cement in case the well is cased (i.e. a steel casing is cemented to the wellbore wall). In such a configuration, the number of unknowns is so large (casing and cement thicknesses and grades, tubing centralization, etc.) that the inverse problem requires the combination of various physics (with nuclear and non-nuclear probes) covering compatible radii of investigation. Exact knowledge of the detectors sensitivity for all nuclear probes is then essential and obtained from nuclear simulations.

The industry standard is to run MCNP with biasing techniques [15][16][17]. However, the fundamental of this assumption needs to be revised when applied to the radially heterogeneous "Cased Hole" domain with independent numerical calculations using MCNP and GEANT4. Prior to validate the concept of sensitivity function for "Cased Hole" environment, discrepancies between both software have been evaluated in this paper without biasing and variance reduction techniques to highlight discrepancies observed between MCNP and GEANT4 and present quantitative results from a comparison in a logging environment. First calculations are performed with a theoretical and simplified spherical model and then in a more realistic "Open Hole" model using simplified logging tools such as Litho-Density, Neutron-Porosity and Carbon/Oxygen.

II. GEOMETRY AND VISUALIZATION

Figure 1 shows the spherical model allowing for a first comparison between the codes, displayed with the MCNP Vised visualization software. It consists of a 22.9 cm radius sphere surrounded by a spherical NaI(Tl) or 3He detection volume with an external radius of 34.9 cm and a thickness of 3 cm. The inner sphere is composed of a pure material commonly encountered in Geoscience problems (carbon ; oxygen ; calcium ; magnesium ; silicon ; hydrogen).

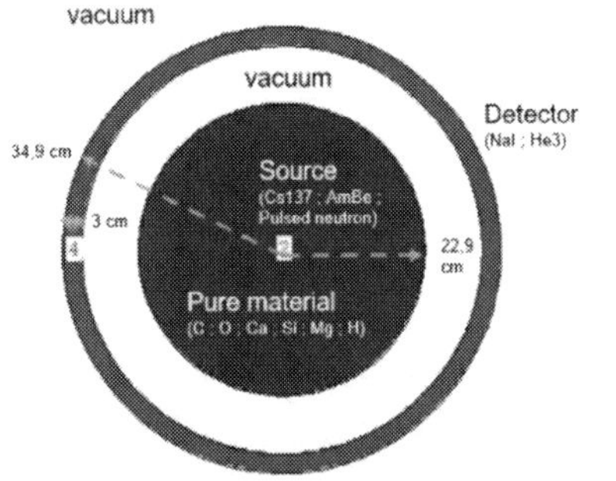

Fig. 1. Spherical model displayed using Vised.

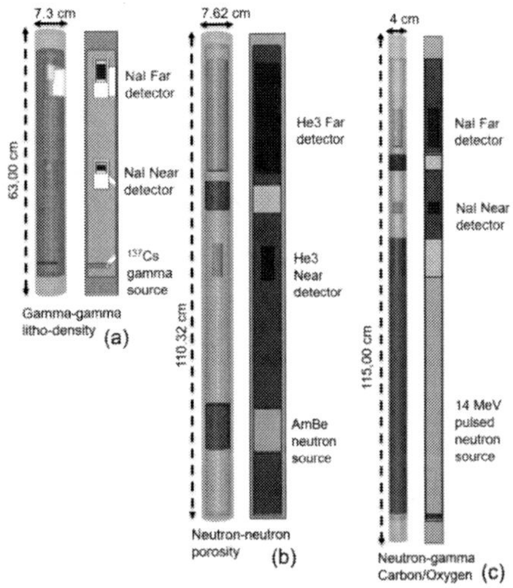

Fig. 2. Gamma-gamma density (a), neutron-neutron porosity (b) and neutron-gamma Carbon/Oxygen (C/O) (c) generic logging tools models displayed using EDGE (left) and Vised (right).

On the GEANT4 side, visualization software such as EDGE can be used to build geometries and export them to GDML format [4][22], to obtain views such as the well logging tool shown in Figure 2. Generic logging tools models are very simplified and supposed to mimic the physical response of commercial logging tools operated by oilfield services companies. These models use main elements of logging tools such as detectors, shields and sources. Key parameters such as source-detectors distances, detectors and shield volumes and tool dimensions are close to real logging tools. Generic logging tools models were developed by the Austin University [14][15].

Figure 3 shows a 3D model of a neutron-gamma C/O logging tool, a pure lithology and a 200 mm water filled borehole. 3D geometry modeling capabilities of MCNP and GEANT4 are comparable and external software such as Vised, or EDGE, can be used to check the model in detail.

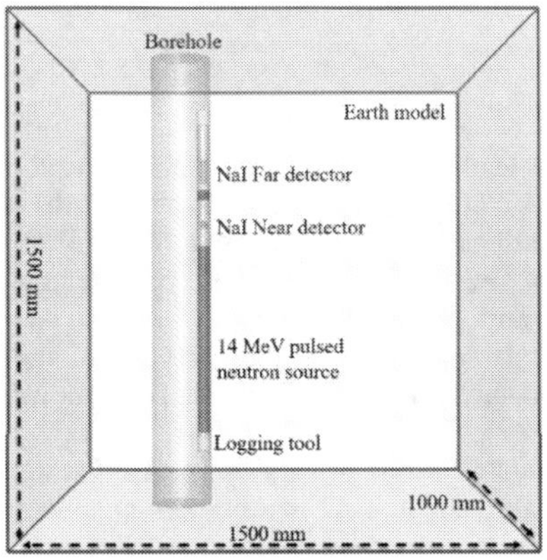

Fig. 3. Neutron-gamma C/O generic logging tools in a water filled borehole model displayed using EDGE. The well is filled with fresh water and the Earth model is composed of a pure lithology at 20 pu (porosity unit) such as sandstone (SiO_2), limestone ($CaCO_3$) or dolomite ($CaMg(CO_3)_2$). Rock porosity is fully saturated with fresh water.

III. GAMMA-GAMMA MEASUREMENT

Gamma-gamma measurement consists in measuring with a NaI(Tl) detector the gamma spectrum originating from a ^{137}Cs radioactive source of mono-energetic 661.7 keV gamma rays, mostly after their Compton scattering in the rock formation [13]. The Compton continuum starting at the initial energy 661.7 keV is measured using a "Hard window" from 540 keV down to about 150 keV, where the Compton interaction dominates. In the Geosciences field, this continuum is used to derive the formation density. On the other hand, the photoelectric effect (PEF) is measured using a "Soft window" between 60 keV and approximately 100 keV. The "Soft window" contains information about the rock mineralogy. To compensate for the photoelectric contribution in the "Hard window", a lithological matrix correction is performed based on PEF.

Gamma-gamma measurements are simulated in MCNP with the F8 pulse height tally, which provides the histogram of photon energy deposits in a detector (i.e. the gamma spectrum), per source particle. The F8 spectrum being normalized to one source particle [2], it is scaled with the number of particles simulated in GEANT4 to allow a quantitative comparison. Gamma-gamma measurements are simulated in GEANT4 by scoring in *"SteppingAction.cc"*, the distribution of energies deposited by gamma rays entering in the NaI detector.

https://doi.org/10.1051/epjconf/202328801002

Figure 4 shows six gamma-gamma measurement spectra in a spherical model for six pure chemical elements.

Tables I and II give the total counts integrated in the Hard and Soft windows for a spherical model and show a maximum discrepancy of 4.0% for calcium.

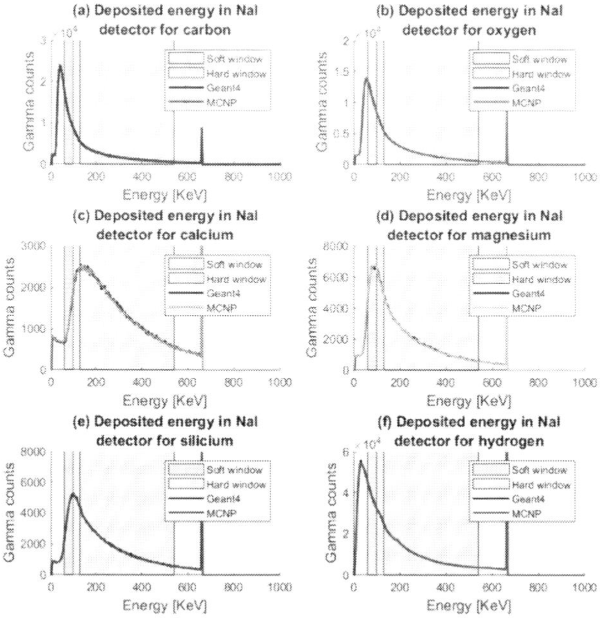

Fig. 4. Simulated gamma-gamma spectra with a spherical model, a ^{137}Cs gamma source, a NaI detector and 6 pure materials. Five materials have a density of 2.71 g/cm3 (C, O, Ca, Mg, Si), and due to its large number of nuclei per volume unit, one material has a density of 0.5 g/cm3 (H).

TABLE I. SOFT WINDOW TOTAL COUNTS OF GAMMA-GAMMA MEASUREMENTS IN A SPHERICAL MODEL FOR SIX PURE ELEMENTS.

Gamma-Gamma density measurements NEAR detector in Soft window			
	Carbon	Oxygen	Calcium
Counts MCNP	147622	128578	14932
Counts GEANT4	151926	131276	14331
GEANT4/MCNP	2,9%	2,1%	4,0%

Gamma-Gamma density measurements NEAR detector in Soft window			
	Magnesium	Silicium	Hydrogen
Counts MCNP	74436	50755	453733
Counts GEANT4	73381	48942	461870
GEANT4/MCNP	1,4%	3,6%	1,8%

TABLE II. HARD WINDOW TOTAL COUNTS OF GAMMA-GAMMA MEASUREMENTS IN A SPHERICAL MODEL FOR SIX PURE ELEMENTS.

Gamma-Gamma density measurements NEAR detector in Soft window			
	Carbon	Oxygen	Calcium
Counts MCNP	192248	190119	150395
Counts GEANT4	194282	192895	152209
GEANT4/MCNP	1,1%	1,5%	1,2%

Gamma-Gamma density measurements NEAR detector in Soft window			
	Magnesium	Silicium	Hydrogen
Counts MCNP	184011	177965	983344
Counts GEANT4	186786	180601	1002273
GEANT4/MCNP	1,5%	1,5%	1,9%

Figure 5 shows six gamma-gamma density measurement spectra of a generic logging tool with two detectors (NEAR and FAR) in a 200 mm diameter borehole and for three rock compositions. On Fig 5, a Gaussian Energy Broadening (GEB) has been performed on each spectrum.

Tables III and IV give the total counts integrated in the Hard and Soft windows for a litho-density logging tool model and show maximum discrepancies of 4.3% for the NEAR detector and 11.3% for the FAR detector.

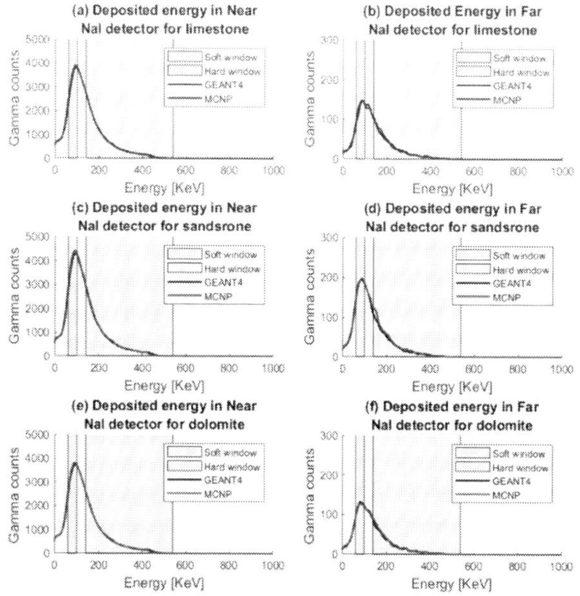

Fig. 5. Simulated gamma-gamma measurements spectra with a lithology-density generic logging tool model, a ^{137}Cs gamma source, two NaI detectors and three different rocks at 20 pu (porosity unit): a 2.368 g/cm3 limestone (a, b), a 2.323 g/cm3 sandstone (c, d), and a 2.496 g/cm3 dolomite (e, f).

Discrepancies are larger between MCNP and GEANT4 using the logging tool model compared to the spherical model, which could be due to the larger number of interactions along the particles random walk. Moreover, the smaller number of counts induce larger calculation statistical uncertainties. Simulations in MCNP and GEANT4 have been performed without biasing techniques to compare the analog gamma transport. Variance reduction techniques could be useful for gamma-gamma density measurements to improve statistics. In Geosciences, the density is mainly based on the FAR detector. Even if the absolute number of counts is lower compared to the NEAR detector, the measurement is more representative of the geological formation because of the larger depth of investigation. NEAR detector is used to correct the density measurement of environmental effects such borehole and mudcake effects. Nevertheless the overall agreement between GEANT4 and MCNP is satisfactory for gamma-gamma physics, as already reported in other benchmarks and shows discrepancies lower than 12% [23].

TABLE III. SOFT WINDOW TOTAL COUNTS OF GAMMA-GAMMA MEASUREMENTS IN A DENSITY GENERIC LOGGING TOOL MODEL WITH TWO DETECTORS IN THREE ROCK COMPOSITIONS AT 20 PU.

Gamma-Gamma density measurements NEAR detector in Soft window			
	Limestone	Sandstone	Dolomite
Counts MCNP	37376	42976	37153
Counts GEANT4	35981	41134	35903
GEANT4/MCNP	3,7%	4,3%	3,4%

Gamma-Gamma density measurements FAR detector in Soft window			
	Limestone	Sandstone	Dolomite
Counts MCNP	1372	1975	1274
Counts GEANT4	1446	1989	1308
GEANT4/MCNP	5,4%	0,7%	2,7%

TABLE IV. HARD WINDOW TOTAL COUNTS OF GAMMA-GAMMA MEASUREMENTS IN A DENSITY GENERIC LOGGING TOOL MODEL WITH TWO DETECTORS IN THREE ROCK COMPOSITIONS AT 20 PU.

Gamma-Gamma density measurements NEAR detector in Hard window			
	Limestone	Sandstone	Dolomite
Counts MCNP	54709	57813	51890
Counts GEANT4	54419	56674	51579
GEANT4/MCNP	0,5%	2,0%	0,6%

Gamma-Gamma density measurements FAR detector in Hard window			
	Limestone	Sandstone	Dolomite
Counts MCNP	2068	2422	1706
Counts GEANT4	2004	2149	1625
GEANT4/MCNP	3,1%	11,3%	4,7%

IV. NEUTRON-NEUTRON MEASUREMENT

The neutron-neutron measurement consists in counting backscattered neutrons originating from an Americium-Beryllium (AmBe) radioactive spectral neutron source with an ^{3}He detector [13]. It is mainly sensitive to the hydrogen content of the rock because of the large elastic scattering and radiative capture cross sections of hydrogen nuclei. The transform of hydrogen content into porosity is strongly linked to the physicochemical nature of fluids or gas (fresh water, brine, hydrocarbon liquid or gas, CO_2). The other chemical elements like C, O, or Ca have a significant impact on neutron slowing down and capture only at very low porosity (i.e. a rock formation porosity close to 0 pu), inducing variations in the total number of counts according to rock density and mineralogy [6][7].

Neutron-neutron measurements are simulated in MCNP with the F4 tally, which is the average flux of neutron particles passing through a cell convoluted with the N°103 nuclear reaction cross section corresponding to the ^{3}He(n,p)^{3}H absorption reaction that creates the signal in the ^{3}He counters. The result obtained with MCNP is multiplied by the detector volume and the number of particles simulated in GEANT4 to allow a quantitative comparison. The neutron-neutron measurement is simulated in GEANT4 by scoring the number of ^{3}He(n,p)^{3}H reactions in *"SteppingAction.cc"*.

Figure 6 shows six neutron-neutron measurement spectra (neutron kinetic energy before inelastic nuclear interaction with helium 3 nuclei) in the spherical model for six pure chemical elements. Results obtained with MCNP are derived from statistical calculations and the abovementioned normalization, hence showing some counts inferior to one, whereas results obtained with GEANT4 are particle tracking calculations and show only counts superior to one.

Table V gives the total number of counts (number of ^{3}He(n,p)^{3}H reactions) for pure elements and shows a maximum discrepancy of 16.6% for hydrogen without molecular effects. Implementation of molecular effects allows reducing deviations of the total number of counts between MCNP and GEANT4 from 9.8% to 0.1% for carbon, and from 16.6% to 2.9% for hydrogen. Therefore, molecular effects on carbon and hydrogen will be implanted in next simulations. The maximum discrepancy is obtained with oxygen (7.0%) and magnesium (7.1%).

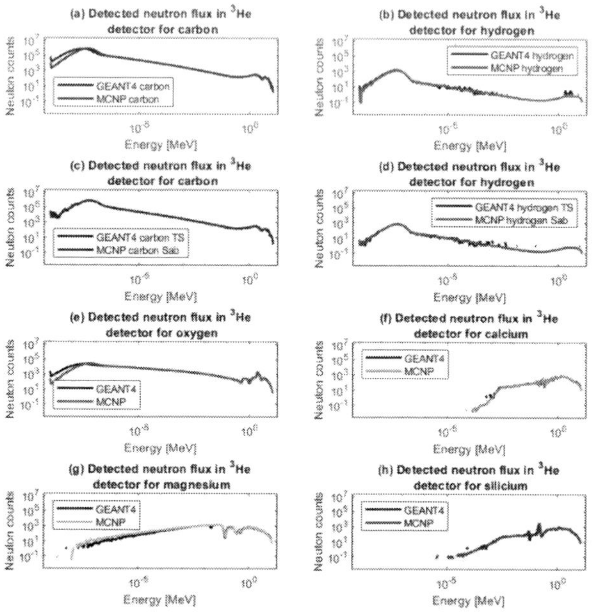

Fig. 6. Simulated neutron-neutron measurement spectra with a spherical model, an AmBe neutron source, a ^{3}He detector and six pure chemical elements. Five materials have a density of 2.71 g/cm3 (C, O, Ca, Mg, Si), and due to its large number of nuclei per volume unit, one material has a density of 0.5 g/cm3 (H). Graphics (c) and (d) show spectra with molecular effects on carbon and on hydrogen "Sab" for MCNP S(α,β) (grph.10t; lwtr.10t) and "TS" for GEANT4 Thermal Scattering (TS_C_of_Graphite; TS_H_of_Water).

TABLE V. TOTAL COUNTS OF NEUTRON-NEUTRON MEASUREMENTS IN A SPHERICAL MODEL FOR SIX PURE ELEMENTS (INCLUDING MOLECULAR EFFECTS FOR CARBON AND HYDROGEN).

Neutron-Neutron porosity measurements NEAR detector				
	Carbon with molecular effects	Carbon	Hydrogen with molecular effects	Hydrogen
Counts MCNP	17628283	18235366	14868	24662
Counts GEANT4	17644604	20019134	14430	28766
GEANT4/MCNP	0,1%	9,8%	2,9%	16,6%

Neutron-Neutron porosity measurements NEAR detector				
	Silicium	Oxygen	Calcium	Magnesium
Counts MCNP	26899	1178892	17635	81882
Counts GEANT4	27472	1261444	18029	76080
GEANT4/MCNP	2,1%	7,0%	2,2%	7,1%

Figure 7 shows six neutron-neutron porosity measurement spectra of a generic logging tool with two detectors (NEAR and FAR) in a 200 mm wellbore and for three rock compositions.

Table VI gives the total number of counts for compound materials and show a maximum discrepancy of 3.8% for the NEAR detector and 24.9% for the FAR detector.

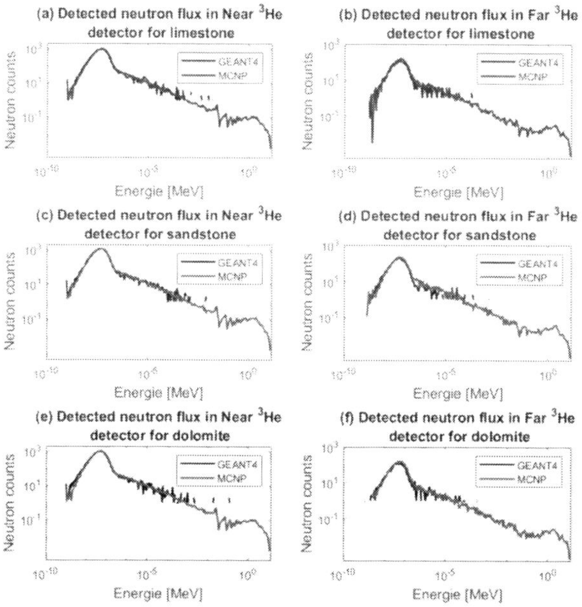

Fig. 7. Simulated neutron-neutron measurement spectra with a neutron-neutron porosity generic logging tool model, an AmBe neutron source, two ^{3}He detectors and three different rocks at 20 pu (porosity unit): a 2.368 g/cm3 limestone (a, b), a 2.3232 g/cm3 sandstone (c, d), and a 2.496 g/cm3 dolomite (e, f).

TABLE VI. TABLE OF TOTAL COUNTS INTEGRATED ON NEUTRON-NEUTRON MEASUREMENTS SPECTRA OF A POROSITY GENERIC LOGGING TOOL WITH TWO DETECTORS IN THREE ROCK COMPOSITIONS AT 20 PU.

Neutron-Neutron porosity measurements NEAR detector			
	Limestone	Sandstone	Dolomite
Counts MCNP	19994	23399	21420
Counts GEANT4	19227	23667	21164
GEANT4/MCNP	3,84%	1,15%	1,20%

Neutron-Neutron porosity measurements FAR detector			
	Limestone	Sandstone	Dolomite
Counts MCNP	3078	4521	3167
Counts GEANT4	2311	3922	2518
GEANT4/MCNP	24,93%	13,25%	20,49%

Using the particle tracking of GEANT4, one can identify nuclei that interacted with neutrons on their path from the Am-Be source to the detectors. In a wellbore model with a 20 pu sandstone lithology, 77% of the counts recorded in the NEAR detector involves interactions on hydrogen nuclei, and 80% for the FAR detector, the other interactions occurring on oxygen and silicon nuclei. In a well model with a 20 pu limestone or dolomite lithology, hydrogen account respectively for 69% and 66% of NEAR detector interactions, 72% and 70% of FAR detector interactions, the other interactions occurring on oxygen, carbon, calcium and magnesium nuclei. The large discrepancies between MCNP and GEANT4 in limestone and dolomite for the FAR detector could be explained by differences in neutron interaction cross sections of elements other than hydrogen. Indeed, for sandstone, in which interactions on hydrogen represent a larger fraction, the difference is smaller. The overall agreement between GEANT4 and MCNP for neutron-neutron physics is nevertheless still acceptable regarding borehole simulation objectives, with discrepancies under 25%.

V. NEUTRON-GAMMA MEASUREMENT

The neutron-gamma measurement consists in measuring the induced gamma spectrum originating from a 14 MeV mono-energetic pulsed neutron generator, with a NaI(Tl) detector [13]. In Geosciences, gamma spectra analysis simply consists in calculating the ratio of counts in two integration windows, the first one from ~ 3.2 MeV up to ~ 4.7 MeV focused on carbon gamma ray at 4.439 MeV (including its escape peaks at 3.928 and 3.417 MeV), the second one from ~ 4.8 MeV up to ~ 7.4 MeV focused on oxygen gamma ray at 6.129 MeV (and escape peaks 5619 and 5108 MeV). Neutron-gamma measurement, also called C/O measurement because it is based on the count ratio between carbon and oxygen windows, is particularly sensitive to quantity of water compared to quantity of oil, and thus to the hydrocarbon rock saturation.

Figure 8 shows four neutron-gamma measurement spectra in a spherical model for a pure carbon material. MCNP flux spectra (red lines) are identical and MCNP deposited energy spectra (blue lines) are also identical. GEANT4 flux and deposited energy spectra are computed with and without a pass band energy filter described below.

In the very particular geometry with a material sphere surrounded by a 4π spherical detector, multiple induced gamma rays generated in the same neutron history by successive inelastic scattering and radiative capture reactions can sum up in the detected gamma spectrum, introducing a single sum peak with an energy corresponding to the sum of the different gamma-ray energies (cf. figure 8, panels c and d, MCNP spectra). A two-step simulation is therefore mandatory to carry out a more realistic neutron-gamma C/O measurement, instead of the above single-step F8 simulation. In the first step, the MCNP F1 current tally scores gamma rays crossing the external surface of the carbon material sphere, the NaI detector cell being here filled with vacuum to avoid photon backscattering. In the second step, using the F1 calculation as a gamma spectral source at the entrance surface of the NaI detector, the F8 pulse height tally scores the energies deposited in the energy bins of the gamma spectrum.

In GEANT4, neutron inelastic scattering on carbon nuclei induces gamma rays over and bellow the main inelastic emission ray at 4.439 MeV that are not experimentally observed (cf. figure 8 a, c). Removing data of emission rays contained in *"G4NDL4.6\Inelastic\Gammas\z6.a12"*, except the ray at 4.439 MeV, does not allow to solve the problem. A pass band energy filter is added in "SteppingAction.cc" for neutron inelastic scattering on carbon nuclei to keep only gamma rays generated between 4 MeV and 5 MeV (cf. figure 8 b, d). Adding such filter allow GEANT4 results to be closer to MCNP [8][12][21]. Moreover, in GEANT4, within the same history as in MCNP, a neutron can create several inelastic gamma rays when scattering on different nuclei. The particle tracking simulation in a spherical geometry yields peaks on the gamma spectrum which are the sum of the energy of the gamma ray generated by capture process and energies of multiple gamma rays generated by inelastic process (cf. figure 8, panel d, GEANT4 spectrum). It does not allow to simulate a realistic neutron-gamma C/O measurement and requires a two-step simulation. First, kinetic energies of gamma rays induced by neutrons entering in a vacuum filled detector are collected in "SteppingAction.cc" to build a spectral gamma source. Then, using the previous current spectrum as a source impinging the entrance surface of a NaI filled detector, deposited energies per energy bins are scored in "SteppingAction.cc".

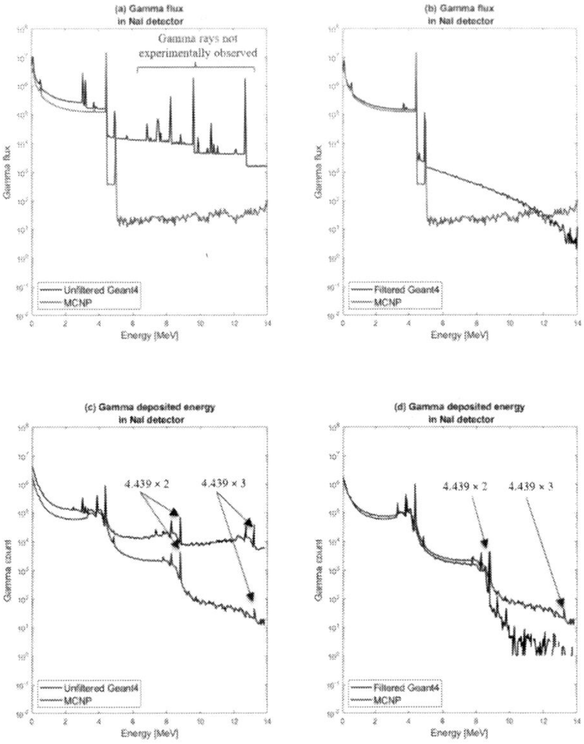

Fig. 8. Simulated neutron-gamma measurement spectra with a spherical model, a 14 MeV pulsed neutron source, an NaI detector and one pure carbon material. Spectra (a) and (b) show flux spectra of gamma rays entering in a NaI detector in cm-2 per 96×10^6 source particles. Spectra (c) and (d) show deposited energy spectra of gamma rays interacting in the NaI crystal.

Figure 9 shows three neutron-gamma measurement spectra in a spherical model for a pure carbon material. Two spectra (black and blue lines) are computed with a two times simulation. One spectrum (brown line) comes from experimental data of a 14 MeV pulsed neutrons source on a well-known sample of graphite [5][19]20].

The associated particle technique used to acquire the experimental spectrum allows recording only the 4.439 MeV inelastic scattering gamma ray of carbon (and its escape peaks also visible on the gamma spectrum). The 4.945 MeV due to radiative capture is not detected (cf. figure 9, brown line). The two-step simulations in MCNP and GEANT4 allows more realistic simulations of the gamma rays induced by fast and thermal neutrons on carbon nuclei.

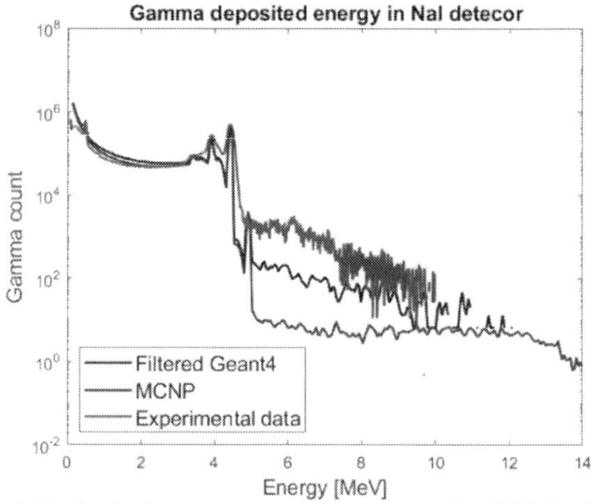

Fig. 9. Simulated and measured neutron-gamma spectra with a 14 MeV pulsed neutron source, a NaI detector and one pure carbon material. The calculated spectra are computed with two steps simulation in MCNP (blue line) and in GEANT4 (black line). The experimental spectrum (brown line) has been normalized to the amplitude of the main inelastic scattering carbon peak at 4.439 MeV of the MCNP spectrum to ease this qualitative comparison.

Figure 10 shows six neutron-gamma measurement spectra in a spherical model for six pure materials. No time decay analysis was implemented. Induced gamma rays by inelastic scattering and capture reactions are summed in the same spectrum.

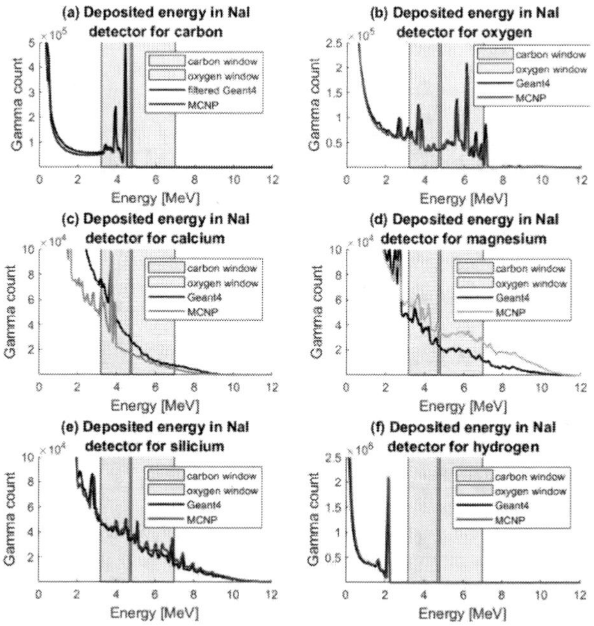

Fig. 10. Simulated neutron-gamma measurement spectra with a spherical model, a 14 MeV pulsed neutron source, an NaI detector and six pure chemical elements. Five materials at density 2.71 g/cm3 (carbon; oxygen; calcium; magnesium; silicon) and because of its large cross section, one material at density 0.5 g/cm3 (hydrogen) have been simulated. Graph (a) shows one MCNP spectrum (in blue) and two GEANT4 spectra, one without modification of the code (dotted black line) and one with a modified code to filter induced gamma rays (see text, black line).

Tables VII and VIII give total counts integrated in the oxygen and carbon windows and show maximum discrepancies of

37.4% with calcium in the carbon window and 42.6% with magnesium in the oxygen window.

MCNP and GEANT4 results for the hydrogen spectrum agree for energies below 2.2 MeV, but the number of counts in the C and O windows of interest is not relevant, similarly as the counts in the oxygen window of the carbon spectrum. Adding a filter on induced gamma creation process after inelastic interaction with carbon yield to decrease the discrepancies between MCNP and GEANT4 from 25.8% to 16.6% in the carbon window on the carbon spectrum. The main count deviations come from calcium and magnesium continuums in both windows which are not as easy to correct as for the carbon spectrum.

TAB. VII. TABLE OF OXYGEN WINDOW TOTAL COUNTS INTEGRATED ON NEUTRON-GAMMA MEASUREMENTS SPECTRA IN A SPHERICAL MODEL FOR SIX PURE ELEMENTS.

Neutron-Gamma C/O measurements NEAR detector in O window				
	Oxygen	Calcium	Magnesium	Silicium
Counts MCNP	2216955	426623	1365249	1221792
Counts GEANT4	2281493	593007	783742	1061198
GEANT4/MCNP	2,9%	39,0%	42,6%	13,1%

TAB. VIII. TABLE OF CARBON WINDOW TOTAL COUNTS INTEGRATED ON NEUTRON-GAMMA MEASUREMENTS SPECTRA IN A SPHERICAL MODEL FOR SIX PURE ELEMENTS.

Neutron-Gamma C/O measurements NEAR detector in C window			
	Carbon filtered	Carbon unfiltered	Oxygen
Counts MCNP	2319258	2319258	1432222
Counts GEANT4	2705439	2917505	1471031
GEANT4/MCNP	16,7%	25,8%	2,7%

Neutron-Gamma C/O measurements NEAR detector in C window			
	Calcium	Magnesium	Silicium
Counts MCNP	1001319	1357499	1217088
Counts GEANT4	1375929	1020289	1145298
GEANT4/MCNP	37,4%	24,8%	5,9%

Figure 11 shows six neutron-gamma C/O measurements spectra of a generic logging tool with two detectors (NEAR and FAR) in a well with 200 mm diameter and for three rock compositions.

Tables IX and X give the total counts integrated in the carbon and oxygen windows, showing maximum discrepancies of 64.8% for the NEAR detector and 110.3% for the FAR detector in the carbon window, 11.5% for the NEAR detector and 9% for the FAR detector in the oxygen window.

TABLE IX. TABLE OF CARBON WINDOW COUNTS INTEGRATED ON NEUTRON-GAMMA MEASUREMENTS SPECTRUMS OF A C/O GENERIC LOGGING TOOL WITH TWO DETECTORS IN THREE ROCK COMPOSITIONS AT 20 PU.

Neutron-Gamma C/O measurements NEAR detector in C window			
	Limestone	Sandstone	Dolomite
Counts MCNP	389	680	395
Counts GEANT4	641	581	597
GEANT4/MCNP	64,8%	14,7%	51,0%

Neutron-Gamma C/O measurements FAR detector in C window			
	Limestone	Sandstone	Dolomite
Counts MCNP	226	439	213
Counts GEANT4	476	411	382
GEANT4/MCNP	110,3%	6,4%	79,1%

TABLE X. TABLE OF OXYGEN WINDOW COUNTS INTEGRATED ON NEUTRON-GAMMA MEASUREMENTS SPECTRUMS OF A C/O GENERIC LOGGING TOOL WITH TWO DETECTORS IN THREE ROCK COMPOSITIONS AT 20 PU.

Neutron-Gamma C/O measurements NEAR detector in O window			
	Limestone	Sandstone	Dolomite
Counts MCNP	448	605	461
Counts GEANT4	420	536	444
GEANT4/MCNP	6,2%	11,5%	3,7%

Neutron-Gamma C/O measurements FAR detector in O window			
	Limestone	Sandstone	Dolomite
Counts MCNP	239	352	242
Counts GEANT4	260	322	236
GEANT4/MCNP	9,0%	8,5%	2,6%

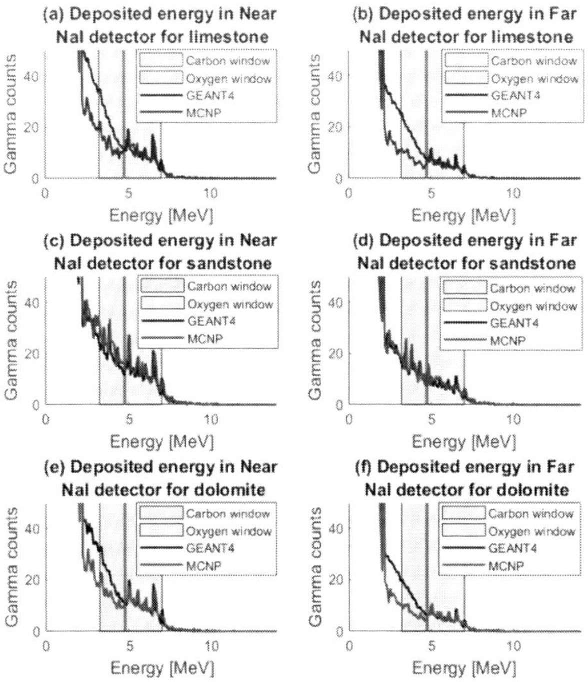

Fig. 11. Simulated neutron-gamma measurement spectra with a neutron-gamma C/O generic logging tool model, a 14 MeV pulsed neutron source, two NaI detectors and three different rocks at 20 pu (porosity unit): a 2.368 g/cm3 limestone (graphics a and b), a 2.323 g/cm3 sandstone (graphics c and d), and a 2.496 g/cm3 dolomite (graphics e and f).

Using the particle tracking of GEANT4, the nuclei that gave birth to the detected gamma rays in the C and O windows of interest can be identified. Induced gamma rays come from interactions between neutrons and calcium or oxygen nuclei, for more than 79% in a 20 pu limestone (cf. figure 11 a, b) and 62% in a 20 pu dolomite (cf. figure 11 e, f). On the other hand, 85% of induced gamma rays come from interactions between neutrons and silicon or oxygen nuclei in a 20 pu sandstone (cf. figure 11 c, d). Calcium appears to be the major element responsible for induced gamma rays in limestone and, to a lesser extent, in dolomite. The larger deviations in the carbon window for the limestone or dolomite lithology, compared to sandstone, is consistent with the observations made with the spherical model that calcium is the element showing the largest discrepancies between MCNP and GEANT4 (cf. figure 10 c and table VIII).

The overall agreement between GEANT4 and MCNP for neutron-gamma physics depends on the chemical composition and ranges of energies. It is excellent for a lithology like sandstone with discrepancies under 15%, but not acceptable for others such as limestone and dolomite in the carbon window.

VI. Conclusions

The comparison of GEANT4 and MCNP for simulating nuclear measurements shows a good agreement for gamma-gamma physics. Neutron-neutron physics also agrees quite satisfactorily between the two codes when molecular effects and proper neutron models are implemented. However, misfits appear for the neutron-gamma physics for some chemical compositions and energy ranges. Some local corrections in GEANT4, such as energy pass band filters on induced gamma rays, can improve the final spectra (e.g. in the case of carbon). The ultimate confirmation would come from experimental measurements that, at the same time, would also validate the concept of sensitivity functions to be used for multiphysics inversion applied to "Cased Hole" wellbore interpretation [14][15][16][17]. Ultimately, simulation results will be compared to experimental measurements in a calibration facility with real logging tools in various wellbore configurations (wells; casing; cement; formation), in view to definitely validate the concept of tool sensitivity functions [3].

References

[1] C.J. Werner, et al., "MCNP6.2 Release Notes", Los Alamos National Laboratory, report LA-UR-18-20808 (2018).

[2] C.J. Werner (editor), «MCNP Users Manual - Code Version 6.2», LA-UR-17-29981 (2017).

[3] Chuilon, P., et al, 2019, From Houston API Calibration Pits… to Artigueloutan Logging Metrological Facility, Society of Petrophysicists and Well-Log Analysts, SPWLA 60th Annual Logging Symposium, June 15–19,SPWLA-2019-GGGGG, DOI: https://doi.org/10.30632/T60ALS-2019_GGGGG.

[4] EDGE—SpaceSuite, Apr. 2019, [online] Available: https://www.space-suite.com/edge/.

[5] El Kanawati, Wassila & Perot, Bertrand & Carasco, C. & Eleon, C. & Valkovic, Vladivoj & Sudac, Davorin & Obhodas, Jasmina & Sannie, G. (2011). Acquisition of prompt gamma-ray spectra induced by 14 MeV neutrons and comparison with Monte Carlo simulations. Applied radiation and isotopes : including data, instrumentation and methods for use in agriculture, industry and medicine. 69. 732-43. 10.1016/j.apradiso.2011.01.010.

[6] Ellis, D. V., Case, C. R. and Chiaramonte, J. M. (2003) 'Tutorial - Porosity from Neutron Logs I - Measurement', Petrophysics - The SPWLA Journal of Formation Evaluation and Reservoir Description, 44(06).

[7] Ellis, D. V., Case, C. R. and Chiaramonte, J. M. (2004) 'Tutorial - Porosity from Neutron Logs II - Interpretation', Petrophysics - The SPWLA Journal of Formation Evaluation and Reservoir Description, 45(01).

[8] FIRESTONE, R. B., & SHIRLEY, V. S. (1996). Table of isotopes. New York, Wiley.

[9] GangLi, GhaoutiBentoumi, ZinTun, LiqianLi, and BhaskarSur. APPLICATION OF GEANT4 TO THE DATA ANALYSIS OF THERMAL NEUTRON SCATTERING EXPERIMENTS. CNL Nuclear Review. 7(1): 11-7. https://doi.org/10.12943/CNR.2017.00002.

[10] GEANT4 - A Simulation Toolkit, S. Agostinelli et al., Nucl. Instrum. Meth. A 506 (2003) 250-303.

[11] Goodyear, G., Sood, A., Andrews, M., Solomon, C.J., Luycx, M., and Torres-Verdín, C., 2018, What's New in Borehole Nuclear Modeling? (A Lot!), Paper AAAA, Transactions, SPWLA 59th Annual Logging Symposium, 2–6 June, London, UK.

[12] Kavetskiy, Aleksandr & Yakubova, Galina & Prior, Stephen & Torbert, H.. (2017). Neutron-Stimulated Gamma Ray Analysis of Soil. 10.5772/68014.

[13] Knoll, G. F. (2010).

[14] Luycx, M. et al. (2020) 'Simulation of Borehole Nuclear Measurements: A Practical Tutorial Guide for Implementation of Monte Carlo Methods and Approximations Based on Flux Sensitivity Functions', Petrophysics - The SPWLA Journal of Formation Evaluation and Reservoir Description, 61(01), pp. 4–36. doi: 10.30632/PJV61N1-2020T1.

[15] Mendoza, A., C. Torres-Verdín, and W. E. Preeg, 2007, Rapid simulation of borehole nuclear measurements based on spatial flux-scattering functions: Presented at SPWLA 48th Annual Logging Symposium, O.

[16] Mendoza, A., C. Torres-Verdín, and W. E. Preeg, 2010a, Linear iterative refinement method for the rapid simulation of borehole nuclear measurements: Part 1 — Vertical wells: Geophysics, 75, no. 1, E9–E29, doi: 10 .1190/1.3267877.

[17] Mendoza, A., C. Torres-Verdín, and W. E. Preeg, 2010b, Linear iterative refinement method for the rapid simulation of borehole nuclear measurements: Part 2 — High-angle and horizontal wells: Geophysics, 75, no. 2, E79–E90, doi: 10.1190/1.3335953.

[18] Mendoza, E., Cano-Ott, D., Update of the Evaluated Neutron Cross Section Libraries for the GEANT4 Code, IAEA technical report INDC(NDS)-0758, Vienna, 2018.

[19] Perot, Bertrand & El Kanawati, Wassila & Carasco, C. & Eleon, C. & Valkovic, Vladivoj & Sudac, Davorin & Obhodas, Jasmina & Sannie, G. (2011). Quantitative comparison between experimental and simulated gamma-ray spectra induced by 14 MeV tagged neutrons. Applied radiation and isotopes : including data, instrumentation and methods for use in agriculture, industry and medicine. 70. 1186-92. 10.1016/j.apradiso.2011.07.005.

[20] Perot, Bertand & Carasco, C. & Bernard, S & Mariani, A & Szabo, J.-L & Sannie, G Valkovic, Vladivoj & Sudac, Davorin & Viesti, G & Lunardon, Matteo & Botosso, C & Nebbia, G. & Pesente, Silvia & Moretto, Sandra & Zenoni, Aldo & Donzella, Antonietta & Moszynski, Marek & Gierlik, Michał & Klamra, W & Salvato, M. (2008). Measurement of 14 MeV neutron-induced prompt gamma-ray spectra from 15 elements found in cargo containers. Applied radiation and isotopes : including data, instrumentation and methods for use in agriculture, industry and medicine. 66. 421-34. 10.1016/j.apradiso.2007.11.011.

[21] Simakov, S P, Pavlik, A, Vonach, H, and Hlavac, S. Status of experimental and evaluated discrete {gamma}-ray production at E{sub n}=14.5 MeV. Final report of Research Contract 7809/RB, performed under the CRP on measurement, calculation and evaluation of photon production data. IAEA: N. p., 1998. Web.

[22] Trouche, *A.*, et al., "EDGE: new GDML CAD tool for GEANT4 based analysis and interoperability bridge for high-energy particle modelling tools", ENSAR2 workshop: GEANT4 in nuclear physics – 24-26 April 2019 – Madrid (Spain).

[23] Velker, Nikolay & Banzarov, Bair & Inanc, Feyzi & Simonov, Nikolai. (2012). GEANT4 for Solving Nuclear Geophysics Problems (Russian). 10.2118/162008-RU.

Detector Development for a Double-Differential Cross Section Experiment with the Emission of Light Charged Particles from High Energy Neutrons

M. Dietz[1,*], R. Beyer[2], A. Junghans[2], R. Nolte[1], E. Pirovano[1], Y. Romanets[3], P. Vaz[3], and the n_TOF Collaboration[4]

[1]Physikalisch-Technische Bundesanstalt (PTB), Bundesallee 100, 38116 Braunschweig, Germany
[2]Helmholtz-Zentrum Dresden-Rossendorf (HZDR), Bautzner Landstraße 400, 01328 Dresden, Germany
[3]Centro de Ciências e Technologias Nucleares (C2TN), CTN - Estrada Nacional 10; 2965-066 Bobadela LRS, Portugal
[4]European Organization for Nuclear Research (CERN), Switzerland

(*) mirco.dietz@ptb.de

Abstract— **Double-differential cross section (DDX) data on the neutron-induced emission of light charged particles are required for assessing the risk of secondary tumors in particle radiation therapy. There are only very few DDX data available for discrete neutron energies close to and above 100 MeV for carbon. A measurement of DDX on carbon is planned at continuous neutron energies from 20 MeV to 200 MeV with particle detector telescopes at n_TOF (CERN). Several detector development criteria and challenges are reported such as coincidence timing and electromagnetic oscillations for high neutron energy events with particle separation.**

Keywords — **n_TOF, Double-Differential Cross Section, Carbon, Time-of-Flight, High Neutron Energy, Particle Detector Telescope, Neutron Dosimetry.**

I. Introduction

HIGH-ENERGY neutrons produced as secondary radiation by cosmic rays or in high-energy accelerators are of importance for radiation protection concerns, radiation treatment planning, and can cause damage in aerospace instrumentation. In particle radiation therapy for cancer, secondary neutrons with energies up to about 200 MeV (proton beams) [1] or 400 MeV (carbon beams) [2] are produced by beam interaction in the treatment head and in the target volume. The risk of secondary tumors induced by these neutrons has received increased interest recently [3,4], and is important in particular for young patients. The risk assessment requires to calculate the absorbed dose outside the target volume from the neutron fluence. The required kerma factors are calculated from double-differential cross sections (DDX) for the emission of light charged particles, e.g., protons, deuterons, tritons and helium particles.

Experimental DDX data for tissue constituents are still rather scarce for neutron energies above 20 MeV and mostly measured for discrete neutron energies at mono-energetic neutron beam facilities. There are only very few DDX data available for discrete neutron energies close to and above 100 MeV for carbon [5,6,7] and oxygen [8,9]. Discrepancies between experimental data on carbon and data calculated using nuclear models can be observed, e.g., around 100 MeV in most reaction channels, especially for (n,dx) and (n,αx).

On the other hand, the calculation of DDX data using statistical, quantum mechanical or intranuclear cascade (INC) models is difficult for low-mass nuclei and composite ejectiles [10]. Due to the large neutron energy range and the relatively small DDX values (usually less than 1 mb/(sr MeV) for $Z = 1$ ejectiles even at forward angles), evaluated data bases cannot be constructed from experimental data alone. Instead, a variety of nuclear model calculations [11] must be used.

Therefore, benchmarking and improvement of codes is important in the energy region between 100 MeV and 200 MeV, where not only experimental data are almost completely missing, but also the statistical model becomes insufficient, and the INC model just starts to become applicable. This requires experimental data at selected forward and backward angles which test the description of the intranuclear cascade and de-excitation phases of the interaction. Data for the neutron-induced emission of helium ions would be of particular importance, as complex ejectiles are not generically described in the INC model [10].

In summary, this motivates a feasibility study [12] of an experiment of double-differential cross section measurements of carbon with the emission of light-charged particles at high neutron energies between 100 MeV and 200 MeV, which is largely unexplored. In contrast of previous work performed at quasi-monoenergetic neutron sources, a measurement at a pulsed neutron source with broad energy distribution such as n_TOF could provide a continuous coverage of the relevant neutron energies. With that, new experimental challenges arise and will be discussed in this article.

II. Experimental setup and tests

In this section, an overview of the experimental setup will be given which was used for the detector development phase of the study. Moreover, the neutron facility n_TOF will be presented. Firstly, the detection principle and experimental demands are outlined.

To measure DDX, particle detector telescopes under different angles are needed to identify ejectile species and to determine energy and angle information on the reaction of interest. The telescopes consist of three stages designed to enable particle identification according to their differential energy loss: Two transmission detectors (also referred as "ΔE detectors") and one stop detector (or "E detector"). As a range of light-charged particles from protons to alpha particles is intended to be measured, the first ΔE detector needs to be thin enough that alpha particles of sufficiently low energy can reach the rear detector. The third detector has to be thick enough or made of high Z material to stop the fastest protons. Additionally, a vacuum chamber should be used to minimize the energy loss between ion production and detection, especially in case of the alpha particles. The principle of operation consists in requiring coincident signals for at least two of the three detectors, to identify the particles emitted by the sample in a given direction. The ΔE-E technique is then used to discriminate protons, deuterons, tritons, and alpha particles.

The range of these different ejectiles presents some major challenges for this experiment. Firstly, the incident neutron energy range to be covered extends from several MeV up to targeted 200 MeV or even higher, which requires fast timing of the detectors using the time-of-flight technique. At the same time, a large energy range of emitted particles has to be detected simultaneously, which challenges the particle identification capabilities. Regarding the alpha particles, low detection threshold detectors are needed to avoid missing information of the pulse height distribution. Furthermore, a linear detector response is needed for calibration purposes and for the detection of high-energy protons, where a characterization of the E detector is desired.

The experimental setup at n_TOF consists of two detector telescopes. The transmission detectors are silicon diodes, quadratic in size with an active area of 900 mm², and their thickness ranging from 50 µm to 1000 µm. EJ-204 plastic scintillators of 8 cm diameter and 10 cm or 15 cm length, and a cylindrical $CeBr_3$ scintillator of 3" diameter and 2" length are used as stop detectors. The scintillators are installed outside the chamber, behind thin windows, to allow easy readout of the scintillation detectors using photomultiplier tubes (PMTs). The light-weight aluminium chamber has a diameter of 40 cm and a height of 15 cm, and it is equipped with eight 47-mm diameter flanges with an angular separation of 45 degrees. Four flanges were equipped with windows, of 50 µm Kapton and additional 200 µm or 100 µm aluminium foil respectively, for the ingoing and outgoing neutron beam, and for two full ΔE_1-ΔE_2-E telescope arms at 45 degrees relative to the beam direction. The chamber was evacuated by a SCROLLVAC pump which produced a residual gas pressure of about 1 Pa to 10 Pa.

A key interest was the readout of the silicon diodes. The resolution of different charge-integrating preamplifiers modules was tested with alpha particles of 5.2 MeV from a ^{239}Pu/^{241}Am/^{244}Cm source at PTB and at n_TOF. Modules like Cividec Cx-L series, Canberra 2004 and Cremat CR110 were used, where the two latter ones were further processed with a fast-filter amplifier Ortec 579. At PTB, the Cividec Cx-L series offered good resolution results for the thicker diodes of about 1% with a fixed signal length in the order of a few hundreds of ns, applicable to time-of-flight experiments. Despite that, the Ortec fast-filter amplification has more flexibility in shaping time, quickly adaptable in respect to the count rate of the studied reaction, but it is worse in terms of resolution by a factor of two to three. Generally, the signals were recorded with the DAQ at n_TOF, which consists of 14-bit flash-ADCs from TELEDYNE SPDevices [13], allowing to record the full pulse shape of each detector signal.

In principle, the neutron time-of-flight facility n_TOF [14] at CERN provides a broad spectrum from thermal up to GeV neutrons, which corresponds to an event time window of about 100 ms for experiments at Experimental Area 1 (EAR1) with a flight path of 185 m, ensuring an excellent neutron energy resolution. The neutrons are produced via spallation reactions of 20 GeV/c proton bunches impinging on a massive lead target. A repetition frequency less than 1 Hz and a high instantaneous rate of about 10^{15} neutrons per bunch lead to a comparable low average fluence of the order of $7 \cdot 10^4$ cm^{-2} s^{-1} at EAR1, which often defines the amount of beam time required to achieve a given statistical uncertainty.

During the prototype test, data were recorded over about 2.5 days with the detector setup shown in Fig. 1. The sample size inside its holder was 47 mm x 47 mm and a graphite sample of 1.5 mm thickness was used. As shown in Fig. 1. , a polyethylene sample of the same dimension can be exploited as reference sample using the well-defined (n,p) recoil reaction as standard.

A final future experiment will need at least ten times more data to achieve similar counting uncertainties as obtained at monoenergetic sources in earlier work [6]. Nevertheless, many experimental challenges were already observed in this time and their solutions are discussed in the following.

III. Discussion of Experimental Challenges and Results

The main goal of the beam time dedicated to detector development was testing silicon diodes of different thicknesses coupled with different models of preamplifiers, to verify if the energy resolution was sufficient to separate hydrogen isotopes and helium isotopes in the ΔE-E two-dimensional histograms. The interesting part of the spectrum in the MeV range arrives at n_TOF EAR1 in the first ten microseconds after an intense γ-flash which requires fast and stable electronics. This initial γ-flash was expected to be the main challenge, as in earlier measurements it was found to partially exhaust the dynamic range of the preamplifier, consequently leading to a loss of valid events due to early saturation.

Fig. 1. The picture shows the detectors and a polyethylene sample mounted inside the vacuum chamber used for the detector test, installed in the EAR1 of n_TOF. The vertical axis of the picture (from the bottom to the top) corresponds to the neutron beam direction. The silicon diodes on the left side of the sample are connected to Cividec Cx-L preamplifiers, which are installed outside the chamber (not captured in this photo). On the right side, the first diode (ΔE_1) is connected to a Cremat CR110 preamplifier, which is mounted on the printed-circuit board frame, directly under the diode. The second diode (ΔE_2) is connected to a Canberra 2004 preamplifier, installed outside the chamber. The two detectors behind the vacuum chamber are a 150 mm long EJ204 plastic scintillator (on the left) and a 76.2 mm long $CeBr_3$ scintillator (on the right).

A. *Experimental Challenges*

With the prototype setup, the γ-flash was not the problem in itself, however it did significantly disturb the data acquisition in the first ten microseconds, which corresponds to neutron energies greater than 1.6 MeV. This was found to be associated to electromagnetic noise that produced oscillations in the preamplifier readout signal with amplitudes comparable to the pulse height of light-charged particle-induced events. The origin of these electromagnetic interferences in the radiofrequency (RF) range, and their relationship to the γ-flash, has not been fully understood but it is also currently under investigation at CERN [15].

Several observations were made from the test of different signal amplification configurations. First of all, excellent shielding against electromagnetic interferences is needed early on in the signal chain to avoid distortions of the baseline. Consequently, the scattering chamber, including the Kapton windows, had to be constructed RF-tight using additional Al foils, and the preamplifier modules also needed additional shielding including all cable connections in between, which were already kept as short as possible. Shielded Cividec Cx-L modules performed well outside the chamber, even though small oscillations in the baseline still occurred but were repetitive enough to handle them with a customized baseline subtraction method during the data analysis. Special capacitance-matched preamplifier modules for the thinnest diodes like Ortec 142C or Mesytec MPR-1 were tested and performed worse than the Cividec Cx-L and Cremat CR110 modules. The advantage of the Cremat module was that it can be used inside the vacuum chamber resulting in the best performance with fully suppressed oscillations in the baseline within their random noise level.

B. *Status of the Development*

The coincidence timing was found to be dependent on the electronic amplification chain and the time difference between ΔE and E detectors was not trivial, as the timing features of the diodes and the scintillator differed significantly. Therefore, the coincidence condition was set rather conservative considering some offset. In a previous experiment [16] using only plastic scintillators for all telescope stages, the identification of coincident events was less difficult because of the higher signal-to-noise ratio.

Two configurations between Cividec Cx-L and Cremat CR110 plus Ortec 579 amplifier on the thin silicon diode are compared in the following as the thinnest diodes were the most sensitive detectors to the challenging experimental conditions at a spallation neutron source.

ΔE_1-ΔE_2 data were produced from a carbon sample using a thin (50 μm) and a thick diode (500 μm or 1000 μm) with the amplification chain outlined in Fig. 1. Due to the short beam time, the data was only analyzed for events covering the full neutron energy range from 5 MeV to 200 MeV. Events with the upper energy limit of 200 MeV were resolved in time-of-flight without distortion from the γ-flash. The data using Cividec Cx-L integrated shaping amplifiers, which had the most promising resolution results at PTB, needed a baseline subtraction procedure at n_TOF which deteriorated the resolution. The results obtained with the Cremat preamplifier inside the vacuum chamber were very promising within the test beam time. Both configurations in this prototype setup delivered good separation between hydrogen and helium ions. Considering the financial aspect and that no special baseline treatment is necessary, the Cremat solution seems favorable for future measurement, where more events will be studied.

C. *Future Work*

Initial simulation work was performed to study the energy deposition of ejectiles in the detector telescopes. The current resolution is not sufficient to resolve protons, deuterons and tritons, but only hydrogen and helium. Therefore, future tests with different improved Cremat (pre)amplifier solutions and some segmented silicon diodes to reduce their capacitance / noise level are contemplated. Additionally, the effect of the energy loss in the graphite sample on the reconstruction of the energy distribution of the ejectiles will be studied experimentally and by performing Monte Carlo simulations.

IV. Conclusions and Outlook

Detector development of particle telescopes for double-differential cross-section measurement of carbon with the emission of light-charged particles was investigated with shielded preamplifiers and preamplifiers inside the vacuum chamber. Excellent shielding against electromagnetic interferences is clearly needed at a neutron spallation source like n_TOF and is a key experience from the experiment with the prototype setup.

For two amplification chains, particle identification of hydrogen and helium ions seems possible from the test data, but more work on the resolution of the different isotopes is required. The planning of a future experiment is ongoing to gain sufficient statistics with a well-adapted setup. Finally, this is a new kind of measurement at n_TOF and offers new possibilities of DDX experiments of tissue elements other than carbon with continuous energy coverage in a largely unexplored energy range.

ACKNOWLEDGMENT

We want to acknowledge the workshop and electrical engineers at the PTB department "Neutron Radiation" for their reliable contributions to this prototype setup.

This project has received funding from the European Union's Horizon 2020 research and innovation programme, Euratom research and training programme 2014-2018 under grant agreement No 847552 (SANDA).

REFERENCES

[1] J. Farah *et al.*, Med. Phys. 42 (2015) 2572-2584
[2] D. Satoh *et al.*, Nucl. Instr. in Phys. Res. A 644 (2011) 59-67
[3] X. G. Xu *et al.*, Phys. Med. Biol. 53 (2008) R191-R241,
[4] U. Schneider, R. Hälg, Front. Oncol. 5 (2015) 235
[5] N. Olsson *et al.*, Nucl. Phys. A 559 (1993) 368-400,
[6] U. Tippawan *et al.* Phys. Rev. C 79 (2009) 064611,
[7] S. Hirayama *et al.*, Progr. Nucl. Sci. and Techn. 1 (2011) 69
[8] U. Tippawan *et al.*, Phys. Rev. C 73 (2006) 034611,
[9] Y. Watanabe *et al.*, Progr. Nucl. Sci. and Techn. 4 (2014) 569-573
[10] J.-C. David, Eur. Phys. J. A 51 (2015) 68
[11] International Commission on Radiation Units and Measurements, *Nuclear Data for Neutron and Proton Radiotherapy and for Radiation Protection*, ICRU Report 63, Bethesda, Maryland, USA (2000)
[12] R. Beyer *et al.*, Letter of Intent INTC-I-221 to the ISOLDE and Neutron Time-of-Flight Committee, September 2020
[13] A. Masi et al. The CERN n TOF facility data acquisition system. JACoW Publishing - ICALEPCS2017 Proceedings, page THPHA195, 2018. URL: https://cds.cern.ch/record/2305649.
[14] C. Guerrero et al., Eur. Phys. J. A 49 (2013) 27
[15] M. Dietz, R. F. Ximenes, *Possible EM interferences at the n_TOF EAR1*, 62nd Impedance Working Group meeting (August 2022), https://indico.cern.ch/event/1190281/
[16] E. Pirovano et al., "A detector system for 'absolute' measurements of fission cross sections at n_TOF in the energy range below 200 MeV", to be published in JINST (2023)

The new readout system for the ALICE Zero Degree Calorimeter in LHC Run 3

Paolo De Remigis[1,*] , Pietro Cortese[1,2], Giorgio Cotto[1,3], Nora De Marco[1], Luca Lombardo[1,4], Carlo Puggioni[5], Mario Sitta[1,2] and Stefan Cristi Zugravel[1,4]
[1]INFN Torino, Italia
[2]Università del Piemonte Orientale, Italia
[3]Università di Torino, Italia
[4]Politecnico di Torino, Italia
[5]INFN Cagliari, Italia
(*) deremigi@infn.it

Abstract —**The Zero Degree Calorimeter (ZDC) was designed to provide the event geometry and luminosity measurements in heavy-ion operation. In order to exploit the potential offered by the LHC's increased luminosity in Run 3, the ZDC upgraded its readout system to acquire all collisions in self - triggered mode without dead time. The purpose of the upgrade was to enable the detector to cope with the increased event rate while preserving its time and charge resolution performance. The ZDC operating conditions in Run 3 Pb – Pb collisions are extremely challenging due to the presence of electromagnetic dissociation processes (EMD). For example when running in self-triggered mode the ZDC system will need to sustain a readout rate of ~2.5 MHz for the channels of the most exposed calorimeters compared to the foreseen hadronic rate of 50 kHz sustained by the other detectors. The previous electronics, based on Charge-to-digital converters (QDCs), with a fixed dead time of ~ 10 μµs, and on readout through VME bus, could not cope with such a high rate. Moreover, a crucial aspect of the ZDC operation in Run 3 is acquiring the events with a reduced bunch spacing of 50 ns (lower than the length of the signal of ~ 60 ns) in the presence of high signal dynamics (from a single neutron to ~ 60 neutrons). The new acquisition chain is based on a 12 bit digitizer with a sampling rate of about 1 GS/s, assembled on an FPGA Mezzanine Card. The signals produced by the ZDC channels are digitized, and samples are processed through an FPGA to extract information such as timing, baseline average estimation and luminosity. The architecture of the new readout system, the auto trigger strategy, the firmware organization and the ZDC performance during 2022 Pb–Pb collisions are presented.**

Keywords —**FPGA readout; ALICE ZDC.**

I. Introduction

The Large Hadron Collider (LHC) is a very big facility at CERN, where it is possible to create interactions between two particle beams, which can be either protons or ions. On that large machine four experimental apparatus are installed to study the fundamental elements of the matter, with each one devoted to a particular aspect. The A Large Ion Collider Experiment (ALICE) is a detector dedicated to heavy – ion physics, particularly designed to study a phase of matter called Quark-Gluon Plasma (QGP) [1] . The ALICE detector consists of different sub detectors, each designed to look for specific types of particle, and the Zero Degree Calorimeter (ZDC) is the one that detects the energy of the spectator nucleons [2] .

It is specifically composed of two identical sets of calorimeters located at 112.5 m from the interaction point. Each set consists of a Neutron (ZN) and a Proton (ZP) calorimeter. The ZN and ZP are quartz – fiber spaghetti calorimeters with silica optical fibres as active material embedded in a dense absorber. The principle of operation is based on the detection of Cherenkov light, produced by the charged particles of the shower in the fibres. The light produced by the hadronic showers is collected by photomultipliers. The main purpose is to provide measurements of the time of the collision, vertex position, centrality, event plane and to measure the luminosity in heavy - ion operations.

The previous readout system for the ZDC detector was based on the commercial Caen module in VME format, such as Charge to Digital Converter (QDC) and Time to Digital Converter (TDC). The goal was charge integration to analyze the signal amplitude, and timing evaluation to detect the arrival time of the nucleons on the calorimeters.

Due to a QDC conversion time of 10 µs the event rate was limited to approximately 100 kHz on average.

The new requirements [3] for the ZDC during the LHC Run 3 is to operate in self – triggered mode without dead time at an average event rate of 2.5 M event/s, preserving the amplitude and time resolution. The proposed readout architecture is FPGA - based to add the possibility of online data processing and managing a system that can be easily reconfigured in case of need. The FPGA is arranged on a VME carrier that hosts two FPGA Mezzanine Card (FMC) ports, see Fig. 1, allowing the possibility to add other hardware functions.

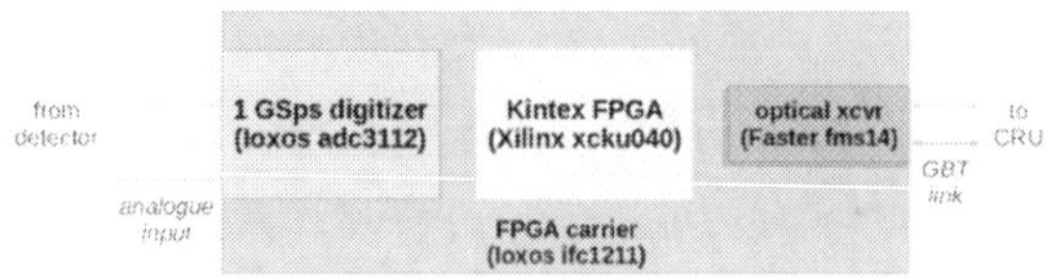

Fig. 1 The readout based on FPGA with two FMC port for the interfacing.

These ports are used to interface with a fast digitizer to acquire the analogue input from the photomultiplier, and for the insertion of a multi – channel optical transceiver to match with the DAQ via optical fibres.

II. Main components

The selected digitizer is the commercial adc3112, from the Ioxos company, in FMC format and equipped with 4 analogue channels able to run at a sample rate of 1 GS/s with a 12 b resolution. The sampling clock is jitter cleaned by a two - stage PLL, and the source can be internal from the VCXO, external from the front panel connector or the FPGA via the FMC connector. The converter circuit features a low – pass filter, which is used for noise reduction, and a 2× decimation filter in order to acquire 12 sample per bunch crossing.

The module is equipped with analogue channels with differential DC coupled input, but the input stage can also be software configured as single - ended mode with 50 Ω input resistance. Direct coupling is crucial for the ZDC because the photomultipliers of the detector produce a unipolar signal that evolves only in the negative voltage domain. So to avoid ADC saturation when a signal becomes very large, a positive voltage offset is inserted to exploit the 1 V_{pp} input range.

The selected FPGA carrier is the commercial ifc1211, from the Ioxos company, that is a module in VME format equipped with a PowerPC processor, a Xilinx FPGA and many other interface ports with two FMC. The embedded FPGA is the ku040 from the Kintex UltraScale family, with more than 500 pin dedicated to custom input / output functions.

That FPGA is used to perform online data processing on the samples coming from the digitizer by implementing a custom trigger algorithm, and to serialize the data by handling some serial data links.

III. The architecture

Each FPGA runs the readout firmware to handle signal samples from the 4 ADC channels. The architecture can be considered flat in the sense that no hierarchically higher entity controls the task of each FPGA, and they do not require to exchange any information with each other. Each FPGA applies a custom trigger algorithm on the data stream coming from each analogue channel, and takes care of the data formatting to comply with the GigaBit Transceiver (GBT) protocol.

The GBT is a custom transmission protocol developed at CERN, based on a serial link running at 4.8 Gb/s, particularly suitable in noisy environments. Each link is a full duplex, meaning it is bi–directional thanks to two optical fibres. The first, in the forward direction, transfers the data, whereas the second, in the reverse direction, carries timing information.

Each readout FPGA, in addition to managing the data stream toward the DAQ, recovers the LHC clock and the trigger messages that are sent on the reverse optical link of the GBT protocol. Then it decodes the information on a secondary communication channel, based on the Single Word Transfer (SWT) protocol, to perform slow control and detector configuration.

During Run 3 the event rate is very high, and the pile-up appears [4] . In other words, it is quite probable that a signal is present on the tail of the previous signal. In such a situation, the selection of the event based on a simple threshold is no longer valid, above all, because to the input dynamics is high up to 1 : 60 considering corresponding signals up to 60 nucleons.

To solve this problem, a custom algorithm was applied to compare close samples with respect to a threshold as in (1), repeated several times:

$$\begin{gathered}(y_i - y_{i+k} > t) \land \\ \land (y_{i+1} - y_{i+k+1} > t) \land \\ \land (y_{i+2} - y_{i+k+2} > t)\,. \end{gathered} \tag{1}$$

In practice, the difference between neighbouring samples y_i and y_{i+k} is calculated and, if it is greater than a threshold value t, the related event is a good candidate for acquisition. When this occurs 3 consecutive times, the event is considered interesting, and the acquisition is triggered. The number of conditions that must be verified to trigger is a parameter. To keep the efficiency high, it is possible to loosen the control from a triple to a double condition. The effect of this custom algorithm is that it behaves like a differential threshold or, in other words, it acts like a threshold on the derivative of the signal.

Considering the typical waveforms produced by photomultipliers and the expected rise time of the signals, the values $k = 4$ and $t = 10$ for the algorithm parameters are very likely. Considering that the signals are matched with the ADC input range of 1 V_{pp} and the digitizer resolution of 12 b, the typical threshold is about one – third the amplitude of a single neutron event.

Once the event is considered interesting due to the auto-trigger algorithm or ALICE central trigger, it is tagged with some information, see Fig. 2, and only those events are pushed in the corresponding channel FIFO. Each FPGA handles data from 4 analogue channels and sends the information over two serial optical links providing the necessary bandwidth, so a multiplexer followed by another FIFO is needed to allow the back – pressure.

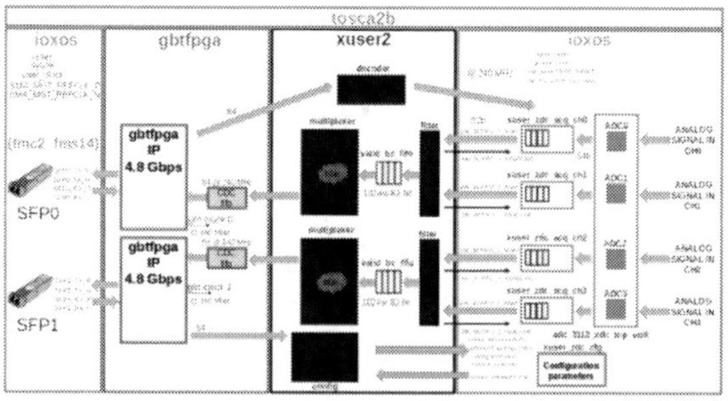

Fig. 2. The firmware architecture implemented managing 4 input channels.

Then the data stream is transferred to the entity implementing the GBT serial protocol at the speed of 4.8 Gbps, where each data word is completed by several redundancy bits for error correction.

Finally, the two serial data streams are sent to the other FMC module equipped with SFP+ optical transceivers; the same links are used to receive the trigger messages or the configuration parameters.

Due to the slight change in frequency after the proton or ion filling of the LHC ring, it can sometimes happen that the sampling clock becomes different from the transmission clock. For this reason, the Clock Domain Crossing (CDC) FIFO was added to the system to allow for smooth data transfer between different clock regions.

IV. Current operation

The new readout system was installed at CERN in the summer of 2022 and was integrated into the ALICE DAQ chain resulting in full compliance with the experiment requirement. Waveforms from a single pp interaction, observed by one ZN, are shown in Fig. 3. This means that the ZDC can acquire data in auto - trigger mode, as required for LHC Run 3, or with triggers provided by the Central Trigger Processor, as it was in the past.

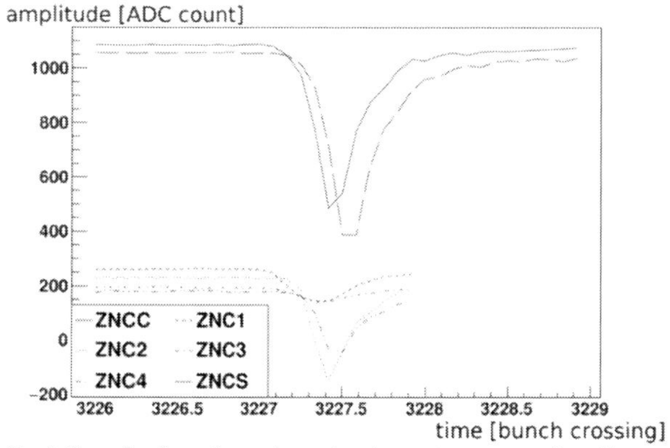

Fig. 3. Example of waveform, where a bunch crossing corresponds to 25 ns.

The commissioning with proton beams was performed last autumn 2022 in special runs at low luminosity, during which ZDC could acquire data thanks to the low value of the crossing angle of the incident beams.

The ZDC then took data together with all other ALICE detectors in november 2022 during the Pb – Pb pilot run.

The standard reference, to assess the quality of the whole ZDC detector, is the amplitude resolution of the peak relative to signals corresponding to events with a single neutron, see Fig. 4. From a preliminary analysis the resolution for the amplitude of the single neutron peak results to be around 15 %, improved with respect to what was achieved in the past Run 2. The peak relating to events with a single neutron is clearly evident and differs from the peak related to other events with 2, 3, 4, 5, .. neutrons [5] .

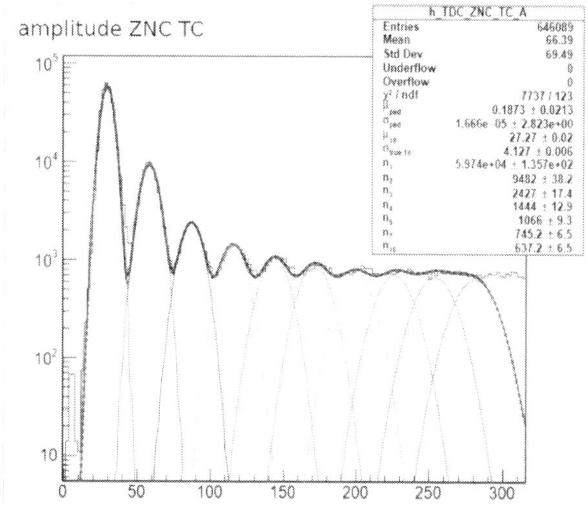

Fig. 4. ZN signal amplitude spectrum in minimum bias Pb – Pb collisions.

V. Conclusions

A new readout system of the ALICE ZDC detector developed for Run 3 can operate in self - triggered mode without dead time at a rate of 2.5 M event/s with an energy resolution of 15 %.

The hardware is FPGA based, equipped with a commercial 1 GS/s digitizer with a 12 b resolution from Ioxos, and implements a custom auto – trigger algorithm.

The system takes data in the global acquisition and complies with the different acquiring modes foreseen in ALICE; furthermore the full integration within the DCS environment has been completed.

The ALICE ZDC detector is expected to be fully operational during the next Pb – Pb data taking in autumn 2023, providing the event geometry and luminosity measurement.

References

[1] K. Aamodt, A. Abrahantes Quintana, R. Achenbach, S. Acounis, D. Adamová, C. Adler, *et al.*, "The ALICE experiment at the CERN LHC," Journal of Instrumentation, Volume 3, s08002, 2008, doi.org/10.1088/1748-0221/3/08/s08002 .

[2] M. Gallio, W. Klempt, L. Leistam, J. De Groot, J. Schükraft, G. Dellacasa, *et al.*, "ALICE Zero Degree Calorimeter (ZDC): Technical Design Report," CERN, alice-tdr-3, cern-lhcc-99-005, 1999, cds.cern.ch/record/381433 .

[3] P. Cortese for the ALICE collaboration, "Performance of the ALICE Zero Degree Calorimeter and upgrade strategy," Journal of Physics, Conference Series 1162, 012006, 2019, doi.org/10.1088/1742-6596/1162/1/012006 .

[4] P. Cortese, M. Arba, G. Cotto, G. Dellacasa, N. De Marco, P. De Remigis, *et al.*, "Performance of the readout system of the ALICE Zero Degree Calorimeter in LHC Run 3," Journal of Physics, Conference Series 2374, 012091, 2022, doi.org/10.1088/1742-6596/2374/1/012091 .

[5] S. C. Zugravel, P. Cortese, G. Cotto, N. De Marco, P. De Remigis, L. Lombardo, *et al.*, "ALICE Zero Degree Calorimeters, the new readout system in LHC Run 3," Journal of Instrumentation, Volume 18, c02009, 2023, doi.org/10.1088/1748-0221/18/02/c02009 .

Trustworthiness of Laser-Induced Breakdown Spectroscopy Predictions via Simulation-based Synthetic Data Augmentation and Multitask Learning

Riccardo Finotello[1,*], Daniel L'Hermite[2], Celine Quéré[2], Benjamin Rouge[2], Mohamed Tamaazousti[1], and Jean-Baptiste Sirven[2]

[1]Université Paris-Saclay, CEA, LIST, Palaiseau, F-91120, France
[2]Université Paris-Saclay, CEA, Service de Physico-Chimie (SPC), Gif sur Yvette, F-91191, France
(*) riccardo.finotello@cea.fr

Abstract—**Laser-induced breakdown spectroscopy is a versatile technique that can be used to quickly measure the concentration of elements in ambient air. We tackle the issues of performance and trustworthiness of the statistical model used for predictions. We propose a method for improving the performance and trustworthiness of statistical models for LIBS. Our method uses deep convolutional multitask learning architectures to predict the concentration of the analyte and additional information as auxiliary outputs. We also introduce a simulation-based data augmentation process to synthesize more training samples. The secondary predictions from the model are used to characterize, quantify and validate its trustworthiness, taking advantage of the mutual dependencies of the weights of the neural networks. As a consequence, these output can be used to successfully detect anomalies, such as changes in the experimental conditions, and out-of-distribution samples. Results on different types of materials show that the proposed method improves the robustness and trueness of the predictions.**

Keywords **—LIBS; trustworthy AI; simulation; multitask; neural networks.**

I. Introduction

Deep Learning (DL) has been adopted with success in many areas of science, and analytical chemistry is no exception. Laser-Induced Breakdown Spectroscopy (LIBS) offers plenty of applications for MVA techniques, especially for hyperspectral imaging applications [1], [2], but the development of DL techniques is quite recent. Basic shallow Neural Networks (NNs) were implemented as soon as the '90s for identification of polymers [3]. The introduction of modern concepts and architectures of DL is however quite recent [4].

In spectroscopic quantitative analysis, we build models relating experimental spectra to the concentration of the species of interest, using calibration samples. Quantitative models are then valid for a given sample matrix and for given experimental conditions. However, for direct analytical techniques like LIBS, unknown samples can have different matrices, and experimental conditions can change. The predicted concentration can thus be biased to a certain extent. Yet, models are usually designed to deliver a concentration but not to estimate to which extent an unknown sample is well represented by calibration ones. In other words, we do not know how reliable the prediction is. Hence, we need models that are able to estimate, if possible quantitatively, the trustworthiness of a prediction. Thanks to their ability to perform different tasks in parallel, MultiTask (MT) NNs are a type of model that could address this issue, by providing, at the same time, the main expected prediction and a characterization of its properties. In turn, these auxiliary data can be used to validate and quantify the model trustworthiness. However, deep NNs can easily overfit the training data, since they require a very large number of spectra to be properly trained: experimentally, time and cost constraints generally do not enable to acquire enough spectra. A possibility to overcome this limitation is to generate calibration data. This can be done by phenomenological modelling of training spectra.

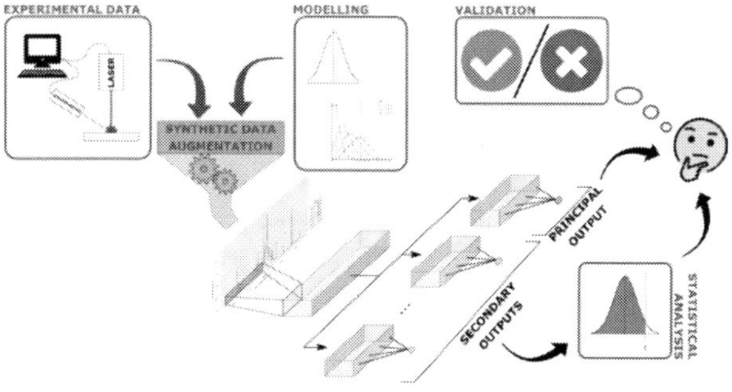

Fig. 1. A robust multitask model is trained on synthetic data to predict a principal variable (e.g. the concentration of the analyte) and several secondary spectral quantities.

We introduce a new approach of quantitative analysis by LIBS and DL, enabling both to generate enough spectra to train a model based on MT Convolutional Neural Networks (CNNs), and to evaluate the confidence in the model predictions. Namely, we propose:

1. a simulation-based synthetic data creation, that is the synthesis of an arbitrary number of new spectra from experimental data to train complex DL architectures;

2. robust deep MT CNNs, capable of processing entire LIBS spectra and increasing the robustness (homoscedasticity) of the model;
3. a measure of the trustworthiness of the predictions of the model through the statistical analysis of the outputs of the MT architecture.

The procedure, presented in Fig. 1, aims at providing a way to increase the robustness of the calibration models, and to benchmark the confidence of the Artificial Intelligence (AI) architecture through its own predictions.

II. Related Works

Univariate and multivariate (MVA) calibration techniques are explored in the LIBS literature [5]. The first are widely adopted for their ability to provide interpretable results quickly and with good precision [6]. In this scenario, calibration standards are used to build a map between the concentration of a given analyte and the information contained in a single measurable variable, such as the integral intensity of an emission line. The model is then inverted during inference to use the measured intensity as predictor of the concentration of the analyte. On the other hand, MVA methods were introduced for their ability to take advantage of more information contained in the input spectra, rather than focusing on a single variable. Techniques based on principal components and multilinear regression have been widely adopted in LIBS [7], [8]. As a MVA technique, DL has also been explored, comparing various types of architectures [4].

A. Data Augmentation

Though DL shows potential for various analyses, time of data gathering and experimental conditions often prevent building large LIBS datasets. Feature engineering and feature selection, for instance through a principal components analysis or a priori expertise, have been employed to reduce the size of the input data in order to be used in smaller or more adapted machine learning (ML) models [9]. At the same time, the production of purely synthetic data, based on local thermodynamical equilibrium, has been explored in LIBS applications [10]. The idea of enriching existing data by means of different representations of the inputs, such as time resolved spectra, was experimented with success [11]. More recently, standard data augmentation techniques were used for the classification of LIBS mapping experiments [4]. Augmentation through synthetic data via the simple addition of random noise to the experimental data is also briefly discussed in the literature [12].

B. Multitask Learning

Though explored at length in ML, MT learning [13] has seen major developments and a wide range of applications recently. Some examples of multi-output algorithms for LIBS analyses were recently explored, based on NNs [11] and on Partial Least Squares (PLS) with two outputs, i.e. PLS2 [14]. In these cases, the response variables are usually the concentrations of multiple elements.

Interestingly, multi-output NN architectures found successful applications for the computation of plasma parameters from LIBS data [15]. MT learning was also introduced for the simultaneous predictions of concentrations of analytes and lithology classes: different NNs were used to process the data with different loss functions, using a latent representation of the input, computed by a common backbone architecture [16].

C. Trustworthy AI

Standard techniques in statistics, such as confidence intervals for mean values and predictions, are usually preferred [17], though they rely on strong assumptions on the type of data analyzed, such as the independence of the residuals from the independent variables. On the other hand, it has been seen heuristically that traditional models do not systematically generalise to unknown data. Some Explainable-AI (xAI) methods have been introduced to analyze the progression of the feature maps in the hidden layers [18], or study the importance of the spectral variables leading to the prediction [19]. This represents indeed a step towards the comprehension of the mechanisms behind NNs. Nevertheless, it does not deal with the confidence of the predictions, or the automatic detection of changes in the distribution of the samples.

III. Methodology

In what follows, we detail the contributions to the end-to-end pipeline of the analysis, from the synthetic data augmentation, to the validation of the model predictions through the statistical analysis.

A. Synthetic Data Augmentation

Training large NNs by optimizing the bias-variance tradeoff may lead to phenomena such as poor stability and bad generalization in inference on unknown samples. Many training samples are usually required to train more complex architectures to overcome the issues. Inspired by usual DL practices, we thus introduce a data augmentation technique for LIBS.

We train the MT model on entire LIBS spectra, without a priori data selection. Given the scarcity of training data, the simple addition of random noise to the experimental spectra may result in a training distribution no longer representative of the use case. Thus, we first proceed to model the distribution of the original spectra, and then to synthetically produce an arbitrary number of spectra. Noise can then be added to each individual channel, provided that its global average effect is negligible, in order to make the synthetic distribution more realistic. The procedure ensures to enlarge consistently the feature space spanned by the synthetic spectra. Specifically, we consider the set of n spectra with p wavelength channels $\{x^{(i)} \in \mathbb{R}^p\}_{i\in[1,n]}$ and the corresponding average spectrum $\overline{\boldsymbol{x}} = (\bar{x}_r)$. For each wavelength channel $r = 1, 2, \dots, p$, we fix the expected value of a random variable $y_r \in \mathbb{R}$ such that $\mathbb{E}[y_r] = \bar{x}_r$. An arbitrary number $m \gg n$ of full spectra can be constructed by similarly proceeding for all wavelength channels. A degree of noise can then be added to each channel, using a multiplicative Gaussian factor $z \sim \mathcal{N}(1, \beta)$, where β is a noise parameter of the synthetic spectra. We can use the

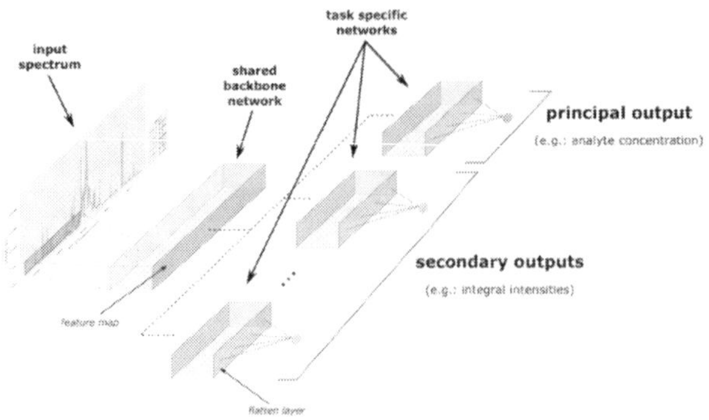

Fig. 2. The MT model is a hard parameter sharing structure, that is a common backbone network and several heads, connected to the shared feature map.

spectra in the training sets to choose appropriate values of β. We observed that, for specific wavelength channels of interest, in the absence of the parameter β, the statistical model representing the spectra is over-confident (i.e. it does not reproduce the full distribution) at low intensity and under-confident at higher intensities (i.e. it creates out-of-distribution intensities), while the opposite occurs for higher values of the noise parameter. This suggests that there is an in-between optimum of the parameter, for which the synthetic distribution covers correctly the variance of the training distribution at a local level. The value of β can be chosen deterministically, for instance, by maximizing the coefficient of determination (R^2) between the ground truths and the synthetic quantiles at a given wavelength of interest.

B. Multitask Convolutional Neural Networks

By definition, MT NNs are a broad class of algorithms, which provide multiple predictions at the same time, using a shared structure of weights, trained simultaneously. This property gives the networks great versatility, as it is capable of using information on one task to improve its generalization. This strategy acts as a regularization and reduces the overfitting of the training data, as the model supposedly learns new representations, which should generalize well on all tasks.

In this analysis, we use a hard parameter sharing implementation of MT learning, with a common set of bottom layers (see the general schematics in Fig. 2). The innermost backbone processes the input spectra and produces a new latent vector representation. The task-specific heads separately use the latent representation as inputs to compute scalar regression outputs (principal and secondary). We use 1D convolutions in the spectral dimension of the data cube as main operation. In what follows, we choose to use the concentration of the analyte as the principal prediction of the network, and the integral intensities of the associated emission lines or bands as secondary outputs of the network. In turn, this helps to stabilize the model and increase its robustness (homoscedastic behavior) during inference. It also provides a set of secondary results, which can be used to validate and quantify the performance of the model and to detect anomalies or out-of-distribution samples.

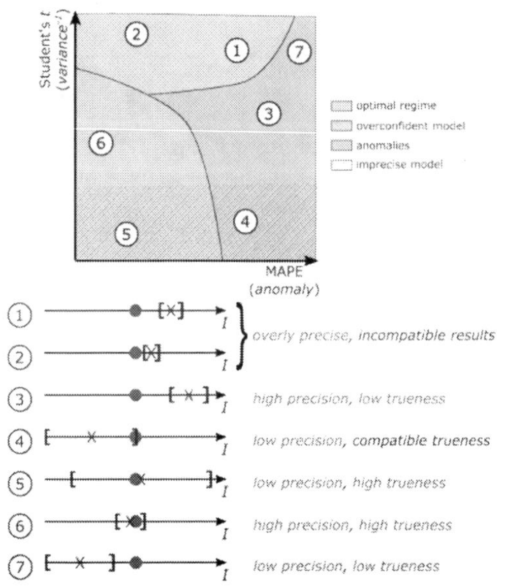

Fig. 3. Outcomes of the MAPE and t variable summarized at the top of the figure. Interpretations are presented schematically at the bottom.

C. Trustworthiness via Validation of the Predictions

In general, the validation of the predictions on new data and the detection of anomalies remain complicated issues in statistics. Given the mutual dependence of the multiple outputs of the MT architecture, we rely on the set of secondary outputs, experimentally measurable on unknown samples. In fact, the integral intensities of emission lines or bands can be extracted from experimental spectra at any given time, even though the concentration of the analyte remains unknown.

When dealing with new data, we compute the predictions of the model on a single spectrum basis, independently. We then average the results per sample, to smooth the influence of defects on the surface. The Mean Absolute Percentage Error (MAPE) gives a measure of the deviation of the data and the performance of the model. For the i-th secondary output of the network and a sample s, let $\boldsymbol{Y}_i^{(s)} = \left|\left(\boldsymbol{I}_i^{(s)} - \hat{\boldsymbol{I}}_i^{(s)}\right)/\hat{\boldsymbol{I}}_i^{(s)}\right|$. We compute the MAPE $M_i^{(s)}$ of the predicted intensities $\boldsymbol{I}_i^{(s)}$ and the corresponding ground truths $\hat{\boldsymbol{I}}_i^{(s)}$, with n samples, that is $M_i^{(s)} := \mathbb{E}\left[\boldsymbol{Y}_i^{(s)}\right]$. This offers a first estimate of the error made by the model, as it describes the trueness of the model. In order to estimate a soft threshold for the quantity, we use a validation set of samples. We compute a confidence interval around the MAPE $\left[M_i^{(s)} - \hat{t}_{1-\alpha}^{n} \frac{\sigma_i^{(s)}}{\sqrt{n}}, M_i^{(s)} + \hat{t}_{1-\alpha}^{n} \frac{\sigma_i^{(s)}}{\sqrt{n}}\right]$, where $\left(\sigma_i^{(s)}\right)^2 = \mathrm{Var}\left(\boldsymbol{Y}_i^{(s)}\right)$, and $\hat{t}_{1-\alpha}^{\nu}$ is the value of the *Student*'s t variable for v spectra, at a confidence level $1-\alpha$.

We then use a Student test on the predicted intensity, given its nature of repeated measurement on the sample. Supposing that the ground truth value of a sample s for the i-th average intensity has a sample variance σ_i^2, we can compute the random variable $t_i^{(s)} = \left|\mathbb{E}\left[\boldsymbol{I}_i^{(s)}\right] - \mathbb{E}\left[\hat{\boldsymbol{I}}_i^{(s)}\right]\right| / \sqrt{\sigma_i^2 + \Sigma_i^2}$, where Σ_i^2 is the sample variance of the predictions. By including the

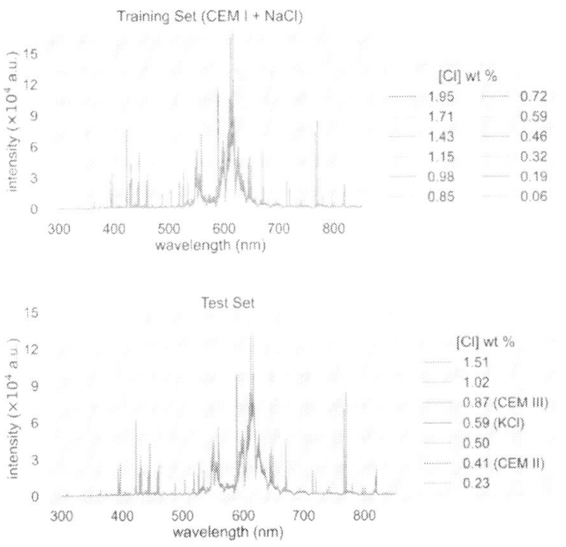

Fig. 4. Average spectra of the cement samples. Samples were fabricated using CEM I, with the addition of NaCl, unless otherwise stated.

dependence on the variance, this test gives a statistical measure of the ability of the model to generalize to unknown data. In order to discriminate possible anomalies, we can use a standard approach by choosing a threshold value $\hat{t}^{\nu}_{1-\frac{\alpha}{2}}$ of a two-tailed t-test at confidence level $1-\alpha$, with ν degrees of freedom, such that the probability $P\left(t_i^{(s)} > \hat{t}^{\nu}_{1-\frac{\alpha}{2}}\right) = \alpha$. This way, we recover a probabilistic interpretation of the result in terms of confidence: models can be compared based on their performance at different values of α on the secondary outputs.

The values of MAPE and of the Student's t variable can be used together to evaluate the trustworthiness of the model and characterize its predictions. We graphically summarize these interpretations in the plane in Fig.3. Though the confidence level of the principal output is not easily computed from the confidence of the secondary outputs, this measure gives an implicit feedback on the main output. Given the dependencies of the MT model parameters, the information determines whether the prediction of the concentration of the analyte is trustworthy.

IV. Experimental Setup

We compare the predictive ability of different algorithms on two types of datasets. We consider 19 cement samples, whose elemental compositions are reported in Fig. 4, and 4 alloy matrices with 4 to 6 samples each, summarized in Fig. 5. The first were built in the framework of an interlaboratory comparison in 2021 [20]. All measurements were carried out in air, at room temperature.

Cement samples were probed using a Nd:YAG laser (*Quantel Brio*) at a wavelength of 1064 nm, 15 mJ pulse energy and 4 ns pulse length. For each sample, 25 spectra were collected, accumulating 40 laser shots (5 pre-ablation shots).

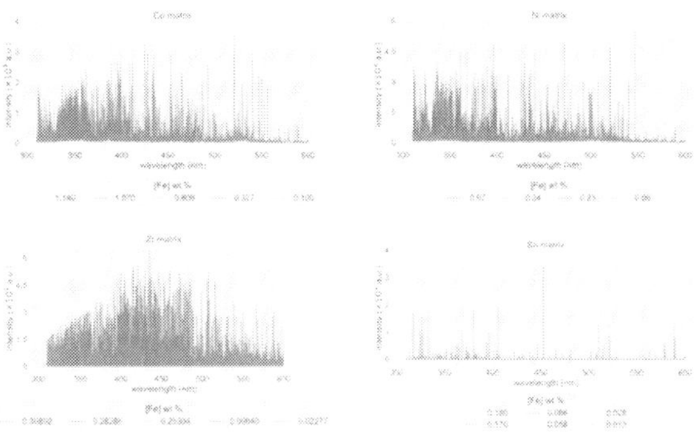

Fig. 5. Different alloy matrices used in the analysis.

We used a *Mechelle* spectrometer with an *Andor iStar* ICCD and a fixed aperture of 50 μm × 50 μm, 10 μs gate delay, and 100 μs gate width in the range 200nm to 975nm (resolving power $\lambda/\Delta\lambda \simeq 4 \times 10^3$ measured at 589.60 nm on a Na peak). The irradiance on the sample surface was 190 GW cm^{-2}.

Data on alloys were collected using a Nd:YAG laser (*Quantel Ultra*) operating at a wavelength of 266 nm, 6 mJ pulse energy, and 4 ns pulse length. The plasma emission was analyzed with a *LTB Aryelle 400* spectrometer with a fixed aperture of 50 μm × 50 μm, equipped with an *Andor* DH740 ICCD, in the range 310 nm to 613 nm (resolving power: $\lambda/\Delta\lambda \simeq 1.8\times10^4$). We used a gate delay of 1 μs and a gate width of 0.5 μs. For each sample, 25 spectra were collected, accumulating 20 laser shots for per crater. The irradiance on the surface of the samples was 76 GW cm^{-2}.

V. Training Methodology

Before entering the discussion of the results, we detail some training techniques used to fit the MT model to the LIBS data.

A. Data Curation

For the analysis of the alloy matrices, we use the intensities of the most intense persistent lines of Fe in the spectral range considered, integrated over 10 wavelength channels, as secondary outputs of the MT model. Specifically, we choose the 8 strongest persistent lines, reported in [21]. Given the small number of samples available, 30% of the spectra for each sample is retained as independent test set, while the rest is used as training set for the baseline models, and as input of the data augmentation for the MT architecture. This ensures, on average, an in-sample inference for the algorithms (the selection of test samples comes from the same samples in the training set). Moreover, it provides the means to verify whether the augmentation technique correctly enhances the training distribution. Notice that this does not automatically translate into a simpler task for the model: spectra in the test set may still differ from the training distribution, due to random and local fluctuations, thus they may represent out-of-distribution data on a spectrum basis.

For the cement samples, we consider two molecular bands centred at 593.46 nm and 617.74 nm, and integrated over 14 channels, as secondary outputs. As a general guideline, choosing many secondary outputs helps the convergence of the network, and preserves a good generalization performance of the model. Samples are separated into training and test sets on a sample basis: 12 matrices are considered as calibration set, while 7 samples are used for inference. In the test set, we insert specifically samples which present a different matrix (type of cement) or manufacturing procedure (different salts added in the mixture) in order to check the out-of-distribution generalization ability of the algorithms, and their ability to recognize possible anomalies.

As preprocessing, outliers are removed from the training set, either experimental or synthetic: at a given wavelength, we define outliers as spectra presenting an intensity outside the interval between the 5th and 95th percentile of the values. The goal is to build performing calibration models without using extreme configurations. In order to test the generalization ability of the models, we retain the outliers in the test set. For the alloy samples, we focus on the Fe line at 373.49 nm (the most intense persistent line), while, for the cement matrices, we consider the molecular band of CaCl at 593.46 nm.

Spectra are normalized using the integral intensity at a given wavelength. Since the procedure is performed independently on each spectrum, we can also safely normalize the spectra in the test sets. For the alloy matrices, we consider the integrated intensity of the most intense emission line of the matrix itself over an interval of 10 wavelength channels. The cement samples have been normalized to the intensity of the CaO molecular band at 615.03 nm, integrated over 20 wavelength channels.

B. *Data Analysis and Methodology*

We compare our results with several algorithms known in the LIBS literature, namely the classical Linear Regression (LR), MVA Linear Regression (MLR), simple Fully Connected NNs (FCNNs), and PLS1. For LR and MLR, no additional validation sets have been considered, as no free parameters are present in the algorithms. For PLS1, we perform a 5-fold cross-validation procedure. In the case of NNs, we use a single holdout validation set made of 20% of the spectra contained in the training set, selected using a stratified strategy to preserve the fraction of spectra for each sample. Hyperparameters are optimized using a tree-structured Parzen estimator [22]. In order to avoid any data leakage, we avoid using the experimental spectra when optimizing the model. We use common regression metrics such as the Root Mean Squared Error (RMSE), the Mean Absolute Error (MAE), and the MAPE to score the results of the algorithms. Results always refer to the independent experimental test set.

In the case of LR, we consider the integral intensity of selected wavelength channels as inputs of the model. Specifically, we consider the integral intensity of the Fe emission line at 373.49 nm, integrated over 10 wavelength channels, for the alloy matrices. For the cement samples, we use the CaCl molecular band at 593.46 nm, integrated over 14 channels. For the MLR and the FCNN, we select several atomic Fe emission lines as inputs, in the case of the alloy matrices. We use the CaCl molecular bands at 593.46 nm and 617.74 nm for

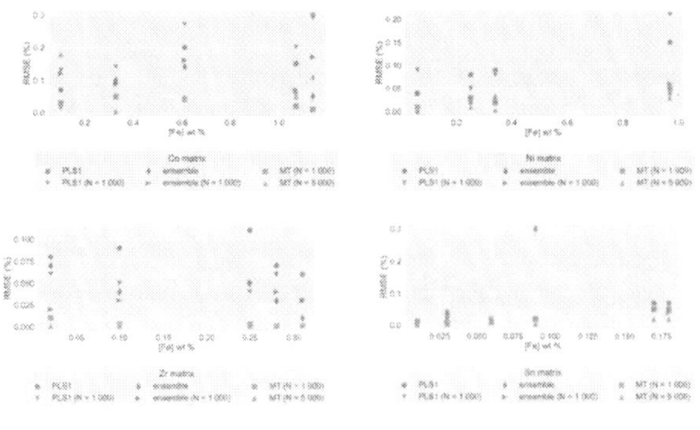

Fig. 6. Prediction uncertainties of the MT model on alloy matrices as a function of training set size.

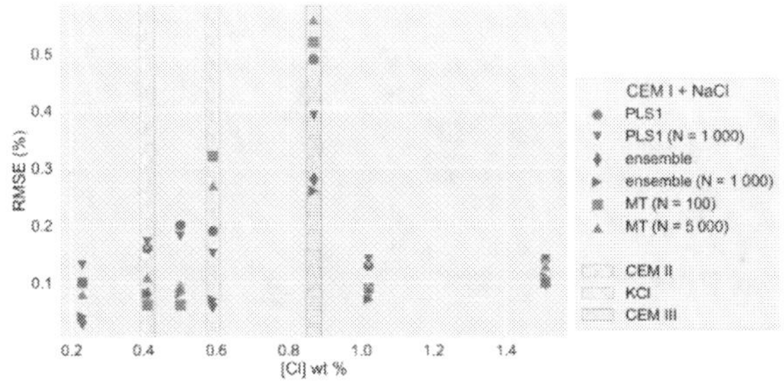

Fig. 7. Prediction uncertainties of the MT model on cement samples as a function of training set size.

the cement samples. On the other hand, for PLS1 and the MT network, we use the entire spectra as input. Finally, we choose a quadratic regression model in the case of LR (without interaction terms), while we consider a simple regression model for MLR.

VI. Results and Discussion

In the analysis, we consider different types of baselines, in order to provide a complete comparison of the proposed technique with the SOTA. Specifically, we show the results of PLS1, which does not require the extraction of information from the experimental spectra, and an ensemble model, with initial feature (emission lines) selection, represented by the best result of LR, MLR, and FCNNs. Our proposed MT model does not require expertise in selecting input data, since the important variables are learnt during training. It requires a degree of knowledge of the emission lines of the analyte to choose the secondary outputs. A simple choice is to consider the full set of persistent lines of the analyte in the available spectral rang, since spectral interference on single lines (or bands) is taken care of by the CNN architecture.

A. *Performance of the Model*

We consider different aspects relating to the nature of the analysis, such as the dependence on the size of the synthetic training set and the choice of the random noise: we train different models, generating a different number of synthetic spectra, with different β parameters, for each sample. The results with a dependence on the synthetic training set are graphically summarized in Fig. 6 for the alloy matrices, and in Fig. 7 for the cement samples. A similar behavior has been observed for the dependence of the noise parameter, with the best results appearing for $\beta = 0.10$ in the case of alloy matrices and $\beta = 0.03$ for the cement samples, as expected from the discussion in Section 3.1.

In terms of robustness, the MT architecture is capable of delivering a homoscedastic performance for all concentrations of the analyte (apart from anomalies, which are discussed in the following). Moreover, the prediction uncertainties are usually comparable with or better than the ensemble model, which already presents good results. The MT architecture is capable of selecting the information contained in the data to base its own predictions. The use of synthetic samples enables to capture the fluctuations at low concentrations, where the PLS1 model struggles to give accurate results. As a general remark, the optimal number of synthetic spectra varies for each matrix, depending on the degree of spectral interference, noise, and sparsity (e.g. 5000 spectra per sample in the case of the Zr matrix, or 1000 spectra per sample for the Cu matrix).

Heuristically, we noticed that the creation of many synthetic spectra impacts on training time for less than 1% of the total training time, while the latter grows linearly. Such behavior makes it usually possible to experiment with a few options, in order to determine, using the validation set or the experimental training samples, the best trade-off between the performance of the model and the computational power available.

B. Validation of the Predictions

For the analysis of the trustworthiness of the model, we focus specifically on the predictions of the MT model on the independent test set of the cement samples. In this analysis, ground truth values of the concentration of the analyte are available for a direct comparison with the predictions of the model. However, in a field application, reference values would not be available. The analysis of the secondary outputs of the MT architecture is a tool to assess the confidence of the predictions and detection of the anomalous samples or modifications in the experimental conditions.

We choose two CaCl molecular bands for the analysis of the trustworthiness of the predictions. This represents an easy choice, as the two bands are the two most intense in the spectral range considered. Following previous sections, we compute the confidence intervals on the experimental training data (unseen by the model, which is trained on synthetic spectra). We consider these values as reference in the analysis of the trustworthiness of the model, since they represent known standards, whose labels are available. *A posteriori*, we notice that the predictions of the MT architecture are all compatible with the respective ground truths, even though some samples present larger uncertainties, which may indicate faulty values. However, in the absence of reference values, predictions alone are not sufficient to measure the trustworthiness of the model.

The predictions of the secondary values show that the band at 593.46 nm displays a pattern which identifies some anomalies in the prediction of the integral intensities (see Table 1). As previously shown (see Fig. 3 for a reference), this pattern of MAPE and t-value is typical of anomalous samples, for which the model does not provide precise predictions. In hindsight, the analysis of the secondary outputs identifies the three out-of-distribution samples present in the dataset (different matrix and salt).

TABLE I
PREDICTIONS ON CEMENT SAMPLES

GROUND TRUTH	PREDICTION	MAPE		T-VALUE	
[Cl] wt %	[Cl] wt %	593.46 nm	617.74 nm	593.46 nm	617.74 nm
0.23	0.25 ± 0.08	1.4	0.7	0.69	0.34
0.41 (CEM II)	0.41 ± 0.11	**4.1**	0.8	0.98	0.32
0.50	0.54 ± 0.09	1.7	0.7	0.48	0.26
0.59 (KCl)	0.33 ± 0.27	**4.2**	0.8	1.29	0.31
0.87 (CEM III)	1.40 ± 0.56	**4.7**	**1.4**	0.77	0.41
1.02	0.97 ± 0.09	1.3	0.4	0.38	0.16
1.51	1.54 ± 0.13	1.5	0.8	0.43	0.30

Trustworthiness of the predictions on cement samples.

To measure the trustworthiness of the predictions, we then use a standard Student's two-tailed test (confidence $1 - \alpha = 0.95$ and 25 degrees of freedom) to assess the predictions of the molecular emission bands. We notice that, although the samples register as anomalies, the variance of the predictions is such to include the true values of the secondary outputs in the error intervals with good confidence (case 4 in Fig. 3). Given the interdependencies of the MT model previously discussed, the confidence on the predictions of the secondary output influences directly the confidence on the main prediction, the concentration of the analyte. The precise quantification of the confidence level is, nonetheless, not trivial because of the large number of parameters involved in the computation. In the case at hand, we can interpret the result by noting that the MT model is still capable of providing trustworthy predictions on the concentration of the analyte. However, its precision is highly affected in the presence of anomalous samples: the predictions of the main output contain the true value, within the uncertainty, with high probability. In other words, in this scenario, the predicted values of the concentration of the analyte can be considered compatible with the reference values, provided by the supplier. Further investigation on three anomalous samples remains necessary.

The procedure provides a way to assess the robustness of the MT architecture on the entire range of variability of the Fe concentration. Notice that the analysis of the secondary outputs is quantitative: for instance, a p-value of the prediction can be computed. Moreover, different choices of the confidence $1 - \alpha$ allow comparing the model as functions of the confidence level. Finally, this analysis is always possible, with any sample, as the information contained in the secondary outputs of the model is directly comparable with the experimental data. This is different from the concentration of the analyte itself, which is known only for standard samples.

C. Calibration Transfer and Anomaly Detection

We test the trustworthiness of the predictions in the presence of a change in the distribution of the samples. Specifically, we

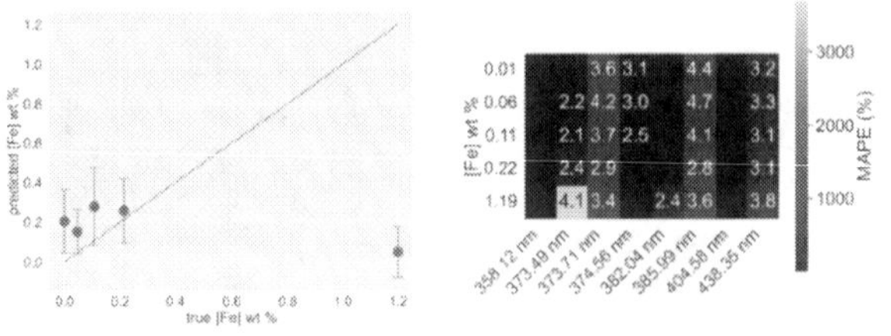

Fig. 8. Comparison of ground truth and predicted values on the Ti matrix, using a model trained on the Co matrix.

consider the MT model trained on the Co matrix, and we perform the inference on the Ti and Zn matrices. In Fig. 8, we show the true and predicted concentrations of the analyte. The a posteriori analysis shows that the model does not generalize among different matrices, hence it is not usable for calibration transfer. However, a real-time application would not reveal the same behavior, as ground truth values would not be available. The Ti matrix presents predictions characterized by large values of the error fractions and by incompatible values of the t variable, whose threshold $\hat{t}^{\nu}_{1-\frac{\alpha}{2}} = 2.06$ for $\nu = 25$ degrees of freedom and a confidence $1 - \alpha = 0.95$. This shows that the predictions of the concentration levels 0.11 wt% and 0.22 wt%, though compatible with their respective ground truths, should not be deemed trustworthy. The MT model is thus capable of detecting anomalies or modifications in the experimental conditions, which is key to assessing correctly the ability of the model to provide trustworthy predictions.

VII. Conclusions

In this work, we use DL techniques to address the quantitative analysis of LIBS data, the prediction of the concentration of an analyte, using a MVA calibration procedure. We focus on complementary aspects: the creation of a synthetic set of spectra as a data augmentation technique to increase the number of samples available for training, the construction of a robust MT learning model based on deep CNNs, and the analysis of the confidence of the predictions. We use the entire experimental emission spectra as inputs, without the need for a preselection of variables or dimensionality reduction. We leverage the robustness and performance of CNNs with the possibility to provide a tool to assess the trustworthiness of the predictions of the model, even for unknown data.

To this end, we introduce a MT learning architecture. The model is capable of predicting the concentration of the analyte and the integral intensities of relevant emission lines (or molecular bands), at the same time. Given the size and complexity of the DL model, we introduce a data simulation technique, to create an arbitrary number of input spectra, statistically representative of the experimental data. The MT architectures display robustness across the range of variation of the analytes. The presence of the secondary outputs, allows us to introduce a statistical analysis, based on the mutual dependencies of the parameters of the AI architecture, which enables the assessment of the trustworthiness of the model. Comparisons of the predicted values with the intensities found in the experimental spectra can be used to study the predictions of the concentration of the analyte, at a given level of confidence of the model. In turn, this grants the ability to assess the extrapolation abilities of the DL model.

Acknowledgements

We acknowledge the financial support of the *Cross-Disciplinary Program on Instrumentation and Detection of CEA*, the French *Alternative Energies and Atomic Energy Commission*. We thank G. Gallou for proposing this collaboration. This publication was made possible by the use of the *FactoryIA* supercomputer, financially supported by the *Ile-De-France Regional Council*.

References

[1] S. Moncayo, L. Duponchel, N. Mousavipak, *et al.*, "Exploration of megapixel hyperspectral LIBS images using principal component analysis," *J. Anal. At. Spectrom.*, vol. 33, 210–220, 2 2018. doi: 10.1039/c7ja00398f.

[2] R. Finotello, M. Tamaazousti, and J.-B. Sirven, "HyperPCA: A powerful tool to extract elemental maps from noisy data obtained in LIBS mapping of materials," *Spectrochim. Acta Part B*, vol. 192, 106418, 2022. doi: 10.1016/j.sab.2022.106418.

[3] R. Sattmann, I. Monch, H. Krause, *et al.*, "Laser-induced breakdown spectroscopy for polymer identification," *Appl. Spectrosc.*, no. 3, 456–461, 1998. doi: 10.1366/0003702981943680.

[4] L.-N. Li, X.-F. Liu, F. Yang, *et al.*, "A review of artificial neural network based chemometrics applied in laser-induced breakdown spectroscopy analysis," *Spectrochim. Acta Part B*, vol. 180, 106183, 2021. doi: https://doi.org/10.1016/j. sab.2021.106183.

[5] V. C. Costa, D. V. Babos, J. P. Castro, *et al.*, "Calibration strategies applied to laser-induced breakdown spectroscopy: A critical review of advances and challenges," *J. Braz. Chem. Soc.*, vol. 31, no. 12, 2439–2451, 2020. doi: 10.21577/0103-5053.20200175.

[6] V. Motto-Ros, S. Moncayo, F. Trichard, *et al.*, "Investigation of signal extraction in the frame of laser induced breakdown spectroscopy imaging," *Spectrochim. Acta Part B*, vol. 155, 127–133, 2019. doi: 10.1016/j.sab.2019.04.004.

[7] N. C. Dingari, I. Barman, A. K. Myakalwar, et al., "Incorporation of support vector machines in the LIBS toolbox for sensitive and robust classification amidst unexpected sample and system variability," *Anal. Chem.*, vol. 84, no. 6, 2686–2694, 2012. doi: 10.1021/ac202755e.

[8] P. Yaroshchyk, D. L. Death, and S. J. Spencer, "Comparison of principal components regression, partial least squares regression, multi-block partial least squares regression, and serial partial least squares regression algorithms for the analysis of Fe in iron ore using LIBS," *J. Anal. At. Spectrom.*, vol. 27, no. 1, 92–98, 2012. doi: 10.1039/c1ja10164a.

[9] T. Takahashi and B. Thornton, "Quantitative methods for compensation of matrix effects and self-absorption in laser induced breakdown spectroscopy signals of solids," *Spectrochim. Acta Part B*, vol. 138, 31–42, 2017. doi: 10.1016/j.sab.2017.09.010.

[10] E. D'Andrea, S. Pagnotta, E. Grifoni, *et al.*, "A hybrid calibrationfree/artificial neural networks approach to the quantitative analysis of LIBS spectra," *Appl. Phys. B*, vol. 118, no. 3, 353–360, 2015. doi: 10.1007/s00340-014-5990-z.

[11] L. Narlagiri and V. R. Soma, "Simultaneous quantification of Au and Ag composition from Au-Ag bi-metallic LIBS spectra combined with shallow neural network model for multi-output regression," *Appl. Phys. B: Lasers Opt.*, vol. 127, no. 9, 135, 2021. doi: 10.1007/s00340-021-07681-y.

[12] T. Chen, L. Sun, H. Yu, *et al.*, "Deep learning with laserinduced breakdown spectroscopy (LIBS) for the classification of rocks based on elemental imaging," *Appl. Geochem.*, vol. 136, 105135, 2022. doi: 10.1016/j.apgeochem.2021.105135.

[13] R. Caruana, "Multitask learning: A knowledge-based source of inductive bias," in Proceedings of the Tenth International Conference on International Conference on Machine Learning, ser. Icml'93, Amherst,

MA, USA: Morgan Kaufmann Publishers Inc., 1993, 41–48. isbn: 1558603077. doi: 10.1016 /b978-1-55860-307-3.50012-5.

[14] R. B. Anderson, J. F. Bell III, R. C. Wiens, *et al.*, "Clustering and training set selection methods for improving the accuracy of quantitative laser induced breakdown spectroscopy," *Spectrochim. Acta Part B*, vol. 70, 24–32, 2012. doi: 10.1016/j. sab.2012.04.004.

[15] F. O. Borges, G. H. Cavalcanti, G. C. Gomes, *et al.*, "A fast method for the calculation of electron number density and temperature in laser-induced breakdown spectroscopy plasmas using artificial neural networks," *Appl. Phys. B*, vol. 117, no. 1, 437–444, 2014. doi: 10.1007/s00340-014-5852-8.

[16] S. Chen, H. Pei, J. Pisonero, *et al.*, "Simultaneous determination of lithology and major elements in rocks using laserinduced breakdown spectroscopy (LIBS) coupled with a deep convolutional neural network," *J. Anal. At. Spectrom.*, vol. 37, no. 3, 508–516, 2022. doi: 10.1039/d1ja00406a.

[17] J.-M. Mermet, "Limit of quantitation in atomic spectrometry: An unambiguous concept?" *Spectrochim. Acta Part B*, vol. 63, no. 2, 166–182, 2008, Honoring Issue A Collection of Papers on Atomic, Molecular and Laser Spectroscopy Dedicated to James D. Winefordner. doi: 10.1016/j.sab.2007.11.029.

[18] W. Zhao, C. Li, C. Yan, *et al.*, "Interpretable deep learning assisted laser-induced breakdown spectroscopy for brand classification of iron ores," Anal. Chim. Acta, vol. 1166, 338574, 2021. doi: 10.1016/j.aca.2021.338574.

[19] X. Zhang, J. Xu, J. Yang, *et al.*, "Understanding the learning mechanism of convolutional neural networks in spectral analysis," *Anal. Chim. Acta*, vol. 1119, 41–51, 2020. doi: 10.1016/j.aca.2020.03.055.

[20] T. Völker, G. Wilsch, I. B. Gornushkin, *et al.*, "Interlaboratory comparison for quantitative chlorine analysis in cement pastes with laser-induced breakdown spectroscopy," *Spectrochim. Acta Part B*, vol. 202, 106632, 2022. doi: 10.1016/j.sab.2023.106632.

[21] J. Sansonetti, Handbook of Basic Atomic Spectroscopic Data, NIST Standard Reference Database 108, 2003. doi: 10.18434/T4FW23.

[22] J. Bergstra, R. Bardenet, Y. Bengio, *et al.*, "Algorithms for hyper-parameter optimization," in Advances in Neural Information Processing Systems, J. Shawe-Taylor, R. Zemel, P. Bartlett, et al., Eds., vol. 24, Curran Associates, Inc., 2011, doi:10.5555/2986459.2986743.

Design of the Angle-Resolving Electron Spectrometer Aboard the PRESET Mission

Benjamin Dyer[1,*], Xingzhi Cheng[1], Andrei Hanu[1,2], and Soo Hyun Byun[1]
[1]McMaster University, Canada
[2]Bruce Power, Canada
(*) dyerbm@mcmaster.ca

***Abstract*—We present a design study for the Pitch REsolving Spectrometer for Electron Transport (PRESET) mission, a CubeSat that is aimed at measuring the pitch angle density spectrum in low-Earth orbit (LEO). While a few missions have measured the pitch angle density spectrum, all have provided low resolution within the loss cone. PRESET will be capable of measuring pitch angle density of electrons with a resolution of 10 degrees or better and an energy range of 0.3-7 MeV filling gaps in both energy range and angular resolution in LEO. The spectrometer is designed to fit within a 10x10x10cm^3 volume (1U) including processing electronics so it can be integrated into a 3U CubeSat to be flown in a polar LEO. To achieve a 10-degree angular resolution, the detector employs a trough shaped collimator with a pin-hole type aperture followed by a single sided silicon strip detector. Aligned coaxially with the strip detector is a stack of 4 silicon detectors. To optimize the spectrometer design, extensive Monte Carlo simulations were carried out. The collimator was optimized by adjusting total length, width and height, aperture size, collimation plate spacing and material. A balance is found between increasing the instrument's geometric factor and reducing the aperture width which directly affects the counting rate and angular resolution of the instrument, respectively. To optimize the stacking geometry of the silicon detectors, simulations were carried out by varying the number and thickness of the silicon detectors, allowing the electron energy resolution and maximum detectable electron energy to be extracted. An optimum design was deduced to accomplish an outstanding performance with a minimum of silicon detectors. Simulation results are verified using a prototype spectrometer and a commercial pulse processing system.**

***Keywords* — silicon telescope, electron collimator, electron spectrometer, pitch angle density**

I. Introduction

The Pitch REsolving Spectroscopy for Electron Transport (PRESET) mission aims to take highly resolved angular measurements of the electron pitch angle density in the Van Allen Belts. The purpose of the mission is to measure the spectrum of electrons entering the Earth's atmosphere from the Van Allen Belts due to pitch angle scattering and other transport mechanisms. To this end the Electron Spectrometer Telescope (EST) is being developed as one of two science instruments aboard the 3U CubeSat. In this paper the design of the EST is explored through the use of Monte Carlo simulation and in lab testing.

II. Background

The Van Allen belts are regions of radiation surrounding the Earth, composed primarily of energetic electrons or protons and a much smaller fraction of heavier ions. Measurement of the spatial and temporal particle flux distributions have always been of great interest due to the negative health effects of energetic particles, strong coupling to space weather, the existence of 'killer' electrons which can disable satellites and understanding atmospheric-ionospheric-magnetospheric interactions (AIM) [1,2]. In more recent years, the loss of Van Allen belt electrons to the upper atmosphere has been of great interest because of the potential link to global warming caused by the depletion of the northern and southern ozone layers.

Research over the past decade has shown that precipitation of medium energy electrons (MEE), that is electrons within the loss cone, leads to production of nitric (NO_x) and hydrogen oxides (HO_x). In turn, these chemicals deplete atmospheric ozone through catalytic reactions [3]. Ozone loss due to NO_x and HO_x production has been predicted using multiple models with varying results. Predictions from the Canadian Middle Atmosphere Model suggest a 60% and 80% ozone loss due to MEE in the northern and southern hemisphere, respectively [4]. Application of the Thermosphere Ionosphere Mesosphere General Circulation Model and the Solar Climate Ozone Links model, suggests a 27% and 33% decrease in ozone, respectively [5,6]. Opposing these models, analysis using the HAMMONIA model suggests no significant effect on ozone due to MEE precipitation [7]. A large factor in discrepancies between models is a lack of accurate MEE precipitation data (i.e. electron spectrum data within the loss cone) leading models to be simulated using approximated data sets with significant uncertainty [8,9].

Measurement of the pitch angle distribution began with the Energetic Particle, Composition, and Thermal Plasma (ECT) suite on board the Van Allen Probes (VAP) [10]. The VAPs launch in 2012 taking the first detailed measurement of the electron pitch angle density with a resolution of 10 degrees [11]. Two spectrometers were used for these measurements, the first being MagEIS a low energy electron spectrometer which uses a slit collimator and magnetic filtering to determine electron energies. The energies detected range from 20keV to 4MeV across three instruments each targeting a different energy range [12]. REPT on the other hand used a cylindrical collimator and a silicon detector stack to measure electrons from 2MeV to 18MeV [11]. While the EST will measure a similar energy range to MagEIS, only a single instrument can be flown so a design more similar to the REPT will be explored.

Following the decommissioning of the VAPs, the ARASE mission began measuring the pitch angle densities of electrons in the energy range of 0.07-2 MeV using the HEP-L and HEP-H instruments. Both HEP-L and HEP-H are based on a pin-hole camera collimator design followed by a position sensitive silicon detector stack. The energies measured by the HEP-L and HEP-H detectors are 0.07-1MeV and 0.7-2MeV, respectively. The pin hole camera design allows high angular resolution measurements to be made with a resolution of 4 degrees in HEP-L and 10 degrees in HEP-H. The major disadvantage to these detector designs is a low geometric factor resulting in low counting rates. To combat this three of each instrument was flown, however this will not be possible on the PRESET mission therefore a modified approach to collimator design will be required. When at non-equatorial latitudes the loss cone increases slightly allowing an additional spectrum containing the edge and region just outside the loss cone [13,14].

The rest of this paper is structured as follows. Section III discusses methods used for simulations and how instrument performance is judged. Section IV discusses experimental verification of simulation results and comparisons between simulated and measured spectra. Section V dives into the instrument development, final design and predicted performance based on simulation data. Section VI concludes the paper and discusses next steps in development of the EST.

III. Simulation Methods

The EST design was carried out using Geant4 v11.0.1. Geant4 was selected as the Monte-Carlo code of choice due to ease of parameterization and high control over data analysis. MCNP was also used in the early stages of the design to simulate idealized collimators. Thanks to MCNPs ease of use, multiple different detector geometries could be explored quickly. However, fine tuning of the detector geometry involved manually rewriting input geometry cards and processing very large PTRAC files which required significant processing time. Geant4 was therefore used for the majority of fine tuning the detector geometry and determining the detector characteristics.

The Geant4 simulations used the EM4 physics package which simulates all relevant physical processes for the EST. Since the silicon detectors can vary in thickness from 50 µm to 1500 µm, the maximum step size for all particles was reduced to 1 µm from the default of 1 mm [15]. This has the added benefit of improving accuracy of forward scattering which is a significant source of uncertainty for the detector's performance.

A. Silicon Detectors

The Electron Spectrometer Telescope (EST) design consists of two major parts, the silicon detector stack and the collimator. The size of the silicon detector stack was determined by simulating a pencil beam of electrons onto a variety of silicon stacks. The energy spectrum of the pencil beam was taken from SPENVIS using the AE-8 solar maxima model in the expected 500km polar orbit. The effectiveness of each silicon stack was judged based on full absorption of electrons up to 4 MeV. Since the upper integral bin will begin at 5 MeV or less the detector does not require high energy particles to fully deposit their energy within the detector stack as long as the instrument can differentiate between electrons and other heavier particles.

B. Collimator

The collimator was designed by describing the geometry parametrically which allowed variations in geometry to quickly be explored. The collimator's effectiveness was judged based on minimizing the number of particles that scattered one or more times before entering the silicon stack, the overall angular resolution of the detector, and the detector's geometric factor.

The geometric factor and number of particles that scattered at least once were calculated using the same simulation. An omnidirectional field was simulated by generating particles uniformly on a sphere with an angular distribution of $\cos^2\theta$ where θ is the polar angle with respect to the sphere's normal vector and $0 \leq \theta \leq \pi/2$. The azimuthal angle was generated uniformly between 0 and 2π. Since these simulations are not focused on the shielding surrounding the detectors, simulation time was significantly reduced by only generating particles on a spherical cap positioned slightly above the entrance to the collimator. The cap was made large enough to ensure all possible angles of particles incident on the collimator were simulated. Based on this simulation set up the geometric factor (G) and its associated uncertainty can be calculated as

$$G = a\,\frac{n}{N}4\pi r^2, \tag{1}$$

$$\sigma_G = a4\pi^2 r^2\left(\frac{n}{N^2}\left(1-\frac{n}{N}\right)\right), \tag{2}$$

where n is the number of particles that are detected, N is the total number of particles, r is the radius of the sphere on which the particles are generated [16,17], and a is a scaling factor. The scaling factor is required since particles are generated only on the spherical cap and is calculated as

$$a = \frac{\text{Area of Cap}}{\text{Area of Sphere}}. \tag{3}$$

The cap is positioned such that all possible particles incident on the top of the collimator will be generated.

The number of particles which scatter before being detected is calculated based on whether the initial direction vector of the particle is the same as when the particle is incident on the first silicon detector. Additionally, the particle must deposit at least 5keV as this is the expected lower detection limit of the first detector. The metric used for evaluating the design was the counting ratio, calculated as the total number of particles that entered the detector over the number of particles that underwent no scattering prior to detection. An ideal collimator would produce a ratio of one, while all non-ideal collimators will produce a ratio greater than one.

The angular resolution of the detector was determined by simulating large area beams of electrons in 5-degree intervals from 0 to 25 degrees. The calculated angle of incidence was determined by measuring the position of the particle incident on the first detector when more than 5keV was deposited. The incident angle is then calculated as

$$\theta = \tan^{-1}\frac{x}{l}, \tag{4}$$

where x is the position where the particle is incident on the detector ($x = 0$) is at the geometric center of the detector) and l is the distance from the detector to the collimator's aperture.

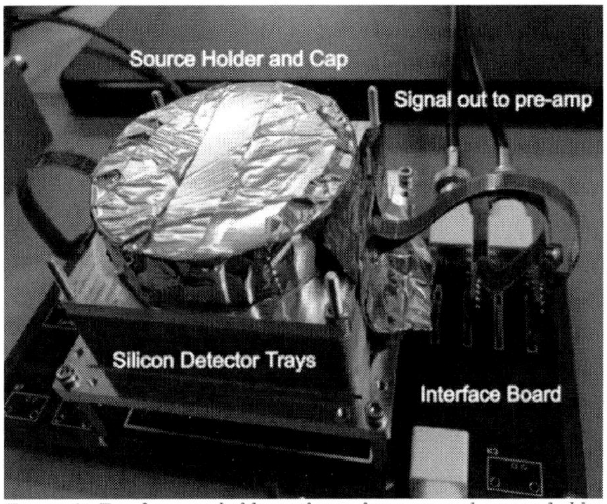

Figure 1: Test housing holding silicon detectors and source holder removed from vacuum chamber and Faraday cage.

IV. Simulation Verification

Verification of the simulations was accomplished using a testing system that could be simulated in Geant4 for result comparisons. The test system, shown in Figure 1, consists of two 1500μm-thick MSD040 silicon detectors produced by Micron Semiconductors Ltd. The detectors are housed in a custom casing which contains an exchangeable source holder for testing. A breakout board was produced that allows the MSD040 detectors to convert its Omnetics connector to a BNC capable of interfacing with the preamplifiers. A CAEN preamplifier module (A1422) with a gain of 45 mV/MeV and a maximum detector capacitance of 1000pF were used and connected to a CAEN DT5724 digitizer. The CAEN preamps were powered using a CAEN DT5523 linear desktop power distributor. The signal between the preamplifier and digitizer were observed on an oscilloscope to ensure correct pulse shape and noise characteristics. High voltage was applied through the preamplifiers using a CAEN DT5534EP high-voltage power supply with a reverse bias voltage of 300V (full depletion of the detectors occurred at 270V and 280V). When high voltage was applied a voltage ramp of 50V/s was used.

The test system was calibrated using a ^{244}Cm alpha source and a 140μm-thick MSD040 silicon detector. The detector was placed in a vacuum chamber which was pumped down to 0.9 Torr to minimize energy loss before reaching the detector. The calibration was validated using a BNC Model PB-5 pulse generator in place of the silicon detector to ensure the system energy response was linear to within manufacturer standards.

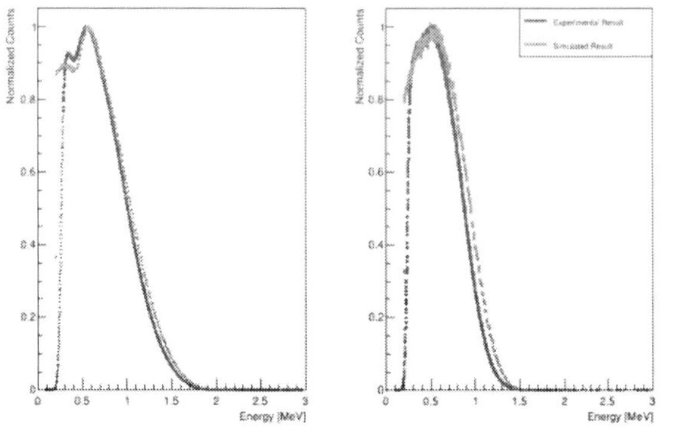

Figure 2: Experimental versus simulated spectrum from a $^{90}Sr/^{90}Y$ source incident on two 1500μm-thick MSD040 silicon detectors in the test housing. First detector is on the left, second detector is on the right.

The test system was also placed in a light tight metal box acting as a faraday cage. A $^{90}Sr/^{90}Y$ source was placed in the source holder and the spectrum in both detectors was collected under atmospheric conditions. The experimental set up was simulated in Geant4 allowing comparisons between results shown in Figure 2. Both the simulation and experimental results have been normalized by their highest count and are rebinned to 10keV/bin to allow for easier comparison. Simulation results show good agreement with experimental data with all bins being within 5%. The largest discrepancy occurs at low energies where the simulation predicts lower counting rates than are observed by 5%. Drop off of detected electron below approximately 250keV can be seen and is caused by system noise making measurement of the lower energies impractical.

V. Instrument Design

The design of the EST is constrained since it must fly on a 3U CubeSat. The total volume of the instrument is limited to 1U (10x10x10 cm^3) and the mass should be minimized. Total power draw of the instrument will also need to be kept below 3W.

A. Collimator

A pin hole camera that allows particles to enter from within a cone was considered. While such a design allows a high geometric factor, the geometric factor as a function of pitch angle is not constant, reaching a minimum for pitch angles of 0 or 180 degrees. This design would therefore reduce the number of particles detected at the center of the loss cone. To ensure the geometric factor remains constant across pitch angles, the aperture of the collimator was changed to a rectangle and the acceptance angle was limited to 5 degrees in the y-axis and kept at 60 degrees in the x-axis. With such a design the geometric factor is constant across all pitch angles, however the geometric factor is severely limited.

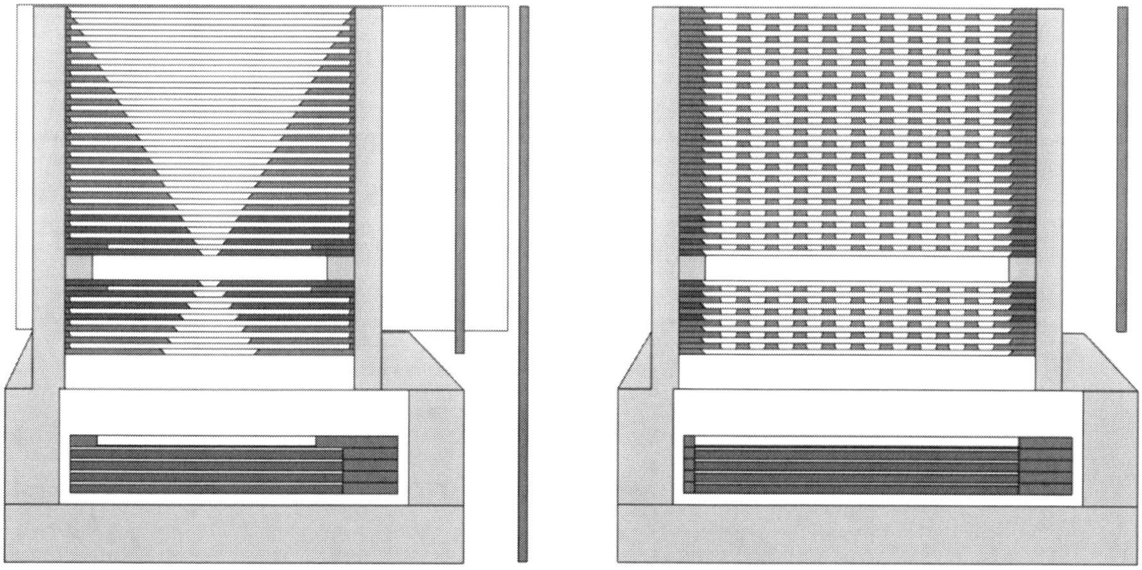

Figure 3: Cross-section views of the instrument in the x-z plane (left) and y-z plane (right). Materials are color coded as Green=FR4, Blue=Silicon, Light Grey = Aluminum, Dark Grey = Tungsten, Orange = Iron or Aluminum.

To solve the reduction in geometric factor, a new collimator design has been developed shown in Figure 3. The collimator is formed using two sets of alternating plates where one set of plates forms the converging-diverging shape of the pinhole camera while the second set of grated plates segment the trough shaped collimator into multiple smaller collimators with a y-axis acceptance angle of 5 degrees. Doing so results increased the geometric factor linearly with the number of smaller collimators formed.

The collimator was designed by varying the length, width, height, material, and edge angle of the collimator plates. The total collimator length is restricted by the need to have a y-axis acceptance angle of 5 degrees. To achieve this the collimator is 60mm in height and each slit in the grated plates are 2.62mm in width resulting in a y-axis acceptance angle of 5.0006 degrees. The number of slits was then varied from 1 to 12 and it was noted that up to 10 slits can be positioned side by side before electrons begin entering the silicon stack after only one scattering event. The total y-axis length of the collimator is restrained to 66mm due to the 1U form factor and the need for support electronics. Therefore, with 8mm of shielding around the collimator the 10 slits result in 2.38mm of material between each slit.

The acceptance angle in the x-axis formed by the converging-diverging plates was predominantly driven by size constraints. Increasing the acceptance angle requires the first silicon detector to be positioned closer to the aperture of the collimator to ensure all electrons can be detected. However, the further away from the first detector is from the aperture the larger the aperture can be for the same angular resolution. The silicon stack was placed 32mm from the aperture as this is the furthest possible without reducing performance of the collimator and staying within the volume constraints. These considerations make the largest possible acceptance angle 65 degrees, with the actual acceptance angle set as 60 degrees to allow for small modifications during manufacturing. Given the 32mm distance from the aperture to the first position sensitive silicon detector, the aperture width in the x-direction was maximized while maintaining acceptable angular resolution. The final aperture size is 2.50mm in the x-axis which when combined with the 10 slits, 2.62mm y-axis width, and 60-degree acceptance angle gives a geometric factor of $1.68(2) \times 10^{-2}\ cm^2 sr$.

The bevel angle on the plates was varied from 0 to 80 degrees in 5-degree increments. As expected, at a 0-degree bevel there is a significant increase in forward scattering resulting in many particles entering the silicon stack after a single scattering event. The collimator performance improved as the bevel angle of the converging-diverging plates was increased but remained fairly constant for angles between 35 and 50 degrees. Bevel angles above 50 degrees also began to decrease collimator performance due to particles penetrating through the much thinner material at the edge of the collimator. The grated plates could not have their bevel angle increased past 15 degrees since the reduction in material reduced the effectiveness of the 5-degree collimation. Therefore, the bevel angle of the grated plates is 15 degrees, and the bevel angle of the converging-diverging plates is 45

 https://doi.org/10.1051/epjconf/202328802001

degrees.

Material choice for the collimator was limited to Aluminum, Tantalum, Tungsten, and Iron as the first three are materials often used for space born electron spectrometers and Iron is both easy to machine and readily available. It was noticed during simulations that there is no significant performance difference between Tantalum and Tungsten, therefore Tungsten was used for most simulations since it is the most likely material to be used in the final collimator. The best performance was provided by a fully Tungsten collimator providing a count ratio of 1.062(3) while the worst performance was provided by the fully aluminum collimator at a count ratio of 1.22(1). A fully Tungsten or Tantalum collimator would be impractical due to the mass budget, so instead a combination of Aluminum and Tungsten plates was tested. It was found that grouping Tungsten plates together at the collimator aperture as seen in Figure 3 has the greatest effect on performance resulting in a counting ratio of 1.13(1). Adding additional tungsten or tantalum plates elsewhere in the collimator does improve performance slightly but at a significant increase in mass that cannot be accommodated. Changing the aluminum plates to Iron does significantly improve performance at a slight increase in mass resulting in a count ratio of 1.084(1). Both collimators with iron and aluminum will be tested before deciding on the final materials based on both total satellite mass budget and real-world collimator performance.

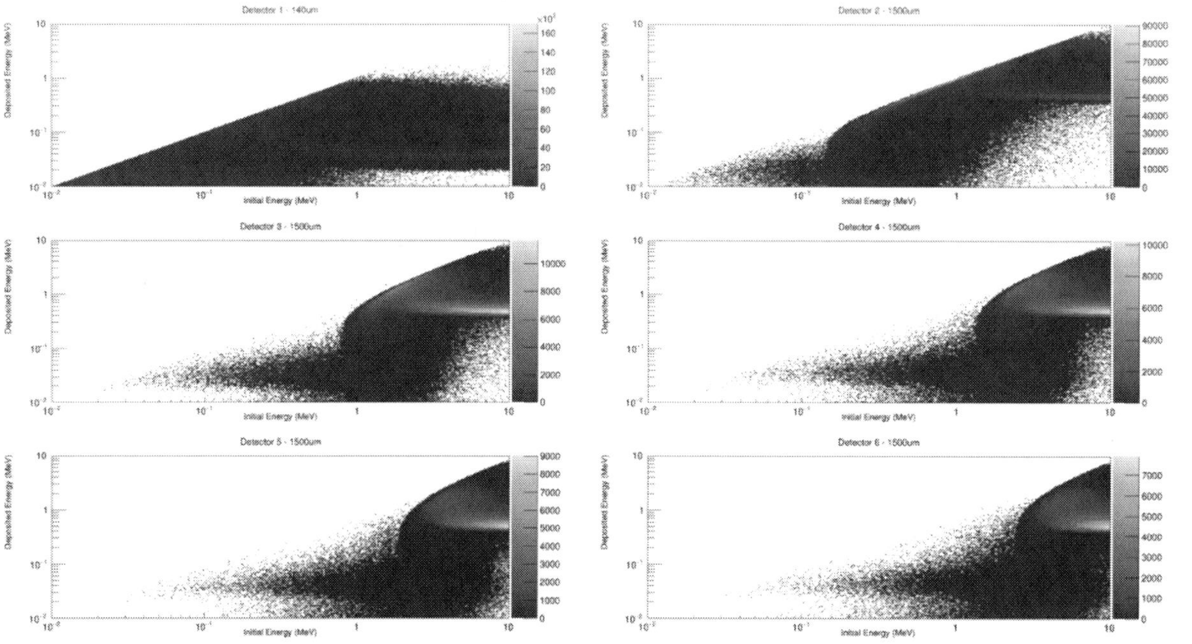

Figure 4: Response matrix of 6 Silicon detectors to electrons. Detector 1 is a 140um thick detector at the front of the detector stack while all other detectors are 1500um thick. Electron energies are distributed as such that each bin along the x-axis (initial energy) has the same number of initial particles. The color scale corresponds to the number of particles detected.

B. *Silicon Stack*

The silicon stack was simulated using a thin front detector followed by 4-7 1500-um thick detectors. The thick detectors were not adjusted as we intend to use the minimum number of detectors possible so the thickest available detectors will be used. The front thin detector will be used as both the initial trigger detector and will determine the position of the particle. The thin detector's thickness was simulated at 50 μm, 140 μm, 300 μm and 500 μm. Using the 300 μm and 500 μm detectors results in an average deposited energy of 110±15 keV and 90±10 keV respectively. Since the detector has no way to distinguish between an electron and a proton that fully deposit their energy in the first detector, minimizing the average energy deposited by a high energy electron will also lower the minimum detectable energy. The 140 μm and 50 μm detectors give an average deposited energy of 30±10 keV and 15±5 keV, both of which allow lower energy electrons to be distinguished from protons. The final design will use either a 50 μm or 140 μm detector depending on real world performance and manufacturability, with preference given towards the 50 μm detector.

The response matrix of the detector stack is shown in Figure 4 for a 140 μm thick front detector followed by five 1500 μm thick detectors. As expected, low energy electrons fully deposit their energy in the front detector while higher energy electrons deposit approximately 30 keV in the front detector. Electrons that pass through the first detector deposit either their entire remaining energy or between 400 keV to 900 keV in each subsequent detector until the electron has lost all energy or exited the detector stack. Electrons with initial energy below 4MeV typically deposit all their energy in the silicon stack while electrons with initial energy greater than 4

MeV fully penetrate the stack. Given the maximum energy deposited is approximately 4 MeV throughout the stack, the upper integral energy bin of the detector will be E>4MeV. The final silicon stack will be composed of one 140 μm thick front detector followed by four 1500 μm thick detectors.

C. Instrument Performance

As previously mentioned, the geometric factor of the instrument is $1.68(2) \times 10^{-2}\ cm^2 sr$. The geometric factor can be converted to an expected counting rate using the SPENVIS AE-8 model [18]. Given a polar orbit between 400 and 500 km, the maximum expected counting rate at the poles is estimated to be 2500 Hz and the maximum counting rate in the South Atlantic Anomaly is estimated to be 9,500 Hz. These counting rates are comparable with previous space spectrometers and should be sufficient for producing useful data sets while not causing excessive dead time in the digital processing.

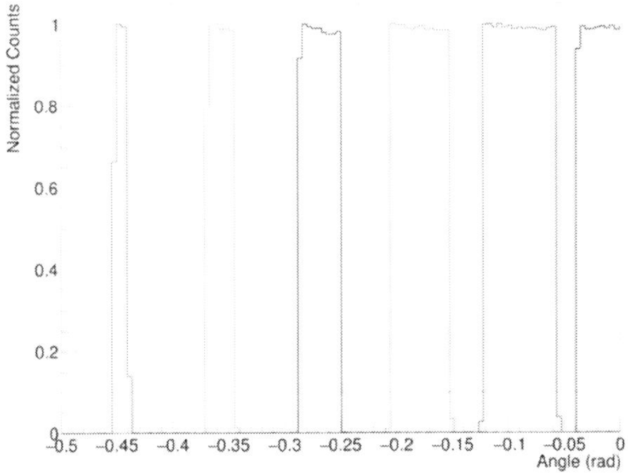

Figure 5: Simulated angular response of the EST to electrons with initial angles of 0,5,10,15,20, and 25 degrees.

Figure 5 shows the angular response of the instrument after removing particles that deposited less than 5 keV in the front detector, normalized by the highest bin counts per angle tested. Note that the angular response is symmetric about 0 degrees therefore the Figure 5 is plotted from -28 to 0 degrees. The particles used in the simulation follow the energy spectrum expected in a polar low earth orbit as extracted from SPENVIS [18]. Using the full width at 10% max gives an angular resolution of 4.5 degrees when the particle is normal to the detector stack and a resolution of 0.75 degrees when the particle is at the edge of the collimator acceptance angle. This performance significantly surpasses the requirement of 10-degree resolution; however it is expected that in real world testing, system noise will reduce the resolution causing the angular distribution to tend towards a Gaussian distribution.

VI. Conclusions

To measure electron pitch angle distribution in the Van Allen Belts a new electron spectrometer has been developed through simulation. Simulations were carried out in Geant4 allowing the instrument geometry to be varied to produce an instrument with excellent performance. The new collimator design allows multiple pin hole camera's to be integrated into a single structure, reducing the number of silicon detector stacks required while providing a high geometric factor of $1.68(2) \times 1.68(s) \times 10^{-2}\ cm^2 sr$. The geometric factor translates to a maximum counting rate of approximately 2,500 Hz at the poles and 9,500 Hz in the South Atlantic Anomaly. The instrument contains a stack of 5 silicon detectors, with the first detector being either 50 μm or 140 μm in thickness and the following detectors each being 1500 μm thick. This results in a minimum detectable energy of 30keV and a maximum discernable energy of 4 MeV. Electrons with energies above 4 MeV will be placed in a high energy integral bin. Using a silicon strip detector with a pitch of 500 μm in conjunction with the collimator results in an angular resolution of 4.5 degrees. It is expected the angular resolution of the final instrument will be worse due to system noise, however it should stay well within the 10-degree resolution limit.

The next steps will be to fabricate the collimator plates out of iron, tungsten, and aluminum so different configurations of materials can be tested and the real-world detector performance can be evaluated. A vacuum system and electron gun are currently being developed to validate the instruments angular and energy resolutions. Custom electronics to process the raw signals and produce histogram data are currently under development and will be compared with the commercial system performance before integration into the PRESET satellite.

References

[1] National Academies of Sciences, Engineering, and Medicine, "Science discoveries and technical advances" in Progress Toward Implementation of the 2013 Decadal Survey for Solar and Space Physics: A Midterm Assessment, Washington, DC, USA: NAP, 2019, Ch.2,pp.34-48.

[2] Zong, Q., Wang, Y., Yuan, C., Yang, B. Wang, C., and Zhang, X., "Fast acceleration of "killer" electrons and energetic ions by interplanetary shock stimulated ULF waves in the inner magnetosphere". *Chin.Sci.Bull.* **56**, 1188-1201 (2011)

[3] Anderson, M.E., Verronen, P.T., Rodger, C.J., Clilverd, and M.A., Seppälä, A., "Missing driver in the Sun-Earth connection from energetic electron precipitation impacts mesospheric ozone. *Nat. Commun.* **5**, 5197 (2014)

[4] Semeniuk, K., Fomichev, V.I., McConnell, J.C., Fu, C., Melo, S.M.L. and Usoskin, I.G., "Middle atmosphere response to the solar cycle in irradiance and ionizing particle precipitation." *Atm.Chem.and Phys.*, **11**(10), 5045-5077 (2011)

[5] Codrescu, M.V., Fuller-Rowell, T.J., Roble, R.G., and Evans, D.S., "Medium energy particle precipitation influences on the mesosphere and lower thermosphere", *J. Geophys. Res.*, **102**, 19977-19987 (1997)

[6] Arsenovic, P., Rozanov, E., Stenke, A., Funke, B., Wissing, J., Mursula, K., *et al.*, "The influence of middle range energy electrons on atmospheric chemistry and regional climate. *J. Atmo. Sol.-Terr.Phys*, **149**, 180-190 (2016)

[7] Meraner, K., and Schmidt, H., "Climate impact of idealized winter polar mesospheric and stratospheric ozone losses as caused by energetic particle precipitation." *Atmo. Chem. and Phys.*, **18**(2), 1079-1089 (2018)

[8] Tyssøy, H.N., Haderlein, H.N., Sandanger, M.I., and Stadsnes, J., "Intercomparison of the POES/MEPED loss cone electron fluxes with the CMIP6 parametrization." *J. Geo.Phys. Res.Space.Phys.*, **124**, 628-642 (2019)

[9] Tyssøy, H.N., Sadanger, M.I., Ødegaard, L.-K.G., Standsnes, J., Aasnes, A., Zawedde, A.E., "Energetic electron precipitation into the middle atmosphere – Constructing the losss cone fluxes from MEPED POES." *J. Geo.Phys.Res.Space.Phys.*, **121**(6), 5693-5707 (2016)

https://doi.org/10.1051/epjconf/202328802001

[10] Stratton, J.M., Harvey, R.J., and Heyler, G.A., "Mission Overview for the Radiation Belt Storm Probes Mission." *Space. Sci. Rev.*, **179**, 29-57 (2013)
[11] Bakr, D.N., Kanekal, S.G., Hoxie, V., Li, X., Jaynes, A.N., Zhao, H., Elkington, S.R., *et al.*, "The relativistic Electron-Proton Telescope (REPT) Investigation: Design, Operational, Properties, and Science Highlights." *Space. Sci. Rev.*, **219**, 68 (2021)
[12] Claudepierre, S.G., Blake, J.B., Boyd, A.J., Clemmons, J.H., Fennel, J.F., Gabrielse, C., *et al.*, "The Magnetic Electron Ion Spectrometer: A Review of On-Orbit Sensor Performance, Data, Operations, and Science." *Space Sci. Rev.*, **217**,80 (2021)
[13] Mitani, T., Takashima, T., Kasahara, S., Miyake, W., Hirahara, M., "High-energy electron experiments (HEP) aboard the ERG (Arase) satellite." *Earth, Planets and.Space.* **70**(77) (2018)
[14] Kasahara, S., Miyoshi, Y., Yokota, S., Mitani, T., Kasahara, Y., Matsuda, S., *et al.*, "Pulsating aurora from electron scattering by chorus waves.", *Nature*, **554**, 337-340 (2018)
[15] Soti, G., Wauters, F., Breitenfeldt, M. Finlay, P., Kraev, I.S., Knecht, *et al.*, "Performance of Geant4 in Simulating Semiconductor Particle Detector Response in the Energy Range Below 1 MeV", *Nucl. Inst. Meth. Phys. Res. Sect. A*, **870**, 64-67 (2018)
[16] Yando, K. Millan, R.M., Green, J.C., and Evans, D.S., "A Monte Carlo simulation of the NOAA POES Medium Energy Proton and Electron Detector instrument." *J.Geo.Phys.Res*, **116**(A10231), (2011)
[17] Sullivan, J.D., "Geometrical factor and directional response of single and multielement particle telescopes." *Nuc.Inst.Meth.*, **95**, 5-11 (1971)
[18] Kruglanski, M., de Donder, E., Messios, N., Hetey, L., Calders, S., Evans, H., *et al.*, "Space Environment Information System (SPENVIS)" in *38th COSPAR Sci. Assem.*, Bremen, Germany, 2010, pp8

https://doi.org/10.1051/epjconf/202328802002

R&D status for an innovative crystal calorimeter for the future Muon Collider

C. Cantone[1], S. Ceravolo[1], F. Colao[2], E. Di Meco[1,3,*], E. Diociaiuti[1], P. Gianotti[1], A. Liedl[1], D. Lucchesi[4], D. Paesani[1,3], N. Pastrone[5], G. Pezzullo[6], A. Saputi[7], I. Sarra[1], L. Sestini[4], D. Tagnani[8]

[1] INFN, Laboratori Nazionali di Frascati, Via Enrico Fermi 54, 00054 Frascati, Italy
[2] Enea Frascati, Via Enrico Fermi 45, 00044 Frascati, Italy
[3] Dipartimento di Fisica, Università degli Studi di Roma Tor Vergata, 00133 Roma, Italy
[4] INFN, Sezione di Padova, Via Francesco Marzolo 8, 35131 Padova, Italy
[5] INFN, Sezione di Torino, Via Pietro Giuria 1, 10125 Torino, Italy
[6] Yale University, New Heaven (CT), USA
[7] INFN, Sezione di Ferrara, Via Saragat 1I, 44122 Ferrara, Italy
[8] INFN, Sezione di Roma Tre, Via della Vasca Navale 84, 00146 Roma, Italy
(*) elisa.dimeco@lnf.infn.it

Abstract: **The measurement of physics processes at new energy frontier experiments requires excellent spatial, time, and energy resolutions to resolve the structure of collimated high-energy jets. Calorimeters, as other detectors, must face this increasing performance demand. In a future TeV-scale Muon Collider, the beam-induced background (BIB) represents the main challenge in the design of the detectors and of the event reconstruction algorithms and can pose serious limitations to the physics performance. However, it is possible to reduce the BIB impact on the Muon Collider calorimeter by exploiting some of its characteristics and by ensuring high granularity, excellent timing, longitudinal segmentation and good energy resolution. The proposed R&D is an innovative semi-homogeneous electromagnetic calorimeter based on stackable modules of lead fluoride crystals (PbF_2) readout by surface-mount UV-extended Silicon Photomultipliers (SiPMs): the Crilin calorimeter (CRystal calorImeter with Longitudinal INformation). The calorimeter should operate in a very harsh radiation environment, withstanding yearly a neutron flux of 10^{14} n_{1MeV} /cm^2 and a dose of 100 krad.**

In this paper, the radiation tolerance measured in several irradiation campaigns and the timing performances evaluated during a test beam at CERN-H2 with 120-GeV electron are discussed. A description of the latest prototype Proto-1, that will be shortly tested, is also provided.

Keywords — **Electromagnetic calorimeters, PbF_2, SiPM, Crystals, High Granularity.**

I. Introduction

NEW High Energy Physics research requires innovative accelerating machines to unlock the multi-TeV range. An interesting option is represented by a Muon Collider facility [1] which, with respect to hadronic machines, would be characterized by an exact knowledge of the initial state (free from parton density functions), by a limited QCD background and, with respect to an electron machine, would reach higher energies because of its very reduced bremsstrahlung losses. Nevertheless, many aspects have hindered the realization of such project: from the point of view of detectors construction one of the main challenges is represented by the Beam Induced Background (BIB). This BIB consists of a pervasive flux of secondary and tertiary particles coming from the interaction of the muons decay products with the machine elements creating a very harsh radiation environment. This means that the whole machine detector interface must be optimized in order to reduce as much as possible this huge contribution.

Concerning the electromagnetic calorimeter barrel the baseline solution consists of 64 million 5×5 mm^2 silicon sensors sampled with 40 layers of Tungsten for a total of 64 million readout channels [2], thus representing a significant technological issue as well as expensive. Indeed, other options are still under investigation. The presented R&D, the Crilin calorimeter [3], is one of the leading alternative choices for this purpose. It consists of a semi-homogeneous electromagnetic calorimeter made of Lead Fluoride Crystals (PbF_2) matrices where each crystal is readout by 2 series of 2 UV-extended surface mount SiPMs. Crilin was indeed optimized in the ambit of the Muon Collider experiment as a candidate design for an electromagnetic barrel calorimeter, because of its fine granularity, excellent timing resolution, good pileup capability and high resistance to radiation.

II. The Crilin calorimeter

The Crilin calorimeter Crilin has a modular architecture based on stackable and interchangeable sub-modules made of matrices of 10 × 10 × 40 mm^3 Lead Fluoride (PbF_2) crystals each readout by two series of two UV-extended 10 μm pixel-size Hamamatsu SiPMs with a surface of 3×3 mm^2 providing a 36% coverage of the crystal area. The main features achievable with this choice are the following:

- High response speed since the Cherenkov response of PbF_2 is instantaneous with respect to the particle passage.
- Narrow signals therefore an excellent ability to resolve temporally close events at high rate leading to a good pileup capability.
- Good light collection that enables a good energy resolution throughout the whole dynamic range.
- Good resistance to radiation needed to withstand the Muon Collider harsh environment.
- Longitudinal segmentation that is crucial to distinguish signal showers from BIB.
- Fine granularity scalable with SiPMs pixel dimensions.

By adapting crystals transversal and longitudinal dimensions to maximize the signal/background ratio and choosing the proper SiPMs the goal is to achieve similar or better performance than the Si-W ECAL baseline solution. Indeed, the simulation framework of the International Muon Collider Collaboration has been employed to compare the performances of the two barrels. With this purpose the W-Si ECAL barrel used in this simulation framework has been substituted with Crilin. The full simulation of b-jets has been used for this purpose, and the beam-induced background was simulated as well. The jet reconstruction efficiency and jet p_T resolutions are presented in Figure 1. It can be noticed that the performances are similar in the two cases but, at the same time, the money cost of Crilin is a factor 10 less.

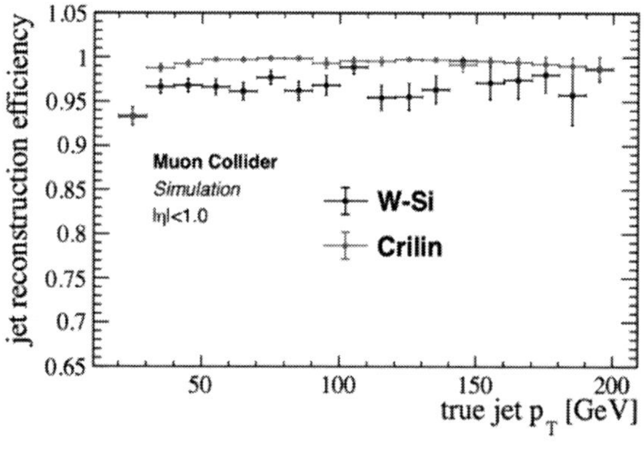

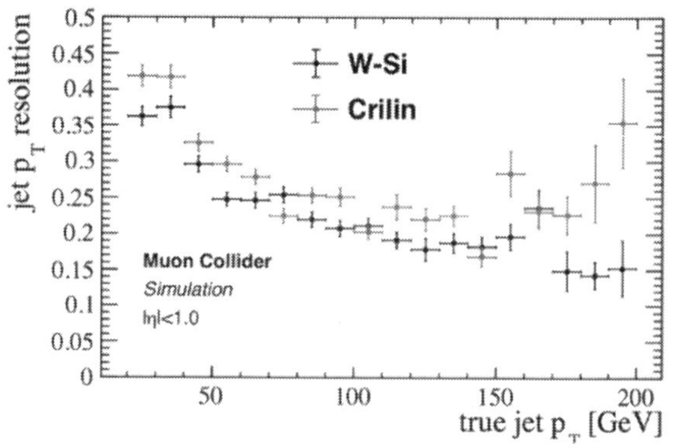

Fig. 1. Top-panel: jet reconstruction efficiency as a function of the jet p_T, obtained by using the Crilin ECAL barrel and the W-Si ECAL barrel. Bottom-panel: jet p_T resolution as a function of the jet p_T, obtained by using the Crilin ECAL barrel and the W-Si ECAL barrel.

A. Radiation hardness

Because of the Beam Induced Background a very harsh environment is expected for the Muon Collider. That is why a FLUKA [4] simulation at $\sqrt{s}$=1.5 TeV has been performed, thus resulting in the Total Ionizing Dose (TID) and neutron fluence levels showed in Figure 2.

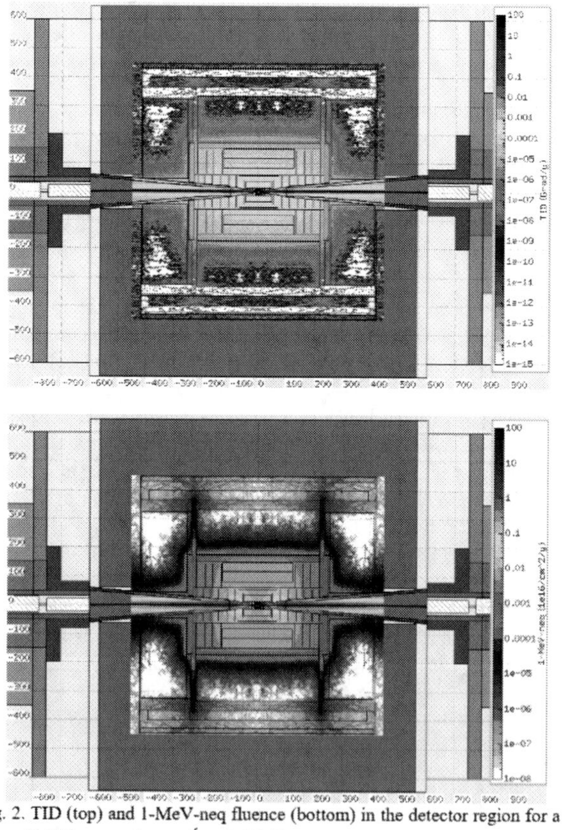

Fig. 2. TID (top) and 1-MeV-neq fluence (bottom) in the detector region for a Muon Collider operating at $\sqrt{s}$ = 1.5 TeV.

Regarding the ECAL barrel region the expected neutron fluence and dose are respectively 10^{14} $n_{1MeV}/(cm^2$ year) and 10^{-4} Grad/year.

Once the radiation levels were known it was necessary to check the radiation hardness of the single components of the calorimeter and to examinate different choices. Starting from crystals, as previously anticipated, the baseline choice for Crilin is represented by PbF_2. At the same time also $PbWO_4$-UF [5] appears to be a valid candidate, indeed this material embeds high density, good light yield, high radiation resistance and fast response speed since it combines both the prompt Cherenkov and a fast scintillation component, yielding a dominant emission with a decay time less then 700 ps.

In order to check the resistance to radiation of both crystals a TID irradiation campaign, after a preliminary one in 2021 [6], has been recently carried out at ENEA-Calliope facility with ^{60}Co. The resulting transmittance spectra after irradiation can be found in Figure 3 for both crystals. After a TID > 35 Mrad no significant decrease in transmittance observed for PbF_2, this happens also for $PbWO_4$-UF up to 100 Mrad.

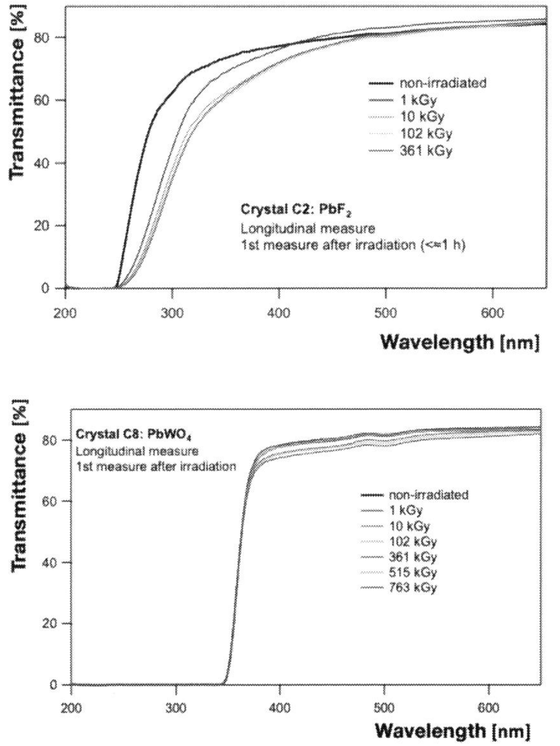

Fig. 3. Transmission spectra obtained in the different irradiation steps for the PbF_2 (top) and $PbWO_4$-UF (bottom) crystal samples.

A second irradiation campaign has been performed to check the radiation hardness of the SiPMs at the Frascati Neutron Generator (FNG-ENEA) facility with 14 MeV neutrons with a fluence up to 10^{14} n/cm^2. Neutron irradiation represents indeed one the main concerns when using these sensors since it leads to a significant increase of dark current. The sample under test were two: one series of 15 μm pixel-size and another one of 10 μm pixel-size Hamamatsu SiPMs. By measuring the amount of dark current after the irradiation at three different temperatures (the results are summarized in Table 1 and 2) it was clear that the right choice of SiPMs pixel-size in order to cope with the Muon Collider radiation environment was represented by the 10 μm once, since they have a more manageable dark rate contribution.

TABLE I
BREAKDOWN VOLTAGES AND DARK CURRENTS FOR 15 μM SIPMS

T [°C]	V_{br} [V]	$I(V_{br}+4V)$ [mA]	$I(V_{br}+6V)$ [mA]	$I(V_{br}+8V)$ [mA]
-10 ± 1	75.29 ± 0.01	12.56 ± 0.01	30.45 ± 0.01	46.76 ± 0.01
-5 ± 1	75.81 ± 0.01	14.89 ± 0.01	32.12 ± 0.01	46.77 ± 0.01
0 ± 1	76.27 ± 0.01	17.38 ± 0.01	33.93 ± 0.01	47.47 ± 0.01

TABLE 2
BREAKDOWN VOLTAGES AND DARK CURRENTS FOR 10 μM SIPMS

T [°C]	V_{br} [V]	$I(V_{br}+4V)$ [mA]	$I(V_{br}+6V)$ [mA]	$I(V_{br}+8V)$ [mA]
-10 ± 1	76.76 ± 0.01	1.84 ± 0.01	6.82 ± 0.01	29.91 ± 0.01
-5 ± 1	77.23 ± 0.01	2.53 ± 0.01	9.66 ± 0.01	37.51 ± 0.01
0 ± 1	77.49 ± 0.01	2.99 ± 0.01	11.59 ± 0.01	38.48 ± 0.01

B. *Timing performances on a single crystal prototype*

During 2022 a preliminary evaluation of the timing performances of Crilin has been carried out by means of a Beam Test at CERN-SPS beamline with 120-GeV electrons. The device under test consisted in a single layer prototype, Proto-0, with only one crystal readout by 2 series of 10 μm pixel-size Hamamatsu SiPMs connected to a custom Front End Electronics (discussed in Sec. III). The goals of this test were [7]: *i*) validate CRILIN new readout electronics and readout scheme *ii*) study systematics of light collection in small crystals with high refractive index *iii*) measure the time resolution achievable with different crystal choices, indeed both PbF_2 and $PbWO_4$-UF were tested.

Regarding the *ii*) some very interesting effects were observed that is why two different orientations of the prototype were studied. The first one was the "front" mode, where the injected electron finds upstream the crystal followed by the photo-sensors. The second one was the "back" mode, here Proto-0 was rotated by 180° degrees, meaning that SiPMs collect only light that has been uniformly reflected inside the crystal, avoiding effects linked to the Cherenkov light directionality. Clear effects of directionality were indeed showed in the "front" runs, especially in the charge sharing variable and time difference as a function of the beam position resulting in two population in the channels time difference distribution as observed in Figure 4-a. This effect is instead negligible in the "back" configuration (Figure 4-b), showing that this is a Cherenkov light propagation systematics, as also confirmed via MC optical simulation [5]. For the "front" run it was possible to perform an event-by-event correction thanks to the almost liner dependence between the asymmetry variable, consisting in the ratio between the two readout channels charge difference and the charge sum. When the correction was applied it was possible to recover a single population time difference distribution, as showed in Figure 4-c.

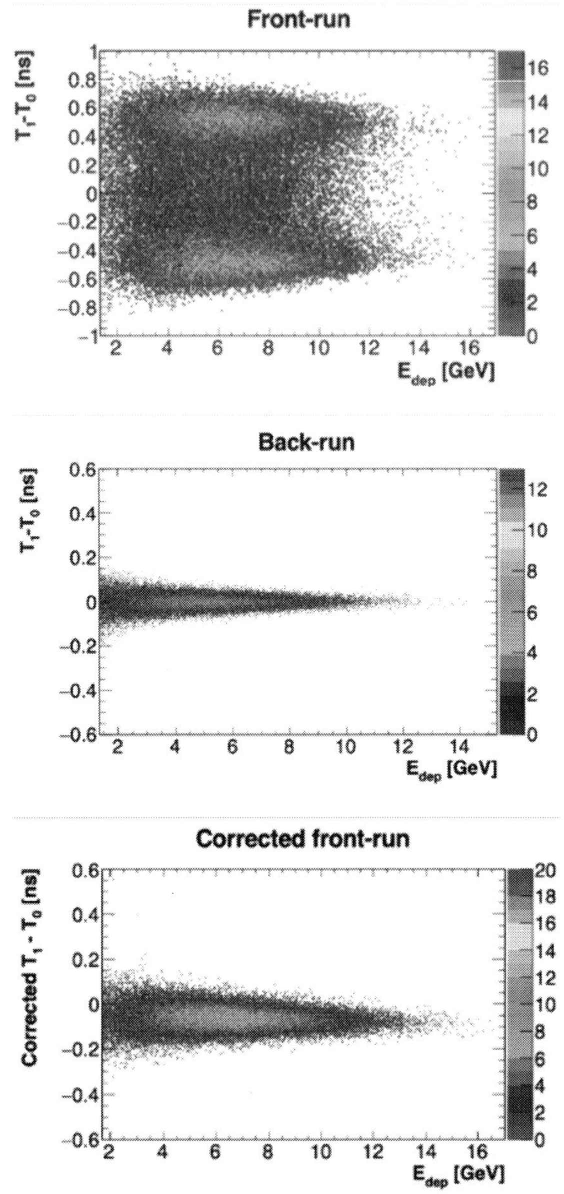

Fig. 4. a) Time difference between the two readout channels as a function of E_{dep}, for front configuration runs. The splitting of the distribution from position-dependent effects is evident. b) Distributions for runs in back configurations. c) Same distribution for front runs right after the event-by-event correction based on the dependence of the time difference from the charge asymmetry.

Starting from these distributions it was possible to estimate the time resolution from a Gaussian distribution fit applied on 10 E_{dep} slices. The results are presented in Figure 5 for all runs. It can be noted that the time resolution as a function of deposited energy was fitted using the function:

$$\sigma_{MT} = \frac{\sigma_{T1-T0}}{2} = \frac{a}{E_{dep}} \oplus b$$

where the subscript MT refers to the resolution obtained for the mean time for the two readout channels of the single calorimeter cell. Despite the higher light yield the timing performances of the front runs are still worse than the back runs, this is still due to some residual position-dependent light-transport effects. It should also be highlighted that, due to the purely Cherenkov nature of the light, the time resolution for PbF_2 is better than that for $PbWO_4$-UF in both configurations, even though the light yield for PbF_2 is only about half of that for $PbWO_4$-UF. Anyhow, the results for both crystals are deeply within the requirements set by the Muon Collider collaboration having indeed a worst-case time resolution for $E_{dep} > 3$ GeV less than 25 ps for PbF_2 and less than 45 ps for $PbWO_4$-UF.

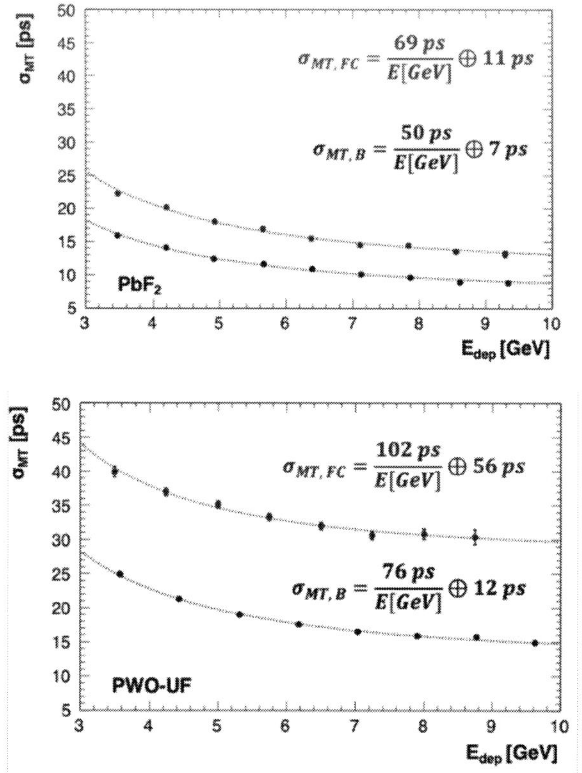

Fig. 5. Mean-time resolution of a single calorimeter cell for PbF_2 (a) and $PbWO_4$-UF (b) as a function of E_{dep} over the range 3-10 GeV. Front-configuration corrected runs ($\sigma_{MT,\ FC}$) are shown in blue, while back-configuration runs ($\sigma_{MT,\ B}$) are shown in black.

III. Proto-1

The novel idea behind the Crilin ECAL is to use multiple layers of PbF_2 crystals and thin surface-mount device (SMD) SiPMs stacked on top to provide longitudinal information of the shower. In order to validate the design choices, a larger prototype was recently built, Proto-1. The Proto-1 design was optimized with the simulation studies starting from transverse and longitudinal dimensions of 0.7 R_M and 8.5 X_0 (~0.3λ) respectively. This size comes from a compromise of an acceptable containment of ~100 GeV electrons showers and cost constraints. Results will be extrapolated to the optimum length of the Muon Collider calorimeter of the order of 20 X_0.

Proto-1, in Figure 6, consist of two layers of 3×3 PbF_2 crystals, each readout by two series of two UV-extended SiPMs, as already done in Proto-0.

Fig. 6. Proto-1 images: a single module during crystal installation, the locking system and the prototype fully assembled.

Two stackable and interchangeable submodules assembled by bolting house the crystals, moreover the light-tight case also embeds the front-end electronic boards and cooling system. Indeed, the on-detector electronics and SiPMs must be cooled during operation, to improve and stabilize the performance of SiPMs against irradiation. This design can remove the heat load due to the increased photosensor leakage current after exposure to the expected 10^{14} n_{1MeV}/cm^2 fluence, along with the power dissipated by the amplification circuitry. The total heat load was estimated as 350 mW per channel. The Crilin cooling system consists of a cooling plant and a cold plate heat exchanger, made of copper, mounted directly over the electronic board. A glycol-based water solution passes through the deep drilled channels and absorbs the heat generated by the SiPMs. It will provide the optimum operating temperature for the electronics and SiPMs at 0/-10 °C.

The protype Front End Electronics (FEE), regulating a single module, is divided in two main parts (Figure 7): a SiPM board and the Mezzanine board. Each SiPM board, Figure 7-a, embeds 36 photo-sensors so that each crystal in the matrix has two separate and independent readout channels, consisting in a series of two 10 µm pixel-size SMD S14160-3010PS SiPMs [8]. These were chosen for their high-speed response, narrow signals and radiation hardness. Four SMD blue LEDs were also placed in between the SiPMs matrices in order to perform in-situ calibration, diagnostic and monitoring. The SiPMs are then connected via 50 Ω micro-coaxial transmission lines to a microprocessor-controlled Mezzanine Board. The Mezzanine Board oversees signal amplification and shaping, along with all slow control functions for all the 18 readout channels of the single layer. The SiPMs biasing is controlled by 12-bit DACs while regulated voltages, bias currents, and the temperature of the SiPM matrix are sensed via dedicated 12-bit ADC channels. The slow control routines are than handled by an onboard Cortex M4 microprocessor.

Fig. 7. FEE parts: the SiPMs board embedding the 36 photosensors and the Mezzanine board that bias, regulate and readout them.

Proto-1 performances will be evaluated in two dedicated Test Beam: the first one with 500 MeV electrons at BTF-LNF, whose analysis will be completed shortly, the second one at CERN-SPS H2 beamline with 120 GeV electrons. Time resolutions and a first estimation of the energy resolutions and the clusterization capability are going to be main interest of these tests.

IV. Conclusions

The Crilin calorimeter represents an innovative and promising choice for future collider experiments. It is a good compromise between homogeneous and sampling calorimeter and is well quoted as alternative solution to W-Si ECAL for future Muon Colliders. Several R&D steps were already made starting from radiation hardness campaigns on the single components and a Beam Test on a single layer prototype that showed very interesting results in terms time resolution. A bigger prototype, Proto-1, composed of two layers of 3×3 crystal matrices was recently developed and assembled. The characterization of this last prototype has already started with a Beam Test at BTF-LNF with 500 MeV electrons, for which the analysis is still ongoing. A second test beam will be also carried out at CERN-SPS H2 beamline with 120 GeV electrons. During these test beams it will be evaluated again the time resolution, given a first estimation of the energy resolution and tested the clusterization capability.
Once the performances of Proto-1 will be stated other irradiation campaigns will be performed with the aim of reaching the expected radiation level of the Muon Collider.
A huge R&D step will happen also during 2023 where a consistently bigger prototype, that will cover 1 Molier Radius, 20 radiation lengths and 1 interaction length, will be designed and built.

Acknowledgment

This project has received funding from the European Union's Horizon 2020 Research and Innovation program under the AIDAinnova project, grant no. 101004761, as well as from INFN projects RD_MUCOL.

References

[1] C. Accettura et al., "Towards a Muon Collider" (2023), https://arxiv.org/pdf/2303.08533.pdf
[2] N.Bartosik et al.,Simulated Detector Performance at the Muon Collider, (2022) https://arxiv.org/pdf/2203.07964.pdf
[3] Ceravolo S, et al., Crilin: A CRystal calorImeter with Longitudinal InformatioN for a future Muon Collider. JINST 17 (2022) P09033. doi:10.1088/1748-0221/17/09/P09033.
[4] A. Ferrari et al., "FLUKA: a multi-particle transport code", CERN-2005-10 (2005), INFN/TC_05/11, SLAC-R-773.
[5] Korzhik M, et al. Ultrafast PWO scintillator for future high energy physics instrumentation. Nucl. Instrum. Meth. A 1034 (2022) 166781. doi:10.1016/j.nima.2022.166781.
[6] Cemmi, A., et al. Radiation study of Lead Fluoride crystals. Journal of Instrumentation 17.05 (2022): T05015.
[7] Cantone C et al. Beam test, simulation, and performance evaluation of PbF2 and PWO-UF crystals with SiPM readout for a semi-homogeneous calorimeter prototype with longitudinal segmentation. Frontiers in Physics 11 (2023). doi:10.3389/fphy.2023.1223183.
[8] Hamamatsu S14160-3010PS datasheet, https://www.hamamatsu.com/content/ dam/hamamatsu-photonics/sites/documents/99_SALES_LIBRARY/ssd/ s14160-1310ps_etc_kapd1070e.pdf

https://doi.org/10.1051/epjconf/202328803001

Modelling of a SiC Based Detector for the Interpretation of 14.1 MeV Neutrons Measurements

Quentin Potiron[1,*], Christophe Destouches[2], Leo Dubus[3], Michael Houry[3], Olivier Llido[2], Abdallah Lyoussi[2], Laurent Ottaviani[1] and Christelle Reynard-Carette[1]

[1]Aix-Marseille Univ, Université de Toulon, CNRS, IM2NP, Marseille, France
[2]CEA, DES, IRESNE, DER, Cadarache, F-13108, Saint Paul-lez-Durance, France
[3]CEA, IRFM, Cadarache, F-13108, Saint-Paul-lez-Durance, France
(*) quentin.potiron@univ-amu.fr

***Abstract*—Wide-bandgap semiconductor-based fast neutron detectors such as silicon carbide (SiC) seem to be a promising concept to meet the implementation requirements as well as the performance specifications for fusion and fission environments such as radiation hardness as well as thermal and mechanical stabilities. Beyond the problem of integration of the device in constrained environments, the issue of the quantitative response of the device in a mixed radiation field must be addressed. Therefore, the characterisation of the detector performances, according to the energy of the neutrons is a key first step needed for neutron detection and monitoring in mixed radiation environments with high levels of gammas and fast and thermal neutron fluxes emissions at large energy scales. This paper presents the interpretation of 14.1 MeV neutron measurements using a numerical model developed with the GEANT4 Monte-Carlo transport code. The methodology used for a first attempt to estimate the fluence rate of incident neutrons is exposed and a calculation versus experiment ratio is determined.**

***Keywords*—Fast neutrons, Semiconductor Detectors, Silicon Carbide, Radiation Detection.**

I. Introduction

WIDE-bandgap semiconductors-based detectors such as Silicon Carbide (SiC) or CDV Diamond had been efficiently used as fast neutron detectors and have shown interesting performances for many applications related to radiation detection in constrained environments [1] [2]. Indeed, neutron detection mechanisms with detectors based on semiconductor materials rely on indirect methods in which neutrons interact with specific atomic nuclei and produce ionising charged particles. These detectors have the advantage of having threshold nuclear reactions with fast neutrons as seen in Table I, while being transparent to thermal neutrons. This characteristic gives them interesting spectrometric properties [3] [4] and they are therefore widely studied in the literature such as for the topic of cross section evaluation [5] [6].

If compared to diamond detectors, SiC-based detectors offer the same benefits along with the added advantage of having further threshold reactions which are of interest in the perspective of using the detector for online spectrometry applications.

A dedicated experiment had been done at the Laboratory of Subatomic Physics and Cosmology (LPSC) in Grenoble in May 2022 in order to evaluate the performance of SiC detectors to monitor fast neutron beam fluence rate. The principle of the experiment consists in several irradiations of SiC sensors in a mono-energetic neutron field. The energy spectrum deposited in the sensor by the charged particles resulting from neutron interactions is thus obtained. The aim is then to interpret this deposited energy distribution in the SiC to evaluate the reaction rate of the $C(n,\alpha)$ interaction and thus estimate the incident neutron fluence rate. It should be noticed that the evaluated nuclear data are incomplete for the detailed description of the sequential decay of the following neutron capture reaction $^{12}C(n,\alpha)^{9}Be^{*}$. But, although the process of generating three alpha particles from the excited states of the beryllium-9 nucleus is complex, the interaction cross section of $^{12}C(n,\alpha_0)^{9}Be^{(g.s)}$ when the beryllium-9 is emitted in the ground state is evaluated [7].

TABLE I
THE THRESHOLD NEUTRON REACTION

Reaction	Reaction threshold energy [MeV]
$^{28}Si(n,n')^{28}Si$	1.8
$^{12}C(n,n')^{12}C$	4.8
$^{28}Si(n,\alpha)^{25}Mg$	2.8
$^{28}Si(n,p)^{28}Al$	4.0
$^{12}C(n,\alpha_0)^{9}Be^{(g.s)}$	6.2

The paper's first part contains comprehensive details about the experimental set-up, measurements, and detector simulation using Monte Carlo transport code. The second part of the paper provides an explanation of the methodology based on the use of the modelisation for the interpretation of the experimental measurements and a calculation versus experiment ratio is determined.

II. Experimental campaign

A. GENESIS facility

The GENESIS platform operated by the CNRS at the LPSC, consists of an electrostatic accelerator which delivers a continuous deuterium ion beam generated by electron cyclotron resonance source, located in a Faraday cage. The source and its platform are then raised to a voltage of 220 kV.

After extraction, the beam of deuterons emitted in a range from $10\,\mathrm{A}$ to $2\,\mathrm{mA}$ is accelerated in a linear accelerator from which it emerges at an energy of $220\,\mathrm{keV}$. Deuterium ions are then guided to the tritiated target with a series of electrostatic quadrupoles to focus the beam [8]. On this line an electromagnetic dipole magnet is used to select ions acting as a mass spectrometer. The target consists of a copper disk $3\,\mathrm{mm}$ thick and $38\,\mathrm{mm}$ in diameter. Tritium alloyed with titanium is deposited on the copper to a thickness of $250\,\mathrm{\mu g\,cm^{-2}}$. The active zone is then a disk of 24 mm in diameter.

B. Experimental set-up

1) Measuring system: The electronic circuit used for signal processing showed in Fig. 1 includes:

- A high-voltage source (HV), marketed by CAEN, providing the positive or negative voltage required to operate the detector and controlled via a laptop.
- A current preamplifier (PA) C2 commercialised by Cividec Instrumentation. The main objective of the PA is to provide an optimised coupling between the detector output and the rest of the counting system. The C2 broadband amplifier with 2 GHz analogue bandwidth and 40 dB gain is also required to minimise noise sources that can alter the signal. To avoid attenuation during signal transmission that could be detrimental to the measurement, the PA is placed at 50 cm from the detector.
- An Acqiris SA230E digitizer with sampling rate of 4 GS/s. This single 14-bit digitizer captures current pulse shapes up to 2 GHz. With a long acquisition memory up to 8 Go, the SA230E includes a Xilinx Kintex UltraScale FPGA offering real-time signal processing capabilities such as waveform averaging. All Acqiris ADC boards and modules implement a low-noise integrated real-time processor which is ideal for applications requiring high bandwidth sampling. This good Signal-to-Noise Ratio (SNR) performance at high frequencies then allows fast and accurate tracking of current pulses.

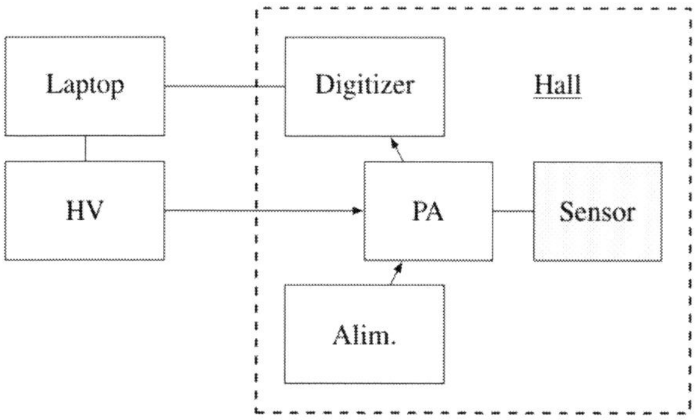

Fig. 1. Measuring system, in the control room and in the experimental hall.

2) Detectors design and irradiation locations: Several types of semiconductors detectors have been tested in the GENESIS installation, presenting different geometries and materials such as CVD diamond and SiC. In the context of this paper, only the experiments performed with the SiC described in Table II will be discussed. The experimental location of the sensor is approximatively 6 cm from the target.

TABLE II
THE CHARACTERISTICS OF THE SIC SENSOR UNDER STUDY

Symbol	Quantity	Value
$W_{sub.}$	substrate thickness	350 µm
$W_{junc.}$	junction thickness	1 µm Al/ 1 µm p^{+}/ 20 µm n/ 100 nm Ti-Ni
S	surface	5 mm × 4 mm
N_A	acceptor atom concentration	1×10^{19} at.cm^{-3}
N_D	donor atom concentration	2×10^{14} at.cm^{-3}

The SiC diode under study were designed and realised in the framework of the European project I_SMART [9].

3) Experimental Protocol: In order to study the different nuclear reaction rates it is first necessary to evaluate the optimal bias voltage to deplete the space charge region (SCR) i. e. the active detection zone. For all measurements presented in this paper, GENEPI delivered $6 \times 10^{9}\,\mathrm{n \cdot s^{-1}}$ in 4π. The neutrons fluence rate is estimated to $1.13 \times 10^{7}\,\mathrm{n \cdot cm^{-2} \cdot s^{-1}}$ at the SiC sensor location. Short irradiations are performed to obtain the saturation curve presented in Fig. 2.

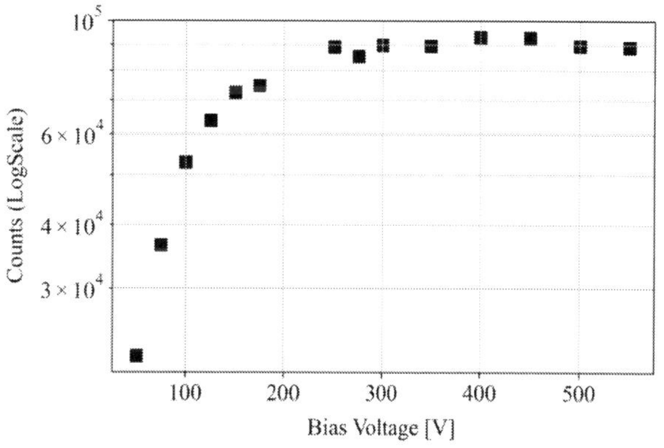

Fig. 2. Saturation curve of the count rate as a function of the applied electric field. The count rate is normalised to the irradiation time.

This curve allows comparison of the count rates obtained with the sensors as a function of the applied reverse bias voltage and therefore as a function of the induced electric field. A reverse bias voltage of $300\,\mathrm{V}$ was therefore chosen at the end of this first characterisation step. Durations were defined in order to achieve a good statistics on the total number of events of interest recorded. In our case, it corresponds to the peak located at 8.4 MeV associated with the maximum energy deposited by the products of $^{12}C(n,\alpha_0)^{9}Be^{(g.s)}$. Since the statistical uncertainties varies according to a Poisson law, acquisitions have been repeated leading to a number of count above 10^4 under the peak of interest.

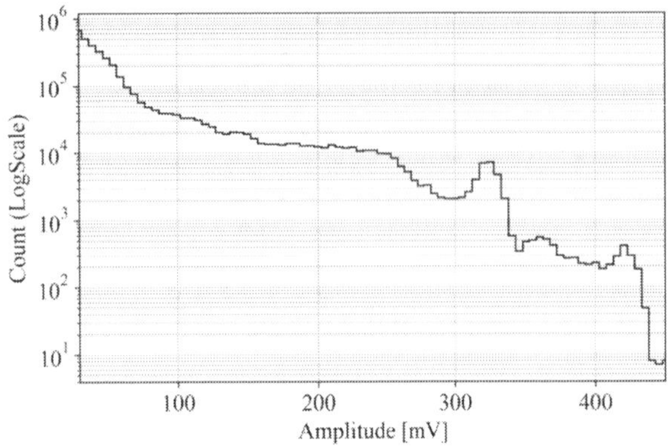

Fig. 3. Distribution of the PHA acquired with one hour irradiation with the detector SiC polarised with a reverse bias voltage of 300 V.

C. Data processing

With the multi-record feature of the digitizer, it becomes possible to acquire a serie of triggered events that take place in rapid succession. Indeed, the SA230E's rapid trigger rearm time is a key attribute in achieving minimal dead time. Additionally, a pre-trigger delay is employed to enhance trigger adaptability, with resolutions down to picoseconds which allows a complete analysis of the shape of the pulses.

The raw data files in Technical Data Management System format (TDMS) recorded by the Human-Computer Interface (HCI) developed on LabVIEW are converted into ASCII format to be post-processed. Each output file corresponds to a single triggering threshold in a previously selected acquisition window on the Acqiris card. Then, in order to analyse individually the pulses collected during this irradiation campaign, the development of automated processing scripts using the Python programming language and associated libraries is performed. Several quantities are extracted from the recorded signals such as the amplitude, determined in relation to the baseline, which is previously determined by averaging the points before the triggering threshold, or the integral representing the area under the peak, determined by the trapezoidal method.

The distribution of the acquired signal parameters are then obtained, like the Pulse Height Amplitude (PHA) distribution in Fig. 3. Then, by utilising the C2 PA internal gain A equal to 156 and the electronically calibrated input impedance Z equal to 55 Ω, the voltage signals are converted to the corresponding collected charges Q such as:

$$Q = \frac{1}{A \cdot Z} \int u(t)\, dt \tag{1}$$

But, the energy deposited in a semiconductor material by a charged particle is equal in a first approach to the product of the number of generated electron-hole pairs N multiplied by the first ionisation energy E_W. Moreover, the number of electron-hole pairs produced is equal to the ratio of the charge collected when they reach the electrodes, by the elementary charge q. It comes thus, the expression of the deposited energy:

$$E_{dep.} = N \cdot E_W = \frac{Q}{q} \cdot E_W \tag{2}$$

The spectrum of energy deposited in the sensor is then obtained. It enables us to interpret in terms of energy, the Fig. 3, which is the distribution obtained experimentally and in particular the peak observed, formed by signals with an amplitude of 310 mV.

III. Detector Simulation

It is necessary to add at this stage that the experimental measurement in pulse mode gives information on the sum of the energy deposited by the two charged particles in the output channel such that the integral on the associated voltage step corresponds to the sum of the charges created by the ionisation of each of the capture products. The aim is therefore to model these interactions in the sensor in order to deconvolute the sensor's energy response and study the reaction of interest, i.e. that of the observed energy peak, to estimate a value for the incident fluence rate.

A. Model assumptions

In this model, we assume the use of the Neutron_HP GEANT4 library for the study of fast neutron interaction with energy under 20 MeV. We also assume the use of intra-nuclear cascade BIC_HP and NRESP package for a better description of threshold neutron interaction with carbon nuclei. Dedicated work has been carried out on the choice of these hypotheses and a full study will be published at a later date. Nevertheless, the results of the numerical calculation compared to the analytical calculation of the kinematics of two-body nuclear interactions have allowed the validation of the assumptions and options implemented in the GEANT4 calculation scheme [10].

B. Dedicated Monte Carlo Calculation

For each neutron interaction in the depleted zone, the transport code gives the energy deposited during the lifetime of the charged particles produced. The distributions of this energy deposited by the different charged particles resulting from the multiple neutron interactions are shown in the Fig. 4. The distribution of the energy deposited by the recoil nuclei is obtained in the Fig. 4a and the distribution of the total energy deposited by all the particles resulting from the interactions on the carbon and silicon atoms is also represented with a dashed line.

It can be noticed a characteristic peak of maximum deposited energy at 8.4 MeV for the events associated with the $^{12}C(n,\alpha_0)^9Be^{(g.s)}$ reaction in Fig. 4b. The latter corresponds to the case where the total kinetic energy of the secondary particles from the reaction $^{12}C(n,\alpha_0)^9Be^{(g.s)}$ is deposited in the SCR. For the observed events where the deposited energy is lower than this maximum value of 8.4 MeV, it can be introduced at this stage the contribution of edge effects in the obtained results.

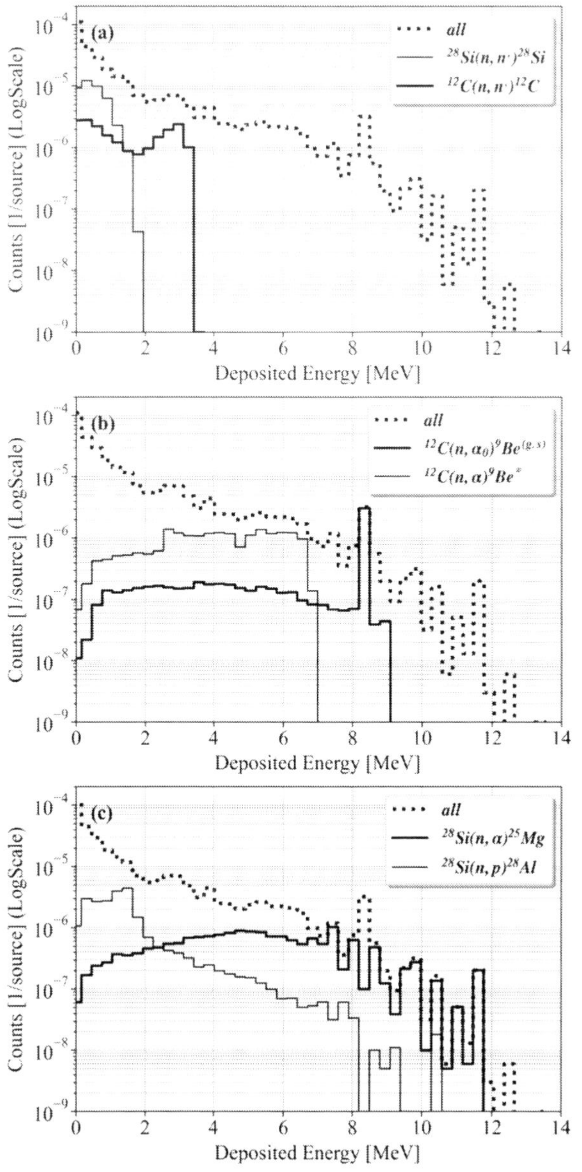

Fig. 4. Deposited energy distributions obtained using Monte-Carlo transport code GEANT4 for neutron captures by carbon nuclei (a), neutron captures by silicon nuclei (b) and inelastic collisions (c). The counts rates are normalised to the number of source neutrons used in the calculation.

Thus, for neutron interactions such as captures, it is found that the events taking place on the edges of the solid state sensor under study, generate charged particles that travel only a very short distance in the SCR. For instance, according to the kinetic energy distribution of the alpha generated in $C(n,\alpha)$ reactions, these particle can deposit energy along their path on 7 to 32 µm. Consequently, depending of the localisation of the interaction, some particles do not deposit all the kinetic energy they have at their creation and these events contribute to the total energy spectrum deposited without populating the characteristic peak of the maximum energy deposited.

IV. Calculation to Experiment Comparison

The estimation of the fast neutron fluence rate from this solid state sensor consists in measuring the reaction rate of the threshold neutron capture $^{12}C(n,\alpha_0)^9Be^{(g.s)}$. The peak energy deposited by the secondary particles of the $^{12}C(n,\alpha_0)^9Be^{(g.s)}$ reaction being well defined and separated from the continuum as presented previously, allows the rate of this reaction to be measured online by exploiting the PHA. To this end, the previously presented GEANT4 model allowed us the evaluation of two important contributions to the analysis and interpretation of the experimental energy spectrum deposited:

- The edge effects i.e. the proportion of secondary particles from $^{12}C(n,\alpha_0)^9Be^{(g.s)}$ that will not deposit the full kinetic energy they have at their creation as shown in Fig. 4b.
- The proportion of secondary particles from $C(n,\alpha_0)$ in the observed peak vs. those originating from $Si(n,\alpha)$ as shown in Fig. 4c.

Thus for a detection thickness of 20 µm the peak observed at 8.4±0.5 MeV is composed of 68% of the energy deposited by the charged particles exiting the $C(n,\alpha_0)$ interactions and 32% of that exiting the $Si(n,\alpha)$ interactions. Moreover, regarding the edge effect, only 46% of secondary particles from $C(n,\alpha_0)^9Be^{(g.s)}$ have deposited all of their available kinetic energy within 20 µm of SiC.

It is therefore possible to estimate the incident neutron fluence rate $\phi_{exp.}$ with this reaction rate R_{exp} since the latter is defined as a function of the interaction cross section σ and the number of target nuclei (4), such that :

$$R_{exp.} = N_C \cdot \sigma \cdot \phi_{exp.} \quad (3)$$

$$N_C = \frac{\mathcal{N}_A \cdot \rho}{M} \cdot V \quad (4)$$

With $\mathcal{N}_A$ the Avogadro number, ρ the density of SiC, V the volume of the sensitive region and M the molar mass. Regarding the evaluation of the effective interaction cross section, the choice of the ENDF/B-VI nuclear data library for the estimation of the fluence rate is not discussed in this paper.

The numerical application enables to obtain an approximate value of the fluence rate of incident neutrons of $0.86 \times 10^7\,\mathrm{n \cdot cm^{-2} \cdot s^{-1}}$. By estimating the relative uncertainty in edge effects and the $C(n,\alpha_0)/Si(n,\alpha)$ ratio in the peak at 8.4±0.5 MeV, a total systematic relative uncertainty of 16% is obtained.

The theoretical fluence rate value $\phi_{ref.}$, on the other hand, taken as a reference is evaluated from the value of the source intensity I_n in 4π reported by the facility. Taking into account uncertainties from d the distance between the target and the diode taking into account uncertainties in the thickness of the target interface; and σ the effective interaction cross section leading to a total systematic uncertainty of 14%. The measurement and calculation results are presented in the Table III.

TABLE III
COMPARISON OF MEASURED AND THEORETICAL NEUTRON FLUENCE RATES

Fluence Rate	Value $[cm^{-2}s{-1}]$	Relative uncertainty at 1σ [%]
$\phi_{exp.}$	0.86×10^7	17
$\phi_{ref.}$	1.13×10^7	14
C/E	1.31	22

As a first approximation, the measured fluence rate value can be considered independent of the reference value, since the source intensity is from an independent determination process using the GENESIS device calibration system.

The two values are coherent within their respective uncertainties at 1σ, showing a good indication that the interpretation process is valid for the SiC sensor regarding the contribution of the edge effects, and reactions differentiation.

V. CONCLUSION AND DISCUSSION

The fast neutron fluence rate in a single energy field has been estimated with encouraging C/E ratio for the validation and interpretation of the SiC detector modelling scheme. The experimental protocol and the acquisition electronics devices dedicated to the pulse measurement have also given satisfaction. Although the C/E obtained in the state with the hypotheses carried out are coherent, a thorough work on the nature of the uncertainties, their propagation, and the reduction of the latter would lead to a reduction of the differences observed between the value of the theoretical fluence rate and that measured.

The validation of this method with an evaluation of a SiC sensor model with GEANT4 Monte-Carlo transport code could firstly allow applying this numerical modelling of the sensor to a distributed neutron fluence rate. Then, the work will continue by studying the variation of several sensor parameters such as the thickness of the junction for the optimisation of the detection performances. This serves as a first step for the definition of SiC based detectors for fast neutron measurements in harsh and constrained environments.

ACKNOWLEDGMENT

This work has been made possible thanks to the CNRS/LPSC GENESIS facility personnel, in particular B.Cheymol, research engineer and executive director of the platform.

REFERENCES

[1] M. De Napoli, "SiC detectors: A review on the use of silicon carbide as radiation detection material," *Frontiers in Physics*, vol. 10, Oct. 2022.
[2] F. H. Ruddy, "Silicon Carbide Neutron Detectors for Harsh Nuclear Environments: A Review of the State of the Art," *IEEE Transactions on Nuclear Science*, vol. 69, no. 4, pp. 792-803 Jan. 2022.
[3] F. H. Ruddy, "Fast Neutron Dosimetry and Spectrometry Using Silicon Carbide Semiconductor Detectors," *Journal of Astm International*, vol. 3, Jan. 2006.
[4] R. W. Flammang, "Fast neutron detection with silicon carbide semiconductor radiation detectors," *Nuclear Instruments and Methods in Physics Research Section A: Accelerators, Spectrometers, Detectors and Associated Equipment*, vol. 579, no. 1, pp. 177-179 Jan. 2007.
[5] C. Weiss, "A CVD Diamond Detector for (n,α) Cross-Section Measurements," Ph.D. dissertation, Atomic Institute of the Austrian Universities, Vienna University of Technology., Geneva, June 13, 2014.
[6] E. Griesmayer, "Neutron Cross Section Measurements with Diamond Detectors," *Engineering of Scintillation Materials and Radiation Technologies*, pp. 187-193, Sep. 2019.
[7] S. A. Kuvin, "Validation of neutron-induced reactions on natural carbon using an active target at neutron energies up to 22 MeV at LANSCE," *Physical Review C*, vol. 104, pp. 177-179 Jul. 2021.
[8] F. Villa, "Accelerator-Based Neutron Irradiation of Integrated Circuits at GENEPI2 (France)," *IEEE Radiation Effects Data Workshop*, vol. 2015, Jul. 2014.
[9] A. Lyoussi, "I_SMART a Collaborative Project on Innovative Sensors for Material Ageing and Radiation Testing: European Innovative Project for SiC Applications in Harsh Media," KIC_Innoenergy, SSCV_SMART Proposal/Exhibit 1, 2012.
[10] S. Agostinelli and *et al.*, GEANT4—a simulation toolkit, *Nuclear Instruments and Methods in Physics Research Section A: Accelerators, Spectrometers, Detectors and Associated Equipment*, vol. 506, pp. 250–303, Jul. 2003.

Characterization of the Fast Neutron Generators for Calibration of Fusion Neutron Diagnostics

T. Kormilitsyn[1,*], A. Pankratenko[1], Yu. Kashchuk[1], S. Obudovsky[1], R. Rodionov[1], D. Fridrikhsen[1], A. Krasilnikov[1], S. Syromukov[2], D. Yurkov[2]
[1]Project Center ITER (ITER RF DA), Moscow, Russia
[2]Dukhov Automatics Research Institute (FSUE VNIIA), Moscow, Russia
(*) t.kormilitsyn@iterrf.ru

Abstract—**Modern magnetic confinement fusion devices increasingly rely on extensive neutron diagnostic configurations to measure a plethora of key plasma parameters. Measurement accuracy for these diagnostics depends heavily on in situ calibration. In order to enable said calibration, we set out to propose a reliable and powerful fast neutron source, to define a characterization plan for this source in terms of yield, flux and energy distributions, propose an optimized set of tools suitable for online monitoring of neutron source performance and its metrological characteristics. In the framework of this research activity with the ultimate aim of ITER tokamak neutron diagnostics calibration we rely on industrial-grade fast neutron generator NG-24 (D-D neutron yield ~10^9 n/s, D-T neutron yield ~10^{11} n/s) with sealed tube and stuffed, titanium target developed by FSUE VNIIA. The well-known analytical expressions for thick target NGs and our measurements using neutron spectrometers utilizing threshold reactions – diamond detector and $LaCl_3$ scintillator for D-T and D-D neutron generators respectively – were found to be significantly coherent. These data are supported through our multi-step forward modelling including ion stopping in target, fusion reaction kinematics modelling and calculating detector response based on modelled neutron spectra. We discuss methods of uncertainty mitigation of neutron spectrometer measurements. Application of both neutron activation analysis and gas-filled neutron flux monitors during source characterization and operation allows for lowing statistical uncertainty of neutron flux measurements to 1% level over 1 minute of time resolution. Over the course of several measurement campaigns the optimal set of measurement tools have been determined including detector dimensions, required acquisition time, calibration methods.**

***Keywords* — neutron generator, neutron spectrometer, fusion product, plasma diagnostics, neutron calibration.**

I. Introduction

The task of fusion neutron diagnostic calibration consists of a continuous set of challenges including factory calibration of the diagnostic equipment, on-site testing and commissioning activities, *in situ* calibration with fast neutron source and cross-calibration during discharges. The objectives of current work revolve around characterization of a chosen source with proven reliability and suitable for the *in situ* neutron calibration activity.

Several aspects limit the choice of the fast neutron source for this task – available space, project schedule, cost, safety considerations and, of course, source neutron spectra. Majority of calibration campaigns of the operating magnetic confinement fusion devices relied on isotope sources, including JET [1], Globus-M2 [2] and TFTR [3]. Only recently the calibration campaign at JET [4] had made use of compact sealed tube neutron generators (NGs) [5] to significant success. The key advantages of the modern sealed tube NGs are:

- neutron spectrum consistent with fusion neutrons for both deuterium and deuterium-tritium plasma scenarios;
- togglable operation without unnecessary irradiation going on during source transportation, storage, handling and between expositions;
- point-like source (Ø2 cm) with yield of D-D and D-T neutrons comfortably reaching the yield of 10^9 and 10^{11} n/s respectively;
- modest weight (<200 kg) and dimensions (<Ø50×115cm) that correspond well to the Type-A package required for transportation and handling of ^{252}Cf source with comparable yield (10^9 n/s).

The main disadvantages of this source type include – neutron yield and energy anisotropy and (for high-yield models) the need for active target cooling. Said disadvantages are an additional source of measurement error for the diagnostics undergoing the calibration, and the goal of characterization of sealed tube NG is elimination of this error.

The NG yield determination for both D-D and D-T cases requires relevant measurement techniques that allow for both time-resolved monitoring and absolute measurement of the number of neutrons produced by the source. The robust tools suitable for monitoring of fast neutron flux include threshold fission chambers [6,7] and gas counters based on boron [8] and helium-3 [9]. Neutron activation analysis (NAA) technique is deemed the fusion industry standard for absolute measurements of neutron yield during discharge [10,11] and can be confidently applied for the task of characterization of a fast neutron source.

Given the nature of fusion reaction kinematics [12] the change in average neutron energy with respect to beam direction will present a significant source of uncertainty. Hence, to maximize the confidence and precision of the calibration we propose to apply for both tasks – source characterization and

 https://doi.org/10.1051/epjconf/202328803002

subsequent monitoring – a set of neutron spectrometers with the goal of determination of neutron energy distribution. For the D-D NG the detector of choice is $LaCl_3$ scintillator [13,14] with fast neutron detection via the $^{35}Cl(n,p)^{35}S_{g.s.}$ reaction, for the D-T NG – diamond detector with its respective $^{12}C(n,\alpha)^9Be$ [15,16] reaction. Both of these instruments allow for pulse-height analysis and feature a classical gaussian-shape response to fast neutrons. This approach allowed us to study the neutron energy distribution at various angle to the ion beam and thus verify the assumptions embedded in the forward modeling of the experiment.

II. Measurement Setup

The measurement setup required for characterization of the fusion neutron source is formed by the source itself – the neutron generator, the neutron spectrometers typically located within 50 cm from the target due to their low sensitivity, neutron monitors – gas-filled ionization chambers – located at a higher distance to avoid pulse pileup. The neutron activation system is constituted of gamma-ray spectrometer and a pneumatic system delivering activation samples to and from the irradiation zone. Coverage of several neutron emission angles with spectrometers is achieved using an automated positioning system operated, together with NG and measurement system I&C from a secure control room.

A. Neutron generator

Considering the dimensions of ITER machine with major radius exceeding 6 m and minor – exceeding 2 m [17], we provision the use of NG-24 neutron generator manufactured by FSUE VNIIA [18], the main technical specifications of this model are listed in Table I.

TABLE I
NG-24 KEY TECHNICAL SPECIFICATIONS

Quantity	Value	
Dimensions, mm	Ø430×1150	
Weight, kg	175	
Sealed tube lifetime, hrs	300	
Ion source, type	Penning	
Ion current, mA	Up to 2	
Accelerating voltage, kV	Up to 250	
Modification	D-T	D-D
Ion beam composition	$D^+, T^+, D_2^+, T_2^+, DT^+$	D^+, D_2^+
Neutron yield, n/s	$\sim10^{11}$	$\sim10^9$
Average neutron energy at 90°, MeV	~14.1	~2.45

The sketch of the NG irradiation unit is illustrated in Several details that constitute the know-how of the manufacturer were omitted. The key element of the NG is the sealed tube and the detailed design around the radiative spot – NG target – has been thoroughly modelled for further research to include the copper cooling circuit, stuffed, titanium target, and sealed tube housing. The supporting structure for the scintillator-based neutron spectrometer to be located in the frontal sphere of the NG has also been designed.

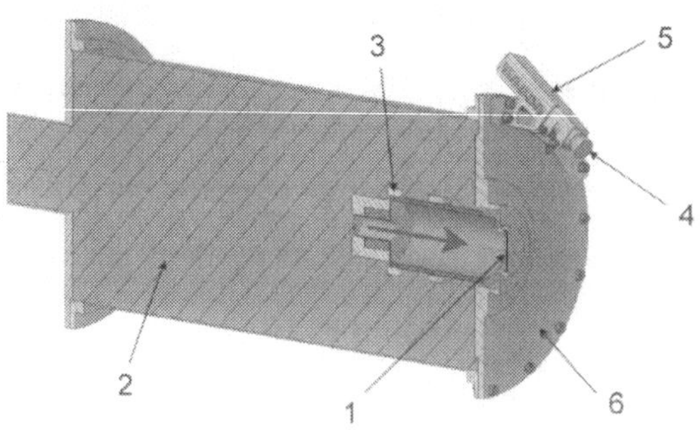

Fig. 1. Vertical cross-section of the simplified neutron generator CAD model. Red arrow shows ion beam direction. 1 – Ti target, 2 – transformer oil (CH_2), 3 – magnets, 4 – sensitive element of neutron spectrometer, 5 – detector support structure, 6 – neutron generator flange.

The duration of expositions necessary for the successful calibration of the key ITER neutron diagnostics lie in the order of hours [19], hence, a 1-minute time resolution for measurements of NG yield and energy we presume sufficient. The tests of the NG-24 model for yield stability measured using laboratory helium-3 neutron counter have yielded good results of only ±2,5% yield change over 1 shift (7 hr 30 min) of continuous operation. Test results further underline the lack of necessity to further minimize the time resolution.

B. Neutron monitors

The task of determination of NG yield starts with determination of neutron flux at detector location. This section comprises a brief overview of achievable results with regard to determination of neutron flux value at several suitable locations around NG irradiation unit.

For this, in D-D NG case operating with the yield of $\sim10^9$ n/s, the boron gas-filled counter is deemed most suitable. Given the Q = 2792 keV the $^{10}B(n,\alpha)^7Li$ reaction with the cross-section at E_n = 2.5 MeV reaching 1 barn, this monitor is highly sensitive to both thermal and D-D neutrons. With a fairly unsophisticated I&C system this sensor constitutes a robust solution achieving $>10^4$ counts per minute at locations near the front on the NG. The sensor mass being as low as 12 g allows designating its location near the back side of the front NG flange on the outer wall of the NG housing. Nevertheless, its sensitivity to thermal neutrons requires mitigation using shielding materials to prevent counts from backscattered fast neutrons. The low abundance of helium-3 [20] and the need increase bias voltage for detector operation (1200 V versus 500 V) rule out the use of said counters for the task of D-D NG characterization and monitoring during operation.

The D-T NG with the yield being 2 orders of magnitude higher ($\sim10^{11}$ n/s) over the D-D modification, allows for use of the threshold fission chambers (FC) with radiators covered by uranium-238 [21]. The FC with 0.5 g of fissile material operating at 300 V of bias voltage is a perfect candidate for NG monitoring purposes due to reaction threshold (~1 MeV) and due to the cross-section increasing with energy up to 2 barn for E_n = 14 MeV. FC location directly behind the back flange of the NG allows reaching count-rates above 10^4 per minute, lowering the statistical uncertainty of flux reconstruction down to 1%.

C. Neutron activation analysis

The more precise way to provide absolute measurements of the neutron yield integrated over time is the use of neutron activation technique also applicable to magnetic confinement fusion devices [10,11,22]. For this, we have utilized several activation samples listed in the Table II with the key characteristics derived based on ENDF-B/V.III [23].

TABLE II
KEY CHARACTERISTICS OF THE MATERIAL SAMPLES USED FOR NAA

Sample	Indium	Cadmium	Aluminum	Teflon
Source	D-D		D-T	
Process	$^{115}In(n,n')^{115m}In$	$^{111}Cd(n,n')^{111m}Cd$	$^{27}Al(n,\alpha)^{24}Na$	$^{19}F(n,2n)^{18}F$
Q, MeV	-1.5	-0.4	-3.132	-10.431
σ,barn	0.327	0.239	0.118	0.057
$T_{1/2}$, hrs	4.486	0.81	14.887	1.829
Eγ, keV / Eff. %	336 / 45.9	245 / 94	1369 / 99.99 2754 / 99.985	511 / 193.46

The D-D neutron energy range ~ 2.5 MeV is covered with only a few threshold reactions occurring in common NAA samples. The two select isotopes feature a gamma-ray energy of the same order of magnitude, but vastly different thresholds, with ^{115}In cross-section more closely matching the required range [24]. In laboratory conditions under the sealed tube D-D neutron generator operating at accelerating voltage of 150 kV, 150 μA of ion current amounting to 10^7 n/s yield the irradiation time required exceeded 10 hours for a 0.2 g sample located at 1.5 cm away from NG target. Increasing the yield to NG-24 operation parameters, increasing the distance to ~50 cm and increasing the weight of the sample tenfold would amount to the same irradiation time needed to achieve the same 3% of statistical uncertainty in neutron fluence determination. This leaves little room for application of NAA technique for detailed multi-angle characterization of the source, but the technique still remains crucial for absolute scaling of measurements by neutron flux monitors and spectrometers provisioned for characterization and monitoring of the D-D NG.

Precise D-T NG yield determination requires less irradiation time and allows the use of samples more fit for purpose of NAA. The pure aluminum sample allows measurements using two gamma-ray energies, further lowering statistical uncertainty. The Teflon sample while having lower cross-section has a more favorable threshold reaching slightly above 10 MeV. The use of these samples with characteristic weight of 1 g under irradiation using NG-24 D-T model operating at 200 kV of accelerating voltage, 2 mA of ion current and producing ~5×10^{10} n/s allowed determination of the local D-T neutron flux with the accuracy within 2% after 1 hour of irradiation.

Activation of each sample has been studied with the laboratory Ø2×2-inch $LaBr_3(Ce)$ scintillator calibrated using a set of standard ^{60}Co, ^{137}Cs and ^{22}Na isotope sources.

The NAA technique highlighted in this section provides an independent and robust measurement of the absolute neutron flux value necessary to minimize the final uncertainty of NG parameters after characterization.

III. NEUTRON SPECTROMETER RESULTS AND DISCUSSION

A. Chlorine-based neutron spectrometer

Recent developments in the field of fast neutron spectrometry have led to a series of scintillating materials being utilized due to availability of $^{35}Cl(n,p)^{35}S_{g.s.}$ (Q=615 keV, σ=0.15 barn for E_n = 2.5 MeV) reaction for measurement of fast neutrons. For the purposes of neutron generator characterization and consequent monitoring we propose the $LaCl_3(Ce)$ scintillator, having in mind the previous multi-step modelling and analysis of experiment with a low-power NG yielding promising results [25]. With the light yield being higher and with response being faster than that of CLYC crystals, this detector looks most promising for operation in the harsh environment – within 50 cm and in direct view of the D-D NG target. The pulse-shape discrimination capability of $LaCl_3(Ce)$ is lower than that of CLYC (Figure-of-Merit = 1.0 versus >2.0) and requires digitization of the signal with sampling rate of at least 500 MSps at 14-bit resolution. The detector energy resolution in our previous work was estimated at 8% [26].

We coupled the Ø2×0.5-inch crystal to the Hamamatsu R6231-100 PMT and conducted a series of expositions at a fixed distance of 43.5 cm from the target center of D-D modification of the NG-24 operating at 220 kV of accelerating voltage and 1 mA of ion current which corresponds to 5×10^8 n/s yield. The only parameter varying in between the expositions was the angle between the detector line of sight and the NG axis. Following the digital pulse-shape discrimination procedure (PSD) [27], we conducted the pulse-height analysis of the detector pulses, associated with slowing down of protons born in the (n,p)-reaction for each of the expositions. The results are illustrated on Fig. 2.

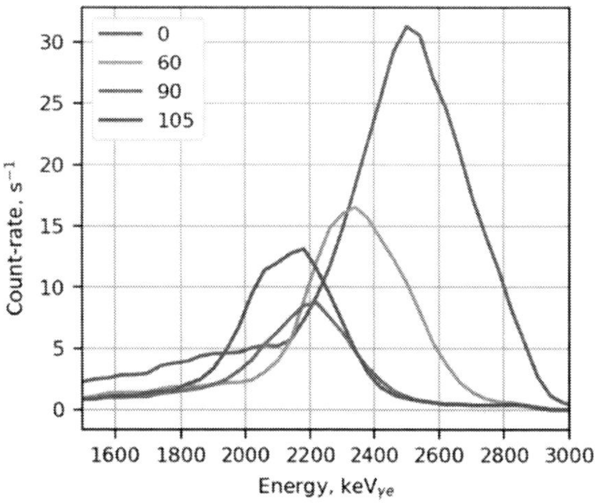

Fig. 2. Pulse-height analysis of proton-associated events in the $LaCl_3(Ce)$ response after PSD plotted for several angles between detector line-of-sight and NG axis, calibrated in the gamma-ray equivalent energy.

The figure-of-merit of current crystal sample was found to be lower and averaged at 0.45 across the expositions, this is primarily due to increased ^{227}Ac contamination [28] of the crystal compared to previously used samples. At the same time, background radiation arising from this contamination in the form of α-decay events was recorded in between each irradiation for the duration of ~5 minutes. No visible deviation has been found in the shape of the pulse-height distribution or the position of the α-decay event peaks. This fortunate circumstance allowed us to self-calibrate the detector using known locations of the peaks, and continue to monitor the energy calibration stability between each exposition. Using the p/β-ratio derived from the previous measurements (~0.75) and (n,p)-reaction kinematics allowed us to infer average neutron energies for each of the irradiation angles. The results of the measurement are summarized in the Fig. 3 and Table III. The agreement between measurement and the data for thick targets by J. Csikai [29] is within 2%.

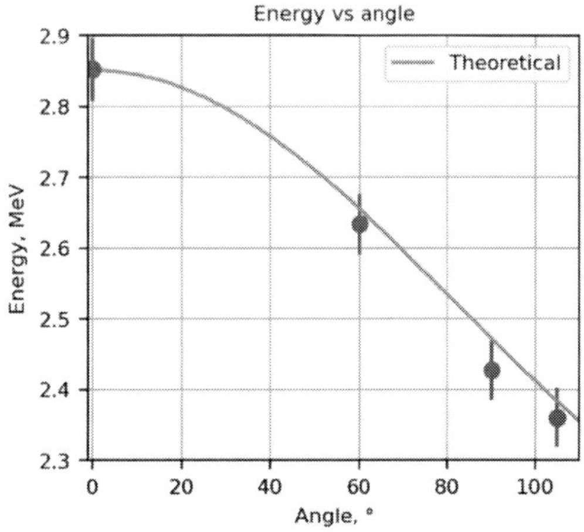

Fig. 3. Neutron energy measured using $LaCl_3(Ce)$ versus the angle plotted against analytical approximation of thick target NG data.

TABLE III
NG NEUTRON ENERGY MEASUREMENT VERSUS THICK-TARGET DATA

Angle, °	Average E_n, Analytical approximation [29], keV	Average E_n, $LaCl_3$ measurement, keV
0	2851	2852±45
60	2656	2634±43
90	2473	2428±42
105	2384	2361±42

The absolute measurements of neutron flux is undermined by discrepancies in the cross-section data for the $^{35}Cl(n,p)$ reactions in the literature [23,30,31]. Experiment data yield incoherent results in terms of absolute yield reconstruction and requires further investigation using a metrological fast neutron source.

Detector performance in terms of relative measurements and average neutron energy determination is on substantial level, allowing it to be used as a primary tool for both characterizing the NG and monitoring its performance during operation. Provisioned placement of $LaCl_3(Ce)$-based detector in the frontal sphere of the NG within 50 cm of its target allows reaching count rates above 10^4 events per minute in the (n,p)-associated peak without significant pulse pileup and PSD degradation, which is effectively aligns with results of the experimental campaign.

B. Diamond neutron spectrometer

The semiconductor detector based on a CVD-diamond mono-crystal is a widely adopted [32-35] method for measurement of D-T neutron energy and flux. Having a high critical fluence of $\sim 10^{14}$ n/cm^2 and impeccable energy resolution ~1% this detector highly benefits form predominantly (except for impurities) carbon composition and a threshold reaction that produces a classical gaussian-like response – $^{12}C(n,\alpha)^9Be$, Q=-5.7 MeV.

Utilizing the 4×4×0.5 mm^3 crystal coupled to a Canberra 2004 preamplifier, Ortec 673 Amplifier and Ortec 926 analyzer we have conducted a set of measurements at 10 cm from the target of the the NG-24 (D-T modification) operating at 200 kV of accelerating voltage and 0.5 mA of ion current, corresponding to the yield of $\sim 2\times10^{10}$ n/s. Detector energy calibration was performed using isotope source of α-particles ^{226}Ra and its analytically estimated sensitivity is $\sim 6\times10^{-5}$ cm^2.

This set of measurements allowed us to study the change in the pulse-height distribution with the angle between detector LOS and NG axis. The PHA of detector for various irradiation angles is illustrated on Fig. 4 with each exposition time being equal to 180 s.

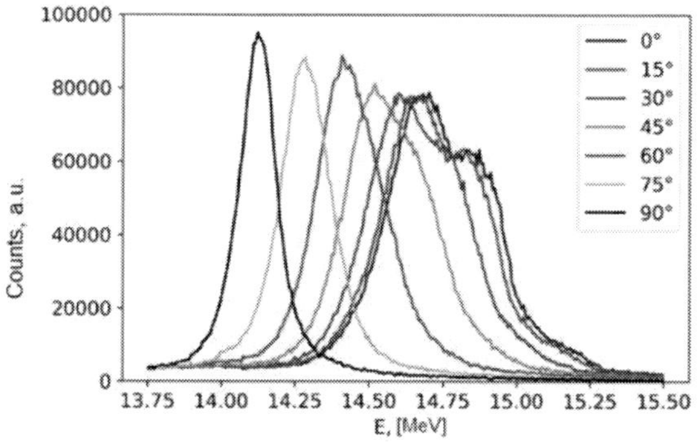

Fig. 4. Diamond detector PHA under D-T NG irradiation plotted for various angles between detector LOS and NG axis, exposition time = 180 s.

Verification of measurement results was done for two critical location – 0 and 90 degrees. For this, we utilized the multi-step modelling approach briefly described in [36]. The neutron generator beam ion composition was presumed 90% molecular ions (45% DT^+, 22.5% D_2^+, 22.5% T_2^+ and 10% single (5% D^+ and 5% T^+). The ranges for stopping of said ions were calculated using SRIM software for the titanium bulk material, each ion had its velocity vector recorded, with that, beam-target reaction kinematics was calculated and the resulting neutron energy and angle distribution was modelled. Beam component-wise energy distribution of the neutrons born in the direction along the NG axis is illustrated on Fig. 5.

The resulting energy distribution convoluted with (n,α)-reaction cross-section was plotted against the results obtained in experiment and shows significant coherence in shape and absolute number of events as illustrated on Fig. 6.

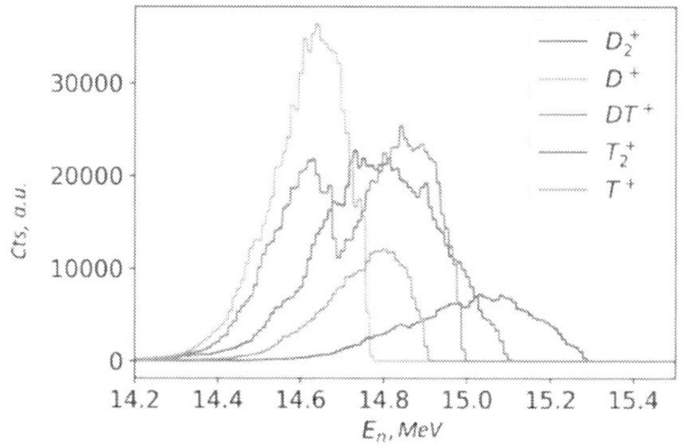

Fig. 5. Diamond detector PHA under D-T NG irradiation plotted for various components of the ion beam for the detector LOS along NG axis, exposition time = 180 s.

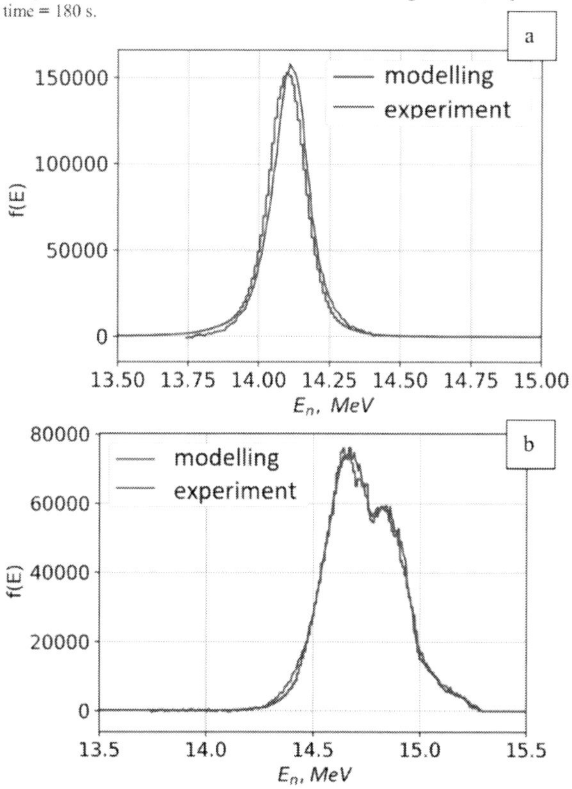

Fig. 6. Comparison between measured detector response and multi-step modelling results for the same experiment. (a) – 0-degree direction, (b) – 90-degree direction.

Diamond detector performance for the purposes of characterizing the D-T NG demonstrates stellar results. With that, its location within 30 cm from target of D-T NG operating at 10^{11} n/s leads to a count rate of the order 5×10^3 per second and its increase would likely constitute a pulse pileup due to significant number of pulses occurring at lower amplitudes simultaneously. The statistical uncertainty in determination of local neutron flux achievable over 1 minute of spectra integration time falls well below 1% in this case nevertheless.

IV. Conclusions and Outlook

In present work we outline the set of key tools necessary for conversion of the industrial-grade sealed tube neutron generator into metrological source fit for calibration and testing of the fusion neutron diagnostics.

Based on extensive experience of operation of sealed tube NGs our recommendation is to use multi-vector approach with simultaneous measurements done by means of neutron activation samples, conventional neutron monitors and unique neutron spectrometers, thus allowing to cover both the absolute neutron fluence measurements, consistent neutron flux monitoring and detailed energy versus angle characterization of the source.

Tests in laboratory conditions allow us to pinpoint the optimal sensor locations for each type of the source provisioned to be used in the ITER calibration campaigns – the sealed tube NG-24 neutron generator. The increase of the number of sensors in each case is a subject to a detailed reliability study.

For D-D source (yield 10^9 n/s) the boron counters are to be located directly behind the front flange of the NG, on the outer surface of the NG body, with careful consideration of the shielding conditions. The neutron spectrometer suitable for both characterization and subsequent monitoring is the lanthanum chloride scintillator, shown able to consistently (<2%) determine the average energy and therefore monitor the stability of NG accelerating voltage. The location in the frontal sphere within 40-50 cm range will allow to drive the statistical uncertainty of neutron flux determination based on this set of two tools well below 1% over each minute of the ongoing exposition.

For D-T source (yield 10^{11} n/s) the uranium-238 fission chambers are a robust tool that may be located behind the back flange of the NG. This location will also require careful design of shielding to prevent backscatter counts. The diamond neutron spectrometer seems a perfect candidate for monitoring of both neutron flux and neutron spectrum from a location in the frontal sphere of the NG. The low dimensions of diamond detector (2-3 cm^3) constitute very little obstruction to the diagnostics undergoing calibration. The cumulative statistical uncertainty of neutron flux determination of this set of tools also reach values well below 1% over 1 minute of irradiation.

The multi-step modelling approach developed and tested over the span of these experiment campaigns provided us with important insight into NG performance and verified a set of measurements of both diamond and chlorine-based neutron spectrometers. The sum total of modelling the ion stopping in the target and the beam-target reaction kinematics allows for analytical definition of the NG source that is both usable in future modelling and is consistent with the measurements.

Acknowledgment

This work has been funded under the Implementing Agreement No. 1 (ITER ref. IO/21/CT/4300002349) between Institution "Project Center ITER" and the ITER International

Fusion Energy Organization "Research and Detailed Design Development of the components needed for Neutron Generators suitable for ITER in-situ neutron calibration''.

ITER is the Nuclear Facility INB no. 174. The views and opinions expressed herein do not necessarily reflect those of the ITER Organization.

Data availability statement

The data that support the findings of this study are available from the corresponding author upon reasonable request.

References

[1] O. Jarvis, G. Sadler, P. Van Belle, T. Elevant (1990) In-vessel calibration of the JET neutron monitors using a 252Cf neutron source: Difficulties experienced. Rev Sci Instrum 61:3172–3174. DOI: 10.1063/1.1141677

[2] O. Skrekel, N. Bakharev, V. Varfolomeev, et al (2022) Calibration of neutron counters at the Globus-M2 tokamak. Tech Phys 92:12. DOI: 10.21883/TP.2022.01.52526.151-21

[3] H. Hendel, R. Palladino, C. Barnes, et al (1990) In situ calibration of TFTR neutron detectors. Rev Sci Instrum 61:1900–1914. DOI: 10.1063/1.1141115

[4] P. Batistoni, S. Popovichev, Z. Ghani, et al (2018) 14 MeV calibration of JET neutron detectors - Phase 2: In-vessel calibration. Nucl Fusion 58:106016. DOI: 10.1088/1741-4326/aad4c1

[5] Yu. Kaschuck, D. Portnov, A. Krasilnikov, et al (1999) Compact neutron generator for diagnostic applications. Rev Sci Instrum 70:1104–1106. DOI: 10.1063/1.1149473

[6] Yu. Kashchuk, V. Shevchenko, V. Frunze, A. Krasil'nikov. (2005) Monitoring the fast-neutron flux density and fluence in a RBMK core using a threshold fission chamber in a screen-absorber. At Energy 98:249–255. DOI: 10.1007/S10512-005-0202-X/

[7] L. Bertalot, R. Barnsley, F. Direz, et al (2012) Fusion neutron diagnostics on ITER tokamak. J Instrum 7:C04012. DOI: 10.1088/1748-0221/7/04/C04012

[8] P. Dighe, D. Prasad, K. Prasad, et al (2003) Boron-lined proportional counters with improved neutron sensitivity. Nucl Instruments Methods Phys Res Sect A Accel Spectrometers, Detect Assoc Equip 496:154–161. DOI: 10.1016/S0168-9002(02)01635-2

[9] R. Batchelor (1952) Neutron energy measurements with a helium 3 filled proportional counter, Proc. Phys. Soc. Sect. A 65:674–675

[10] R. Prokopowicz, B. Bienkowska, K. Drozdowicz, et al (2011) Measurements of neutrons at JET by means of the activation methods. Nucl Instruments Methods Phys Res Sect A Accel Spectrometers, Detect Assoc Equip 637:119–127. DOI: 10.1016/j.nima.2011.01.128

[11] C. Barnes, M. Loughlin, T. Nishitani (1997) Neutron activation for ITER. Rev Sci Instrum 68:577–580. DOI: 10.1063/1.1147657

[12] H. Brysk (1973) Fusion neutron energies and spectra. Plasma Phys 15:611–617. DOI: 10.1088/0032-1028/15/7/001

[13] T. Kormilitsyn, S. Obudovsky, Yu. Kashchuk, et al (2021) Application of the LaCl3(Ce) Scintillator to Fast Neutron Measurements. Phys Part Nucl Lett 18:75–81. DOI: 10.1134/S154747712101009X

[14] P. Vuong, H. Kim, N. Luan, S. Kim (2021) Neutron spectroscopy using pure LaCl3 crystal and the dependence of pulse shape discrimination on Ce-doped concentrations. Nucl Eng Technol 53:3784–3789. DOI: 10.1016/J.NET.2021.05.020

[15] A. Krasilnikov (1998) Natural Diamond Neutron Spectrometer. In: Diagnostics for Experimental Thermonuclear Fusion Reactors 2. Springer US, Boston, MA, pp 439–448

[16] A. Krasilnikov, V. Amosov, P. Van Belle, et al (2002) Study of d-t neutron energy spectra at JET using natural diamond detectors. Nucl Instruments Methods Phys Res Sect A Accel Spectrometers, Detect Assoc Equip 476:500–505. DOI: 10.1016/S0168-9002(01)01497-8

[17] K. Ioki, C. Choi, E. Daly, et al (2012) ITER Vacuum Vessel design and construction. Fusion Eng Des 87:828–835. DOI: 10.1016/j.fusengdes.2012.02.023

[18] S. Syromukov, V. Stepnov, R. Dobrov, et al (2015) NG-24 Neutron Generator for Nuclear Medicine and Thermonuclear Research. At Energy 119:68–71. DOI: 10.1007/s10512-015-0031-5

[19] T. Nishitani, M. Ishikawa, ML.-S. Cheon, et al (2023) Neutronic simulations of in-vessel neutron calibrations for ITER neutron diagnostics by using simplified ITER model. Fusion Eng Des 191:113548. DOI: 10.1016/j.fusengdes.2023.113548

[20] R. Kouzes, A. Lintereur, E. Siciliano (2015) Progress in alternative neutron detection to address the helium-3 shortage. Nucl Instruments Methods Phys Res Sect A Accel Spectrometers, Detect Assoc Equip 784:172–175. DOI: 10.1016/j.nima.2014.10.046

[21] S. Obudovskii, A. Batyunin, V. Sevast'yanov, et al (2016) Metrological Assurance of Thermonuclear Neutron Flux Density Measurements. Meas Tech 59:288–292. DOI: 10.1007/s11018-016-0960-y

[22] M. Hoek, T. Nishitani, Y. Ikeda, A. Morioka (1995) Neutron yield measurements by use of foil activation at JT-60U. Rev Sci Instrum 66:885–887. DOI: 10.1063/1.1146527

[23] D. Brown, M. Chadwick, R. Capote, et al (2018) ENDF/B-VIII.0: The 8th Major Release of the Nuclear Reaction Data Library with CIELO-project Cross Sections, New Standards and Thermal Scattering Data. Nucl Data Sheets 148:1–142. DOI: 10.1016/j.nds.2018.02.001

[24] C. Duniiu, Y. Baosheng, L. Hanlin, Z. Wenrong, 115In(n,n')115mIn cross section evaluation, INDC(CPR) 024/L, (1991), IAEA

[25] T. Kormilitsyn, S. Obudovsky, R. Rodionov, et al (2021) Novel LaCl3(Ce)-based spectrometer for deuterium plasma neutron diagnostics. Rev Sci Instrum 92:043528. DOI: 10.1063/5.0042394

[26] V. Pankratenko, T. Kormilitsyn, Yu. Kashuck, et al (2023) Analysis of the LaCl3(Ce) scintillator response function to fast neutrons. Nucl Instruments Methods Phys Res Sect A Accel Spectrometers, Detect Assoc Equip 1052:168282. DOI: 10.1016/j.nima.2023.168282

[27] A. Pankratenko, T. Kormilitsyn, S. Obudovsky, et al (2022) Digital Pulse Shape Discrimination Method for D–D Neutron Spectrometry Using the LaCl3(Ce) Scintillator. Phys Part Nucl Lett 19:66–76. DOI: 10.1134/S1547477122010095

[28] W. Wolszczak, P. Dorenbos (2017) Shape of intrinsic alpha pulse height spectra in lanthanide halide scintillators. Nucl Instruments Methods Phys Res Sect A Accel Spectrometers, Detect Assoc Equip 857:66–74. DOI: 10.1016/j.nima.2017.02.041

[29] J. Csikai (1987) CRC handbook of fast neutron generators; 240 p; CRC Press Inc; Boca Raton, FL (USA)

[30] S. Kuvin, H. Lee, T. Kawano, et al (2020) Nonstatistical fluctuations in the Cl 35 (n,p) S 35 reaction cross section at fast-neutron energies from 0.6 to 6 MeV. Phys Rev C 102:024623. DOI: 10.1103/PhysRevC.102.024623

[31] J. Batchelder et al (2019). Possible evidence of nonstatistical properties in the Cl 35 (n, p) S 35 cross section, Physical Review C, 99(4). DOI: 10.1103/PhysRevC.99.044612

[32] L. Bertalot, J. Adams, M. Angelone, et al (2005) ITER relevant developments in neutron diagnostics during the JET Trace Tritium campaign. In: Fusion Engineering and Design. North-Holland, pp 835–839, DOI: 10.1016/j.fusengdes.2005.06.213

[33] G. Ericsson G. (2019) Advanced Neutron Spectroscopy in Fusion Research. J Fusion Energy 38:330–355. DOI: 10.1007/s10894-019-00213-9

[34] C. Cazzaniga, M. Nocente, M. Rebai, et al (2014) A diamond-based neutron spectrometer for diagnostics of deuterium-tritium fusion plasmas. Rev Sci Instrum 85:11E101. DOI: 10.1063/1.4885356

[35] B. Esposito, D. Marocco, G. Gandolfo, et al (2022) Progress of Design and Development for the ITER Radial Neutron Camera. J Fusion Energy 41:22. DOI: 10.1007/s10894-022-00333-9

[36] T. Kormilitsyn, S. Obudovsky, Yu. Kashchuk, et al (2021) Application of the LaCl3(Ce) Scintillator to Fast Neutron Measurements. Phys Part Nucl Lett 18:75–81. DOI: 10.1134/S154747712101009X

Concept of calorimetry system for ITER

Lukasz Tomkow[1,*], Eleonore Geulin[1], Daniel Iglesias[1], Matthew Clough[1], George Vayakis[1]
[1]ITER Organization, France
(*) lukasz.tomkow@iter.org

Abstract—

The ITER tokamak is expected to produce up to 500 MW of fusion power. Other energy inputs will contribute up to 100 MW. All of this power ultimately converts to heat and the Tokamak Cooling Water System (TCWS) will remove most of it. To ensure safety, diagnostics methods, including calorimetry, will be used.
Thermal methods used in calorimetry are not suitable for online control but rather provide time-averaged or integrated values. The ITER calorimetric system aims to calculate the integrated fusion power after each pulse by determining the total tokamak power using data from the TCWS and other systems involved in the energy transfers. The total fusion power will be then determined by analyzing the thermal balance of the machine. As most of the energy transfers are realised by cooling loops, a general forward model has been developed for their analysis.
The system is designed to measure net fusion power production. At approximately 200MW, the accuracy is around 10%, and at about 500MW, the accuracy improves to 3%. The expected time constant of the measured instantaneous value is in order of 40s (the nominal ITER burn times are in the range 300 - 3000s). System will be most useful in cross-calibration with fast time resolution neutron systems. The developed system will also serve as a model for future tokamak-based fusion power plants, where its accuracy will be even higher. The paper gives a general description of the system concept, with main inputs required and the expected outputs.

***Keywords* —ITER, fusion power, fusion energy, calorimetry, thermal analysis.**

I. INTRODUCTION

In the pursuit of controlled fusion energy ITER is a landmark project developing the tokamak. ITER will be equipped with a pressurized water cooling system to ensure efficient cooling of all surfaces exposed to the plasma. The effective removal of heat is essential during stationary long discharges to protect the machine. Fusion power generated in ITER is limited by regulatory requirements and capped at 700 MW although power in most scenarios does not exceed 500 MW [1].
Instantaneous control is provided by neutron flux monitors that have ms level response time, but accuracy limited to ≈ 10% suffering from incomplete knowledge of the local flux distribution as a function of the global power, calibration and geometrical errors [2]. Calorimetry can reduce the overall error, calibrate the fast neutron measurements and improve the confidence in the output power used in extrapolation to future reactors.
The total energy of the pulse can be accurately determined by calorimetric methods [3,4]. Various studies have demonstrated the accuracy of calorimetry as an informative source for quantifying energy or power deposition [5,6]. Additionally, this method is successfully applied in fission power plants [7]. Therefore, calorimetry can be used to calibrate neutronic diagnostics by comparing the obtained pulse energy with the integrated neutron flux. Calorimetric measurement in ITER can achieve high accuracy thanks to high fusion power, good coverage by the cooling system and excellent thermal insulation. The majority of heat from ITER is removed by the Tokamak Cooling Water System (TCWS) [8], which consists of several cooling loops and is accompanied by a number individual component cooling systems. Estimation of the heat extracted by a cooling loop can be performed using the sensors installed at the inlet and outlet of the loop. By combining data from all the cooling loops, one can analyse the global energy balance within the system.

The focus of this study is to develop an algorithm capable of determining an integrated value of fusion power - pulse energy. Understanding thermal fluxes in the cooling system offers insights into the dynamics of fusion reactions within the ITER reactor. This algorithmic approach provides a valuable tool for understanding the temporal behaviour of fusion power and contributes to the broader goal of optimising fusion reactor performance.

II. CALCULATION OF HEAT BALANCE

Various energy sources contribute to thermal balance of the tokamak. During power operation, the main contributors to the balance are the thermal energy resulting from fusion reactions (with its thermal flux denoted as P_{fusion}), and the energy supplied by the plasma heating system ($P_{heating}$). Minor contributing sources include the energy delivered by auxiliary devices such as lasers ($P_{auxiliary}$) and the enthalpy change in the pump (P_{pump}).
Heat from the tokamak is removed mostly by the cooling system which captures and transfers the combined energy from various sources. This is divided into two sections, tokamak (P_{TCWS}) and component (P_{CCWS}) cooling system. Some heat is captured by the cryogenic system, however, the values are quite small (in the order of watts), and will therefore be overlooked in this study.
There are additional thermal transfers that are not directly managed, including convection to and from the cryostat structure, heat conduction through the supports, and thermal

radiation from the machine (P_{losses}). Analysing these transfers is a complex task due to the non-uniform temperature distribution of the cryostat and the multi-directional heat flow.

The overall tokamak power balance can be expressed as:

$$P_{fusion} + P_{heating} + P_{auxiliary} + P_{pump} = P_{TCWS} + P_{CCWS} + P_{losses} \quad (1)$$

The power fluxes are visualised in Fig. 1. Table I describes the typical values of these flows.

TABLE I
TYPICAL THERMAL FLOWS VALUES

Symbol	Components	Value [MW]
P_{fusion}	Fusion power	500
$P_{heating}$	Induced current, ECH, ICH, NBI	100
$P_{auxiliary}$	Laser, control coils	O(1)
P_{pump}	Pumps	10
P_{TCWS}	Tokamak cooling water system	-560
P_{CCWS}	Component cooling water system	-50
P_{losses}	Convection, Conduction and radiation to and from the cryostat external surface	O(1)

P_{TCWS} and P_{CCWS} comprise the vast majority of heat being transferred away from the tokamak. The heat transfer in this system is realised by water cooling loops. In terms of measurements and analysis all the cooling loops can be treated similarly. The thermal energy carried by the cooling water is computed with formula (2). The experimental inputs to the formula are the temperatures and pressures measured at both the entrance and exit of a loop.

$$P = \dot{\mathbf{m}} \cdot \left(h_{out}(\mathbf{T_{out}}, \mathbf{p_{out}}) - h_{in}(\mathbf{T_{in}}, \mathbf{p_{in}})\right) \quad (2)$$

Where h is the enthalpy and subscript "in" and "out" denote the inlet and outlet of the control volume, $\dot{\mathbf{m}}$ the mass flow, **T** the temperature and **p** the pressure. In this formula the bold characters ($\dot{\mathbf{m}}, \mathbf{T_{in|out}}, \mathbf{p_{in|out}}$) are directly measured in the cooling system. In the considered thermodynamic regime the variation of enthalpy is dominated by temperature, pressure has minor effect.

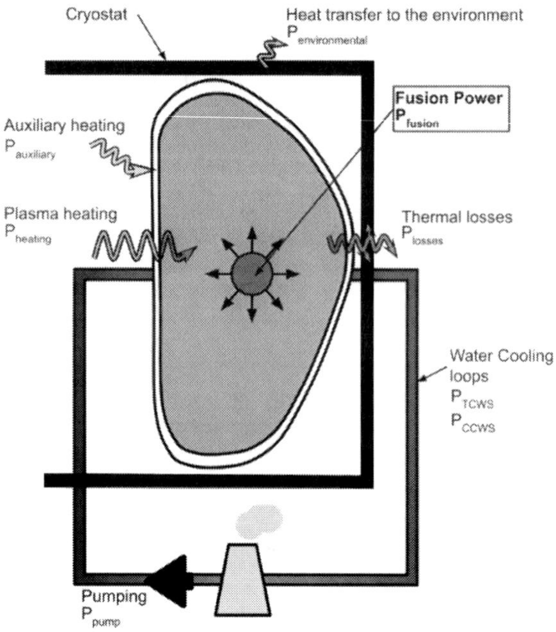

Fig. 1. Power balance schematic showing inputs ($P_{heating}$, $P_{auxiliary}$, P_{pump}), outputs (P_{TCWS}, P_{CCWS}, P_{losses}), and fusion-generated power (P_{fusion}). Tokamak Cooling Water System (TCWS) loop depicted with blue for cold water and red for hot water. Includes representation plasma heat interaction.

III. FORWARD MODEL

Since the cooling systems are still in their preliminary phase, there is a lack of experimental data to draw from. Consequently, the creation of the calorimetry system must rely on mathematical modeling and insights gained from past experiences. Developing this system necessitates the analysis of a substantial variety of pulse and operational scenarios. To generate the synthetic results allowing for the behaviour of the calorimetric system to be predicted, a forward model of the cooling loops has been created. A more complex model already exists such as RELAP5 [9]. However, the complexity of the existing model is not needed for the system development, as from the point of view of the calorimetry system the entire tokamak is treated as a single volume. The synthetic data from the forward model is compared with existing complex model results for validation (Fig. 2.).

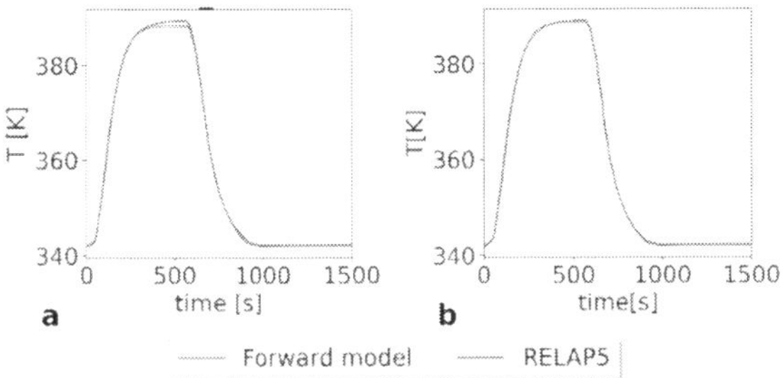

Fig. 2. Comparison between results from RELAP5 and forward model described in this paper obtained from different scenarios.

The principal outputs of the model used for testing the calorimetry diagnostic are inlet and outlet temperatures ($\mathbf{T_{in|out}}$) of a fluid while the internal dynamics of heat exchange in the tokamak are simplified. The internal values of temperature of the tokamak obtained in the model do not directly represent the temperature of the actual tokamak structures.

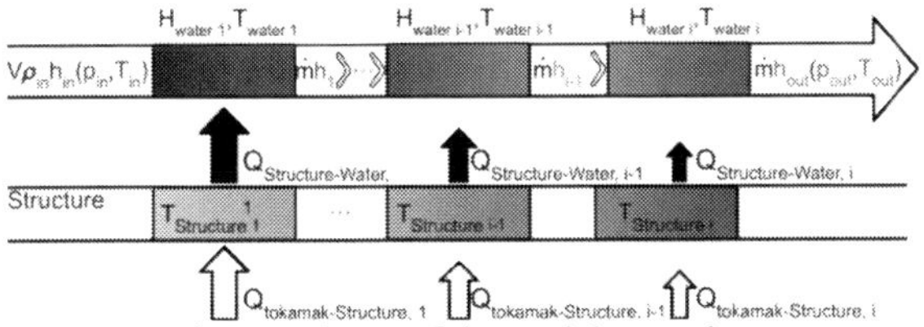

Fig. 3. Scheme of the model operation

The forward model has been implemented in Python. The modelled system is divided into two domains that represent cooling water flowing through the pipes, and a homogenised tokamak structure. The structure domain represents the first wall, the vacuum vessel, the cryostat and all other solid elements of the tokamak. This system could be modelled as a simple heat exchanger, but it was observed that it is not sufficient to align with RELAP5 model.

Thus, the domains are partitioned into multiple sections, as depicted in Fig. 3. The tokamak's structure captures the heat generated within it, which is subsequently removed by the water used as a working fluid in most loops.

The model is time dependent. At each time step heat balance Q_{bal} is calculated for each section, with (3) at the structural sections, and (4) at the water sections.

$$Q_{bal_i} = \tau \cdot \left(Q_{tokamak-structure_i} - Q_{structure-water_i}\right) \quad (3)$$

$$Q_{bal_i} = \tau \cdot \left(\dot{m} \cdot (h_{i-1} - h_i) + Q_{structure-water_i}\right) \quad (4)$$

In the TCWS model, the structure domain heat balance is a difference between the heat coming from the tokamak and the heat removed by a cooling fluid. For the water domain mass flow is conserved, and the heat balance is found as the sum of heat coming from the structure, and a difference between inlet and outlet enthalpies h. Thus, the heat transfer depends on the state of a section i and the preceding section $i-1$ (judging by the mass flow direction). An inlet mass and energy flux with predefined temperature, pressure and mass flow is imposed on the first water domain section.

Heat transfer between the respective sections of water and structure domain $Q_{structure-water}$ is calculated with formula (5).

$$Q_{structure-water_i} = (T_{water} - T_{structure}) \cdot k \quad (5)$$

The parameter k is calculated using formula (6), which includes a coefficient γ to consider the geometry of the tokamak. As the determination of the actual heat exchange parameters is a complex task beyond the scope of the system, the coefficient γ is selected to match the pre-existing tokamak analyses performed with more complex models [10,11]. Parameter k is dependent on temperature through the parameters of dynamic viscosity (κ), thermal conductivity (μ), and Prandtl number (Pr) which are calculated using the PropSI library.

While (6) is often utilized for more intricate, temperature-dependent systems, it retains its effectiveness even for our less complex model. In fact we successfully managed to fit the power calculated from the complex model to ours, which validated its applicability (Fig. 2).

$$k = \gamma\, \kappa\, \mu^{-\frac{4}{5}} \mathrm{Pr}^{-\frac{2}{5}} \quad (6)$$

All the structure domain sections are affected by a Neumann boundary condition of temperature-independent prescribed heat flux $Q_{tokamak-structure}$, representing the heat load from plasma and other sources. It is assumed that heat transfer by conduction between the sections of structure domains is negligible compared to the heat load from other sources.

The temperature change in a section of structure domain is found with formula (7).

$$T_{i_\tau} = T_{i_{\tau-1}} + \frac{Q_{bal_i}}{C_i\left(T_{i_{\tau-1}}\right)}, \quad (7)$$

C is a temperature-dependent parameter representing the thermal capacity of the structure. Temperature-dependance is assumed to be proportional to that of the stainless steel described with formula (8) [12].

$$C \propto 6.7 + 0.05 \times T + 80.7 \times \ln(T) \quad (8)$$

Similarly to k it is based on matching the results with the pre-existing calculations used to design the TCWS system.

Calculations in the water domain are done using CoolProp [13], with the thermodynamic state defined by enthalpy and pressure. Pressure in each section is pre-defined and assumed to drop linearly along the modeled system between the inlet and outlet values. It is also assumed to be independent from the flow conditions. In the actual system the inlet pressure is maintained by a pressurizer.

Initial conditions assume a constant temperature of 70°C across the modeled system, pre-defined mass flow and no heat input from the tokamak. During the model operation, scenarios with different heat input, mass flow and thermal inertia are analysed. The model in its current form calculates heat removal from the structure by a single loop. Works are on-going to include parallel operation of several loops and to mathematically close the loop by the inclusion of the heat rejection system, pump and a pressurizer.

IV. Results and discussion

Fig. 4. illustrates the temperatures of the water passing through the cooling system at the inlet and outlet of the tokamak. These temperatures are expected to be measured in operational mode. For the purpose of the paper, they are calculated by the forward model. In this simple approach it assumed that the entire internal power is removed by a single cooling loop.

In this example the inlet pressure is set at 4 MPa, while a pressure loss of 1.35 MPa is assumed to occur linearly along the sections. Throughout the calculation, the inlet mass flow is consistently maintained at 3300 kg.s^{-1}. The cooling tower system is considered perfect, giving a constant inlet temperature at 70°C. The simulated calorimetry system measures the power input into the cooling system based on the observed temperature and related enthalpy changes using (2).

The integrated pulse energy value is the primary focus of the results analysis. The imposed input power registers a total energy of 250.5 GJ, while the system reads it at 250.4 GJ. This near-perfect alignment between the input power and the system's measured integrated pulse energy is an expected outcome. It serves as a reassurance that the code is functioning as designed, with no identifiable bugs.

There is a shift between the instantaneous input power and the recalculated value, as shown by Fig. 5. This is due to the moderating effect of the tokamak structure thermal capacity. The shift does not has a significant effect on the integrated value of the pulse energy. The time constant of the power measurement can be estimated from this curve as 40s.

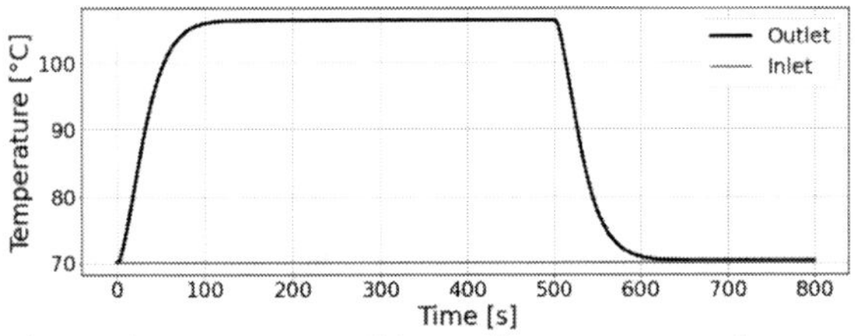

Fig. 4. The temperature of the cooling system over time. inlet (blue) and outlet (black) the tokamak.

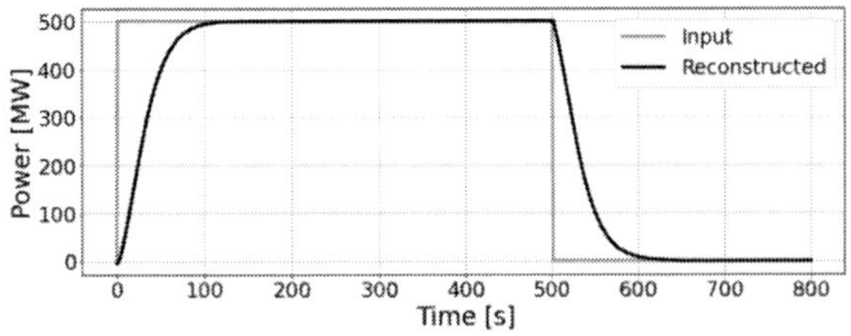

Fig. 5. Power comparison over time: original input (Orange) vs. recalculated (Black).

Because ITER is equipped with multiple cooling loop systems, a future step is to consider the individual inputs to the thermal balance of the tokamak. Additionally, the acquisition of results from various points across the device may enhance the temperature distribution reconstruction within the tokamak structure and lead to a better understanding of plasma power outputs.

A future aspect of this research involves reconstructing the initial and final stages of the calculated power, which will rely on the development of an algorithm. It will enable performing correlation analyses between the calorimetry and neutronics diagnostics, allowing for cross-calibration and actively contributing to anomaly detection. Correlation between integrated power and integrated neutron output will be analysed.

Based on the accuracy of sensors, the expected accuracy of the system has been estimated. The indicative results are given in the Table II, the exact assessment is ongoing.

Table II
Estimated accuracy

Nominal Power [MW]	500s pulse energy [GJ]	Estimated error [GJ]	Accuracy [%]
100	50	+/- 6.7	13.4
250	125	+/- 6.8	5.4
500	250	+/- 7.1	2.8
700	350	+/- 7.4	2.1

If the temperature difference between input and output is higher, the accuracy can be even better. This will be the case in the future power generating fusion plants. In these facilities calorimetry system will play a crucial role in measuring the power output of the tokamak in the power cycle.

V. Conclusions

The calorimetry system is used to determine fusion power from temperature, mass flow and pressure measurements. It provides independent albeit slow measurement of the total pulse energy on its own but can be cross-utilised with the neutron systems to give a fast response.

This paper describes the operation of simulated calorimetric system using ideal sensors as inputs. An excellent agreement is reached between the simulated input pulse energy and the reading of the system. The accuracy of the actual system will be affected by the accuracy of the sensors. Nevertheless, since the result value is integrated, the accuracy will remain very high after identification and elimination of systematic errors. The development of analysis algorithms for this purpose is ongoing. Also, further investigation is needed using the forward model to quantify the actual accuracy. Calorimetry has a major role to play in the future fusion power plants not only as an offline system measuring pulse energy but also as an online system measuring actual thermal power output of the machine.

In the current state of the model, it is possible to estimate the change in temperature and enthalpy of the water in the cooling system based on the injected power. From this information, the model allows for the reconstruction of the power that was applied. Further accuracy studies are needed. The model will be improved to take into account the inlet temperature changes over time and to incorporate the parallel operation of the different cooling circuits. A reconstruction algorithm to enable "real-time" analysis of the calorimetry diagnostic will be developed.

REFERENCES

[1] Decret n° 2012-1248 ,autorisant l'Organisation internationale ITER à créer une installation nucléaire de base dénommée « ITER » sur la commune de Saint-Paul-lez-Durance (Bouches-du-Rhône) JORF 0262 (2012)
[2] ITER Organization, ITER Research plan within the staged Approach, Report ITR-18-003 (2018)

[3] J.-C. Vallet *et al*, "Calorimetry in Tore Supra. : An accurate tool and a benchmark for ITER", Journal of Nuclear Materials, vol. 313–316, 2003, pp. 706-710
[4] Y. K. Liu *et al*, "Power balance investigation in long-pulse high-performance discharges with ITER-like tungsten divertor on EAST", Nucl. Fusion, vol. 60, 2020, pp. 096019
[5] T. Richter *et al*, "The cooling water calorimetry system at ASDEX upgrade tokamak", Review of Scientific Instruments, vol. 65, 1994, pp. 1613-1615
[6] K. Hanada *et al*, "Calorimetric measurement of heat load in full non-inductive LHCD plasmas on TRIAM-1M", Journal of Nuclear Materials, vol. 363-365, 2007, pp. 1425-1428
[7] J. Coble, "Calorimetric Analysis to Infer Primary Circuit Flow in Integral and Pool- Type Reactors", IEEE Transactions on nuclear science, vol. 64, 2017, pp. 837
[8] G. Dell'Orco *et al*, "Design modification and optimization of the ITER cooling water system", Fusion Engineering and Design, vol. 86, 2011, pp. 15-19.
[9] J.J. Carbajo *et al*, "Thermal Design of the ITER Vacuum Vessel Cooling System", 60, 2011, (150-155)
[10] L. Vicini *et al*, "Thermal Design of the ITER Vacuum Vessel Cooling System", Fusion Engineering and Design, vol. 173, 2021, pp. 112906
[11] B. Guo *et al*, Thermal behaviour analysis on ITER component cooling water system loop 2B", Fusion Engineering and Design, vol. 100, 2015, pp. 249-253
[12] D.P. Bentz, "Thermal Performance of Fire Resistive Materials I. Characterization with Respect to Thermal Performance Models" NIST, NISTIR 7401, Feb. 2007.
[13] Bell *et al*, "Pure and Pseudo-pure Fluid Thermophysical Property Evaluation and the Open-Source Thermophysical Property Library CoolProp", vol. 53, 2014, pp. 2498-2508

Detection of fast neutrons using 4H-SiC radiation detectors

Bohumír Zaťko[1*], Andrea Šagátová[2], Eva Kováčová[1], Andrej Novák[2], Rudolf Sýkora[3], Zdeněk Kohout[3,4] and Štepán Polansky[5]

[1]Institute of Electrical Engineering, Slovak Academy of Sciences, Dúbravská cesta 9, SK-814 04 Bratislava, Slovak Republic

[2]Faculty of Electrical Engineering and Information Technology, Slovak University of Technology, SK-812 19 Bratislava, Slovak Republic

[3]Institute of Experimental and Applied Physics, Czech Technical University in Prague, Husova 240/5, CZ 110 00 Prague 1, Czech Republic

[4]Faculty of Mechanical Engineering, Czech Technical University in Prague, Technická 4, CZ 160 00 Prague 6, Czech Republic

[5]Advacam, U Pergamenky 12, 17000 Prague 7, Czech Republic

(*) bohumir.zatko@savba.sk

***Abstract*—The particle detector based on a low concentration 4H-SiC epitaxial layer shows promising properties for the detection of various types of ionizing radiation. The wide bandgap energy of the 4H-SiC semiconductor material (3.23 eV at room temperature) allows the detector to operate reliably at room temperature and at elevated temperatures up to several hundred degrees Celsius. The 80 μm thick 4H-SiC epitaxial layer grown on a 350 μm 4H-SiC substrate was used for detector preparation. The active area of the detector was defined by a 3 mm Schottky contact. The current-voltage measurement shows a reverse current of less than 30 pA up to 1000 V. Capacitance-voltage measurements show that the epitaxial layer is completely depleted at bias voltages above 250 V. The detector has been tested with neutrons of energies from 370 keV up to 14.9 MeV. Neutrons were produced by three nuclear reactions p-T, d-D and d-T. The measured spectra clearly identified the elastic and inelastic scattering at silicon and carbon atoms of detector material.**

***Keywords* —4H-SiC detector, low concentration SiC epitaxial layer, fast neutron detection.**

I. Introduction

In recent years, neutron detectors have become an essential part of many areas of science and research. These are currently: fundamental physics research, cosmology research, neutron imaging as a complementary technique to X-rays and magnetic resonance, small angle scattering for materials research, radiation safety and security, monitoring of transport, nuclear energetics, and others. Neutrons do not directly ionize the material they pass through, so their detection is most often possible through elastic collisions or nuclear reactions that produce secondary radiation (charged particles, photons, etc.). A typical way of describing neutron detection methods is in terms of two broad categories of neutrons, namely of slow thermal neutrons and fast neutrons. Standard neutron detectors can be categorized as gaseous, semiconductor, scintillation and passive [1]. We have focused on semiconductor detectors, which have a number of advantages such as fast response, compact size and high operational stability. Since neutrons have no charge, a conversion layer is usually required for proper detection. The neutron-converting material can be incorporated into the detector as an external layer placed close to the detector. For the detection of thermal neutrons, the ^{10}B or ^{6}Li conversion layer is most commonly used because of its high cross section, and when the neutron is captured, it produces charged particles that are easy to detect. The most effective is ^{3}He, but due to its scarcity and very high price other alternatives are in demand. Fast neutrons can be detected by elastic scattering from light nuclei. Here we can use a High Density PolyEthylene (HDPE) layer, which contains hydrogen and converts neutrons into protons. Other possibilities are endoergic reactions, which are threshold processes. There is a minimum energy below which the reaction is energetically impossible.

We have many years of experience with 4H-SiC (silicon carbide) detectors and they also show good potential as neutron detectors. For thermal neutron detection, the conversion layer is required and 4H-SiC detectors show excellent performance [2-5]. In the case of fast neutrons, SiC detectors are capable of direct detection because they consist of light atoms such as silicon and carbon. Fast neutrons can be detected by means of the elastic scattering and threshold nuclear reactions. In our previous studies we detected fast neutrons from 16.0 MeV to 18.3 MeV with 4H-SiC detectors [6-8].

In this paper we have tested 4H-SiC detectors over a wide range of neutron energies from 370 keV to 14.90 MeV. Fast neutrons were produced by p-T (proton – tritium), d-D (deuterium – deuterium) and d-T (deuterium – tritium) reactions at the Van de Graaff (VdG) accelerator.

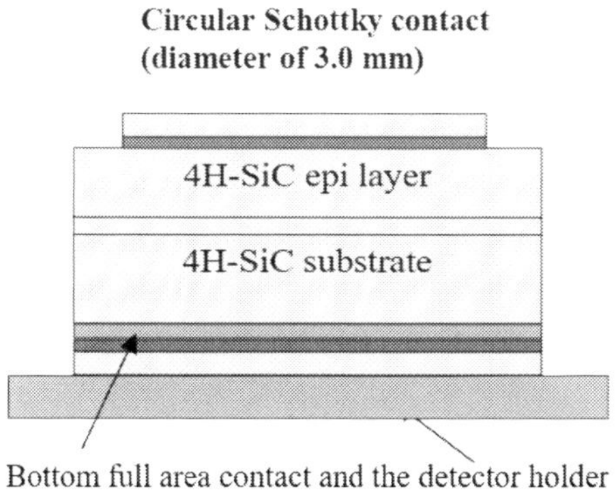

Fig. 1. The cross-section of fabricated 4H-SiC radiation detector.

II. Detector structure

The detector structure has been fabricated from an 80 μm thick nitrogen-doped 4H-SiC layer (with a donor doping level of about 7×10^{13} cm^{-3}) grown by liquid phase epitaxy (LPE) on a fragment of 4-inch 4H-SiC wafer (with donor doping level ~2×10^{18} cm^{-3}, thickness 350 μm) by growing a 0.5 μm thick n++ 4H-SiC buffer layer with a donor concentration of 1×10^{18} cm^{-3} between them. The radiation detectors were produced by evaporation of a double layer of Ni/Au with a thickness of 15/25 nm on the top side in a high vacuum chamber. The Schottky barrier contacts of Ni/Au with a diameter of 3.0 mm are formed through a metal contact mask, while a full area contact of Ti/Pt/Au (15/30/80 nm) was deposited on the other side (substrate). The cross-section of detector is in Fig. 1.

III. Experiment and discussion

First, the current-voltage characteristics of the 4H-SiC Schottky barrier detectors were measured at room temperature. The reverse current is below 30 pA up to 1000 V and the typical characteristic is shown in Fig. 2. The capacitance–voltage characteristic depicted in Fig. 3. was then measured to determine the full depletion reverse bias of the detector. The capacitance (blue curve) does not change with a reverse bias higher than 250V. The black curve shows the calculated depletion width using the following formula:

$$W_D = \frac{\varepsilon_r \varepsilon_0 S}{C_D} \quad (1)$$

where ε_r is the relative permittivity of 4H-SiC (9.66 [9]), ε_0 is the permittivity of free space and C_D is the measured capacitance of the detector and S is the detector contact area. The low concentration epitaxial layer of the detector is completely depleted at bias voltages above 250 V.

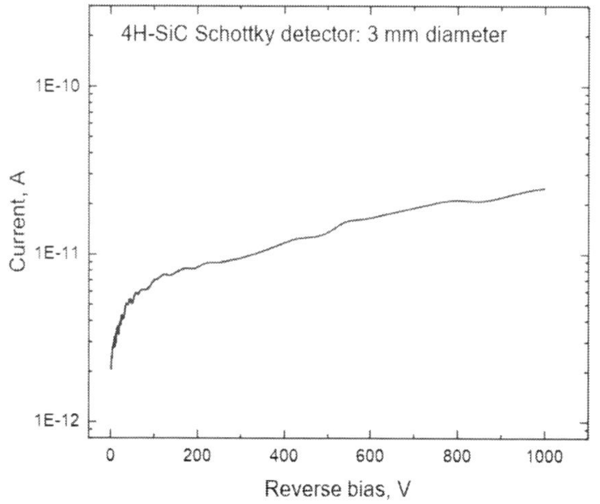

Fig. 2. The typical reverse current–voltage characteristic of 4H-SiC Schottky radiation detector.

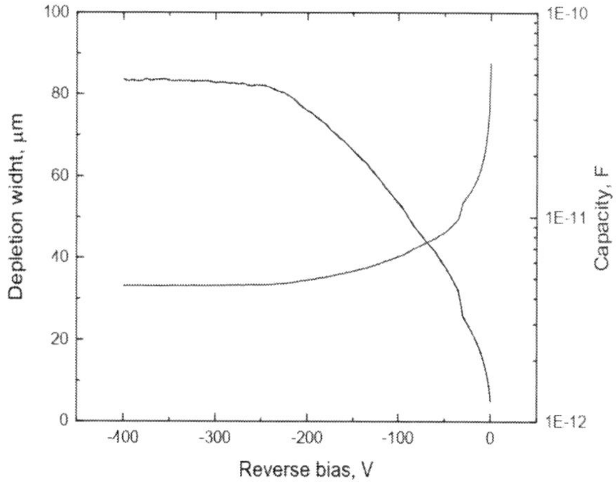

Fig. 3. The capacitance-voltage characteristic of 4H-SiC Schottky radiation detector and the calculated depletion width.

The fast neutron detection was performed at the Van de Graaff Laboratory of the Institute of Experimental and Applied Physics of the Czech Technical University in Prague. Three nuclear reactions were used to produce fast neutrons in a wide range of energies. The applicable nuclear reactions are as follows: p-T (proton-tritium), d-D (deuterium-deuterium) and d-T (deuterium-tritium). The detector has been tested with neutron energies from 370 keV up to 14.9 MeV. As mentioned above, the SiC detector is capable of detecting fast neutrons by elastic scattering in light atoms of the detector material ^{12}C(n,n')^{12}C or ^{28}Si(n,n')^{28}Si and the nuclear reactions presented in Table I [10]. There are two reactions on ^{12}C and ^{28}Si atoms that produce α-particles and protons. The Q-value (released energy during reaction) is negative so its absolute value represents the amount of energy that is absorbed during the nuclear reaction and reduced for the resulting products.

TABLE I
NEUTRON REACTIONS ON SIC AND THE CORRESPONDING ENERGY THRESHOLDS.

Nuclear reaction	Branch	Q-value (MeV)
$^{12}C(n,n')^{12}C$	Ground state	0
$^{12}C(n,n')^{12*}C$	First excited state	-4.4389
$^{12}C(n,\alpha)^{9}Be$	Ground state α_0	-5.7012
	1st excited state α_1	-7.3859
	2nd excited state α_1	
$^{12}C(n,p)^{12}B$	Ground state p_0	-12.5865
	1st excited state p_1	-13.5396
$^{28}Si(n,n')^{28}Si$	Ground state	0
$^{28}Si(n,n')^{28*}Si$	First excited state	-1.779
$^{28}Si(n,\alpha)^{25}Mg$	Ground state α_0	-2.6537
	1st excited state α_1	-3.2387
	2nd excited state α_2	-3.6284
	3rd excited state α_3	-4.2655
	4th excited state α_4	-4.6183
$^{28}Si(n,p)^{28}Al$	Ground state p_0	-3.8599
	1st excited state p_1	-3.8905
	2nd excited state p_2	-4.8323

The 4H-SiC Schottky barrier detector was connected to the spectrometric chain consisting of the charge sensitive preamplifier Amptek A250CF, the digital multi-channel analyzer CAEN Hexagon with a high voltage source providing the reverse bias to the detector and a personal computer. The detection spectra of fast neutrons of different energies are shown in Fig. 4. The most complex spectrum was obtained by detecting 14.9 MeV monoenergetic neutrons produced by the d-T reaction. In the detected spectrum, the elastic scattering of neutrons on Si and C atoms is clearly visible. The peaks from the main nuclear reactions were then identified. According to Table I, the peak corresponding to an energy of 9.19 MeV is produced by the nuclear reaction $^{12}C(n,\alpha)^{9}Be$. The second clearly visible peak corresponds to the nuclear reaction between neutrons and silicon, $^{28}Si(n,\alpha)^{25}Mg$. The energy of the peak is 12.27 MeV according to the input fast neutrons. Smaller peaks with lower energies represent higher excited states. In the case of the 3.5 MeV neutrons produced by the d-D reaction, the peak from the silicon reaction is identified with a corresponding energy of 846 keV. Neutrons with energies of 370 keV and 1.1 MeV were produced in the p-T reaction. As the neutron energy is below the nuclear reaction threshold, only elastic scattering occurs in their detection. The measurements presented demonstrate that the 4H-SiC detector is suitable for the detection of a wide range of fast neutrons.

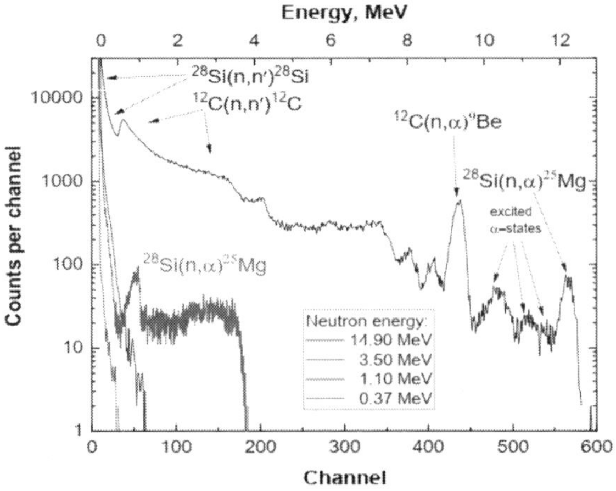

Fig. 4. The detected spectra of fast neutrons with different energies (370 keV, 1.1 MeV, 3.5 MeV and 14.9 MeV) using 4H-SiC detector.

IV. CONCLUSIONS

The 4H-SiC epitaxial Schottky barrier detector was fabricated and studied as a fast neutron detector. The thickness of the active layer was 80 μm. Firstly, the electrical characterization (current–voltage and capacitance–voltage measurements) of a Schottky contact detector was carried out to find the optimum operating bias. The detector is fully depleted at a reverse bias greater than 250 V. The detector was then used for the detection of monoenergetic fast neutrons with energies from 0.37 MeV to 14.9 MeV produced by the three nuclear reactions (p-T, d-D, d-T). The measured neutron spectra revealed elastic and inelastic neutron scattering from Si and C atoms, as well as nuclear reactions that produce α-particles. The results show that the 4H-SiC-based detectors are capable of detecting fast neutrons over a wide range of energies.

ACKNOWLEDGMENT

This work was partially supported by grants of the Slovak Research and Development Agency No. APVV-SK-CZ-21-0116, APVV-22-0382 and of the Scientific Grant Agency of the Ministry of Education of the Slovak Republic and the Slovak Academy of Sciences No. 2/0084/20. The measurements took place in the Van de Graaff accelerator laboratory supported by the LM2018108 grant of MEYS, Czech Republic.

REFERENCES

[1] A. Pietropaolo *et al.*, "Neutron detection techniques from μeV to GeV", Physics Report 875 (2020) 1-65.
[2] B. Zaťko *et al.*, "Radiation detector based on 4H-SiC used for thermal neutron detection", Journal of Instrumentation 11 (2016) C11022.
[3] K. Sedlačková *et al.*, "Comparison of semi-insulating GaAs and 4H-SiC-based semiconductor detector covered by LiF film for thermal neutron detection", Applied Surface Science 461 (2018) 242-248.

[4] A. R. Dulloo *et al.*, "The thermal neutron response of miniature silicon carbide semiconductor detectors", NIM in Phys. Res. A 498 (2003) 415-423.
[5] Haili Huang *et al.*, "Design and spectrum calculation of 4H-SiC thermal neutron detectors using FLUKA and TCAD", NIM in Phys. Res. A 822 (2016) 192-198.
[6] B. Zat'ko *et al.*: Detection of fast neutrons from D-T nuclear reaction using 4H-SiC radiation detector. Inter. J. Modern Phys.: Conf. Ser. 44 (2016) 1660235.
[7] K. Sedlačková *et al.*: MCNPX Monte Carlo simulations of particle transport in SiC semiconductor detectors of fast neutrons. J. Instrument. 9 (2014) C05016.
[8] B. Zat'ko *et* al., Detection of fast neutrons using detectors based on semi-insulating GaAs. J. Instrument. 6 (2011) C12047.
[9] http://www.ioffe.ru/SVA/NSM/Semicond/SiC/bandstr.html#Basic
[10] F. H. Ruddy *et al.*: The fast neutron response of silicon carbide semiconductor radiation detectors, IEEE Nuclear Science Symposium Conference Record, Vols. 1-7. (2004) pp.4575-4579.

Digitizer hardware for magnetic data acquisition on COMPASS-U

J. Oliveira[1*], A. Torres[1,2], A.J.N Batista[1], J. Sousa[1], B. Carvalho[1], A. Havranek[2], H.Fernandes[1]
[1]Instituto de Plasmas e Fusão Nuclear, Instituto Superior Técnico, Lisbon, Portugal
[2]Institute of Plasma Physics of the CAS, Prague, Czech Republic
(*) joao.s.oliveira@tecnico.ulisboa.pt

Abstract—**This paper presents the design and development of a modular digitizer prototype tailored for magnetic sensor data acquisition within the COMPASS-U tokamak facility, an upcoming experiment in Prague that will operate under conditions relevant to future experiments like ITER and DEMO. The magnetic diagnostic data serves a dual role, both for scientific analysis and real-time control. Due to the latter, the data integrity and fidelity is paramount for the plasma performance and overall safety of the fusion experiment. The digitizer prototype showcases a contemporary adaptation of prior board designs deployed in large fusion experiments such as JET and W7X. Leveraging the advanced capabilities of the Xilinx K26 Zynq Ultrascale+ System-on-Module, this digitizer interfaces seamlessly with ADC modules, facilitating real-time data pre-processing. The digitizer sends data to the real-time control system by a high-speed PCI Express interface. A distinctive facet of the design lies in its configurable clock architecture, allowing versatile distribution of clocks and triggers to accommodate diverse experimental scenarios. The inherent modularity and dynamic reprogramming of the hardware mean the system can find applications beyond its intended role. These include data acquisition for other diagnostics or application in different experiments. The modularity and flexibility and scalability of this design position it as invaluable candidate for high-volume data acquisition systems with a high number of isolated channels at a low cost per channel. As the scientific community endeavors towards advancing fusion technologies, the presented modular digitizer design stands poised to make significant contributions in magnetic diagnostics and real-time control applications.**

Keywords—**Data acquisition, Fusion reactors, Plasma measurements, CODAC, integrator, magnetic diagnostics**

I. Introduction

Research in magnetic fusion experiments is steadily moving towards longer pulse lengths which places increased requirements on their control, data acquisition and communication systems (CODAC) [1]. The largest experiments have a hazardous operation environment due to the high neutron flux caused by the fusion reaction and must be run remotely.

COMPASS Upgrade (COMPASS-U) is a new tokamak experiment (R = 0.894 m, a = 0.27 m, B_t = 5 T,I_p = 2 MA), under commission in Prague, Czech Republic [2, 3]. Its scientific goals are relevant to support the operation of ITER and DEMO such as plasma exhaust under extreme heat loads and advanced confinement modes. Additionally the hot wall operation places constraints on the design of the diagnostic systems of the machine[4].

One of the fundamental diagnostics in magnetic confinement fusion experiments is the magnetic diagnostic. A multitude of plasma parameters can be measured with magnetic sensors, among them: plasma current, plasma position, plasma shape and magnetic equilibrium. While some of these parameters, such as the plasma position and shape are crucial for the safe and reliable operation of the machine, others like the magnetic equilibrium are essential for other's plasma parametric analysis [5].

In fusion experiments, most of sensors used for the magnetic diagnostic system are inductive sensors. Since they use the principle of magnetic induction to measure magnetic fields, the signal they output is proportional to the time derivative of the magnetic field. As such, to obtain the magnetic field the induction sensor outputs need to be integrated. In the context of magnetic fusion experiments there are a number of integration methods that can be split into two major ones: analog integration and digital integration. Both of these methods are susceptible to integration drift: over time, noise with DC or low frequency components in the input signal accumulate and result in an error in the output of the integrator. As pulse lengths increase this becomes one of the major challenges that must be overcome in order to obtain accurate magnetic field measurements.

The data acquisition (DAQ) system is an integral part of any diagnostic. In the particular case of the magnetic diagnostic, while the bandwidths demanded by plasma shape and position control are quite low (up to the tenths of KHz) the measurement of MHD modes and more advanced instability control algorithms demand higher bandwidths (over 300KHz up to 1MHz in some experiments) [5, 6]. Combined with the high number of sensors present on large devices, and real-time latency constraints, rapidly puts the demands placed on the data acquisition system outside of those offered by off-the-shelf hardware. This implies the design of custom data acquisition systems tailored to the particular needs of magnetic data acquisition.

Due to the low latency, high data volume constraint of magnetic data acquisition systems most of the DAQ systems feature a Field Programmable Gate Array (FPGA) as its main processing element [7–9]. Moreover, their highly parallel nature and low latency FPGA's are particularly well suited to the needs of magnetic data acquisition. Recently System on a Chip (SoC) hardware has become increasingly popular. SoC's combine a traditional processor with a programmable logic fabric, providing the benefits of both kinds of hardware. The

 https://doi.org/10.1051/epjconf/202328803005

ease of development and familiar environment of a CPU gets combined with the low latency and highly parallel nature of an FPGA.

In this paper we present the design of the COMPASS-U magnetic data acquisition system. It leverages the capabilities of SoC FPGA devices to implement a flexible hardware design that can be readily reprogrammed to as the needs of the data acquisition system evolve.

II. System design

The design of the data acquisition system as a whole was guided by previous designs already proven to be effective[7, 9, 10]. Since then, these have become obsolete as the electronic components they depend on are no longer manufactured. A major aim of this design is then to harness the capabilities of newer components to satisfy the requirements of the magnetic diagnostics of COMPASS-U at a reduced cost per acquisition channel. A second objective was to create a sufficiently modular architecture that could also be used in other low latency, high channel count and high data volume applications.

The acquisition hardware is part of the overall magnetic diagnostic acquisition system. It stands between the magnetic sensors and the real-time plasma controllers. Its function is to digitize and pre-process the signals detected by the sensors, the acquired signals are then made available to the real-time plasma controllers and are also stored in a central database for offline data analysis.

A. Constraints and requirements

The hardware we developed targets the COMPASS-U magnetic acquisition system and as such, there are a set of constraints and requirements it must meet. Some of these constraints arise from the nature of the magnetic sensors selected for use in COMPASS-U, while others come from the design of the CODAC systems of the machine. The environment of COMPASS-U (up to 500 °C wall temperature and high vacuum) places several constraints on the design of the magnetic sensors themselves, mainly in the selection of materials for their construction [4]. Most of the sensors that will be used in COMPASS-U magnetic diagnostic will be induction coils. Since induction sensors present floating voltage readings they must be digitised with differential input analog to digital converters to prevent common mode errors. Furthermore, it was shown that isolation of the inputs is important to reduce crosstalk and mitigate integrator drift [11, 12]. In the acquisition system this is done with the modules described in [13], which our hardware is specifically designed to interface. The ADC modules are based on the digital integration with modulation design described in [12], that has been previously used at ISTTOK [14] and W7X [15].

From the design of the COMPASS-U magnetic diagnostic we will need at least 350 synchronous acquisition channels [16], [2]. The per channel bandwidth requirements imposed by the magnetic sensors (up to 1 MHz) require a sampling rate of at least 2 MSPS. With a bit depth of 24 bit per sample (the maximum under consideration for COMPASS-U [17]), each channel requires a digital bandwidth of 48 Mb/s which corresponds to a total system bandwidth requirement of 16.8 Gb/s. The CODAC system also places other constraints on the design of the magnetic data acquisition system. The real-time control loop will have a period of 50 μs which constrains the maximum latency on the data acquisition system. There is also the requirement of galvanic isolation between the acquisition hardware and the control hardware. Given the operating environments, bandwidth and latency requirements, this points towards the use of an optical connection for data transfer between the real-time control and data acquisition systems.

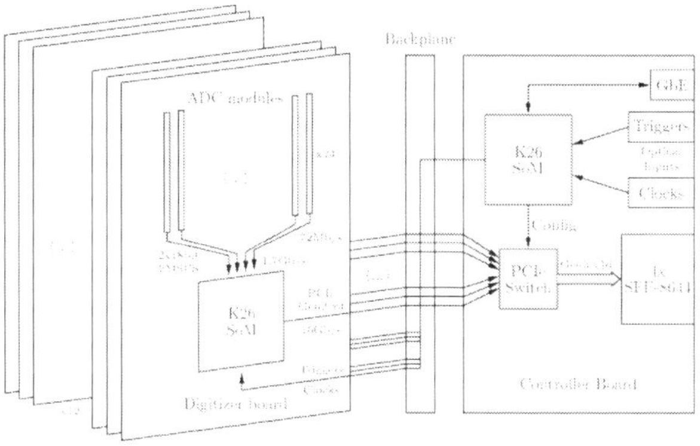

Fig. 1. High level schematic of the magnetic data acquisition system. The bandwidths shown in the digitizer boards are calculated for dual channel 18-bit 2MSPS ADC modules. We see the 1.8 Gbps demanded by 24 of these modules comfortably fit within the 16 Gbps available on the PCIe link.

A third constraint that was, unfortunately, also present was the low availability of electronic components on major distributors during the beginning of the design process circa 2021. This greatly limited component choice, in particular for the programmable logic components which we found in short supply in major electronic distributors. Since one of the aims of the design was to achieve a low cost per channel, the availability of electronic components on major markets is a major consideration.

B. Acquisition system architecture

The acquisition system we designed is based on the Advanced Telecommunications Computing Architecture (ATCA) standard like previous systems [7, 8, 15]. We chose ATCA, a telecommunications industry standard proven in its reliability and availability archived through the use of redundancy. The form factor allows for high channel densities in a compact chassis. We also considered the mTCA.4 standard since its widely used on other physics experiments[18–20], but due to its smaller board size it could not achieve the same channel densities as ATCA. The ATCA standard defines a backplane which a number (from two to sixteen) of boards utilize to communicate with each other through point to point multi Gigabit serial links. The standard offers multiple connection typologies: full mesh, where each board has a link to every board, and dual star topology where two redundant central boards have a link to all the other node boards. Since our design is focused on data acquisition we chose a dual star topology for ease of implementation.

The high level design of the system can be seen in Fig. 1. The system is composed by a central switch board and multiple (up to twelve) node boards. Each node board interfaces up to 24 dual-channel ADC modules, with isolated differential inputs, 18-bit precision and an acquisition rate up to 2 MSPS. With this each board will have 48 input channels which translates in 576 channels per ATCA chassis (assuming a 19" chassis with 14 slots), optical clock and trigger input ports and a set of four SFF-8644 ports for communication with the real-time control system. This communication is established with a PCIe link. The PCIe expansion bus was chosen because it is capable of providing sufficient bandwidth to carry the required data volumes to the real-time control system with low transfer latencies and because it was used successfully on previous data acquisition systems. The SFF-8644 ports can carry a PCIe 3.0 x16 link over optic cables, providing a bandwidth of 16GB/s with sub microsecond latencies and ensuring galvanic isolation between both ends of the connection.

The analog signals go into the the node boards which function as a digitizer. These boards host the ADC modules responsible for digitizing the incoming signals. The modules all interface to an FPGA which gathers the data from the ADC's, numerically integrates the signals in real time and forwards the resulting data to the central board through a low latency, high-speed serial PCIe connection carried over the fabric channel in the ATCA backplane. The central board's primary function is a data aggregator. It has a PCIe switch that is used join the PCIe Gen2 x4 links from each digitizer board into a single x16 Gen 3 up-link to the real-time control system. The central board also has the clock and trigger inputs to the acquisition system which are distributed to the node boards over the ATCA backplane synchronization clock interface.

Both the node and central boards are based on the same design. The central board is a modified version of the node board with the ADC module interface removed and a PCIe switch added. This enables the reuse of most of the board design, considerably lowering the design effort. Both boards contain the same SoM too, reducing the effort of maintaining the software stack.

III. Digitizer design

The design of the digitizer board is centered around the AMD Kria K26 System on a module. This SoM contains a Zynq Ultrascale+ SoC that combines programmable logic (PL), like the fabric of a conventional FPGA, with a multi-core ARM processor and a series of peripheral interface hard blocks. Its programmable logic capabilities are well suited to the massively parallel real time computations that a real-time data acquisition system requires. The processor coupled with the peripheral interfaces form a processing system that provides a familiar Linux based programming environment that is useful for system management, remote operation and supervision. The system on a module integrates many of the power regulators and peripherals required for SoC operation such as DRAM, non-volatile storage and firmware storage which considerably reduces development effort. This particular SoM was chosen due to its manufacturer support, availability on major distributors, low cost and high number of I/O pins. The SoM has two types of I/O pins connected to the PL: high performance I/O, which support differential signaling standards at data rates up to 1.25 Gbps and high density I/O pins, which mostly support single ended signaling standards at rates under 250 Mbps. A scheme of the design can be seen in Fig. 2. It illustrates all major subsystems of the design. These are: the ADC interface, the clock and trigger distribution circuits, the high speed serial interfaces and the processor system peripherals.

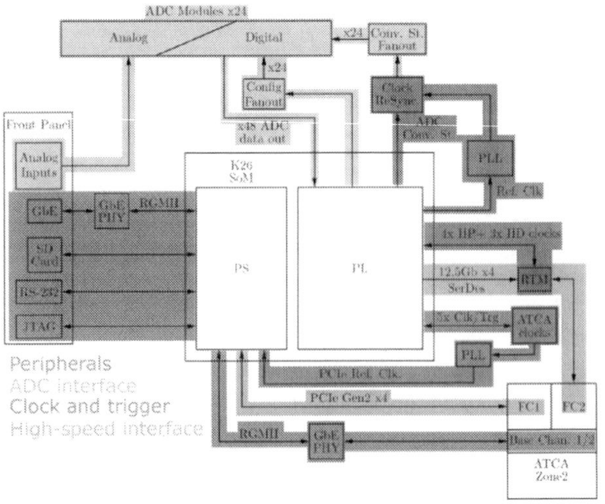

Fig. 2. Schematized design of the acquisition board, illustrating the connections between all major components.

The use of K26 module as the processing element of the digitizer board brings us some benefits to the overall architecture of the acquisition system. The on-board RAM and storage provide us with and extra layer of redundancy during the data acquisition since its is possible to buffer the acquired data in RAM and later store it on the digitizer board to non volatile storage. This prevents data loss in case of a fault in the data streaming interfaces.

Taking into account the form factor (130 mm x 20 mm x 9.5 mm) of ADC modules and the space available on the board, we have found that 48 channels per board is the maximum density that can be achieved. The ADC modules are mounted on to the main board at a right angle and they are oriented parallel to the airflow on the ATCA crate for temperature homogeneity. Since there are two channels per module there are 24 per board. The analog signals enter the board through the front panel on DD-50 connectors. The signals are routed to the modules' analog connectors using a stripline structure with dedicated shielding planes on both sides of the differential pair carrying the signal. This ensures crosstalk between acquisition channels is minimized while maintaining electrical isolation up to 1500 V between each channel. The modules connect to the main board with two right-angle connectors; one of them carries analog signals and the other carries digital signals. The digital connector is a high-density connector for high-speed signals; the connector on the analog size is of a different type and lower pin density, it was selected taking into consideration the electrical isolation requirement. The use of different types of connectors for the ADC module interface adds the obstacle of finding compatible connectors, but allows us to choose the best type for each situation with less compromises.

The digital interface to the ADC modules is based on SPI with a dedicated acquisition trigger signal. The SPI control signals are carried over LVDS to the modules to minimize crosstalk and interference. Since the ADC control signals are the same for every module, they are distributed using fan-out buffers which greatly reduces the demand for I/O pins on the SoC and also simplifies the board layout and routing. The acquisition trigger for the ADC is also distributed to every module using precision clock distribution fan-out buffers that guarantee low jitter on their outputs. To avoid lowering the ENOB of the ADC's we targeted a acquisition clock jitter under 1 ps which is sufficient for 18-bit precision at 1 MHz (The upper limit of COMPASS-U's high speed inductive sensors' bandwidth [16]). Since the mixed-mode clock managers (MMCM) available on the SoC can only obtain a jitter of 200 ps the design features an external PLL to reduce the jitter of the clock signal to acceptable levels. Because the trigger to the ADC is not a symmetric pulse we use a clock and data synchronizer IC to re-time the acquisition trigger pulse coming from the SoC to the low jitter clock synthesized by the PLL.

Each node board can send or receive up to six independent clock or trigger signals from the ATCA backplane. One of the clocks is reserved as the PCIe reference clock leaving the rest of them to be used to synchronize the data acquisition across all node boards. Apart from the PCIe clock which is routed to a PLL to reduce its jitter, all of the clocks are routed to the SoC where they are used to synthesize the acquisition start signal. The clock distribution circuit is shown on Fig. 3. With the use of bi-directional MLVDS buffers and the programmable logic on the SoC we can see that the design is quite flexible and offers several possibilities for clock and trigger distribution to suit the needs of a particular data acquisition system.

The SoC features two sets of four lanes of high-speed serial multi-gigabit transceivers (MGT). One of the sets is controlled by the processor system and is capable of speeds up to 6 Gbps per lane while the other is within the programmable logic section and is capable of speeds of 12.5 Gbps per lane. The processor system MGT lanes are restricted to a small set of communication protocols, notably PCIe Gen2 which we use for data output. The programmable logic MGT lanes allow for a much larger number of communication protocols and as such we opted to have them available on the board's expansion port.

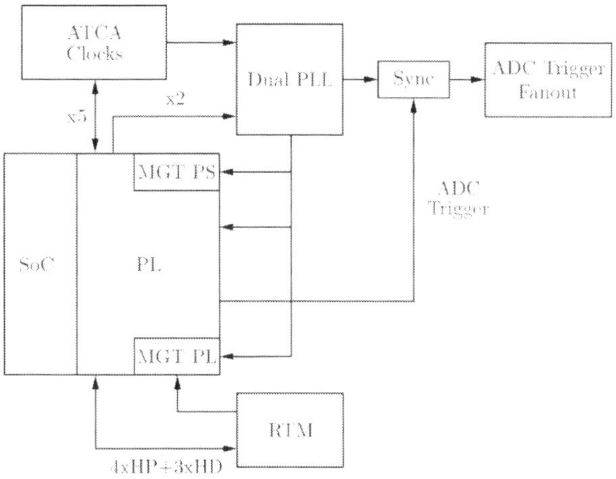

Fig. 3. Scheme of the clock/trigger distribution circuit. PL refers to the programmable logic fabric of the SoC. MGT refers to the multi-gigabit transceivers available on the K26 SoM. RTM is the rear transition module, used for extra user defined I/O there are 7 clock inputs from the RTM: three are connected to High Performance I/O pins and four are connected to high density I/O pins on the SoC. We use a dual PLL to clean jitter in the incoming PCIe reference clock and to reduce the jitter of the ADC conversion start signal.

The board features an expansion port that can be used to augment its functionality with a rear transition module (RTM) should that need arise. The ATCA specification provides for an RTM as a means to add used defined I/O to a board should the need arise. Aside from the programming logic's MGT lanes there are twenty-six 250 Mbps high density I/O pins plus three 1.25 Gbps high performance clock compatible differential pairs routed to the RTM connector. Together with the MGT lanes this can be used to extend and alter the functionality of the digitizer board should a different architecture of the data acquisition system be required. The availability of expansion I/O is one of the main factors contributing towards the flexibility and adaptability of the design.

There are multiple processing system peripherals present on the SoC. Of all the options available we chose to use two Ethernet controllers one SD card controller and a serial port. While none of these are required for the operation of the data acquisition system, they can be used to significantly extend its functionality. The Ethernet connections are convenient for system management, remote operation and supervision. There are two Ethernet ports on the board, one of them is on the front panel and is meant for debugging during firmware development and testing. The other one is connected to the ATCA backplane on its base channel and is meant for remote operation and management during production use. The SD card controller is connected to a micro SD card slot that can be used to extend the amount of storage available on the digitizer boards; it can also serve as an alternate SoC boot device should the internal storage get corrupted or erased. Although the write speed to the SD card is too slow to store experiment data in real time it nevertheless could be useful as a redundant, local archive of experiment data. Finally, the serial port is useful to debug the operating system running on the processor system of the SoC.

IV. Conclusions

We presented the design of a digitizer board optimized for the acquisition of magnetic signals in fusion experiments. Our design ,in particular, was tailored to the requirements of COMPASS-U. The design of the digitizer revolves around the use of dual channel isolated ADC modules that were purposely developed for digitizing signals from inductive probes while integrated in an acquisition system using digital integration. The hardware we designed allows for an acquisition system with real-time processing capabilities, with up to 576 channels in a single ATCA chassis that is galvanically isolated from the real-time control system.

We manged to harness the capabilities of modern FPGA hardware to design a flexible modular system that satisfies the requirements of COMPASS-U for channel count, bandwidth, latency and isolation. also can easily be reconfigured to satisfy the requirements of other high channel count and data volume data acquisition systems at a low cost per channel. The key behind this achievement is the use of the K26 SoM as the main processing element which allowed us to reduce the design effort as the module integrates the SoC along with power converters DRAM and storage at an accessible price point.

Acknowledgements

This work has been carried out within the framework of the project COMPASS-U: Tokamak for cutting-edge fusion research (No. CZ.02.1.01/0.0/0.0/16_019/0000768) and co-funded from European structural and investment funds and by MEYS project LM2018117. J. Oliveira was supported by the Fundação para a Ciência e Tecnologia through the Instituto de Plasmas e Fusão Nuclear, a research institute of Instituto Superior Técnico, under the Ph.D. grant PD/BD/150408/2019.

References

[1] G. Vayakis, C. Walker, I. I. Team, and P. Teams, "Magnetic diagnostics for iter/bpx plasmas," *Review of scientific instruments*, vol. 74, no. 4, pp. 2409–2417, 2003.

[2] P. Vondracek, R. Pánek, M. Hron, J. Havlicek, V. Weinzettl, T. Todd, D. Tskhakaya, G. Cunningham, P. Hacek, J. Hromádka, *et al.*, "Preliminary design of the compass upgrade tokamak," *Fusion Engineering and Design*, vol. 169, p. 112490, 2021.

[3] R. Panek, T. Markovic, P. Cahyna, R. Dejarnac, J. Havlicek, J. Horacek, M. Hron, M. Imrisek, P. Junek, M. Komm, *et al.*, "Conceptual design of the compass upgrade tokamak," *Fusion Engineering and Design*, vol. 123, pp. 11–16, 2017.

[4] V. Weinzettl, J. Adamek, P. Bilkova, J. Havlicek, R. Panek, M. Hron, O. Bogar, P. Bohm, A. Casolari, J. Cavalier, *et al.*, "Constraints on conceptual design of diagnostics for the high magnetic field compass-u tokamak with hot walls," *Fusion Engineering and Design*, vol. 146, pp. 1703–1707, 2019.

[5] E. Strait, E. Fredrickson, J.-M. Moret, and M. Takechi, "Chapter 2: Magnetic diagnostics," *Fusion Science and technology*, vol. 53, no. 2, pp. 304–334, 2008.

[6] A. Sushkov, A. Belov, G. Igonkina, A. Kachkin, E. Khayrutdinov, A. Melnikov, and M. Sokolov, "Design of inductive sensors and data acquisition system for diagnostics of magnetohydrodynamic instabilities on the t-15md tokamak," *Fusion Engineering and Design*, vol. 146, pp. 383–387, 2019.

[7] A. Batista, C. Leong, V. Bexiga, A. Rodrigues, A. Combo, B. Carvalho, P. Carvalho, J. Fortunato, B. Santos, P. Carvalho, *et al.*, "Control and data acquisition atca/axie board designed for high system availability and reliability of nuclear fusion experiments," *Fusion Engineering and Design*, vol. 88, no. 6-8, pp. 1332–1337, 2013.

[8] B. Carvalho, A. Batista, M. Correia, A. Neto, H. Fernandes, B. Gonçalves, and J. Sousa, "Reconfigurable atca hardware for plasma control and data acquisition," *Fusion Engineering and Design*, vol. 85, no. 3-4, pp. 298–302, 2010.

[9] D. Valcárcel, A. Neto, J. Sousa, B. Carvalho, H. Fernandes, J. Fortunato, A. Gouveia, A. Batista, A. Fernandes, M. Correia, *et al.*, "An atca embedded data acquisition and control system for the compass tokamak," *Fusion Engineering and Design*, vol. 84, no. 7-11, pp. 1901–1904, 2009.

[10] I. S. Carvalho, P. Duarte, H. Fernandes, D. F. Valcárcel, P. J. Carvalho, C. Silva, A. S. Duarte, A. Neto, J. Sousa, A. J. Batista, *et al.*, "Real-time control for long ohmic alternate current discharges," *Fusion engineering and design*, vol. 89, no. 5, pp. 576–581, 2014.

[11] P. Spuig, P. Kumari, M. Moreau, P. Moreau, A. Le-luyer, and P. Malard, "Enhanced integrators for west magnetic diagnostics," *Fusion Engineering and Design*, vol. 96, pp. 966–969, 2015.

[12] A. Werner, "W7-x magnetic diagnostics: Performance of the digital integrator," *Review of scientific instruments*, vol. 77, no. 10, 2006.

[13] A. Torres, B. Carvalho, T. Markovic, A. Batista, A. Havranek, V. Weinzettl, and H. Fernandes, "Data acquisition with real-time numerical integration for compass-u magnetic diagnostics," *Fusion Engineering and Design*, vol. 191, p. 113580, 2023.

[14] D. Corona, N. Cruz, J. J. E. Herrera, H. Figueiredo, B. B. Carvalho, I. S. Carvalho, H. Alves, and H. Fernandes, "Design and simulation of isttok real-time magnetic multiple-input multiple-output control," *IEEE Transactions on Plasma Science*, vol. 46, no. 7, pp. 2362–2369, 2018.

[15] B. Carvalho, A. Batista, M. Zilker, K. Rahbarnia, *et al.*, "Design, implementation and commissioning of atca based high-speed multichannel data acquisition systems

for magnetic diagnostics in w7-x," in *11th IAEA Technical Meeting on Control, Data Acquisition, and Remote Participation for Fusion Research*, 2017.

[16] V. Weinzettl, P. Bilkova, I. Duran, M. Hron, R. Panek, T. Markovic, M. Varavin, J. Cavalier, K. Kovarik, A. Torres, *et al.*, "Development of the diagnostic tools for the compass-u tokamak and plans for the first plasma," *Fusion Engineering and Design*, vol. 191, p. 113545, 2023.

[17] A. Havránek, O. Bogár, D. Fridrich, I. Hanak, T. Majer, J. Burant, J. Havlíček, P. Vondráček, M. Komm, M. Hron, *et al.*, "Conceptual design of the compass-u control systems," *Fusion Engineering and Design*, vol. 170, p. 112550, 2021.

[18] J. Sousa, M. Correia, R. Pereira, A. Rodrigues, P. Carvalho, A. Combo, B. Santos, B. Carvalho, A. Batista, R. Dias, *et al.*, "Mtca control and data acquisition platform for plasma diagnostics," *Journal of Instrumentation*, vol. 14, no. 11, p. C11025, 2019.

[19] J. Zhang, H. Li, and X. Jiang, "The microtca fast control board for generic control and data acquisition applications in hep experiments," *IEEE Transactions on Nuclear Science*, vol. 66, no. 7, pp. 1169–1173, 2019.

[20] T. Lenzi, "A micro-tca based data acquisition system for the triple-gem detectors for the upgrade of the cms forward muon spectrometer," *Journal of Instrumentation*, vol. 12, no. 01, p. C01058, 2017.

https://doi.org/10.1051/epjconf/202328803006

Evaluation of the energy calibration of diamond detectors for fast neutrons applications

Giada Gandolfo[1,*], Luigi Lepore[1], Nadia Cherubini[1], Ermanno Piccinelli[1], and Giuseppe Augusto Marzo[1]

[1]ENEA, Fusion and Technology for Nuclear Safety and Security Department, Rome, Italy

(*) giada.gandolfo@enea.it

Abstract—**The radiation hardness, the chemical resistance, and the capabilities to operate at high temperature conditions make diamond detectors a good option for carrying out fast neutron measurements on fusion plasma experiments or facilities using accelerator-driven neutron sources. A correct energy calibration of pulse-height spectra acquired through diamond detectors allows to perform fast neutron spectrometry. As a general rule, energy calibration of diamond detectors is performed using an alpha source, e.g. ^{239}Pu, ^{241}Am, ^{244}Cm, whose characteristic emission energies are in the range from 5 to 6 MeV. Calibration at higher energies, such as those related to charged particles (about 8.4 MeV) produced by the ^{12}C(n,alpha)^{9}Be induced by fast neutrons coming from D-T fusion reaction, is traditionally extrapolated with the hypothesis of linearity for the energy calibration curve. In this work an evaluation of the diamond detector energy calibration based on alpha source emissions is performed at higher energies by means of a compact D-T neutron generator, able to produce neutrons within a broad energy range, by changing the accelerator voltage and the neutron emission angle. A relative deviation less than 2% between experimental and theoretical energies was observed, showing that the energy calibration through alpha sources could be still valid for fast neutrons coming from D-T fusion reactions.**

***Keywords* — diamond detectors, fusion experiments, energy calibration, fast neutrons**

I. Introduction

Diamond detectors are widely used in high energy physics and nuclear fusion experiments because of their resistance to hostile radiation environments [1]. For instance, they are part of the diagnostic designed for the International Thermonuclear Experimental Reactor (ITER) [2] and they are currently used in Joint European Torus (JET) [3]. Additionally, they can be used as point-like probe for characterizing the neutron field emitted by accelerator-driven neutron sources, as compact D-T neutron generators involved in detection of hidden explosives [4] or fissile material [5] and in almost every field of industry and research [6].

In a diamond detector, neutron detection occurs by means of charged particles produced by nuclear reactions involving the carbon nuclei. In the particular case of fast neutrons with energies higher than 5.702 MeV, the pulse-height spectrum resulting from neutron-induced reactions shows a well isolated peak corresponding to the ^{12}C(n,alpha)^{9}Be reaction byproducts, and directly proportional to the incident neutron energy [7]. A correct energy calibration of pulse-height spectrum allows to perform fast neutron spectrometry. As a general rule, energy calibration of diamond detectors is performed using an alpha source, whose characteristic emission energies lie in the range from 5 to 6 MeV. Calibration at higher energies, such as those related to charged particles produced by the ^{12}C(n,alpha)^{9}Be induced by fast neutrons from D-T fusion reaction, is traditionally extrapolated with the hypothesis of linearity for the energy calibration curve obtained by considering the alpha source [3,7-9].

Usually, in radiation spectrometry (e.g., gamma or alpha spectrometry) the energy calibration is performed using sources emitting radiation with energies that cover the whole energy interval of interest, with experimental points included in such energy range. If possible, linear extrapolation should not be used to extend the energy calibration outside the range where the measured points of the calibration source actually lie.

In this paper, an experimental evaluation of the energy calibration of a diamond detector obtained by means of an alpha source to higher energies by using a D-T neutron generator is carried out, being the neutron generator able to produce neutrons within a broad energy range. This will build confidence in adopting the alpha calibration within conditions closer to real application of diamond detectors.

II. Experimental setup

A. Single crystal diamond detector

A small size diamond detector, i.e. 4.7 mm x 4.7 mm x 0.5 mm, obtained by chemical vapor deposition [10] was used as fast neutrons detector (Fig. 1).

Fig. 1. Diamond detector.

The diamond detector working principle is similar to that of a solid-state ionization chamber. A charged particle with an energy greater than its band-gap produced ionization through the diamond, and consequently generates electron-hole couples. Charges are then collected by means of an electric field. Neutrons, lacking electrical charge, do not cause direct ionization within the detector and can be detected only through secondary charged particles produced by conversion reactions. Focusing on 14 MeV neutrons emitted by D-T reaction, the best candidate reaction channel for neutron spectroscopy is the (n,alpha) reaction, $^{12}C(n,alpha)^{9}Be$, whose characteristic peak in the measured pulse height spectrum (PHS) corresponds to the total energy of reaction products, ^{9}Be and alpha particle, as shown in Fig. 2. The well-isolated peak is centered at the energy $E_n - 5.702$ MeV, where E_n is the energy of the incident neutron, and 5.702 MeV is the threshold over which the reaction between neutrons and ^{12}C can occur.

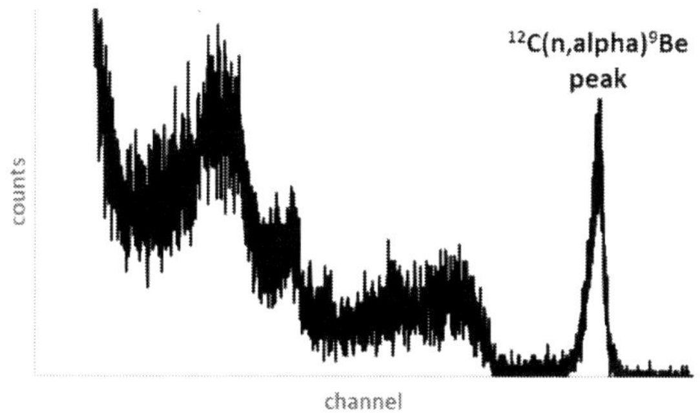

Fig. 2. Measured pulse height spectrum for 14 MeV neutrons interacting with a diamond detector.

B. *Calibration sources*

For the experiments two types of calibration sources were used: certified alpha sources and a compact D-T neutron generator.

To collect calibration points in an adequate energy range, two certified alpha sources were used, one containing ^{237}Np, ^{241}Am, and ^{244}Cm and the other one containing ^{239}Pu, ^{241}Am, and ^{244}Cm. Characteristic alpha energies emitted by the aforementioned radionuclides are listed in Table I.

TABLE I
CHARACTERISTIC MOST PROBABLE ENERGIES EMITTED FROM ALPHA SOURCES

Radionuclides	Energy for the most probable alpha emissions (MeV)
^{237}Np	4.788
^{239}Pu	5.157
^{241}Am	5.486
^{244}Cm	5.805

Since the experiments were performed in air, an appropriate correction was introduced for estimating the actual energy of the alpha particle on the detector, taking into account the loss of energy due to the air thickness between detector and alpha source.

A D-T compact neutron generator was used as alternative source of secondary charged particles through neutrons interactions. Inside the generator a beam of deuterons is accelerated and hits a metallic target where tritium is adsorbed on. The D-T fusion reaction $^{3}H(d,n)^{4}He$ produces fast neutrons emitted in the whole solid angle. The theoretical energy of the emitted neutron E_n depends on its emission direction with respect to incident deuteron direction θ as well as on the deuteron energy E_d, directly linked to the accelerator voltage, following kinematic of the reaction:

$$\sqrt{E_n(\theta, E_d)} = A + \sqrt{A^2 + B} \quad (1)$$

where

$$A = \frac{\sqrt{M_d M_n E_d}\cos\theta}{M_\alpha + M_n} \quad (2); \quad B = \frac{M_\alpha Q + E_d(M_\alpha - M_d)}{M_\alpha + M_n} \quad (3)$$

with M_d, M_α, M_n the masses of deuteron, alpha particle, and neutron, respectively. While Q is the reaction Q-value which, for the fusion nuclear reaction $^{3}H(d,n)^{4}He$ has a value of 17.59 MeV [11]. Since the accelerator voltage is selectable between 40 kV and 90 kV, it offers the possibility to produce neutrons within a broad energy range around 14 MeV. In Fig. 3, the neutron energy as a function of its emerging angle is shown, produced in the case of a deuteron acceleration voltage of 40 kV and 90 kV.

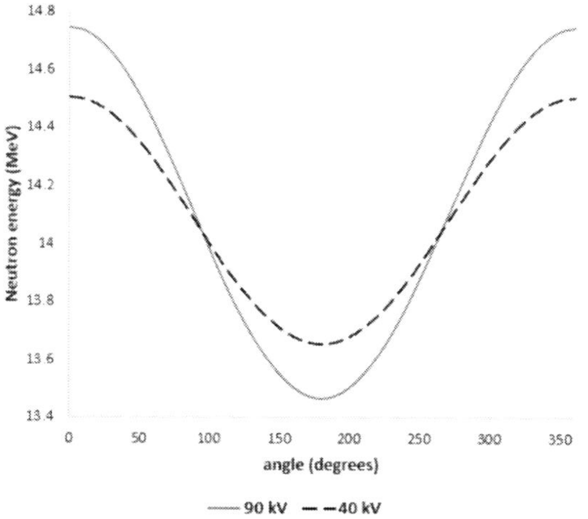

Fig. 3. Neutron energy as a function of its emerging angle from D-T neutron generator. The solid line shows the energy variation for an accelerator voltage of 90 kV, the dotted line corresponds to an accelerator voltage of 40 kV.

Using the D-T neutron generator with different accelerator voltages it is possible to produce neutrons within the range 13.466 and 14.748 MeV. Since the well-isolated peak in the diamond measured PHS is centered at $E_n - 5.702$ MeV, these neutrons interacting with the diamond detector could produce secondary charged particles with a total energy deposition inside the diamond crystal between 7.764 MeV and 9.046 MeV, thus over the energy range covered by the alpha sources.

C. *Standard calibration procedure*

The certified point alpha sources were placed one at a time on the detector and the two PHS shown in Fig. 4 were acquired and recorded.

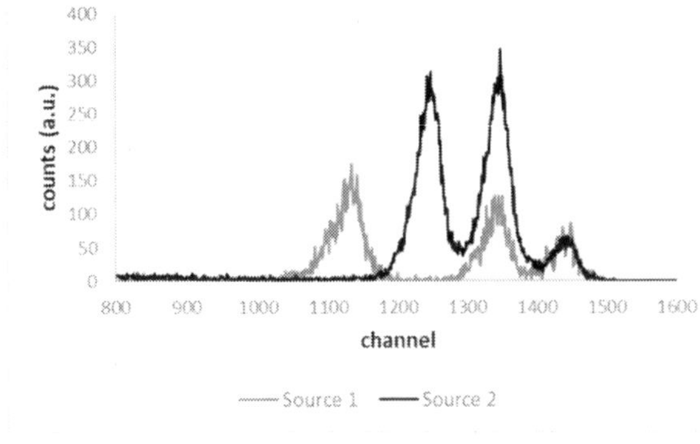

Fig. 4. Measured PHS for certified alpha sources. In grey Source 1 (^{237}Np, ^{241}Am, and ^{244}Cm), in black Source 2 (^{239}Pu, ^{241}Am, and ^{244}Cm).

A standard calibration of the PHS was performed, associating the channels of the peak centroids to the corresponding alpha energies and performing a least squares regression. The goodness of the regression was evaluated through the R-square that resulted 0.9998. Usually in radiation spectrometry (i.e., gamma or alpha spectrometry) the energy calibration is performed using sources emitting radiation with energies that cover the whole energy interval of interest, with experimental points included in such energy range.

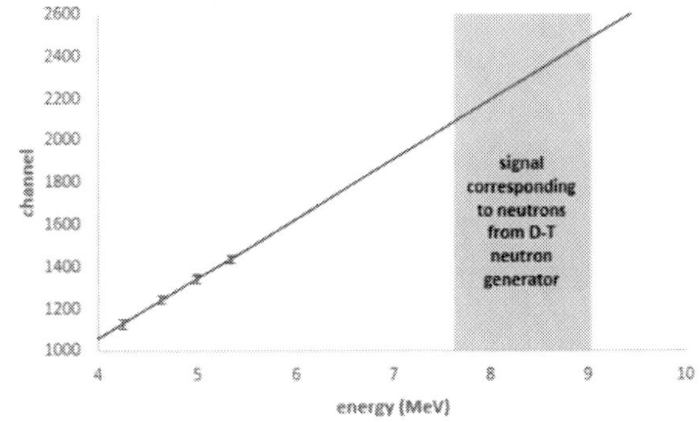

Fig. 5. Regression line corresponding to the standard calibration procedure (R-square = 0.9998). On the left the experimental points corresponding to alpha sources emissions, on the right in grey the region related to neutrons from D-T neutron generator.

In this case, the regression line is based on experimental points corresponding to energies between 4.246 MeV and 5.331 MeV, relatively far away, in terms of energy, from the region related to the signal coming from D-T neutrons, highlighted in grey in Fig. 5.

For this reason, an experimental evaluation of the standard calibration through alpha sources was performed using the compact D-T neutron generator.

D. Experimental evaluation of standard calibration for 14 MeV neutrons

To record PHS related to different neutron energies the diamond detector was mounted on a metal arm solidly connected to the generator and able to rotate around the center of the target (Fig. 6). Such a configuration makes it possible to acquire the neutron energy spectra at different angular positions with respect to the deuteron ions direction.

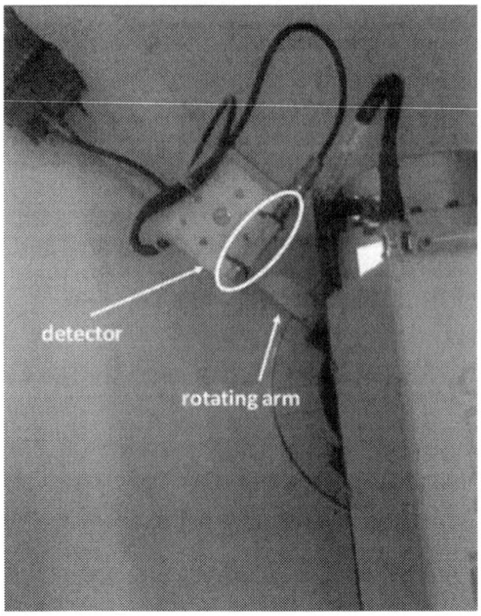

Fig. 6. Experimental setup. Diamond detector mounted on a metal rotating arm solidly connected to the neutron generator.

To maximize the flux and consequently minimize the measurement time, the smallest allowed detector-to-target distance was chosen. Given the overall dimensions of the neutron generator, such a distance is 17 cm. A semicircle with a radius of 17 cm has been sampled in 8 positions, illustrated in Fig. 7.

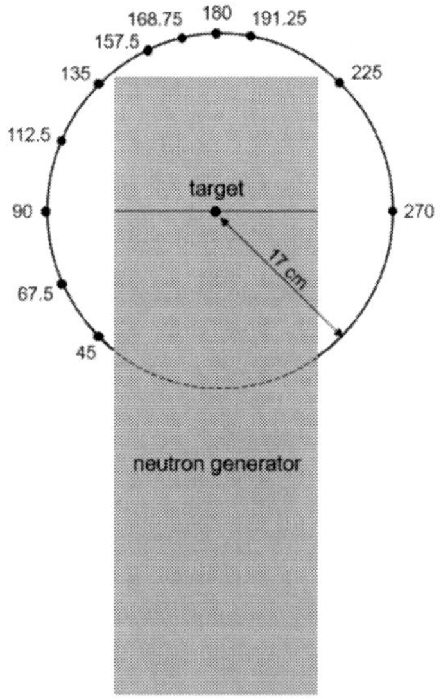

Fig. 7. Diamond detector experimental positions around the compact D-T neutron generator.

Additional measurements were carried out to verify the irradiation symmetry. Measurements with $\theta < 45°$ was not investigated because of the overall dimensions of the device, preventing the detector to be placed at the minimum selected distance for those angles. Three sets of measurements were performed corresponding to three different accelerating voltages, 50 kV, 80 kV, and 90 kV. The 80 kV measurements set investigated all the 11 positions, the other two sets focused on 6 of them (45°, 135°, 157.5°, 168.75°, 180°, and 270°).

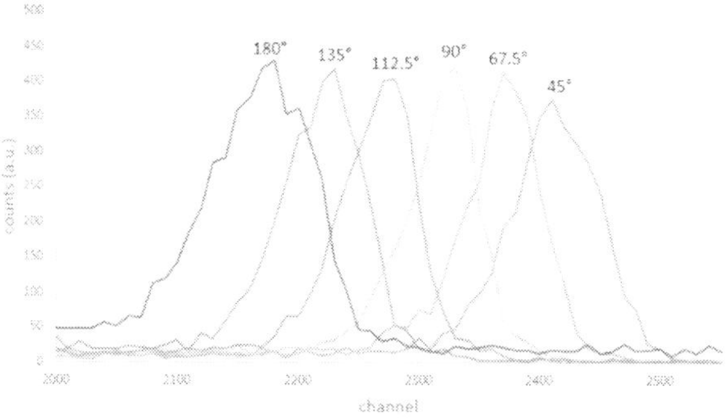

Fig. 8. $^{12}C(n,alpha)^{9}Be$ reaction peak at different detector positions for an accelerating voltage of 80kV.

Focusing on the position of $^{12}C(n,alpha)^{9}Be$ reaction peak, the expected shift in channels –corresponding to the variation in energy of neutrons emitted at different angles– was recorded, as illustrated in Fig. 8.

The peak shape depends on the incident particle beam composition within the neutron generator that in general can be a mixture of D/T ions causing the presence of DT^+, D_2^+, T_2^+ T^+, and D^+ within the beam [12]. Since in this case study the accelerated beam is made of deuterons impinging on a tritiated target, the main contribution to the peak shape is expected to be due to D_2^+ and D^+ ions. The simultaneous presence of these two types of ions could cause a double-peak structure for angles near to 0° and 180°, depending on the proportion between the two types of ions. Since the ratio of D_2^+ and D^+ ions is an unknown parameter, a simplified approach considering the incident particle beam made by only one type of ions has been considered. In particular, the analysis has been performed for each case (i.e., only D^+ ions and only D_2^+ ions). In the case considering only D_2^+ ions, the molecular ion will produce two particles with half energy respect to the case considering only D^+ ions and the neutron theoretical energy will vary accordingly.

For each detector position and accelerating voltage, the characteristic peak was fitted with a Gaussian distribution function to determine the mean value (μ) and the related standard deviation (σ). With the hypothesis that neutrons interacting with diamond detector at 17 cm from the target position and causing a signal in the well isolated peak are uncollided neutrons, each mean value was associated with the respective theoretical energy (E_{th}), and the experimental points, whose coordinates in an Energy-channel reference system are (E_{th}, $\mu\pm\sigma$), were compared with the predicted values corresponding to points (E_{th}, $A+B\cdot E_{th}$), where A and B are the coefficients of the calibration regression line.

Fig. 9-10 show the experimental points related to neutrons produced at different accelerating voltages for an incident particle beam made respectively only by D^+ ions and D_2^+ ions.

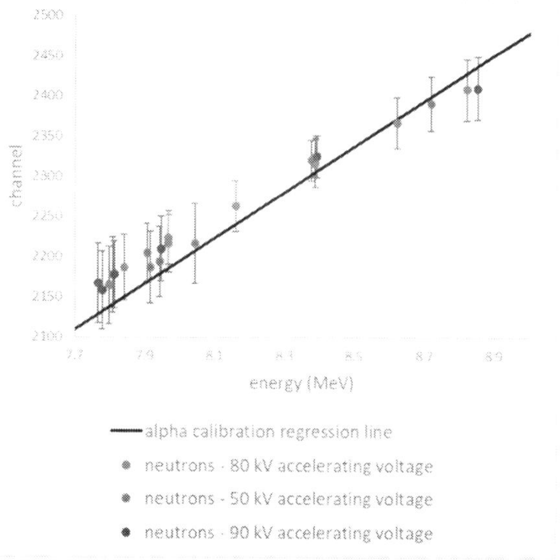

Fig. 9. Evaluation of alpha calibration regression line in the energy region 7.764 MeV - 9.046 MeV through experimental points from neutrons produced by D-T neutron generator. The incident particle beam is considered made only by D^+ ions.

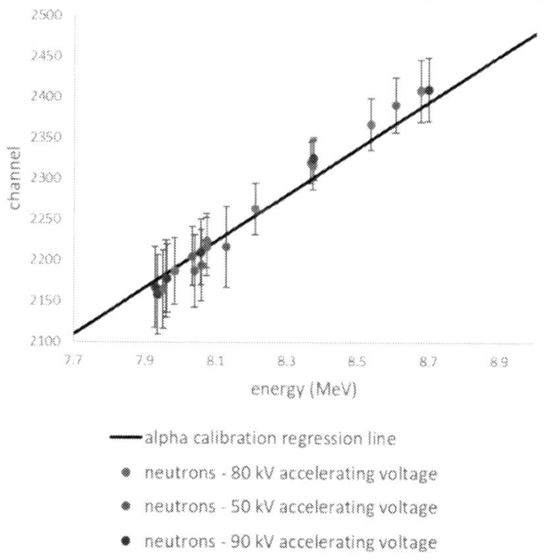

Fig. 10. Evaluation of alpha calibration regression line in the energy region 7.764 MeV - 9.046 MeV through experimental points from neutrons produced by D-T neutron generator. The incident particle beam is considered made only by D_2^+ ions.

For both hypotheses, the experimental results are consistent with predicted values within the uncertainties (assumed equal to standard deviations, σ) (Fig. 9-10), providing evidence that that diamond detector energy calibration based on alpha sources and extrapolated at higher energies is still consistent for 14 MeV neutrons produced by a D-T neutron generator.

Each centroid (μ) was associated to the corresponding energy value E_{exp} calculated via the alpha calibration, the ratio respect to theoretical value E_{th} was then evaluated.

As shown in Fig. 11-12, the relative deviation was always less than 2% for D^+ hypothesis and less than 1.05% for D_2^+ hypothesis, confirming the good agreement between experimental and theoretical values. Since the actual incident particle beam is made by a mixture of the two considered types of ions, the agreement will be still consistent.

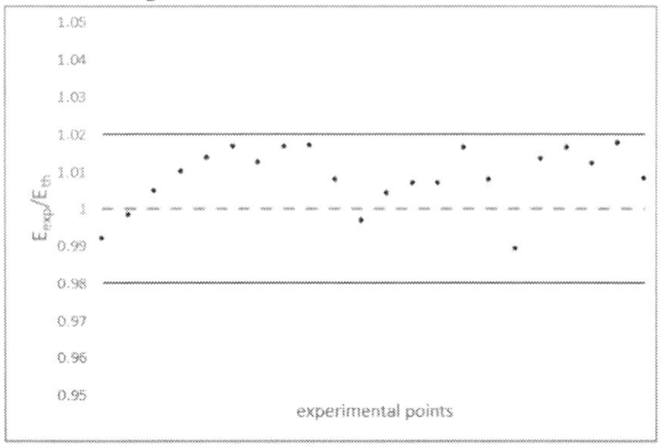

Fig. 11. E_{exp}/E_{th} ratio evaluated for the experimental points from neutrons produced by D-T neutron generator. Red lines show the 2% confidence interval. The incident particle beam is considered made only by D^+ ions.

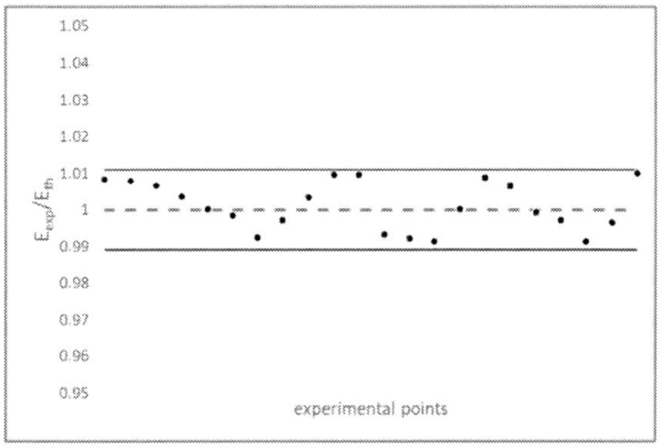

Fig. 12. E_{exp}/E_{th} ratio evaluated for the experimental points from neutrons produced by D-T neutron generator. Red lines show the 2% confidence interval. The incident particle beam is considered made only by D_2^+ ions.

III. Conclusions

In this work an evaluation of the diamond detector energy calibration based on alpha source emissions (5-6 MeV) is performed at higher energies by means of a compact D-T neutron generator. Inside the generator, a beam of deuterons is accelerated to a selectable accelerator voltage (between 40 keV and 90 kV) and impacts on a tritiated target to produce 14 MeV neutrons by means of the fusion reaction T(d,n)alpha. Because of the kinematics of the reaction, the energy of emitted neutrons is not 14 MeV in all conditions. It depends in fact on neutron emission direction with respect to incident deuterons direction as well as on the deuteron energies (i.e., directly linked to the accelerator voltage). Changing the accelerator voltage offers the possibility to produce neutrons within a broad energy range, 13.466 MeV - 14.748 MeV.

For different selected accelerator voltages, several measurements with a 4.7 mm x 4.7 mm x 0.5 mm diamond detector were performed at different angular positions around the compact D-T neutron generator. The PHS corresponding to different conditions were acquired by means of a Multi Channel Analyzer and calibrated in energy through alpha sources. At the same time, a simple theoretical model is used for calculating the energy and the relative uncertainty of emitted neutrons in each case, in order to evaluate the effectiveness of the diamond detector energy calibration extrapolation at energies higher than those used for the traditional alpha calibration process. The analysis has been performed for two different hypothetical compositions of the incident particle beam. A relative deviation less than 2% between evaluated and theoretical energies was observed recorded showing that the PHS energy calibration of a diamond detector through alpha sources and extrapolated at higher energies could be still consistent for fast neutrons coming from D-T fusion reactions.

References

[1] J. Dankowski, K. Drozdowicz, A. Kurowski, U. Wiącek, T. Nowak, Y. Zabila, Diamond detectors for spectrometric measurements of fusion plasma products, Diamond and Related Materials, Volume 79, 2017, Pages 88-92, ISSN 0925-9635, DOI. 10.1016/j.diamond.2017.08.016.

[2] B. Esposito, D. Marocco, G. Gandolfo, et al. Progress of Design and Development for the ITER Radial Neutron Camera. J Fusion Energ 41, 22 (2022). DOI. 10.1007/s10894-022-00333-9.

[3] C. Cazzaniga et al., "Single crystal diamond detector measurements of deuterium-deuterium and deuterium-tritium neutrons in joint European torus fusion plasmas", Rev. Sci. Instrum., vol. 85, no. 4, pp. 43506, 2014, DOI. 10.1063/1.4870584.

[4] S. Blagus, D. Sudac, V. Valkovic, 2004. Hidden substances identification by detection of fast neutron induced rays using associated α particle technique. Nuclear Instruments and Methods in Physics Research Section B: Beam Interactions with Materials and Atoms 213, 434–438. DOI. 10.1016/S0168-583X(03)01586-6.

[5] K. Jordan, T. Gozani, 2007. Pulsed neutron differential die away analysis for detection of nuclear materials. Nuclear Instruments and Methods in Physics Research Section B: Beam Interactions with Materials and Atoms 261, 365–368. DOI.10.1016/j.nimb.2007.04.294.

[6] V. Valkovic, 2016. 14 MeV Neutrons Physics and Applications. CRC Press, London.

[7] M. Pillon, M. Angelone, A.V. Krasilnikov, 14 MeV neutron spectra measurements with 4% energy resolution using a type IIa diamond detector, Nuclear Instruments and Methods in Physics Research Section B: Beam Interactions with Materials and Atoms, Volume 101, Issue 4, 1995, Pages 473-483, ISSN 0168-583X, DOI. 10.1016/0168-583X(95)00507-2.

[8] M. Rebai et al., Diamond detectors for fast neutron measurements at pulsed spallation sources, Journal of Instrumentation, Volume 7, May 2012, DOI 10.1088/1748-0221/7/05/C05015.

[9] V. D. Kovalchuk, V. I. Trotsik, V. D. Kovalchuk, Diamond detector as a fast neutron spectrometer, Nuclear Instruments and Methods in Physics Research Section A: Accelerators, Spectrometers, Detectors and Associated Equipment, Volume 351, Issues 2–3, 1994, Pages 590-591, ISSN 0168-9002, DOI. 10.1016/0168-9002(94)91393-5.

[10] M. Angelone, G. Aielli, S. Almaviva, R. Cardarelli, D. Lattanzi, M. Marinelli, 2009. Neutron detectors based upon artificial single crystal diamond. Nuclear Science, IEEE Transactions 56, 2275– 2279. DOI. 10.1109/TNS.2009.2025177.

[11] R. Mandal, G. Pansare, D. Sengupta, V. Bhoraskar, 2012. Angular distribution of neutron flux around the tritium target of 14 MeV neutron generator. Journal of the Physical Society of Japan 81, 4006–. DOI. 10.1143/JPSJ.81.104006.

[12] D. Rigamonti et al, 2018, Neutron spectroscopy measurements of 14 MeV neutrons at unprecedented energy resolution and implications for deuterium-tritium fusion plasma diagnostics, Measurement Science and Technology, 29, 045502, DOI 10.1088/1361-6501/aaa675.

A novel smart rad-hard fast detection system for Radioactive Ion Beam Tagging and Diagnostics

Luis Acosta[1], Carmen Altana[2], Giuseppe Cardella[3], Andrea Castoldi[4,5], Michele Costa[2], Enrico De Filippo[3], Elena Geraci[6,3], Brunilde Gnoffo[3], Chiara Guazzoni[4,5,*], Cettina Maiolino[2], Nunzia Simona Martorana[3], Andrea Naggi[4,5], Angelo Pagano[3], Emanuele Vincenzo Pagano[2], Sara Pirrone[3], Giuseppe Politi[6,3], Fabio Risitano[7,3], Francesca Rizzo[6,3], Antonio Domenico Russo[2], Paolo Russotto[2], Marina Trimarchi[7,3], Salvo Tudisco[2]

[1]Instituto de Física, Universidad Nacional Autónoma de México, Mexico
[2]LNS-INFN, Italy
[3]INFN, Sezione di Catania, Italy
[4]Politecnico di Milano, Dip. Elettronica, Informazione e Bioingegneria, Italy
[5]INFN, Sezione di Milano, Italy
[6] Dipartimento di Fisica e Astronomia, Universita' degli Studi di Catania, Italy
[7]Dip. di Scienze MIFT Universita' di Messina, Italy

(*) chiara.guazzoni@mi.infn.it

***Abstract*— Radioactive Ion Beams (RIBs) of large intensity (10^6 pps or higher) are at the frontier in nuclear physics. We designed a novel detection system for RIBs diagnostics and tagging based on Silicon Carbide detectors and on custom frontend electronics ready to be coupled with a Real Data Management Unit. The full detection system is designed to measure the spatial distribution of the beam intensity and trajectory with sufficient spatial resolution (of the order of 1-2 mm). In addition, the detection system has to determine the RIB composition that can be obtained from the joint measurement of the energy loss (ΔE) of the ions passing through the sensors and the time of flight between two sensors or with respect to a given reference signal as the Radio-Frequency signal of a Cyclotron. In this paper we present the full design of the proposed system together with the results of the first experimental qualification of the first mini-prototype. The paper also shows the steps towards the final detection system, housed in a DN160 spherical cross and able to cover an active area of 30 mm × 60 mm.**

***Keywords*— Radioactive Ion Beams, Particle Tagging, Beam Monitor, Fast Electronics**

I. Introduction

RADIOACTIVE Ion Beams (RIBs) [1], [2], [3] of large intensity (10^6 pps or higher) are at the frontier in nuclear physics and are available since one decade in different facilities [4], [5], [6] and new ones are presently under construction [7] to deliver intense RIBs to the nuclear physics community and to produce radiopharmaceuticals.

The exotic nuclei (short-lived nuclear species) are a unique tool to investigate properties of nuclei and nuclear matter under extreme conditions and have proven to be very effective in unveiling the fundamental properties of the still not well-known nuclear force. RIBs allow the study of reactions of fundamental importance for the nucleosynthesis of elements in the universe.

Radioactive isotopes produced in RIBs facilities find a relevant application in the medical field as radiopharmaceutical and innovative techniques foresee the use of radioactive isotopes, such as ^{11}C, in hadron therapy [8].

The tuning and transport of the radioactive beam are of critical importance to deliver high-quality beams, but they represent a time-consuming process and require dedicated diagnostics and tagging devices measuring different RIBs features in combination with skilled personnel. The same devices should be capable of operation in radioactively activated environments because of the expected high direct and background irradiation and to sustain several experiments per year.

To this aim, we are developing a dedicated instrument [10] for the diagnostics and the tagging of RIBs along the transport line based on an array of Silicon Carbide (SiC) diodes [11], readout by an optimized fast frontend electronics and ready to be coupled with a smart DAQ with Data Real-Time Management capabilities and a dedicated software layer implementing Artificial Intelligence and machine learning techniques.

The system is originally designed to cope with the characteristics of a new fragment separator, FRAgment In-fligth SEparator (FraISe) [7], [12], [13], [14], being built at Laboratori Nazionali del Sud of INFN in Catania (Italy), as part of the upgrade of the facility [15] to push the rate of production of high luminosity stable and radioactive beams.

The facility will produce RIBs exploiting the in-flight technique based on the fragmentation of the primary beam that impinges on light Be or C targets. FraISe makes use of light and medium mass primary beams, having power limited to ≈ 2–3 kW for radioprotection issues, leading to RIBs, whose intensities vary in the range of ≈ 10^3–10^7 pps [14], for nuclei far from and close to the stability valley, respectively. FraISe aims at providing high-intensity and high-quality RIBs for nuclear physics experiments, also serving to interdisciplinary research areas, such as medical physics. The improvement in beam intensity opens the way to more precise studies on nuclei close to stability and allows the extension of the investigations also to nuclei far from stability.

Several beam diagnostic/tagging devices are needed within FraISe: i) and ii) within the fragment separator in two points (degrader & exit slits), where there is the need during beam optimization of a point-to-point measurement of the cocktail intensity, of its relative composition, of the energy distribution, of the 2D profile, and of the angular distribution; and during data taking of monitoring the beam properties, of measuring the start time for the event-by-event ToF/energy measurement for particle identification; iii) in the final set-up point (measurement chamber), where there is the need of event-by-event tagging of the cocktail beam and of trajectory measurements.

We chose a one device-fits-all approach. The modularity and the flexibility of the system make it suitable both for beam diagnostics and for particle tagging not only for the aforementioned needs in FraISe but also in different fragment separators and facilities.

In this paper we present the full design conception of the proposed system together with the results of the first experimental qualification of the first mini-prototype. The paper is organized as follows. Section II illustrates the architecture of the full tagging system. Section III describes the first developed workhorse, called mini-prototype and section IV is devoted to the first experimental qualification of the first mini-prototype. Section V ends with conclusions and gives an .

II. Architecture of the Full Tagging System

The full detection system [9], [10] is designed to be housed in a DN160 spherical cross to be inserted along the beam path. Eight monolithic detector tiles of 1.5 cm × 1.5 cm 4H-SiC diodes will be placed side-by -side to cover the detection active area of about 30 mm × 60 mm as required to reconstruct typical RIBs profiles in the high dispersion point of the fragment separator. SiC has been chosen due to its superior properties with respect to Si [16]. The detector thickness is about 100 μm, as a compromise among mechanical strength, epitaxial layer growable thickness and readout node capacitance. SiC detectors are segmented in square pads of 5 mm side as a tradeoff between particle rate, power budget and required resolution, resulting in 72 readouts elements. Fig. 1a shows a simplified CAD view of the final tagging system housed in its spherical DN160 cross to be located along the beam path. The beam is coming orthogonal to the figure. Fig. 1b shows a closer view of the detection area. Wire bondings connect each individual pad to the corresponding readout channel. Less than 11mm difference in the bonding wires length cause a negligible dispersion of the parasitic capacitance do not cause a relevant dispersion in the output node capacitance.

(a)

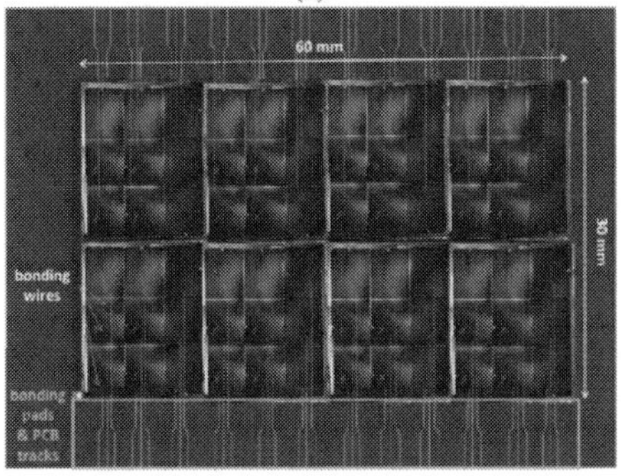

(b)

Fig. 1. a) CAD view of the final tagging system housed in its spherical DN160 cross to be located along the beam path. b) closer view of the detector active area, where 8 monolithic detector tiles are placed side-by-side to cover the 30 mm × 60 mm detection area.

Each detection unit is readout by a dedicated frontend in charge preamplifier configuration with fast decay to cope with the foreseen particle flux up to 10^7 pps on the full detection area. A selected, high-performance operational amplifier plays the role of forward gain stage. A fully differential amplifier is responsible of line driving and additional filtering if needed. Ref. [17] contains a detailed description of the frontend electronics architecture.

III. The First Mini-Prototype Structure

In order to experimentally qualify the system performance, we assembled a mini-prototype of the final tagging system, featuring 1 cm^2 active area. The detector carrier is able to house a 1 cm^2 single die segmented in 4 detection units or up to 4 dies, with an active area of 25 mm^2 per single unit. Fig.2a and Fig.2b show the photo of two of the different detector prototypes available, a 25 mm^2 detection unit and a monolithic 1 cm^2 active area detector segmented in four individual detection units.

We designed two different substrates to be used as detector carriers (motherboards) and to house the frontend electronics. Fig.2c and Fig.2d shows the two different motherboards. The first developed carrier is the safest one and allows to easily glue and bond even non-monolithic arrays. It is designed with a dedicated stack-up to reduce the thickness in the detector area to the minimum (300 μm) so as to minimize the material budget along the beam path. The second one is closer to the final design as it has a milled area in correspondence of the detector active area to remove all the material budget and work in full transmission mode. The gluing bridge in the center of the transmission window features the minimum width for the technology that is 600 μm and will be used also to probe its impact on the quality of the identification matrices as it is unavoidable also in the final system to mount the individual dies.

Fig. 2. a) Photo of one 5 mm × 5 mm active area detector prototype. b) Photo of one monolithic detector prototype 1cm^2 active area, segmented in 4 independent detection units. c) Photo of one detector prototype 5 mm × 5 mm active area mounted on the motherboard with minimized material budget along the beam path. d) Photo of the motherboard designed to allow the mounting of different detectors prototypes (up to 1 cm^2 die size) in transmission mode.

A 4-channel custom frontend board with the same configuration of the final system is directly coupled with the detector motherboard by means of the 0.50mm Pitch SlimStack Plug visible at the lowest side of the motherboard in the middle of the two fixing holes. The signal coming from the frontend electronics in charge preamplifier configuration is driven to the outer world by means of a fully-differential line driver through a dedicated high-integrity Samtec HSEC8-113-01-S-DV-A-K connector. Fig. 3 shows the photo of the 4-channel frontend board. Individual U.FL-R-SMT-1(10) coaxial connectors allow probing also the single-ended output of the charge preamplifier.

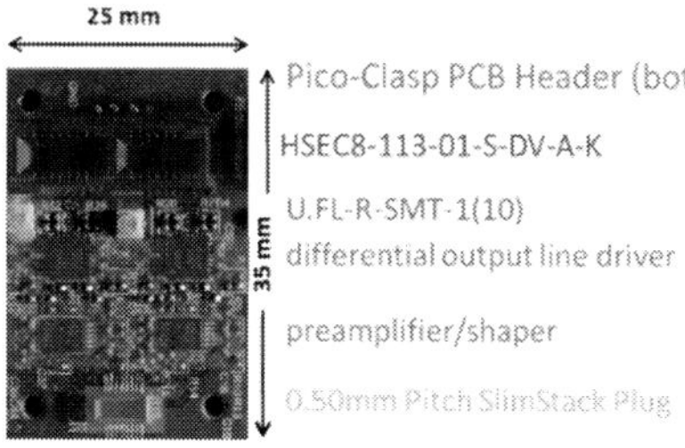

Fig. 3. Photo of the custom 4-channels frontend board with the main components highlighted.

IV. Experimental Qualification of the First Mini-Prototypes

We performed a detailed qualification of the individual system components prior to assembling the first mini-prototypes. Ref. [10] and [17] present and discuss the results of this qualification. We report here the results of the first qualification of one mini-prototype.

In order to qualify the system response, we probed the output waveform as a function of the applied reverse bias upon electrical injection exploiting a dedicated injection line through a 1 pF injection capacitance connected to the charge preamplifier virtual ground. Fig. 4a shows the corresponding output waveforms. Fig. 4b shows the maximum value, minimum value and amplitude of the output waveform as a function of the applied reverse bias voltage. The vertical line indicates the full depletion voltage as resulting for the considered prototype from independent CV measurements [10]. As it can be noticed above 200 V reverse bias, when the detector capacitance drops well below 50 pF [10], 6the output response follows the expected behavior with no overshoot, however the output waveform maximum, minimum and amplitude saturate to the target value (indicated by the corresponding horizontal line), only above depletion.

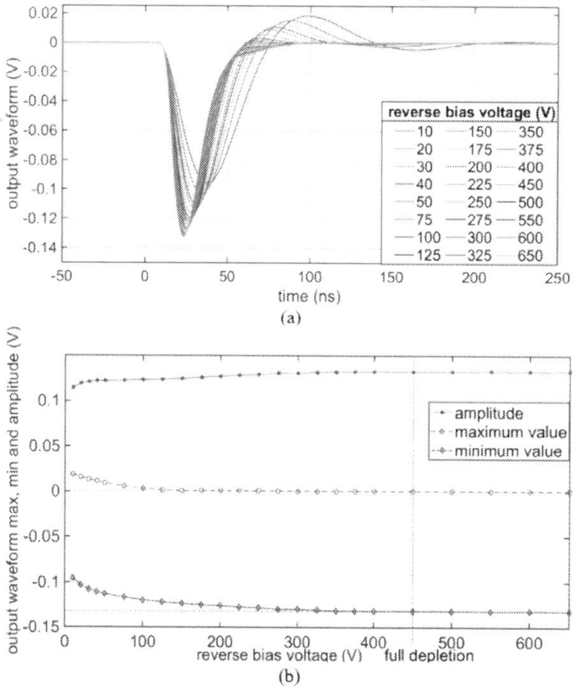

Fig. 4. a) Output waveforms upon electrical injection as a function of the applied reverse bias voltage in the case of the setup with minimized material budget. b) Maximum value, minimum value and amplitude of the output waveform as a function of the applied reverse bias voltage. The vertical line indicates the full depletion voltage as resulting for the considered prototype from independent CV measurements [10].

We illuminated the detector with a 3kBq mixed-nuclei α source (^{239}Pu, ^{241}Am, ^{244}Cm). Also in this case we acquired the output waveforms as a function of the applied reverse bias in order to better investigate the system performance, including also the charge collection process. Fig. 5 shows the corresponding output waveforms. Output waveforms as a function of the applied reverse bias in the case of the setup with minimized material budget in the case of alpha particle detection from the mixed nuclei source. We chose always the same alpha energy, corresponding to the ^{244}Cm alpha energy (5806 keV).

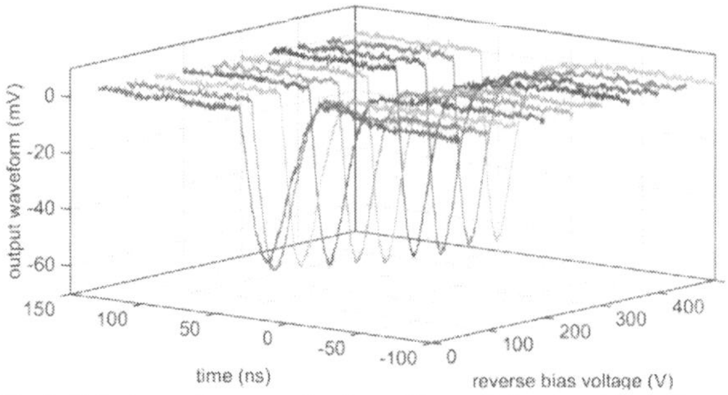

Fig. 5. Output waveforms as a function of the applied reverse bias in the case of the setup with minimized material budget in the case of alpha particle detection from the mixed nuclei source. We chose always the same alpha energy, corresponding to the 244Cm alpha energy (5806 keV).

We directly digitized the output waveforms with a High Resolution Oscilloscope from Teledyne LeCroy (12 bits vertical resolution). From the acquired waveforms we performed a preliminary qualification of the spectroscopic performance of the system, limited by operation in air. Fig.6 shows the measured spectrum in air when the waveforms are filtered with an 11 taps digital filter. The quality of the spectrum is affected by operation in air, however the measured resolution at the pulser line (not spoiled by operation in air) is 60 keV FWHM well within required specifications.

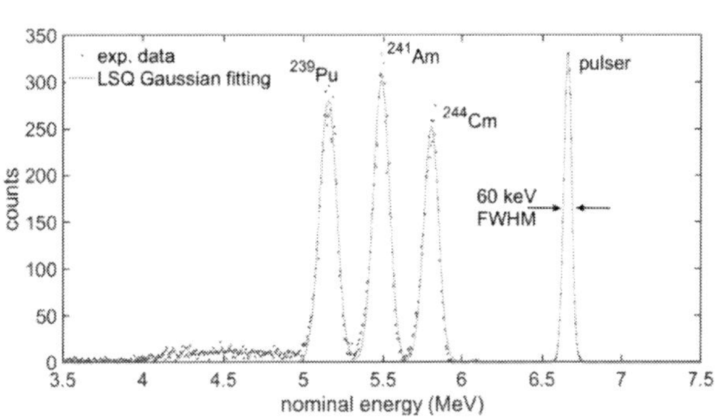

Fig. 6. Measured spectrum in air when the detector is illuminated by a mixed-nuclei α source. The measured resolution at the pulser line is 60 keV FWHM. The quality of the spectrum is degraded by operation in air.

V. Conclusions

This paper presented the results of the first experimental qualification of the first mini-prototype of the designed tagging system for the new FraISe facility now in its final phase of construction. The mini-prototype features a 1 cm^2 active area and allows probing all the feature of the final system. The opamp based fast frontend is fully developed and qualified. The system performance was experimentally assessed in the lab with electrical injection and with a mixed nuclei alpha radioactive source. The measured pulse rise time is sufficiently faster, below 10 ns for the 10%-90% rise time and subject to further improvement. The measured INL is below 0.2% up to 50 MeV input signal and we measured an energy resolution of 60 keV FWHM in air with the fast electronics.

Given the positive and promising results of the qualification in the lab the mini-prototype will be qualified with stable beams in the next months, pending the availability of the first radioactive beams of FraISe. This test will give precious results for the design and assembly of the final system. The mini-prototype will be populated also with the monolithic array of 2 × 2 SiC diodes in order to investigate charge sharing and cross-talk.

Acknowledgment

This work was supported by INFN, Istituto Nazionale di Fisica Nucleare in the framework of the CHIRONE experiment and by Politecnico di Milano, with the visiting scientists initiative.

The authors gratefully acknowledge the precious contribution of Mattia Pozzi (Politecnico di Milano) in the PCB design and the valuable qualified professionalism of Sergio Masci in the detector gluing and bonding.

References

[1] K. Hagino, I. Tanihata and H. Sagawa, Exotic Nuclei Far from the Stability Line, In: Henley, E.M., Ellis, S. D. (ed.) 100 Years of Subatomic Physics. World Scientific, Singapore, vol. 231, 2013.
[2] I. Tanihata, "Nuclear structure studies from reaction induced by radioactive nuclear beams", Prog. Part. Nucl. Phys., vol. 33, pp. 505-573, 1995.
[3] EPJA Focus Point on Rewriting Nuclear Physics textbooks: 30 years with radioactive ion beam physics, Eur. Phys. J. Plus 131, 2016.
[4] Y. Blumenfeld, T. Nilsson, P. van Duppen. "Facilities and methods for radioactive ion beam production", Physica Scripta, pp.14023, 2013.
[5] K.-H. Schmidt, E. Hanelt, H. Geissel, G. Münzenberg, J.P. Dufour, "The momentum-loss achromat — A new method for the isotopical separation of relativistic heavy ions", Nucl. Instrum. Meth. A, vol. 260, pp. 287-303, 1987.
[6] R. Anne and A.C. Mueller, "LISE 3: a magnetic spectrometer—Wien filter combination for secondary radioactive beam production", Nucl. Instrum. Meth. B, vol. 70, pp. 276-285, 1992.
[7] P. Russotto et al., "Status and Perspectives of the INFN-LNS In-Flight Fragment Separator", Jour. of Phys.: Conf. Ser., vol. 1014, 012016, pp. 1-12, 2018.
[8] L. Penescu et al., "Technical Design Report for a Carbon-11 Treatment Facility", Front. Med., vol. 8, Apr. 2022, DOI: 10.3389/fmed.2021.697235.
[9] C. Altana et al., "Feasibility study of a smart rad-hard fast detection system for Radioactive Ion Beam Tagging and Diagnostics", 2021 IEEE Nuclear Science Symposium and Medical Imain Conference Records, IEEE, 2021, pp 1-5.
[10] L. Acosta et al., "Design of a novel smart rad-hard fast detection system for Radioactive Ion Beam Tagging and Diagnostics", 2022 IEEE Nuclear Science Symposium, Medical Imaging Conference and Room Temperature Radiation Detectors Conference Conference Records, November 5-12, 2022, Milano (Italy), pp. 1-4.
[11] S. Tudisco et al., "SiCILIA—Silicon Carbide Detectors for Intense Luminosity Investigations and Applications", *Sensors*, vol. 18(7), 2289, pp. 1-16, 2018.
[12] A.D. Russo, L. Calabretta, G. Cardella and P. Russotto, "Preliminary design of the new FRAgment In-flight SEparator (FraISe)", Nucl. Instrum. Meth. B, vol. 463, pp. 418-420, 2020, DOI: 10.1016/j.nimb.2019.04.037
[13] N. S. Martorana, "Status of the FraISe facility and diagnostics system", Il nuovo cimento C, vol. 44 (1), pp. 1-10, 2021, DOI: 10.1393/ncc/i2021-21001-2
[14] N. S. Martorana et al., "Radioactive ion beam opportunities at the new FRAISE facility of INFN-LNS", Front. Phys., Vol. 10, Dec. 2022, article n. 1058419, pp. 1-13, 2022, DOI: 10.3389/fphy.2022.1058419.
[15] https://potlns.lns.infn.it/it/

[16] J. M. Rafi' et al, "Electron, Neutron, and Proton Irradiation Effects on SiC Radiation Detectors", *IEEE Trans. Nucl. Sci.* Vol. 67, no 12, 2020, pp. 281-2489.
[17] A. Castoldi et al., "Feasibility study of the use of operational amplifiers as forward gain stages in charge preamplifiers and shaping filters for radiation detectors", IEEE Trans Nucl Sci., July 2023, DOI:10.1109/TNS.2023.3277223.

Absolute neutron and photon flux characterization in the CNESTEN's TRIGA Mark II research reactor

H. Ghninou[1,*], A. Gruel[1], A. Lyoussi[1], C. Reynard-Carette[2], C. El Younoussi[3], B. El Bakkari[3], Y. Boulaich[3]

[1]CEA, DES, IRESNE, DER (SPESI/LP2E), Cadarache F-13108 Saint-Paul-Lez-Durance, France
[2]Aix Marseille Univ, Université de Toulon, CNRS, IM2NP, Marseille, France
[3] CNESTEN/CENM, POB 1382, Rabat, Morocco
(*) hamza.ghninou@cea.fr

Abstract— **The National Center for Nuclear Energy, Sciences and Technology (CNESTEN) located in Rabat, Morocco, operates a 2 MW TRIGA Mark II research reactor. This type of reactor is specially designed to effectively implement the various fields of nuclear research such as neutron activation analysis, neutron radiography, detectors testing, radioisotopes production as well as education and training. In the last few years, a collaboration between the French Atomic Energy and Alternative Energies Commission (CEA) and the CNESTEN was established to expand the utilization of the TRIGA computational model by carrying out new in-situ measurements in order to characterize neutron and photon fields within and beyond the TRIGA reactor core. These new measurements will consolidate the knowledge of neutron and photon fluxes in different irradiation and instrumentation channels. The results of these experiments will also be used to extend the experimental validation of the new developed TRIPOLI-4® computational model of the reactor and to quantify the uncertainties and biases. This paper focuses on the neutron and photon flux characterization of two irradiation channels inside the TRIGA reactor core. Neutron measurements are ensured by activation dosimetry whereas photon measurements are ensured by thermo-luminescent detectors (TLD400 – CaF_2: Mn). Based on these techniques, the experiments were carried out during an experimental campaign conducted in June 2022. Preceding the implementation of these measurements, two experimental devices were specifically designed and manufactured to ensure a reproducible positioning of the detectors in the selected irradiation channels. This paper presents the experimental results analysis and the associated uncertainty quantification. These results will be then compared to the calculation ones obtained by the computational models of the TRIGA reactor.**

Keywords **—In-core measurements, TRIGA reactor, Activation detectors, TLD detectors, Ionization chamber.**

I. Introduction

The TRIGA Mark II research reactor at the National Center for Energy, Sciences and Technology (CNESTEN) is a typical 2 MW TRIGA reactor which achieved its initial criticality on 02 May, 2007. It is a light water pool-type research reactor cooled by natural convection. Schematic drawings of the reactor core configuration are displayed in Fig. 1 and Fig. 2. The reactor core is immerged into an 8 m high and 2.52 m diameter tank filled with water and surrounded by a 21 cm thick cylindrical graphite block. The reactor core assembly consists of 101 FEs (including 5 fuel-follower control rods), 17 graphite elements, a central thimble (CT), a pneumatic transfer system (PTS) and a new irradiation channel located in the G25 location (cf. Fig. 2). The reactor is also equipped with a rotary specimen rack (RSR) containing 40 irradiation positions, a thermal column and four beam channels penetrating the concrete shield, the aluminum tank, and the reactor tank. Beam port NB1 is a tangential port, whereas beam ports NB2 – 4 are radially oriented. The TRIGA reactor fuel consists of a uniform mixture of uranium (8.5 %wt, enriched to 19.7% of ^{235}U) and zirconium hydride and it is contained in stainless steel cylindrical elements of approximately 73.2 cm in length and 3.6 cm in diameter. Similar to many other research reactors, the CNESTEN's TRIGA is designed to effectively implement a wide range of applications, including the production of radioisotopes for medical and industrial purposes [1] [2], qualification, under irradiation, fuel materials, inert materials and sensors [3] [4] [5] [6], validation and qualification of modelling and simulation tools [7] [8] [9] [10] [11] [12]. The implementation of such applications requires a reliable characterization of the neutron and photon fluxes in the reactor core. Indeed, the knowledge of neutron and photon fluxes is crucial for planning and carrying out experimental programs and for validating and qualifying computational models of the reactor. In fact, the first validation of the CNESTEN's TRIGA computational model was done by comparing the MCNP [13] calculated multiplication factor and control rod reactivity worth with the criticality benchmark experiments carried out in 2007 [14]. Up until then, this had been the only available validation of the computational model of the TRIGA reactor. Therefore, a new set of measurements was carried out in the TRIGA reactor core as part of the bilateral collaboration between CEA Cadarache, CNESTEN, and the joint laboratory LIMMEX (Aix-Marseille University, CEA, and CNRS). The primary objective of this collaboration is to expand the experimental validation of the TRIGA reactor by performing, analyzing, and interpreting experiments for the characterization of the irradiation and instrumentation channels. These measurements consist of three types of detectors: activation detectors for neutron characterization, thermo-luminescent detectors and a miniature ionization chamber for photon characterization. This decision was based on several criteria, including feedback from the use of these detectors in nuclear research reactors, such as detector dimensions, measurement uncertainties, and analysis and interpretation methods. It also considered the unique features of the TRIGA reactor and the associated irradiation constraints, such as the dimensions of the irradiation channels and the levels of neutron and photon flux.

The present paper describes the neutron and photon flux characterization of two in-core irradiation channels of the CNESTEN's TRIGA Mark II research reactor. For each type of measurement, the obtained results are analyzed and the associated uncertainties are evaluated.

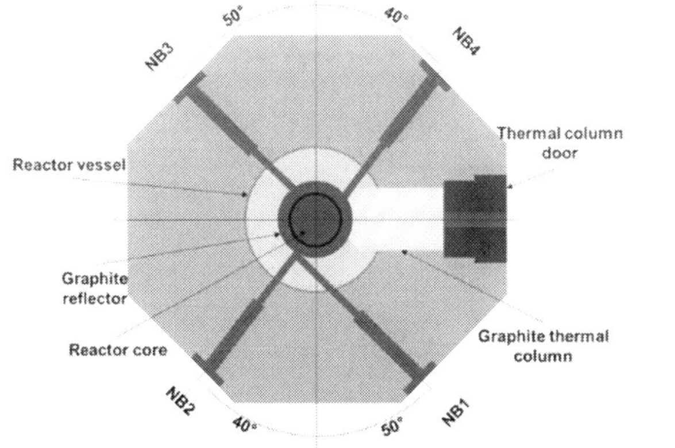

Fig. 1. Radial view of the CNESTEN's TRIGA reactor.

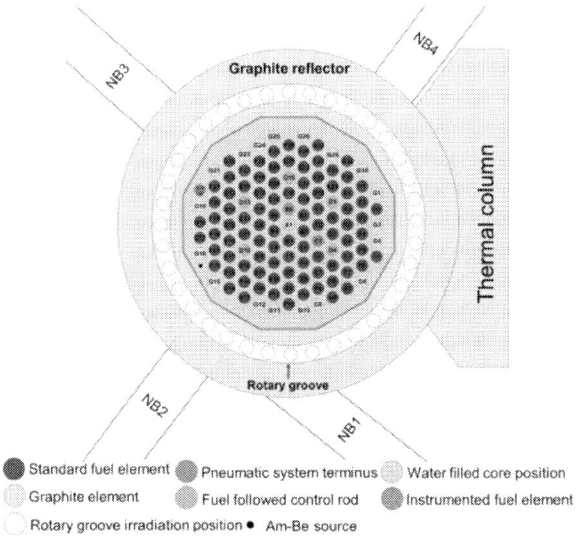

Fig. 2. Radial view of the TRIGA reactor core in the current operating configuration.

II. Measurements design

A. Experimental set-up

As mentioned above, the TRIGA reactor is equipped with different irradiation channels that have different characteristics in terms of neutron flux levels and energy spectra. For this study, our aim is to characterize two in-core irradiation channels:

- CT: a water-filled irradiation channel located in the core center. It is an aluminum tube with, respectively, 3.2 and 3.4 internal and external diameters and 8 m length.
- G25: a new water-filled irradiation channel located at the core periphery, which has approximately the same diameter as the CT.

Three types of detectors were used to characterize the neutron and photon in different axial position in the irradiation channels: activation detectors, thermos-luminescent detectors and ionization chamber (cf. B). Carrying out of such measurements requires dedicated experimental devices to ensure the reproducibility of the measurements in the reactor. Schematic drawings of the experimental devices used in this study are presented in Fig. 3 and Fig. 4.

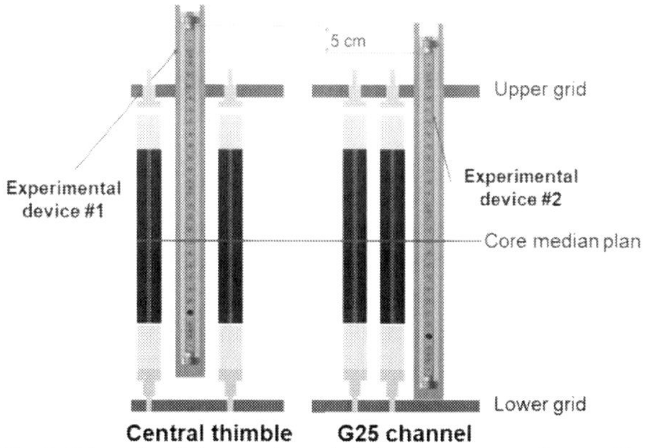

Fig. 3. Schematic of the experimental devices that were used to position activation and TLD dosimeters in the CT and G25 irradiation channels.

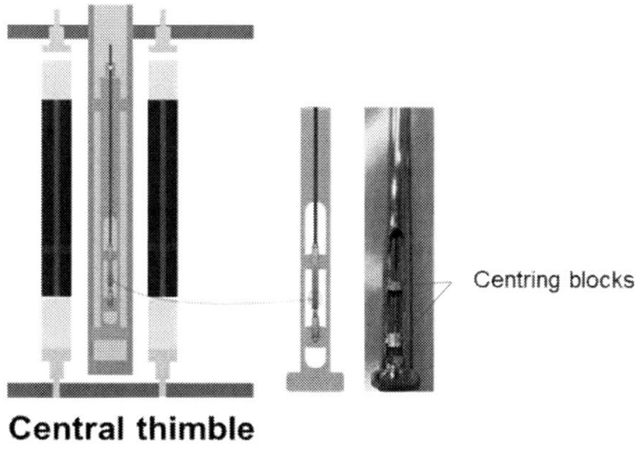

Fig. 4. Schematic of the positioning system used to place the MIC in the CT and G25 irradiation channels.

B. Experimental design

In the following paragraphs, the characteristics of the different detectors used in this study are described.

1) Activation detectors

For neutron flux measurements, we used two types of activation detectors to cover different energy regions of the neutron spectra: $^{59}Co(n,\gamma)$ and $^{58}Ni(n,p)$. The selection of these dosimeters is based on different criteria such as the cross section of the neutron reaction of interest, energy range of response, energy and intensity of decay γ rays' emission. The experimental design and the preliminary calculations are presented in [15]. At the end of irradiation, the analysis of the irradiated dosimeters is performed by means of the γ spectrometry technique using the CNESTEN's HP-Ge detector (cf. Fig. 5).

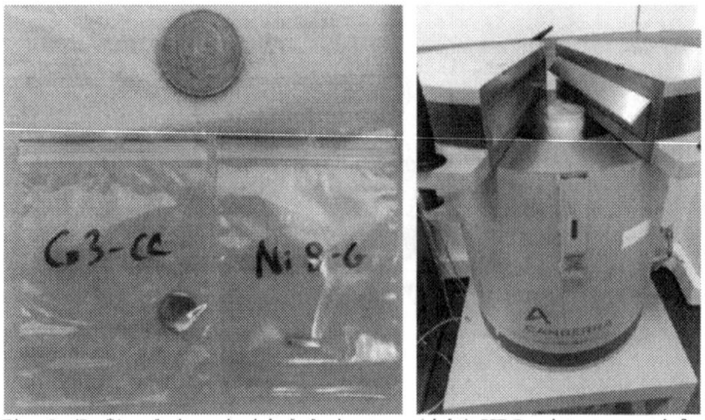
Fig. 5. (Left) cobalt and nickel dosimeters, (right) HPGe detector used for spectrometry γ measurements.

Two irradiations at two steady state power levels, 10 kW and 20 kW, were carried out. Table I presents the characteristics of the irradiations performed in the CT and G25 irradiation channel.

TABLE I
IRRADIATIONS CARRIED OUT IN CT AND G25 CHANNELS.

Date of irradiation	IC[2]	Power level	Irradiation time	Irradiated samples
09/12/2020	CT	20 kW	15 min	9 Ni
10/06/2022	CT G25	10 kW	15 min	19 Co 10 Co and 9 Ni

[2] IC : Irradiation Channel

2) Thermo-luminescent detectors

For gamma dose measurement, in the range of few kGy, we used thermo-luminescent detectors of type TLD400 (CaF_2: Mn) obtained from Thermo-Fisher Scientific. The TLD400 post-irradiation reading protocol is performed with a Harshaw Model 3500 TLD reader (cf. Fig. 6), allowing the TLDs to be heated up to 400°C. The charge integrated in the TLD detector is read using the following sequence: preheating phase at 150 °C for 5 s, followed by a ramp-up to 350 °C at a rate of 10 °C.s^{-1} and maintaining this temperature for 30 s.

Prior to the irradiations, the TLD detectors were calibrated, at a dose range of several kGy, using a reference gamma field generated by the PAGURE and POSEIDON ^{60}Co sources at CEA Saclay. This calibration establishes the link between the integral charge Q (expressed in nC or counts) of the luminescent signal read post irradiation and the dose to which the detectors are exposed, which is represented by the reference quantity of Kerma air (in Gy). The obtained results were adjusted with rational model, defined as ratios of polynomials.

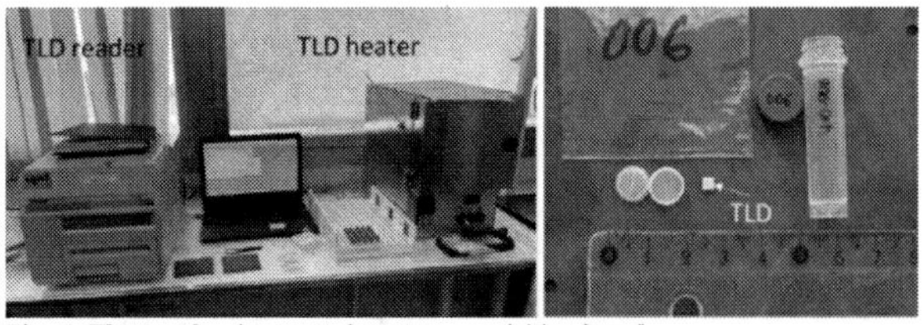

Fig. 6. Thermo-luminescent detectors acquisition bench.

As shown in Table II, two irradiations, 5 kW – 10 min (irradiation 1) and 5 kW – 5 min (irradiation 2), were performed in the CT and G25 irradiation channels. The TLD measurements were carried out in 3 phases: measurement of the background gamma (reactor shutdown), measurement during transients (divergence up to 5 kW then rod-drop) and finally a measurement during a constant power level of 5 kW. To determine the net dose integrated over constant power level, we subtract the dose integrated during the background and transient measurements.

TABLE II
TLD IRRADIATIONS IN BOTH CT AND G25 CHANNELS.

Date of irradiation	EXPERIMENT	Power level	Irradiation time	IC
24/06/2022	Background	0 W	1 h	5 axial positions in the CT and G25
27/06/2022	Divergence	5 kW	-	
24/06/2022	**1st irradiation**	5 kW	10 min	
27/06/2022	Background	0 W	2 h 49 min	
28/06/2022	**2nd irradiation**	5 kW	5 min	

Prior to the measurements, preliminary simulations, using the TRIPOLI-4® [16] computational model of the TRIGA reactor, were carried out to design the TLD measurements. Hence, the duration and power level were adjusted to obtain an integrated dose within the calibration and linear operating range of the TLD detectors (in the range of Gy to few kGy).

3) Miniature ionization chamber

To assess photon flux in the two-irradiation channels, we used a miniature ionization chamber (MIC) manufactured by PHOTONIS. It is a gas-filled (Argon at 5 bar) detector, with 3 mm diameter and 10 mm sensitive length, that is sealed to a mineral cable (cf. Fig. 7).

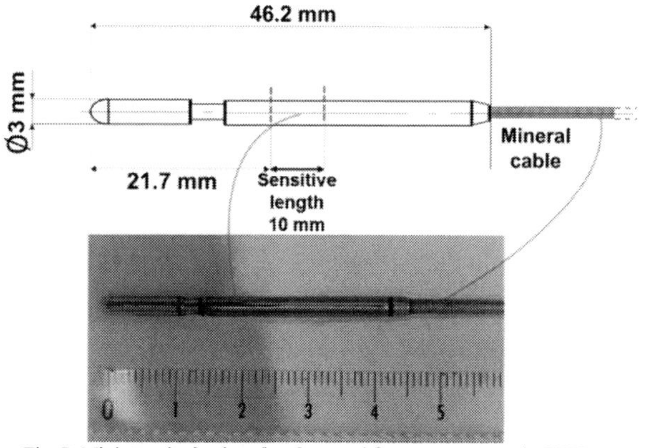

Fig. 7. Miniature ionization chamber used for measurements in TRIGA.

For current measurement and high voltage supply, a Keithley 6517B electrometer, controlled by a LabVIEW-based data acquisition application, was used. The bias voltage was set to 150 V (manufacturer's specified value) and the current was measured with a sampling rate of 1 Hz.

Different measurements with the MIC were carried out in the CT and G25 channels. At first, was positioned at the core center in order to verify the linearity of the measured current as a function of the reactor power. To do this, the power was gradually increased in steps ranging from 0 to 1800 kW, while allowing the power to stabilize at each step to verify any drifting in the measured signal. Then, the MIC was used to characterize the axial profile of the photon flux in the two-irradiation channels.

III. Results and discussions

The following sections are dedicated to the analysis of the experiments' results. Each type of measurements will be treated and analyzed and the associated uncertainties will be evaluated.

A. Activation measurements

At the end of the irradiation, the specific activity (in unit of $Bq.mg^{-1}$) of the radioisotopes formed during the irradiation process is evaluated using γ spectrometry measurements. By determining the number of counts under the total absorption peak, the activity A_j of a radioisotope is calculated by the following formula:

$$A_j = \frac{N(E_j)}{m I_\gamma(E_j)\varepsilon(E_j)T(E_j)t_a}\prod_i C_i \quad (1)$$

where $N(E_j)$ is the number of counts in the absorption peak of energy E_j, m is the mass in g of the measured dosimeter, $I_\gamma(E_j)$ is the intensity of the photon emission, $\varepsilon(E_j)$ is the detector's efficiency for the photon with the energy E_j, $T(E_j)$ is the yield transfer factor for the photon with energy E_j, t_a the acquisition (counting) duration in s and $\prod_i C_i$ is the product of the correction coefficients of dead time, decay and coincidences. The activity profiles of the cobalt and nickel dosimeters irradiated in CT and G25 channels are presented in Fig. 8 and Fig. 9. The overall relative uncertainty associated to the activity at the end of irradiation is the quadratic combination of the relative uncertainties on each term of equation (1) assuming that the variables are independent of each other. A summary of these uncertainties (at 1σ) is presented in the following:

- The uncertainty in the determination of the net peak area combines both the uncertainties of counting rate and the background. This uncertainty is directly given by the acquisition software used and is of the order of 1% for all the dosimeters irradiated in the CT and G25.
- The uncertainty of the detector's efficiency is evaluated from the covariance data associated with the fourth-order polynomial fit of experimental data [15]. These uncertainties are around 1 to 2%.
- The uncertainties associated with the photon emission probabilities and decay uncertainties for cobalt and nickel dosimeters are generally very well-known and are around 0.04% for the radioactive constants and around 0.06% for the photon emission intensities [17].
- The coincidence correction was evaluated with the ETNA software developed by Laboratoire National Henri Becquerel (LNHB). The uncertainty on the coincidence correction is estimated at 10% of the corrective term (LNHB recommendation) which corresponds to a relative uncertainty of 0.1%.
- The yield transfer factor, which corrects the difference between the dosimeter measurement geometry and the calibration geometry, was estimated by TRIPOLI-4 calculations. The associated uncertainty is around 0.5% for different calibration configuration.

Finally, the relative uncertainty on the activity at the end of irradiation varies from 1% to 3%.

The axial profiles of the cobalt and nickel dosimeters present a maximum around the reactor core mid-plane and decrease symmetrically as moving away from the core center toward the graphite reflectors at the top and bottom of the FE.

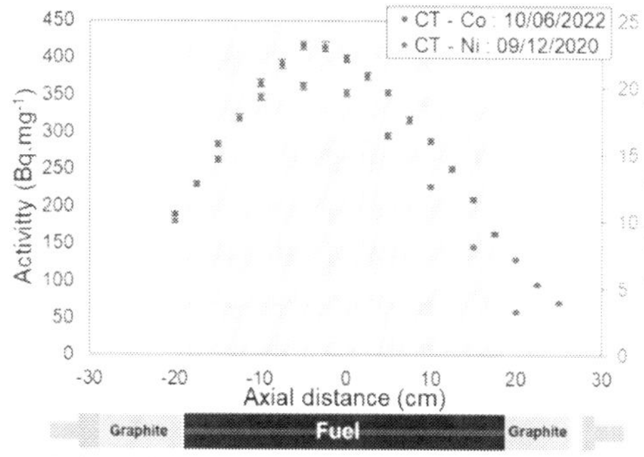

Fig. 8. Cobalt and nickel activities measured in the CT irradiation channel. The area marked in blue corresponds to the active fuel region of the standard FE.

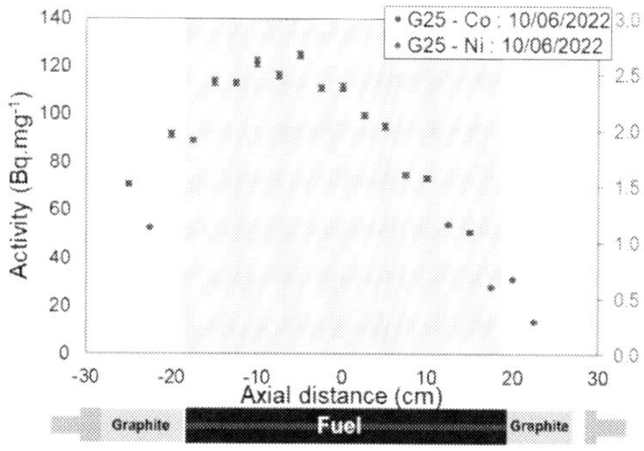

Fig. 9. Cobalt and nickel activities measured in the G25 irradiation channel.

B. TLD measurements

The net integrated dose, by the TLDs, over a constant power level and corrected for background and transient is obtained by the following formula:

$$D_{net} = D_{irr} - D_{div} - D_{Bd}\left(\frac{t_{irr} - t_{div}}{t_{Bd}}\right) \quad (2)$$

where D_{net} is the net dose integrated during the constant power level (kGy), D_{irr} is the integrated dose over the entire irradiation duration, D_{div} is the dose integrated during the transient phase, D_{Bd} is the dose integrated during the background measurement, t_{irr}, t_{div} and t_{Bd} are respectively the irradiation, the transient and the background durations (see Table II).

In order to be qualitatively compared, the net doses are normalized to the average energy (in $W.s^{-1}$) released during the irradiation process. This normalization is calculated as follows:

$$D_{i,norm} = \frac{D_{i,net}}{\overline{P_{irr}} \times t_{irr}} \quad (3)$$

where $D_{i,net}$ is the net dose at the position "i" in kGy, $\overline{P_{irr}}$ and t_{irr} are respectively the power level and the irradiation duration.

Fig. 10 and Fig. 11 summarize the results of the net dose rates normalized to the irradiation power. From these results, one can conclude that:

- The maximum dose rate in the CT is around (6.44 ± 0.75) $\times 10^{-7}$ kGy.W^{-1}.s^{-1} for 10 min irradiation at 5 kW.
- For the G25, the maximum dose rate is around (1.40 ± 0.14) $\times 10^{-7}$ kGy.W^{-1}.s^{-1}.
- The uncertainties associated with the dose measurement vary from 7 to 15% (1σ) and are primarily due to the uncertainties associated with the calibration of the TLDs.

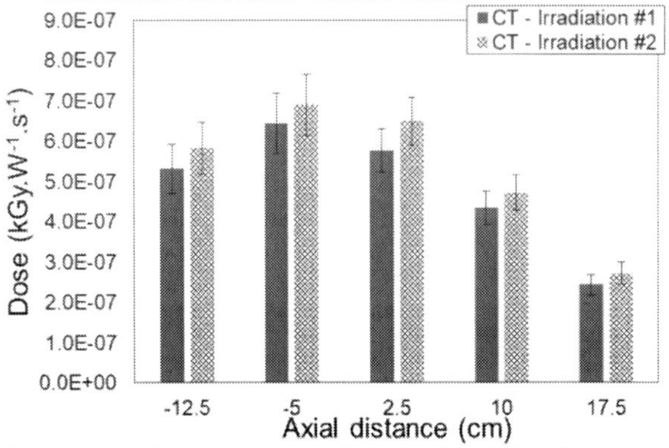

Fig. 10. Results of doses measurements in the CT channel, the solid and dotted lined bars stand for each irradiation.

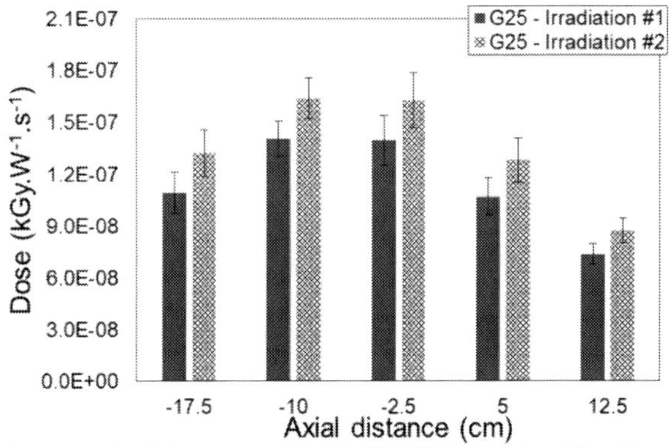

Fig. 11. Results of doses measurements in the G25 channel, the solid and dotted lined bars stand for each irradiation.

C. *MIC measurements*

The axial current profiles measured by the MIC are presented in Fig. 12 and Fig. 13. For each channel, two irradiations at the same power level were performed in order to evaluate the reproducibility of the measurement. The uncertainties associated with these measurements are directly estimated from the variation of the measured current and are around 1% (k=1). Comparing these profiles, one can observe that:

- The profiles in the CT and G25 have maxima at -6 cm below the mid-plane of the reactor core and decrease away from this maximum.
- The measurements are reproducible and consistent given the associated uncertainties.

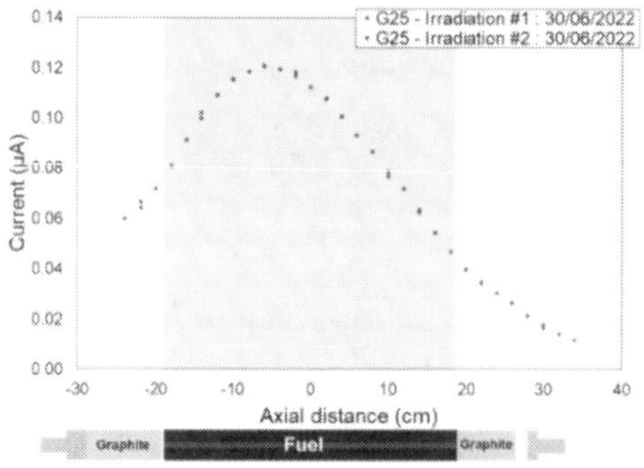

Fig. 12. Results of MIC current measurements in the CT channel.

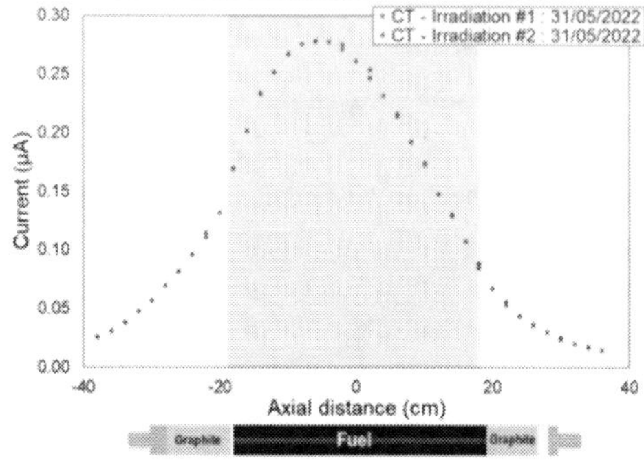

Fig. 13. Results of MIC current measurements in the G25 channel.

D. *Inter-comparison of measurements*

To compare the neutron and photon measurements with each other, the axial profiles were normalized to the maximum value in each channel. Fig. 14 and Fig. 15 illustrate the normalized neutron and photon profiles. The results show that:

- For the CT, the neutron and photon profiles show a maximum between -6 cm and 2.5 cm below the core mid-plane.
- For the G25, the maximum is between -10 cm and -2.5 cm below the core mid-plane.
- The different profiles, around the core mid-plane, are consistent with each other.

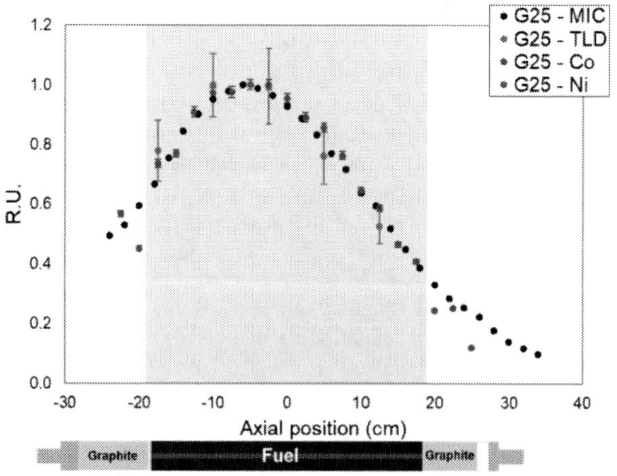

Fig. 14. Inter-comparison of measurements in the CT channel.

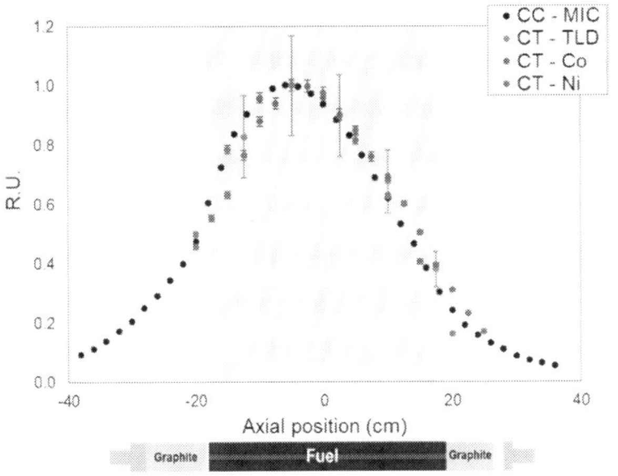

Fig. 15. Inter-comparison of measurements in the G25 channel.

IV. Conclusion and perspectives

In this paper, the experimental characterization of two in-core irradiation channels of the CNESTEN's TRIGA Mark II reactor is presented. Three different types of detectors (activation detectors, TLD and MIC) were used to assess neutron and photon flux measurements. For each type of measurement, the results are analyzed and the associated uncertainties are evaluated and quantified. Indeed, the inter-comparison of measurements show a qualitatively good consistency within the associated uncertainties at 1 or 2σ.

This work should be continued by a thorough interpretation of these measurements using Monte-Carlo calculations. The comparison between the calculated and measured values will allow the evaluation of discrepancies and biases and will extend the experimental validation basis of the TRIGA computational models (developed with TRIPOLI-4® and MCNP codes).

References

[1] B. El Bakkari, B. Nacir, T. El Bardouni, C. El Younoussi, Y. Boulaich, et H. Boukhal, « Feasibility analysis of I-131 production in the Moroccan TRIGA research reactor », *Annals of Nuclear Energy*, vol. 78, p. 140-145, avr. 2015, doi: 10.1016/j.anucene.2014.11.044.

[2] A. Fllaoui *et al.*, « Validation of a New Design of Tellurium Dioxide-Irradiated Target », *Nuclear Engineering and Technology*, vol. 48, n° 5, p. 1273-1279, oct. 2016, doi: 10.1016/j.net.2016.05.004.

[3] T. Žagar, M. Božič, et M. Ravnik, « Long-lived activation products in TRIGA Mark II research reactor concrete shield: calculation and experiment », *Journal of Nuclear Materials*, vol. 335, n° 3, p. 379-386, déc. 2004, doi: 10.1016/j.jnucmat.2004.07.047.

[4] A. Räty *et al.*, « Characterization measurements of fluental and graphite in FiR1 TRIGA research reactor decommissioning waste », *Nuclear Engineering and Design*, vol. 353, p. 110198, nov. 2019, doi: 10.1016/j.nucengdes.2019.110198.

[5] M. Ferrari, A. Zenoni, Y. Lee, et Y. Hayashi, « Characterization of a polyphenyl ether oil irradiated at high doses in a TRIGA Mark II nuclear reactor », *Nuclear Instruments and Methods in Physics Research Section B: Beam Interactions with Materials and Atoms*, vol. 497, p. 1-9, juin 2021, doi: 10.1016/j.nimb.2021.03.021.

[6] M. Cagnazzo, A. Borio di Tigliole, H. Böck, et M. Villa, « Fission products detection in irradiated TRIGA fuel by means of gamma spectroscopy and MCNP calculation », *Applied Radiation and Isotopes*, vol. 135, p. 123-130, mai 2018, doi: 10.1016/j.apradiso.2018.01.027.

[7] M. A. Allaf, S. Ş. Lüle, et Ü. Çolak, « The development of TM2-RIA code for TRIGA type research reactors », *Annals of Nuclear Energy*, vol. 145, p. 107545, sept. 2020, doi: 10.1016/j.anucene.2020.107545.

[8] M. R. Omar, J. A. Karim, et T. L. Yoon, « The development of a multigroup Monte Carlo code for TRIGA reactors », *Nuclear Engineering and Design*, vol. 342, p. 99-114, févr. 2019, doi: 10.1016/j.nucengdes.2018.11.039.

[9] K. Ambrožič et L. Snoj, « JSIR2S code for delayed radiation simulations: Validation against measurements at the JSI TRIGA reactor », *Progress in Nuclear Energy*, vol. 129, p. 103498, nov. 2020, doi: 10.1016/j.pnucene.2020.103498.

[10] E. Chham *et al.*, « Fuel reloads optimization for TRIGA research reactor using Genetic Algorithm coupled with neutronic and thermal-hydraulic codes », *Progress in Nuclear Energy*, vol. 133, p. 103637, mars 2021, doi: 10.1016/j.pnucene.2021.103637.

[11] V. Radulović, R. Jaćimović, A. Pungerčič, I. Vavtar, L. Snoj, et A. Trkov, « Characterization of the neutron spectra in three irradiation channels of the JSI TRIGA reactor using the GRUPINT spectrum adjustment code », *Nuclear Data Sheets*, vol. 167, p. 61-75, juill. 2020, doi: 10.1016/j.nds.2020.07.003.

[12] B. El Bakkari, T. El Bardouni, O. Merroun, Ch. El Younoussi, Y. Boulaich, et E. Chakir, « Development of an MCNP-tally based burnup code and validation through PWR benchmark exercises », *Annals of Nuclear Energy*, vol. 36, n° 5, p. 626-633, mai 2009, doi: 10.1016/j.anucene.2008.12.025.

[13] X-5 Monte Carlo Team, « MCNP – A general Monte Carlo N-Particle Transport Code, Version 5, LA-UR-03-1987. », 2004.

[14] B. El Bakkari *et al.*, « Monte Carlo modelling of TRIGA research reactor », *Radiation Physics and Chemistry*, vol. 79, n° 10, p. 1022-1030, oct. 2010, doi: 10.1016/j.radphyschem.2010.04.016.

[15] H. Ghninou *et al.*, « Characterization of Neutron Reaction Rates in Different Irradiation Channels of the CNESTEN's TRIGA Mark II Research Reactor », *IEEE Transactions on Nuclear Science*, vol. 69, n° 7, p. 1806-1814, juill. 2022, doi: 10.1109/TNS.2022.3174673.

[16] E. Brun *et al.*, « TRIPOLI-4®, CEA, EDF and AREVA reference Monte Carlo code », *Annals of Nuclear Energy*, vol. 82, p. 151-160, août 2015, doi: 10.1016/j.anucene.2014.07.053.

[17] « NUCLÉIDE-LARA on the web (2022) ». http://www.nucleide.org/Laraweb/index.php (consulté le 25 août 2022).

Assessment of the gamma and neutron dose field around the closed-water activation loop at the JSI TRIGA reactor

Domen Kotnik[1,2,*], Luka Snoj[1,2], and Igor Lengar[1]

[1]Reactor Physics Department, Jožef Stefan Institute, Slovenia

[2]Faculty of mathematics and physics, University of Ljubljana, Slovenia

(*) domen.kotnik@ijs.si

Abstract — A closed-water activation loop is being built at the Jožef Stefan Institute TRIGA reactor in Slovenia to serve as a well-defined and stable source of high-energy gamma rays and neutrons. The radial piercing port, which penetrates the graphite reflector and touching the reactor core was chosen for the installation of the closed-water loop due to the high neutron flux and favourable shielding conditions of the surrounding concrete bioshield. The main objective of this work is to assess the neutron and gamma dose field outside the port to obtain important details for the final design of the inner part of the irradiation facility and to assess the background noise for the detectors. The main part of the work consists of the design of the shielding plugs and the construction of a detailed MCNP model of the whole irradiation facility. The dose field calculations were performed with a two-step hybrid transport approach using ADVANTG for variance reduction and MCNP for particle transport. Such deep penetration and shielding calculations are challenging and computationally intensive. The results showed that the dose rate using shielding plugs is more than 7 orders of magnitude lower compared to an empty open port. To reduce the computational uncertainty, further optimisation of the ADVANTG input is essential. The final design of the shielding plugs is described. Additional lead shielding blocks will be added outside the port afterwards if needed.

***Keywords* —TRIGA, water activation, MCNP, ADVANTG.**

I. Introduction

The cooling water in a fusion reactor is crucial to the performance of the reactor, but as a source of radiation it is still poorly understood. In fusion reactors, activated water is one of the main sources of radiation during operation, resulting in radiation exposure to personnel and equipment. The threshold energy for the main water activation reaction, i.e. $^{16}O(n,p)^{16}N$, is about 10.5 MeV. Neutrons in fusion reactors lead to water activity that is 5 orders of magnitude higher than in fission reactors of similar power [1]. Numerous computational analyses of the water activation process have been performed for ITER and DEMO. However, the results are subject to uncertainties and are therefore of poor quality due to the lack of experimental nuclear data, inaccurate computational methods/codes and experimental facilities to validate the methodology experimentally, specifically considering time- and spatial dependent radiation sources (such as the flow of activated water in the cooling system [2]). Consequently, the calculated dose rates are subject to larger uncertainties.

On this basis, a closed¬- water activation loop is currently being built at the Jožef Stefan Institute (JSI) TRIGA Mark II reactor in Slovenia, which will serve also as a well-defined and stable 6 MeV – 7 MeV gamma-ray and 0.4 MeV - 1.2 MeV neutron source. Such a high-energy and well-characterised irradiation facility, i.e. a closed-water activation loop, will enable to perform benchmark experiments based on water activation to be performed, e.g. shielding experiments using ITER relevant materials, investigation of the response of detectors to high-energy gamma radiation, investigation of short-lived moving radiation sources, etc., and will also serve as a benchmark facility for the validation of computational methods and computer codes for the modelling of time- and spatial-dependent radiation sources. Currently, basic water activation experiments are already routinely performed for educational purposes at the JSI TRIGA reactor, using a simplified open water loop on the reactor platform. Even with such a simplified experimental setup, the main peaks of ^{16}N can be observed with a High Purity Germanium (HPGe) detector [3].

An important part of the overall irradiation facility is to ensure adequate protection from gamma rays and neutrons originating from the reactor core through the RPP and from the activated water of the closed-water loop. It is necessary to ensure that dose rates in the reactor hall are sufficiently low during operation of the reactor at full power (250 kW) and that personnel are not exposed to high radiation doses, as are the detectors (background noise) that will be located around the water activation loop. The influence of the reactor core and the influence of the water activation can be considered separately.

The aim of this work is to build a detailed MCNP model and to assess the neutron and gamma dose field outside the Radial Piercing Port (RPP), considering only the contribution of the reactor operation. In this way, the efficiency of the proposed main shielding inside the RPP, i.e. the neutron and gamma shielding plugs, is assessed.

The paper is structured as follows: Section 2 presents a

description of the JSI irradiation facility focusing on the inner part of the closed- water activation loop and Section 3 presents the analysis performed, focusing on the computational methods and tools, the creation of the detailed MCNP model and the dose rate calculations.

II. Closed-water activation loop at JSI TRIGA

The JSI TRGIA is a pool-type reactor operated by natural convection with a thermal output of 250 kW [4]. A special feature of this reactor is its ability to reach briefly a power of 1 GW for a few milliseconds in pulsed mode [5]. The reactor core contains several irradiation channels, including vertical channels at the edge of the core and three channels embedded in the core. In addition, there are three horizontal irradiation channels – two radial and one tangential – that penetrate the concrete bioshield of the reactor. One of the radial ports ends outside the graphite reflector, while the other, called the Radial Piercing Port (RPP), also penetrates the graphite reflector and practically reaches the reactor core. A visual representation of the JSI TRIGA reactor core with its irradiation channels can be seen in a horizontal cross-section in Fig. 1 in a horizontal cross-section view. Previous assessments [6] led to the choice of the RPP (highlighted in red) for the installation of a closed-water loop due to the high neutron flux and the favourable shielding conditions of the surrounding reactor concrete bioshield.

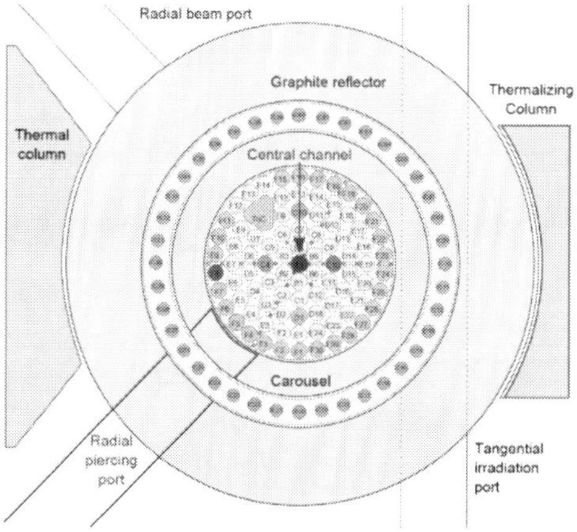

Fig. 1. Horizontal cross-section of the JSI TRIGA reactor core with the irradiation channels. The RPP shown in red was chosen for the installation of the closed-water loop.

The central idea behind the proposed closed- water activation loop system, highlighted in red in Fig. 2, is to set up a pipe loop that is inserted into the RPP. This conceptual framework consists of three key components: first, the "inner activation part," which is located inside the RPP in close proximity to the reactor core. This is where most of the water activation process takes place. Secondly, the "transport pipes" are responsible for transporting the activated water to the "outer observation part" of the circuit. Around this outer part is a series of advanced gamma and neutron detectors, i.e. HPGe, LaBr, TLD and He-3 detectors. The expected length of the transport segment of the loop (including the entire circuit pathway, but excluding the inner activation and outer observation parts) is estimated to be about 12 m. However, due to the constraints imposed by the fixed dimensions of the RPP, approximately 6 metres of this length is predetermined.

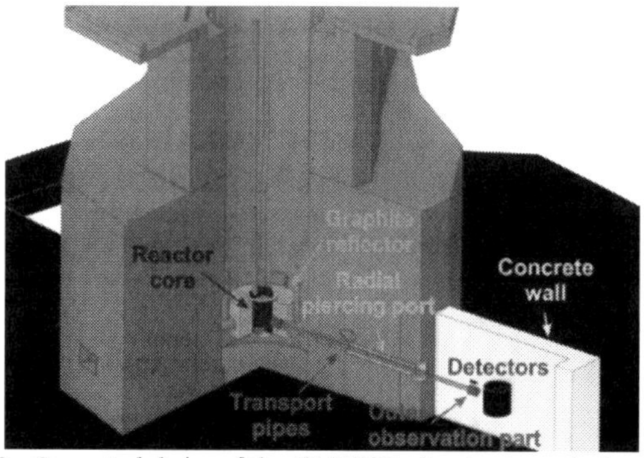

Fig. 2. Conceptual design of the JSI TRIGA closed-water activation loop, which consist of three main components, i.e. the inner activation part, the transport pipes and the outer observation part, using the RPP.

It is important to note that all necessary equipment, including essential elements such as pumps, flow metres, temperature and pressure sensors, cooling systems and mechanisms for refilling water, are located outside of the RPP. To effectively control radiation exposure inside the reactor hall, an additional 90 cm thick concrete wall will surround the entire irradiation facility [7]. It serves as a physical barrier and separates the experimental area from the rest of the reactor hall. In addition, it serves the dual purpose of reducing the contribution of dose rate from the reactor core and, most importantly, providing sufficient protection from high-energy gamma radiation from the activated water. The largest contribution is due to the outer observation part with its large volume of water compared to the narrow transport pipes.

A. *Inner part of the closed-water loop*

The inner set-up of the closed- water loop consists of the inner activation part, the narrow transport pipes and the appropriate shielding, i.e. gamma and neutron shielding plugs. The proposed design is shown in Fig. 3. The major size constraint of the whole setup is the RPP itself, which consists of two parts, an inner narrow part (length 150.3 cm and diameter 15.4 cm) and an outer wider part (length 151.5 cm and diameter 20.5 cm). The final design of the inner activation part was chosen based on a previous optimisation analysis [8] and includes a complex snail-shape design, with the highest effective water volume, which achieved the highest activities of the main water-activated isotopes (^{16}N, ^{17}N and ^{19}O) with respect to other candidate designs. The complex designed snail was fully 3D printed with aluminium. As a result, the cylindrical snail shape with a diameter of 14.8 cm and a length of 21.9 cm with 3 mm thick aluminium walls has an effective water volume of ~3 l.

The primary gamma-shielding plug consists of a 5 mm thick aluminium shell and is filled with (from left to right in Fig. 3) 19.2 mm thick Boral plates to minimise material degradation by neutrons, ~56 cm of concrete, ~62 cm mixture of concrete and lead shot, and a subsequent 12 cm long stainless-steel part (SS), which has three wheels for easier insertion into the RPP and a cavity for mounting a handling tool. The neutron shielding plug, located behind the gamma shield, is made of borated polyethylene (BP). The BP plug, with a boron mass fraction of 5 % enables much more effective shielding against neutrons than the previous wooden plug because it contains the isotope 10B, which has a very large absorption cross-section for thermal neutrons. Similar to the gamma plug, a 19.2 mm thick Boral plate is placed in front of it.

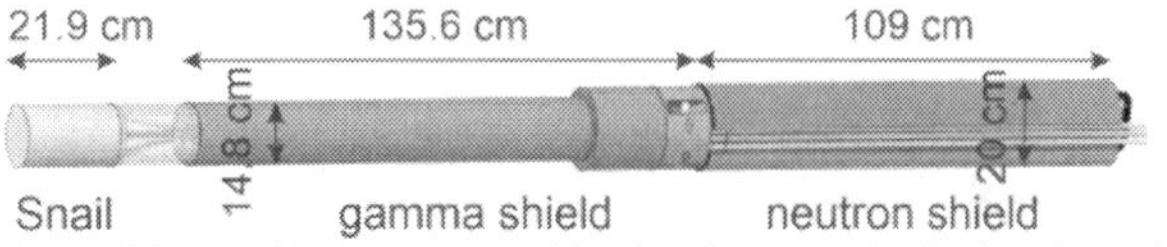

Fig. 3. Scheme of the inner set-up of the closed- water activation loop located inside of the RPP at the JSI TRIGA reactor consisting of three main components, i.e. inner activation part (snail), aluminium-concrete gamma shielding plug and the neutron shielding plug made of borated polyethylene. The water inlet and outlet tubes are shown in blue and the central hollow instrumentational tube in red.

In addition, three pipes lead from the snail through the aforementioned shielding plugs: an inlet and an outlet pipe for the water circuit (shown in blue in Fig. 3) and a central hollow instrumentational pipe (shown in red) for the insertion of detectors and cables during reactor operation, e.g. a fission cell for measuring the neutron flux inside the snail (inner activation part) to monitor reactor power or using activation foils to determine the neutron spectrum along the snail. The above tubes are also embedded in the concrete plug and bent to minimise the streaming of direct (non-scattered) neutrons and gamma rays from the reactor core.

III. Calculations

One of the most important components of the irradiation facility with closed-water activation loop as a whole is to ensure adequate shielding for the personnel and instruments (detectors). For this reason, an assessment of the neutron and gamma dose field outside the RPP has been performed, excluding the contribution of the activated water in the whole circuit at this stage. In this way, the shielding efficiency of the proposed inner set-up is assessed.

A. Computational methods and tools

The calculations for the assessment of neutron and gamma dose rates were performed using MCNP6 v2.0, while the variance reduction parameters, more specifically the weight window (WW) parameters, were prepared using ADVANTG [10]. The use of a variance reduction technique is essential as deep penetration/shielding calculations are challenging and extremely computationally intensive. This two-step hybrid transport approach was chosen because large speedups in terms of computational time (more than a factor of 1000) were observed for a variety of shielding and radiation geometries compared to analogue MCNP calculations [11-15]. Due to the ADVANTG limitation, a fixed SDEF source based on a kcode calculation was prepared. A deterministic ADVANTG calculation used a 27n19g nuclear cross-section library, a variable geometric mesh with $13.7 \cdot 10^6$ voxels that varied in size depending on the geometry, i.e. finer mesh in the areas of the reactor core and RPP (a few centimetres) and coarser mesh in the surroundings (from several centimetres to one metre), a P_3 scattering- angle expansion and an S_4 angular approximation.

The ANSYS SpaceClaim programme [16] was used to create CAD models of the entire closed- water activation loop. It allowed detailed modelling and easier 3D visualisation.

In addition, the open code GEOUNED [17] was used to convert the models from CAD to MCNP input files. This step was essential to conserve as much details as possible, which is otherwise practically impossible to achieve manually.

B. Computational model

The computational model for the MCNP calculation was created in several steps. In the first step, the detailed CAD model, shown in Fig. 3, was simplified because some surfaces, e.g. splines, torus, extruded, rotated, etc., are not supported in MCNP. Specifically, 2 spline, 30 torus and 115 extruded/rotated surfaces were manually simplified as planes, cylinders, cones or spheres. In the second step, the GEOUNED code was used to convert the simplified CAD model, shown in Fig. 4a, into an MCNP input file. In general, the conversion was smooth, with less than 3 cells requiring subsequent manual correction. To avoid errors during the conversion, complex cells were preferably split into smaller simpler geometries, i.e. the most problematic parts were curved pipes and the area around them (Fig. 4b).

Last but not least, the generated MCNP input file of the entire inner set-up of the closed-water activation loop was implemented into the detailed MCNP input of the JSI TRIGA reactor considering the current core configuration #244 (Fig. 4c).

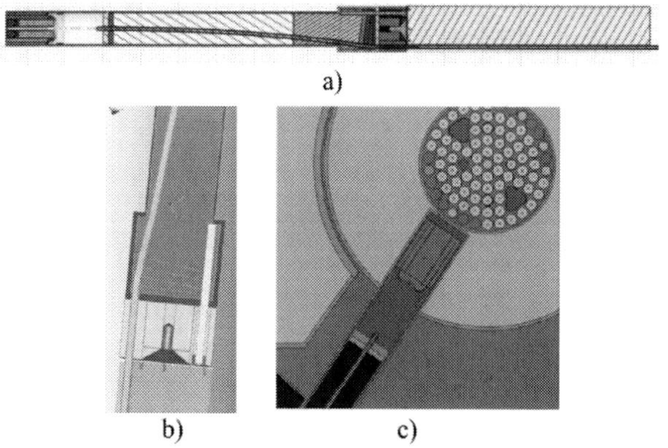

Fig. 4. a) Simplified CAD model of the inner set-up of the closed-water activation loop. b) Detailed MCNP model generated with the GEOUNED code. c) Final computational model.

C. Dose rate calculations

Neutron and gamma dose rate values were calculated as ambient dose equivalent H*(10) using NCRP-38 [18] and ICRP-21 flux- to- dose- rate conversion factors, respectively. The measurement position (MP) was chosen at the entrance of the RPP (outer boundary of the reactor bioshield), about 320 cm from the centre of the reactor core. The overall efficiency of the proposed shielding plugs was assessed by comparing the dose rate values at the MP with the entire inner set-up of the closed-water loop and without it (empty RPP). In both scenarios, the reactor was considered at full power (250 kW).

The results of the neutron and gamma dose rate calculations, with a relative statistical uncertainty of 1 sigma, are shown in Fig. 5 at the entrance of the RPP for different scenarios. The results for the reference scenario, the empty RPP, are very large, i.e. ~400 Sv/h for neutrons and ~13 Sv/h for gamma rays. Of course, this scenario cannot be reproduced experimentally, but it shows that any shielding used in the RPP should reduce the dose rate by several orders of magnitude. Specifically, the dose rate limitation in the reactor hall is 10 μSv/h and up to several mSv/h within the experimental area. Considering the proposed inner set-up of the closed-water activation loop, the dose rate values decrease significantly. In particular, they amount to ~82 μSv/h for neutrons and ~11 μSv/h for gammas. The absolute values are quite promising, but should be considered with caution as they are currently of lower quality due to the high (~40 %) statistical uncertainty even when using a variance reduction technique. We would like to point out that such deep penetration/shielding calculations are extremely computationally intensive and the analogue MCNP calculation did not yield any result at all. The statistical uncertainty could be reduced by generating more effective WW. Further work on optimising the ADVANTG parameters is essential.

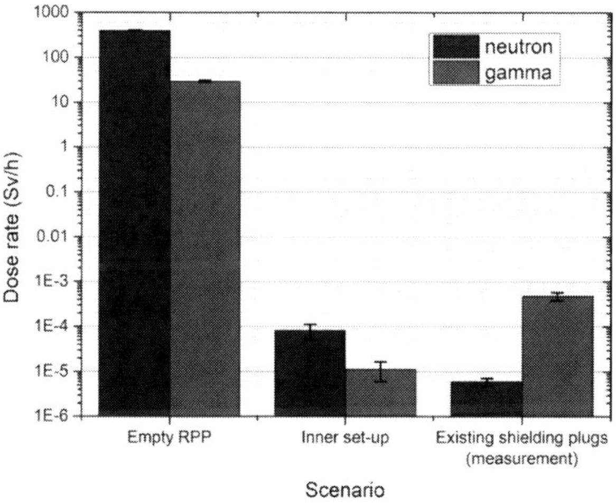

Fig. 5. Neutron and gamma dose rate values at the entrance of the RPP for empty RPP and inserted inner set-up of the closed-water loop. Due to the lower quality of the latter, experimental results of using similar existing shielding plugs are also shown for orientation.

To provide a more tangible comparison of the shielding effectiveness of the proposed design of the inner set-up, the experimental values of using similar existing shielding plugs are also shown in the graph. Existing shielding plugs currently used inside the RPP were the basis for developing new plugs for the inner set-up of the closed-water loop. Although a difference of more than an order of magnitude can be observed, the experimental results are only indicative. However, they indicate promising shielding properties of the proposed new shielding plugs. The most important comparison will be the experimental validation after the whole set-up is built and tested.

IV. Conclusions

A unique irradiation facility for high-energy gamma rays (6 MeV – 7 MeV) and neutrons (0.4 MeV - 1.2 MeV) is being built at the research reactor JSI TRIGA in Slovenia, using a closed-water loop for experiments based on water activation. The main focus of this work was to build detailed MCNP model and to assess the neutron and gamma dose rate outside the radial piercing port of the closed-water activation loop, with particular emphasis on the shielding efficiency of the proposed inner set-up that is inserted inside of the port. In this way, the efficiency of the proposed main shielding inside the port, i.e. the neutron and gamma shielding plugs, is assessed.

It turned out that the GEOUNED code is very useful for converting CAD models into MCNP input files. This step was essential to obtain as much detail as possible, which is otherwise practically impossible to do manually.

The results show that the dose rate using shielding plugs is more than 7 orders of magnitude lower than for a empty open port and in good agreement with the experimental results of existing similar shielding plugs. While the calculated dose rate values appear promising, they are currently subject to high statistical uncertainties. The deep penetration/shielding calculations are extremely computationally intensive and the analogue MCNP calculation gave no result at all. Further optimisation of the ADVANTG input parameters is foreseen to optimise the weight windows and thus improve the accuracy of the results.

The proposed design of the shielding plugs and the whole internal set-up of the closed-water loop has been confirmed. The most important confirmation will be the experimental validation after the whole set-up is built and tested. After that, any additional lead shielding blocks will be placed outside the port if needed. The commissioning of the entire irradiation facility and first experiments are scheduled by the end of the year.

Acknowledgment

The authors acknowledge the support of the Slovenian Ministry of Education, Science and Sport (project codes P2-0405 Fusion technologies; P2–0073 Reactor physics; PR-09769 Training of young researchers). This work has been carried out within the framework of the EUROfusion Consortium, funded by the European Union via the Euratom Research and Training Programme (Grant Agreement No 101052200 — EUROfusion). Views and opinions expressed are however those of the author(s) only and do not necessarily reflect those of the European Union or the European Commission. Neither the European Union nor the European Commission can be held responsible for them.

References

[1] A. Žohar, L. Snoj, On the dose fields due to activated cooling water in nuclear facilities, Progress in Nuclear Energy, Volume 117, 2019, 103042.
[2] M. De Pietri, J. Alguacil, A. Kolsek, G. Pedroche, N. Ghirelli, E. Polunovskiy, et. al., Integral modelling of the ITER cooling water systems radiation source for applications outside of the Bio-shield, Fusion Engineering and Design, Volume 171, 2021, 112575, ISSN 0920-3796.
[3] A. Žohar, V. Radulovic, A. Jazbec, I. Lengar, S. Rupnik, L. Snoj, Water Activation Experiments and Calculations at JSI TRIGA Research

Reactor, 29th International conference Nuclear Energy for New Europe, 302.1, Portoroz, 2020.

[4] A. R. Jeraj, M. Ravnik, U(20)-Zirconium Hydride Fuel Rods in Water with Graphite Reflector, International Handbook of Evaluated Criticality Safety Benchmark Experiments, NEA/NSC/DOC(95)03, OECD-NEA, 2010.

[5] I. Vavtar, A. Pungercic, L. Snoj, Utilisation of JSI TRIGA Pulse Experiments for Testing of Nuclear Instrumentation and Validation of Transient Models, EPJ Web Conf. 225 04027, 2020.

[6] A. Žohar, A. Pungercic, K. Ambrozic, V. Radulovic, A. Jazbec, S. Rupnik, et al., Conceptual Design of Irradiation Facility with 6 MeV and 7 MeV Gamma Rays at the JSI TRIGA Mark II Research Reactor, EPJ Web Conf. 225 04014, 2020.

[7] D. Kotnik, L. Snoj, I. Lengar, Overview of the closed water activation loop at JSI TRIGA research reactor, The European research reactor conference RRFM, Antwerp, 2023, ISBN - 978-92-95064-39-3

[8] D. Kotnik, A. K. Basavaraj, L. Snoj, I. Lengar, Design optimization of the closed-water activation loop at the JSI irradiation facility, Fusion Engineering and Design, Volume 193,2023, 113632, ISSN 0920-3796, https://doi.org/10.1016/j.fusengdes.2023.113632.

[9] C. J. Werner, et al., "MCNP6.2 Release Notes", Los Alamos National Laboratory, report LA-UR-18-20808 (2018)

[10] Mosher, S. W., Johnson, Seth R., Bevill, Aaron M., Ibrahim, Ahmad M., Daily, Charles R., Evans, Thomas M., Wagner, John C., Johnson, Jeffrey O., and Grove, Robert E., ADVANTG An Automated Variance Reduction Parameter Generator, Rev. 1. United States: N. p., 2015. Web. doi:10.2172/1210162.

[11] D. Kotnik, A. Cufar, B. Kos, L. Snoj, Validation and evaluation of the ADVANTG hybrid code on the ICSBEP labyrinth benchmark experiment, Ann. Nucl. Energy 114 (2018) 464–481, https://doi.org/10.1016/j.anucene.2017.12.011. ISSN 0306-4549.

[12] D. Kotnik, B. Kos, A. Cufar, S.W. Mosher, Validation and evaluation of the ADVANTG code on the ICSBEP skyshine benchmark experiment, Ann. Nucl. Energy 125 (2019) 249–260.

[13] D. Kotnik, B. Kos, D. Calic, L. Snoj, Use of ADVANTG to analyse skyshine γ-dose, rates around a silo type LILW repository, Ann. Nucl. Energy 145 (2020), 107585, https://doi.org/10.1016/j.anucene.2020.107585. ISSN 0306-4549.

[14] A. Jazbec, B. Kos, K. Ambrozic, L. Snoj, Dose rate calculations at beam tube no. 5 of the JSI TRIGA mark II research reactor using Monte Carlo method, Appl. Radiat. Isot. (2020), 109510, https://doi.org/10.1016/j.

[15] D. Kotnik, B. Kos, I. Lengar, A. Cufar, C. Bachmann, L. Snoj, Assessment of sky-shine in DEMO during breeding blanket maintenance, Fusion Engineering and Design, Volume 167, 2021, 112348, ISSN 0920-3796, https://doi.org/10.1016/j.fusengdes.2021.112348.

[16] ANSYS, 2022. https://www.ansys.com/products/3d-design/ansys-spaceclaim#tab1-3 (4. 10. 2022)

[17] J. Garcia, J. P. Catalan, J. Sanz, Development of the automatic void generation module in GEOUNED conversion tool, Fusion Engineering and Design, Volume 168, 2021, 112366, ISSN 0920-3796, https://doi.org/10.1016/j.fusengdes.2021.112366 https://github.com/jpcatalanUNED/GEOUNED_0.9.8.2

[18] NCRP Scientific Committee 4 on Heavy Particles, H. H. Rossi, chairman, "Protection Against Neutron Radiation," NCRP-38, National Council on Radiation Protection and Measurements (January 1971).

[19] ICRP Committee 3 Task Group, P. Grande and M.C. O'Riordan (April 1971). Data for Protection Against Ionizing Radiation from External Sources: Supplement to ICRP Publication 15, ICRP-21, International Commission on Radiological Protection, Pergamon Press.

Characterization of a new reduced-height CALORRE differential calorimeter for CALOR-I Irradiation in MITR

A. Volte[1,*], S. Hauptman[2], D. Carpenter[2], M. Carette[1], A. Lyoussi[3], G. Kohse[2], C. Reynard-Carette[1]

[1]Aix Marseille Univ, Université de Toulon, CNRS, IM2NP, Marseille, France
[2]Massachusetts Institute of Technology, Nuclear Reactor Laboratory, Cambridge, Massachusetts, USA
[3]CEA/DES/IRESNE/DER, Section of Experimental Physics, Safety Tests and Instrumentation, Cadarache, F-13108, Saint Paul-lez-Durance, France

(*) adrien.volte@univ-amu.fr

***Abstract*— This paper presents the preliminary characterization of a new reduced-height CALORRE differential calorimeter designed and fabricated for the irradiation in the MITR within the framework of the CALOR-I program. The paper begins by focusing on the preparation of the irradiation campaign, providing a brief and concise description of the MITR core and the newly fabricated differential calorimeter assembly. Next, the paper presents the preliminary experimental characterization conducted under laboratory conditions. This section includes a detailed presentation of the updated experimental set-up and the key metrological characteristics of the sensor response obtained from the first experimental results. Furthermore, the paper estimates and highlights the responses of the calorimeter calculated by means of a 3-D numerical thermal model under real conditions considering local heat sources determined by the NRL of the MIT using MCNP code. In conclusion, the paper offers some final remarks and prospects to realize the irradiation campaign successfully.**

***Keywords* —Calorimeter, Nuclear Heating Rate, On-line Measurements, Calibration, Irradiation campaign.**

I. Introduction

The physical knowledge of the behavior of materials and fuels in extreme environments is an essential contribution to the development of new systems, whether for nuclear fission or fusion purposes. Thus, research works are carried out using major facilities, such as Material Testing Reactors (MTRs) in the field of fission, to address issues such as safety, security, and the lifetime of current power reactors, the testing of new concepts for future reactors, and the optimization of fuel and radioactive waste. To conduct required experimental studies, MTRs produce relevant conditions including high neutron and gamma fluxes as well as strong displacements per atom.

Nevertheless, the European research reactor park is aging (more than 50 years) and the number of MTRs has been decreasing over the years. For instance, the OSIRIS reactor (at CEA Saclay Center, France) was shut down in late 2015 after its first criticality in 1966. A new facility with high performances was required in Europe to deal with the in-core experimental facilities needs for irradiating material under realistic controlled conditions (temperature, neutron and photon fluxes, etc.). Therefore, the Jules Horowitz Reactor (JHR) construction started in 2007 at the CEA Cadarache Center in the south of France [1, 2]. This new 100 MW_{th} nominal power MTR will offer unequaled research and development possibilities in Europe thanks to a high fast neutron flux of $5.5x10^{14}$ $n.cm^{-2}.s^{-1}$ (from 1 MeV) leading to high accelerated aging (up to 16 dpa/year) and a high nuclear absorbed dose rate (up to 20 $W.g^{-1}$ in aluminum).

The JHR program initiated new collaborative research work between Aix-Marseille University (AMU) and the CEA (within the framework of the joint laboratory LIMMEX - Laboratory for Instrumentation and Measurement in Extreme Environments) and its IN-CORE program (Instrumentation for Nuclear Radiations and Calorimetry on-line in REactor) in 2009. The main objectives are to enhance on-line instrumentation and advanced measurement methods for the quantification of crucial nuclear parameters such as neutron and photon fluxes and high nuclear absorbed dose rate, also called nuclear heating rate [3]. This latter quantity is measured on-line by means of specific sensors: non-adiabatic calorimeters. These sensors must be preliminary studied under laboratory conditions by out-of-pile characterization and calibration and then require qualification and/or validation under real conditions thanks to several irradiation campaigns [4-7].

As a result of this latter statement, a new research program, called CALOR-I (compact-CALORimeter Irradiations inside the MIT Research Reactor) began in 2020. This program is funded by A*Midex Foundation. It is led by AMU and involves two partners: the Nuclear Reactor Laboratory of the MIT and the CEA. The first two aims of this program are the design of a new compact prototype of the CALORRE calorimeter and the study under various real conditions. The CALORRE calorimeter was patented by AMU and the CEA in 2015 [8, 9]. A first prototype was tested successfully for the first time in November 2015 in the MARIA reactor [6, 7]. The third objective of this program is to map a dedicated in-core water loop of the MIT Research Reactor (MITR) in terms of nuclear heating rate using the new reduced-size prototype of CALORRE developed and optimized thanks to coupled thermal and neutronic modeling.

This paper will present the new prototype of the CALORRE differential calorimeter designed by taking feedbacks into account in particular from the previous irradiation campaign in the MARIA reactor [6, 7] and with a comprehensive approach coupling experimental, theoretical, and numerical work from laboratory conditions to nuclear environments.

The first section of the paper will briefly describe the MITR core conditions and outline the latest design of the CALORRE prototype and its main features will be discussed. This new prototype corresponds to the smallest differential calorimeter with two superimposed cells (88.5 mm from the nose to the watertight passage). The second section will be dedicated to the experimental characterization of the prototype response under laboratory conditions. The metrological characteristics: sensitivity, linearity, response time, and reached absolute temperatures will be detailed and compared to previous CALORRE configurations. The last section will present new 3-D numerical thermal calculations under real conditions to estimate the behavior and the response of the sensor for future irradiation in the MITR (resolution with COMSOL Multiphysics code). The 3-D thermal model takes local heat sources into account for each part of the calorimeter assembly. These local heat sources correspond to the nuclear heating rates and were preliminary determined through particle transport calculations (photon-electron and neutron-photon-electron MCNP modes) by the MIT and for different core fuel loadings, starting random seeds and calorimeter positions [10].

II. Preparation of an irradiation campaign inside the MIT reactor

The aim of the campaign is threefold: the mapping of the nuclear heating rate in the MITR core by using two sensors (CALORRE prototype and a commercial sensor), the study of a new CALORRE prototype for the in-core water loop by testing two measurement methods and by determining the influence of several conditions (the convection intensity, the power of the reactor). In the next sub-section, the reactor core will be briefly described, then the new calorimeter prototype will be presented.

A. *The MITR core*

The MIT Reactor (MITR), located in the center of the MIT Campus (Cambridge, MA, USA), is a 6 MW_{th} tank-type nuclear reactor that operates with light-water cooled and moderated while utilizing heavy-water and graphite reflectors.

The MITR provides experimenters with a maximum flux of $1.2x10^{14}$ $n.cm^{-2}.s^{-1}$ (E > 1 MeV) and $6.0x10^{13}$ $n.cm^{-2}.s^{-1}$ for thermal flux. The core of the MITR has dimensions of 56 cm in height and 38 cm in diameter. Within the core, there are a total of 27 positions, primarily occupied by fuel elements like the one shown in position C-9 (cf. Figure 1). The remaining three positions (A1, A3, and B3) are either occupied by a solid aluminum dummy element or dedicated to hosting an In-Core Experiment [11, 12].

The CALOR-I Irradiation in MITR will take place in the in-core water loop (located at B3 in Figure 1). This loop allows the control of both fluid flow temperature and velocity.

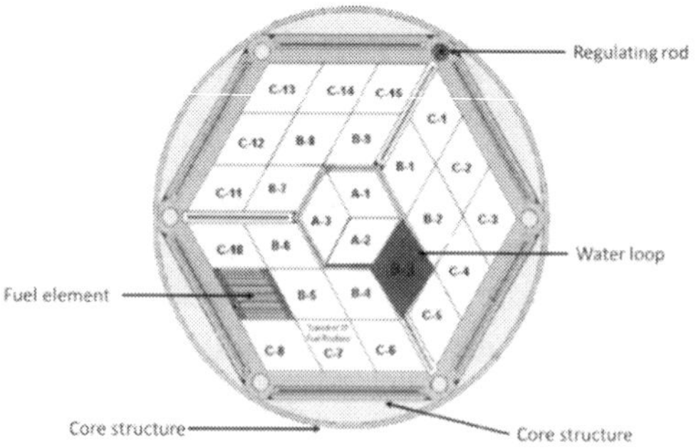

Fig. 1. Simplified diagram of the MITR core.

Considering the calibration requirements under laboratory conditions [5-7], the temperature of the in-core water loop will be set at approximately 50 °C. However, the fluid flow velocity will be varied for a specific range. The impact of this boundary condition is important and was studied carefully previously to determine the optimal heat transfer coefficient [7] to ensure efficient heat dissipation regarding safety and security demands.

B. *The new differential calorimeter prototype*

As presented previously, the prototype was designed thanks to the results of previous irradiation campaigns, recent research work, the results of six additional calorimeter cells, and the nuclear heating rate predicted in the MITR [10]. The prototype was defined using 3-D numerical thermal simulations with the objectives to eliminate contact thermal resistances by simplifying assembly and components, and to reduce the overall height of the calorimeter [7].

A modified configuration was chosen to achieve our objectives, with a reduced-height calorimetric cell measuring 11.55 mm. The new calorimetric cell is composed of a stainless steel half-horizontal fin arrangement (8 metal sectors and 8 empty sectors with the same angle equal to 22.5°) and an increased vertical fin (31.55 mm instead of 11.55 mm). Notably, the sample and head structure were integrated into a single block made of stainless steel, simplifying the assembly. Furthermore, the vertical fins of both the sample and reference cells were welded together (cf. Figure 2), eliminating the need for spacers and creating a unified calorimeter structure filled with di-nitrogen.

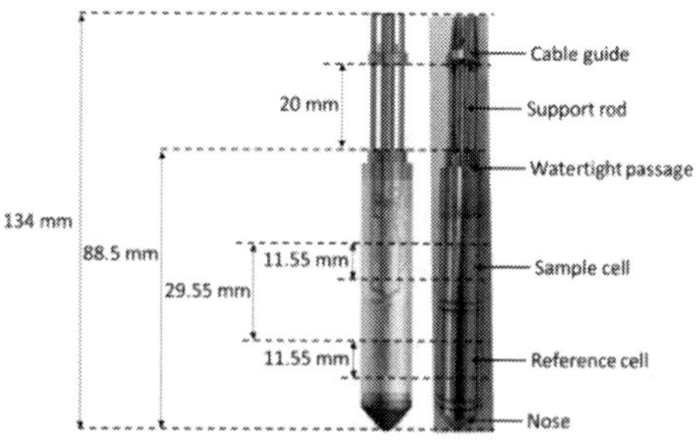

Fig. 2. 3-D scheme (left-hand section) and picture (right-hand section) of the CALORRE differential calorimeter.

Figure 2 gives a 3-D scheme and a picture of the complete CALORRE differential calorimeter manufactured for MITR irradiation. The calorimeter is made up of two superimposed calorimetric cells with an inter-cell space of 18 mm, resulting in an inter-sample space of 29.55 mm (as against 95 mm for the MARIA campaign) [7]. In addition, the heating elements are wired in a 4-wire assembly with Nickel-Chromium wires, allowing the necessary range of electrical power to be injected during calibration to be representative of the targeted nuclear heating rate range. There are always two K-type thermocouples per cell to measure temperatures at specific points (same location as previous configurations). A total of 9 cables need to be pulled up (with a length of 20 m each).

The assembly is composed, from bottom to top, of the nose, the reference cell, the sample cell, the watertight passage, the support rod, and the cable guide. The total height from the bottom of the nose to the top of the support rod is 134 mm and only 88.5 mm from the bottom of the nose to the top of the watertight passage (as against 222.5 mm for CALORRE-MARIA, CARMEN, and CALMOS differential calorimeters). The external diameter remains the same at 18 mm.

III. Experimental characterization under laboratory conditions

This section outlines the experimental characterization of the newly manufactured calorimeter under laboratory conditions. Due to the extended cable length and calorimeter assembly requirements, an updated experimental set-up was developed. This set-up will be detailed before presenting the first calibration results of the prototype and the corresponding metrological characteristics obtained during the experiments.

A. The experimental set-up

As previously mentioned, the experimental set-up had to be adapted to take the specific characteristics of the final assembly into account as they differ from those of previous laboratory studies (only one calorimetric cell). It is presented in Figure 3.

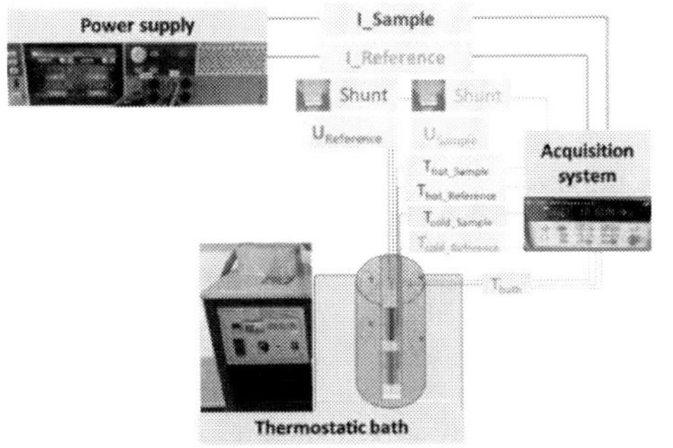

Fig. 3. A schematic representation of the new experimental set-up including photographs.

The composition of the experimental set-up still follows a conventional design: a power supply, an acquisition system, a computer, a precision resistance (a high accuracy burster 0.1 Ohm), and a thermostatic bath. However, to ensure appropriate current injection during the calibration of the sample cell and reference cell simultaneously, the implementation of a new 2-channel programmable power supply was necessary. This will facilitate the controlled supply of current to both cells, in relation to the length of cables increasing the total line resistance.

In addition, a newly designed support structure has been developed to facilitate the insertion of the assembly into the thermostatic bath, providing exact alignment in terms of height and radial position (compared to previous studies for different configurations). This support structure ensures precise positioning and stability of the assembly throughout the experimental calibration or parametrical studies. It was made in PLA using a 3-D printer. Three thermocouples were also inserted to measure the temperature of the bath around the calorimeter at the inter-cell space level and check the bath homogeneity.

B. The metrological characteristics of the CALORRE prototype

The CALORRE prototype was studied for a fluid temperature of 33 °C and a Reynolds number of 1607 (consistent with PLA and previous studies). Successive values of electric current are simultaneously injected (from 0 to around 1.7 A) for 20 minutes into the two heating elements so that the power is between 0 and 6 W for the sample cell and the reference cell. This electrical power range corresponds to a preliminary tested range in order to do comparisons with previous studies. Considering the mass of the head of the reference cell (2.85 g) and the sample cell (5.65 g), the nuclear heating rate ranges simulated by this 6 W power range are 2.1 $W.g^{-1}$ and 1.1 $W.g^{-1}$ for the reference and sample cells, respectively.

The temporal response obtained is presented in Figure 4.

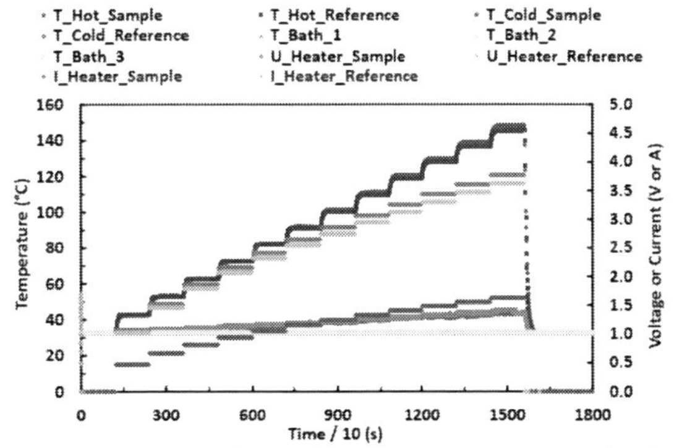

Fig. 4. Temperatures, voltage and current measurements versus time for the sample and reference cells, a fluid temperature of 33 °C and a Reynolds number of 1607.

By analyzing the temporal response, the absolute temperatures can be examined, and the response time of the sensor evaluated. The maximum hot and cold temperatures obtained for an injected electrical power of 6 W are 148.3 °C and 145.4 °C and 43.3 °C and 43.4 °C for the sample cell and the reference cell respectively. The three bath temperatures are homogeneous and equal to 32.8 °C. The previous results obtained under laboratory conditions, for only one calorimetric cell (half-height and half-horizontal fin arrangement made of stainless steel, inserted in an individual envelope), were 157 °C and 55.5 °C for the hot and cold temperatures, respectively.

The mean response times (at 3τ) from 0 to 6 W are: 211 s and 191 s for T_{Hot} for sample and reference cells and 164 s and 144 s for T_{Cold} for sample and reference cells respectively. The response time obtained previously with only one calorimetric cell (as described above) was 174 s for the hot temperature.

Finally, the calibration curve of each cell, which represents the difference of the mean steady temperatures ($T_{Hot} - T_{Cold}$) versus the mean electrical power, is obtained by analyzing the steady states.

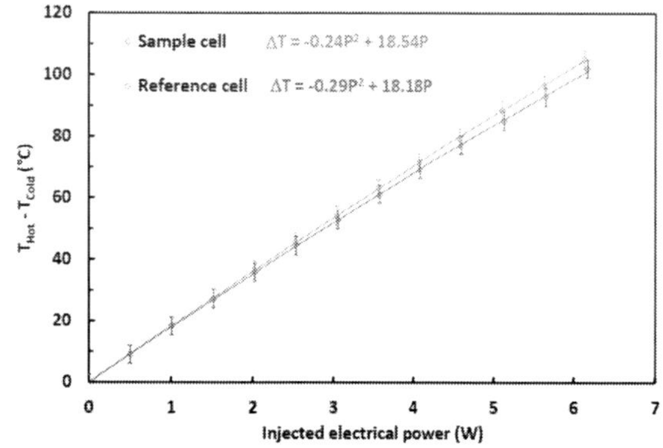

Fig. 5. Calibration curve of the sample and the reference cells obtained for a fluid temperature of 33 °C and a Reynolds number of 1607.

Figure 5 shows the calibration curves obtained for the sample and reference cells (second-order coefficients equal to -0.24 $°C.W^{-2}$ and -0.29 $°C.W^{-2}$ for the sample and reference cell respectively). The sensitivity for an electrical power of 6 W is 15.7 $°C.W^{-1}$ for the sample cell as against 14.7 $°C.W^{-1}$. The sensitivity obtained for only one calorimetric cell (as described above) was 15.1 $°C.W^{-1}$ [7].

IV. 3-D NUMERICAL THERMAL CALCULATIONS UNDER REAL CONDITIONS

The performance of the new compact differential calorimeter was studied under real conditions focusing on the two cells, the nose, watertight passage, support rod, and cable guide, and excluding the cables and thermocouples. The 3-D numerical study employed a validated 3-D thermal model [6, 7], which was solved using COMSOL Multiphysics. The 3-D thermal model included thermal conductive transfers (taking the thermal conductivity of stainless steel and di-nitrogen as a function of temperature into account) as well as thermal radiative transfers (assuming a constant emissivity of 0.25 for stainless steel). Local heat sources corresponding to the local nuclear heating rate, calculated using MCNP codes [10] for various positions of the calorimeter and material densities were considered. Convective boundary conditions were applied to the external surfaces of the calorimeter, assuming turbulent fluid flow with a heat transfer coefficient of 10000 $W.m^{-2}.K^{-1}$ and a coolant fluid temperature of 50 °C.

Figure 6 shows the numerical response of the differential calorimeter corresponding to the difference of the two temperature differences ($\Delta\Delta T = \Delta T_{Sample} - \Delta T_{Reference}$) versus the distance to the core center line for a reactor power of 6 MW_{th}. The obtained response validates the ability of the new prototype to measure the peak nuclear heating rate, which reaches approximately 2 $W.g^{-1}$ at 6 MW_{th} for a stainless steel sample with a sensor sensitivity of ~56.5 $°C.g.W^{-1}$.

The results also confirm the possibility of mapping the B3 channel. Additionally, the maximum temperature inside the calorimeter and the maximum wall temperature obtained are around 300 °C and 56 °C respectively for the nuclear heating rate peak.

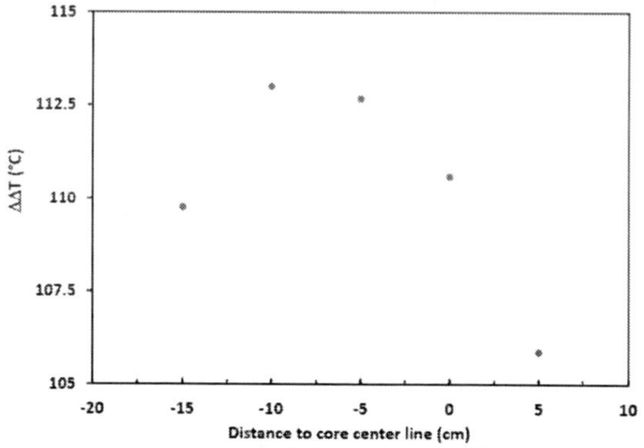

Fig. 6. 3-D numerical response of the new compact differential calorimeter as a function of the distance to core center line (h=10000 $W.°C^{-1}.m^{-2}$ and T_f=50 °C) for a reactor power of 6 MW_{th}.

V. CONCLUSIONS AND OUTLOOKS

A new compact differential calorimeter was fabricated by considering previous 3-D numerical parametrical studies. It was preliminarily characterized under laboratory conditions by means of an updated experimental set-up. The key metrological characteristics of the sensor were presented (Maximum temperatures, time responses, calibration curves, and sensitivity) and compared to previous calorimetric cells. Then, the calorimeter prototype was studied under real conditions thanks to the validated 3-D thermal model by taking local heat sources into account and corresponding to the nuclear heating rates preliminary determined through particle transport calculations by the MIT for different calorimeter positions. The 3-D numerical response of the new compact differential calorimeter as a function of the distance to the core center line was determined for a heat transfer coefficient of 10000 $W.m^{-2}.K^{-1}$ and a coolant fluid temperature of 50 °C. A right behavior was observed (maximum sensor temperature and wall temperature).

The experimental outlooks are further characterizations under laboratory conditions (repeatability, reproducibility, etc.) by increasing the range of injected electrical power for the sample cell to simulate the nuclear heating expected in the MITR and by testing two experimental protocols (individual and/or simultaneous current injection). The numerical outlooks will be new 3-D numerical thermal simulations with the final multi-sensor device used for the future irradiation campaign and composed of a thermocouple (to measure the fluid temperature inside the in-core water loop) and a gamma thermometer (to compare the nuclear heating rate values obtained).

ACKNOWLEDGMENT

The CALOR-I project leading to this publication has received funding from the Excellence Initiative of Aix-

Marseille University - A*Midex, a French "Investissements d'Avenir programme".

References

[1] G. Bignan, X. Bravo and P.M. Lemoine, "The Jules Horowitz Reactor: A new high Performances European MTR (Material Testing Reactor) with modern experimental capacities: Toward an International Centre of Excellence", Proceedings of Conferences RRFM and IGORR 2012, pp. 21-25, Prague, Czech Republic, 18-22 March 2012.

[2] A. Lyoussi, D. Fourmentel, J-F. Villard, J-Y. Malo, P.Guimbal, H. Carcreff, C. Gonnier, G. Bignan, J-P. Chauvin, C. Reynard-Carette, J. Brun, O. Merroun, M. Carette, M. Muraglia, A. Janulyte, Y. Zerega and J. Andre, "Advanced methodology and instrumentation for accurate on line measurements of neutron, photon and nuclear heating parameters inside Jules Horowitz MTR Reactor", Proceedings of Conferences RRFM and IGORR 2012, pp. 21-25, Prague, Czech Republic, 18-22 March 2012.

[3] C. Reynard-Carette, G. Kohse, J. Brun, M. Carette, A. Volte, A. Lyoussi, "Review of nuclear heating measurement by calorimetry in France and USA", Proc. ANIMMA 2017, Jun. 2017, EPJ Web of Conferences 170, 04019 (2018).

[4] H. Carcreff, L. Salmon and F. Malouch, "Recent developments in nuclear heating measurement methods inside the OSIRIS reactor", Nucl. Instrum. Methods Phys. Res. A. Accel. Spectrom. Detect. Assoc. Equip., vol. 942, pp. 1-19, Oct. 2019.

[5] A.Volte, C. Reynard-Carette, A. Lyoussi, J. Brun, M. Carette, "Study of the Response of a New Compact Calorimetric Cell for Nuclear Heating Rate Measurements", IEEE Trans. Nucl. Sci., vol. 65, no. 9, pp. 2461-2470, Sept. 2018, doi: 10.1109/TNS.2018.2827084.

[6] A.Volte, J. Brun, A. Lyoussi, M. Carette, C. Reynard-Carette, "Qualification of a New Differential Calorimeter Configuration Dedicated to Nuclear Heating Rates up to 20 W.g-1", IEEE Trans. Nucl. Sci., vol. 67, no. 11, pp. 2405-2414, Nov. 2020, doi: 10.1109/TNS.2020.3027913.

[7] A.Volte, M. Carette, A. Lyoussi, G. Kohse, J. Rebaud, V. Valero, C. Reynard-Carette, "Review of CALORRE Calorimeter Characterizations Under Laboratory and Irradiation Conditions", IEEE Trans. Nucl. Sci., vol. 69, no. 4, pp. 840-848, April 2022, doi: 10.1109/TNS.2022.3150148.

[8] M. Carette, A. Lyoussi, J. Brun, C. Reynard-Carette, J.F. Villard, « Eprouvette pour mesure d'échauffements nucléaire dans un réacteur nucléaire, et cellule calorimétrique comprenant au moins une telle éprouvette », FR1553136A, 2015-04-10.

[9] M. Carette, A. Lyoussi, J. Brun, C. Reynard-Carette, J.F. Villard, P. Guimbal, "Sample holder for measuring nuclear heating in a nuclear reactor, and calorimetric cell including at least one such sample holder", CEA/AMU, US patent, US10755823B2, 2020-08-25.

[10] S. Hauptman, A. Volte, G. Kohse, M. Carette, C. Reynard-Carette, "Flux and Nuclear Heating Rate Calculations for CALORRE Irradiation in MITR", ANIMMA conference, Lucca, Italy, 2023.

[11] S. Kim, D. Carpenter, G. Kohse, L. Hu "Hydride fuel irradiation in MITR-II: Thermal design and validation results", Nucl. Eng. Des., 277 (2014), pp. 1-14.

[12] A. J. Dave, K. Sun, L. Hu, E. Wilson, S. Pham, D. Jaluvka, "Thermal-hydraulic analyses of MIT Reactor LEU Transition Cycles," Progress in Nuclear Energy 118 (2020):103-117.

Defected fuel rods identification in TRIGA Reactors: the experience at the ENEA Casaccia TRIGA RC-1 reactor

L. Lepore[1,*], L. Falconi[1], V. Fabrizio[1], A. Roberti[1], D. Formenton[1], M. G. Iorio[1], and L. Sperandio[2]
[1] ENEA, Research Nuclear Reactor Laboratory, Casaccia Research Centre, Rome, Italy
[2] ENEA, Laboratory for radiological and environmental surveillance, Casaccia Research Centre, Rome, Italy
(*) luigi.lepore@enea.it

***Abstract*—Experience in running TRIGA reactors all around the world has shown that fuel cladding failures can occur. Fission products, especially in gaseous physical form, can exit the defected fuel rods being dispersed within the water primary coolant. Suspects of a cladding failure event can be confirmed by detection of short-lived fission products, i.e., Krypton, Xenon, and even Iodine isotopes in primary water, or within the ionic-exchange resins tank installed for the purification of the primary coolant loop. The magnitude of the release of those 'key-indicators' from the defected fuel element(s) is the driving method by which the defected rod(s) can be identified. 'Significant' releases can be detected with a direct online sampling of the water from the top of the suspected fuel rod with reactor at power, leading this water to an online high-resolution gamma spectrometry analysis system. By considering delays due to the water velocity in tubes and decay time of radionuclides identified within the gamma spectrum, it is possible to calculate the concentration of those radionuclides just emerging from the inspected fuel rod. Releases lower than the minimum detection capabilities of the previous online experimental configuration push to modify the detection method with an indirect identification of the release. This is the case when the 'normal' radioactive background of the activated water, when reactor is on power, is the dominant component in the gamma spectrum of the sampled water, and fission gases (even produced) are not identified promptly, i.e. a relationship to a specific fuel rod by the sampling circuit before could not be identified. The paper describes the experience carried out at the Italian ENEA TRIGA RC-1 reactor, deepening the technical aspects and solutions applied to solve the issue. In particular, a 'significant' release has been found in the instrumented fuel rod within ring B, i.e. the inner fuel ring, exposed to the maximum neutron flux of the reactor. The leaking element was found within a week, in two days of operation, being the sampling system designed on detection of minutes-short-lived fission gases (no need to 'cool' down the primary loop by waiting decay of hours-lived radionuclides). After the removal of the instrumented fuel rod in ring B, further days in searching other 'significant' release with the sampling circuit before have reported nothing detected. But in samples of water taken after reactor shutdown, some iodine emerged after decaying of 'normal' radioactive background of the activated water. This was sufficient evidence of another defected fuel rod in the pool. Identification of the latter defected fuel element is still ongoing, being based on: 1) selective removal of rod(s) from the reactor core, 2) run power of the reactor, 3) take samples of water after shutdown and measure iodine after decaying of 'normal' radioactive-background of the activated water, 4) identify the rod responsible of the remaining leakage when no further iodine is detected.**

***Keywords* — Fuel defects, Fission Gases Leakage, Gamma Spectrometry Characterization.**

I. Introduction

CLADDING defects in fuel elements of the General Atomic TRIGA pool-type reactor is a well-known phenomenon, and it is normally expected during the reactor lifetime.

The cladding failure involves, in relation to the entity of the crack itself, the transfer of radioactivity from inside the defected fuel rods to the moderating and cooling demineralized water, being gases the most likely species able to escape the cladding. This event is emphasized by the heating of the fuel element, since the increase in its temperature produces an increase of the internal pressure of the rod, promoting the release.

Several evidence of the occurrence of this kind of event are found in literature since the 1970s, and many techniques have been developed to solve this issue, i.e. identifying the fuel element(s) responsible of the release of fission products, with the following removal from the reactor core [1-5].

Identification of a cladding defect could be monitored by gamma spectrometry characterization of water samples taken from the primary water loop of the reactor, or by monitoring the ionic-exchange resins installed for primary water purification. Identification of short-lived fission gases (Krypton and Xenon isotopes) and Iodine-131 is a straightforward and sufficient evidence of fuel cladding defects. The identification of the defected fuel rod(s) should rely on direct measurement of short-lived fission products e.g., in samples of water taken just on the top of the suspected fuel element(s), while reactor is on power.

This paper describes all the activities carried out to address the issue in the case of the TRIGA RC-1 reactor of the ENEA Casaccia Research Centre, showing the experimental solutions identified.

II. The ENEA TRIGA RC-1 Reactor

The TRIGA RC-1 at the ENEA Casaccia Research Centre is a thermal neutron spectrum, pool-type, water-cooled research reactor, with MARK III type containment, originally having a thermal power of 100 kW when built in 1960, increased to 1 MW based on an in-house developed ENEA design in 1967.

Fig. 1. A picture of the ENEA TRIGA RC-1 Reactor Hall.

The current reactor core configuration has 111 fuel elements, each one with a different degree of burn-up. The cladding of the fuel elements is made of stainless steel, with a uniform thickness of 0.5 mm. The reactor cooling system is configured with a primary water loop directly exchanging heat with the reactor core by natural convection. The water in the pool is circulated by means of pumps to exchange heat with a physically separated secondary water loop (two shell-and-tube exchangers in cascade).

The first evidence and suspect of a cladding failure event came in 2017, identifying the presence of Cs-137 within the ion exchange resins tank of the primary water purification system. The presence of this radionuclide is consistent with the cladding failure because Cs-137 could come from the short-lived fission gas Xe-137 decay, with the Cs-137 accumulating in the resins tank with water circulation.

III. Materials And Methods

In 2017, during the substitution of the primary water purification ion exchange resins tank, high resolution gamma spectrometry detected Cs-137. It is a long-lived 'contaminant' (half-life in ~30 years), it is not straightforward evidence of an effective and current cladding failure. From 2017 to 2019, several main technical maintenances have limited the normal operation of the reactor. In July 2019 an experimental setup was realized with a HPGe gamma detector to an online monitoring of the water circulating in the reactor primary water purification loop, while the reactor was on power. The aim of the test was to detect short-lived gaseous fission products, to confirm the release from the fuel rods. Krypton and Xenon isotopes were searched for. Among those with the highest fission yield, it is possible to distinguish three macro-groups:

A. Radioisotopes with half-lives > 30 seconds and < 1 minute: Kr-90, Xe-139;

B. Radioisotopes with half-lives > 1 minute and < 10 minutes: Kr-89, Xe-137;

C. Radioisotopes with half-lives > 10 minutes: Kr-87, Kr-88, Xe-138, Xe-135m and Xe-135.

The ability to determine the isotopes classified depends on the observation point chosen, substantially. Contaminants have a travel time from the generation point to the HPGe detector location, and more "far" in space and time the observation position is, the less is the possibility of measuring the radionuclides of groups A and B due to their decay. On the contrary, radionuclides of group C will be more easily measurable since they are longer-lived, but they produce an increasing radioactive "pollution" of the reactor water.

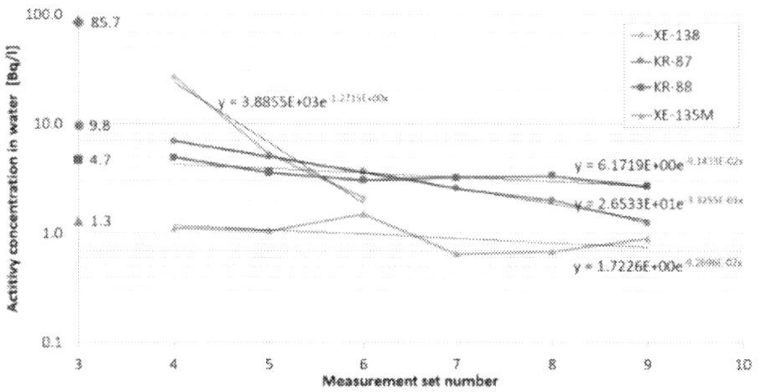

Fig. 2. Evidence of presence of short-lived fission gases dissolved in TRIGA RC-1 Reactor primary water. Activity concentration evolution vs. time of radionuclides in group C (Kr87, Kr-88, Xe-135m, Xe-138) in water flowing in a branch of the primary water purification loop, July 2019.

In the experience conducted in July 2019, the radiological characterization of the water in the primary circuit of the TRIGA RC-1 reactor was carried out during the operation of the reactor at different power levels, and a few hours after the shutdown. The measuring position was close to the flowmeter installed on a branch of the primary water purification loop, in the pump room below the main floor of the reactor building. From the analysis of the spectrum at 1000 kW thermal power, acquired for about 30 minutes, the following noble fission gases are clearly visible: Kr-87, Kr-88, Xe-135m, Xe-138, Kr-89, Xe-137. Concentrations of radionuclides in water have been measured also in water decay, after reactor shutdown. By means of exponential extrapolation, it was possible to calculate the concentration values for the species studied at the instant of reactor shutdown. The values are: Kr-87, 10 Bq/l; Kr-88, 5 Bq/l; Xe-138, 86 Bq/l, supposing a homogeneous dilution of contaminants in the whole primary water. The detection of those radionuclides was unequivocal evidence of cracking in the cladding of one or more fuel elements. With this study, no information regarding which of the fuel rod loaded in the core was responsible for the release.

To identify the element responsible for the leaking, an online monitoring water sampling loop was built.

Among the hypotheses available in the literature [1-5], the one concerning the sampling and direct measurement of the

https://doi.org/10.1051/epjconf/202328804005

primary water in which the fission gases are dissolved is of particular interest, because: i) it is the most direct and simple method from the point of view of the sampling circuit to be implemented, ii) it represents the potentially fastest solution.

The figure shows a conceptual scheme of the sampling circuit.

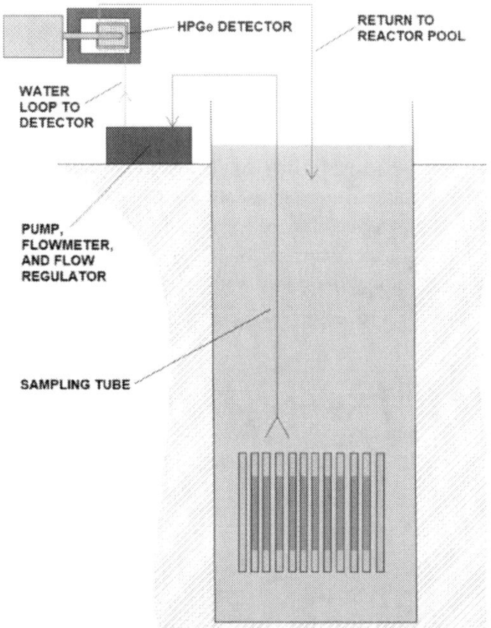

Fig. 3. Conceptual sketch of the water loop circuit for sampling water on the top of each fuel element while reactor on power..

A sampling circuit relying on radionuclides in group A and B would allow to search for the defected fuel rod(s) even while the water contamination is increasing, avoiding any needs for cooling-periods for "radiological clean-up" of the reactor water.

IV. The Measurement Campaign

The online loop system design was made by the end of December 2019. The operation of the circuit was not possible till May 2021 due to the COVID-19 pandemic disruption.

On 2021 May 11th, prior to the beginning of the activities, the Radiation Protection Expert in charge for the TRIGA RC-1 reactor requested a full day operation at 1 MW to assess radiation protection conditions with the reactor at full power [6]. The operation of the reactor was also used to test and tune the online monitoring water sampling loop, as to adjust all operational parameters.

It was decided to analyze spare elements in the reactor core (black circled in Fig. 4), and 9 fuel elements were investigated in about 2 hour of operation. When moving the aluminum sampling tube toward the core center, it was discovered an increase in count rates detected for Kr-89 (221.0 keV) and Xe-137 (455.5 keV), as shown in Fig. 4.

A repetition of the measurement, on a "far" fuel rod already measured, demonstrated that the count rate spikes was effectively belonging to a localized position, and not from the general contamination of the water. Immediately the reactor operation was stopped.

The identification of an increase of very short-lived fission gases near the core center opened the investigation about the possible fuel element responsible for the leakage. It should be noted that at 1 MW thermal power the reactor primary pumps were circulating water, promoting convection phenomena in the reactor pool: because of that the localization of the Kr-89 and Xe-137 count rate spikes could be just representative of a region, and not of a single fuel element. However, the oldest fuel element in the inner fuel ring (element #7208 in position B01) was extracted for a visual inspection, immediately. No significant defects were found, even using an Olympus industrial endoscope for a closer inspection. It was decided to keep the element #7208 outside the reactor core, at the decay storage grids inside the reactor pool.

A second irradiation day was requested by the Radiation Protection Expert. On 2021 May 18th, without introducing any new element to substitute the #7208 in position B01, reactor power was raised at 100 kW. In 4 hour operation, 10 positions were investigated, deepening the search in the core center. At 100 kW thermal power, reactor pumps were taken off. A direct measurement at the empty position B01 showed nothing significant. The sampling aluminum tube was moved around the element #TC3931 (instrumented fuel element for the measurement of fuel temperature) in position B03. Spikes in Kr-89 and Xe-137 count rates emerged once again, and several measurements confirmed the localization of the leakage around the element #TC3931 in position B03. A repetition of the measurement of a "far" fuel rod (position D11) demonstrated that the count rate spikes was effectively belonging to a localized position, and not from the general contamination of the water.

As for the element #7208 in position B01, even #TC3931 in position B03 was extracted and put at the decay storage grids inside the reactor pool, for a visual inspection. A defect on the lateral cladding of the element was found by visual inspection with the Olympus endoscope: the defect was identified by an intermittent bubbling of gaseous species from the stainless-steel surface of the element.

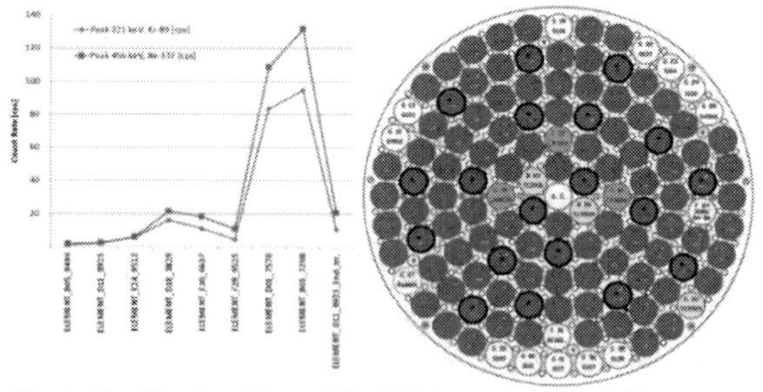

Fig. 4. Identification of the position B03 instrumented fuel element as the principal responsible of fission gases leaking in the reactor primary water loop.

On February 2022, operations with the online loop detection system were resumed. During this timeframe, it has been developed an in-house Python-based routine controlling the HPGe detector for acquisition, spectral analysis, and contextual plotting to ease and speed up all search operations.

Having removed the element #TC3931 and restored the element #7208 in position B01, a complete (5 hours) full power (1 MW) irradiation day was realized on 2022 February 22nd, with the online loop sampling water about at the center of the reactor core, continuously. No fission gases were detected from water samples taken, confirming that the element #TC3931 was a leaking fuel rod, once again.

Unfortunately, in running primary water spectral analysis during the radiological decay of water in the days following, traces of Xe-135 (249.8 keV) and I-131 (364.5 keV) were found, confirming the presence of another leaking fuel element, with a minor leakage emission-rate, probably.

On 2022 March 14th, a new search campaign was conducted, with the reactor power at 50 kW and the online sampling loop. Thirty fuel elements were investigated (about a quarter of the reactor core) in about 4 hours irradiation. No fission gases were detected once again, but in spectral analysis of water during the radiological decay in the days following, traces of Xe-135 (249.8 keV) and I-131 (364.5 keV) were found again.

The remaining leaking element was supposed to have an emission-rate producing water contamination below the minimum detectable activity concentration of the online sampling loop, and an alternative approach to solve the issue was proposed after discussion with the reactor staff.

The delayed detection in water of fission gases and iodine, once enough time has passed for the Ar-41 and Na-24 decay (from 2 to 5 days), makes not possible an online identification of the remaining 'minor' defected element by the sampling loop conceived.

Any element in the core could be responsible of the leakage, ideally, and each element could be tested by unloading it from the core, run reactor at power, and following delayed spectral analysis of the water contamination. Considerations should be done about "the most likely element(s) responsible for the leakage". Several criteria could be followed, based on the experiences in the literature and on the reactor history. From the literature, e.g., it is known that instrumented fuel rods are delicate elements, and defects of those one with respect to normal fuel elements seems to be more likely. The same concept could be applied to control rods with the fuel-follower, since the reactor SCRAM makes those elements falling rapidly into the reactor core, hitting the support structures. From the reactor history, e.g., the history of each fuel element (i.e., burn-up) and its usage could be a driving criterion, or a defect identified in a fuel element could be replicated on other fuel rods from the same batch too.

V. Conclusions

The paper has resumed the approach conceived at the ENEA TRIGA RC-1 reactor to solve the cladding defects identified in fuel elements.

In the available literature, there are several evidence of the occurrence of this kind of event since the 1970s, and many techniques have been developed to solve the issue, i.e. identifying and removing the fuel element(s) responsible of the release of fission products.

The identification of the defected fuel rod(s) relying on direct measurement of short-lived fission products has been acknowledged to be the fastest solution e.g., in samples of water taken just above the suspected fuel element(s) while reactor is on power, with a continuous sampling water loop and an online gamma spectrometry analysis.

At the TRIGA RC-1 reactor, such analysis loop has been realized and operated in a configuration based on the short-lived radionuclides research, Kr-89 and Xe-137 particularly, being the corresponding gamma energy lines free from increasing contamination of the water while reactor is on power. In about ten hours operation of the analysis loop, the major defected fuel element, an instrumented one, was found.

After the unloading of the last one, no other online detection of fission gases happened while reactor on power. Evidence of presence of a remaining, weaker, leakage came from the identification in primary water spectral analysis of traces of Xe-135 (249.8 keV) and I-131 (364.5 keV) during the water radiological decay in the days following the irradiation (from 2 to 5 days, for Ar-41 and Na-24 to decay). The remaining leaking element was supposed to have an emission-rate producing water contamination below the minimum detectable activity concentration of the online sampling loop, and an alternative approach to solve the issue was proposed.

Currently, the reactor 'shutdown state' due to administrative constraints has stopped all activities. Search operations will be resumed as soon it will be possible.

References

[1] G. D. Bouchey and S. J. Gage, "Detection and Location of Leaking Triga Reactor Fuel Elements," Nuclear Technology, vol. 10, no. 2, 2017.
[2] J. S. Bennion, et al., "IDENTIFICATION OF LEAKING TRIGA FUEL ELEMENTS," University of Utah Nuclear Engineering Laboratory, Salt Lake City, Utah, 1987.
[3] A. B. d. Tigliole, M. Cagnazzo, F. Lana, et al. "IDENTIFICATION OF A LEAKING TRIGA FUEL ELEMENT AT THE REACTOR FACILITY OF PAVIA," Laboratorio Energia Nucleare Applicata (L.E.N.A.) of University of Pavia, Pavia, Italy, 2004.
[4] R. R. Rodrigues, et al., "Designing a system to detect leaking in fuel elements in Brazilian Triga research reactor," Int. J. Nuclear Energy Science and Technology, vol. 12, no. 3, 2018.
[5] A. Z. Mesquita and R. R. Rodrigues, "SYSTEM TO DETECT LEAKING IN FUEL ELEMENTS IN TRIGA NUCLEAR RESEARCH REACTOR," in Proceedings of the 1st Iberic Conference on Theoretical and Experimental Mechanics and Materials, 11th National Congress on Experimental Mechanics, Porto/Portugal 4-7 November 2018, 2018.
[6] L. Sperandio, L. Lepore, et al. "Identification of a defected fuel rod in the TRIGA R.C.-1 facility of the ENEA Casaccia Research Center: radiation protection aspects" Accepted paper to The European Physical Journal Plus.

Design and fabrication of an axial neutron flux profile measurement assembly for the Advanced Test Reactor Critical Facility

M. A. Reichenberger[1,*], M. R. Holtz[1], D. Nichols[2], B. Harris[1], R. D. Ball[1], H. Rollins[1], and C. Downey[1]

[1]Idaho National Laboratory, United States of America

[2]Kansas State University, United States of America

(*) Michael.Reichenberger@INL.gov

***Abstract*— Real-time characterization of irradiation facilities improves the utilization of the core capabilities of test nuclear reactors. The ability to observe how the local neutron flux (level and spectrum) changes as control elements and experiments change will fundamentally transform our understanding of the underlying physical phenomena that govern the operation of present and advanced nuclear reactors, ultimately providing valuable information for the nuclear energy industry. The objective of this research was to demonstrate how advanced sensors could be used to significantly reduce the time and cost of experiments, improve our understanding of experimental environments, and enable verification and validation of simulation and modeling methods. This was accomplished by designing and fabricating a dedicated real-time instrument test train for the Advanced Test Reactor Critical (ATR-C) facility. The first year of this project focused on the design and modeling of real-time axial neutron flux monitors, leveraging proven technologies pioneered at the Idaho National Laboratory, to characterize the transient that occurs in the Small-B positions at the Advanced Test Reactor and the Advanced Test Reactor Critical Facility. We found that the flux amplitude in those positions can fluctuate as much as 380% depending on the outer shim control cylinder position. The engineering design of the test fixture and flux monitor instrumentation was the objective of the 2nd project year. New capabilities were established to electrodeposit enriched uranium for fission chamber development at the Idaho National Laboratory and trials were begun to characterize the process. The final year included the fabrication of the test fixture and instruments for Advanced Test Reactor Critical Facility. The fabrication process was delayed by supply chain and personal availability caused by the COVID-19 pandemic. However, we were still able to deliver this unique capability to Advanced Test Reactor Critical Facility that will enable future instrument testing and scientific experiments.**

***Keywords* — ATR-C, Fission Chamber, MPFD, Real-Time Neutron Flux Measurement.**

I. INTRODUCTION

The Advanced Test Reactor (ATR) located at the Idaho National Laboratory (INL) in the USA is a prominent thermal test reactor designed to facilitate a wide range of nuclear experiments. With a typical operating full power of approximately 100 MWth, the ATR operates using Outer Shim Control Cylinders (OSCCs) to vary lobe powers, illustrated in Fig. 1. This versatility accommodates an array of high fluence tests that can simulate end-of-life conditions for reactor component materials (including fuel and structural materials) after only a few years of sustained irradiation.

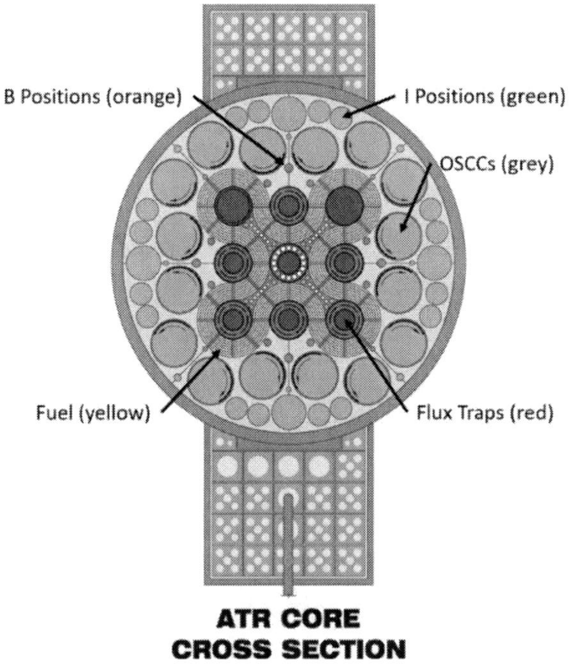

Fig. 1. The ATR core power can be tailored by rotation of the OSCCs (grey) to enhance or suppress neutron flux in different regions of the reactor [1].

The ATR is renowned for its numerous test positions, each offering unique capabilities, including in-pile tubes (red), A- (white), B- (orange), and I- (green) positions illustrated in

Fig. 1. These heavily utilized positions provide essential insights into the behaviors of materials in nuclear environments. The neutron environment within the ATR is highly heterogenous, illustrated in Fig. 2. Furthermore, the neutron flux distribution is not static throughout the nominal 60-day operating cycle, posing a challenge to reactor operators who aim to maintain stable power levels during experiments and experimenters as they interpret irradiation results.

One critical aspect of ATR experiments is burnup, the depletion of fuel over time. The reactor also accommodates a host of experiments simultaneously, spanning diverse areas of nuclear research. To manage and optimize these experiments, precise control mechanisms are essential. Currently, the ATR relies primarily on integral dosimetry measurements to determine neutron fluence rates during irradiations, a practice that has some limitations. Dosimetry is employed throughout the reactor, although it is often not specific to the experiment at hand. In the interpretation of experiment results, most teams adopt this conventional approach. However, real-time neutron flux monitors can also be used. Calibration of conventional real-time monitors is typically conducted within a thermalized neutron field, compared to the integral fluence in the field that is determined using dosimetry techniques. This approach yields results "per nv" a unit typically valid for similarly thermalized fields. This methodology, while widely used due to its applicability in various scenarios, presents a challenge. The response of most sensors is intricately linked to the neutron energy spectrum, which can significantly vary (especially in ATR).

The ATR's unique characteristic lies in its heterogenous nature, particularly in positions of utmost interest (in-pile tubes). These positions, characterized by the highest neutron flux, tend to also be the least thermalized. This inherent property complicates the straightforward application of conventional dosimetry methods. As a result, experiment teams working within the ATR must grapple with the intricacies of interpreting results and accounting for the complex neutron energy spectral shifts both in space and time.

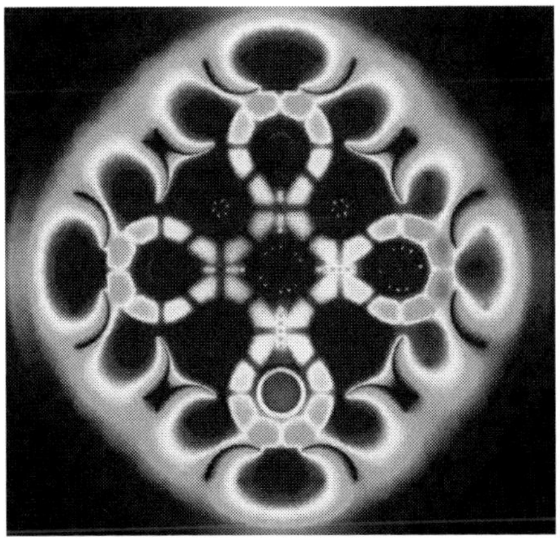

Fig. 2. The unique design of ATR creates an extremely heterogenous neutron flux distribution throughout the core, illustrated in this reproduced heat-map [2].

II. Irradiation and Test Plan

The present research proposed an intriguing experiment conducted within the Advanced Test Reactor's Critical (ATR-C) facility. An irradiation test train was designed and constructed to utilize a Small-B position in ATR-C, strategically situated near the fuel and the OSCCs. The potential impact of OSCC rotation on the thermal neutron field within the Small-B position was investigated through modeling and simulation. These models characterized the influence such rotations wield on experiments which manifest dramatic shifts in both amplitude and spectral shape of the neutron flux. Specialized real-time neutron flux monitors were constructed based on prior research to be installed alongside conventional neutron flux monitors and passive dosimetry. A test plan was developed but operational and fabrication delays spurred by the COVID-19 pandemic prohibited the execution of the irradiation.

A. Irradiation Plan

The critical configurations of the ATR-C core were modeled to determine the expected neutron flux levels for the irradiation experiment. An ATR-C model was developed in MCNP [3], that could facilitate different OSCC. Critical core configurations were determined which had OSCCs adjacent to the Small-B position of interest at the nominal core critical position of 58.55° as well as the two most extreme positions which are allowed under the operating envelop of 5.73° and 150.1°, illustrated in Fig. 3. These represent positions where the OSCCs are turned completely inward and outward such that the hafnium absorbers have the greatest and least impact respectively on the irradiation position.

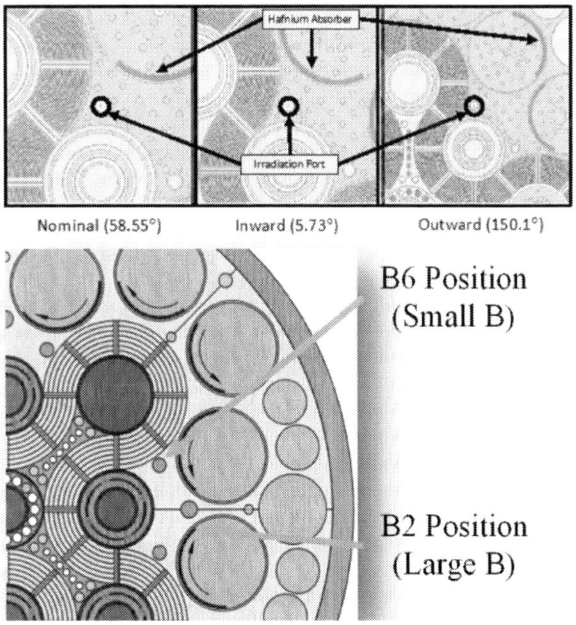

Fig. 3. The Small-B positions in ATR-C experience similar transients from OSCC rotation (illustrated above) as the Large-B positions (illustrated below) [4].

https://doi.org/10.1051/epjconf/202328804006

Neutron energy spectra were captured for each of 70 axial zones at these 3 positions. The normalized neutron energy spectra, comparing the neutron flux at the centerline axial position for each OSCC position, are shown in Fig. 4. Simulations revealed that even a slight shift from an inward position to the nominal setting brings about a discernible change in the thermal neutron environment. However, the most remarkable revelation stems from the transition from the nominal to an outward OSCC position, where a significant and pronounced transformation in the thermal neutron field is observed, shown in Fig. 4. These findings underscore the intricate interplay between reactor geometry and neutron behavior, shedding light on the dynamic and complex nature of nuclear systems. The variation in neutron flux can be explained by the greater local absorption of thermal neutrons when the outer shim control cylinder is turned inward (the 5.73° position) coupled with the greater neutron multiplication from the beryllium reflector when the outer shim control cylinder is turned out (the 150.1° position). These results confirm one hypothesis that was the basis of this research, that the neutron energy spectrum amplitude and shape are substantially perturbed by the position of the local outer shim control cylinders, demonstrating that there are penalties to experiments in 'B' positions that necessarily limit irradiation capabilities.

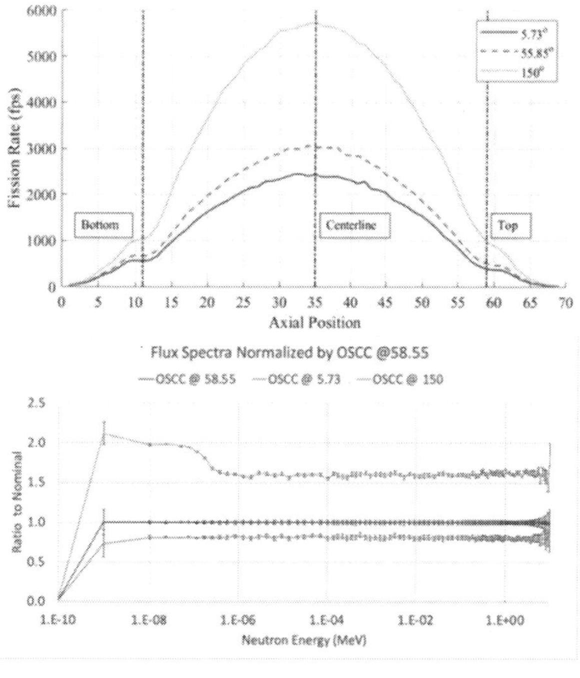

Fig. 4. Simulations of the neutron energy amplitude (above) and energy profile (below) demonstrate the dramatic shift that can occur nearby the OSCCs as they rotate from inward to outward positions [4].

The objective of the present research was to observe these spectral shifts in real-time by the installation of numerous neutron flux monitors in the Small-B position in ATR-C.

B. Sensor Selection, Design, and Fabrication

A tiered approach was used in the selection of neutron flux measurement sensors for the experiment. First, passive dosimetry was desired. This conventional approach is sensitive to the integral neutron fluence in a single position during the experiment. Therefore, an established real-time sensor was also selected to provide a scaling factor to translate the integral signal to the time-domain. Two Self-Powered Neutron Detectors were selected with hafnium emitters to provide sufficient sensitivity for the low-power ATR-C experiment. However, even SPNDs only provide a single point measurement. A multi-nodal instrument was also desired for the present research so that the axial distribution of neutrons in the experiment could be observed. Micro-Pocket Fission Detectors (MPFDs) have been developed previously which offered advantages for this experiment [5] [6] [7] [8]. Typical parameters of MPFDs were used to construct a 3-D model to assess the electric field within the sensors in an array of 4 detectors, shown in Fig. 5 (Top and Middle). Then, MPFDs were simulated in a characteristic neutron flux from the reactor-scale models. The response of sensors with various loadings of fissile ^{235}U were determined, shown in Fig. 5 (Bottom). The required fissile layer mass was determined to be between 0.5 to 1.0 µg of ^{235}U [4].

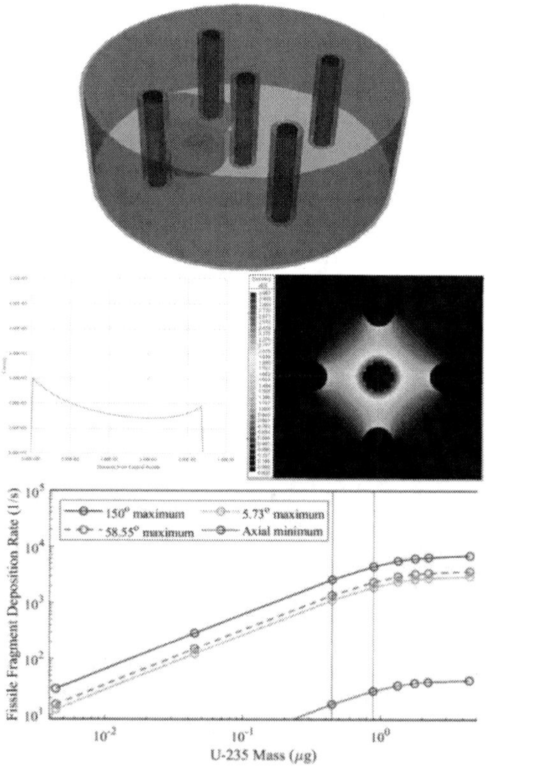

Fig. 5. A 3-D model of an MPFD (top) was developed to model the electric field characteristics (middle) within the sensor. These models were used to determine the response rate of MPFDs in the desired irradiation location (bottom) [4].

The high enrichment required for the present effort prohibited sub-contracting outside the national laboratory complex for the electroception of MPFD components. A method of electrodeposition of enriched uranium onto the small (~250 micro-meter) diameter electrodes was developed for the present research. The chemical preparation of the electrolytic solution was optimized, starting from the method developed previously [9]. Improvements to the consistency of the chemical preparation of the solution and the electrodeposition process dramatically improve reproducibility and reduced material waste compared to prior research [10]. A high-concentration enriched uranium solution was prepared from the enriched uranium dioxide source material. The chemistry and electrodeposition were performed at the Radioanalytical Chemistry Laboratory (RaCL) at ATR. The RaCL has appropriate work controls and support to facilitate safe work with both the powder uranium dioxide and liquid electrolytic solutions containing highly enriched uranium. The source solution was then diluted to prepare an appropriate amount of working solution that was pH balanced using precision pipetting for a highly reproducible system, shown in Fig. 6 (Top). The electrolytic solution was fixed under a microscope to allow the chemists to make electrical contact with the working electrode. Two electrodeposition methods were performed to deposit fissile material on the platinum substrate. Although both cyclic voltammetry and bulk electrolysis methods yielded electrodeposited uranium, the bulk electrolysis appeared to deposit more material. After the electrodeposition, the samples were cleaned, and the alpha-particle activity of the samples was measured using an alpha-particle spectrometry system at RaCL. A sample spectrum is shown in Fig. 6 (Bottom).

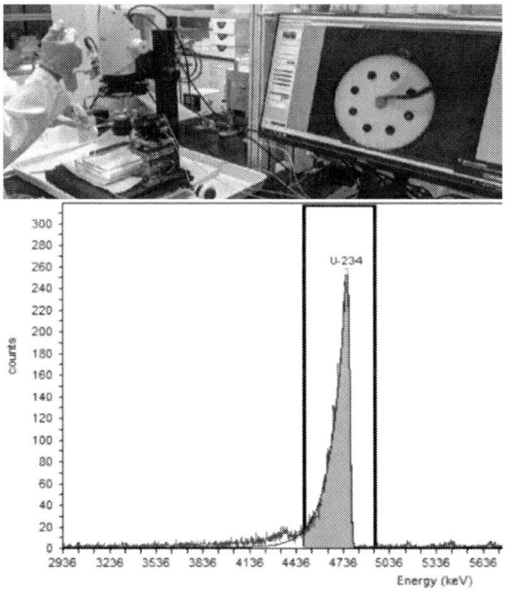

Fig. 6. An electrodeposition system (top) was developed that improved on prior research [9] [10]. The amount of U-235 was determined by measuring the activity of U-234 on the samples by alpha-particle spectroscopy.

Schematics were developed by which MPFD arrays could be constructed with proper spacing to ensure placement within the ATR-C core to align with the desired measurement locations, shown in Fig. 7 (Top). Then two MPFD arrays were fabricated using parts that were produced by laser waterjet cutting. The gas chamber, shown in the top-left of the bottom of Fig. 7, provides space for ionization from fission fragments in the MPFD. Silica spacers were used to ensure proper spacing between MPFD nodes, and the 2-ft. long MPFD array was attached to a mineral insulated extension cable by micro-tig welding inside an Ultra-High Purity Argon glove box at approximately 1 atm and a laser weld was used to seal any imperfections.

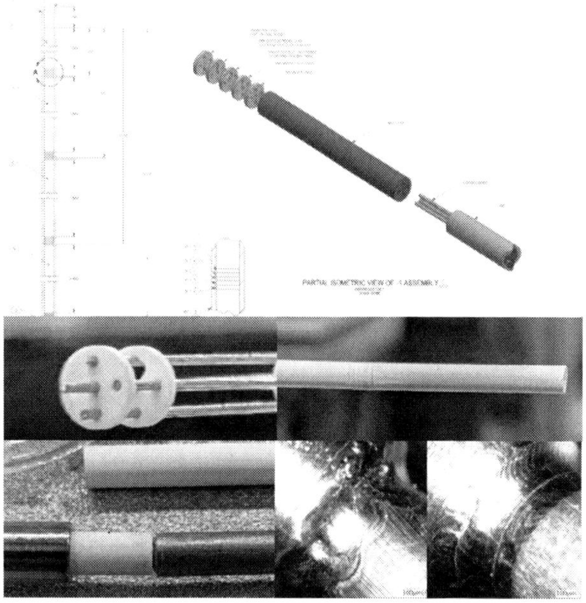

Fig. 7. MPFD arrays were constructed with 4 measurement locations separated by silica insulation and encapsulated in an Inconel sheath.

C. Test Train Design

The experiment test-train was designed to situate all the measurement devices in the proper location within the irradiation position in ATR-C. Illustrated in Fig. 8, each of the 5 sensors (2 SPNDs, 2 MPFD arrays, and 1 central dosimetry holder) were suspended in the test train with symmetry around the center of the position.

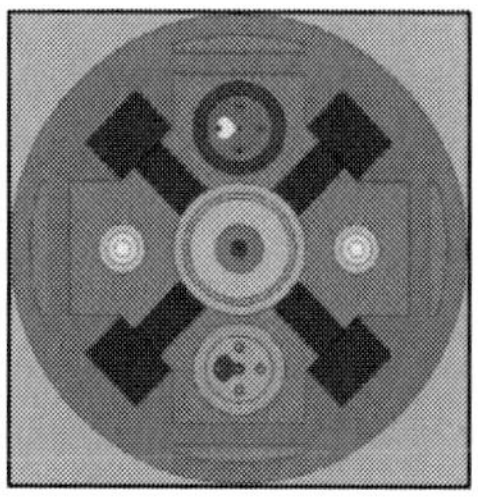

Fig. 8. The test train included support structures for two MPFD arrays (top and bottom), two SPNDs (left and right), and a central dosimetry package.

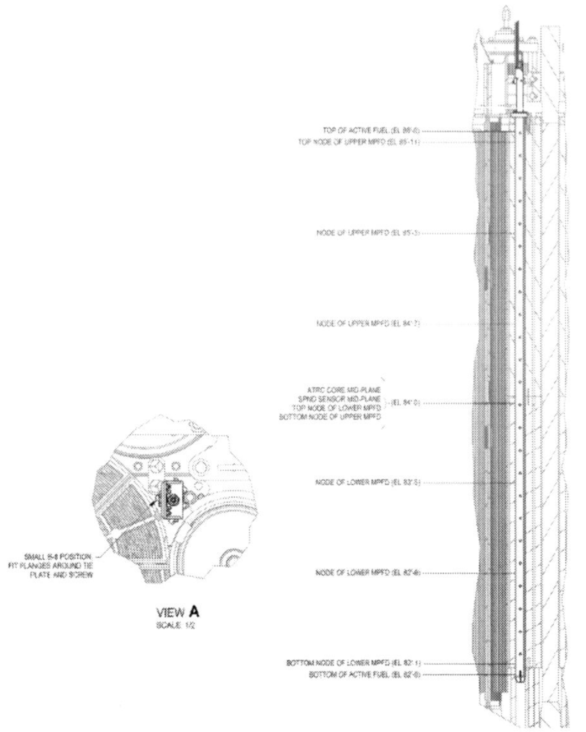

Fig. 9. The test train was designed to be inserted into the B-6 irradiation position at ATR-C without interfering with other structures while maintaining axial spacing of the neutron flux monitors.

The two MPFDs were each composed of 4 nodes. The bottom of one array overlapped with the top of the second array such that 7 independent axial positions were able to be monitored. The overlapping position, located at the core center plane, corresponded with the signals from the SPNDs and the dosimetry package where were also located on the core center plane, shown in Fig. 9. The outer sleeve of the test-train and the central support tube were constructed of 6061-T6 aluminum. However, Polyether ether ketone (PEEK) holders were fabricated to support each instrument to isolate them electrically from the reactor components. A special cruciform was designed to center the test train in the irradiation position, while a custom bale assembly was necessary to allow for the experiment to be lowered into the position and rotationally fixed. The structural integrity of the bale was assessed [11]. These efforts culminated in the publication of eight internally released engineering drawings at INL (DWG-607812 through DWG-607819). These drawings document the technical details of each component and were not only necessary to ensure the proper fabrication and assembly of components, but also as a tool to improve access to similar experiments in the future. The quality level determination for this irradiation was determined such that similar experiments could utilize many of the same features to reduce their preparation costs. Similarly, the drawings of the MPFD arrays can be further leveraged by other irradiation programs for production of these specialized instruments.

III. Conclusions

Although the irradiation associated with this project is still pending, several significant research outcomes were achieved. Future irradiations will be able to utilize the new test fixture for ATR-C to observe the difference in neutron flux in the Small-B positions at ATR-C due to different outer shim control cylinder configurations will be assessed. These results will improve irradiation capabilities by reducing uncertainties associated with experiments in these positions, improving INL's capability to meet irradiation missions. Furthermore, the new test fixture will enable experiments at ATR-C that were previously not possible, enhancing the utilization of this unique facility and producing new scientific results. The instrument design research and results developed new researcher skills by supporting two Ph.D. students and an early career scientist. This improved our capabilities to continue to meet the needs for in-core instrumentation in the future. The added capacity to perform thin-layer uranium compound electrodeposition at INL improved sensor fabrication capabilities enabling future ATR experiments to provide progressively higher fidelity measurements that will accelerate fuel cycle and advanced materials research. This research was well-aligned with the Nuclear Energy Enabling Technologies Advanced Sensors and Instrumentation (NEET ASI) real time in-core instrumentation work package that implements research and development activities to develop advanced sensors that address critical technology gaps for monitoring and controlling existing and advanced reactors and supporting fuel cycle development. The new capability to perform enriched uranium electrodeposition at ATR will further benefit fission chamber development efforts at INL. Finally, the unique test fixture at ATR-C can also be utilized for simulation and modeling validation in transient reactor conditions.

IV. Acknowledgment

This work, presented at the Advancements in Nuclear Instrumentation Measurement Methods and their Applications (ANIMMA) 2023 conference, June 12-16, 2023, was supported by the United States Department of Energy Contract Number: DE-AC07-05ID14517. This information was prepared as an account of work sponsored by an agency of the U.S. Government. Neither the U.S. Government nor any agency thereof, nor any of their employees, makes any warranty, expressed or implied, or assumes any legal liability or responsibility for the accuracy, completeness, or usefulness, of any information, apparatus, product, or process disclosed, or represents that its use would not infringe privately owned rights. References herein to any specific commercial product, process, or service by trade name, trademark, manufacturer, or otherwise, does not necessarily constitute or imply its endorsement, recommendation, or favoring by the U.S. Government or any agency thereof. The views and opinions of authors expressed herein do not necessarily state or reflect those of the U.S. Government or any agency thereof.

V. References

[1] Idaho National Laboratory, "ATR Core Cross Section Diagrams, DWG-606000," 2013.

[2] Leppänen, J., et al. , "Serpent a Continuous-energy Monte Carlo neutron and photon tranposrt code," 24 3 2023. [Online]. Available: https://serpent.vtt.fi/serpent/gallery/atr.htm. [Accessed 2023 8 8].

[3] J. A. Kulesza, et al, "MCNP® Code Version 6.3.0 Theory & User Manual," Los Alamos National Laboratory Tech. Rep. LA-UR-22-30006, Rev. 1, Los Alamos, NM, USA, 2022.

[4] D. M. Nichols, M. A. Reichenberger, A. D. Maile, M. R. Holtz and D. S. McGregor, "Simulated Performance of the Micro-Pocket," *Nuclear Science and Engineering,* no. https://doi.org/10.1080/00295639.2021.1898922, pp. 1098-1106, 2021.

[5] M. A. Reichenberger, "Micro-Pocket Fission Detectors: Development of Advanced, Real-Time, In-Core, Neutron -Flux Sensors," Kansas State University, Manhattan, KS, 2017.

[6] M. A. Reichenberger, D. M. Nichols, S. R. Stevenson, T. M. Swope, C. W. Hilger, T. C. U. D. S. McGregor and J. A. Roberts, "Fabricaton and testing of a 4-Node Micro-Pocket Fission Detecotr Array for the Kansas State University TRIGA Mk. II research Nuclear Reactor," *Nucl. Inst. and Meth. A,* pp. 8-17, 2017.

[7] M. A. Reichenberger, D. M. Nichols, S. R. Stevenson, T. M. Swope, C. W. Higer, R. G. Fronk, J. A. Geuther and D. S. McGregor, "Fabrication and Testing of a 5-Node micro-Pocket Fission Detector Array for Real-Time Spatial, Iron-Wire Port Neutron-Flux Monitoring," *Annal of Nuclear Energy,* pp. 995-1001, 207.

[8] D. Nichols, M. Reichenberger, S. Stevenson, C. Hilger, T. Swope, K. Kellogg, J. Hewitt, J. Roberts and D. McGregor, "TRIGA Pulse Tracking Utilizing a Multi-Node Micro-Pocket Fission Detector," in *Transactions of American Nuclear Society, Vol. 119*, Orlando, FL, 2018.

[9] M. A. Reichenberger, D. M. Nichols, K. Tsai, S. R. Stevenson, T. M. Swope, C. W. Hilger, J. D. Hewitt, K. E. Kellogg, J. A. Roberts, C. Chen and D. S. McGregor, "Electrodeposition of Low-Enriched Uranium onto Small Platinum Electrodes," *Nuclear Instruments and Methods in Physics Research A,* vol. 941, 2019.

[10] M. A. Reichenberger, T. Ito, P. B. Ugorowski, B. W. Montag, S. R. Stevenson, D. M. Nichols and D. S. McGregor, "Electrodeposition of Uranium and Thorium onto Small Platinum Electrodes," *Nuclear Instruments and Methods in Physics Research A,* vol. 812, pp. 12-16, 2016.

[11] M. A. Reichenberger, D. M. nichols, M. R. Holtz and W. F. Bauer, "Engineering Calculation and Analysis Report (ECAR) 5410 "Evaluation of Bail Loop for Real-Time Axial Neutron Flux Profile Measurement at The Advance Test Reactor Critical Facility"," Idaho National Laboratory, Idaho Falls, ID, 2021.

Design of an Instrumented Fuel Rod for Intra-pin Neutron Flux Measurements in the CROCUS Experimental Reactor

Tom Mager[1,*], Jean-Baptiste Valentin[1], Vincent Lamirand[1,2], Thomas Ligonnet[1], Won Dong Shin[3], Jean-Marie Fürbringer[4], Mathieu Hursin[2], Andreas Pautz[1,2]

[1]Laboratory for Reactor Physics and Systems Behaviour, Ecole Polytechnique Fédérale de Lausanne, Switzerland
[2]Nuclear Energy and Safety Research Division, Paul Scherrer Institute, Switzerland
[3]Laboratory of Intelligent Systems, Ecole Polytechnique Fédérale de Lausanne, Switzerland
[4]Section of Physics (SB SPH-GE), Ecole polytechnique Fédérale de Lausanne, Switzerland
(*) tom.mager@epfl.ch

***Abstract*—In response to the need for validating high-fidelity deterministic neutronics solvers capable of pin-resolved neutron flux distributions, an instrumented fuel rod was designed for the experimental reactor CROCUS, operated by EPFL's Laboratory for Reactor Physics and Systems Behaviour. This rod aims at obtaining intra-pin data using activation dosimetry techniques. Placed in the outer region of the reactor core, the rod utilizes spaces between four metal uranium cigars (25 cm each) to house various disk dosimeters, targeting different neutron energy ranges and allowing retrieval of radial and azimuthal intra-pin neutron reaction rates. A design of experiment (DOE) study, aided by Serpent2 Monte Carlo calculations, facilitated the selection of dosimeter material and rod design, ensuring detection of within-pin neutron flux variations while adhering to mechanical and regulatory constraints. Azimuthal divisions of activation dosimeters were irradiated at the core center and subsequently subjected to activity determination using an HPGe gamma spectrometer. The objectives were met, detecting expected azimuthal variations within dosimeters and confirming uncertainties on reconstructed activity to be smaller than the amplitude of the observed variations. A permit from the Swiss nuclear regulatory authority ENSI-IFSN was obtained, allowing installation and usage of the instrumented fuel rod in CROCUS, known as the NECTAR experiments.**

Keywords—**Instrumented fuel rod, Intra-pin measurements, High-fidelity, CROCUS**

I. Introduction

In recent years, significant efforts have been devoted to advancing nuclear reactor modeling and simulation capabilities, encompassing aspects such as neutron transport, reactor thermal hydraulics, fuel performance, and material behaviors. These advancements have played a decisive role in addressing operational and safety challenges in light-water reactors, including pellet-clad interaction (PCI), departure from nucleate boiling (DNB), and corrosion product deposition (CRUD). To support these endeavors, nuclear programs worldwide have initiated the development of three-dimensional (3D) full-core time-dependent methods with pin-resolved details [1]. These high-fidelity deterministic neutronics codes can predict the steady-state neutron flux distribution directly at the pin or sub-pin scale [2]–[4], offering detailed insights into reactor behavior and performance.

However, successful utilization of such advanced modeling techniques heavily relies on the availability of high-resolution experimental data measured in a reactor. Particularly, accurate measurements of intra-pin reaction rates have been identified as crucial validation requirements for high-fidelity neutronics codes that aim to resolve local phenomena. Unfortunately, the availability of such measurements remains limited. The lack of access to locations of interest, such as within the fuel lattice and inside fuel pins, is a significant limitation for localized measurements in operating NPPs, and when such measurements exist their access is restricted. In this context, zero-power research reactors play a decisive role in generating high-quality in-core validation data. The availability and flexibility of operation in research reactors allow for investigating neutronics effects at a local scale with inter-pin and intra-pin measurements.

Some efforts have been made over the years to produce such data, including the indirect measurements of intra-pin reaction rates performed along the radial and azimuthal directions of a BWR assembly at the PROTEUS reactor of the Paul Scherrer Institute [5]. Regarding PWRs, the only available experimental intra-pin data were conducted on the IPEN/MB-01 reactor facility, where radial distributions of reaction rates were obtained from gamma ray spectroscopy and collimation techniques on irradiated fuel disk samples [6]. More recently, highly-localized inter-pin data was acquired in the CROCUS experimental reactor thanks to miniature Li-6 scintillators [7].

Despite these advancements, the validation of high-fidelity modeling codes for pin-resolved neutron flux distributions remains scarce, and to address this validation gap, the Laboratory for Reactor Physics and Systems Behavior at EPFL took the initiative to design an instrumented fuel rod for the CROCUS experimental reactor. With its double-lattice core, CROCUS poses modeling challenges, making it a demanding test for high-fidelity codes that aim to model new generations of complex cores. The instrumented fuel rod, strategically positioned in the outer region of the reactor core, is expected to capture significant azimuthal and radial intra-pin neutron flux variations thanks to direct or indirect detection technics.

This paper presents the outcomes of an extensive design study, incorporating design of experiment (DOE) methodologies that utilize Serpent2 Monte Carlo calculations to optimize the detector system and instrumented fuel rod design. Additionally, we discuss the successful irradiation and analysis of some activation dosimeter samples with non-regular shapes at the core center of CROCUS. Finally, we present the final design of the instrumented fuel rod, for the so-called NECTAR experiments in the CROCUS zero-power reactor.

II. Description of the Project

A. The CROCUS Experimental Reactor

CROCUS is a zero-power research reactor located on the EPFL campus, with a maximum allowed power of 100 W. It is an ideal tool for research and education due to the acceptably low dose rates within the shielding after shutdown. The core is located in an aluminum vessel measuring 1.3 m in diameter and 1.2 cm in thickness. Demineralized light water is used as both the moderator and reflector. The active core region is cylindrical, 1 m high and approximately 60 cm in diameter, consisting of two fuel zones with square lattices of different pitches: an inner uranium oxide region containing 336 rods enriched to 1.806% with a pitch of 1.837 cm, and an outer uranium metal region with uo to 180 rods enriched to 0.947% with a pitch of 2.917 cm. The fuel loading pattern is illustrated in Fig.1.

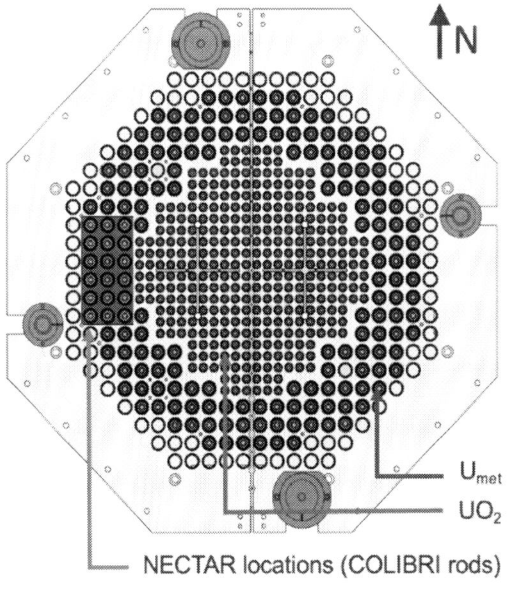

Fig. 1. CROCUS top-view of the fuel loading pattern with the two fuel lattices and the 18 locations available for the NECTAR fuel rod.

While UO_2 rods consist in a 1-m stack of cylindrical pellets with aluminum cladding, U_{met} rods are formed of four cigars, each measuring 25 cm, also encased with aluminum. The rods are held in position by two octagonal aluminum grid plates spaced 1 meter apart. Each plate contains a 1-mm-thick cadmium layer to limit axial thermal flux and neutron doses in the reactor cavity, with the active zone of the fuel located between the two cadmium layers. Reactivity is controlled by the water level, which is controlled via a spillway, and B_4C control rods.

B. Instrumented Fuel Rod

The instrumented fuel rod will be composed of four U_{met} cigars obtained from the LRS reserve, thereby positioning it in the outer core region of the reactor. To achieve intra-pin neutron flux distributions, detectors or dosimeters will be strategically placed in the space between the two central cigars. To facilitate manipulation and prevent contamination of individual cigars and dosimeters, each cigar will be individually encased. In consideration of structural integrity and water tightness, an outer cladding may be added to encompass the four cigars and the detector holder. If this arrangement is adopted, the rod will need to be positioned in the COLIBRI [8] fuel rod oscillator region (highlighted in green in Fig. 1) to take advantage of the enlarged holes in the aluminum grid plates. A representation of the potential design for the instrumented fuel rod is illustrated in Fig. 2.

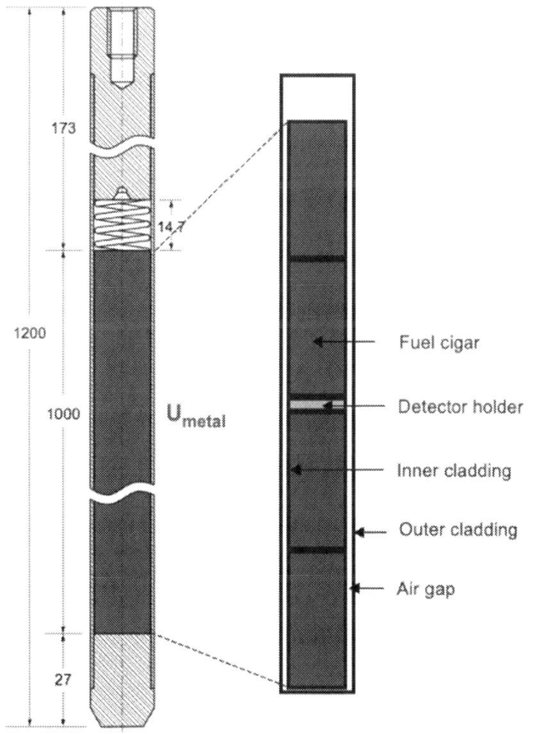

Fig. 2. Schematics of a CROCUS uranium metal fuel rod, along with a sketch of the instrumented fuel rod basic principle.

III. Design of Experiments

In light water reactors, fission "fast" neutrons are emitted with an average energy of 2 MeV. To induce fissions again, these neutrons need to be thermalized down to approximately 0.025 eV. Accordingly, we classify neutrons into two groups based on their energies: the "thermal" neutrons with energies around 0.025 eV and the "fast" neutrons with energies above the cutoff of approximately 0.625 eV. Consequently, the "thermal" neutrons are more prevalent in the moderator surrounding fuel rods, while the "fast" neutrons are predominantly present within the fuel. This spatial distribution results in a depression of thermal neutron flux within the fuel rod, while the fast neutron flux peaks in the fuel. These variations from the moderator to the center of the fuel rods are of particular interest for code validation purposes. Therefore, an instrumented fuel rod design that maximizes the detection of these radial variations within the fuel is desired.

To achieve this objective, we will employ Design Of Experiments (DOE) methodologies [9] to establish the relationship between the fuel rod design and the amplitude of the measured intra-pin neutron flux variations. The strategy employed is presented hereafter and will be elaborated upon in individual subsections.

Objective: The primary objective of this study is to design an instrumented fuel rod for the CROCUS reactor, specifically aimed at conducting precise intra-pin neutron flux measurements for the validation of high-fidelity deterministic codes. Therefore, the focus is on identifying which instrumented rod design maximizes the detection of the radial neutron flux depression within the fuel.

Factors: Four factors influencing the flux depression have been identified:

- the cladding material;
- the axial thickness of the inner cladding which is directly in contact with the detector holder;
- the presence or absence of an outer cladding, depending on the detector system selected;
- the detection system among gold dosimeters sensible to thermal neutrons, nickel dosimeters sensible to fast ones, or a miniature fiber-coupled scintillators developed at EPFL [10]. Dosimeter materials have been selected based on the experience gained during the PETALE program [11]–[13], the main criteria being reaction half-life, incident neutron energy and material capability for easy cutting and handling.

Response: Thermal and fast neutron fluxes: ^{6}Li(n,α), ^{197}Au(n,γ), ^{58}Ni(n,p) reaction rates computed with Serpent2.

Model: A constant coefficient model without interactions will first be considered to describe the causal relations from the factors to the response.

Strategy: A full factorial design could be considered, but Graeco-latin squares [14] will be used to reduce the number of Serpent2 simulations to be performed.

A. Procedure

The stochastic Monte Carlo neutron transport code Serpent2 [15] is used to model the instrumented control rod. The model consists of a 3 by 3 arrangement of outer core fuel rods with the instrumented rod positioned at the center (Fig. 3a). Reflective boundary conditions are applied in all three dimensions. Tracklength detectors are employed to tally reaction rates along a radial traverse from the moderator to the fuel center (Fig. 3c).

For the Serpent calculations, the neutron population was set at 10^6 neutrons per cycle, with 200 inactive followed by 10000 active cycles. Fig. 3c illustrates an example of some neutron flux profiles obtained by simulating the ^{6}Li(n,α) reaction detector response. A comparison is made among various axial cladding thicknesses and a reference fuel rod profile. Notably, an increase in the axial cladding thickness corresponds to a reduction in the neutron flux depression within the instrumented rod.

To evaluate the success of the simulations, the relative amplitude of the flux depression, denoted as "r", is used as a parameter. "r" is defined as the ratio between the flux values at the fuel center and at the cladding internal radius. This parameter serves as a metric in assessing the accuracy of the model in capturing neutron flux variations within the fuel assembly.

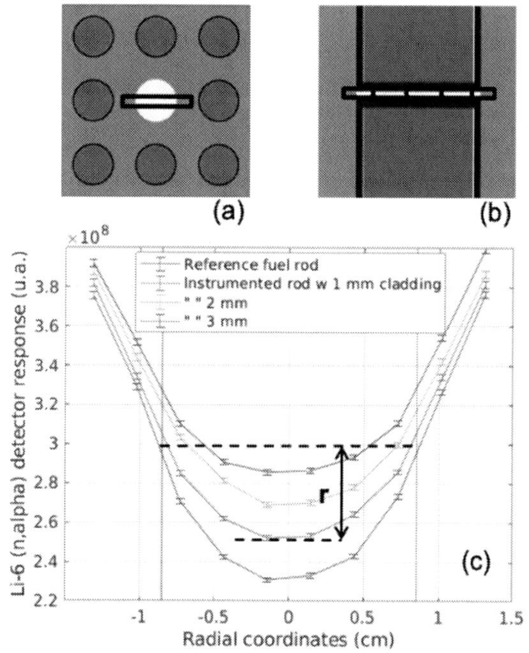

Fig. 3. (a) Axial section of the Serpent model. (b) Radial section of the instrument fuel rod with schematics of the tracklength detectors. (c) Example of Li-6 detector response distribution along the instrumented fuel rod radial profile.

B. Matrix of Simulations

as stated previously, four factors were considered to have a possible influence on the design. The three possible levels for each factor are presented in Tab. I.

A full factorial design would necessitate 81 Serpent2 simulations, requiring extensive computation time to obtain sufficient statistics on the detector response. To address this computational burden, a Graeco-Latin square design was adopted

TABLE I
SELECTED FACTORS INFLUENCING THE DESIGN OF THE INSTRUMENTED FUEL ROD.

Factors	Levels		
Axial clad thickness (mm)	1	2	3
Outer radial clad	no	yes w/o air	yes w/ air
Material	Al	Zr	SS
Detector	Li-6	Au-197	Ni-58

(Tab. II), effectively reducing the simulation matrix from 81 to 9 entries.

In the Graeco-Latin square design, the axial cladding factor is represented over the rows, while the presence of outer cladding is represented over the columns. Moreover, the arrangement of materials and detectors is carefully ordered to ensure that each element of material and each element of detectors appear exactly once in every row and column, and no two cells contain the same ordered pair. This design strategy optimizes the efficiency of the simulations while maintaining a representative dataset for analysis and validation.

TABLE II
GRAECO-LATIN SQUARE DESIGN.

	no	yes w/o air	yes w/ air
1 mm	Al & Li-6	Zr & Au-197	SS & Ni-58
2 mm	Zr & Ni-58	SS & Li-6	Al & Au-197
3 mm	SS & Au-197	Al & Ni-58	Zr & Li-6

Instead of tallying just one detector response per case, we made modifications to the Serpent 2 inputs, enabling us to obtain three virtual detector responses for each configuration. Detectors were therefore not explicitly modeled. As a result, we were able to generate an extended dataset consisting of 27 simulations. To organize these simulations efficiently, they can be grouped into three Graeco-Latin squares, mirroring the format of the initial set of 9 simulations (Tab. III). The first three rows represent the first set of 9 simulations, while the 6 additional rows were obtained by conducting permutations over the three detector levels.

TABLE III
GRAECO-LATIN SQUARE DESIGN.

	no	yes w/o air	yes w/ air
1 mm	Al & Li-6	Zr & Au-197	SS & Ni-58
2 mm	Zr & Ni-58	SS & Li-6	Al & Au-197
3 mm	SS & Au-197	Al & Ni-58	Zr & Li-6
1 mm	Al & Au-197	Zr & Ni-58	SS & Li-6
2 mm	Zr & Li-6	SS & Au-197	Al & Ni-58
3 mm	SS & Ni-58	Al & Li-6	Zr & Au-197
1 mm	Al & Ni-58	Zr & Li-6	SS & Au-197
2 mm	Zr & Au-197	SS & Ni-58	Al & Li-6
3 mm	SS & Li-6	Al & Au-197	Zr & Ni-58

C. Model

In DOE, constant coefficient models, also known as simple linear models, are often employed to approximate the relationship between factors and the response variable [14]. This type of models assumes that the effects of the factors are constant across the entire experimental domain. They are relatively simple and easy to interpret, especially when dealing with a small number of factors. The experimental design and data analysis process is easy to implement as the assumption of linearity allows the use of standard linear regression techniques for parameter estimation. Utilizing a simple model facilitates the identification and interpretation of significant effects of the factors on the response variable. However, constant coefficient models may not accurately capture complex relationships between factors and the response variable, especially in cases where the true relationship is nonlinear or contains interactions between factors. Constant coefficient models are also not suitable for extrapolation beyond the range of the experimental data.

Although we have interaction between factors, especially between the choice of detector and the presence of an outer cladding, we started our analysis with the following constant coefficient model for the sake of simplicity:

$$r_{otmd} = \mu + \gamma_{oc} + \tau_{thick} + \omega_{mat} + \zeta_{det} + \epsilon \quad (1)$$

where r_{otmd} represents the relative amplitude of the flux depression that we want to maximize, μ is the total average of the simulations, γ, τ, ω and ζ are respectively the effects of the four considered factors oc: outer cladding, $thick$: cladding thickness, mat: material and det: detectors. The residuals are considered in the coefficient ϵ. For each of the 27 configurations described in Tab. II, the simulated relative amplitude of the detector response depression r_{otmd} observed between the fuel center and its boundary has been reported in Tab. IV.

TABLE IV
RELATIVE AMPLITUDE OF THE DETECTOR RESPONSE DEPRESSION (%).

	no	yes wo air	yes w air
1 mm	11.53 ± 0.098	4.7 ± 0.16	5.32 ± 0.14
2 mm	8.38 ± 0.11	14.90 ± 0.19	3.28 ± 0.11
3 mm	9.04 ± 0.038	5.99 ± 0.0081	6.63 ± 0.0055
1 mm	3.46 ± 0.16	8.05 ± 0.22	16.64 ± 0.11
2 mm	9.72 ± 0.0079	7.93 ± 0.42	4.49 ± 0.0058
3 mm	3.81 ± 0.0053	4.97 ± 0.0040	4.10 ± 0.019
1 mm	9.99 ± 0.14	10.27 ± 0.0080	8.42 ± 0.47
2 mm	4.27 ± 0.19	4.59 ± 0.15	7.82 ± 0.0063
3 mm	19.12 ± 0.0017	2.03 ± 0.0090	4.05 ± 0.0053

From this set of 27 simulations, we were able to extract the constant coefficients displayed in Eq. 2.

$$\begin{aligned}\widehat{r_{otmd}} \approx 7.54 + &\begin{Bmatrix} 1.27 \\ -0.49 \\ -0.79 \end{Bmatrix}_{oc} + \begin{Bmatrix} 1.17 \\ -0.27 \\ -0.90 \end{Bmatrix}_{thick} \\ + &\begin{Bmatrix} -1.59 \\ -0.85 \\ 2.44 \end{Bmatrix}_{mat} + \begin{Bmatrix} 3.75 \\ -2.29 \\ -1.46 \end{Bmatrix}_{det}\end{aligned} \quad (2)$$

Fig. 4 provides a visual representation of the effect of each factor, helping the reader in better understanding their impact. The centerline, depicting a 0% effect, represents the average response of the 27 simulations. As our objective is to identify the optimal design maximizing the detector response depression, it is crucial to focus on the levels with higher effects. Notably, the selection of the detector and cladding material holds greater importance than the thickness of the axial cladding and the presence of an outer cladding.

The n-way analysis of variance (ANOVAN) in Fig. 5 further supports this observation, revealing that the p-values for the factors corresponding to material and detector are both lower than 0.05. Hence, these two factors are the only ones with a statistically significant effect, representing a 5% risk of concluding that an effect exists when, in reality, there is no actual effect.

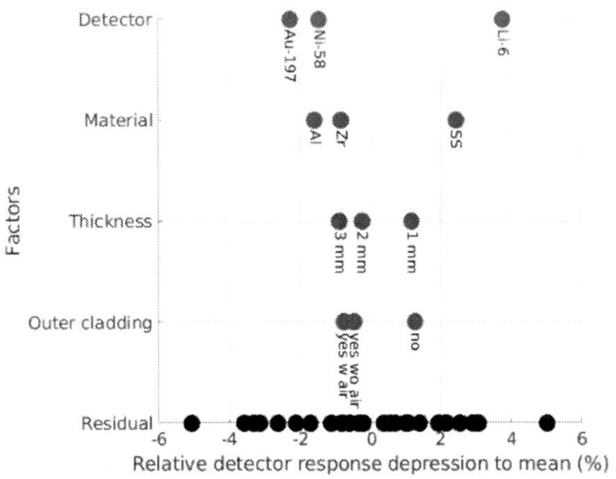

Fig. 4. Effects of each factor's level on the detector response depression in the fuel.

Source	Sum Sq.	d.f.	Mean Sq.	F	Prob>F
Outer cladding	22.39	2	11.1948	1.38	0.276
Thickness	20.306	2	10.1529	1.26	0.3088
Material	82.63	2	41.3151	5.11	0.0175
Detector	193.106	2	96.5532	11.94	0.0005
Error	145.6	18	8.0889		
Total	464.032	26			

Fig. 5. N-way analysis of variance (ANOVAN).

D. Discussion

Based on the constant coefficient model and its analysis of variance, the choice of the detection system and the cladding material appears to be the only factors significantly impacting the response. Initially, the seemingly favorable solution for the instrumented rod design would involve a single stainless steel cladding with an axial thickness of 1 mm, coupled with Li-6 scintillators as the detection system. However, it is important to consider that Li-6 scintillators are connected to optical fibers, requiring the presence of an outer cladding with sufficient space to accommodate them from the center of the fuel rod to the acquisition system. Consequently, the level of the "outer cladding" factor will be set to "yes w/space," with little impact on the response, as evident from the ANOVAN results.

Another material constraint is the choice of cladding material. Despite stainless steel being the optimal choice, Al-6060 has been selected to ensure compatibility with all CROCUS internals, leveraging our workshop's expertise, and simplifying the process of obtaining a permit from the Swiss regulator, ENSI.

As a result, the instrumented fuel rod will be produced using Al-6060, featuring the smallest possible inner cladding axial thickness and an outer cladding. With these design choices, the constant coefficient model predicts the radial reaction rate depression within the fuel to be $10.11 \pm 0.13\%$ with Li-6 scintillators, $4.9 \pm 0.21\%$ with Ni-58 dosimeters, and $4.0 \pm 0.23\%$ with gold dosimeters. In order to check the validity of the model, we proposed to fix 3 factors arbitrarily and to run simulations on the 3 levels of the last one. We therefore fixed the outer cladding on "yes w/ space", the thickness on "1 mm", the material on "Aluminum", and we ran 3 simulations varying the detector on "Li-6", "Au-197" and "Ni-58". Relative errors between the flux depressions predicted by the model and the simulated ones are respectively 13, 21 and 23%. However, these first estimates offer valuable insights into the expected performance of the instrumented fuel rod under various conditions and highlight the importance of appropriate material selection and detector system in achieving the desired response. A DOE model taking interactions between factors into account would be needed to improve our predictions.

IV. The Instrumented Fuel Rod

Considering the results of the Design of Experiments (DOE) and the various constraints presented in Section III-D, the following design for the instrumented rod was adopted and submitted to the Swiss nuclear regulatory authority ENSI-IFSN for examination, with the goal of obtaining a permit for the installation and usage of the rod in CROCUS.

The instrumented rod, which can be placed alternatively in one of the 18 COLIBRI slots (see Fig. 1), taking the position of the existing rod at the desired measurement location, is constructed as follows (refer to Fig. 6):

The rod comprises four U_{met} fuel cigars from the LRS reserve. These cigars are identical to those used in CROCUS U_{met} rods and are enriched to 0.947% in ^{235}U. Each cigar measures 25 cm in length and has a diameter of 17 mm. They are individually clad with an Al-6060 sheath, sealed using O-rings with a thickness of 1.5 mm, resulting in an outer diameter of 20.3 mm.

An Al-6060 outer sheath with a 22 mm internal diameter and a 25 mm external diameter forms the second layer. This layer has a wall thickness of 1.5 mm and is sealed with an aluminum plug screwed at its lower end, ensuring its tightness with an O-ring. The plug also serves as a centering piece in the lower grid.

Centering devices for dosimeters are placed between each clad cigar, enabling precise and reproducible positioning in terms of centering, height, and angle.

A fixing piece is attached to the upper end of the outer sheath, facilitating the manual or remote retrieval of the rod. At the upper part, the instrumented rod is positioned using a centering piece fixed in the passage hole of the upper grid and the moving plate of COLIBRI.

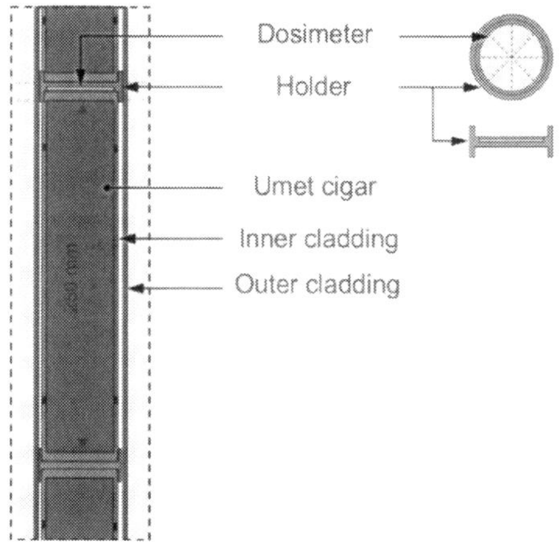

Fig. 6. Schematic diagram of the NECTAR instrumented fuel rod.

V. Dosimetry Testing

The last step in the design process involves conducting reactor dosimetry testing to verify the observation of reaction rate variations within dosimeters. For this purpose, nickel dosimeters with azimuthal sections are irradiated at the core center of CROCUS, with two distinct objectives in mind. Firstly, this testing aims to gain experience in HPGe gamma spectroscopy measurement and post-processing techniques. Secondly, it seeks to demonstrate the detection of neutron flux variations within pin-size dosimeters and to estimate experimental uncertainties.

The nickel dosimeter, measuring 17.3 mm in diameter and 0.5 mm in thickness, will feature 8 azimuthal sections and will be placed at the CROCUS core center, as schematically shown in Fig. 7a. Upon irradiation, the ^{58}Ni(n,p) reaction will produce for one third $^{58}Co^m$ with a half-life of 9.04 hours and for two thirds ^{58}Co, which, through electron capture, will yield ^{58}Fe with a half-life of 70.86 days [16]. To ensure sufficient counts, the dosimeter will be irradiated at 60 W for 1 hour, aiming for at least 80000 counts per section for the 811 keV gamma peak of the ^{58}Ni(n,p)^{58}Co reaction, which will be measured using the LRS high-efficiency gamma-ray spectrometer.

Post-irradiation, a waiting period of 4 days is required to allow the quantity of parasitic $^{58}Co^m$ in the dosimeter to decrease to a negligible amount.

The full nickel dosimeter activity was initially measured before being cut into its 8 azimuthal sections. As shown in Fig. 7b, the specific activity of each of the 8 azimuthal sections is compared to the activity of the full dosimeter measured before cutting. As expected, the mass-weighted average of the 8 slices matches the full dosimeter's activity, serving as proof of principle for the observation of neutron flux variations within the dosimeter.

The two-sigma error bars encompass uncertainties related to statistics, HPGe efficiency at 810 keV, ^{58}Co gamma emission probability, auto-absorption, and activation cross-section. The uncertainty on the detector efficiency at 810 keV was determined by the efficiency calibration of the HPGe spectrometer. The ^{58}Co gamma emission probability is known from literature [16]. The activation cross-section along with its uncertainty was computed from IRDFF microscopic cross section data [17] and neutron flux tallied in the dosimeter with Serpent2 simulations. Auto-absorption probability and respective uncertainty were retrieved from Serpent2 simulations with point source and exact detector geometry.

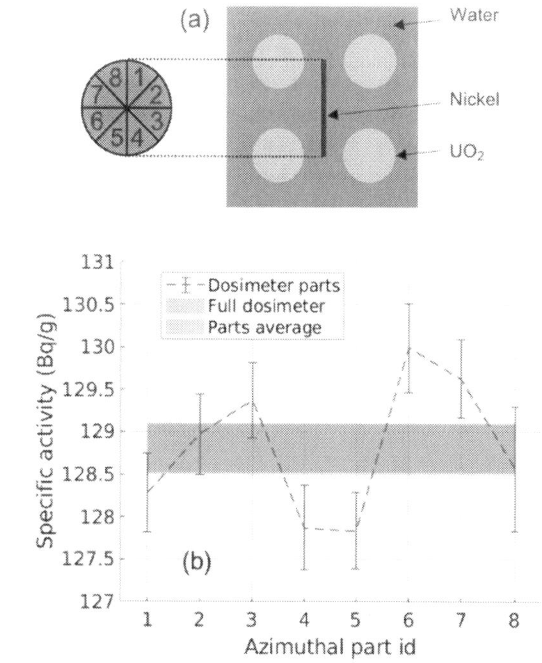

Fig. 7. (a) Schematics of the irradiation location. (b)Mass weighted activity of the dosimeters.

Regarding the azimuthal variations, it can be observed that sections 2, 3, 6, and 7, which are closer to the fuel pins, exhibit higher specific activities compared to the other sections. This outcome confirms our capability to detect azimuthal neutron flux variations within the dosimeter, as ^{58}Ni is more sensitive to fast neutrons, which are more abundant in the fuel than in the moderator. However, the maximum relative amplitude of these variations is only 1.7%. While uncertainties on HPGe efficiency at 810 keV, ^{58}Co gamma emission probability, auto-absorption, and activation cross-section are correlated from one section of dosimeter to the other, the primary contributor to the error bars is the counting uncertainty, which could be mitigated through higher power irradiation and longer HPGe times, while respecting to the power limit authorization of 100 W.

The observed 1.7% amplitude falls below the 4% predicted by the constant coefficient model for within-pin variations, which can be attributed to the fuel pins' fast neutron flux distribution differing from that in the moderator region.

VI. Conclusion

In this study, we successfully designed and manufactured an instrumented fuel rod tailored for intra-pin neutron flux measurements within the CROCUS reactor. Employing design of experiments (DOE) methodologies, we systematically investigated the key parameters affecting our response, specifically the amplitude of neutron flux variations within the fuel pin. The constant coefficient model used and its analysis of variance revealed that the choice of cladding material and detection system were the primary factors influencing the observed response.

Adhering to safety and regulatory constraints, we were constrained in our selection of cladding material, leading us to opt for Al-6060. To maximize data acquisition capabilities, we incorporated multiple detection systems, including Li-6 scintillators, Ni-58, and Au-197 dosimeters. The DOE study highlighted Li-6 fiber-coupled scintillators as the most favorable detection system, but considering practical aspects, we conducted reactor dosimetry testing with Nickel scintillators. The results confirmed that our design enabled the observation of within-pin variations, although with smaller amplitudes compared to Li-6.

In light of the testing outcomes and to ensure future flexibility, we prioritized the use of dosimeters while ensuring our design could accommodate fiber-coupled Li-6 scintillators when suitable. In parallel, we sought and successfully obtained a permit from the Swiss nuclear regulatory authority ENSI-IFSN, authorizing the installation and operation of the instrumented fuel rod in the CROCUS reactor.

Standing as the second intrapin neutron flux measurements in a LWR performed worlwide, this experiment should yield rare and highly valuable data. Our primary objective is to complement the existing CROCUS IRPhE benchmark in order to make the experimental dataset available to the community and facilitate high-fidelity neutronics solver validation.

Acknowledgment

I would like to express my sincere gratitude to the mechanics from the EPFL workshops, who played indispensable roles in the design, testing and development, of the instrumented fuel rod for the CROCUS nuclear reactor. Their unwavering support and expertise were essential throughout the entire project.

I extend my heartfelt thanks to:

- Gilles Grandjean, the workshop supervisor, for his essential guidance and initial assistance in getting started with the design of the instrumented fuel rod.
- Daniel Clément and Luc Chevalley, for their invaluable contributions in producing the first prototypes and conducting the crucial initial tests.
- Virgile Cavin, for his dedication and skill in overseeing the entire production process of the fuel rod, ensuring its successful realization.
- Claude Amendola, responsible for the production and laser pre-cut of dosimeters, a critical component of the rod.
- Daniel Godat, our lab mechanics, who provided great support in the final adjustments to ease handling and integration of the rod into the reactor core.

This project would not have been possible without their remarkable collective efforts and commitment.

References

[1] P. J. Turinsky and D. B. Kothe, "Modeling and simulation challenges pursued by the consortium for advanced simulation of light water reactors (CASL)," *Journal of Computational Physics*, vol. 313, pp. 367–376, may 2016.

[2] MPACT Team, "MPACT Theory Manual," Oak Ridge National Laboratory, Technical Report CASL-U-2019-1874-001, November 2019.

[3] N. Choi, H. Park, H. G. Lee, S. Jae, S. Jeon, and H. G. Joo, "Recent capability and performance enhancements of the whole-core transport code ntracer," *EPJ Web of Conferences*, vol. 247, p. 06033, 2021.

[4] T. Mager, M. Hursin, C. Fiorina, and A. Pautz, "Development of a high fidelity model of the crocus experimental reactor," in *PHYSOR 2022 - International Conference on physics of reactors.*, 05 2022.

[5] K. Macku, F. Jatuff, M. Murphy, M. Plaschy, P. Grimm, O. P. Joneja, and R. Chawla, "Radial and azimuthal 235u fission and 238u capture distributions in BWR UO2 pins: CASMO-4 and MCNP4c versus activation foil measurements," *Nuclear Science and Engineering*, vol. 155, no. 1, pp. 96–101, jan 2007.

[6] A. D. Santos, *U(n,f) and 238 U(n,gamma) Reaction Rates Across the Fuel Pellet Radius of the IPEN/MB-01 Reactor,*. International Handbook of Evaluated Reactor Physics Benchmark Experiments, NEA/NSC/DOC/(2006)1, Organisation for Economic Co-operation and Development, Nuclear Energy Agency, 2018, ch. IPEN(MB01)-LWR-RESR-019.

[7] F. Vitullo, "Miniature and minimalistic neutron detectors for online high-resolution experiments in the zero-power reactor crocus," p. 171, 2022. [Online]. Available: http://infoscience.epfl.ch/record/296483

[8] V. Lamirand, P. Frajtag, D. Godat, O. Pakari, A. Laureau, A. Rais, M. Hursin, G. Hursin, C. Fiorina, and A. Pautz, "The colibri experimental program in the crocus reactor: characterization of the fuel rods oscillator," *EPJ Web of Conferences*, vol. 225, p. 04020, 01 2020.

[9] G. E. P. Box, W. G. Hunter, and J. S. Hunter, *Statistics for Experimenters: Design, Innovation, and Discovery*, 2nd ed. Wiley-Interscience, 2005.

[10] F. Vitullo, V. Lamirand, J.-B. Mosset, P. Frajtag, O. Pakari, G. Perret, and A. Pautz, "A mm3 fiber-coupled scintillator for in-core thermal neutron detection in crocus," *IEEE Transactions on Nuclear Science*, vol. 67, no. 4, pp. 625–635, apr 2020.

[11] A. Laureau, A. Gruel, V. Lamirand, T. Ligonnet, A. Sardet, and A. Pautz, "Analysis of the preliminary campaign for the petale program," *EPJ Nuclear Sciences Technologies*, vol. 9, 06 2023.

[12] Lamirand, Vincent, Laureau, Axel, Rochman, Dimitri, Perret, Gregory, Gruel, Adrien, Leconte, Pierre, Blaise, Patrick, and Pautz, Andreas, "An experimental programme optimized with uncertainty propagation: Petale in the crocus reactor," *EPJ Web Conf.*, vol. 211, p. 03003, 2019. [Online]. Available: https://doi.org/10.1051/epjconf/201921103003

[13] T. Ligonnet, "Preliminary analysis of the petale program and comparison to simulations," p. 85, 2021. [Online]. Available: http://infoscience.epfl.ch/record/289130

[14] D. Montgomery and C. St, *Design and Analysis of Experiments, 9th Edition*, 07 2022.

[15] J. Leppanen, "Development of a new monte carlo reactor physics code," *978-951-38-7018-8*, 2007.

[16] W. C. Morgan, "Long-term neutron activation products of nickel-58." [Online]. Available: https://www.osti.gov/biblio/10172408

[17] A. Trkov, P. Griffin, S. Simakov, L. Greenwood, K. Zolotarev, R. Capote, D. Aldama, V. Chechev, C. Destouches, A. Kahler, C. Konno, M. Koštál, M. Majerle, E. Malambu, M. Ohta, V. Pronyaev, V. Radulović, S. Sato, M. Schulc, E. Šimečková, I. Vavtar, J. Wagemans, M. White, and H. Yashima, "Irdff-ii: A new neutron metrology library," *Nuclear Data Sheets*, vol. 163, pp. 1–108, 2020. [Online]. Available: https://www.sciencedirect.com/science/article/pii/S0090375219300687

Experimental and numerical results for a new single-cell calorimeter dedicated to nuclear heating rate measurement

J. Rebaud[1,*], A. Volte[1], M. Carette[1], A. Lyoussi[2], C. Reynard-Carette[1]

[1]Aix Marseille Univ, Université de Toulon, CNRS, IM2NP, Marseille, France
[2]CEA/DES/IRESNE/DER, Section of Experimental Physics, Safety Tests and Instrumentation, Cadarache, F-13108, Saint Paul-lez-Durance, France

(*) jeremy.rebaud@univ-amu.fr

***Abstract*— This paper presents the study of a new innovative heat flow single-cell calorimeter dedicated to the measurement of the nuclear heating rate in research reactors. Before irradiating this type of sensor, it is necessary to characterize it under laboratory conditions to determine its metrological properties.**

A general description (design, geometry and materials nature) of this new calorimeter will be given, focusing on the motivations that led to its development. This part will be followed by a presentation of the experimental set-up and operating protocol used to characterize this sensor. The sensor sensitivity and response time will be quantified and discussed with a comparison of those obtained with previous calorimeters. In parallel with these experimental characterizations, parametric 3-D numerical simulations will be carried out using COMSOL Multiphysics software. These results, obtained both experimentally and numerically, will be compared to present a complete calorimeter study under laboratory conditions.

***Keywords* — Calorimetry, Calibration, Nuclear heating rate, 3-D numerical simulations.**

I. Introduction

Material Testing Reactor (MTR) is a type of research reactor with multiple objectives. From the production of medical radioisotopes to the study of materials under high photon and neutron fluxes, these nuclear facilities are crucial tools for research in nuclear fission field.

The design and interpretation of experiments carried out in the core and the reflector require the accurate knowledge of specific quantities. Among these quantities, the absorbed dose rate is a key parameter. The origin of this quantity lies in the energy deposition generated by the interaction of radiation with matter expressed per unit of time and mass. In the reactor core, this mass energy deposition rate is so intense that it results in a local rise in temperature and consequently the absorbed dose rate is also known as the nuclear heating rate in $W.g^{-1}$. Thus, this quantity can be measured by using specific calorimeters. These sensors are inserted into experimental reactor channels for on-line measurements. These calorimeters are non-adiabatic sensor, by evacuating their energy with the external environment in order to ensure the sensor integrity and realize on-line measurements. The literature distinguishes two families of heat flow calorimeter dedicated to measuring nuclear heating rate:

- Differential calorimeter: with at least two calorimetric cells, including a reference cell, and incorporating heating elements in their design [1-5],
- Single-cell calorimeter: with just one calorimetric cell, in most cases without a heating element [4-9].

Since 2009, research work on the development of these calorimeters has been carried out by the LIMMEX laboratory (a joint AMU-CEA-CNRS laboratory), with the main aim of continuously improving these sensors. Various areas of improvement are targeted, such as the design, miniaturization, metrological characteristics (sensitivity, operating range, response time, selectivity etc.). The modularity of these sensors enables to meet a wide range of needs, including a prototype for measuring the 20 $W.g^{-1}$ in aluminum expected in the Jules Horowitz Reactor [3]. In 2015, a new differential calorimeter (named CALORRE) comprising two compact calorimetric cells using predominantly radial heat transfer was patented by AMU and CEA [10] and successfully qualified in the Polish MARIA research reactor [11]. Recently, this calorimeter has been improved as part of the CALOR-I project to achieve a smaller size of 89.5 mm axial height (instead of 222.5 mm previously) and an innovative design to eliminate contact thermal resistances [12]. This new prototype will shortly be qualified in the MIT reactor in-core water loop.

In parallel, a new prototype of a single-cell calorimeter called Mono-CALO is currently being studied. This research is part of the MICRO-CALOR project, and follows on a previous PhD thesis [13] leading to a first prototype of Mono-CALO and an AMU-CEA-CNRS patent [14] in 2020. The first prototype corresponds to a design of a single-cell calorimeter incorporating a thin-film deposited heating element. The aim of this innovative design is to keep the advantages of both calorimeter families by offering a single-cell calorimeter integrating a heating element.

The first section of this paper will describe the assembly of the first prototype of the Mono-CALO calorimeter with its various components, and present the experimental set-up and the protocol used to characterize it. The second section will discuss the results obtained both experimentally and numerically, and compare the metrological characteristics of this new prototype with those of other calorimeters developed in the laboratory. Finally, the last section will explain the main results and open up perspectives on future actions in order to develop a second prototype with intrinsic characteristics compatible with an irradiation campaign.

II. Mono-CALO calorimeter and experimental set-up with operating protocol

A. Mono-CALO calorimeter

The first prototype of Mono-CALO calorimeter has a parallelepiped shape (H = 25 mm, w = 18 mm, t = 10.2 mm) and features the characteristic elements of previous calorimeters: at its center there is a 4-wire heating element that also incorporates a temperature sensor (T-type thermocouple). This temperature sensor is referred to a hot spot and corresponds to the temperature at the center of the calorimeter. Around this heating element, two samples of material are placed on either side. These two samples are held in the center of the calorimeter by two spacers. The entire calorimeter is encapsulated in a jacket and cap, making it watertight. All materials structure used in the calorimeter are AISI 316L, with the exception of the heating element, which is made of Kapton as insulating material and Constant and Copper for the resistive element and T-type thermocouple. The calorimeter assembly and the identification of the various components are presented in Figure 1.

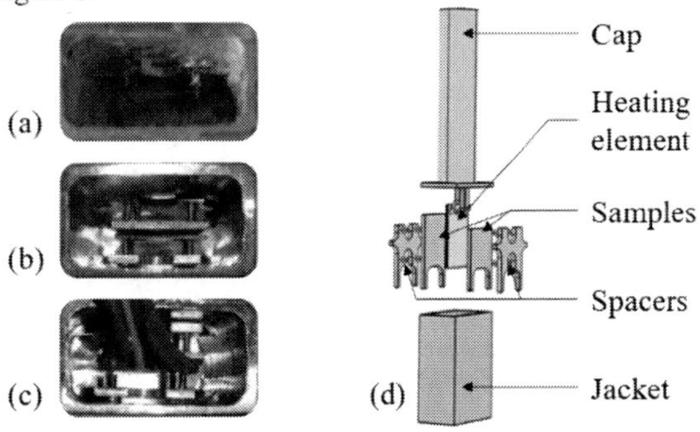

Fig. 1. Assembly of the Mono-CALO calorimeter with pictures on the left-hand section (jacket + spacers (a), + samples (b), + heating element (c)), and a diagram in the right-hand section (identification of elements (d)).

In order to limit the occurrence of contact resistance under laboratory conditions, thermal grease is applied to all calorimeter components. Another K-type thermocouple is used to measure a cold temperature, located outside the calorimeter, in contact with the cooling fluid.

B. Experimental set-up

A specific bench is used to characterize this sensor under laboratory conditions. The calorimeter is immersed in a thermostatic bath using water as the heat transfer fluid. The temperature and rising speed of this coolant fluid flow can be controlled by applying setpoints.

Thanks to the presence of a heating element at the center of the Mono-CALO calorimeter, the injection of a current through the resistive element creates a Joule effect. These currents are injected via a programmable power supply. The Joule effect created inside the calorimeter thus enables laboratory simulation of the energy deposition generated by the radiation-matter interactions that the calorimeter undergoes when irradiated in the reactor core. The energy deposited by the heating element will lead to a local increase in temperature.

To complete the test bench, a precision resistance is connected in series with the heating element electrical circuit to determine the value of the current injection due to limitations in the data acquisition system. All quantities are measured by a data acquisition system with a sampling time of 10 seconds. A monitoring software program plots the evolution of all these parameters in real-time. All these elements are shown in Figure 2.

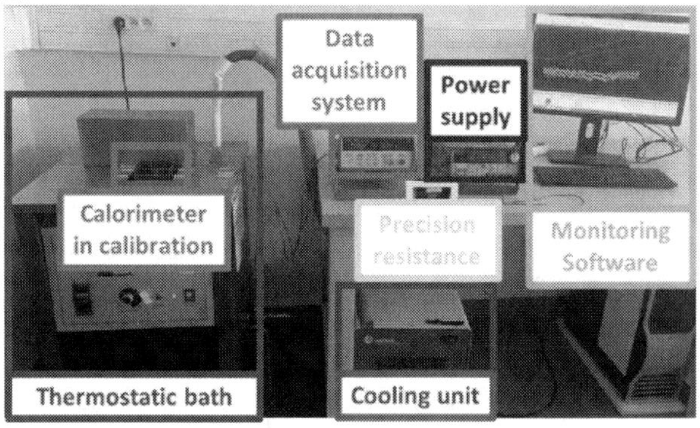

Fig. 2. Pictures of the various components of the experimental set-up.

Thanks to this bench, five quantities are measured:

- Hot temperature: at the center of the calorimeter, in the immediate vicinity of the heating element
- Cold temperature n°1: outside of the calorimeter, in direct contact with the heat transfer fluid
- Cold temperature n°2: outside of the calorimeter, in direct contact with the heat transfer fluid
- Voltage n°1: at the heating element terminals
- Voltage n°2: at the precision resistance terminals

C. Operating protocol

The aim of the study is to determine the calibration curve of the sensor. This curve corresponds to the difference in the mean steady temperatures between the hot spot, inside the calorimeter, and the cold spot corresponding to the temperature of the fluid, as a function of electrical power inputs in steady state. In this way, an increasing increment of the electrical current is injected into the heating element to determine the sensor response over a defined range of electrical power from 0 W to 0.8 W. As the power increment is equal to 0.05 W, 18 current injection steps are required (including the 0 W value at the start and end of the test). Each step lasts 30 minutes. This time is necessary to ensure that the system reaches a steady state. Figure 3 shows the temporal evolution of the temperatures and the generated electrical power.

 https://doi.org/10.1051/epjconf/202328804008

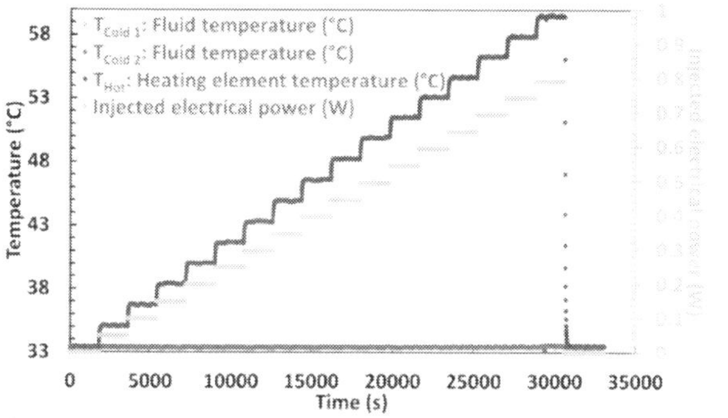

Fig. 3. Temperatures of the Mono-CALO calorimeter and fluid flow, and of the injected electrical power as a function of time for the following parietal conditions: T_f = 33 °C and Re = 1746.

The time curve described in Figure 3 can then be used to obtain the sensor calibration curve. That involves determining the mean steady temperatures and then calculating the difference of the mean steady temperatures between the hot and cold points of the sensor, and finally plotting this temperature difference versus the electrical power injected (cf. Fig. 4). Each mean temperature is calculated over the last 10 minutes of each step.

Thanks to this experimental protocol, various parietal conditions (coolant fluid flow temperature and velocity) could be applied to quantify and study their influence on the sensor response. The experimental study in this paper was carried out with a coolant fluid flow temperature of 33 °C and a coolant fluid flow velocity corresponding to a Reynolds number of 1746 around the calorimeter.

III. Experimental and numerical results

Characterization results for the new Mono-CALO prototype were obtained experimentally using the experimental set-up described above, and by coupling these results with 3-D numerical thermal studies. The metrological characteristics obtained were then compared with those of other calorimeters.

A. Experimental results

Among the metrological characteristics that need to be determined in the laboratory before an irradiation campaign, the calibration curve coefficients are of major interest. These will be used during the irradiation to process the on-line measured data and then to determine the nuclear heating rate.

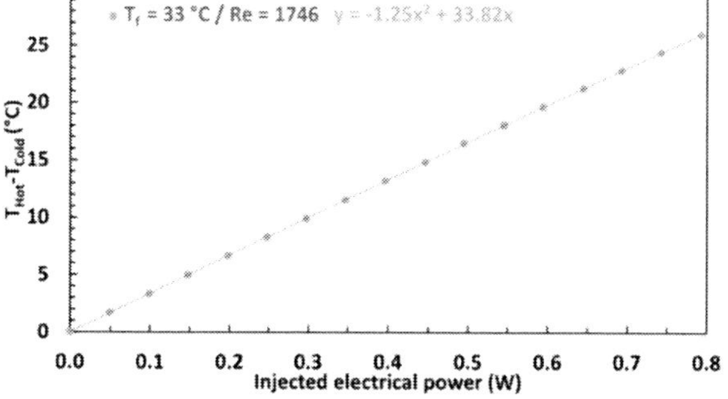

Fig. 4. Mono-CALO calibration curve obtained for the following parietal conditions: T_f = 33 °C and Re = 1746.

The calibration curve of the first prototype of the Mono-CALO calorimeter is given in Figure 4. It corresponds to a second-order polynomial curve. This behavior could be explained by the temperature dependence of the thermal properties of the materials. Consequently, the Mono-CALO calorimeter sensitivity is not constant (cf. Table 1). It decreases with increasing the value of injected electrical power.

In addition to this metrological characteristic, it is also important to determine the calorimeter response time. At the start of each power injection step, the calorimeter has a phase of energy accumulation before reaching a thermal equilibrium (steady state). The time required to reach this thermal equilibrium corresponds to the calorimeter response time. This response time is specific to each calorimeter and depends on its design, its mass, its materials, and the external boundary conditions. Each response time is evaluated at 3τ (95% of the final response variation). The average of the Mono-CALO calorimeter response time obtained for all steps is 121 s.

B. 3-D numerical results

A complementary 3-D numerical study was carried out using COMSOL Multiphysics software. This software, using a finite element calculation method, allows the estimation of the response of the sensor under laboratory conditions as close as possible to those achieved during the experimental study. To do that, a physics coupling the solving of the Heat and Maxwell equations was considered for this new prototype. This physics was never used for previous calorimeters. This method allows the calculation of the heat source induced by the Joule effect inside the heating element as well as the temperature field in the whole sensor. The conductive and radiative heat transfers inside the calorimeter are considered. The thermal conductivity depends on the temperature and the emissivity is taken constant and equal to 0.25. The results of this simulation in terms of calibration curve and sensitivity are compared with those obtained experimentally in Table I.

TABLE I
COMPARISON OF EXPERIMENTAL AND NUMERICAL RESULTS FOR THE MONO-CALO CALORIMETER

Studies	Equation of the calibration curve $\Delta T = T_{hot}-T_{cold}$ in °C with P in W	Sensitivity $°C.W^{-1}$	Sensitivity for 0.8 W $°C.W^{-1}$
Experiments	$\Delta T = -1.25*P^2+33.82*P$	-2.5*P+33.82	31.82
Simulations	$\Delta T = -0.13*P^2+35.32*P$	-0.26*P+35.32	35.11

The experimental and numerical results are consistent. However, there is a difference between the second-order coefficients. This difference can be explained by the value of the emissivity which is not well known. Nevertheless, these results provide a validation of the 3-D numerical model coupling the solving of the Heat and Maxwell equations.

C. Comparison of metrological characteristics

The main metrological characteristics (sensitivity and response time) of this new Mono-CALO calorimeter can be compared in Table II with those of previous prototypes developed in the laboratory. The comparison is realized with experimental results from two CALORRE calorimetric cells using the same structural material (AISI 316L) [15].

TABLE II
COMPARISON OF THE SENSITIVITY AND THE RESPONSE TIME FOR THREE CALORIMETERS

Calorimeter	Equation of the calibration curve	Sensitivity	Response time
	ΔT in °C with P in W	°C.W^{-1}	s
Mono-CALO	ΔT = -1.25*P²+33.82*P	-2.5*P+33.82	121
CALORRE (quarter-horizontal fin)	ΔT = -0.82*P²+30.03*P	-1.64*P+30.03	498
CALORRE (half-horizontal fin)	ΔT = -0.54*P²+22.65*P	-1.08*P+22.65	291

On the one hand, the sensitivity of the Mono-CALO sensor is relatively close to that of the CALORRE calorimetric cell with a quarter-horizontal fin design and a height equal to 23.1 mm and higher than that of the CALORRE calorimetric cell with a half-horizontal fin design and a calorimetric cell height of 23.1 mm. On the other hand, the response time of the new Mono-CALO prototype (121 s) is lower than that of two CALORRE calorimetric cells (4 times lower than that of the CALORRE configuration with a quarter-horizontal fin design and a height of 23.1 mm). This metrological characteristic is of prime importance during irradiation campaigns, since the protocol, the irradiation time, and consequently, the sensor ageing depend on it.

IV. CONCLUSIONS AND OUTLOOKS

Laboratory characterization of the first prototype of a new single-cell calorimeter, called Mono-CALO, incorporating a thin heating element allowed the determination of its main metrological characteristics (calibration curve, sensitivity, and response time). This new sensor has a non-linear calibration curve as previous sensitive CALORRE calorimetric cell configurations. By against, the response time of this new single-cell calorimeter is lower than that of CALORRE calorimetric cells having a similar height and made of stainless steel. Its response time is up to 4 times lower than that of the CALORRE calorimetric cell with a quarter-horizontal fin configuration. A sensitivity of 31.82 °C.W^{-1} for an electrical power injection of 0.8 W and an average response time of 121 s are found. In addition, the use of 3-D numerical simulations coupling the solving of Heat and Maxwell equations led to results consistent with those obtained experimentally. The validation of this new numerical simulation model is important in order to carry out future parametrical studies to modify and optimize the sensor design and response.

Following the characterization of this first prototype in the laboratory, a 3-D parametric study under laboratory and irradiation conditions will be carried out. The aim of this new study will be to design a second prototype with a greater sensitivity, in order to carry out an irradiation campaign in a TRIGA-type research reactor. In addition to changes in the sensor design, the heating element used to calibrate needs to be modified. At present, the Kapton used as the insulating material for the resistive element is the limiting point for injecting greater electrical power, due to its low melting point. A change in the manufacturing method of the thin heating element is being studied, in order to use nuclear-hardened materials and to withstand higher temperatures. In this way, the future prototype will offer greater sensitivity and enable metrological characterization over a wider operating range.

ACKNOWLEDGMENT

This research work was made possible thanks to the LIMMEX joint laboratory and the IM2NP micro-mechanics workshop for the manufacture of the Mono-CALO calorimeter.

REFERENCES

[1] H. Carcreff, L. Salmon, F. Malouch, "Recent developments in nuclear heating measurement methods inside the OSIRIS reactor", Nuclear Instruments and Methods in Physics Research, Volume 942, 2019.
[2] J. Brun et *al.*, "Comparison of the thermal response of two calorimetric cells dedicated to nuclear heating measurements during calibration", IEEE TNS, Vol. 61, no. 4, pp. 2254-2261, Aug. 2014.
[3] A. Volte, J. Brun, A. Lyoussi, M. Carette, C. Reynard-Carette, "Qualification of a New Differential Calorimeter Configuration Dedicated to Nuclear Heating Rates up to 20 W.g−1", IEEE TNS, Vol. 67, no. 11, pp. 2405-2414, Nov. 2020.
[4] D. Fourmentel et *al.*, "Nuclear Heating Measurements in Material Testing Reactor: A Comparison Between a Differential Calorimeter and a Gamma Thermometer", IEEE TNS, Vol. 60, no. 1, pp. 328-335, Feb. 2013.
[5] C. Reynard-Carette, G. Kohse, J. Brun, M. Carette, A. Volte, A. Lyoussi, "Review of nuclear heating measurement by calorimetry in France and USA", Proc. ANIMMA 2017, Liege, Belgique, 2017.
[6] R. Van Nieuwenhove, L. Vermeeren, "Nuclear Heating Measurements by Gamma and Neutron Thermometers", IEEE TNS, Vol. 67, no. 9, pp. 2073-2080, Sept. 2020.
[7] M. Tarchalski, "Nuclear heating measurements in MARIA reactor and implementation of neutron and photon calculation scheme", Aix-Marseille University thesis, 2016.
[8] M-S Kim, B-G Park, "Confirmation of Nuclear Heating Rate for Installation of Cold Neutron Source at HANARO", Proc. of ANIMMA 2019, Portoroz, Slovenia, 2019.
[9] M. Alqahtani, "On-line measurement and simulation of the in-core gamma energy deposition in the McMaster nuclear reactor", Nuclear Engineering and Technology, Vol. 54, Issue 1, pp. 30-35, 2022.
[10] M. Carette, A. Lyoussi, J. Brun, C. Reynard-Carette, J.-F. Villard and P. Guimbal, "Éprouvette pour mesure d'échauffement nucléaire dans un réacteur nucléaire, et cellule calorimétrique comprenant au moins une telle éprouvette", Patent N° 10755823 deposited on 8 April 2016.
[11] J. Brun, A. Volte, C. Reynard-Carette, M. Carette, A. Lyoussi, D. Fourmentel, L. Barbot, J.-F. Villard, P. Guimbal, "Nouveau capteur compact pour la mesure de l'échauffement nucléaire : Conception par simulation, Etudes en laboratoire, Qualification en réacteur de recherche", Proc. SFT, Marseille, France, May 2017.
[12] A. Volte, M. Carette, A. Lyoussi, G. Kohse, J. Rebaud, V. Valero, C. Reynard-Carette, "Review of CALORRE Calorimeter Characterizations Under Laboratory and Irradiation Conditions", IEEE TNS, Vol. 69, no. 4, pp. 840-848, April 2022.
[13] A. Volte, "Calorimètre miniaturisé innovant pour la mesure de l'énergie déposée par les interactions rayonnements nucléaires/matière en réacteur de recherche : de sa conception à son étude en laboratoire", Aix-Marseille University thesis, 2019.
[14] C. Reynard-Carette, A. Volte, A. Lyoussi, M. Carette, "Noyau de mesure pour la mesure d'échauffement nucléaire en réacteur nucléaire et capteur calorimétrique intégrant un tel noyau de mesure", Patent N° FR 2013804 deposited on 21 December 2020.
[15] A. Volte, C. Reynard-Carette, A. Lyoussi, J. Brun, M. Carette, "Study of the response of a new compact calorimetric cell for nuclear heating rate measurements", IEEE TNS, Vol. 65, no .9, pp. 2461-2470, Sept. 2018.

Exploring the ENEA Casaccia and ENEA Frascati irradiation capabilities: Status and perspectives

Michele Croia[1,*], Nunzio Burgio[1], Alessandro Ampollini[2], Maria Denise Astorino[2], Giulia Bazzano[2], Barbara Bianchi[1], Mateo Cesaroni[1], Luca Falconi[1], Salvatore Fiore[2,3], Luigi Lepore[1], Augusto Nascetti[4], Paolo Nenzi[2], Antonino Pietropaolo[2], Concetta Ronsivalle[2], Alfonso Santagata[1], Antonino Ratto[1], Pierpaolo Ricci[1] and Luigi Scaramuzzo[1]

[1]ENEA Casaccia research center, Rome, Italy
[2] ENEA Frascati research center, Frascati, Italy
[3]CERN, Geneva, Switzerland
[4]School of Aerospace Engineering, Sapienza University of Rome, Italy
(*) michele.croia@enea.it

***Abstract*—A collaboration between different facilities in Casaccia and Frascati ENEA research centers has recently opened the possibility of performing irradiation experiments using different kinds of particles such as protons, neutrons, and electrons. The facilities involved in the project are the TAPIRO fast nuclear research reactor, the TOP-IMPLART proton accelerator (71 MeV), the REX electron accelerator (5 MeV) and the FNG neutron facility (2.5/14 MeV). This suite of plans represents a distributed irradiation facility using different particles and spectra, offering various irradiation capabilities for experiments in many fields, such as: nuclear physics, accelerator physics, aerospace, science material, medicine, detectors, radiation diagnostics, etc. Complete numerical models of all the facilities have been implemented to perform start-to-end simulations before experimental irradiation using beam dynamics and nuclear transport simulation codes. In addition, our distributed irradiation facility makes it possible to recreate an essential part of the Van Allen Belt radiation allowing us to check the FLUKA quantitative calibration regarding this specific simulation, facilitating the disentanglements of the uncertainties. Finally, a comparison of the estimated silicon 1 MeV neutron equivalent fluence (SI1MEVNE) of the radiation damage imparted by the different facilities is presented.**

***Keywords* —Nuclear research reactor, Particle accelerator physics, Neutron generator, Irradiation Facility, Aerospace, Science material, Beam dynamics, Monte Carlo simulations.**

I. Introduction

Historically, ENEA nuclear irradiation facilities were involved, during the years, in activities related to material sciences, Gen IV nuclear reactors, Accelerator Driven Systems (ADS), Nuclear Fusion, Medical Isotope production and radiotherapy, qualification of radiation-hardened electronic components, nuclear detector calibration, radio-dosimetry. Up to date, these laboratories have worked on the above-cited activities in parallel into separate tasks, without needing to be directly connected. However, the need to design complex components and test them against a complex mixed radiation field, such as aerospace systems, poses the question of ascertaining the representativeness limits of the ENEA facilities of the radiative spatial condition. An example of the practical representation of the Van Allen Belt complex orbital radiation field at 5830 km to be used in design activities is given in [1].

All experiments or tests that must be conducted in these environments require as much as possible the creation of the same multi-particle radioactive environment, covering as many energy spectra as possible.

The task of artificially reproducing these radioactive environments is entrusted to irradiation facilities (nuclear research reactors, particle accelerators, radioactive sources), which allow for the irradiation of experimental samples subjected to a limited number of types of radiation and energy spectra. Regarding the neutron energy spectrum, nuclear reactors can generate a thermal spectrum or a fast spectrum essentially depending on the moderator used [2]. For particle bunches or ions produced via particle accelerators the final spectra depend on the accelerator itself i.e., accelerator length and accelerating gradient.

In some facilities it is possible to reproduce an environment consisting of multi-particle radiation. Typically, this is a bunch of primary particles impinging on an appropriate target to produce secondary radiation. This is what happens in the neutron and muon sources [3-6].

Our collaboration between different facilities within the same institute has created a Distributed Facility for multi-particle irradiation, which allows irradiation experiments with different types of radiation (neutrons, protons, and electrons) and different energy spectra. One of the main advantages of the distributed facility is therefore the simplicity of access for the user who needs to irradiate a sample to be subjected to a radioactive environment using particles of different nature.

The aim of this work is to introduce the design of a simple irradiation experiment having the scope to demonstrate the feasibility of the ENEA distributed facility for aerospace investigation. In the following, after the presentation of each single irradiation facility participating in the distributed system, we present a simple one-unit CubeSat layout that has a dosimeter as payload. Based on the simulation outcomes, we

https://doi.org/10.1051/epjconf/202328804009

establish that the distributed facility has the capability to cover the trapped particles and the solar energetic particle emission energy range with the exclusion of the Galactic Cosmic Rays (GCR) contribution. The simulation tuning, based on the MCNP and FLUKA codes, has been performed using as input the outcomes from beam dynamics codes for the particle accelerators and nuclear properties of the research nuclear reactor. As a reference, we used the experimental data obtained from the ABCS satellite that was launched into orbit in the Van Allen belt (5860 km of altitude) in 2022 [1].

II. The distributed irradiation facility

The ENEA Distributed Irradiation facility is composed by four different devices: the RSV TAPIRO nuclear research reactor, the Frascati Neutron Generator (FNG), the TOP-IMPLART proton linac accelerator and the REX electron linac accelerator.

A. The RSV TAPIRO fast nuclear research reactor

RSV TAPIRO (Reattore Sorgente Veloce - TAratura PIla Rapida a Potenza 0 - Fast Pile Calibration at 0 Power) is a source of neutrons generated in the fission process in a highly enriched Uranium-Molybdenum alloy. Depending on the irradiation position inside the nuclear reactor, different neutrons' energy distributions, and intensities are used to satisfy the experiments' needs. The almost fission energy spectrum is available in the core center with a neutron flux intensity of 4×10^{12} cm^{-2}/s^{-1} at the maximum power of 5 kW. In the other irradiation channels, less energetic neutrons are usable up to epithermal and thermal neutrons in the Thermal Column Cavity (TCC). A detailed description of the reactor and possible applications are in [7]. Fig. 1 shows the RSV TAPIRO schematic layout, with experimental channels indicated.

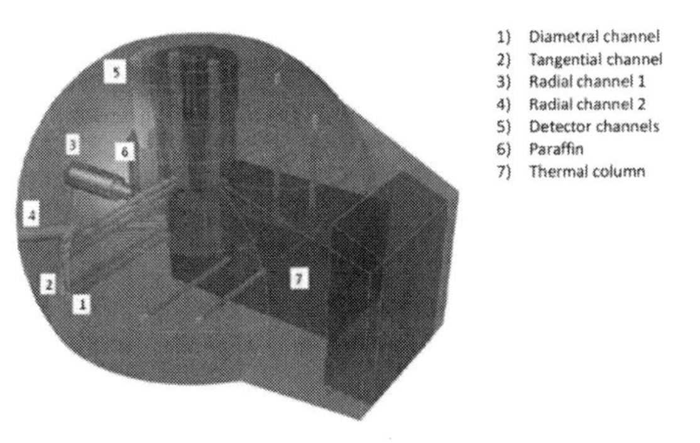

Fig. 1 Schematic layout of the RSV TAPIRO nuclear research reactor and its experimental channels.

B. The Frascati Neutron Generator – FNG

The Frascati Neutron Generator (FNG) is a compact accelerator driven neutron source installed and operating at the ENEA Frascati Research Centre, thoroughly described in [8,9,10]. It relies on both Deuteron-Deuteron (D-D) and Deuteron-Tritium (D-T) fusion reactions, in turn producing almost monochromatic 2.5 MeV and 14 MeV neutrons with a maximum neutron emission rate up to 10^9 s^{-1} and up to 10^{11} s^{-1}, respectively. Fig.2 shows a schematic 3D representation of FNG.

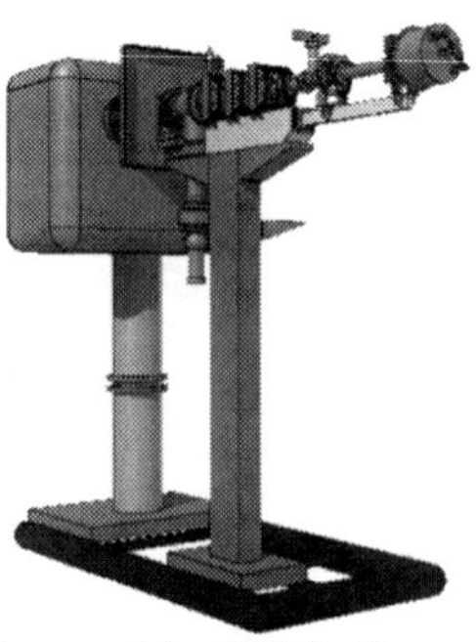

Fig. 2 Schematic layout of the FNG facility.

In D-T mode, the source neutron emission rate is determined by means of an absolute measurement, the so-called associated alpha particle technique [11]. Simulations, benchmarked by means of experimental measurements, provide a neutron yield and flux determination within an uncertainty of 3% [8,9,10]. In D-D mode the 2.5 MeV neutron emission rate is determined using foil activation technique, where a ^{115}In is irradiated and the gamma activation following the ^{115}In (n,n') $^{115}In_m$ reaction channel is measured by means of an absolutely energy calibrated High-Purity Germanium detector. The D-D neutron emission is measured at ±7% uncertainty.

C. The TOP-IMPLART proton linac

The TOP-IMPLART particle accelerator [12] is a technology demonstrator of a pulsed, fully linear machine for proton therapy under development at ENEA Frascati Research Centre in collaboration with the Italian Institute of Health (ISS) and the Oncological Hospital Regina Elena-IFO. It is a unique and compact accelerator, exploiting S-band accelerating structures from energies as low as 7 MeV, and one of the few proton sources above 60 MeV available in Italy. It comprises a commercial injector, by ACCSYS-Hitachi, and 8 Side Coupled DTL structures, patented by ENEA, accelerating the proton beam to a maximum energy of 71 MeV and maximum intensities of 4.5×10^8 protons per pulse. A set of two scanning magnets allow irradiation of an area up to 10 cm × 10 cm. Fig. 3 shows a schematic layout of the accelerating line, from the injector up to reach the final energy of 71 MeV.

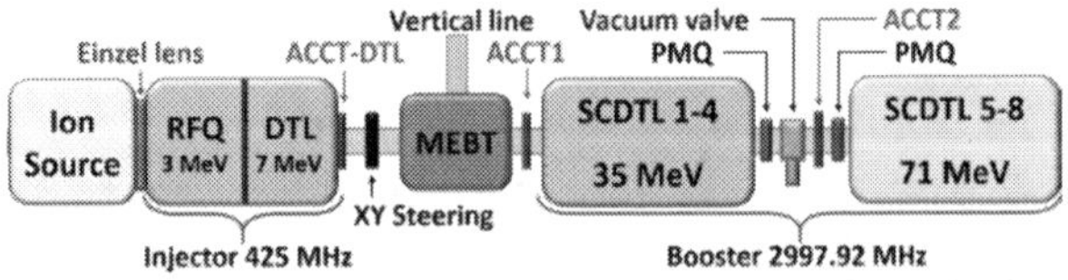

Fig. 3 Layout of the TOP-IMPLART accelerating line. The first extracting point is in the vertical line (7 MeV.) At the end of the entire line (71 MeV) are located the two scanning magnets.

Table I reports the main accelerators and beam parameters at the end of the linac, before the two scanning magnets.

TABLE I
TOP-IMPLART FINAL PARAMETERS

Quantity	Final values
Pulse length	2.4 μs
Pulse repetition frequency	25 Hz
Protons per pulse (max)	4.5×10^8
Pristine energy on target	61 – 70 MeV
Pencil beam spot size (FWHM)	17 mm
Maximum proton flux on a 10 cm × 10 cm area	$1.1 \times 10^8\ cm^{-2}s^{-1}$

Main parameters for the irradiations at the end of the accelerator line.

D. The REX electron linac.

The REX irradiation facility [13] operating at ENEA Frascati Research Center is based on a 5 MeV S-band on-axis coupled electron linear accelerator. It can provide two types of radiation: electrons and photons. It is equipped with a removable conversion unit for secondary X-Ray emission, generated by bremsstrahlung from the 5 MeV electron beam. The irradiation chamber (dimensions: 40cm x 40cm x 80cm) can be equipped with a remote positioning system to scan the specimens for complete exposure to the beams. Fig.4 shows the schematic layout of the accelerator and the irradiation chamber.

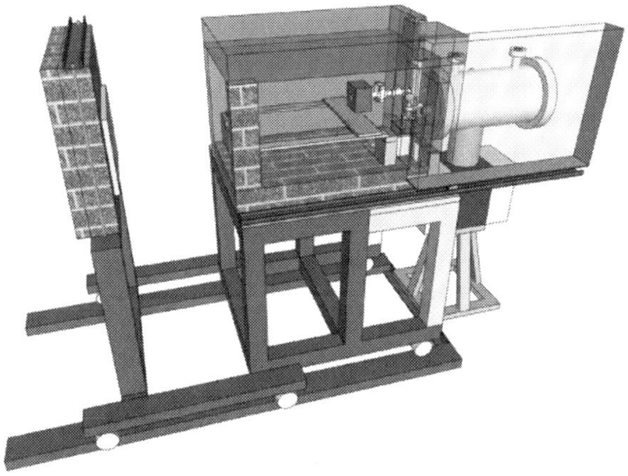

Fig. 4 Schematic layout of the REX accelerator and of the irradiation chamber.

In Table II are reported the main REX beam parameters.

TABLE II
REX FINAL PARAMETERS

Quantity	**Final values**
Pulse duration	3 μs
Electron per pulse (max)	2.95×10^{12}
Pristine energy on target	5 MeV
Spot size 5cm from the linac exit (FWHM)	20 mm
Pulse current (max)	150 mA
Average current	9 μA
Electrons per second (max)	$5.6 \times 10^{13}\ s^{-1}$

Main parameters at the end of the accelerator line.

III. Irradiation of a satellite in the distributed facility to partially recreate the Van Allen belt radiation conditions

The Astro Bio CubeSat (ABCS) was launched in the Van Allen belt [1] in July 2022, in a period of Solar maximum activity that triggered a Coronal Mass Ejection, causing the arrival of high flux Solar Energetic Particles (SEP) in the proximity of the Earth. After seven days of the mission, the satellite data transmission to Earth suddenly ceased, and the causes are currently under investigation.

However, the first two mission days completed all the experimental activities concerning the onboard multi-analytical platform. Also, an exhaustive data flow on the dose measures within the payload was received. The measured cumulative dose time evolution within the payload obtained by the onboard metal-oxide-silicon Field-Effect Transistors radFET [14], oscillates but its trend increases almost linearly, fitting the data it is possible to quantify the dose in 7.7 rad/day [14], that seems to confirm that the satellite was only marginally irradiated by the SEP.

Comparing the measured integral dose with the one simulated using FLUKA (FLUktuierende KAskade) Monte Carlo code [15], assuming a very poor SEP contribution to the dose along the orbit, we found an initial difference of 27% [16] that refining the calculation was lowered to 1.89% [17]. To verify, the simulated radFET answer, we will measure the dosimeter response using the distributed irradiation facility. The program for the irradiation consists firstly in the implementation of a CubeSat-like unit with a payload shell hosting the same radFET dose counter used in the ABCS mission, a VARADIS VT02 [18]. We already built the satellite shell. Fig. 5 shows a copy of the CubeSat that was produced by T.S.C [19], recreating the same technical features used in the ABCS mission.

Fig. 5 One of the copies that we realized for the irradiations. Inside it is visible a housing for placing the electronic board containing the radFET. A 2mm thick hole was made in one of the walls to allow for the possible passage of cables during the irradiation.

By means of FLUKA simulations, we evaluated the material activation during and after the irradiations in all the facilities. The shell, as in the mission, was realized in an anticorodal alloy, this will allow to extract and handling in a relatively short time the CubeSat unit, even after the irradiation with neutrons.

Currently, we are implementing the electronic board to insert in the CubeSat unit. The information collected during the irradiation will be also useful to establish the general irradiation protocol of a full replica of the ABCS.

The second step of the program before the irradiations foresees complete Start-to-End (S2E) simulations for each facility. The last step will be the irradiations to measure the components' response comparing experimental results with simulations outcomes.

IV. Simulations of the irradiations

The facility covers an interval of energies with different kinds of particles (neutrons, protons, electrons, and gammas) and can be used in parallel to investigate the impact of mixed field radiations. In Table III there are the flux intensities, their abundance, and mean energies of the energy distribution of the primary orbital source term in ABCS orbit. The values show that we are almost representative of the Van Allen belt radiative environment.

TABLE III
PARAMETERS IN THE ABCS ORBIT

	Flux intensity ($cm^{-2}s^{-1}$)	Worth (%)	Flux weighted energy (MeV)
Trapped protons	5.9×10^6	6.31	2.884×10^{-1}
GCR protons	8.8	9×10^{-6}	576.71
SEP protons	5.7×10^5	6.03×10^{-1}	1.501
Trapped electron	8.7×10^7	93.09	1.409×10^{-1}
Total	9.4×10^7	100	

Trapped protons and SEP protons have flux weighted energies covered by the 70 MeV protons of TOP_IMPLART while REX electron accelerator covers the trapped electron mean flux. GCR protons flux weighted energy outperforms the TOP_IMPLART energy range.

A. Simulated proton beam irradiation @ TOP_IMPLART

To perform S2E simulations we simulated the beam from the injector up to the interaction point. The simulated beam parameters at the interaction point are a spot size of $\sigma_{x,fwhm}$=1.7 cm, a medium (peak) current of $\bar{I}$ (I) = 1.2 μA (20 mA), a final beam energy of E $\approx$ 70 MeV.

To simulate the interaction between the proton bunch and the CubeSat we used FLUKA, in which we modeled the CubeSat structure, and we used as input the proton beam parameters from the beam dynamics simulations. Fig. 6 shows a cross-section of the CubeSat geometry having superimposed the integral mesh of the estimated energy density in GeV cm^{-3} nA^{-1}, normalized at 1 nA obtained in the simulated irradiation at TOP_IMPLART, in the central part of the satellite is located the card containing the radFET.

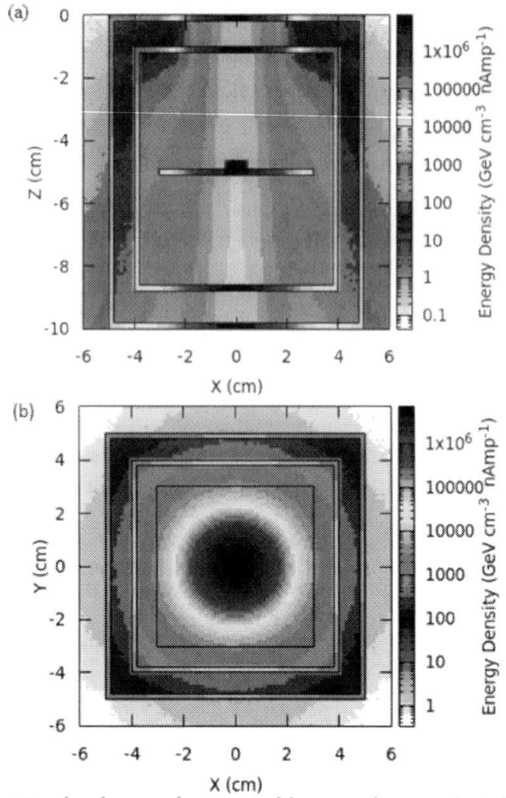

Fig. 6 Simulated energy deposition of the proton beam on the CubeSat. 6. (a) Axial view of the irradiation, in the upper part of the figure the beam from TOP_IMPLART penetrates the satellite. The loss of energy and the widening of the beam is clearly visible. 6. (b) Radial view of the energy deposition.

B. Simulated electron beam irradiation @REX

Using the same procedure, we simulated the interaction between the electron bunch from the REX accelerator and the CubeSat. Using beam dynamics simulation codes, we obtained the beam parameters at the interaction point: a spot size of $\sigma_{x,fwhm}$=3.8 cm, a medium (peak) current of $\bar{I}$ (I) = 9 μA (150 mA), a final beam energy of E $\approx$ 5 MeV. Implementing the beam parameters in FLUKA we simulated the interaction between accelerated electrons and the dummy satellite. Fig. 7 shows the estimated axial and radial deposited energy densities normalized per μA (GeV cm^{-3} μA^{-1}).

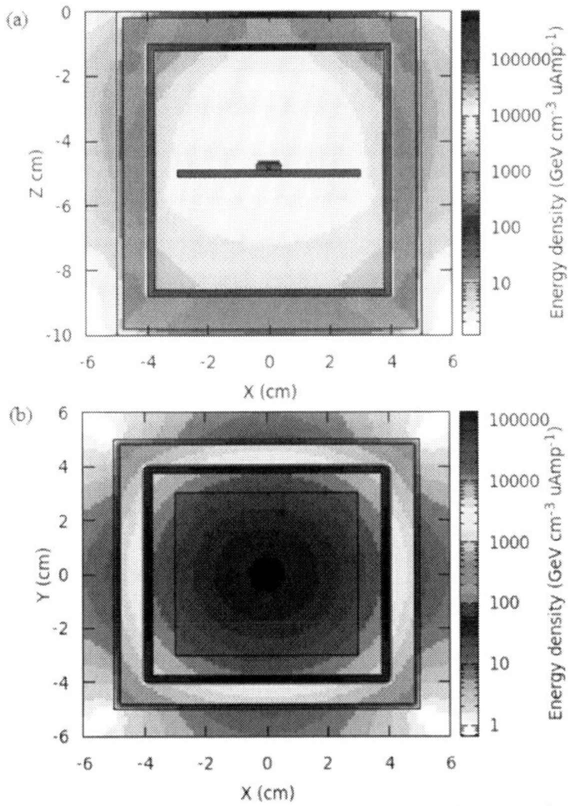

Fig. 7 Simulated energy deposition of the electron beam on the CubeSat. Fig 7. (a) Top view of the irradiation, in the upper part of the figure the beam from REX penetrates the satellite. The loss of energy and the widening of the beam is clearly visible. 6. (b) Front view of the irradiation. The view shows the integral of the simulated interaction.

C. *Irradiation with neutrons @TAPIRO*

Fig. 9 reports the geometrical cross-section of the MCNP TAPIRO model [20] showing the irradiation position inside the TCC in which we simulated the irradiation of the CubeSat-like structure within the TCC intending to estimate its performance in terms of equivalent damage (S1MEVNE). By comparing such simulation outcomes with the equivalent damage estimated for the orbital source, we can estimate the effectiveness of the neutrons as a source of damages. A similar estimation has been already carried out in [1] for the whole ABCS structure.

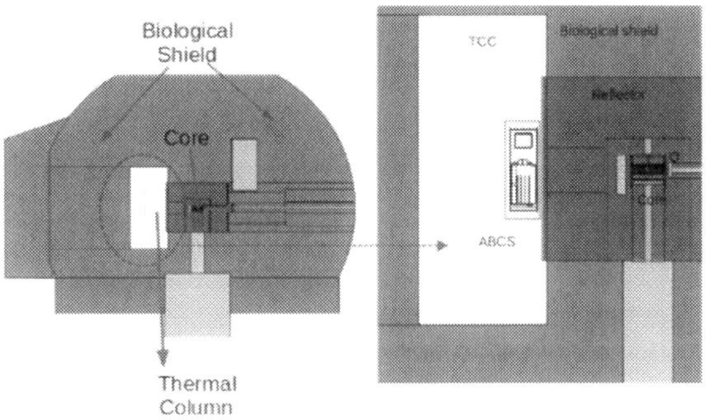

Fig. 9 Cross-section of the TAPIRO MCNP model layout. The left part of the figure reports the Thermal Column Cavity (TCC), the neutron reflector, the core, the radial channel, and the biological shield. The right part reports a magnification of the TCC, where is reported a section of the full ABCS model.

D. *Irradiation with neutrons @FNG*

For the scope of the present FLUKA simulation, we considered only the forward neutron flux generated by FNG in D-T mode (source emission rate 1×10^9 s^{-1}). Fig. 8 reports the distribution of the energy deposited along the CubeSat structure. More detailed modelling of the D-T neutron source will be implemented in the next phase of experimental design.

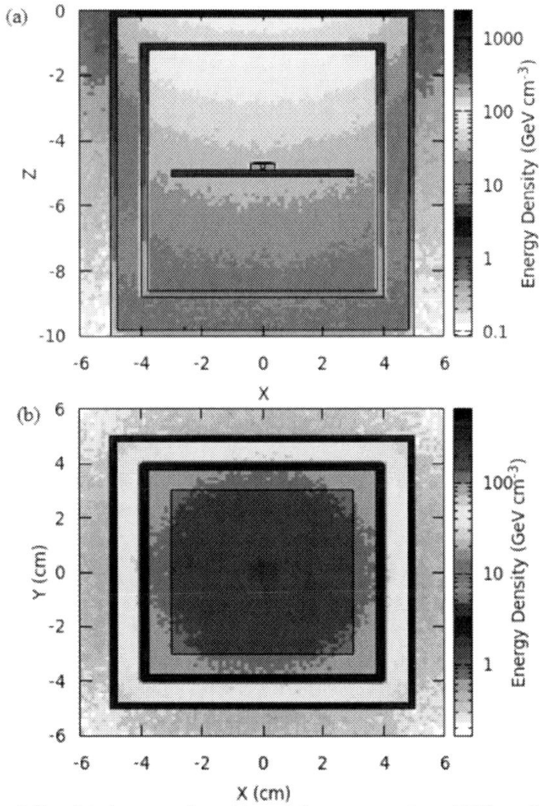

Fig. 8 Simulated energy deposition of the neutrons from FNG on the CubeSat. Fig. 8. (a) shows the axial interaction of the neutrons (coming from the upper part of the figure) with the satellite. Fig. 8. (b) is a sketch of the integral dose in the radial direction.

 https://doi.org/10.1051/epjconf/202328804009

E. *Comparison of the simulated response*

The SI1MEVNE fluence allows comparing damages level imparted from different kinds of particles. Using the distributed facility, we can reproduce equivalent damages of days or years (according to the selected beam) of exposure to the orbital conditions in one hour of irradiation. The response difference between doses on radFET and the PCB card depends on the irradiation geometry (positions and beam spot sizes) that are not controlled in the orbital conditions. Tab. IV reports a comparison between the Van Allen dose rate and the simulated ones acquired during the irradiation in the distributed facility. To compare the damages from different particles the values of the corresponding SI1MEVNE fluences are reported.

TABLE IV
COMPARISON OF THE SIMULATED IRRADIATIONS

Radiative source	**Dose Rate to radFET active part**	**Dose Rate on PCB card**	**SI1MEVNE fluence on PCB Card after 1-hour irradiation**	**Years of equivalent damage in orbital condition to the PCB card**
	rad/day	rad/day	p/cm^2	
TOP_IMPLART @ 1 nA	962.57	145.75	7.67×10^{10}	*2.32 years*
REX @ 1 μA	14544	1494	6.64×10^{13}	*2005 years*
FNG (1 x 10^9 s^{-1})	4.7×10^{-4}	3.2×10^{-3}	7.07×10^{9}	*77.93 days*
TAPIRO (1kW)	155.4	0.366	8.83×10^{11}	*26.67 years*
Van Allen belt	7.7	9.3×10^{-4}	3.78×10^{6}	

The estimated dose rates to the radFET grow proportionally to the intensity of the particle fluxes and the equivalent damages fluences grow accordingly.

As reported in Table IV, it is apparent that all the considered facilities greatly outperform the equivalent damages fluences of the orbital source term imparted to the PCB card. The greater effect is obtained using the electrons from the REX accelerator. To confirm the estimated trend work is in progress to realize the homologous experiments in each facility. Probably, besides the radFET the use of dedicated sensors for measuring the Non-Ionizing Energy Loss (NIEL) could be planned. In this manner, we will have direct and indirect confirmation of the facilities' performance.

V. CONCLUSIONS

This work was the first approach to consider the possibility of using ENEA irradiation facilities suite as a distributed laboratory to investigate the effect of mixed radiation fields.
The irradiation for a CubeSat-like structure is the occasion to test the potential of such a distributed facility. The simulations show the possibility of obtaining damages equivalent to years of exposure to space radiation in our facility with a few hours of irradiation. Examining the energy range covered by our electron and proton accelerator, we found that the irradiations represent Low Earth Orbit radiation source terms and up to the Van Allen belt, where the contribution of GCR is not dominant. Future work will be addressed to refine some simulation models like the FNG one. The second experimental phase has begun with the production of the prototypes of the satellites which will soon be irradiated. As a long-term activity, we will foresee the design and execution of similar experiments to European facilities with higher energetic particle output, and based on the experience gained, we will begin studying a national neutron spallation source and heavy ions accelerator to be located in one of the ENEA research centers.

REFERENCES

[1] N. Burgio, et al. "Modelling the interaction of the Astro Bio Cube Sat with the Van Allen's Belt radiative field using Monte Carlo transport codes." *Radiation Detection Technology and Methods* 6.2 (2022): 262-279.
[2] IAEAb, 2017. Research Reactors for the development of materials and fuels for innovative nuclear energy systems, IAEA NUCLEAR ENERGY SERIES NO. NP-T5.8, International Atomic Energy Agency, Vienna, Austria.
[3] J. W. G., Thomason, "The ISIS spallation neutron and muon source—the first thirty-three years." Nuclear Instruments and Methods in Physics Research Section A: Accelerators, Spectrometers, Detectors and Associated Equipment 917 (2019): 61-67.
[4] J. Kohlbrecher., and W. Wagner. "The new SANS instrument at the Swiss spallation source SINQ." Journal of applied crystallography 33.3 (2000): 804-806.
[5] T. E., Mason, et al. "The Spallation Neutron Source in Oak Ridge: A powerful tool for materials research." Physica B: Condensed Matter 385 (2006): 955-960.
[6] Maekawa, Fujio, et al. "First neutron production utilizing J-PARC pulsed spallation neutron source JSNS and neutronic performance demonstrated." Nuclear Instruments and Methods in Physics Research Section A: Accelerators, Spectrometers, Detectors and Associated Equipment 620.2-3 (2010): 159-165.
[7] https://www.enea.it/it/Ricerca_sviluppo/documenti/nucleare/TAPIRO.pdf
[8] M. Martone, et.al. J. Nucl. Mat. 212-215, 1661 (1994)
[9] M. Angelone et al., Rev. Sci. Instr. **67**, 2189 (1996).
[10] A. Pietropaolo et al. "The Frascati Neutron Generator: A multipurpose facility for physics and engineering", IOP Conf. Series: Journal of Physics: Conf. Series 1021 (2018) 012004, doi:10.1088/1742-6596/1021/1/012004.
[11] E. Rhodes, C.E. Dickerman, A. De Volpi, C.W. Peters, IEEE Trans. Nucl. Sci. 39, 1041 (1992).
[12] V. Surrenti et al. Proc. 14th International Particle Accelerator Conference, 2394-2397, TUPM092 (2023), doi:10.18429/jacow-ipac2023-tupm092
[13] M. Vadrucci et al., Nuclear Inst. and Methods in Physics Research, A 930 (2019) 126–131.
[14] A. Holmes-Siedle, and L. Adams. "RADFET: A review of the use of metal-oxide-silicon devices as integrating dosimeters." International Journal of Radiation Applications and Instrumentation. Part C. Radiation Physics and Chemistry 28.2 (1986): 235-244.
[15] G. Battistoni, et al. "Overview of the FLUKA code." Annals of Nuclear Energy 82 (2015): 10-18.
[16] N.Burgio, "Radiometric impact assessments and shielding of CubeSat class satellites in interaction with orbital radiation fields and representativity of the calibration procedure and radiation damage tests in the ground facility." PhD thesis, DIAEE, Sapienza University of Rome, Rome, Italy, 2023.
[17] N. Burgio, private communication, April 2023.
[18] https://www.varadis.com/products/
[19] http://www.tscsrl.net/Home.html
[20] N. Burgio, et al. "Monte Carlo simulation analysis of integral data measured in the SCK-CEN/ENEA experimental campaign on the TAPIRO fast reactor. experimental and calculated data comparison." Nuclear Engineering and Design 273 (2014): 350-358.

Flux and Nuclear Heating Rate Calculations for CALORRE Irradiation in MITR

S. Hauptman[1,*], A. Volte[2], G. Kohse[1], M. Carette[2], A. Lyoussi[3], C. Reynard-Carette[2]

[1]Massachusetts Institute of Technology, Nuclear Reactor Laboratory, Cambridge, Massachusetts, USA
[2]Aix Marseille Univ, Université de Toulon, CNRS, IM2NP, Marseille, France
[3]CEA/DES/IRESNE/DER, Section of Experimental Physics, Safety Tests and Instrumentation, Cadarache, F-13108, Saint Paul-lez-Durance, France

(*) hauptman@mit.edu

***Abstract*— This work summarizes neutronic calculations performed for the CALORRE differential calorimeter specifically designed by Aix Marseille University to inform the irradiation campaign planning for testing in MITR within the framework of the CALOR-I research program. MCNP software was used to predict the neutron and gamma flux spectrum, and total nuclear heating rate to the components of the calorimeter in a variety of positions. Results were compared to evaluate spatial bias, core loading effects, optimization for axial position, and provide input data for thermal multi-physics modelling.**

***Keywords* — Calorimeter, Nuclear Heating Rate, Neutronics.**

I. Introduction

The CALOR-I research program, (compact-CALORimeter Irradiations inside the MIT Research Reactor) began in 2020 within the framework of the joint LIMMEX laboratory between Aix Marseille University (AMU), the CEA and the CNRS. This program is funded by A*Midex Foundation. It is led AMU and involves two partners: the Nuclear Reactor Laboratory at MIT and the CEA. The first two aims of this program focused on design of a new compact prototype of the CALORRE calorimeter [1] and performance of realistic condition testing in laboratory thanks to calibration protocol, both of which have been successfully completed. The final goal is to test a dedicated in-core water loop multi-sensor device in the MIT Research Reactor (MITR) to verify measurement performance and validate the predicted nuclear heating rate using the new reduced-size prototype of CALORRE by imposing several conditions (reactor power levels, convection intensity) and by applying different measurement methods.

In order to support the completion of this project, accurate predictions of the in-core environment and the effects on components are needed to inform irradiation campaign planning and instrumentation design. These predictions investigate the energy spectrum for both the neutron flux and gamma flux in a representative volume, the total heating rate expected in the materials, and the potential for spatial bias and core loading effects to influence the measurement.

II. Description

A. MITR Water Loop Facility

The Massachusetts Institute of Technology (MIT) Research Reactor (MITR), is a light-water cooled and moderated tank-type reactor, licensed for operation up to 6 MW_{th}. The MITR has several experimental facilities, including three in-core positions for irradiation over a 10-week operating cycle. The position of interest for the CALOR-I project is the B3 position, in the middle ring of the core which is designated for high-temperature water loop (HTWL) facilities. A mockup of the core tank with the HTWL seated in the upper and lower grid plate is shown in Figure 1.

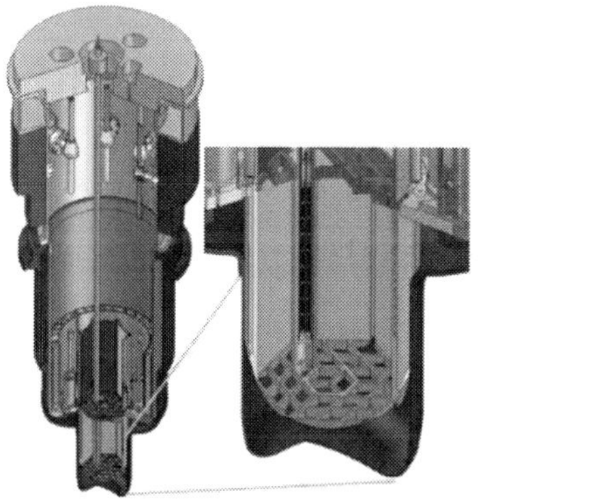

Fig. 1. Diagram of installed HTWL facility in B3 position. Fuel elements are omitted for clarity.

The HTWL allows for accurate control of in-core experiment temperature and flow conditions while exposing the installed sample components to high neutron and gamma flux. A unique feature that makes MITR an ideal choice for testing is the similar flux spectrum and power density to a commercial pressurized water reactor (PWR) core. The independent water loop provides the control functions to tailor the coolant chemistry, temperature, flow, and pressure to desired experimental conditions without adversely affecting MITR operation [2].

B. Calorimeter

The CALORRE capsule is composed of two calorimetric cells (a sample cell and a reference cell). Each cell has a central alumina and nichrome heating element, a structural holder, and

8 equally spaced fins connecting the cell to the outer wall of the calorimeter capsule. The sample cell has the stainless-steel material (SS316L) of interest between its heater and fin ring, and the reference cell leaves a gap around the heating element to be filled by di-nitrogen. The cells are superimposed and are 29.55 mm apart. The various thermocouples (2 per cell) and connection wiring (4-wire heater) will be routed through the inside of the capsule and out the top to connect back to the measurement system outside the core. The calorimeter capsule will be placed thanks to a specific device in the water loop autoclave which is monitored by its own MITR controllers and indications [3].

III. Model Parameters and Methodology

A. Model Assumptions

Nominal STP densities were used for the modeling of the capsule components. Figure 2 shows the comparison between the mechanical model and the visualization of the geometry in the MCNP input deck. The smaller instrumentation masses of the wires and thermocouples were neglected from the neutronic input deck as they are unlikely to impact the tally results of the larger components. The fixtures and structural supports holding the capsule itself in place in the HTWL were also neglected due to their distance from the region of interest. The top plug of the capsule and the nose cone have non-negligible mass by comparison and were included, the shape of the capsule and more specifically the displacement of the water volume within the HTWL was preserved.

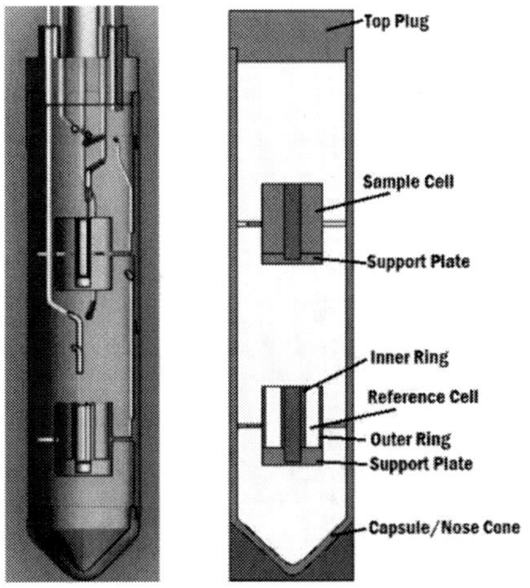

Fig. 2. CAD drawing (left-hand section) created by A. Volte, converted into MCNP input geometry (right-hand section) with labels corresponding to component designations for nuclear heating rate results.

B. Neutronic Problem Settings

The problems were all run in MCNP5 v1.60 using ENDF-VII nuclear data libraries [4, 5]. Comparison checks were done with MCNP6.2 using the same data resources to confirm agreement and compare different tally methodologies. MCNP6 was also used to verify the energy component due to electron deposition can be reasonably assumed using the default local deposition settings. For a coupled neutron-photon problem, MCNP5 was selected due to the reduced runtime. F6 neutron and gamma energy deposition tallies were used in order to separate the heating contributions.

Each problem was run with 800 active cycles and 800,000 neutron histories per cycle to obtain tally statistics with low uncertainties, particularly for the fin geometries. The CALORRE capsule components are geometrically very small and the fins relatively thin, which required more simulation time to satisfactorily converge. Additional random seeds were used and the resulting tallies merged for further variance reduction.

C. Normalization

All tally results were normalized to a reactor power of 6 MW. Measurements during the irradiation campaign may be performed at various power levels but the correction to lower power is linearly proportional and does not require separate simulation results. MCNP does not have the capability to track energy deposition from the delayed gamma component. However, the total gamma heating rate for MITR in-core irradiations can be calculated using established coefficients following (1) [6].

$$\gamma_{total} = \gamma_{prompt} + \gamma_{delayed} = 1.53 * \gamma_{prompt} \quad (1)$$

D. Problem Setup

The axial flux profile across the MITR core is not sharply peaked but has a changing gradient. To account for differences as a function of height, pairs of problems were run centering first the sample cell and then the reference cell at a fixed axial point. Five locations were chosen 5 cm apart ranging from 15 cm below the core centerline to 5 cm above the core centerline. These comparison predictions allow for a best estimate of the correction factor that may need to be applied during the actual measurement to obtain the net difference in measurements between the two cells of the calorimeter.

It was also investigated to see if local fuel loading had a notable effect on the calculated heating rates. Unlike most other reactors, MITR does not have a fixed fuel loading or shuffling pattern and both the burnup and orientation in the elements that border the B3 experimental position can vary between operating cycles. Four core loading patterns that were representative of typical loadings and included the different potential configurations of the border elements were used. The power output for each nodalized section of the closest fuel plates was calculated and compared to the resulting total nuclear heating rates.

The other area of investigation concerned the orientation of the fins. The B3 position is not directly in the center of the core and although it receives a relatively uniform radial flux, both neutron and gamma, if there is a directional component influencing the total nuclear heating rate it could necessitate adding rotation to the facility design and irradiation campaign. The nuclear heating rates were calculated in each individual fin for two capsule positions: one in the peak flux position with the sample cell centered at 5 cm below the core centerline, and one in a low flux position with the sample cell centered 15 cm below the core centerline. For these two axial positions, a problem run

https://doi.org/10.1051/epjconf/202328804010

was tallied with five different starting random seeds and four different core loading files.

IV. Results

A. Flux Profiles

The neutron and gamma flux profiles seen by each of the capsule components, matches closely with the known flux profiles for MITR in-core positions. The total neutron flux, shown in Figure 3, has a peak around 2.4 x10^{14} n. $cm^{-2}.s^{-1}$ and drops to 1.2 x10^{14} n. $cm^{-2}.s^{-1}$at the higher core planes. The higher the total flux, the more tightly the values for each component are clustered. This again follows the known core profile trends where the flux gradient is lowest across the peak region, and increases towards the bottom and top of the active core.

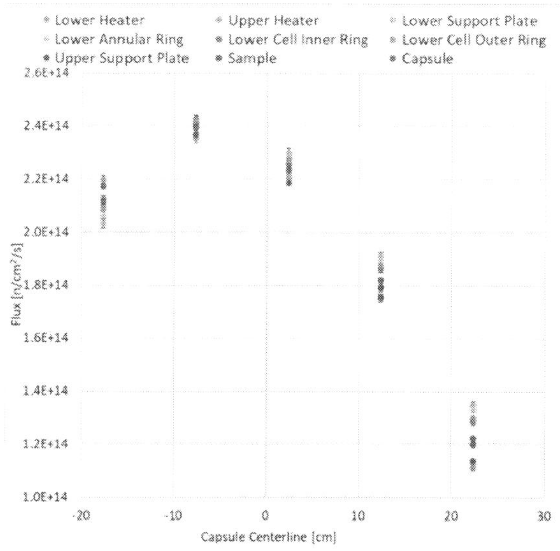

Fig. 3. Neutron flux in capsule components over several axial positions with 1σ error bars.

In Figure 4, the gamma flux also peaks at the same core height and displays a similar clustering and gradient behavior, but is overall less peaked and more uniform throughout the core.

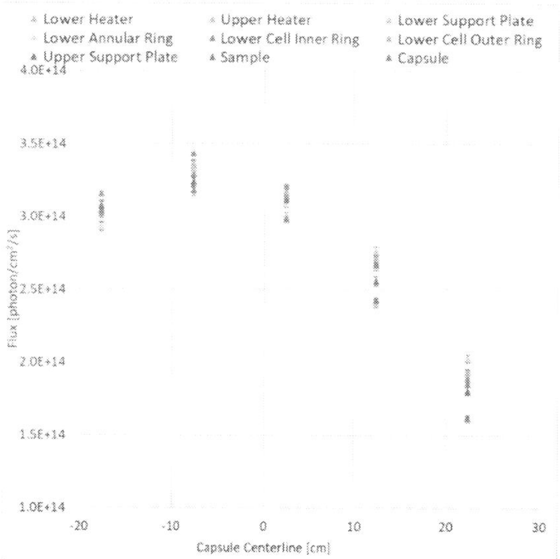

Fig. 4. Gamma flux in capsule components over several axial positions with 1σ error bars.

The gamma spectrum was also tallied at four quadrant locations on the exterior of the capsule to investigate any local shielding effects that might occur and how that could influence the placement of a second independent measurement to be performed by a commercial gamma thermometer. The spectrums were nearly identical which supported no physical limitation on radial positioning as the gamma flux is negligibly perturbed by the calorimeter.

B. Nuclear Heating Rates

The total nuclear heating rates across all non-fin SS316L components and positions for the statistical combination of tallies, ranged from 1.59 ± 0.02 $W.g^{-1}$ to 2.01 ± 0.02 $W.g^{-1}$. All non-fin heating rates as a function of capsule position are shown in Figure 5. The relative order of which component saw the highest heating also was fairly consistent above the peak flux point and below it. The expected inversion across the plane of peak flux, about 5 cm below the core centerline, was observed as the reference and sample cells switch which one is closer to the higher flux point (i.e. in the top half of the core it is anticipated that the reference cell will see more heating, whereas in the bottom half of the core the sample cell is expected to see more heating).

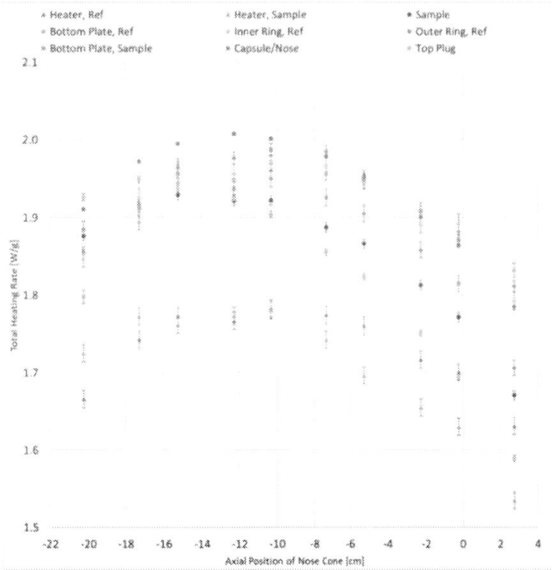

Fig. 5. Total nuclear heating rates in each component over several axial positions. Nuclear heating rates have been calculated from the statistical combination of three random seed runs and normalized to 6 MW.

C. *Bias Checks*

Figure 6 shows the fin designation with respect to core position as well as the positions bordering B3. Despite the different node power outputs and orientations between the neighboring element positions in core configurations 238, 239, 240, and 242. An example of the variation in power profile is demonstrated by the 10th axial nodes for core configurations #240 and #242 in Figure 7. No impact was observed on the calculated nuclear heating rates tallied in the fins. Particularly of note is the core loading of #240 with the C5 element in a different posture and an inverted power distribution throughout the plates did not impact the local heating rate.

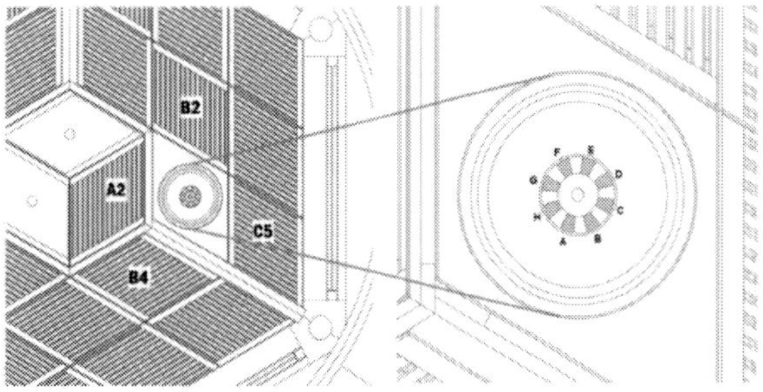

Fig. 6. Cross section of CALORRE capsule in B3 core position (left-hand section) with zoomed in image (right-hand section) showing sample cell fins and label designation for orientation. Positions of neighboring elements are also labelled.

240				242			
A2	B2	B4	C5	A2	B2	B4	C5
1489	1383	1517	1789	1568	1432	1466	1077
1411	1346	1472	1450	1508	1337	1407	1055
1363	1353	1468	1249	1468	1335	1389	1016
1380	1319	1422	1133	1463	1328	1360	988
1383	1299	1405	1086	1425	1310	1346	1020
1390	1307	1412	1043	1440	1322	1345	1007
1366	1311	1423	1012	1421	1320	1313	992
1361	1322	1389	1034	1432	1355	1341	1008
1339	1315	1374	1025	1451	1323	1296	1005
1352	1343	1367	1052	1438	1347	1297	1044
1344	1358	1318	1023	1452	1376	1294	1016
1402	1390	1346	1081	1470	1408	1322	1019
1390	1380	1393	1098	1463	1422	1317	1057
1393	1427	1445	1152	1466	1447	1352	1039
1454	1522	1521	1232	1557	1514	1427	1033

Fig. 7. Power profiles of node 10 of each neighboring fuel element to B3, for two of the four analyzed core loadings (#240 and #242). The 10th node borders the sample cell with the capsule in the high flux position, encompassing the -7.104 cm to -3.552 cm axial section. Power output values for each of the 15 fuel plates are listed in watts and normalized to nominal operating power of 5.7 MW.

The nuclear heating rate results of the two flux positions chosen for fin bias checks are given in Figures 8 and 9, for the high flux and low flux positions respectively. In each, the highest predicted nuclear heating rates are shown in red and the lowest in blue.

	Reference Cell				Sample Cell			
Core	238	239	240	242	238	239	240	242
A	1.95	1.96	1.96	1.96	1.94	1.92	1.97	1.98
B	1.93	1.92	1.96	1.99	1.92	1.89	1.95	1.94
C	1.94	1.96	1.97	1.95	1.92	1.91	1.96	1.96
D	1.91	1.95	1.97	1.95	1.91	1.92	1.94	1.96
E	1.96	1.97	1.98	2.00	1.95	1.94	1.97	1.98
F	1.96	1.99	1.99	1.99	1.96	1.97	1.98	1.97
G	1.98	1.95	2.01	2.00	1.97	1.96	1.99	2.00
H	1.95	1.98	1.96	1.98	1.95	1.96	1.99	2.00

Fig. 8. Fin total nuclear heating rate results for high flux position (5cm below core centerline) in four core loadings. Highest nuclear heating rates are shown in red and lowest in blue. All rates are a statistical combination of five random seed runs and are given in $W.g^{-1}$. 1σ uncertainties are 0.07 $W.g^{-1}$ for all fins.

	Reference Cell				Sample Cell			
Core	238	239	240	242	238	239	240	242
A	1.74	1.72	1.74	1.75	1.80	1.80	1.85	1.83
B	1.73	1.71	1.73	1.73	1.80	1.80	1.83	1.80
C	1.70	1.74	1.72	1.74	1.80	1.82	1.81	1.83
D	1.73	1.71	1.73	1.73	1.76	1.78	1.82	1.79
E	1.73	1.70	1.75	1.74	1.81	1.77	1.83	1.85
F	1.74	1.74	1.74	1.75	1.83	1.82	1.83	1.85
G	1.74	1.73	1.74	1.75	1.82	1.81	1.84	1.84
H	1.74	1.74	1.75	1.76	1.79	1.82	1.85	1.87

Fig. 9. Fin total nuclear heating rate results for low flux position (15cm below core centerline) in four core loadings. Highest nuclear heating rates are shown in red and lowest in blue. All rates are a statistical combination of five random seed runs and are given in $W.g^{-1}$. 1σ uncertainties are 0.07 $W.g^{-1}$ for all fins.

While the highest rates were observed in the tallies for the G and H fins in both cells in each position, the difference between the highest and lowest observed heating rate is within one standard deviation of the tally. Therefore, there is no statistically significant support for a spatial bias in the fins that might be observed over random variation. The same reasoning extends to the fins with the lowest observed heating rates, although the minimum values appeared across a larger set of fins B, C, D, and E.

V. Conclusions

The alignment of flux profiles and nuclear heating rate trends supports the early predictions of in-core response and affirms the reliability of the model results. The consistency within each problem run of nuclear heating rates, relative to the axial positioning also implies that the resulting nuclear heating rates are accurate inputs for the subsequent 3-D thermal calculations realized by AMU thanks to COMSOL Multiphysics software. It also suggests that with data measurements at a few points of interest, the shape of the resulting nuclear heating rates could be accurately estimated with interpolation. The magnitudes of the nuclear heating rates are all within the expected bounds of what has previously been measured and modelled for similar materials in similar irradiation conditions of MITR.

The lack of influence from core loading significantly simplifies the project analysis as it removes the necessity for re-performing calculations with the actual core loading at the time of irradiation or having to adjust the irradiation schedule or core plan to avoid unnecessary local effects. Likewise, the lack of evidence supporting a spatial bias in the fin orientation indicates that a rotation of the calorimeter during irradiation is not strictly necessary to obtain even heating and thermal transfer. These verifications increase confidence in the irradiation facility design and establishes a flexible envelope for obtaining different test measurements for calorimeter validation and verification.

Acknowledgment

The CALOR-I project leading to this publication has received funding from the Excellence Initiative of Aix-Marseille University - A*Midex, a French "Investissements d'Avenir programme".

References

[1] A.Volte, M. Carette, A. Lyoussi, G. Kohse, J. Rebaud, V. Valero, C. Reynard-Carette, "Review of CALORRE Calorimeter Characterizations Under Laboratory and Irradiation Conditions", IEEE Trans. Nucl. Sci., vol. 69, no. 4, pp. 840-848, April 2022, doi: 10.1109/TNS.2022.3150148

[2] MIT Nuclear Reactor Laboratory, "In-Core Experiments," Facilities, https://nrl.mit.edu/facilities/in-core (accessed July, 2023).

[3] A. Volte, S. Hauptman, D. Carpenter, M. Carette, A. Lyoussi, G. Kohse, C. Reynard-Carette, "Characterization of a new reduced-height CALOREE differential calorimeter for CALOR-I Irradiation in MITR", ANIMMA conference, Lucca, Italy, 2023.

[4] X-5 Monte Carlo Team, MCNP – A General Monte Carlo N-Particle Transport Code, Version 5, Volume II: User's Guide. April 24, 2003.

[5] M.B. Chadwick, P. Obložinský, M. Herman, N.M. Greene, R.D. McKnight, D.L. Smith, P.G. Young, R.E. MacFarlane, G.M. Hale, S.C. Frankle, A.C. Kahler, T. Kawano, R.C. Little, D.G. Madland, P. Moller, R.D. Mosteller, P.R. Page, P. Talou, H. Trellue, M.C. White, W.B. Wilson, R. Arcilla, C.L. Dunford, S.F. Mughabghab, B. Pritychenko, D. Rochman, A.A. Sonzogni, C.R. Lubitz, T.H. Trumbull, J.P. Weinman, D.A. Brown, D.E. Cullen, D.P. Heinrichs, D.P. McNabb, H. Derrien, M.E. Dunn, N.M. Larson, L.C. Leal, A.D. Carlson, R.C. Block, J.B. Briggs, E.T. Cheng, H.C. Huria, M.L. Zerkle, K.S. Kozier, A. Courcelle, V. Pronyaev, S.C. van der Marck, "ENDF/B-VII.0: Next generation evaluated nuclear data library for nuclear science and technology", Nucl. Data Sheets 107(2006)2931.

[6] K. Sun, "Nuclear Heating Rate of Bounding Case FS-2 Design" Internal MIT Nuclear Reactor Laboratory Document. April 2014.

Fuel Irradiation Devices Test of Feedthroughs Equipped With Optical Fibers in support of the development of innovative instrumentation.

Authors: S.Gaillot (1),*, C.Destouches (2), G.Cheymol (3), J.Brinster and Al (1)

(1) CEA, DES, IRESNE, Nuclear Technology Department
(2) CEA, DES, IRESNE, Reactors Studies Department
(3) CEA, DES, ISAS, Chemical & Physical Department
(1,2) CEA, French Alternative Energies and Atomic Energy Commission, Division of Energies, CEA Cadarache.13108 Saint Paul lez Durance. France.
(3) Université Paris-Saclay, CEA, Service d'Études Analytiques et de Réactivité des Surfaces, 91191, Gif-sur-Yvette, France

(*) stephane.gaillot@cea.fr

***Abstract*: As part of the activities for developing experimental devices for research reactors and associated test loops, some concern the development and qualification of instrumentation.**
The on-board instrumentation makes it possible to monitor online the evolution of physical parameters of samples and/or components subjected to different stresses (thermal, hydraulic, chemical and nuclear, etc.).
In addition to the so-called "classic" instrumentation implemented in this type of equipment (pressure, temperature, flow, elongation), R&D actions are in progress in order to propose innovative instrumentation, able to improve the experimental offer and describe the physical phenomena with more accuracy.
This is the case of optical measurements which present many advantages (compactness, insensitivity to EM waves, complex measurements by optical interrogation of a single fiber).
In addition, actions are underway for the selection of optical fibers that can be used for applications in a nuclear reactor (in particular resistance to the irradiation).
The tests presented in this paper relate to the behavior of sealed feedthroughs for optical fibers to the thermal hydraulics conditions of light water reactors.
These tests are part of the safety demonstration in so-called degraded operating cases i.e. corresponding to a configuration where the optical fiber is placed directly in the process fluid (pressurized water). One note that in normal operation, the optical fiber is isolated from the process fluid by a jacket consisting of a metal capillary.
After an introduction and a brief presentation of the use of optical fibers in experimental equipment, the paper describes the tightness feedthroughs used, the tests carried out, indicate the main results obtained and opens up some perspectives for future development phases.

***Keywords:* Experimentation, Thermal Hydraulic tests, Feedthroughs, Optical Fiber**

I. Introduction

To carry out technological irradiations in Materials Research Reactors (MTR), experimental devices are implemented in the core or in the reflector.

These devices make it possible to reproduce on a small scale the physical stresses on material or fuel samples. The stresses applied to the samples can be of different types: thermal, thermal-hydraulic, nuclear, chemical, etc.
These stresses come from the operating conditions of the reactor (neutron flux, gamma heating) or can be linked to the operations of the device (thermo-hydraulics, thermal, chemistry of the coolant, mechanical stress on the sample).

It is in this context that the resistance tests of tightness feedthroughs with optical fibers are identified.

These tests are carried out under light water reactor (LWR) conditions.

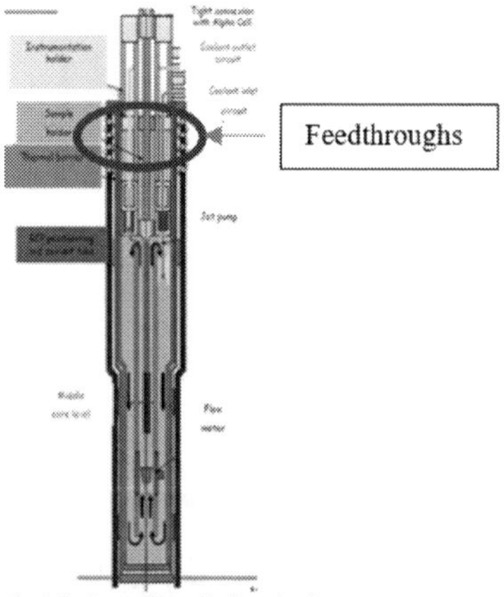

Fig. 1. Typical design of irradiation device

II. CONTEXT REMINDER

The realization of the experimental devices in a MTR is the result of a long and complex design process [1] taking into account the experience feedback on other reactors, the thermal hydraulics parameters in normal and degraded conditions, the safety analysis defining the number of constructive provisions (number of barriers) and management (operating area) in order to ensure reliable and long-lasting operation of the device as well as taking into account the current regulations (in particular the Nuclear Facilities Quality Order (INB) as well as French Nuclear Pressure Equipment Order ESP(N).

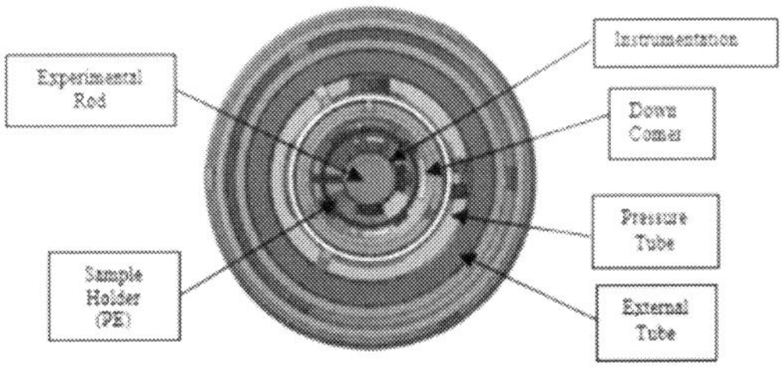

Fig. 2. Typical design of irradiation device cross section (CEA-Preliminary design) [7]

III. THE TIGHTNESS FEEDTHROUGHS

As mentioned above, the irradiation device constitutes a closed and sealed enclosure to accommodate the samples and also makes it possible to reproduce the experimental conditions necessary for irradiation. In order to communicate with the outside, the device is equipped with tightness feedthroughs of the fluid and electrical type, instrumentation allowing the test to be controlled and the evolution of the experimental parameters to be monitored online [2].

Depending on the safety analysis carried out, the sealed passages constitute a single or even double barrier connection between the process fluid (area where the test sample is placed) and the surrounding environment (reactor pool).

Another objective concerning the leak tight passages of devices in the reactor is to be able to maximize the integration of connections in a given geometry. This generally involves multiple watertight passages in order to increase the number of connections.

Generally, watertight passages allow different types of connections to be made:

- ✓ Fluid: pressure measurement, gas gaps, etc.
- ✓ Electrical: heating power supply,
- ✓ Instrumentation: flow meter, thermocouples, LVDT, nuclear measurements,

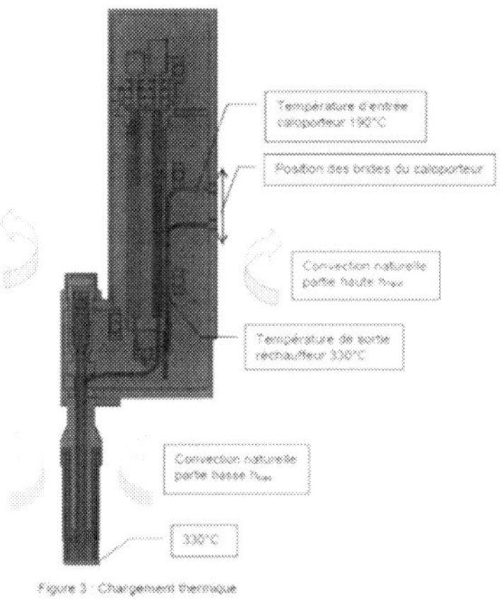

Fig. 3. Head of irradiation device
Typical configuration (illustration) [7]

Fig. 4. Example of device feedthroughs (illustration)

IV. OPTICAL MEASUREMENTS

In addition to the classic instrumentation, progress in R&D actions now makes it possible to consider other types of instrumentation. Optical measurement is part of this panel of innovative instrumentation that we seek to implement in irradiation devices.

Its advantages are numerous:

Geometry: an optical fiber has a diameter of about 200microns compared to 1 mm or more for a standard thermocouple. A hardened Fabry-Pérot type fiber optic elongation sensor for MTR application has a diameter of 2 mm compared to more than 10 mm for an LVDT [3].

FOS (Fiber Optic Sensors) allow measuring numerous parameters like vibrations, temperature, pressure, elongation, displacement, strain. Chemical measurements also possible.

Multipoint response by interrogation of a single fiber equipped with Bragg gratings [4] or response all along a fiber by analyzing Raman, Brillouin, Rayleigh backscattering. Immunity to electromagnetic interference

Progress in resistance to irradiation in particular through the appropriate choice of fiber material (cf. TESCA program with CEA & BR2 foreseen in 2023).

Hardened FOS or measurement using optical fibers for irradiation rig have already been developed (for temperature [ref 2 ter], elongation [3]) and tested under irradiation (ATR, MITR, BR2,...) up to more than 10E20 n/mc² and GGy. One studied application correspond to a sensor for measuring displacement / swelling. The physical principle is based on interferometric measurement of the Michelson type [5]. Moreover a non-contact measurement is also in development [6].

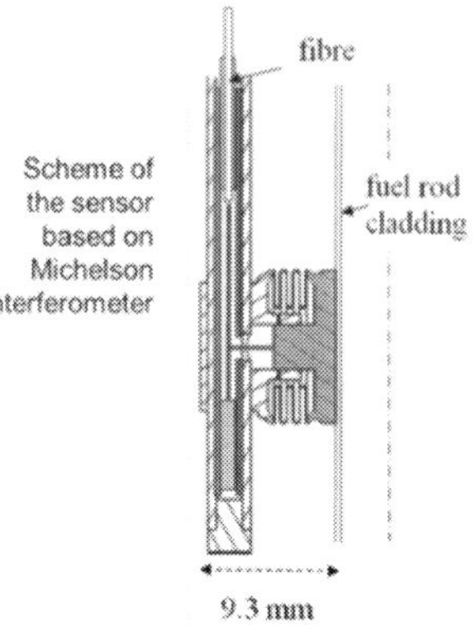

Fig. 5. Cladding deformation sensor (simplified scheme)

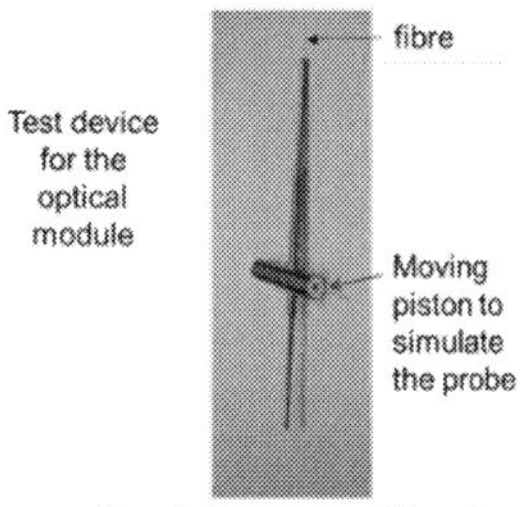

Fig. 6. Prototype with a piston simulating the feeler piece (CEA, R&D) [7].

V. OPTICAL FEEDTHROUGHS EQUIPPED WITH OPTICAL FIBERS

To implement optical measurements in a research reactor, specific sealed feedthroughs have been studied, produced and tested according to different experimental scenarios.

These watertight passages consist of mini metal tubes of diam. ext.1/16'' containing the Single Mode(SM) optical fiber coated with Al or Cu.

The sealing between the mini tube and the fiber is achieved by the use of special glue and manufactured by SEDI Company.
The watertight passages thus formed are fixed to a metal flange which isolates the experimental enclosure (the device) from the outside (reactor pool).

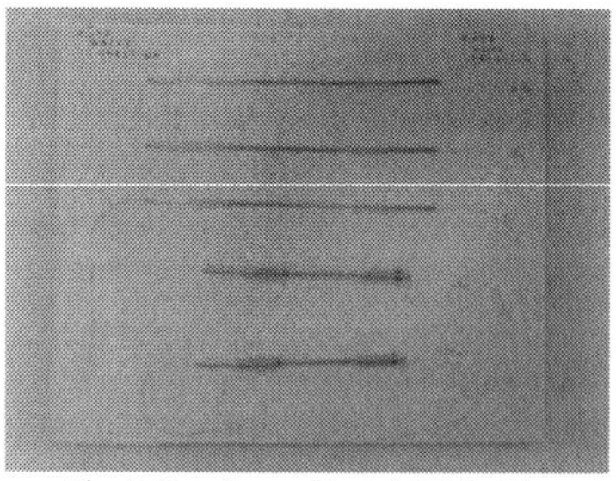
Fig. 7. Typology of tested OF feedthroughs [7]

The "standard implementation configuration" of an optical fiber in a device can be described as following:

The fiber is located in a metal capillary closed at its end (jacket) which physically isolates it from the surrounding environment (in this case the process fluid). This configuration prevents the degradation of the fiber's silica by dissolution in pressurized water. The mini metal tube protecting the fiber constitutes the first isolation barrier in terms of safety.

The watertight passage through the flange constitutes the second barrier in terms of safety.

The objective of the resistance tests of airtight passages with optical fiber is to test a so-called "degraded" configuration corresponding to the consequences of a degradation of the first barrier constituted by the mini-tube and protecting the fiber. This case falls within the field of investigation of "defence in depth" in the sense of security.

In this case, the fiber is in direct contact with the process fluid and the purpose of the tests is to verify the behavior of the sealed passage (second safety barrier) under these experimental conditions.

VI. EXPERIMENTAL TESTS CONFIGURATION

The configuration identified for the tests is described as follows:
• The watertight passages are fixed to the flange by various processes: welding, compression seal feating.

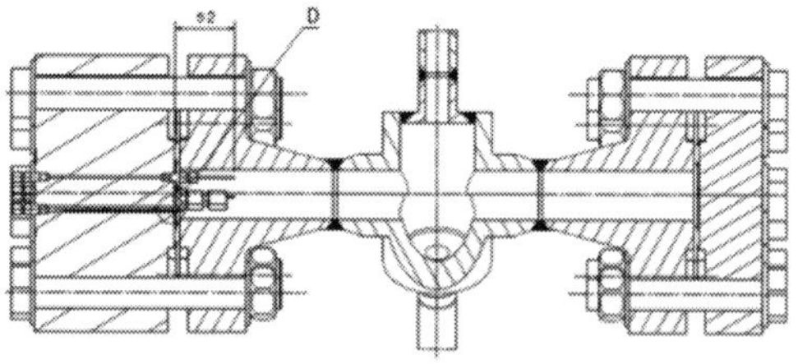

Fig. 8. Experimental test section. Cross section

https://doi.org/10.1051/epjconf/202328804011

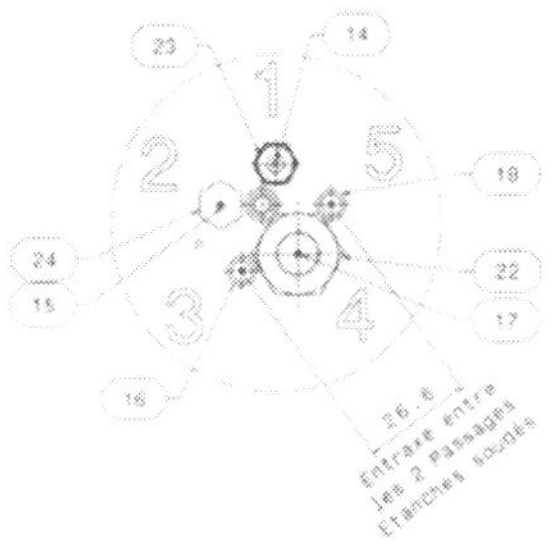

Fig.9: Experimental test section
Local view of optical feedthroughs

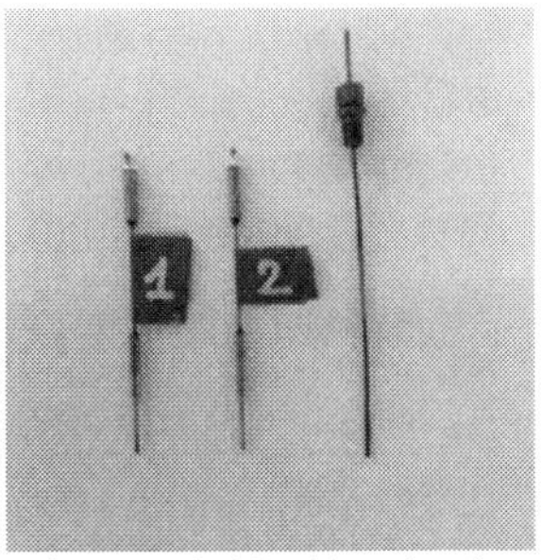

Fig.10: Optical Feedthrough to be welded (1) (2)
or mounted on a compression seal fitting

Note: For these tests, the screwed connections are positioned in the fluid in the enclosure (device). This configuration makes it possible to monitor the tightness of the sealed passage during the test. This configuration is specific to these tests and does not correspond to the "classic" configuration: seal fittings outside the flange.

Note: monitoring of the leak tightness of the leak tight passage during the tests is carried out by monitoring the pressure on capillaries fixed downstream of the leak tight passage.

Fig.11: BIKINI experimental loop (HP-HT)
Operator CEA-DTN-STCP-LETH [7]

Fig.12: Experimental
test section flanges

VII. EXPERIMENTAL WORKPLAN

The configuration identified for the tests is described as follows:

• The watertight passages are fixed to the flange by various processes: welding, compression seal feating.

In order to check the behavior of airtight passages with optical fiber in "degraded configuration", various tests were carried out:

The first test corresponds to the configuration of an ADELINE V0 [2] type device which would be equipped with optical fibers included in metal capillaries. For example, the optical fiber could figure FBG to produce temperature profiles.

The test carried out on the sealed passage corresponds to a situation where it is considered that the first seal (represented by the capillary) is lost or is degraded. In this case, the tightness of the leaktight passage is checked (second leaktightness).

For this test, the fiber is in contact with the primary fluid. The duration of the test is set at one week, typically corresponding to the duration of an ADELINE type test.

The thermo-hydraulic conditions of the test correspond to those prevailing at the level of the sealed passage at the head of the device, i.e. P=155b and fluid temperature =100°C.

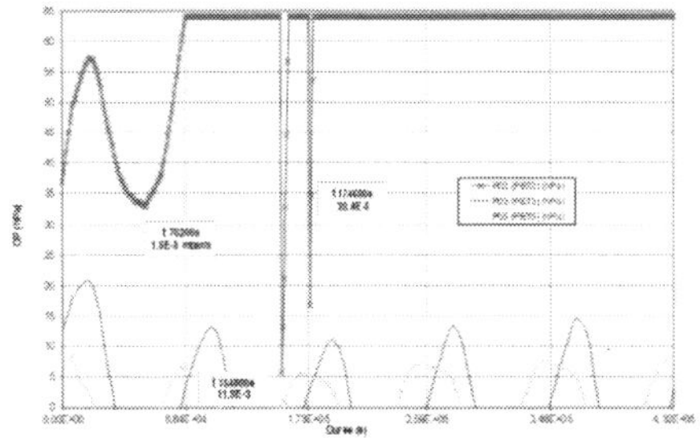

Fig.13: Control of the evolution of the pressure of individual passages tested
legend: yellow - red: (welded Pet) / blue (screwed Pet) [7]

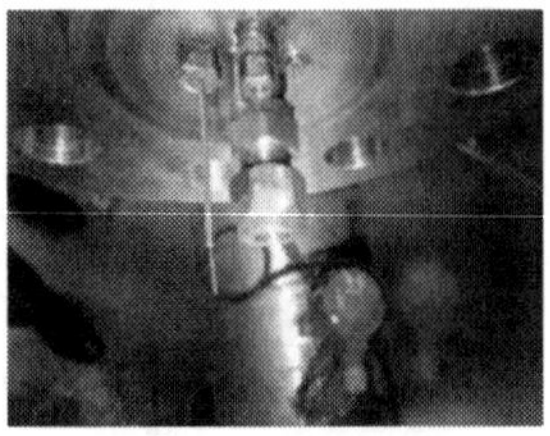

Fig.14: view of the feedthroughs after the 1st test campaign

The second test carried out corresponds to the resistance of the leak tight feedthrough to the following thermal-hydraulic conditions: 150b, 280°C.

For this test, the duration was set at 24 hours.

Finally, the 3rd test corresponds to the resistance of the leak tight feedthrough to the conditions of LWR reactors (155b, 320°C), also over a period of 24 hours.

For this test, water characteristics have been recorded:

pH=6, 23 at T =21°C.

VIII. FIRST RESULTS OBTAINED

At the end of the three tests campaigns carried out, the following results can be noted:

The technology of tightness feedthroughs with mini metal tube and special glue in order to create the seal between the optical fiber and the capillary is validated.

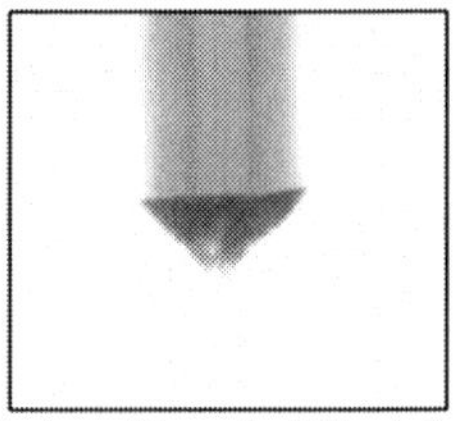

Fig 15: view of the optical fiber in the feedthrough

The assembly technology by welding between the sealed passage and the flange is validated.

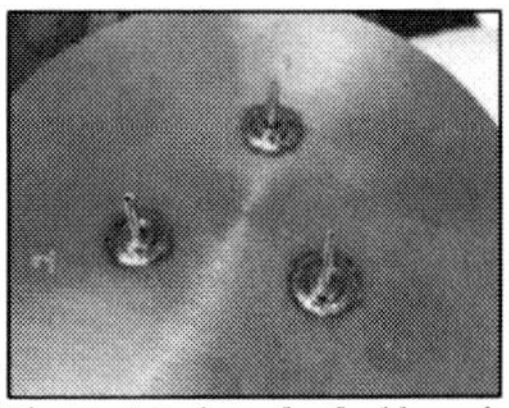

Fig.16: Mock-up for feedthrough welding qualification [R4]

The compression seal fitting technology remains to be confirmed.

Nevertheless, subsequent tests carried out at CEA-Saclay showed the feasibility of implementing and using this type of assembly.

Other tests have also been carried out to demonstrate that tightening the screws beyond what is necessary for sealing does not affect the optical signal to be transmitted.

Fig 17: Screwed connexion with optical feedthrough

IX. POST TESTS EXAMINATIONS

At the end of the tests carried out, the sealed passages were removed from the test flange. X-ray controls have been carried out in an attempt to determine the length of silica dissolved by the pressurized water during the tests.

Then, the watertight passages will be, if possible, cut out in order to carry out some final visual examinations.

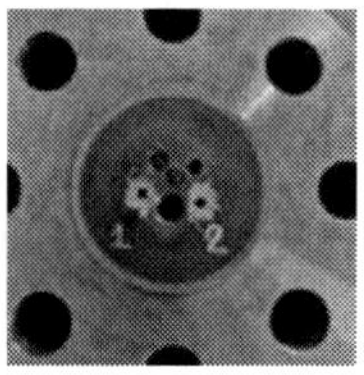

Fig 18a: Visual inspection Pressure cover with feedthroughs holes

Fig 18b: OF feedthrough (with Cu layer)

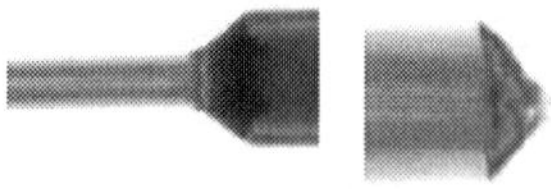

Fig 18c: XRays examination OF feedthrough Intermediate & lower parts

X. FIRST CONCLUDING REMARKS & PERSPECTIVES

The use of irradiation devices in research reactors requires the implementation of instrumentation to allow online monitoring of the physical parameters of the experiment.

In addition to the use of conventional sensors, the use of optical fiber sensors is promising.

It is in this context that watertight feedthrough resistance tests were carried out.

They made it possible to validate a tight passage technology and thus to have initial results in support of the associated safety demonstration.

On notes, that experiment has also been conducted to validate the behavior of these feedthroughs in the PWR conditions.

In the light of the first results obtained, the objective today is to continue the development of optical measurement in devices.

The next step will aim to study, produce and test, in representative thermo-hydraulic conditions, an instrumented sample holder equipped, among other things, with several optical lines.

This instrumented sample holder will make it possible to validate in particular the operations of multiple integration of optical fibers via the head plugs.

These actions will aim to consolidate the use of optical fibers for carrying out experimental measurements in research reactors.

XI. LIST OF REFERENCES

[1] PHYSOR 2014. Jules Horowitz Reactor. FRANCE DEVELOPMENT OF AN experimental loop INTEGRATING AN optimized IRRADIATION PROCESS.S.GAILLOT, S.VITRY & T.DOUSSON
CEA, DEN, CEA CADARACHE, F-13108 Saint-Paul-lez-Durance, France.

[2] IGORR 2009. JHR. The ADELINE irradiation loop in the Jules Horowitz MTR. Testing a LWR fuel rod up to the limits with a high quality level. Experimental process.
S.GAILLOT, D.PARRAT, G.LAFFONT, C.GONNIER, C.GARNIER
CEA, DEN, CEA CADARACHE, F-13108 Saint-Paul-lez-Durance, France.

[3] G. Cheymol et al., "High-Temperature Measurements With a Fabry–Perot Extensometer," in IEEE Transactions on Nuclear Science, vol. 67, no. 4, pp. 552-558, April 2020, doi: 10.1109/TNS.2020.2975707.

[4] G. Laffont et al., Temperature Resistant Fiber Bragg Gratings for On-Line and Structural Health Monitoring of the Next-Generation of Nuclear Reactors. Sensors 2018, 18, 1791. https://doi.org/10.3390/s18061791

[5] IGORR 2013 JHR Project. Fuel irradiation device. Innovative instrumentation proposal for experimental real time measurement.
S.GAILLOT, G.CHEYMOL
CEA, DEN, CEA CADARACHE, F-13108 Saint-Paul-lez-Durance, FRANCE
CEA, DEN, CEA SACLAY, F-91191 Gif sur Yvette, FRANCE.

[6] M. Agoyan et al., « Confocal chromatic sensor for displacement monitoring in research reactor", ANIMMA 2021 Conference, Jun 2021, Prague, Czech Republic.

[7] IGORR-RRFM 2019. INSNU Project. Fuel irradiation devices
Test of sealed passages with optical fibres in support of the development of innovative instrumentation.
S.Gaillot, G.Cheymol, J.Brinster, D.Picard, G.Chichereau , N.Housse, P.Degousee , C.Destouches

The CABRI fast neutrons Hodoscope: CABRI model and signal-to-mass conversion charts

Vincent Chevalier[1,*] and Jacques Di Salvo[1]
[1]IRSN, PSN-RES/SEREX/L2EP, Cadarache, F-13108 Saint-Paul-Lez-Durance, France

(*) vincent.chevalier@irsn.fr

Abstract— **The CABRI experimental pulse reactor, located at the Cadarache nuclear research center, southern France, is devoted to the study of Reactivity Initiated Accidents (RIA). The hodoscope, installed in the CABRI reactor, is a unique online fuel motion monitoring system, operated by IRSN. This equipment is dedicated to the measurement of the fast neutrons emitted by the tested rod, in real time (with a rate of 1ms), during the power pulse. It is one of the distinctive features of the CABRI reactor facility, which is operated by CEA. To support the experimental task around CABRI reactor, by the experimenters who work on the Hodoscope, a Monte Carlo model, using the MORET code, is used by IRSN. This paper presents the main outcomes obtained during the reactor commissioning tests functioning, using Hodoscope results compared to MORET calculations, which proves the validity of the CABRI MORET model. Furthermore, we show how MORET code is used to build the signal-to-mass conversion charts of the Hodoscope.**

Keywords **—CABRI, Hodoscope, RIA transient, fast neutron detection system, MORET simulations.**

I. Introduction

FOR enhancing safety of Nuclear Power Plants (NPP), the French Institute for Radiological Protection and Nuclear Safety (IRSN) carries out some experimental programs in order to improve the understanding of the fuel behavior under severe accident conditions. One of them is the CABRI International Program, managed and funded by IRSN under an OECD/NEA agreement. The program is devoted to the study of Reactivity Initiated Accidents (RIA) in representative Pressurized Water Reactor (PWR) conditions [1] .

For this purpose, the renovation of the CABRI facility has been conducted by the French Alternative and Atomic Energies (CEA), its operator. This pool type reactor is designed to submit a test rod placed in the center of the core to a RIA. A Pressurized Water Loop, to reproduce the thermal hydraulic conditions of a Pressurized Water Reactor (PWR), for testing the cladding of a fuel rod under prototypical conditions of RIA in PWRs, has been implemented in the center of the core. This so-called test cell is filled with the experimental device during a CABRI transient. It can also receive a system maintaining different dosimeters, with or without water filling the loop during these kinds of experiments.

The CABRI facility is equipped with two nondestructive measurement systems operated by IRSN teams:

- The IRIS facility, for performing X-ray radiography and tomography imaging before and after a power transient thanks to a linear electron accelerator, as well as quantitative gamma scanning analyses [2] .
- The Hodoscope, an online fuel motion measurement system, which aims at analyzing the fuel motion deduced from the detection of fast neutrons emitted by the tested rod, in real time (with a time step of 1ms) during the transient [3] .

In Section II of this article, a brief description of the hodoscope measuring system will be given. In Section III, comparison between dosimetry and Hodoscope's detectors measurements are shown. Then, section IV, these experimental results will be compared to numerical Monte Carlo simulations. These calculations are thereafter used to build signal-to-mass conversion charts, that are presented in section V. Finally, conclusion and perspectives of work are given.

II. General description of the system

The CABRI reactor has been equipped with the hodoscope since 1977. This system measures in real time the behavior of the test rod in the center of the CABRI core during the transient. The hodoscope is able to quantify the amount of fuel ejected in the milliseconds following failure, time-dependent axial fuel mass distributions and to follow the fuel clusters after failure. In addition, the initial state of the test rod is precisely known thanks to the hodoscope measurement as well as the fissile length and the axial power profile.

The hodoscope centerpiece is a 3 m steel collimator (in green in Fig. 1) of 3 columns and 51 rows, so 153 channels. Behind each one of these line-of-sights, a ^{237}Np Fission Chamber detector (FC) and a Proton Recoil proportional detector (PR) measure the fast neutrons coming from the core and test rod. Two different technologies of detectors are used in CABRI in order to follow all the experiments from low power (~50 kW) up to 20 GW. Proton recoil counters are better suited for low power (up to ~1 GW which means 2.5 million impulsions per second), thanks to a higher efficiency and no discrimination of the γ noise, and ^{237}Np fission chambers are used for higher power. They have low gamma sensitivity, and an energy threshold of 650 keV. The signal acquisition rate may be adjusted between 1 ms and 10 s according to the count rate.

The hodoscope collimator has three different degrees of freedom, for adjusting its position in front of the test rod. The distance from the collimator nose to the core axis may be adjusted manually in translation, while two Direct Current Motors are used to move the collimator in verticality and rotation. Each one of the 306 detectors is linked by a µ-metal cable to its electronic chain placed 30 m far from the detector. These electronics have been renewed taking benefit from the renovation period [4] .

To keep the signal-to-noise ratio as low as possible [5] , the CABRI driver core is traversed radially by the irradiation channel. In that way, none of the fuel rods of the core are in direct view of the hodoscope detectors. Nevertheless, the signal-to-noise ratio remains low and requires specific data treatment algorithms.

The collimator was installed to make a pixelization of the test rod placed in the center of the reactor. Each detector looks at a slice of the test rod. In steady state conditions, the signals measured by all detectors are stable, whereas in transient conditions (i.e. a RIA pulse) they may vary according to a fuel displacement, a fuel densification (yielding to an increasing signal), or a fuel dilution or ejection (yielding to a decreasing signal.

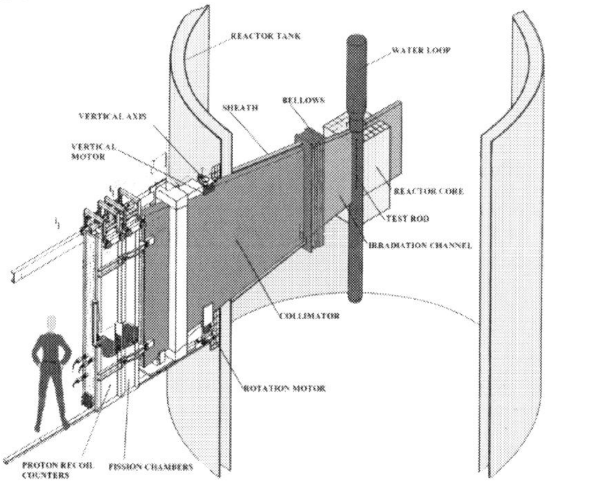

Fig. 1. Hodoscope detector system.

III. Hodoscope experimental results compared to dosimetry measurements

The CABRI commissioning campaign included several experiences with dosimetry measurements [6] .

Wire dosimeters were placed at several known elevations in the center of the test cell to measure the axial power profile of the CABRI core. Two irradiation campaigns have been performed: one with the test cell in air and one with the test cell filled with water. These results were compared to those obtained with the hodoscope PR detectors, in order to determine their qualitative response. Fig. 2 and Fig. 3 respectively present the axial profile measured with the PR detectors of the hodoscope and the dosimeter counting rates, for water or air filling the loop. The power plateau for these experiences were set at 30 kW (respectively 3 kW).

For this comparison, new sensitivity coefficients, estimated during the calibration campaign of hodoscope detectors, performed in 2018, are used [7] .

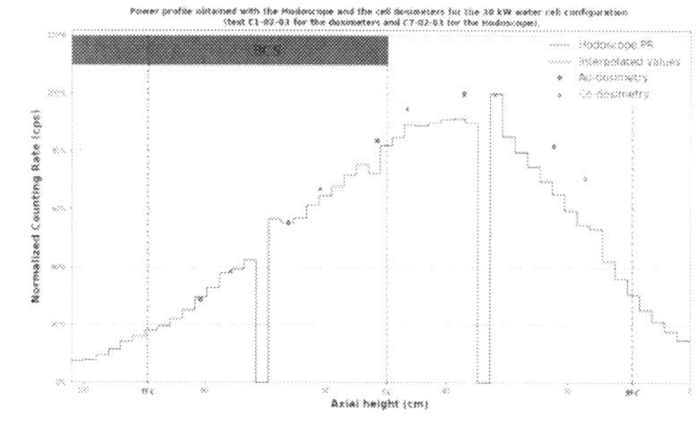

Fig. 2. Comparison between the counting rate measured by the Hodoscope and the "water cell" dosimetry.

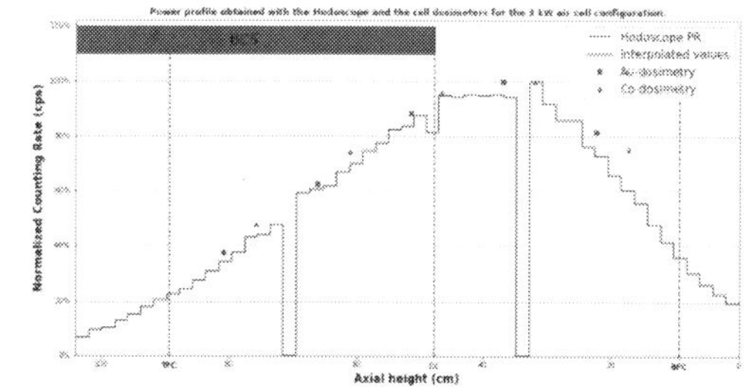

Fig. 3. Comparison between the counting rate measured by the Hodoscope and the "air cell" dosimetry.

An important aspect to observe in these graphs is that hodoscope PR detectors measure fast neutron flux coming from the core, whereas dosimeters measure thermal and epithermal fluxes of the core. Consistency between these curves is due to a constant ratio between thermal and fast flux during steady state conditions. This assumption is also verified in transient conditions.

The position of the Control and Safety Rods (BCS) has been plotted, this allows a better understanding of the profile of the neutron flux.

Indeed, these rods are made of hafnium which is a neutron absorber, the fast neutrons from the fuel rods are therefore less present in the areas of the test cell located at the same height as these rods. This then distorts the theoretical flux shape of the parallelepipedic core, following a "cosine" function. Transient bars filled with helium-3 also play the role of neutron absorber on CABRI in order to quickly reduce the power of the reactor, however in the various tests carried out during these campaigns, these bars did not contain helium-3.

In conclusion, the profile obtained with the new sensitivity coefficients is more suited with the dosimetry measurements, compared to previous results presented in [3] .

IV. Comparison with numerical simulation

In order to solve the Boltzmann Equation, for CABRI reactor, Monte Carlo MORET code [8] , developed by IRSN, has been used to determine some neutronic parameters.

In this section, results obtained with MORET 5C, using JEFF3.1.1 library [9] , are compared to experimental results. Monte Carlo codes give results in steady state conditions. The MORET model of the CABRI core is shown on Fig. 4.

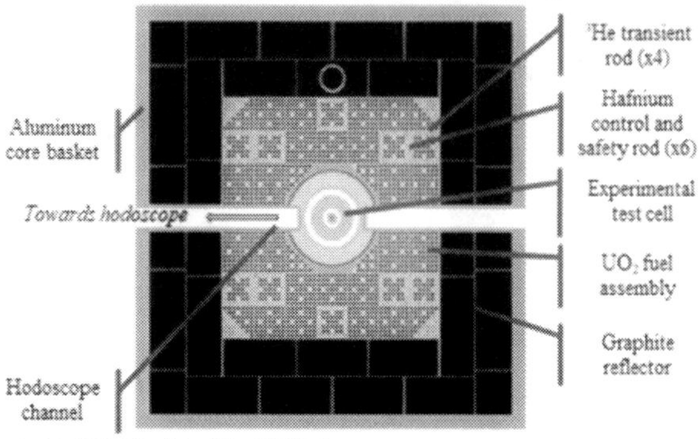

Fig. 4. MORET Model of the CABRI core.

A. MORET calculation compared to hodoscope results

A comparison has been carried out with hodoscope experimental results obtained during a 23 MW power plateau.

In order to present these results, the same 2018 sensitivity coefficients have been applied to the Hodoscope measurements. The MORET simulation was made by representing a lead rod inside the experimental device, in the center of the test cell. This dummy rod has been cut into cylinders of 2cm-high (representing the projections of the Hodoscope channels on the test rod).

In this configuration, the BCS are less inserted into the core, because the reactor power being higher (23 MW against a few tens of kW previously). At higher power level, the fuel is hotter, leading to a broadening of the resonances and therefore of the increase of core anti-reactivity. The critical position of the control rods obtained is then greater than the previous one. This can also be seen in the shape of the curves, which is much more symmetrical than those at low power (figures in the previous section).

Considering the 2 curves on Fig. 5, a difference at the extremities can however be noticed: the calculations seem to underestimate the flux in these places. This can be partly explained statistically because the neutron flux being lower, the uncertainty on the associated values is higher.

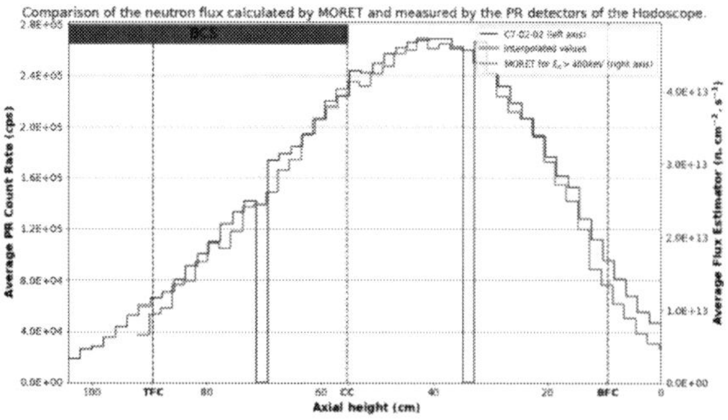

Fig. 5. MORET 5 compared to PR profile steady-state conditions.

B. MORET calculation compared to dosimetry experiments

Another step in the validation of our model consists in simulating the operation of the dosimeters within our modeling. This simulation requires adding several elements within the CABRI core. The experimental conditions of the test corresponding to "air cell" dosimetry, and it was necessary to precisely know the geometry of the dosimeters used and their locations.

10 wired dosimeters were used, including 5 in gold (Au197) and 5 in cobalt (Co59). These wires are extremely thin, 5 mm long and 0.1 mm in diameter (gold) to 0.25 mm in diameter (cobalt).

Fig. 6. Example of dosimeter under quartz (wire made of iron).

They were therefore deposited in quartz capsules 1 mm in radius and about 1 cm in height (Fig. 6). These 10 dosimeters were then placed in an aluminum dosimeter holder tube according to the following diagram (Fig. 7).

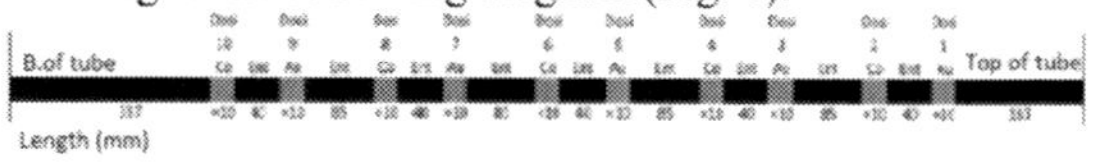

Fig. 7. Dosimeter's layout diagrams.

This 6 mm diameter of the holder tube replaces the entire test device in the center of the cell and is fixed by two aluminum centralizers located at its ends.

A MORET model was therefore carried out by reproducing the following various elements as exactly as possible: the tube, the centering devices, the quartz capsules, and the gold and cobalt wires. After configuring the different chemical compositions of the elements used, reaction rates (n, γ) were requested as output for each volume of gold and cobalt wires. As the volume of the wire is tiny, the reaction rate is low, and the uncertainties remain therefore substantial (between 10% and 15%). However, it can be notice in the following Figures, that the relative profiles calculated by the MORET code are in good agreement with cobalt dosimeter measurements and show some limited differences with gold dosimeter measurements.

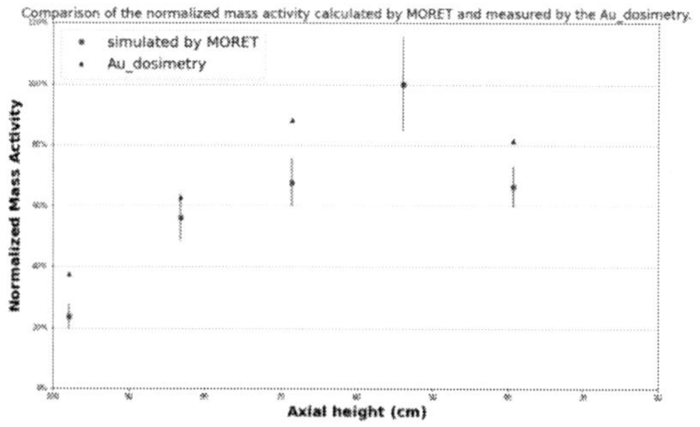

Fig. 8. Comparison between measured and computed specific activities of "cell air" for dosimeters in gold.

https://doi.org/10.1051/epjconf/202328804012

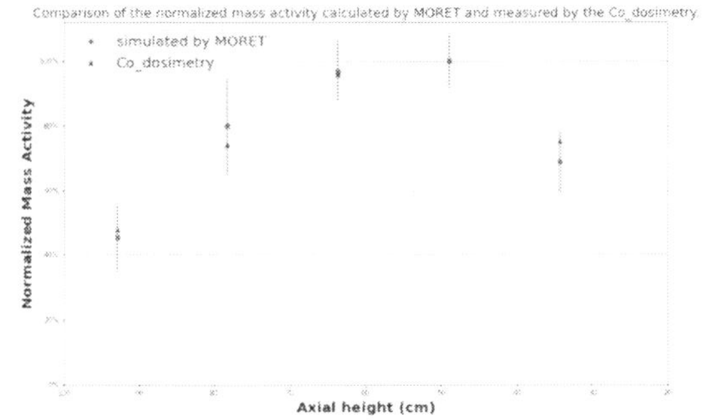

Fig. 9. Comparison between measured and computed specific activities of "cell air" for dosimeters in cobalt.

Despite these relative differences, we can consider a rather good ability of our CABRI model with MORET, to account for physical reality regarding flux variations. Moreover, this model is used for relative calculations, as it will be explained in the next section.

V. Signal-to-mass conversion charts

A. Axial distribution signal

Thanks to the Hodoscope software it's possible to obtain the axial distribution of the signals recorded by FC during a test. This is one of the most important hodoscope measurements, performed during the transient. The axial distribution is made up of the ratio of the "useful signal" (neutrons coming from the rod, in the line of sight of the corresponding row, monitored for each of the Hodoscope line) to the "noise" (neutrons from the core) measured by the detectors over time (SNR(t)), divided by the value of this same ratio recorded during the power plateau, before the transient, (SNR_0).

The experimentalist then evaluates, on this axial distribution, the different periods of time where a fuel motion seems to have been occurred. This is evidenced by the value of the aforementioned report. Indeed, if $SNR(t)/SNR_0 > 1$, this means an increase of the mass of fuel on the line and during the time interval considered. Conversely, if $SNR(t)/SNR_0 < 1$, this reflects a mass deficit.

As an example, from a real test, the $SNR(t)/SNR_0$ ratio, displayed in Fig. 10 for the hodoscope rows 19 to 25, describes fuel motions during a power transient. For a given row, a ratio greater (respectively lower) than one is represented in green (respectively red) on the Fig. 10.

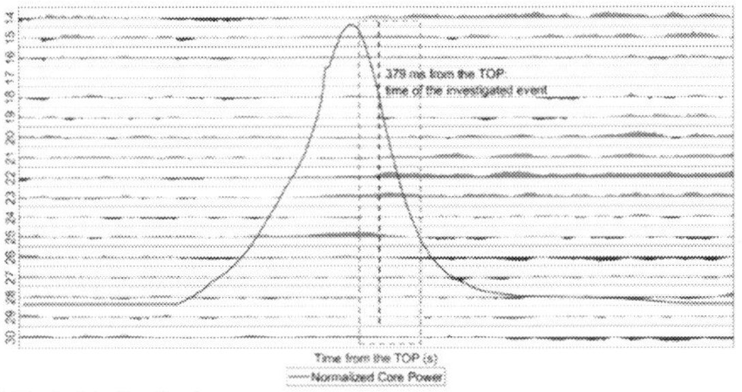

Fig. 10. Axial distribution.

The grey dashed rectangle (around T0+379 ms) delimits the time interval where a significative fuel motion started: a fuel accumulation for the rows 22 and 23, followed later by a lack of fuel for the rows 25 to 30.

B. Signal-to-mass conversion charts

In order to be able to quantify these mass movements highlighted by the axial distribution signal, the experimentalist uses the signal/mass conversion charts.

The value of $SNR(t)/SNR_0$ is linked to the ratio of fission rate of the test rod during the transient, to the fission rate of this same intact rod (before the transient) [10].

These fissions rates are computed for different ideal degradation configurations that the rod could have during the transient, as well as for the intact rod configuration, to obtain the normalized ratio of fission rates. These calculations are realized using the CABRI model, performed with the IRSN MORET code, for a given critical state configuration. It is important to note that this normalized ratio ($\sum\Phi/\sum_0\Phi$) is independent of the axial profile.

To build signal/mass conversion charts, 6 configurations of the test rod were imagined in addition to that corresponding to the intact test rod. By noting m_0 the fissile mass of the initial fuel in the rod section considered and m, the fissile mass of the fuel in the considered configuration, we can characterize these different models by the ratio m/m_0.

It is important to note that even if these configurations are ideal, they could represent a real case. Anyway, they are to be considered as points of reference to build the charts. They are presented on the Fig. 11.

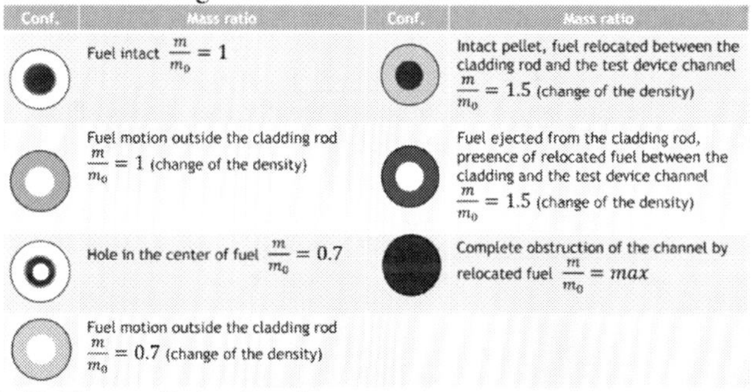

Fig. 11. The seven ideal configurations.

The charts resulting from the 7 MORET calculations on these configurations are plotted in the Fig. 12.

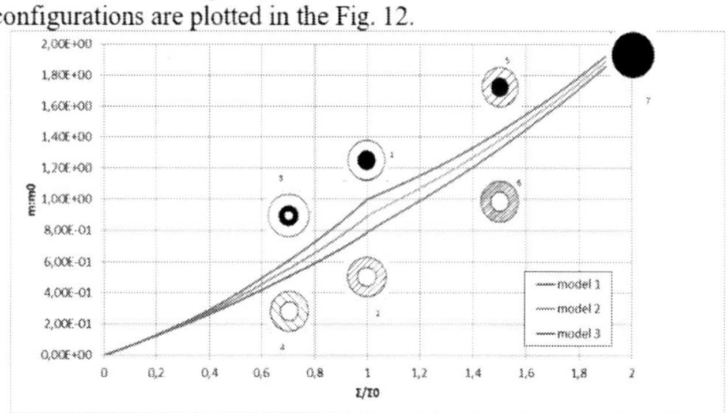

Fig. 12. Signal-to-mass conversion charts.

The different configurations are:

- Configuration 1 is the reference, the m/m_0 ratio is 1 and all the fuel is inside the cladding.
- Configurations 2, 4 and 6 correspond to a situation where the fuel has left the clad and is located between it and the channel test device (corresponding to the so-called "model 1" in the Fig. 12. They respectively have a ratio $m/m_0 = 1$, 0.7 and 1.5.
- Configuration 3 corresponds to core meltdown, i.e., the fuel is located inside the clad but the central part has been destroyed, giving a ratio $m/m_0 = 0.7$.
- In configuration 5, the fuel pellet is still intact, but additional fuel coming from upstream has been added, between the clad and the channel test device with a ratio $m/m_0 = 1.5$. The three cases 1,3,5 constitute the so-called "model3" in the Fig. 12.
- For the last configuration, case 7, the inner section of the channel test device is completely filled with fuel, which gives $m/m_0 = 3.5$.

The "model 2" is made up by taking the intermediate cases between those of "model 1" and those of "model 3".

To carry out these seven simulations, the experimentalist must know the isotopic composition of the rod as well as that of its cladding, then modify the mass concentrations of the constituent elements of the fuel scattered around the cladding (for cases 2, 4, 5 and 6).

The MORET model can take into account the influence of the spatial self-shielding effect by modeling 3 slices of axial degradation of the test rod.

C. *Application to a real case*

The axial distribution signal illustrated in Fig. 10 is first transformed to a signal histogram, by averaging the values in the time period mentioned above, for each slice of the test rod (Fig. 13). One can notice the line without counting rate, which is full of stainless steel in order to assure the mechanical support of the collimator. Then, the signal-to-mass conversion charts is applied to transform the signal histogram to mass histogram (Fig. 14).

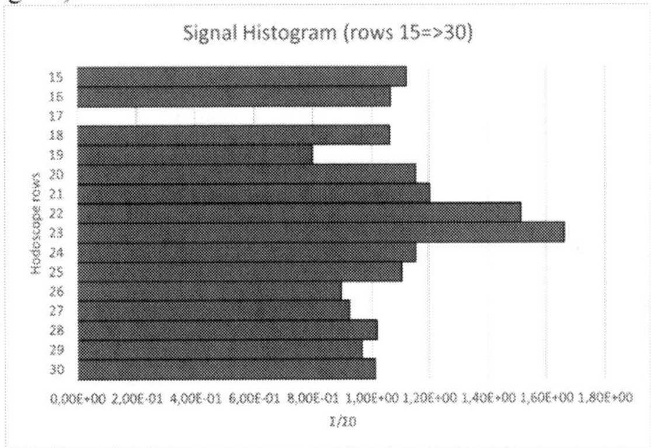

Fig. 13. Signal histogram.

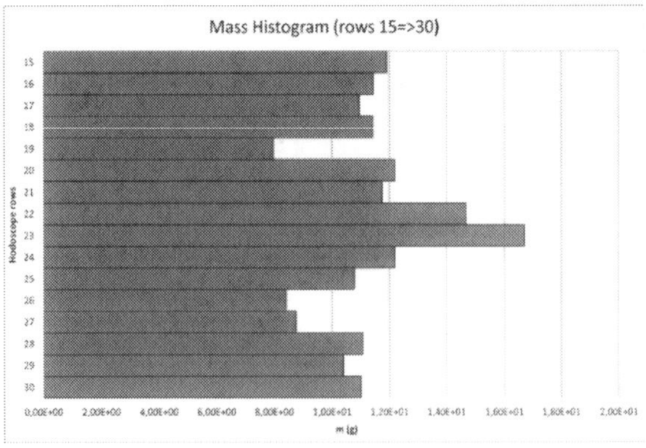

Fig. 14. Mass histogram.

Reproducing this methodology for each time period, it is possible to obtain the fissile mass motion during the whole transient. Hence, to completely analyze the degradation of a test rod, it is required to create a suited scenario, combining all the events spotted in the axial distribution to the corresponding degradation configurations.

VI. Conclusions and perspectives

This paper presents the experimental validation of a MORET model, which is used to analyze the hodoscope measurements monitored during a reactivity transient in the CABRI reactor.

The MORET model has been compared to results coming from the hodoscope calibration campaign, as well as from a dedicated neutron activation dosimetry characterization.

The neutron flux profile, calculated by MORET at the center of the CABRI reactor, is consistent with the hodoscope detectors measurements. For the comparison to the specific activities of the gold and cobalt dosimeters, the MORET calculations show some limited differences to the measurements, despite high calculations uncertainties. As this model is used for fission rates ratio calculations, this validation has been considered sufficiently satisfying.

Different configurations, considering different degradations of the test rod, have then been simulated with the model, in order to obtain signal-to-mass conversion charts. Thanks to these charts, the quantitative mass motion of the test rod can be directly deducted from signal histograms, acquired by hodoscope detectors all along the transient, and requiring the creation of a scenario of degradation.

The definition of this scenario is a manual combination of the choice of the degradation models, for each line of the hodoscope. Further developments could be studied, to automate this combination. Besides, a complementary work of the estimation of the resolution of the Hodoscope, has been undertaken, in order to be able to use it in the case of fuel relocation during or after the test

Acknowledgment

The authors thanks J. Andries, who made its internship in the laboratory, for its contribution to this work.

 https://doi.org/10.1051/epjconf/202328804012

References

[1] B. Biard et al, "Reactivity Initiated Accident transient testing on irradiated fuel rods in PWR conditions: The CABRI International Program," Annals of Nuclear Energy, Vol **141**, 15 June 2020, 10.1016/j.anucene.2019.107253.

[2] Q. Grando et al, "CABRI test events monitoring through three measurement systems," in Proc. ANIMMA, Prague, Czech Republic, 2021. EPJ Web of Conferences, Vol **253**, 04013 (2021), 10.1051/epjconf/202125304013.

[3] V. Chevalier et al, "The CABRI fast neutron Hodoscope: Renovation, qualification program and first results following the experimental reactor restart," in Proc. ANIMMA, Liège, Belgium, 2017. EPJ Web of Conferences, Vol **170**, 04003 (2018), 10.1051/epjconf/201817004003.

[4] S. Mirotta et al, "Qualification and characterization of electronics of the fast neutron Hodoscope detectors using neutrons from CABRI core," in Proc. ANIMMA, Liège, Belgium, 2017. EPJ Web of Conferences, Vol **170**, 04016 (2018), 10.1051/epjconf/201817004016.

[5] J. Di Salvo et al, "Signal-to-Noise Ratio of CABRI hodoscope: Monte Carlo calculation versus experiments," IEEE Trans. On Nuclear Science, Vol **69**, n°4, pp. 648-655, April 2022, 10.1109/TNS.2022.3150069.

[6] J. Lecerf et al, "Analysis of the dosimetry campaign performed during the CABRI commissioning tests," in Proc. ANIMMA, Liège, Belgium, 2017, hal-02418142.

[7] V. Chevalier et al, "The CABRI fast neutron Hodoscope: calibration campaign results," in Proc. RRFM, Helsinki, Finland, 2020, hal-03080772.

[8] B. Cochet et al, "Capabilities overview of the MORET5 Monte Carlo code," Annals of Nuclear Energy, Vol **82**, pp. 74-84, August 2015, 10.1016/j.anucene.2014.08.022.

[9] A. Santamarina et al, "The JEFF-3.1.1 Nuclear Data Library. Validation results from JEF2-2 to JEFF-3.1.1,", JEFF Report 22, ISBN 978-92-64-99074-6. © OECD 2009.

[10] K. Baumung et al, "Quantitative fuel motion determination with the CABRI fast neutron hodoscope: Evaluation methods and results," Nucl. Technology, Vol **96**, pp. 302-311, June 1991.

Identified Opportunities in Utilization of the European Research Reactor Fleet as a Part of the TOURR Project

Anže Pungerčič[1,2], Roberta Cirillo[3], Joanna Walkiewicz[4], Evžen Novák[5], Jacek Gajewski[4], Laszlo Szentmiklósi[6], Lisanne Van Puyvelde[7], Jörg Starflinger[8], Daniel Cano[9], Georg Pohlner[8], Renata Mikolajczak[4], Gabriel Pavel[3], Luka Snoj[1,2,*]

[1]"Jožef Stefan" Institute (JSI), Slovenia
[2]Faculty of Mathematics and Physics, University of Ljubljana, Slovenia
[3]European Nuclear Education Network (ENEN), Belgium
[4]National Centre for Nuclear Research (NCBJ), Poland
[5]Centrum Výzkumu Řež (CVR), Czech Republic
[6]Centre for Energy Research (EK-CER), Hungary
[7]Studiecentrum Voor Kernenergie, Centre D'etude de L'energie Nucleaire (SCK-CEN), Belgium
[8]Institute of Nuclear Technology and Energy Systems, University of Stuttgart (USTUTT), Germany
[9]Centro de Investigaciones Energeticas, Medioambientales y Technologicas-Ciem (CIEMAT)

(*) luka.snoj@ijs.si

Abstract—This paper discusses the current status of the European research reactor (RR) fleet and identified opportunities for its utilization. The data for this analysis was collected through a specific questionnaire from which a database of the European RR fleet was created. The questionnaire was designed to assess the degree of exploitation of different RR applications and to identify gaps and opportunities for future utilization. The results indicate that the European RR fleet is older compared to the world average, with no new research reactors built in Europe since 1992. The majority of RRs reported low levels of exploitation across all applications, and a desire to expand utilization was expressed by 78% of respondents. Lack of manpower, finance, and customers were identified as the main obstacles preventing expansion, while the need for more nuclear engineers was identified as a clear opportunity to attract people to the field. The findings of this study emphasize the need for actions to be taken to combat future needs and to improve the utilization of the European RR fleet. Overall, this study provides valuable insights for policymakers, industry professionals, and researchers working in the field of nuclear energy.

***Keywords* —Research Reactor, TOURR, European Reactors, Utilization, Research Reactor Applications.**

I. Introduction

In recent years, a significant number of research reactors in Europe have stopped its operation due to financial constraints or the aging of reactor facilities. Consequently, it has become essential to evaluate the present state of the research reactor (RR) fleet. The TOURR project has been initiated in 2019 to address this challenge by coordinating the optimization of available research reactors in Europe.

The primary goal of the TOURR project is to devise a comprehensive strategy for research reactors in Europe and lay the groundwork for its implementation. The first phase of the project includes specific objectives, one of which is to assess the current status of the European RR fleet [1]. Based on this assessment, a strategy will be developed to modernize the reactor fleet, aimed at identifying existing gaps and potential opportunities in RR activities. This analysis will form the foundation for a comprehensive strategy to optimize irradiation time and ensure uninterrupted accessibility for experimental, educational, and production purposes.

Recently, several documents have been published that explore the use of research reactors, their future, and potential opportunities for scientific and technological applications. One such document, titled "Strategic Planning for Research Reactors" [2], primarily focuses on enhancing the use of existing research reactors and offers guidance for developing and implementing a strategic plan for new research reactor projects. Another document, "Neutron scattering facilities in Europe, Present status and future perspectives" [3], provides a detailed overview of the current status of neutron scattering applications using RR and outlines various scenarios and perspectives for the future. Notably, the document identifies a crucial gap: the need for a reactor-based source in addition to the existing spallation source in Europe. Furthermore, the document "Neutron Users in Europe: Facility-Based Insights and Scientific Trends" [4] includes a similar questionnaire analysis to the one presented in this paper, focusing on the role of each RR in shaping the European Spallation Source (ESS) policy for innovation and access. This document also highlights gaps and opportunities related to ESS and its use in research and development.

The analysis of the current European research reactor fleet, as presented in this paper, is organized into three main sections. The first section outlines the methodology used for the analysis and explains how gaps were identified through the intended questionnaire. The second section provides an overview of the current status of the research reactor fleet and presents statistical analyses applicable to different reactor applications. The final section centers on the identification of opportunities in present and future RR activities. In conclusion, the main gaps and opportunities identified throughout the TOURR project are listed.

II. Methodology

In order to obtain data needed for this analysis, we constructed a specific questionnaire, which enabled us to create a comprehensive database of the European research reactor (RR) fleet [5]. The questionnaire was designed with a gap analysis in mind, aiming to assess each RR's application in Research and Development (R&D), Education and Training (E&T), and isotope production and its related medical applications. Through this assessment, we aimed to identify clear gaps, which would indicate potential opportunities in the specific application or field. In addition, identification of application which might become obsolete in the future could be performed.

To identify these gaps and opportunities, we relied on the expertise of both project members and authors involved in this study, in addition to the data from the questionnaire. We performed a statistical analysis of the responses obtained, searching for definitive indications of gaps. Some conclusions, especially concerning gap identification and potential opportunities, were drawn from the collective expertise and knowledge in this domain. Additionally, we obtained general information about the research reactors from the existing IAEA RR database [6].

The degree of exploitation for various RR applications was determined using the questionnaire, and initial data from the questionnaire is available in [5]. Experts and project partners identified 18 different research reactor applications during the questionnaire's creation. To determine the degree of exploitation for each application, we asked the question: "To what degree do you think this application is exploited at the RR of your institute?". Subsequently, respondents were asked a follow-up question: "Would your institute like to expand the utilization of your RR for this application?", with the option to provide further explanations for their "yes" or "no" responses.

Gaps and opportunities analysis utilized a questionnaire, combined with the expertise of project members and authors, and involved a statistical assessment of responses to identify gaps and opportunities in existing research reactor activities across various applications. The findings, presented in next Section III, will help improve the overall utilization and strategic planning of research reactors in Europe.

III. EU Research reactor fleet status

A. Age of EU research reactors

The IAEA RR database [6] provides publicly available general information about research reactors, including details such as power, year of construction, and type. To assess the age of the European research reactor (RR) fleet, we compared it to the world's average, as shown in Figure 1. Notably, no new research reactors have been constructed in Europe since 1992, and the majority of existing research reactors were built during the 1960-1970 period. Consequently, the average age of European RRs is 56 years, with a median age of 62 years. This indicates that the European RR fleet is relatively older than the global average, emphasizing the need for proactive measures to address future requirements. Research reactor average and median age for Europe and the world is listed in Table 1.

Table I
Research reactor age in Europe and the world

Analyzed parameter	Europe	World
# of operational RR	30	197
Average age (years)	56	48
Median age (years)	62	51
Year last new RR was built	1992	2016

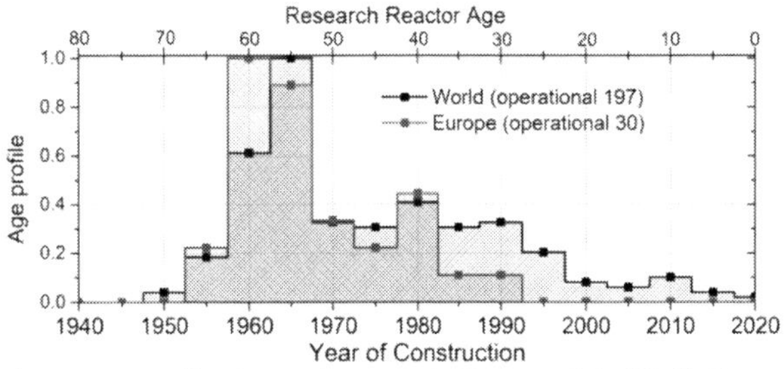

Fig. 1. Age profile of research reactors in Europe and the World (Europe included). Statistical representation of the age profile curve is presented in Table I.

B. Degree of RR applications exploitation

The questionnaire responses were analyzed, grouping all answers regarding RR exploitation without limiting the scope to specific applications. A total of 162 responses were received, and the findings are summarized in Figure 2. The analysis showed that 43% of research reactors reported low levels of exploitation across all applications. For 31% of the RRs, the exploitation level is considered medium, while the minority of 27% indicated a high level of exploitation.

Furthermore, an average of 78% of respondents expressed a desire to expand the utilization of their respective RRs, irrespective of whether they reported low, medium, or high exploitation levels. No correlation between the current level of exploitation and the willingness to expand was observed.

In conclusion, the analysis of the IAEA RR database and general questionnaire responses, indicated the aging of European RR fleet and the need for future measures for increasing. The findings underscore the importance of optimizing RR utilization across various applications and planning for the future demands of research reactors in Europe.

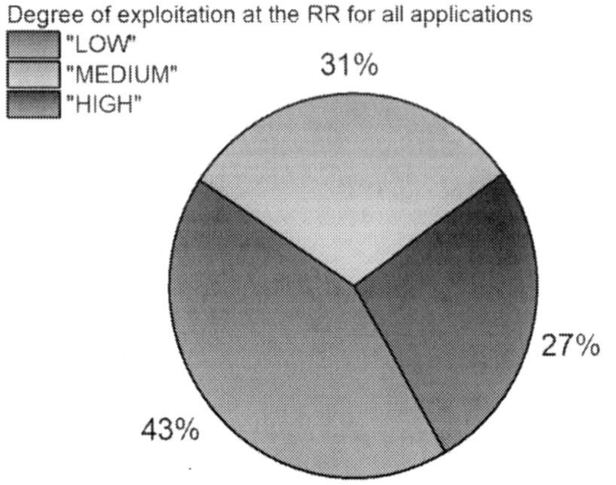

Fig. 2. Relative representation of the degree of exploitation of research reactors. Total number of answers was N=162.

Furthermore, we analyzed the average or major reason why their institute wouldn't want the expansion of their RR for individual application. Most common reason was that the reactor is already 100 % utilized in that application and no further expansion is required. Another frequent answer was that the expansion is not possible as it is limited by the reactor properties (not high enough flux, limited number of neutron ports, etc.).

C. Identification of Gaps in RR utilization

Large portion of conclusions can be obtained from the analysis of research reactors who answered "yes" to the question: "Would your institute like to expand the utilization of your RR for this application", where a follow-up question was asked: "If yes, what obstacles are preventing your institute to do so:". Multiple answers were possible, such as "Lack of manpower", "Lack of customers", etc. At first, an analysis of the results for all applications combined is presented in Figure 3. It can be observed that out of all answers the main reason for not expanding the utilization is in lack of manpower, followed by lack of finance and lack of customers.

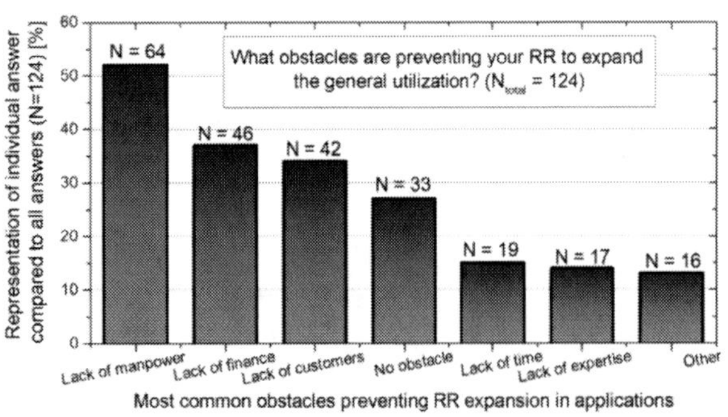

Fig. 3. Relative representation of answers to what is the reason preventing your RR to expand the utilization of all applications together. The columns within one application do not sum to 100 %, as multiple-option answer was possible.

The analysis of what is preventing the RR to expand the utilization can be performed on the level of individual application. Individual applications in R&D were analyzed. For most applications similar to the average answer, lack of manpower was the main reason, resulting in the main gaps of RR utilization to be:

- **Lack of manpower – need for more nuclear engineers and scientists:** The „Lack of manpower" was the most common answer to reasons for not expanding the general RR utilization. This opportunity is related to Education & Training and is further explained in Deliverable 1.4 [7].
- **Lack of long-term financing:** The "Lack of finance" was the second most common reason for not expanding RR utilization, showing a clear gap in long-term investment for operation of current RR.

There were some outliers, such as gemstone coloration and geochronology, where lack of customers was the main obstacle, marking such activity to be obsolete in the future. Another obstacle worth mentioning is the lack of financing, which is the biggest obstacle preventing the expansing in actinide transmutation, neutron scattering, neutron radiography and boron capture theraphy. A clear gap of higher financing in many of the activities is observed. Another obstacle that clearly shows the gap is the „lack of time", meaning there are not enough reactor hours to be used. One such example is the radiochemistry, however similar portion of answers was the lack of customers, meaning a more general gap can be observed:

- **Better communication to bridge the gap between customers and reactor operators:** It was observed that the number of answers for „Lack of time" and „Lack of customers" was the same and this issue would be fixed with communication and distribution of work.

IV. Identified opportunities in RR utilization

Based on the presented gaps, observed from the analysis of the questionnaire answers, opportunities in RR utilization can be devised. In addition, state-of-the-art research was analyzed to determine general opportunities for new future activities ranging from radiation hardness testing, nuclear fusion to nuclear-driven chemical production and processing.

Based on the identification of gaps in RR utilization three opportunities can be determined:

- **Opportunity to attract people to the nuclear field** – The analysis showed a large gap, where new possibilities for nuclear engineers and scientist are possible (e.g. NAA scientist are aging, need for transfer of knowledge). High need for employment among research reactors in Europe.
- **Opportunity for long-term financing** – With stable financing high portion of EU research reactors will be able to afford the expansion of utilization in various RR applications.
- **RR communication platform** – In the scope of the TOURR project a platform will be launched, which will make the communication between

customer and institute, operating the research reactor, easier and more direct. The platform is being launched in the final year of the TOURR project in 2023 [8].

Based on the analysis in the state-of-the-art, the identified opportunities in future RR R&D utilization are:

- Nuclear-driven production and processing of chemicals (chemical transmutations in research reactors) [9].
- Radiation Hardness Testing.
- Fusion applications (e.g. Water activation loop in research reactors [10]).
- Neutron scattering – In the future reactor-based source will be needed.
- Research in the field of fuel cells, hydrogen storage and batteries.
- Increase of silicon doping due to the increase of renewable energy.

A. *Actions to implement in the long-term*

One of the conclusions of our analysis is that: it can be inferred that RRs are a great resource in terms of 'teaching' and allowing students and young professionals to get first-hand experience. This is in contrast with the fact that the number of RRs in the EU is decreasing.

Having fewer RRs means losing opportunities for the E&T sector including a high risk for human capacity in terms of qualified RR operators and/or employees. The RR fleet should be kept numerous and equipped with state-of-art installations to be attractive for the new generation of nuclear workforce. This leads to the conclusion that having a larger pool of nuclear-educated people will represent a resource for the RRs community in need of a larger workforce to expand their E&T activities. The following actions summarized in the table are encouraged to fill the gap between the current scenario and the ideal one. Table II summarizes and prioritizes the main actions to be implemented in the long-term scenario.

TABLE II
ACTIONS TO BE IMPLEMENTED IN THE LONG-TERM SCENARIO

1	• Continuous promotion of nuclear education
2	• Retention of nuclear-educated people in the nuclear field
3	• Retention and attraction of nuclear-educated people specifically **in Europe** and to EU RRs
4	• Continuous exchange with other nuclear networks and nuclear projects
5	• Modernize the EU RR fleet
6	• New RR for testing new technologies (electricity generation with high % of renewables)

In order for the EU RR fleet to be attractive for the young generations it is advisable to take into account the option of building new RRs, more modern and more accessible, focusing on cutting-edge research and being an E&T opportunity in the same research fields. A new research reactor could serve as well to test new technologies for future electricity generation using nuclear energy, together with a higher percentage of renewables. This reactor could be a micro-reactor (or SMR) in development by several companies around the world, the concepts of which have not yet been tested in real experiments (e.g. being built and operated).

This aspect will be analysed in more detail in D3.4 - Recommendations for planning refurbishment of existing research reactors or construction of new ones.

The TOURR consortium recommends that new Research Reactor focused projects should be started under the coordination of the European Union level, where all EU states will have access to a vast number of RR applications. We recommend supporting national initiatives regarding the building of new reactors (e.g. PALLAS or projects in central Europe countries). Euratom has already supported the construction of JHR and similar approach should be explored for supporting new projects. This support does not necessarily have to be financial but it can also consist in creating the adequate framework for this cooperation to take place.

These recommendations aim at recovering and enlarging the capacity to cover a large number of applications mentioned and discussed in this report. A good example is a medium-power research reactor flexible for almost all domestic applications of Science and Technology (S&T) (e.g. NAA, nuclear data measurements, neutron, and gamma irradiation,) and education & training (E&T) by being capable of performing unique experiments for future European nuclear engineers. Another is a zero-power RR, dedicated mainly to the research of reactor physics and benchmark experiments but that can also be employed for E&T.

V. CONCLUSIONS

The European research reactor (RR) fleet faces challenges due to financial constraints and aging equipment, resulting in many reactors ceasing operations recently. The TOURR project seeks to modernize the European RR fleet and optimize research reactor utilization. It aims to identify gaps and opportunities in existing RR activities, forming the basis for a comprehensive strategy ensuring optimized use for experimental, educational, and production purposes.

Data for the analysis were collected through a specialized questionnaire, creating a database of the European RR fleet. Statistical analysis revealed that the European RR fleet is relatively older compared to the global average, with no new reactors built in Europe since 1992.

In conclusion, the TOURR project is an important initiative that aims to develop a comprehensive strategy for the modernization of the European RR fleet, optimizing the exploitation of available research reactors. The analysis of the current status of the European RR fleet showed that there is a significant need for modernization and improvement. The identification of gaps and opportunities in existing activities will form the basis for a comprehensive strategy that will define the framework for optimized use of irradiation time and uninterrupted accessibility for experimental, educational, and production purposes. With this strategy in place, it is hoped that the European RR fleet can be maintained and developed for the benefit of scientific research and development in Europe.

ACKNOWLEDGMENT

This project has received funding from the Euratom research and training programme 2019-2020 under grant agreement No 945 269

LIST OF TOURR PROJECT CONTRIBUTORS

- European Nuclear Education Network, ENEN, Belgium: Gabriel Lazaro Pavel, Roberta Cirillo, Francisco Suárez Ortiz
- Centrum Výzkumu Řež, CVR, Czech Republic: Evžen Novák, Vlastimil Juříček, Ján Milčák, Jaroslav Šoltés, Lukáš Veselý, Miroslav Hrehor
- Energiatudományi Kutatóközpont, EK, Hungary: László Szentmiklósi, Péter Juhász
- Narodowe Centrum Badań Jądrowych, NCBJ, Poland: Renata Mikołajczak, Jacek Gajewski, Grzegorz Krzysztoszek, Iliana Chwalińska, Małgorzata Kot, Joanna Walkiewicz
- Belgian Nuclear Research Centre SCK CEN: Michèle Coeck, Lisanne Van Puyvelde
- Institut "Jožef Stefan", JSI, Slovenia: Luka Snoj, Bor Kos, Anže Pungerčič, Vladimir Radulović, Anže Jazbec, Saša Škof, Tanja Klopcic
- Evalion SRO, EVALION, Czech Republic: Petr Koran, Michaela Velckova, Jakub Heller, Jana Peroutkova
- Universität Stuttgart, USTUTT, Germany: Georg Pohlner, Joerg Starflinger
- Centro de Investigaciones Energeticas, MedioAmbientales y Tecnologicas, CIEMAT, Spain: Daniel Cano, Enrique Gonzalez

REFERENCES

[1] B. Kos, R. Cirillo, A. Pungercic, G. L. Pavel, and L. Snoj, "Gathering of data on the european research reactor fleet as part of the tourr project," in Proc. International Conference Nuclear Energy for New Europe (NENE), Bled, Slovenia, 2021.

[2] IAEA Nuclear Energy Series No. NG-T-3.16, "Strategic Planning for Research Reactors", IAEA, Vienna, 2017

[3] "Neutron scattering facilities in Europe: present status and future perspectives", ESFRI Physical Sciences and Engineering Strategy Working Group, 2016

[4] BrightnESS - "Neutron users in Europe: Facility-Based Insights and Scientific Trends", European Spallation Source ERIC, 2018

[5] R. Cirillo, TOURR Project Deliverable 1.1, "Data Base of European RR fleet", 2021

[6] IAEA' Research Reactor Database – RRDB, available on webpage: https://www.iaea.org/resources/databases/research-reactor-database-rrdb.

[7] R. Cirillo, TOURR Project Deliverable 1.4, "Current utilization and possible trends in RR's utilization for E&T", 2021

[8] R. Cirillo, TOURR Project Deliverable 3.3, "Online platform for optimized use of research reactors in Europe", 2023

[9] Plant, Arran George, et al. "Nuclear-driven production of renewable fuel additives from waste organics." Communications chemistry 4.1 (2021): 132.

[10] Kotnik, Domen, et al. "Design optimization of the closed-water activation loop at the JSI irradiation facility." Fusion Engineering and Design 193 (2023): 113632.

In-core measurement methodology with optical fibers using Cherenkov radiation for transient induced power measurement in the CABRI experimental reactor

Georges Gay[1*], Adrien Gruel[1], Romain Boffy[1], Grégoire De Izarra[1], Christian Jammes[5]
[1]CEA, DES, IRESNE, DER, SPESI, Cadarache, F-13108 Saint-Paul-lez-Durance, France
[2]CEA, DRF, IRFM, STEP, GEAC, Cadarache, F-13108 Saint-Paul-lez-Durance, France
(*) georges.gay@cea.fr

Abstract— **This paper presents our experimental work to assess the capability to estimate the transient-induced power distribution in the CABRI experimental reactor using Cherenkov radiation. The CABRI reactor is designed to produce a power transient up to 21 GW within a time less than 100 ms in order to irradiate a test fuel pin in condition representative of a Reactivity insertion Accident in pressurized water reactors. The large response range and short response time required to follow the flux evolution during a complete transient makes classical means of detection, such as ionization or fission chamber, inoperative. For that purpose, we suggest to measure Cherenkov light produced within optical fibers. Indeed, Cherenkov light emission is linked to the local electron production, which is proportional to the local gamma flux through the Compton or pair production cross-section, the intensity of Cherenkov radiation is related to the photon flux intensity. The knowledge of the fission photons emitted by the reactor gives direct insight on the fission rate, hence a spatial power density distribution could be reconstructed thanks to the measure of the Cherenkov light at different point in the reactor.**

Keywords — **CABRI, power Pulse, transient, Detector, Cherenkov, Optical fibers.**

I. Introduction

THE purpose of this document is to show the main results obtained by measuring the Cherenkov light in optical fibres during a CABRI power transient. The CABRI reactor is designed to produce pulses to irradiate a fuel pin at the center of the core. The short duration and the high amplitudes of the CABRI transient poses a particular problematic regarding the mean of detection available to follow 5 decades of variation in less than 100 ms with a size allowing in-core measurement. In order to detect how the spatial fission distribution evolves during a reactor transient. Indeed, the evolution of the fission distribution has a direct impact on the variation of the coupling factor [1] during the transient which quantify the energy released in the experimental fuel pin.

The mean of measurement such as ionization or fission chamber are not appropriate due to their overall size and the presence of static space charged in the detector due to the high neutron and photon flux. Another option would be the Optical Neutron Detector [2], which is still under development.

As optical fibers have the right size to allow in core measurement, we decided to explore the possibility to use them in order to monitor the reactor power during the CABRI power transient.

II. Presentation of CABRI

The CABRI facility at CEA Cadarache is a "pool reactor" with the core at the bottom of an 8m-deep water pool. The core is made up of 1,487 UO2 fuel rods, enriched to 6% and clad in stainless steel.

The reactor is equipped with 138 hafnium control rods divided into 6 groups of 23 elements symmetrically arranged in the core. Reactivity is injected by depressurizing 96 helium-3 rods, divided into 4 groups of 24 elements in the core. The speed at which the helium valves open and the number of valves open (1 or 2) will determine the shape and amplitude of the pulse. The reactivity insertion can be up to 4 $ at a rate of 40 $/s.

In stationary regime, the reactor has a power of between 100 kW and 25 MW. During the power transients, the power can reach 21 GW. The main detectors used to monitor CABRI's reactor power is given in TABLE I.

TABLE I
DETECTORS FOR CABRI'S CORE POWER MONITORING

Chamber	type	Power range
BN1/BN2	*Fission chamber (^{235}U)*	*0-200 W*
HN1/HN2	*Ionization chamber (^{10}B)*	*0.2-100 kW*
G2-1/G2-2	*Ionization chamber (^{10}B)*	*0.100-450 MW*
G311/G312/G34	*Ionization chamber (^{10}B)*	*0.450-21 GW*

CABRI pulses are used to irradiate a test rod located in a pressurised water loop that passes through the centre of the reactor. The rod is in PWR condition and is subjected to a pressure of 155 bar and a temperature of 300°C.

III. Cherenkov Effect In Optical Fibers

The Cherenkov effect is due to charged particles which velocity are faster than the phase velocity of light in a given transparent media [3]. In the case of optical fibers, Cherenkov light is generated in the core and cladding of a fibers and propagates all along the fibers. The particle detected by the mean of Cherenkov effect in a nuclear reactor are mainly electrons produced by Compton interaction of gamma-rays with matter [4]. Unlike most of the detectors, optical fibers play the roles of both particle detectors and signal transmitters.

IV. Description of the Experiments

A. Irradiated materials

The material irradiated during the experiments consisted in an optical fiber from artPhotonics with the following characteristics:

- *Core diameter :* 400 µm
- *Cladding outer diameter :* 440 µm
- *Coating outer diameter :* 560 µm
- *OH content :* Low
- *Numerical aperture :* 0.22
- *Coating material :* Aluminum

The fiber is set up in a waterproof container next to CABRI's reactor core as it is shown in Fig. 1. A waterproof pipe is connecting the tank to the reactor hall so that the pipe and tank are in air.

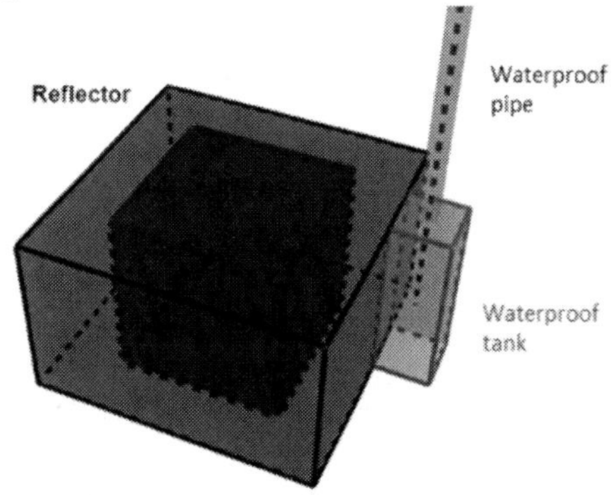

Fig. 1. Schematic view of the optical fiber position next to the reactor core (not to scale). The dashed line represents the optical fiber.

B. Optical system and electronics

The optical system aims at filtering the Cherenkov light intensity so that the photodetectors does not saturate during the transient. To do so, we mounted a band-pass optical filter centered at 800 nm with a bandwidth of 40 nm as it is shown in Fig. 2.

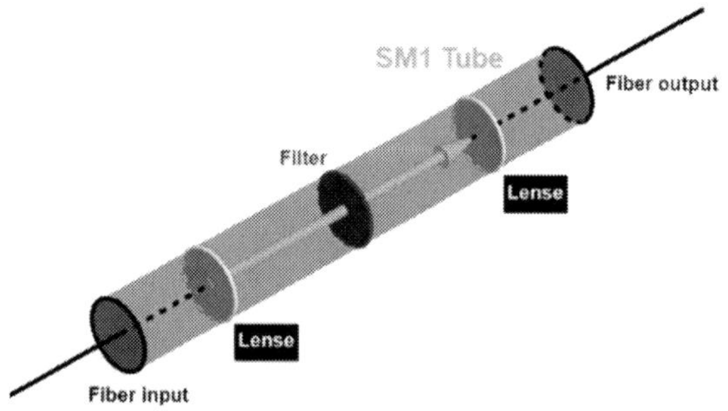

Fig. 2. Schematic view of the optical system (not to scale). The optical filter and the lenses are inserted in a SM1 optical tube. The white arrow represents the direction of light through the system.

The filter is placed between two lenses which focal lengths are 2 cm. The first lense makes the light beam from the optical fiber parallel to the optical axis, for it to be filtered, and the second lenses focuses the light on the output optical fiber. The SM1 lens tube and optical filter are designed by ThorLabs ®.

The filtered light is measured by a Single Photon Avalanche Detector (SPAD C1300-01 from Hamamatsu ®) which linearity and quantum efficiency are given in Fig. 3.

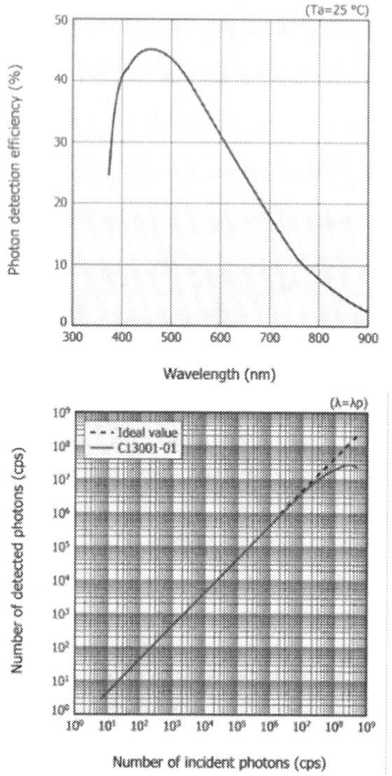

Fig. 3. (a) Quantum efficiency of the photosensitive surface and (b) linearity of the SPAD (C13001-01) at 450 nm [5].

The SPAD generates TTL pulses with 20 ns width. The TTL are counted by the mean of oscilloscopes (Agilent ®) and a FPGA (raspberry pi ®). The oscilloscopes record the full signal generated by the SPAD as the raspberry pi counts the TTL pulses on a 100 µs time interval. Both devices are triggered by an external signal controlled by the reactor operators.

V. Results of the Experiments

The results of the experiments were obtained during two reactor transients. Due to reactor maintenance, the waterproof tanks had to be replaced for each transients so that the same fiber has been placed differently in the tank from one transient to the other. The shape of the power transient is described on Fig. 4. We notice the steep increase in power next to the helium depressurization followed by a rapid drop due to the Doppler feedback, the control rods are then dropped to shutdown the reactor.

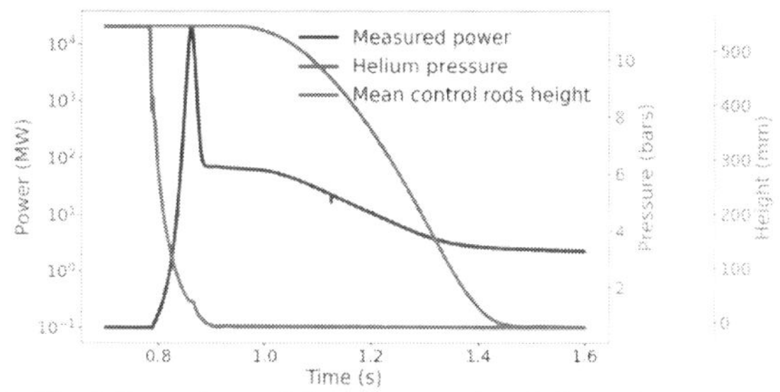

Fig. 4. The different phases of the reactor transient, helium release, power increase and control rods drop.

A. First Transient Experiment

For the first transient we counted the TTL with the FPGA only, the results is shown on Fig. 5. The high deviation of the counting rate at low power is due to the short time interval (100µs) over which the low counting rate is calculated.

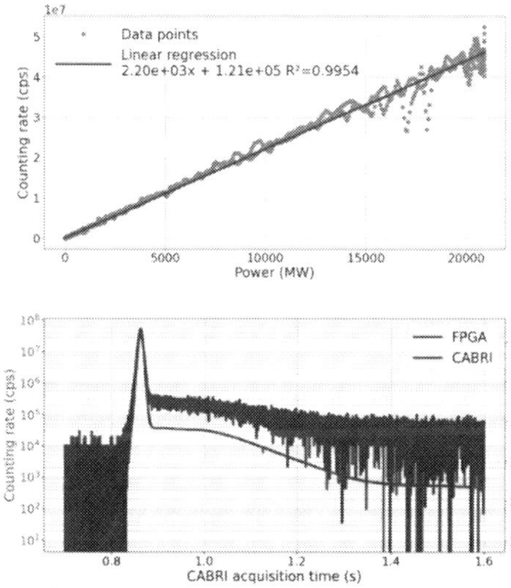

Fig. 5. (a) Linear regression between the counting rate measured by the FPGA and the reactor power and (b) the time evolution of the counting rate and the CABRI monitoring system.

B. Second Transient Experiment

For the second transient we counted the TTL with the FPGA and the oscilloscope with the raw data displayed on Fig. 6.

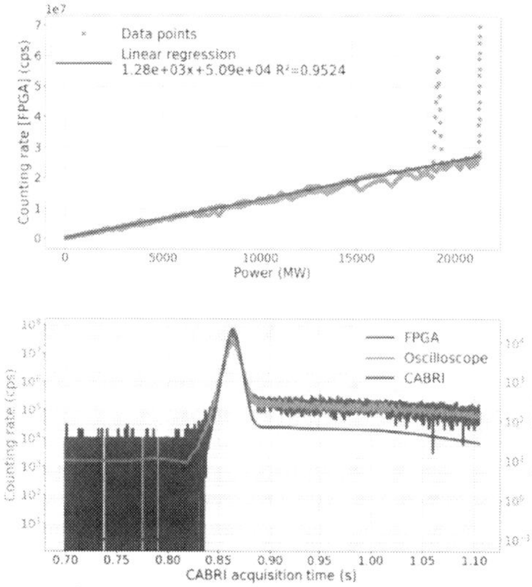

Fig. 6. (a) Linear regression between the counting rate measured with the FPGA and the reactor power and (b) the time evolution of the signals from the oscilloscope and the FPGA.

The measurement with the oscilloscope gives us an estimation of the counting rate at low power of 1 10^3 cps (Fig. 6 (b)). The correlation between the counting rate from the oscilloscope and the FPGA is given on Fig. 7. We can see that the slope is close to unity, with some overshoot on the FPGA, which confirm the reliability of the measurement with the FPGA.

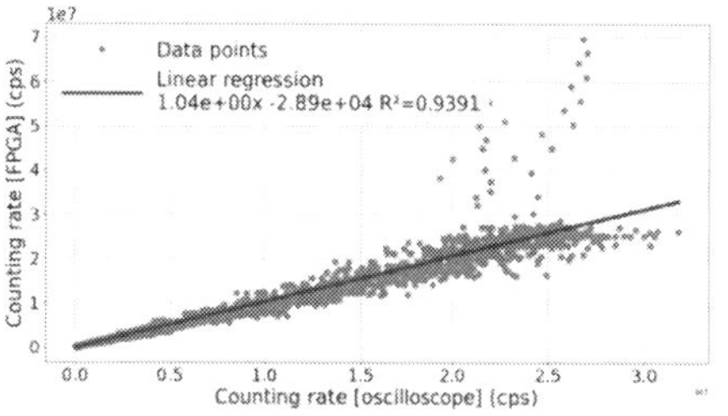

Fig. 7. Linear regression between the counting rate of the oscilloscope and the FPGA.

The comparison between the FPGA measurement from the first and second transient is shown in Fig. 8.

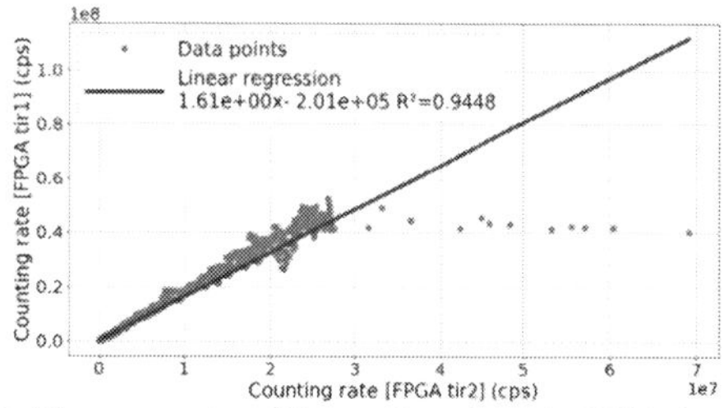

Fig. 8. Linear regression of the counting rate on the FPGAs between the first and second transient.

In this case, the slope is different from unity due to the change of the fiber position in the waterproof container from the first to the second transient.

In each transients we notice that the Cherenkov signal, unlike the measured power on CABRI is not varying through 5 decades.

Indeed the counting rate is varying between 1 10^3 cps up to 3-4 10^7 cps as the signal from CABRI is varying from 100 kW to 21 GW. In addition, the estimated WHM from the raw data shows that there is relative deviation of 6-9% from the WHM of each transient. This lead us to think that the linearity of the SPAD at high counting rate is not exactly described by Fig. 3.

TABLE II
WHM estimated on both transient

Transient	WHM	Deviation from the WHM on CABRI[a]
1st Transient	*9.55 ms*	*9%*
2nd Transient	*9.28 ms*	*6%*

[a]CABRI's WHM is 8.75 ms for each transient.

VI. Conclusions

The measurement of the Cherenkov light filtered at 800 nm allows us to measure the counting rate variation on a Single Photon Avalanche Detector over 4 decades. The measurement on both the oscilloscope and FPGA gives us insight of the counting rate for the reactor at low power and at maximum power. However The 5 decades covering of the CABRI's transient with a SPAD would require further work on the optical system to set the counting rate at a given value before the transient. Further more, the position of the fiber in the waterproof tank is not strictly defined which leads to fairly different counting rate from one transient to the other.

Finally, the use of other kind of photodetector, such as Avalanche Photodiode, could be more suitable for our experiments especially for in-core measurements.

Acknowledgment

I would like to thank the team of the LDCI especially Grégoire De Izarra who provided suitable optical fiber and FPGAs and Marc Pouradier who helped me to configure the experiments. I would also like to thank the CABRI operators who made this experiments possible by handling all the equipment with great care and for their rigorous work on the preparation of the reactor transients.

References

[1] J. Bartos, A. Gruel, C. Vaglio-Gaudard, et C. Coquelet-Pascal, « Some Considerations on the Energy Deposition During a RIA Transient Based On Monte Carlo Simulations », *EPJ Web Conf.*, vol. 253, p. 04005, 2021, doi: 10.1051/epjconf/202125304005.

[2] M. Lamotte, G. De Izarra, et C. Jammes, « Pulse-reactor core monitoring with an innovative optical neutron detector », *Nucl. Instrum. Methods Phys. Res. Sect. Accel. Spectrometers Detect. Assoc. Equip.*, vol. 995, p. 165086, avr. 2021, doi: 10.1016/j.nima.2021.165086.

[3] L. D. Landau et E. M. Lifshitz, *Electrodynamics of continuous media 2n edition*. 1984.

[4] B. Brichard, A. F. Fernandez, H. Ooms, et F. Berghmans, « Fibre-optic gamma-flux monitoring in a fission reactor by means of Cerenkov radiation », *Meas. Sci. Technol.*, vol. 18, n° 10, p. 3257-3262, sept. 2007, doi: 10.1088/0957-0233/18/10/S32.

[5] Hamamatsu, « SPAD module 13001-01 ». 2022. [En ligne]. Disponible sur: https://www.hamamatsu.com/content/dam/hamamatsu-photonics/sites/documents/99_SALES_LIBRARY/ssd/c13001-01_kacc1207e.pdf

Investigation of organic scintillators for neutron-gamma noise measurements in a zero power reactor

F.B. Darby[1,2,*], O. Pakari[1,2,3], M.Y. Hua[1], V. Lamirand[2,3], S.D. Clarke[1], A. Pautz[2,3], and S.A. Pozzi[1]

[1] Department of Nuclear Engineering and Radiological Sciences, University of Michigan, Ann Arbor, 48109 MI, USA
[2] Laboratory of Reactor Physics and Systems Behaviour, Ecole Polytechnique Fédérale de Lausanne, 1015 Lausanne, Switzerland
[3] Nuclear Energy and Safety Division, Paul Scherrer Institut, 5232 Villigen PSI, Switzerland
(*) fdarby@umich.edu

Abstract—**Noise measurements in light water reactor systems aid in generating validation data for integral point kinetic parameter predictions and monitoring parameters for reactor safety and safeguards. The CROCUS zero-power reactor has been used to produce both data types to date, using thermal neutron detectors to observe neutron noise and inorganic scintillators to observe gamma noise. Also, the cross-correlation of gamma and neutron noise has been investigated at CROCUS with separate gamma and neutron detectors. Organic scintillators can be used to cross-correlate gamma and neutron noise with only one detector type, within a single detector volume, and provide nanosecond timing resolution for time-correlated measurements. Dual-particle measurements require particle-type discrimination and are hence possible with organic scintillators since such detectors have the property of presenting statistically different pulse shapes for gamma rays and fast neutrons. The fine timing precision increases the signal-to-noise ratio relative to moderated thermal neutron detectors for correlated measurements and the dual-particle sensitivity allows for multiple modalities of estimating the prompt neutron decay constant. In this work, we present data obtained with 5.08 cm-length by 5.08 cm-diameter trans-stilbene cylindrical detectors set in the water reflector of CROCUS. Preliminary results estimate the prompt neutron decay constant to be (155 ± 5) s^{-1} at delayed critical.**

Keywords—**power spectral density, organic scintillators, pulse-shape discrimination**

I. Introduction

NOISE measurements offer a powerful non-invasive technique to observe a given fissile system's kinetic parameters. The kinetic parameters hereby refer to the coefficients of the differential equations describing the temporal behavior of a neutron population in a neutron multiplying medium. These equations are often derived using the so-called point kinetics assumptions, leading to a set of equations that can be solved cost-efficiently and, if the system indeed allows for the used assumptions, offer a precise predictor for time-dependent phenomena of neutron populations [1]. Noise measurements typically refer to a method to estimate the prompt decay constant, $\alpha = (\beta_{\text{eff}} - \rho)/\Lambda$ [2], [3] - where β_{eff} is the effective delayed neutron fraction, and Λ is the mean neutron generation time - via the analysis of a time series signal from a detector set close to a fissile system. More advanced noise measurements can also include the determination of β_{eff} and Λ [4], [5]. These measurements can then be used for code validation [6], integral parameter databases [7], and potentially for nuclear data assimilation [8].

In previous work, estimates of the prompt neutron decay constant from neutron noise measurements of the CROCUS zero power reactor at critical were completed with thermal neutron detectors and inorganic scintillators [4], [9]–[11]. Power spectral density analysis (also known as Cohn-α [2]) was used to estimate α using the gamma-ray sensitive inorganic scintillators ($CeBr_3$) and neutron sensitive thermal neutron detectors (^{235}U fission chambers and ^{3}He). Additionally, the two detector types were cross-correlated to obtain (γ,n) correlations [12]. The estimates compared well to simulated values using iterative fission probability (IFP) predictions in Serpent 2 [13], where the highest accuracy per unit measurement time was found with (γ,γ) correlations [14].

This work uses organic scintillator measurements of the CROCUS zero-power reactor to leverage dual-particle sensitivity in a single detector volume for combined (γ,n) noise analysis. We set a single cylindrical 5.08-cm-diameter by 5.08-cm-length trans-stilbene detector [15], [16] in the water reflector of CROCUS, and analyzed time stamp data with the power spectral density technique to estimate α.

II. Methods

A. The CROCUS research reactor

The CROCUS zero-power research reactor is a two-region, water-moderated uranium core operated by the Laboratory for Reactor Physics and Systems Behaviour (LRS) at the Swiss Federal Institute of Technology Lausanne (EPFL). It is housed in a concrete shielding of about 1.3m thickness, see Figure 1. It is a zero-power reactor, with a maximum power of up to 100 W. The reactor core is an approximately cylindrical configuration with a diameter of about 58 cm and a height of 100 cm, consisting of two fuel zones (see Figure 2). The central zone is loaded with 336 UO_2 fuel rods (1.806 wt.%-enriched), set in a square lattice with a pitch of 1.837 cm. The peripheral zone is loaded with up to 176 thicker, U_{met} fuel rods (0.947 wt.%-enriched) with a pitch of 2.917 cm, also in a square lattice. The core is brought to criticality by introducing water from below via pumps, with an excess reactivity at

Fig. 1. Top-down view of the CROCUS reactor from above the concrete containment. The top wall may be opened when the reactor is shut down.

the maximum water level being 200 pcm. The water level is controlled by a spillway that allows for 0.1 mm accuracy on the control of the water level (equivalent to about 0.4 pcm). The core is located in an aluminum water tank, its diameter is 130 cm and its thickness is 1.2 cm [9].

B. Organic scintillators

Organic scintillators are a dual-particle sensitive detector. The hydrocarbon volume is sensitive to gamma rays mainly via Compton scattering and neutrons mainly via proton elastic scattering. The Compton scattering interaction provides the information for light-output calibration via a measured, monoenergetic source. The neutron light output relative to gamma-ray light output is quenched, non-linear, and delayed. For this reason, the detectors are only sensitive to fast neutrons (as low as 700 keV) with a detection threshold of about 50 keV electron equivalent (keVee).

The incident neutron and gamma ray radiation may be discriminated on-the-fly with a charge integration based pulse-shape discrimination. By quantifying the delayed light-output of neutrons in comparison to gamma-rays, we can discriminate the two types of radiation [16].

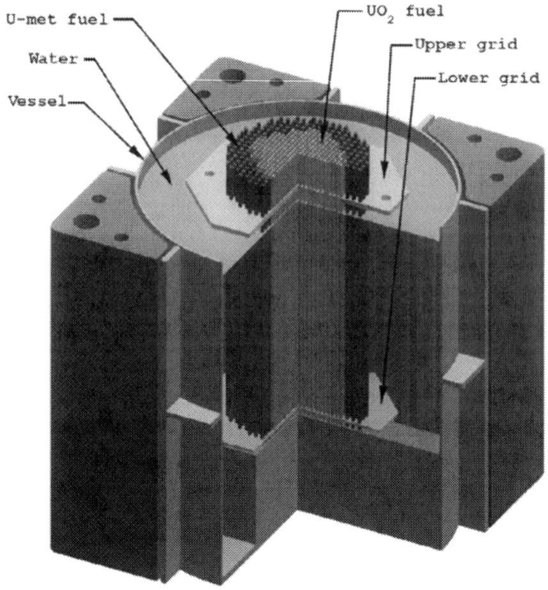

Fig. 2. Schematic view of the CROCUS reactor, showing the vessel, fuel grids, fuel elements, and water level when in operation.

C. Neutron-gamma noise experimental setup

We placed one 5.08-cm-diameter by 5.08-cm-length trans-stilbene detector in the water moderator offset 20 cm from the edge of the U_{met} zone, as detailed in Figure 3, with the active volume of the stilbene centered about the mid-height of the active fuel volume. A sealed, clear plastic tube fastened to the grid held the detector in position and protected the assembly from water. The detector was connected to a CAEN DT5730S 500-MHz 14-bit digitizer and a CAEN DT1470ET high-voltage unit in the reactor control room through diagnostic channels connecting the control room and containment while maximizing shielding [16]. After a gradual search for the critical water level, we measured the reactor for 60 minutes at critical.

D. Power spectral density technique

The time dependent behavior of the neutron flux $n(t)$ can be described by the point kinetics approximation [1]:

$$\frac{dn(t)}{dt} = \frac{\rho(t) - \beta_{\text{eff}}}{\Lambda} n(t) + \sum_i \lambda_i c_i + S, \qquad (1)$$

with the quantities ρ, the reactivity, β_{eff}, the effective delayed neutron fraction and Λ, the prompt generation time. The beta-delayed neutron precursor concentrations c_i and decay constants λ_i are derived via a balance equation, also known as the Bateman equation. The prompt neutron decay constant is defined as

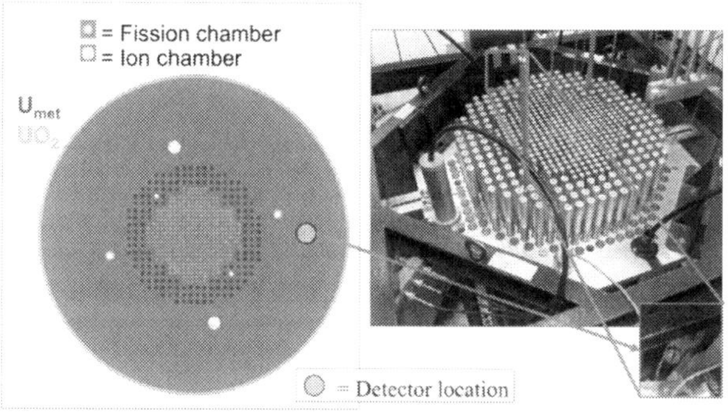

Fig. 3. The axial stilbene detector placement in the CROCUS reflector.

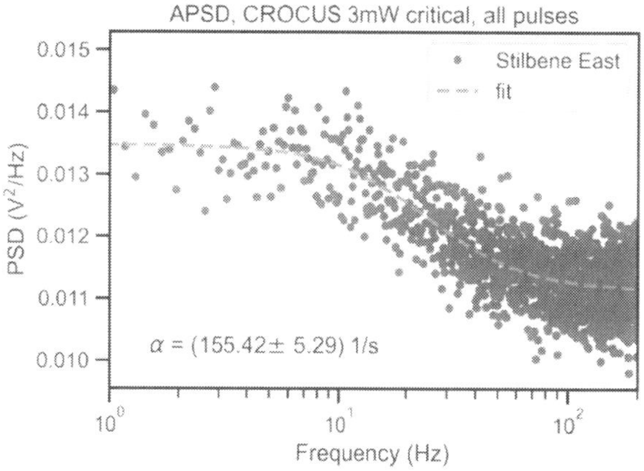

Fig. 4. Auto power spectral density (APSD) of the stilbene detector signal pulse time stamps.

$$\alpha = \frac{\beta_{\text{eff}} - \rho}{\Lambda}. \tag{2}$$

Using [17], the autocorrelation of a point-like reactor can be written as:

$$P_{ii}(\tau) = \frac{1}{2} Y \alpha e^{-\alpha|\tau|}, \tag{3}$$

with Y being

$$Y = \frac{\epsilon D_\nu}{(\beta_{\text{eff}} - \rho)^2}, \tag{4}$$

where ϵ is the detector efficiency in counts (C_i) over fission rate (F_0), and D_ν is the Diven factor [18]. When accounting for the overlap of uncorrelated fission chains, an autocorrelation term $\delta(\tau)$ needs to be added, here neglecting delayed neutrons for illustrative purposes [14]:

$$P_{ii}(\tau) = \epsilon F_0 \ (\frac{1}{2} Y_1 \alpha e^{-\alpha|\tau|} + \delta(\tau)). \tag{5}$$

This formulation, called the "Rossi-α method", is a common noise analysis technique using time interval distributions [3], [19]. It describes the correlation of a given detection with one later in time - which evidently shows a relation that is linked through the prompt fission chain, i.e., detecting a neutron now will have a high likelihood of resulting in another neutron detected as the fission chain continues to propagate. The time correlation is hereby exponential in nature: the fission chain will eventually die out - thus the name prompt decay constant for the coefficient of the exponent.

The auto power spectral density method (APSD) is an arguably more (unwanted) noise-resistant method to estimate α. The APSD of a detector is found by Fourier-transforming the Rossi-α function, yielding a Lorentzian [14]:

$$G_{ii}(\omega) = \int_{-\infty}^{\infty} dt e^{-i\omega t} P_{ii}(t) =$$
$$\epsilon_i F_0 + \frac{\epsilon_i^2 F_0 D_\nu}{(\beta_{\text{eff}} - \rho)^2} \frac{1}{1 + \omega^2/\alpha^2}. \tag{6}$$

By Fourier transforming the time series of a detector signal, one may directly fit this expression to obtain α, in this case not an exponential decay, but the Fourier transform of it: a Lorentzian bell curve with cut-off frequency.

III. Result

The resultant APSD distribution yields a precise estimate of $\alpha = 155.42 \pm 5.29$ when fit with equation 6 (see Figure 4). The total data for both neutron and gamma-ray detection time stamps were analyzed together without discrimination.

IV. Conclusions and Future Work

The result presented in this work represents a precise estimate of the prompt neutron decay constant of CROCUS at critical in a relatively brief, 60-minute measurement time. This measurement time was shorter and compares well to previous measurements and simulations [4], [10], [12], [13]. Exact simulation modeling of this reactor configuration will be needed to support this estimate of α due to the unique perturbation caused by this detector configuration. A successful comparison would be an additional benchmark for code validation.

Analysis utilizing pulse-shape discrimination and multiple trans-stilbene detectors will be analyzed to increase modalities and improve precision. Small volume, 6-mm edge, cubic organic scintillators placed next to the reactor U_{met} (less than 1 cm offset) that exhibit lower relative magnitudes of pulse pile-up will be analyzed for the same purpose.

Acknowledgment

This work was supported in part by the Consortium for Monitoring, Technology, and Verification under the Department of Energy National Nuclear Security Administration award number DE-NA0003920. We thank J-P. Abegg for the logistical support in getting the sensitive detection equipment from EPFL to GVA.

REFERENCES

[1] K. O. Ott and R. J. Neuhold, *Introductory nuclear reactor dynamics*. Amer Nuclear Society, 1985.

[2] C. E. Cohn, "Determination of reactor kinetic parameters by pile noise analysis," *Nuclear Science and Engineering*, vol. 5, no. 5, pp. 331–335, 1959, Publisher: Taylor & Francis.

[3] M. Y. Hua, F. B. Darby, J. D. Hutchinson, G. E. McKenzie, S. D. Clarke, and S. A. Pozzi, "Validation of the two-region rossi-alpha model for reflected assemblies," *Nuclear Instruments and Methods in Physics Research Section A: Accelerators, Spectrometers, Detectors and Associated Equipment*, vol. 981, p. 164535, Nov. 21, 2020, ISSN: 0168-9002. DOI: 10.1016/j.nima.2020.164535. [Online]. Available: https://www.sciencedirect.com/science/article/pii/S0168900220309323.

[4] O. Pakari, V. Lamirand, G. Perret, P. Frajtag, and A. Pautz, "Kinetic parameter measurements in the CROCUS reactor using current mode instrumentation," *IEEE Transactions on Nuclear Science*, 2018, Publisher: IEEE.

[5] B. Geslot, F. Mellier, A. Pepino, *et al.*, "Kinetic parameters of the GUINEVERE reference configuration in VENUS-f reactor obtained from a pile noise experiment using rossi and feynman methods," in *Advancements in Nuclear Instrumentation Measurement Methods and their Applications (ANIMMA), 2015 4th International Conference on*, IEEE, 2015, pp. 1–6.

[6] O. Pakari, D. Mancusi, O. Petit, A. Zoia, V. Lamirand, and A. Pautz, "Towards the validation of noise experiments in the CROCUS reactor using the TRIPOLI-4 monte carlo code in analog mode," in *PHYSOR 2020*, 2020.

[7] M. Salvatores, G. Palmiotti, G. Aliberti, *et al.*, "Methods and issues for the combined use of integral experiments and covariance data: Results of a NEA international collaborative study," *Nuclear Data Sheets*, vol. 118, pp. 38–71, 2014, Publisher: Elsevier BV. DOI: 10.1016/j.nds.2014.04.005.

[8] D. J. Siefman, "Development and application of data assimilation methods in reactor physics," Ph.D. dissertation, Lausanne, EPFL, 2019. DOI: 10.5075/EPFL-THESIS-7525.

[9] O. Pakari, V. Lamirand, G. Perret, L. Braun, P. Frajtag, and A. Pautz, "Current mode neutron noise measurements in the zero power reactor CROCUS," in *EPJ Web of Conferences*, vol. 170, EDP Sciences, 2018, p. 04017.

[10] O. V. Pakari, V. Lamirand, T. Mager, P. Frajtag, and A. Pautz, "High accuracy measurement of the prompt neutron decay constant in CROCUS using gamma noise and bootstrapped uncertainties," *Annals of Nuclear Energy*, vol. 170, p. 109011, Jun. 1, 2022, ISSN: 0306-4549. DOI: 10.1016/j.anucene.2022.109011. [Online]. Available: https://www.sciencedirect.com/science/article/pii/S0306454922000469 (visited on 05/25/2023).

[11] O. Pakari, V. Lamirand, B. Vandereydt, *et al.*, "Design and simulation of gamma spectrometry experiments in the CROCUS reactor," in *ANIMMA, Portoroz, Slovenia*, 2019.

[12] T. Mager, O. Pakari, V. Lamirand, and A. Pautz, "The LEAF system and gamma detection applications in CROCUS," Master's Thesis, École polytechnique fédérale de Lausanne (EPFL), Grenoble Phelma, and Politecnico di Torino, 2020.

[13] J. Leppänen, M. Aufiero, E. Fridman, R. Rachamin, and S. v. d. Marck, "Calculation of effective point kinetics parameters in the serpent 2 monte carlo code," *Annals of Nuclear Energy*, vol. 65, pp. 272–279, Mar. 2014, Publisher: Elsevier BV. DOI: 10.1016/j.anucene.2013.10.032.

[14] O. V. Pakari, "Experimental and numerical study of stochastic branching noise in nuclear reactors," Ph.D. dissertation, Lausanne, EPFL, 2020. DOI: 10.5075/EPFL-THESIS-8336.

[15] N. Zaitseva, A. Glenn, L. Carman, *et al.*, "Scintillation properties of solution-grown trans-stilbene single crystals," *Nuclear Instruments and Methods in Physics Research Section A: Accelerators, Spectrometers, Detectors and Associated Equipment*, vol. 789, pp. 8–15, Jul. 21, 2015, ISSN: 0168-9002. DOI: 10.1016/j.nima.2015.03.090. [Online]. Available: https://www.sciencedirect.com/science/article/pii/S0168900215004635.

[16] F. Darby, M. Hua, O. Pakari, S. Clarke, and S. Pozzi, "Multiplicity counting using organic scintillators to distinguish neutron sources: An advanced teaching laboratory," *American Journal of Physics, Accepted July 20, 2023, In Pre-proof*, 2023.

[17] M. M. R. Williams, *Random processes in nuclear reactors*. Elsevier, 2013.

[18] B. Diven, H. Martin, R. Taschek, and J. Terrell, "Multiplicities of fission neutrons," *Physical Review*, vol. 101, no. 3, p. 1012, 1956, Publisher: APS.

[19] V. Lamirand, G. De Izarra, A. Krása, *et al.*, "Intercomparison of neutron noise measurement systems in the CROCUS reactor," in *Proc. PHYSOR*, 2018.

New experiment for measuring the delayed neutron yields of ^{238}U fission in the range 1MeV to 19 MeV

D.Belverge[1,*], P.Leconte[1], B.Geslot[2], G.Kessedjian[1], P.Mutti[3], F.Rodiac[4], E.Pirovano[5], B.Lutz[5], X.Ledoux[6], L.Mathieu[7], O.Meplan[8]
[1]CEA, DES, IRESNE, DER, SPRC, LEPh, Cadarache, F-13108 Saint Paul Lez Durance, France
[2]CEA, DES, IRESNE, DTN, SMTA, LMN, Cadarache, F-13108 Saint Paul Lez Durance, France
[3]Institut Laue-Langevin, Grenoble, France
[4]CEA, DES, IRESNE, DER, SPESI, LEXIC, Cadarache, F-13108 Saint Paul Lez Durance, France
[5]PTB, Braunschweig, Germany
[6]GANIL, Caen, France
[7]CNRS, IN2P3, LP2i, Bordeaux, France
[8]CNRS, IN2P3, LPSC, Grenoble, France
(*) dorian.belverge@cea.fr

***Abstract*— In nuclear reactors operating in nominal conditions, delayed neutrons (DN) drive the kinetic behavior of the neutron population. Predicting this behavior requires therefore the knowledge of the DN yield per fission, average lifetime, and kinetic of their emission, which is described by precursors groups abundances. Characterizing the uncertainties of these macroscopic data is essential, especially as they are used to design the safety margins of reactors.**

Even for major fissile isotopes such as ^{235}U and ^{239}Pu, discrepancies still appear between the main nuclear data library. In 2018, CEA launched the ALDEN project (Average Lifetime of DElayed Neutrons), gathering several laboratories (CEA/DES and DRF, CNRS/LPSC, LP2i and LPC Caen, ENSICAEN, and University of Caen). This collaboration aims at producing new precise DN macroscopic data for reactor applications.

Three experimental campaigns were conducted in 2018, 2019 and 2021 on the thermal fission of ^{233}U, ^{235}U and ^{239}Pu at Institut Laue Langevin in Grenoble (France), giving high quality results. In February 2023, a new campaign was performed at the PTB Ion Accelerator Facility in Braunschweig (Germany), focusing on the fast fission of ^{238}U. In particular, the DN yield was measured for eight energies between 1.5 and 19 MeV. This paper presents the experimental setup, that was adapted to the new facility and neutron spectrum. The methodology of analysis is also detailed, and some preliminary delayed neutron decay curves are showed.

***Keywords* — Delayed neutrons, ^{3}He proportional counters, uranium 238, fast neutron induced fission**

I. Introduction

Delayed neutrons (DN) are produced during the beta decay of particular fission products, called "precursors". Emitted with a mean energy of 500 keV and between 1 ms to 1 minute after the fission, they play a key role in predicting the kinetic behavior of nuclear reactors, even though they represent less than 1 % of the neutrons emitted after a fission . In particular, knowing the average number of DN produced after a fission (called DN multiplicity or yield) and the kinetics parameters (precursors groups abundances) associated to their emission is needed to estimate the effective DN fraction (βeff, fraction of fissions induced by DN), which also defines the limit of criticality accident in nuclear reactors.

For isotopes of main interest for reactor applications, discrepancies still exist between nuclear data libraries (such as JEFF-3.3 and ENDF/B/VIII) [1]. In the case of ^{238}U in particular, very few experimental data are also available, and are usually associated with uncertainties higher than 5% [2]. In 2018, the ALDEN collaboration was therefore settled, gathering laboratories working in the nuclear data field (CEA/DES and DRF, CNRS/LPSC, LP2i and LPC Caen, ENSICAEN and University of Caen). The objective of the collaboration was to produce new experimental DN data with reduced uncertainties, focusing at first on the thermal fission of fissile isotopes.

An experiment was designed, based on an efficient fast neutron detection system. The principle of the measurement was the following: a sample of the investigated isotope is irradiated inside the detector in order to induce fissions and produce precursors. After a certain period of time, the irradiation is stopped and the DN emitted during the precursors decay are counted. Cycles of irradiation/DN counting are repeated to increase the statistics, the duration of both phases being adjusted depending on the DN data that is targeted. The experimental setup was used three times at the PF1b cold neutron beam at ILL (Institut Laue Langevin, Grenoble, France) in September 2018 [1], June 2019 and March 2021 [3]. Very precise measurements (uncertainty lower than 1 % on the DN yield) were produced for the thermal fission of ^{233}U, ^{235}U and ^{239}Pu [4].

After the success of these measurements, it was decided to investigate the fast fission of ^{238}U, and especially the energy dependency of the DN multiplicity. The experimental setup was then adjusted to be used with fast monoenergetic neutrons at the PTB Ion Accelerator Facility (PIAF, Braunschweig, Germany). After being tested on several occasions at the GENESIS facility (Grenoble, France, in February and October 2022) and at PIAF (in December 2022), it was finally used during a two weeks long campaign conducted at PIAF in February 2023), during which eight energies were investigated between 1.5 and 19 MeV.

The present paper focuses on the campaign performed in February 2023 at PIAF. After presenting the experimental setup used, the measurements performed are detailed. The methodology used to analyze the data is then described. Finally, preliminary results are shown and discussed.

II. Experimental Setup

A. PTB Ion Accelerator Facility

The measurements took place in the "Low-scatter hall" of the PTB Ion Accelerator Facility at Braunschweig (Germany). Monoenergetic neutrons flux were produced through the interaction of a solid tritium or gaseous deuterium target and a beam of protons or deuterons delivered by a Tandem accelerator. The solid tritium targets were composed of a titanium layer loaded with tritium and deposited on either a silver or an aluminum backing. The deuterium gas target was 30 mm in length and had a pressure of around 1.6 bar. A molybdenum foil (5 μm in thickness) was placed at its entrance. Table 1 summarizes the characteristics of the targets used during the campaign.

A steering magnet was mounted on the beam line before the experimental hall. It was used to deflect the incident charged particles beam in order to perform the irradiation/decay cycles required for the measurements.

Table 1
Characteristics of the targets used to produce the incident monoenergetic neutron flux

E_n MeV	Reaction	E_{proj} KeV	Target	Thickness μg/cm^2	T/Ti mass ratio
1.5	T(p, n)^{3}He	2375	Solid	1831	1.25
2.5	T(p, n)^{3}He	3319	Solid	999	1.18
3	T(p, n)^{3}He	3840	Solid	1831	1.28
4	D(d, n)^{3}He	1740	Gas	-	-
5	D(d, n)^{3}He	2419	Gas	-	-
6.5	D(d, n)^{3}He	3705	Gas	-	-
14.8	T(d, n)^{4}He	215	Solid	1026	1.25
19	T(d, n)^{4}He	2674	Solid	1871	1.25

E_n and E_{proj} are respectively the energies of the neutron produced and of the incident charged particle.

B. Uranium sample

The uranium sample was a cylinder of metallic depleted uranium (0.22% in ^{235}U), with 15 mm in diameter, 9.7 mm in height and a mass of 29.5481 g. To ease its manipulation and for radioprotection issues, it was encapsulated in an aluminum container (1 mm in thickness) prior to the campaign (see Fig. 1). A passive gamma counting of the sample was performed in the beginning of the campaign with the HPGe detector of the facility to verify its isotopic composition.

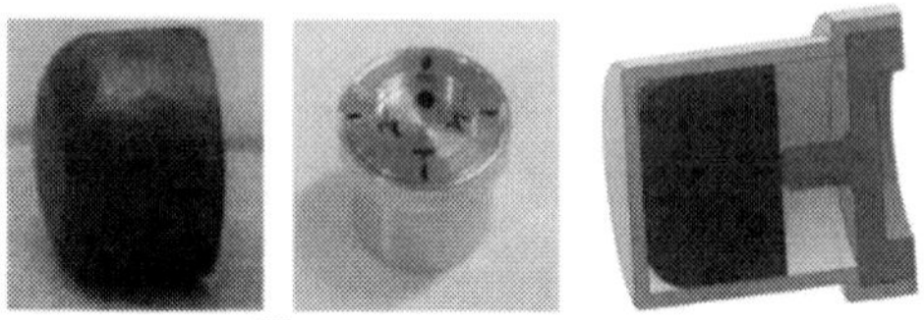

Fig. 1. From left to right: ^{238}U sample before and after encapsulation, scheme of the encapsulation (red: sample, grey: aluminum container, blue: aluminum plug).

C. Detection system

The detector LOENIEv2 was already used in the previous ALDEN measurements performed at ILL [1]. It was made of sixteen ^{3}He proportional counters (PCs) at a pressure of 10 bar, polarized at a high voltage around 2100 V and placed on three concentric rings in a cylinder of high-density polyethylene (HDPE, 46 cm in diameter, 34 cm in length, see Fig. 2). The center of the detector was left free to allow placing the sample under investigation. Delayed neutrons emitted in the sample were thermalized in the HDPE before inducing ^{3}He(n, t) reactions in the PCs. The global geometry of the detector, including the position of each PC, has been optimized in order to ensure an almost flat efficiency (2% relative variation) in the range 0.1 to 1 MeV, the energy range of delayed neutrons emitted inside the sample. Finally, boron rubber sheets covered the inner and outer faces of the HDPE cylinder, absorbing thermal neutrons. These sheets played a key role in reducing the detection of room-returned neutrons as well as limiting the fission of ^{235}U inside the sample from incident neutrons thermalized in the HDPE.

LOENIEv2 was calibrated for delayed neutrons emitted at its center in December 2021 at the NPL facility in England [5]. For the measurements performed at PIAF however, the sample was placed at some distance from the center of the detector (see section D). Therefore, a relative calibration was performed between these two positions with an AmBe source during the campaign.

Fig. 2. Picture of LOENIEv2 during the installation. ^{3}He PCs are inserted from the back face of the detector. Inner and outer faces of the PEHD cylinder are covered with boron rubber sheets (in black).

D. Source, sample and detector geometry

To maximize the fission rate inside the sample and be able to count delayed neutrons as soon as the irradiation stops, LOENIEv2 was placed at a few centimeters from the neutron production target, with the sample inside. The detector was aligned thanks to a set of sights, used in association with the lasers and telescopes available in the experimental hall and allowing visualizing the horizontal and vertical axes of the beam line. A specific tool was developed based on a depth gauge to measure the distance separating LOENIEv2 from the target.

The encapsulated sample was placed in a three wings aluminum support ensuring its radial centering. The support was inserted in an aluminum guiding tube placed at the center of the detector. An aluminum stop placed at the end of the tube provided a repeatable and precise positioning of the sample, at around 4 cm from the front face of the detector (see Fig. 3).

Fig. 3. Placement and alignment of LOENIEv2 (in red) at proximity of the neutron production target (in green). Positioning of the 238U sample inside the detector (in blue).

E. Data Acquisition System

The acquisition system was designed by the ILL instrument control group in 2018 for the first AL33DEN campaign. It was operated using the NOMAD software also developed at ILL [6].

The signals of the sixteen ^{3}He PCs were shaped using a sixteen channels charge sensitive preamplifier card. They were then processed by two synchronized digitizers (CAEN V1724, 8 channels, 14 bits 100 MS/s Flash ADC) running a real time digital pulse processing program (CAEN DPP-HPA). This program allowed detecting the events and obtaining their arrival time and energy (pulses amplitudes). Data were stored in binary files in "list mode" before being further processed. The signal of a neutron monitor placed at 97° and around 2.5 m from the target, already shaped by the facility, was recorded as well using an additional histogram card (recording in histogram mode, no energy determination). Finally, the acquisition system communicated with a monitoring computer through an Ethernet link, allowing transferring a large data rate from the experimental hall to the command room of the facility.

To avoid saturating the system during irradiations (detection of a large amount of incident, prompt and delayed neutrons), the acquisition of the ^{3}He PCs was inhibited during the main part of it, the rising and falling edges left apart. A low frequency generator was therefore used to generate an inhibition signal. It was synchronized with the cycles performed by the steering magnet cutting the neutron production off, thanks to a trigger signal provided by the facility. As the measurements required to record complete cycles of irradiation/decay (especially for group abundances measurements), a synchronized trigger signal was also generated using the same low frequency generator. This allowed launching automatically acquisitions at the beginning of the irradiations.

During irradiations, ^{3}He PCs were subject to a very high reaction rate. Tests have therefore been conducted prior to the campaign in February 2022 at the GENESIS facility in Grenoble (France) to verify that such reactions rates would not affect the efficiency of the counters after the end of the irradiation and that the system was able to recover after the beam stop in less than a few milliseconds.

III. The experimental campaign of February 2023

A. Overview of the delayed neutrons measurements

The experimental campaign was conducted over two weeks, during which DN multiplicity was measured for eight energies between 1.5 – 6.5 and 14.8 – 19 MeV by performing short cycles. The durations of these cycles (2/0.2 s, 2.1/0.1 s, 4/0.3 s or 6/1 s) were adjusted depending on the energy due to the variations of the beam cut off duration (see section IV.B). The DN group abundances were also measured for three of these energies (see Table 2) by performing long cycles allowing to count the DN emitted by each group (50/300 s).

For 19 MeV, an additional DN multiplicity measurement was performed with a blank target (same target without tritium) to correct for the neutrons produced by deuteron break-up reactions due to incident deuterons implementing in the target backing.

TABLE 2
LIST OF DN MEASUREMENTS PERFORMED DURING THE CAMPAIGN

Day	Energy MeV	Cycle	Objective
5	14.8	2 / 0.2 s	DN multiplicity
5	14.8	50 / 300 s	Groups abundances
6	4	2.1 / 0.1 s	DN multiplicity
7	5	4 / 0.3 s	DN multiplicity
7 and 8	5	50 / 300 s	Groups abundances
11	2.5	4 / 0.3 s	DN multiplicity
11	2.5	50 /300 s	Groups abundance
12	1.5	4 / 0.3 s	DN multiplicity
12	3	4 / 0.3 s	DN multiplicity
13	6.5	4 / 0.3 s	DN multiplicity
13	6.5	6 / 1 s	DN multiplicity
14	19	6 / 1 s	DN multiplicity

B. *Sample fission rate estimation*

Two methods were used for each energy to estimate the neutron flux seen by the uranium sample and determine the fission rate inside it during DN measurements. The first one comes from the knowledge that the facility has of the emitted neutron flux. For each energy, two neutron monitors (at 16° and 97 ° and around 2.5 m from the target) were calibrated using a reference long counter. The operation was repeated after placing LOENIEv2 close to the target to account for incident neutrons scattered in the detector. The counting rates inside the neutron monitors were then recorded during each DN measurement, allowing estimating the neutron flux emitted by the source and seen by the sample, thanks to a Monte-Carlo simulation of the experimental setup.

The second method consists in characterizing the neutron flux by dosimetry. Indium, Cobalt, Nickel or Magnesium dosimeters (2 mm in thickness, 15 mm in diameter) were encapsulated in an aluminum container and irradiated during a separate run at the position of the uranium sample. They were then measured by gamma spectrometry either on site using the HPGe detector of the facility (Indium and Magnesium dosimeters mainly) or at the Low-Activity Laboratory at LPSC Grenoble (France) for the radionuclides of high half-life. The fission rate will then be estimated by simulating the irradiation of both the dosimeters and the uranium sample (Monte-Carlo calculations).

A third method was employed for the first energy of the campaign (14.8 MeV). It consisted in activating the not-yet irradiated sample through a thirty minutes long continuous irradiation before estimating the reaction rates of some fission products (^{92}Sr and ^{135}I) by gamma spectrometry. The fission rate was then deduced using the cumulative fission yields of the measured fission products. As this method does not require to simulate the experimental setup, it will be used as a reference to validate the two previously presented methods at 14.8 MeV.

C. *Optimization and characterization of the background and beam cut-off*

Both the background and beam cut off durations varied significantly with the neutron energy investigated. As these are key points in estimating the DN multiplicity and groups abundances, a consequent amount of time and efforts was dedicated to their optimization and characterization for each energy.

The optimization phase was performed with a dummy container similar to the one encapsulating the uranium sample. The background optimization was conducted by deflecting the incident charged particles beam using the steering magnet (configuration used during the delayed neutrons counting) and acting on the settings of the magnet and the beam line entrance window while monitoring the counting rate inside the ^{3}He PCs. The same procedure was applied for the beam cut off optimization, except this time cycles of irradiation/decay were performed (similar to the ones used for DN multiplicity measurements).

At the end of the optimization phase, the beam cut off was characterized using a dummy container and performing cycles, whereas the background was measured with the uranium sample in place and with the incident charged particles beam deflected.

IV. DATA ANALYSIS STRATEGY AND FIRST RESULTS

A. *Methodology of analysis*

Each "run" of acquisition produced a binary file containing the data associated to every event recorded by each ^{3}He PCs during the acquisition. Runs performed for DN multiplicity measurement contained several cycles whereas one run corresponded to one cycle for groups abundances measurement. For DN multiplicity measurements, the first runs (around ten minutes) were left out of the analysis, as the population of precursors had not reached its equilibrium yet.

The first step of the analysis process is to convert every binary file into a readable format compatible for instance with the MATLAB environment. Then several steps are taken in order to produce the DN decay curves that will be used to estimate the DN data of interest:

1. PHA spectra are built for each ^{3}He PC, in order to select the energy interval corresponding to neutron detection.
2. Time histograms of the counting rate inside the ^{3}He PCs are generated for each run.
3. Counting rates of each ^{3}He PC is corrected for dead time, by applying a non-extendable model of parameter τ, which was estimated in previous tests performed in February 2022 at the GENESIS facility (Grenoble, France).
4. For each run, the cycles are separated based on a threshold applied to the counting rate summed over all ^{3}He tubes, allowing identifying the beginning and the end of the irradiation.
5. The "DN counting only" time interval is then identified for each cycle (between the end of an irradiation and the beginning of the next one).

6. Finally, the obtained counting rates inside the time interval of interest are summed over all ^{3}He PCs and all cycles.

During these steps, the actual duration of the cycles performed are also estimated.

The next steps consists in estimating the background observed during the DN counting, the fission rate inside the sample during irradiations (using the three methods introduced in section 0) and the efficiency of LOENIEv2 towards DN emitted inside the sample (based on the relative calibration performed with an AmBe source). Once these steps are completed, the DN multiplicity can be deduced using (1):

$$c(t) = b_{off} + \nu_d F_0 \sum_k a_k \varepsilon_{d,k} \frac{(1-e^{-\lambda_k t_{irr}})e^{-\lambda_k t}}{1-e^{-\lambda_k t_m}}, \quad (1)$$

Where $c(t)$ is the counting rate at t ($t = 0$ corresponding to the end of irradiation), b_{off} the background during DN counting, ν_d the DN multiplicity, F_0 the fission rate inside the sample, a_k the abundance of group k, $\varepsilon_{d,k}$ the efficiency of LOENIEv2 for DN of group k, λ_k the decay constant associated to group k, t_{irr} the duration of the irradiation and t_m the duration of the irradiation/decay cycle.

B. Preliminary results: DN decay curves, signal to noise ratio and veto time

Fig. 4 presents two example of DN decay curves obtained for DN multiplicity measurements at 14.8 and 2.5 MeV. DN counting rates varied a lot between energies, as could be expected due to the variations of fission cross section, DN multiplicity and flux emitted by the targets. The measured background also showed some variations with the neutron energy, leading to signal over noise ratios ranging from 12 to 163 (see Table 3).

Substantial variations were also observed during the campaign concerning the beam cut off duration, depending on the energy investigated, as illustrated in Fig. 4. During this period of time, not only delayed neutrons are detected, but also remaining incident neutrons. Therefore, a veto time has to be applied after the identified end of irradiation, during which the neutron detections are not taken in consideration for the estimation of the DN multiplicity. Although it is most likely insignificant compared to the whole irradiation time, the impact of the additional fissions that are induced during this time on the model described by (1), which considers perfectly squared irradiations, will have to be investigated.

Table 4 presents the preliminary fission rates estimated using the methods introduced in section III.B, for 14.8 MeV. A good agreement is observed between the results given by the activation of the ^{238}U sample and the flux calibration using dosimeters with high-energy threshold reactions. The method based on the PIAF neutron monitors calibration and monitoring to determine the neutron flux however leads to an apparent underestimation of the fission rate. More analysis is required to understand where this discrepancy comes from.

TABLE 3
PRELIMINARY RESULTS OBTAINED FOR EACH ENERGY: COUNTING RATE AFTER THE VETO TIME FOLLOWING THE END OF THE IRRADIATION (INCLUDING DN AND BACKGROUND, BKG), BACKGROUND MEASURED WITH THE SAMPLE IN PLACE AND THE BEAM DEFLECTED, SIGNAL TO NOISE RATIO OBTAINED AND VETO TIME ESTIMATED.

Energy MeV	DN + bkg counting rate (s^{-1})	Bkg counting rate (s^{-1})	Signal to noise ratio	Veto time ms
1.5	54	0.7	76	70
2.5	76	0.7	108	70
3	134	9.3	13	70
4	101	0.8	125	60
5	148	0.9	163	73
6.5	184	4.8	37	100
14.8	450	16	27	15
19	38	2.9	12	80

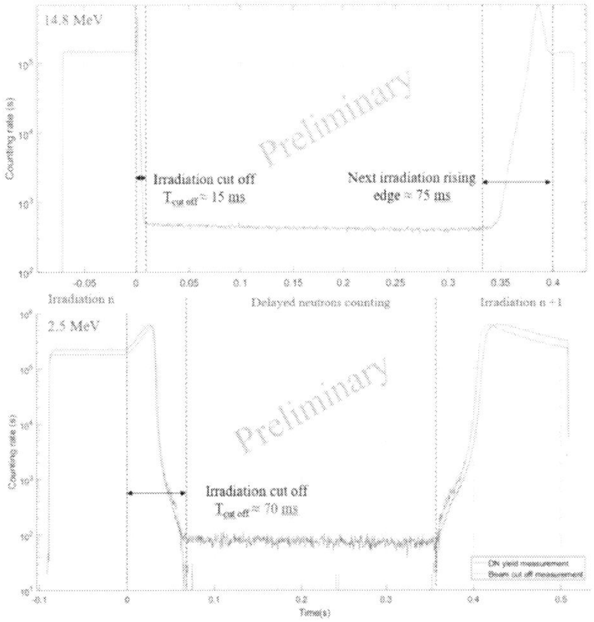

Fig. 4. Decay curves obtained for 14.8 (up) and 2.5 MeV (bottom). Irradiation cut off starts at $t = 0$. Inhibition of the ^{3}He PCs acquisition stops approximately 0.1 s before the beginning of the cut off (counting rate increasing from 0 to more than 10^5 counts/s. For 2.5 MeV, the counting rate obtained during the beam cut off characterization with a dummy container is also displayed in red.

TABLE 4
PRELIMINARY ESTIMATIONS OF FISSION RATE INSIDE THE SAMPLE DURING IRRADIATIONS AT 14.8 MEV USING THE THREE METHODS PRESENTED IN SECTION III.B

Method	Reaction (Threshold, MeV)		Fission rate (x 10^5 fissions/s)
Flux monitoring (PIAF)	-		1.784 (3.5 %)
Dosimetry	^{115}In (n, n') ^{115m}In	(≈0.340)	1.408 (1.2 %)
	^{59}Co (n, p) ^{59}Fe	(≈3)	2.016 (3.0 %)
	^{60}Ni (n, p) ^{60}Co	(≈4)	2.195 (10.1 %)
	^{24}Mg (n, p) ^{24}Na	(≈6)	1.920 (1.1 %)
	^{115}In (n, 2n) ^{114m}In	(≈9)	1.969 (3.2 %)
	^{59}Co (n, 2n) ^{58}Co	(≈11)	2.072 (3.1 %)
^{238}U sample activation	-		2.026 (2.3%)

V. CONCLUSIONS

Since 2018, CEA has conducted new measurements of DN macroscopic data such as multiplicity and groups abundances, in the framework of the ALDEN collaboration. An experimental setup was designed to count the DN emitted from an irradiated sample placed at the center of an efficient neutron detector. This setup has been used three times over the last years at ILL on the cold neutron beam PF1b, with fissile samples (^{233}U, ^{235}U and ^{239}Pu).

Started in 2021, a new project inside the collaboration focused on measuring the DN data for the fast fission of fertile isotopes. The setup was adapted to be used at the PTB Ion Accelerator Facility in Germany. Two weeks of measurements were performed in February 2023, during which DN data associated to the fast fission of ^{238}U were measured for eight incident neutron energies between 1.5 and 19 MeV.

Overall and despite the change of installation and incident neutron spectrum, the campaign was a success, as data were acquired for all energies planned. The analysis is still in progress, but preliminary results suggest that the measurements performed were of good quality.

Finally, a second campaign is planned in September 2023, focusing on energies between 7 and 14 MeV. It will benefit from the feedback of the measurements conducted in February, including for example the need to better characterize the beam cut off profile at the end of each irradiation.

ACKNOWLEDGMENT

This work was partly funded by I3P Institute gathering CEA, EDF and Framatome. It also received funding from the Euratom research and training program 2014-2018 under grant agreement No 847594 (ARIEL).

The authors would like to gratefully thank the PTB Ion Accelerator Facility staff members for their involvement in preparation of and throughout the campaign.

REFERENCES

[1] D.Foligno, "New evaluation of delayed-neutron data and associated covariances", PhD thesis, N°2017AIXM0001/001ED62. Aix Marseille Université, France, Oct. 2019.

[2] IAEA Reference Database for Beta-Delayed Neutron Emission. Produced by IAEA Coordinated Research Project (2013-2018). [Online]. Available: https://nds.iaea.org/beta-delayed-neutron/database.html, Accessed on: August. 10, 2023

[3] B.Geslot, A.Sardet, P.Casoli, P.Leconte, G. De Izarra, A.Chebboubi *et al.*, "Measuring the delayed neutrons multiplicity and kinetic parameters for the thermal induced fission of ^{235}U, ^{239}Pu and ^{233}U", ANIMMA Conference, Prague, Czech Republic, 2021.

[4] P.Leconte, B.Geslot, A.Sardet, P.Casoli, T.Kooyman, D.Belverge, "Measurement of the delaed-neutron yield in the thermal neutron induced fission of ^{239}Pu", EPJ Web of Conf. Volume 284, 2023. Doi: https://doi.org/10.1051/epjconf/202328408008

[5] P. Leconte, B. Geslot, T. Kooyman, "Validation of the monte-carlo efficiency calculation of the LOENIEv2 long counter for delayed neutron measurements", proceedings of the wonder2023 workshop, Aix-en-Provence, 5-9 June 2023.

[6] P. Mutti, et al., "NOMAD –More than a simple sequencer". Proceedings of 13th Int. Conf. on Accelerator and Large Experimental Physics Control Systems (ICALEPCS), Grenoble, France, 10-14th October 2011.

Optimization of the magnetic core of a Linear Variable Differential Transducer

Rudi Van Nieuwenhove*,
Belgian Nuclear Research Institute, SCK CEN, Belgium
(*) rvnieuwe@sckcen.be

Abstract— **Research into materials and fuels for nuclear power reactors is carried out in so-called research reactors where various types of fuels and materials can be monitored online and in real time by in-core instruments. At the Halden reactor in Norway, online measurements such as cladding elongation, inner fuel rod pressure, fuel swelling, material creep and stress relaxation were based on Linear Variable Differential transducers (LVDT).**
The magnetic core of the standard LVDT core is made of silicon iron, encapsulated by an Inconel 600 or AISI 316L tube for corrosion protection. In the present study, we consider FeCrAl as magnetic core material. This material, which is currently also being investigated as accident tolerant fuel cladding material, is corrosion resistant in the chemical environment of PWR or BWR reactors and therefore doesn't required an extra corrosion protection tube. As a result, the magnetic core diameter can be made larger, resulting in an increase of the LVDT sensitivity. In order to verify the various dependencies, the LVDT sensitivity was measured and compared to finite element calculations.
The sensitivity of an LVDT is influenced by the magnetic permeability of the magnetic core, as well as by eddy current losses in this magnetic core. When the temperature changes, the eddy current losses in the core also change, resulting in a slight change of the LVDT sensitivity. It is shown that this effect is reduced when using FeCrAl as magnetic core material, instead of SiFe. On the other hand, the change in magnetic permeability with temperature for FeCrAl is larger than for SiFe. The overall combined effect results in a higher dependence of sensitivity for the FeCrAl core.
Finally, using a solid magnetic core (without encapsulation) allows operation of the LVDT at higher gamma heating in the core of a reactor because of improved cooling by the surrounding water. This feature is relevant for operation in the Jules Horowitz Reactor, where the gamma heating in some locations can be up to 20 W/g.

Keywords **—LVDT, core, magnetic, eddy current**

I. INTRODUCTION

In the Halden LVDT [1], the magnetic core is made out of Si(2.5%)Fe. The dimensions of this core are shown in Fig. 1. This core has a length of 24 mm and a diameter of 2 mm. Since SiFe is highly corrosive in high temperature water, it is encapsulated by a housing in Inconel 600 with an outer diameter of 3.3 mm and an inner diameter of 2.5 mm. The gap between the magnetic core and the housing (0.25 mm) is filled by helium at a pressure of 3 bar. The core housing fits inside the bore of the LVDT (inner diameter 4 mm), such that there is gap for water (or liquid metal) between the core holder and the LVDT of 0.35 mm. This gap is necessary for the correct cooling of the LVDT core.

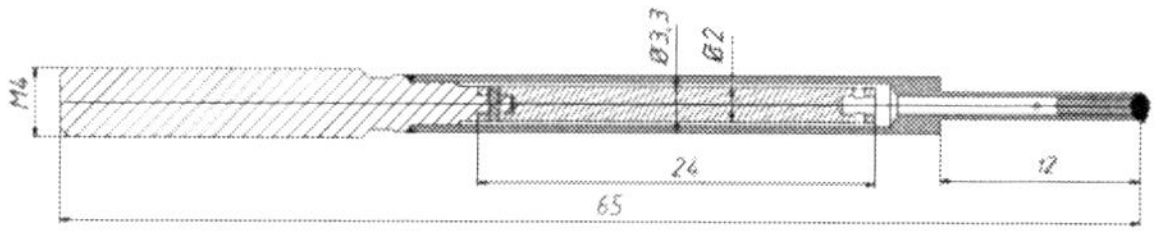

Fig. 1. LVDT core holder (Halden type)

In order to fabricate this type of core, some difficult steps are required:

1. Fabrication (by electrical wire discharge cutting of a 2 mm SiFe bolt out of a larger diameter (10-20 mm) bolt. This is a time consuming step.
2. Heat treatment of the SiFe core in order to obtain the required magnetic properties (involves heating in vacuum up to 850 °C).
3. Filling the gap region by a mixture of He (90 %) and Ar (10 %).
4. Seal welding the gas region.

Besides these mechanical steps, it can be recognized that the magnetic core assembly resembles very much a gamma thermometer (GT). Although He has a rather good thermal conductivity, the temperature of the magnetic core can nevertheless reach a high temperature at high gamma heating, such that the Curie point is reached. At this point, the LVDT stops functioning.
It is clear that if a material could be found which is both magnetic and corrosion resistant (in BWR or PWR conditions) that the design of the core could be simplified considerably because the core housing could be omitted in this case.

II. DESIGN OF THE NEW MAGNETIC CORE

Such a material actually exists, namely FeCrAl. Kanthal is the trademark for a family of iron-chromium-aluminium (FeCrAl) alloys used in a wide range of resistance and high-temperature applications. Kanthal FeCrAl alloys consist of mainly iron, chromium (20–30%) and aluminium (4–7.5 %). The first Kanthal FeCrAl alloy was developed by Hans von Kantzow in Hallstahammar, Sweden. The alloys are known for their ability to withstand high temperatures and having intermediate electric resistance. As such, it is frequently used in heating elements. The trademark Kanthal is owned by Sandvik Intellectual Property AB. In recent years, FeCrAl is being investigated as cladding material for accident tolerant fuel rods in nuclear power reactors [2]. A FeCrAl rod (diameter 3.2 mm, length 200 mm) was ordered from Goodfellow. From Goodfellow's website, the following properties are retrieved:

Composition (%): Fe 72.8, Cr 22, Al 5 %, Y 0.1, Zr 0.1
Physical properties:
Density (g/cm^3): 7.22
Melting point (°C): 1380-1490
Electrical properties:
Electrical resistivity (µWcm): 134
Temperature coefficient (K^{-1}): 0.0001
Thermal properties:
Coefficient of thermal expansion @20-100 °C : 11.1 x10^{-6} K^{-1}
Thermal conductivity @23 °C (W/mK): 16
Magnetic properties:
Curie temperature (°C): 600

The magnetic permeability could not be found. This is probably due to the fact that this material is normally not used for its magnetic properties.
Note that the electrical resistivity of FeCrAl (134 µWcm) is a factor 2.8 larger than that of Si(3%)Fe (47.2 µWcm). Because of this, eddy current generation within FeCrAl core will be even smaller than in SiFe and this should result in even better linearity.
The drawings of this prototype core are shown in Figs. 2,3,4.

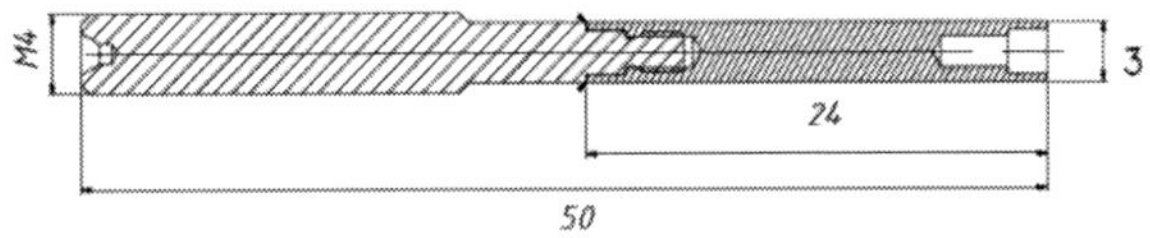

Fig. 2. Assembly of the FeCrAl core (right part). The left part is made in AISI 316L stainless steel.

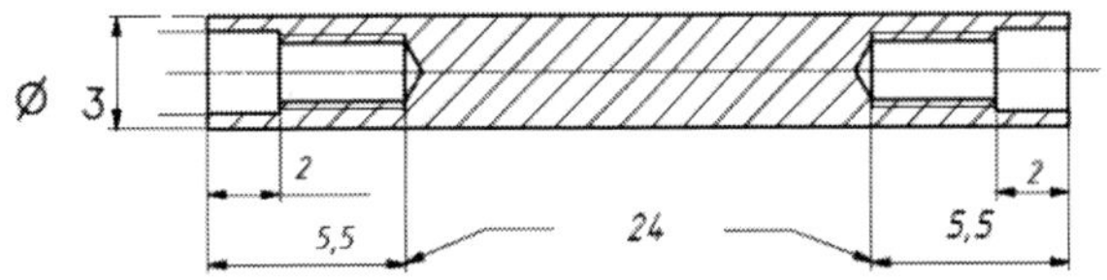

Fig. 3. Drawing of the FeCrAl core.

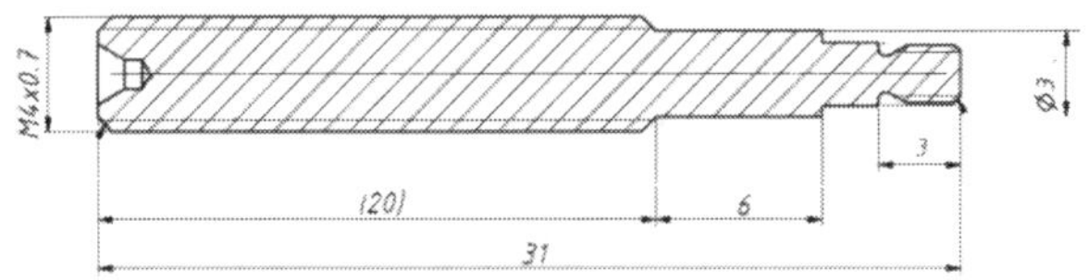

Fig. 4. Drawing of the core suspension part (AISI 316L).

Because no housing is needed, the magnetic core material can now have a diameter which is as large as the outer diameter of the non-magnet housing for the the SiFe core. In this prototype, the diameter was however slightly less; 3 mm instead of 3.3 mm. This larger diameter of the magnetic core will definitely increase the sensitivity of the LVDT. On the other hand, the magnetic permeability of FeCrAl is likely lower than that of SiFe.

III. CALCULATIONS

Calculations of the sensitivity of the LVDT with the new core were made with the finite element software package QuickField.

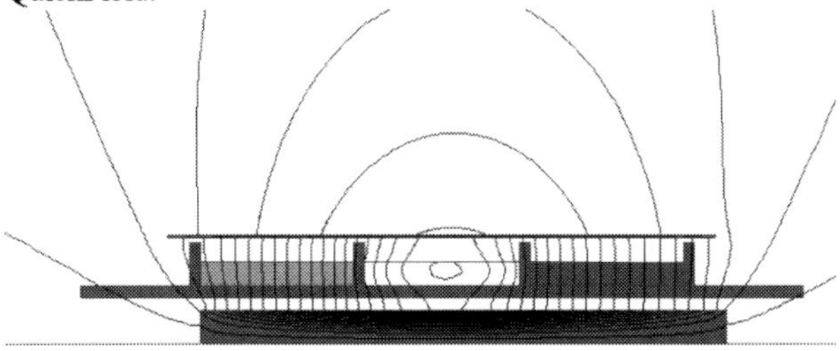

Fig. 5. Result of finite element calculations (QuickField) for the FeCrAl core, showing field lines.

The magnetic permeability of FeCrAl is not known, but it is likely lower than that of SiFe. The measurements (see further) show a total increase in sensitivity of 26 %. If the magnetic permeability of FeCrAl and SiFe (assumed relative permeability of 400) would have been the same, the finite element calculations (by QuickField) predict an increase in sensitivity of 43 %. In order to arrive at the experimentally observed improvement of 26 % (see later), the relative magnetic permeability of FeCrAl should be about 220 (as based on the calculations), thus about a factor 2 lower than that of SiFe. For the simulations, the core was displaced by 1 mm, the driving frequency was 1000 Hz and the driving current amplitude was 50 mA. The number of turns in each coil was 450. When changing the temperature of the FeCrAl core, the calculations showed only a very small increase in LVDT sensitivity with temperature. From 20 °C to 320 °C, the calculations show that the LVDT sensitivity was found to increase by only 0.2 %. For the same core geometry, the change in sensitivity for a SiFe core would be 13 %. In these calculations, the magnetic permeability was assumed to be independent of temperature. The measurements (see next section) indicate however that the latter effect is dominant.

IV. MEASUREMENTS

In order to verify the performance of the new FeCrAl core relative to the standard Halden LVDT core, a Halden LVDT type 5 was used, while using respectively a standard Halden core (based on SiFe inside a housing) and a FeCrAl core.

A. Measurements at room temperature

The LVDT and core were mounted in a room temperature calibration set-up which was designed for this purpose. Instead of using the standard Halden LVDT conditioning unit, a newly developed electronic unit was used (based on digital technology instead of the analog technology used in Halden). This new unit uses a coil driving current of 50 mA at an excitation frequency of 1000 Hz (whereas the Halden units work at 400 Hz). This higher frequency increases the sensitivity of the LVDT by a factor 1000/400 = 2.5. Here we will focus only on the relative sensitivities and the signal linearity.

Both the ac output of the secondary coils (measured by a Fluke 87V True RMS multimeter) and the DC output after demodulation (measured by a Fluke 787 process meter) were measured simultaneously for different positions of the magnetic core.

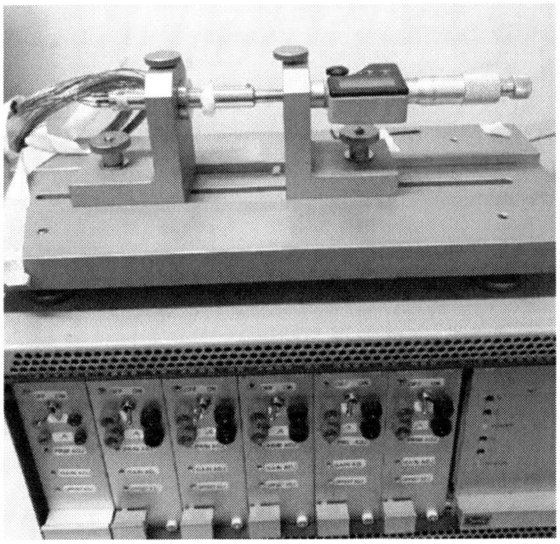

Fig. 6. Electronic units and calibration unit (on top) with mounted LVDT.

As shown in Fig. 7 to Fig. 10, the LVDT works very well with the new, simplified FeCrAl core. The sensitivity is even higher by 26 % as compared to the Halden core (see Fig. 9 and Fig.10). For both cores, one observes a slight deviation from linearity at the zero crossing. The deviation from linearity is slightly lower for the FeCrAl core.

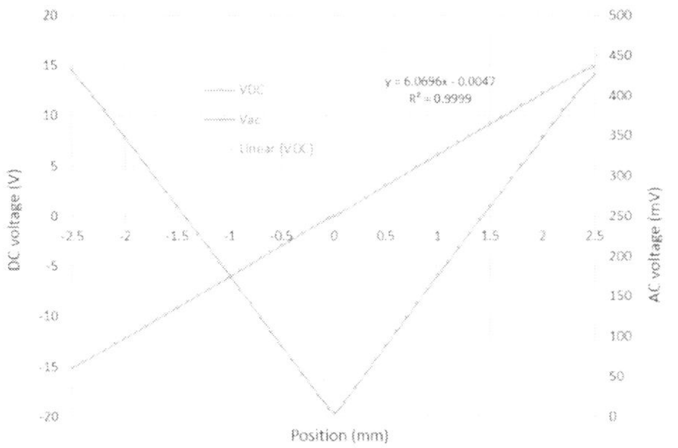

Fig. 7. Response of the LVDT with FeCrAl core.

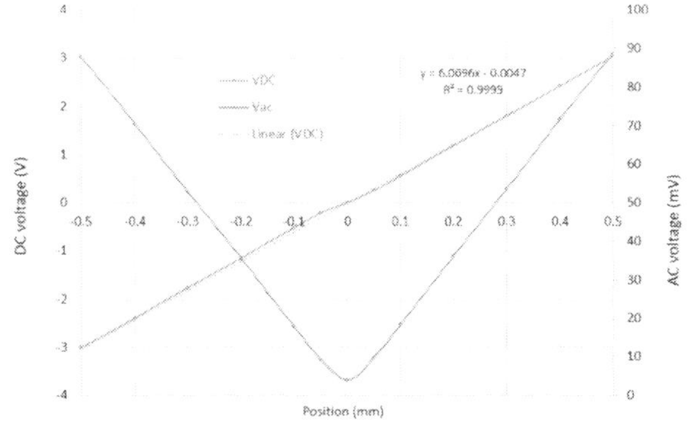

Fig. 8. Response of the LVDT with FeCrAl core showing a slight deviation from linearity of the DC signal at the zero crossing.

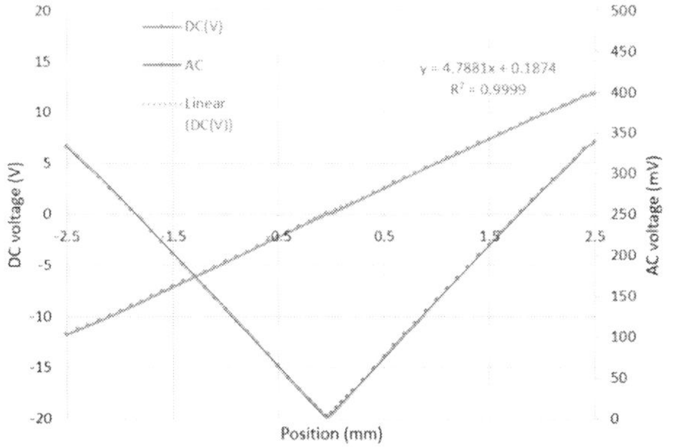

Fig. 9. Response of the LVDT with the Halden (SiFe) core.

https://doi.org/10.1051/epjconf/202328804017

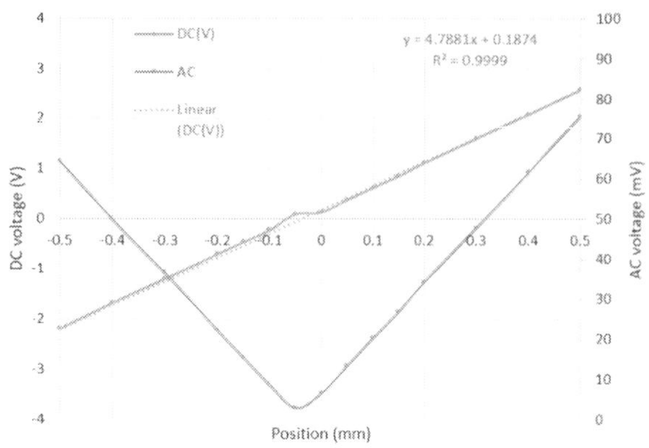

Fig. 10. Response of the LVDT with Halden core (SiFe) showing a slight deviation from linearity of the DC signal at the zero crossing.

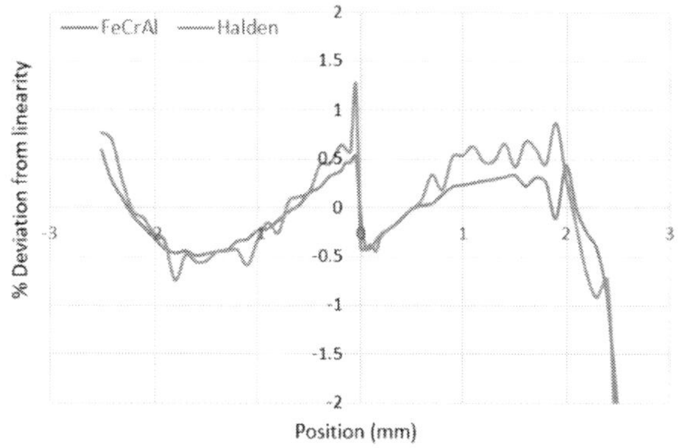

Fig. 11. Comparison of deviation from linearity between FeCrAl core and Halden core.

Fig. 12. High temperature LVDT calibration system.

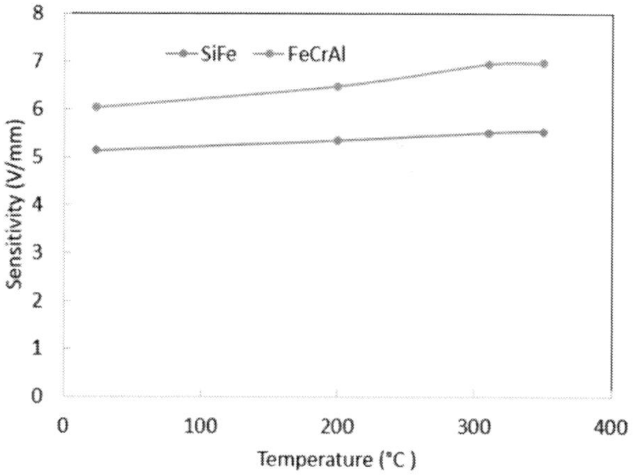

Fig. 13. Comparison of the LVDT sensitivity with SiFe and FeCrAl core.

B. Measurements at high temperature

A newly developed system for calibrating LVDTs at high temperature was used next (see Fig. 12). For the standard Halden LVDT and standard core (type 5), the sensitivity was measured with the newly developed conditioning unit, using a driving frequency of 1000 Hz. The sensitivity was found to increase from room temperature to 260 °C by 6.7 %. At a driving frequency of 400 Hz, the corresponding increase was 1.3 %. Thus the temperature dependence at 1000 Hz is a factor 5.15 higher than at 400 Hz. This factor is of the order of the ratio of the square of the frequency (6.25) and is in line with the fact that eddy current losses scale as the square of the frequency. The temperature dependence of the sensitivities at 400 Hz and 1000 Hz were found to be in good agreement with the calculated sensitivities (using QuickField), namely 1.3 % (400 Hz) and 6.07 % (1000 Hz), while assuming a constant magnetic permeability. Next, the LVDT sensitivities are compared with the SiFe and FeCrAl cores at different temperatures (see Fig. 13). The relative increase of the sensitivities for the two cores, relative to the sensitivities at room temperature are compared in Fig. 14.

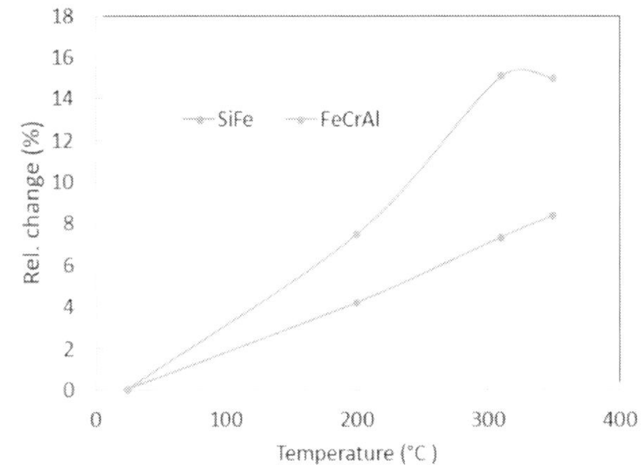

Fig. 14. Relative increase of sensitivity with temperature for the SiFe and FeCrAl cores.

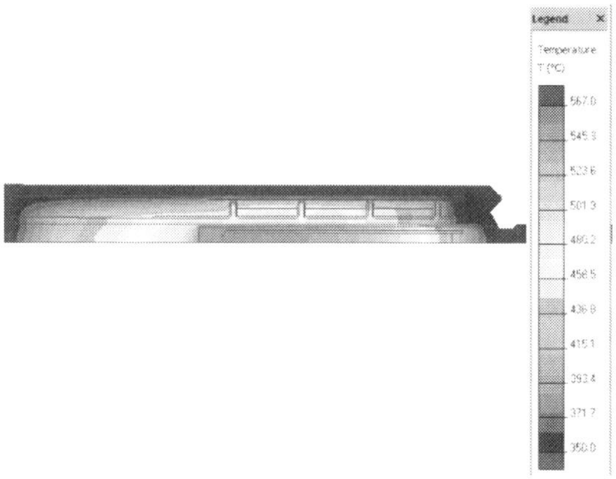

Fig. 15. Temperature distribution in an LVDT with standard core at a (stagnant) water temperature of 350 °C and a gamma heating of 5 W/g.

The higher dependence on temperature for the FeCrAl core cannot be explained by a change in eddy current losses but has to be due to a significant increase of the magnetic permeability of FeCrAl with temperature. According to the finite element calculations, the relative magnetic permeability should change from 220 at room temperature to 340 at 300 °C, thus corresponding to an increase of 54 %. If one would use the 5-wire electronic set-up, the relative increase of sensitivity from room temperature to 300 °C would be reduced from 15 % to 5 % (based on the finite element calculations).

V. OPERATION AT HIGH GAMMA HEATING

The Halden type LVDT was designed to operate in the core of a reactor with flowing water at a maximum temperature of 350 °C (PWR conditions). The design is such that there is still some gap around the magnetic core for water to flow and there is an entrance and exit hole for this water at opposite sides of the magnetic core. In some special cases, a combination of high gamma heating and low water flow velocity can lead to overheating of the LVDT. An example of zero flow and a gamma heating of 5 W/g is shown in Fig. 15.
In this case, the water is converted to steam in the thin gap between magnetic core and the LVDT bore, resulting in a strongly reduced thermal conductivity in this region.

In this case, the temperature of the coils reaches 476 °C and the magnetic core reaches 566 °C. The Curie temperature of the SiFe core is around 700 °C, so the temperature of the core is still below it. However, the temperature of the coil is at the limit for the ceramic insulated nickel clad copper wire (used in almost all the experimental rigs in the Halden reactor). In the newer design, the wire is made of a ceramic insulated silver palladium alloy which allows continuous operation up to 500 °C. Nevertheless, such high temperatures are undesirable and the internal sealing of the mineral insulated wire can also be damaged at this temperature.
Another aspect to consider is that the temperature of the core is not uniform over its length. This is because the He-filled encapsulated core effectively behaves as a gamma thermometer. Even with sufficient flow, there is (for this case), a temperature difference between the tip and the base of the core of about 50 °C. This in turn will lead to a small variation of the magnetic permeability which will induce an imbalance of the signals generated in the two secondary coils, thus effectively mimicking a small displacement of the core.
Finally, at a gamma heating of 20 W/g (expected gamma heating for the JHR [3]), the finite element calculations show that the core temperature would reach 760 °C (above Curie temperature) in the case of water flow around the core and the LVDT would thus not function.
For the FeCrAl core, this effect can be avoided entirely since the core is not thermally insulated by a gas gap.

A further improvement could be obtained by making the core hollow. One could use a tube (3 mm outer diameter and 2 mm inner diameter), such that also water could flow through (and around) the core. This will allow a higher flow of water around the core resulting in a better internal cooling. This, in combination with a lower heated mass (as compared to a solid core) will allow operation at higher gamma heating. Detailed thermal and fluid calculations would be required to determine the actual temperature distribution for a particular situation.

VI. Effect of irradiation

In the case of the Halden LVDT, the sensitivity drops by about 3 % during the first days of irradiation. This drop in sensitivity is thought to be due to the effect of fast neutrons on the magnetic permeability of the magnetic core. For an LVDT with a FeCrAl core, the effect of neutron irradiation has never been studied. The 3 % decrease for the SiFe core was observed using the 4-wire LVDT conditioning unit at a driving frequency of 400 Hz. With a 5-wire conditioning unit, the drop in sensitivity would be only 0.9 %, as based on the finite element simulations. Even if some drop in sensitivity upon irradiation would be observed for the FeCrAl core, the impact of it could likewise be made much smaller by operating the LVDT in 5-wire mode.

VII. Conclusions

In the standard Halden LVDT, an encapsulated SiFe magnetic core is used, leading to excellent performance in terms of linearity and temperature sensitivity. Making the magnetic core of unprotected FeCrAl however, allows 1) to fabricate the core in a much simpler and cheaper way, 2) to increase slightly the linearity of the LVDT, 3) to use the LVDT at higher gamma heating up to 20 W/g (expected for the JHR).
The sensitivity of the LVDT with FeCrAl core increased from room temperature to 300 °C by 54 %, due to a significant increase of the magnetic permeability with temperature. Whatever the degree of sensitivity on temperature, it is shown that this can be reduced significantly by operating the LVDT in 5-wire mode. Finally, the LVDTs are always calibrated using a high temperature calibration set-up so that the sensitivity is determined in function of temperature.
Further, it is shown that changes in sensitivity due to neutron irradiation (during the first days of irradiation in a research reactor) can also be reduced significantly by operating the LVDT in 5-wire mode.

ACKNOWLEDGEMENT

The author wants to acknowledge Guido Engelen for support for setting up the high temperature calibration unit and for doing the LVDT calibrations.

References

[1] R. Van Nieuwenhove, S. Solstad, "In-Core Fuel Performance and Material Characterization in the Halden Reactor", IEEE Trans. Nucl. Sci., 57(5), pp. 2683-2688, Oct. 2010, 10.1109/TNS.2010.2050701.
[2] B.A. Pint, K.A. Terrani, Y. Yamamoto and L.L. Snead, Material Selection for Accident Tolerant Fuel Cladding, Metallurgical and Material Transactions E, 2, 190-196 (2015).
[3] D. Blanchet, M. Antony, H. Carcreff, et al., "Assessment of Iradiation Performance in the Jules Horowitz Reactor (JHR) using the CARMEN Measuring Device", EPJ Web of Conferences, 253, 04010 (2021), ANIMMA 2021, https://doi.org/10.1051/epjconf/202125304010 .

POLLEN: A Pile-Oscillator for the BLOOM Experimental Program

Thomas Ligonnet[1,*], Nicolas Weiss[2], Vincent Lamirand[1,3], Axel Laureau[4], and Andreas Pautz[1,3]
[1]LRS, EPFL, CH-1015 Lausanne, Switzerland
[2]IPHYS, EPFL, CH-1015 Lausanne, Switzerland
[3]NES, PSI, CH-5232, Switzerland
[4]Univ. Grenoble Alpes, CNRS, Grenoble INP*, LPSC-IN2P3, 38000 Grenoble, France

(*) thomas.ligonnet@epfl.ch

***Abstract*—In the scope of its project on the assimilation and reproduction of experiments for the evaluation of stainless-steel nuclear data HARVEST-X, the LRS launches a pile-oscillation experimental program in the CROCUS reactor: BLOOM. For this program, an oscillator called POLLEN originally developed to be used as a vibrating absorber to compensate the fuel rods oscillator COLIBRI was reworked as a standalone pile-oscillator called. The oscillator operates with an arbitrary periodical shape, an amplitude of 1 m and a frequency ranging from the mHz to the Hz. An emphasis was put in the development of an interface for pseudo-square oscillations for BLOOM. This interface also allows independent adjustment of the ramping time and dwell time of the pseudo-square. The qualification of POLLEN was performed by video analysis of sinusoidal oscillations, using a 4K 30fps camera. With the current system, a precision of 0.2 mm was achieved during slow sinusoidal oscillation and a precision of 0.17 mm was obtained for pseudo square oscillations with a 500 g load. Whereas the results are satisfactory with respect to the current system and fulfill the requirement of the BLOOM program, it is planned to upgrade the system with the addition of a mechanical reference in the system. It is also in consideration to upgrade the acquisition card and controller to 16 bits systems to allow the use of POLLEN in setups where larger displacements are needed. The first in-core oscillation experiments are planned for early 2024.**

***Keywords* —Pile-Oscillation, POLLEN, BLOOM, HARVEST-X, CROCUS, Nuclear data, Stainless-Steel.**

I. INTRODUCTION

THE BLOOM (*Bayesian-aided Local Oscillation for Optimized Modulation*) program is an incoming experimental program in the CROCUS reactor at EPFL. BLOOM is part of the HARVEST-X project (*Heuristic Assimilation, Reproduction, and Validation of Experiments for stainless-STeel Cross-sections*), the EPFL follow-up to the EPFL-CEA PETALE (*Programme d'Etude en Transmission de l'Acier Lourd et Eléments*) experimental program [1–3]. The BLOOM program consists in performing pile-oscillations in the CROCUS reactor with an open-loop methodology, similarly to the experiments performed during the CEA MAESTRO-SL campaign [4–8]. Like PETALE, the program will primarily focus on the study of samples related to nuclear grade stainless-steel.

This paper focuses on the developments conducted on the pile-oscillator POLLEN (*Pile-Oscillator for Localized and Low Effect Noise*). Originally developed to extract samples for the study of transients in CROCUS [9], and adapted for oscillations in the CORTEX project campaigns [10], POLLEN was especially used as a secondary oscillator in synchronization with the fuel rods oscillator COLIBRI [11], oscillating a cadmium sample in the reactor core center. In view of a standalone usage for the BLOOM program, the controls and commands were upgraded, and a system qualification was performed in-house by video analysis.

II. BLOOM IN THE CROCUS REACTOR

A. The CROCUS reactor

CROCUS is a zero-power light water research reactor. It operates at ambient pressure, at a temperature of 20 °C, and is licensed for a maximum thermal power of 100 W. The reactivity is primarily controlled by adjusting the vessel water level using a spillway. This adjustment affects fuel rods immersion, which in turn affects neutron thermalization. It can also be controlled by two optional B_4C control rods. The water level can be adjusted up to ±0.1 mm, equivalent to a reactivity worth of ±0.4 pcm around criticality, and the control rods up to ±0.5 mm, equivalent to a minimum reactivity worth of about ±0.2 pcm based on the rods S-curve [3,12,13].

The core consists of two types of fuel rods distributed in two interlocking zones: an inner zone and an outer zone separated by a water gap. The inner zone contains 336 uranium dioxide fuel rods enriched at 1.806 wt.%, while the outer zone has metallic uranium fuel rods enriched at 0.947 wt.%. The number of rods in the outer zone can be adjusted depending on the specific experimental configuration, with currently 180 metallic uranium rods for compensating the SAFFRON array [14]. CROCUS is equipped with four monitors: two boron coated compensated ionization chambers for safety monitoring (east and west) and two fission chambers (north and south) for monitoring and recording the reactor's power history. The fission chambers operate in pulse mode with an adjustable dwell time, typically of 1 second, enabling real-time monitoring of the reactor, and are typically used to reconstruct the reactor power history.

B. The BLOOM pile-oscillation program

The BLOOM program is a pile-oscillation program set to take place in the CROCUS reactor. As a follow-up to the EPFL-CEA PETALE program on stainless steel [1–3], BLOOM aims at producing experimental results usable for Bayesian Monte Carlo data assimilation [15], and comparable to PETALE. In this sense BLOOM is a reproduction of the criticality experiments of PETALE but with a different methodology, one of the objectives being to identify potential methodological biases. Indeed, whereas PETALE had criticality experiments (and transmission experiments), the BLOOM program is composed of a type of modulation experiments called pile-oscillations. In these experiments, the periodic perturbation of the local neutron flux, due to the insertion and extraction of a sample, is measured.

The chosen methodology for the BLOOM program is the open-loop pile-oscillations [4,5], in opposition to closed-loop ones. With this method the oscillations of macroscopic samples are performed in the reactor core without compensation for the periodic change in reactivity. Therefore, it is necessary to not only measure the local flux perturbation but also the general evolution of the neutron population to correct for the induced periodic power drift. This method was already successfully tested by CEA during the MAESTRO-SL campaign [7,8].

In the CROCUS reactor, it is planned to take advantage of LRS recent developments of novel miniature neutron detectors called MiMi detectors to measure the local flux perturbation [16], while the SAFFRON array of 160 detectors will be used to monitor the general and spatial behaviour of the reactor [14].

The primary focus of the BLOOM oscillation program is stainless steel and its components. As in the PETALE program, the reference stainless-steel will be 304L. Its main components iron, nickel, and chromium, will also be taken separately with elemental samples of high purity [1,2]. Especially, it is planned to recut samples from the spare materials of the PETALE program to limit the number of sources of uncertainty during future analysis and comparisons of the results. In addition, it is planned to oscillate other samples such as reference elements D_2O and gold, and possibly other alloys or elements of interest such as stainless-steel 316L, tungsten, and possibly EUROFER.

III. POLLEN DESCRIPTION AND CALIBRATION SETUP

A. Mechanics and Controls

The POLLEN oscillator mechanics is fairly direct. The vertical displacement of a sample is produced by the rotation of a wheel around which a thread, attached to the sample, is lapping, and the verticality of the displacement is assured by gravity. In CROCUS, the oscillation will be performed in an air-filled guide tube at atmospheric pressure. The friction in the guide being low, the maximum downward acceleration and speed are constrained by the gravity as the motor can only slowdown the displacement when going down. A picture of POLLEN with its main elements is displayed in Fig. 1.

The different elements of POLLEN are listed below:

- A wheel of 10 cm in diameter
- A Kevlar thread for sample support
- A Brushless DC motor with 3000 steps and a gear ratio to the wheel of 30
- A 12-bits controller
- A 14-bits data acquisition card (DAQ) with an absolute voltage precision of 26 mV and a voltage range of ±10 V
- An interface convertor ethernet-RS232
- A Windows computer with LabVIEW-2021 and the motor drivers

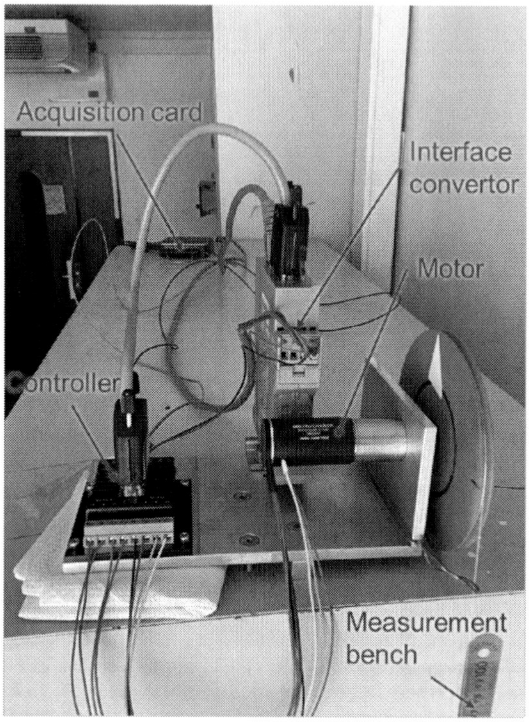

Fig. 1. Picture of the POLLEN pile-oscillator above the measurement bench used for calibration and testing.

The programming of a displacement is as follows:

First the limits in the motor angular speed and acceleration, and the motor PD settings are programmed into the controller using the motor driver and Eternet-RS232 interface convertor, represented with dashed arrows in Fig. 2. These settings are programmed once with generic values, which avoid the thread jumping out of the wheel, and are independent of the oscillation shape and the LabVIEW interface.

Second, the diameter of the wheel, the maximum range of displacement, and the acquisition card voltage range are set in the LabVIEW interface. For the BLOOM program those values are set to 1 m for the maximum displacement, ±10 V for the acquisition card voltage range, and 10 cm for the wheel diameter before calibration. At the start of the communication between the LabVIEW interface and the controller, the conversion factor between voltage and motor step is computed from the set parameters and calibration results, and then programmed into the controller.

 https://doi.org/10.1051/epjconf/202328804018

Third, once an oscillation is defined, the LabVIEW interface converts it to a corresponding voltage table and programs it in the data acquisition card. Following it, the data acquisition card will loop on the table and send in real time the corresponding voltage to the controller. The controller will then rotate the motor to move the sample to the expected height, represented by plain blue arrows in Fig. 2. The rotation of the motor is then done based on the previously mentioned driver settings. The acquisition card will continue to loop on the table until a manual or programed trigger request it to stop at the end of the current period.

Lastly, the position of the motor according to its internal encoder and the voltage measured by data acquisition card internal voltmeter are sent back to the computer LabVIEW interface (orange checked arrows of Fig. 2).

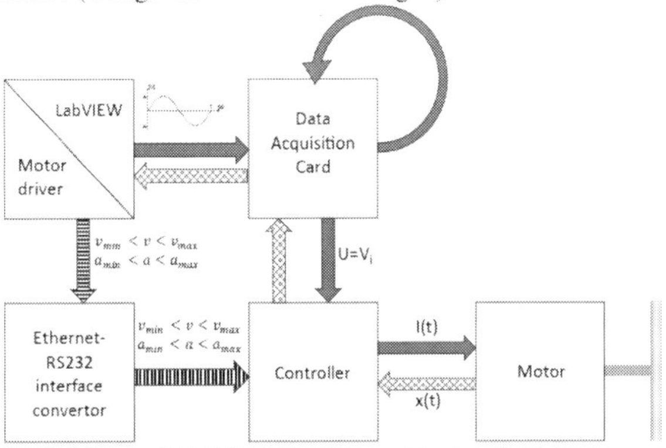

Fig. 2. Scheme of POLLEN control systems. The dashed arrows represent the motor and controller presettings. The blue full arrows are the path of the oscillation commands from the LabVIEW to the motor. The orange plaid arrows represent the feedbacks sent from the motor and data acquisition card to the LabVIEW.

B. *POLLEN functionalities*

At its inception with less advanced controls, POLLEN was limited to sample unique displacement and extraction. With its first *vibrating absorber* version for CORTEX, it oscillated in response to COLIBRI using an associated trigger signal: thus, without direct user control and at a frequency up to a few Hz. The latest developments aim at a standalone usage for pile-oscillation purposes.

Its first functionality is the possibility to perform single, non-periodic displacements. The firmware developments allowed correcting the loss of reference previously observed in going home. POLLEN can perform displacement to a targeted position, or of a given length relatively to its current position. The possibility to perform single and non-periodic displacement is necessary for the BLOOM program as it allows to obtain a first estimation of the sample reactivity worth in CROCUS.

The second functionality is the possibility to perform periodic displacement of arbitrary shape. The shapes are defined in ASCII files read by the LabVIEW interface. The user then defines the total amplitude of the oscillation as well as its period. This functionality was reworked from the version produced for CORTEX, the interface is created to ease the use and modification of the parameters of oscillation.

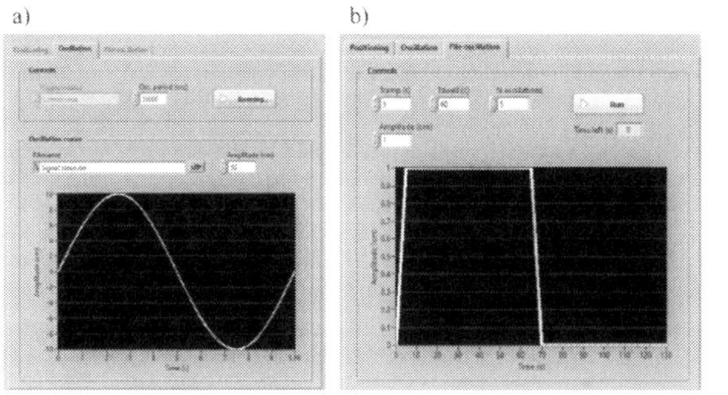

Fig. 3 a) LabVIEW interface for arbitrary oscillation shapes. b) Special interface for the pile-oscillation program.

The last functionality has been developed especially for the BLOOM program. It consists in a preconfigured interface for pseudo square oscillation such as the ones that will be performed during the pile-oscillation program. In this interface the user can define the amplitude of oscillation and, instead of the period, separately the ramping time between positions and dwell time at the positions. Screen shots of the two oscillation interfaces are displayed in Fig. 3.

C. *Calibration setup and methodology*

The displacements of POLLEN are slow, large, and one dimensional. Thus, the methodology chosen to calibrate and qualify the oscillator is based on video analysis of the displacement. The calibration setup consists of a 1 m long metallic ruler (class II), a 30 fps 4k camera, and an object of appropriate mass and size to displace (in fact, a second metallic ruler), see Fig. 4.

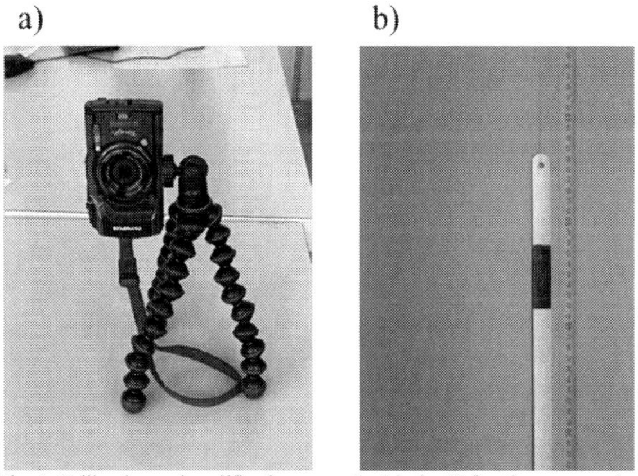

Fig. 4. Calibration and qualification setup. In a) is the 30 fps 4k camera and in b) the object and ruler used for the displacement measurements.

The prerequisites are an object darker than the video background and enough luminosity for the contrast of the ruler graduation. The methodology is the following. The video of the moving object is analyzed frame by frame. First, each frame is converted to a relative grey scale, then in the frames the columns of pixels are swept for an abrupt drop in the pixel brightness, example in Fig. 5.

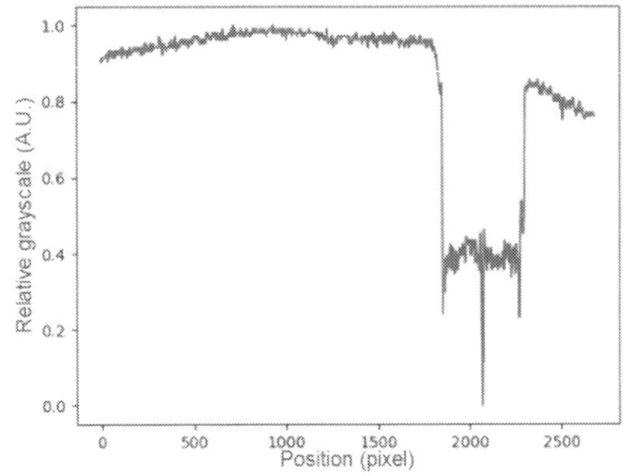

Fig. 5. Example of a swept column of pixel. The presence of the object is causing a sharp drop in the relative greyscale.

Once the position of the object is found, the image is cropped around it, and the column of pixel of the ruler is similarly swept to identify its graduation and determine locally the conversion factor between pixels and mm. Then, the number on the cropped frame are read using the Tesseract OCR [17], and the corresponding graduation and the related pixels are determined to be used as reference. The position in mm of the object is then computed as the reference position plus, the distance in pixel multiplied by the conversion factor between pixels and mm. The movement of the object is then reconstructed from all measured positions.

The theoretical precision of the oscillator is driven by the 12-bits of the controller. The amplitude of the displacement being 1 m the theoretical uncertainty on the displacement is 0.12 mm. In order for the characterization to be consistent with this value, the total amplitude of the displacement for the characterization is 40 cm and the field of view of the camera is fixed with a width of 50 cm allowing an average resolution of 0.125 mm/pixel. The period of the oscillation was fixed at 10 s to have velocities close to the ones planned for pile-oscillations.

IV. Calibration results

The characterization of the oscillator displacement was done with a sinusoidal oscillation of 20 cm in amplitude and a period of 10 s. The measured signal obtained with the methodology described in Section III.C is displayed in Fig. 6.

The measured main frequency of the measured signal is 10.0 ± 0.2 s. From the oscillation signal amplitude, it is measured that the effective diameter of the wheel is 103.5 ± 0.1 mm. The observed uncertainties at top and bottom positions along the oscillation is 0.2 mm, thus on the same order of the pixel resolution and the controller precision. The highest measured standard deviation for the position in a time bin is 2.2 mm at time 7.43 ± 0.03 s. This value is consistent with the time resolution of the camera as the velocity per frame at this time is 4.3 mm/frame. Tests were also performed with pseudo-square oscillations with ramping times of 5 s, dwell times of 60 s, and a load of 500 g. The measured standard deviation on the position during the dwell times at top and bottom positions is 0.17 mm.

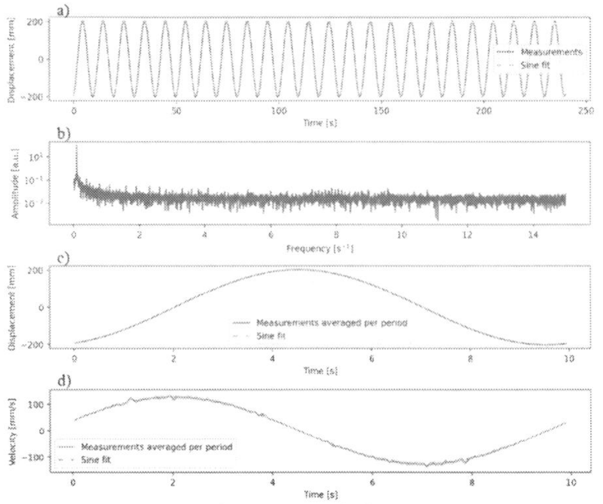

Fig. 6. Video analysis of a 20 cm amplitude sine oscillation with a 10 s period. Subplot a) shows the displacement of the object during the 4 min long video. Subplot b) shows the FFT of the signal. Subplot c) is the average on periods of the displacements. Subplot c) is the averaged measured velocity of the oscillator.

V. Future developments of POLLEN

Although the tests have proven conclusive, additional developments are planned. The first one will be the addition of an absolute reference of the position of the system. Indeed, the current version of POLLEN relies solely on the motor internal encoder to position itself and perform displacements. Thus, an unknown offset may be present at each power up of the system and currently it is required to manually verify the state of POLLEN before each experiment, which is impractical for a reactor use. In addition, an external encoder would be added to the motor to ensure that no drift is happening during and between oscillations.

A second upgrade consists in the replacement of the acquisition card for a model with a better precision on the output voltage. This precision directly affects the sample displacement precision, and may therefore be necessary as mechanical uncertainties are driving the total uncertainty of such experiments. Similarly, it is envisaged to upgrade the controller and motor to 16-bits systems, for greater precision, especially in case POLLEN is used for larger displacements.

VI. Conclusion

The POLLEN oscillator has been updated to work as a standalone pile-oscillator for the CROCUS reactor. Unlike its original version, which operated in response to another device, COLIBRI, the new version has been qualified for performing accurate single displacements, enabling sample reactivity worth to be estimated prior to the oscillations. The new interface still allows to choose arbitrary oscillation shapes from ASCII files, to adjust its period, and oscillation amplitude up to a maximum of 1m. In addition, a dedicated interface was developed for the BLOOM pile-oscillation program. It allows to perform pseudo-square oscillations with adjustable ramping time and dwell time.

Tests and calibrations of the oscillator were performed using video analysis with a 4K 30 fps camera. The results of the tests show an uncertainty on the top and bottom position of the oscillation of the same order than the controller of the motor precision. Similarly, the motor was observed to be stable during pseudo-square oscillations with a 500 g load.

Updates of POLLEN are planned for the near future with the addition of an absolute reference for the position and later on with upgrades to 16 bits systems. The current version is ready for in-core testing with the first oscillation of the BLOOM program planned for early 2024.

References

1. V. LAMIRAND et al., "An Experimental Programme optimized with Uncertainty Propagation: PETALE in the CROCUS Reactor," EPJ Web Conf. **211**, O. Serot and A. Chebboubi, Eds., 03003 (2019); https://doi.org/10.1051/epjconf/201921103003.
2. T. LIGONNET, Ed., *Preliminary analysis of the PETALE program and comparison to simulations* (2021).
3. V. LAMIRAND et al., *Future experimental programmes in the CROCUS reactor* (2016); https://doi.org/10.13140/RG.2.1.2309.6728.
4. W. K. FOELL, "Small-sample reactivity measurements in nuclear reactors," TID-26511, American Nuclear Society, Hinsdale, IL (1972).
5. B. A. BAKER, "Comparison of open loop and closed loop reactivity measurement techniques on the ISU-AGN-201 reactor," Ph.D., Idaho State University (2013).
6. P. LECONTE et al., "MAESTRO: An ambitious experimental programme for the improvement of nuclear data of structural, detection, moderating and absorbing materials - First results for natV, 55Mn, 59Co and 103Rh," in 2013 3rd International Conference on Advancements in Nuclear Instrumentation, Measurement Methods and their Applications (ANIMMA), pp. 1–9 (2013); https://doi.org/10.1109/ANIMMA.2013.6728101.
7. B. GESLOT et al., "Innovative Hybrid Pile Oscillator Technique in the Minerve Reactor: Open Loop Versus Closed Loop," IEEE Trans. Nucl. Sci. **65** 11, 2767 (2018); https://doi.org/10.1109/TNS.2018.2874690.
8. "A hybrid pile oscillator experiment in the Minerve reactor | Elsevier Enhanced Reader;" ; https://doi.org/10.1016/j.anucene.2017.04.036; (current as of Dec. 15, 2022).
9. A. RAIS, "Performance assessment of a 3-D steady-state and spatial kinetics model for the CROCUS reactor," EPFL (2018); https://doi.org/10.5075/epfl-thesis-8248.
10. V. LAMIRAND et al., "Experimental report of the 3rd campaign at AKR-2 and CROCUS," Public Deliverables 2.4, CORTEX, p. 38 (2021).
11. V. LAMIRAND et al., "ANALYSIS OF THE FIRST COLIBRI FUEL RODS OSCILLATION CAMPAIGN IN THE CROCUS REACTOR FOR THE EUROPEAN PROJECT CORTEX," EPJ Web Conf. **247**, 21010, EDP Sciences (2021); https://doi.org/10.1051/epjconf/202124721010.
12. U. KASEMEYER et al., "Benchmark on Kinetic Parameters in the CROCUS Reactor," 978-92-64-99020-3, Nuclear Energy Agency of the OECD (NEA), p. 93 (2007).
13. V. LAMIRAND et al., Eds., "Power calibration methodology at the CROCUS reactor," Adv. Nucl. Instrum. Meas. Methods Their Appl. Animma 2019, E D P SCIENCES, Cedex A (2020); https://doi.org/10.1051/epjconf/202022504022.
14. F. VITULLO, "Miniature and Minimalistic Neutron Detectors for Online High-Resolution Experiments in the Zero-Power Reactor CROCUS," EPFL (2022); https://doi.org/10.5075/epfl-thesis-9492.
15. A. LAUREAU et al., "Bayesian Monte Carlo assimilation for the PETALE experimental programme using inter-dosimeter correlation," EPJ Web Conf. **239**, 18004, EDP Sciences (2020); https://doi.org/10.1051/epjconf/202023918004.
16. F. VITULLO et al., "A mm3 Fiber-Coupled Scintillator for In-Core Thermal Neutron Detection in CROCUS," IEEE Trans. Nucl. Sci. **67** 4, 625 (2020); https://doi.org/10.1109/TNS.2020.2977530.
17. J. OOMS, *tesseract: Open Source OCR Engine* (2023).

Qualification of the Industrialized Libera MONACO 3 Digital Acquisition System dedicated to Fission/Ion Chamber Measurements in Research Reactors

Loic Barbot[1,*], Gregoire De Izarra [1], Christophe Destouches[2], Danilo Bisiach[3], Manuel Cargnelutti[3], Sebastjan Zorzut[3], Matjaž Škabar[3], Aleš Bardorfer[3], Aurora Makovac[3], Jure Trnovec[3], Peter Paglovec[3]

[1]CEA, DES, IRESNE, DER, Instrumentation Sensors and Dosimetry Laboratory, France
[2]CEA, DES, IRESNE, Reactor Studies Department, France
[3]Instrumentation Technologies, Slovenia
(*) loic.barbot@cea.fr - danilo.bisiach@i-tech.si

***Abstract*— The CEA Sensor, Dosimetry and Instrumentation Laboratory (LDCI) has been working since 2011 on an integrated acquisition system called MONACO, *'Multichannel Online Neutron Acquisition in Campbell mOde'*, for fission chamber measurements. Taking into account the feedback acquired up to the CEA TRL7 version, the LDCI Lab and the Instrumentation Technologies company (Solkan – Slovenia) initiated in 2020 a two years collaboration agreement to build the Libera MONACO 3 industrialized version required for reactor use and mass production. After a final qualification performed in the Slovenian TRIGA Mark II reactor in March 2023, the system will be available on the nuclear instrumentation market by the end of 2023.**

Keywords* — *Fission Chamber, Ion Chamber Digital Acquisition System, Neutron Flux, Pulse Mode, Campbell Mode, Current Mode.

I. Introduction

THE CEA Sensor, Dosimetry and Instrumentation Laboratory (LDCI) has been developing since 2011 an integrated acquisition system called MONACO which stands for *'Multichannel Online Neutron Acquisition in Campbell mOde'* for fission chamber (FC) measurements. The spread use of such online detectors were suffering from the lack of up-to-date and integrated signal acquisition system on the nuclear instrumentation market. While building its new Jules Horowitz Reactor (JHR), CEA was seeking a maintainable and standardized measurement system.

The MONACO system is unique as it integrates all the required features in one module to find out optimum parameters for FC operations and calibration prior to absolute online measurements. This module is also unique as it makes accessible the overlapping analysis of FC operating modes using a sole acquisition system.

The LDCI Lab and the Instrumentation Technologies company (Solkan – Slovenia) agreed in 2020 on a 30 months collaboration to develop the MONACO 3 version with a maximum level of technological maturity as well as all validations and certifications required for reactor use and mass production.

This joint development is based on the existing Instrumentation Technologies digital board, called Libera Digit 500, taken as-is. This also includes a fully redesigned electronic architecture, a redeveloped firmware containing all the data processing algorithms including the implementation of a true PHA acquisition mode, an innovative development for current measurement electronics with an improved bandwidth of 0 to 1 kHz and a new graphical user interface development carried out by Instrumentation Technologies based on the CEA existing interface.

After comprehensive lab test series with synthetic signals to ensure overall operation and performances, the two MONACO 3 prototypes completed a final qualification performed in the TRIGA Mark II reactor at the Jožef Stefan Institute (Ljubljana – Slovenia) in March 2023, using CEA made miniature fission and ion chambers.

The "Libera MONACO 3" version will be available on the nuclear instrumentation market by late 2023. It is meant to ease the implementation of fission and ion chambers as part of in-core instrumentation, gathering all required data acquisition systems in one with four channels, encouraging multiplication of neutron flux core mapping and neutron/gamma field characterization of experimental locations in research reactors.

II. Context

Fissions chambers (FCs) are online neutrons detectors widely used in nuclear reactors, often as part of the reactor command control system. In this case, FCs are commonly excore massive detectors with a strong sensitivity to neutrons and operated in pulse mode for reactor start-up measurements. They are frequently part of a displacement device for nominal power measurements while used in current mode. Both pulse and current mode acquisition electronics used in command control systems are specifically validated for nuclear safety, with a strict usage dedicated to the reactor operation and fixed automatic data processing.

Based on the same gaseous/ion detector technology, CEA developed from the 1960s some miniature (Ø3 mm) and subminiature (Ø1.5 mm) fission chambers [1] with a mineral integrated cable for incore neutron flux measurements in research reactors. These FCs are deployed in reactor core for local neutron measurements close to irradiation locations. Using dedicated FC calibration methods [2] made absolute neutron flux measurements achievable. It is also possible to get neutron spectrum with spectral indices techniques, miniature FC sensitivity being easily adapted to a neutron energy range with appropriate fissile coatings. These FCs are often used either at low neutron flux range in zero-power reactors [3] in pulse mode or at high neutron flux range in material testing reactor in current mode using both dedicated acquisition systems mainly derived from command control systems.

With the start-up of new research reactors such as the Jules Horowitz Reactor (JHR) at CEA Cadarache, miniature FCs are expected to be part of most instrumented experimental devices for technological irradiations or neutron field characterization. FCs will play the role of local neutron flux monitors giving access to online accurate neutron flux levels/spectra over a wide reactor power range, during either stabilized or transient reactor operations. This more intensive use of FCs requires new FC acquisition systems with enhanced performances.

Such experimental constraints for neutron flux measurement were first met at CEA, within the bilateral collaboration project with ANSTO (Australia), during the preparation of the neutron flux mapping of fuel coolant channels of the OPAL reactor [4]. Based on the previous CEA IRINA (Instrumentation RES Interne Neutronique Automatisée) project initiated in 2002 [5], the CEA LDCI Lab launched in 2011 the development of an integrated acquisition system in support to FC comprehensive measurements, called MONACO (Multichannel Online Neutron Acquisition in Campbell mOde) [6]. From the beginning, the MONACO system was designed to perform FC measurements over several decades of neutron flux, being capable of running FCs in both pulse and current modes as well as in the intermediate mode, called Campbell or fluctuation mode.

In the 2011-2019 period, the LDCI Lab, along with the CEA LIST Sensors and Electronics Lab (LCAE) worked on various implementations of the MONACO system, ending up in 2019 to the MONACOv2 TRL7 version in 2019 [7, 8]. This multichannel acquisition system, dedicated to neutron (and gamma using ion chambers) measurements was unique as it was integrating all the following features for FC measurements in one module: automatic generation of saturation curves, automatic generation of pulse discrimination curves, detector pulse characterization using the embedded oscilloscope module, pulse mode acquisitions in count rate or pulse height analysis, fluctuation mode and current mode acquisitions. FCs plugged to a single connector and the implemented operating modes run constantly in parallel, with a minimum acquisition time of 1 ms.

Taking into account the expertise and feedback acquired during CEA labs development, the LDCI Lab and the Instrumentation Technologies company (Solkan – Slovenia) initiated in 2020 a 30 months collaboration agreement to build the Libera MONACO 3 industrialized version.

III. Libera MONACO 3 design and performances

The following sections provide an overview of the key components of the Libera MONACO 3 system. The instrument's main blocks are described in the schematic diagram illustrated on Fig. 1.

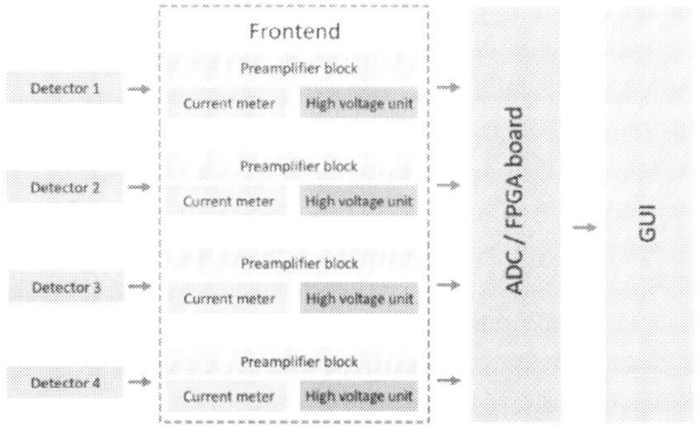

Fig. 1. Libera MONACO 3 block structure description.

The signals at the outputs of the detectors are received through 4 separate HN connectors. These signals are subsequently directed into a frontend consisting of a transimpedance preamplifier block [9], a current meter unit, and a High Voltage (HV) unit. A digitizer unit, based on the Libera Digit 500 DC (developed by Instrumentation Technologies), is acquiring the data, is performing Digital Signal Processing (DSP), and is transmitting the acquired data to a Graphical User Interface (GUI) running on a local computer.

A. Front end

The frontend is described in Fig. 2:

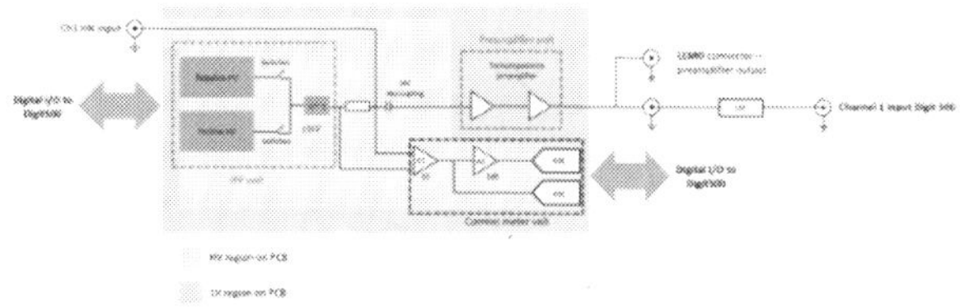

Fig. 2. Libera MONACO 3 frontend block description.

The current signal entering the HN input connector undergoes decoupling from the HV component using a capacitor. Within the transimpedance preamplifier, the current signal is amplified and converted into a voltage signal. The amplifier has a total bandwidth greater than 20 MHz and a maximum output noise of 40 mVpp when connected to an instrument output with a cable having a total capacitance exceeding 8 pF. The transimpedance gain of the preamplifier is approximately 3100 V/A.

A high voltage block is providing both negative and positive high-voltage polarization within the range of -1 kV to +1 kV, with a maximum precision of 1% on the target value. The total ripple in this block falls within the range of 50 mV.

The current measurement unit is designed to measure the current supplied to the detector and can operate in two separate

 https://doi.org/10.1051/epjconf/202328804019

ranges:

- The high gain range spans from 0 nA to ±10 uA.
- The low gain range spans from 0 to ±1 mA.

Both range measurements run concurrently, utilizing two separate Analog Digital Converters (ADC) with a sampling rate of 1 kS/s each for precise current measurement and integration of lower currents.

The high voltage and current measurement blocks on the frontend boards are controlled through an I/O bus interface that is directly connected to the FPGA on the digitizer.

B. Digitizer unit

Four ADCs on the Libera Digit 500 DC unit convert the four output signals from the transimpedance amplifiers into digital form. The ADCs operate at a sampling frequency of 491.52 MS/s, offering a resolution of 14 bits and a noise floor of 90 dB. Within the analog RF chains, there are user-controllable variable attenuators with a range of 31 dB. These attenuators are used to control the input signal intensity and to enhance the overall dynamic range of acquired pulses.

The digitized data are then transmitted to the Xilinx 7035 System On Chip (SOC) unit, which integrates both the Field Programmable Gate Array (FPGA) and the Advanced Risc Machine (ARM) processing unit. The data acquisition trigger enables the storage of data in the memory buffer chunk, while parallel DSP calculations are performed simultaneously.

C. DSP

The DSP chain incorporated within the digitizer unit is depicted by the blocks illustrated in Fig. 3.

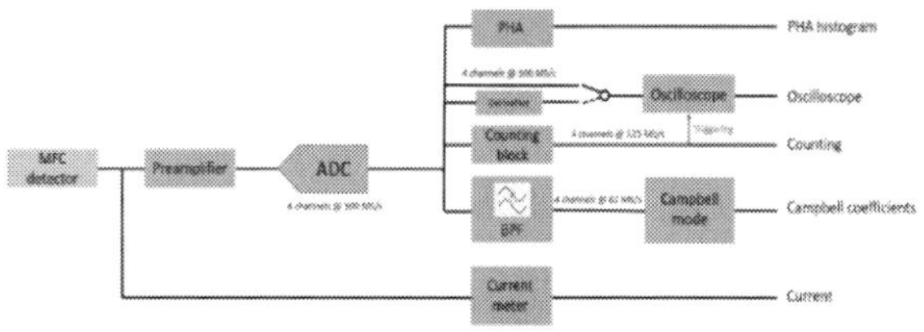

Fig. 3. DSP block schematics.

The DSP block, implemented within the FPGA, encompasses various sub-blocks that serve different functions:

- Pulse Height Analysis (PHA): this block computes the charge associated with the pulse population and presents data as a histogram. It also includes a pile-up rejection mode.
- Oscilloscope: this block is acquiring and visualizing pulse signals. It functions as an oscilloscope, allowing the user to monitor and collect pulse signals.
- Counting mode block: this block is dedicated to pulse discrimination and counting. Individual pulses are counted when their amplitude exceeds a predetermined threshold. The output of the counting block is a count rate in counts per second (Cps). The implemented counting method employs high-pass filtering in conjunction with a Schmitt trigger. This approach enables low dead time and reduced sensitivity to pile-up while maintaining a lower signal-to-noise ratio (SNR).
- Campbell mode block: the acquired ADC data pass through a bandpass filter with a frequency range of 400 to 800 kHz and a data rate of 61.44 MHz. Subsequently, the K2, K3, and K4 coefficients are calculated.
- Current meter: the data recorded by the current meter block in the integrated frontend is transmitted to the LD500 digitizer via the SPI bus.

D. GUI

The GUI serves as primary user interface through which the operator can access and acquire data from the detectors. It utilizes Python 3.10, PyQT4 graphic framework, and runs on Windows 10™. A screenshot example of the GUI is shown on Fig. 4.

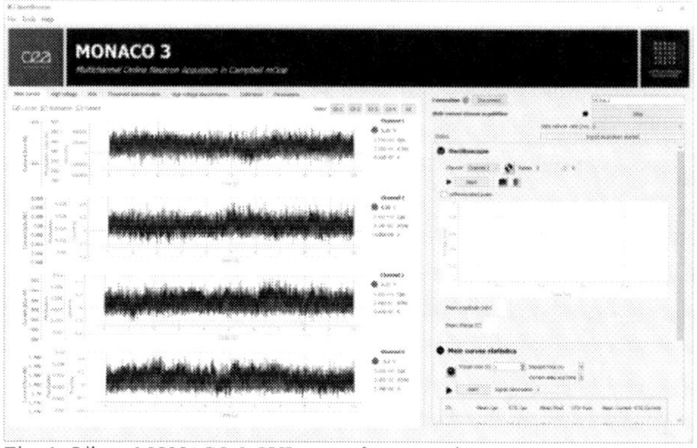

Fig. 4. Libera MONACO 3 GUI screenshot example.

The GUI development is built upon the ergonomic principles of the previous MONACOv2 system, and it incorporates additional modules to perform statistical analysis and calculations on the acquired data.

The communication infrastructure with the Libera Digit 500 unit operates over Ethernet and utilizes an HTTP adapter, enabling the exchange of large volumes of compressed data as streams of binary data.

In Fig. 4, the left hand part displays plots of the signals from the four channels, which consist of counting mode, fluctuation mode, and current modes. Data decimation can be adjusted by the user, allowing for a maximum resolution of 1 ms. For long-term reactor campaigns, data acquisition can be performed with reduced data rates.

The GUI integrates essential functions for instrument control, including:

- High voltage adjustment for the detector polarization,
- Automatic threshold discrimination curve,
- Automatic saturation (HV) curve-,
- Parameter settings for DSP functions.

IV. Qualification

Libera MONACO 3 qualification was completed through reactor tests carried out during 2 days in early March 2023, in the Slovenian TRIGA Mark II reactor. The fully characterized irradiation conditions helped to validate the fully operability of the Libera MONACO 3 in representative neutrons and gamma flux conditions.

A. TRIGA Mark II Reactor

The JSI TRIGA Mark II reactor is a pool type research reactor (250 kW – first criticality in 1966) cooled by natural water convection. The uranium zirconium-hydride fuel mixture designed by General Atomics has unique properties that make the reactor inherently safe and suitable for training, research and isotope production. The JSI reactor physics division and the LDCI Lab have a 15 years collaboration in radiation measurements and instrumentation qualifications [10], which helped to fully characterize different experimental locations within the TRIGA Mark II reactor.

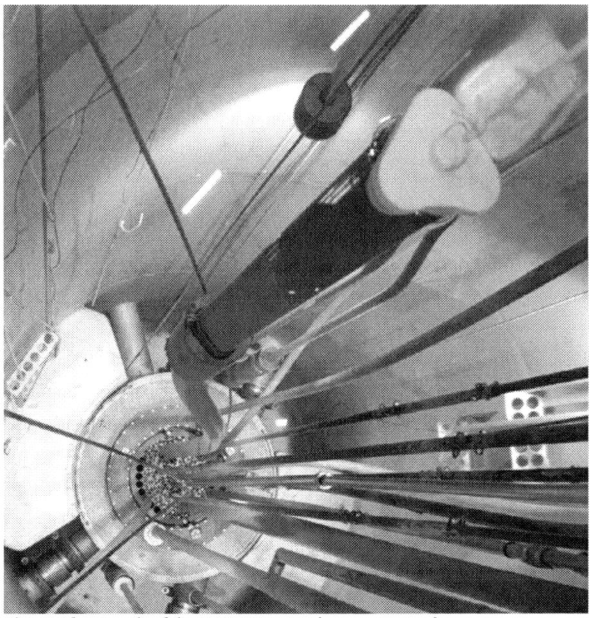

Fig. 5. Photograph of the JSI TRIGA Mark II reactor pool

B. Experimental Set-Up.

The full range of the three operation modes of the Libera MONACO 3 system were qualified thanks to two CEA Ø3 mm miniature FCs (respectively with 10 µg of uranium-235 and 100 µg of uranium-238) placed incore in the MP17 and MP25 central locations. The current mode upper range was confirmed using one Centronic (Croydon, UK) FC (0.165 g of uranium-235) usually used for excore measurements, kindly loaned by JSI, and placed in the reactor triangular channel (yellow cap on Fig. 5).

In complement, a CEA Ø3 mm miniature ion chamber and a rhodium emitter SPND were deployed to confirm the possibility to use Libera MONACO 3 with such detectors.

C. Reactor Tests

The TRIGA reactor was operated with several stabilized power steps from 0.1 W to 250 kW. Libera MONACO 3 operating modes were successfully tested with the following results:

- Pulse mode linearity is demonstrated up to 3.10^6 c.s^{-1}; it should be extended to >1.10^6 c.s^{-1} providing a fine-tuning of pulse mode parameters.
- Campbell mode is verified to be linear over more than 5 decades of neutron flux.
- Current is confirmed operational within the target 1 nA – 1 mA current range (See Fig. 7).

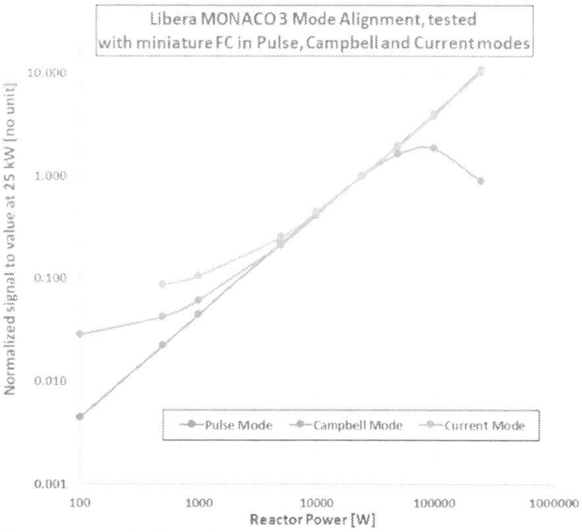

Fig. 6: Tests results of the Libera MONACO 3 mode alignment, using CEA miniature U235 FC.

As it was one of the most challenging part of this industrialization process, Fig. 7 is a focus on the current mode. Thanks to an artificial "reactor power equivalent U235 fission rate" x-axis, CEA U235 miniature FC and Centronic FC measurements are plotted on the same graph; presented here along with IC, Rh SPND and Keithley electrometer currents, the latter taken here as reference.

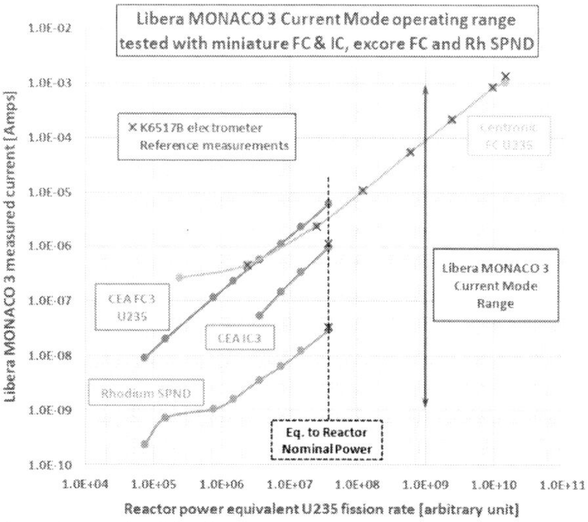

Fig. 7: Tests results of the Libera MONACO 3 current mode range, using CEA FC and IC, Centronic FC and a Rh SPND.

The 1 nA – 1 mA current range is highlighted in red on Fig. 7, its linearity is validated using the two different FC measurements as well as current absolute values in comparison to the electrometer measurements. Fig. 7 also shows the possible use of Libera MONACO 3 with IC detectors and even with SPNDs (HV set to 0 V) which confirm the versatility of the system for different kind of reactor measurements.

V. Conclusions

The Libera MONACO 3 system will be available on the nuclear instrumentation market at Instrumentation Technologies by the end of 2023.

Acknowledgment

The authors acknowledge the financial support from the nuclear instrumentation project (INSNU) of the CEA Gen2&3 program for this 30 months bilateral and fruitful collaboration between the LDCI Lab and the Instrumentation Technologies company.

References

[1] M. Guéry, "Improvements in or Relating to a Miniature Neutron-Fission Ionisation Chamber", Patent GB 938, 594 (1963).

[2] V. Lamirand, B. Geslot, J. Wagemans, L. Borms, E. Malambu, P. Casoli, X. Jacquet, G. Rousseau, G. Grégoire, P. Sauvecane, D. Garnier, S. Bréaud, F. Mellier, J. Di Salvo, C. Destouches, P. Blaise, "Miniature fission chambers calibration in pulse mode: interlaboratory comparison at the SCK•CEN BR1 and CEA CALIBAN reactors", *ANIMMA 2013, June 23–27, 2013 Marseille, France.*

[3] N. Thiollay, J. Di Salvo, C. Sandrin, M. Soldevila, S. Bourganel, C. Fausser, C. Destouches, P. Blaise, C. Domergue , H. Philibert, J. Bonora, A. Gruel, B. Geslot, V. Lamirand, A. Pepino, A. Roche, O. Méplan and M. Ramdhane, "FLUOLE-2: An Experiment for PWR Pressure Vessel Surveillance", *ISRD 2015, May 18-23, 2015, Aix-en-provence, France.*

[4] L. Barbot, C. Domergue, J.F. Villard, C. Destouches, G. Braoudakis, D. Wassink, B. Sinclair, J. Osborn, H. Wu, C. Blandin, "On line neutron flux mapping in fuel coolant channels of a research reactor", *ANIMMA 2013, June* 23-27, 2013, *Marseille, France.*

[5] L. Barbot, S. Normand, P. Pasdeloup, B. Lescop, "Final qualification of an industrial wide range neutron instrumentation in the OSIRIS MTR reactor", *ANIMMA 2009, June* 7-10, 2009, *Marseille, France*

[6] M. Thevenin, L. Barbot, G. Corre, R. Woo, C. Destouches and S. Normand, "Digital Real-Time Multiple Channel Multiple Mode Neutron Flux Estimation on FPGA-based Device", *ISRD 2015, May 18-23, 2015, Aix-en-provence, France.*

[7] L. Barbot, D. Fourmentel, G. De Izarra, C. Destouches, J-F. Villard, Y. Moline and H. Hamrita, "MONACO v2: Multipurpose and Integrated Data Acquisition System for On-line Neutron and Gamma Measurements", *ANIMMA 2019, June 17-21, 2019, Portorož, Slovenia.*

[8] Y. Moline, A. Dabat-Blondeau, H. Hamrita, S. Klasa, G. Corre, J-M. Bourbotte, K. Boudergui, L. Barbot, C. Destouches, "MONACO: A Case Study of a New System on Chip and Stream Based Architecture for Nuclear instrumentation", *ANIMMA 2019, June 17-21, 2019, Portorož, Slovenia.*

[9] D. Bisiach, L. Barbot, G. De Izarra, C. Destouches, M. Cargnelutti and S. Zorzut, "Development of a Wideband Current Amplifier dedicated to Fission Chamber Measurement", *ANIMMA 2021, June 21-25, 2021, Prague, Czech Republic.*

[10] C. Destouches, L. Snoj, L. Barbot, V. Radulović, D. Fourmentel, N. Thiollay, B. Geslot, G. De Izarra, A. Gruel, G. Grégoire, C. Domergue, K. Ambrožič, I. Lengar, T. Goričanec, G.Žerovnik, I. Kodeli, A. Trkov, "Update of the Review of more than 15 years of JSI – CEA Fruitful Collaboration on Nuclear Instrumentation Development", *to be presented at the NENE 2023, September 11-14, 2023, Portorož, Slovenia.*

Radiation measurements in reactor pulse mode at the JSI TRIGA reactor – Power meter based on Cherenkov light intensity measurements

Julijan Peric[1,*], Vladimir Radulović[1], Loïc Barbot[2] and Grégoire De Izarra[2]
[1]Jožef Stefan Institute, Reactor Physics Division, Slovenia
[2]CEA, DES, IRESNE, DER, Instrumentation Sensors and Dosimetry Laboratory, Cadarache, F-13108 Saint Paul-lez-Durance, France
(*) julijan.peric@ijs.si

***Abstract*—The Cherenkov power meter developed at the Jožef Stefan Institute (JSI) is an independent, reliable and cost-effective measurement system based on Cherenkov radiation detection, used to measure fast power transients during pulse operation at the JSI TRIGA research reactor. It is based on a closed tube in order to avoid interference from external light sources and radiation damage experienced by optical fibers. The tube is placed in the reactor core periphery, the bottom part of the tube is filled with water, serving as the source of Cherenkov light. The measurements obtained in the framework of an extensive experimental campaign in collaboration with the French Atomic and Alternative Energy Commission (CEA) focused on reactor pulse operation using the Cherenkov power meter show excellent agreement with existing nuclear instrumentation (TRIGA pulse recorder), especially for high peak power pulses. However, the Cherenkov power meter outperforms the TRIGA pulse recorder in accurately recording low peak power pulses. The power vs. time behavior measured with the Cherenkov power meter is in accordance with measurements using miniature fission and ionization chambers developed at the CEA. The possibility of reactor pulse characterization based on measurements with the Cherenkov power meter combined with neutron dosimetry measurements sets the basis for enabling irradiation during pulse operation at JSI TRIGA research reactor in the future.**

***Keywords* — Cherenkov radiation, Cherenkov power meter, Silicon photomultiplier, SiPM, Reactor pulse, Irradiation, Dosimetry, Miniature fission chamber, CEA, JSI, TRIGA research reactor.**

I. INTRODUCTION

The Jožef Stefan Institute (JSI) TRIGA research reactor is used to test electronic components and systems in a well-characterized radiation field during steady-state operation. It is known as a well-established irradiation facility. The reactor can produce a maximum neutron flux of about 2×10^{13} n cm^{-1}s^{-1} at a thermal power of 250 kW. However, the neutron flux level relevant for testing of instrumentation for use in material testing reactors (MTRs) or nuclear power plants (NPPs) is in the range between 10^{14} and 10^{15} n cm^{-1}s^{-1}. On the other hand, in the JSI TRIGA research reactor, such neutron flux levels are achievable in pulse operation mode, albeit only for short periods of time. The height and duration of the pulses depends on the inserted reactivity. Irradiations in pulse operation mode represent a possibility to extend the experimental capabilities of the JSI TRIGA research reactor.

Cherenkov radiation is present in most water-cooled nuclear reactors, with the exception of zero-power reactors. It is caused by energetic charged particles traveling faster than the speed of light in a dielectric medium. In open-pool reactors, Cherenkov radiation can be observed as a blue glow around the reactor core. Since the intensity of the Cherenkov light produced in the reactor cooling water is in principle proportional to the neutron flux during a reactor pulse, Cherenkov light intensity measurements as an alternative method for measuring the time dependence of reactor power were investigated, implemented, and tested at the JSI TRIGA research reactor.

II. REACTOR PULSE OPERATION

The JSI TRIGA research reactor is equipped with a pneumatic system that enables pulse mode operation by allowing for the rapid ejection of the transient control rod. After a rapid ejection of the transient control rod, the reactor becomes prompt supercritical, and the power starts to increase exponentially. The change in power and consequently in the fuel temperature causes a decrease in reactivity, since the reactivity coefficient of the fuel is prompt negative, therefore the reactor rapidly and efficiently reaches a new equilibrium state. The decrease in reactivity slows or stops the chain reaction, resulting in a drop in power.

The peak power of the pulse is in the range between 1 MW and several 100 MW (depending on the core configuration), and the total energy released in the pulse is relatively small (usually a few MJ) due to the short pulse duration. In the TRIGA reactor, the temperature reactivity coefficient of the fuel is the strongest and most important feedback effect of the reactor state on reactivity. The temperature feedback mechanism of the TRIGA reactor is due to the unique composition of its fuel, which is a homogeneous mixture of 20 % enriched uranium and zirconium hydride with a Zr-H ratio of 1.6. In this design, the hydrogen in the zirconium hydride acts as a moderator, with most of the moderation occurring in the fuel elements themselves and only a small fraction in the surrounding water. As a result, changes in the fuel temperature are rapidly reflected on the moderator within the fuel, which results in an immediate effect on the reactivity of the core.

A. Fuchs-Hansen theoretical model

The time dependence of a pulse can be calculated using the point kinetics equations by considering only the contribution of the prompt neutrons to the neutron population and assuming that the reactivity decrease during the pulse is proportional to the energy released - adiabatic process. The model derived by Fuchs and Hansen is also referred to as the "adiabatic model" because of the second assumption used in its derivation [1].
From the theoretical model, important dependencies of pulse parameters can be calculated. In particular, the pulse height is proportional to the square of the prompt reactivity ($P_{peak} \propto \rho'^2$), the energy released is proportional to the prompt reactivity ($E_{rel} \propto \rho'$), and the pulse width is inversely proportional to the prompt reactivity ($FWHM \propto 1/\rho'$). The prompt reactivity (ρ') is equal to the inserted reactivity (ρ_i) reduced by the delayed neutron fraction (β) [1-2].

III. Cherenkov light

The electromagnetic radiation known as Cherenkov light is emitted when a charged particle moves in a dielectric medium at a speed greater than the speed of light in that medium [3]. From Einstein's special theory of relativity, we know that the speed of light in a vacuum is a universal physical constant c_0 = 299792458 m/s, and this is the upper limit on the speed at which conventional matter can move through space. On the other hand, when light travels through transparent material, its speed is lower. Using the refractive index, we can easily formulate the condition for the occurrence of Cherenkov light in a transparent medium:

$$\frac{v}{c_0} = \beta > \frac{1}{n}, \quad (1)$$

where v is the speed of charged particle, and β is the ratio between the speed of the particle v and the speed of light in vacuum, c_0 [3-4].

Any charged particle with a velocity in the range of $0.75 \cdot c0 < v < c0$ causes the emission of visible Cherenkov light in water. For pure water at 20°C, the refractive index is $n = 1.33$, which results in a speed of light of $c = 0,75 \cdot c_0$ [5]. The threshold kinetic energy for electrons moving in water and emitting Cherenkov light is about 250 keV. Since most Cherenkov photons are emitted at lower wavelengths, it is important to note that water is not transparent at wavelengths in the UV range, and photons of these wavelengths are immediately absorbed. In nuclear reactors, there are several sources of ionizing radiation that cause the production of Cherenkov light. Heavier isotopes can undergo α-decay, but they are not expected to contribute significantly to the production of Cherenkov light, since α-particles are easily stopped within the fuel material or fuel cladding [6]. One important source of radiation are the electrons produced in the β-decay process. These electrons often have energies greater than 250 keV, therefore they can produce Cherenkov light in water. However, they are also effectively stopped in materials (e.g. fuel, cladding, reactor structural materials) and therefore their contribution to the overall Cherenkov light emission is low. During reactor operation, the fuel and other materials in the reactor are an intense source of prompt and delayed gamma radiation, which is not attenuated as strongly as alpha particles or electrons and can generate energetic electrons in the reactor cooling water.

The prompt γ-rays represent the largest contribution to the intensity of the Cherenkov light in a nuclear reactor during operation. A much smaller contribution is due to the γ-rays from the fission and activation products and the β-particles. When a nuclear reactor is operated at power (the fission power is dominant compared to the residual power), the intensity of the Cherenkov light is linear with the reactor power. When the fission power is no longer dominant in a nuclear reactor, linearity is no longer valid [6-7].

IV. Experiment at the JSI TRIGA reactor

An extensive experimental campaign was carried out at the JSI TRIGA reactor from 17 January 2022 to 28 January 2022 as part of the bilateral collaboration between the Jožef Stefan Institute and the French Atomic and Alternative Energy Commission (CEA). The aim of the campaign was to explore the feasibility of utilizing the high neutron flux that can be achieved during reactor pulse mode for testing nuclear instrumentation detectors at the JSI TRIGA research reactor. Additionally, the campaign sought to evaluate the performance of CEA miniature fission with ultra-low fissile coatings and ionization chambers, as well as the MONACO v2 data acquisition system [8-9] at high neutron flux levels.

In the campaign, a Cherenkov pulse recorder was used as an independent measurement technique in conjunction with miniature fission (FCs) and ionization chambers (ICs) operated by the CEA-developed MONACO v2 data acquisition system, and different dosimetry foils. During the campaign 128 pulses were recorded at the JSI TRIGA research reactor and 159 dosimetry foils were irradiated during different pulses.

A. Experimental setup

The Cherenkov pulse recorder system consists of a measurement channel, Cherenkov light radiator, a light sensor, amplifier, and a data acquisition system. The measurement channel was an aluminum tube with an inner diameter of 36 mm and approximately 6 m in length. The bottom end of the tube was blanked off with an aluminum plate, making it watertight. The channel was designed to fit into the standard fuel element positions in the JSI TRIGA reactor core. The channel was inserted in the periphery of the reactor core, in position F-18. 0.65 l of water was poured into the channel (filling the height of 0.65 m in the tube), which served as the source of Cherenkov radiation. The top end of the channel reached the reactor platform level. The channel had a slight bend in order to limit radiation streaming from the reactor core to the reactor platform. Due to the bend, there was no direct line-of-sight from the top end of the channel to the water in the bottom end. However, the inner surface of the channel was sufficiently smooth to allow for some light reflection from the water in the bottom end to the top end.

The sensor consisted of a silicon photomultiplier - SiPM (KETEK PM3315- WB) selected for its peak photon detection efficiency at around 450 nm, a readout circuit (KETEK PEPCB-EVAL-MCX-RESP) and a plastic holder. The SiPM and the readout circuit are manufactured by the German company KETEK. The detector was connected to the data acquisition system with a BNC cable through a Thorlabs amplifier (AMP220) and lowered approximately 2.5 m inside the measurement channel. The top end of the measurement channel was taped up in order to minimize light interference from external sources.

A Red Pitaya [10] module was used as the data acquisition system and was controlled by a script implemented in Python language for pulse recording in the form of a GUI - Graphical User Interface. The script triggers data acquisition when the measured signal is above a threshold level ($U_{threshold} = 0.015$ V). The signal is measured at a sampling rate of 1.8737 kS/s for 8.59 s and stored in a csv file.

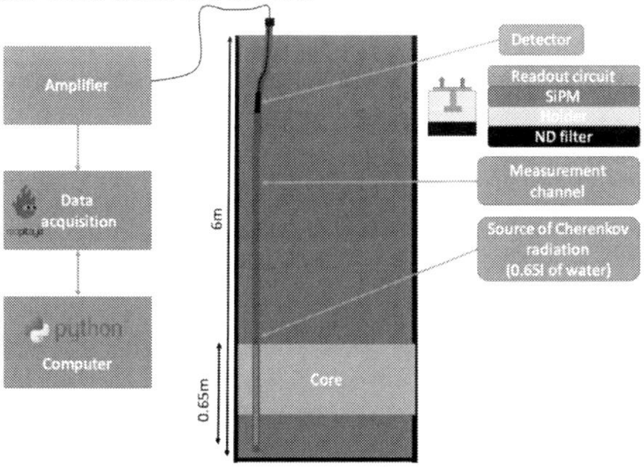

Fig. 1. Cherenkov pulse recorder experimental setup scheme [11].

In pulse operation mode, the power of the nuclear reactor can vary from low power levels (below 1 kW) to the GW range in a short amount of time. Therefore, the Cherenkov pulse recorder requires a large dynamic range, made possible by the Red Pitaya module. However, for high peak power pulses, neutral density filters (ND) were used in front of the SiPM to reduce the Cherenkov light intensity, as otherwise the signal generated by the sensors would have exceeded the Red Pitaya input range.

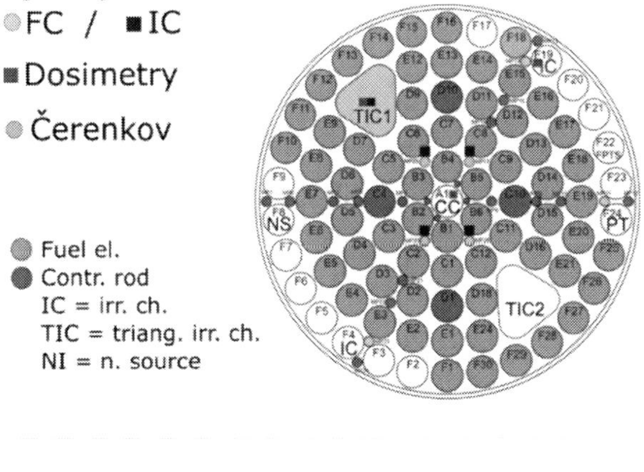

Fig. 2. Experimental reactor core arrangement scheme.

In addition to the Cherenkov pulse recorder, in the campaign 10 miniature fission and ionization chambers were arranged in various measurement positions in the reactor core and 3 irradiation positions were used for dosimetry experiments. The arrangement of the measurement and irradiation positions during the experiment can be seen in Figure 2 and the positions are listed in Table 1.

TABLE I
MEASUREMENT AND IRRADIATION POSITIONS

Location	Detector	Dosimetry
MP17	FC	
MP20	FC	
MP25	IC	
MP26	IC	
MP23	FC	
MP11	FC	
MP14	IC	
TIC1	FC + IC	Dosimetry
CC		Dosimetry
F19		Dosimetry
F4		Dosimetry
F18	Cherenkov	
MP2	FC	

B. *Reactor pulse shape measurements*

To enable the use of the JSI TRIGA reactor in pulse operation mode and thus to achieve higher neutron flux levels than in steady-state operation, it is critical to characterize the time behavior of the reactor power during each individual pulse. Pulses with a low inserted reactivity have a longer duration and a lower peak power, whereas pulses with a high inserted reactivity have shorter duration and a high peak power. In practical terms, pulses with a peak power in the range between 2.5 MW and 25 MW correspond to a peak neutron flux range between 2×10^{14} n cm^{-1}s^{-1} and 2×10^{15} n cm^{-1}s^{-1}. The corresponding inserted reactivities are 1.3 \$ - 1.55 \$. Therefore, the primary focus is on lower and longer pulses with inserted reactivities from just over 1 \$ to around 1.55 \$. Higher peak power pulses are also of interest; however their quick duration may represent an important constraint for applications.

Pulse operation of the JSI TRIGA reactor is monitored by a dedicated instrumentation channel – the pulse channel, based on an uncompensated B-10 lined ionization chamber, the signal being acquired and processed by a pulse recorder. The pulse recorder calculates quantities of interest for each reactor pulse, i.e. the peak power, duration (FWHM) and energy released. It is designed to cover all the pulse range, up to peak power levels of 2 GW. The pulse recorder produces reliable results for high peak power pulses, however at the very low end of the range ($\rho < 1.2\$$) the results are unreliable as the pulse shape and characteristic times change significantly. This shortcoming was one of the main drivers in the development of the Cherenkov pulse recorder.

Since the Cherenkov pulse recorder is based on a different principle of reactor power measurement than the TRIGA pulse recorder, which uses an uncompensated ionization chamber as detector, a signal comparison is necessary. All measurements used in this analysis were taken during the first week of the campaign. Measurements were performed again in the second week with a different arrangement of fission and ionization chambers. Figures 3 and 4 display a representative set of pulse shapes obtained, respectively, with the Cherenkov pulse recorder (in units of voltage, proportional to the Cherenkov light intensity) and the TRIGA pulse recorder (in units of power). The comparison of the pulse shapes between Figure 3 and Figure 4 shows that the signals of the two systems are practically identical and confirms that the Cherenkov pulse recorder produces reliable pulse shapes.

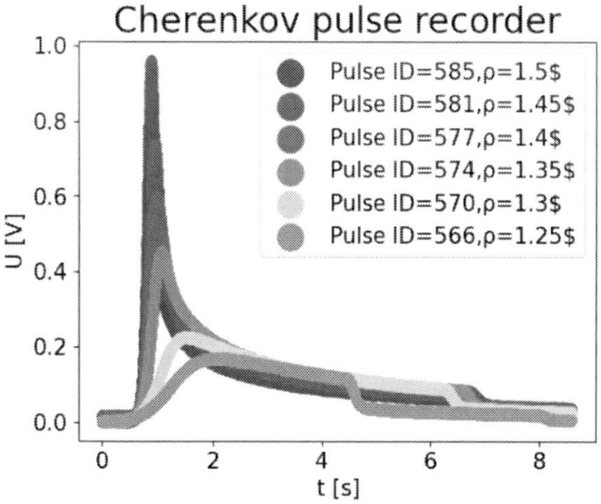

Fig. 3. Selected pulse signals measured with Cherenkov pulse recorder.

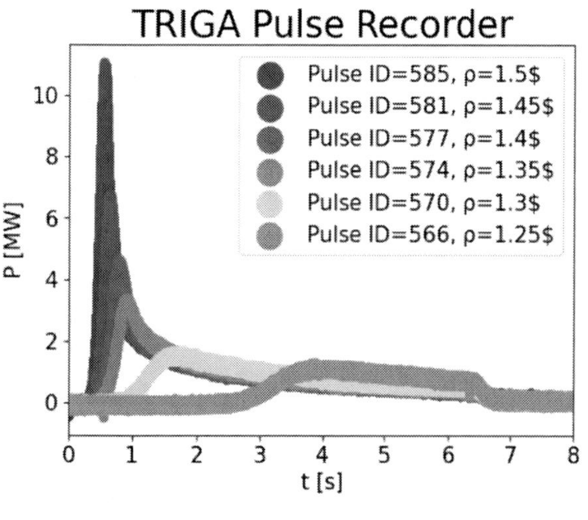

Fig. 4. Selected pulse signals measured with TRIGA pulse recorder.

It should be noted that between the two systems the trigger signal is different, because, as mentioned earlier, signal acquisition in the Cherenkov pulse recorder is triggered by the magnitude of the signal, whereas in the TRIGA pulse recorder signal acquisition is triggered by the "fire" signal, which initiates the pulse. Each time a pulse is initiated, an automatic reactor "scram" (shutdown by rapid insertion control rods) occurs after a pre-programmed time interval. This is why in Figure 4 the scram point is common to all the pulses at time 6.5 s after the "fire" signal. The width of the measured signal is similar for both systems at higher pulses ($\rho > 1.4\$$), but at the smaller pulses, the advantage of the Cherenkov pulse recorder is evident, providing a more accurate measurement because it is optimized for the entire pulse range.

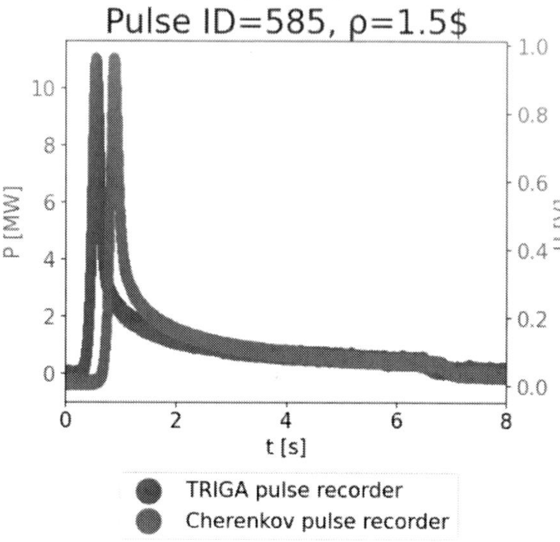

Fig. 5. Combined pulse (ID =585) signals measured with TRIGA and Cherenkov pulse recorder.

Figure 5 shows combined measured pulse signals with both systems for one specific 1.5 \$ pulse, and the qualitative agreement is excellent. The shape of the pulse is the same for both systems, indicating that the Cherenkov pulse recorder operates as well as commonly used nuclear instrumentation.

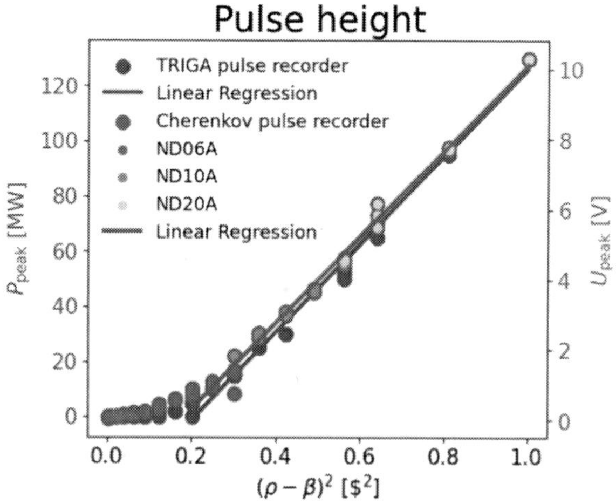

Fig. 6. Pulse height vs. the square of the prompt reactivity.

Figure 6 shows the measured pulse height as a function of the square of the prompt reactivity for both systems. It is evident that both measurements follow a linear trend resulting from the Fuchs-Hansen model, as described in the previous section. It can be seen that linearity holds for pulses with inserted reactivity greater than $\rho = 1.4\$$. The Fuchs-Hansen model is not suitable for describing smaller pulses because the assumptions used for its derivation do not hold in this range.

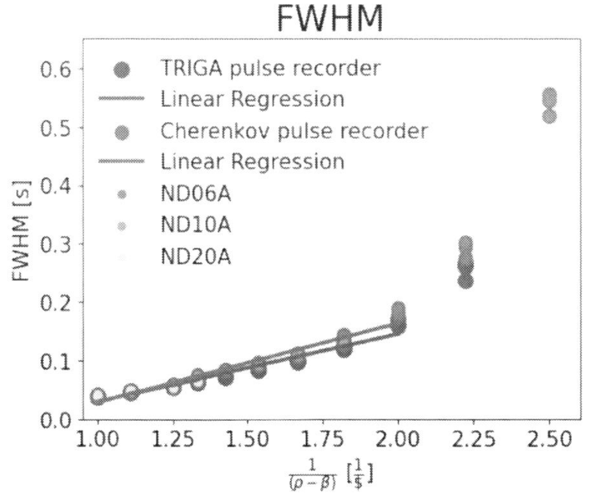

Fig. 7. Pulse full width at half maximum vs. the inverse prompt reactivity.

The full width at half maximum (FWHM) of the pulses is an important pulse parameter that allows us to assess whether the measured pulse is consistent with the assumptions used in deriving the theoretical model. Figure 7 shows the measured FWHM values vs. the inverse of the prompt reactivity. For higher pulses ($\rho > 1.4\$$) the FWHM dependency agrees with the theoretical model. On the other hand, the FWHM for lower pulses begins to increase nonlinearly as the lengthening pulse tail is included in the determination of the FWHM. In the tail of the pulse, the assumption of adiabatic heating is no longer valid. The alignment of FWHM measurements for both systems means that the shape of the measured pulses is in excellent agreement, although the measurement principle is completely different. This shows that the Cherenkov pulse recorder is more than suitable for this kind of application.

The energy released during the pulse can be calculated using various methods. For this analysis, the pulse shape was integrated in an interval corresponding to 1000 measurement points before and 1000 measurement points after the pulse peak. In terms of the integration time this selection is consistent with the theoretical model assumptions. However, these values do not correspond to the actual energy released during the pulse, which is crucial for TRIGA pulse operation for irradiation of different samples. Figure 8 displays the energy released versus prompt reactivity from both systems and shows that both systems produce results in accordance with the Fuchs-Hansen theoretical model. Figure 8 also shows that the TRIGA pulse recorder did not perform as well as the Cherenkov pulse recorder in measuring the energy released at lower pulses ($\rho < 1.3\$$) due to the low signal-to-noise ratio at its operating limit.

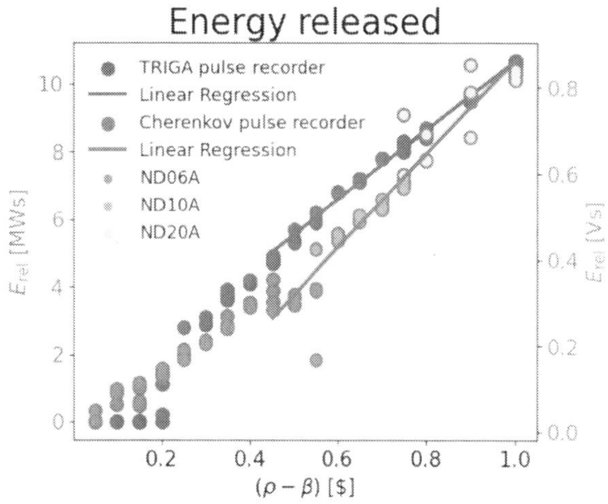

Fig. 8. Pulse energy released vs. the prompt reactivity.

With the exception of the FWHM comparison, all other comparisons were performed in a relative sense, since the Cherenkov pulse recorder is not absolutely calibrated. There are several options for calibration. The most promising are calibration to the peak power of the TRIGA pulse recorder or calibration to the total energy released during the pulse, since it has been shown in this article that the shape of the measured pulse is the same as in the nuclear instrumentation commonly used to measure reactor flux. For calibration of the total energy released, dosimetry measurements can be used to evaluate the total neutron fluence during the pulse in the irradiation channel where the dosimetry foils are irradiated. Once the Cherenkov pulse recorder is calibrated, the total neutron fluence delivered to irradiated samples can be derived from the measured signal.

TABLE II
LINEAR REGRESSIONS – PULSE PARAMETERS

Pulse parameter (Measurement system)	k	y_0
Peak power (TRIGA)	$159.0 \cdot (1 \pm 0.01) MW/\$^2$	$-32.9 \cdot (1 \pm 0.04)\ MW$
Peak power (Cherenkov)	$12.2 \cdot (1 \pm 0.02) V/\$^2$	$-2.1 \cdot (1 \pm 0.05)\ V$
FWHM (TRIGA)	$0.1 \cdot (1 \pm 0.04)\ s\$$	$0.1 \cdot (1 \pm 0.08)\ s$
FWHM (Cherenkov)	$0.1 \cdot (1 \pm 0.04)\ s\$$	$0.1 \cdot (1 \pm 0.1)\ s$
Energy released (TRIGA)	$10.3 \cdot (1 \pm 0.02)\ MWs/\$$	$0.4 \cdot (1 \pm 0.27)\ Vs$
Energy released (Cherenkov)	$1.1 \cdot (1 \pm 0.06)\ Vs/\$$	$0.1 \cdot (1 \pm 0.19)\ Vs$

The linear regression parameters for peak power and released energy cannot be compared between the two systems because they are calibrated differently. However, the comparison of their computational errors shows that the calculated linear regression parameters for both systems are consistent with the proposed theoretical adiabatic model described previously. On the other hand, the FWHM linear regression parameters are the same with similar uncertainties, which means that the measured signals have a pronounced similarity in their waveform properties.

C. Reactor pulse dosimetry measurements

Neutron activation dosimetry is the reference method in the determination of the neutron flux or fluence, and are included in the present work as a reliable experimental technique to obtain measurements of the total neutron fluence which can be delivered in pulse operation mode to samples of interest (e.g. neutron detectors, etc.). A series of test irradiations was performed in reactor pulse mode in which samples of certified reference materials Al-0.1%Au, Al-0.1%Co and pure Ni were irradiated within polyethylene capsules in the Central Channel (CC), F-19 Irradiation Channel (F19) and Triangular Irradiation Channel 1 (TIC1) of the JSI TRIGA reactor. The induced activities in the samples were subsequently measured using two absolutely calibrated High-Purity Germanium (HPGe) detectors [12].

Fig. 9 displays the measured specific end-of-irradiation (EOI) activities (in Bq per mg of sample material) vs. the pulse prompt reactivity. The induced activities are proportional to the total neutron fluence received during the irradiations, which is in turn proportional to the released energy in the pulse. According to the Fuchs Hansen model [11] for reactor pulse operation, the energy released in a pulse is proportional to the prompt reactivity. The observed dependencies clearly exhibit a linear behavior, taking into consideration the experimental uncertainties. In this work, these are relatively low (of the order of 1.5-1.8 %) in the context of reactor dosimetry measurements [9]. It is evident that the measured specific activity of the samples irradiated during pulse No. 629 with inserted reactivity ρ_i=1.15$ is lower than expected. The comparison of the measured power versus time signals for pulses with the same inserted reactivity shows that the pulse was interrupted earlier (scram), consequently the samples were activated less and are removed from the calculation of the linear regression parameters.

TABLE III
LINEAR REGRESSIONS – DOSIMETRY PARAMETERS

Position	$k\ [Bq/mg\$]$	$y_0\ [Bq/mg]$
CC	$5.15 \cdot 10^5 \cdot (1 \pm 0.05)$	$9.20 \cdot 10^4 \cdot (1 \pm 0.15)$
F-19	$2.65 \cdot 10^5 \cdot (1 \pm 0.06)$	$5.19 \cdot 10^4 \cdot (1 \pm 0.16)$
TIC 1	$3.20 \cdot 10^5 \cdot (1 \pm 0.05)$	$5.28 \cdot 10^4 \cdot (1 \pm 0.15)$

The dosimetry measurements serve as a reference for the absolute normalization of the time-dependent signals measured with the Cherenkov power meter and thus enable sample irradiation during pulse operation at the JSI TRIGA research reactor.

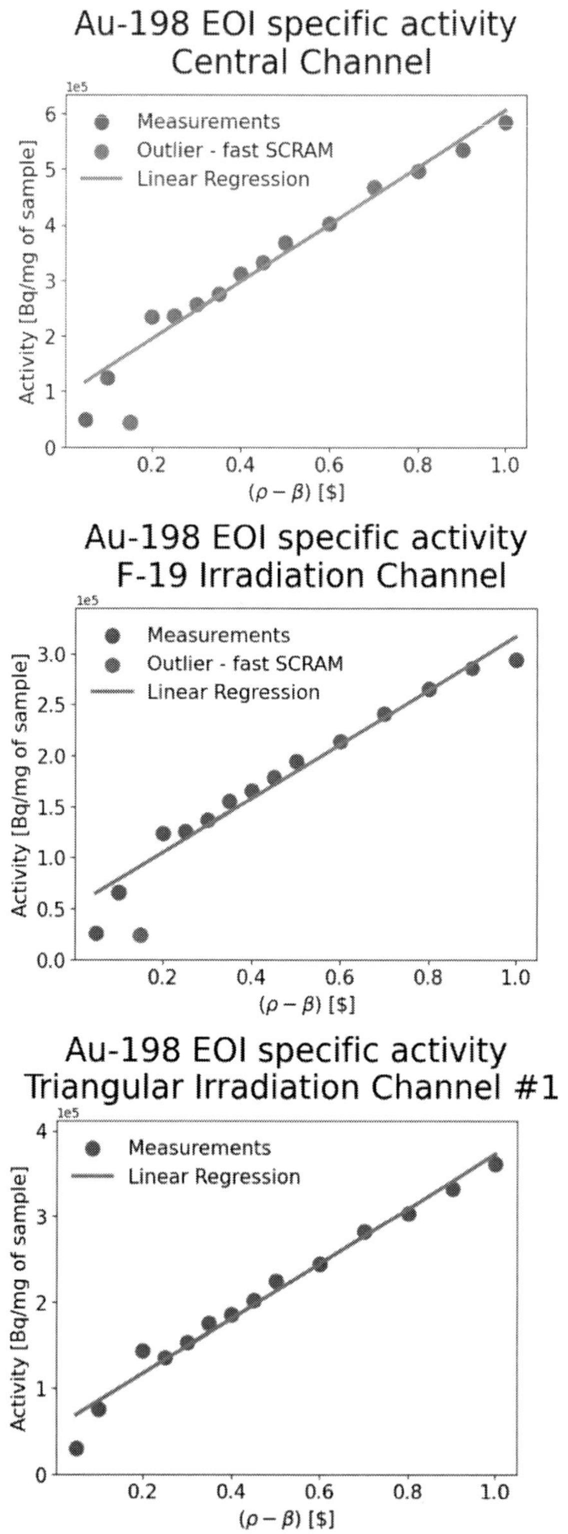

Fig. 9. Measured specific end-of-irradiation activities [Bq per mg of sample material] for Au-198, originating from the Au-197(n,γ) reactions, as a function of the pulse prompt reactivity.

V. Conclusions

This work demonstrates the use of the Cherenkov power meter for irradiation measurements during TRIGA pulse operation. The presented results clearly demonstrate that the Cherenkov power is a capable experimental technique for measurements of rapid transients in nuclear reactors. The present research is aimed at extending the experimental capabilities of the JSI TRIGA research reactor, making use of the high neutron flux levels achievable in reactor pulse operation. Cherenkov detector could be envisaged to complement the online nuclear instrumentation fleet such as fission and ionization chambers and SPNDs. It is shown that Cherenkov power meter in conjunction with dosimetry measurements enables sample irradiation during pulse operation at the JSI TRIGA research reactor. Nuclear reactor power meters based on Cherenkov radiation detection offer great potential for increased redundancy, adaptability, and diversity of instrumentation and control in modern nuclear applications. It should be taken into account that the developed Cherenkov power meter at the research reactor JSI TRIGA is more than a hundred times cheaper than the implemented instrumentation, which provides the same results and, in some cases, exceeds the performance of the implemented solution.

Acknowledgment

The authors acknowledge the financial support from the Slovenian Research Agency (research core funding No. P2-0073 – Reactor Physics, research project No. NC-0011 - Absolute radiation measurements at very high neutron flux levels in reactor pulse mode) and the instrumentation project of the CEA Gen2&3 program.

References

[1] I. Vavatar, A. Pungerčič, L. Snoj, "Utilisation of JSI TRIGA Pulse Experiments for Testing of Nuclear Instrumentation and Validation of Transient Models." EPJ Web of Conferences. Vol. 225, EDP Science, 2020.
[2] I. Vavtar, L. Snoj. "Evaluation of approximations in reactor pulse." EPJ Web of Conferences, Vol. 247, EDP Sciences, 2021.
[3] J. V. Jelley, "Cerenkov Radiation and Its Applications." London: Pergamon Press, 1958.
[4] J. V. Jelley, "Cerenkov radiation and its applications." British Journal of Applied Physics", 6(7, 1955), pp. 227–232.
[5] E. Hecht, Optics. Addison-Wesley, 2002, ISBN 978-0-321-18878-6.
[6] S. E. Rippon, "Cherenkov detectors for the measurement of reactor power." Nuclear Instruments and Methods, Volume 21, 1963, pp. 192-196.
[7] M. Arkani, M. Gharib, "Reactor core power measurement using Cherenkov radiation and its application in Tehran Research Reactor." Annals of Nuclear Energy, Volume 36, Issue 7, 2009, pp. 896-900.
[8] L. Barbot, D. Fourmentel, G. De Izarra, C. Destouches, J-F. Villard, Y. Moline and H. Hamrita, "MONACO v2: Multipurpose and Integrated Data Acquisition System for On-line Neutron and Gamma Measurements", ANIMMA 2019, June 17-21, 2019, Portoroz, Slovenia.
[9] Y. Moline, A. Dabat-Blondeau, H. Hamrita, S. Klasa, G. Corre, J-M. Bourbotte, K. Boudergui, L. Barbot, C. Destouches, "MONACO: A Case Study of a New System on Chip and Stream Based Architecture for Nuclear instrumentation", ANIMMA 2019, June 17-21, 2019, Portoroz, Slovenia
[10] https://www.redpitaya.com/
[11] J. Peric, "Using Cherenkov radiation to measure power of nuclear reactor during steady-state and pulsed operation (in Slovene)," M.S. thesis, FMF, Ljubljana Univ., Ljubljana, Slovenia, 2022.
[12] V. Radulović, L. Barbot, G. De Izarra, J. Peric, I. Lengar, "Reactor Pulse Operation for Nuclear Instrumentation Detector Testing–Preparation of a Dedicated Experimental Campaign at the JSI TRIGA Reactor." In EPJ Web of Conferences (Vol. 253, p. 04019). EDP Sciences, 2021.

Reactivity modulation experiments for nuclear data in CROCUS within a CEA-EPFL collaboration

Benoit Geslot[1*], Vincent Lamirand[2], Yifeng Jiang[1], Pierre Leconte[2], Oskari Pakari[2], Daniel Godat[2], Laurent Braun[2], Pavel Frajtag[2], Laurent Pantera[1] and Andreas Pautz[2]
[1] CEA, DES, IRESNE, DER, SPESI, LP2E, Cadarache F-13108 Saint-Paul-Lez-Durance, France
[2] Laboratory for Reactor Physics and Systems Behaviour (LRS), Ecole polytechnique fédérale de Lausanne (EPFL), 1015 Lausanne, Switzerland
(*) benoit.geslot@cea.fr

Abstract—**In nuclear research reactors, integral experiments are powerful tools to measure integral core parameters, such as the delayed neutron fraction. Within the scope of the point kinetic approximation, reactivity modulation experiments can be used for probing the reactor transfer function and then infer integral parameters of the core.**

In this context, Commissariat à l'Energie Atomique et aux Energies Alternatives (CEA) and Ecole Polytechnique Fédérale de Lausanne (EPFL) have been collaborating for developing a probe device (PISTIL) and measurement setup adapted to the CROCUS zero power research reactor operated by EPFL (Lausanne, Switzerland).

Despite some mechanical limitations of PISTIL, its maximum reactivity worth was measured with a good precision and repeatability using different methods (8.82 ± 0.07 pcm), and its value is found rather close to the simulated one using TRIPOLI-4 (9.4 ± 0.4 pcm with JEFF-3.3). Above 1 Hz, the shape of the used modulation is pseudo-sinusoidal, with only a few well defined harmonics of excitation. The strongest harmonic only was analyzed using standard signal processing algorithms such as the Fourier transform and the Bartlett estimator. Twelve data points were produced in the range 0.5 Hz to 200 Hz, with uncertainty ranging from 1 % to 15 %. The prompt decay constant was measured at 150 ± 3 rad/s. Below 1 Hz, step-wise modulations were used with pseudo-random time sequences, which allowed exciting at once a large number of frequencies. Around 150 data points were produced in this particularly interesting frequency domain, between 1.6 mHz and 0.75 Hz, thanks to the use of three distinctive sequences with different base frequencies and overlapping ranges. The amplitude and phase of the RTF were measured satisfactorily, with uncertainties below 1 % for the strongest harmonics. The shape of the RTF was found consistent with the predictions of both JEFF-3.3 and ENDF/B-VII.1 libraries.

Keywords —**Reactor physics, reactor transfer function, CROCUS, delayed neutron groups**

I. Introduction

INTEGRAL measurements in nuclear research reactors are used to provide information on core integral parameters, such as the delayed neutron fraction, as well as to probe the kinetics of delayed neutrons (DN). For this purpose, reactivity modulation experiments are employed to excite the kinetic response of the core, which is then recorded by neutron detectors. This method allows obtaining precise information on the reactor transfer function (RTF). In the point kinetic approximation, the amplitude and phase of the RTF are dependent on the DN kinetic parameters.

The Commissariat à l'Energie Atomique et aux Energies Alternatives (CEA) and Ecole Polytechnique Fédérale de Lausanne (EPFL) have been collaborating towards conducting modulation experiments in the CROCUS zero power research reactor operated by EPFL (Lausanne, Switzerland). The objective was to measure the RTF of CROCUS, its power and phase, on a broad frequency range and with the same device in order to acquire information on the DN (range 1 mHz - 1 Hz) but also on prompt neutrons (range 1 Hz - 100 Hz).

The study was conducted within a PhD work [1], which focused first on designing the modulation device, named PISTIL [2]. An experimental campaign was performed in CROCUS in June 2021. A significant part of the experiment was dedicated to characterizing the reactivity modulation profile.

This article gives an overview of the experimental setup and results obtained so far and gives an outlook onto the comparison with the point kinetic predictions using TRIPOLI-4 with JEFF-3.3 and ENDF/B-VII.1 libraries [3]–[5]. Conclusions are drawn for further experiments on RTF and specifically DN to be scheduled in CROCUS.

II. Probing the Reactor Transfer Function with Reactivity Modulation Experiments

A. Expression of the point kinetic transfer function

In the framework of the point kinetics model [6], the neutron population in the reactor can be treated as a group, without taking into account the energy and position of each neutron. Only the kinetic behavior of the neutron population is considered $n(t)$. The neutron population is related to several integral parameters, namely the core's reactivity ρ, the effective delayed neutron fraction β_{eff}, the prompt neutron generation time Λ. By "effective", one refers to the fact that all fissile actinides in the core account for a part of the delayed neutron fraction. The production of neutrons is also related to the concentrations of the DN groups of precursors $Ci(t)$, where i is the index of the group, and i=1 refers to the longest lived DN precursors.

For discussing reactivity modulation, it is convenient to express the model in the frequency domain. By considering a small modulation of reactivity $\rho(t) = \rho_0 + \delta\rho(t)$ which gives rise to a modulation in the neutron population $N(t) = N_0 + \delta N(t)$, one can express the reactor transfer function (RTF) of the reactor by the following equation :

$$H(j\omega_0) = \frac{1}{N_0}\frac{\delta N(j\omega_0)}{\delta\rho(j\omega_0)} = \left(j\Lambda\omega_0 + \beta_{eff}\sum_i^G \frac{j\omega_0 a_i}{j\omega_0 + \lambda_i} - \rho_0\right)^{-1}$$

In the previous equation, the notations are the following:

- j is the unit imaginary number;
- ω_0 is the pulsation (in radian);
- ρ_0 and N_0 are the time-averaged values of reactivity and neutron population
- a_i and λ_i are the delayed neutron abundance and time constant of the ith DN group. The sum of abundances is normalized to unit (i.e. $\sum_i^G a_i = 1$)
- G is the number of DN groups.

The RTF is a complex function which can be represented by calculating its modulus $\|H\|$ and argument $arg(H)$ as a function of the modulation frequency. The kinetic parameters of the CROCUS reactor were calculated using TRIPOLI-4 associated to two nuclear data libraries: JEFF-3.3 and ENDF/B-VII.1 (see table I) and then fed into the RTF model described above. The results are plotted in figure 1. In the range of the prompt neutrons (frequency above 1 Hz), a good agreement is observed with a difference as low as a 2 % between the predictions. This is not the case below 1 Hz, in the frequency range of DN, where a difference of around 15% can be observed between predictions with JEFF-3.3 and ENDF/B-VII.1.

TABLE I
CROCUS KINETIC PARAMETERS COMPUTED WITH TRIPOLI-4

Parameter	JEFF-3.3	ENDF/B-VII.1
Effective DN fraction (pcm)	758.7 ± 1.4	737.2 ± 1.1
Generation time (μs)	47.703 ± 0.001	47.502 ± 0.001
DN average period (s)	12.3 ± 0.24	10.55 ± 0.13
Prompt neutron decay constant (s^{-1})	159.03 ± 0.29	155.20 ± 0.23

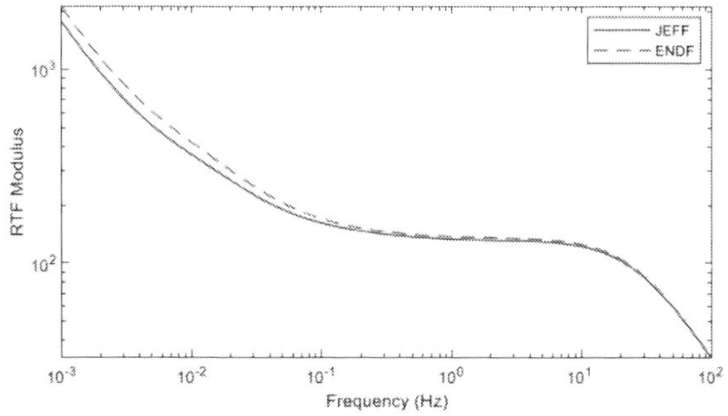

Fig. 1. Modulus of CROCUS RTF calculated with TRIPOLI-4 associated to JEFF-3.3 or ENDF/B-VII.1 nuclear data libraries.

B. *PISTIL, a device for reactivity modulation in CROCUS.*

CEA and EPFL co-developed a reactivity modulation device to meet the purpose to provide feedback on kinetic parameters [7] and the constraints of the CROCUS environment. PISTIL (*Periodic reactivity Injection System for Transients Induced Locally*) is an in-core fixture made of a thin aluminum tube in which a rotary modulator is inserted (see figure 2). The modulator is made of a stator and a rotor. The stator is an aluminum tube with two rectangular cadmium foils stuck on its outer surface with mirror symmetry. Inside of it is the rotor: a polyethylene rod with two cadmium foils facing the ones of the stator. The distance between the rotor and the stator is 0.1 mm.

The rotor is attached to a brushless motor thanks to a thin metallic string. When the motor is rotating, the cadmium foils are alternatively put face to face (when the motor angle is 0° or 180°) and at perpendicular directions (when the motor angle is 90° or 270°).

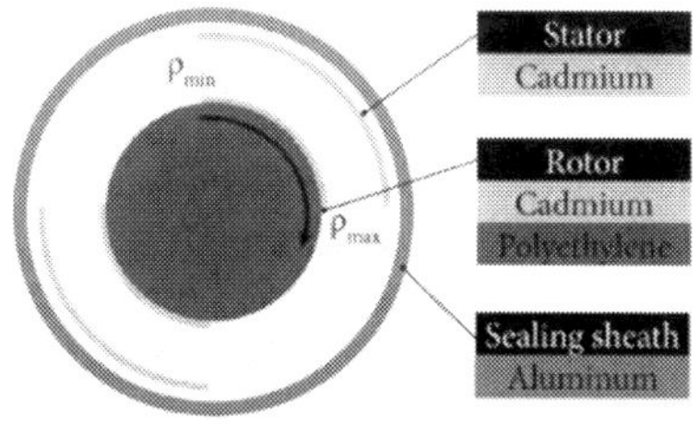

Fig. 2. Principle of PISTIL. Two cadmium sheets are fixed on a stator and a rotor, respectively. The reactivity worth depends on the respective positions of the foils, i.e. the rotation angle.

Because of the high cross sections $^{113}Cd(n,\gamma)$ with thermal neutrons, the local thermal neutron flux is depleted around the cadmium foils. The total reactivity worth of PISTIL, which was calculated with TRIPOLI-4 around -100 pcm, is loosely dependent on the position of the foils. The flux depletion is minimized when the foils are facing each others and maximized otherwise. The maximum modulation was calculated at 8.3 pcm, with an uncertainty of 4 pcm (at 1 sigma).

III. ENDF/B-VII.1 CROCUS REACTOR TRANSFER FUNCTION: RESULTS AND DISCUSSION

A. *The CROCUS experimental nuclear reactor*

CROCUS is a zero-power teaching and research reactor operated by EPFL. This pool-type reactor has a maximum nominal power of 100 W, but it is standardly operated around 1 W [8], [9]. Its core is 1 m in height and about 60 cm in diameter. It consists of two interlocked fuel zones of UO_2 (1.806 wt.% enriched) and metallic uranium (0.947 wt.%) rods, with square lattices of different pitches. The reactor is operated at atmospheric pressure and the moderator temperature is controlled. The power is adjusted either by changing the water level in the reactor vessel, or by moving two optional B_4C control rods. It has been used for numerous studies on neutron noise in recent years, both intrinsic and modulation [10], [11].

B. *Experimental setup*

For the purpose of the modulation experimental campaign that took place in June 2021, PISTIL was installed in the core center position. It was operated thanks to a numerical command (TrioMotion) linked to a servo-controller (XtrapulsePac) to command the brushless motor's position. A National Instruments Labview program communicates with the numerical command through Ethercat connection. The acquisition frequency was set to 1 kHz.

For monitoring the reactivity modulation experiments, four fission chambers were installed in the reflector (Photonis CFUL-01, thermal neutron sensitivity around 1 c/s/(cm²/s), bias voltage 600 V). Their signals were fed to four EPFL-developed current-to-voltage amplifiers [12]. A National Instruments c-RIO mini-crate with a NI-9223 acquisition card was used to record the fission chambers signals.

The PISTIL monitoring software allowed for two operating modes: rotation at constant angular velocity or step-by-step motion. The first mode was used for modulation experiments above 1 Hz. In this mode, the modulation signal contained only a few harmonics, the main one, at the double of the rotation frequency, containing more than 90% of the signal power. For these experiments, the main harmonic (at a frequency twice the rotation frequency) was analyzed and produced one estimate of the RTF.

In the second operating mode of PISTIL, a time sequence of 90 ° angular steps was fed to the monitor software. The initial angular position was chosen so that the reactivity change between two steps is maximum. The time intervals between steps were chosen according to a modified 5-bits Maximum Length Sequence (MLS) given in reference [13]. In theory, by generating reactivity steps according to this pseudo-random series samples in time, ones can induce a pseudo-random signal associated to numerous harmonics of modulation. The length of the 5-bits sequence is equal to $2^{5+1} - 2 = 62$, which corresponds to a sequence of 62 s if the sampling time dt is chosen equal to 1 s. The first harmonics of the modulation signal is then 1/62 = 16 mHz. In practice, three time series were used by choosing dt equal to 1 s, 2 s and 10 s. The first harmonic of these series are respectively at 16 mHz, 8 mHz and 1.6 mHz. During the experiments, the modulation was repeated multiple times in order to acquire sufficient data for Fourier analysis and peak estimation.

C. Differential reactivity worth calibration

The estimation of the shape and intensity of the reactivity modulation is crucial to correctly analyze data and produce correct estimates of the RTF. Indeed, the power spectrum of the modulation signal is needed to get $\delta\rho(\omega_0)$.

The differential reactivity worth is a function that links the induced reactivity versus PISTIL angular position. By construction, the average of the differential reactivity worth is zero. Its peak amplitude gives the maximum reactivity worth achievable with PISTIL and the shape of the function allows calculating its power distribution, i.e. the power in each significant modulation harmonic.

A specific experiment was conducted to estimate the differential reactivity worth by rotating PISTIL at an angular speed of 4°/s (~11 mHz). First, by applying an inverse kinetics algorithm on the fission chamber signals, the reactivity versus time was obtained. Second, by using the angular position of PISTIL, it was possible to cut and average the many of the time signal into portions of 360° each, then average them to produce the experimental differential reactivity worth function.

From the shape of the curve (Fig. 4), one can see that the modulation signals are not perfectly sinus and not even symmetrical: the positive peaks are narrower than negative ones. This is possibly because of slight differences in the mass of the cadmium foils (the relative difference is about 1 %). Because of this characteristic, the power spectrum of the modulation is associated with several odd harmonics. A five harmonics ARMA model was used to fit the reactivity worth and obtain the power distribution in the harmonics.

From the peak amplitude of the curve, a maximum reactivity worth of about 1.1 ¢ was estimated (1 ¢ = β_{eff}/100 pcm). As expected, there is a significant difference between the JEFF and ENDF predictions (cf. table II). This can be explained by the gap between the predicted RTF at the modulation frequency, as illustrated in figure 1).

One conclusion to be drawn is that estimated reactivity worth by this method cannot be used to normalize the experimental RTF. On the contrary, the shape of the reactivity modulation, which is independent on the reactivity scale, can indeed be used in order to predict the power spectrum and the distribution of harmonics induced by PISTIL. Based on the assumption that the power distribution does not change with the rotation frequency, the modulation experiments at any frequency was analyzed and rescaled using the reactivity spectrum. The authors refer to [1] for more explanations on the data processing.

Fig. 3. View of PISTIL inserted in the central channel the CROCUS reactor.

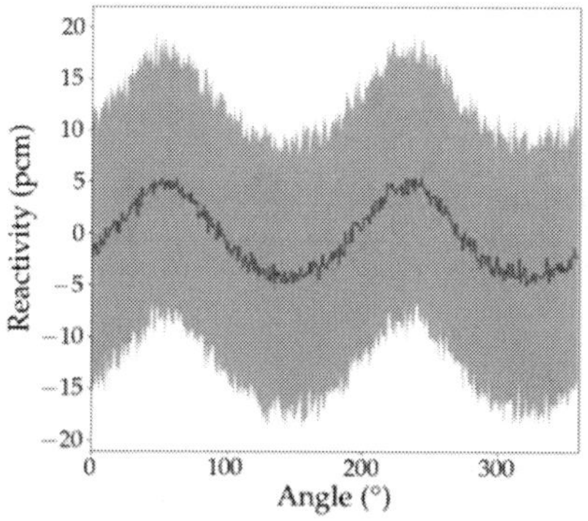

Fig. 4. Differential reactivity worth of PISTIL using an inverse kinetics method. The solid blue line is the curve obtained by averaging the many repetitions (background grey line).

TABLE II
PISTIL REACTIVITY WORTH OBTAINED WITH JEFF-3.3 AND ENDF/B-VII,1.

	ENDF/B-VII	JEFF3.3	Difference
Reactivity worth in ¢	1.06	1.15	7.8 %
Reactivity worth in pcm	7.84	8.76	10.5 %

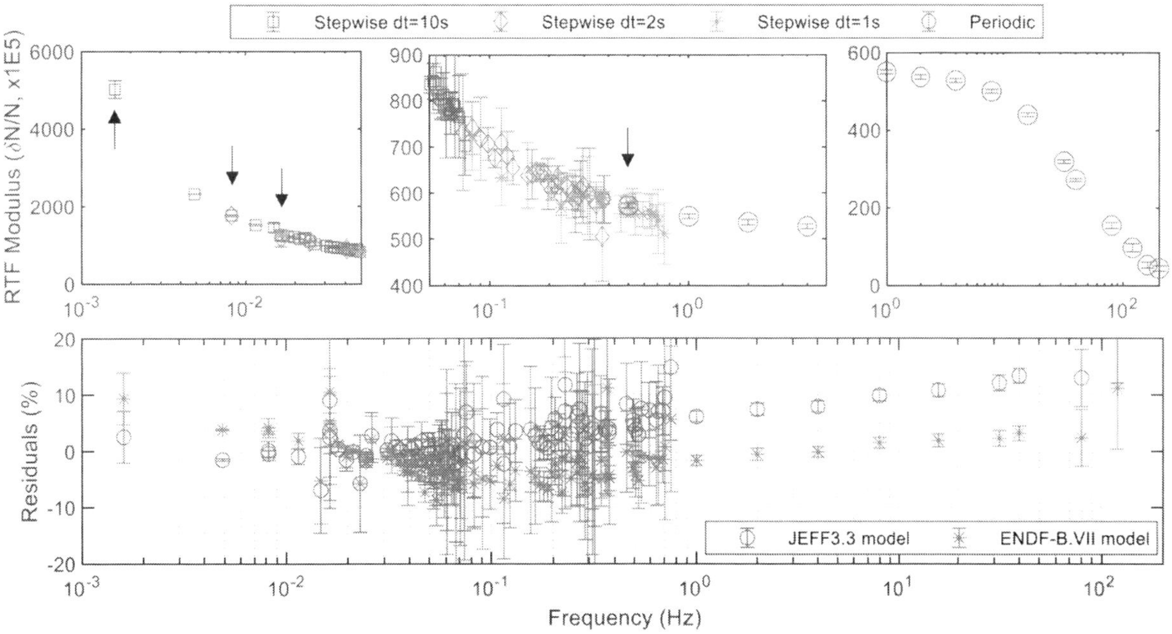

Fig. 5. Experimental estimates of CROCUS modulus on three frequency ranges (upper graphs). Relative residuals from the comparison with models (bottom).

D. Experimental estimates of CROCUS reactor transfer function.

The experimental estimates of CROCUS RTF are presented in Fig. 5 (upper graphs). Because the scale of the data is large, it was split into three sub-graphs.

In the prompt neutron range (right, above 1 Hz), each data point corresponds to an experiment with periodic modulation. In the low frequency range (left, below 50 mHz), three stepwise experiments are represented. The data point of the first harmonic of each experiment is pointed by an arrow. In the intermediate range (middle, between 50 mHz and 5 Hz), the four experiments are represented. One can see that there is an excellent agreement between the various estimates. This is especially the case at 0.5 Hz with a difference of less than 0.1 % between the periodic and stepwise modulations.

E. First comparison with point kinetics models.

To compare the experimental data points with JEFF-3.3 and ENDF/B-VII.1 models, it was necessary to rescale the models. This is because it was not possible to calibrate the reactivity worth of PISTIL independently of a nuclear data library (see section III.C). Rescaling the models was done by multiplying the prediction by the reactivity worth of PISTIL, expressed in \$, estimated by inverse kinetics with each nuclear data library. This step has the drawback of possibly introducing a bias coming from the reactivity estimate at low frequency.

The difference between data and models was computed and then divided by the data to obtain relative residuals. They are plotted on figure 5 (bottom graph). One can see that there the two models exhibit different behaviors at low and high frequency. On the one hand, the 6-groups model associated to ENDF/B-VII.1 show a better agreement with the data in the prompt neutrons energy range (above 1 Hz). This tends to indicate that the DN fraction of CROCUS is best predicted by ENDF/B-VII.1.

On the other hand, the 8-groups model associated to JEFF-3.3 show a better agreement with the data in the DN neutron frequency range (below 0.1 Hz). This is particularly the case below 0.01 Hz, which tends to indicate that JEFF-3.3 predicts the behavior of the DNs better than ENDF/B-VII.1.

F. Least square fit of the prompt decay constant.

We chose to analyze the periodic modulation experiments on the frequency range between 2 Hz and 120 Hz, with a simplified RTF model, which has often been used to analyze branching noise experiments [10]:

$$\|H\| = \frac{\delta\rho}{\beta}\frac{1}{|1 + j\omega/\alpha_p|}$$

Using a weighted non-linear least square fit, the objective was first to obtain an estimate of the prompt decay constant:

$$\alpha_p = 147.2 \pm 3.4 \text{ s}^{-1}.$$

This value compares rather well with the calculated values (see table I), with a relative difference of -5 % as compared to ENDF/B-VII.1, as well as with the reference value previously published [12].

From the amplitude factor, it was also possible to estimate the reactivity worth of PISTIL, expressed in \$: $\delta\rho / \beta = 1.07\times10^{-2} \pm 10^{-4}$. It is interesting to note that this estimate is independent from the nuclear data libraries and falls in the very middle of the interval of values from inverse kinetics (see section III.C and table II), but with a much lower uncertainty.

IV. Conclusions and outlook

In the context of improving DN nuclear data, CEA and EPFL have collaborated towards measuring the RTF of the CROCUS reactor by making use of reactivity modulation. PISTIL, a dedicated reactivity modulation device was co-developed for that purpose and used for a first campaign in June 2021. The RTF was estimated over a wide frequency range, from 1.6 mHz to 200 Hz. More than 150 data points were produced, most of them in the DN frequency range. These data can be used to estimate integral parameters of CROCUS, such as the prompt decay constant and the DN abundances.

Based on the results presented here, it is envisaged to go further in estimating the RTF of CROCUS. Indeed, more precise measurements are needed to estimate the DN abundances, within the frame of the point kinetics equation. In the future, and following the CORTEX European project [14], PISTIL could also be used for improving the detection and localization of spurious perturbations in nuclear reactors.

Acknowledgment

This work was funded by the French research institute regrouping CEA, EDF and Framatome. The authors would like to thank the EPFL and the operating team of CROCUS for their support.

References

[1] Y. Jiang, B. Geslot, V. Lamirand, and P. Leconte, "Experimental determination of the transfer function of the zero-power reactor CROCUS with reactivity modulation," Université Grenoble-Alpes, 2022.

[2] Y. Jiang *et al.*, "PISTIL, a reactivity modulation device to probe the transfer function of the nuclear reactor CROCUS," *EPJ Web Conf.*, vol. 253, p. 04007, Nov. 2021.

[3] E. Brun *et al.*, "Tripoli-4®, CEA, EDF and AREVA reference Monte Carlo code," *Ann. Nucl. Energy*, vol. 82, pp. 151–160, 2015.

[4] A. J. M. Plompen *et al.*, "The joint evaluated fission and fusion nuclear data library, JEFF-3.3," *Eur. Phys. J. A*, vol. 56, no. 7, pp. 1–108, 2020.

[5] M. B. Chadwick *et al.*, "ENDF/B-VII.1 nuclear data for science and technology: Cross sections, covariances, fission product yields and decay data," *Nucl. Data Sheets*, vol. 112, no. 12, pp. 2887–2996, 2011.

[6] K. O. Ott and R. J. Neuhold, "Introductory Nuclear Reactor Dynamics," 1985.

[7] Y. Jiang, B. Geslot, V. Lamirand, and P. Leconte, "Review of kinetic modulation experiments in low power nuclear reactors," *EPJ Nucl. Sci. Technol.*, vol. 6, p. 55, 2020.

[8] J. M. Paratte, R. Früh, U. Kasemeyer, M. a. Kalugin, W. Timm, and R. Chawla, "A benchmark on the calculation of kinetic parameters based on reactivity effect experiments in the CROCUS reactor," *Ann. Nucl. Energy*, vol. 33, no. 8, pp. 739–748, May 2006.

[9] U. Kasemeyer, R. Früh, J. M. Paratte, and R. Chawla, "Benchmark on Kinetic Parameters in the CROCUS Reactor," in *International Reactor Physics Experiments Handbook (IRPhE)*, no. 4440, OECD, Ed. 2007, p. 94.

[10] O. V. Pakari, V. Lamirand, and A. Pautz, "Experimental and numerical study of stochastic branching noise in nuclear reactors," EPFL, Lausanne, 2020.

[11] V. Lamirand *et al.*, "Analysis of the first COLIBRI fuel rods oscillation campaign in the CROCUS reactor for the European project CORTEX," *EPJ Web Conf.*, vol. 247, p. 21010, Feb. 2021.

[12] O. Pakari, V. Lamirand, G. Perret, P. Frajtag, and A. Pautz, "Kinetic Parameter Measurements in the CROCUS Reactor Using Current Mode Instrumentation," *IEEE Trans. Nucl. Sci.*, vol. 65, no. 9, pp. 2456–2460, Sep. 2018.

[13] D. D. Rife and J. Vanderkooy, "Transfer-Function Measurement with Maximum-Length Sequences," *J. Audio Eng. Soc.*, vol. 37, pp. 419–444, 1989.

[14] C. Demazière, P. Vinai, M. Hursin, S. Kollias, and J. Herb, "Overview of the CORTEX project," in *Proceedings of PHYSOR 2018*, 2018, pp. 2971–2980.

Reactor power measurement using N16 and N17 production rate in MARIA reactor

Justyna Cybowska[1,*]
[1] National Centre for Nuclear Research, Poland
(*) justyna.cybowska@ncbj.gov.pl

Abstract — **In the water surrounding the reactor core, especially in the water flowing through the cooling system of the fuel channels, the (n,p) reactions on ^{16}O and ^{17}O occur producing ^{16}N and ^{17}N isotopes. The decay products of ^{16}N and ^{17}N may be used to regain information about the reactor power and the neutron flux density in the core. The aim of the study was the simulation of the activity of the ^{16}N and ^{17}N isotopes during the water flow through the fuel channels' cooling system in the MARIA reactor, checking the possibility of the improvement of the currently performed measurements and the determination of the most optimal placement for the activity measurements. The simulations were performed taking into account the power generated by the individual fuel rods and the information on their burnout. The calculations were made for various configurations of the reactor core over the last 3 years. To check the reliability of the simulation, the activation measurement of the high-purity Au foil was performed. The Au have been activated at the outlet pipeline of the cooling system during the whole reactor cycle (4 days) and after that, the activity of the produced ^{198}Au was measured. The ^{198}Au is the product of the (n, gamma) reaction on the ^{197}Au and its activity can be the measure of the neutron flux density in the foil, which in turn gives the information of the ^{17}N content in the cooling water. For better neutron flux density estimations, the models of the fuel element and the model of the outlet pipe with the Au foil were made in the MCNP simulation code. The Au foil activation was also simulated in the Fispact-II code system which allowed estimating the neutron flux density needed to obtain the measured foil activity. This work shows the results of the studies together with the possible sources of uncertainties and the estimation of their influence on the final result.**

Keywords — **reactor power measurement, neutron activation, simulation**

I. Introduction

The reactor power measurement can be performed using many different techniques. The most popular, among others, are thermal power or neutron flux measurements. In this study, the technique based on the water activation was considered. In the water surrounding the reactor core, especially flowing through the cooling system of fuel elements, reactions involving ^{16}O and ^{17}O take place, resulting in a) γ radiation with an energy of 6.13 MeV ((*1*)),

$$^{16}O + n \rightarrow {}^{16}N + p$$
$$^{16}N \xrightarrow{\beta-} {}^{16}O + \gamma(6.13\ MeV) \tag{1}$$

and b) neutrons with an energy of approximately 1 MeV ((*2*)).

$$^{17}O + n \rightarrow {}^{17}N + p$$
$$^{17}N \xrightarrow{\beta-} {}^{17*}O \rightarrow {}^{16}O + n(\approx 1MeV) \tag{2}$$

The activity of ^{16}N and ^{17}N can be used for the reactor power measurement.

Reactor MARIA is a 30MW, open-pool research reactor, cooled with H_2O and moderated with H_2O and Beryllium. It uses ≈19.75 % enriched fuel in the form of concentric tubes and its core is surrounded with a graphite reflector. The power and power changes measurement is mainly based on the neutron flux measurements (with the use of ionization and fission chambers) as well as the temperature difference between the fuel channel outlet and the inlet header. There is also ^{16}N activity measurement performed as an additional method. The aim of this study was to improve the existing measurement by finding the best possible spot for the detector (if such exists) and to extend the measurement for the ^{17}N measurement.

II. Methods

Several frameworks were used to simulate a water flow through the cooling system of the fuel channels. The cooling system consists of 33 inlet and outlet pipelines of fuel channels and inlet and outlet header. Inlet and outlet pipelines are connected to the fuel channels with the fuel element inside, linked with the so-called "clips" or closed with a valve, depending on the current number of fuel elements and the core configuration. The flow rates in each fuel channel and inlet/outlet pipelines depend on the generated power and are equal to 25 m^3/h or 30 m^3/h for power less than or more than 1.64 MW respectively. The total flow rate in the inlet/outlet header is equal to $(n \cdot 25 + m \cdot 30)$ m^3/h, where n and m are the numbers of fuel elements generating power less than or more than 1.64 MW respectively. The neutron flux density (ϕ_{fuel}) in each fuel channel is calculated taking into account the burnup of the fuel element and the generated power, see (3).

$$\phi_{fuel} = \frac{P}{\sigma_{fiss} \cdot E_{fiss} \cdot N_{^{235}U} \cdot \left[\frac{J}{eV}\right]}, \tag{3}$$

where: P – power generated in the fuel channel (MeV), σ_{fiss} – cross section for ^{235}U(n,fission) reaction for given neutron energy spectrum (cm^2), $E_{fiss} = 193$ – energy produced by one fission (MeV), $N_{^{235}U}$ – number of ^{235}U atoms, including the fuel element burnup. The top view of the fuel cooling system in the MARIA reactor is presented in Fig. 1.

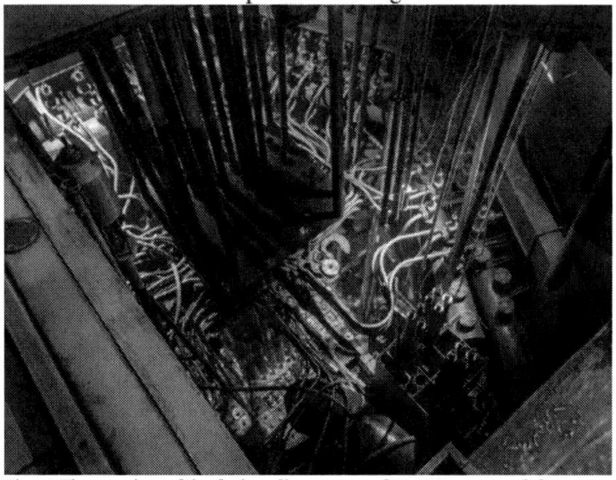

Fig. 1 The top view of the fuel cooling system of MARIA reactor *[1]*.

The production of ^{17}N and its propagation through the fuel cooling circuit was calculated using python ver. 3.8.6 [2]. The ^{17}N activity was calculated using formula (4) obtained from solving the differential equations.

$$A = e^{-\lambda \cdot t_c} \cdot \frac{\lambda \cdot N_0 \cdot RR_1}{\lambda + RR_2 - RR_3} \cdot \left(e^{-RR_3 \cdot t_a} - e^{-(\lambda + RR_2) \cdot t_a} \right) \tag{4}$$

where, λ – decay constant of ^{17}N (1/s), t_a – activation time [s], t_c – cooling time (s), N_0 – initial number of ^{17}O atoms, RR_1 – reaction rate for ^{17}O(n,p) (1/s), RR_2 – reaction rate for ^{17}N(n,total) (1/s), RR_3 – reaction rate for ^{17}O(n,total) (1/s).

The MCNP6.2 (called further MCNP) was used to model the irradiation of the ^{197}Au foil and the water irradiation in fuel. In the first case a steel pipe, filled with water and surrounded with air, with the ^{197}Au foil at the top of it was simulated, see Fig. 2, and the neutron flux density (ϕ_n) and the reaction rates (RR) in the foil were calculated. Knowing these quantities, the activity of the ^{198}Au in foil ($A_{^{198}Au}$) can be calculated with the same formula as for ^{17}N, but with the following arguments: λ – decay constant of ^{198}Au (1/s), N_0 – initial number of ^{197}Au atoms, RR_1 – reaction rate for ^{197}Au(n,γ) (1/s), RR_2 – reaction rate for ^{198}Au(n,total) (1/s), RR_3 – reaction rate for ^{197}Au(n,total) (1/s). In the second case, the neutron energy spectrum in the cooling water in the fuel element, see Fig. 3, was obtained. The geometry in both cases was modeled in SpaceClaim™ [3] and then the .stp file was converted to MCNP input script in SuperMC ver. 3.3.0 [4].

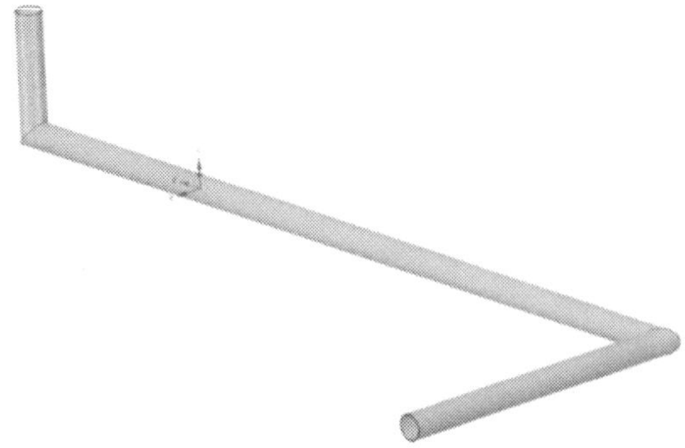

Fig. 2 The 3D model of an outlet header with the Au foil on the top, in the duct under the reactor pool.

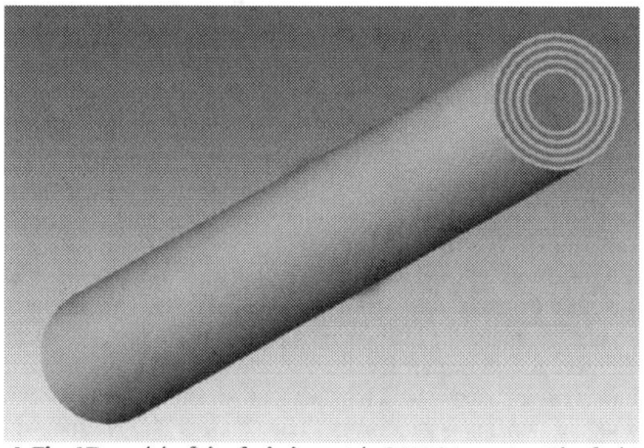

Fig. 3 The 3D model of the fuel element in MARIA reactor. The fuel tubes are marked with yellow, the water is marked with green.

Two high-purity Au foils were used for the measurement of the neutron flux density at the outlet header, in the duct under the reactor pool, see Fig. 4. The dimensions of the foils were equal to 50 mm × 50 mm × 0.005 mm („small foil") and 50 mm × 80 mm × 0.01 mm („large foil"), and their masses – 244 mg and 742.5 mg respectively. They were irradiated for the full reactor cycle, meaning 4 days, and then their activity was measured with the germanium spectrometer.

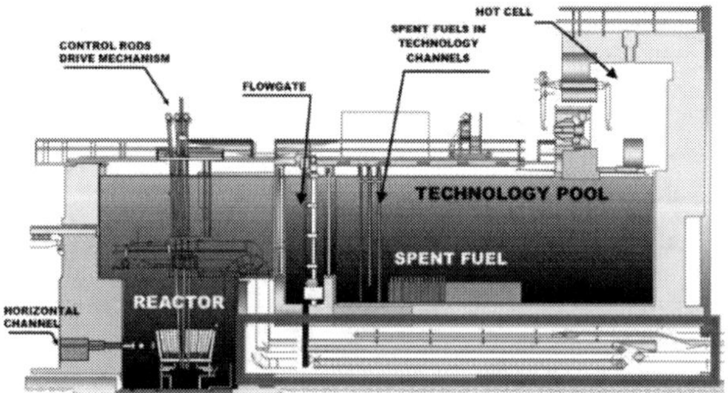

Fig. 4 The MARIA reactor side view. The duct under the reactor pool, where the measurement was performed is marked with the red rectangle *[1]*.

The Fispact-II code was used for the estimation of the neutron flux density needed to activate the Au foil to the measured activity, given the foil mass. For all calculations, the cross-sections from ENDF/B-VIII.0 were used.

III. Discussion

The procedure of estimating the neutron flux density in the Au foils consists of calculating the ^{198}Au activity for different neutron flux densities. Then, the linear function was fitted to the obtained results, and the expected neutron flux was calculated. In Fig. 5, the ^{198}Au activity in the function of the neutron flux density in the small foil is presented.

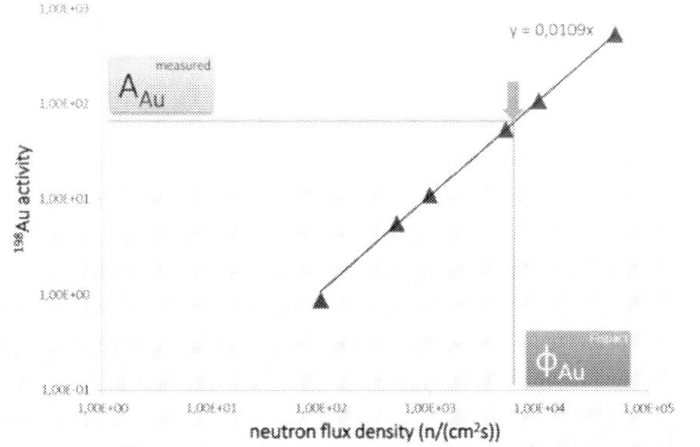

Fig. 5 The ^{198}Au activity in the function of the neutron flux density for the small foil. The black line the line fitted to the simulation points. The blue lines and the arrow marks the values of interest.

The measured activity of the ^{198}Au in foils ($A_{^{198}Au}$) and the Fispact-II neutron flux density ($\phi_{n\,Fisp}$) in foils estimation are presented in Table 1.

TABLE 1
RESULTS OF THE ^{198}AU ACTIVITY MEASUREMENTS AND THE NUETRON FLUX DENSITY IN THE AU FOIL SIMULATIONS MADE IN FISPACT-II.

	measured: $A_{^{198}Au}$ (Bq)	Fispact-II: $\phi_{n\,Fisp}$ $(\frac{n}{cm^2s})$
Small foil	66.89(6.81)	6137(625)
Large foil	183.63(22.82)	5547(689)

The measured ^{198}Au activities were compared with the ^{198}Au activities calculated (4) based on reaction rates obtained from MCNP simulation of the measurement setup (Fig. 2) and ^{17}N activity in the outlet header in the duct. The ^{17}N activity (Fig. 6) was used for normalization of the reaction rates as a number of neutrons emitted from the source per secondFig. 2 The 3D model of an outlet header with the Au foil on the top, in the duct under the reactor pool.. The $\phi_{n\,Fisp}$ was compared with the neutron flux density calculated in MCNP and normalized with the ^{17}N activity ($\phi_{n\,MCNP}$). Both $A_{^{198}Au\,MCNP}$ and $\phi_{n\,MCNP}$ are presented in Table 2.

TABLE 2 THE ^{198}AU ACTIVITY AND THE NEUTRON FLUX DENSITY IN THE AU FOIL CALCULATED BASED ON MCNP SIMULATIONS AND ^{17}N ACTIVITY CALCULATION IN THE FUEL COOLING CICRUIT.

	MCNP: $A_{^{198}Au\,MCNP}$ (Bq)	MCNP: $\phi_{n\,MCNP}$ $(\frac{n}{cm^2s})$
Small foil	7.27(0.37)	353(16)
Large foil	21.53(0.42)	

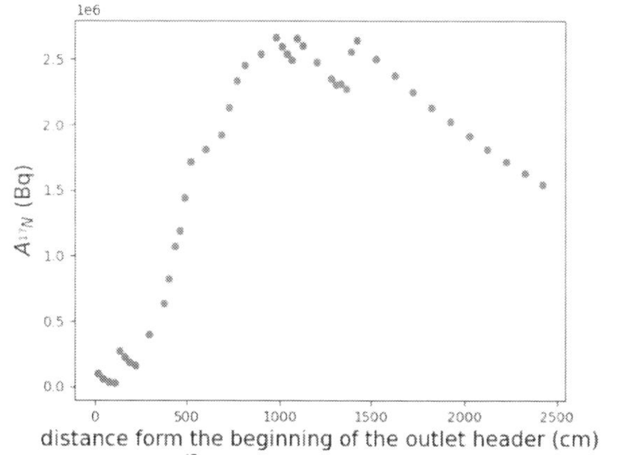

Fig. 6 The activity of ^{17}N during the water flow through the outlet header in single loop. The activity increases when 1 liter of water activated in the fuel element enters the header. For the later calculations the final activity obtained from this algorithm was divided by the number of fuel elements to get the normalised ^{17}N activity per 1 liter of water.

A comparison of $A_{^{198}Au}$ and ϕ_n from Table 1 and Table 2 shows a big discrepancy between the values. The measured and simulated $A_{^{198}Au}$ differ by a factor of ≈ 9. The measured and simulated ϕ_n differ by a factor of ≈ 15.

They may be several potential sources of such large discrepancies between measurement and simulation, such as underestimated neutron flux densities in the fuel element or in the duct under the reactor pool where the Au foil was measured. Possible other neutron sources, contributing to the background in the measurement place, should be also considered. Another possible source of the large discrepancy is the uncertainty of the cross section of the $^{17}O(n,p)^{17}N$ reaction. The large difference between the cross section in JEFF-3.3 database and the cross section in TENDL-2019 and ENDF/B-VIII.0 databases is presented in Fig. 7.

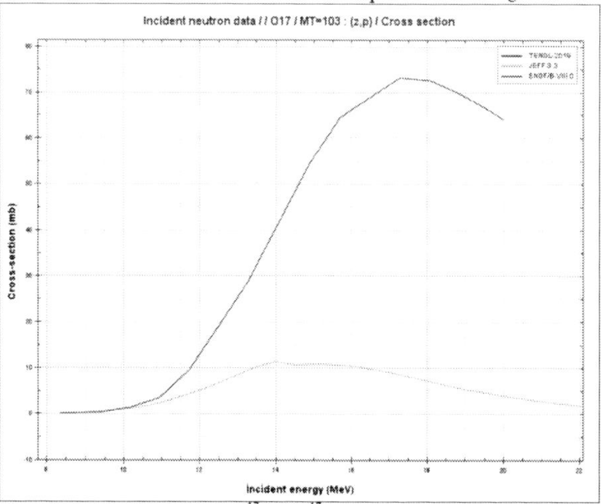

Fig. 7 Cross sections for the ^{17}O(n,p)^{17}N reaction in TENDL-2019 (blue), JEFF-3.3 (green) and ENDF/B-VIII.0 (red) databases.Source: JANIS [5].

Another step in the investigation of this problem could be using a different kinds of foils for the neutron flux measurement in the duct under the reactor pool (e.g. In foil) and measuring also the activity of ^{16}N, which is produced in the reaction (1).

IV. Uncertainties

All numerical values of uncertainties presented in this study are in the form of the standard deviation assuming the Gaussian shape of the distribution of the given quantity.
The uncertainty of the activity measurement $A_{^{198}Au}$ (σ_{meas}) includes both statistical (σ_{stat}) and systematic (σ_{sys}) part. The statistical part is the result of the low activity of the sample and the detector time resolution, which influenced the collected data statistic. The systematic uncertainty is a result of the detector calibration and is estimated as ≈10% of the measured quantity. The final uncertainty was calculated as a sum of square roots of systematic and statistical uncertainty, see (5).

$$\sigma_{meas} = \sqrt{\sigma_{stat}^2 + \sigma_{sys}^2} \tag{5}$$

The uncertainty of the ϕ_n calculated in Fispact-II ($\sigma_{\phi_n\,Fisp}$) is propagated from the measurement uncertainty, as in (6),

$$\sigma_{\phi_n\,Fisp} = \frac{\sigma_{meas}}{a}, \tag{6}$$

where a is the slope of the linear fit (see Fig. 5).
The uncertainty of the simulated $\phi_{n\,MCNP}$ and $A_{^{198}Au\,MCNP}$ is propagated from the uncertainties of the MCNP simulation (obtained from the MCNP output file) and the uncertainty of the ^{17}N activity. The uncertainty of the ^{17}N activity was sampled by running the Python code in a loop for $5 \cdot 10^4$ times, each time changing the following properties of the cooling circuit according to the Gaussian distribution with a mean equal to the given quantity obtained from the documentation and standard deviation listed below:

- Lengths of inlet and outlet pipelines: $\sigma = \frac{8}{\sqrt{3}}$ cm,
- Water flow rate: $\sigma = \frac{0.1 \cdot \mu}{\sqrt{3}}$, where μ is the average flow rate in the given pipeline, obtained from the documentation.

The distribution of the ^{17}N activity uncertainty is presented in Fig. 8. The final uncertainty of the ^{17}N activity is taken as the standard deviation of the given distribution. The final uncertainties of $\phi_{n\,MCNP}$ and $A_{^{198}Au\,MCNP}$ are calculated using the Law of Propagation of Uncertainty for uncorrelated variables (since variables obtained via MCNP simulation and Python code are uncorrelated).
The uncertainty of the power measurement in the fuel channels is not included. It is estimated to have a considerable impact on the uncertainty overall so it is planned to be implemented in the next step of this study.

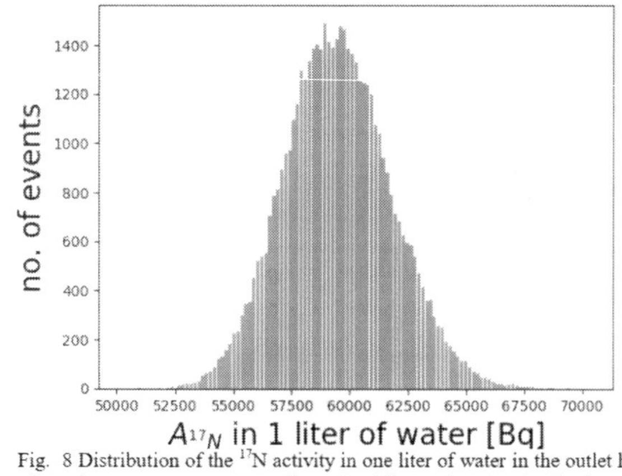

Fig. 8 Distribution of the ^{17}N activity in one liter of water in the outlet header where the measurement was performed, while varying the pipeline length and the water flow rate. Number of samples: $5 \cdot 10^4$.

V. Conclusions

The study shows that the results of the measurement and the simulation show large discrepancies, but the source of them is still unknown. However, there is still much space for improvement in both cases, which has been discussed in this paper. Both measurement techniques, algorithm, and uncertainty estimation needs to be revised carefully. Repeating the measurement using different foil materials or techniques may provide additional information about the source of the error.

REFERENCES

[1] National Centre for Nuclear Research, *Photo archive,* 2023.

[2] *Python Software Foundation. Python Language Reference, version 3.8.6.,* Available at http://www.python.org (accessed: 26th June 2023).

[3] "Ansys SpaceClaim. 3D Modeling Software," [Online]. Available: https://www.ansys.com/products/3d-design/ansys-spaceclaim. [Accessed 11 July 2023].

[4] Y. Wu, J. Song, H. Zheng, G. Sun, L. Hao, P. Long and L. Hu, "CAD-based Monte Carlo program for integrated simulation of nuclear system SuperMC," *Annals of Nuclear Energy,* pp. 82:161-168, 2015.

[5] N. Soppera, M. Bossant and E. Dupont, "JANIS 4: An Improved Version of the NEA Java-based Nuclear Data Information System," *Nuclear Data Sheets,* vol. 120, pp. 294-296, 2014.

Simultaneous inter-calibration of 160 MiMi neutron detectors

Fanny Vitullo[1,*], Vincent Lamirand[1,2], Daniel Godat[1], and Andreas Pautz[1,2]
[1]École polytechnique fédérale de Lausanne (EPFL), CH-1015 Lausanne, Switzerland
[2] Paul Scherrer Institute (PSI), CH-5232, Switzerland

(*) fanny.vitullo@epfl.ch, vincent.lamirand@epfl.ch

***Abstract*—The 160 MiMi neutron detectors of SAFFRON, a 3D full-core mapping system developed at EPFL for the CROCUS zero-power reactor, have been simultaneously inter-calibrated in preparation for their in-core installation. An experimental setup was built to distribute up to 180 MiMi detectors radially at 15 cm around the Pu-Be neutron source of the CARROUSEL facility. An efficient inter-calibration methodology is presented: first, the azimuthal shape of the relative source strength is characterized by rotating 18 MiMi detectors distributed every 20° around the Pu-Be, followed by the simultaneous determination of the relative sensitivity of 160 MiMi detectors, distributed all around the Pu-Be source.**

***Keywords*—Neutron detector, MiMi detector, Inter-calibration, Pu-Be neutron source, CARROUSEL, Relative sensitivity, Source factor, CROCUS.**

I. Introduction

THE development of a three-dimensional (3D) full-core mapping system, called SAFFRON [1], was initiated in 2020 for its installation in the core of the CROCUS zero-power research reactor [2], [3] operated at the Ecole Polytechnique de Lausanne (EPFL), Lausanne, Switzerland. SAFFRON consists of 149 miniature and minimalistic (MiMi) neutron detectors [4], [5] distributed at different radial and axial inter-pin locations of the CROCUS core (plus 11 mobile MiMi detectors), the detectors' supporting structure, and custom digital readout electronics. SAFFRON, with its large number of neutron detectors distributed in between fuel rods and covering the entire core volume, targets the investigation of space-dependent neutronics phenomena in CROCUS and enables the online measurement of thermal neutron flux maps in both a static and time-dependent fashion.

Among SAFFRON objectives, the experimental reconstruction of the 3D thermal neutron flux shape of the CROCUS core is envisaged for future comparison with high-fidelity reactor physics codes results [6]. However, each of the SAFFRON detectors is a unique hand-made instrument, with peculiar counting capabilities, exposed to a different thermal neutron flux dictated by its core location. For this reason, the knowledge of the relative sensitivity ε between MiMi neutron detectors is essential to rescale their count rate readings and obtain a detector-independent shape of the CROCUS thermal flux map.

In this work, the 160 MiMi neutron detectors of SAFFRON are inter-calibrated prior to their in-core installation with the Pu-Be source installed in the CARROUSEL facility, operated at EPFL. The innovative inter-calibration methodology proposed in this work aims at retrieving efficiently and in a short time the relative sensitivities between such a large number of MiMi detectors while reducing experimental uncertainties.

II. MiMi neutron detectors

The MiMi neutron detectors [4], [5] have been developed at EPFL since 2018 targeting a variety of zero-power reactor applications ranging from the investigation of localized neutronics phenomena in CROCUS, i.e., azimuthal flux dependencies [7] and lateral fuel rod displacements [8], [9], to their application to 3D full-core mapping systems such as SAFFRON [1].

The MiMi technology features an inorganic $ZnS(Ag){:}^{6}LiF$ scintillator screen [10] positioned at the front-end of a plastic optical fiber [11] (see Fig. 1) transporting the scintillation light from the detector's active volume to a silicon photomultiplier (SiPM) [12] to count thermal neutron interactions after proper signal processing with either analog or digital readout electronics.

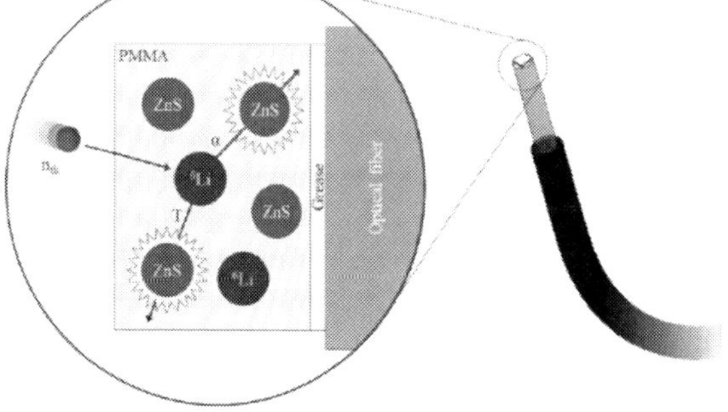

Fig. 1. Schematic of the front-end of a MiMi neutron detector.

A. *Upgrade to SAFFRON version*

The latest version of MiMi neutron detectors used in the SAFFRON system is composed of a 0.225-mm thick $ZnS(Ag){:}^{6}LiF$ screen cut into 0.66×0.66 mm squares using a precise wire-cutting technique. The screen is positioned at the

https://doi.org/10.1051/epjconf/202328804023

tip of the sanded and uncovered tip of the 1-mm diameter optical fiber PMMA core (2.2 mm of total external diameter) with a layer of optical grease in between the two. The fiber tip is covered with an aluminum cap having the same external diameter as the optical fiber. The cap is glued to the optical fiber with an epoxydic adhesive (Araldite), used due to its excellent resistance in the water, fast setting, and good transparency to prevent both unwanted movements or loss of the screen in the CROCUS core.

A crucial upgrade for the extension of the applicability of MiMi detectors to full-core mapping systems was the transition from an analog readout system to a custom-made six-module multi-channel processing electronics system, coupled with FPGA-based digital readout [1]. The combination of the two systems allows the simultaneous processing and acquisition of up to 160 MiMi detector signals.

B. *Detector sensitivity*

Each individual MiMi neutron detector is unique, as it was manufactured first-hand and in series directly at EPFL. In fact, different MiMi detectors measuring in the same neutron field, in terms of neutron flux level and spectrum, show variations in their thermal neutron counting capabilities. These disparities can arise from several factors:

- *Different amounts of ^{6}Li converter*: inhomogeneities in the mixing of ^{6}LiF and ZnS in the screen, or slight variations of screen thickness or size (after cutting) may affect the neutron count rate;
- *Photon yield variations*: variations in the ZnS concentration, misalignments in the positioning of the scintillator screen at the tip of the optical fiber, and infiltrations of filtering materials as glue or dirt may reduce the light collection efficiency causing detector spectrum shifts and, consequently neutron count rate variations;
- *Efficiency of processing electronics*: the SiPM photon detection efficiency and the capabilities of the processing electronics to transform SiPM signals into photon counts might differ due to manufacturing. In this work, each scintillator screen is always paired with its own fixed SiPM and processing electronic channel, and no cross-connections are performed.

The quantity describing the counting capabilities of a single neutron detector is the intrinsic counting efficiency ϵ, or sensitivity, which relates the detector output with the amount of radiation incident on the active detector volume [13]. Instead, the relative sensitivity $\varepsilon_{i,j}$ between two detectors i and j is defined as:

$$\varepsilon_{i,j} = \frac{\epsilon_i}{\epsilon_j} \qquad \text{Eq. 1}$$

and is a characteristic parameter of the detector's duo. $\varepsilon_{i,j}$ can be considered independent of the experimental setup and the irradiation conditions in which it is measured, as long as the two detectors experience a similar neutron spectrum and they will be further used in neutron fields with a similar spectrum. In the particular case of the MiMi neutron detectors, it is also important to maintain the same connection between the ZnS(Ag):^{6}LiF screen and its SiPM to obtain reproducible values of $\varepsilon_{i,j}$.

III. The CARROUSEL facility

CARROUSEL is a nuclear facility operated at EPFL and consisting of a Pu-Be neutron source installed at the center of a cylindrical stainless steel (SS) tank filled with water. Neutrons are produced via the (α, n) reaction between an α particle generated by the decay of ^{239}Pu and a ^{9}Be nucleus as:

$$^{9}_{4}Be + \alpha \rightarrow {}^{12}_{6}C^{*} + n \rightarrow {}^{12}_{6}C + n + \gamma \qquad \text{Eq. 2}$$

The Pu-Be source of CARROUSEL emits 8.9 × 10^{6} neutrons/s [14] with an energy spectrum of 3.5 MeV average and 10.6 MeV maximum energy. Although different shapes of the neutron emission energy spectrum of Pu-Be sources are available in literature [15], [16], the spectrum of the CARROUSEL source is not characterized. However, it was shown that the neutron spectrum becomes fully thermal after 10 cm of distance from the source center [17]. In addition, the 4.4-MeVgamma rays originating from the de-excitation of $^{12}_{6}C^{*}$, make CARROUSEL a mixed neutron-gamma field facility.

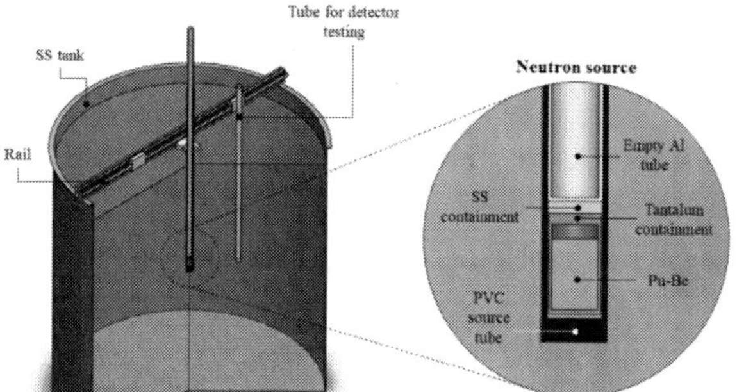

Fig. 2. Schematic of the CARROUSEL facility operated at EPFL.

As shown in Fig. 2, the Pu-Be source has a cylindrical shape, with a double-wall container made of tantalum and welded stainless steel. The source is connected to an aluminum tube for facilitating a convenient and safe handling. On occurrence, the source, along with the handling tube, is inserted into a Polyvinyl chloride (PVC) tube of 40 mm of internal diameter and 3.2 mm thickness positioned in the central vertical axis of the water tank. The PVC tube, designed so that the Pu-Be source is located at mid-height of the water tank, is held in position by being attached to a horizontal rail installed above the tank. The same rail is also used to install and move plastic measurement tubes where different neutron detectors can are installed at will for teaching and research purposes. Finally, the water tank is filled with industrial water at approximately 12 °C, and gradually reaching thermal equilibrium with the controlled area's temperature of 20 °C.

IV. Inter-calibration methodology

The inter-calibration procedure presented in this work aims at finding the relative sensitivity $\varepsilon_{i,j}$ between MiMi detectors in a well-known thermal neutron field in order to successively re-apply those factors for rescaling experimental flux shape measurements in CROCUS as:

$$\frac{\phi_{th,i}}{\phi_{th,j}} \propto \frac{CR_i}{CR_j} \cdot \frac{1}{\varepsilon_{i,j}} \quad \text{Eq. 3}$$

where $\phi_{th,i}$ and $\phi_{th,j}$ are the thermal neutron fluxes in two core locations, and CR_i and CR_j are the count rates of the MiMi detectors at the corresponding locations.

The CARROUSEL facility was selected to perform the MiMi detectors' inter-calibration, as readily available at EPFL. The classic neutron detector calibration methodology used in CARROUSEL consists in measuring the relative sensitivity between two detectors by positioning them alternatively in the same measuring tube, located at a fixed radial distance from the source, and compare their count rate in constant irradiation conditions. This method would allow to retrieve relative efficiencies independently from possible Pu-Be source anisotropies. However, due to the large number of detectors to be inter-calibrated for the SAFFRON system, such procedure would be time-consuming and cumbersome, resulting in larger experimental uncertainties. Therefore, an innovative methodology was developed ad-hoc to simultaneously inter-calibrate up to 160 MiMi neutron detectors in the CARROUSEL facility. This methodology includes two steps: first, the experimental characterization of the Pu-Be azimuthal source factor, and then the simultaneous inter-calibration of 160 MiMi detectors.

A. *Mathematical formulation*

Considering a generic neutron source located in the center of a non-multiplying media, the intrinsic sensitivity ϵ of a neutron detector, positioned at a radial distance r from the source with an azimuthal angle θ and a polar angle φ, within a neutron energy spectrum χ, can be defined as:

$$\epsilon = \frac{CR(m,g,\chi,r,\theta,\varphi)}{S(g,\chi,r,\theta,\varphi)} \quad \text{Eq. 4}$$

where:

- $CR(m,g,\chi,r,\theta,\varphi)$ is the detector's count rate. It depends on the detector's properties, i.e., material composition m and detector geometry g, the neutron energy spectrum χ, and the detector position (r,θ,φ);
- $S(g,\chi,r,\theta,\varphi)$ is the source factor and represents the number of neutrons with a χ spectrum incident on the detector's geometry g when placed at (r,θ,φ) coordinates.

If the radial distance r and the polar angle φ at which the detector is positioned with respect to the point source are fixed, and the neutron spectrum is constant at different azimuthal coordinates θ, Eq. 4 can be simplified as:

$$\epsilon = \frac{CR(m,g,\theta)}{S(g,\theta)} \quad \text{Eq. 5}$$

The relative sensitivity $\varepsilon_{i,j}$ between two neutron detectors i and j with similar geometrical properties $g_i \approx g_j$, but different material composition $m_i \neq m_j$, positioned respectively at the azimuthal angles θ_i and θ_j is defined as:

$$\varepsilon_{i,j} = \frac{\epsilon_i}{\epsilon_j} = \frac{CR_{i,\theta_i}}{CR_{j,\theta_j}} \cdot \frac{S_{\theta_j}}{S_{\theta_i}} \quad \text{Eq. 6}$$

In case the two detectors are positioned alternatively at the same azimuthal position $\theta_i = \theta_j = \bar{\theta}$, or if the source factor shows an isotropic azimuthal profile, the relative sensitivity $\varepsilon_{i,j}$ between two detectors becomes simply the ratio of their count rates:

$$\varepsilon_{i,j} = \frac{CR_i}{CR_j} \quad \text{Eq. 7}$$

On the contrary, if the azimuthal profile of the neutron field $S(\theta)$ is unknown, the prior estimation of the relative source factor $s_{\theta_i,\theta_j} = S_{\theta_j}/S_{\theta_i}$ is essential to derive the relative sensitivity between geometrically similar detectors, as per Eq. 6, without having to perform their inter-calibration by positioning them one-by-one in constant irradiation conditions $S(\bar{\chi},\bar{r},\bar{\theta},\bar{\varphi})$.

B. *Source factor characterization*

The source factor characterization in the CARROUSEL facility consists in finding the $S(\theta)$ profile with respect to a reference azimuthal position, at fixed radial $\bar{r}$ and polar $\bar{\varphi}$ coordinates, and in a constant neutron spectrum $\bar{\chi}$.

1) *Simple case methodology*

Two detectors of similar geometries, called detector 0 and detector 1, are located respectively at the azimuthal positions α and β, and both are at the same fixed $\bar{r}$ and $\bar{\varphi}$ around a neutron source. The count rates measured by the two detectors, having similar geometrical properties, are:

$$\begin{aligned} CR_{0,\alpha} &= \epsilon_0 \cdot S_\alpha \\ CR_{1,\beta} &= \epsilon_1 \cdot S_\beta \end{aligned} \quad \text{Eq. 8}$$

If the azimuthal position of the two detectors is switched, the successive measurement yields to:

$$\begin{aligned} CR_{0,\beta} &= \epsilon_0 \cdot S_\beta \\ CR_{1,\alpha} &= \epsilon_1 \cdot S_\alpha \end{aligned} \quad \text{Eq. 9}$$

By summing the two independent experimental output obtained with different detectors at the same azimuthal coordinate, the following quantities are obtained:

$$\begin{matrix} CR_{0,\alpha} + CR_{1,\alpha} = (\epsilon_0 + \epsilon_1) \cdot S_\alpha \\ CR_{0,\beta} + CR_{1,\beta} = (\epsilon_0 + \epsilon_1) \cdot S_\beta \end{matrix} \rightarrow \frac{S_\alpha}{S_\beta} \quad \text{Eq. 10}$$

The ratio between the right-hand terms of Eq. 10 leads to the estimation of the relative source factor $s_{\alpha,\beta} = S_\alpha / S_\beta$. In a similar way, the ratio between the sum of count rates measured at different azimuthal angles with the same detector returns the relative detector efficiency $\varepsilon_{0,1} = \epsilon_0/\epsilon_1$:

$$\begin{matrix} CR_{0,\alpha} + CR_{0,\beta} = \epsilon_0 \cdot (S_\alpha + S_\beta) \\ CR_{1,\alpha} + CR_{1,\beta} = \epsilon_1 \cdot (S_\alpha + S_\beta) \end{matrix} \rightarrow \frac{\epsilon_0}{\epsilon_1} \quad \text{Eq. 11}$$

The strength of this methodology lies in the possibility to assess at the same time both the relative source term and the relative detector efficiency, independently from each other. In this simple case with two detectors, only two experimental measurements are required to obtain independent relative parameters. However, if the number of detectors to be inter-calibrated becomes large, the number of independent measurements required grows linearly, leading once again to a cumbersome procedure. For instance, 160 measurements would be required to perform a full inter-calibration of 160 MiMi neutron detectors with this approach. As a result, this methodology is employed in this work solely with the intent of reconstructing the source factor's shape in CARROUSEL by using a limited number of MiMi detectors.

2) Extension to 18 detectors

The simple methodology described in the previous paragraph was extended to the use of 18 MiMi neutron detectors evenly spaced around the CARROUSEL Pu-Be source. The 18 MiMi detectors, assumed to have a comparable active volume's geometry, are positioned at 15 cm of radial distance from the Pu-Be neutron source of CARROUSEL with an azimuthal spacing of 20 degrees and at the same level of the source mid-height, giving a zero polar coordinate. The radial distance of 15 cm, considerably larger than the neutron slowing-down length of Pu-Be neutrons in water, was chosen to ensure that the scintillator screens are in a thermal neutron spectrum. By re-positioning each detector at each azimuthal position, for a total of 18 measurements, it becomes possible to delineate the relative source term profile $S(\theta)$ with respect to the zero-angle reference position with a 20 degrees resolution. Such azimuthal resolution was deemed appropriate to find a global shape of the source term around the neutron source making use of the number of MiMi neutron detectors available at the time of measurement. A visual illustration of the method extended to 18 detectors is shown in Fig. 3.

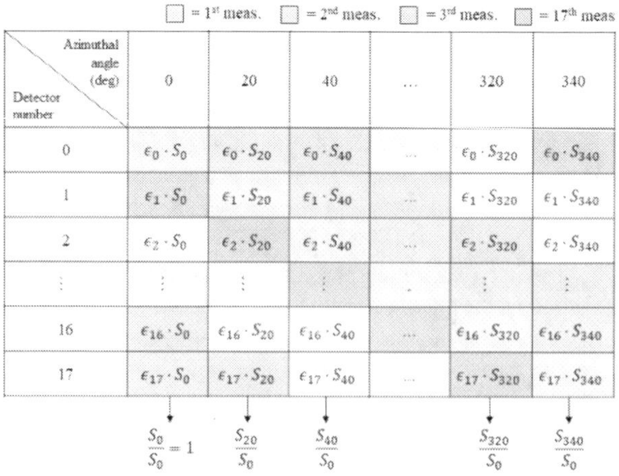

Fig. 3. Visual illustration of the source factor characterization methodology performed in CARROUSEL with 18 MiMi neutron detectors.

C. Relative sensitivity measurement with 160 detectors

Knowledgeable of the relative source factor of the Pu-Be source of CARROUSEL, Eq. 6 is used to rescale the ratio of the count rates between MiMi detectors at different azimuthal positions, but distributed at 15 cm around the Pu-Be source of CARROUSEL with a 2 degrees resolution. In this way, the relative detector efficiencies $\varepsilon_{i,j}$ of 160 MiMi detectors are simultaneously estimated.

V. Experimental setup

The MiMi detectors' positioning around the neutron source is performed with a dedicated plastic structure designed and built at EPFL for testing and inter-calibrating all 160 MiMi detectors of SAFFRON. The structure, shown in red in Fig. 4 for visualization purposes, consists of two black POM-C disks of 30.7 cm diameter and 3 cm thickness. The disks, inserted around the CARROUSEL source tube, are held together by three vertical plastic rods dimensioned to position the detectors' active volume at the source mid-level. At the same time, the rods force the upper and lower disk to rotate in solidarity around the source tube. The lower POM-C disk is machined with 180 blind holes of 2.3 mm diameter and 2.2 cm depth, distributed every 2 degrees at 15 cm from the disk center. Each blind hole can accommodate one of the MiMi detectors, safely held in position by a rubber O-ring through a later aperture, as visible in Fig. 4.

A. Setup for source factor characterization

For the source factor characterization experiment, the 18 detectors are positioned one every five holes to respect the desired 20 degrees of spacing. The upper POM-C disk serves as a reference for the azimuthal position of the lower disk. The 36 through holes of 5 mm diameter distributed every 10 degrees around the upper disk define the unit of a rotational step that the structure can perform. A plastic piece attached to the rail running over CARROUSEL allows for the precise selection of the azimuthal coordinate via the insertion of a metallic pin bounding the piece itself and the upper disk through one of its

https://doi.org/10.1051/epjconf/202328804023

holes. At the initial conditions, the reference MiMi detector is positioned in correspondence with the piece fixating the rotation, oriented towards the west and defining the 0° azimuthal position. The rest of the detectors are distributed counter-clockwise in the lower POM-C disk.

The front-end of each of the 18 MiMi detectors is inserted in the lower disk and is connected via the optical fiber to its own counting channel, composed of a SiPM and a processing board, both part of a multi-channel stand-alone module developed at EPFL for the SAFFRON experiment [1]. In this particular case, the processing channels from ch.65 to ch.82 of module D are used to process the scintillation light. Module D is stored, together with the other five operational modules, in the Memmert IPP750eco incubator [18], kept at a stable temperature of 20 ± 0.1 °C to guarantee stable SiPM photon counting efficiency. SiPMs are operated with a high voltage of 56.66 V, and the conversion of the SiPM output into photon counting is performed with a threshold of 1 V. The photon counting is digitally processed by the CAEN V2495 FPGA [19] with a proprietary firmware based on moving sum differentiation (MSD) algorithm [20]. The resulting thermal neutron counting per channel is sent to the acquisition computer via the CAEN V1718 VME bus [21].

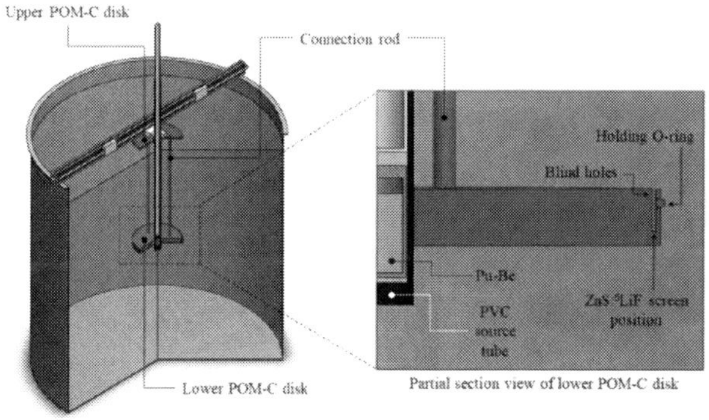

Fig. 4. Experimental setup in CARROUSEL for inter-calibration procedure.

The holding structure is installed and equipped with the 18 MiMi detectors while the water tank of CARROUSEL is empty and the Pu-Be source is not present in the tube. After the installation, the tank is filled with industrial water flowing at around 12 °C of temperature. Three heaters are installed on the side of the water in order to speed up the attainment of the thermal equilibrium between the industrial water and the controlled area environment, kept constant at around 20°C. Once the water equilibrium temperature is reached, the series of 18 measurements by rotation of the structure is started. Each counting measurement is named as the azimuthal position of det. #0. The selection of the azimuthal position is optimized to obtain preliminary results and, for this reason, it is not performed following the increase of the azimuthal angle with respect to the initial position (corresponding to the reference detector at position 0°) linearly. All the measurements are performed continuously, with only interruptions of a few minutes to manually rotate the holding structure and the 18 MiMi detectors together with it. In this way, potential drift due to external conditions, such as the water temperature, can be taken into account and quantified. The measuring time at each azimuthal position was chosen to have an uncertainty of at least 0.5% on the measured count rate of each MiMi detector at every azimuthal position. The minimum measuring time at each azimuthal position is about 2 hours, while five sub-measurements of approximately 20 minutes are automatically run by the digital readout. Longer measurements – performed with continuous sub measurements during nights and weekends – contribute to lowering the final uncertainty of the source factor profile.

B. Setup for the inter-calibration

The CARROUSEL tank was then emptied after properly storing the Pu-Be source, and the same setup was equipped with 160 MiMi neutron detectors in preparation of the measurement to simultaneously inter-calibrate them. The 160 MiMi detectors are labeled with numbers going from det.#0 to det.#159. The detectors from #0 to #155 are installed in the lower POM-C disk occupying the azimuthal positions from 0° to 310° with a 2° spacing. The remaining 4 MiMi detectors are spread in the last 50° to cover the entire range of azimuthal coordinates. Hence, these detectors are positioned every 10° and are accompanied by two small fictive detectors in the neighboring azimuthal positions to reproduce potential self-shielding effects between detectors as in the other 156.

The five stand-alone processing modules A, B, D, E, and F are installed inside the Memmert IPP750 eco incubator set at 20 ± 0.1 °C, together with the digital read-out electronics composed of the FPGA and the VME bus from CAEN. Each optical fiber is connected to one of the counting channels of the multi-channel processing electronics. The connection follows the detector numbering in a way that det.#0 is connected to ch.0, det. #1 is connected to ch.1, and so on. The same connection between the MiMi active volume and its processing channels must be maintained when installing the detectors in-core with the SAFFRON system. All the MiMi detectors are inter-calibrated with respect to the det.#0, positioned in the reference 0° azimuthal position of CARROUSEL. The duration of the counting experiment is dictated by the need for low statistical uncertainties in the counting. A total acquisition time above 40h was estimated sufficient to obtain uncertainties in the order of 0.1% on the single detector count rate of approximately 7.5 cps. The 42-h acquisition is divided into consecutive automatic sub-measurements of approximately 15 minutes each with the intent to observe potential trends over time and split the result data files.

At the starting time, the heaters installed in CARROUSEL are left intentionally powered after already two hours of functioning to observe if there are effects related to the water temperature increase. They are tuned off after about four hours, and the water is mixed at the same time, reaching the temperature of 21°C at approximately the mid-height level of the tank water. The data acquisition from the next 16 hours is discarded to let the CARROUSEL tank reach equilibrium with the environment, followed by the 42-h acquisition.

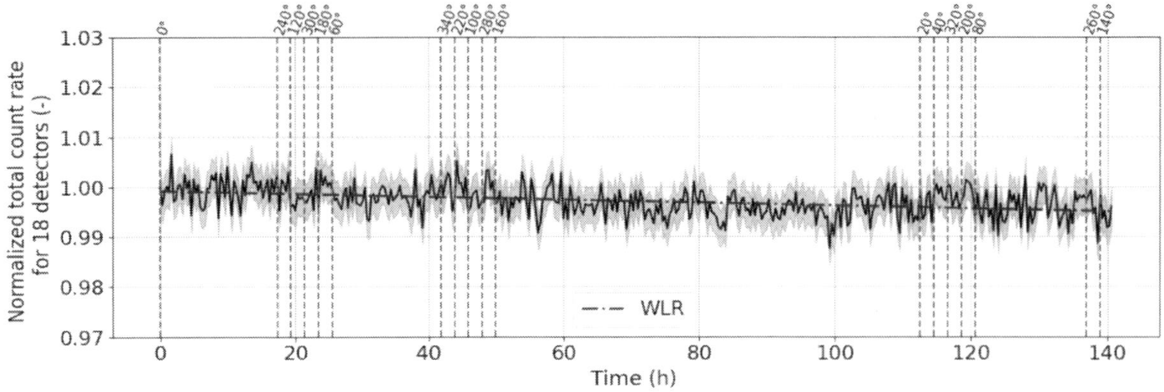

Fig. 6. Normalized total count rate for the 18 MiMi detectors used for source characterization as a function of the measuring time in CARROUSEL. Vertical red lines represent a change in the detector position, indicated as the azimuthal coordinate of det. #0. The trend from linear regression of data is in blue.

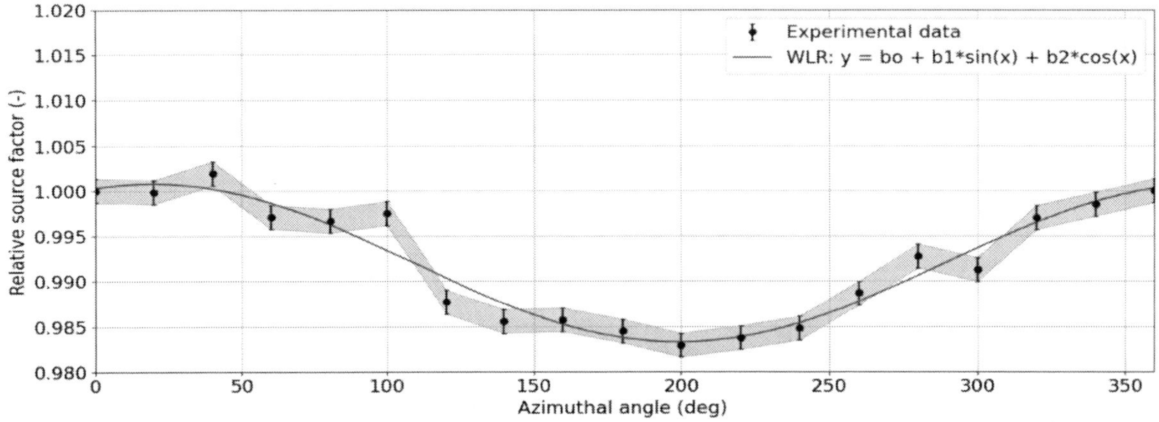

Fig. 5. Relative source factor profile at 15 cm from the Pu-Be source in CARROUSEL.

VI. Inter-calibration results

A. *Relative source strength*

The trend of the total count rate for the 18 MiMi detectors during the whole measuring time is shown in Fig. 5, normalized to its value at the first 20-min sub-measurement. Observing the trend of total count rate instead of those of single detectors allows to reduce the magnitude of the statistical counting uncertainties. The normalized total count rate for the ensemble of 18 MiMi detectors presents a slight negative drift. The drift magnitude was estimated with a weighted linear regression (WLR). A t-test for the regression slope was performed, resulting in the conclusion that a statistically significant drift is present in the total count rate of 18 MiMi detectors. Such drift causes a reduction of the total count rate of -0.48 ± 0.05 % at the end of the measuring time with respect to the initial conditions. The causes of this drift are yet to be investigated and clarified, although it is possible that the drift is linked to an observed drift in the CARROUSEL water temperature from 19°C to 22°C between the beginning and the end of the measurement. It was estimated that such a temperature variation has a negligible impact on the moderation properties of water and plastic. Hence, the suspect is that a water temperature variation is influencing either the optical properties of the ZnS(Ag):^{6}LiF screen, increasing its opacity, thus reducing the amount of photons emerging from the screen and transported to the SiPM, or respective optical indices between screen, optical grease and fiber. Nevertheless, it was estimated that a correction of the experimental data for the drift, under the assumption that it is constant and equal among all 18 detectors, induces variations in the order of 0.001% in the shape of the source factor and can therefore be neglected.

The source factor shape for the uncorrected data is presented in Fig. 6. The uncertainty on the relative source factor is in the order of 0.14%. A reduction of -1.5% in the source factor shape at 15 cm from the source corresponding to the 200 degrees azimuthal angle is visible in Fig. 6. Assuming that the detectors were perfectly positioned and aligned, which was continuously verified along setup and testing, the causes for such reduction could be:

- An anisotropy of the Pu-Be source of CARROUSEL, attenuated by the water volume and the lower disk plastic at 15 cm;
- A miss-centering towards the azimuthal direction of 200° of the Pu-Be source in its dedicated tube.

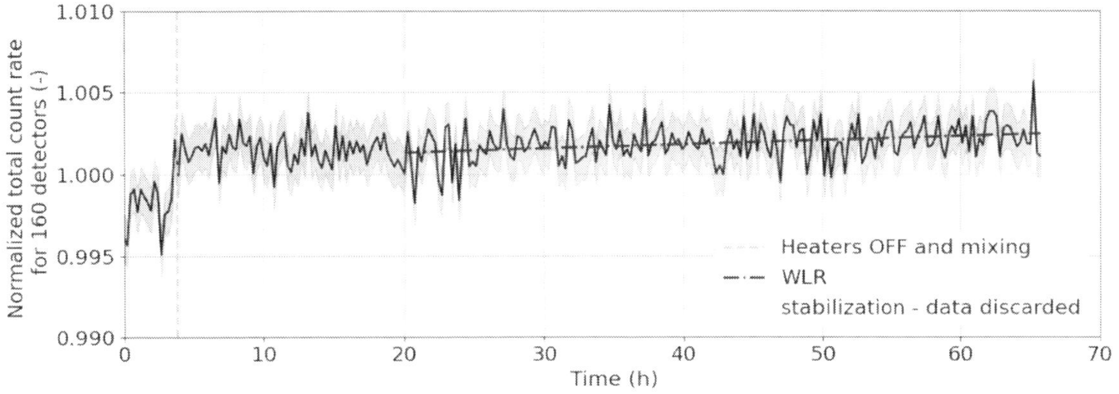

Fig. 8. Normalized total count rate for 160 MiMi detectors as a function of the acquisition time during inter-calibration.

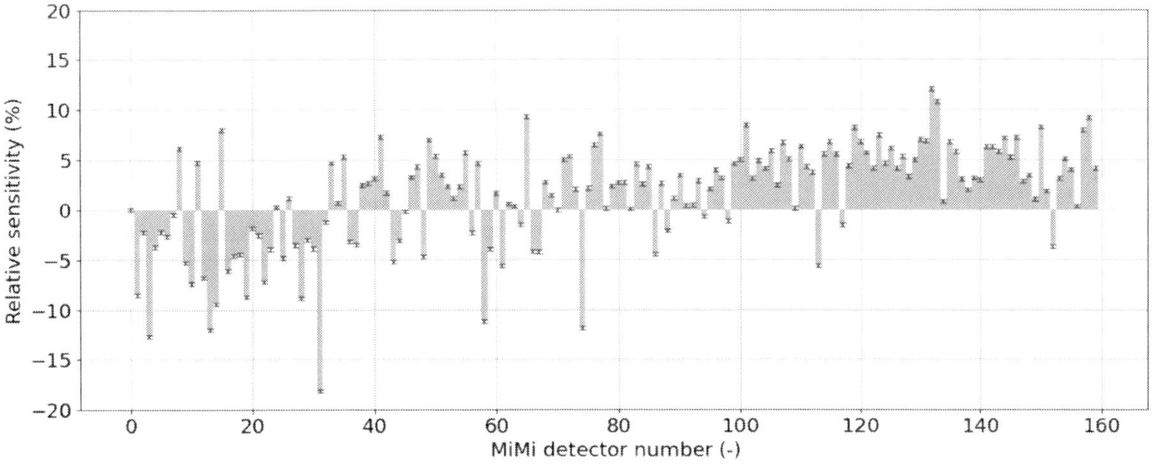

Fig. 7. Relative sensitivity for 160 MiMi detectors with respect to det. #0. The values shown are for data corrected by the total count rate drift.

Considering the known difficulties in the manufacturing and characterization of Pu-Be sources, and the low mechanical play between source and tube (mm-range), as well as its probably low incidence, we rather favor the first option as an explanation.

The experimental data are fitted with a sine function having a 2π period, linearized as $y = b_0 + b_1 \cdot \sin(x) + b_2 \cdot \cos(x)$ with the regression parameters given in Table I.

TABLE I
WLR PARAMETERS FOR THE LINEARIZED SINE FUNCTION DESCRIBING THE RELATIVE SOURCE FACTOR IN THE CARROUSEL FACILITY.

Regression parameter	*Value*	*Standard error*
b_0	0.992	< 0.001
b_1	0.003	0.001
b_2	0.008	0.001

B. *Relative MiMi detectors sensitivities*

The trend in time of the total number of counts acquired during the 42-h inter-calibration measurement is reported in Fig. 7. At the time of water mixing, an increase in the total count rate in the order of 1% is clearly visible. This represents an ulterior proof of the influence of the water temperature on the detector counting capabilities. However, further investigations are necessary to conclude how the direction of the change is caused by an increase in the water temperature. After the stabilization time, the normalized total count rate for the 160 MiMi detectors shows as well a slight positive and statistically significant drift.

The count rate for each MiMi neutron detector acquired over the entire acquisition time is first corrected by the drift, under the assumption that it is constant and equal for all the detectors, and then normalized by the same value for the det. #0 positioned at the 0° reference position of CARROUSEL. Following Eq. 6, each normalized count rate is divided by the value of the linearized sine function with the parameters in Tab. 1, representing the relative source factor at 15 cm in CARROUSEL. The final values obtained are the relative sensitivities of the 160 MiMi neutron detectors with respect to det.#0. The maximum uncertainty of 0.13% measured on the relative source factor azimuthal points is added to the final computation of uncertainties for conservative purposes. The relative detector sensitivities $\varepsilon_{i,0}$ for corrected data are shown in Fig. 8.

The 160 MiMi neutron detectors show differences in the

relative sensitivity between -18.2 ± 0.2 % and +12.1 ± 0.2%. The average uncertainty is 0.2%. The difference between corrected and uncorrected data is calculated to be within the final uncertainty, including the uncertainty of the source factor, and indicates that the total count rate drift can be neglected in the estimation of the relative sensitivities.

VII. CONCLUSIONS

A methodology to simultaneously inter-calibrate 160 MiMi neutron detectors, built for being installed in the CROCUS zero-power reactor as part of the SAFFRON 3D full-core mapping system, was developed in this work. The proposed methodology was performed in the CARROUSEL facility and consists in two-steps: first the relative azimuthal source factor of the Pu-Be source was experimentally estimated with 18 MiMi neutron detectors, and successively the relative sensitivity of 160 MiMi detectors was measured by installing them all at 15 cm around the Pu-Be source. The advantages of this procedure are a shortening of the measuring time and a reduction of the experimental uncertainties. The experimental results highlighted an anisotropy of -1.5% in the shape of the Pu-Be source factor at 15 cm from the source, while the MiMi detectors' relative sensitivities showed variations between -18.2 ± 0.2 % and +12.1 ± 0.2% with respect to the reference detectors. Such variations in the relative sensitivity might be caused by differences in the amount of sensitive material, in the detector mounting, and in the dedicated processing electronics. The obtained relative sensitivities will then be used to re-scale the count rates of the same MiMi detectors while installed in CROCUS as part of the SAFFRON system, and measure the absolute 3D shape of the thermal neutron flux map in CROCUS.

REFERENCES

[1] F. Vitullo *et al.*, "Design of a 150-miniature detectors 3D core-mapping system for the CROCUS reactor," *EPJ Web Conf.*, vol. 253, 2021.

[2] Nuclear Energy Agency (NEA), "IRPhE Handbook 2020." International Reactor Physics Evaluation Project Handbook (database), 2020.

[3] V. Lamirand, "Installation nucléaire CROCUS - Rapport de sécurité," Techical report RT-100-104-17-48, 2017.

[4] F. Vitullo *et al.*, "Developing and testing a miniature fiber-coupled scintillator for in-core neutron counting in CROCUS," in *EPJ Web Conferences*, 2020.

[5] F. Vitullo *et al.*, "A mm³ Fiber-Coupled Scintillator for In-Core Thermal Neutron Detection in CROCUS," *IEEE Trans. Nucl. Sci.*, vol. 67, no. 4, pp. 625–635, Apr. 2020.

[6] F. Vitullo, "Miniature and Minimalistic Neutron Detectors for Online High-Resolution Experiments in the Zero-Power Reactor CROCUS," Ph.D. thesis TH9492, École polytechnique fédérale de Lausanne (EPFL), Lausanne, 2022. [Online]. Available: https://infoscience.epfl.ch/record/296483?ln=en

[7] F. Vitullo, V. Lamirand, P. Frajtag, G. Perret, and A. Pautz, "Highly localized azimuthal measurements in the CROCUS reactor towards the validation of high-fidelity neutronics codes," in *EPJ Web of Conferences*, 2021, p. 08014.

[8] F. Vitullo *et al.*, "Millimetric fuel rod displacements: an experimental study of the impact on local thermal neutron flux in the CROCUS reactor," in Proceedings of PHYSOR2022, Pittsburgh, PA, USA, 2022.

[9] V. Lamirand *et al.*, "Local and high distance neutron and gamma measurements of fuel rods oscillation experiments," *EPJ Web Conf.*, vol. 253, p. 04024, 2021.

[10] Scintacor, "Neutron screens," 2020. [Online]. Available: https://scintacor.com/wp-content/uploads/2015/09/Datasheet-Neutron-Screens-High-Res.pdf

[11] Mitsubishi Chemical Co., "ESKA optical fibers." https://www.pofeska.com/pofeskae/product/02/index.html

[12] Hamamatsu Photonics K.K., "MPPC S13360 sereis datasheet." 2016. [Online]. Available: https://www.hamamatsu.com/content/dam/hamamatsu-photonics/sites/documents/99_SALES_LIBRARY/ssd/s13360_series_kapd1052e.pdf

[13] G. F. Knoll, *Radiation detection and measurement*, 4th ed. Hoboken, N.J: John Wiley, 2010.

[14] NUMEC-USA, "Shipping document Pu-Be source." 1961.

[15] L. Stewart, "Neutron Spectrum and Absolute Yield of a Plutonium-Beryllium Source," *Phys. Rev.*, vol. 98, no. 3, pp. 740–743, May 1955.

[16] M. E. Anderson and W. H. Bond, "Neutron spectrum of a plutonium-beryllium source," *Nucl. Phys.*, vol. 43, pp. 330–338, May 1963.

[17] E. L. Brunetto, "Validation of the Serpent 2 model for the CARROUSEL facility," École polytechnique fédérale de Lausanne (EPFL), Semester project, 2020.

[18] Memmert Gmbh, "DS_Memmert Peltier-cooled incubator IPP750eco." [Online]. Available: https://www.memmert.com/products/incubators/peltier-cooled-incubator/IPP750eco/

[19] CAEN, "V2495/VX2495 VME Programmable Logic Unit." 2019.

[20] J.-B. Mosset *et al.*, "A 16-ch module for thermal neutron detection using ZnS:6LiF scintillator with embedded WLS fibers coupled to SiPMs and its dedicated readout electronics," *Nucl. Instrum. Methods Phys. Res. Sect. Accel. Spectrometers Detect. Assoc. Equip.*, vol. 845, pp. 494–498, Feb. 2017.

[21] CAEN, "V1718/VX1718 - VME - USB 2.0 Bridges." 2020.

Specially developed LR-0 reactor graphite environment for gen IV reactor support and cross-section measurement

Tomas Peltan[1, 2, *], Eva Vilimova[1], Tomas Czakoj[2], Zdenek Matej[3], Filip Mravec[3], Frantisek Cvachovec[3], Jan Simon[2], Vlastimil Juricek[2], Michal Kostal[2]

[1]University of West Bohemia in Pilsen, Czech Republic
[2]Research Centre Rez, Czech Republic
[3]Masaryk University in Brno, Czech Republic
(*) peltan@fel.zcu.cz

Abstract—**This paper is focused on the development of the experimental environment connected to reactor graphite. Regarding its very good neutronic and mechanical properties, graphite will be very important in some new reactor designs, such as high-temperature or molten salt SMR reactors. These new reactor concepts require a new experimental environment as support for further research. In the laboratories of the Research Centre Řež and at the LR-0 reactor, the new experimentally validated graphite environment was created. This large graphite insertion is the largest graphite mono-block, which is possible to assemble at the LR-0 reactor. Sets of experiments for measuring reaction rates of different activation detectors for neutron field mapping were performed. This approach was used for thermal and epithermal region descriptions. For the fast neutron spectrum evaluation, the stilbene scintillation detector was used. All parameters, such as criticality height of moderator level, neutron spectrum, and other parameters for all experiments, were performed using Monte Carlo neutronic codes Serpent and MCNP. The obtained results were finally compared to the measurement of neutron leakage spectra from the graphite cube and graphite cylinder. These specially developed graphite-shaped neutron fields, reactor insertions, and external cube and cylinder with Cf neutron source can be used in the future for validation of not only materials used in SMR reactors but for arbitrary cross-section verification.**

Keywords **— LR-0 reactor, a research reactor, graphite, cross-section, neutron leakage spectra, californium leakage spectrum**

I. Introduction

The reactor graphite has become a more popular material in Gen IV reactors. It can be used as a neutron reflector or a moderator in nuclear reactors due to very good neutron properties and mechanical behavior during irradiation and thermal stresses. Graphite is a very important material for a wide range of common nuclear reactor types such as Gas Cooled Reactors (GCR) and Light Water Graphite Reactors (LWGR) and even for newly developed reactors such as Molten Salt Reactors (MSR) or other currently very popular Small Modular Reactors (SMR).

Regarding the promising future of SMR reactors, a new adequate experimental environment and research facilities should be designed.

Although the behavior and neutron properties of the graphite are well known, there are still some discrepancies in graphite microscopic cross-section [1] , which can play a non-negligible role in new reactor development.

Moreover, currently worldwide is a lack of well-characterized neutron fields in graphite that can be used for experimental purposes. This article deals with a preliminary characterization of the graphite neutron field in a special large graphite insertion in the LR-0 experimental reactor. With the unique graphite insertion, the LR-0 reactor core could serve as a new reference neutron field for various experiments requiring a precise thermal and epithermal neutron spectrum.

This paper presents the results of a neutron distribution mapping performed using a set of activation foils selected according to their suitable microscopic cross-sections for thermal and epithermal neutron field reconstruction to characterize the neutron field of a large graphite prism. Then the paper analyses a fast neutron spectrum measured by the stilbene detector above 1 MeV and compares the experimental data from measurement in the LR 0 reactor with calculations performed in Monte Carlo code MCNP.

A. Experimental reactor LR-0

Reactor LR-0 is a zero power, light water pool type reactor located in the Research Centre Rez (Czechia). The LR-0 reactor was designed to research pressurized water reactor (PWR) physics (mainly for VVER-type cores). Its special design allows multi-purpose use. The technological equipment allows the carrying out of experiments on VVER-1000 or VVER-440 type assemblies.

The criticality of the reactor is reached by pumping a moderator to the reactor tank to the critical height, which is preliminarily calculated and latterly experimentally derived. As a moderator, pure demineralized water or a mixture of light water and boric acid up to 12 g/kg can be used. The second way how to reach criticality is by using control rods. The absorbing elements are boron carbide pellets in stainless-steel tubes.

Fuel assemblies are mockups of shortened VVER-1000

elements, which means a Zr cladding tube with an outer diameter of 9.15 mm filled with sintered UO_2 pellets. The active length of the fuel pin is 125 cm, and the total length is 135.7 cm. Other parameters are the same in the radial sense as the standard fuel assemblies VVER-1000. The characteristics of the reactor enable a wide range of experiments and different core arrangements.

Continuous nominal thermal power is 1 kW with thermal neutron flux $\approx 10^{13}$ m^{-2} s^{-1}. Fig. 1 shows the photo of the reactor core arrangement presented in this research.

Fig. 1. LR-0 reactor core arrangement with central graphite insertion

II. Experimental and calculation methods

A. *Reactor core arrangement*

Twelve fuel assemblies surround a special dry experimental module, corresponding to seven fuel assemblies in size and shape. The dry experimental module is made of pure aluminum walls and assembled with stainless steel bottom weights and a support socket located in the lower part of the whole module. The dry module support socket fits to fuel construction supports of the reactor. The complex geometry of this special module allows the placement of the graphite blocks into the dry module in very tight geometry without significant air gaps, which keeps the module dry during all phases of the experiment. This feature, together with the addition of central graphite cylinder plugs, ensures a very "clean" neutron field dependent only on the embedded material because there is no additional excess structural material or water gaps between the separate experimental modules. The visualization of the reactor core, with developed large graphite insertion in the dry experimental module, can be seen in Fig. 2.

The experimental dry module is filled with seven identical graphite blocks. Each block consists of six smaller trapezoidal parts with a central cylinder in tight geometry. These seven blocks were installed on a special aluminum base for easier handling. The height of each graphite block is 60 cm with a hexagonal key dimension of 21.65 cm. The graphite insertion used in these experiments has a density of 1.72 ± 0.02 g/cm^3, a concentration of impurities below 0.2 ppm of boron equivalent. These properties meet the nuclear graphite limits [2] .

In this case, fuel assemblies have a nominal enrichment of 3.59 % and 3.60 % of ^{235}U. The fuel pins have a pin pitch of 1.275 cm, and fuel assemblies have a pitch formed of 23.6 cm, the same as in VVER-100 fuel geometry. The experimentally determined critical level of the moderator in the reactor was determined based on three independent measurements to H_{cr} = 39.133 ± 0.005 cm. Fuel assemblies are surrounded by nine dry aluminum installed 4 cm away from the core in a light water reflector. These channels serve for neutron monitoring regarding the reactor operation.

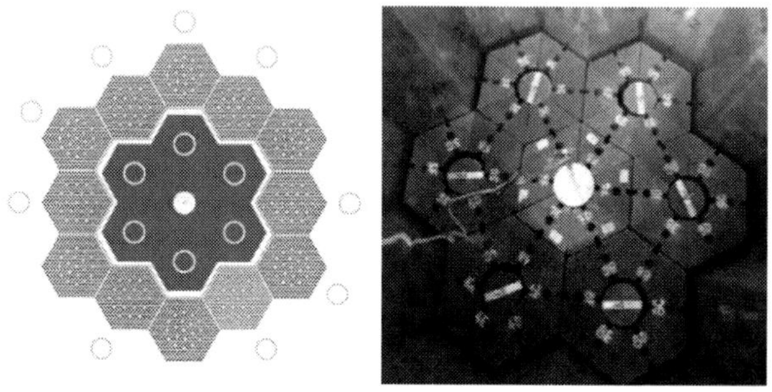

Fig. 2. Schematic view on reactor core arrangement (left), photo of real graphite insertion in the reactor core (right)

B. *Activation detectors and reactor core arrangement*

The specially developed activation foil holder is a single pure aluminum rod with a diameter of 0.8 cm and a special spatial displacer to ensure the precise position in the center of the graphite prism. Activation detectors were attached to this special holder. The distance between each activation material was manually set to 5 cm in the axial direction, which gives us seven axial positions in this case. This distance sufficiently ensures no neutron field disruption and interface effects between detectors. The Au, Cu, Fe, and Mn activation detectors were chosen for these purposes. The size of all activation detectors was 3 x 3x 0.1 mm. Only the manganese detector was spherical in diameter of less than 1 mm.

Irradiation of the activation foils was performed in two days irradiation, taking 9 hours each day to reach sufficient activity of detected radionuclides. Reactor power during irradiation was approximately 5 W.

C. *Graphite cube measurement with ^{252}Cf neutron source*

The measurement with a ^{252}Cf neutron source in a graphite cube geometry was performed at a special laboratory connected to the LR-0 facility [3] . The dimension of the whole laboratory is 5 x 5 x 7.2 m (height). The graphite cube was hung 2 m above the floor, which minimizes the neutron scattering from the wall and floor (the so-called room effect). The length of the edge of the graphite cube is 30 cm, and the cube is constructed from four smaller cubes tied together. In the center of the cube is a 2.5 cm diameter cavity for placing the flexo-rabbit system for the neutron source movement.

The neutron leakage spectra were measured only by the stilbene detector for the purposes of this paper. The detector was placed 1 m away from the cube side, and the center of the crystal was set to the center of the neutron source in the central part of the cube. The graphite used in this experiment has a 1.72 g/cm^3 density, and the purity of this graphite is also very high.

This experiment was carried out by two independent measurements, one for fast neutron measurement, and the second with the special shielding cones. Measuring with the cones helps us evaluate the room effect and correct measured results regarding this measurement. The schematic diagram of the experiment can be found in Fig. 3.

Fig. 3. Measuring the neutron leakage spectra in special laboratory [3]

D. HPGe detector measurement and reaction rate determination

After the irradiation in the reactor core, the activity of irradiated activation detectors was measured using a well-defined HPGe coaxial vertical detector (Ortec GEM35). The HPGe detector efficiency was calculated using MCNP6 code in previous work, see [4] . The precise computational model for the detector was compiled using experimentally determined dimensions by a special radiogram of the crystal REF, and precisely measured dead layer. The efficiency curve of the used HPGe detector can be found in Fig. 4 compared to different models developed in the MCNP code.

Using the HPGe detector, the Net Peak Areas (NPA) of irradiated activation foils were measured. The calculations with a fixed source of neutron spectrum were used to determine the resonance self-shielding factors. Other necessary correction factors and parameters were obtained based on other MCNP calculations. The reaction rate of radioisotopes originating during non-constant irradiation power level in the reactor core is derived using Equations (1) and (2), while the capture reactions of the produced radioisotopes are neglected. Coincidence summing effect correction was calculated based on the method presented in previous work, see [7] .

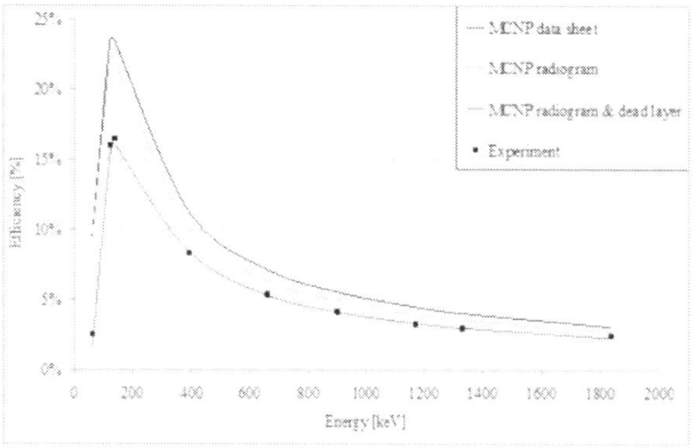

Fig. 4. Calculated and experimentally evaluated efficiency of the mentioned HPGe detector

$$the\ \frac{A(\underline{P})}{A_{Sat}(\underline{P})} = \sum P^{i}_{rel} \times \left(1 - e^{-\lambda . T^{i}_{ir}}\right) \times e^{-\lambda . T^{i}_{end}} \quad (1)$$

$$q(\underline{P}) = \left(\frac{A(\underline{P})}{A_{Sat}(\underline{P})}\right)^{-1} \times NPA(T_M) \times \frac{\lambda}{\varepsilon \times \eta \times N} \times \frac{\frac{t_{real}}{t_{live}}}{(1 - e^{-\lambda . T_m})} \times \frac{1}{e^{-\lambda . \Delta T}} \times \frac{1}{k_{CSEF}} \times k_{SSEF} \quad (2)$$

where:

$\frac{A(\underline{P})}{A_{Sat}(\underline{P})}$ is relative portion of saturated activity induced during irradiation experiment,

P^{i}_{rel} is relative power on the i-th day of irradiation, $P^{i}_{rel} = \frac{P^i}{\underline{P}}$,

$q(\underline{P})$ is reaction rate of activation foil during power density$\underline{P}$,

T^{i}_{ir} is irradiation time on the i-th day of irradiation,

T^{i}_{end} is time from the end of i-th day of irradiation to end of all irradiations,

λ is the decay constant of corresponding material,

T_m is the time of activation foil measurement by HPGe,

ΔT is the time between the end of irradiation and the start of HPGe measurement,

$NPA(T_m)$ is the measured number of counts,

ε is gamma branching ratio of activation material – depending on the material,

η is detector efficiency – the result of MCNP calculation,

N is number of target isotope nuclei in activation foil,

t_{real} is the real-time of counting system of the HPGe (= T_m),

t_{live} is the live time of the counting system of the HPGe (< t_{real}),

k_{CSEF} is the coincidence summing effect correction,

k_{SSEF} is the resonance self-shielding effect correction determined by MCNP.

E. Stilbene detector measurement

The neutron flux up to 1 MeV was measured with a two-parameter multichannel analyzer, which is fully digitized, and can process up to 300 000 impulses per second [8] . The input analog signal from the photomultiplier is divided in the preamplifier into two branches. Each branch is differently amplified in ratio 1:8 and digitized by separate A/D converts with 12 bits resolution. The stilbene scintillator of cylindrical geometry with dimensions ø 10 mm × 10 mm coupled to the photomultiplier RCA 8575 was used as the neutron detector. Since stilbene is an organic scintillation detector, it is sensitive not only to neutrons but also to gamma. The separation between neutron and gamma pulses is realized by means of pulse shape discrimination (PSD) of the measured response. Pulse Shape Discrimination parameter (D) is derived by an integration algorithm, which principle lies in the comparison of area limited by part of a trailing edge of the measured response (*Q1*) with area limited by the whole response (*Q2*). The areas *Q1* and *Q2*, as integrals over time, are expressed in Equation (3), and their illustration is shown in Fig. 5.

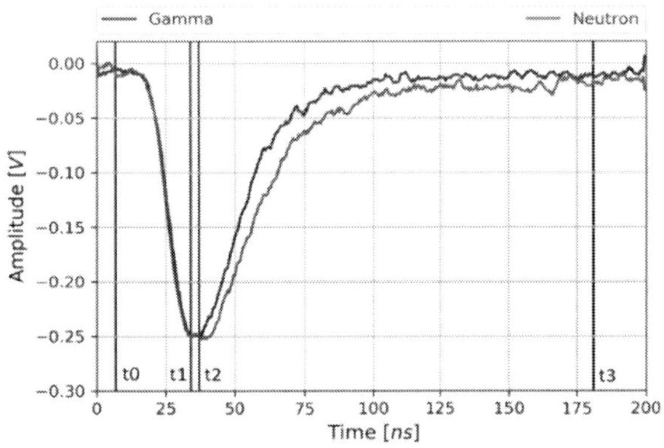

Fig. 5. Comparison of real neutron and gamma pulses from Stilbene scintillation detector with marked examples of a separation boundary for integration algorithm

The time offset t_2 is set for the optimal discrimination properties (namely, the most significant possible difference in the discrimination parameter for neutrons and gammas) to about 1/10 to 1/3 of the trailing edge. This parameter is different regarding the scintillation detector type.

Charge Q_1 is determined by an area limited by the response course within a time interval (t_2, t_3). The charge Q_2 is determined by an area limited by the response course within firmly defined times t_0 and t_3. Times t_0 and t_3 depend on the parameters of the measuring apparatus, and time t_3 is defined as the end of the response.

Using PSD, energy-dependent recoil proton responses $S(E_p)$ are evaluated. The neutron fluxes are then evaluated by deconvolution according to Equation (4). The response matrix of the crystal $K(E_N, E_P)$ was determined employing Monte Carlo code NEU-7 [9] [10] [11] .

$$Q_1 = \int_{t_2}^{t_3} i(t)dt, \quad Q_2 = \int_{t_0}^{t_3} i(t)dt, \quad D = \frac{Q_1}{Q_2} \tag{3}$$

$$S(E_p) = \int K(E_N, E_P)\phi(E_N)dE_N \tag{4}$$

The stilbene detector is very suitable for measuring higher neutron energies. The ability to accurately measure stilbene at energies above 1 MeV was demonstrated by a specially designed experiment on the LVR-15 reactor. The neutron beam in one radial port was filtered by a 100 cm thick plug of high-purity silicon monocrystals. This filter appropriately shapes the neutron spectrum, making it possible to accurately calibrate the apparatus [12] . The comparison between simulation and real measurement can be found in Fig. 6

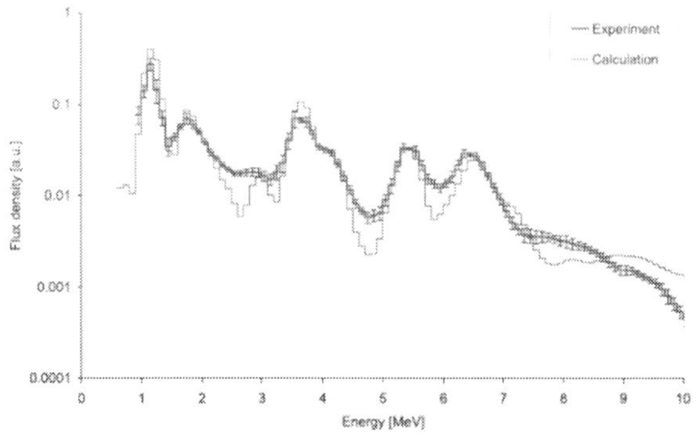

Fig. 6. Calculated and experimentally measured neutron flux on Si filtered neutron beam with stilbene detector [12]

Stilbene was used to measure the neutron leakage spectrum in the range of 1 MeV to 10 MeV in the group structure with energy steps of 100 keV. The total uncertainty of measurement in both experiments was estimated between 1.9% and 10.5%, depending on energy and position in the experiment.

F. Calculations

Reactor experiment

The neutron transport and neutron spectra simulations were carried out using the MCNP 6.2 [13] Monte Carlo code with ENDF/B-VII.1 [14] nuclear data library and corresponding TSL (Thermal Scattering Library). The whole reactor core was modeled in accurate dimensions and shapes, the only simplification being the neglect of a small air gap in each graphite block. The simplification was done by homogenizing the air cavity and graphite blocks by changing the density of the whole graphite module to ensure the same amount of graphite in each module and reactor core.

Neutron spectra were calculated in each seven axial positions in the activation foil holder. The calculation was performed at a fixed critical level of the moderator obtained experimentally, critical height was set to H_{cr} = 39.133 cm. The number of simulated neutrons was 40,000 neutrons per cycle in 585,000 active cycles with 50 inactive cycles. The statistical uncertainty of the calculated neutron spectrum, reached in the central position 3 of the holder, is below 2% in each energy group in the energy range from 1×10^{-8} MeV to 3 MeV. In higher and lower energy regions, the statistical uncertainty in each energy group slightly rises. Obtained neutron spectrum serves as a defined neutron spectrum for reaction rate calculation in activation materials described in the sections above.

Reaction rates of all activation foils were calculated separately with a fixed neutron spectrum source and the exact shape of the activation detector in the separate MCNP calculation. In each calculation step, 1×10^9 source neutrons were simulated with an energy spectrum corresponding to the precise position in the activation foil holder. The statistical uncertainty in the calculated reaction rate was between 0.3% and 0.7%, depending on the activation material and size. Finally, the correction factors and self-shielding factors were calculated for all positions and all materials.

Reaction rates of all foils used for neutron mapping in the activation foil holder and comparison with the experiment were manually calculated using the scalar multiplication of the calculated neutron spectrum in a defined position and microscopic cross-section obtained from ENDF/B-VIII.0 [15] nuclear data library. This technique determines the reaction rate per atom in each activation foil.

MCNP was also used to calculate a wide range of the neutron spectrum. The ENDF/B-VIII.0 nuclear data library was used with a corresponding TSL matrix with 30% graphite porosity, consistent with previous research on graphite calculations and experiments performed on the LR-0 reactor. The calculation was carried out with 200,000 neutrons per cycle in 20,000 active and 50 inactive cycles. The detection volume was the same size and position as a stilbene detector volume placed in the center of the height of the moderator level. The uncertainty of the calculated neutron spectrum is below 0.8% in energy ranges from 1×10^{-8} MeV to 3 MeV.

Graphite cube experiment.

As in the previous section, the MCNP Monte Carlo code with ENDF/B-VII.1 nuclear data library was used to calculate the neutron leakage spectrum from the graphite cube. In this case, Mannhart's [16] neutron spectrum for ^{252}Cf simulation was used. As graphite, the same composition as in the reactor experiment was used. The energy grid structure of the simulated detector was set to the same energy resolution as in the stilbene measurement.

III. Results

A. Reaction rate measurement in LR-0 graphite core

Regarding the methodology described above, the activity of the activation detectors was measured on the HPGe detector and then calculated based on Equations (1) and (2). Estimated reaction rates based on measurement were normalized using a scaling factor. The scaling factor is a calculated constant representing the neutron emission in the reactor core. For the scaling factor calculation, the average calculated values based on Ta and $_{1\%}$Au activation foils were used (axial positions 2, 3, and 4 in the activation foil holder). The Ta and $_{1\%}$Au were used only like a power monitor during irradiation, and the reaction rate of these foils is not a part of this work. After that, the scaling factor was multiplied by the self-shielding correction factor and divided by the calculated reaction rate from the MCNP calculation. The results of all mentioned calculations lead to the experimentally determined scaling factor to be 4.245×10^{11} with 0.86% statistical uncertainty. The thermal power of the reactor during irradiation can be estimated based on the number of neutrons per fission, 2.447 neutrons, with fission energy released of 180.9034 MeV. All these parameters were obtained from the MCNP output file. From these values, the irradiation power was calculated to be approximately 5 W.

Table I shows experimentally determined reaction rates of the selected activation foils. The statistical uncertainty of measurement has been quantified with a combination of measured geometry uncertainty and uncertainty of HPGe detector below 0.79 % for all observed activation detectors.

Table I
Experimentally determined reaction rates of activation foil per one atom in the exact axial position on the foil holder - central rod. The unit [1/s]

Pos.	Au	Cu	Fe	Mn
1	7.650E-27	1.872E-28	5.406E-29	5.311E-28
2	8.441E-27	2.040E-28	5.990E-29	5.733E-28
3	9.171E-27	2.047E-28	5.996E-29	6.122E-28
4	9.489E-27	2.083E-28	5.866E-29	6.290E-28
5	8.990E-27	2.005E-28	-	6.077E-28
6	8.896E-27	1.912E-28	5.167E-29	5.511E-28
7	7.877E-27	1.698E-28	4.664E-29	4.830E-28

For a better comparison of obtained experimentally determined results with MCNP calculation, the C/E-1 (calculation/experiment-1) ratio was estimated, see Table II.

Table II
Calculated C/E-1 for presented activation detectors

Pos.	Au	Cu	Fe	Mn
1	-0.01%	-9.64%	-8.88%	-5.22%
2	5.49%	-4.88%	-5.87%	-0.05%
3	5.24%	0.10%	-0.86%	-1.01%
4	1.31%	-2.68%	0.57%	-4.41%
5	0.85%	-6.07%	-	-7.90%
6	-10.58%	-13.98%	-7.45%	-11.55%
7	-17.03%	-21.08%	-16.61%	-18.04%

An excellent agreement between measurement and calculation can be observed in positions 3 and 4, position 5 is sometimes better no, which is caused by its position in the axial center of the activation holder. Position 5 is close to the spacing grid of fuel, which can slightly affect the measured results. In contrast, positions 1, 2, and 6, 7 are underestimated for almost all detectors. Position 1 and 2 are placed in the bottom part of the holder, which caused inhomogeneity in the neutron field caused by construction parts of a special experimental channel where the graphite prism is placed. Under the channel, the stainless-steel weight and more water below the fuel can be found, which can negatively affect the neutron flux by these materials, which seems like heavy reflector with shifted neutron spectrum. At the same time, other boundary phenomena, such as the end of the fuel column in this area, may play a non-negligible role in neutron spectra shaping.

The situation is slightly different on the other side of the foil holder. Position 6 is in the axial position, corresponding to the moderator-air interface. Position 7 is placed above the moderator water level but still in graphite blocks. Position 7 shows the highest inconsistency of all detectors examined. As the difference between the moderator level and the position of the activation foil increases, the underestimation rises. This phenomenon was previously observed and described in different type of experiment with a partially flooded reactor core, see [17] .

The central positions are the most suitable for future cross-section testing or for another irradiation experiment, which means axial positions 3, 4, and 5. In these positions, the neutron flux is the most homogeneous. The precise positioning will depend on the material properties, size, and the requirement for the shape of the neutron spectrum.

B. Fast neutron spectra measurement in LR-0 graphite core

In this step, the main goal was to evaluate the neutron spectrum in the fast energy region. Activation detectors presented in this work are sensitive only in thermal and epithermal energy regions, so the stilbene was chosen for fast neutron mapping.

Mapping the fast neutron flux using activation detectors case appears to be very complicated because a minimal amount of the fast neutrons reaches this central well-moderated position. The second issue is that the threshold reactions commonly used for these purposes have a small cross-section, which complicates the measurement.

The preliminary assumption about the well-thermalized neutron spectrum has been verified based on activation detector measurement and MCNP calculation. For better imagination of the graphite shaping influence, the three different reactor core neutron spectrum calculation by MCNP is shown in Fig. 7.

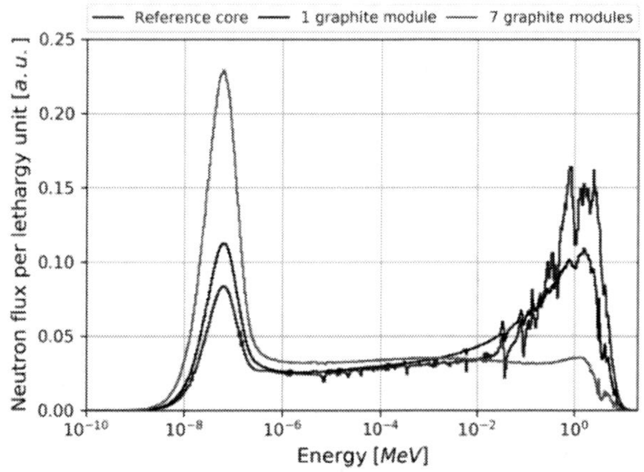

Fig. 7. Calculated neutron spectrum for reference core (blue curve), one graphite module in the centre of the reference core (black curve) and proposed seven graphite reactor core (red curve)

In contrast to the reference LR-0 core [18] , the red curve shows significant thermalization occurring in the proposed reactor core with a thermal neutron ratio of almost 25% and a fast neutron ratio of less than 4% above 1 MeV. It can be noticed that a small amount of graphite (case with only one block in the center of the reference core) slightly increases the number of thermal neutrons in the detector position but hardly decrease the number of fast neutrons, see the comparison of blue and black curve. The epithermal region in large graphite case is nearly constant, which can be helpful for cross-section evaluation.

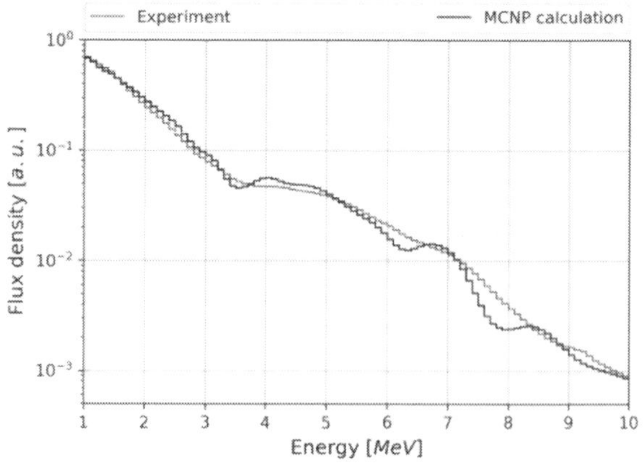

Fig. 8. Comparison of the experimentally measured fast neutron spectrum by stilbene detector and spectrum calculated by MCNP code – LR-0 reactor experiment

The last part of the modeled neutron spectrum from Fig. 7 was compared with the real measurement provided by the stilbene detector. The measurement was carried out from 1 MeV to 10 MeV. For this purpose, the calculated and measured spectrum was normalized to 1 in the range from 4 MeV to 10 MeV. This normalization is advantageous due to ensure that the gamma counts do not contribute to neutron spectra. The calculation was finally modified by the Gaussian broadening function based on experimentally determined parameters for stilbene detector. The comparison between calculation and measurement can be seen in Fig. 8.

Due to the resolution and measured energy grid structure of the stilbene crystal, the sharp resonances cannot be recognized. The agreement between calculation and measurement is very well in the range from 1 MeV to 5.8 MeV. With increasing energy of the neutrons, the deviation starts rising except for 7 MeV. Maximal disagreement can be observed around the 7.8 MeV.

C. Fast neutron leakage spectra measurement with graphite cube and ^{252}Cf neutron source

The fast neutron leakage spectrum was measured with a ^{252}Cf neutron source. Using the californium neutron source instead of ^{235}U and comparing them to each other is legit because the ^{252}Cf neutron source is often mentioned as the neutron PFNS standard [3] . The differences between uranium and californium are very small, so they can be compared without any complications.

The results from the measurement on the graphite cube compared to the calculation in MCNP code can be found in Fig. 9. The comparison of measurement and MCNP calculation provides the same trend in neutron spectra shape [19] . Up to 6.5 MeV, the agreement between measurement and experiment is perfect. Only one problematic region, slightly similar behavior as in Fig. 8, was observed between 3.1 MeV and 4 MeV. In this case, probably the resolution of the stilbene detector plays the main role. In graphite cross-section, a very sharp resonance exists in this area, which can cause the deformation of a few surrounding energy bins. In higher energies up to 6.5 MeV, the discrepancy between measurement and calculation rises. In energies up to 9 MeV, the agreement is again perfect.

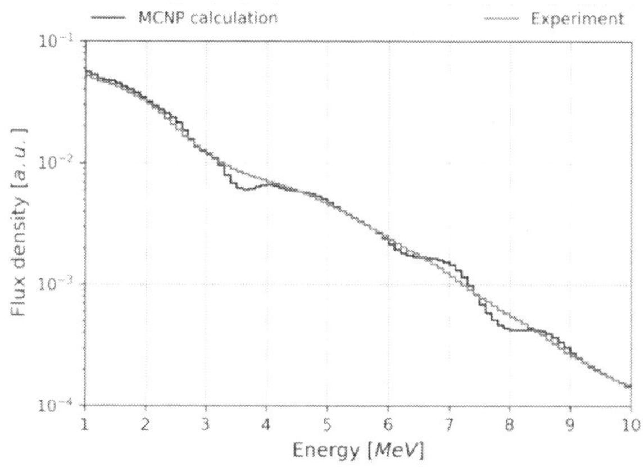

Fig. 9. Comparison of the experimentally measured fast neutron spectrum by stilbene detector and spectrum calculated by MCNP code – graphite cube with ^{252}Cf neutron source

For better imagination and better comparison between both experiments and the calculated results, the C/E-1 (calculation/experiment-1) chart shows the behavior of this phenomena in one chart, see Fig. 10.

It can be noticed that even if the experimental setup of the graphite cube is quite different compared to the LR-0 core, the same trends in the behavior of the fast neutrons neutron spectra shape can be observed in Fig. 8 and Fig. 9.

In the LR-0 experiment setup, the neutrons go from outside to the middle of the graphite geometry. In the case of the graphite cube, the neutrons transport oppositely from the source in the center to the outside.

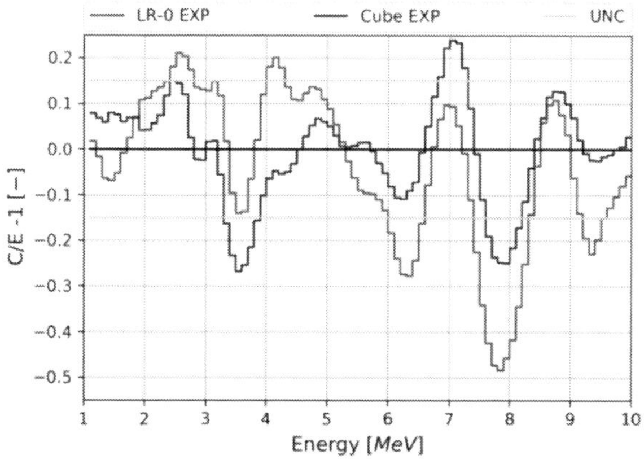

Fig. 10. Comparison of both experiments in C/E-1 methodology

In Fig. 10, the same behavior explained in the upper section is more evident. The yellow lines serve to imagine the maximal possible uncertainty caused by the methodology of the experiment and stilbene measurement. One can notice that in both cases, the most problematic energetic region is from 6 MeV to 8.5 MeV. This effect is probably multiplied by the amount of filtering graphite. With the increasing amount of graphite in the experiment, the energies higher than 6 MeV are in much worse agreement than the cube and the LR-0 experiment. The amount of graphite and the different behavior of neutron flux was previously observed and investigated in [20] [21] [22] .

More significant disagreement in both experiments is located around 7.8 MeV.

IV. CONCLUSIONS

In this work, a comparison of two independent experiments with rector graphite was performed. The first experiment in the LR-0 core provides very good results in thermal and epithermal energies. Another benefit is the fact that graphite filter effectively decreases the number of fast neutrons to a very low share. The epithermal region is almost flat and approx. 25% of all neutrons are in thermal energy. This core configuration can be used in the future for various types of experiments regarding validation of the cross-section in thermal and epithermal energies with a cadmium filter, for example.

On the other hand, the fast neutron spectrum filtered by graphite shows some deficiencies in the description of graphite nuclear data. The biggest discrepancy between measurement and calculation was observed in the energy region from 6 MeV to 9.5 MeV. This discrepancy may be caused by inaccuracy in the graphite cross-section around the 6-8 MeV energy region, probably a combination of the elastic and inelastic cross-section.

Finally, these results lead to the development of new reactor core arrangements with a special arrangement of graphite block to confirm or refute the finding mentioned higher in this work.

ACKNOWLEDGEMENT

The presented results were obtained using the CICRR infrastructure, which is financially supported by the Ministry of Education, Youth and Sports – project number LM2023041, supported by Energy Storage TK02030069 project and special project SGS-2021-018.

REFERENCES

[1] Brown, D.A. 'ENDF/B-VIII.0: The 8th major release of the ENDF/B library containing Cielo-project evaluations, new standards and thermal scattering data', 20th Topical Meeting of the Radiation Protection and Shielding Division, RPSD 2018, 148, pp. 1–142

[2] Bolewski, A. et al. 'A practical method for measuring the boron equivalent of graphite impurity', Nuclear Instruments and Methods in Physics Research, Section B: Beam Interactions with Materials and Atoms, 237 (3–4), 2005 pp. 602–612. Available at: https://doi.org/10.1016/j.nimb.2005.03.002

[3] Schulc M. et al. 'Application of ^{252}Cf neutron source for precise nuclear data experiments', Applied Radiation and Isotopes, 2019,Vol 151, 187-195, Available at: https://doi.org/10.1016/j.apradiso.2019.06.012

[4] Košťál, M. et al. 'Measurement of various monitors reaction rate in a special core at LR-0 reactor', Annals of Nuclear Energy, 112, 2018, pp. 759–768. Available at: https://doi.org/10.1016/j.anucene.2017.10.036

[5] Dryak, P. and Kovar, P. 'Experimental and MC determination of HPGe detector efficiency in the 40-2754 keV energy range for measuring point source geometry with the source-to-detector distance of 25 cm', Applied Radiation and Isotopes, 64(10–11), 2006, pp. 1346–1349. Available at: https://doi.org/10.1016/j.apradiso.2006.02.083.

[6] Boson, J., Ågren, G. and Johansson, L. 'A detailed investigation of HPGe detector response for improved Monte Carlo efficiency calculations', Nuclear Instruments and Methods in Physics Research, Section A: Accelerators, Spectrometers, Detectors, and Associated Equipment, 587 (2–3), 2008, pp. 304–314. Available at: https://doi.org/10.1016/j.nima.2008.01.062

[7] Tomarchio, E. and Rizzo, S. 'Coincidence-summing correction equations in gamma-ray spectrometry with p-type HPGedetectors', Radiation Physics and Chemistry, 80(3), 2011, pp. 318–323. Available at: https://doi.org/10.1016/j.radphyschem.2010.09.014.
[8] Veskrna et al. 'Digitalized two parametric system for gamma/neutron spectrometry' 18th Topical Meeting of the Radiation Protection & Shielding Division of ANS, Knoxville, TN USA, 2014.
[9] Bureš, Z. et al. 'Multiparameter Multichannel Analyser System for Characterisation of Mixed Neutron – Gamma Field in the Experimental Reactor Lr-O', in Reactor Dosimetry in the 21st Century: Proceedings of the 11th International Symposium on Reactor Dosimetry Brussels, 2003, pp. 194–201. Available at: https://doi.org/10.1142/9789812705563_0025
[10] Cvachovec, J. et al. 'Neutron response function for BC-523A scintillation detector in the energy range 0.5 MeV to 20 MeV', Journal of ASTM International, 5(5), 2008. Available at: https://doi.org/10.1520/JAI101149.
[11] Veškrna, M. et al. 'Digitalized two parametric system for gamma/neutron spectrometry', in 18th Topical Meeting of the Radiation Protection and Shielding Division of ANS, RPSD, 2014. Knoxville, TN USA, pp. 216–219.
[12] Košťál M. et al. 'Measurement of neutron spectra in a silicon filtered neutron beam using stilbene detector at the LVR-15 research reactor', Applied Radiation and Isotopes 128, 2017, 41-48, Available at: https://doi.org/10.1016/j.apradiso.2017.06.026
[13] Werner, C.J. 'MCNP6TM User's Manual - Code Version 6.2'. Los Alamos: Los Alamos National Laboratory, 2017.
[14] Chadwick, M.B. et al. 'ENDF/B-VII.1 nuclear data for science and technology: Cross sections, covariances, fission product yields and decay data', Nuclear Data Sheets, 112(12), 2011, pp. 2887–2996. Available at: https://doi.org/10.1016/j.nds.2011.11.002
[15] Brown, D.A. 'ENDF/B-VIII.0: The 8th major release of the ENDF/B library containing Cielo-project evaluations, new standards and thermal scattering data', 20th Topical Meeting of the Radiation Protection and Shielding Division, RPSD, 2018, 148, pp. 1–142.
[16] Capote, R. et al. 'Updating and Extending the IRDF-2002 Dosimetry Library' Journal of ASTM International, Vol. 9, No. 4, 2012, pp. 1-9, https://doi.org/10.1520/JAI104119. ISSN 1546-962X
[17] Košťál, M., Švadlenková, M., et al. 'Determining the axial power profile of partly flooded fuel in a compact core assembled in reactor LR-0', Annals of Nuclear Energy, 90, 2016, pp. 450–458. Available at: https://doi.org/10.1016/j.anucene.2015.12.028.
[18] Košťál, M. et al. 'A reference neutron field for measurement of spectrum averaged cross sections', Annals of Nuclear Energy, 2020, 140. Available at: https://doi.org/10.1016/j.anucene.2019.107119.
[19] Peltan, T. et al. 'Validation of graphite cross section in various integral experiments', in International Conference on Nuclear Engineering, Proceedings, ICONE, 2019. Available at: https://doi.org/10.1299/jsmeicone.2019.27.2083.
[20] Czakoj, T., Losa, E. and Košťál, M. 'Design of a special core for large graphite insertion studies in the LR-0 reactor', in PHYTRA4 – The Fourth International Conference on Physics and Technology of Reactors and Applications. Marrakesh, Morocco, 2018, pp. 238–249
[21] Vilímová E., et al. 'Position evaluation of ex-core neutron flux measurement in new type graphite reactors', The European Physical Journal Conferences, 2021, vol 253, Available at: ttps://doi.org/10.1051/epjconf/202125305005
[22] Košťál, M., Rypar, V., et al. 'Study of graphite reactivity worth on well-defined cores assembled on LR-0 reactor', Annals of Nuclear Energy, 87, 2016, pp. 601–611. Available at: https://doi.org/10.1016/j.anucene.2015.10.010.

 https://doi.org/10.1051/epjconf/202328804025

Validation of Monte Carlo simulations by experimental measurements of neutron-induced activation in cyclotrons

Jonathan COLLIN[1,*], Jean-Michel HORODYNSKI[2], Nicolas ARBOR[1], Massimo BARBAGALLO[3], Federico CARMINATI[3], Giuliana GALLI CARMINATI[3], Luca J TAGLIAPIETRA[3] and Abdel-Mjid NOURREDDINE[1]

[1]Institut Pluridisciplinaire Hubert CURIEN (IPHC), CNRS, Université de Strasbourg, FRANCE
[2]ingénierie Radioprotection Sûreté Démantèlement (iRSD), CNRS, Université Paris-Saclay, FRANCE
[3]Transmutex SA, Suisse
(*) jonathan.collin@iphc.cnrs.fr

Abstract— **Nuclear activation is the process of production of radionuclides by irradiation. This phenomenon concerns particle accelerators used in various fields, from medical applications to industrial ones, both during operation and at the decommissioning phase. For more than three decades, the possibility of using cyclotrons for nuclear power generation and nuclear waste reduction has also been discussed, i.e. in the case of Accelerator-Driven Systems [1]. The radioprotection and dismantling issues of accelerator facilities, that have been raised recently, is even more potent for such installations.**

In our study, we are particularly interested in the activation due to secondary neutrons produced by (x,n) reactions, mostly (p,n) occurring in the accelerator's components. This work focuses on the study of the radioactivity induced in various materials (V, Sc, Tb, W, Ta) irradiated by fast and thermal neutrons, in two different scenarios: through direct irradiation -with an AmBe source- and around an operating cyclotron at the CYRCé facility (Strasbourg). A broad Monte Carlo study including FLUKA, GEANT4, PHITS and MCNP simulation has been performed, with and without a FISPACT-II coupling, to estimate the reaction rates and to trace the induced radioactivity in samples of known composition. The results of the simulations are compared with the values extracted in two dedicated experimental campaigns in which activated samples underwent high resolution gamma-ray spectrometry.

Keywords —**Neutron activation, decommissioning, ADS, Fluka, PHITS, Geant4, FISPACT-II, MCNP6, HP-Ge spectrometry.**

I. Introduction

THE use of particle accelerators has expanded beyond the scientific field to include applications in the medical and industrial sectors, such as hadrontherapy and sterilization. The IAEA accounts for more than 577 accelerator facilities in the world [2], a number progressively increasing, in particular for particle therapy facilities, as seen on Figure 1.The dismantlement plans of such facilities were put on agenda at the beginning of XXIst century [3], but no related standards were established accordingly. Since then, the IAEA has been calling for stricter controls of the life-cycles of such facilities, with the idea to converge towards the standard of nuclear installations. The IAEA recommends the establishment of dismantling plans prior to any new commissioning of particle accelerator facilities. Such plans, that must be revised regularly (and at any upgrade or modifications to the facility), consists in a funding plan, an inventory of techniques and procedures for the decommissioning. The inventory of expected (and estimated) radionuclides produced during the life-cycle of facilities is of the utmost importance [4].

The relevance of such recommendations is more pertinent for the Accelerator Driven Systems (ADS) facilities, as they are both particle accelerator facilities and nuclear facilities. Indeed, ADS couples a nuclear reactor to a high-energy proton accelerator, allowing to produce energy as well as to transmute nuclear waste, in a safer subcritical regime [5]. Several ADS are currently being designed or constructed, such as the Transmutation Experimental Facility (TEF) (proton beam of 100 MeV -max 400 MeV/250 kW) [6], the Multipurpose Hybrid Research Reactor for High-tech Applications (MYRRHA) (600 MeV with 4 mA) [7], the Chinese Initiative Accelerator Driven Systems (CiADS) (250 MeV with 10 mA -max 1 GeV) [8] and the Subcritical Transmuting Accelerated Reactor Technology (START) from Transmutex (800 MeV with 5mA) [9].

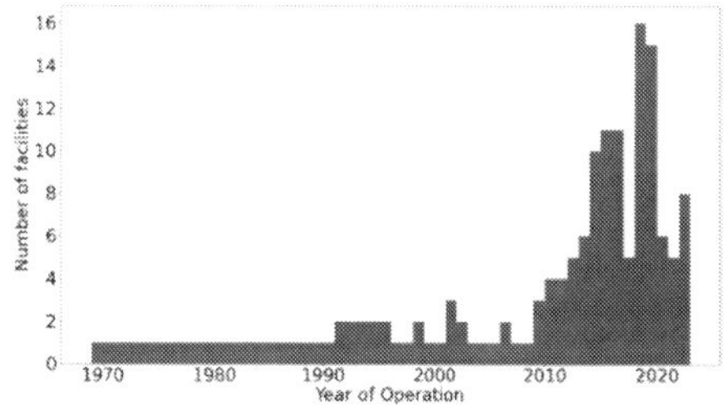

Figure 1 Particle therapy facilities in clinical operator since 1970 recorder by PTCOG *[10]*

The components of such systems will be activated by direct or indirect irradiation (secondary particles). Depending on the half-life of the radionuclides produced, an inventory is required for the radiation protection and the future dismantling plans. Hence the need for a quantitative radiological inventory, for which thermalized secondary neutrons are the main responsible for activation of components, primarily via (p,n) reactions [4], but (γ,n) reactions [11] and to some extent (α,n) reactions contribute as well to the inventory, in particular in the case of high energy cyclotrons, such the START system

Monte Carlo (MC) simulations of radiation-matter interactions are a powerful tool to estimate the induced activities and dose rates, as they can manage a great range of scenarios (fluence, irradiation/cooling time, …).

The objective of this work is to build a tool capable of accurately evaluate the radionuclides inventory for the START project. An exploration of which MC codes would be the most suitable for such application is necessary. This work is a first step to make an informed choice. We focused our development to activation studies for beam energy below 20 MeV, using the proton cyclotron CYRCé (Cyclotron for Research and Education) [12], available at the IPHC laboratory. This value corresponds, indeed, to physics models threshold in main Monte Carlo codes. The goal is to validate the methodology used here and apply it again to explore higher energy, where models may have greater discrepancies.

Neutron activation and (p,n) reactions will be discussed in this paper through the comparison of different MC codes, with an association to an analytical code (for activation computation). In the first section, the details on the codes used and the methodology adopted are reported, then the experimental setups (direct/indirect) and the codes used are presented. Finally, the simulation and the experimental results are compared.

II. Methodology

We will compare the four main MC tools for neutron capture and activation computation i.e., Fluka [13], [14], Geant4 [15], PHITS [16] and MCNP6 [17]. If most of them can do a radionuclides inventory natively, the access to an activity value is limited. Hence, the analytical code FISPACT-II [18] will be systematically used in association to the neutron spectra obtained via the mentioned MC tools. The comparison will be validated by experimental measurements, using high-resolution gamma spectrometry.

In 2020, the IAEA [4] suggested a list of materials, that are the most activated components in the accelerators environment, among them one can find iron, zinc, copper, aluminium, tungsten, steel, graphite, concrete, plastics, resins.

Using this list as a starting point, we selected V, Sc, Tb, Ta, W and Au samples to expose them to the two different types of neutron fields available. These samples were chosen according to the following criteria:

- Preferably only one stable nucleus.
- Materials available in foils and at high purity.
- Activation products half-life are compatible with gamma spectrometry and measures of radioprotection (access to irradiation zone, …)

Spectrometry using High Purity Germanium (HPGe) semiconductor detectors is one of the most widely used qualitative and quantitative multi-elemental analysis techniques in the field of radiation measurements. This technique allows the assessment of all γ-emitting radionuclides in a single analysis. The physical phenomena of radiation-matter interaction, for a given geometry, are integrated into a transfer function that provides the precise estimate of the sample's activity [19].

In this work, the measurements were performed with a HPGe LABSOCS from Canberra. The sensitivity of the setup used spans from 40 keV to 2650 keV with a relative efficiency of about 40%, while the resolution is 1.9 keV at 1.33 MeV. The system had been previously characterized using an ^{152}Eu source.

The activities were computed using the software Genie2000®, considering the decay scheme compiled by the Laboratoire National Henri Becquerel (LNHB) [20]. An example of measured spectra is provided Figure 3.

III. Experiments and Models

Two types of experiments have been performed, one with direct irradiation of neutron emitted by an AmBe source and another using the secondary neutron field produced by the interaction of the proton beam accelerated by the cyclotron and impinging on suited targets.

A. Direct neutron field

Two samples were irradiated with the AmBe source: Au and V. The isotropic ^{241}Am-Be neutron source used has an activity of 3.7×10^{10} Bq and a nominal neutron emission rate of 2.24×10^6 n/s [21]. The AmO_2-Be powder is double-encapsulated within welded stainless-steel cylinders, which are inside of a Pb cylinder. The source was placed inside a paraffine (CH_2) barrel of 75 cm diameter and height, with 5 extruded cylinders, allowing to guide safely the source in the middle of the barrel, see Figure 2.

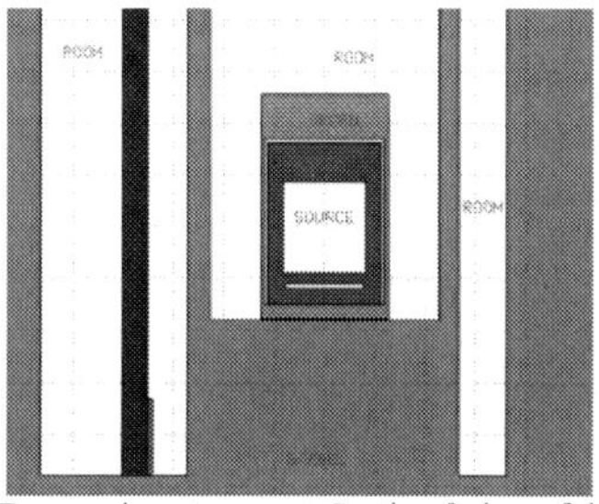

Figure 2. Am-Be experiment concept. Sectional view of the center of the paraffine barrel (teal) with PVC cane (dark blue) holding the sample (violet) close to the source (grey: lead shell, red: stainless steel)

The V (1.5x3x0.1cm^2) and Au (Ø1.2x0.1cm^2) samples were placed for respectively 32 and 6027 minutes inside the barrel and gamma spectrometry analysis were performed. One of the gamma spectra acquired is shown in Figure 3.

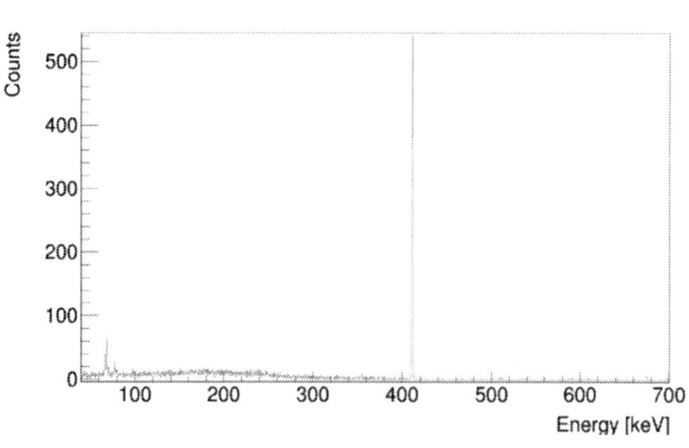

Figure 3. Gamma spectrometry of the Au sample irradiated using the AmBe neutron field measured by HP-Ge

B. Secondary neutron field

CYRCé is a TR24 from Advanced Cyclotron System Inc (ACSI), built in 2010 and installed in 2012. The experiment was conducted inside the cyclotron room, in the radioisotope line, which allows to produce ^{18}F, ^{89}Y and ^{64}Cu, with respectively a liquid target of enriched water (H_2O_{18}) and two solid targets (deposit of ^{89}Zr or ^{64}Ni). The targets are controlled by a selector that can align the devices with the beam. In this paper, only the secondary field from ^{18}F was studied.

Two runs were performed, using the 1 mL liquid target of enriched water (with 16.5 MeV primary protons), see

Table I and Table II for the detailed characteristics of the irradiations.

The standard dimensions of the samples, or activation detectors, are of 2.5x2.5 cm^2 and of thickness respectively 2 mm for Ta, 1 mm for Tb, 0.55 mm for Sc and 1 mm for Au. They were placed where high neutron flux was expected: inside the biological shield (where the target is located), on the biological shielding walls, next to the cyclotron yoke and on the wall in the periphery of CYRCé. The positions are shown in Figure 4.

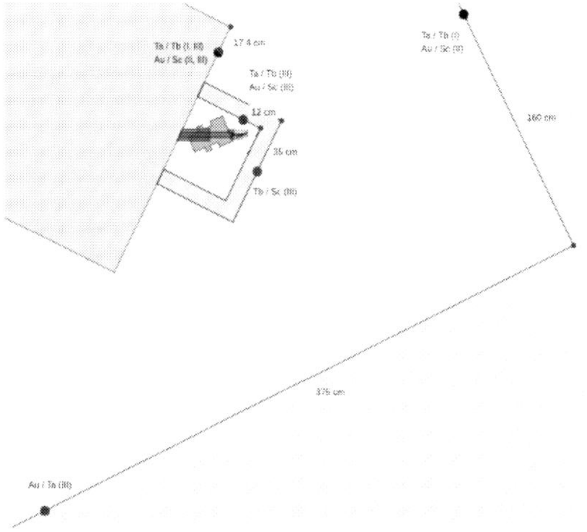

Figure 4. Historic of the locations for irradiated samples for the different runs (black dots for Run#1, red for the second one)

TABLE I
CHARACTERISTICS OF RUN#1

Primary beam	
Particle	Proton
Energy on target	16.5 MeV
Irradiation profile	
Irradiation [min]	Current [μA]
13	32
2880	0
23	34
1440	0
13	32
8640	0
14	30

TABLE II
CHARACTERISTICS OF RUN#2

Primary beam	
Particle	Proton
Energy on target	16.5 MeV
Irradiation profile	
Time length [min]	Current [μA]
17	35
1486	0
17	35
1354	0
19	35
8622	0
15	35

C. Model assumptions

1) Physics choices

To conduct this study, the options used for the MC tools will be detailed in the following:

- Fluka 2021.2.9

Neutrons are managed group-wise below 20 MeV (with 260 groups). ENDF/B-VIII.0 [22] cross-sections is used for all materials.

- FlukaCERN 4-3.2

In this case, the neutron point-wise cross-sections JEFF-3.3 library [23] is used for all materials.

- Geant4 11.0.3

The high precision models and cross-sections for neutron below 20 MeV is used (for elastic and inelastic scattering, capture and fission) through the physics lists ShieldingLEND [24] and QGSP_BIC_HP [25].

- PHITS 3.32

The photon transport EGS is activated. The library JENDL-4.0 [26] is used for neutron cross-sections.

- MCNP6.2

The neutron, proton and photon mode are activated in this simulation. The library ENDF/B-VII.1[27] were privileged. For the first irradiation run, TENDL-21 [28] neutron cross-section library was mainly used, excepted for ^{nat}C (ENDF/B-VII.1), ^{nat}Mg, ^{nat}S (ENDF/B-VI.8) and ^{nat}Si (ENDF/B-VI). TENDL-19 proton cross-section library is used for all materials instead of physics models excepted for H, ^{nat}C, ^{nat}Mg, ^{nat}S and ^{nat}Si.

- FISPACT-II

Is used in a two-step process, using the average neutron fluxes in the volume of interest. All computations were done with TENDL-19 library.

2) Geometrical choices

In the case of the direct irradiation, only the barrel and the surrounding air were modelled in the simulations. And for the AmBe, the ISO8529-1:2021 (small source) spectrum was assumed [29].

As far as the indirect irradiation is concerned, the concrete (Portland) walls, ground and floor close to the cyclotron were implemented, the cyclotron itself was simplified as a simple parallelepiped in stainless steel while a finer model was built for the target system components (target selector, target support, ...)

IV. DATA AND ANALYSIS

The value of interest for these simulations were the incoming neutron spectra on the sample, as well as the production yield of nuclides inside the samples. This last value has been compared to experimental values acquired via gamma spectrometry and computed through FISPACT.

A. AmBe experiment

The neutrons produced by the Am-Be source are thermalised inside the paraffine barrel, maximizing the neutron capture rate in the vanadium and gold samples. The energy differential neutron flux calculated by various codes is reported in Figure 5. We found an agreement of all simulation for the energy spectrum of the neutron arriving on the sample, but for Geant4-ShieldingLEND which seems to have a different thermalisation behaviour. At thermal energy, the LEND (Low Energy Nuclear Data) model deviates of 43% from the MCNP value, when Fluka, PHITS and the High Precision (HP) from Geant4 stays below 8% of difference. For the fast neutron component, the standard deviation is within 5% for all models.

The four codes allow scoring the yield of the radioactive nuclei of interest. The simulation results are synthetised in the Table III. Fluka, PHITS and MCNP well agree within 1% (3% for Au). As for Geant4, while QGSP_BIC_HP stays within 3.9% (and 1% for Au), ShieldingLEND gives the most discrepancies with as much as 14% of difference for gold. We note that when using the physics list QGSP_BIC_HP the computation's speed is a factor 10 slower.

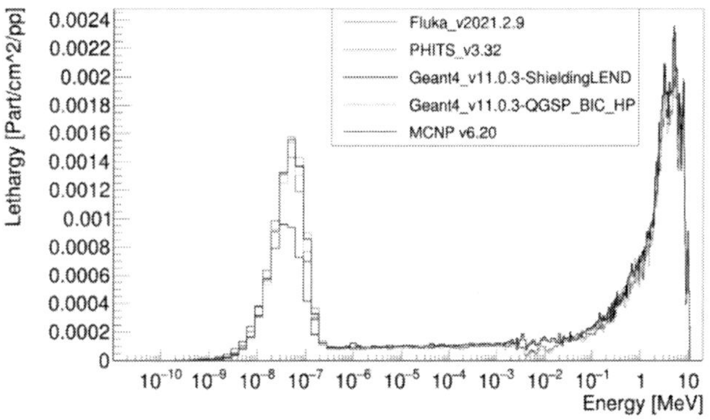

Figure 5. Track-length estimation of the neutron fluence inside the vanadium sample irradiated by the direct neutron field

TABLE III
COMPARISON OF EXPECTED YIELD IN THE IRRADIATED SAMPLES

Vanadium	
Code	Production rate [$\times 10^{-4}$ ^{52}V/pp]
Fluka	4.08 $\pm$ 0.002
Geant4 (ShieldingLEND)	3.78 $\pm$ 0.036
Geant4 (QGSP_BIC_HP)	4.24 $\pm$ 0.146
PHITS	4.05 $\pm$ 0.020

Gold	
Code	Production rate [$\times 10^{-3}$ ^{198}Au/pp]
Fluka	0.99 $\pm$ 0.002
Geant4 (ShieldingLEND)	0.87 $\pm$ 0.002
Geant4 (QGSP_BIC_HP)	1.02 $\pm$ 0.002
PHITS	1.02 $\pm$ 0.003

The same agreement has been found when considering FISPACT-II, as ashown in Figure 6. Native Fluka and Fluka coupled with FISPACT-II give same results within the given statistical uncertainties as well.

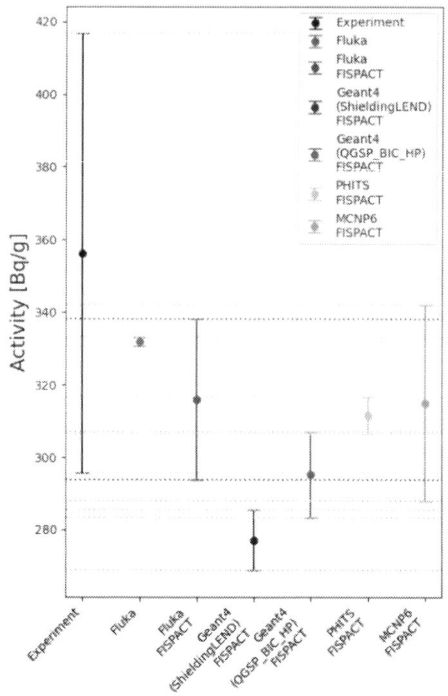

Figure 6. Compilation of measured vs estimated activities for the direct field experiment with the vanadium sample

Experimental data points were taken with V and Au, the samples were respectively activated at 357 $\pm$ 61 Bq/g and 318 $\pm$ 21 Bq/g. Considering uncertainties, the simulations are validated by the experimental data points.

B. CYRCé experiment

1) 1 mL target

Figure 7 shows the comparison between the activity value extracted using FLUKA-CERN one-step process and using FISPACT-II with fluence estimated with FLUKA-CERN (USRTRACK) and MCNP6.2 (f4). The measured values are also reported therein.

MCNP6/FISPACT-II results are in very good agreement with the experimental data. In most cases, FLUKA results are higher than experimental data, but large discrepancies seem to occur especially for Ta activation foils due to high uncertainties in the thermal neutron fluence range (Figure 8). Variance reductions techniques used with MCNP6 improve the statistics of the neutron fluence scored inside the activation foils (DXTRAN). Then, FLUKA-CERN model need to be improved to reduce statistical uncertainties without increasing the number of primary particles. Statistical uncertainties for neutron fluence estimated with FLUKA-CERN are larger than MCNP6 ones (Figure 8) although a higher number of primary particles than MCNP6 (7,4e7 for FLUKA-CERN, 4e7 for MCNP6) was used. Models and variance reduction set up in FLUKA-CERN model need to be improved to reduce statistical uncertainties without increasing the number of primary particles.

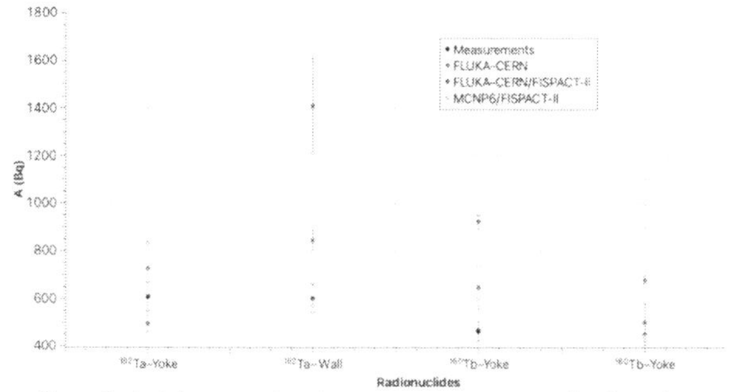

Figure 7. Activity comparison between measurements and estimated activities (indirect irradiation) using FISPACT-II, MCNP6 and FlukaCERN

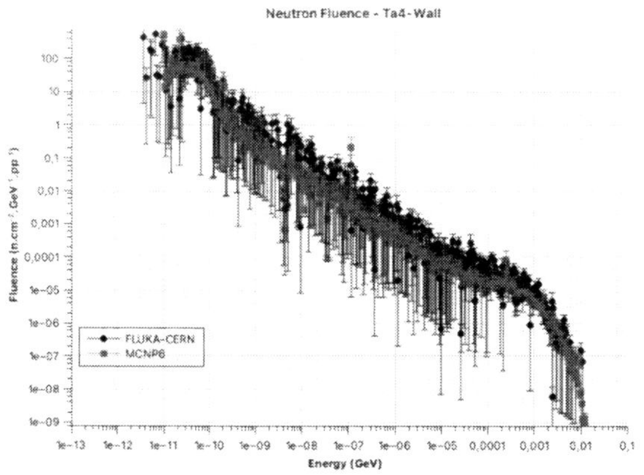

Figure 8. Neutron fluence for Ta foils located on the wall of the CYRCé bunker estimated with FLUKA-CERN and MCNP6

V. Conclusions and Perspectives

The objective of this work was to compare several MC codes, including a coupling with FISPACT, for two experimental scenarios: neutrons from AmBe source and secondary neutrons generated in a low energy cyclotron environment irradiated a set of samples and induced activations have been compared.

For the AmBe experiment, there is a good agreement between all the codes, and experimental values as well. While the Fluka, MCNP and PHITS agree within less than 3%, Geant4 and its ShieldingLEND physics list strays to 15%. This difference could be explained by a difference in the thermalisation process of neutron (Neutron managed with geant4 physics lists, not marked with '_HP' i.e., not with high precision models are not able to thermalise).

As for the irradiations at CYRCé cyclotron, MCNP6 code gives the best results, also thanks to the fact that the statistical uncertainties are reduced using efficient variance reduction techniques, especially for small regions like activation foils used in this study. FLUKA-CERN model needs to be improved to reduce statistical uncertainties for the neutron fluence. As for the experimental values, we plan increasing the concerned statistics as it is difficult to draw conclusions with the current uncertainties.

If one can compare codes through the yield of nuclei per primary atoms, to confront to experimental value with ease, FISPACT-II remains essential. As well as for the intercomparison of codes, this analytical allows to compare radionuclides inventories with the neutron spectra as the only input, restraining the number of influences.

To further the study, the irradiation campaign is ongoing, and it features other kinds of target (liquid and solid) and new locations for activation foils. Comparisons will be made with Solid-State nuclear track detectors. Geant4 and PHITS code will be included in the panel of MC code in the cyclotron studies.

Transmutex is also developing its own MC tool, called TMX-MC, based on Geant4 to simulate its nuclear and particle accelerator installation, as well as the dismantling of the last one. This code will join the panel of MC tools compared, for radiological inventories and radiation protection studies.

To extend this work, we are planning a systematic intercomparison of codes, in particular for neutron capture in the full range from thermal neutrons to ultrafast. A complete survey of (x,n) reactions of interest in proton accelerators up to GeV will also be carried out to address the problem of ADS decommissioning.

Acknowledgment

The authors would like to thank the CYRCé technical team and in particular M. PELLICIOLI, as well as T. FOEHRENBACHER for the spectrometry measurements.

References

[1] C. Rubbia *et al.*, "Conceptual design of a fast neutron operated high power energy amplifier", *CERN*, 1995.

[2] "Cyclotron Master List - cyclotrons_view_2020", Jun. 08, 2023. https://nucleus.iaea.org/sites/accelerators/lists/cyclotron%20master%20list/cyclotrons_view_2020.aspx#InplviewHashb4c61c57-92e3-4919-885c-7bc3bba78de7=Paged%3DTRUE-p_ID%3D271-PageFirstRow%3D151 (accessed Jun. 08, 2023).

[3] Université libre de Bruxelles, Ed., *Evaluation of the radiological and economic consequences of decommissioning particle accelerators: final report*. in EUR Nuclear safety and the environment, no. 19151. Luxembourg: Off. for Off. Publ. of the European Communities, 1999.

[4] I. A. E. Agency, "Decommissioning of Particle Accelerators", International Atomic Energy Agency, Text, 2020. Accessed: Jun. 02, 2023. [Online]. Available: https://www.iaea.org/publications/12371/decommissioning-of-particle-accelerators

[5] H. Nifenecker, S. David, J. M. Loiseaux, and O. Meplan, "Basics of accelerator driven subcritical reactors", *Nucl. Instrum. Methods Phys. Res. Sect. Accel. Spectrometers Detect. Assoc. Equip.*, vol. 463, no. 3, pp. 428–467, May 2001, doi: 10.1016/S0168-9002(01)00160-7.

[6] T. Sasa, "Design of J-PARC Transmutation Experimental Facility", *Prog. Nucl. Energy*, vol. 82, pp. 64–68, Jul. 2015, doi: 10.1016/j.pnucene.2014.07.031.

[7] "MYRRHA: Preliminary front-end engineering design - ScienceDirect". https://www-sciencedirect-com.scd-rproxy.u-strasbg.fr/science/article/pii/S036031991500717X (accessed Aug. 01, 2023).

[8] Z. Wang *et al.*, "Materials for Components in Accelerator-driven Subcritical System", *Strateg. Study Chin. Acad. Eng.*, vol. 21, no. 1, Art. no. 1, doi: 10.15302/J-SSCAE-2019.01.006.

[9] "TECHNOLOGY | Transmutex_technology", *Transmutex*. https://www.transmutex.com/technology (accessed Aug. 01, 2023).

[10] "PTCOG - Facilities in Operation". https://ptcog.site/index.php/facilities-in-operation-public (accessed Aug. 01, 2023).

[11] S. Hu, K. Fan, L. Zhang, Z. Mei, X. Li, and Z. Zeng, "Beam loss

at HUST", *Nucl. Instrum. Methods Phys. Res. Sect. Accel. Spectrometers Detect. Assoc. Equip.*, vol. 911, pp. 87–93, Dec. 2018, doi: 10.1016/j.nima.2018.09.126.
[12] E. Bouquerel *et al.*, "Design and commissioning of the first two CYRCé extension beamlines", *Nucl. Instrum. Methods Phys. Res. Sect. Accel. Spectrometers Detect. Assoc. Equip.*, vol. 1024, p. 166034, Feb. 2022, doi: 10.1016/j.nima.2021.166034.
[13] A. Ferrari, P. R. Sala, A. Fasso, and J. Ranft, "FLUKA: A Multi-Particle Transport Code", SLAC-R-773, 877507, Dec. 2005. doi: 10.2172/877507.
[14] T. T. Böhlen *et al.*, "The FLUKA Code: Developments and Challenges for High Energy and Medical Applications", *Nucl. Data Sheets*, vol. 120, pp. 211–214, Jun. 2014, doi: 10.1016/j.nds.2014.07.049.
[15] J. Allison *et al.*, "Recent developments in Geant4", *Nucl. Instrum. Methods Phys. Res. Sect. Accel. Spectrometers Detect. Assoc. Equip.*, vol. 835, pp. 186–225, Nov. 2016, doi: 10.1016/j.nima.2016.06.125.
[16] T. Sato *et al.*, "Features of Particle and Heavy Ion Transport code System (PHITS) version 3.02", *J. Nucl. Sci. Technol.*, vol. 55, no. 6, pp. 684–690, Jun. 2018, doi: 10.1080/00223131.2017.1419890.
[17] C. J. Werner *et al.*, "MCNP Version 6.2 Release Notes", Los Alamos National Lab. (LANL), Los Alamos, NM (United States), LA-UR-18-20808, Feb. 2018. doi: 10.2172/1419730.
[18] J.-Ch. Sublet, J. W. Eastwood, J. G. Morgan, M. R. Gilbert, M. Fleming, and W. Arter, "FISPACT-II: An Advanced Simulation System for Activation, Transmutation and Material Modelling", *Nucl. Data Sheets*, vol. 139, pp. 77–137, Jan. 2017, doi: 10.1016/j.nds.2017.01.002.
[19] ISO20042, "Measurement of radioactivity - Gamma-ray emitting radionuclides - Generic test method using gamma-ray spectrometry", Aug. 2021.
[20] "Nucléide – Lara Application – Laboratoire National Henri Becquerel". http://www.lnhb.fr/nuclear-data/module-lara/ (accessed Aug. 02, 2023).
[21] K. Amgarou *et al.*, "Characterization of the neutron field from the 241Am-Be isotopic source of the IPHC irradiator", *Radiat. Meas.*, vol. 50, pp. 61–66, Mar. 2013, doi: 10.1016/j.radmeas.2012.11.015.
[22] D. A. Brown *et al.*, "ENDF/B-VIII.0: The 8th Major Release of the Nuclear Reaction Data Library with CIELO-project Cross Sections, New Standards and Thermal Scattering Data", *Nucl. Data Sheets*, vol. 148, pp. 1–142, Feb. 2018, doi: 10.1016/j.nds.2018.02.001.
[23] A. J. M. Plompen *et al.*, "The joint evaluated fission and fusion nuclear data library, JEFF-3.3", *Eur. Phys. J. A*, vol. 56, no. 7, p. 181, Jul. 2020, doi: 10.1140/epja/s10050-020-00141-9.
[24] "Shielding — PhysicsListGuide 11.1 documentation". https://geant4-userdoc.web.cern.ch/UsersGuides/PhysicsListGuide/html/reference_PL/Shielding.html#shielding (accessed Aug. 10, 2023).
[25] "QGSP_BIC — PhysicsListGuide 11.1 documentation". https://geant4-userdoc.web.cern.ch/UsersGuides/PhysicsListGuide/html/reference_PL/QGSP_BIC.html#qgsp-bic (accessed Aug. 10, 2023).
[26] K. SHIBATA *et al.*, "JENDL-4.0: A New Library for Nuclear Science and Engineering", *J. Nucl. Sci. Technol.*, vol. 48, no. 1, pp. 1–30, Jan. 2011, doi: 10.1080/18811248.2011.9711675.
[27] M. B. Chadwick *et al.*, "ENDF/B-VII.1 Nuclear Data for Science and Technology: Cross Sections, Covariances, Fission Product Yields and Decay Data", *Nucl. Data Sheets*, vol. 112, no. 12, pp. 2887–2996, Dec. 2011, doi: 10.1016/j.nds.2011.11.002.
[28] A. J. Koning, D. Rochman, J.-Ch. Sublet, N. Dzysiuk, M. Fleming, and S. van der Marck, "TENDL: Complete Nuclear Data Library for Innovative Nuclear Science and Technology", *Nucl. Data Sheets*, vol. 155, pp. 1–55, Jan. 2019, doi: 10.1016/j.nds.2019.01.002.
[29] ISO8529-1, "Neutron reference radiations fields - Part 1 : characteristics and methods of production", Dec. 2021.

Ten springs of experiments in CROCUS

Vincent Lamirand[1,2,*]
[1]École polytechnique fédérale de Lausanne (EPFL), CH-1015 Lausanne, Switzerland
[2] Paul Scherrer Institute (PSI), CH-5232, Switzerland
(*) vincent.lamirand@epfl.ch

***Abstract*—Around a decade ago, a new team rebuilt experimental research at EPFL's nuclear facilities, and in particular in the CROCUS nuclear reactor. After a broad investigation and open discussions with colleagues from the nuclear community, a number of experiments and research directions were selected. They range from reactor physics to nuclear data, with a focus on instrumentation. We present here the variety of experiments carried out and how instrumentation has been instrumental in these perspectives. Developments on branching or intrinsic reactor noise were made possible thanks to extensive theoretical investigation coupled with the developments of pulse and current modes neutron detection systems, as well as the LEAF gamma detection array. With regard to modulation or perturbation reactor noise, a study of fuel rods vibration was carried out thanks to the unique COLIBRI in-core fuel rods oscillation device; in the VOID experiments, a method was tested to measure the void coefficient with neutron modulation; in the APRHODITE project, the PISTIL rotating absorber, or absorber of variable strength, was used to determine the zero power reactor transfer function, in order to obtain feedback on kinetics nuclear data, particularly delayed neutrons. On the topic of nuclear data, the PETALE programme consisted on criticality and transmission experiments for the study of stainless steel, using instrumented metal reflectors. New dosimetry methodologies were developed for consistent and complete propagation of uncertainties, which also enabled experiments to be optimized. It will be continued in the HARVEST-X project, and its pile-oscillation program, BLOOM. Last but not least, developments in dosimetry as well as for a novel miniature neutron scintillation technology (MiMi detectors) allowed for inter- and intrapin (NECTAR) neutron measurements. 160 MiMi detectors have recently been distributed throughout the CROCUS core in a unique 3D detector array called SAFFRON, paving the way for novel high-resolution neutronics.**

***Keywords*—Reactor experiments, Nuclear data, Neutron detector, CROCUS, COLIBRI, CORTEX, PETALE, HARVEST-X, BLOOM, NECTAR, MiMi detector, SAFFRON.**

I. Introduction

EXPERIMENTAL programmes in the CROCUS reactor were rejuvenated a decade ago, following the appointment of a new professor at EPFL, and the assembling of a new team. CROCUS is a zero-power light water reactor, which was previously used mainly for teaching purposes. It is continuously employed for EPFL students in Physics in their second and third year of Bachelor, as well as for students in the joint EPFL/ETH Zurich Swiss Master programme in Nuclear engineering, in close collaboration with the Paul Scherrer Institute. It also hosts internships, Master semester projects and Master theses.

Although CROCUS lacks flexibility in fuel type and configuration, its low technological uncertainties, operation precision and stability, reflector/interpin space and general accessibility revealed it to be a perfect candidate for reference, but also atypical experimental endeavours. The core consists in two interlocked fuel zones with different fuel and pitches which, despite being an additional complexity, are also a challenge of interest for deterministic codes, towards their validation. Over the past decade, an expertise and a unique set of experimental results were obtained, along and thanks to developments in nuclear instrumentation.

The present article introduces both the experiments and associated developments, their salient outcomes, as well the conditions that enabled them. Four main types of conducted integral experiments are presented successively. The first two concerns the study of fluctuations in time of the neutron population, also called neutron noise: due to the intrinsic stochastic behaviour of the fission chains propagation, i.e. branching noise; and following a perturbation introduced in the system, also called neutron modulation. Another type of experiments concerns the study of selected nuclear data. In the case of the PETALE program, and its continuation the HARVEST-X project, we targeted the cross sections of stainless steel elements. Last but not least, following instrumentation developments, unique high-resolution experiments, SAFFRON and NECTAR, are conducted in CROCUS for the constitution of data sets aiming at the validation of high-fidelity neutronic codes.

All the experiments were made possible thanks to the continued support of CROCUS and experiments at laboratory level, but also at the level of the Faculty of Basic sciences and the Physics Institute. Given the small size of the facility and experimental teams, an effort was needed to cross-fertilize projects, for the sake of creativity, relevance, and feasibility.

II. Branching reactor noise

The relaunch began in 2013 with a project supported by swiss*nuclear* that involved building a data acquisition system in pulse mode to measure neutron noise [1], [2]. This method consists in the acquisition and analysis of fluctuations of neutron detections in time that are due to branching in the fission process in a reactor. Also called zero power reactor noise, it is conventionally used for determining reactor kinetic parameters according to the following formulation:

$$G_{ij}(\omega) = \frac{2C_iC_jD_\nu}{\beta^2 F_0}\frac{1}{1+\left(\frac{\omega}{\alpha}\right)^2} + \sqrt{C_iC_j}\delta_{ij} \qquad \text{Eq. 1}$$

where C_i and C_j are the count rates in the detectors, D_ν the Diven factor, β the delayed neutron fraction, F_0 the total fission rate, $\omega = 2\pi f$ the angular frequency, α the prompt decay constant, and δ_{ij} the Kronecker delta.

The initial project was followed by an intercomparison exercise with CEA and SCK CEN colleagues [3], and eventually by a broader experimental investigation as part of a PhD thesis [4], [5]. Further developments involved increasing detection efficiency for improved statistics, by shifting to the use of large fission chambers (1 g deposit of ^{235}U) requiring the acquisition of their signal in current mode, performed with in-house amplifiers. The obtained setup allowed improved and reference measurements of CROCUS kinetic parameters [6], [7]. This instrumentation was used to investigate theoretical considerations, regarding spatial dependences for instance [8].

More significantly, it led to a re-interpretation of noise as a study of time correlations where neutron detections are time-tagging means only. First, this led to the development of a new methodology for complete and reduced uncertainty budget based on bootstrapping (see Figure 1, top) [9]. This allowed as well contributing to the development of new Monte Carlo computational approaches in collaboration with CEA and IRSN, using Serpent2 and TRIPOLI-4 in analog fission [10]. This finally opened up to the study of branching correlations using gamma detectors for unprecedented precision (see Figure 1, bottom) [9], [11]. Incidentally, the gamma detection array developed for this purpose [12], called LEAF , allowed the first measurement of in-core gamma spectra [13], and the determination of the delayed gamma fraction in CROCUS [14].

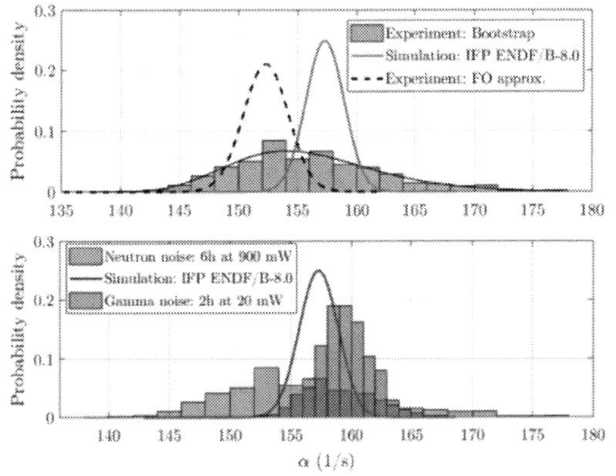

Figure 1. Two comparisons for the determination of the prompt decay constant α. On top, methodology comparison between first-order approximation and bootstrap uncertainty quantification, based on reference experiments with large fission chambers and current mode instrumentation, compared to simulation results. On the bottom, comparison between neutron noise (same experiments) and gamma noise, with simulation results again. [5]

III. Modulation noise

The second type of experiments consists in modulation experiments: a perturbation is introduced in the reactor, leading to a response in neutron population. Following the point kinetics approximation, the same perturbation applied at different frequencies would allow reconstructing the transfer function of the zero power reactor noise mentioned above. For each perturbation frequency, a peak is observed in the frequency domain on top of the branching noise, with a gain which is depending on the amplitude of the perturbation, being it in reactivity or neutron source emission [15], [16].

A. COLIBRI: fuel rods displacement and oscillation

The COLIBRI programme aims at investigating reactor noise related to fuel vibrations. The main motivation was the increased amplitudes in the neutron noise distributions recorded in ex- and in-core detectors that have been observed in Siemens pre-Konvoi type of pressurized water reactors. Among others, changes in fuel assembly or pin vibration patterns, due to recent modifications of assembly structural designs, were pointed out as a possible cause [17].

The experiments consists in the lateral oscillation of one to 18 fuel rods in the outer fuel zone of CROCUS, to measure the induced noise and its eventually spatial dependence over the core. It required the development, qualification and licensing of a unique in-core device for fuel displacement [18]–[20]. It aims at the validation of the noise simulation codes, and was used in particular in three experimental campaigns [21]–[26] within the framework of the Horizon2020 project CORTEX [27]–[29].

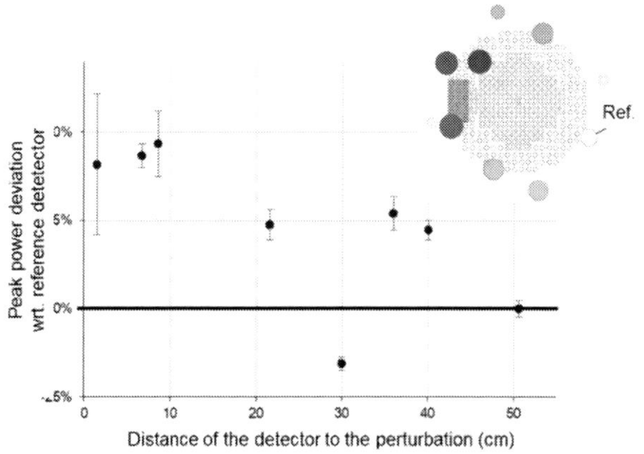

Figure 2. Relative difference in noise levels with respect to the reference detector, as a function of distance to COLIBRI (main plot), and on the CROCUS map (miniature, blue to red scale). In the main plot, the 0% value (black line) correspond the absence of space dependence: it is consequently crossed by the reference detector at ~50 cm, but the difference for other detectors should be noted. [20]

The methodological developments in noise analysis using bootstrapping that were started for branching noise, were pursued and adapted for modulation noise [30]–[33]. The bootstrapping as well as a ratio methodology allowed the definitive observation of spatial dependences with reduced and appropriate uncertainties (see Figure 2) [20]. Last but not least, thanks to the LEAF gamma detection array, as well as the development of MiMi detectors (see section V.B), it has been possible to measure the noise directly at rod level (moving rod and adjacent rod), and at up to 7 meters from the core with gamma rays [34].

B. VOID: void fraction determination

In the case of the VOID experiments, we demonstrated the possibility of measuring in-core the void fraction in an experimental channel using neutron detectors and noise analysis. In this particular case, target modulation is not linked to core neutronics, but to the passage of neutrons through an air stream (i.e., bubbles). Whereas the theory was developed by collaborators at University of Chalmers, it was tested experimentally under CROCUS clean conditions, and by comparison with an out-of-pile setup for airflow characterization, and corresponding simulations [35]–[38]. For now, the results are qualitative, but developments have been initiated to achieve quantitative capabilities [39].

C. APHRODITE and PISTIL: delayed neutrons

In the case of APHRODITE, the goal is opposite to COLIBRI: a clean and well characterized perturbation allows providing feedback on the nuclear data for kinetic parameters. Carried out within a PhD thesis in collaboration between CEA and EPFL [40], a rotating oscillator – or *oscillator of variable strength* – called PISTIL was developed to induce a reactivity modulation over a large frequency range, from 1 mHz to 200 Hz [41]. Whereas branching noise allows only limited feedback at low frequencies, focusing on the integral β_{eff} parameter at most because of a lack of detection statistics, the gain observed in modulation noise allows overcoming this limitation. Hence the zero power transfer function (ZPTF) of CROCUS was determined in power and phase, with a focus on delayed neutrons, which is of most interest for nuclear data, especially for the determination of respective abundances of precursors (see Figure 3) [40], [42]. In addition to typical continuous rotations inducing unique frequency modulations, stepwise displacements allowed triggering a large range of frequencies at once, which was key in the resolution of low frequencies.

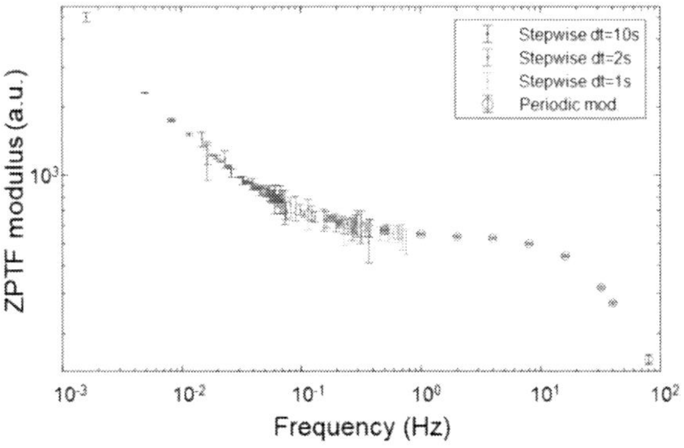

Figure 3. Experimental estimates of the zero power transfer function modulus of CROCUS. [42]

IV. Stainless steel nuclear data

With regard to nuclear data, other types of experiments have been carried out to contribute to the validation of stainless steel nuclear data: first the PETALE programme in CROCUS, and its continuation, the HARVEST-X project.

A. PETALE in CROCUS

Within a collaboration with CEA, the PETALE programme consisted in both critical and neutron transmission semi-integral experiments [43]. The programme featured movable metal reflectors, which replaced part of the water reflector in the reactor vessel (see Figure 4). Four different reflectors were used to study the properties of nuclear grade stainless steel, and its main elements – namely iron, nickel, and chromium. The elemental – or semi-integral – nature of the experiments allows removing material ambiguity, and thus compensation effects between elemental nuclear data. The impact of the reflectors was assessed by conducting critical experiments, while neutron transmission was measured using activation dosimeters.

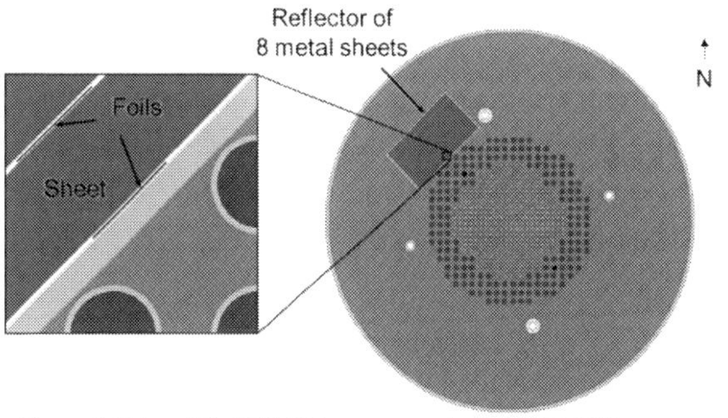

Figure 4. Setup of the PETALE programme in CROCUS. [42]

1) Uncertainty propagation and experimental design

In preparation for the transmission experiments, a large effort was devoted to developing tools for reactor dosimetry, including a new gamma spectrometry platform, CERVIN, and in-house analysis scripts for full propagation of all uncertainties [44], [45]. Uncertainties with covariance include reactor power history, HPGe calibration, and dosimetry nuclear data using the specialized library IRDFF [46], [47]. A series of measurement campaigns was conducted for validating the tools, and deriving C/E comparison for several nuclear data libraries [48].

From these tools, a unique method was developed to optimize the experimental programme and to ensure feedback on nuclear data [49]. The stainless steel nuclear data uncertainties were propagated in a preliminary design of the experiments using a Monte Carlo methodology with the TENDL-2017 nuclear data library and correlated sampling for reducing the computational cost [50], [51]. The propagation resulted in expected reaction rates for a list of considered dosimetry reactions, with their corresponding distributions and covariance, for each measurement position in each metal reflector. These were compared to accessible dosimetry uncertainties, enabling the appropriate reactions to be selected easily and efficiently (see Figure 5). Moreover, the complete schedule of the programme was optimized for minimal uncertainties and facility load [52].

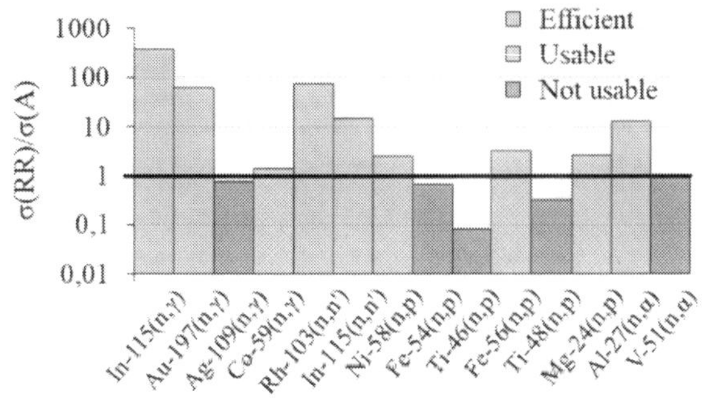

Figure 5. Optimization for the pure iron reflector: comparison between the expected distribution of dosimetry reaction rates $\sigma(RR)$ due to uncertainties of the iron cross sections, and the accessible uncertainties of activation dosimeters $\sigma(A)$. A ratio above 1 indicates usability of the reaction. [42]

2) On-going analysis and preliminary results

PETALE was initiated to support the community needs for integral experiments on stainless steel, hence it was planned from the outset to include it in the CROCUS IRPhE benchmark [53], [54]. Following the end of dosimetry measurements, the joint analysis of CEA and EPFL is on-going, and preliminary results were published on the EPFL side (see Figure 6) [55]. In particular, a high resolution Monte Carlo model of all experiments was recently produced in order to take into account all setup details, in preparation of the benchmark [56].

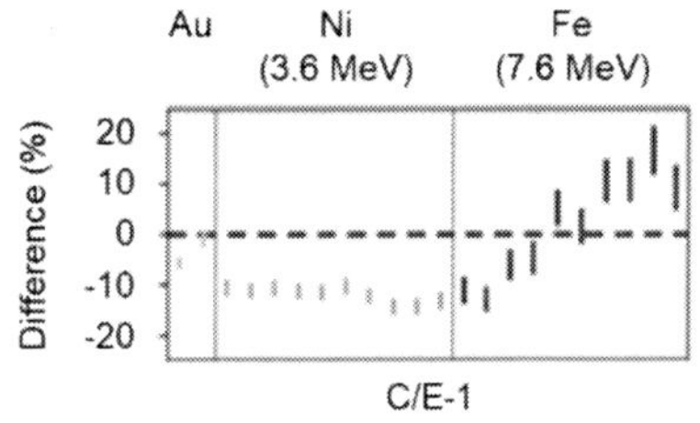

Figure 6. Preliminary comparison between computation (C) and experimental results (E) of Ni and Fe reaction rates for the chromium reflector using JEFF-3.3 [52], [55]. For each reaction, dosimeters are ordered from left to right by reflector depth, from core to periphery. At the time of publication, an offset due to power normalization was not corrected yet, but a clear trend was already visible in the case of the iron dosimeter (right), i.e. higher energies.

B. *HARVEST-X, BLOOM and GRAPE*

The continuation of PETALE, HARVEST-X, consists firstly in the analysis (validation) and use (assimilation) of PETALE [57], but also in its reproduction for bias identification. Using multiple facilities and measurement methods, we aim to provide a consistent set of reliable data for nuclear data feedback and evaluation purposes. The first two sets of experiments are BLOOM and GRAPE. The BLOOM program consists of pile-oscillation experiments in CROCUS to repeat PETALE critical experiments, but with a different methodology [58]. As a modulation program, it will use the developments presented in section III, and the development of the POLLEN pile-oscillator [59]. The GRAPE campaign was allocated beam time at the GELINA facility in the Horizon2020 project ARIEL. It consists of reproducing PETALE transmission experiments with the same metal blocks, but using GELINA's "white" neutron field.

V. High-resolution neutronics

Last but not least, thanks to the unique design of CROCUS, its core accessibility, and operation flexibility, high resolution experiments were carried out for the constitution of unique data sets aiming at validation of high-fidelity neutronic codes. It first made use of existing instrumentation, in particular the TRAX device [60]. Following developments in activation dosimetry but as well miniature neutron scintillators called MiMi detectors, inter- and intra-pin measurements were and are carried out within two PhD theses [61].

A. *TRAX experiments: thermal flux profiles*

Initially for teaching purposes, CROCUS is equipped with a miniature fission chamber set at the end of a vertical shaft. Thanks to a cut in the superior grid, essentially consisting of two half-grids, and appropriate motors and controls, the TRAX device can move the fission chamber axially and radially along the mid-cross section of the core. Precise axial but especially radial relative neutron flux profiles were measured and compared to Monte Carlo simulations, with the interesting observation of the transition between the two core lattices [62]. In addition, using a preliminary version of the POLLEN linear oscillator, the axial extraction of a Cd sample was measured locally with the fission chamber [60].

B. *MiMi detectors: high-resolution interpin thermal flux*

Initiated in collaboration with the Paul Scherrer Institute [63], [64], a new type of neutron detector has been developed [65]–[69]. These advanced miniature and minimalistic (MiMi) neutron detectors have an active volume smaller than 1 mm^3 but a relatively high efficiency in the thermal range, which make them perfect tools for in-core use. They consist of a ZnS(Ag):^{6}LiF scintillator screen connected to an optical fiber. The photon signal is collected using SiPM technology, and processed using in-house electronics, initially with analog electronics, then with FPGA digital acquisition and processing. With the latest design, their sensitivity is about 5 kcps/W.

The millimetric size allowed measuring local interpin azimuthal gradients in an experimental channel [70], as well as the impact of fuel displacement using the COLIBRI fuel rods oscillator [71] (see Figure 7). They were also applied to measurements at the SUR-100 reactor of the University of Stuttgart, and for the second CORTEX modulation campaign at the AKR-2 reactor of TU Dresden [23], [72], [73].

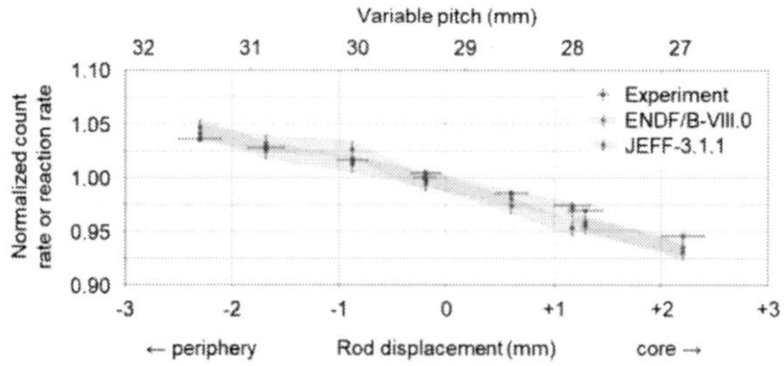

Figure 7. Comparison of MiMi measurements and Serpent2 calculations of the impact on neutron flux of displacing a fuel rod in the U_{met} lattice of CROCUS [71], in effect corresponding to a local variation of pitch.

C. SAFFRON: thermal flux core mapping

Thanks to the scalability of MiMi detectors using digital acquisition, a core-mapping array called SAFFRON was developed, licensed, and finally loaded in CROCUS in 2022 [74]. It consists of 149 detectors distributed radially in both inner and outer lattices of CROCUS, and axially on three main levels, i.e. about 50 detectors per level, as well as 11 supplementary ones for a total 160 detectors (see Figure 8). With this unique 3D instrumentation we aim at the study of static and dynamic neutron flux maps. For this purpose, each individual MiMi detector was intercalibrated in the CARROUSEL facility, which consists of a PuBe neutron source set in a water tank [75]. First results were obtained for an absolute flux map of CROCUS, i.e. comparing each MiMi detector with one another, as well as for the difference in flux between two critical configurations: critical with water level only, and critical with one control rod inserted to compensate the maximum water level [61]. Recent and further experiments include the studies of branching and modulation noise, as well as kinetics experiments.

Figure 8. Photograph of the core of CROCUS instrumented with SAFFRON, during the core loading procedure. One can see the red horizontal rods of the three main axial levels, as well as the two intermediate ones instrumenting at positions close to the control rods only.

D. NECTAR: intrapin neutron flux profiles

Most recently, the NECTAR experiments consist in intrapin neutron flux profiles using activation dosimetry for code validation purposes. To this end, an instrumented fuel rod was designed, licensed and built to use the enlarged grids holes of the COLIBRI region [76]. The rod is made up of four 25 cm-long U_{met} cigars, which are individually encased, and assembled in a second outer cladding. The cigars come from the facilities reserve and are equivalent to those from the rest of U_{met} rods. The activation foils are set in-between the cigars, in holders that act as centering elements, including angular reference. The foils are laser pre-cut, which allows to measure the pieces in order to reconstruct the neutron flux profiles, azimuthally and radially. For the experiments, one COLIBRI fuel rod is replaced by the NECTAR instrumented rod, which enabled irradiating it at a variety of core locations, from the vicinity of the water gap between the two lattices, to the core periphery. The first experiments were conducted in 2023, and are being published.

VI. CONCLUSIONS

A decade ago, following the appointment of a new professor, a rejuvenated team was assembled at EPFL to relaunch research activities. In particular, emphasis was placed on experimental programmes to make use of local nuclear facilities, especially the now only Swiss experimental reactor CROCUS, with a view to maintaining and developing competences. A wide-ranging investigation between EPFL, PSI, and historical and new collaborators led to the selection of a number of research directions, ranging from reactor physics to nuclear data. The results were then presented for scientific external evaluation, and to the nuclear community [77]–[83]. Based on the unique design of CROCUS, the accessibility of its core and its clockwork technological qualities, we oriented our activities towards instrumentation-based and reference programmes.

Beyond purely instrumention developments [84]–[89], each individual prospect was made possible by redefining the tools and methods in relation to the purpose and theory, without limitation in the first investigation phase. It is worth noting that the projects used a variety of frameworks, ranging from internal laboratory and School funding, to informal and contractual collaborations funded in-kind, as well as Swiss and European projects, both academic and industrial. The underlying logic was: whatever works, people and ideas first, funding second. Accordingly, all the experiments mentioned above were made possible thanks to a continued support at the laboratory level, but also at the level of the Faculty of Basic Sciences and the Physics Institute. On average, facilities and experiments were covered at EPFL by the work of around 4 scientists (about 2 full-time), 2 technicians, and 2-3 PhD and Master thesis students and postdoctoral fellows. Given the small size of the facilities and experimental teams, a major effort was required to cross-fertilize projects, in the interests of creativity, relevance, and feasibility.

The content and results of the experiments have been published throughout the projects, although a significant part is still hidden in theses and project reports, and is being analyzed for the most complex projects, especially COLIBRI and PETALE, both of which are nearing completion and publication. The next steps will be to establish benchmarks for making data available to the wider community. With regard to new experiments, both high-resolution SAFFRON and NECTAR experiments are still underway, with dynamic experiments for the former [90], particularly in conjunction with COLIBRI. Within the HARVEST-X project, PETALE has entered its analysis phase, and is followed by the BLOOM and GRAPE experiments. New developments have recently been undertaken concerning modern detection methods using scintillators [91].

ACKNOWLEDGEMENTS

The research leading to these results has received funding from the Euratom research and training programme 2014-2018 under grant agreement No 754316.

The author would like to thank all contributors over the years, and especially the CROCUS operation team and the academic collaborators who have made this journey a path of friendship.

References

[1] V. Roland, G. Perret, G. Girardin, P. Frajtag, and A. Pautz, "Neutron noise measurement in the CROCUS reactor," in *Proc. IGORR 2013*, 2013, pp. 2–7.
[2] G. Perret, G. Girardin, P. Frajtag, and M. Hursin, "Decay constant and delayed neutron fraction measurements in CROCUS," TM-41-14-02, 2014.
[3] V. Lamirand *et al.*, "Intercomparison of neutron noise measurement systems in the CROCUS reactor," in *Proceedings of PHYSOR 2018*, 2018.
[4] O. V. Pakari, V. Lamirand, G. Perret, and A. Pautz, "Development of Current and Fast Neutron Noise Measurements in CROCUS," MSc thesis ETHZ-EPFL, 2016.
[5] O. V. Pakari, V. Lamirand, and A. Pautz, "Experimental and numerical study of stochastic branching noise in nuclear reactors," EPFL, Lausanne, 2020.
[6] O. Pakari, V. Lamirand, G. Perret, L. Braun, P. Frajtag, and A. Pautz, "Current Mode Neutron Noise Measurements in the Zero Power Reactor CROCUS," *EPJ Web Conf.*, vol. 170, p. 04017, Jan. 2018.
[7] O. Pakari, V. Lamirand, G. Perret, P. Frajtag, and A. Pautz, "Kinetic Parameter Measurements in the CROCUS Reactor Using Current Mode Instrumentation," *IEEE Trans. Nucl. Sci.*, vol. 65, no. 9, pp. 2456–2460, Sep. 2018.
[8] O. Pakari *et al.*, "Investigation of spatial effects on neutron noise measurements in the zero power reactor CROCUS," in *Proceedings of PHYSOR 2018*, 2018, p. 12.
[9] O. V. Pakari, V. Lamirand, T. Mager, P. Frajtag, and A. Pautz, "High accuracy measurement of the prompt neutron decay constant in CROCUS using Gamma Noise and bootstrapped uncertainties," *Ann. Nucl. Energy*, vol. 170, p. 109011, Jun. 2022.
[10] O. Pakari, D. Mancusi, O. Petit, A. Zoia, V. Lamirand, and A. Pautz, "Towards the Validation of Noise Experiments in the CROCUS Reactor Using the TRIPOLI-4 Monte Carlo Code in Analog Mode," *EPJ Web Conf.*, vol. 247, p. 04007, Feb. 2021.
[11] O. Pakari, V. Lamirand, T. Mager, P. Frajtag, and A. Pautz, "Gamma noise to monitor future nuclear reactors for safeguards , nonproliferation , and safety," *Submitt. to Nat. Sci. reports*, 2023.
[12] O. Pakari *et al.*, "Design and Simulation of Gamma Spectrometry Experiments in the CROCUS Reactor," *EPJ Web Conf.*, vol. 225, p. 04016, Jan. 2020.
[13] O. Pakari, V. Lamirand, T. Mager, A. Laureau, P. Frajtag, and A. Pautz, "First in-core gamma spectroscopy experiments in a zero power reactor," *EPJ Web Conf.*, vol. 253, p. 04022, Nov. 2021.
[14] O. Pakari, T. Mager, V. Lamirand, P. Frajtag, and A. Pautz, "Delayed gamma fraction determination in the zero power reactor CROCUS," *EPJ Nucl. Sci. Technol.*, vol. 7, p. 16, Sep. 2021.
[15] Y. Jiang, B. Geslot, V. Lamirand, and P. Leconte, "Review of kinetic modulation experiments in low power nuclear reactors," *EPJ Nucl. Sci. Technol.*, vol. 6, p. 55, 2020.
[16] Y. Jiang, "Mesure de la fonction de transfert du réacteur CROCUS à l'aide d'un dispositif de modulation de réactivité," in *Atelier NEEDS, 8 December 2021, Aix-en-Provence (France)*, 2021.
[17] D. Chionis, "Development of Advanced Methodologies for Monitoring and Modelling of Neutron Noise in Modern LWR Cores," 2020.
[18] V. Lamirand *et al.*, "The COLIBRI programme in CROCUS: development and licensing of a fuel rods oscillator," in *RRFM/IGORR 2019*, 2019.
[19] V. Lamirand *et al.*, "The COLIBRI experimental program in the CROCUS reactor: characterization of the fuel rods oscillator," *EPJ Web Conf.*, vol. 225, p. 04020, Jan. 2020.
[20] V. Lamirand *et al.*, "Analysis of the first COLIBRI fuel rods oscillation campaign in the CROCUS reactor for the European project CORTEX," *EPJ Web Conf.*, vol. 247, p. 21010, Feb. 2021.
[21] V. Lamirand *et al.*, "Neutron noise experiments in the AKR-2 and CROCUS reactors for the European project CORTEX," *EPJ Web Conf.*, vol. 225, p. 04023, Jan. 2020.
[22] V. Lamirand *et al.*, "Experimental report of the 1st campaign at AKR-2 and CROCUS," CORTEX Deliverable 2.1, 2018.
[23] V. Lamirand *et al.*, "Experimental report of the 2nd campaign at AKR-2 and CROCUS," CORTEX Deliverable 2.2, 2020.
[24] V. Lamirand, A. Knospe, K. Ambrožič, F. Vitullo, and C. Lange, "Experimental report of the 3rd campaign at AKR-2 and CROCUS," CORTEX Deliverable 2.4, 2021.
[25] V. Lamirand *et al.*, "Modulation campaigns at AKR-2 & CROCUS for the validation of neutron noise codes," *Submitt. to Ann. Nucl. Energy*, 2023.
[26] V. Lamirand, "COLIBRI: Fuel rods oscillation experiments in CROCUS," in *NEEDS workshop - Perturbations in experimental reactors*, 2021.
[27] C. Demazière, P. Vinai, M. Hursin, S. Kollias, and J. Herb, "Overview of the CORTEX project," in *Proceedings of PHYSOR 2018*, 2018, pp. 2971–2980.
[28] A. G. Mylonakis *et al.*, "CORE SIM+ simulations of COLIBRI fuel rods oscillation experiments and comparison with measurements," *EPJ Web Conf.*, vol. 247, p. 21006, Feb. 2021.
[29] M. Hursin *et al.*, "Modeling noise experiments performed at AKR-2 and CROCUS zero-power reactors," *Ann. Nucl. Energy*, vol. 194, 2023.
[30] A. Rais *et al.*, "Towards the validation of neutron noise simulators: Qualification of data acquisition systems," in *International Conference on Mathematics and Computational Methods Applied to Nuclear Science and Engineering, M and C 2019*, 2019, pp. 2638–2647.
[31] K. Ambrožič *et al.*, "Noise analysis techniques of in-core modulation experiments of the European CORTEX project," in *29th International Conference Nuclear Energy for New Europe*, 2020, vol. 2, no. September, pp. 1–8.
[32] K. Ambrožič, V. Lamirand, and A. Pautz,

"Characterization of High Harmonic Frequencies in Reactor Noise Experiments Within the CORTEX Project," *IEEE Trans. Nucl. Sci.*, vol. 69, no. 4, pp. 825–831, Apr. 2022.

[33] K. Ambrožič *et al.*, "Postprocessing of experimental time series to determine reliable mean and uncertainties," *Submitt. to Ann. Nucl. Energy*, 2023.

[34] V. Lamirand *et al.*, "Local and high distance neutron and gamma measurements of fuel rods oscillation experiments," *EPJ Web Conf.*, vol. 253, p. 04024, Nov. 2021.

[35] V. Dykin and I. Pazsit, "Simulation of in-core neutron noise measurements for axial void profile reconstruction in boiling water reactors," *Nucl. Technol.*, vol. 183, pp. 354–366, 2013.

[36] M. Hursin *et al.*, "Measurement of the gas velocity in a water-air mixture in CROCUS by neutron noise technique," in *International Conference on Mathematics and Computational Methods applied to Nuclear Science and Engineering (M&C 2019)*, 2019.

[37] M. Hursin *et al.*, "Measurement of the Gas Velocity in a Water-Air Mixture in CROCUS Using Neutron Noise Techniques," *Nucl. Technol.*, vol. 206, no. 10, pp. 1566–1583, Feb. 2020.

[38] M. Hursin *et al.*, "Validation of axial void profile measured by neutron noise techniques in CROCUS," *EPJ Web Conf.*, vol. 247, p. 08004, Feb. 2021.

[39] S. Lapaire, "Design of a conductivity detector for Void measurement in a two component mixture," EPFL, 2022.

[40] Y. Jiang, B. Geslot, V. Lamirand, and P. Leconte, "Experimental determination of the transfer function of the zero-power reactor CROCUS with reactivity modulation," Université Grenoble-Alpes, 2022.

[41] Y. Jiang *et al.*, "PISTIL, a reactivity modulation device to probe the transfer function of the nuclear reactor CROCUS," *EPJ Web Conf.*, vol. 253, p. 04007, Nov. 2021.

[42] B. Geslot *et al.*, "Reactivity modulation experiments for nuclear data in CROCUS within a CEA-EPFL collaboration," in *ANIMMA 2023*, 2023.

[43] V. Lamirand *et al.*, "Design of separated element reflector experiments in Crocus: Petale," *ASTM Spec. Tech. Publ.*, vol. STP 1608, pp. 1–14, 2018.

[44] Y. Jiang, A. Laureau, V. Lamirand, P. Frajtag, and A. Pautz, "In-core dosimetry for the validation of neutron spectra in the CROCUS reactor," *EPJ Web Conf.*, vol. 225, p. 04021, Jan. 2020.

[45] A. Laureau, V. Lamirand, A. Gruel, P. Frajtag, and A. Pautz, "DOSIMETRY MODELING AND EXPERIMENTAL VALIDATION FOR THE PETALE PROGRAM IN THE CROCUS REACTOR," *EPJ Web Conf.*, vol. 247, no. September, p. 08015, Feb. 2021.

[46] E. M. Zsolnay, R. C. Noy, H. J. Nolthenius, and A. Trkov, "Summary description of the new international reactor dosimetry and fusion file (IRDFF release 1.0)," vol. 14, no. July 2011, pp. 1–78, 2012.

[47] R. Capote, K. I. K. I. Zolotarev, V. G. V. G. Pronyaev, and A. Trkov, "Updating and Extending the IRDF-2002 Dosimetry Library," *J. ASTM Int.*, vol. 9, no. 4, p. 104119, 2012.

[48] A. Laureau, A. Gruel, V. Lamirand, T. Ligonnet, A. Sardet, and A. Pautz, "Analysis of the preliminary campaign for the PETALE program," *EPJ Nucl. Sci. Technol.*, vol. 9, p. 25, 2023.

[49] V. Lamirand *et al.*, "An Experimental Programme optimized with Uncertainty Propagation: PETALE in the CROCUS Reactor," *EPJ Web Conf.*, vol. 211, p. 03003, Jun. 2019.

[50] A. Laureau, V. Lamirand, D. Rochman, and A. Pautz, "Uncertainty propagation based on correlated sampling technique for nuclear data applications," *EPJ Nucl. Sci. Technol.*, vol. 6, p. 8, 2020.

[51] A. Laureau, V. Lamirand, D. Rochman, and A. Pautz, "Uncertainty propagation for the design study of the PETALE experimental programme in the CROCUS reactor," *EPJ Nucl. Sci. Technol.*, vol. 6, p. 9, Apr. 2020.

[52] V. Lamirand *et al.*, "PETALE in CROCUS - Conduction in COVID times," in *17th International Symposium on Reactor Dosimetry (ISRD17)*, 2023.

[53] J. M. Paratte, R. Früh, U. Kasemeyer, M. a. Kalugin, W. Timm, and R. Chawla, "A benchmark on the calculation of kinetic parameters based on reactivity effect experiments in the CROCUS reactor," *Ann. Nucl. Energy*, vol. 33, no. 8, pp. 739–748, May 2006.

[54] U. Kasemeyer, R. Früh, J. M. Paratte, and R. Chawla, "Benchmark on Kinetic Parameters in the CROCUS Reactor," in *International Reactor Physics Experiments Handbook (IRPhE)*, no. 4440, OECD, Ed. 2007, p. 94.

[55] T. Ligonnet, A. Laureau, V. Lamirand, and A. Pautz, "Preliminary analysis of the PETALE program and comparison to simulations," EPFL, 2021.

[56] T. Ligonnet and A. Laureau, "High Resolution Modelling of the PETALE Program without Slowdown," in *M&C 2023 International Conference on Mathematics & Computational Methods Applied to Nuclear Science & Engineering*, 2023.

[57] A. Vasiliev *et al.*, "Assessment of representativity of the PETALE experiments for validation of Swiss LWRs ex-core dosimetry calculations," *EPJ Web Conf.*, vol. 239, p. 22001, Sep. 2020.

[58] T. Ligonnet, A. Laureau, P. Simsbler, A. Pautz, and V. Lamirand, "Design of an Open-Loop Pile-Oscillation Program in the CROCUS Reactor," *Submitt. to IEEE Trans. Nucl. Sci.*, 2023.

[59] T. Ligonnet, N. Weiss, V. Lamirand, A. Laureau, and A. Pautz, "POLLEN: A Pile-Oscillator for the BLOOM Experimental Program," in *ANIMMA 2023*, 2023.

[60] A. Rais, "Performance Assessment of a 3-D Steady-State and Spatial Kinetics Model for the Crocus Reactor," Ecole Polytechnique Fédérale de Lausanne (EPFL), 2017.

[61] F. Vitullo, "Miniature and Minimalistic Neutron Detectors for Online High-Resolution Experiments in the Zero-Power Reactor CROCUS," EPFL, Lausanne, 2022.

[62] A. Rais, M. Hursin, G. Perret, and A. Pautz,

"Experimental validation of control rod reactivity worth and fission rate distributions for the CROCUS Reactor," *Phys. React. 2016, PHYSOR 2016 Unifying Theory Exp. 21st Century*, vol. 6, pp. 1–10, 2016.
[63] J. B. Mosset *et al.*, "A 16-ch module for thermal neutron detection using ZnS:6LiF scintillator with embedded WLS fibers coupled to SiPMs and its dedicated readout electronics," *Nucl. Instruments Methods Phys. Res. Sect. A Accel. Spectrometers, Detect. Assoc. Equip.*, vol. 845, pp. 494–498, 2016.
[64] A. Wolfertz, R. Adams, and G. Perret, "First tests of a gamma-blind fast neutron detector using a ZnS:Ag-epoxy mixture cast around wavelength-shifting fibers," *Nucl. Instruments Methods Phys. Res. Sect. A Accel. Spectrometers, Detect. Assoc. Equip.*, vol. 971, no. January, p. 164003, 2020.
[65] F. Vitullo, V. Lamirand, and A. Pautz, "A Novel Neutron Detection System for Highly Localized In-core Measurements in the CROCUS Reactor," PhD candidacy exam, 2019.
[66] F. Vitullo *et al.*, "Developing and testing a miniature fiber-coupled scintillator for in-core neutron counting in CROCUS," *EPJ Web Conf.*, vol. 225, p. 04018, Jan. 2020.
[67] F. Vitullo *et al.*, "A mm 3 Fiber-Coupled Scintillator for In-Core Thermal Neutron Detection in CROCUS," *IEEE Trans. Nucl. Sci.*, vol. 67, no. 4, pp. 625–635, Apr. 2020.
[68] V. Lamirand, F. Vitullo, K. Ambrožič, O. Pakari, L. Braun, and D. Godat, "Development of fibre-based neutron scintillators," CORTEX Deliverable 2.3 (D2.3), 2021.
[69] R. Colin, F. Vitullo, V. Lamirand, and A. Pautz, "Extension of miniature neutron scintillators features in CROCUS : novel acquisition scheme and extended flux range," EPFL SB IPHYS LRS, 2021.
[70] F. Vitullo, V. Lamirand, P. Frajtag, G. Perret, and A. Pautz, "Highly localized azimuthal measurements in the CROCUS reactor towards the validation of high-fidelity neutronics codes," *EPJ Web Conf.*, vol. 247, p. 08014, Feb. 2021.
[71] F. Vitullo *et al.*, "Millimetric fuel rod displacements : an experimental study of the impact on local thermal neutron flux in the CROCUS reactor," in *Proceedings of PHYSOR 2022*, 2022, pp. 2632–2641.
[72] E. L. Brunetto, F. Vitullo, and V. Lamirand, "Measurements and Monte Carlo simulations with miniature neutron detectors for flux characterization in the SUR-100 zero power reactor," Politecnico di Torino, 2020.
[73] E. L. Brunetto *et al.*, "High resolution measurements with miniature neutron scintillators in the SUR-100 zero power reactor," *EPJ Web Conf.*, vol. 253, p. 04029, Nov. 2021.
[74] F. Vitullo *et al.*, "Design of a 150-miniature detectors 3D core-mapping system for the CROCUS reactor," *EPJ Web Conf.*, vol. 253, p. 04023, Nov. 2021.
[75] F. Vitullo, V. Lamirand, D. Godat, and A. Pautz, "Simultaneous inter-calibration of 160 MiMi neutron detectors," in *ANIMMA 2023*, 2023.
[76] T. Mager *et al.*, "Design of an instrumented fuel rod for intra-pin neutron flux measurements in the CROCUS experimental reactor," in *ANIMMA 2023*, 2023.
[77] V. Lamirand, "Selection of experimental research directions for LRS," CR-100-113-15, 2015.
[78] V. Lamirand *et al.*, "Experimental neutronics: Outlook," 2015.
[79] V. Lamirand, M. Hursin, G. Perret, P. Frajtag, and A. Pautz, "Proposal for a new CROCUS experiment : Qualification of reflector materials for GEN-III reactors (PETALE)," RT-100-113-15-16, 2015.
[80] V. Lamirand, M. Hursin, G. Perret, P. Frajtag, and A. Pautz, "Proposal for a new CROCUS experiment : Investigation of mechanical noise induced by fuel rod vibrations (COLIBRI)," RT-100-113-15-15, 2015.
[81] V. Lamirand, "New experiments at the LRS facilities within the VEP collaboration," CR-100-113-15-14, 2015.
[82] V. Lamirand, M. Hursin, G. Perret, P. Frajtag, O. Pakari, and A. Pautz, "Future experimental programmes in the CROCUS reactor," in *RRFM/IGORR 2016, Berlin (Germany) 13-17 March*, 2016, no. 02–2016, pp. 284–292.
[83] V. Lamirand, "Experiments in a ZPR in the 2010s : Harvesting data in CROCUS," in *ANIMMA 2019 - Workshop N°1 "The coupling of experiments and modelling to enhance experiments in research reactors,"* 2019.
[84] A. Theler, V. Lamirand, B. Geslot, and A. Pautz, "Method development for discrimination of anodic and cathodic signals in miniature fission chambers for reactor physics applications," MSc thesis EPFL, 2017.
[85] M. Hursin *et al.*, "Testing of a sCVD diamond detection system in the CROCUS reactor," *Eur. Phys. J. A*, vol. 54, no. 5, p. 82, May 2018.
[86] C. Kong *et al.*, "Modeling the response of a diamond detector in the zero power reactor CROCUS," *Eur. Phys. J. Plus*, vol. 137, no. 1, 2022.
[87] P. Andričević *et al.*, "Halide Perovskites: Kilogram-Scale Crystallogenesis of Halide Perovskites for Gamma-Rays Dose Rate Measurements (Adv. Sci. 2/2021)," *Adv. Sci.*, vol. 8, no. 2, p. 2170009, Jan. 2021.
[88] P. Andričević *et al.*, "Kilogram-Scale Crystallogenesis of Halide Perovskites for Gamma-Rays Dose Rate Measurements," *Adv. Sci.*, vol. 8, no. 2, pp. 1–9, 2021.
[89] P. R. Andricevic, "Carbon/Hybrid Perovskite Interfaces: From Light Emission to Radiation Detection," 9801, 2019.
[90] M. Saliba, O. Pakari, V. Lamirand, W. Monange, and A. Pautz, "Temporal and spatial correlation in the CROCUS reactor with the SAFFRON 3D detector array," Ecole Polytechnique Fédérale de Lausanne (EPFL), 2023.
[91] F. B. Darby *et al.*, "Investigation of organic scintillators for neutron-gamma noise measurements in a zero-power reactor Introduction : Neutron Noise," in *ANIMMA 2023*, 2023.

Characterization and performance of an acoustic sensor for fission gas release characterization devoted to JHR environment measurements

F. Baudry[1,**], E. Rosenkrantz[1], P. Combette[1], D. Fourmentel[2], C. Destouches[2], PV. Bonzon[2], K. Paumel[2], J.Y. Ferrandis[1,*]
[1] IES, University of Montpellier, CNRS, Montpellier, France
[2] CEA, DES, IRESNE, Cadarache, Saint-Paul-Lez-Durance, France
(*) Jean-yves.Ferrandis@umontpellier.fr - (**) florian.baudry@ies.univ-montp2.fr

For over a decade, the IES laboratory has been working in collaboration with the CEA on the development of acoustic instrumentation in the nuclear field. As part of this collaboration, the IES acoustics team is developing a miniaturized gas composition sensor for in situ measurements of gas composition in a fuel rod. The first experimental use of an acoustic sensor to measure gas composition dates back to 2010 with the REMORA 3 experiment, which estimated the release of fission gas in an experimental fuel rod. Unlike the sensor tested during REMORA 3, this new sensor should be able to operate at 300°C, a performance made possible by the use of a piezoelectric element based on bismuth titanate (NBT). The material is screen-printed onto an alumina substrate. The manufacturing process and initial characterization were presented in the previous version of ANIMMA 2021 [1].
In this article, we will focus on the performance of this type of sensor by characterizing different gas mixtures representative of the gases inside a fuel rod, in order to estimate relative deviations under laboratory conditions.

***Keywords* —Acoustic sensor, composition, screen printing**

I. Introduction

The research presented in this article is a direct follow-up to the work presented in 2021 at the ANIMMA conference [mettre ref]. This work detailed the study and design of an experimental gas composition sensor, dedicated to measurements in an experimental fuel rod of the Jules Horowitz Reactor (RJH).
The laboratory developing these devices is the Institut d'Électronique et des Systèmes (Montpellier, France), and more specifically the acoustics team, which in part develops acoustic technologies dedicated to NDT operations in harsh environments. The sensor dedicated to measuring gas composition will be subjected to environmental constraints similar to those that might be encountered in the vessel of a PWR, namely:

- A maximum temperature of around 350°C
- A radiative flux of up to 1.175E13 (Thermal+Epithermal+Fast) [2].
- A volume of less than 3cm3

The sensor consists of one or more piezoelectric elements, which can operate in reflection or transmission mode. This piezoelectric element generates an acoustic wave in an acoustic cavity whose volume is of the order of 1 cm3. By measuring the time of flight of the acoustic wave, it is then possible to estimate the speed of the acoustic wave. The speed of the wave depends, among other things, on the atomic composition of the gas. The theoretical principle of sensor operation was demonstrated experimentally in 2010 in the OSIRIS reactor (CEA Saclay, Paris) by a sensor named CACP-1 [3], [4]. The principle of the measurement was to monitor gaseous releases from fuel pellets, which are either MOX or UO2, during reactor operation. This experiment was carried out with a radiative flux similar to that which might be found in the vessel of a power generation reactor, but at a temperature of 150°C, a far cry from that envisaged for operation of the RJH. That's why, starting in 2015, our work focused on building a sensor capable of continuously measuring gas composition up to at least 350°C. To achieve this, we had to rethink the sensor manufacturing process and the materials used in its manufacture. We have abandoned lead zirconate titanate (PZT) in favor of bismuth niobate titanate (NBT), which has a Curie temperature of 650°C. And the piezoelectric element is no longer a solid material, but a layer of NBT deposited on an alumina ceramic substrate, which is relatively insensitive to oxidation and expansion compared with stainless steel, for instance. This manufacturing process avoids bonding problems and parallelism defects, as the adhesion between the substrate and the NBT is a direct bond.
The article will be divided into two parts, the first of which will deal with the principle of operation and manufacture of the sensor. The second part will focus on the sensor's initial results under laboratory conditions.

II. Fabrication and Principle of working of the sensor

The innovative aspect of the sensor comes in part from the materials used in its manufacture and the manufacturing process [5]–[8]. The sensor is made exclusively of ceramic and is divided into two parts: The piezoelectric elements which are shown in Fig 1 A), these elements are disc-shaped and are assembled at the top and bottom of the cavity and bonded

https://doi.org/10.1051/epjconf/202328805001

with alumina cement. Fig 1 B) shows an assembled sensor with the gold connecting wires that enable the sensor to operate.

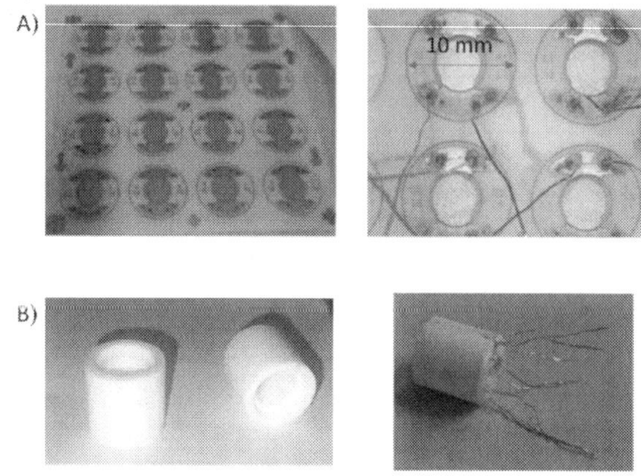

Fig 1: A) piezoelectric elements plate of 16 elements and a focus on 4 piezoelectric elements with gold wires soldering onto gold pads B) Acoustic cavity and sensor finalized with gold wire connectors

Piezoelectric elements are manufactured by successively depositing a first layer of a conductive material, in our case gold, on an alumina substrate. This is followed by a multitude of layers of the piezoelectric element, and finally a layer of gold to form the top electrode. This stack of layers is then fired at a temperature of 850°C to fuse the gold and sinter the ceramic, producing a ready-to-use piezoelectric element. The manufacturing method is similar to the low-temperature co-firing technique (LTCC). The layers are deposited using an ink that is a mixture of organic binder and the material to be deposited in fine powder form, in our case either gold for the electrodes or NBT for the active layer with piezoelectric properties. The ink is deposited *via* screen printing, a method that uses a doctor blade to force ink through a steel mesh stencil to reproduce the stencil pattern on the substrate. The sensor manufacturing process and the technical choices made to produce this sensor are discussed in more detail at the ANIMMA 2021 conference [1] and in this series of articles [5], [6], [8], [9].

The principle of the sensor is to measure the time-of-flight of an acoustic wave, and therefore to measure the wave velocity, and then, using an analytical model, to estimate the gas composition of the medium in which the wave is moving. The sensor can operate in two modes, the first using a single piezoelectric element and the second using two piezoelectric elements facing each other. If the sensor operates with a single piezoelectric element, it works in "reflection" mode, with the same piezoelectric element transmitting and receiving the wave. The second operating mode used for our sensor is the transmission mode, which uses two piezoelectric elements, one transmitting and the other receiving. This operating mode was chosen for several reasons:

- The use of two piezoelectric elements means that only one of the two elements can be electrically excited, which means that the second piezoelectric element is more sensitive. This is impossible in reflection mode, as the same sensor emits and receives at the same time.
- When the sensor operates with two piezoelectric elements, the first echo received only travels one way into the cavity, so acoustic transmission losses are limited. On the other hand, in reflection mode the first echo necessarily makes one outward and one return travel.
- The advantage of having two piezoelectric elements is redundancy. In fact, there are two active elements per sensor, reducing the chance of global sensor failure.

The operating principle of the sensor is illustrated in Fig 2.

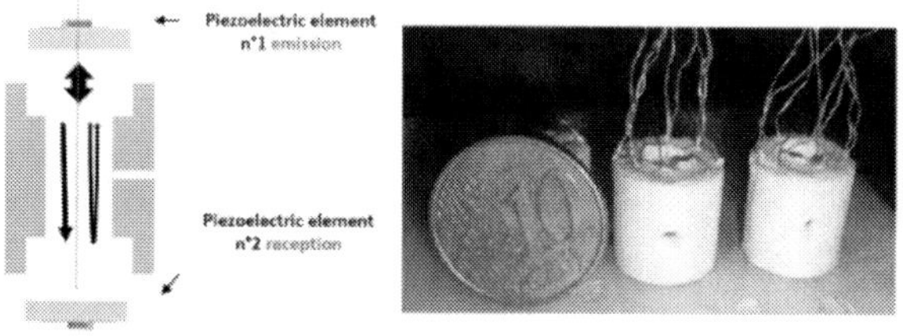

Fig 2 : Transmission sensors used for the experiments

In fact, the echoes emitted by the piezoelectric element are reflected several times in the resonant cavity. This phenomenon is shown on the tempograph in Fig 4, which is a measurement of the sensor. We can estimate the speed of sound by measuring the time between the various echoes, and knowing the height of the cavity. In a perfect gas, the acoustic wave velocity can be written as a function of the gas density and temperature [3]:

$$C = \sqrt{\frac{\gamma RT}{M}} \quad (1)$$

With:
γ: Laplace coefficient
R : Universal perfect gas constant
T : Gas temperature
M : gas molar mass

In the study of fission gas release, the gas present in the fuel rod can be assimilated to a diatomic gas mixture. The helium initially present in the fuel rod and the gaseous species produced, which are xenon and krypton. So the molar mass of the gas in the fuel rod can be written as the sum of the molar mass of the helium and the molar mass of the fission gases named M_{MFG}.

$$\mathrm{M} = xM_{MFG} + (1-x)M_{He} \quad (2)$$

The molar mass of M_{MFG} fission gases is a known ratio, depending exclusively on the type of fuel used. It's a mixture of xenon and krypton, in a proportion of $\frac{x_{Xe}}{x_{Kr}}$=17 [3].

At the first order, the equation of state does not take pressure into account when calculating velocity. By including Virial equation of state at least to first order, it is possible to include pressure in the estimate. With β_{mix} and γ_{mix} related to Virial equation of state coefficient, and P the pressure of gas:

$$C^2 = \frac{\gamma RT}{xM_{MFG}+(1-x)M_{He}}(1+\beta_{mix}P+\gamma_{mix}P^2+\cdots) \quad (3)$$

So, knowing all these parameters; the gas temperature being estimated by a thermocouple and the pressure being known by a manometer, we can estimate the ratio x which corresponds to the molar fraction of fission gases produced in the fuel rod.

III. Experimental measurement

A. Material and method

The first tests are carried out at room temperature in a pressurized chamber, an image of which is shown in Fig 3. The sensor is placed inside, and sealed contacts are used to power the sensor and retrieve data. The sensor is connected to a TEKTRONICS TDS3032 oscilloscope and a Panametrics 5800 pulse (Dirac) generator. Two types of gas will be tested: firstly, pure helium, so as to calibrate the sensor and secondly, a mixture of helium, Xenon and Krypton, representative of the gas that might be found in a fuel rod.

Fig 3: Experimental pressure test bench

The sensor under test is a so-called transmission sensor with two piezoelectric elements facing each other as shown in Fig 2.

B. Gas characterization

First, the sensor is tested in pure helium at a pressure of 50 bar. The test chamber in which the sensor is housed is kept at a temperature of 25 degrees. Fig 4 shows two echograms of the same transducer, the different peaks representing multiple reflections inside the transducer's acoustic cavity. Two measurements are made, the first at 50 bar and the second at 25 bar. As the pressure is lower, echo attenuation is greater, as the medium is less dense. However, as long as two echoes are distinguishable from the noise, it is possible to measure the time of flight of the acoustic wave and thus obtain an estimation of the velocity.

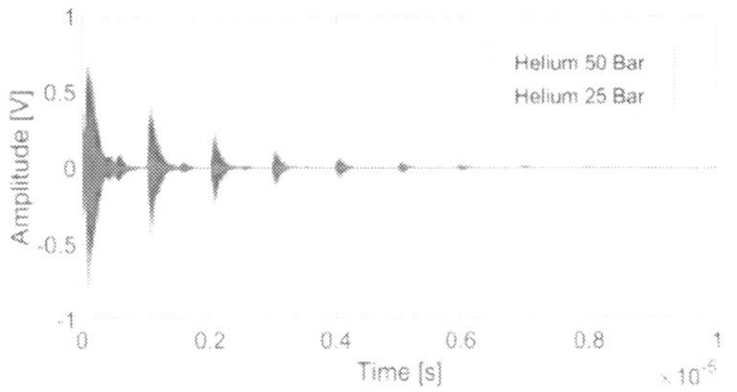

Fig 4: Echograms from the acoustic sensor tested at 50 and 25 bar in helium

For the estimation of the acoustic velocity, there is an uncertainty in the height of the acoustic cavity, which is machined to a height of 10 millimeters. But the machining uncertainties and the bonding process lead to uncertainties in its actual height, which is specific to each sensor. Given that we know the nature of the gas, which is helium, and its thermodynamic parameters (pressure and temperature), we can accurately determine the theoretical speed of sound in the gas from NIST data (Table I) [10];

TABLE I
Speed of sound in helium

	Speed of sound (m/s) at 25°C and 50 bar	Uncertainty
Data from NIST	1038.2	0.1%

For the sensor tested in this way, the real height of the cavity was estimated at h=10.2 mm; from the echogram it is possible to find the speed of sound in helium, which comes from the NIST data. Then Fig 5 shows the echogram of the sensor, this time in a mixture of helium, xenon and krypton, representative of the type of mixture that might be encountered when monitoring gas release in a fuel rod.

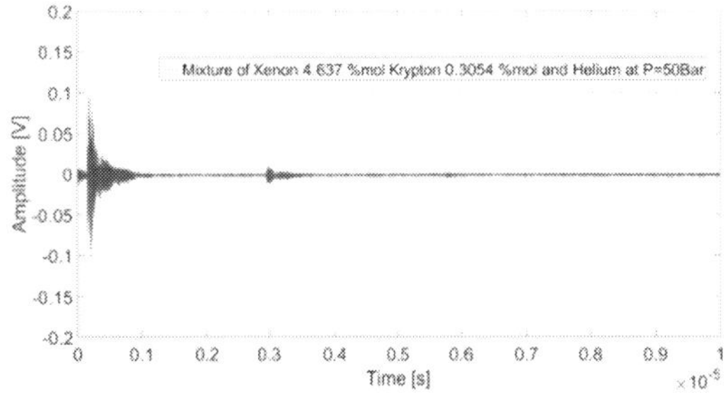

Fig 5: Echograms from the acoustic sensor tested at 50 for a mixture of Helium, Xenon and Krypton

The Table II shows the composition of the gas supplied by the manufacturer, together with its uncertainty. The same table shows the composition estimated by the Virial equation from the velocity measured by our sensor (Fig 5).

TABLE II
Comparison between gas composition estimated by the manufacturer and Sensor measurements

DATAS FROM GAS SUPPLIER (%MOL)		Uncertainty from gas supplier (%mol)		Measurement of the sensor (%mol)	Relative deviation for each gas (%)
Xenon	**4.637**	Xenon	0.046	**4.313**	7.5
Krypton	**0.305**	Krypton	0.003	**0.284**	7.3
Helium	**95.057**	Helium	0.951	**95.403**	0.36

C. *Discussion*

Under laboratory conditions, initial measurements show a difference between the experimental and theoretical results, i.e. the manufacturer's data and our measurements, of between 7 and 8% (Table II). This accuracy is sufficient for our applications, but these measurements are carried out at room temperature and at a pressure of 50 bar, which gives a sufficient signal-to-noise ratio, as shown in Fig 5. It should be noted that the experimental rods in RJH will be at a temperature of around 300°C and a pressure of between 75 and 155 bar. The next stage of the tests is to combine the characterization of a gaseous mixture at a higher temperature, at least up to 300°C. According to the studies in [9], we should expect a decrease in acoustic performance. But at high temperature the pressure should be far more than 50 bar, we hope that this increase of pressure will compensate the diminution of acoustic performances.

IV. CONCLUSION

In this article, we have essentially presented the first experimental tests of our gas sensor. The materials used for the sensor's manufacturing make it an innovative sensor, unique at present in the state of the art. The sensor was able to characterize a mixture of three different gases with an accuracy of less than 10%, on the first sensor tested. Currently, a series of more than a dozen sensors similar to this one is being tested. And in the coming months, we plan to carry out further tests at a temperature of 300°C to validate the operation of our prototypes at high temperatures.

ACKNOWLEDGMENT

This work was financed by a partnership between IES and the CEA center of Cadarache and more specifically the LISM laboratory and the INSNU project. The authors would like to thank the CTM (Centrale de technologie en micro et nanoélectronique de Montpellier).

REFERENCES

[1] F. Baudry, E. Rosenkrantz, P. Combette, D. Fourmentel, C. Destouches, et J. Y. Ferrandis, « Design of an acoustic sensor for fission gas release characterization devoted to JHR environment measurements », *EPJ Web Conf.*, vol. 253, p. 04028, 2021, doi: 10.1051/epjconf/202125304028.

[2] Jy. Ferrandis *et al.*, « Acoustic instrumentation of the new generation of MTR: effect of nuclear radiation on modified Bismuth Titanate piezoelectric elements », *EPJ Web Conf.*, vol. 225, p. 04012, 2020, doi: 10.1051/epjconf/202022504012.

[3] E. Rosenkrantz, J. Y. Ferrandis, F. Augereau, T. Lambert, D. Fourmentel, et X. Tiratay, « An Innovative Acoustic Sensor for In-Pile Fission Gas Composition Measurements », *IEEE Trans. Nucl. Sci.*, vol. 60, n° 2, p. 1346-1353, avr. 2013, doi: 10.1109/TNS.2013.2252624.

[4] D. Fourmentel, J. F. Villard, J. Y. Ferrandis, F. Augereau, E. Rosenkrantz, et M. Dierckx, « Acoustic Sensor for In-Pile Fuel Rod Fission Gas Release Measurement », *IEEE Trans. Nucl. Sci.*, vol. 58, n° 1, p. 151-155, févr. 2011, doi: 10.1109/TNS.2010.2090899.

[5] F. Very *et al.*, « Acoustic Sensors for Fission Gas Characterization in MTR Harsh Environment », *Phys. Procedia*, vol. 70, p. 292-295, 2015, doi: 10.1016/j.phpro.2015.08.157.

[6] D. Lapeine, J.-Y. Ferrandis, O. Gatsa, F. Pascal, et P. Combette, « Study and optimization of screen-printed PZT transducers for acoustic applications », in *2016 IEEE International Conference on Dielectrics (ICD)*, Montpellier, France: IEEE, juill. 2016, p. 253-256. doi: 10.1109/ICD.2016.7547592.

[7] O. Gatsa, E. Rosenkrantz, D. Fourmentel, C. Destouches, P. Combette, et J.-Y. Ferrandis, « Validation of the first prototype high temperature ultrasonic sensor for gas composition measurement », présenté à 2017 ICU Honolulu: Sixth International Congress on Ultrasonics, Honolulu, Hawaii, USA, 2017, p. 030001. doi: 10.1121/2.0000679.

[8] O. Gatsa, « Development of acoustic sensors for the extension of measurements to high temperature in the experimental reactors », p. 177.

[9] O. Gatsa, P. Combette, D. Fourmentel, C. Destouches, et J. Y. Ferrandis, « High temperature ultrasonic sensor for fission gas characterization in MTR harsh environment », p. 9.

[10] W. Wagner et A. Pruß, « The IAPWS Formulation 1995 for the Thermodynamic Properties of Ordinary Water Substance for General and Scientific Use », *J. Phys. Chem. Ref. Data*, vol. 31, n° 2, p. 387-535, juin 2002, doi: 10.1063/1.1461829.

Experimental analysis of natural convection in a differentially heated cavity

Tommaso Carlesi[1,2,3,*], Delphine Laboureur[1], Philippe Planquart[1], Julio Pacio[3], Katrien Van Tichelen[3] and Patrice Mégret[2]

[1]von Karman Institute for Fluid Dynamics, Belgium
[2] University of Mons, Belgium
[3] Belgian Nuclear Research Centre (SCK CEN), Belgium

(*) tommaso.carlesi@vki.ac.be

Abstract— **Turbulent heat fluxes (THFs) estimation is of paramount importance in the determination of heat transfer in fluids. Numerical models for low Prandtl number fluids are still unreliable and their experimental evaluation is a challenging task since it requires simultaneous measurement of fast velocity and temperature fluctuations. In nuclear applications, a better understanding of THFs in liquid metals could lead to more precise predictions of the primary coolant temperature for the evaluation of nominal operation (forced convection regime) and accidental conditions (mixed and/or natural convection regime). The first part of this work focuses on the selection of the measurement techniques suitable for water, GaInSn and LBE and the thorough literature review required. Tests in different setups led to the choice of sheathed type K thermocouples and fiber Bragg gratings for temperature measurements and Ultrasound Doppler Velocimetry and Hot Wire Anemometry for velocity measurements. The comparison carried out among the different techniques underlines advantages and limitations of each of them. Calibration of each technique is performed and cross-effects of temperature and velocity are evaluated. Uncertainty analyses are also carried out. To conclude, first results obtained in a differentially heated cavity made of stainless steel 316L with an edge of 60 mm are presented. DNS numerical simulations are performed to know the ranges of the quantities to be measured and to have results available for comparison with experiments.**

Keywords **—Natural Convection, Differentially Heated Cavity, Turbulent Heat Flux, Nuclear Power Reactors and Nuclear Fuel Cycle**

I. Introduction

One of the biggest challenges to be faced nowadays is to fulfill the increasing energy demand while meeting the requirements of carbon neutrality imposed by the Paris agreement [1]. Most of the successful scenarios foreseen by the International Energy Agency see an increase in the use of nuclear energy for electricity production [2] and consequently in the amount of nuclear waste to be treated. To tackle the waste problem, transmutation performed in generation-IV nuclear reactor systems seems a valuable solution.

This paper is conceived in the framework of the design and construction of such a system: MYRRHA (Multi-purpose hYbrid Research Reactor for High-tech Applications), a pool-type fast-spectrum research reactor under design at SCK CEN (the Belgian nuclear research center). In particular, this work aims at the understanding of the thermal hydraulic behavior of its primary coolant, the liquid metal Lead Bismuth Eutectic (LBE). For the design and safety evaluation of MYRRHA, the temperature of the primary coolant and of the reactor core components must be predicted in nominal operation (forced convection regime) as well as in accidental conditions (mixed and/or natural convection regime). For this assessment, the turbulent heat flux (THF) as it appears in the Reynolds Averaged Navier-Stokes equations (RANS) is a term of paramount importance. It describes the convective heat transfer mechanism related to fast and simultaneous turbulent fluctuations of the local fluid velocity (U) and temperature (T). Their evaluation is usually performed with a combination of numerical simulations and experiments [3], leading to the availability of reliable numerical models for common fluids like water or air. On the other hand, for lower Pr number fluids (e.g. liquid metals) the available validation data are still scarce and unreliable, especially in the natural convection regime and therefore new experiments are needed for the development of numerical models [4,5,6].

Measurement of THF is challenging as it requires simultaneous measurement of velocity and temperature fluctuations. Some attempts have been made in air over a vertical backward facing step with Laser Doppler Velocimetry (LDV) and cold wire anemometry, used to measure the velocity and temperature fluctuations respectively [7]. Other papers address the same issue in water, performing measurements of the turbulent natural convection boundary layer on a vertical plate using hot films and thermocouples [8] or in a closed cavity with an isothermal wall using Particle Image Velocimetry (PIV) and Laser Induced Fluorescence (LIF) [9]. Despite of the difficulties already encountered in these works, these campaigns do not present the additional challenges that must be faced in liquid metals (LMs), especially with respect to high temperatures, opacity and material compatibility. Nowadays techniques for the individual measurement of velocities and temperatures in LMs exist but simultaneous and fast measurements of these two quantities in such environment have not been published yet. In fact, most of the work addressing liquid metal measurements are focused on average quantities [10], finding correlations in a particular setup [11] or large-scale dynamics [12].

https://doi.org/10.1051/epjconf/202328805002

An application more similar to the current project is the turbulence measurement performed in Galinstan (GaInSn) in a non-isothermal vertical confined backward facing step [13]. Both forced and mixed convection regimes have been studied. Velocity measurements have been performed with a newly developed Permanent Magnet Probe (PMP) probe while temperature measurements, useful also for velocity corrections, have been made with standard type K thermocouples. Despite of the high acquisition frequency (5000 Hz with a cutoff frequency of 200 Hz), a full turbulent heat flux measurement has not been performed because the probes were sampled by two different measuring systems. For this reason, synchronization of the two measurements was impossible to achieve and consequently only first order statistics have been calculated.

In this context, the focus of this Ph.D. is twofold:

- Develop a turbulent heat flux sensor (THFS) suitable for THF measurements in liquid metals (EPICURUS, ExPerImental teChnique for tURbUlence in liquid metalS)
- Provide reference data for the development and validation of numerical models for THF in liquid metals

The present paper will give an overview of the selection of the techniques for the measurement of temperature and velocity and will focus on the description of the experimental setup presenting and discussing the first results obtained in differentially heated cavity with water as working fluid.

II. Methods

This section describes the methodology adopted in the paper, starting with a description of the requirements of the sensors and the consequent choices made. At the end, a description of the experimental setup is carried out.

A. Requirements of the sensors

The selected sensors must satisfy specific requirements in terms of spatial and temporal resolution, temperature and velocity measurement range and capability of withstanding high temperature and corrosive environment. All these characteristics are summarized in Fig.1 and translated in physical specifications that the sensor must meet.

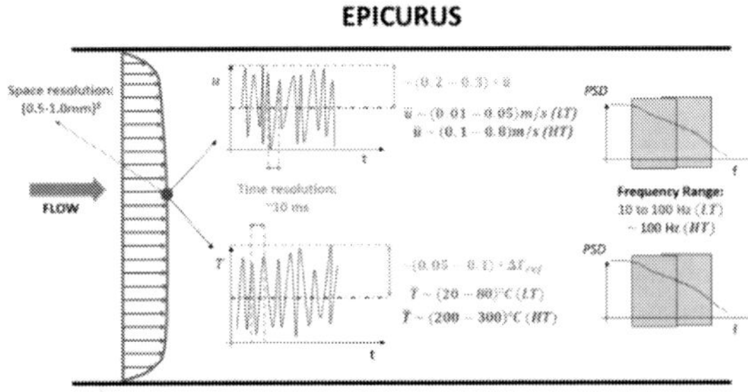

Figure 1 - Sensor requirements with distinction of the low temperature and high temperature cases of the Ph.D. project

Many of these choices are intrinsically related to the physics of the liquid metals. As the experiments presented in this report utilize water, the emphasis of this section will be on discussing the sensor characteristics that are applicable to its use in water. On the other hand, GaInSn experiments are also part of the Low Temperature (LT) phase of the project, therefore the more strict sensor requirements given by the liquid metals are also considered.

The temperature and velocity range needed by the sensor is determined by the characteristics of the turbulent fluctuations. In particular, the natural convection flow regime establishes the oscillations frequencies in the order of 10-100 Hz [9,14,15,16,17] increasing for lower Pr number.

Requirements for water are generally less strict but typical sampling frequencies found in literature for turbulence measurements are in the range 1-95 Hz [9,16,17] and for this reason liquid metals constraints are still taken as a reference here. As a consequence, very strict time response requirements are established for the sensor (ideally lower than 10 ms) to be able to follow the fluctuations even in the fastest case. In general, this will translate in a small dimension of the sensor itself meeting also the constraints given by the spatial resolution. To avoid aliasing effects and alter the frequency content of the signal, it is also advisable to set the target frequency as double of the frequency of the physical phenomenon. Nevertheless, this issue does not affect the time signal, which is of primary interest in this work.

At the same time, an appropriate acquisition system must be chosen. As a rule of thumb, the acquisition frequency should be roughly 10 times higher than the target frequency. A wise choice of commercial systems available on the market should be sufficient in this sense.

In general, the requirements on sensor response time are supposed to be less strict in the LT phase of the project than in the HT. Nevertheless, it is important to consider that every coating or adaptation of the sensor to high temperature is expected to increase its dimension and consequently its response time.

Similar considerations are valid for the velocity magnitude range. Most probable velocities in natural convection are expected in the range of cm/s in GaInSn [18] and mm/s [19] in water. Table 1 reports a summary of the fluctuations characteristics retrieved from the DNS simulation in GaInSn [18].

Table 1 - Fluctuations characteristics

Variable	Max Frequency	Amplitude
Temperature	60 Hz	15%
Velocity	30Hz	16%

In the experiments presented in this paper, the use of water as working fluid could lead to lower fluctuations amplitudes (for a given Gr number) that might be more challenging to be measured.

The smallest scale of the turbulence that is interesting to measure determines the spatial resolution needed. In terms of instrumentation, this is directly related to the size of the sensor. The Kolmogorov scale has been estimated to be in the order of

10μm for GaInSn in natural convection [18] and it is reasonable to think that a similar value can be obtained also in the LBE setup. These values are far smaller than the best achievable with experiments. Moreover, it is not meaningful to reach such small scales since no useful information about the dissipation rate could be deduced anyway. For this reason, numerical results and literature studies suggest that the best compromise would be to resolve turbulent structures with a 1 cm dimension meaning that the probes should have a spatial resolution in the order of 0.5-1mm. Once again, liquid metals requirements are taken as a reference also for water experiments.

B. *Choice and characterization of the sensors*

According to the requirements expressed in the previous Section, some suitable measurement techniques were selected. Regarding the velocity measurements, optical techniques like PIV or LDV were discarded because of the opacity of liquid metals and therefore the lack of optical access in the setup even for the water phase of the experiments. Other techniques like LFV, CIFT, X-ray or neutron imaging were also discarded mostly for their scarce spatial or temporal resolution. Among the more reasonable techniques considered, the Pitot tube was considered not suitable for the measurements, mainly for a matter of spatial resolution and instrusivity in the flow. Indeed, is hard to keep the probe smaller than 1 mm

The **Hot Wire Anwmometry (HW)** and the **Ultrasound Doppler Velocimetry (UDV)** were both selected for the project. The former has proved to give very good performances when measuring turbulence and applications in liquid metals (even if not very recent) have been found in literature [20]. The intrusiveness given by the support and the prongs can be overcome with an accurate design of the facility and does not really affect the spatial resolution that is instead given by the dimensions of the wire. The probe was purchased by DANTEC and its main characteristic is to have the sensitive part (a nickel film smaller than 1μm) sputtered around a core in quartz and covered by another quartz layer that ensures electrical insulation and protection.

The UDV has proven to be one of the best technique for velocity measurements when optical access is not available. For this reason, its use in liquid metals is strongly supported by many literature studies [21, 22, 23]. Moreover, the maximum axial spatial resolution of the chosen commercial sensor by Signal Processing is 0.126mm while its volumetric spatial resolution depends on the aperture of the US cone and therefore on the emitting frequency. The temporal resolution can arrive up to few ms. The main limitations will be given by low velocities and working environment and cannot be precisely predicted before. Despite of that, there is no reason to think that they cannot be overcome.

Regarding temperature measurement techniques, **thermocouples** have been used in liquid metals for years and often with good results. Despite the difficulty of finding the best type, there is no reason why they should not be tested in GaInSn and LBE. For GaInSn and water measurements, a sheathed K-type thermocouple with a predicted time response around 30 ms will be tested [24]. The HBM QuantumX MX440B acquisition system was chosen allowing acquisition frequencies up to 5000 Hz thanks to its integrated cold junction compensation.

Fiber Bragg Gratings (FBGs) were also selected for temperature measurements. Preliminary tests [24,25] showed that the time response of bare FBGs could be even better than the one of unsheathed standard K-type thermocouples presenting similar dimensions and therefore intrusiveness and spatial resolution. Another big advantage is that, being made of glass, no particular problem should arise in terms of chemical corrosion in the liquid metal stages. Moreover, applications of coated FBGs up to 500 °C is reported in literature and acquisition frequencies in the order of 1000 Hz are possible with the BaySpec interrogator. Once again it is important to underline that the above mentioned techniques can also be adapted to LBE to withstand the higher temperatures and the chemical corrosion.

C. Experimental setup

In this section, a detailed description of the experimental setup used for the investigation is provided, including some design choices and some procedures followed for its usage. By understanding the experimental setup, we can better interpret and analyze the results obtained from the experiments and draw meaningful conclusions about the behavior of the system under study.

Natural convection was chosen as flow regime to be focused on this first stage, not only because the turbulent frequencies are lower and the facility will be easier to operate [26] but mostly because validated turbulence models for natural convection are not available and experimental data for this flow regime are strongly needed.

Because most of the studies in literature focus on the Rayleigh-Bénard convection [18, 27, 28, 29] the differentially heated cavity setup was chosen for the purpose of this experiment to fill this knowledge gap. A schematic of the setup is shown in Fig. 2.

In differentially heated cavities, a constant temperature difference is maintained between two opposite side walls while all other sides of the cavity are insulated, i.e. ideally adiabatic. This configuration usually gives a flow with similar characteristics to the isothermal vertical plate on its active walls and for this reason it is also called vertical convection (VC) [11] Inside the cavity, the imposed temperature difference triggers a natural convection loop rotating in the direction from the warmer side towards the colder side [30]. The main features of the flow change according to the working fluid but in all cases the main non-dimensional parameters that control the flow and the turbulence phenomena inside the cavity are the Prandtl number (Pr), the Grashof number (Gr) and the Rayleigh number (Ra = Gr*Pr).

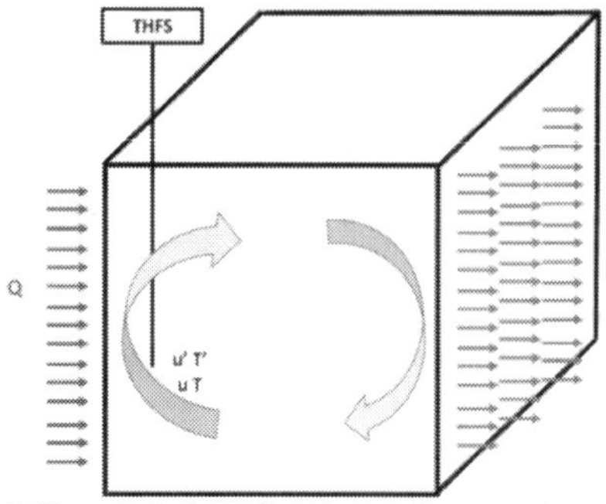

Figure 2 - Differentially heated cavity used as experimental setup for the experiments

In particular, increasing the Gr (Ra) number promotes instabilities leading to fully turbulent flows. Previous studies have shown a critical Grashof number of about 10^8 where the transition to turbulence in buoyant flows in cavities is expected [15,27]. Other studies place the transition further, e.g. Ra from 10^{12} to 10^{14} [8].

The tests performed for this project were conducted in a cubic cavity made of 316L steel with an (inner) edge of 60 mm. The thickness of the walls was 10 mm.

Inside natural convection was established through heated/cooled Peltier elements from Laird Thermal Systems capable of reaching a maximum heating power of 800 W and a maximum cooling power of 300 Wsupplied by an EA-PSI 9000 2U. Full power was not required for the purpose of this experiment. The two elements were positioned oppositely on each side of the cavity in contact with two copper walls inserted in place holders in the cavity.

The copper walls were coated with a 1μm carbon coating deposited through magnetron sputtering at the University of Namur to ensure material compatibility during future tests with liquid metals.

Two fans were added in contact with the Peltier elements to avoid overheating. On the hot wall a fan supplied with 12 V in DC was installed while on the cold wall a fan of 24 V was cooling the hotter face of the Peltier element.

The working fluid (water in the present case) was fed through an opening on the top wall controlled by a ball-valve directly linked to a container filled with the fluid. Another ball-valve placed on the bottom of the cell was instead used to empty the cavity. One last ball-valve can be used to feed inert gas used to work in low-oxygen environment during the liquid metal experiments. All the non-used ingresses on the top needed as temperature instrumentation access are kept closed with a plastic cap screwed during operations.

As already mentioned in this Section, the four walls that are neither heated nor cooled are ideally adiabatic. To ensure the minimum amount of heat loss from these walls, an insulation made of a combination of Insulfrax and aluminium foil was made.

Due to the many irregularities on the cavity surfaces, e.g., valves and openings for instrumentation, it was not possible to place effectively the insulation on every wall. As a result, the adiabaticity of the cavity during the operations was not fully ensured and a global averaged heat transfer coefficient h = 20 W/(m^2 K) was experimentally retrieved.

An important point for the design of the experiment was the access for the instrumentation. In fact, in such a small setup it is not trivial to find the best compromise between the available space and the optimal locations of the sensors. Access for the instrumentation was accurately designed according to previous Direct Numerical Simulations (DNS) performed in GaInSn [18]. The simulations showed that the higher intensities in terms of THFs are concentrated close to the copper plates (i.e. the heated/cooled walls), in particular in the upper part of the hot plate and and the lower part of the cold plate. The same distribution can be retrieved for the temperature fluctuations. A final picture of the real cavity with the main elements highlighted is reported in Fig. 3.

Figure 3 - Natural convection setup: 1) Moving chariot moved by an electrical Maxon Motor 110209; 2) TC + FBG access; 3) Valve to feed the fluid; 4) Valve to feed the inert gas; 5) Cavity; 6) Fan; 7) UDV access; 8) Valve to empty the cavity; 9) Support; the whole setup is mounted on an optical table to ensure stability and precision during the measurements. The non-heated walls are therefore insulated with a layer of Insulfrax®.

In general, the most interesting zones where the sensors should be placed are close to the walls. Here the turbulent phenomena are more accentuated and the higher amplitudes reached by temperature and velocity will be easier to measure.

The x and z coordinates of the measurements are limited to 4 combinations (x = 4 mm; 56 mm and z = 12 mm; 48 mm) while the y direction is in principle fully accessible by all the sensors. The access for the temperature sensors (TC and FBG) was guaranteed from the top, while the UDV was flush-mounted on the bottom wall. In this way, the UDV could acquire the profile of the y-component of the velocity while the other sensors could move along the vertical line through the moving chariot. At the same time, the UDV is placed such that the axis of the cylindrical sensor (i.e. the center of the ultrasonic beam) is in correspondence with the temperature sensor tip so that the last meaningful point of the velocity profile measured by the UDV coincides with the point where the temperature sensors acquire data. In this way, the theoretical need of measuring temperature and velocity "at the same point" to obtain the THF was satisfied the best way possible.

III. Results

In this section, the results obtained in the experimental setup previously described are presented and commented. As the hot wire was not tested in the cavity, a separate section is dedicated to the preliminary results obtained regarding this technique. For more information about the preliminary results obtained with each techniques in other setups and presented at the conference, please contact the author.

A. Hot wire results

The experimental results obtained during the calibration of a hot wire for utilization within a differentially heated cavity are shown in Fig. 4.

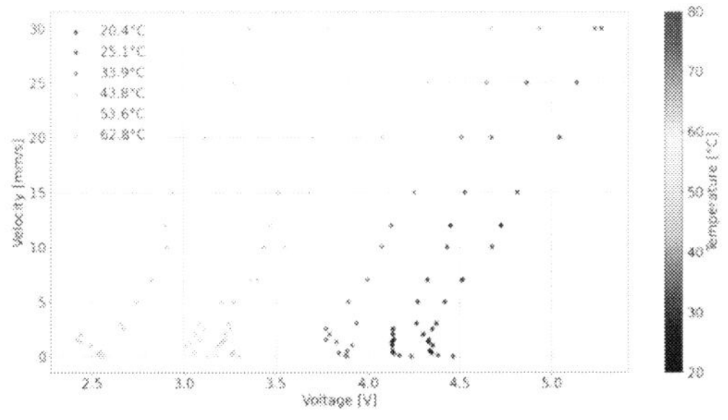

Figure 4 - Hot wire calibration for different temperatures

A distinct correlation emerges between lower temperatures and higher voltage outputs, showcasing a remarkable sensitivity of 50 mV/°C to temperature changes. This high sensitivity, while indicative of the precision of the measurements, provides a significant perspective on the system's thermal dynamics. However, in this specific case, excessive sensitivity to temperature proves to be less advantageous, since my primary interest lies in measuring velocity, and the temperature sensitivity becomes an undesirable and bothersome secondary aspect. Moreover, the tests display an encouraging level of repeatability, validating the robustness of the methodology employed. However, the behavior of the system becomes somewhat ambiguous when operating below 5 mm/s, indicating the need for further investigation in this regime. As the temperature climbs, the trend in the data tends to diminish in clarity, suggesting the emergence of additional influencing factors. These findings collectively emphasize the importance of considering temperature effects, repeatability, and velocity regimes when interpreting the hot wire calibration outcomes for applications within a differentially heated cavity.

B. Results in the differentially heated cavity

Experimental results are therefore available in different points along the vertical lines of the positions shown in Fig 4 to respectively represent the flow behaviour close to the hot and cold wall.

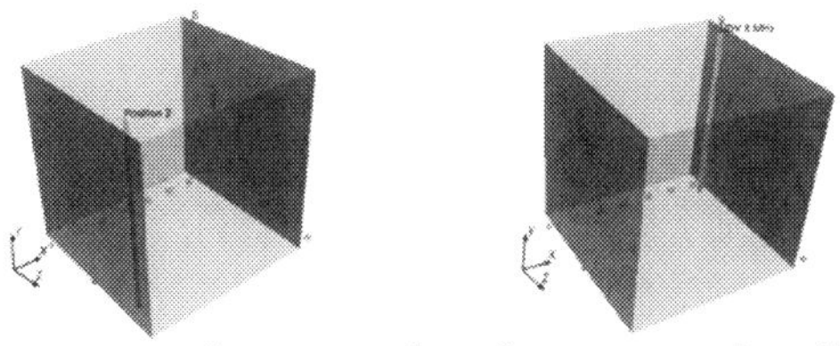

Figure 5 - Vertical positions where the experimental profiles are measured

The results are also presented and compared with Direct Numerical Simulations (DNS) results previously performed in the same setup. More information about how these tests are carried out can be found in the literature [19].

The numerical simulations are performed keeping the cold wall at T = 20°C and the hot wall at T = 80°C, with an average ΔT = 60 °C. On the other hand, due to practical constraints and limitations, the experimental tests were carried out keeping the cold wall at T = 21°C and the hot wall at T = 74°C, i.e., with an average ΔT = 53 °C. Anyway, these conditions still correspond to a Grashof number of $2x10^8$, i.e., to the limit of the transition to turbulence according to some of the studies previously cited [8,15] therefore keeping the results in the same regime of the ones obtained numerically.

The drift of the temperature difference was 0.16 °C /min leading to a maximum change of 0.25°C during the time of the acquisition which corresponded to 50 flow-around times like in the numerical simulations. The imposed ΔT can therefore be considered constant during the measurements. First of all, the average velocity profile along the lines is reported in Fig. 6 and Fig. 7. The experimental data are presented with the uncertainty calculated according to the Guide of of expression of Uncertainty in Measurement [31].

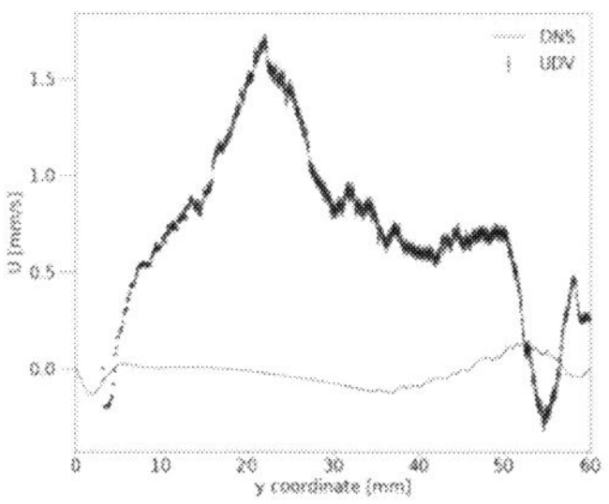

Figure 6 - UDV velocity profile in position 2 (hot wall) compared with numerical results

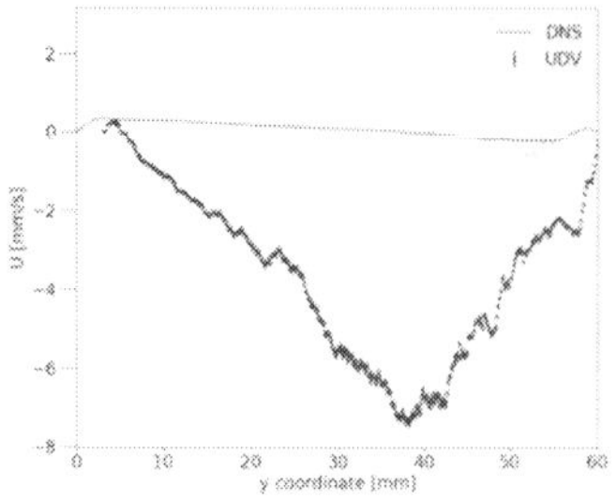

Figure 7 - UDV velocity profile in position 3 (cold wall) compared with numerical results

To establish a meaningful comparison between the experimental and numerical profiles, a careful averaging process was employed to account for the spectral weight of the available DNS data. The velocity data for each point in the UDV profile was obtained by averaging measurements from particles within a cone volume with a 5° aperture angle. While the velocity profiles exhibit the expected direction, aligning with the flow towards the sensor near the cold wall (corresponding to downward motion along the y-axis) and vice versa near the hot wall, there exists a notable order of magnitude difference between the experimental and numerical results. This unexpected divergence in velocities can be attributed to the experimental setup's design for liquid metals, potentially causing limitations in the UDV sensor's ability to accurately measure such low velocities.

Furthermore, distinct features present in the experimental profile are absent in the DNS simulations, potentially stemming from inherent characteristics of the experimental technique. The initial segment of the experimental profile, situated in close proximity to the UDV sensor, is influenced by the near field, often leading to unreliable measurements. To enhance the signal strength farther along the profile, a higher emitting power was employed, exacerbating this near-field effect. Additionally, a discrepancy arises between experimental and DNS results at position 2 near the upper wall (y = 60 mm), where a slight velocity decrease and change in direction occur. This phenomenon, potentially linked to recirculation induced by the impinging jet effect of the horizontal flow against the wall, contrasts with DNS predictions.

Contrarily, the same phenomenon does not manifest in the profile obtained at position 3, warranting further investigation into its physical or artificial nature. Considering the experimental constraints, attempts to attribute physical interpretations to the velocity profile, particularly regarding rms values, yielded inconclusive outcomes due to the risk of electronic noise overpowering genuine turbulent fluctuations at such low velocities. Consequently, exercising caution in drawing conclusions from incompletely understood phenomena is imperative at this stage. Notably, the vast disparity between DNS predictions and experimental measurements, likely attributable to the sensor's limitations, underscores that achieving direct alignment between the two results will not be pursued in the current phase where water serves as the working fluid.

Further considerations regarding the flow in the cavity can be made for temperature profiles. The ΔT imposed to the cavity has already been commented at the beginning of this section. The average temperature profile along the position 2 and 3 inside the cavity is shown respectively in Fig. 8 and Fig. 9.

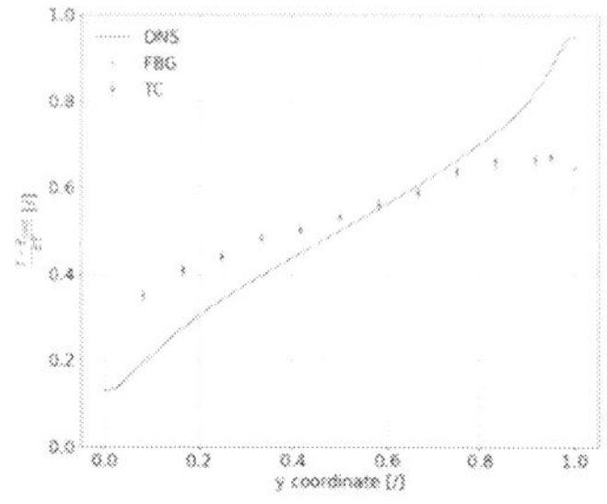

Figure 8 - Temperature profile along position 2 (hot wall) measured with thermocouple and Fiber Bragg gratings compared with numerical results

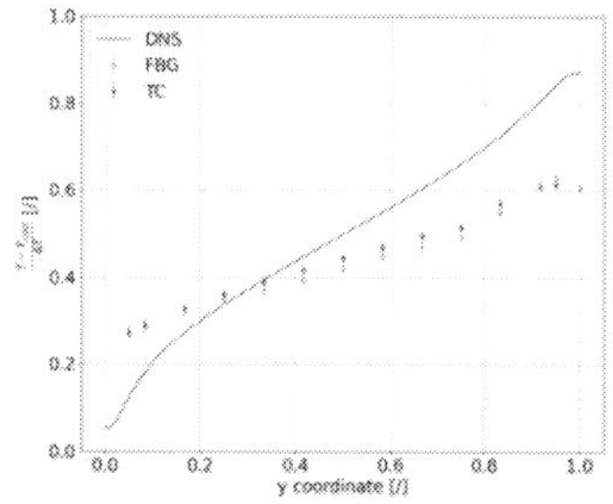

Figure 9 - Temperature profile along position 3 (cold wall) measured with thermocouple and Fiber Bragg gratings compared with numerical results

The comparison between FBG and TC measurements reveals good agreement, accounting for uncertainties [19]. Conversely, it becomes evident that the experimental data generally do not align with the numerical temperature profile, even when considering uncertainties. While the temperature magnitudes are consistent, the slope of the numerical profile is notably steeper. Unlike the velocity measurements, this discrepancy cannot be attributed to sensor sensitivity issues, indicating the reliability of experimental temperature outcomes and enabling certain deductions.

Several factors might contribute to the observed discrepancies. Firstly, the slight difference in imposed ΔT (temperature difference) - $\Delta T_{exp} = 53$°C and $\Delta T_{DNS} = 60$°C - could potentially explain the steeper numerical profile. Furthermore, inherent heat losses within the experimental setup, attributed to non-perfectly adiabatic insulated walls, influence the temperature distribution. To account for these losses, measurements were conducted on five out of six cavity walls, excluding the back wall used for support. A global heat transfer coefficient estimation of 20 W m−2 K−1, reasonable for air in natural convection, was obtained through tests filling the cavity with hot water. These heat losses, negligible compared to the Peltier power input (approximately 100 W), were factored into a Biot number calculation (Bi = 0.007), suggesting that the stainless steel wall's thickness allows for uniform temperature assumption and neglect of conduction losses.

Although neither the numerical nor the experimental analyses reveal distinctive temporal characteristics at a specific point, post-processing of data is omitted due to this lack of differentiation. Further exploration is needed to comprehend the divergence between numerical and experimental results and to conduct meaningful liquid and thi experiments. The discrepancies between experimental and simulated temperature profiles persist, even when accounting for error bars. Multiple potential causes contribute to this disparity, including the influence of heat losses, discrepancies in the temperature difference, and non-uniform temperature distribution on the cavity walls. To address these issues, further analyses are warranted, and symmetric measurements at the cold wall could provide additional insight into the flow behavior within the experimental setup.

IV. Conclusion and future works

This study addresses the intricate challenge of quantifying turbulent heat fluxes (THFs) within liquid metals, characterized by the rapid and simultaneous fluctuations of fluid velocity and temperature. To unravel this complexity, the research focuses on an initial experimental phase, meticulously crafting the experimental framework and adopting water as the primary working fluid. Commencing with an insightful survey of cutting-edge techniques for velocity and temperature measurements, the paper judiciously evaluates the suitability of each method for application in liquid metals. In this deliberation, Ultrasonic Doppler Velocimetry and Hot Wire Anemometry emerge as the optimal choice for velocity measurements, while temperature measurements find their precision through the utilization of thermocouples and Fiber Bragg Gratings. The spotlight then turns toward the design of the differentially heated cavity, meticulously selected as the platform for investigating natural convection. This section provides an exposition of the cavity's architectural intricacies, accompanied by a lucid exposition of the underlying rationale guiding the chosen configurations. Presenting a significant milestone, the paper presents preliminary outcomes stemming from water-based experimentation, thoughtfully juxtaposed against the backdrop of available Direct Numerical Simulations (DNS) results. These comparisons enable a preliminary assessment of the methodology's viability and provide a foundation for future exploration. The paper's achievements are underscored by the successful development of an innovative experimental apparatus, tailor-made for unraveling the complexities of turbulent heat flux across diverse fluid media. The meticulously devised cavity facilitates accurate measurements, with the initial water-based tests serving as a prudent step to preclude any unforeseen intricacies before embarking upon experiments involving GaInSn or LBE. It is pertinent to note that a comprehensive turbulent analysis is reserved for subsequent stages of the research, as the current phase, constrained by time, has yet to unveil substantial turbulence. This calculated progression epitomizes the paper's commitment to methodical and rigorous scientific inquiry. However, the comparison between experimental and numerical results shows a significant discrepancy at this stage. The authors refrain from drawing major conclusions about the mass and heat transfer phenomena in the cavity until the cause of this mismatch is clarified. They propose and analyze possible reasons for the disparity in the results in the corresponding section which will be crucial for future improvements. Importantly, the mismatch cannot be attributed to design or measurement technique selection errors, as the cavity is designed for GaInSn experiments, and water may not yield identical outcomes. Further investigations are required to fully comprehend the source of the discrepancies. Future work will focus on identifying the main reasons for the mismatch and addressing them. The researchers plan to conduct meaningful experiments with water, comparing the results with DNS data. Additionally, they intend to test the Hot wire technique in the same setup to measure velocity at high frequencies. With synchronized acquisition systems, calculations of THFs will also be possible. After completing this phase, experiments with GaInSn will be carried out, providing THFs data in liquid metals to validate numerical models.

References

[1] Working Group, I. I. I. contribution to the, W. G. I. I. I. Sixth, and Assessment Report. Climate change 2022 mitigation of climate change. Technical report, Intergovernmental Panel on Climate Change, 2022.

[2] IEA International Energy Agency. Global energy and climate model documentation. Technical report, International Energy Agency, 2021.

[3] Katrien Van Tichelen, Graham Kennedy, Fabio Mirelli, Alessandro Marino, Antonio Toti, Davide Rozzia, Edoardo Cascioli, Steven Keijers, and Philippe Planquart. Advanced liquid-metal thermal-hydraulic research for MYRRHA. Nuclear Technology, 206(2):150–163, 2020.

[4] A. Shams, F. Roelofs, I. Tiselj, J. Oder, Y. Bartosiewicz, M. Duponcheel, B. Niceno, W. Guo, E. Stalio, D. Angeli, A. Fregni, S. Buckingham, L. K. Koloszar, A. Villa Ortiz, P. Planquart, C. Narayanan, D. Lakehal, K. van Tichelen, W. Jäger, and T. Schaub. A collaborative effort towards the accurate prediction of turbulent flow and heat transfer in low-prandtl number fluids. Nuclear Engineering and Design, 366:110750, sep 2020.

[5] H.I. Abu-Mulaweh, B.F. Armaly, and Chen T.S. Turbulent mixed convection flow over a backward-facing step. International Journal of Heat and Mass Transfer, 2001.

[6] Tohru Takeshita, Takehiko Segawa, James A. Glazier, and Masaki Sano. Thermal turbulence in mercury. Phys. Rev. Lett., 76:1465–1468, February 1996.

[7] H.I. Abu-Mulaweh, B.F. Armaly, and Chen T.S. Turbulent mixed convection flow over a backward-facing step. International Journal of Heat and Mass Transfer, 2001.

[8] G. C. Vliet and C. K. Liu. An experimental study of turbulent natural convection boundary layers. Journal of Heat Transfer, 1969.

[9] M. E. Negretti T Caudwell, J.-B Flòr. Convection at an isothermal wall in an enclosure and establishment of stratification. Journal of Fluid Mechanics, 2016.

[10] A. Cramer, C. Zhang, and S. Eckert. Local flow structures in liquid metals measured by ultrasonic Doppler velocimetry. Flow Measurement and Instrumentation, 15(3):145–153, 2004. Ultrasonic Flowmetering.

[11] Lukas Zwirner, Mohammad S. Emran, Felix Schindler, Sanjay Singh, Sven Eckert, Tobias Vogt, and Olga Shishkina. Dynamics and length scales in vertical and convection of liquid metals. Journal of Fluid Mechanics, 2022.

[12] T. Zurner, F. Schindler, T. Vogt, S. Eckert, and J. Schumacher. Combined measurement of velocity and and temperature in liquid metal convection. J. Fluid Mech. (2019), vol. 876, pp. 1108–1128, 2019.

[13] Thomas Schaub, Frederik Arbeiter, Wolfgang Hering, and Robert Stieglitz. Forced and mixed convection experiments in a confined vertical backward facing step at low-prandtl number. Experiments in Fluids, 63(1), dec 2021.

[14] Tohru Takeshita, Takehiko Segawa, James A. Glazier, and Masaki Sano. Thermal turbulence in mercury. Phys. Rev. Lett., 76:1465–1468, February 1996.
[15] S. Cioni, S. Ciliberto, and J. Sommeria. Experimental study of high-Rayleigh-number convection in mercury and water.
[16] D. Auliano, M. Auliano, and E. Næss. Near wall temperature measurements and turbulent features in a waterflow at transition reynolds numbers in a square heated asymmetric cavity channel. Thermal Science and Engineering Progress, 2023.
[17] A. Aubert, S. Poncet, P. Le Gal, S. Viazzo, and M. Le Bars. Velocity and temperature measurements in a turbulent water-filled taylor-couette-poiseuille system. International Journal of Thermal Sciences, 2015.
[18] Mohammed Alsailani. Numerical simulation of transition in liquid metals. Master's thesis, Von Karman Institute for Fluid Dynamics, 2021.
[19] J. Oder, M. Alsailani, L. Koloszar, W. Munters, and D. Laboureur. Direct numerical simulation of differentially heatedcavity at low prandtl numbers. In Eleventh International Conference on Computational Fluid Dynamics, 2022.
[20] David G. Malcolm. Thermo-anemometry in magneto-hydrodynamics. Phd thesis, University of Warwick, 1968.
[21] Marc Dierckx. Feasibility of ultrasonic inspection techniques for use in the MYRRHA reactor. Phd thesis, Katholieke Universiteit Leuven, 2014.
[22] S. Eckert, S. Franke, T. Gundrum, G. Gerbeth, and J.-C. Willemetz. Applications of Ultrasonic Doppler Velocimetry to flow measurements in hot liquid metals. In 8th International Conference on Electromagnetic Processing of Materials, Oct 2015, Cannes, France, 2015.
[23] Yoshitaka Ueki, Yuya Noguchi, Juro Yagi, Teruya Tanaka, Takehiko Yokomine, Masaru Hirabayashi, Kuniaki Ara, Tomoaki Kunugi, and Akio Sagara. Ultrasonic Doppler Velocimetry experiment of lead-lithium flow with Oroshhi-2 loop. Fusion Science and Technology, pages 1–7, June 2017.
[24] Clemens Naumann. Unsteady temperature measurement in hot/cold heated cavity filled with liquid metal. Master's thesis, TU Ilmenau, 2022.
[25] T. Carlesi, D. Laboureur, P. Planquart, J. Pacio, K. van Tichelen, and P. Megret. Experimental analysis of the dynamic behaviour of fbg temperature sensors. In 25th Annual Symposium of the IEEE Photonics Benelux Chapter, 2021.
[26] NEA. Handbook on lead-bismuth eutectic alloy and lead properties, materials compatibility, thermal-hydraulics and technologies. Nuclear Energy Agency, 2 edition, 2015.
[27] J. A. Glazier and A. Naert. Evidence against ultrahard thermal turbulence at very high Rayleigh numbers. Nature, 398:307–310, 1999.
[28] Janet D. Scheel and Jörg Schumacher. Global and local statistics in turbulent convection at low Prandtl numbers. Journal of Fluid Mechanics, 802:147–173, 2016.
[29] S. Paolucci. The differentially heated cavity. Sadhana, 1994.
[30] R. J. A. Janssen and R. A. W. M. Henkes. Influence of prandtl and number on and instability mechanisms and transition and in a differentially heated square cavity. Journal of Fluid Mechanics, 1994.
[31] Evaluation of measurement data — Guide to the expression of uncertainty in measurement.

Experimental validation of a $CeBr_3$ gamma-ray logging probe MCNP model

Thomas Marchais[1,*], Bertrand Pérot[1], Pierre-Guy Allinei[1], Hervé Toubon [2], Youcef Bensedik[2], Romain Mieszkalski[2], Sebastien Hocquet [2], Dragomir Savov [2], Matthew Schubert [3], Bruno Legros [3], Patrick Van Eyll [3], Michaël Pernette[3]

[1]CEA, DES, IRESNE Institute, DTN, SMTA, Nuclear Measurement Laboratory, France
[2]Orano Mining, France
[3]Advanced Logic Technology, Luxembourg
(*) Thomas.MARCHAIS@cea.fr

Abstract— Orano Mining has relied on Nuclear Measurement Laboratories for several years to estimate calibration factors for borehole radiometric probes. The total gamma count rate recorded with a NaI(Tl) scintillation detector (NGRS probe) is converted into equivalent uranium grade using a calibration coefficient in $s^{-1}.ppm_U^{-1}$ units, estimated thanks to different calibration blocks ranging from 0 to 10,000 ppm of uranium, at Orano CIME calibration facility in Bessines, France. Recently, Orano Mining embarked on an update of its measurement means by associating gamma spectrometry directly in the wellbore. Therefore, Orano Mining has recently tested a $CeBr_3$ probe designed and developed by ALT (Advanced Logic Technology) to qualify a patented borehole spectroscopic method based on energy bands recently developed by CEA and Orano. In this context, we have implemented a fine Monte Carlo modelling of this new probe in order to correct the calibration coefficients as a function of different parameters such as the drilling diameter, the presence of a casing, the density of the mineralization, the distance of the probe with respect to the well or casing walls, etc. In the present work, we have simulated the $CeBr_3$ probe in the concrete calibration blocks of Orano CIME. The MCNP model of the probe has been validated through a comparison with experimental data. The calibration coefficient determined by simulation is 11.6 $s^{-1}.ppm_U^{-1}$ ± 10 % in total gamma counting, which is in good agreement with the one measured in Bessines, i.e. 10.6 $s^{-1}.ppm_U^{-1}$. The calculation vs. experiment agreement is also satisfactory (i.e. within 10 %) for spectroscopic energy bands used to characterize the uranium grade in case of disequilibrium in the uranium chain, when total count rate does not apply. This method recently developed for samples and boreholes will be tested in future work with this new $CeBr_3$ probe.

Index Terms— $CeBr_3$(Ce) probe for uranium borehole logging, MCNP simulation, experimental calibration

I. Introduction

The development of the $CeBr_3$ probe was entrusted by Orano Mining to the company Advanced Logic Technology (ALT [1]). The $CeBr_3$ crystal has dimensions of 2.0 cm in diameter and 9.6 cm in length. The $CeBr_3$ probe was simulated with the MCNP6 computer code [2] using data provided by ALT (see Fig.1).

The main emissions of uranium ore come from ^{238}U, ^{235}U and ^{232}Th decay chains shown in Fig. 2, and from ^{40}K. The majority of the gamma signal comes from ^{214}Bi and ^{214}Pb located at the end of the ^{238}U decay chain.

It should be noted that the decay chains may be subject to possible radioactive imbalances resulting from geochemical phenomena (such as differential leaching of uranium and radium in roll-fronts) or from physicochemical treatments on the ore (radon evaporation during crushing or drilling). Indeed, uranium is a mobile element in the ore and its mobility is influenced by the chemical conditions of the environment. The main imbalance considered during mining prospection is between uranium and radium, denoted U/Ra (activity ratio). As the majority of gamma emissions come from ^{214}Pb and ^{214}Bi, the uranium content estimated in total count rate, such as the NGRS probe [3] (Natural Gamma Ray Probe), might be erroneous without the knowledge of the imbalance between the top and bottom activities of the ^{238}U decay chain. Therefore, we developed new gamma spectroscopy approaches for a $LaBr_3$(Ce) probe prototype [4], which will be now tested with the new industrial $CeBr_3$ probe.

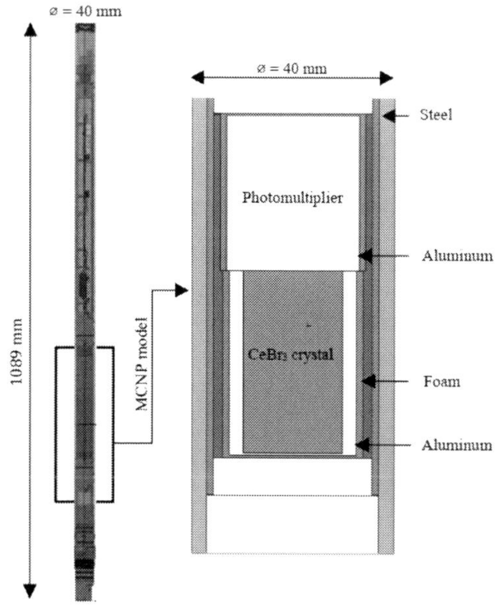

Fig. 1. QL40-SGR-2G probe from Advanced Logic Technology (ALT) and its associated MCNP model.

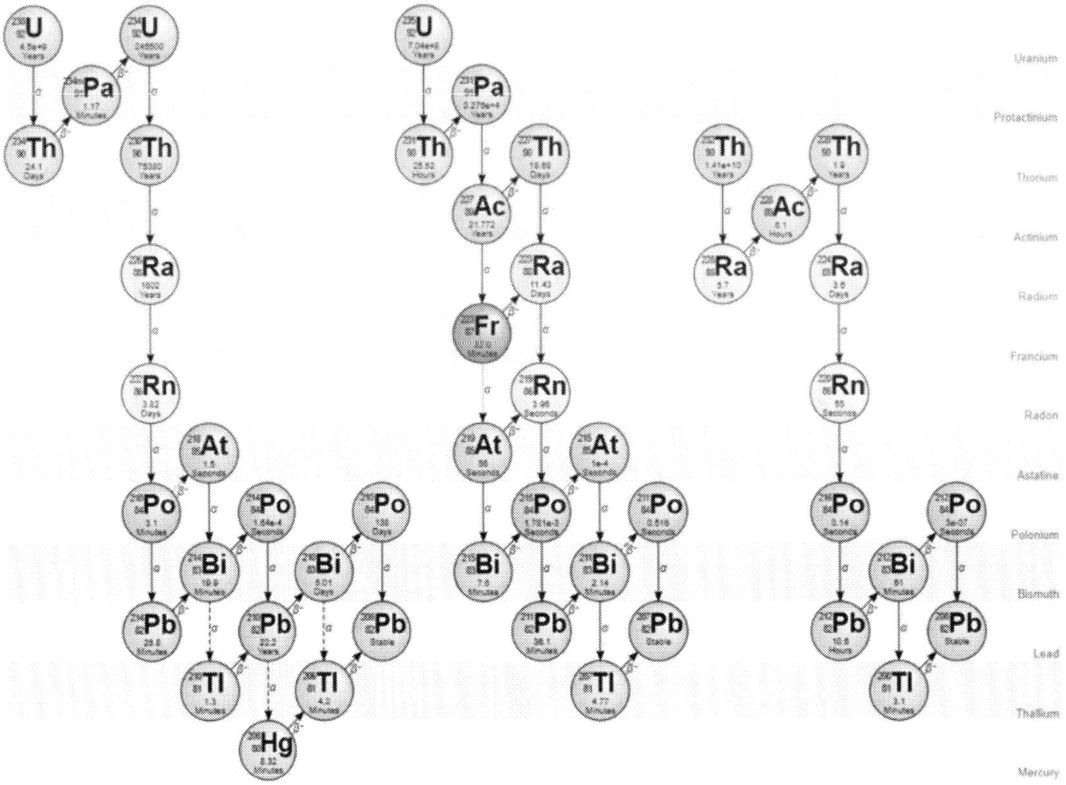

Fig. 2. Decay chains of ^{238}U, ^{235}U and ^{232}Th and associated radioactive periods [5].

II. $CeBr_3$ probe simulation in calibration blocks

Foremost, the experimental calibration of this probe in the Sosvurt block (located in Kazakhstan) and in the blocks of Orano CIME calibration station (in Bessines, France) made it possible to acquire gamma spectra with uranium contents ranging from 500 to 9700 ppm_U.

The $CeBr_3$ probe was initially simulated in the Sosvurt block with a content of 1000 ppm_U, see Figure 3. The thorium and potassium contents of this block were estimated by Orano Mining at respectively 20 ppm and 3.7 % (mass fractions). The spectrum obtained by MCNP simulation is compared to the spectrum obtained experimentally, see figure 4.

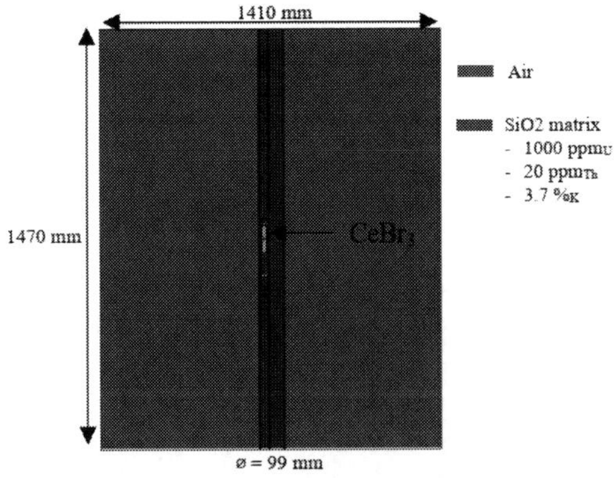

Fig. 3. MCNP modelling of $CeBr_3$ probe in Sosvurt block.

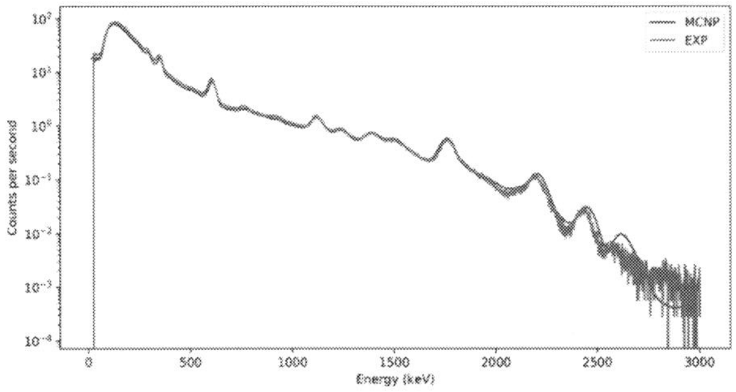

Fig. 4. Simulated and measured spectra of $CeBr_3$ probe in Sosvurt block.

A good agreement is observed, with only a 5.0 % overestimation of the experimental total count rate (9770 s^{-1}) by numerical simulation (10240 s^{-1}) for an energy larger than the electronics low-energy threshold of 80 keV.

The experiment-calculation comparison was then performed with gamma spectra measured in the calibration blocks of Bessines, with a uranium content ranging from 500 to 9700 ppm_U. The thorium and potassium contents of these blocks were estimated by Orano Mining at 15 ppm_{Th} and 3.6 % (mass fractions), respectively. Figure 5 shows the MCNP model of the $CeBr_3$ probe in a Bessines block and Figure 6 shows the spectra obtained by numerical simulation and experiment for the 2900 ppm_U block.

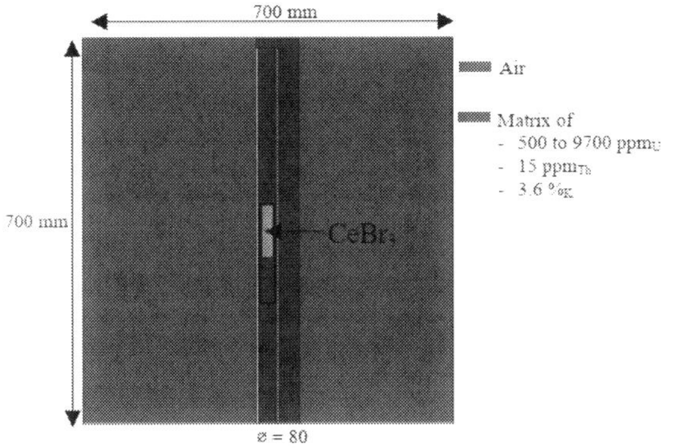

Fig. 5. MCNP modelling of $CeBr_3$ probe in Bessines block.

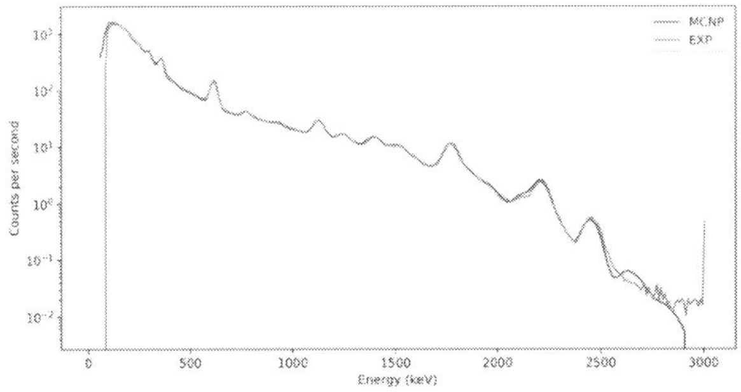

Fig. 6. Simulated and measured spectra of CeBr3 probe inside Bessines block with a 2900 ppm_U grade.

Table I reports, for each of the six Bessines blocks, the difference between experiment and simulation on the total counts recorded above the low-energy threshold of 80 keV. An average relative difference of 7 % is observed between numerical simulation and experiment. This value is similar to that of the Sosvurt reference block (+ 5 %).

TABLE I
TOTAL COUNTING RATES, C_{TOT}, OBSERVED EXPERIMENTALLY AND BY NUMERICAL SIMULATION FOR ENERGY GREATER THAN 80 KEV

Block	U grade (ppm)	C_{TOT}^{EXP} (s^{-1})	C_{TOT}^{MCNP} (s^{-1})	$\frac{MCNP - EXP}{EXP}$
B2	500	5356	5878	+ 9.7 %
B3	1000	10620	11515	+ 8.4 %
B4	1900	21155	21452	+ 1.4 %
B5	2900	30169	32188	+ 6.7 %
B6	4800	48772	51789	+ 6.2 %
B7	9700	89872	98191	+ 9.3 %

Fig. 7, shows the total counting rate C^{TOT} as function of the uranium grade of Bessines blocks.

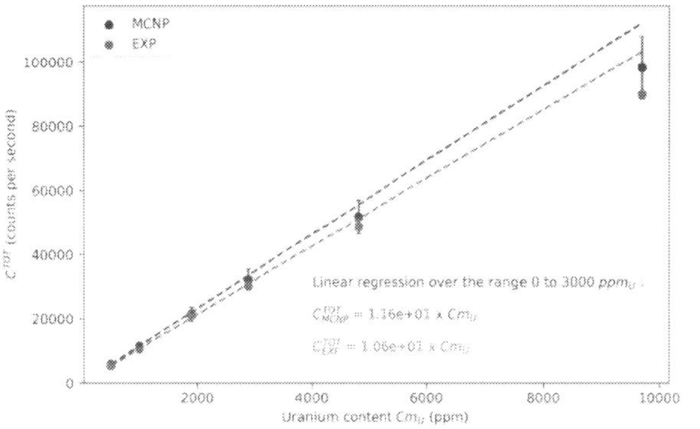

Fig. 7. Total counting rate C_{TOT} observed experimentally and by numerical simulation for energy greater than 80 keV as function of uranium grade (ppm). The linear regression fits only includes the data from 0 to 3000 ppm_U to avoid the saturation effects mentioned in the text.

The uncertainty plotted for MCNP simulation (10 %) includes those due to detector modelling, statistical uncertainty of MCNP calculations and knowledge of the standard blocks [6]. A beginning of saturation is observed both in MCNP simulation, due to gamma self-absorption in uranium [3], and in the experiment due to self-absorption but also electronics losses in the highest count rates.

In order to use the K_{APC} indicator of uranium content developed in previous study [4] [5] and described in section III, we also compare in Table II the experimental and calculated count rates in two energy areas (120-170 keV and 960-1040 keV), which evidences a slight distortion of the spectra: MCNP globally overestimates the low-energy signal but underestimates the high-energy count rate for most of the blocks.

TABLE II
COUNT RATES OBSERVED EXPERIMENTALLY AND BY NUMERICAL SIMULATIONS BETWEEN 120-170 KEV AND 960-1040 KEV

	Count rates between 120-170 keV			Count rates between 960-1040 keV		
Block	EXP (s^{-1})	MCNP (s^{-1})	$\frac{MCNP - EXP}{EXP}$	EXP (s^{-1})	MCNP (s^{-1})	$\frac{MCNP - EXP}{EXP}$
B2	1361	1466	+ 7.7 %	32	34	+5.1 %
B3	2690	2855	+ 6.1 %	66	67	+0.8 %
B4	5256	5242	- 0.3 %	136	125	- 8.0 %
B5	7326	7756	+ 5.9 %	198	190	- 4.2 %
B6	11438	12144	+ 6.2 %	334	313	- 6.2 %
B7	19202	21632	+ 12.7 %	693	631	- 9.0 %

III. Indicators of the uranium grade

The traditional approach based on the total count rate uses a CC_{TOT} calibration coefficient, which is the slope of the curves of Fig. 7. This coefficient links the total count rate above the low-energy threshold of 80 keV, C_{TOT}, and the uranium content:

$$Cm_U = \frac{C_{TOT}}{CC_{TOT}} \quad (1)$$

Their values are CC_{TOT}^{EXP} = 10.6 $c.s^{-1}.ppm_U^{-1}$ and CC_{TOT}^{MCNP} = 11.6 $c.s^{-1}.ppm_U^{-1}$ (see Fig. 7), with an uncertainty estimated to 10 %. These uncertainties include those due to detector modelling, statistical uncertainty of MCNP calculations, knowledge of the standard blocks [6] (uranium

grade, density, concrete composition), and also the precision on the energy of the electronics threshold.
On the other hand, the new spectroscopic approach described in [4] and [5] uses another indicator K_{APC} defined as the ratio between the counting rates in the 120 – 170 keV and 960 – 1040 keV areas. K_{APC} is related to the uranium content through the following formula:

$$Cm_U = \frac{\beta - K_{APC}}{\alpha} \quad (2)$$

with:

- $K_{APC} = \frac{C_{[120-170]}}{C_{[960-1040]}}$, $C_{[120-170]}$ and $C_{[960-1040]}$ being the number of counts in the abovementioned energy bands,
- α and β the parameters of the linear relationship between K_{APC} indicator and Cm_U uranium content, established with the data of Table II, see Fig. 7.

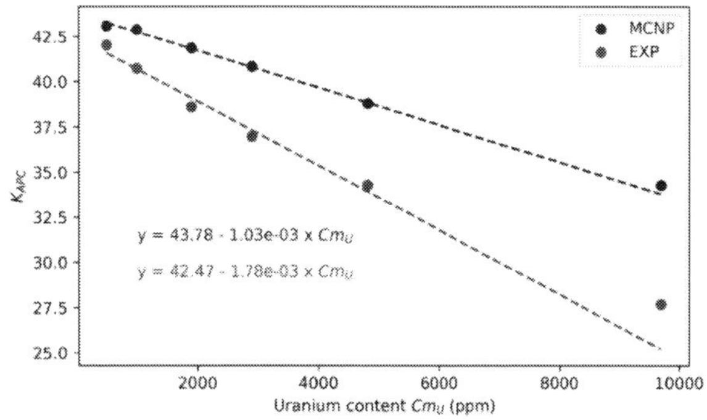

Fig. 8. K_{APC} indicator as function of uranium content in Bessines blocks.

As in Fig. 7 for the total count rate indicator, the linear fits reported on Fig. 8 only use the points below 5000 ppm$_U$. Indeed, regardless a possible spectrum distortion between the low and high energy areas at high count rate, the attenuation of the Compton continuum in the low-energy band, which is due to gamma self-absorption in uranium, starts saturating above 5000 ppm$_U$. The β coefficients are quite similar in MCNP and experiment, but the slopes are quite different, probably due to the gamma spectrum distortion that can be observed in the data of Table II. Therefore, the new $CeBr_3$ probe will be sent to the Nuclear Measurement Laboratory of CEA to investigate the effect of increasing count rates on the shape of the gamma spectrum and on count losses.

IV. Conclusion

The new $CeBr_3$ probe developed by ALT was simulated by CEA with MCNP6 computer code and its model was qualified using experimental data provided by Orano Mining. The probe modelling in the Sosvurt block and in the Bessines calibration blocks made it possible to observe the influence of the uranium content (from 500 to 9700 ppm$_U$) on the gamma spectra and on two uranium indicators: on the one hand the total count rate above the 80 keV low-energy threshold, and on the other hand the K_{APC} ratio between two energy bands of the spectrum (120-170 keV and 960-1040 keV, respectively). We observe a good agreement between simulation and experiment on the total count rate, but the expected self-attenuation of the Compton continuum in the 120-170 keV band seems to be underestimated in numerical simulations. In fact, we rather suspect count losses at low energy when the total count rate increases along with uranium content. This hypothesis will be investigated in further studies when the $CeBr_3$ probe will be sent by Orano Mining to the Nuclear Measurement Laboratory of CEA for a detailed characterization, both to refine the modelling of the crystal and its probe environment, but also to investigate count-rate related spectrum distortions and losses.

References

[1] "ALT site web," *ALT*. https://www.alt.lu/ (accessed Feb. 22, 2023).

[2] Los Alamos National Laboratory, "MCNP6, User's manual – Version 1.0 - LA-CP-13-00634, Rev. 0 – May 2013 – Denise B. Pelowitz, editor." 2013.

[3] C. Carasco, B. Perot, J.-L. Ma, H. Toubon, and A. Dubille-Auchere, "Improving Gross Count Gamma-Ray Logging in Uranium Mining With the NGRS Probe," *IEEE Trans. Nucl. Sci.*, vol. 65, no. 3, Art. no. 3, Mar. 2018, doi: 10.1109/TNS.2018.2800909.

[4] Bertrand Pérot, Thomas Marchais, Pierre-Guy Allinei, Hervé Toubon, Youcef Bensedik, Romain Goupillou, Alexandre Berland, "Development of Gamma Spectroscopic Tools for Uranium Ore Samples and Borehole Exploration", 2022 IEEE Nuclear Science Symposium, 5 – 12 November 2022, Milano, Italy.

[5] T. Marchais, B. Perot, C. Carasco, H. Toubon, and Y. Bensedik, "Method for assessing the uranium content using gamma spectrometry in a wellbore and associated device," WO2021122772 (A1), Jun. 24, 2021 Accessed: Oct. 25, 2021. [Online]. Available: https://worldwide.espacenet.com/publicationDetails/biblio?FT=D&date=20210624&DB=EPODOC&locale=fr_EP&CC=WO&NR=2021122772A1&KC=A1&ND=4

[6] T. Marchais *et al.*, "Detailed MCNP Simulations of Gamma-Ray Spectroscopy Measurements With Calibration Blocks for Uranium Mining Applications," *IEEE Trans. Nucl. Sci.*, vol. 65, no. 9, pp. 2533–2538, Sep. 2018, doi: 10.1109/TNS.2018.2797312.

Towards the enhancement of nuclear reactor safety: physical phenomena involved in the measurement accuracy of boron concentration online monitoring systems

Nouhaila Tabti*, Adrien Sari, and Dominique Tromson

Université Paris-Saclay, CEA, List, F-91120 Palaiseau, France
(*) nouhaila.tabti@cea.fr

Abstract— **In Pressurized Water Reactors (PWRs), boron-10 (^{10}B) is added to the primary fluid as boric acid to regulate criticality and power distribution within the core. Indeed, ^{10}B possesses a significant neutron absorption cross-section of approximately 3845 barns for thermal neutron energy (around 25.3 meV). Thus, under certain circumstances, even a slight variation in boron concentration could lead to criticality issues. This paper addresses the problems regarding the accuracy and real-time measurement of Boron Concentration Online Monitoring Systems (BCOMS) in PWRs. First, the paper focuses on presenting BCOMS's key performance and qualification criteria to ensure a real-time, accurate response during normal operation, operational transients, and accidental events. Then, we evaluate the influence of different dimensioning parameters on the overall measuring performance through Monte Carlo simulations using the MCNP6.2 code. Finally, we investigate, combining theoretical knowledge and simulation results, the physical phenomena involved in the measurement accuracy of BCOMS. This analytical approach not only fosters a deeper comprehension of the underlying phenomena but also facilitates the identification of potential areas for improving measurement accuracy, thereby enhancing the safety of nuclear reactors.**

Keywords —**Pressurized Water Reactor (PWR), nuclear safety, primary fluid, boron, online monitoring, neutron absorption, Monte Carlo simulation.**

I. Introduction

BORON is added to the primary fluid of Pressurized Water Reactors (PWRs) as boric acid (H_3BO_3) to adjust reactivity and power distribution within the core. Indeed, boron-10 (^{10}B) has a significant neutron absorption cross section following the $^{10}B(n, \alpha)^7Li$ reaction, of approximately 3845 barns for a thermal neutron energy of 25.3 meV. During both normal operation and operational transients, boron is introduced into the primary fluid to offset fuel burnup and xenon oscillations. Moreover, the addition of boron becomes essential in the event of specific faults that might lead to a reactivity excursion in the core, in order to prevent criticality issues.

Therefore, it is of paramount importance to maintain the appropriate level of boron concentration to avoid any unintended criticality and ensure safe and reliable operation throughout the reactor's operational lifecycle. The implementation of robust monitoring and control measures for boron levels is crucial to effectively mitigate the potential hazards associated with boron dilution events. These measures play a crucial role in safeguarding the reactor's integrity and ensuring the safety of personnel and the surrounding environment.

There are different methods used to measure boron concentration [1], amongst which two ways are commonly used in nuclear reactors industry: periodic chemical sampling and Boron Concentration Measuring Systems (BCMS), also referred to as the "boron meters". It is well known that chemical sampling provides highly precise measurements, but it produces radioactive waste and necessitates a considerable turnaround time. Conversely, BCMS enables continuous measurement of boron concentration; however, its margin of error is higher than that of chemical titration. BCMS can be categorized into different types based on their operational mode and level of intrusiveness. They can be classified as either online or offline, referring to whether the measurement is performed in real-time or at discrete intervals, respectively. Additionally, boron meters can be further classified as either intrusive [2-7] or non-intrusive [8-12], depending on whether they require direct physical contact with the primary fluid or operate through non-contact methods.

This paper focuses on non-intrusive Boron Concentration Online Monitoring Systems (BCOMS) mounted on the Chemical and Volume Control System's discharge line (CVCS), since they can easily equip the actual operating reactors without a significant material modification in the design basis of the circuit. First, we highlight the essential performance and qualification criteria relative to boron meters. Second, we provide a description of the evolution of the overall performance following the variation of certain dimensioning parameters. This study is carried out by Monte Carlo simulation using the MCNP6.2 code [13]. Finally, we discuss results combining theoretical knowledge and some analytical calculations in order to gain deeper insights into the physical phenomena that affect the measuring accuracy of BCOMS. This comprehensive and analytical approach aims to enhance our understanding of boron meter current performance and contributes to the advancement of their design and application in nuclear reactors.

II. BCOMS Performance and Qualification Criteria

To establish accurate real-time measurements of boron concentration, a comprehensive examination of design and safety requirements is essential in determining performance and qualification criteria for BCOMS. Nuclear Power Plants (NPPs) are complex systems with intricate interconnections between different phenomena, such as neutronics, thermal hydraulics, mechanics, and thermomechanics. By studying these diverse phenomena, a set of crucial criteria can be formulated to assess the overall efficiency of the system. We divide the selected criteria into two categories: performance criteria, which directly assess accuracy and response time, and qualification criteria, which are essential to ensure compliance with safety measures. By categorizing the criteria in this manner, we can effectively evaluate both the operational effectiveness and the adherence to safety standards required for nuclear reactors systems.

We have identified three main performance criteria: counting rate (CR), measuring resolution ($\mathfrak{R}$), and neutron sensitivity (S_n) compared to thermal sensitivity (S_T). The counting rate (CR) criterion reflects the measurement statistics, thus influencing the system's response time. Measuring resolution, as described in (1), quantifies the variation in CR relative to a reference value (CR_{ref}). It serves as a crucial metric for evaluating the system's ability to discern and resolve minute differences in boron concentration (C_B). By assessing measuring resolution, we can effectively gauge the system's precision in distinguishing subtle variations in C_B. The comparison between neutron and thermal sensitivities, described as in (2) and (3), is to ensure that S_n is greater than S_T by at least one order of magnitude. This criterion ensures that the BCOMS provides a relatively stable response across temperature variation.

$$\mathfrak{R} \stackrel{\text{def}}{=} \frac{\Delta CR}{CR_{ref}} \quad (\%) \tag{1}$$

$$S_n \stackrel{\text{def}}{=} \frac{\Delta CR}{\Delta C_B} \quad (\text{cps/ppm}) \tag{2}$$

$$S_T \stackrel{\text{def}}{=} \frac{\Delta CR}{\Delta T} \quad (\text{cps/°C}) \tag{3}$$

Additionally, to ensure the safety and reliability of BCOMS, three key qualification criteria have been established: radiation protection, mechanical qualification and redundancy. Boron meters are equipped with a neutron source, it is therefore necessary to mitigate radiation exposure to system operators and surrounding environments. The mechanical qualification criterion assesses the structural integrity and durability of the system under various operational conditions and accidental situations, ensuring its stability and longevity. Lastly, as per the redundancy criterion, it is imperative to implement a dual-channel acquisition line spanning from the neutron detector to the processing unit. This redundancy measure ensures system integrity by providing redundant data pathways, enhancing reliability and safeguarding against potential failures along the entire acquisition and processing chain. In this paper, we focus our study on evaluating the performance of BCOMS based exclusively on the criteria of CR and measuring resolution ($\mathfrak{R}$) in a redundant non-intrusive BCOMS.

III. Design of a Non-Intrusive BCOMS

Prior to conducting the performance assessment of BCOMS, this section provides a comprehensive review of the measuring principle and key characteristics associated with these systems.

A. *Measuring principle*

The measuring principle of boron meters, depicted in Fig. 1, remains consistent, irrespective of whether they are intrusive or not. It revolves around quantifying the attenuation and absorption of a neutron flux. Neutrons, suitably slowed down by a neutron moderator, undergo capture within the primary fluid through the (n, α) reaction. Subsequently, the remaining neutrons are detected by a dedicated neutron counter, establishing an anticorrelation between the detector count and boron concentration. This anticorrelation is determined through calibration techniques, which involve finding a fitting function capable of converting the measured CR (in cps) to C_B (in ppm). Multiple studies have put forth various solutions aiming at improving calibration accuracy regarding fitting functions [2, 4, 7, 11].

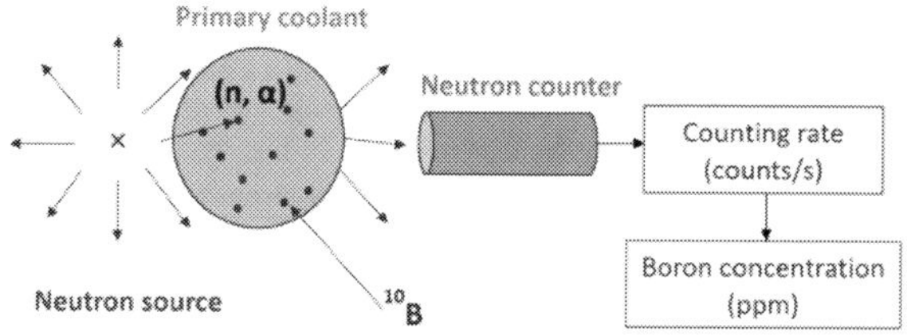

Fig. 1. Boron meter measuring principle.

B. *Neutron source*

The borated water is interrogated by neutrons emitted from a neutron source such as Am-Be or ^{252}Cf. In the frame of this work, we study the performance of a BCOMS equipped with an Am-Be neutron source, since this latter is characterized with a long half-life, of roughly 432 years, which alleviates the need for frequent recalibration. Indeed, ^{241}Am decays via α emission to ^{237}Np. Neutrons are then generated through the reaction ^{9}Be(α, n)^{12}C.

C. *Neutron counter*

Based on specific properties regarding neutron sensitivity to thermal neutrons and sensitive length, we select a CPNB48 boron-lined proportional counter [14], manufactured by PHOTONIS. Neutrons interact with the boron coating on the inside wall of the tube following the ^{10}B(n, α)^{7}Li reaction. Since reaction products (i.e., ^{7}Li and α particles) are oppositely directed, only one reaction product ionize the gas per (n, α) reaction. The ionized particles are then converted to a signal by the applied voltage, inducing the gas multiplication process of electrons near the anode wire.

D. Neutron moderator

Efficient neutron moderation is crucial to enhance the sensitivity of neutron detectors towards thermal neutrons. This moderation process primarily involves elastic and inelastic scattering. In this study, we compare the measuring performance of three commonly used neutron moderators: high-density polyethylene (HDPE), graphite and amorphous carbon.

IV. Performance Assessment

In this section, we examine the impact of various dimensioning parameters on the selected performance criteria by Monte Carlo simulation using the MCNP6.2 code. This rigorous analysis allows us to gain insights into how different parameter configurations affect the identified performance metrics. By elucidating these relationships, we can enhance our understanding of the system's behavior and inform effective design and optimization strategies.

Fig. 2 presents the MCNP6.2 model depicting the preliminary design of a non-intrusive BCOMS. In this work, we investigate the influence of the detector position, spacing and moderator material. Note that the study of the detector position and spacing was performed in the case of the use of HDPE as an external neutron moderator, with a radius of 20 cm and a height of 50 cm.

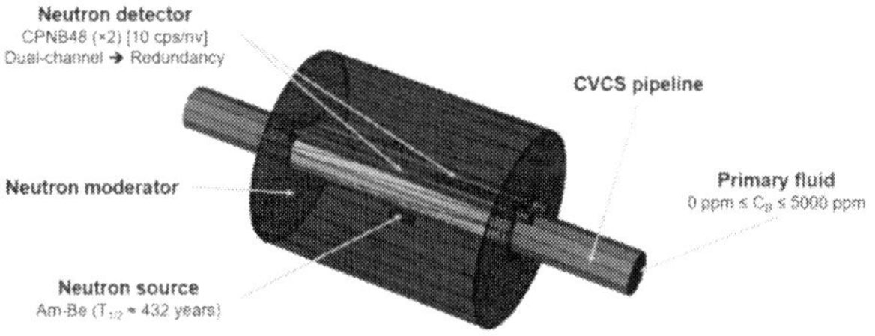

Fig. 2. 3-D visualization of the MCNP6.2 model of a non-intrusive BCOMS mounted around the CVCS pipeline.

A. Detector position

In order to assess the impact of detector position relative to the neutron source, we conducted an evaluation around the CVCS pipeline, as depicted in Fig. 3. The objective was to compare two configurations: direct attenuation (on the left) and backscatter geometry (on the right). Various angular distributions of detectors were simulated around the pipeline, ranging from 30° (for direct attenuation) to 330° (for backscattering).

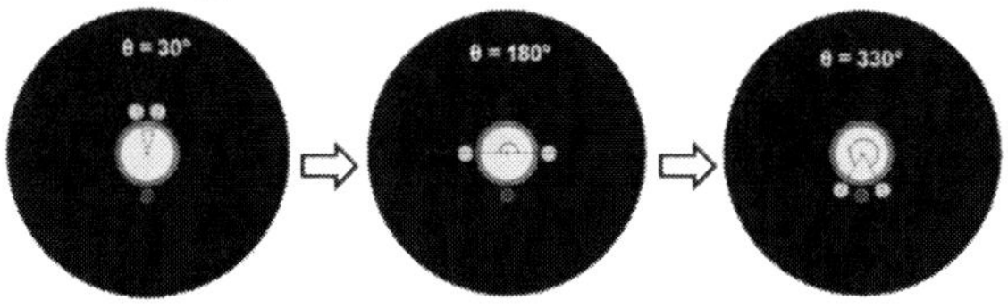

Fig. 3. 2-D view of the simulated detector position configurations.

Fig. 4 presents the plotted evolution of CR (at 0 ppm) and measuring resolution. It is important to note that CR correlates with the (n, α) reaction rate occurring in the boron deposition layer of the counter tube. The (n, α) reaction rate was calculated using the *F4 tally* to estimate the detection efficiency.

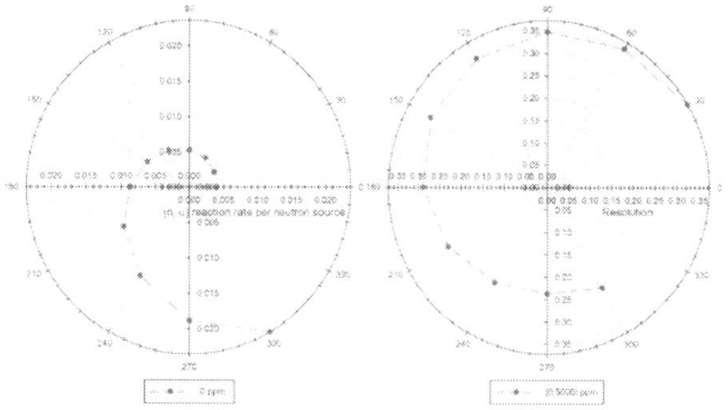

Fig. 4. Impact of detector position. MCNP6.2 simulation results.

Observations indicate that as the system transitions towards a backscatter geometry, there is a notable increase in measuring statistics while encountering a decrease in measuring resolution. In this context, selecting a high sensitivity detector helps minimize constraints related to CR, making resolution a more crucial performance criterion.

Based on this analysis, a direct attenuation measuring configuration is chosen, enabling the best measuring resolution of approximately 35% while maintaining a sufficiently fast response time. This decision is made considering the intensity of the neutron source emitting approximately 10^6–10^7 n.s^{-1} in 4π sr.

B. Detector spacing

Building upon the previously selected direct attenuation configuration, we proceed to evaluate the impact of detector spacing. In this assessment, we examine the evolution of CR and $\Re$ across a range of spacing options between the two measuring channels, spanning from 1 to 30 cm, as illustrated in Fig. 5.

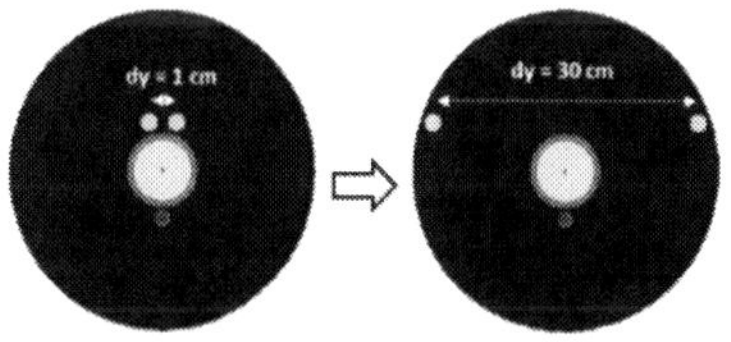

Fig. 5. 2-D view of the simulated detector spacing configurations.

Simulation results are presented in Fig. 6, revealing important insights into the influence of detector spacing. A notable observation is that at a detector spacing of approximately 4 cm, there is a slight increase in CR. However, this increase comes at the expense of a 10% decrease in $\Re$; when compared to the minimum detector spacing that yields the highest resolution, approaching 35%. Based on these findings, we confidently select the measuring configuration with the minimum detector spacing, ensuring optimal measuring resolution.

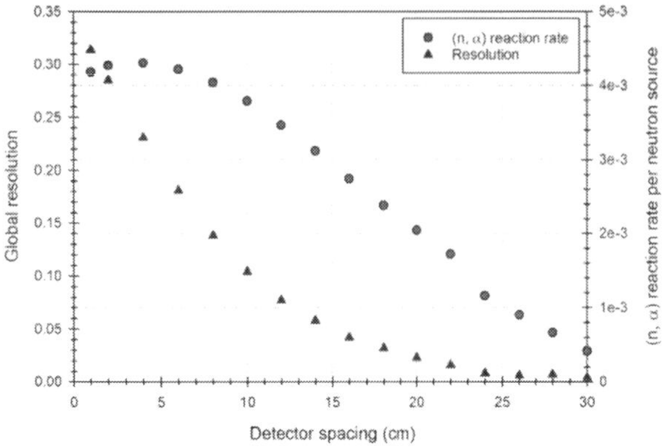

Fig. 6. Impact of detector spacing, MCNP6.2 simulation results.

C. *Neutron moderator material*

In the final phase of our analysis, we conducted a comprehensive comparison of neutron moderator materials, specifically HDPE, graphite and amorphous carbon. The simulation results, depicted in Fig. 7 and summarized in Table 1, underscore the profound influence of the neutron moderator on overall performance. In fact, carbon-based moderators, such as graphite and amorphous carbon, exhibit a remarkable enhancement in measuring resolution, reaching nearly 60% compared to the 35% achieved by HDPE. This notable improvement enables better discrimination of C_B, particularly for higher values where we observe a drastic decrease $\Re$. Furthermore, it is noteworthy that the higher density assigned to amorphous carbon contributes to a simultaneous increase of 34% in CR, while graphite exhibits a 16% decrease, compared to HDPE.

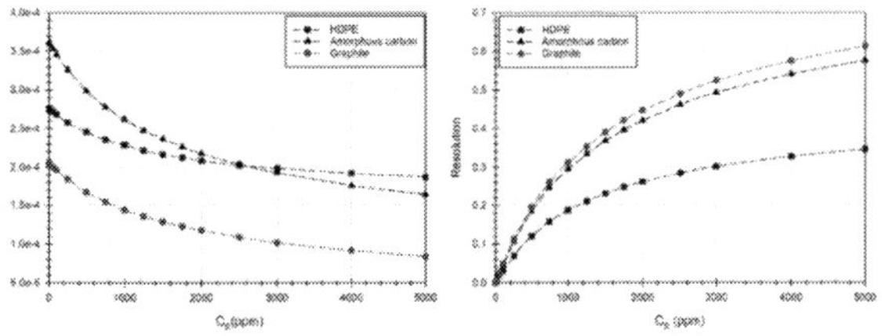

Fig. 7. Influence of neutron moderator material, MCNP6.2 simulation results.

TABLE I
COMPARISON OF CR AND MEASURING RESOLUTION AMONG HDPE, GRAPHITE, AND AMORPHOUS CARBON

	HDPE	Graphite	Amorphous carbon
$\Delta CR_{/HDPE}$ (0 ppm)	–	- 16%	+ 34%
Global resolution ($\Re$)	35%	61%	58%
$\Re_{[0-1250]\ ppm}$	21%	35%	33%
$\Re_{[1250-2500]\ ppm}$	9%	21%	19%
$\Re_{[2500-5000]\ ppm}$	9%	24%	21%

V. PHYSICAL PHENOMENA

The primary focus of this section is to gain insights into neutron-induced phenomena that affect the overall performance of the studied BCOMS. Specifically, our aim is to understand the underlying physical mechanisms contributing to the decrease in measuring resolution, clearly observed in Table 1, as boron concentration increases. Furthermore, we seek to explore the influence of different materials on the performance criteria, unraveling the intricate relationship between neutron moderator properties and system performance.

The behavior of neutron population is governed by the Boltzmann transport equation establishing a delicate balance between mechanisms that generate and deplete neutron population [15]. Solving the Boltzmann equation presents substantial challenges due to its reliance on seven independent variables (three in space, two in direction, energy and time). In response to this complexity, two main approaches have been developed. On the one hand, various numerical methods, such as deterministic methods and the Monte Carlo method, have been utilized [16]. On the other hand, the diffusion theory approximation offers a practical and efficient solution, providing an analytical approach under specific assumptions regarding the system's geometry and neutronics. This approximation has garnered recognition for its effectiveness in addressing a wide range of challenges encountered in reactor design. By simplifying the complex Boltzmann equation, the diffusion approximation presents a more manageable formulation, as in (4), where "Δ" is the Laplacian operator, parameter K combines the macroscopic transport (Σ_{tr}) and absorption (Σ_a) cross-sections, and D denotes the diffusion coefficient.

$$\Delta\phi - K^2\phi = 0\,;\ K^2 = \frac{\Sigma_a}{D} = 3\Sigma_a\Sigma_{tr} \quad (4)$$

This study endeavors to attain a deep comprehension of the fundamental physical phenomena contributing to the observed decline in measuring resolution within the BCOMS. Hence and to simplify the analysis, we adopt a one-dimensional Cartesian coordinate system to solve (4) in both the borated water region and the neutron moderator (graphite) region, while incorporating appropriate boundary conditions. Then, we compare the analytical solution with the MCNP6.2 calculations. Results are presented in Fig. 8, where we observe a coherence in the behavior of the two curves. However, there are discrepancies between them, which can be attributed to the geometric and neutronic assumptions and simplifications made within the diffusion theory approximation.

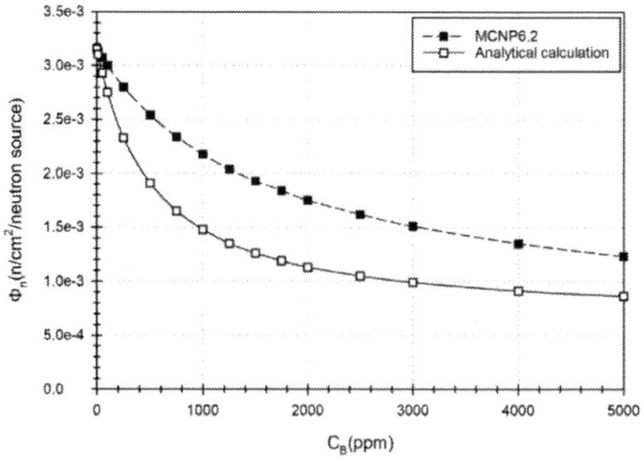

Fig. 8. Comparison between the analytical solution and the MCNP6.2 calculation of the neutron flux distribution in the detector region.

Remarkably, the analytical solution of the neutron flux distribution (ϕ_n) offers an insightful explanation for the decrease in measuring resolution, which is observed through the flattening tendency in calibration curves as the C_B increases. The solution of (6) in a Cartesian coordinate system involves a combination of exponential functions that yield an exponential-like shape to the neutron flux distribution. Specifically, the solution includes exponential functions with arguments of Kx and –Kx, where x represents the distance traveled along the considered axis.

Additionally, the neutronic characteristics of the moderator play a crucial role in shaping the neutron flux population and energy distribution within both the borated water and the detector region. Notably, HDPE demonstrates significantly greater absorption rates compared to graphite and amorphous carbon, thus exerting a substantial impact on the neutron flux population. Furthermore, HDPE exhibits a higher macroscopic scattering cross-section than amorphous carbon and graphite, thereby influencing the energy distribution of neutrons. These contrasting properties contribute to the observed disparities in terms of CR and $\Re$.

VI. Conclusions and Outlook

This study presents a comprehensive performance analysis of non-intrusive Boron Concentration Online Monitoring Systems (BCOMS) installed around the CVCS pipeline of a PWR. We evaluate, using the MCNP6.2 code, the impact of three critical dimensioning parameters on the counting rate and measuring resolution. The results indicate that a direct attenuation configuration with minimized detector spacing yields improved accuracy. Furthermore, our investigation into the influence of external neutron moderators reveals that carbon-based materials such as graphite and amorphous carbon exhibit a significant increase in measuring resolution compared to HDPE.

However, it is worth noting that a substantial decrease in resolution occurs as boron concentration increases across the various configurations. This intriguing phenomenon motivates our focus on understanding the underlying physical mechanisms responsible for the observed flattening behavior in calibration curves. Through an analytical study employing the diffusion theory approximation, we establish a correlation between the neutron's behavior within the boron meter and the observed resolution. This finding aligns with the results obtained from MCNP6.2 calculations. Our conclusion is that the flattening behavior is an intrinsic phenomenon inherent to borated water and cannot be eliminated. Consequently, the best approach is to identify an appropriate external neutron moderator that offers optimal measuring resolution. Currently, carbon-based moderators exhibit the most favorable performance, with a resolution of nearly 60%. However, the selection of the neutron moderator is still under investigation, as it is crucial to consider other design and safety requirements.

Moreover, we aim to enhance the analytical calculations by solving the equation (6) in a three-dimensional coordinate system, enabling the derivation of a formula that links resolution to the moderator's properties. In addition, an expanded study of various advanced moderators is planned. Ultimately, our objectives include manufacturing an optimized prototype of BCOMS to experimentally calibrate the device and assess overall performance, including thermal stability.

References

[1] R. N. Sah and P. H. Brown, "Boron Determination—A Review of Analytical Methods," *Microchemical Journal*, vol. 56, no. 3, pp. 285-304, 1997/07/01, doi: 10.1006/mchj.1997.1428.

[2] C. Kong, H. Lee, T. Tak, D. Lee, S. H. Kim, and S. Lyou, "Accuracy Improvement of Boron Meter Adopting New Fitting Function and Multi-Detector," (in English), *Nucl Eng Technol*, vol. 48, no. 6, pp. 1360-1367, Dec 2016, doi: 10.1016/j.net.2016.06.012.

[3] S. H. Sung and H. R. Kim, "Development of accuracy enhancement system for boron meters using multisensitive detector for reactor safety," (in English), *Nucl Eng Technol*, vol. 52, no. 3, pp. 538-543, Mar 2020, doi: 10.1016/j.net.2019.08.004.

[4] S. H. Sung and H. R. Kim, "Experimental characterization of the accuracy of multidetector boron meter for operational safety of reactors," (in English), *Int J Energ Res*, vol. 42, no. 8, pp. 2701-2709, Jun 25 2018, doi: 10.1002/er.4059.

[5] S. H. Sung, U. J. Lee, J. S. Min, and H. R. Kim, "Experimental Setup for Boron Meter Accuracy Analysis in Nuclear Reactor Environment," (in English), *Materials in Environmental Engineering*, pp. 1201-1207, 2017, doi: 10.1515/9783110516623-118.

[6] J. Bulkowski, W. Fletcher, A. Opitz, V. Rajagopal, and D. Whyte, "DEVELOPMENT OF A BORON CONCENTRATION METER. Final Report," Westinghouse Electric Corp., Pittsburgh, Pa. Nuclear Energy Systems Div., 1969.

[7] Y. Jo, C. Kong, J. Yu, D. Lee, and S. H. Kim, "High accuracy boronometer design developed for light water reactors," *Ann Nucl Energy*, vol. 110, pp. 25-30, 2017.

[8] P. M. Dighe, "Design simulation studies of pipe mounted boron concentration meter," (in English), *Ann Nucl Energy*, vol. 134, pp. 414-423, Dec 2019, doi: 10.1016/j.anucene.2019.07.021.

[9] Framatome. "Boronline®, A new generation of boron meter. Technical Sheet No: A3022." (accessed April 15, 2022).

[10] Mirion Technologies. "BM501™, Boronmeter. Technical Sheet No: 155780EN-F." https://www.mirion.com/products/bm-501-boron-meter (accessed November 23, 2021).

[11] P. Pirat, "Boronline, a new generation of boron meter," in *2011 2nd International Conference on Advancements in Nuclear Instrumentation, Measurement Methods and their Applications*, 2011: IEEE, pp. 1-2.

[12] M. Jimenez and P. Pirat, "CONTINUOUS MEASUREMENT OF BORON CONCENTRATION IN NUCLEAR POWER PLANTS: WHY IS IT ESSENTIAL FOR SAFE AND EFFICIENT OPERATION?," 2017.

[13] C. J. Werner *et al.*, "MCNP6.2 Release Notes", Los Alamos National Laboratory, report LA-UR-18-20808, 2018.

[14] PHOTONIS. "CPNB48 Specifications." https://www.photonis.com/system/files/2019-03/CPNB48-specifications.pdf (accessed April 28, 2022).

[15] J. J. Duderstadt and L. J. Hamilton, *Nuclear reactor analysis*. Wiley, 1976.

[16] M. Coste-Delclaux, C. Diop, A. Nicolas, and B. Bonin, *Neutronics*. CEA Saclay; Groupe Moniteur, 2015.

A mobile radiation detection system for security inspection and monitoring

Luís Marques[1,*], Rui Coito[2], Sofia Fernandes[2], Tiago Costa[2], Alberto Vale[3], and Pedro Vaz[4]

[1]Centro de Investigação da Academia da Força Aérea, Instituto Universitário Militar, Portugal
[2]Faculdade de Ciências, Universidade de Lisboa, Portugal
[3]Instituto de Plasmas e Fusão Nuclear, Instituto Superior Técnico, Universidade de Lisboa, Portugal
[4]Centro de Ciências e Tecnologias Nucleares, Instituto Superior Técnico, Universidade de Lisboa, Portugal

(*) lumarques@academiafa.edu.pt

Abstract— **In Nuclear Security scenarios, traditional radiation portal monitors are normally fixed, large, and expensive equipment. This study introduces an innovative, cost-effective solution: a lightweight, compact, mobile gamma-ray/beta and neutron detection system, comprising two plastic scintillators with silicon photomultipliers. Designed for inspecting maritime containers, a real-time gamma-ray source search algorithm operates during perimeter scans. Using maximum likelihood estimation, the algorithm computes radiation source coordinates. To optimize the detection system's path and enhance survey quality (informed path), a profit function is employed. This function is grounded in either gamma-ray count analysis or source position estimations, providing dual approaches to improve source position accuracy. Evaluating this approach, simulations used a ^{137}Cs source inside a container, with gamma-ray counts adjusted for geometric detection efficiency and material attenuation. The informed path generated by the source search algorithm is compared to predefined paths, revealing substantial improvements in source height estimation accuracy and reduced confidence intervals. The use of profit functions to refine the initial path enhances system reliability and streamlines inspection time.**

Keywords — **Plastic scintillators, silicon photomultiplier, neutron and gamma-ray detection, source localization, mobile radiation detection system, Security scenarios.**

I. Introduction

In order to detect special nuclear materials (SNM) and other radioactive materials and sources, large and expensive radiation portal monitors (RPMs) are often employed at some seaports and international borders to scan maritime containers cargo and vehicles [1]. Nonetheless, ensuring the smooth flow of goods during inspections poses a significant challenge, demanding innovative approaches and alternative radiation detection systems. For instance, the distance between the inspected cargo and the RPMs, along with the radiation attenuation caused by cargo materials and potential shielding, hamper the ability to detect and precisely locate such radioactive sources and SNM [2]. Furthermore, following an RPM alarm, secondary inspections conducted using handheld equipment may consume up to 20 minutes, while physical cargo inspections might extend up to three hours [3] [4]. Mobile radiation detection systems can also be utilized, but their weight and large dimensions often require them to be mounted on cars, trucks, or trailers. While dual-mode detection systems and gamma cameras mounted on multirotors (a.k.a. drones) and unmanned helicopters have been described in literature, they suffer from limitations such as a small sensitive volume and long acquisition times. To achieve simultaneous detection and identification of most gamma-ray emitters radionuclides, detection systems with spectroscopy capability are commonly used, particularly based on semiconductors and inorganic scintillators. However, semiconductors (e.g., CZT) have volume limitations and are expensive, while inorganic scintillators offer limited energy resolution, limited crystal volume size and shape, and some of them are hygroscopic (increasing casing weight) [2]. Moreover, considering neutron detection systems, the worldwide shortage of helium procurement dictates the need to find alternatives to helium-3 based detectors´ technology [5].

Therefore, the development of a low-cost and highly efficient mobile radiation detection system capable of detecting both gamma-rays and neutrons is of paramount importance. In this study, a novel radiation detection system combining two independent detectors was developed: a 110 mm diameter and 30 mm thick EJ-200 plastic scintillator (for gamma and beta detection) and a thermal neutron detector EJ-426HD plastic scintillator embedded on a modular and compact high-density polyethylene (HDPE) moderator [6] [7]. For this study, only the EJ-200 detection system will be considered.

Compared to other radiation detection systems, plastic scintillators offer several advantages, including fast response, good neutron and beta sensitivity, much higher gamma-ray detection efficiency than gas-filled detectors (e.g., Geiger Muller), availability in different shapes and sizes which allows for increased geometric detection efficiency (e.g., pancake shape), non-hygroscopic nature, ruggedness, and low maintenance requirements. The use of silicon photomultipliers allows for lightweight, compact, low power consumption, and magnetic interference-free detection systems [2].

Detecting and localizing hidden radiative sources is typically conceived as a path optimization problem, dependent on the source search algorithm [8] and the chosen detection system path [9] [10] [11]. Although predefined paths can be used to create a radiation map of the area of interest (e.g., parallel sweeping search pattern) [9] [10], advantages can be found in utilizing real-time (or near real-time) radiation data gathered by the detectors and sensors to optimize the detection system path (informed path) and survey time [1]. Several radioactive hot-spot localization algorithms have been described in literature [2] [12], such as the use of a maximum likelihood estimation (MLE) algorithm [13] [14], the use of a radiation detector network [15] [16] and the application of deep learning approaches [17].

This study aims at investigating the influence of the flight path of a mobile radiation detection system, which can be easily integrated on a platform with weight restrictions (e.g., drone) or used as handheld equipment, on its ability to localize a hidden gamma-ray source inside a maritime container. Different paths around the container were considered, as well as the influence of radiation detection system counts and source position estimates on the path (informed path) to optimize the detection and localization of a gamma-ray source. Using a radiation detection simulator, a MLE algorithm is used to estimate the source position and corresponding confidence intervals at each iteration (detector position). The developed simulator considers a point source of ^{137}Cs, placed inside a standard maritime container, and calculates the counts obtained in the EJ-200 detector while it performs complete paths around the container. Comparing the obtained results with the ones of predefined paths, improvements in the estimation of the source position (particularly its height) and survey time were observed when profit functions were used to change the radiation detection path based on the counts and estimates.

II. Materials and Methods

To calculate the gamma-ray counts of the EJ-200 detector, a radiation detection simulator was developed using the Python language. The simulation software performs assuming the following parameters:

- Spherical detector approximation with a diameter of 11 cm (equivalent to the cylindrical diameter of the EJ-200 detector described in Chapter I).
- A 4 MBq ^{137}Cs point source placed at the center and at the corner of a maritime container with internal dimensions of 5.94 m × 2.34 m × 2.39 m [18] was considered. A similar radiation source was also considered in field tests by the authors in [5]. The source emits an average number of 662 keV gamma-rays (according to its activity and branching ratio).
- A radiation detection system dwell time of 1 second (acquisition time) was considered.
- The radiation detection system performs complete paths around the container at a horizontal speed of 0.2 m/s. This speed was also considered in [6], allowing sufficient radiation data points for the source position estimation according to the acquisition rate of 1 Hz.
- A safety distance of 50 cm between the detection system and the container walls is kept constant during the container inspection. The choice of the mobile platform for transporting the detection system will determine the appropriate safety distance to prevent potential collisions with the container.
- Gamma-ray attenuation due to air and 3 mm of iron, which represents the mean thickness of the container walls.
- Background noise is considered as a Poisson distribution with 26 counts based on measurements performed.

The MLE algorithm is used to estimate the source position and consists of the following steps:

- It starts running after four radiation data points above a decision threshold given by the critical level L_C [19] given by equation (1), and then at each data point. The standard deviation s_B of the background distribution is $\sqrt{26} \approx 5\ counts$.

  $$L_C = 1.645\sqrt{2s_B^2} = 2.33\ s_B \qquad (1)$$
- For each source position estimation, the confidence intervals given by the Fisher information are obtained. Fisher information assesses MLE-based parameter estimate quality and enhances comprehension of efficiency, precision, and limitations in estimation [20].
- The initial guess for the source activity was set to 10^6 Bq (intermediate value of the activity bounds interval considered) and updated at each MLE calculation, with bounds between 10^4 and 10^8 Bq (to consider a wide range of activity values).
- The initial guess for the source position is set at the container center and is updated in every MLE algorithm iteration.

To investigate the benefits arising from using the detection system data for path optimization and for source position estimation and confidence intervals improvement, different paths were considered:

- Predefined paths: The detection system performs complete turns around the container, and each radiation data point is analyzed by the MLE algorithm. The path height remains fixed during the container inspection. Either a single path at one-third the container´s height and a double path at one-third and at two-thirds the container´s height was considered. The paths' heights were chosen according to previous work performed by the authors [6], however other heights could be considered (which would not change the conclusions of this work).
- Informed paths: The detection system initially performs complete turns around the container based on a predefined path. The data is then analyzed at each second using the MLE algorithm along with a profit function. The initial predefined path can change dynamically based on the detection system counts and source position estimation at each MLE iteration, optimizing the radiation field information.

To take into consideration statistical variations due to background (Poisson distribution) five complete simulations were performed for each path and source position considered.

Fig. 1 illustrates the proposed algorithm for the radioactive source search considering an informed path.

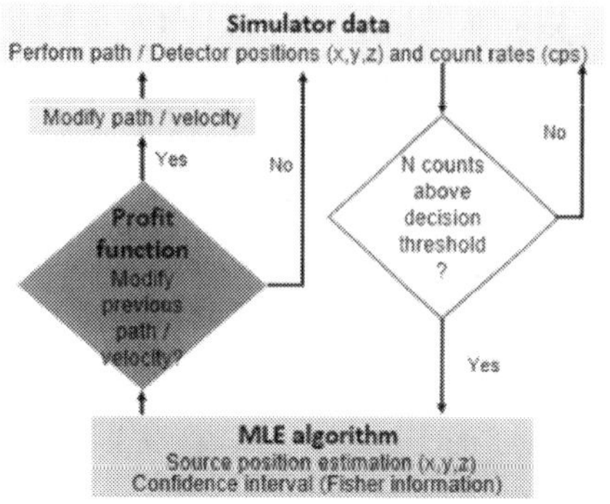

Fig. 1. Proposed source search algorithm for informed path planning (N=4).

Employing a profit function allows for a comparison between the detection system's position and the estimated source position (step profit function), or the detection of gamma-ray count increments as the detection system approaches the source position (up and down profit function). As a result, this triggers modifications to the previous path to gather supplementary information regarding the source position.

The profit functions used in the simulations are as follows:

- Step profit function (Fig. 2a): The detection system starts by following a predefined horizontal path along the container at one-half of the container's height. At each iteration of the MLE algorithm, the detection system adjusts its vertical velocity component to converge its position to the z-component of the radioactive source's estimated position.
- Up and down (UD) profit function (Fig. 2b): The detection system initiates its predefined horizontal path along the container at one-third of the container's height. When a gamma-ray count peak is detected, the system comes to a stop (horizontal speed is zero), and it performs a UD movement with a vertical speed of 0.2 m/s. The purpose of this movement is to identify a gamma-ray count peak in the upward or downward direction. If a peak is found in the vertical direction, the detection system proceeds to move horizontally around the container at the peak height. If another peak is detected in the horizontal direction, the procedure is repeated.

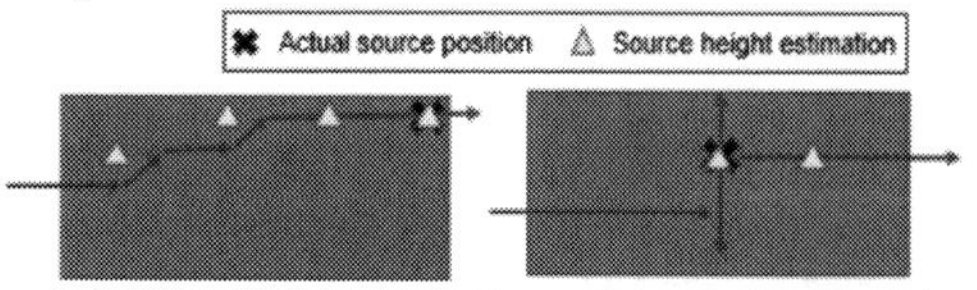

Fig. 2. Proposed profit functions to improve source position estimation by changing detection system path according to: a) the source height estimation at each MLE iteration; b) the maximum detector counts registered.

III. Results and Discussion

A. Predefined paths

To estimate the source position (x, y, and z coordinates) of a 4 MBq ^{137}Cs point source located both at the center and at the corner of the container, an MLE algorithm was used. The position source estimates, considering a single path and two paths around the container, are displayed in Table I and Table II, respectively. Additionally, examples of the detection system paths and their corresponding counts for a centered source are shown in Fig. 3 (single path) and Fig. 4 (double path).

Table I
Simulated values for a single path (one-third the container height) around the container - predefined path

Estimated parameters	Source at the center[a]	Source at the corner[b]
x (m)	2.97±0.02	0.13±0.01
y (m)	-1.17±0.01	-0.14±0.01
z (m)	0.80±0.05	0.8±0.02
d (m)	0.4	0.8

[a] (x=2.97 m, y=-1.17 m, z=1.195 m). [b] (x=0.1 m, y=-0.1 m, z=0.1 m). Confidence intervals are given by the Fisher information.

Table II
Simulated values for Two paths (One-third and two-thirds the container height) around the container - predefined paths

Estimated parameters	Source at the center[a]	Source at the corner[b]
x (m)	2.97±0.01	0.13±0.01
y (m)	-1.17±0.01	-0.13±0.01
z (m)	1.20±0.05	0.19±0.01
d (m)	< 0.01	< 0.01

[a] (x=2.97 m, y=-1.17 m, z=1.195 m). [b] (x=0.1 m, y=-0.1 m, z=0.1 m). Confidence intervals are given by the Fisher information.

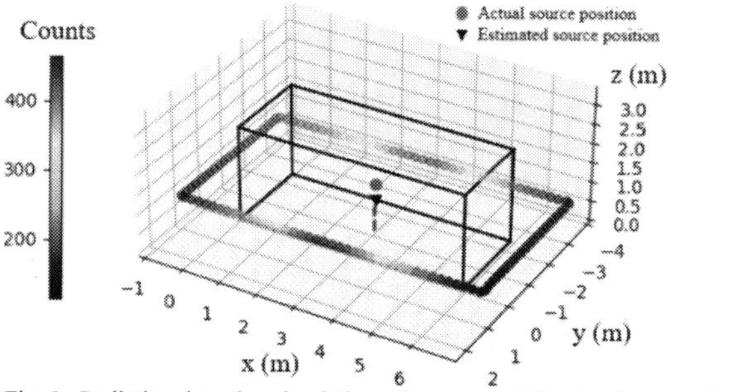

Fig. 3. Radiation detection simulation using one predefined path at one-third the container's height – actual source position at the container center: (x=2.97 m, y=-1.17 m, z=1.195 m).

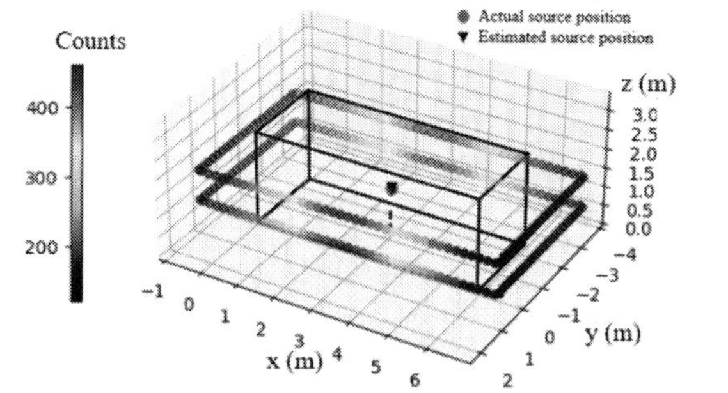

Fig. 4. Radiation detection simulation using two predefined paths – actual source position at the container center: (x=2.97 m, y=-1.17 m, z=1.195 m).

According to Table II, the z component estimation of the source position (source height estimation), and consequently, the distance between the estimated and actual source position, improve when compared to a single path inspection (Table I). The poor source height estimation in the single path inspection is attributed to the lack of information (gamma-ray counts) along the z-axis, as the path is performed only in the x and y axes. However, it should be noted that using two paths will take twice the time (206 s) compared to a single path inspection (103 s). Therefore, it becomes necessary to optimize the information gathered by the detection system during a single path inspection around the container.

B. Informed paths

Table III and Table IV display the simulation results of the source position estimations, considering the Step and UD profit function paths, respectively. Figures 5 to 8 illustrate some examples of the informed paths, considering both the source in the center and at the corner of the container.

In the Step profit function path, the detection system maintains a constant velocity around the container, resulting in the same time as the single predefined path (103 s). This is due to the fact that the detection system needs to stop at the gamma-ray counts peak position to vertically scan the z-axis (source height). On the other hand, the UD profit function path takes 137 s for a source placed at the center and 145 s for a source placed at the corner of the container.

Despite the longer time spent by the UD profit function path, it significantly improved the source position estimation (lower distance *d* and z component of the source position closer to the simulated one) compared to a single predefined trajectory and yields better results compared to the Step profit function. Moreover, the UD profit path takes about one minute less to perform than two predefined paths, which yielded the best results.

A third profit function (not addressed in this study), which combines the two previously analyzed profit functions, could likely enhance both the source position estimation and inspection time.

TABLE IIII
SIMULATED VALUES FOR A STEP PROFIT FUNCTION PATH

Estimated parameters	Source at the center[a]	Source at the corner[b]
x (m)	2.98±0.02	0.15±0.01
y (m)	-1.17±0.01	-0.16±0.01
z (m)	1.25±0.03	0.53±0.01
d (m)	0.05	0.4

[a] (x=2.97 m, y=-1.17 m, z=1.195 m). [b] (x=0.1 m, y=-0.1 m, z=0.1 m). Confidence intervals are given by the Fisher information.

TABLE IIIV
SIMULATED VALUES FOR A UD PROFIT FUNCTION PATH

Estimated parameters	Source at the center[a]	Source at the corner[b]
x (m)	2.97±0.02	0.12±0.01
y (m)	-1.18±0.01	-0.13±0.01
z (m)	1.22±0.03	0.22±0.01
d (m)	0.03	0.1

[a] (x=2.97 m, y=-1.17 m, z=1.195 m). [b] (x=0.1 m, y=-0.1 m, z=0.1 m). Confidence intervals are given by the Fisher information.

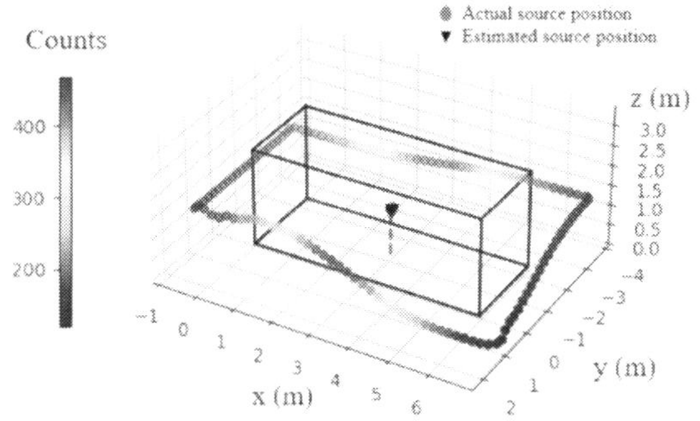

Fig. 5. Radiation detection simulation using the Step profit function – Source centered (x=2.97 m, y=-1.17 m, z=1.195 m).

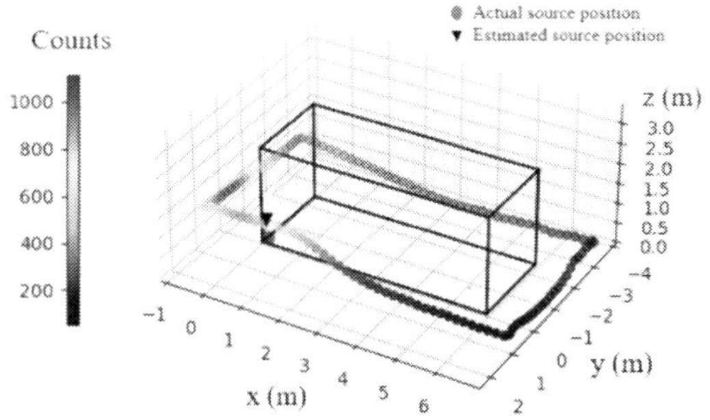

Fig. 6. Radiation detection simulation using the Step profit function – Source at the corner (x=0.1 m, y=-0.1 m, z=0.1 m).

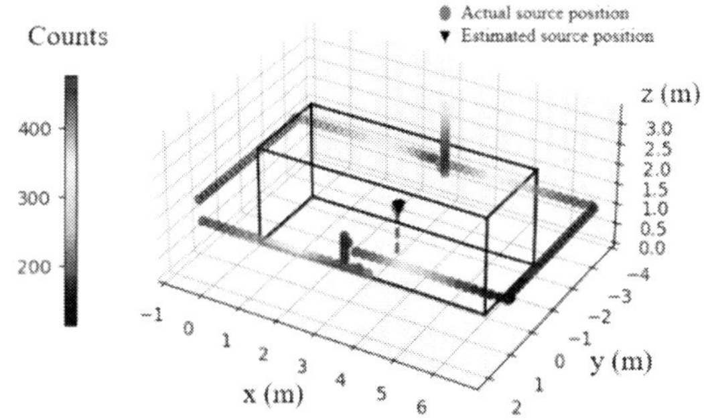

Fig. 7. Radiation detection simulation using the UD profit function – Source centered (x=2.97 m, y=-1.17 m, z=1.195 m).

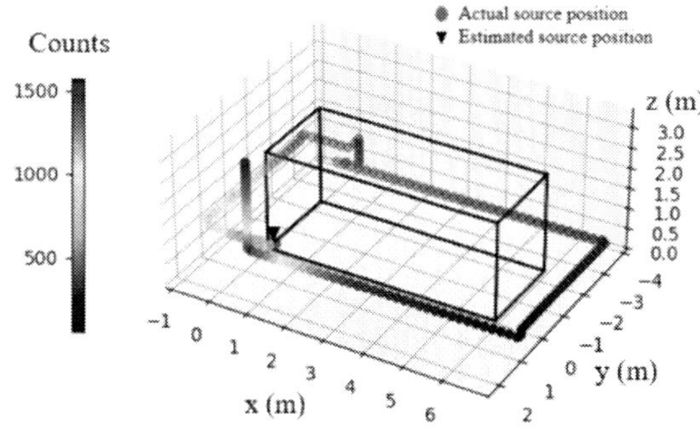

Fig. 8. Radiation detection simulation using the UD profit function – Source at the corner (x=0.1 m, y=-0.1 m, z=0.1 m).

IV. Conclusions

A three-dimensional radioactive source search algorithm for security inspection and monitoring was developed, encompassing radiation detection data simulation, an MLE algorithm with Fisher information, and profit functions.

When comparing the use of two predefined paths to a single path, the algorithm showed a better estimation of the source height (z coordinate) and consequently, an improvement in the estimation of the global source position. However, it should be noted that implementing two trajectories requires doubling the inspection and monitoring time. The incorporation of profit functions into the source search algorithm resulted in a significant enhancement in the source height estimation and confidence intervals compared to a single predefined path, with the UD profit function showing superior results compared to the step profit function. Although two predefined paths showed the best results, the use of profit functions will significantly reduce the container inspection time.

As future work, the authors plan to validate the simulated data with field tests and to address the impact of the detector position uncertainty on the results. Other issues that need to be addressed are multiple sources scenario and improve the realism of the movement in the UD profit function. Finally, a composite profit function, merging the two profit functions analyzed in this study, will be implemented, and assessed for its impact on localization accuracy and reduction of inspection time.

Acknowledgment

One of the authors (Luís Marques) thanks the Portuguese Air Force Academy for all the financial support provided.

References

[1] International Atomic Energy Agency, "Combating Illicit Trafficking in Nuclear and Other Radioactive Material," IAEA, Vienna, Austria, IAEA Nuclear Security Series No. 6 Tech. Guidance STI/PUB/1039, Dec. 2007.

[2] L. Marques, A. Vale, and P. Vaz, "State-of-the-Art Mobile Radiation Detection Systems for Different Scenarios," *Sensors*, vol. 21, no. 4, Feb. 2021, Art. no. 1051, DOI. 10.3390/s21041051.

[3] OECD/European Union Intellectual Property Office, Misuse of Containerized Maritime Shipping in the Global Trade of Counterfeits, Illicit Trade. OECD Publishing: Paris, France, 2021.

[4] R. Downes, C. Hobbs, and D. Salisbury, "Combating nuclear smuggling? Exploring drivers and challenges to detecting nuclear and radiological materials at maritime facilities," *Nonprolif. Rev.*, vol. 26, nos. 1-2, pp. 83-104, June 2019, DOI. 10.1080/10736700.2019.1610256.

[5] R.T. Kouzes, J.H. Ely, L.E. Erikson, W.J. Kernan, A.T. Lintereur, E.R. Siciliano, *et al.*, "Neutron detection alternatives to 3He for national security applications," *Nucl. Instrum. Meth. Phys. Res. A*, vol. 623, no. 3, pp. 1035–1045, Nov. 2010, DOI. 10.1016/j.nima.2010.08.021

[6] L. Marques, L. Félix, G. Cruz, V. Coelho, J. Caetano, A. Vale, *et al.*, "Neutron and Gamma-Ray Detection System Coupled to a Multirotor for Screening of Shipping Container Cargo," *Sensors*, vol. 23, no. 1, Dec. 2022, Art. no. 329, DOI. 10.3390/s23010329.

[7] L. Marques, A. Vale and P. Vaz, "Development of a portable neutron detection system for Security and Defense applications," in *Developments and Advances in Defense and Security, Smart Innovation, Systems and Technologies*, vol. 328, Á. Rocha, C.H. Fajardo-Toro, J.M. Riola, Eds. Singapore: Springer, 2023, pp. 283-293, DOI. 10.1007/978-981-19-7689-6_24.

[8] Y. Brouwer, A. Vale, D. Macedo, B. Gonçalves, and H. Fernandes, "Radioactive Hot-spot Detection Using Unmanned Aerial Vehicle Surveillance," *EPJ Web Conference*, vol. 225, pp. 1-8, January 2020, Art no. 06005, DOI. 10.1051/epjconf/202022506005.

[9] A.H. Zakaria, Y.M. Mustafah, J. Abdullah, N. Khair, and T. Abdullah, "Development of Autonomous Radiation Mapping Robot," *Procedia Comput. Sci.*, vol. 105, 2017, pp. 81–86, DOI. 10.1016/j.procs.2017.01.203.

[10] H. Carvalho, A. Vale, R. Marques, R. Ventura, Y. Brouwer, and B. Gonçalves, "Remote inspection with multi-copters, radiological sensors and SLAM techniques," *EPJ Web Conf.*, vol. 170, Jan. 2018, Art. no. 07014, DOI. 10.1051/epjconf/201817007014.

[11] T. Lazna, P. Gabrlik, T. Jilek, and L. Zalud, "Cooperation between an unmanned aerial vehicle and an unmanned ground vehicle in highly accurate localization of gamma radiation hotspots," *Int. J. Adv. Robot. Syst.*, vol. 15, no. 1, pp. 1–16, Jan. 2018, DOI. 10.1177/1729881417750787.

[12] E.L. Connolly, and P.G. Martin, "Current and prospective radiation detection systems, screening infrastructure and interpretive algorithms for the non-intrusive screening of shipping container cargo: a review," *J. Nucl. Eng.*, vol. 2, no. 3, pp. 246–280, Aug. 2021, DOI. 10.3390/jne2030023.

[13] M.K. Sharma, A.B. Alajo, H.K. Lee, "Three-dimensional localization of low activity gamma-ray sources in real-time scenarios," *Nucl. Instrum. Methods Phys. Res. Sect. A Accel. Spectrometers Detect. Assoc. Equip.*, vol. 813, pp. 132–138, March 2016, DOI. 10.1016/j.nima.2016.01.001.

[14] H.E. Baidoo-Williams, "Maximum likelihood localization of radiation sources with unknown source intensity," arXiv 2016, arXiv:1608.00427, DOI. 10.48550/arXiv.1608.00427.

[15] C.Q. Wu, M.L. Berry, K.M. Grieme, S. Sen, N.S. Rao, R.R. Brooks, G. Cordone, "Network Detection of Radiation Sources Using Localization-based Approaches," *IEEE Trans. Ind. Inform.*, vol. 15, no. 4, pp. 2308–2320, Jan. 2019, DOI. 10.1109/TII.2019.2891253.

[16] N.S.V. Rao, S. Sen, N.J. Prins, D.A. Cooper, R.J. Ledoux, J.B. Costales et al., "Network algorithms for detection of radiation sources," *Nucl. Instrum. Methods Phys. Res. Sect. A Accel. Spectrometers Detect. Assoc. Equip.*, vol. 784, pp. 326–331, June 2015, DOI. 10.1016/j.nima.2015.01.037.

[17] F. Mendes, M. Barros, A. Vale, and B. Gonçalves, "Radioactive hot-spot localisation and identification using deep learning," *J Radiol Prot.*, vol. 42, no. 1, Jan. 2022, DOI. 10.1088/1361-6498/ac1a5c.

[18] Secure container solutions, "Everything You Need To Know About Sea Container Dimensions." https://www.securecontainer.ca/everything-need-know-sea-container-dimensions/ (accessed June 05, 2023).

[19] E.W. Abelquist, J.P. Clements, A.M. Huffert, D.A. King, T.J. Vitkus, and B.A. Watson, "Minimum Detectable Concentrations with Typical Radiation Survey for Instruments for Various Contaminants and Field Conditions," Office of Nuclear Material Safety and Safeguards, U.S. Nuclear Regulatory Commission, Washington, D, USA, NUREG-1507 Revision 1, Aug. 2020.

[20] C.R. Rao, "Information and the Accuracy Attainable in the Estimation of Statistical Parameters," in: *Breakthroughs in Statistics. Springer Series in Statistics*. S. Kotz, N.L. Johnson, Eds. New York: Springer, 1992, DOI. 10.1007/978-1-4612-0919-5_16.

A portable aerosol monitor with a novel linear deconvolution algorithm for fast and accurate detection of alpha emitters

Luca F. Ferrante Vero[1,*], Giacomo P. Manessi[1], Giuseppe A. Marzo[2], and Gabriele Zorloni[1]
[1]ELSE NUCLEAR Srl, Italy
[2]ENEA C. R. Casaccia, Italy
(*) luca.ferrante@elsenuclear.com

***Abstract*—Radiochemistry laboratories and nuclear power plants are required to install aerosol monitors for real-time detection of alpha-particle emitters in air. However, common radioactive aerosol monitors are unsuited for rapid emergency response, since they are neither portable nor easily deployable. Moreover, their spectrometric capabilities are usually limited, showing particularly poor performance of radioisotope identification, which makes it difficult to correctly assess the alpha-emitter concentration in air. We developed a rugged and compact radioactive aerosol monitor, fully battery powered, which can be quickly deployed in field, as well as used as a fixed station for routine monitoring. The monitor is equipped with a silicon detector, and the acquired alpha-particle spectra are analyzed employing a patented linear deconvolution algorithm. A prototype of this device was tested to evaluate its performance of radioisotope identification by using artificial alpha-emitters. Initial encouraging tests provide evidence for alpha-emitters identification with multi-line mixed-isotope alpha sources, and in energy-degraded low-counting-statistics conditions.**

***Keywords* —Alpha-particle spectroscopy; Radioactive aerosol monitor; Linear deconvolution algorithm**

I. Introduction

ALPHA-particle emitters are commonly found at radiochemistry laboratories and nuclear power plants, during either normal operation or decommissioning. These nuclides pose a serious risk to the workers, and general public, in the case of accidents [1]. In fact, even though their low penetration capability makes somehow negligible the risk due to external exposure, in the case of ingestion or inhalation the committed effective dose might be significantly relevant even in the case of low activity concentrations in air [2]. The installation of aerosol monitors is therefore often mandatory in those workplaces where alpha-emitters are manipulated, and inside and around nuclear sites [3,4].

Several aerosol monitors are available on the market, generally designed as fixed stations devoted to routine monitoring. Even though largely appropriate during normal operation, their usability during an accident is limited. First, they are generally mains powered, and show low Inner Protection (IP [5]) grade; thus, in the case of emergency, they are neither portable nor easily deployable. Second, their usually poor energy resolution and peak deconvolution algorithms, generally with limited capabilities of isotopic recognition, make it difficult to correctly assess the alpha-emitter concentration in air [6]. An accurate isotopic recognition algorithm is also required during routine monitoring, to subtract the contribution of background Naturally Occurring Radioactive Material (NORM), mostly coming from the decay of ^{222}Rn, ^{220}Rn, and related daughter nuclei.

To tackle the aforementioned limitations, a new aerosol monitor is under development. The device is designed to be both a routine monitoring system as well as a readily usable emergency device. It is characterised by a rugged and compact design, fully battery powered, and it is easily deployable in-field. It is equipped with a silicon detector sensitive to alpha- and beta-particles. The software devoted to the analysis is based on a patented linear deconvolution algorithm.

The initial tests performed with a first prototype of the device are discussed herein, focusing on its spectrometric capabilities. The instrument was tested with some multi-line alpha-emitters to simulate accidental contamination from artificial radionuclides.

II. Materials and Methods

A. Hardware

The sensitive component of the prototype aerosol monitor is a 450 mm^2 active area silicon detector. The signal from the probe is directly fed to a Charge Sensitive Amplifier (CSA) followed by an analogue shaping stage (1 μs semi-Gaussian shaping). The signal is then digitized by a custom-made Multichannel Analyser (MCA) based on a 125 Ms/s 14-bit Field Programmable Gate Array (FPGA). The amplitude of the digitized pulse is eventually sampled and saved in a histogram. The system allows sampling on different types of filters depending on the specific application. It uses Mixed Cellulose Ester (MCE) Membrane or polytetrafluoroethylene polymer (PTFE) circular filters (standard diameter 47 mm) as standards.

An industrial Single Board Computer (SBC) manages the data acquisition, the real time elaboration, the visualization on a local display and the remote communication of the instrument (Wi-Fi, Bluetooth Low Energy – BLE, 4G). The system uses a low-voltage-powered diaphragm gas pump with a sampling flow rate on the filter of about 3 m^3/h. It is powered by a lithium ion battery pack which guarantees an operating autonomy of approximately 8 hours without recharging. The dimensions of the monitor are 60 cm × 40 cm × 30 cm, with an overall weight < 25 kg. Fig. 1 shows a 3D rendering of the prototype aerosol monitor at the current stage of development.

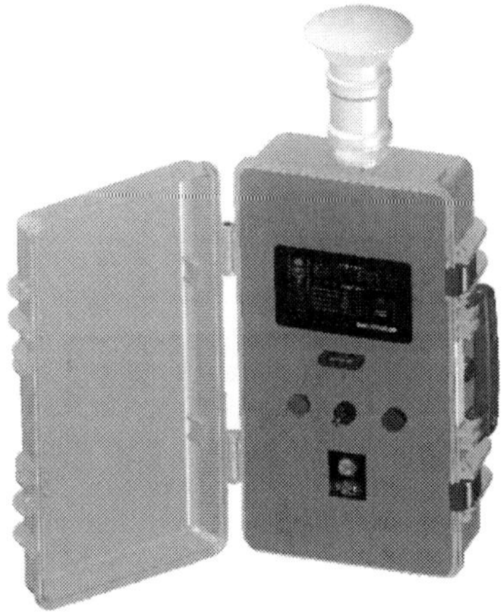

Fig. 1: 3D rendering of the aerosol monitor.

B. Deconvolution algorithm

Alpha-decaying isotopes emit alpha-particles with discrete energies. When a spectroscopy measurement is carried out using a silicon detector, the observed shape of the energy spectrum is determined by the stochastic elastic and inelastic collisions of the alpha-particles in air and in the detector lattice itself. The determination of the radionuclide activity in the measured sample therefore requires radioisotope identification by deconvolving either a multi-peak spectrum or, in general, an energy-degraded low-resolution left-tailed single peak. Therefore, often multiple radioisotopes are identified as candidates to be associated to a specific alpha peak (see e.g. ref. [7]). The patented ENEA approach [8] is adopted to overcome the common problem of ambiguity in the identification of radioisotopes. The algorithm is based on the following calculation steps:

(1) The response function corresponding to the specific silicon detector and the measurement geometry is determined from a reference alpha-particle spectrum;

(2) A library of synthetic alpha-particle spectra is calculated using appropriate modified Gaussian functions [9,10] based on the parameters retrieved in step (1);

(3) Steps (1) and (2) are due once: further measurements will take advantage of this initial effort;

(4) The measured alpha-particle spectrum is then deconvolved by minimizing its difference with a linear combination of the synthetic alpha-particle spectra calculated at step (2).

The goodness of the fitting procedure is evaluated based on the trend of the residuals R, i.e.

$$R = \frac{C_{exp} - C_{fit}}{\sqrt{C_{exp}}} \quad (1)$$

where C_{exp} are the experimental counts and C_{fit} are the ones deconvolved by the fitting algorithm.

The algorithm has proven efficient in laboratory conditions in providing unequivocal identification of radioisotopes in samples, showing a more accurate and sensitive evaluation of the sample activity with respect to traditional approaches [8].

C. Experimental setup

Tests were performed at the Casaccia Research Centre laboratories of the Italian National Agency for New Technologies, Energy and Sustainable Economic Development (ENEA). The laboratory hosts several certified multi-line alpha-sources. Among them, the following have been selected to carry out the monitor tests:

- ^{237}Np-^{241}Am-^{244}Cm (150 Bq, 98 Bq, 71 Bq, respectively).
- ^{239}Pu-^{241}Am-^{244}Cm (5546 Bq, 5320 Bq, 2020 Bq, respectively).
- ^{241}Am (213 Bq).

All tests were performed by placing the radioactive source at 2 mm from the silicon detector in air. In the case of the ^{237}Np-^{241}Am-^{244}Cm source, tests were repeated at source-to-detector distances equal to 2 mm, 5 mm and 10 mm, in order to investigate the effects of the distance, i.e. different spectral resolution and counting statistics, on the deconvolution algorithm.

The acquisition time was set to obtain at least about 1000 counts in the most populated spectral bin, with the exception of the test performed at 10 mm from the ^{237}Np-^{241}Am-^{244}Cm source. In this case, the measurement was stopped when the most populated bin reached about 200 counts, in order to investigate the algorithm performance in the case of poor spectral resolution due to low counting statistics.

III. Results and Discussion

Prior to the measurement with the radioactive sources, a background measurement was launched for 10 minutes, obtaining an average count rate lower than one count per minute over the integral MCA spectrum. Therefore, the monitor is almost blind to background radiation and it is not affected by electronic disturbances.

Fig. 2 shows the counts obtained with the pure ^{241}Am source (red dots) and the reconstructed peak shape calculated by the algorithm (solid line). The peak is correctly recognised by the software. The fit is satisfactory both quantitatively and qualitatively, by visual inspection of the data distribution vs fit function, and residuals distribution vs zero line (discrepancies within ± 10%), without any significant under-/overestimation trend.

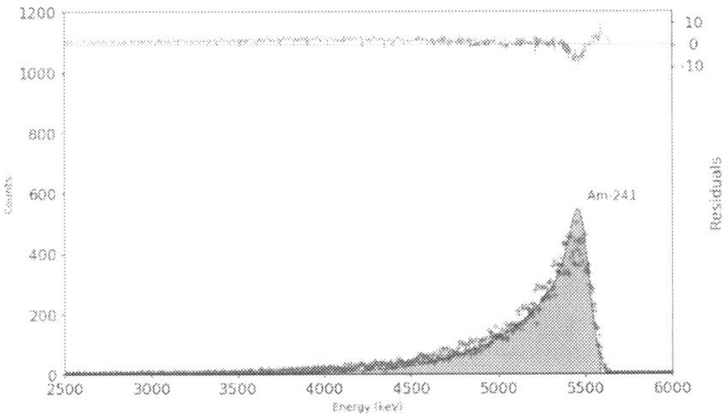

Fig. 2: Experimental counts (dots) and reconstructed peak (dashed line) in the case of the pure ^{241}Am source. The isotope identified by the software is labelled just above of the reconstructed peak. Residuals are compared with the zero line on the top of the graph.

Encouraging results were obtained while measuring a multi-peak source where spectral band overlapping is significant, such as ^{239}Pu-^{241}Am-^{244}Cm (Fig. 3). In this case, the three nuclides of the mixture are correctly deconvolved, represented as solid lines in the graph. Above each peak, the software correctly labels the corresponding isotope: 5.157 MeV peak for ^{239}Pu, 5.486 MeV peak for ^{241}Am, and about 5.8 MeV for ^{244}Cm (differently from the other isotopes of the source, the ^{244}Cm peak is given by the superposition of two emission lines which are almost overlapped, i.e. 5.805 MeV (76.9%) and 5.763 MeV (23.1%)). The spectrum reconstruction (dashed line) is satisfactory both quantitatively and qualitatively, by visual inspection of the data distribution vs fit function, and residuals distribution vs zero line (discrepancies within ± 10%), without any significant under-/overestimation trend. It is worth noting that the algorithm is capable of recognising (and reconstructing) the contribution of the left tail of the ^{241}Am peak, even though about 50% of its area is overlapped with the ^{239}Pu peak.

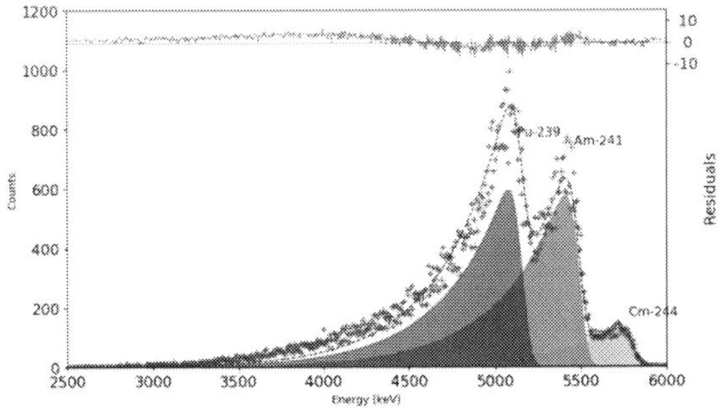

Fig. 3: Experimental counts (dots) and reconstructed peaks (solid lines) in the case of the mixed ^{239}Pu-^{241}Am-^{244}Cm source. The isotope identified by the software is labelled just above of the reconstructed peak. The final convolved spectrum reconstructed by the algorithm is shown as a dashed line. Residuals are compared with the zero line on the top of the graph.

Fig. 4 shows the counts obtained with the mixed ^{237}Np-^{241}Am-^{244}Cm source (red dots), the reconstructed peaks (solid lines and labels), and the reconstructed convolved spectrum (dashed lines), for increasing source-to-detector distances, namely 2 mm (a), 5 mm (b), and 10 mm (c). The spectrometric capabilities of the device are satisfactory even at the largest source-to-detector distances. The full shape of the spectrum is always correctly deconvolved, and the radionuclides are correctly identified for each source-to-detector distance, and for different measurement times, the latter corresponding to distinct counting statistics. The spectrum reconstruction (dashed line) is satisfactory both quantitatively and qualitatively, by visual inspection of the data distribution vs fit function, and residuals distribution vs zero line, without any significant under-/overestimation trend, with the possible exception of the 2 mm distance spectrum, where the largest deviations of the residuals are observed, even though the variations vs zero line still fall between ± 15%. Those deviations are due to chance coincidence between alpha-particles and beta-rays, interactions of x-rays, scattering effects, etc. [11,12,13], which are not directly included in the fit function. However, such effects do not prevent the deconvolution algorithm to correctly identify the alpha-emitters occurring in the spectrum. For the two other cases (5 mm and 10 mm distance), such effects are not appreciable. In fact, the increased source-to-detector distance slightly improves the spectral resolution, since only particles emitted along narrower angles can be detected, i.e. limiting the peak broadening due to alpha-particle collisions in air. Moreover, the lowered counting efficiency due to the increased source-to-detector distance causes a significant drop in the probability of chance coincidence between alpha-particles and beta-/x-rays.

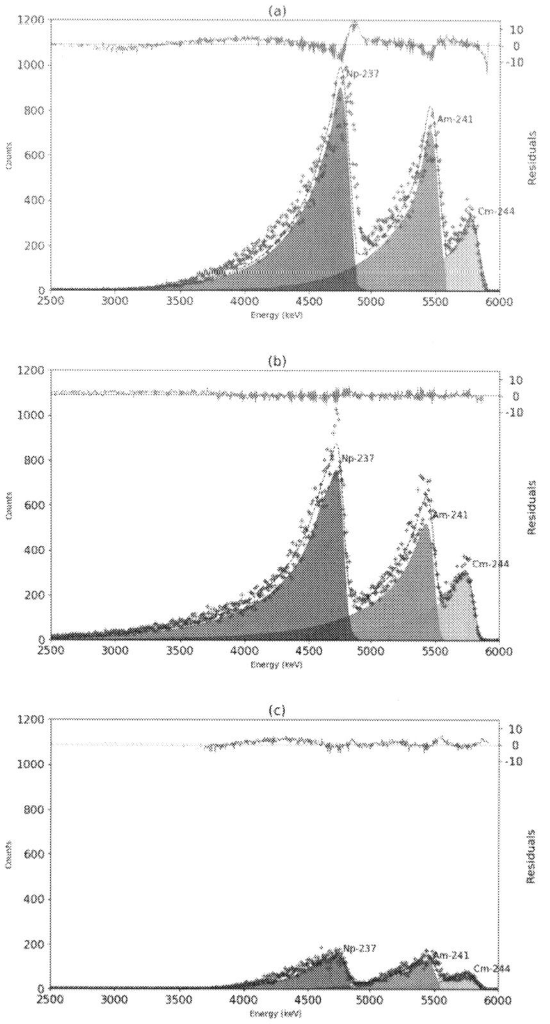

Fig. 4: Experimental counts (dots) and reconstructed peaks (solid lines) in the case of the mixed ^{237}Np-^{241}Am-^{244}Cm source at different source-to-detector distances, namely, 2 mm (a), 5 mm (b), and 10 mm (c). The isotope identified by the software is labelled just above of the reconstructed peak. The final convolved spectrum reconstructed by the algorithm is shown as a dashed line. Residuals are compared with the zero line on the top of the graphs. In the case of 2 mm distance (graph (a)), the relatively large deviations of the residuals (still included between ± 15%) do not prevent the radioisotope identification.

IV. Conclusions and Outlooks

A novel alpha-beta aerosol monitor is under development. The device was developed to be either a fixed station, as well as an easily deployable device for emergency use. The implemented algorithm for isotopic recognition was tested with multi-peak alpha sources, showing appropriate and reliable spectrometric performances, both in terms of isotopic identification and spectrum reconstruction (residuals always within ± 10% without showing any significant under-/overestimation trend), also in condition of extremely poor peak resolution and low counting statistics. The device then proved to be equivalent (or superior) to conventional fixed aerosol monitoring stations for isotopic recognition, with the advantage of being designed also for emergency use.

Further steps of the research will involve the investigation of the spectrometric properties of the device for NORM recognition, i.e. background subtraction, and employing the final measurement geometry, i.e. measuring contaminated and uncontaminated filters. This will allow investigating the capability of the system of quantitatively measuring the concentration of artificial radionuclides (or NORM) in the ambient air.

References

[1] INTERNATIONAL COMMISSION ON RADIOLOGICAL PROTECTION, "Recommendations of the International Commission on Radiological Protection, Publication 60", Pergamon Press, Oxford and New York (1991).

[2] INTERNATIONAL COMMISSION ON RADIOLOGICAL PROTECTION, "Dose Coefficients for Intakes of Radionuclides by Workers, Publication 68", Pergamon Press, Oxford and New York (1994).

[3] EUROPEAN UNION, "Council Directive 2013/59/Euratom of 5 December 2013 laying down basic safety standards for protection against the dangers arising from exposure to ionising radiation, and repealing Directives 89/618/Euratom, 90/641/Euratom, 96/29/Euratom, 97/43/Euratom and 2003/122/Euratom", *Off. J. Europ. Comm.* Document 32013L0059 (2013).

[4] INTERNATIONAL ATOMIC ENERGY AGENCY, Occupational Radiation Protection, Safety Standards Series No. GSG-7, IAEA, Vienna (2018).

[5] Degrees of protection provided by enclosures (IP Code), IEC standard 60529, 2019.

[6] M. Blaauw, E. García-Toraño, S. Woods and S. Fazinic, "The 1997 IAEA intercomparison of commercially available PC-based software for alpha-particle spectrometry", *Nuclear Instruments and Methods in Physics Research A*, 428:317–329, 1999.

[7] E. García-Toraño. "Current status of alpha-particle spectrometry", *Applied Radiation and Isotopes*, 64:1273–1280, 2006.

[8] G.A. Marzo and N. Cherubini, "Metodo di scomposizione lineare di spettri energetici e relativo apparato" Italian Patent 102018000007610, July 30, 2018.

[9] G. Bortels and P. Collaers, "Analytical function for fitting peaks in alpha-particle spectra form Si detectors", *Applied Radiation and Isotopes*, 38:831-837, 1987. http://dx.doi.org/10.1016/0883-2889(87)90180-8.

[10] G.A. Marzo, "A comparison of different peak shapes for deconvolution of alpha-particle spectra *Nuclear Instruments and Methods in Physics Research A*, 832:191-201 (2016). https://doi.org/10.1016/j.nima.2016.06.111

[11] M. Marouli, S. Pommé, V. Jobbágy, R. Van Ammel, J. Paepen, H. Stroh and L. Benedik, "Alpha-particle emission probabilities of 236U obtained by alpha spectrometry", *Applied Radiation and Isotopes*, Volume 87, 2014, Pages 292-296, ISSN 0969-8043, https://doi.org/10.1016/j.apradiso.2013.11.020.

[12] S. Pommé, E. García-Toraño, M. Marouli, M.T. Crespo, V. Jobbágy, R. Van Ammel, J. Paepen and H. Stroh, "High-resolution alpha-particle spectrometry of ^{238}U", *Applied Radiation and Isotopes*, Volume 87, Pages 315-319, ISSN 0969-8043, 2014. https://doi.org/10.1016/j.apradiso.2013.11.075.

[13] S. Pommé and B. Caro Marroyo, "Improved peak shape fitting in alpha spectra", *Applied Radiation and Isotopes*, 96:148–153, 2015. http://dx.doi.org/10.1016/j. apradiso.2014.11.023.

An Extended List-Mode MLEM Algorithm for 3D Compton Image Reconstruction from Multi-View Data

Nhan Le[1,*], Hichem Snoussi[1], Zied Hmissi[2], Alain Iltis[2], Guillaume Lebonvallet[1], and Ghislain Zeufack[2]
[1]Computer Science and Digital Society laboratory, Troyes University of Technology, France
[2]Damavan Imaging, 2 rue Gustave Eiffel, 10430 Rosières-Près-Troyes, France
(*) thi-ai-nhan.le@utt.fr

Abstract—**The lack of parallax in the measurements is a big challenge in 3D image reconstruction for handheld Compton cameras. Our solution to this issue is to extend the conventional list-mode maximum-likelihood expectation-maximization (LM-MLEM) 3D reconstruction algorithm to allow the simultaneous use of multi-view Compton data seeking parallax improvement. It involves building a new list-mode simultaneous data space from multi-view Compton events, formulating the associated probabilistic models for the system response matrix and sensitivity, and developing an extended LM-MLEM algorithm. For the performance assessment, we experiment the extended 3D reconstruction algorithm on real multi-view data conducted with a handheld CeBr3 Compton camera developed by Damavan Imaging and a punctual 0.2 (MBq) ^{22}Na source. Various comparative studies with different view numbers, source locations and energy ranges confirm the outperformances of our extended LM-MLEM algorithm.**

Keywords **—3D reconstruction, handheld Compton camera, LM-MLEM algorithm, multi-view Compton events, real data.**

I. Introduction

3D Compton image reconstruction allows to visualize gamma ray sources in a given space by measuring a Compton scatter and a photoelectric absorption for each radioactive decay in terms of positions and deposited energies. This technique covers nowadays a wide range of applications in nuclear industry, medical imaging, homeland security and astronomy. In principle, the radioactive source related to a measured event is located on a cone surface issued from the Compton kinematics with the measured information. By drawing and overlaying multiple cones given from different measured events, we can retrieve the source location. Based on this idea, the literature has developed numerous reconstruction methods in two main streams: (*i*) *analytical inversion* [1] and (*ii*) *iterative model-based* [2]. The first stream exploits the relationship between a function and its line integrals. Although it often leads to fast reconstruction algorithms such as filtered back-projection [3], the accuracy of reconstructed images is limited, as it requires approximations in the line-integral model while eliminating uncertainties inherent in photon-limited detection. In contrast, the second stream enables more accurate reconstruction algorithms by explicitly modeling statistical variations in the photon detection process and progressively refining reconstructed images through repetitive calculations. The price of this accuracy is the significantly higher computational cost. This is why the current state-of-the-art has put much effort into improving the computation time for statistical iterative reconstruction algorithms [4].

The class of maximum-likelihood expectation-maximization (MLEM) algorithms has been the leading statistical iterative reconstruction algorithms for several years [5]. Being applicable for Poisson Compton data in both the bin-mode [6] and list-mode [7], the conventional MLEM algorithm exhibits attractive properties including consistent and predictable convergence behavior, guaranteed non-negativity, and count preservation at each iteration. Nevertheless, since the algorithm is relatively slow, several variants have been proposed to speed up its convergence. The main improvements reside in using grids of different sizes to reduce processing time [8], extrapolating and increasing the magnitude of the change made at each iteration [9], updating each pixel individually with a series of different complete-data spaces [10], or in increasing the number of updates by using only part of the data in each update [11]. Despite their abundance, most MLEM algorithms only yield good reconstructed images when the parallax in the measurements is large enough. In other words, we cannot apply these algorithms to handheld Compton cameras that are becoming popular in nuclear engineering applications due to their convenience and economic advantages [12]. This fact prompted us to improve MLEM algorithms to address the lack of parallax inherent in handheld Compton cameras.

A solution is to measure potential radioactive sources with a handheld Compton camera deployed at different view angles, then combine the gathered data to increase the parallax [13]. Indeed, using *in turn* events recorded by an ultra-compact Compton camera from one view to another, Kishimoto *et al.* successfully locate radioactive sources in their experiments [14] with a conventional list-mode MLEM (LMMLEM) algorithm [7]. It is worth noting that the conventional LM-MLEM algorithm still works for single-view Compton data because the authors deployed the cameras very near to potential sources so that the parallax for each camera remains enough, albeit small. The major challenge in this current work is the *severe lack of parallax in single-view measurements*, which makes the existing LM-MLEM algorithms inapplicable. To overcome this challenge, we propose using multi-view Compton data *simultaneously* to build a so-called *list-mode simultaneous data space* where each element is a set of independent events arbitrarily given from all the view angles. This new list-mode data leads to new probabilistic models for the system response matrix and sensitivity. Using the probability theory, we find that these models are the sum of all response matrices and sensitivity models of individual cameras deployed at different view angles. They are thus flexible and can inherit the state-of-the-art models previously proposed for single-view Compton data (see e.g., [15,16] and references therein). Following a similar formulation process as in [17], we arrive at an extended LM-MLEM algorithm adapted to the new list-mode simultaneous data. This extended version shares the same iteration equation form as in the conventional LM-MLEM algorithm, but with different meaning. To assess the performance of the proposed 3D reconstruction algorithm, we setup an experimental system with a punctual 0.2 (MBq) ^{22}Na source and a handheld CeBr3 Compton camera developed by Damavan Imaging [18] to obtain 1-view, 2-view and 3-view Compton data. Using an extension of the well-known Maxim *et al.*'s probabilistic model [19] as the basic camera response matrix and sensitivity, we apply the extended list-mode MLEM to determine the radioactive source location. Various numerical results with different numbers of views and different locations of the punctual source confirm the outperformances of our extended LM-MLEM 3D reconstruction algorithm.

We structure the remainder of the paper as follows. In Section II, we present the list-mode simultaneous data space, the associated model for the system response matrix and sensitivity, and an extended version of the conventional LM-MLEM algorithm under the assumption of the Poisson data model. We dedicate Section III to numerical experiments and analyses. Finally, we conclude the paper and give some perspectives for the future works in Section IV.

II. Extended LM-MLEM 3D Reconstruction Algorithm

Let consider a system consisting of a radioactive source located in an image volume $\boldsymbol{V}$ and a handheld Compton camera deployed at K different view angles around the volume $\boldsymbol{V}$ as in Fig. 1. Our goal is to reconstruct the radioactive source image from the multi-view Compton events measured by the camera.

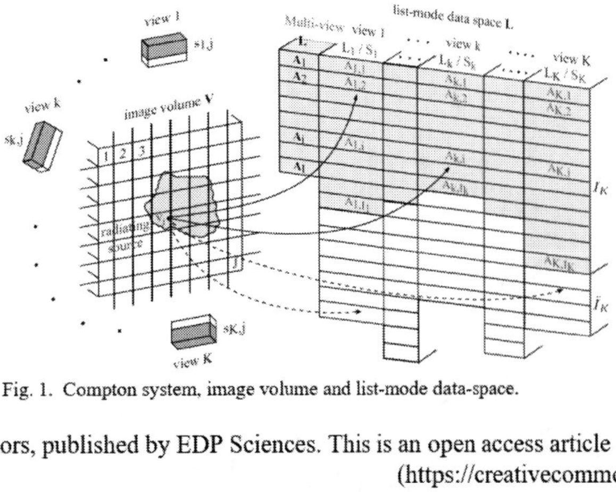

Fig. 1. Compton system, image volume and list-mode data-space.

A. List-Mode Data Space, Response Function and Sensitivity

Let divide the image volume $\boldsymbol{V}$ in to $J \in N^*$ voxels v_j, with $j \in \mathcal{J} = \{1, \cdots, J\}$. A photon emitted from v_j and detected by the k-th view angle camera, with $k \in \boldsymbol{K} = \{1, \cdots, K\}$, is recorded as event $A_{k,i}$, $i = 1,2, \cdots$ of a list L_k. Each recorded event $A_{k,i}$ consists of 4 key attributes $\{P_{k,i,1}, E_{k,i,1}, P_{k,i,2}, E_{k,i,2}\}$ as illustrated in Fig. 2, where $P_{k,i,1}, P_{k,I,2} \in R^3$ are respectively the positions of the first and second interaction on the scatterer and absorber, and $E_{k,i,1}$ and $E_{k,i,2}$ are the associated energy. We

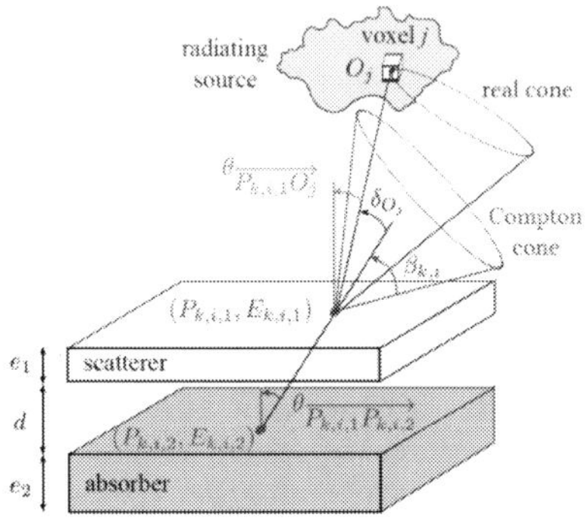

Fig. 2. Acquisition principle of the Compton camera at the k-th view angle.

call $A_{k,i}$ a valid event if $E_{k,i,1} + E_{k,i,2}$ satisfies a preset energy range. Among events recorded in list L_k, we assume I_k valid events constituting the unordered set $S_k \triangleq \{A_{k,1}, \cdots, A_{k,I_k}\}$. Each valid event generates a Compton cone, and the itersection of several resulting Compton cones can return an image of the radioactive source [20]. Unfortunately, the lack of parallax inherent in Compton cameras does not allow an efficient image reconstruction when using only events from a single view (i.e., it returns a direction toward the radioactive source rather than its location). That is why we should coordinate events from multi-views seeking parallax improvement and ultimately a better reconstruction result. Unlike Kishimoto *et al.* [14] who consider Compton event in turn from one view to another, we propose using events from all view angles simultaneously. It allows us to deal with the severe lack of parallax for which the 3D reconstruction from a certain view angle definitively fails. More precisely, we built a new list-mode data-space $\boldsymbol{L} = \{A_1, \dots, A_i, \dots, A_I\}$, where $I = \min\{I_1, \dots I_k, \dots, I_K\}$ denotes the length of $\boldsymbol{L}$ and $A_i \triangleq \cup_{k=1}^{K} A_{k,i}$, $i \in \mathcal{I} \triangleq \{1, \dots, I\}$, is a set of K independent valid events arbitraily given from K view angles. We give in Fig .1 an illustration of the data list $\boldsymbol{L}$.

Using $\boldsymbol{L}$ for 3D reconstruction requires an associated system response matrix and sensitivity. Let denote t_{ij} an element of the *system* response matrix $\boldsymbol{T}$ indexed on the event A_i and on voxel v_j, it corresponds to the probability that photons emitted from v_j are detected by the system and recorded as A_i. Mathematically, we formulate t_{ij} as

$$t_{ij} = P(A_i, D|v_j) = P(A_i|D, v_j)s_j, \qquad (1)$$

where D stands for the event "*detected by the system*", and $s_j \triangleq P(D|v_j)$ is known as *system sensitivity*. Denoting D_k the event "*detected by the camera employed at the k-th view angle*", we obtain $D = \cup_{k=1}^{K} D_k$ and further express t_{ij} by

$$t_{ij} = P\left(\cup_{k=1}^{K} \{A_{k,i}, D_k\} | v_j\right) = \sum_{k=1}^{K} t_{ijk}, \qquad (2)$$

where $t_{ijk} \triangleq P\left(A_{k,i}, D_k | v_j\right)$ can be seen as an element of the camera response matrix $\boldsymbol{T}_k$ at the k-th view angle. Similarly, we obtain the system sensitivity s_j as

$$s_j = P\left(\cup_{k=1}^{K} D_k | v_j\right) = \sum_{k=1}^{K} s_{jk}, \qquad (3)$$

where s_{jk} stands for the sensitivity of the k-th view angle camera. The sum over k in both the expressions (2) and (3) confirms that combining multi-view Compton events as in $\boldsymbol{L}$ can significantly improve the detection probability by the system, especially when a number of view angles becomes higher. It is worth noting that the expressions (2) and (3) constitutes a general framework for the system response and sensitivity. The models of camera response t_{ijk} and sensitivity s_{jk} have been extensively developed in the litterature. Despite their abundance, the camera response models share the same development method initially proposed by Wilderman *et al.* in [17]. Their differences reside mostly in the introduction of various kinds of measurement uncertainties (i.e., detector spatial resolution [19], detector energy resolution [21], Doppler broadening effect [22], etc.) to improve the model accuracy. Similarly, various models have been proposed for the camera sensitivity such as ones summarized by Muñoz *et al.* in [15].

B. eLM-MLEM Algorithm

We formulate here an extension of the conventional LM-MLEM algorithm for 3D reconstruction using the list-mode data space $\boldsymbol{L}$. For convenience, we first conduct the formulation under the bin-mode data as in [17], and then derive the model of interest by considering list-mode data as a limiting case of bin-mode when the bin widths tend to zero. This approach is motivated by the same information contained in the likelihood functions of both kinds of data albeit their different in format, so that the convergence properties of the corresponding LM-MLEM algorithms are identical [5]. By this way, we can define, for the bin-mode data, a vector of measurement $\boldsymbol{g}$ where each element g_i, $i \in \mathcal{I}$, is a random variable describing the number of counts in a virtual bin i corresponding the event A_i. When $g_i \to 1, \forall_i \in \mathcal{I}$, then the bin-mode data return to the list-mode ones [23]. Besides, we can also find in [7] a direct formulation for the list-mode data.

Let denote $\boldsymbol{f}$ the source distribution in the considered image volume $\boldsymbol{V}$, each element $f_j, \forall j \in \mathcal{J}$, of $\boldsymbol{f}$ characterizes the emission intensity of the voxel v_j in $\boldsymbol{V}$. Assuming the Poisson event-counting model [2], we obtain the conditional probability mass function of observing g_i given $\boldsymbol{f}$ as

$$p(g_i|\boldsymbol{f}) = e^{-\mu_i \frac{\mu_i^{g_i}}{g_i!}} \qquad (4)$$

where $\mu_i = \sum_{j=1}^{J} t_{ij} f_j$ denotes the mean value of g_i. Reconstruct 3D image of a radioactive source is equivalent to find the estimate that maximizes the log-likelihood function $\mathcal{L}(\boldsymbol{g}|\boldsymbol{f})$

$$\hat{\boldsymbol{f}} = \arg\max_{\{\boldsymbol{f} \geq 0\}} \mathcal{L}(\boldsymbol{g}|\boldsymbol{f}), \qquad (5)$$

where

$$\mathcal{L}(\boldsymbol{g}|\boldsymbol{f}) = \sum_{i=1}^{I} \left(-\sum_{j=1}^{J} t_{ij} f_j + g_i \ln\left(\sum_{j=1}^{J} t_{ij} f_j\right) - \ln(g_i!) \right). \qquad (6)$$

By following the same procedure in [24] for the bin-mode data, we can characterize the E-step at the k-the iteration by the function

$$Q_b(\boldsymbol{f}|\hat{\boldsymbol{f}}^{(k)}) = \sum_{i=1}^{I} \sum_{j=1}^{J} \left(\frac{g_i t_{ij} \hat{f}_j^{(k)}}{\sum_{s=1}^{J} t_{is} \hat{f}_s^{(k)}} \ln(t_{ij} f_j) - t_{ij} f_j \right), \qquad (7)$$

where $\hat{\boldsymbol{f}}^{(k)}$ is the maximum likelihood estimate of $\boldsymbol{f}$ at the k-th iteration, and $\hat{f}_j^{(k)}$, $\forall j \in \mathcal{J}$, are elements of $\hat{\boldsymbol{f}}^{(k)}$. Approaching $g_i \to 1$, $\forall i \in \mathcal{I}$, we obtain the associated function for the list-mode data

$$Q_l(\boldsymbol{f}|\hat{\boldsymbol{f}}^{(k)}) = \sum_{i=1}^{I} \sum_{j=1}^{J} \left(\frac{t_{ij} \hat{f}_j^{(k)}}{\sum_{s=1}^{J} t_{is} \hat{f}_s^{(k)}} \ln(t_{ij} f_j) - t_{ij} f_j \right). \qquad (8)$$

In M-step, we maximize $Q_l(\boldsymbol{f}|\hat{\boldsymbol{f}}^{(k)})$ by merely vanishing its partial derivative with respect to f_j, $\forall j \in \mathcal{J}$, and obtain

$$f_j = \frac{\hat{f}_j^{(k)}}{\sum_i^I t_{ij}} \sum_{i=1}^{I} \frac{t_{ij}}{\sum_{s=1}^{J} t_{is} \hat{f}_s^{(k)}}. \qquad (9)$$

Setting $s_{jk} \triangleq \sum_{i=1}^{I} t_{ijk}$, and using (2) and (3), we arrive to the estimate of $\boldsymbol{f}$ at the $(k+1)$-st iteration as

$$\hat{f}_j^{(k+1)} = \frac{\hat{f}_j^{(k)}}{s_j} \sum_{i=1}^{I} \frac{t_{ij}}{\sum_{s=1}^{J} t_{is} \hat{f}_s^{(k)}}, \quad \forall j \in \mathcal{J}, \qquad (10)$$

where s_j is given by (3). Repeating (8) until verify a certain convergence criterion, we obtain the final maxim likelihood estimate $\hat{\boldsymbol{f}}$.

We can remark that the update equation (10) shares exactly the same form as the one of the conventional LM-MLEM algorithm. However, they do not have the same meaning. With t_{ij} and s_j defined by (2) and (3), ours takes into account simultaneously multi-view Compton events, and degenerates into the classical ones when only a single-view data are used. It is thus more generalized, flexible and can inherit the state-of-the-art models previously proposed for t_{ijk} and s_{jk}. We call therefore the proposed algorithm extended LM-MLEM (eLM-MLEM).

III. Numerical Experiments and Analyses

We present in this section an experimental system that can generate various configurations of real multi-view Compton data in function of the view number, the source location and the energy range. These rich data allow assessing thoroughly the performance of the proposed eLM-MLEM 3D reconstruction algorithm.

A. Experiments Setting and Data

To obtain real data for performance assessment, we setup an experimental system consisting of a punctual 0.2 (MBq) ^{22}Na source and a handheld CeBr3 Compton camera developed by Damavan Imaging [18]. We deploy sequentially the camera at three different view-angles as in Fig. 3: *the two first views are set perpendicularly on a same horizontal plane* ***H*** *of the punctual source, while the third view is placed vertically above.*

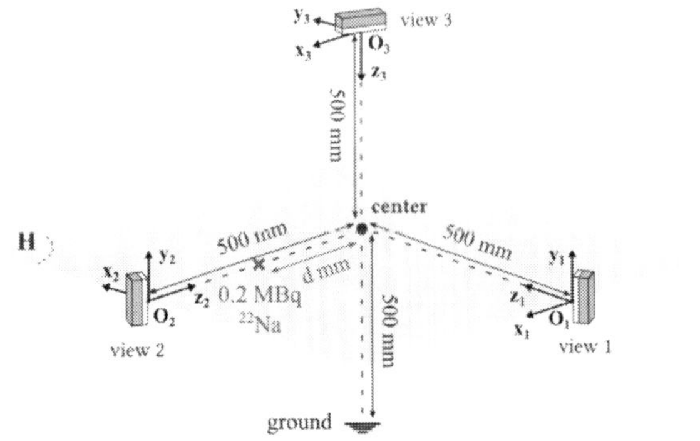

Fig. 3. Experiment setting with one punctual source and three view-angles.

The distance between the camera and the center is 500 (mm), and all the system is located 500 (mm) from the ground. We locate the punctual 0.2 (MBq) ^{22}Na source on the plane **H** at d (mm) from the center. For each location of the radioactive source, we conducted the data acquisition within 15 minutes per view-angle. Table I shows the number of Compton events recorded from three view-angles for different energy ranges depicted in Fig. 4. The events satisfying the low energy range

TABLE I
NUMBER OF COMPTON EVENTS IN FUNCTION OF SOURCE LOCATION, ENERGY RANGE AND VIEW ANGLE

d (mm)	Energy range (MeV)	Number of Compton events / view		
		view 1	view 2	view 3
	full range	1339	1398	1308
0	[1.15, 1.38]	140	148	140
	[0.48, 0.54]	344	335	335
	full range	1951	5132	987
250	[1.15, 1.38]	96	279	56
	[0.48, 0.54]	288	541	104

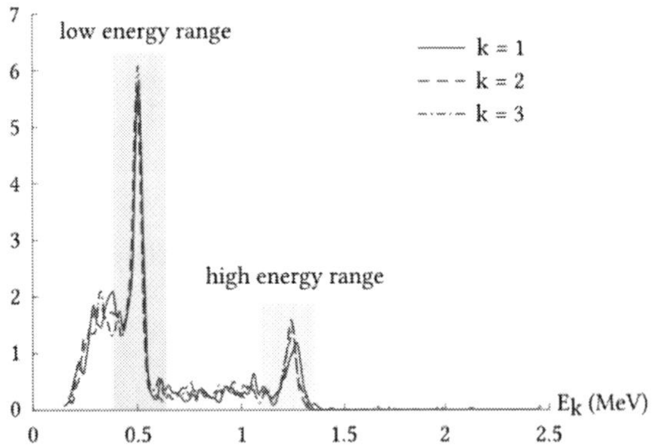

Fig. 4. Spectra of energies.

[0.48, 0.54] and the high energy range [1.15, 1.38] are selected to constitute different 1-view, 2-view and 3-view datasets.

B. Experimental Results and Analyses

The eLM-MLEM 3D reconstruction algortihm requires a probabilistic model for the camera response t_{ijk} and sensitivity s_{jk}. Among various choices (see Subsection II.A), we adopt the model t_{ijk} proposed by Maxim *et al.* in [19] and add different measurement uncertainties related to the spatial and energy resolution of detectors, as well as the Doppler broadening effect to make the model more flexible. This extended model leads us to set the camera sensitivity as uniform $s_{jk} = 1$ because uncertainties have been taken into account in t_{ijk}.

Using 3-view datasets at high range energy, we show in Fig. 5 and Fig. 6 how the reconstructed 3D image evolves in function of iteration number for two locations of the radioactive souces $d = 0$ and $d = 250$ (mm). The eLM-MLEM algorithm

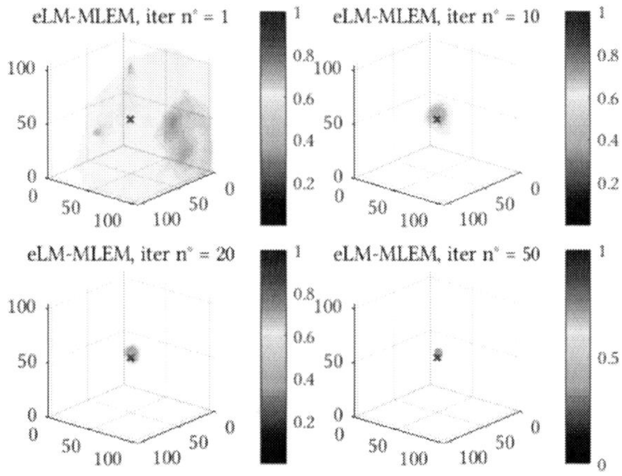

Fig. 5. Convergence of the eLM-MLEM algorithm when $d = 0$ (mm).

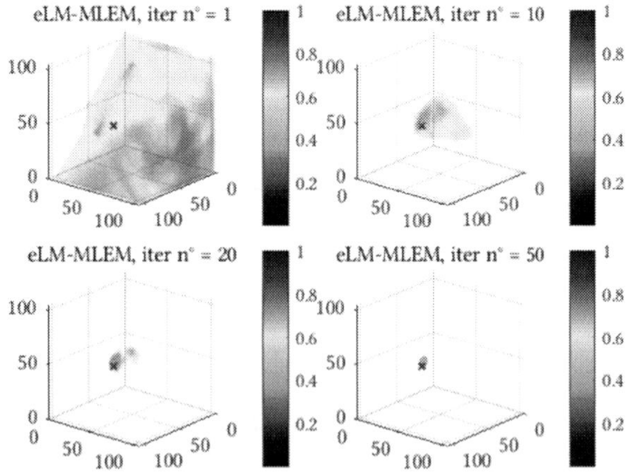

Fig. 6. Convergence of the eLM-MLEM algorithm when $d = 250$ (mm).

converges since 50th iteration. Consider now the datasets at high range energy of the radioactive source $d = 0$ (mm), we give in Fig. 7 the recontructed results given by the well-kown back-projection and the eLM-MLEM algorithm, as well as the X-Y, Y-Z and X-Z profiles for different multi-view datasets. When the 1-view data are used, the eLM-MLEM algorithm returns to

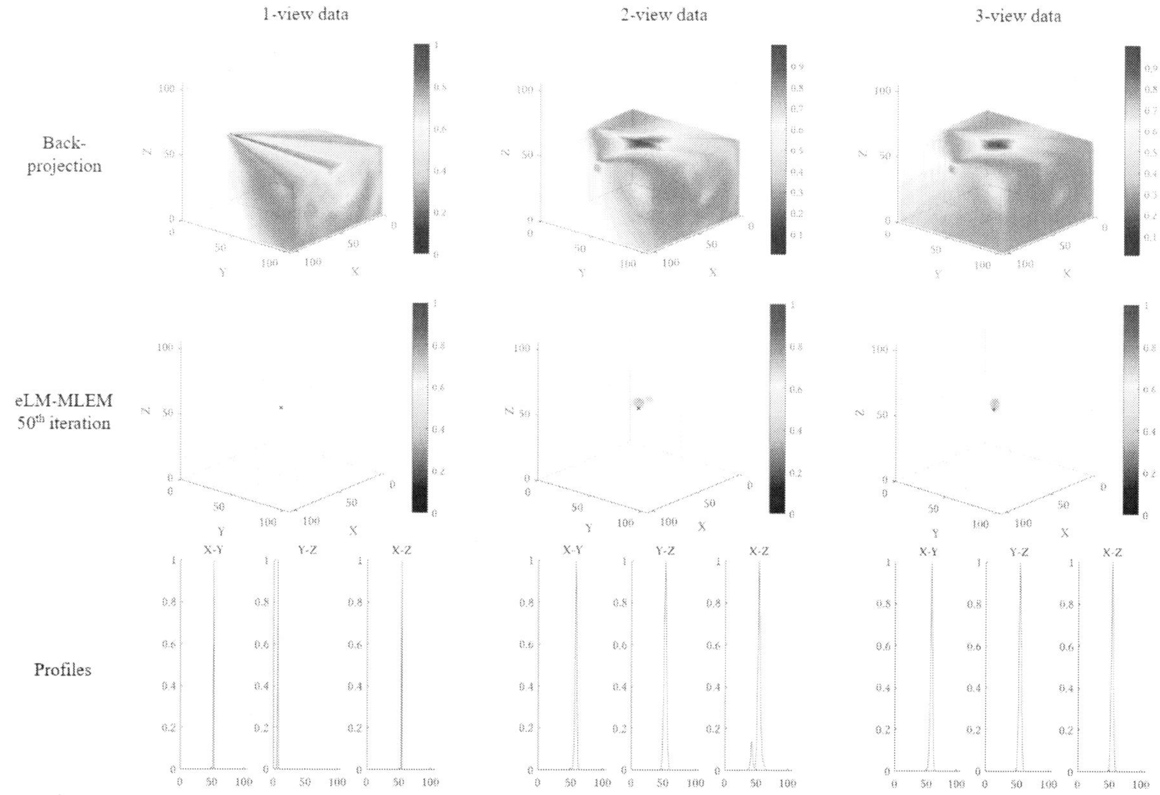

Fig. 7. Recontructed results given by the well-kown back-projection and the eLM-MLEM algorithm, as well as the X-Y, Y-Z and X-Z profiles.

the conventional LM-MLEM algorithm. Clearly, the reconstruction fails in the case of 1-view data and succeeds in the cases of 2-view and 3-view data. This confirms the outperformance of the eLM-MLEM algorithm over the conventional one. Looking at the reconstructed image at the 50th iteration and their profiles, we find that the higher number of views allow a better result.

For a quantitative assessement, we propose using the so-called *sum of weighted distances* (SWD) between the reconstructed point cloud and the true source as a criterion for the quality assessment at the k-th iteration

$$SWD^{(k)} = \sum_{j=1}^{J} \hat{f}_j^{(k)} \left\|P_j - P_S\right\|_2,$$

where $\hat{f}_j^{(k)}$ is given from (10), P_j and P_S are respectively the coordinate vectors of the voxel v_j and the punctual source, and $\|\cdot\|_2$ denotes the Euclidean norm. The lower the value of $SWD^{(k)}$, the more the reconstructed 3D image is near to the true source. We show in Table II the $SWD^{(k)}$ scores when the

TABLE II
SUM OF WEIGHTED DISTANCES BETWEEN THE RECONSTRUCTED POINT CLOUD AND THE TRUE SOURCE LOCATION

Data	Low energy range		High energy range	
	$d = 0$	$d = 250$	$d = 0$	$d = 250$
2-view	6943.2	9199.2	7846.9	3588.4
3-view	5271.2	4728.5	4046.9	1587.7

2-view and 3-view data at the low and high energy range are used to reconstruct both the radioactive sources $d = 0$ and $d = 250$ (mm). The lower scores obtained by the 3-view data justify the relevance in using higher number of view-angles for the 3D image reconstruction.

IV. CONCLUSIONS AND PERSPECTIVES

We have developed in this work an extended LM-MLEM 3D reconstruction algorithm adapted to multi-view Compton data. Based to real multi-view data of punctial radioactive source, we prove that the algorithm is effective for the serious lack of parallax in measurements. Espectially, the higher the number of view-angles in the Compton data, the more the algorithm return better results.

Albeit very promising results, we can remark that the algorithm still converges slowly. A way to speed up it is to apply the variants of the conventional LM-MLEM algorithm introduced in Section I to the eLM-MLEM algorithm. However, such a solution always requires a large number of Compton events for noise reduction. To overcome this problem, we think about a 3D reconstruction algorithm in the Bayesian framework with a prior reflecting the smoothness in the expected image. It is the main orientation of our future works.

ACKNOWLEDGMENT

This research was financially supported by ANR RED-7D project and BPI PIA4 DreamScanner project.

REFERENCES

[1] P. E. Kinahan, M. Defrise and R. Clackdoyle, "Analytic image reconstruction methods", in *Emission Tomography: The Fundamentals of PET and SPECT*, Elsevier Academic Press, 2004, pp. 421-442.

[2] D. S. Lalush and M. N. Wernick, "Iterative image reconstruction", in *Emission Tomography: The Fundamentals of PET and SPECT*, Elsevier Academic Press, 2004, pp. 443-472.

[3] V. Maxim, "Filtered backprojection reconstruction and redundancy in Compton camera imaging", *IEEE Transactions on Image Processing*, vol. 23, no. 1, 2014, pp. 332-341.

[4] B. F. Hutton, "Recent advances in iterative reconstruction for clinical SPECT/PET and CT", *Acta Oncologica*, vol. 50, no. 6, 2011, pp. 851-858.

[5] J. Qi and R. M. Leahy, "Iterative reconstruction techniques in emission computed tomography", *Physics in Medicine & Biology*, vol. 51, no. 15, 2006, pp. R541.

[6] L. A. Shepp and Y. Vardi, "Maximum Likelihood Reconstruction for Emission Tomography", *IEEE Transactions on Medical Imaging*, vol. 1, no. 2, 1982, pp. 113-122.

[7] L. Parra and H. H. Barrett, "List-Mode Likelihood: EM Algorithm and Image Quality Estimation Demonstrated on 2-D PET", *IEEE Transactions on Medical Imaging*, vol. 17, no. 2, 1998, pp. 228-235.

[8] M. V. Ranganath, A. P. Dhawan and N. Mullani, "A multigrid expectation maximization reconstruction algorithm for positron emission tomography", *IEEE Transactions on Medical Imaging*, vol. 7, no. 4, 1988, pp. 273-278.

[9] L. Kaufman, "Implementing and accelerating the EM algorithm for positron emission tomography", *IEEE Transactions on Medical Imaging*, vol. 6, no. 1, 1987, pp. 37-51.

[10] J. Fessler and A. O. Hero, "Space-alternating generalized expectation-maximization algorithm", *IEEE Transactions on Signal Processing*, vol. 42, no. 10, 1994, pp. 2664-2677.

[11] H. M. Hudson and R. S. Larkin, "Accelerated image reconstruction using ordered subsets of projection data", *IEEE Transactions on Medical Imaging*, vol. 13, no. 4, 1994, pp. 601-609.

[12] Y. Sato, Y. Terasaka, W. Utsugi, H. Kikuchi, H. Kiyooka and T. Torii, "Radiation imaging using a compact Compton camera mounted on a crawler robot inside reactor buildings of Fukushima Daiichi Nuclear Power Station", *Journal of Nuclear Science and Technology*, vol. 56, no. 9-10, 2019, pp. 801-808.

[13] A. Kishimoto, J. Kataoka, T. Nishiyama, T. Taya and S. Kabuki, "Demonstration of three-dimensional imaging based on handheld Compton camera", *Journal of Instrumentation*, vol. 10, no. 11, 2015, pp. P11001.

[14] A. Kishimoto, J. Kataoka, T. Taya, L. Tagawa, S. Mochizuki, S. Ohsuka, T. Nagao, K. Kurita, M. Yamaguchi, N. Kawachi, K. Matsunaga, H. Ikeda, E. Shimosegawa and J. Hatazawa, "First demonstration of multi-color 3-D in vivo imaging using ultra-compact Compton camera", *Scientific reports*, vol. 7, no. 1, 2017, pp. 2110.

[15] E. Muñoz, J. Barrio, J. Bernabéu, A. Etxebeste, C. Lacasta, G. Llosá, A. Ros, J. Roser and J. F. Oliver, "Study and comparison of different sensitivity models for a two-plane Compton camera", *Physics in Medicine & Biology*, vol. 63, no. 13, 2018, pp. 135004.

[16] C. Wu, S. Zhang and L. Li, "An accurate probabilistic model with detector resolution and Doppler broadening correction in list-mode MLEM reconstruction for Compton camera", *Physics in Medicine & Biology*, vol. 67, no. 12, 2022, pp. 125017.

[17] S. J. Wilderman, N. H. Clinthorne, J. A., Fessler and W. L. Rogers, "List-mode maximum likelihood reconstruction of Compton scatter camera images in nuclear medicine", in *IEEE Nuclear Science Symposium Conference Record*, vol. 3, 1998, pp. 1716-1720.

[18] M. Z. Hmissi, A. Iltis, C. Tata, G. Zeufack, L. Rodrigues, B. Mehadji, C. Morel and H. Snoussi, "First images from a CeBr 3/LYSO: Ce Temporal Imaging portable Compton camera at 1.3 MeV", in *2018 IEEE Nuclear Science Symposium and Medical Imaging Conference Proceedings*, Sydney, NSW, Australia 2018, pp. 1-3.

[19] V. Maxim, X. Lojacono, E. Hilaire, J. Krimmer, E. Testa, D. Dauvergne, I. Magnin and R. Prost, "Probabilistic models and numerical calculation of system matrix and sensitivity in list-mode MLEM 3D reconstruction of Compton camera images", *Physics in Medicine & Biology*, vol. 61, no. 1, 2016, pp. 243.

[20] L. C. Parra, "Reconstruction of cone-beam projections from Compton scattered data", *IEEE Transactions on Nuclear Science*, vol. 47, no. 4, 2000, pp. 1543-1550.

[21] B. Mehadji, M. Dupont, Y. Boursier and C. Morel, "Extension of the List-Mode MLEM Algorithm for Poly-Energetic Imaging with a Compton Camera", in *2018 IEEE Nuclear Science Symposium and Medical Imaging Conference*, 2018, pp. 1-8.

[22] Y. Feng, J. M. Létang, D. Sarrut and V. Maxim, "Influence of Doppler broadening model accuracy in Compton camera list-mode MLEM reconstruction", *Inverse Problems in Science and Engineering*, vol. 29, no. 13, 2021, pp. 3509-3529.

[23] H. H. Barrett, T. White and L. C. Parra, "List-mode likelihood", *Journal of the Optical Society of America A*, vol. 14, no. 11, 1997, pp. 2914-2923.

[24] K. Lange and R. Carson, "EM Reconstruction Algorithms for Emission and Transmission Tomography", *Journal of Computer Assisted Tomography*, vol. 8, no. 2, 1984, pp. 306-316.

EPJ Web of Conferences **288**, 06004 (2023) https://doi.org/10.1051/epjconf/202328806004
ANIMMA 2023

Development of a new method for the detection of illicit materials based on the Active Photon Interrogation Method and photo-neutron spectrometry

C. Besnard-Vauterin[1,*], V. Blideanu[1], and B. Rapp[1]
[1]Université Paris-Saclay, CEA, List, Laboratoire National Henri Becquerel (LNE-LNHB)
F91120, Palaiseau, France
(*) clement.besnardvauterin@cea.fr

Abstract—**The detection of illicit materials is a critical task in the field of homeland security, as international trade has contributed to the increase of smuggling activities. Non-intrusive on-site inspections are crucial in this context, but the current active interrogation methods have limitations. Neutron-induced reactions have been used, but the measurement of gamma spectra is complex due to background noise. Active photon interrogation methods have also been overlooked but they are currently limited to actinides detection using photo-fission reactions. This work presents a novel method for the detection of illicit materials based on active photon interrogation and photo-neutron spectrometry. This approach extends the application of active photon interrogation by including the detection of conventional explosives, narcotics, and chemical weapons based on the use photo-nuclear reactions to determine the content of light elements such as nitrogen, oxygen, and carbon. Monte-Carlo codes are the main tool used to simulate this process for the application need. However, because of the lack in the present literature of measured neutron spectra, the experimental validation of the simulations is not straight forward. At present, benchmarking the Monte-Carlo codes seems to be the sole option for testing their ability to accurately simulate photo-neutron production and spectra for the nuclear reactions of interest. Simultaneously, we designed a test bench based on a linear accelerator to generate photons, induce photo-nuclear reactions, and acquire photo-neutron spectra. This study on the fundamental aspects of photo-neutron production lays the groundwork for a promising new detection method for illicit materials.**

***Keywords*—Homeland security, neutron spectrometry, photo-nuclear, photo-neutron, Monte Carlo simulation, CBRNE.**

I. Introduction

In the field of material detection by non-intrusive on-site inspection, the existing techniques can be divided into two families depending on the exploring particles used. Neutron-induced reactions can be used to detect illicit materials in containers (the so-called active neutron interrogation method), using in particular the associated particle technique (APT) [1, 2]. In this case, detection is based on the measurement of gamma radiation produced by neutron-induced reactions (namely inelastic scattering, capture). However, the precise measurement of gamma spectra is generally complex due to the experimental conditions, being often characterized by a high level of gamma background. In the case of the use of photons, the method is called active photon interrogation (API) and is currently used only for threats related to nuclear materials based on the actinide detection using photo-fission reactions and spectrometry of gammas emitted subsequent to the decay of radioactive fission fragments [3].

Our work extends the applicability of traditional API techniques by the development of a novel method, allowing for the first time the detection of light elements such as nitrogen, oxygen and carbon present in illicit materials (conventional explosives, narcotics and chemical weapons) (see Fig. 1), based on active photon interrogation associated to photo-neutron spectrometry. This new method overcomes the difficulties related to gamma-ray spectrometry on which the neutron interrogation is based [4]. In addition, compared to the active neutron interrogation method, this new technique has the advantage that photons are much more penetrating than neutrons for most materials and therefore less subject to scattering and attenuation.

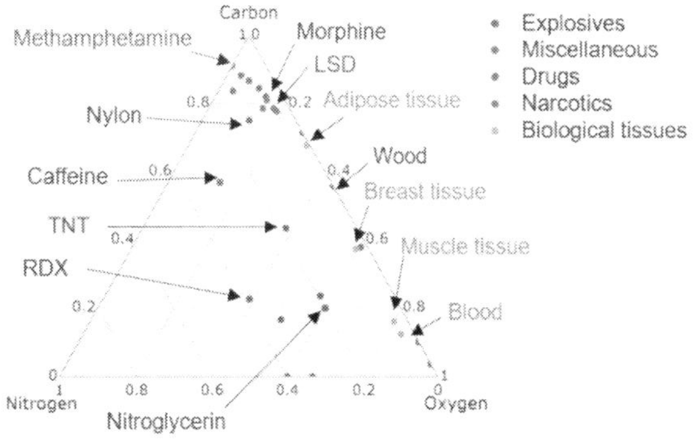

Fig. 1. Triangle of normalized proportions of carbon, nitrogen and oxygen for different materials.

The method presented in this paper for the detection of illicit materials based on the active photon interrogation and photo-neutron spectrometry involves inducing photo-nuclear reactions on light nuclei using ideally mono-energetic photons, with energies above the reaction threshold of about 10 MeV for nitrogen isotopes. This leads to nuclear excitation and subsequent emission of neutrons. The measurement of their spectra allows the identification of the emitting nuclei, and potentially determine their quantity. The existence of energy levels in nuclei and their low density for light nuclei, gives to the energy spectra of the emitted photo-neutrons the particularity of presenting distinctive peaks at well-known energies.

Among the existing photon sources suitable for this purpose, compact single and dual energy sources with energies above 10 MeV can be obtained namely based on low-energy proton induced resonance reactions, as for example $^7Li(p,\gamma)^8Be$ at 441 keV . These sources are however multi-directional with low effective photon intensities in a given solid angle thus significantly degrading the signal over noise ratio. Another approach, which avoids this limitation while being simpler to set up given the higher availability of the required equipment, consists of using a source of photons produced by an electron linear accelerator (LINAC). However, the bremsstrahlung photons produced in this way do not meet the conditions mentioned above relating to their energy spectrum, which is in this case continuous. Therefore, the implementation of this method requires a special treatment in order to select in the total spectrum of the detected neutrons generated in photo-nuclear reactions by all photon energies, the events associated with a given energy of the photon. This approach, known as the photon tagging method, can be implemented experimentally by using additional equipment but this requires the deployment of a complex and non-compact system for electron spectroscopy and electron-neutron coincidence electronics. Our new approach intends to overcome these limitations. The selection of contribution from photons in a well-defined energy range can be achieved by subtracting two neutron spectra induced by bremsstrahlung photon beams created by electrons of close energies. Furthermore, a deep learning-based method has been developed and successfully tested for the identification of the signatures from specific light elements in the simulated complex photo-neutron spectra from irradiation of compound materials and to extract their respective contributions.

II. Methods

As there is a large reliance on Monte-Carlo methods for the above-mentioned application, especially for establishing the training data sets for the neural networks, the capabilities of the existing codes to accurately simulate the photo-nuclear reactions and a good understanding of physics involved becomes crucial and justifies the need for intercomparisons and benchmarks. The first step of the work described in the present paper was dedicated to the comparison of the photo-neutron energy spectra calculated by the main Monte-Carlo codes used by the community: FLUKA [5], GEANT4 [6], MCNP6.2 [7] and PHITS3.27 [8]. On the other hand, due to the lack of measured neutron spectra in the existing literature, experimental validation of these simulations is challenging. Therefore, benchmarking Monte-Carlo codes appears to be the only viable option to accurately test their capabilities in simulating photo-neutron production. In this study, we analyzed the agreement between Monte-Carlo codes tested for calculated photo-neutron spectra from different nuclei covering a wide range of mass numbers. Based on assumptions derived from photo-nuclear reaction theory, our analysis allowed us to thoroughly test the performance of the codes for this physics case. In order to focus the study on the physics of the photo-nuclear reactions, a simple simulation geometry has been used to calculate with Monte-Carlo codes the photo-neutron spectra and production rates. The geometry consists of a vacuum sphere containing a cylindrical target of 1 mm thickness and 1 cm in diameter. The 1 mm thickness has been chosen as a good compromise in order to increase the rate of photo-nuclear reactions occurring in the target while minimizing the probability of the secondary reactions induced by the particles primary produced by the impinging beam. The target is uniformly irradiated with a parallel 20 MeV photon beam with 1 cm diameter. This photon energy has been chosen in order to ensure a significant photo-nuclear reaction cross-section for most of the elements. Considered materials ranges from beryllium to lead and include carbon, oxygen and nitrogen as elements of particular interest for the application of illicit material detection.

The second step consists in the development of a test bench based of a Varian TrueBeam® medical electron LINAC available in our laboratory to study photo-neutron spectra from the reactions induced on different materials by photons with a continuous energy spectrum up to 20 MeV. A preliminary feasibility and optimization study based on simulations has been performed for the test bench from the point of view of the target, collimation, and detector shielding, considering the fully modeled geometry of the medical accelerator. Neutron spectra from the irradiation of tungsten were measured using a passive Bonner sphere spectrometry system associated to dedicated unfolding techniques. Tungsten has been selected in order to allow quick testing of the simulation capabilities, since this material has a high cross-section for photo-nuclear reaction and consequently the photo-neutron production rate is higher than in the case of light elements. Reasonably low statistical uncertainties for both measured and simulated neutron spectra can therefore be achieved in relatively short times. Comparison of the measured spectra with the results from Monte-Carlo simulations, accurately reproducing experimental conditions, was performed to strengthen the validation of the Monte-Carlo simulation code selected as main tool for further development of the new illicit materials detection method.

III. Results and Discussions

A. Simulation of photo-neutron emission by light nuclei

1) Monte-Carlo codes benchmark for photo-neutron energy spectra

As there are no usable experimental neutron spectra available in the literature allowing the comparison with the Monte-Carlo simulations, the energy conservation aspect is the only indicator available to check the consistency of the simulated spectra. According to the energy conservation law, the expected energy of the photo-neutron can be calculated from Eq. 1 for each excited energy level of the residual nucleus left after the emission of one neutron.

 https://doi.org/10.1051/epjconf/202328806004

$$E_n = \frac{E_\gamma - E_{thres} - E_{state}}{1 + A_{residual}^{-1}} \quad (1)$$

Where E_γ is the incident photon energy, E_{thres} is the reaction threshold, E_{state} is the energy of the excited state of the residual nucleus (being equal to 0 for the ground state, hereafter abbreviated to g.s.) and A its mass number.

In the present study, carbon was considered as one of target materials, with natural isotopic composition (98.9% of ^{12}C and 1.1% of ^{13}C) and a density of 2 g/cm^3. Figure 2 displays the emitted photo-neutron spectrum from natural carbon when irradiated by 20 MeV photons, calculated by the Monte-Carlo codes considered in this study. A series of well-defined peaks are observed. For the channel $^{12}C(\gamma,n)^{11}C_{g.s.}$, a photo-neutron peak is expected at around 1.24 MeV (see Table I) and is accurately reproduced by all codes considered here, with the exception of PHITS3.27. The first excited level of the residual ^{11}C is at 2 MeV ± 0.4 keV. The channel $^{12}C(\gamma,n)^{11}C_{1st}$ therefore requires a photon energy of at least 20.74 MeV when considering the reaction threshold of 18.72 MeV, which exceeds the incident photon energy considered in this study. Thus, no signature from excited states of the residual ^{11}C is found in the simulated photo-neutron spectra. Peaks at higher energies are signature of various $^{13}C(\gamma, n)^{12}C$ channels, leaving the residual ^{12}C in the ground-state or in the first few low-lying excited states. The corresponding excitation energies are reported in Table II, along with the associated neutron peak energies. FLUKA, GEANT4 and MCNP6.2 all reproduce the peak corresponding to the ground state as well as the first and second excited states of the residual nucleus, although peak intensities may differ by up to an order of magnitude. These peak intensities are roughly a factor of 100 lower than those for $^{12}C(\gamma, n)^{11}C$ reflecting the much lower abundance of ^{13}C in natural carbon, even though the cross-section of the photo-nuclear reaction in the case of ^{13}C is 2.4 higher. The photo-neutron peak at 3.88 MeV is only present in the spectra calculated by MCNP6.2 and GEANT4. Additionally, MCNP6.2 seems to predict two other peaks at around 6 MeV and 8.5 MeV that cannot be associated to the $^{13}C(\gamma, n)^{12}C$ reaction.

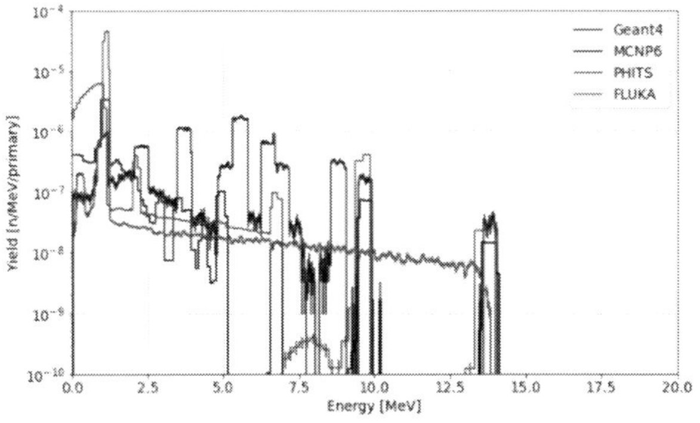

Fig. 2. Simulated energy spectra of neutrons emitted from the irradiation of natural carbon target with 20 MeV photons.

TABLE I
EXPECTED PHOTO-NEUTRON ENERGIES FROM THE $^{12}C(\gamma, n)^{11}C$ REACTION

Energy levels of ^{11}C (MeV± keV)	Photo-neutron energy (MeV)
g.s.	1.24

TABLE II
EXPECTED PHOTO-NEUTRON ENERGIES FROM THE $^{13}C(\gamma, n)^{12}C$ REACTION

Energy levels of ^{12}C (MeV± keV)	Photo-neutron energy (MeV)
g.s.	14.06
4.439 ± 0.21	9.79
7.654 ± 0.19	6.83
9.641 ± 5	4.99
9.870 ± 60	4.78
10.847 ± 4	3.88
11.836 ± 4	2.97
12.710 ± 6	2.16
13.316 ± 20	1.60
14.019 ± 5	0.90

Special attention has been given to the neutron spectrum from nitrogen irradiation as this element can be used as marker allowing to detect conventional explosives, narcotics and drugs and some chemical weapons with, notably, the emergence of new compounds related to opioid attacks. In this case also, natural nitrogen composition with 99.6 % of ^{14}N and 0.4 % of ^{15}N was used. The density was chosen to be of 1 g/cm^3 in order to speed-up the calculations by artificially increasing the neutron production rate. Both ^{14}N and ^{15}N have very close neutron binding energies of respectively 10.55 MeV and 10.83 MeV. For ^{13}N (residual nucleus from the $^{14}N(\gamma,n)^{13}N$ reaction), the ground state predicts a photo-neutron at 8.92 MeV and the first four excited levels photo-neutrons at 6.58 MeV, 5.52 MeV, 5.48 MeV and 2.87 MeV as reported in Table III. Again, the spectra calculated by all codes but PHITS3.27 are showing peaks at these expected energies. It is worth noting that MCNP6.2 also predicts a photo-neutron peak at around 3.75 MeV which can be potentially explained as corresponding to the $^{14}N(\gamma,pn)^{12}C$ reaction channel. The MCNP6.2 peaks found at energies higher than 9.5 MeV appear to be beyond the maximum energy allowed by the kinematics of (γ,n) reaction.

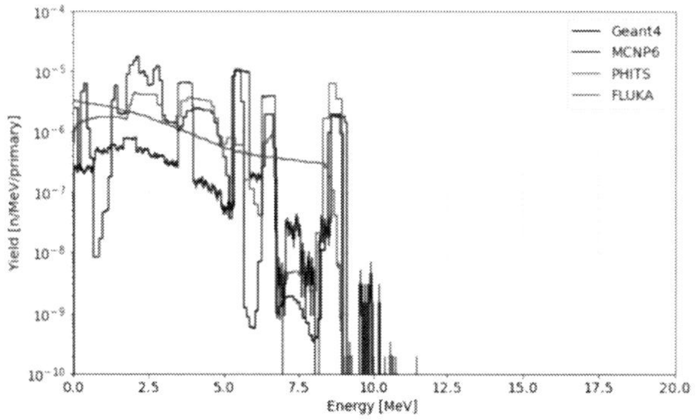

Fig. 3. Simulated energy spectra of neutrons emitted from the irradiation of natural nitrogen target with 20 MeV photons.

 https://doi.org/10.1051/epjconf/202328806004

TABLE III
EXPECTED PHOTO-NEUTRON ENERGIES FROM THE $^{14}N(\gamma, n)^{13}N$ REACTION

Energy levels of ^{13}N (MeV± keV)	Photo-neutron energy (MeV)
g.s.	8.92
2.3649 ± 0.6	6.58
3.502 ± 2	5.52
3.547 ± 4	5.48
6.364 ± 9	2.87
6.886 ± 8	2.38
7.155 ± 5	2.13

These results show that the shape of the photo-neutrons spectra is strongly influenced by nuclear levels structure for light target elements. Significant differences in how this effect is handled by the various codes were identified, with an extreme behavior in the case of PHITS3.27 for which physics models and data libraries available at the time of the calculations in the distribution package do not account at all for this effect.

2) Intercomparison on total neutron production cross-sections using several Monte Carlo codes and TALYS

While experimental neutron differential distributions in energy and/or angle are rather scarce in the literature, integrated cross section data are more abundant. Thus, while a comparison against experimental data in terms of double-differential cross section is not feasible, a direct comparison against integrated cross sections is possible. To this end, experimental integrated cross section data available through the EXFOR [9] database have been compiled and compared with those extracted from the simulations performed in this work. In addition to the cross-section values estimated from the calculations performed with the Monte-Carlo codes considered here and the experimental data from EXFOR, cross sections calculated with TALYS [10] have been included in the comparison. TALYS is a deterministic code using specially developed and continuously updated reaction models to calculate particle emission probabilities for various reactions. It is used to generate the TENDL nuclear data library which is one of the libraries extensively used by the broad nuclear physics community.

The results are presented in the Figure 4, which displays the cross section as a function of the mass number for the elements considered in this study.

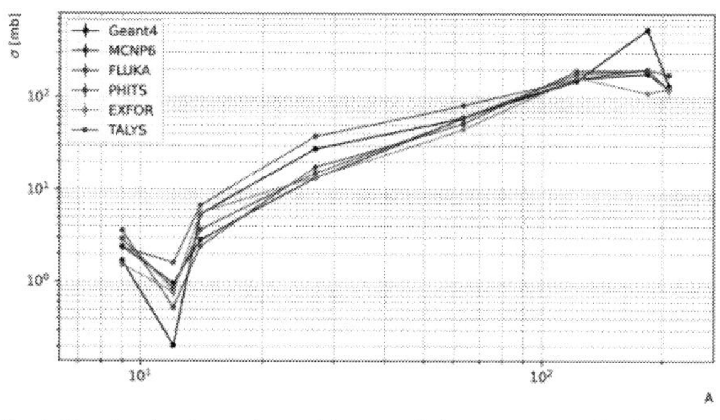

Fig. 4. Simulated and experimental integrated photo-neutron production cross-sections as a function of the atomic mass.

The general trend observed is characterized by an increase of the photo-neutron production cross-section with the mass of the target nucleus, reaching a plateau for heavy elements since no significant increase is observed between tungsten and lead. An exception is to be noted for light elements with the cross-section for carbon being lower than for beryllium. This trend is reproduced by all the Monte Carlo codes tested and by TALYS, however MCNP6.2, FLUKA and PHITS3.27 show better general agreement with the data while GEANT4 and TALYS produce higher cross-sections for light and intermediate mass nuclei. For tungsten all the calculated cross-sections are overestimated. An interesting finding concerns PHITS3.27, which despite showing an inaccurate simulation of the energy spectra of photo-neutrons as discussed in the previous section, is able to reproduce the same trend as the other codes for integrated cross sections. This fact strengthens the importance of the approach of this study, consisting in benchmarks not only in terms of integrated cross-section and production rates, but including also energy differential cross section allowing to better assess the energy spectrum features.

Figure 5 provides a detailed comparison of the cross-section values used by MCNP6.2 and PHITS3.27 concerning photo-neutron production for ^{12}C target nucleus, as a function of the incident photon energy. MCNP6.2 and PHITS3.27 values are in a very good agreement for the cross-section trends for (γ, n) reactions. In addition, the observed trends closely match the experimental data from B. C. Cook *et al.* ([11]) confirming their agreement with measured values.

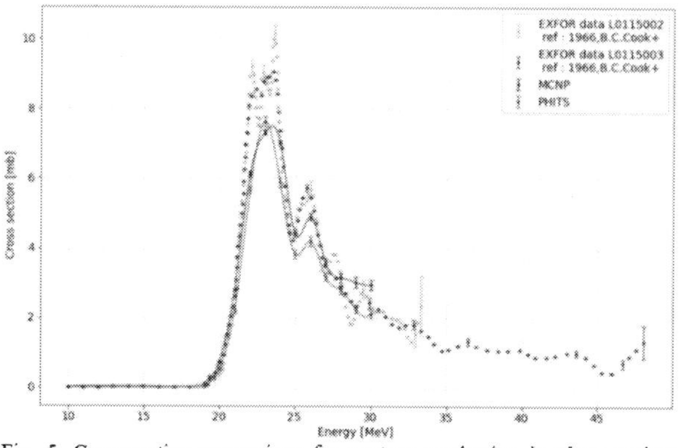

Fig. 5. Cross-section comparison for neutron production in photo-nuclear reactions on ^{12}C between MCNP6.2 and PHITS3.27 as a function of incident photon energy.

Figure 6 highlights contrasting findings in the cross-sections calculated by MCNP6.2 and PHITS3.27 for ^{14}N. While MCNP6.2 results agrees with the cross-sections measured by B. L. Berman et al. ([12]), PHITS3.27 results are closer to the data from J. D. King et al. ([13]). This can be explained by the fact that these two experimental data sets differ significantly in terms of physical process measured, as data from J. D. King refers only to the cross-sections for the (γ,n) channel, while B. L. Berman et al. provides cross-sections including contributions from both (γ, n) and (γ,pn) channels.

Moreover, as data from B. L. Berman et al. suggest, fact also confirmed by other works, the (γ,pn) process is dominant in the decay of the giant resonance in ^{14}N. This could be an explanation for the difference found in the reaction cross-sections calculated by MCNP6.2 and PHITS3.27 (a factor of ten for 23 MeV photons).

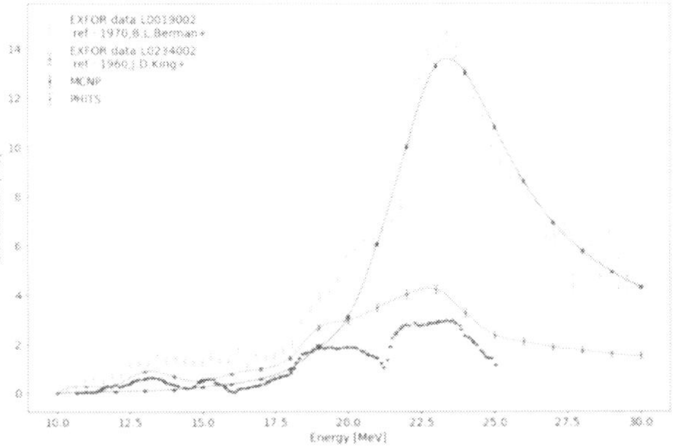

Fig. 6. Cross-section comparison for neutron production in photo-nuclear reactions on ^{14}N between MCNP6.2 and PHITS3.27 as a function of incident photon energy.

The observed discrepancies motivate the importance to be assigned to the underlying neutron emission mechanisms for accurate interpretation and use of cross-section values (either from evaluated libraries or model calculated) in Monte-Carlo simulations. In particular, the observation that MCNP6.2 appears to be in agreement with measured values in terms of total cross-section must be taken with caution since this is achieved without activating the proton physics in the calculations. This leads to the conclusion that the total cross section is calculated by MCNP6.2 without a complete picture of the physics involved but only based on cross-section evaluation using a particular data set. Additional investigation on this matter is essential to delve deeper into this observed phenomenon and gain a comprehensive understanding.

3) MCNP6.2 simulated photo-neutron spectra from molecules of interest for illicit material detection

Based on the findings from the initial study concerning the simulation of photo-neutron energy distributions for natural elements, one can conclude that MCNP6.2 allow a reasonably good accuracy for the purpose of our application in predicting neutron energy values in the outgoing spectra for the materials of interest. This code has been therefore selected for further simulations.

In this context, our focus lies on examining the photo-neutron spectra from various molecules of interest in the context of CBRNE (Chemical, Biological, Radiological, Nuclear, Explosive) risks, when irradiated by 20 MeV photons. Figure 7 illustrates the photo-neutron energy distribution for trinitrotoluene, methamphetamine, and tabun. These molecules correspond respectively to a conventional explosive, a narcotic substance, and a neurotoxic gas hazardous through inhalation or skin contact. Despite the different associated risks, their compositions are very similar, being made of same elements (carbon, nitrogen, and oxygen) only in different proportions.

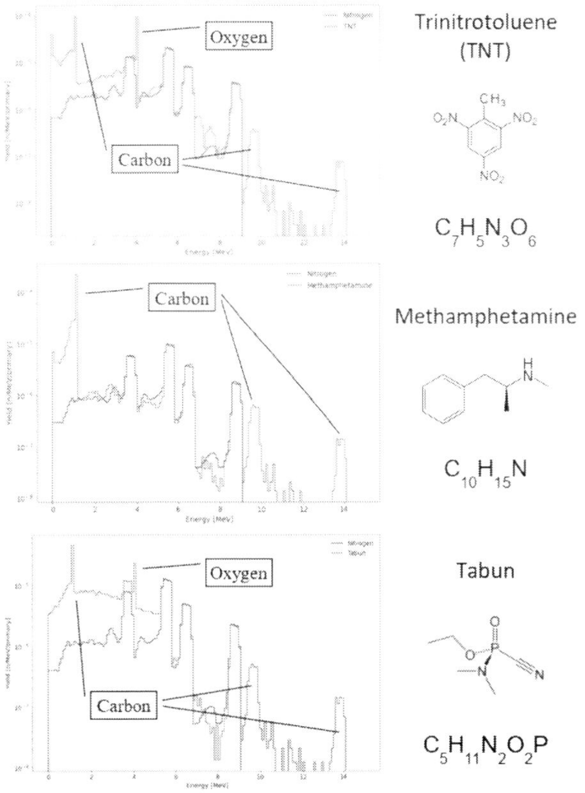

Fig. 7. MCNP6.2 simulated photo-neutron spectra from different molecules when irradiated with 20 MeV photons. The right side of each spectrum provides information on the structure and composition of each molecule.

The blue spectrum on each plot corresponds to the contribution of nitrogen in terms of photo-neutrons being produced. When present, carbon and oxygen contributions are marked with an arrow mark.

One can clearly observe in the simulated spectra the photo-neutron peaks expected at 1.24 MeV and 4.16 MeV, assigned respectively to the ^{11}C and ^{15}O residual nuclei left in their ground states subsequent to (γ,n) reactions on these target nuclei. These contributions can be easily separated from the photo-neutrons emitted in (γ,n) on nitrogen isotopes and thus providing evidence that specific shapes of photo-neutron spectra can be used as signatures for the detection and quantification of illicit materials containing light nuclei such as carbon, nitrogen, or oxygen.

B. Preliminary feasibility studies for the new illicit materials detection method using simulations and measurements with a LINAC based test bench

1) Modeling of the Varian TrueBeam® medical LINAC

As highlighted through intercomparisons of energy spectra and integrated cross-sections, the substantial lack of experimental data makes the development and validation of Monte-Carlo codes for photo-nuclear reactions very challenging. To address this critical gap, the Laboratoire National Henri Becquerel at CEA-Saclay is presently undertaking the construction of a test bench based on a photon source produced by a medical linear electron accelerator. The chosen accelerator, a TrueBeam® model from Varian, has the capability to accelerate electrons at various energies up to a maximum of about 22 MeV.

A comprehensive geometry model of the TrueBeam® accelerator's head and its surroundings has been developed with MCNP6.2 as shown in Figure 8, enabling accurate replication of experimental conditions in Monte-Carlo simulations.

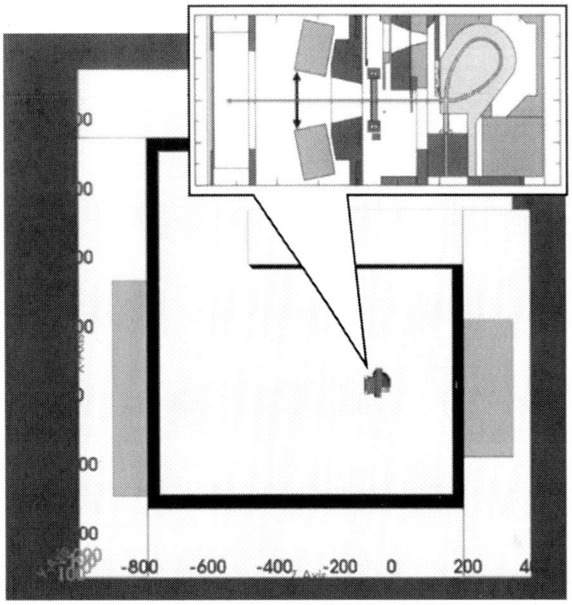

Fig. 8. Global view of the model showing the bunker of the Varian TrueBeam accelerator and a visualization of the accelerator's head geometry highlighting the level of details integrated in the simulation model.

Figure 9 is an illustration of the experimental protocol set for photo-neutron measurements. A 20 MeV electrons beam from the LINAC is stopped in the target, generating photons via the bremsstrahlung process. Subsequent interactions of photons with nuclei inside the target can occur through photo-nuclear reactions, resulting in the production of photo-neutrons. Thus, by using this optimized configuration the same target acts as producer of the photon source and as interacting media allowing photo-nuclear reactions to occur.

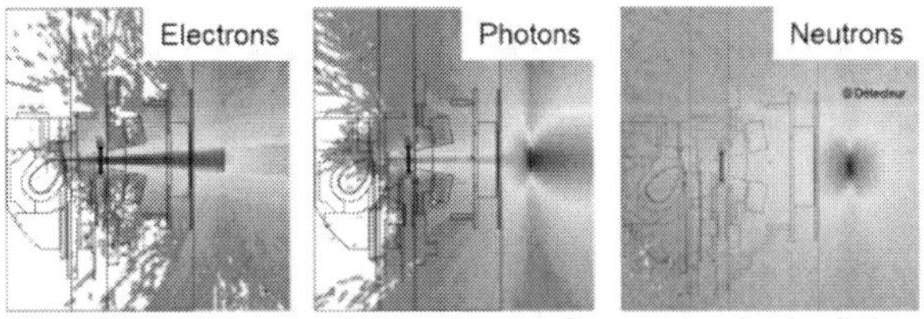

Fig. 9. Simulated electron, photon and neutron fluence maps when irradiating a 10 x 10 cm^2 wide and 1 cm thick tungsten target with a 20 MeV electron beam produced by the Varian TrueBeam® accelerator.

2) Bonner sphere based neutron spectrometry

One of the first experiments performed with the new set-up was dedicated to the measurement of photo-neutron spectrum emitted subsequent to the irradiation of a 10 x 10 cm^2 wide and 1 cm thick sample of tungsten placed at a distance of 70 cm from the beryllium vacuum window. The neutron spectrometry was conducted using a passive Bonner sphere system [14], well suited for measuring neutron spectra in pulsed and mixed n-γ fields with a strong high-energy photon component. The system comprises ten polyethylene spheres with densities of 0.95 g/cm^3. The selected sphere diameters are 7.62 cm, 8.89 cm, 10.16 cm, 11.43 cm, 12.70 cm, 15.24 cm, 17.78 cm, 20.32 cm, 25.40 cm, and 30.48 cm. Neutron detectors are placed in the center of each sphere being made of ^{197}Au foils (99.99% chemical purity, density of 19.3 g/cm^3) with a diameter of 15 mm and thickness of 0.25 mm.

The response matrix of the Bonner sphere spectrometer (BSS) was computed using MCNP6.2 simulations to cover an extended neutron energy range from 10^{-9} MeV to 20 MeV. As a result of the ^{197}Au(n,γ)^{198}Au radiative capture reaction, ^{198}Au radionuclides are created, which subsequently decay with a half-life of approximately 2.7 days through a β/γ cascade, primarily emitting 412 keV γ-rays. These γ-rays were detected and quantified using a hyper pure germanium detector (HPGe) at the LNHB activity metrology laboratory.

The specific ^{198}Au saturation activities obtained for each sphere diameter were analyzed using two custom unfolding algorithms: GRAVEL (an iteration algorithm based on SAND-II) and MLEM (a maximum likelihood expectation maximization unfolding algorithm). These unfolding algorithms are using a default spectrum, assumed to be flat in this study. In addition, a third method based on Markov chain Monte Carlo (MCMC) custom algorithm performing a Bayesian estimation of a parameterized spectrum was used to derive the final neutron spectra. The use of different unfolding algorithms allowed for the validation of results' robustness and confirmed their stability. Those algorithms were validated with the datasets given for the EURADOS comparison exercise on neutron spectra unfolding in Bonner spheres spectrometry during the 14th EURADOS Webinar [15].

Furthermore, the experimental conditions were simulated with the MCNP6.2 Monte-Carlo code using the detailed TrueBeam® geometry model described in section B-1. The calculated photo-neutron spectra were compared to the unfolded experimental spectra and the outcomes are presented in Figure 10.

As shown in the figure, the unfolded spectra using different algorithms consistently agree for neutron energies up to 6 MeV. For higher energies the agreement between GRAVEL and MLEM algorithms is preserved beyond 6 MeV, but it is essential to highlight that the shape of the neutron spectra is not physically meaningful for energies beyond around 10 MeV according to the kinetic energy constraint for primary (γ,n) reactions on natural tungsten. The continuous nature of the response matrix used by these two algorithms results in an energy spectrum artificially extended at high energies since no restriction is applied to constrain the shape of the spectrum. In contrast, the unfolded spectrum based on the MCMC method shows a trend which is more physically accurate and agreeing with the shape of the Monte-Carlo simulated photo-neutron spectrum. However, it is important to note the significant disagreement in amplitude, as high as a factor of three, between the experimental and simulated spectra. This discrepancy has been further investigated in terms of simulation parameters to achieve improved consistency with the measurements.

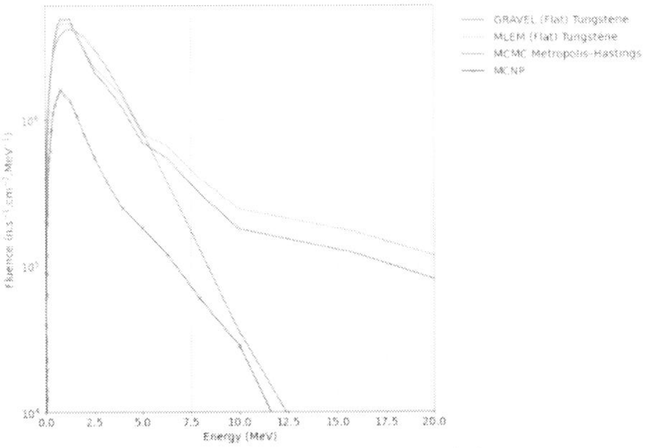

Fig. 10. Comparison between measured and simulated photo-neutron spectra from tungsten sample irradiation with 20 MeV electrons at the Varian TrueBeam® accelerator.

Consequently, additional optimization and modifications of the accelerator have been performed based on electron beam size measurements using Gafchromic® EBT-XD [16] films at a distance of 14 cm form the beryllium vacuum window. The Gafchromic® film readings are displayed in Figure 11 and allowed to conclude that, due the higher-than-expected size of the beam spot at this location, the peripheral electrons were impinging on a steel plate leading to an unwanted additional production of photons (and subsequently neutrons). This effect was not account for in the simulations which considered only the neutrons produced by the interaction of the electron beam with the tungsten sample, explaining the neutron production underestimation by the simulations observed in Figure 10.

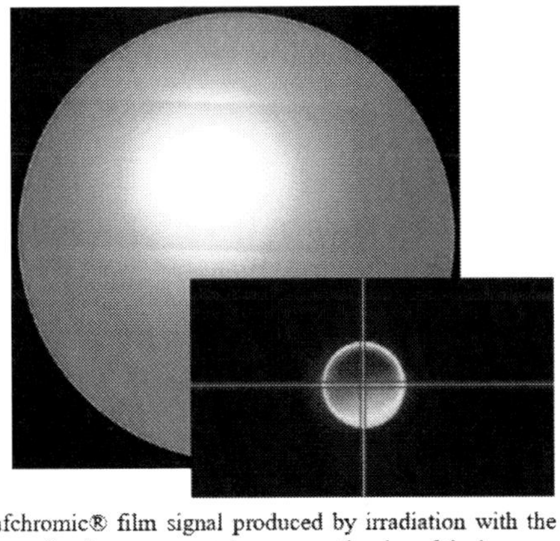

Fig. 11 Gafchromic® film signal produced by irradiation with the 20 MeV electron beam allowing to accurately measure the size of the beam spot.

Based on these findings, the configuration of the accelerator has been further optimized in order to reduce the size of the beam spot and thus to avoid additional interactions with elements other than the sample being studied. As shown in Figure 12, the electron beam spot size at the sample location (70 cm away from the beryllium vacuum window) has been decreased from the initial diameter of 29 mm to 21 mm.

Beam size measurements have been performed in addition at a distance of 14 cm, and the values obtained together with the ones measured at 70 cm allowed to calculate the divergence angle of the beam which was found to be 1.5°.

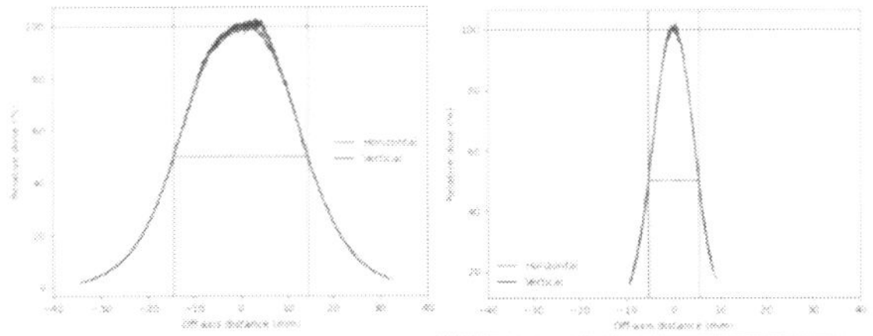

Fig. 12. Profile of the 20 MeV electron beam, before and after the modifications made to the accelerator (left and right side of the figure respectively).

With this new information in hand, an improved more accurate model of the Varian TrueBeam® accelerator for Monte-Carlo simulations has been set allowing calculations to be performed in a configuration as close as possible to the real-life experimental conditions.

These developments strongly contribute to the reliability and validity of further experiments to be performed for the establishment of an inspection method based on the active photon interrogation technique associated to the spectroscopy of photo-neutrons to be used for the detection of illicit materials.

IV. Conclusions and Perspectives

The lack of experimental energy and/or angle distributions of photo-neutrons is a serious challenge in the development and validation of Monte-Carlo codes for the simulation of applications involving photo-nuclear reactions. Current data available from the EXFOR database, which are in most cases related to total cross-section values integrated over energy and angle, does not provide enough information to rigorously test and validate the accuracy of Monte-Carlo simulations in realistically reproducing the differential distributions of photo-neutrons. This is particularly true in the case of photo-nuclear reactions on light elements, characterized by a pronounced influence of nuclear level structures on the shape of photo-neutron spectra as highlighted by the present work. Moreover, we identified significant differences when comes to the method used by the different Monte-Carlo codes tested to handle this effect, resulting in noticeable discrepancies in the results obtained namely for photo-nuclear reactions on carbon and nitrogen.

Nevertheless, our investigation revealed that very nucleus-specific, easy to identify structures in the photo-neutron spectra can serve as signatures for the detection and quantification of illicit materials containing light nuclei, such as carbon, nitrogen, or oxygen. Consequently, there is a crucial need for new measurements dedicated to experimental photo-neutron energy spectra, allowing to enhance the understanding of physics related to photo-nuclear reactions in general. This need is strengthened by the preliminary results presented in this work, showing evidence that these specific reactions are an excellent tool in the frame of homeland security applications and that the good knowledge of the related physics is crucial in the development and optimization process.

In response, the Laboratoire National Henri Becquerel at CEA-Saclay started the construction of a test bench, based on a photon source produced by a medical linear electron accelerator and specifically optimized for photo-neutron measurements.

This pioneering setup is a major milestone as it will not only enrich the existing nuclear data on photo-nuclear reactions but will also result in a technical platform allowing to assess, optimize and finally validate the proposed innovative method for detection of illicit materials. It therefore represent a significant contribution to strengthening the reliability and accuracy of Monte-Carlo simulations when addressing photo-nuclear reactions and consequently bolstering homeland security efforts through the development of an first of its kind inspection method based on the active photon interrogation technique and the spectroscopy of photo-neutrons.

Acknowledgment

The authors express their gratitude to the IRSN laboratory for neutron metrology, particularly Nelson Magalotti and Quentin Ducasse, for their contributions to the neutron spectra measurements and for the loan of their passive Bonner sphere spectrometry system. Special thanks are also addressed to Marie-Christine Lepy for the support in the analysis of activated gold foils. Finally, the authors would like to acknowledge the financial support provided by the French National Laboratory of Metrology and Testing.

References

[1] Y. Zhao, T. Cui, Y. Yang, "The design of a photo-neutron source for the narcotic drugs detection in a large-truck" in *Proc. 2017 IEEE Nuclear Science Symposium and Medical Imaging Conference (NSS/MIC 2017)*, Atlanta, Georgia, October 21–28 (2017).

[2 C. L. Fontana, A. Carnera, M. Lunardon, F. Pino, C. Sada,*et al.*, "Detection system of the first rapidly relocatable tagged neutron inspection system (RRTNIS), developed in the framework of the European H2020 C-Bord project", Physics Procedia, 90, pp. 279-284 (2017).

[3] M. Gmar, E. Berthoumieux, S. Boyer, F. Carrel, D. Doré *et al.*, "Detection of nuclear material by photon activation inside cargo containers", in *Proc. SPIE Defense and Security Symposium*, Orlando, Florida, April 17–21 (2006).

[4] K.W. Habiger, J.R. Clifford, R.B. Miller, W.F. McCullough, *et al.*, "Explosives detection with energetic photons", *Nuclear Inst. and Methods in Physics Research B*, vol. 56-57, no. 2, pp. 834-838 (1991).

[5] The Official CERN FLUKA website, https://fluka.cern (2022).

[6 S. Agostinelli, J. Allison, K. Amako, J. Apostolakis, H. Araujo, *et al.*, « GEANT4—a simulation toolkit », *Nuclear Inst. and Methods in Physics Research A*, vol. 506, no.3, pp. 250–303 (2003).

[7] C. J. Werner, MCNP users manual - code version 6.2, LA-UR-17-29981.

[8] T. Sato,Y. Iwamoto, S. Hashimoto, T. Ogawa, T. Furuta, *et al.*, "Features of particle and heavy ion transport code system (PHITS3.27) version 3.02", *Journal of Nuclear Science and Technology*, vol. 55, no. 6, pp. 684–690 (2018).

[9] N. Otuka, B. Pritychenko, N. Otuka, E. Dupont, V. Semkova, *et al.*, "Towards a More Complete and Accurate Experimental Nuclear Reaction Data Library (EXFOR): International Collaboration Between Nuclear Reaction Data Centres (NRDC)", *Nucl. Data Sheets*, vol. 120, no. 272 (2014).

[10] A. Koning, D. Rochman, "Modern nuclear data evaluation with the TALYS code system", *Nuclear Data Sheets*, vol. 113, no. 2841 (2012).

[11] B. C. Cook, J. E. E. Baglin, J. N. Bradford, J. E. Griffin, "^{12}C (γ,n) ^{11}C Cross Section to 65 MeV", *Phys. Rev.* vol. 143, no. 724 (1966).

[12] B. L. Berman, S. C. Fultz, J. T. Caldwell, M. A. Kelly, S. S. Dietrich, "Photoneutron Cross Sections for Ba-138 and N-14", *Phys. Rev. C*, vol. 2, no. 2318 (1970).

[13] J. D. King, R. N. H. Haslam, R. W. Parsons, "The gamma-neutron cross section for N-14", *Canadian Journal of Physics*, vol. 38, no. 2 (1960).

[14] K. Amgarou, V. Lacoste, A. Martin, B. Asselineau, L. Donadille, "Neutron Spectrometry With a Passive Bonner Sphere System Around a Medical LINAC and Evaluation of the Associated Unfolding Uncertainties", *IEEE Transactions on Nuclear Science*, vol. 56 no. 5, pp. 2885-2895 (2009).

[15] J. M. Gómez-Ros, R. Bedogni, C. Domingo, J. S. Eakins, N. Roberts, *et al.*, "International comparison exercise on neutron spectra unfolding in bonner spheres spectrometry: problem description and preliminary analysis", *Radiation Protection Dosimetry*, vol. 180, no. 1-4, pp. 70–74 (2018).

[16] http://www.gafchromic.com/

Development of a prototype of monitor for alpha and beta radiation in water using new silicone-based contamination-safe detectors

Marcello Lunardon[1,*], Paol o Sartori[1], Pierino De Felice[2] and Luca Stevanato[1]
[1]Dept. of Physics and Astronomy, University of Padova and INFN, Italy
[2]National Institute of Ionizing Radiation Metrology, ENEA, Casaccia - Rome, Italy

(*) marcello.lunardon@pd.infn.it

Abstract **— We present here the main results of an experimental activity aimed at the realization of a prototype of monitor for alpha and beta radioactivity in water exploiting a novel type of contamination-safe scintillation detector developed by our research group.**

Due to the short path-length of alpha and beta particles in water and the low detection limits needed to be compliant with the international legislations in matter of radiation safety for water, the detectors to be used for such a kind of instruments should have in general large area, very low intrinsic background and avoid any kind of window between the monitored water and the active volume of the detector. Even an extremely low contamination of the surface can therefore destroy the detector performance, while protecting the detector with a layer of passive material will reduce the detection efficiency, in particular for alpha particles, and the passive layer can get contaminated itself, making necessary its substitution and representing an important limitation to the realization of a radioactivity monitor that should work continuously for years.

The novel detectors are large-area silicone-based scintillators developed and produced in our laboratories. The technology we propose is a significant step forward in the direction of the realization of radioactivity monitors for water with high sensitivity and reasonable costs, to be used to improve security and quality of the water distributed to European and worldwide citizens.

The detection performance for alpha and beta particles was first tested in air with standard sources, showing results very close to similar detectors commercially available. Later a number of tests were carried out in the ENEA-INMRI laboratories with the aim to characterize the contamination and decontamination properties. Initial activity of each detector foil was checked by low background radioactivity measurements, then the detectors were immersed for fixed times and constant procedure in a variety of radioactive reference liquid solutions with different radionuclides in different chemical forms. After rinsing with distilled water the detector foils were again measured for residual radioactivity adsorbed on the foils surface.

After this preliminary campaign of contamination measurements carried out on small size samples, several large area detectors were realized and used to build a small demonstrator device for continuous monitoring of the radioactivity in water. The monitor was tested with distilled water (blank) and radioactive aqueous solutions, demonstrating the possibility to realize a new generation of radiation monitors for water with high sensitivity and with the possibility to easily manage the contamination issues.

Keywords **— radioactivity in water, windowless plastic scintillation detector, decontamination.**

I. Introduction

In recent history we have been witnessing several important accidents and disasters that led to a large spread of radioactive material in water. Let's remember, as an example, the recent accident at Fukushima's Daiichi reactor following the devastating tsunami in Japan (2011), or the large release of radioactive liquids and solid wastes in the environment at Church Rock in New Mexico (1979), or again the contamination events in the Ontario Lake in the last decade (2009 and 2011 the last ones).

Such events, together with the increasing terroristic threat of possible attacks involving RN materials, has pushed the interest of international community in monitors for radioactivity in water (MRW) intended for human consumption, directly (drinking water) or indirectly (water for crops irrigation, wasted water that can impact on the environment, etc…).

Currently water radioactivity screening procedures are based on periodic samplings followed by laboratory analyses. It's a well-established procedure, but has several deep limitations on the possibility of performing a continuous real-time monitoring, due to the significant time and cost needed to complete a single analysis that should be replicated continuously many and many times. For this reason a number of alternative water monitoring systems based on different technologies are being implemented and are facing the market, most of them taking advantage of the recent advances on scintillation detectors technology. There are a few commercially available MRW for continuous monitoring based on scintillation detectors. They usually declare medium-high Minimum Detection Activities (MDA) between several hundreds of Bq/L down to several tens of Bq/L for reference elements like Cs-137, K-40, Sr-90 and U-238 or Po-210, two order of magnitude or more larger than the WHO Guidelines detection limits. See as an example the NEX-GEN-SSS and NEX-BETA-ABG systems [1]. Typically no clear information is provided about effects of contamination in the sensors.

Nevertheless several key aspects need still to be further improved, before this technology can become the reference for MRWs. Among them, the problem of contamination of sensors in contact with water to be analyzed is definitely one of the most significant. For this reason many commercial MRW are based on detection of the gamma radiation emitted by contaminants, that is more penetrating and allows protecting the gamma sensor with thick and solid shielding (typically metallic). But gamma detection is not adequate to detect all possible radioactive contaminants in water, e.g. pure beta emitters like Sr-90 (see e.g. the SARA Water system [2]). This kind of devices does not cover properly all possible RN contaminants in water and therefore they are more often used in dedicated applications, as an example the measure of Cs-137 in sea water (see e.g. [3]).

Current radioactivity limits in legislations all around the world refer, in fact, to "gross alpha and beta activities" as first level of screening (see e.g. [4]) and direct alpha and beta radiation detection in water is therefore recommended. Detecting such low activities (1 Bq/L or less) of beta and, even more, of alpha radiation in water with scintillators requires large detection areas, since the mean free path of these particles in water is short (a few tens of microns for alphas). Consequently, even very low levels of contamination per unit of area can destroy the monitor performance. On the other hand, the need of large detection areas makes economically impractical the use of existing robust but expensive scintillators like YAP:Ce crystals.

II. DESCRIPTION OF THE RESEARCH PROGRAM

The main objective of the research activity described in the work is the development of a novel detector for alpha-beta detection in water, which can overcome the limitations of the existing devices as reported above.

This activity is as a follow-up of the EU project TAWARA_RTM (FP7-SEC-2012-1, Grant n. 312713), where our research group developed a high-sensitivity monitor for radioactivity in water intended for human consumption [5]. The core of such system, the Real Time Monitor, used a number of large area commercial scintillation phoswich EJ-444 as sensors for alpha and beta detection [6].

The TAWARA_RTM system, after being extensively tested and calibrated at the INMRI laboratories of ENEA in Casaccia (Rome), was finally installed and operated for the demonstration campaign at the North Water Treatment Plant of the Warsaw Waterwork Company (MPWiK). Despite its notable results in terms of sensitivity and overall features for water radioactivity monitoring, which make it still a state-of-the-art instrument in this context, the characterization campaign showed several aspects that could be improved, the main one being the effect of surface contamination of the radiation detectors, in terms of both deterioration of the sensitivity and possibility to efficiently clean-up the system to recover the original performance.

After the end of this project we continued the research on this topic by focusing on the problem of the detector contamination on the surface directly exposed to water with a radioactive content.

A very promising solution was identified in the use of a novel category of scintillation detectors, consisting in a large area phoswich made of ZnS crystal powder embedded in a matrix of active silicone-based scintillator [7]. The silicone physical properties, in particular its hydrophobicity, chemical resistance and mechanical robustness to water exposure, make it particularly interesting for long-term use in water. Surface can be further functionalized in order to achieve a higher level of robustness, making it contamination-safe when exposed to aqueous radioactive solutions.

The basic material of the new detector we are proposing is a silicone-based scintillation material, namely phenyl-polysiloxane (PPS), non-toxic, non-dangerous, easy to be manufactured and handled, working in a wide range of temperatures (-100, +250 C), much wider than in case of common plastic scintillators. A number of possible surface functionalization (e.g. sylanization processes) can be applied in order to maximize its resistance against RN contaminants. These detectors can be easily shaped in thin foils (~ 1 mm) with large active areas, of the order of 20 cm x 20 cm or more. Efficient alpha-beta discrimination can be obtained by pulse shape analysis techniques applied to signals.

The particular surface properties described above make this silicone-based detector also a good sensor for RMW operating with raw water from rivers, lakes or sea.

The expected features of this novel detector are:

- good detection efficiency for alpha and beta radiation
- high hydrophobicity, chemical resistance and long-term stability of the surface
- negligible contamination level after exposition to radioactive aqueous solutions
- possibility to clean up the surface from possible residual contaminations using chemical cleaning agents without damaging the detector
- low production cost

After realizing and characterizing the new detectors, a simple single-module radioactivity monitor was built, with the aim of demonstrate the usability of this technology to make a new generation of contamination-safe high-sensitivity WRM, better than the state-of-art instruments.

III. CONTAMINATION TESTS ON SMALL AREA DETECTORS

Many small samples of the new silicone-based detectors have been produced at the chemical labs of the Department of Physics and Astronomy of the Padova University (UNIPD).

They have first optically and mechanically characterized and then tested with alpha (Po-210) and beta (Sr/Y-90) sources in air. Finally they have been exposed to different aqueous radioactive solutions at ENEA-INMRI and the final contamination has been evaluated.

A. Production and test of small samples at UNIPD

The production starts by mixing liquid silicone (polydimethylsiloxane Gelest DMS-V21) with the cross-linking agent (polydiphenyl-co-dimethylsiloxane Gelest PDV-2331) and the scintillation dopants (2,5-diphenyl oxazole + Lumogen F Violet 570) [7,8].

A ZnS:Ag scintillating powder (EJ-600 by Eljen Technology) is added to the silicone and the compound is poured on the mold to have a uniform layer 1.0 - 1.2 mm thick.

A powder sedimentation of about 12h is used to create an active bottom layer about 50 um thick with 60%wt of ZnS:Ag.

The silicone is finally cross-linked in oven at 70 °C for 1h to create a solid elastomeric slab.

A surface functionalization with nanometric fluorinated silane film is created by vapor phase deposition on several samples. Other samples were made using passive PDMS silicone (only alpha detection capability).

The small bottom layer is the "alpha active layer" where alpha particles coming from radioactive decays are stopped transferring most of their energy to the ZnS:Ag crystals that produce the typical delayed light flash. Beta particles, on the contrary, typically loose their energy in a much longer path inside the active silicone body, producing the typical prompt flash of the phoswich detector.

The final result is shown on Fig. 1 : on the left a cross-section at 40x optical transmission microscope showing the 50 um alpha active layer and the silicone beta active body. On the right a picture of a typical 5 cm x 5 cm detector sample.

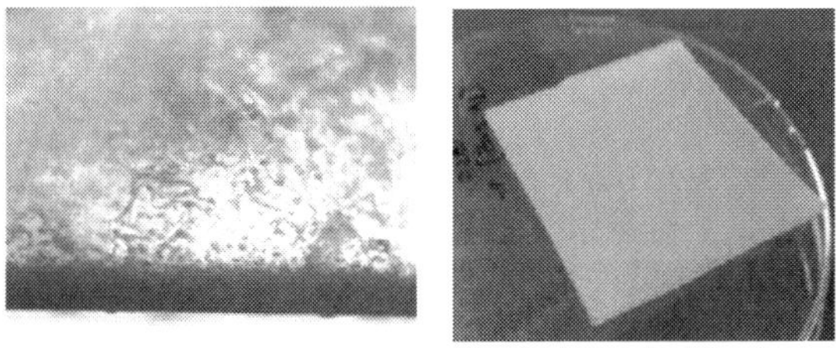

Fig. 1. Left panel: a cross-section at 40x optical transmission microscope showing the 50 um alpha active layer and the silicone beta active body. (right panel: a picture of a typical 5 cm x 5 cm detector sample.

The samples have been tested as radioactivity detectors using standard alpha (Po-210) and beta (Sr/Y-90) sources in air. The silicone detector was placed on a planar wavelength-shifting light guide (EJ-280) inside a light-proof box. The scintillation light was collected by two PMTs Hamamatsu R3550A 1" diameter acquired in coincidence using a CAEN DT5725B digitizer.

Alpha and beta identification was achieved by standard pulse shape discrimination technique. The alpha and beta spectra were compared to equivalent spectra taken using commercial reference phoswich detector EJ-444/EJ-440 by Eljen Technology. An example of raw alpha and beta spectra are shown in Fig. 2.

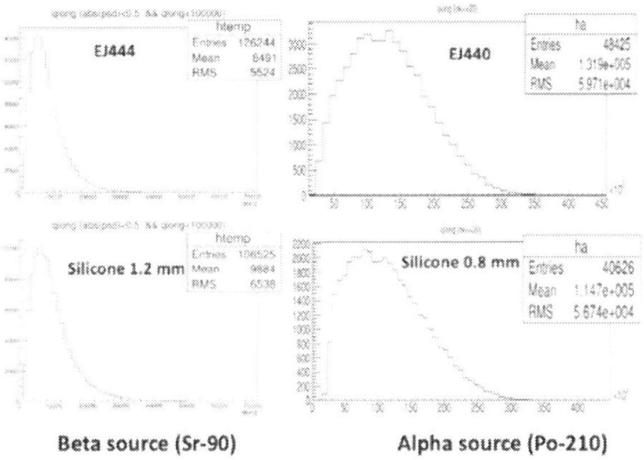

Fig. 2. Example of raw alpha (on the right) and beta (on the left) spectra taken with the reference EJ-444/440 detectors and with the new silicone-based detectors.

A measured efficiency of 80% to 85% relative to EJ-444 detector was measured for the new silicone-based detectors. The corresponding light output amplitude was very similar. For alpha particles, a relative efficiency between 85% to 90% was found, with a slightly lower light output (90% relative to EJ-440).

B. Contamination tests at ENEA-INMRI

Many detectors were produced at UNIPD and moved to the ENEA-INMRI labs in order to test the contamination properties.

Several bunches of 8 sample pairs were hooked to a dedicated fork-like support (Fig. XX, left panel). Each support hosted detectors with different features: new silicone detectors with or without surface silanization, made with scintillating silicone or passive silicone and reference EJ-440 detectors. To be noticed than the external surface of the EJ-444 detectors is the same of EJ-440, changing only the lower plastic support.

The supports were immersed in the aqueous radioactive solutions by means of a moving arm of a dip-coater (Fig. 3, right panel).

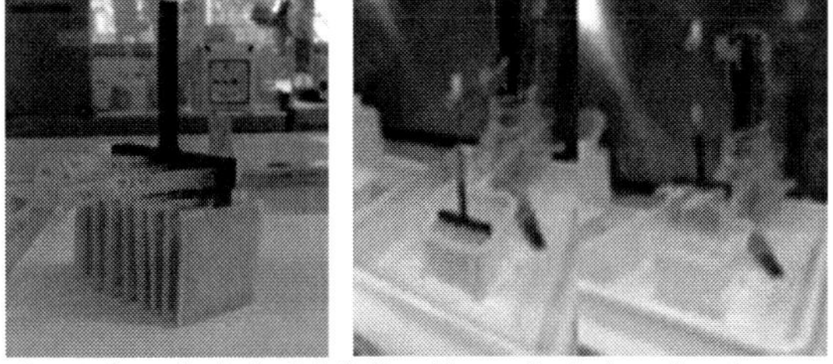

Fig. 3. set of samples in the fork (left) and immersion in the radioactive aqueous solutions. See text for details.

The contamination was obtained therefore by immersion and motion inside the aqueous solution, to simulate the flowing of the water inside the real instrument.

Cs-137 and Co-60 solutions with specific activities of about 1, 10 and 100 kBq/L were used. The bath time was varied from 30 min to 15 h.

After contamination, the samples were rinsed with distilled water, dried in air and measured with the certified HPGe gamma spectrometers available at INMRI in order to quantify the residual activity.

Finally we tested several decontamination procedures to further reduce the residual contamination:

1. immersion in a detergent solution of CONTRAD 2000 for 30 min;
2. immersion in an acid solution of HCl 0.5M for 30 min;
3. immersion in CONTRAD 2000 for 30 min + rubbing with a soft towel;
4. immersion in a solution of HCl 1.5M for 30 min.

Steps 1+2 are labelled with "soft cleaning" in the following, while 3+4 are labelled as "strong cleaning". The first procedure was thought as possible cleaning of a sealed device with only a short temporary stop in the monitoring, while the second would require a deeper maintenance intervention.

The main results are summarized in Fig. 4, where the residual contamination of Cs-137 is shown as a function of the specific activity for different samples.

During a first test run we found that the residual contamination saturated after a few minutes, no significant differences were found between shortest and longest immersion times. Furthermore an almost linear increase of with the specific activity used, at least in the considered activity range.

The reference samples (EJ-440) showed the largest residual contamination (magenta points). Moreover, due to the structure of the outer active layer, no cleaning procedure after contamination, either "soft" or "strong", could be used.

The new silicone-based samples, without cleaning procedure, showed in general a contamination level about 10 times lower than the reference samples (blue points). A further reduction of contamination of about a factor 2 was observed for samples with surface functionalization (red points).

A second test run was performed with only large specific activity solutions. In this run we first evaluated the samples with no cleaning procedure, obtaining results similar to the first run (light blue points), then we tried the cleaning procedure.

After the "soft cleaning" the residual contamination was reduced of about one order of magnitude (violet points). The "strong cleaning" reduced further the contamination below the minimum detectable activity of the gamma spectrometer, about three order of magnitude lower than what measured with the reference detectors.

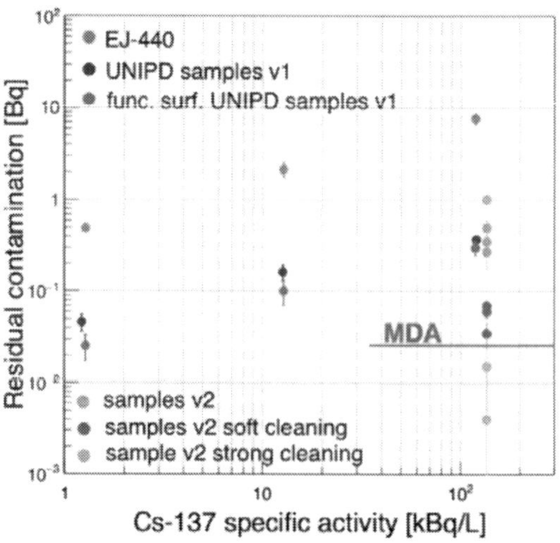

Fig. 4. Residual contamination as a function of the specific activity for different samples in different configurations. See text for details.

IV. Demonstrator Monitor Device with Large Area Detectors

A. *Large area detector production and test*

Several large-area 20 cm x 20 cm alpha-beta detectors were produced at UNIPD. The production procedure was the same described above for small samples. Only one difference was introduced in order to improve the sedimentation of the ZnS:Ag powder that was initially found not enough homogeneous: a high static electrical field (several kV/cm) was used to push the powder toward the bottom taking advantage of the small electrostatic charge of the powder observed during manipulation. The thickness of the alpha active layer was finally found consistent with the small samples over the whole area.

The homogeneity of the detection efficiency was measured using standard alpha (Po-210) and beta (Sr/Y-90) sources in air. In Fig. 5 we show the detection efficiency relative to a small-size EJ-444 reference detector for a large area detector for alpha (left panel) and beta (right panel) particles. The experimental setup was the same used for the efficiency measurements of the small detectors described above. The efficiency is slightly higher in the upper rows where the PMTs where placed, allowing therefore for a better light collection. This effect was seen also in the homogeneity tests of the TAWARA_RTM device [9], where EJ-444 detectors were used. Considering the efficiency values at the same distance from the PMTs, the spread of the values is limited to a few percent.

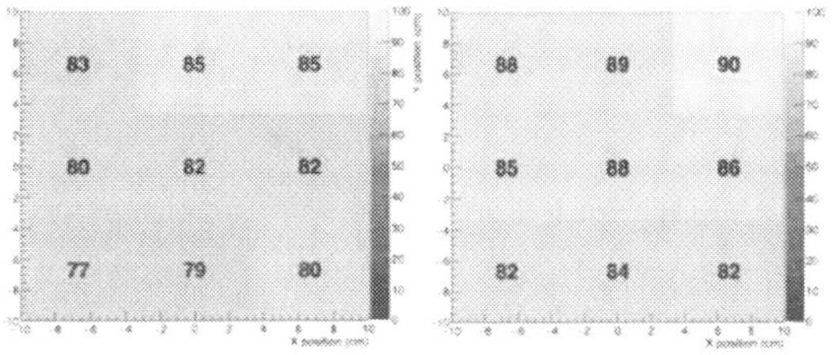

Fig. 5. Detection efficiency relative to a small-size EJ-444 reference detector for a large area silicone-based detector for alpha (left panel) and beta (right panel) particles. Numbers are percentages.

B. Demonstrator device design and realization

After validating the functioning of the new large-area detectors, two of them were mounted on a demonstrator device to show the possibility to realize an alpha-beta monitor system for radioactivity in water with very low intrinsic contamination and decontamination capability.

The system is composed by a wavelength shifting (WLS) light guide EJ-280 of size 21 cm x 21 cm mounted inside a water-proof INOX chamber (Fig. 6, left panel). On the top and the bottom of the WLS two large area detectors are placed (Fig. 6, right panel). After mounting and sealing an active area of about 19 cm x 19 cm was left in each side, for a total of about 2 x 350 cm^2.

Fig. 6. Design of the demonstrator chamber (left panel) and a picture of the silicone-based detectors placed on the WLS during the assembly phase (right panel). See text for details.

The scintillating light was collected on one side of the WLS by means of two PMTs ET 9266 2" manufactured by Scionix. Electric signals were acquired in coincidence by a CAEN DT5725B digitizer. The full waveforms were acquired on a PC and processed in real time in order to discriminate alpha, beta and residual background signals and display the counting rates.

C. Demonstrator characterization: blank

The demonstrator was tested at UNIPD using demineralized water to characterize the blank and using NORM to evaluate the detection performances. Unfortunately, due to logistic problems related to the post-Covid period, at the moment of writing the calibration tests and finale contamination tests with aqueous solutions of standard radioactive isotopes (like Cs-137) could not be performed.

The blank test was realized with demineralized water by measuring the alpha and beta rates for long periods. In Fig. 7 we show a typical output of a 5 days measuring run.

The observed average beta rate was about 10 counts per second, in line with what was measured in Padova with the TAWARA_RTM detectors (between 6 to 18 cps). The rate was quite stable, except for a small non stochastic day-night fluctuation due probably to temperature, varying of about 5-6 degrees between night and day. In fact, the known effect of temperature on the PMTs was not compensated at the time of writing.

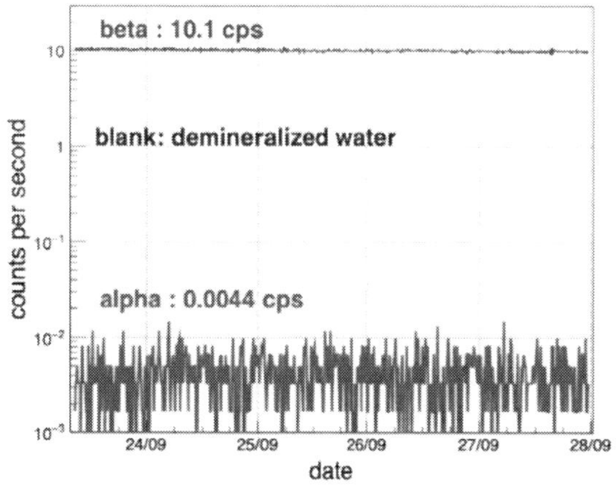

Fig. 7. Differential (a) and integral probability of interaction as a function of angle of incidence. Note that "Fig." is abbreviated. There is a period after the figure number, followed by two spaces.

The observed alpha rate was about 0.004 counts per second. Again, this value is consistent with what measured in the past for the TAWARA_RTM detectors (between 0.003 and 0.008 cps). In this case the temperature effect is negligible and the distribution of counts in 10 min is reasonably poissonian, as expected. We show in Fig. 8 the frequency plot together with the poissonian fit.

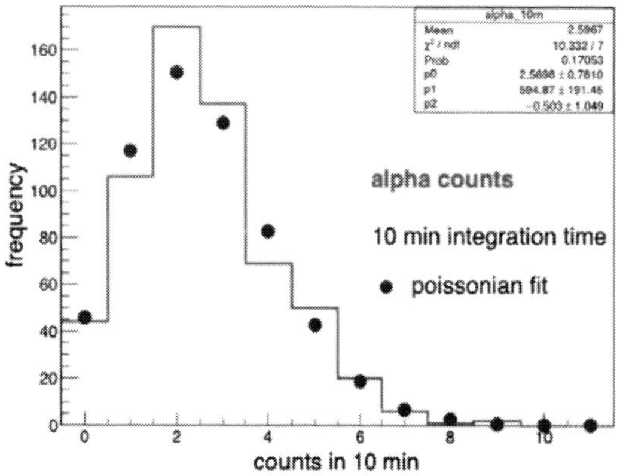

Fig. 8. Frequency plot of the alpha counts in 10 minutes together with the poissonian fit. The experimental distribution follows reasonably the expected behaviour.

D. Demonstrator characterization: test with NORM

Finally the detection efficiency was tested using NORM. In particular for the beta efficiency an aqueous solution containing

K-40 was used. The plot of Fig. 9 shows the increment of the beta rate following the insertion of KCl inside the water circuit. The final specific activity of the solution was about 200 Bq/L.

The measured rate increment was about 2.5 counts per second. This number can be directly compared to the efficiency measured and certified at ENEA-INMRI for the TAWARA_RTM units [10]. In that case, a calibration coefficients of 0.1 to 0.15 Kg^{-1} were measured for the three units when K-40 is used. Each unit were composed of 8 modules with active area of 2 x 400 cm^2 each. By rescaling for the number of modules and for the active area, an expected rate increment between 2.2 and 3.3 counts per second with 200 Bq/L of K-40 can be calculated. The measured number fits well with the calculated range of expectation.

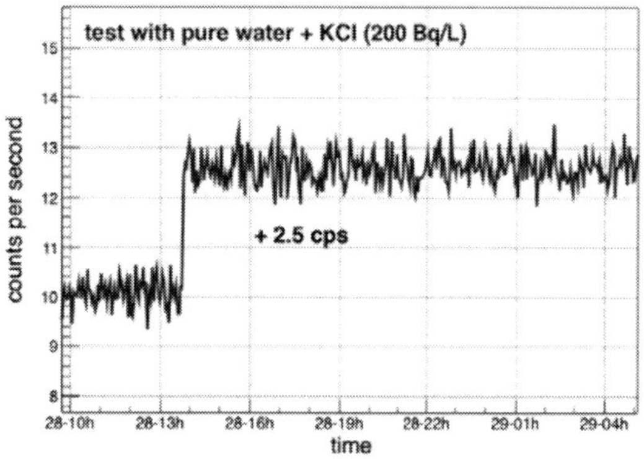

Fig. 9. Beta counts rate as a function of time for the beta efficiency test. The step of +2.5 cps corresponds to the injection of 200 Bq/L of K-40 in the water circuit.

The alpha efficiency was verified only qualitatively by measuring the Radon melted in the Padova tap water. The Padova tap water comes from underground and contains a small amount of Radon, of the order of 5 to 10 Bq/L [11]. This number can vary during the years for natural reasons and, even more important, it depends strongly on the way the water is treated before and during the measurement, therefore ii has to be taken only as an indicative value.

Radon was observed and measured during the tests of the TAWARA_RTM detectors. The typical measured values (maximum value before the Radon decay) were ranging from about 0.2 to 0.3 counts per second per module.

In this case we measured a corresponding maximum value of about 0.7 counts per second [Fig. 10]. This value, even if a direct comparison is not possible, is much larger than the one measured with the TAWARA_RTM detectors and suggests the possibility that, as we expect, the alpha efficiency of these detectors is higher due to the absence of passive protection layers on the surface of the detectors.

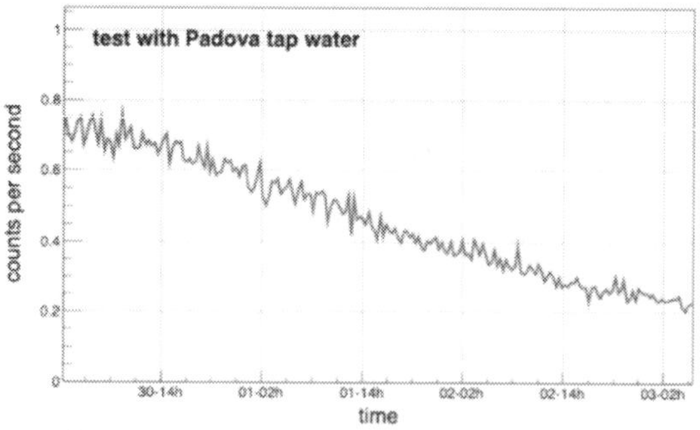

Fig. 10. Alpha rate as a function of time for a run with Padova tap water. The alpha counts are due to the Radon contained in the tap water, coming from the underground aquifer.

V. CONCLUSIONS

In this work we showed the results of a research program aimed at develop and characterized a new type of phoswich detectors for alpha and beta radioactivity in water.

These detectors, thanks to their silicone-based bulk composition, are demonstrated to have lower residual contamination than analogue commonly used commercial detectors. Moreover, thanks to their improved mechanical and chemical resistance, they can undergo simple and effective decontamination procedures that can reduce further the residual contamination up to two order of magnitude.

These new detectors can be used to build a new generation of monitors for alpha and beta radioactivity in water. A simple single-module demonstrator has been built and tested. At the moment of writing, even if the foreseen experimental characterization is not fully completed, it is showing encouraging results.

REFERENCES

[1] Continental Hydrodyne Systems Inc., Ohio, US http://www.hydrodynesystems.com/radiationwater.htm

[2] Scienta Envinet / ENVINET GmbH, Germany, https://scientaenvinet.com/en/products/sara.html

[3] A. Naumenko et al., “Autonomous NaI(Tl) gamma-ray spectrometer for in situ underwater measurements”, Nucl. Inst. and Meth. A, vol. 908, pp. 97-109, Nov. 2018, 10.1016/j.nima.2018.08.001

[4] World Health Organization, "WHO Guidelines for drinking-water quality, 4th edition", 2011, **ISBN:** 9789241548151

[5] P. Carconi et al., " Characterization and calibration of a novel detection system for real time monitoring of radioactive contamination in water processed at water treatment facilities", Appl. Rad. and Iso., vol. 126, pp. 138-145, Aug. 2017, 10.1016/j.apradiso.2017.01.023

[6] E. Bodewits et al., "", IEEE Trans. on Nucl. Sci., Vol. 63, Issue 3, pp. 1565-1569, Jun. 2016, 10.1109/TNS.2015.2495207

[7] M. Dalla Palma et al., " Non-toxic liquid scintillators with high light output based on phenyl-substituted siloxanes", Opt. Mat., Vol. 42, pp. 111-117, Apr. 2015, 10.1016/j.optmat.2014.12.028

[8] A. Quaranta et al., " Doping of polysiloxane rubbers for the production of organic scintillators ", Opt. Mat., Vol. 32, pp. 1317-1320, 10.1016/j.optmat.2010.04.021

[9] TAWARA_RTM project, Deliverable D2.4, private communication, 2015.

[10] ENEA-INMRI, Certificate of calibration N. 661/ARN, Jun. 30, 2016

[11] ARPAV Vicenza, Italy, private communication, 2014.

Field tests of the MICADO monitoring detectors in real radwaste storages

Paolo Finocchiaro[1 *], Luigi Cosentino[1], Fabio Longhitano[2], Gaetano Elio Poma[1], Simone Amaducci[1], Bernard Dumercq[3], Christopher Helbert[3], Lionel Tondut[3],
[1]INFN Laboratori Nazionali del Sud, Italy
[2]INFN Sezione di Catania, Italy
[3]ORANO, France
(*) finocchiaro@lns.infn.it

Abstract— **In the framework of the MICADO project, funded by Euratom H2020 program, a system of gamma ray and neutron counters was developed. These counters are meant to be deployed as a monitoring grid around radwaste drums in low and/or intermediate waste storage sites, even though they could be usefully employed also for high level waste. A prototype of the MICADO gamma and neutron monitoring grid was installed and tested around real radwaste drums in two storage sites on the ORANO and NUCLECO premises, at La Hague and Casaccia respectively. A modular support structure, was assembled in four units in order to accommodate four radwaste drums. Each support unit hosted four gamma ray counters and four neutron counters. In both cases the units were installed around four radwaste drums also containing small or tiny quantities of actinides, and data were acquired. The results are quite promising as the system behaved as expected from previous tests and from the nominal radiological info about the drums.**

Keywords —**radwaste monitoring; neutron counter; gamma counter.**

I. Introduction

THE easiest way to perform external radiological monitoring of radioactive waste drums (radwaste, in short) is to count the outcoming gamma rays and neutrons, as they represent the only useful form of penetrating radioactivity that can reach out of the drums. Indeed alphas, beta particles and fission fragments are basically stopped inside the drums and cannot be directly detected in a non-destructive fashion. In the framework of the MICADO H2020 project [1] a set of gamma ray and neutron counters was developed (Fig. 1). These counters are meant to be deployed as a monitoring grid around radwaste drums in low and/or intermediate waste storage sites, even though they could be usefully employed also for high level waste.

The gamma counter, named SciFi, is based on a scintillating optical fiber readout at both ends by SiPM photodetectors. Whenever a gamma ray interacts with the fiber, mainly via Compton scattering, a tiny and fast flash of light is produced that can be detected by the two SiPMs (Fig. 2). A suitable signal amplitude threshold together with the coincidence between the two signals suppresses the spurious background noise [2]. The neutron counter, named SiLiF, is based on a double-sided silicon diode sandwiched between two ^{6}LiF neutron converters. When a neutron is captured by ^{6}Li two high energy particles are produced, namely an alpha (2.05 MeV) and a triton (2.73 MeV) emitted back to back, that can be detected by the silicon diode (Fig. 3). A suitable threshold on the deposited energy strongly suppresses the possible gamma background contribution down to about 10^{-10} [3].

Fig. 1. The set of 32+32 SciFi and SiLiF detectors developed for gamma rays and neutrons. One of the SiLiF detectors is shown as allocated inside its own polyethylene moderator.

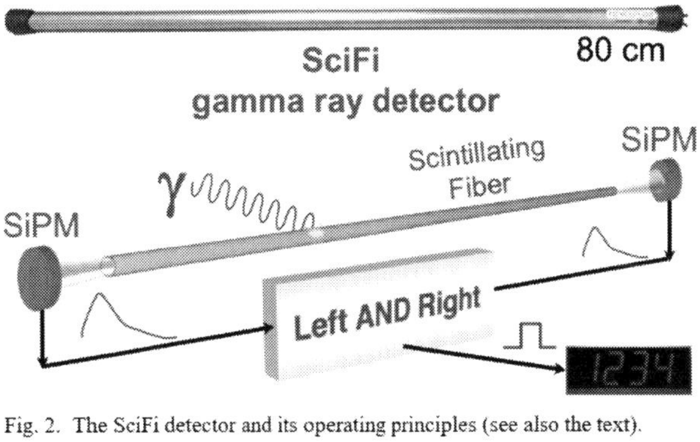

Fig. 2. The SciFi detector and its operating principles (see also the text).

A prototype of such a monitoring grid was initially tested with a mock-up drum and laboratory sources at ENEA Casaccia near Rome, Italy, then was installed and tested around real radwaste drums in a storage site on the ORANO premises at La Hague, France, and in another one at NUCLECO in Casaccia. A modular support structure, based on Bosch Rexroth aluminum profiles, was assembled in four units in order to accommodate four radwaste drums [4]. Each support unit hosted four SciFi and four SiLiF counters, for a total of 16+16 detectors. In Fig. 4 we show a high resolution spectrum of one

https://doi.org/10.1051/epjconf/202328806006

of the two SiPM photosensors of a SciFi detector exposed to a ^{137}Cs gamma source. The low energy background is indicated in blue, whereas the contribution due to the source, in red, shows a series of peaks testifying the intrinsic capability of the SiPM to detect discrete number of photons.

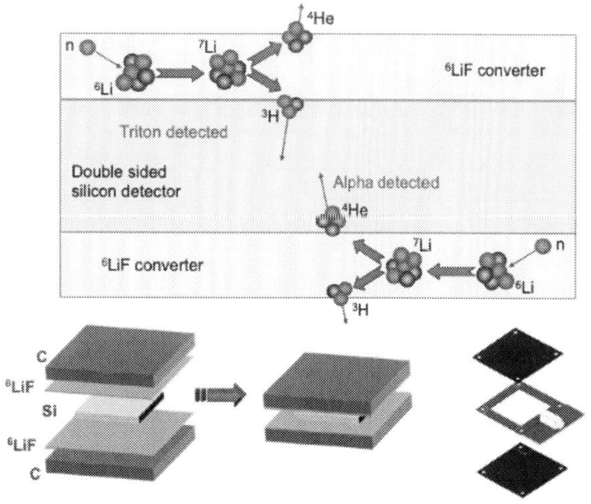

Fig. 3. The SiLiF detector and its operating principle (see also the text).

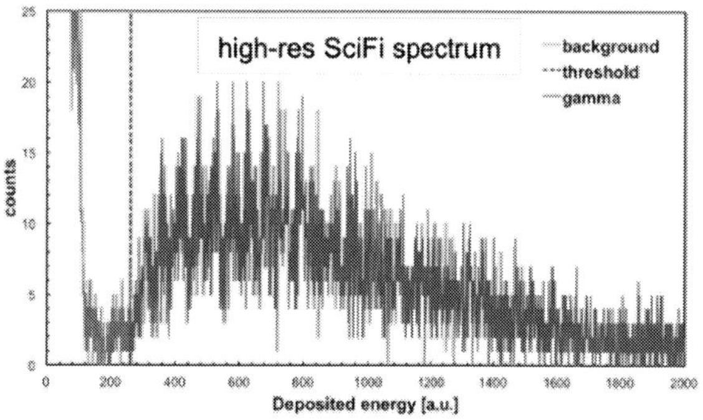

Fig. 4. High resolution spectrum of one of the two SiPM photosensors of a SciFi detector exposed to a ^{137}Cs gamma source.

In the real operating conditions a low resolution digitization is faster and more convenient. In Fig. 5 we show such a spectrum, acquired while installed close and far from a radwaste drum, with the indication of the chosen threshold and the contribution of the cosmic muons. Similarly, an example of a low resolution spectrum from a SiLiF detector installed close to a radwaste drum containing tiny amounts of actinides is shown in Fig. 6. The intrinsic detection efficiency of a SciFi for gamma rays between 200 and 2500 keV is between 1 and 2%, and the neutron detection efficiency of a SiLiF is typically 5%. In Fig. 7 we show a sketch and a picture of one modular support structure to be placed around the radwaste drums in order to host the monitoring detectors.

The preliminary test done with a suitable mockup drum (Fig. 8) provided clear indications of the feasibility of the monitoring system, even though one SiLiF and one SciFi did not work properly due to a strong electromagnetic interference from other equipment in the position where they had been installed in the laboratory. The count rates measured by the 3+3 gamma and neutron detectors were reported in Fig. 9. The same data were reported in Fig. 10 in two so-called radar plots also showing the position of the gamma and neutron radioactive sources.

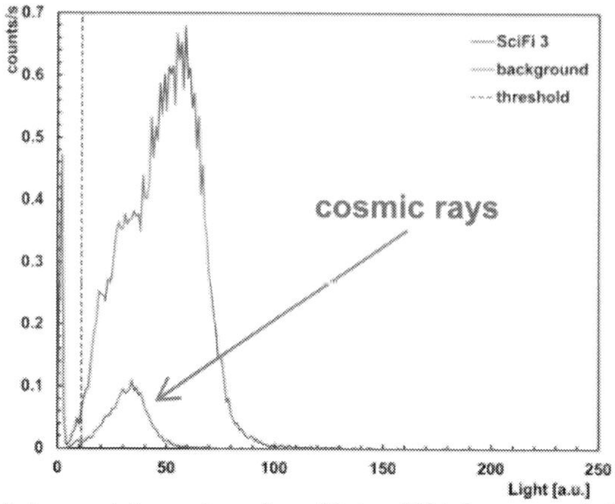

Fig. 5. Low resolution spectrum of one of the two SiPM photosensors of a SciFi detector close and far from a radwaste drum.

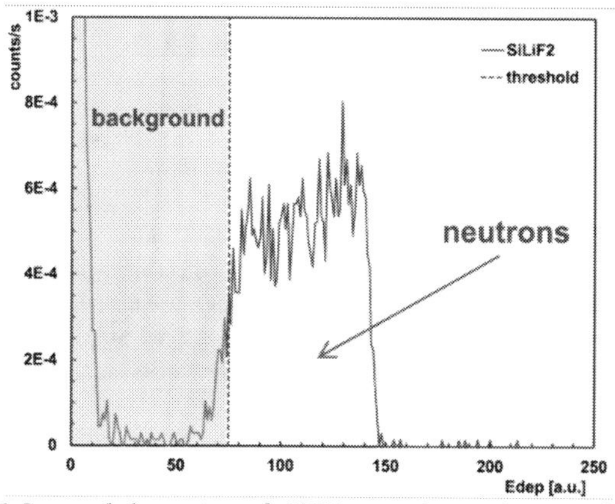

Fig. 6. Low resolution spectrum of a SiLiF neutron detector close to a radwaste drum containing tiny amounts of actinides.

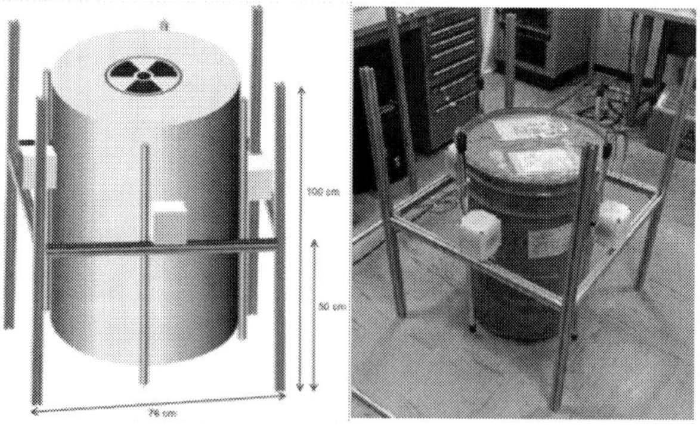

Fig. 7. The modular support structure to be placed around the radwaste drums to host the monitoring detectors.

II. Test at ORANO La Hague

The test was done on four 120-liter drums of Intermediate Level Waste containing between 13 and 25 grams of plutonium and 0.6 to 1 g of americium. The overall alpha activity content was of the order of 10^{11} Bq per drum. The nominal neutron emission ranged from ≈10000 to ≈40000 n/s, even though the neutron dose rate in contact, as measured by the local radiation protection experts using handheld devices, was declared << 1 μSv/h (basically below the instrument sensitivity). As for the gamma dose rate, it was measured between 50 and 200 μSv/h.

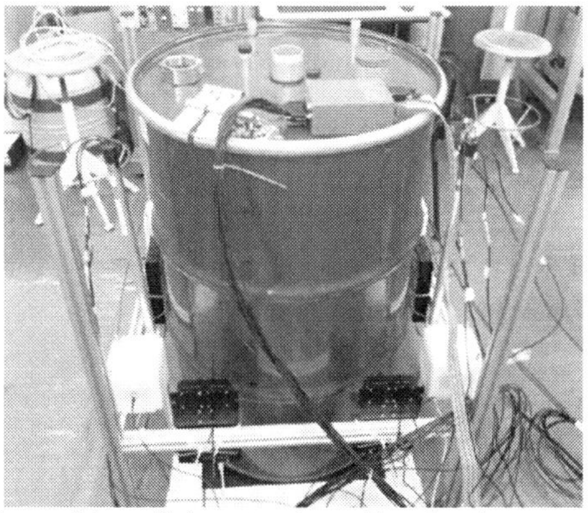

Fig. 8. Setup of the preliminary test done with radioactive sources inside a mockup drum.

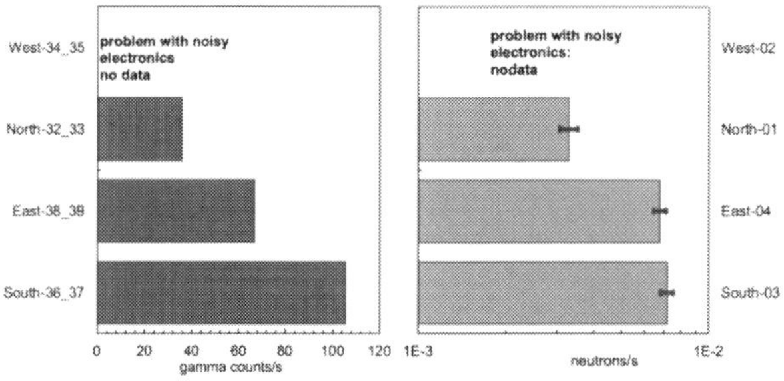

Fig. 9. Count rates measured around the mockup by the 3+3 SciFi and SiLiF.

Four support structures, hosting 16 SiLiF and 16 SciFi detectors in total, were assembled and arranged around the four drums, as shown in Fig. 11. After coping with several logistic difficulties, due to the limited time allowed inside the storage room because of safety issues, the data taking went on throughout 18 hours. The measured count rates are reported in the plots of Fig. 12 and Fig. 13. Interestingly, both plots show a remarkable anisotropy in the emission suggesting a relevant role of the matrix inside the drums which was generically indicated as "plastic". Moreover, the declared radioactivity content was mainly ^{238}Pu, ^{239}Pu, ^{240}Pu, ^{241}Pu, ^{242}Pu, ^{241}Am, with activity between 10^7 and 10^{11}, and the only relevant gamma emission came from ^{241}Am with ≈ 10^{11} Bq in each drum. The corresponding radar plots, not shown here, would further testify the mentioned anisotropic emission. Finally, in Fig. 14 we show as an example the gamma count rate timeline covering the duration of the test for one drum. A similar plot did not make sense for neutrons due to their very low count rate.

III. Test at NUCLECO Casaccia

The test was done on four 220-liter drums of (Very) Low Level Waste containing tiny amounts of actinides producing a much lower neutron flux than observed in the previous test. A weak gamma emission was to be expected mainly due to small amounts of ^{137}Cs, ^{241}Am and ^{235}U.

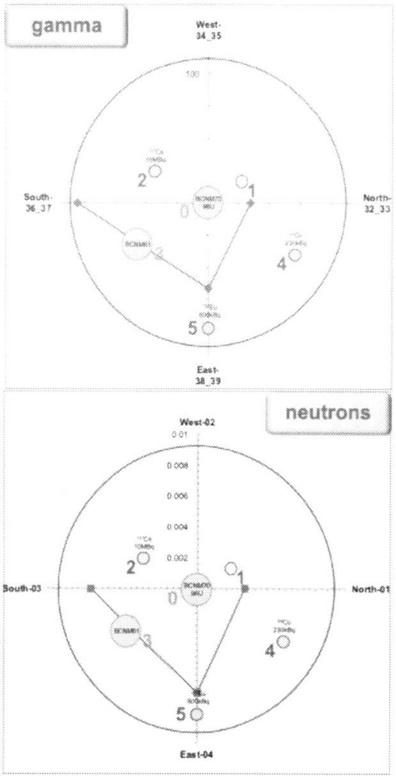

Fig. 10. Radar plots indicating the position of the sources (small blue/gray circles for gamma, big orange/gray circles for neutrons) and the count rates of SciFi and SiLiF reported in Fig. 9.

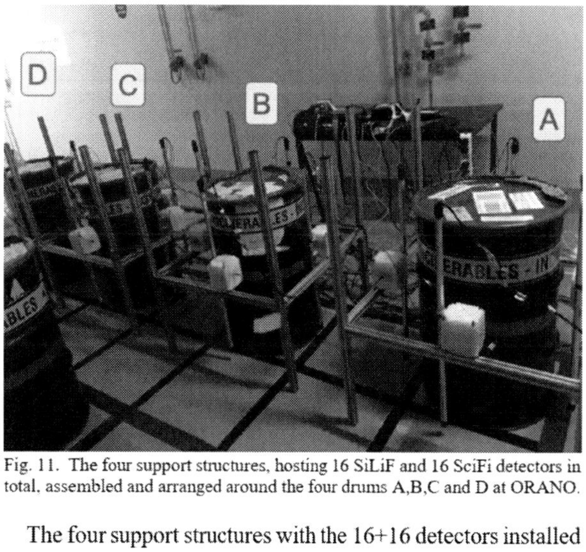

Fig. 11. The four support structures, hosting 16 SiLiF and 16 SciFi detectors in total, assembled and arranged around the four drums A,B,C and D at ORANO.

The four support structures with the 16+16 detectors installed around the drums at NUCLECO are shown in Fig. 15. The test went on quite stable throughout one month, and the average measured count rates for gamma and neutrons are reported in Fig. 16 and Fig. 17 respectively. We remark that the count rate observed with the SciFi sensors far from the drums, mainly due to cosmic rays, was < 0.5 counts/s. In Fig. 18 we show as an example the gamma count rate timeline covering the duration of the test for the drum B.

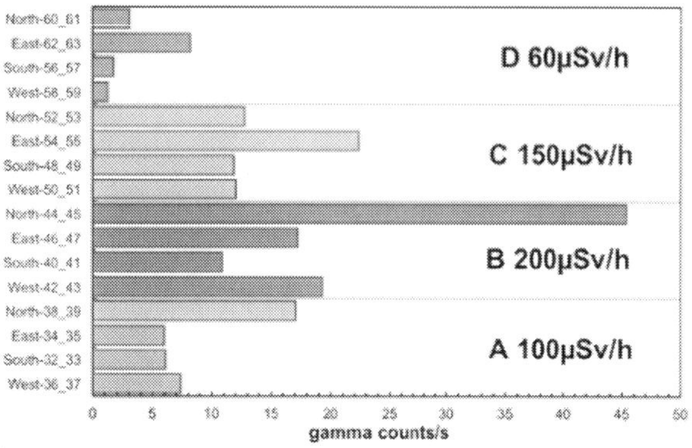

Fig. 12. Gamma count rates measured around the four ILW drums at ORANO.

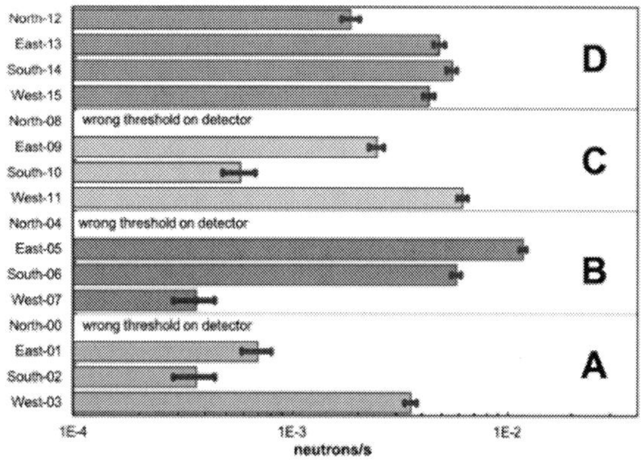

Fig. 13. Neutron count rates measured around the four ILW drums at ORANO.

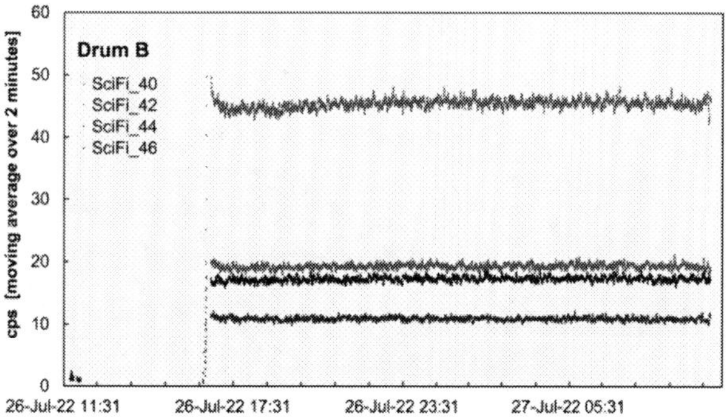

Fig. 14. Gamma count rate timeline around drum B at ORANO.

During the one-month a few ordinary activities were performed in the area close to the drums, including the activation of a huge X-ray machine and operations with a forklift around and above our four drums. Hints of these activities were reported by the SciFi sensors.

IV. Conclusions

Tests of the MICADO monitoring detectors were done on a mockup drum and sources, on four ILW drums at ORANO and on four (V)LLW drums at NUCLECO. Due to the small sensitive area and intrinsic detection efficiency of the detectors, namely 24 cm^2 with 1.8% efficiency for SciFi and 9 cm^2 with 5% efficiency for SiLiF, and the long possible counting time, the system can be employed in a wide dynamical range of radwaste activity, as proved by the obtained results. The detectors showed to be stable during the one-month test at NUCLECO, and the SciFi also proved to be sensitive to slight changes in the surrounding environment, possibly extending its perspective of application from safety to security.

Fig. 15. The four support structures, hosting 16 SiLiF and 16 SciFi detectors in total, assembled and arranged around the four drums A,B,C and D at NUCLECO.

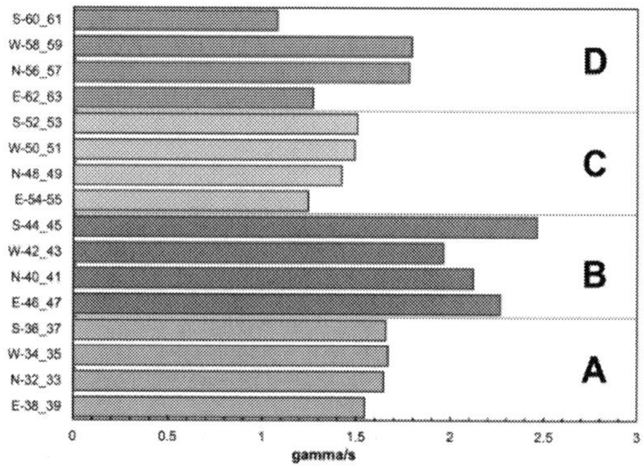

Fig. 16. Gamma count rates measured around the four (V)LLW drums at NUCLECO.

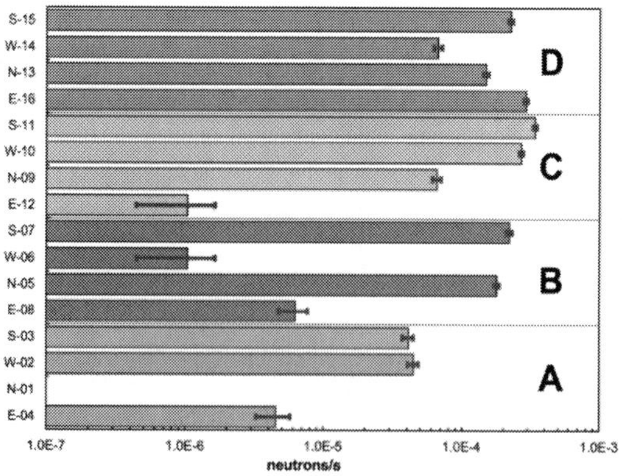

Fig. 17. Neutron count rates measured around the four (V)LLW drums at NUCLECO.

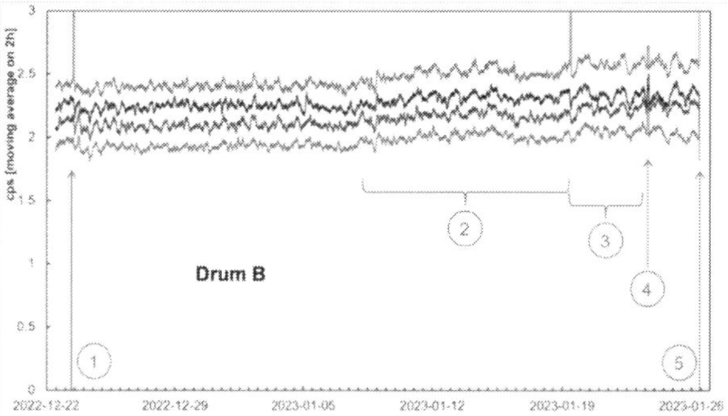

Fig. 18. Gamma count rate timeline around drum B at NUCLECO. (1): measurement start. (2): people operating with forklift around and above the drums. (3): X-ray machinery close to the drums in operation. (4): the console and the electronics were moved, live, close to the door. (5): measurement end.

ACKNOWLEDGMENT

The MICADO project (Measurement and Instrumentation for Cleaning and Decommissioning Operations) was funded within the framework of the European Union's Horizon 2020 research and innovation program under grant agreement No 847641. The authors are grateful to all the partners of MICADO, in particular Massimo Morichi, Erica Fanchini, Eric Breuil, Quentin Ducasse. A special thank goes to Marco Ripani, Abdallah Lyoussi and Paolo Peerani.

REFERENCES

[1] MICADO Project. Available online: https://www.micado-project.eu/ (accessed on 30 July 2023).
[2] Cosentino,L.;Giuffrida, M.; Lo Meo, S.; Longhitano, F.; Pappalardo, A.; Passaro, G.; Finocchiaro, P. Gamma—Ray Counters to Monitor Radioactive Waste Packages in the MICADO Project. *Instruments* 2021, 5, 19. https://doi.org/10.3390/ instruments5020019.
[3] Cosentino, L.; Ducasse, Q.; Giuffrida, M.; Lo Meo, S.; Longhitano, F.; Marchetta, C.; Massara, A.; Pappalardo, A.; Passaro, G.; Russo, S.; et al. SiLiF Neutron Counters to Monitor Nuclear Materials in the MICADO Project. *Sensors* 2021, 21, 2630. https://doi.org/10.3390/s21082630
[4] Bosch Rexroth profiles, https://www.boschrexroth.com

Hot-spots finding with modular gamma-ray system for sort and segregate activities

Gaetano Elio Poma[1*], Luigi Cosentino[1], Fabio Longhitano[2] and Paolo Finocchiaro[1]

[1]INFN Laboratori Nazionali del Sud, 95123 Catania, Italy
[2]INFN Sezione di Catania, 95123 Catania, Italy

(*) elio.poma@lns.infn.it

Abstract—**In the framework of the PI3SO project (Proximity Imaging System for Sort and Segregate Operations) funded by INFN-Energy project, aimed at the low and intermediate level radioactive waste (radwaste) classification, conditioning and characterization, a scanning system based on a set of 128 gamma-ray detectors based on CsI(Tl) crystal scintillators readout by a silicon photomultiplier have been built and characterized. These are arranged in linear arrays grouped in modules, each one being made of sixteen detection elements compose a sensitive unit, named module, and eight modules define the full-scintillators scanning system. The combination of four modules, arranged in a linear configuration, defines a sliding linear array named detection unit. Placed one above a scanning table and the other parallel to it underneath the table, the two co-moving detection units detect gamma-rays from sparse objects on table improving the detection of partially gamma sources and they will identify the hot-spot. The described modules are suitable for the sorting of nuclear materials and can be used as scanner of radioactive objects placed on a large area surface. Each unit was tested with point-like laboratory gamma-ray ^{137}Cs and ^{22}Na source: the results are promising in view of the integration with the mechanical system which will move the detection units scanning the radwaste. Next step are planned tests in a real environment.**

***Keywords* — radwaste; hot-spot; sort and segregate; SiPM; CsI; gamma-ray scanner.**

I. Introduction

The radioactive materials have to be handled in several fields, for instance radiation waste (radwaste) sorting and segregation [1,2], nuclear medicine [3,4], nuclear physics research and inspections [5,6]. Well defined prescriptions are given concerning material handling, radiation protection and safety. In the framework of radioactive material handling, like in radwaste sorting and segregating operations, the availability of an efficient and easy-to-handle tool to quickly detect gamma radiation hot-spots can be very useful. Furthermore, spectroscopic features, even if with moderate energy resolution, could provide significant benefits. The radwaste danger decreases with time, therefore the worldwide handling strategy is seal-and-isolate for as long as needed to reduce the activity, and this time can range from a few hours to tens of thousand years.

The IAEA radwaste classification defines 6 classes of radwaste based on overall activity, type and decay time, and a proper storage should be available at all stages in waste processing, for isolation and environmental protection as well as to facilitate retrival and monitoring [1]. In particular low (LLW) and intermediate (ILW) level waste represent the vast majority of the radwaste worldwide, that are stored into near-surface disposals [1].

As recommended by Innovative UK in conjunction with the Nuclear Decommissioning Authority, the development of innovative solutions i.e. integrated, autonomous toolkits which sort and segregate radwaste generated from nuclear decommissioning activities into optimized containers represents a priority [7]. Each adopted solution requires a tool to monitor the waste radioactivity with the priority in recycling over its disposal, to reduce sorting and segregation process carried out by humans and to increase productivity.

Many tools are commercially available to monitor the radioactivity, in particular gamma-rays which are the most ubiquitous form of radioactivity. Most of the current detection systems are environmental detectors or hand-held detectors. Newly developed compact low-cost radiation detectors, to be directly applied to the radwaste drums, could monitor nuclear materials in place and/or during transportation, detecting possible diversion.

A few projects are currently being developed for the automatic manipulation of nuclear waste, like ROMANS [8] and VIRERO [9] projects, that reduce the waste volume by disassembling and re-sorting radioactive waste material on the basis of the dose-rate levels. Based on system like these, a new compact gamma-ray scanning system named PI3SO (Proximity Imaging System for Sort and Segregate Operations) has been proposed and it is currently under construction.

II. PI3SO precursors

On the wave of existing projects for sorting and segregation of radwaste, two prototype detectors have been built at INFN-LNS [10] as demonstrators for the quick detection of hotspots. The first one is a 60x60 cm^2 table, named RADSIEVE, featuring an array of scintillating fibers, coupled to Silicon PhotoMultiplier (SiPM) at each end, whose counting rates can tag the presence of hot-spots (fig. 1). RADSIEVE demonstrator has limited performances due to the low detection efficiency of the fibers, poor space resolution, and the ambiguity in position reconstruction in case of more hot-spots.

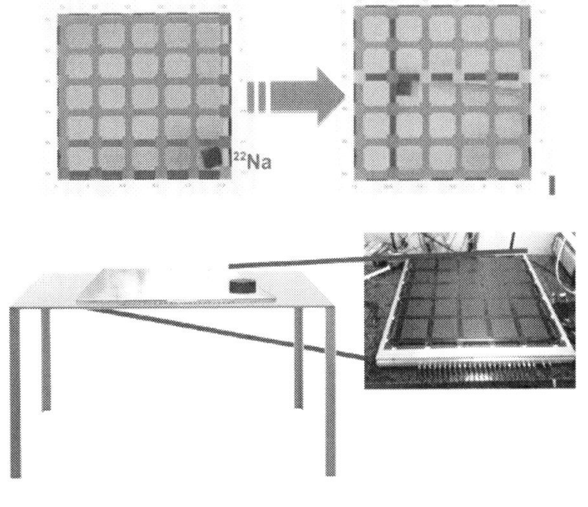

Fig. 1. RADSIEVE precursor: online gamma-ray counting by moving a ^{22}Na source (top); detector sketch and photo (bottom).

A second demonstrator, named SCINTABLE, was built as a proof-of-principle in order to overcome the RADSIEVE limitations: it consists of a hundred CsI(Tl) scintillators 1 cm^3 with spectroscopic capability, arranged in a 10x10 cm^2 array and read-out by a 10x10 SiPM array [11].

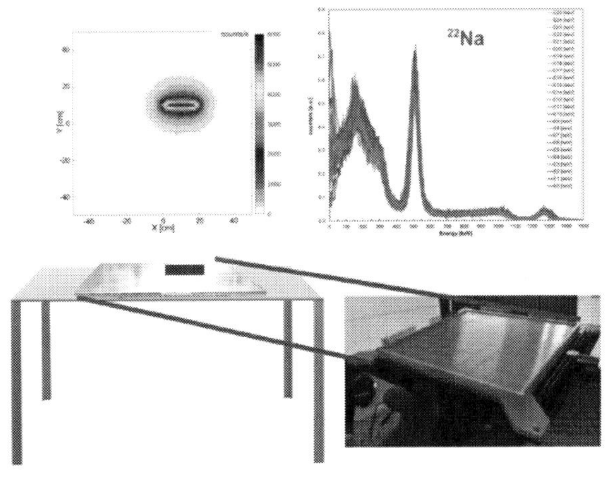

Fig. 2. SCINTABLE precursor: gamma-ray imaging (left, Monte Carlo simulation) and spectra (right) with ^{22}Na source (top); detector sketch and photo (bottom).

Based on Monte Carlo simulations and acquired data (fig. 2, top) to characterize the performances of proximity imaging, counting and spectroscopic capabilities [11], SCINTABLE proved the feasibility of the technology and paved the way to the innovative PI3SO system.

PI3SO system has been proposed as SCINTABLE-like detector with the aim of speed up the sorting process of the radwaste material and to reduce operations with human exposure.

The technology explored at INFN-LNS was proved suitable to build a modular gamma detection system as part of a work desk. Realistic configurations are needed as a trade-off between complexity, detection efficiency, counting capability and spectroscopic features.

III. LINEAR GAMMA-RAY ARRAY

Designed as a sensitive table for reclassification of radwaste, PI3SO system is equipped with a reasonable operational 84x128 cm^2 desk surface: this would imply that to make the whole surface sensitive to the radiation, a thousand of 1 cm^3 scintillators are required, which is something too expensive in terms of costs, DAQ and time consuming.

In order to quadratically reduce the complexity of the system and build a near-final device that can be easily employed in real radwaste handling operation, a simplified linear array of scintillators is proposed (fig. 3, 4).

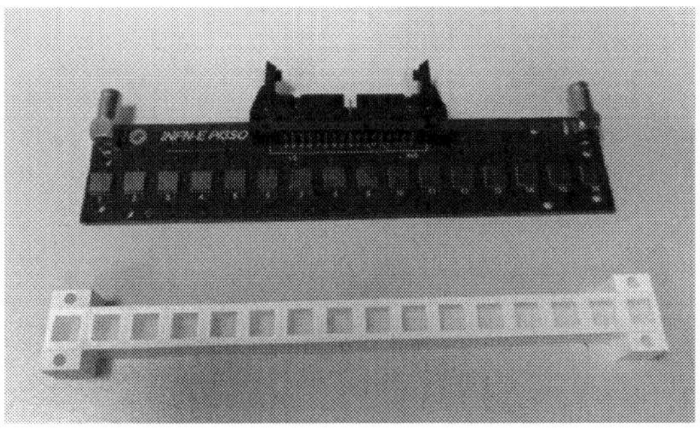

Fig. 3. A PI3SO printed circuit board equipped with 16 SiPMs (top) and 16 CsI(Tl) crystal unit (bottom, white array) compose a module: four modules compose one linear array or detection unit.

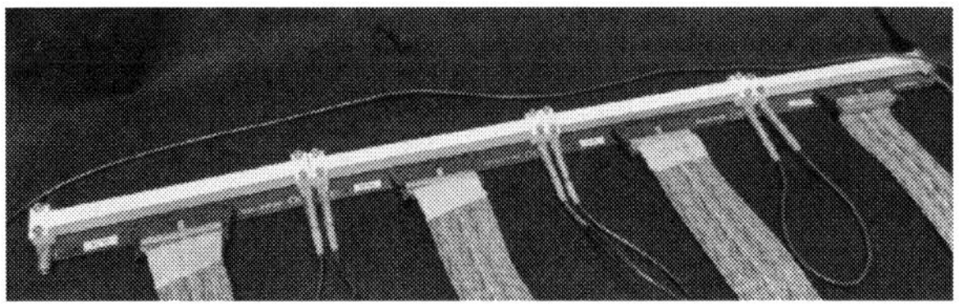

Fig. 4. A detection unit with four flanked modules, arranged for DAQ and characterization tests.

Specifically, the tool is composed of two linear arrays named detection units, top and bottom, composed each by 1x64 channels (CsI(Tl)+SiPM) and a work table for the radwaste support: the bottom array is be placed beneath the table and will stay fixed, while the upper one hovering will be moved up and towards the materials placed on the work table, guided by a led-based proximity sensor.

These arrays scan the total table length in 1 cm steps by means of two handling engines which provide also the vertical translation of the linear array in order to avoid possible collisions with the deposited materials. Based on statistical requirements, the acquisition times can be suitably set.

The sensitive element of each detection unit is a CsI(Tl) 1 cm^3 crystal [12]. Each crystal is coupled to a 6x6 mm^2 S14160-6050HS Hamamatsu SiPM by means of optical grease [13], and an array of 16 elements compose a module (fig. 5). The chosen size is a trade-off between detection efficiency, position resolution and compatibility with the SiPM.

The CsI(Tl) has a very good light yield of about $6x10^4$ photons/Me, with a spectral emission centered around the wavelength of 550 nm, and is negligibly hygroscopic. The spectral match between the scintillation light and the SiPM sensitivity gives rise to an overall PDE of about 20%.

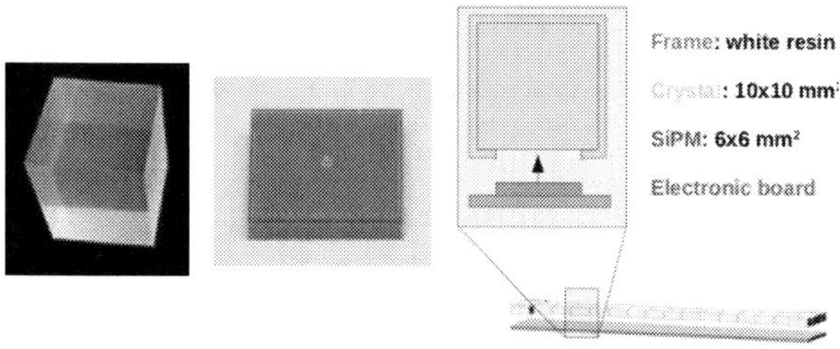

Fig. 5. Arrangement of elements for a detection unit made by a CsI(Tl) 1 cm^3 crystal [12] and a 6x6 mm^2 S14160-6050HS SiPM Hamamatsu [13]. A replica of 16 elements makes a detection module, fig. 3.

The CsI(Tl) high density of 4.51 g/cm^3 guarantees a good intrinsic detection efficiency. The 1 cm thickness provides a reasonable interaction probability of about 30% (20%) for 600 (1500) keV gamma rays.

SiPMs are biases at low voltage, typically 41.3 V (+2.8 V overbias above breakdown), and the overall 128 output signals are directly connected to two CAEN VX2745 64-channel digitizers operating in charge integration mode [14]. Each channel samples its input waveform at 125 MS/s 16 bits and the data are processed by onboard FPGA, which evaluates the pulse height of the detected signal.

Measurements with gamma sources have been done in order to evaluate its performances in terms of single photon detection capability of SiPM, spectroscopic response, energy resolution, counting rate and peak-to-total spectral efficiency [10].

The measured Full-Width Half-Maximum (FWHM) energy resolution, referred to the 511 and 1275 keV (^{22}Na) peaks, is 9.7% and 5.8% respectively, in line with typical values reported in literature as well as the 23% peak-to-total efficiency. Due to the decay time of the scintillation light (3 μs), the maximum counting rate permitted in the CsI(Tl) is of the order of 30-50 kHz [10]. This could represent a limitation in case of very high activity because of pile up effects, therefore if necessary future configurations with faster scintillators will be take into account.

PI3SO feature three different and dependent operational modalities:

- as an overall gamma radiation counter, where the global counting rate is the sum of the individual counting rates;
- as a proximity imaging system (1 cm^2 pixels);
- as an array of spectroscopic detectors.

In the second and third mode the data from individual channels can be summed, if needed, to improve the statistics at the price of a worse space resolution. The three modes are meant to be used in sequence: first, a coarse and quick counting measurement followed, if a predefined threshold is overcome, by a slower hot-spot search; finally, in case hot spots are found, a finer spectroscopic investigation to look for specific isotopes and/or to perform an energy selective imaging based on full energy gamma detection with reduction of Compton background.

IV. Radwaste scanner

The PI3SO detection units described above, will face each other with an operational table in between, in order to locate the gamma hot-spots, reconstruct their energy spectra and compute the isotope-related images (fig. 6). The need of two detectors arises from the possibility that flat metal objects could be activated on one side only or could shield other objects.

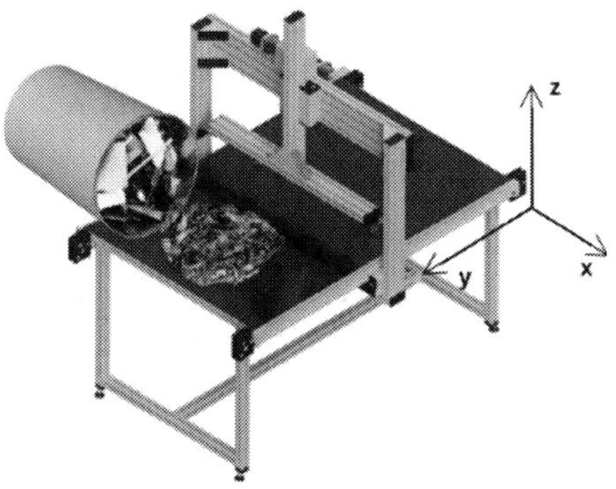

Fig. 6. Sketch of PI3SO scanner: spread radioactive objects, placed on large surface, are scanned by top and bottom linear arrays in XY plane, approaching radwaste during scan along Z direction.

The main application for PI3SO is in the framework of radwaste sorting and repackaging, that involves the opening of drums and sorting their content in order to reduce the volume to be stored. In this application the waste would be placed on top of the work table and the two detector arrays would scan the objects assessing their activity (fig. 6). This would drive the decision on whether an object can be released or should be repacked and stored.

In fig. 6 the configuration of the PI3SO radwaste scanner

based on a fixed table is sketched, with a CCD camera on top and two linear arrays of detectors. The bottom array, almost in contact with the table, is moved in the horizontal direction, synchronized with the upper one. The latter is installed on an up-down actuator in order to be placed as close as possible to the materials to be inspected, according to the indications of a proximity sensor.

The scan is performed in steps of 1 cm and the data are acquired for predefined time intervals. In alternative the acquisition time at each step could be defined depending on the counting rate. In case a hot-spot is detected, the operator can decide is perform a finer spectroscopic analysis.

In fig. 7 a sketch of the radwaste scanner that is going to be developed by a private company is shown. The scanning area is 84x128 cm^2 and the table is made of carbon fiber (1-2) mm thick for a gamma-rays absorption as low as possible. The table is watched by a CCD visible camera, to overlap the visible light image with the gamma-ray image for improving the hot-spot identification. The Anti Collision System (ACS) allows a step by step vertical movement of the upper array, depending on the height of the objects onto the table. It is necessary for reducing as much as possible the distance of the detectors eliminating the collision risks.

A customized software using DAQ digitizers libraries is going to be developed starting from PSD firmware of VX2745 CAEN digitizer [14].

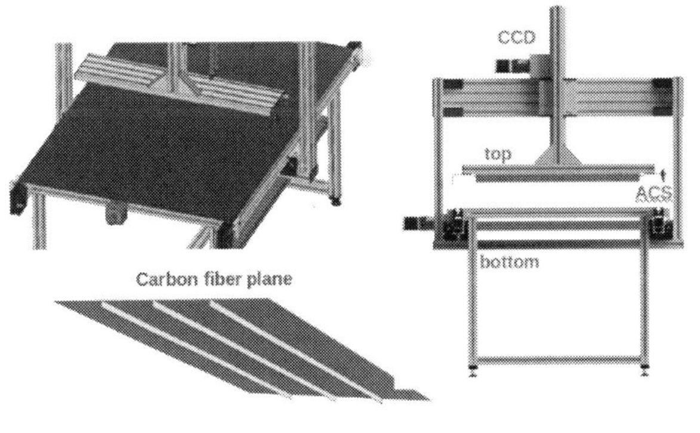

Fig. 7. PI3SO radwaste scanner by SIATEL. 3D view (top-left), carbon fiber plane (bottom-left), side view (right). A CCD camera, top and bottom detectors and the anti-collision system (ACS): it prevents the hitting of top line with the objects placed on the table.

The full development of the detector is carried out at INFN-LNS, including the Monte Carlo simulations in FLUKA [15] required to design a complete model of the system. The SIATEL company [16] has in charge the development of mechanics and controls of the motorized stages, included managing software (fig. 6, 7).

V. Preliminary results

A test campaign with gamma-ray sources together with FLUKA simulations allowed to study the performances of PI3SO as proximity imaging system. In details, the aim of the tests is to characterize the detection elements in terms of energy resolution, hot-spot reconstruction and signal-to-noise ratio (S/B ratio) as a function of the channel threshold.

In fig. 8 the scanner implemented in FLUKA framework is shown with the two linear arrays marked in green. In order to speed-up the simulation, a 64x120 cm^2 matrix detector has been implemented as the corresponding of a single linear 64x1 cm^2 detector (fig. 9). In this way, one simulation by N_{tot} particles represents a complete scan i.e. 120 steps and $N_{tot} = N_{single}$ x 120 primary particles simulated.

Simulations consider a single gamma source of 662 keV with activity 10 kBq, 1 cm diameter, isotropically emitted. Source-to-detector distance was varied from 1 to 20 cm, with and without 1 mm carbon fiber in order to evaluate its contribution in terms of gamma-rays absorption and image degradation. Detectors top (det-2) and bottom (det-1) are shown in fig. 9.

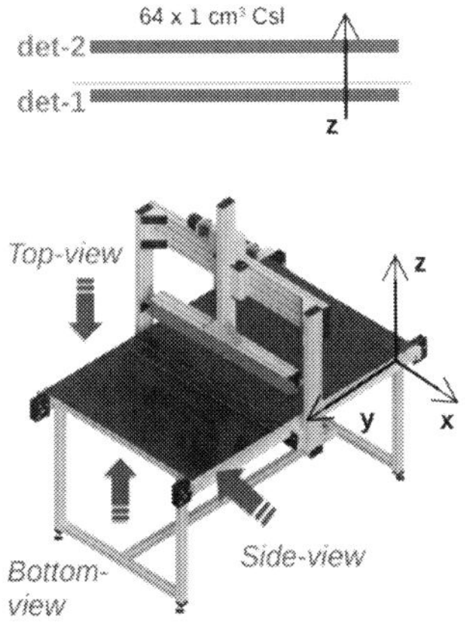

Fig. 8. PI3SO radwaste scanner implemented in FLUKA framework: linear array top (det-2) and bottom (det-1) on top (in green); scanner views on bottom.

By varying the source-to-detector distance (1-20 cm), the hot-spot image reconstruction was evaluated in terms of FWHM (fig. 11) showing the role of the source-to-detector distance to localize the hot-spot. As shown in fig. 10, although no collimator has been used, at a distance of 1 cm the system is able to localize and operate a rough imaging of the hot-spot by employing 20 minutes of scanning time i.e. 10 seconds for each step.

This demonstrates that a reduced source-to-detector distance is very important to localize the hot-spot and identify it. The contribution of few mm carbon fiber plane in terms of image degradation and gamma-rays reduction is negligible.

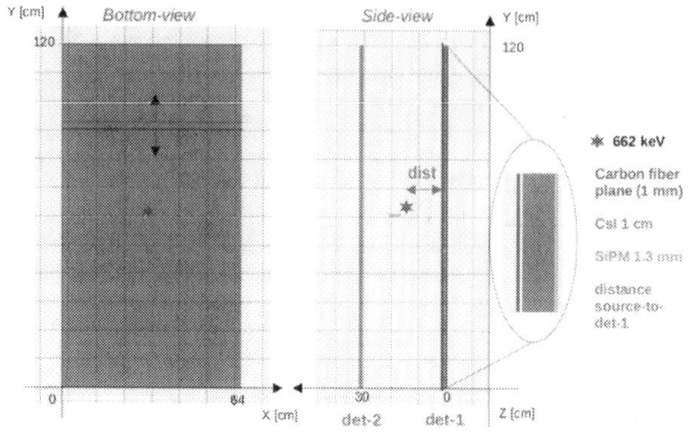

Fig. 9. Sketch on PI3SO scanner implemented in FLUKA framework in bottom (left) and side (center) views, and a zoom of detector with its proprieties (right).

As shown in fig. 11, which report the values of the FWHM of the hot-spot as a function of the source-to-detector distance, hot-spot is well-reconstructed at small distance, worsting at increased distance, which testified the importance to approach objects at each step during scanning.

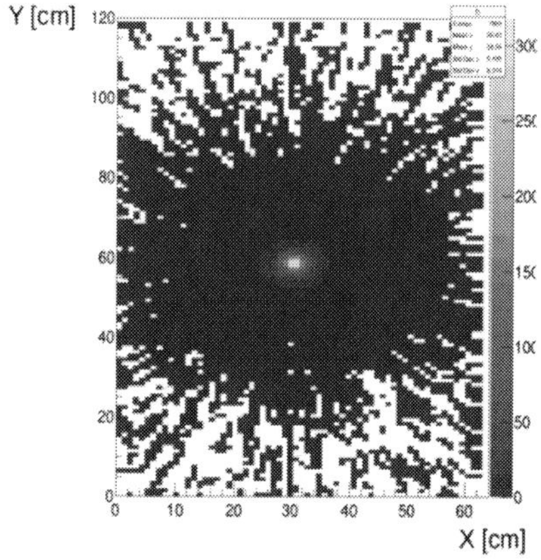

Fig. 10. Example of hot-spot image reconstruction in proximity imaging, by simulating a 10 kBq 662 keV ^{137}Cs source (1 cm diameter sphere, iso distributed) for 10 sec at each step (1 cm along Y), and a total 20 minute scan. Source-to-detector distance is 1 cm.

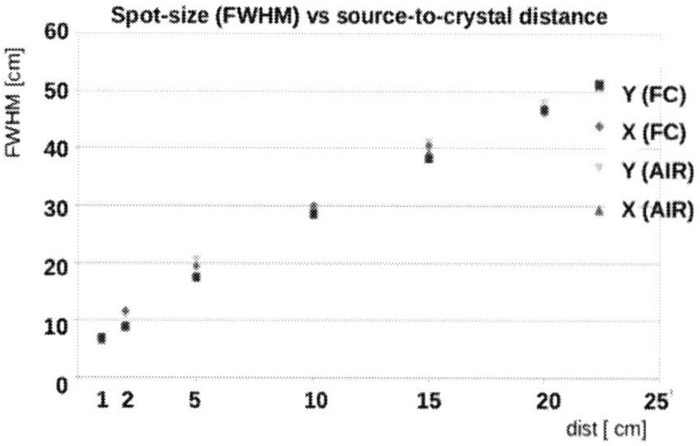

Fig. 11. Hot-spot reconstruction as FWHM as function of source-to-detector distance, by varying the support plane (FC fiber carbon, AIR no plane).

In order to study the system performance in terms of energy calibration and S/B ratio, CsI(Tl)+SiPM characterization has been done in laboratory using two gamma-ray sources using two gamma-ray source: ^{22}Na (662 keV photons) and ^{137}Cs (511, 1275 keV photons), both with 20 kBq activity. S/B ratio represents the quality parameter of a peak and is proportional to the square root of the number of scans.

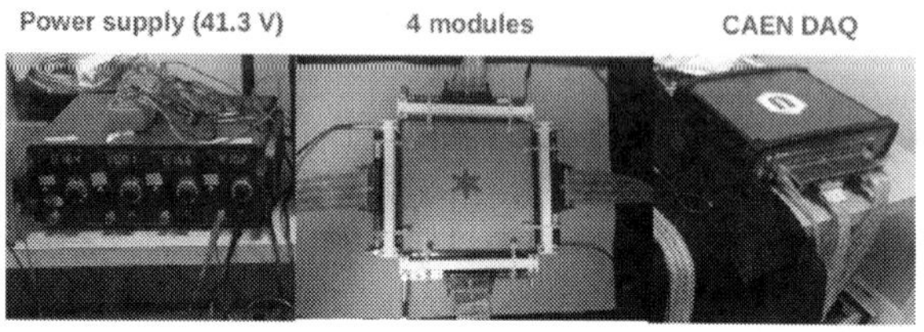

Fig. 12. Setup for energy calibration's test. A power supply (left) at 41.3 V feed 64 channels at a time of four modules (center), that are connected to digitizer for acquisition.

By using the setup shown in fig. 4, but with the modules arranged in squared configuration, some tests have been done for two set of modules (4 at a time). In particular, the energy calibration has been done for all 128 channels by means of the setup shown in fig. 12, placing both sources at the center of the square. Background spectra have been also acquired to evaluate the detector performance in terms of S/B ratio.

Due to the high number of channels and considering the presence of three peaks in the energy spectrum obtained with from ^{22}Na and ^{137}Cs, an automatic energy calibration procedure for peaks finding and energy reconstruction has been implemented. In fig. 13 an example of the acquired spectrum in channel and energy are show, with the corresponding energy resolutions of the peaks.

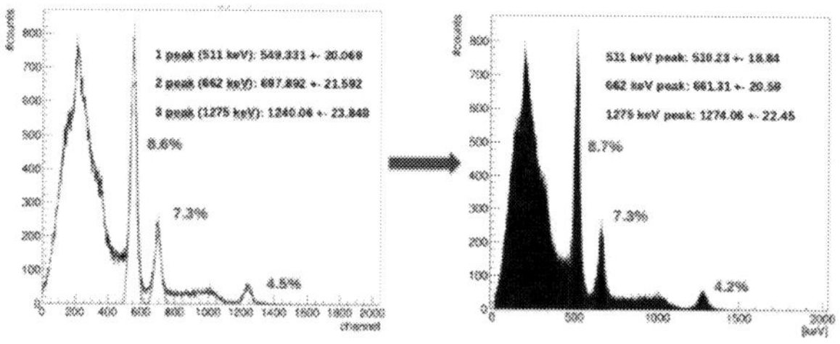

Fig. 13. Energy calibration: automatic peaks finding (left), reconstructed calibrated spectrum (right) for one channel.

In particular, the 662 keV peak resolution is 7.3 %, in line with the expectation from existing literature [11].
In order to evaluate detector performance, the energy spectrum quality has been studied for all the channels in terms of S/B ratio as a function of the energy threshold.

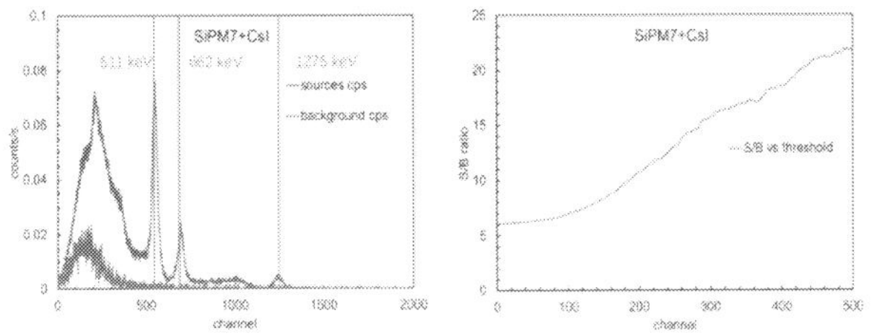

Fig. 14. Energy spectrum (left) for source and background runs; S/B ratio as function of channels by varying threshold in [0, 500] channels (right).

In fig. 14 left, the spectra acquired with the source and without the source (background) is reported, considering the S/B ratio from 6 to 22 as function of channel threshold.
Fig. 14 right shows the S/B ratio as a function of the channel threshold, ranging from 6 to 22. By choosing a threshold close to 200, S/B ratio is close to 10 which assures well-resolved small peaks.

Fig. 15 shown the energy resolution of the 662 keV peak for all the SiPMs.

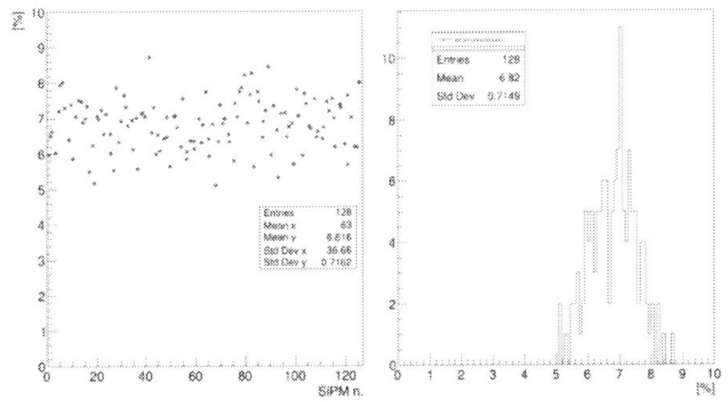

Fig. 15. 662 keV relative energy resolution for all 128 SiPMs: its behavior as a function of # SiPM (left), 1D distribution (right).

The mean value is 6.82 % with sigma 0.71 %. These values evidence that the automatic fitting procedure and crystals response are reliable.

VI. Conclusions

A modular and portable gamma-ray system was described as radwaste scanner for radioactive objects characterization, proximity imaging and selective energy hot-spot findings.

Tests of the PI3SO scanning detectors were done by means of Monte Carlo simulations and experimental tests in order to study its ability to reconstruct hot-spots, proximity images and energy spectra in terms of energy resolution and signal-to-noise ratio.

Due to the large sensitive area and intrinsic detection efficiency of the detectors, as well as reasonable short scanning time for large surface area, the system can be employed in a wide range of radwaste size, shape and radioactive nature, as proved by the obtained preliminary results. The system could also be extended in future with complementary alpha/beta detectors.

PI3SO system can be used in several ways, due to its versatility and modularity, as:

- radwaste sorting and segregation system;
- inspector of large surfaces, floors or walls in decommissioning and/or decontamination of nuclear installations;
- safeguards inspections by IAEA to detect and localize out of control and not declared gamma-emitting materials, as well as unintentional spilling or spread of contaminated materials;
- contamination monitor in nuclear facilities or laboratories;
- supplementary system to the safety baggage scanner in airports.

Acknowledgment

The PI3SO project (Proximity Imaging System for Sort and Segregate Operations) was approved and funded by INFN-Energy. The authors are grateful to all the components.

References

[1] IAEA International Atomic Energy Agency. Available online: https://www.iaea.org/topics/processing (accessed on 18 May 2022).
[2] Status and trends in spent fuel and radioactive waste management. IAEA nuclear energy series, NW-T-1.14 (Rev. 1), IAEA, Vienna, 2022.
[3] D.L. Bailey, D.W. Townsend, P.E. Valk, M.N. Maisey. Positron-Emission Tomography: Basic Sciences; Springer: Berlin, 2005.
[4] A. Granov, L. Tiutin, T. Schwarz, Positron Emission Tomography; Springer: Berlin, 2013.
[5] Handbook for regulatory inspectors of nuclear power plants. IAEA-TECDOC-1867, IAEA, Vienna, 2019.
[6] Functions and processes of the regulatory body for safety. IAEA Safety Standard, General Safety Guide, GSG-13, IAEA, Vienna, 2018.
[7] Guidance "Sort and segregate nuclear waste: specification", 27 July 2020, https://www.gov.uk/government/publications/sort-and-segregate-nuclear-waste-specification/sort-and-segregate-nuclear-waste-specification
[8] Romans project, https://cordis.europa.eu/project/id/645582
[9] VIRERO project, https://www.faps.fau.eu/curforsch/virero-virtual-remote-robotics-for-radiometric-sorting
[10] LNS-INFN, https://www.lns.infn.it/it/
[11] F. Longhitano, G. E. Poma, L. Cosentino and P. Finocchiaro. 2022. "A Scintillator Array Table with Spectroscopic Features" Sensors 22, no. 13: 4754. https://doi.org/10.3390/s22134754
[12] Hilger crystals, CsI(Tl), https://www.hilger-crystals.co.uk/materials/thallium-doped-caesium-iodide/
[13] Hamamatsu, MPPC S14160-6050HS, https://www.hamamatsu.com/eu/en/product/optical-sensors/mppc/mppc_mppc-array/S14160-6050HS.html
[14] CAEN digitizer VX274, https://www.caen.it/products/vx2745/
[15] FLUKA software, http://www.fluka.org/fluka.php
[16] SIATEL company, https://www.linkedin.com/in/paolo-nicotra-24500188/?originalSubdomain=it
[17] Rossi, F.; Cosentino, L.; Longhitano, F.; Minutoli, S.; Musico, P.; Osipenko, M.; Poma, G.E.; Ripani, M.; Finocchiaro, P. The Gamma and Neutron Sensor System for Rapid Dose Rate Mapping in the CLEANDEM Project. *Sensors* **2023**, *23*, 4210. https://doi.org/10.3390/s23094210

Neutron coincidence counters for disarmament verification: experiments and modeling

Alessandro Borella[1,*], Yurdunaz Çelik[1] and Riccardo Rossa[1]
[1] Belgian Nuclear Research Centre SCK CEN, Boeretang 200, B2400 Mol (Belgium)

(*) alessandro.borella@sckcen.be

***Abstract*—A measurement campaign in support of nuclear disarmament verification was carried out in 2019 on the premises of the Belgian Nuclear Research Centre, SCK CEN. Well-characterized MOX sources with different types of shielding materials were used during the campaign. The relative content of Pu was up to 14%$_{wt}$ and the isotopic abundance of ^{239}Pu was up to 93%$_{wt}$. One of the technologies that were deployed was neutron coincidence counting with ^{3}He based slab counters.**

The obtained background and dead time corrected total and reals rates were used to determine the mass of the assayed sample by using the so-called point model equations. The obtained masses were systematically underestimated by 15 % in average. A calibration approach based on Monte Carlo simulations where the mass is estimated from the calculated number of reals per spontaneous fission event was also used. A Monte Carlo model of the measurements setup was therefore developed to account for the finite source size, multiplication, moderation and absorption in the detector and scattering in the environment, including the measurement room walls, floor and ceiling. By using a Monte Carlo based efficiency for the reals the masses were slightly overestimated by typically few percent. However, outliers were observed.

In an effort to understand the obtained results, additional calculations were carried out to estimate the impact due to both the neutron emission by (α,n) reactions and spontaneous fission decay on the measured observables. The associated total and reals counts were obtained for various configurations and compared with the experimental data. The obtained results indicate the importance of (α,n) reactions when neutron multiplication is not negligible.

***Keywords* —Neutron coincidence counting; Monte Carlo modeling; Point model; Disarmament verification**

I. Introduction

WITHIN the framework of the activities of the International Partnership for Nuclear Disarmament Verification (IPNDV) [1], an international measurement campaign took place on the premises of the Belgian Nuclear Research Centre (SCK CEN) in the fall of 2019 [2]. The aim of the exercise was testing and comparing different non-destructive assay (NDA) techniques to be potentially used for the verification of nuclear material in the framework of the dismantlement of nuclear weapons. The measurements were carried out with well-characterized Mixed Oxide (MOX) fuel with different compositions, mass and shielding configurations. The gathered data would allow to draw conclusions on the capability of the considered NDA techniques to: verify the presence or absence of nuclear material originating from a nuclear weapon; distinguish weapon grade from civil grade nuclear material. One of the techniques of interest is neutron coincidence counting. Passive neutron counting is one of the techniques envisaged within the new START treaty for absence measurements [3].

In this work we report about the results obtained with ^{3}He based slab counters for the considered MOX fuel assemblies. We first introduce neutron coincidence counting and how it can be used to determine the mass of material in the assayed material either by using the so-called point model approach or a Monte Carlo based approach. We will then present the measurement campaign by describing the used detectors and electronics and the measured MOX fuel assemblies. The analysis of the data with the proposed models follows, including a description of the used Monte Carlo model. A comparison of the obtained masses with the nominal ones is given as well as additional simulations that were carried out to interpret the results.

The uncertainties in this paper are quoted at a 68% confidence limit. We quote the uncertainties between parentheses as recommended in paragraph 7.2.2 of Ref. [4].

II. Neutron Coincidence and Multiplicity Counting

The measurement of time correlated neutrons is a non-destructive technique that allows assaying material containing radionuclides undergoing spontaneous fission and has applications in the field of safeguards [5], waste and disposal [6] and disarmament verification [7]. Following a spontaneous or induced fission event, one or more neutrons are emitted simultaneously. Correlated neutron counting takes advantage of the simultaneous emission of fission neutrons to separate them from neutron emitted by other reactions such as (α,n) reactions where only one neutron is emitted.

Neutron coincidence counting allows estimating the mass of a sample containing material undergoing spontaneous fission provided that the radionuclide vector is known [8]. The technology measures the time correlation between events measured with neutron detectors. In addition to determine the total counts *T*, the so-called *R+A* (reals plus accidentals) and *A* (accidentals) are determined from the number of detected neutrons in two time windows or gates that are opened every time a neutron is detected. In this work, the gates for the calculation of the *R+A* and *A* events have the same duration of 64 μs. The first gate opens 4.5 μs after the trigger and there is a time interval (delay) of 1000 μs between the gates. The concept is illustrated in Fig. 1., where the so-called Rossi-Alpha distribution (RAD), which represents the time distribution of neutron detections after an arbitrary neutron detection, is shown for a ^{252}Cf source.

In addition to measuring the total counts in the two gates (neutron coincidence counting), the multiplicity distributions in the two gates can also be measured with dedicated systems; in this case one talks of neutron multiplicity counting. From the multiplicity distributions, one can derive the moments of the distribution. The moment of order zero represents the total counts, while the *R+A* and *A* terms corresponds to the moments of order one of the multiplicity distributions in the two gates, respectively.

The rate of reals or double *D=(R+A) –A* is a measure of the amount of material emitting neutrons by spontaneous fission. In this work, to determine the mass of the assayed sample we apply the so-called point model and a calibration approach, where the calibration factor is determined by Monte Carlo simulations.

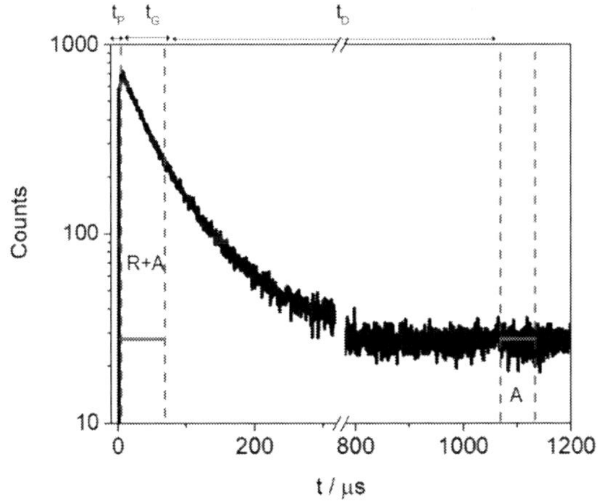

Figure 1. Rossi-alpha distribution from the measurement of a ^{252}Cf source.

A. Point Model approach

Starting from basic principles, theories were developed [9,10,11] that express the observables of neutron coincidence or multiplicity measurement as function of the mass of neutron emitting radionuclides and other parameters. This model, called the point model, can be applied in case the assayed sample has point geometry, no neutrons return from the detector to the sample, the time response function of the moderator detector assembly is a pure exponential function with decay constant, and there are no dead time losses of signals. Other conditions also apply but we refer to the references for the full description.

A generalization of the point model equations to a generic source of n radionuclides can be found in [12]. In absence of multiplication, the theoretical estimates from the point model (PM) of Hage [11] are the following for the totals and reals:

$$T_{PM} = t\varepsilon F_s \nu_{s(1)}[1+\alpha] \tag{1}$$

$$R_{PM} = t\varepsilon^2 F_s \nu_{s(2)} \tag{2}$$

where ε is the detection efficiency, α is the ratio between the neutron emission due to (α,n) reaction and spontaneous fission, F_S the number of fission per seconds, ν_{s1} and ν_{s2} are the 1st and 2nd factorial moments of the distribution for the neutron emission through spontaneous fission, *t* the measurement time.

The corresponding experimental observables from the recorded pulse train are the total count rate *T'* and the measured double rate *D'* corrected by the *f* factor that accounts for the finite duration of the trigger gate and the presence of a predelay:

$$\frac{(R+A)-A}{tf} = \frac{D'}{f} \tag{3}$$

Where

$$f = \int_{t_p}^{t_p+t_g} \frac{e^{-t/\tau}}{\tau} dt = e^{-t_p/\tau}\left(1 - e^{-t_g/\tau}\right) \tag{4}$$

and *t* is the measurement time. The given expression for *f* holds only for an exponential RAD.

B. Monte Carlo based calibration approach

The mass determination of the assayed sample can be also done by using a calibration approach, where the double rates are related to the mass of assayed sample (typically expressed as $^{240}Pu_{eq}$) by a calibration coefficient [13]. This coefficient can be determined by using reference materials. If the geometry or chemical composition of the assayed samples differ from the ones of the reference materials correction factors need to be introduced, for example to account for the multiplication or hydrogen content [13]. The calibration coefficient can also be obtained from Monte Carlo simulations as explained in [8, 14]. One of the main advantages of this approach is that the impact of the geometry of the sample, the gate occupation factor *f* and, if the mass is known, the multiplication factor are accounted for in the simulations. Some knowledge about the composition of the source is however required in order to develop the model.

We have implemented a calibration approach where the number of doubles per fission event ε_D is determined by Monte Carlo simulations [8,12]. This term implicitly accounts for the occupation factor associated to the experimental conditions as well as the multiplicity distribution associated to the spontaneous fission in the source. The composition of the radionuclide source must be known to determine ε_D.

Therefore one can write:

$$D' = \varepsilon_D\, m \sum_j f_{m,j} F'_{S,j} \tag{5}$$

Where $F'_{S,j}$ is the spontaneous fission rate per gram and $f_{m,j}$ is the mass fraction of the radionuclide j, and m mass of the sample.

Since the source composition may not be known at the moment when the Monte Carlo calculation is done, one could use a reference radionuclide (e.g. ^{240}Pu) for which the $\varepsilon_{D,REF}$ is calculated.

One can write then

$$D' = \varepsilon_{D,REF}\ m'_{REF,eq} F'_{S,REF} \tag{6}$$

Where $m'_{REF,eq}$ represents the mass of the reference radionuclides yielding the same doubles rate.

From Equations 5 and 6 of $m'_{REF,eq}$ can be expressed as:

$$m'_{REF,eq} = m \frac{\varepsilon_D}{\varepsilon_{D,REF}} \left(\sum_j f_{m,j} \frac{F'_{S,j}}{F'_{S,REF}} \right) \tag{7}$$

The quantities $\varepsilon_{D,REF}$ and ε_D depend on the multiplicity and neutron energy distributions associated to the source and can be very different e.g. for a ^{252}Cf or a ^{240}Pu source and not accounting for such a difference can be a source of systematic effects.

If we neglect the impact of the neutron energy distributions, in good approximation the number of doubles per fission event is proportional to the average number of pairs per fission generated by the sources and we can therefore approximate the efficiency ratio as the ratio of the average number of pairs per fission:

$$\frac{\varepsilon_D}{\varepsilon_{D,REF}} \approx \frac{\dfrac{(\sum_j Q_j f_{m,j})}{(\sum_j F'_{S,j} f_{m,j})}}{\dfrac{Q_{\ REF}}{F'_{S,REF}}} = \frac{\dfrac{(\sum_j Q_j f_{m,j})}{Q_{\ REF}}}{\dfrac{(\sum_j F'_{S,j} f_{m,j})}{F'_{S,REF}}} \tag{8}$$

which provides a way to account for the actual radionuclide composition of the source and determine m from Eq. 5 and using $\varepsilon_{D,REF}$ which may be easier to calculate.

III. Experiments

A. Materials

The assayed material was unirradiated mixed oxide (MOX) fuel. The plutonium (Pu) is mixed with uranium (U), chemically in the form of oxide. The total plutonium content is up to 14%$_{wt}$ with a relative ^{239}Pu amount up to 96%$_{wt}$. The MOX fuel pins were arranged in a compact hexagonal configuration as indicated in Fig. 2 to maximize the average density of the item to be assayed.

The measurement campaign allowed investigating the influence on the detector response due to the amount of nuclear material, the type of nuclear material, and the type of shielding material. For example, by performing measurements on samples with various amounts of plutonium mass and a fixed isotopic composition, one could study the sensitivity of the methods to the amount of nuclear material.

Figure 2. Top view of a 19-pin configuration (Source: SCK CEN. Used by permission).

Five assemblies with given composition were available for the experiments, with neutron emission rates indicated in Table I calculated from their nominal composition. The hexagonal bundle was hosted in a 3 mm thick square container of stainless steel to reduce the gamma-ray dose rate and facilitate the manipulation. The dimensions of the horizontal cross-section of the container were 10 cm × 10 cm. On the container different types of shielding material could be installed such as: lead 5 mm and 10 mm thick; cadmium 1.1 mm thick; polyethylene 50 mm thick; and/or a combined shielding of polyethylene (50 mm thick on the inside) and lead (10 mm thick on the outside).

The assemblies *62–1, 62–19, 62–61* were fuel pins with a plutonium content of 12.6%$_{wt}$ and with a ^{239}Pu amount of 61%$_{wt}$. The number of rods were 1, 19, or 61; these numbers allowed to keep a hexagonal arrangement and to have a ^{239}Pu mass range from 0.1 to 2.6 kg. The uranium enrichment was 0.4%. These assemblies could be used for measurements dedicated to the study of the mass determination or sensitivity.

The assemblies *62–19, 79–19* and *96–19* had a ^{239}Pu relative content in the PuO$_2$ of 61%$_{wt}$, 79%$_{wt}$, and up to 96%$_{wt}$ respectively. The *96–19* was an axially heterogeneous assembly ^{239}Pu relative content of 79% at the top and bottom and 96%$_{wt}$ at the mid of the assembly. Among these assemblies, the plutonium content ranged between 4%$_{wt}$ and 12%$_{wt}$ and the uranium enrichment was between 0.4% and 2.0%.

TABLE I
Neutron emission rates for the considered assemblies (values expressed in 10^3 neutron/s).

Assembly	Spontaneous Fission			(α,n) reaction		
ID	U	Pu	Am	U	Pu	Am
79-19	51	0.1	<0.1	17	0.5	21
96-19	21	0.1	<0.1	10	0.5	8
62-1	23	<0.1	<0.1	13	<0.1	14
62-19	439	0.1	<0.1	241	0.7	262
62-61	1,410	0.3	<0.1	774	2.1	841

B. Detector and setup

Two Canberra WM3400 slab counters [15] were deployed. A picture of the detectors during the measurement of MOX assembly is shown in Figure 3.

Each WM3400 counter measures 104 cm (height) × 41 cm (width) × 16 cm (depth) and has an approximate weight of 50 kg. The WM3400 counter contains six ^{3}He tubes (active length of 91.44 cm and diameter of 2.54 cm) which are positioned in-line and embedded in a polyethylene slab. The polyethylene slab is then surrounded by 1-mm Cd sheet that covers five sides of the slab. During a measurement the WM3400 should be placed such that the side without the Cd cover is facing the sample being measured.

A dedicated electronic system [16] was used that includes a dedicated preamp for each ^{3}He tube, the possibility to adjust for gain differences between the ^{3}He tubes, and an optimised shaping time for ^{3}He detectors. The logical OR of the signals coming from both slab counters was processed with a MCA-527 multi-channel analyzer equipped with time stamp processing firmware [17]. By considering the two detectors as a whole one limits the impact of possible asymmetrical detector arrangements. Such a system allowed saving the time stamps of the neutron detections for off-line data with a dedicated software to determine *T*, *R+A*, and *A* events, the multiplicity distribution, as well as to remove system instabilities due to high voltage and fluctuations or interferences.

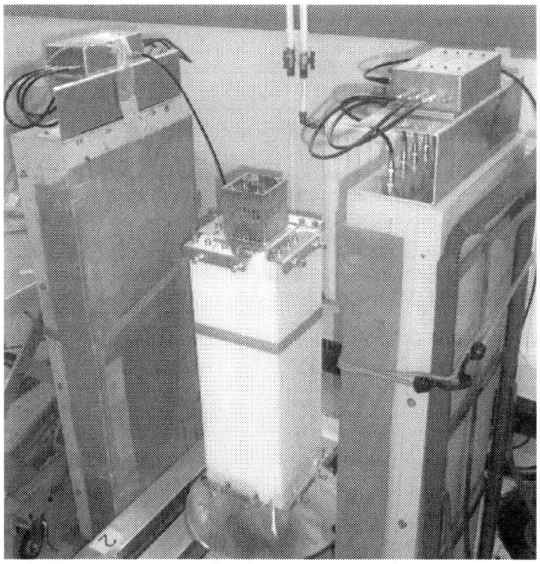

Figure 3. Measurement setup for the neutron coincidence counters.

C. Measured samples

All the five assemblies (*96–19, 79–19, 62–1, 62–19, 62–61*) were measured. All measurements were carried out with 1.1 mm Cd around the fuel assembly. For fuel types *79–19* and *96–19* the measurements were also carried out without Cd. The fuel type *96–19* was also measured with other two configurations: the first one with 5 cm CH_2 around the assembly and 1.1 mm thick Cd sheet on the detector (shown in Figure 3) and the second one with 5 mm Pb around the assembly and a 1.1 mm thick Cd sheet on the detector. Measurements were done at two distances 26 cm and 96 cm from the outer surface of the assembly. The measurement time depends on the neutron emission of the sample and was either 1800 s or 3600 s. In addition to measurements with MOX fuel assemblies, also background measurements were carried out.

IV. DATA ANALYSIS

The background subtracted and dead time corrected totals and reals rate values for the considered assemblies are shown in Table II. The uncertainty due to counting statistics is also given.

The time stamps recorded with the MCA527 were analyzed by means of Python [18] scripts, in the Spyder environment version 4 [19] . The scripts first carried out stability checks of the data by monitoring the total count rate over 10 seconds interval. A dead-time correction was carried out by implementing the dead time correction algorithm with paralysable model as described in Dytlewski [20] and Vincent [21]. The dead-time was estimated to be 1.0 μs with a corresponding maximum dead time correction was for the case 62-61 at 26 cm, where the totals correction was about 11 % and the reals correction was about 34 %, respectively.

Then the data were analyzed to determine the totals and the reals rate for each measurement. The background subtraction was then carried out and the net total and net real count rates, indicated with *T'* and *D'* respectively, were determined.

TABLE II

Background subtracted totals and reals rate values for the different measurement configurations. (*) denotes the fact that the Cd sheet was placed on the detector. DTCF indicate the dead time correction factor.

Assembly	*Shielding*	*d*	*T'*	T_{DTCF}	*D'*	D_{DTCF}
		cm	s^{-1}	%	s^{-1}	%
62-61	Cd	96	30435 (5)	3.1	104.6 (93)	9.4
62-1	Cd	96	431.8 (2)	0.1	1.1 (1)	0.2
62-61	Cd	26	100089 (13)	10.6	1115 (46)	34.3
62-19	Cd	26	30415 (5)	3.1	303.9 (93)	9.3
62-19	Cd	96	8848.8 (16)	0.9	23.9 (17)	2.6
62-1	Cd	26	1505.5 (10)	0.2	14.5 (5)	0.5
96-19	Pb,Cd*	26	1480.3 (10)	0.2	21.7 (5)	0.6
96-19	PE,Cd*	26	1044.5 (6)	0.1	12.4 (2)	0.4
96-19	Cd	26	1558.0 (10)	0.2	21.8 (5)	0.5
96-19		26	1558.4 (7)	0.2	21.7 (3)	0.6
79-19		26	3772.8 (15)	0.4	55.7 (11)	1.2
79-19	Cd	26	3731.5 (12)	0.4	55.7 (8)	1.2

The data in Table II for the assemblies 79-19 and 96-19, indicate that both the totals and reals rate are almost insensitive to the Cd absorber indicating that the counted neutrons are the mainly due fast neutrons emerging from the assembly. For the *96–19* assembly, both totals and reals were almost insensitive to Cd+Pb (5% reduction on the totals), with PE+Cd the totals were attenuated by 30% and the reals by 43%.

In all the cases the uncorrelated background was relatively high with the ratio A/(R+A) ranging from 88.4% (96–19 with PE+Cd shielding) to 99.8% (*62–61* with Cd, at 1 m distance). In the worst case the reals rate could be determined with a relative uncertainty of 9%. This is in line with the data of Table I, from which one would expect that for *96–19 and 79-19* assemblies a lesser impact due to (α,n) reaction when compared to the 62 type assemblies.

The obtained data were used to determine the mass of the assayed sample by applying two methodologies, one based on the point model approximation and one based on Monte Carlo modeling. The nuclear data used in the data analysis such as decay constants, (α,n) emission probabilities, spontaneous fission half-lives, factorial moments of the neutron emission distribution by spontaneous fission used in the data analysis were taken from a compilation of different sources e.g. [22-31], with atomic masses taken from [32].

A. Point model approach

The equations of the Hage's point model equations in absence of multiplication were solved by calculating the gate occupation factor f from Eq. 4 and using a die-away of the system, measured with a ^{252}Cf source (55.3±1.0 μs). The α term in the Eq. 1 is determined from the composition and nuclear data of the fuel. We derived the calculated F_S source term; from the F_s one can determine the mass of the sample by dividing by F'_s , the number of spontaneous fission per second and per gram.

$$m_c = \frac{F_s}{F_s'}$$

The obtained ratio between the calculated mass of heavy metal and the nominal one from the fuel specification for different cases and as a function of the mass of heavy metal mass is shown in Fig. 4.

The error bars represents the 68% confidence limit due to counting statistics and include a 1% uncertainty of the gate occupation factor f.

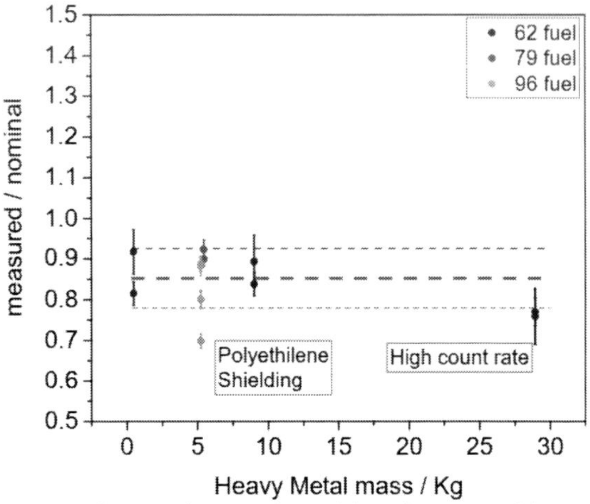

Figure 4. Ratio between the mass obtained from the point model approach and nominal masses as a function of the mass for the different measurement configurations.

The results show an average underestimation of the mass by about 15%. The fact that the multiplication factor is not accounted for can explain the observed bias as well as the fact that the conditions to apply the point model were not fully met. The main outlier is represented by the case with polyethylene shielding for which a larger correction for that multiplication in the assembly is needed when solving the point model equations.

B. Monte Carlo based calibration approach

The Monte Carlo code MCNP6 [33] was used to carry out a full scale modelling of the measurement setup. Following the results of previous modeling efforts [8], the model included the detectors and the fuel assembly, shielding, as well as the walls, ceiling and floor of room in which the measurements took place. Since the specifications of detectors and assemblies were very well known, the geometry and material composition could be modelled with a high degree of fidelity. A cross section of an example of a modeled geometry is shown in Fig. 5.

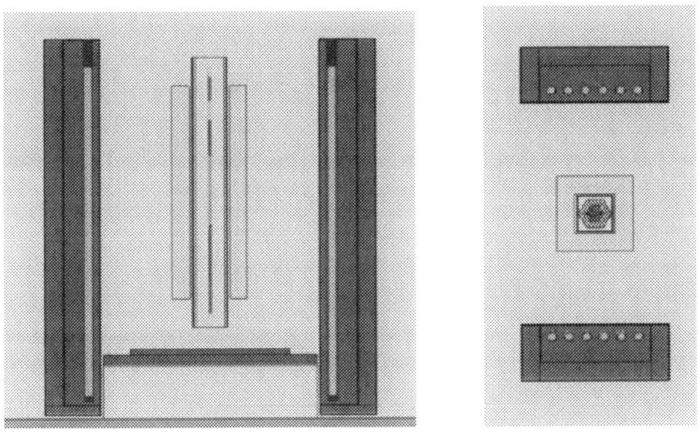

Figure 5. Vertical and horizontal cross sections of the Monte Carlo model of the measurement setup for the 96-16 assembly with PE shielding and the slab counters. The slab counters are indicated in blue.

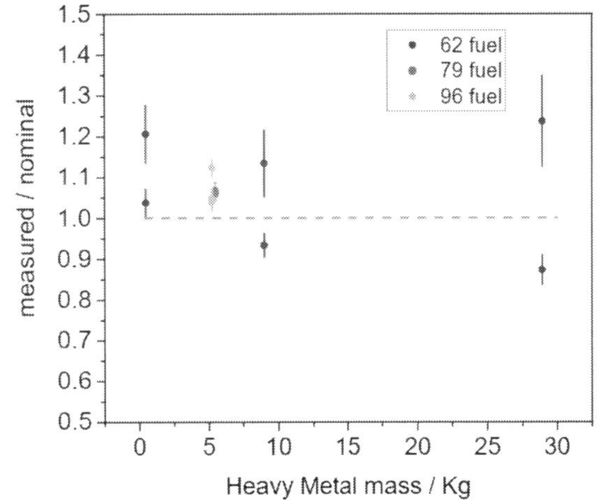

Figure 6. Ratio between the mass obtained from the Monte Carlo based calibration approach and nominal masses as a function of the mass for the different measurement configurations.

MCNP6 was run in the so-called analog mode i.e. tracking event-by-event the interactions of neutrons rather than their average behaviors. A tally of the type F8 with the CAP special treatment [34] was used. With such a tally the neutron multiplicity is scored in two time intervals both of duration of 64 μs corresponding to the experimental ones: the first one called *R+A gate* and starting after 4.5 μs from the spontaneous fission event; the second one called *A gate* and starting 1,000 μs after the end of *R+A gate*. The simulations implicitly account for the pre-delay and finite duration of the gate and therefore a correction for the *f* factor, such as the one used in the point model approach, is not necessary. The simulations only account for neutrons emitted by spontaneous fission and therefore do not account for reals induced by (α,n) reaction which are present only in the presence of a multiplying sample. A source term corresponding to the spontaneous fission of the source radionuclides was used. The nuclear data library JEFF 3.3 was used.

The obtained masses m_c with the previously outlined procedures were compared with the nominal masses m_n of the measured samples are shown in Figure 6. The obtained ratios m_c/m_n for the considered cases are given in Table III for the point model and Monte Carlo approaches. The quoted uncertainties result from counting statistics on the net reals rate and on the reals per spontaneous fission event from Monte Carlo simulations, which varied between 0.2% and 0.4%, and have been obtained by first-order Taylor expansion-based error propagation..

With the exception of the data for the 62-19 and 62-61 assemblies measured at 26 cm distance, are overestimated between 2% and 24%. As previously stressed, the Monte Carlo simulation do not account for the effect of (α,n) reactions. By not accounting for (α,n) reactions, the share of the reals due (α,n) reactions are attributed to spontaneous events and this may result in a systematic effect on the estimated mass. We notice that for the two outliers for the 62-19 and 62-61 assembly at 26 cm distance fuel for which the dead time correction of the real was higher than for most of the other cases.

In an effort to understand the role of (α,n) reactions on the obtained results additional MC calculations were carried out.

TABLE III
Ratio between the mass obtained from the point model, the Monte Carlo based calibration approach and nominal masses for the different measurement configurations.

Assembly	*Shielding*	*d*	m_c/m_n	m_c/m_n
		cm	*Point Model*	*MC calibration*
62-61	Cd	96	0.76(7)	1.24(11)
62-1	Cd	96	0.92(6)	1.21(7)
62-61	Cd	26	0.77(3)	0.87(4)
62-19	Cd	26	0.84(3)	0.93(3)
62-19	Cd	96	0.89(7)	1.13(8)
62-1	Cd	26	0.82(3)	1.04(3)
96-19	Pb,Cd*	26	0.80(2)	1.04(2)
96-19	PE,Cd*	26	0.70(2)	1.13(2)
96-19	Cd	26	0.88(2)	1.05(2)
96-19	---	26	0.89(2)	1.05(2)
79-19	---	26	0.92(2)	1.07(2)
79-19	Cd	26	0.90(2)	1.06(2)

C. *Impact of the (α,n) reactions*

The Monte Carlo approach proposed in previous section assumes that only spontaneous fission affect the observed reals rate. As indicated in Table in addition to spontaneous fission a non negligible amount of neutron are emitted by (α,n) reactions. In absence of multiplication, due to the fact that the emission of such neutrons is not time correlated, they do not contribute to the net real rate *D'*. Given the mass of fissile material, the assumption of absence of multiplication is questionable and therefore we carried out additional simulations to account also for the (α,n) component.

 https://doi.org/10.1051/epjconf/202328806008

The energy dependency of the neutrons stemming from the (α,n) reaction was deduced from the code SOURCE-4C [35] based on the radionuclide composition of the source. The results of the MC simulations are total and reals rate per fission event or per neutron emitted by (α,n) reaction. By multiplying these data by the corresponding source strength we obtain the total and reals rate due to spontaneous fission and (α,n), respectively.

We then determine the ratio between the dead time and background corrected experimental data and the simulated reals rate in Table IV. This ratio should be compared with the mass ratio m_c/m_n from MC calibration in Table III. The calculated mass m_c was obtained from Eq. 5 by using the ε_D term, that is the number of simulated reals per spontaneous fission event, and the experimental data. The change in ratio due to the presence of (α,n) is an indication of the impact of accounting for the (α,n) in the mass determination.

We observe that accounting for (α,n) systematically reduces the ratio; this is due to the fact that with the first approach part of the reals were ascribed to spontaneous fission source and this resulted in an overestimation of the mass. For most of the cases there is an agreement within the uncertainties between calculated and measured values. An overestimation is still present on all the measurement data at 96 cm distance. The data for the 62-61 and 62-19 at 26 cm distance are clearly underestimated possibly due to insufficient dead time correction. The data for the 62-61 assembly at 96 cm probably suffer from both effects and have a relatively large uncertainty.

The quoted uncertainty accounts for the statistic uncertainty only. For completeness we also include the ratio between the experimental and calculated totals rates, the share of reals due to spontaneous fission (D_{SF}/D) and multiplication factor (M) obtained from MC simulations.

TABLE IV
Ratio between the experimental and calculated reals and totals rate for the different measurement configurations.

Assembly	*Shielding*	*d*	*D'*	*T'*		
		cm	*E/C*	*E/C*	D_{SF}/D	*M*
62-61	Cd	96	1.07(10)	1.02	0.86	1.09
62-1	Cd	96	1.17(7)	1.05	0.97	1.01
62-61	Cd	26	0.75(3)	0.94	0.86	1.09
62-19	Cd	26	0.85(3)	0.99	0.91	1.05
62-19	Cd	96	1.05(8)	1.04	0.92	1.05
62-1	Cd	26	1.01(3)	1.00	0.97	1.01
96-19	Pb,Cd*	26	0.98(2)	0.94	0.95	1.04
96-19	PE,Cd*	26	0.99(2)	0.90	0.88	1.12
96-19	Cd	26	0.99(2)	0.98	0.95	1.04
96-19		26	0.99(2)	0.98	0.95	1.04
79-19		26	1.01(2)	1.02	0.95	1.04
79-19	Cd	26	1.01(2)	1.01	0.95	1.04

V. CONCLUSIONS

We carried neutron coincidence measurements with ^3He based slab counters in the framework of a measurement campaign in support of nuclear disarmament verification with well-characterized MOX sources

The obtained background and dead time corrected total and reals rates were used to determine the mass of the assayed sample by using the so-called point model equations in absence of multiplication. The obtained masses were systematically underestimated by 15 % in average.

A calibration approach based on Monte Carlo simulations where the mass is estimated from the calculated number of reals per spontaneous fission event was also used. The obtained masses were slightly overestimated and few outliers, up to 20%, were observed.

In an effort to better understand the obtained results, additional calculations were carried out to estimate the impact due to both the neutron emission by (α,n) reactions and spontaneous fission decay on the measured observables. The associated total and reals counts were obtained for various configurations and compared with the experimental data. The obtained results indicate a very good agreement between experimental and calculated data. The importance of (α,n) reactions when neutron multiplication is not negligible. The remaining discrepancy are most likely due to the dead time correction.

REFERENCES

[1] https://www.ipndv.org, Last accessed on 2023/05/05.
[2] https://www.ipndv.org/reports-analysis/results-of-the-sck-cen-exercise-for-disarmament-verification-technologies/, Last accessed on 2023/05/05.
[3] https://media.nti.org/documents/new_start_article_analysis.pdf, Last accessed on 2023/05/05.
[4] Evaluation of measurement data — Guide to the expression of uncertainty in measurement, JCGM 100:2008, First edition September 2008, available at https://www.iso.org/sites/JCGM/GUM/JCGM100/C045315e-html/C045315e.html?csnumber=50461
[5] Safeguards techniques and equipment. – 2011 edition. International Atomic Energy Agency, 2011, IAEA/NVS/1/2011 (Rev. 2), ISBN 978-92-0-118910-3
[6] Baeten, P, Bruggeman, M & Carchon, R 1997, *An investigation of the neutron die-away time in passive neutron waste assay systems*. SCK•CEN Reports, no. BLG-732, Studiecentrum voor Kernenergie, Mol, Belgium.
[7] L.R. Avens, J.E. Doyle, and M.F. Mullen, *The fissile material transparency technology demonstration*, LA-UR-01-3570, Los Alamos National Laboratory (2001).
[8] A. Borella, et al., Neutron Coincidence Measurements and Monte Carlo Modelling of Waste Drums Containing Reference Nuclear Material, EPJ Web Conf. 253 07001 (2021), DOI: 10.1051/epjconf/202125307001
[9] R. Dierckx & W. Hage (1983) Neutron Signal Multiplet Analysis for the Mass Determination of Spontaneous Fission Isotopes, Nuclear Science and Engineering, 85:4, 325-338,
[10] W. Hage & D.M. Cifarelli Nucl. Sci & Eng. 89 (1985) 159-176
[11] D.M. Cifarelli & W. Hage Nucl. Instr. & Meth. A251 (1986) 550-563
[12] A. Borella, R. Rossa, 2022. A Data Analysis Suite for Neutron Time Correlation Measurements. BLG-2940 SCK CEN report
[13] Ensslin E., 1991. "Principles of Neutron Coincidence Counting" in Passive Nondestructive Assay of Nuclear Materials, NuREG/CR-5550, LA-UR-90-732.
[14] Q. Ducasse et al. *Design of MICADO advanced passive and active neutron measurement system for radioactive waste drums*, Nuclear Instruments and Methods A 1005 (2021) 165398, https://doi.org/10.1016/j.nima.2021.165398.

[15] https://www.mirion.com/products/wm3400-passive-neutron-slab-counter. Last accessed 28/04/2021
[16] E. Boogers et al., "Improved electronics for 3He based neutron counters." EPJ Web of Conferences 225 (2020): 05002. http://dx.doi.org/10.1051/epjconf/202022505002.
[17] https://www.gbs-elektronik.de/media/download_gallery/mca527_neutron.pdf. Last accessed 28/04/2021
[18] https://www.python.org/
[19] https://docs.spyder-ide.org
[20] N. Dytlewski, "Dead-time Corrections for Multiplicity Counters," Nucl. Instr. Meth. A305, 492–494 (1991)
[21] C.G. Vincent "Type II dead-time losses between detections occurring at random times within a gate-open time", Nuclear Instruments and Methods in Physics Research Section A, Accelerators, Spectrometers, Detectors and Associated Equipment, 269(1), 335-337.
[22] N. Ensslin et al., Application Guide to Neutron Multiplicity Counting, 1998, LA-13422-M
[23] J.W.Boldeman, M.G.Hines, Prompt Neutron Emission Probabilities following Spontaneous and Thermal Neutron Fission, Nucl.Sci.Eng. 91, 114 (1985)
[24] W. Hage K. Caruso, Safeguards and Fissile Materials Management, a Preliminary procedure for the treatment of time correlation analyser measurement data for non-destructive assay of Pu, Technical Note I.07.C3.86.115, 1987
[25] Norman E. Holden & Martin S. Zucker Prompt neutron multiplicities for the transplutonium nuclides, Radiation Effects: Incorporating Plasma Science and Plasma Technology, 96:1-4, 289-292, 1986, DOI: 10.1080/00337578608211755
[26] Handbook Of Nuclear Data For Safeguards: Database Extensions, August 2008, INDC(NDS)-0534
[27] S. Simakov, M. Verpelli, N. Otsuka, Update of the nuclear data for the neutron emissions for actinides of interest in safeguards, Nuclear Data Section, IAEA, 2015, https://www-nds.iaea.org/sgnucdat/SF_n-Yield_20150313.pdf
[28] Nuclear data sheets, NDS 57,558 (1989), NDS 59,545 (1990), NDS 59,507 (1990) retrieved from http://nucleardata.nuclear.lu.se/toi/
[29] https://www.nndc.bnl.gov/nudat2/
[30] P. Santi & M. Miller (2008) Reevaluation of Prompt Neutron Emission Multiplicity Distributions for Spontaneous Fission, Nuclear Science and Engineering, 160:2, 190-199, DOI: 10.13182/NSE07-85
[31] Zucker, M.S., & Holden, N.E. (1984). Parameters for several plutonium nuclides and 252Cf of safeguards interest (BNL--34804-Draft). United States
[32] Coursey, J.S., Schwab, D.J., Tsai, J.J., and Dragoset, R.A. (2015), Atomic Weights and Isotopic Compositions (version 4.1). [Online] Available: http://physics.nist.gov/Comp [year, month, day]. National Institute of Standards and Technology, Gaithersburg, MD.
[33] MCNP6 Users Manual - Code Version 6.1.1 beta, LA-CP-14-00745, June 2014.
[34] M. T. Swinhoe, J. S. Hendricks, and D. R. Mayo, "MCNPX for Neutron Multiplicity Detector Simulation," Los Alamos National Laboratory report LA-UR-04-8025 (2004).
[35] Wilson, William B., Perry, Robert T., Shores, Erik F., Charlton, William S., Parish, Theodore A., Estes, Guy P., Brown, Thomas H., Arthur, Edward Dana, Bozoian, Michael, England, T.R., Madland, D.G., & Stewart, James E. (Jan 2002). SOURCES 4C: a code for calculating (α,n), spontaneous fission, and delayed neutron sources and spectra (LA-UR--02-1839). United States

On the use of free code tools to simulate the propagation of radiation following dirty bomb explosions in sensible contexts

Riccardo Quaranta[1,3,*], Gian Marco Ludovici[1], Guglielmo Manenti[2], Pasqualino Gaudio[1] and Andrea Malizia[2]

[1]Department of Industrial Engineering, University of Rome Tor Vergata, Italy
[2]Department of Biomedicine and Prevention, University of Rome Tor Vergata, Italy
[3]The American University of Rome, Italy

(*) quaranta@ing.uniroma2.it

Abstract— **The current geopolitical situation suggests an increasing possibility of radiological dispersal device attacks on sensitive targets. Consequently, understanding the transport of radiation over great distances in a short time can help first responders and decision makers in effectively managing emergencies. By utilizing open-source computational codes, intentional releases of radioactive material and their transmission from person to person can be simulated to provide first responders and decision makers with a rapid tool to facilitate their work. In this study, the HotSpot code was employed to simulate two releases of Cs-137 resulting from the detonation of a dirty bomb in a major city and near an aqueduct waterworks. Additionally, the STEM code was used to simulate radiation propagation from the initially affected individuals, drawing comparisons to the vectors of viral infections. This approach allowed to compare the outcomes of a scenario involving many individuals in an urban setting with another scenario having fewer individuals but posing the risk of contaminating critical infrastructure. The results showed that both scenarios had similar relatively mild health consequences for the population, despite their considerable differences and variations in the analyzed timelines. However, both scenarios present numerous challenges in emergency management. In the first case, the incident generates widespread panic and media frenzy. In the second case, the dissemination of radiation and potential public unawareness must be taken into account. Addressing these considerations needs the involvement of multiple stakeholders, including police, firefighters, healthcare professionals, journalists, politicians, and others, in emergency management efforts.**

Keywords — **Dirty bomb, HotSpot, radiological release, radiation epidemiology, RDD, simulations, STEM.**

I. Introduction

CURRENTLY, mathematical modeling proved to be a useful tool to understand the dynamics of the spread of infectious diseases [1,2]. Scientific literature proposes mathematical models for studying epidemics, outbreaks and also the improvement of any interventions [3]. These mathematical models are extremely useful for the analysis of risks, even if they could be used by personnel not properly trained in both the use of the specific software and the basic knowledge of epidemics [4,5]. For this reason, the development of user-friendly software for application and display of epidemic models stirs great interest in the field of risk assessment and emergency management [3]. In recent years a branch of medicine known as Radiation Epidemiology, a field of epidemiology that deals with the health effects of radiation exposure, whose aspects range from health to public safety is gaining in importance [6]. The combination of mathematical models and radiation epidemiology form the basis of this work as it correlates simulations of dispersion of radiation by radiological dispersal devices (RDDs) and both immediate and long-term effects that radioactive sources might have on the population. The objective of this study is to combine radiation atmospheric dispersion models with epidemiological models to investigate immediate and long-term effects that radioactive sources might have on the population. This is done by using open-source computational codes to simulate intentional releases of radioactive material (HotSpot) and their transport from person to person (STEM) in order to provide an expeditive tool to first responders and decision makers to support the management of the emergency.

II. Methodology

To understand the spread of the radiological contamination two case studies in different conditions were simulated:

- A radiological dispersal device (RDD) explosion in the vicinity of the main station of an important city in central Italy (case study 1).
- The same RDD of case 1 used in a rural environment, in the vicinity of an aqueduct supplying a large city (case study 2).

The analysis of both case studies was split in two steps:

1) The radiological release was simulated by using the Lawrence Livermore National Laboratory code HotSpot to evaluate the extension of the radiation dispersion in terms of the ground deposition.
2) The results of Step 1, in terms of number of people involved in the event, were fed as initial data in the Eclipse

foundation code Spatio-Temporal Epidemiological Modeler (STEM) using a SEIR epidemic model STEM in a manner congruent with those of a common flu (Orthomyxovirus).

The HotSpot code generally provides a conservative estimate of the radiation effects associated with the atmospheric release of radioactive materials [7,8,9] and it is designed for short-range, less than 10Km, and short-term, less than a few hours, predictions [7,10].

STEM is an open-source tool that allows users to create spatial and temporal models of emerging infectious diseases [11,12]. It comes with many existing compartment models, as Susceptible/Infectious (SI), Susceptible-Infectious-Recovered (SIR) and Susceptible-Exposed-Infectious-Recovered (SEIR) models pre-coded with both deterministic and stochastic engines and allows users to extend or modify existing projects or create new ones [11,12]. These models can aid in understanding and potentially preventing the spread of such diseases and help counter their spread [11,13]. In this study, the radiation that, after the initial release, is deposited on the clothing and objects of people present in the area of the event, is then transported by these people, who behave as viral vectors and can spread radiological contamination in a manner similar to a viral transmission.

For the HotSpot simulations, a RDD loaded with a source of Cs-137 in an insoluble form was employed [14,15,16]. The source was selected because it is commercially available and the amount of Cs-137 chosen is comparable to this type of commercial sources [9].

The other parameters used for the HotSpot simulation, shown in Table I, were adapted from [17] and were chosen with the aim to maximize the area involved by the radiological release.

TABLE I
PARAMETERS USED FOR THE HOTSPOT SIMULATIONS

Parameter	Value	Unit
Isotope	Cs-137	Slow absorption
Activity	15	TBq
Explosive	1	Kg TNT equivalent
Wind	1	m/s
Wind input height	10	m
Pasquill Stability Class	C	-
Damage ratio (DR)	100	%
Leakpath factor (LF)	100	%
Deposition velocity (DV)	0.30	cm/s
Airborne fraction (ARF)	100	%
Respirable fraction (RF)	20	%
Receptors height	1.5	m
Non-respirable deposition velocity (NRDV)	40	cm/s
Breathing rate (BR)	4.17e-4	m^3/s
Sampling time	10	minutes
Holdup time	0	minutes
Terrain	City	

Wind direction was selected to maximize the impacts on the surrounding areas, based on the location selected for the release. The simulations in HotSpot were carried out using the Dose Conversion Factor library from the Federal Guidance Report (FGR) 13.

The same parameters were used for the radiological dispersion simulation in both case studies, with the only difference being the terrain set to rural to reduce the influence of buildings surrounding the blast and the wind direction in Case Study 2.

The simulation in HotSpot allowed for the definition of the area covered by the ground deposition of the radiation dispersed by the explosion. From this information it was possible to assess the number of people involved that could act as vectors of the radiation dispersion and this data was fed into STEM.

Given the different impacts of the event analyzed in the two scenarios, in terms of people directly involved, the simulations in STEM had to be carried out considering different timeframes. In the first case study, the large number of people directly involved was taken into consideration and the STEM simulation ran for 7 days. In the second case study, the few people directly involved and the contamination of the water in the aqueduct were both taken into consideration and the STEM simulation ran for 30 days.

III. RESULTS

A. *Case study 1 – Step 1*

The results of the case study 1 simulation in HotSpot, in term of Ground Deposition Contour (GDC), can be seen in Fig. 1. The limits of the three areas were set at 3,200 kBq/m^2, 32,000 kBq/m^2 and 320,000 kBq/m^2 for the inner, middle, and outer zone respectively for a good visualization. As seen in Fig. 1 (a), which uses maps from Google Earth Pro, the area in the proximity of the blast presents elevated GDC values, while the radiation deposited on the ground decreases with the distance from the blast.

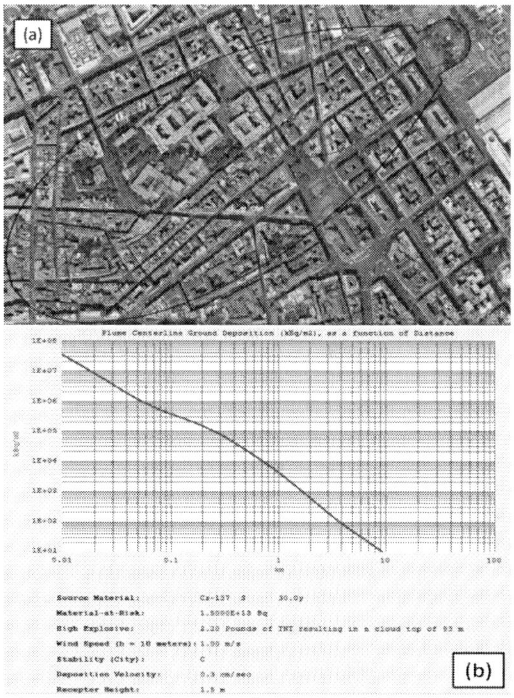

Fig. 1. Ground deposition contours for Case Study 1 (a) and Plume Centerline Ground Deposition in kBq/m^2, as a function of distance (b).

Considering the geometry of the area, the extension of the plume, the maximum distance from the blast and the population density in the city it was possible to estimate the number of people who are indirectly affected by the radioactivity deposited on the ground and thus become unaware carriers of radiation, at 200. Although this value seems small given the large area covered by the GDC, it is important to highlight that everyone in the proximity of the blast will be aware of the event and will therefore seek immediate medical attention and probably undergo rapid decontamination.

B. Case Study 2 – Step 1

The results of the case study 2 simulation in HotSpot, in term of Ground Deposition Contour (GDC), can be seen in Fig. 2. The limits of the three areas were set as before. As seen in Fig. 1 (a), which uses maps from Google Earth Pro, the area in the proximity of the blast presents again elevated GDC values, while the radiation deposited on the ground decreases with the distance from the blast.

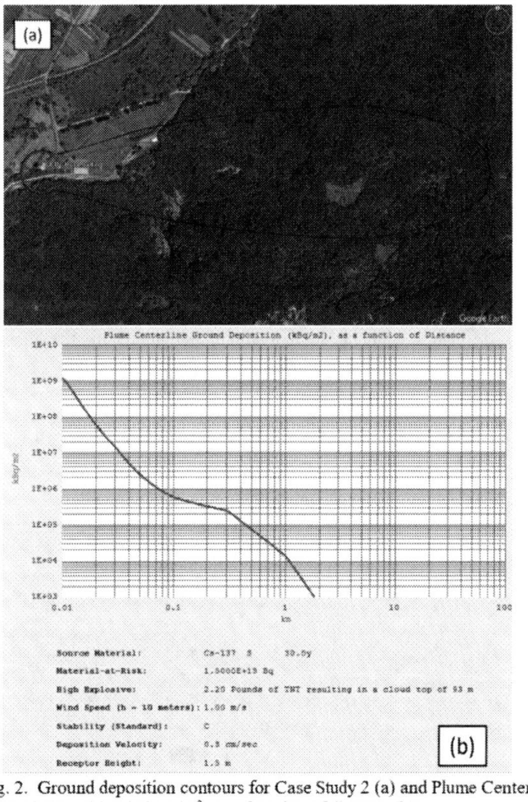

Fig. 2. Ground deposition contours for Case Study 2 (a) and Plume Centerline Ground Deposition in kBq/m^2, as a function of distance (b).

In this case, although it is very unlikely that many people are in the vicinity of the blast when it happens, it can be assumed that over a longer period, such as the 30 days for which the STEM simulation ran, many people can get in contact with the radiation released through ground deposition and deposition in the water feeding the aqueduct. For this reason, 200 people were assessed to be a reasonable input to be fed to the calculations with STEM.

C. Case Study 1 – Step 2

As seen in Fig. 3, starting from central Italy, where the target city is located, the dispersion of the radiation released by the event reached several contiguous regions in 7 days.

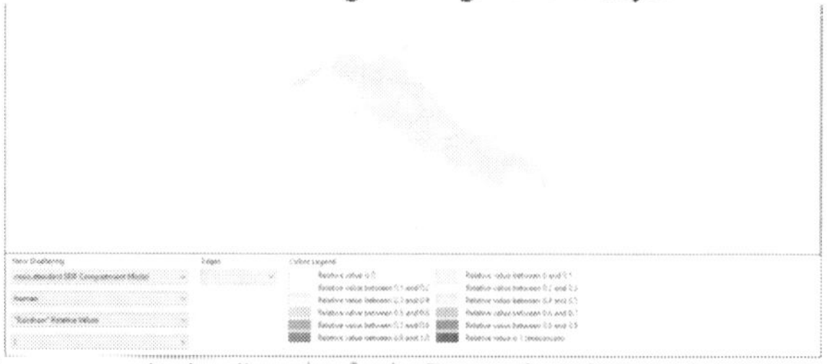

Fig. 3. Contamination dispersion for the Case Study 1 after 1 week.

In Case Study 1, the area closest to the event is the most affected, due to the rapid dispersal of people, but at the same time, as the event is media-impactful, the maximum distance from the source is relatively limited because the visibility of the event makes the people involved aware of the risk. In this case, a large incidence but a limited territorial propagation can be seen.

D. Case Study 2 – Step 2

As seen in Fig. 4, for Case Study 2 the dispersion of the radiation generated by the event reached farther than Case Study 1, in 30 days.

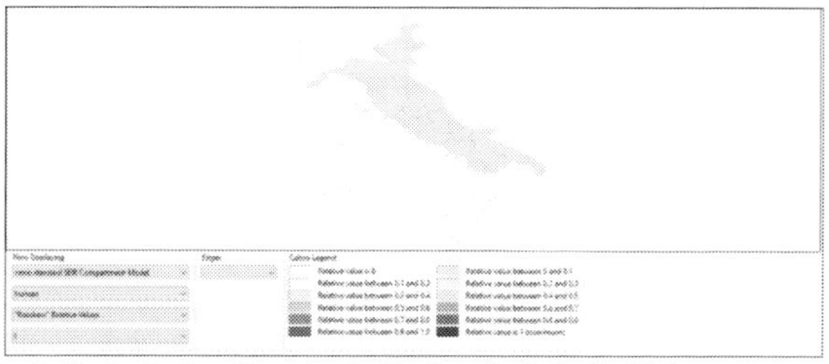

Fig. 4. Contamination dispersion for the Case Study 2 after 30 days.

In Case Study 2, on the other hand, the silence of the attack leads to the public not being fully aware of the risk, thus causing a lower incidence but a larger territorial spread than in Case Study 1.

E. Discussion

Both scenarios analyzed present similar results, despite being significantly different. The main differences regard the challenges that arise in the management of the emergency. A RDD in a crowded area will likely have a strong impact in terms of generalized panic and media coverage but not as much in terms of human health effect of the radiation. However, when the target of such an incident is a critical infrastructure, even in the case that not many people are directly involved, the challenges that arise could significantly increase the complexity of the emergency management. These differences likely lead to having to deal with the two emergencies in different ways and require different approaches.

IV. Conclusions

In this work, radiation epidemiology studies following an intentional release of a Cs-137 source for two case studies, a RDD in a crowded area of an important city and the same RDD in the vicinity of an aqueduct supplying a large city, were carried out by means of simulations using free open-source codes.

The combined use of free open-source codes has demonstrated able to describe rapidly and realistically releases of radioactive material and their transport from person to person. The method proposed is a promising expeditive tool for first responders and decision makers to support the management of the emergency. It is however necessary to highlight that other, more accurate, systems and codes exist, but they require higher computational resources and costs that are not compatible with an effective rapid response.

It should be noted that considering the radioactive material transported by unaware people following such an event is an innovative approach that was sought to be developed. The results obtained with HotSpot and STEM combined show how, very quickly, this radioactivity can spread on a par with a viral infection.

From the results obtained in this study, it will be possible to refine the models and algorithms used to obtain increasingly realistic and comparable simulations of possible events of this type. Different epidemiologic and dispersion model must be studied as well. The ultimate goal is to obtain further insights in order to improve the management of emergency events of a radiological nature.

This first attempt of combining the two open-source codes yielded promising results, but at the same time the models employed will have to be further studied and improved in order to perform simulations that are as truthful as possible and that are meaningful in possible real-world scenarios.

References

[1] S. Y. Del Valle, J. M. Hyman, N. Chitnis, "Mathematical models of contact patterns between age groups for predicting the spread of infectious diseases" *Mathematical biosciences and engineering: MBE*, vol. *10*, no. *5-6*, pp. *1475–1497*, 2013, DOI. 10.3934/mbe.2013.10.1475.

[2] A. Din, K. Shah, A. Seadawy, H. Alrabaiah, D. Baleanu, "On a new conceptual mathematical model dealing the current novel coronavirus-19 infectious disease", *Results in Physics*, vol. *19*, 2020, 103510, ISSN 2211-3797, DOI. 10.1016/j.rinp.2020.103510.

[3] C. J. E. Metcalf, D. H. Morris, S. W. Park, "Mathematical models to guide pandemic response", *Science (New York, N.Y.)*, vol. *369*, no. *6502*, pp. *368–369*, July 2020, DOI.10.1126/science.abd1668.

[4] Y. Mohamadou, A. Halidou, P. T. Kapen, "A review of mathematical modeling, artificial intelligence and datasets used in the study, prediction and management of COVID-19", *Applied intelligence (Dordrecht, Netherlands)*, vol. *50*, no. *11*, pp. *3913–3925*, July 2020, DOI.10.1007/s10489-020-01770-9.

[5] O. Cenciarelli, S. Pietropaoli, A. Malizia, M. Carestia, F. D'Amico, A. Sassolini, D. Di Giovanni, S. Rea, V. Gabbarini, A. Tamburrini, L. Palombi, C. Bellecci, P. Gaudio, "Ebola virus disease 2013-2014 outbreak in west Africa: an analysis of the epidemic spread and response", *International journal of microbiology*, 769121, March 2015, DOI. 10.1155/2015/769121.

[6] W. J. Lee, "Lessons from radiation epidemiology", *Epidemiology and Health*, vol. *40*, e2018057, November 2018, DOI. 10.4178/epih.e2018057.

[7] S. Biancotto, A. Malizia, M. Pinto, G. M. Contessa, A. Coniglio, M. D'Arienzo, "Analysis of a dirty bomb attack in a large metropolitan area: simulate the dispersion of radioactive materials", *Journal of Instrumentation*, vol. *15*, Feb. 2020, DOI. 10.1088/1748-0221/15/02/P02019.

[8] A. Malizia, A. Chierici, S. Biancotto, M. D'Arienzo, G. M. Ludovici, F. d'Errico, G. Manenti, F. Marturano, "The hotspot code as a tool to improve risk analysis during emergencies: Predicting I-131 and Cs-137 dispersion in the Fukushima nuclear accident", *International Journal of Safety and Security Engineering*, vol. *11*, no. *4*, pp. *473-186*, August 2021, DOI. 10.18280/ijsse.110421.

[9] D. Di Giovanni, *et al.*, "Two realistic scenarios of intentional release of radionuclides (Cs-137, Sr-90) - the use of the HotSpot code to forecast contamination extent", *WSEAS Transactions on Environment and Development*, vol. *10*, pp. *106-122*, January 2014.

[10] R. Gallo, *et al.*, "Development of a georeferencing software for radiological diffusion in order to improve the safety and security of first responders", *Defence S&T Tech. Bulletin*, vol. *6*, no. *1*, pp. *21-32*, February 2013.

[11] J. V. Douglas, *et al.*, "STEM: An Open Source Tool for Disease Modeling", *Health security*, vol. *17*, no. *4*, pp. *291–306*, Jul/Aug 2019, DOI. 10.1089/hs.2019.0018.

[12] F. Baldassi, *et al.*, "Testing the accuracy ratio of the Spatio-Temporal Epidemiological Modeler (STEM) through Ebola haemorrhagic fever outbreaks", *Epidemiology and infection*, vol. *144*, no. *7*, pp. *1463–1472*, Dec. 2015, DOI. 10.1017/S0950268815002939.

[13] R. K. Meentemeyer, S. E. Haas, T. Václavík, "Landscape epidemiology of emerging infectious diseases in natural and human-altered ecosystems", *Annual review of phytopathology*, vol. *50*, pp. *379–402*, June 2012, DOI. 10.1146/annurev-phyto-081211-172938.

[14] A. Malizia, *et al.*, "The free license codes as decision support system (DSS) for the emergency planning to simulate radioactive releases in case of accidents in the new generation energy plants", *WSEAS Transactions on Environment and Development*, vol. *10*, no. *1*, pp. *453-464*, January 2014.

[15] A. Abate, *et al.*, "The local effects of a global disaster: Case study on the Fukushima radiological emergency management in Italy", *Defence S&T Tech. Bulletin*, vol. *9*, no. *2*, pp. *126-133*, 2016.

[16] I. Cacciotti, *et al.*, "Simulation of Caesium-137 (137CS) local diffusion as a consequence of the Chernobyl accident using HOTSPOT", *Defence S&T Tech. Bulletin*, vol. *1*, no. *1*, pp. *18-26*, 2014.

[17] A. Rump, S. Eder, C. Hermann, A. Lamkowski, P. Ostheim, M. Abend, M. Port, "Estimation of radiation-induced health hazards from a "dirty bomb" attack with radiocesium under different assault and rescue conditions", Military Medical Research, vol. 8, no. 1, Dec. 2021, DOI. 10.1186/s40779-021-00349-w.

Pulse shape discrimination of CLYC, CLLBC and EJ-276 with SiPM readout

Bjoern Seitz[1,*], David Bennett[1], and Frank Thomson[1]
[1]SUPA School of Physics and Astronomy, University of Glasgow, Glasgow G12 8QQ, U.K.
(*) Bjoern.Seitz@glasgow.ac.uk

Abstract— **Solid state detectors for the detection of thermal and fast neutrons find widespread applications. They often operate in a high gamma-radiation field or are required to discriminate between gamma-radiation and neutrons. Organic and in-organic scintillation materials are proposed as detector materials. These materials often exhibit different pulse shapes in their light output, which allows a variety of pulse shape analysis (PSA) techniques to be used to distinguish the two species of radiation. The current maturity of silicon based single photon counters (SiPM) provides a viable visible photon detector alternative to conventional vacuum based photo multiplier tubes (PMT). However, their impact on PSA has not been deeply studied. Three solid state scintillation materials, CLYC, CLLBC and EJ-276 are coupled to an array of SiPM and exposed to neutron and gamma radiation. Their response is characterized using a variety of PSA algorithms and quantified in a Figure of Merit. Conventional charge comparison algorithms perform well for all materials, while a Fourier component analysis shows particular strength for the in-organic materials tested.**

Keywords —**neutron detection, pulse shape analysis, SiPM, Fourier analysis, scintillation detectors**

I. Introduction

THE detection of neutrons and the discrimination between neutrons and gamma-radiation finds widespread applications across a variety of fields, from evaluating additional dose in radiotherapy to scanning cargo to the monitoring of nuclear installations and the calibration of neutron sources or nuclear emitting radio-isotopes. Scintillation counters exhibiting different scintillation light pulse shapes, depending on the type of radiation detected are a promising solution meeting these requirements. The analysis of these pulse shapes (Pulse Shape Analysis (PSA)) provides the identification of each particle species, with several methods being available. In addition to the well established liquid scintillation materials, recent organic and in-organic solid state scintillation materials show great promise in detecting neutrons and discriminating these from gamma radiation.

In parallel, solid state visible light photon counters based on SPAD arrays have been developed and provide a rugged and compact alternative to conventional vacuum based Photo Multiplier Tubes (PMT). These Silicon Photo Multipliers (SiPM) show a high photon detection efficiency, fast timing and acceptable levels of noise [1]. However, SiPM will exhibit non-linearities with increasing incident light level, raising questions about their suitability for use in pulse shape analysis applications.

The following text describes tests performed with two inorganic and one organic scintillator coupled to anSiPM array and exposed to thermal and fast neutrons in the presence of a gamma radiation field. The detector response was studied on a wave-form level and analyzed using a variety of pulse shape analysis algorithms. Section II describes the materials and experimental setup. The PSA and performance metrics used to compare the different materials is detailed in section III. Section IV summarizes the results and provides a conclusion.

II. Experimental setup and methods

The detector system tested consists of the scintillator material under test coupled to a SiPM array via a custom light guide and read out via a custom front-end board and a commercial waveform digitizer to record the full pulse shape for each event. An example of the constructed detector system is shown in fig. 1 below.

Fig. 1. Photograph of the detector setup used for these tests. The grey cylinder houses the scintillation material. Above it, the custom pulse shape preserving light guide is visible, followed by the electronics board housing the SiPM array.

The three materials tested were two inorganic crystals, CLYC (Cs_2LiYCl_6:Ce) [2,3] and CLLBC ($Cs_2LiLa(Br,Cl)_6$:Ce), and one organic material, EJ-276. The inorganic crystals provide a good sensitivity to thermal neutrons through capture reactions on ^{6}Li and ^{35}Cl. Their average density is comparable with NaI:Tl and they are expected to perform better than this in terms of gamma detection efficiency and resolution. Both inorganic crystals are expected to exhibit clear pulse shape differences between neutron and gamma radiation. In contrast, E-276 is not expected to provide a good response to gamma radiation, but an excellent response to fast neutrons from collisions with the material's protons [4,5].

Each scintillator was constructed as a rect-cylinder with 51 mm diameter and 51 mm height. All three detector materials are packaged and sealed in a cylindrical aluminum housing with reflective coating on the inside and a quartz or perspex window for the inorganic and organic scintillators respectively. A pulse shape conserving light guide made from PMMA was constructed to couple the circular exit surface of the scintillator to a square 8x8 array of SensL J 30035 SiPM. The individual SiPM have a size of 3mm x 3mm. Each individual SiPM consists of 5676 3.07μm x 3.07μm pixels. The maximum photon detection efficiency is given as 38% at 420 nm with an overvoltage of V=2.5V and 50% at V=6V. The array was operated at a common voltage of 30V. This corresponds to a setting of 5V above the established breakdown voltage. The analog signals of the individual SiPM were summed on a custom PCB and fed to a CAEN DT5730B waveform digitizer. A schematic of the setup is shown in fig. 2. It should be noted that throughout the experiment, the source configuration, light transport, light detection and read-out electronics have been kept constant. Coupling a different scintillation crystal was the only change.

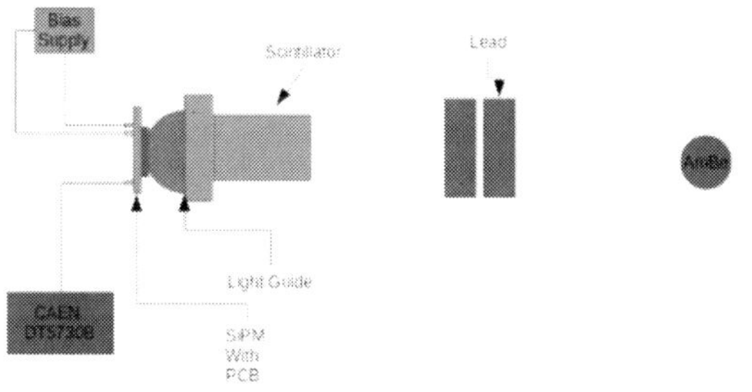

Fig. 2. Schematics of the test setup depicted as a photograph in fig.1. The SiPM array is shown coupled to the pulse-shape conserving light guide coupled to the scintillator under test. A lead shield shield the detector from gamma radiation emitted from the AmBe neutron source. The detector is powered by a LV bias supply and read out by a CAEN DTS5730B waveform digitizer.

Neutrons were generated using an AmBe source through α−capture on ^{9}Be. This results in the emission of a fast neutron and the emission of a gamma ray with an energy of E_γ = 4.4 MeV. The branching ratio for this reaction is 55%. The emitted neutrons have a maximum neutron energy E_n of E_n=11 MeV, with a mean unmoderated energy between 4 MeV < E_n < 5 MeV. Paraffin was used to moderate neutrons to thermal energies when desired and substantial lead shielding was employed to limit the flux of gamma radiation into the detector when needed.

The data were collected from the CAEN DT5730B waveform digitizer and stored directly in waveform format for further analysis.

III. Pulse shape analysis and performance metric

All scintillator prototypes were examined with identical experimental setups. Three different methods of PSA were used and a figure of merit calculated to allow a direct comparison of the different materials and algorithms.

A. Charge comparison method

One of the longest established method which was even realized in analog electronics is the charge comparison method. As illustrated in Fig 3, taken from a measurement using stilbene as scintillator material, the scintillation pulse created by gamma radiation has a similar rise time, but much faster fall time compared to the scintillation pulse created by neutrons.

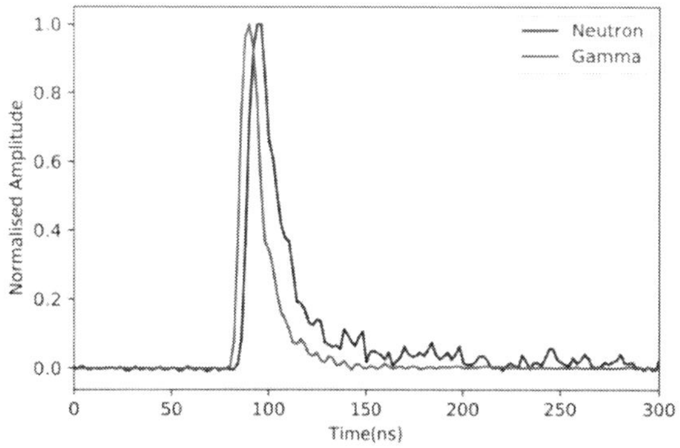

Fig. 3. Recorded waveforms measured using a Stilbene crystal exposed to radiation from an AmBE source. A clear distinction in pulse rise time and pulse fall time is visible in the signal from both radiation species. These pulse shape differences can be exploited in PSA to identify each particle species..

A measured pulse will be integrated over two different time periods, the short gate and the long gate. In the case of a short, fast pulse, these two pulse integrals will be very similar, while in the case of a pulse with a slow decay time, the integral over the longer gate will yield a significantly larger value. Defining a value *PSD* using the integrated charge over the short gate Q_1 and the long gate Q_2, a discrimination variable between the two particle species can be defined as

$$PSD = 1 - \frac{Q1}{Q2}.$$

When this value is plotted as a function of the total energy deposited inside the scintillator volume, a separation between the two particle species is clear visible, as shown in Fig. 4 taken from a measurement of Stilbene exposed to thermal neutrons from an AmBe source.

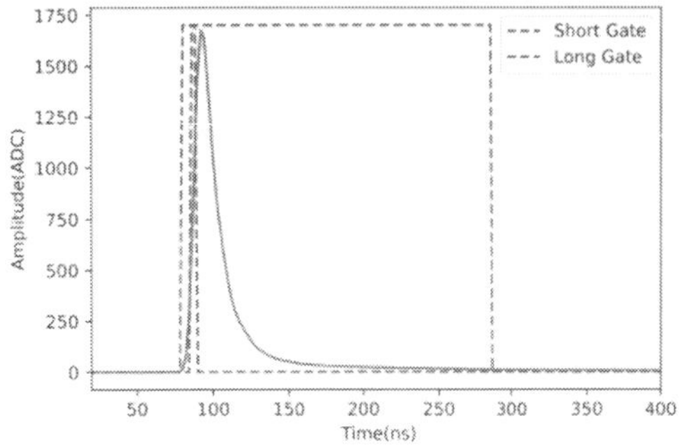

Fig. 4. Illustration of the charge comparison method for a pulse emitted by a stilbene scintillator irradiated by neutrons. The boxes indicate the integration regions for short and long pulse gates providing the charge integral Q_1 and Q_2 used for PSA.

B. *Fourier component analysis*

The use of a waveform digitizer provides access to the full waveform, allowing a wider form of analysis to be performed. Most PSA algorithms are performed in the time domain. The full access to the waveform allows PSA to be performed in the frequency domain as well [6,7]. This type of method offers several advantages over time domain based methods, as it mitigates many of the effects of noise common in scintillation signals by reducing the dependence on the timing aspects of the signal.

One particular method in the frequency domain is provided by a simple Fourier transform of the signal, greatly simplifying the deconvolution of the signal for an affordable computational cost. In many cases, a simple comparison of the amplitudes in the high frequency (HF) and low frequency (LF) domains for the Fourier transform may be sufficient. As the zeroth component is equal to the integral of the entire signal in the time domain, the comparison of the zeroth and first order component can be considered equivalent to the charge comparison methods. As such, the PSD factor is given as the ratio of the sum over all Fourier components starting with the first order component over the sum of all Fourier components starting at zeroth order, yielding a range of $0<PSD<1$.

C. *Risetime method*

The risetime method is a variant of the charge comparison method in that it tries to exploit not the fall time differences, but differences in the risetime of the scintillation light pulse due to the different scintillation mechanisms for neutrons and gamma rays [8]. In this method, the time it takes a pulse to reach a set fraction of the maximum pulse height is measured. The date presented here compare the risetime between 20% and 80% of the pulse. Cheers This method could also be realized in analog electronics, while this study uses a full software implementation of the algorithm. A PSD value is defined from the fractions of the leading edge in a CFD algorithm f_1 and f_2 as

$$PSD = -1/2\left(\left(\frac{t1}{log(1-f1)}\right) + \left(\frac{t2}{log(1-f2)}\right)\right).$$

D. *Figure of Merit*

In order to compare different detector materials and different PSA algorithms, a metric has to be introduced to measure their relative performance on an equal footing. As outlined above, the scintillation crystal was the only variable during the experiment. The metric used in this comparison follows the well established definition of a Figure of Merit (FoM) for two variables with reasonably distinct (ideally Gaussian) distributions. It is assumed that the distribution of the aforementioned PSD factors for the two particle species is Gaussian in shape. The FoM is then defined as the ratio of the separation S of the mean value of the two distributions divided by the sum of their individual full width at half maximum, d_1 and d_2:

$$FoM = \frac{S}{d1 + d2}.$$

The larger the FoM value, the better the detector performance, with values FoM<1 signifying a strong overlap of the two distributions and values of FoM>3 showing a complete separation. A FoM of 1 is considered a good separation [9].

IV. Results

Three materials were exposed to fast and moderated neutrons emitted from an AmBe source. The light collection, detection and read-out was kept the same for all experiments. The three described PSA algorithms were applied to the recorded waveforms.

A. *CLYC:Ce*

CLYC:Ce is an inorganic scintillation material with strong pulse shape differences between thermal neutrons and gamma radiation. Given that most neutron detection relies on capture reactions, it's performance when detecting fast neutrons is expected to be poor. Fig. 5 shows the conventional PSA using charge comparison when CLYC:Ce is exposed to thermal neutrons in the presence of a gamma radiation field.

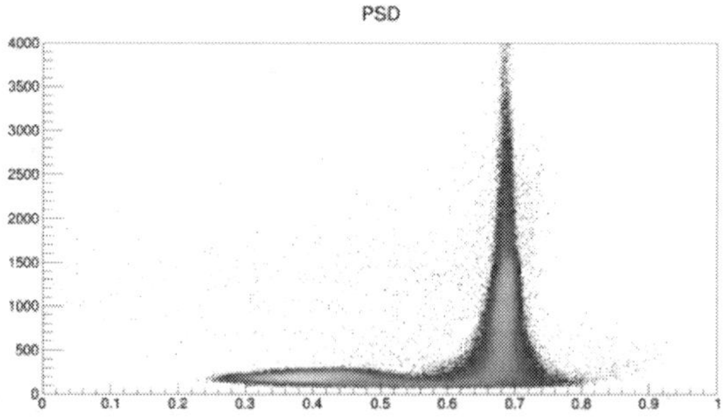

Fig. 5. Heatmap showing the distribution of the PSD variable on the x-axis versus the distribution of uncalibrated ADC channels representative of the total energy deposited in the scintillation volume on the y-axis for a CLYC:Ce crystal exposed to radiation from an AmBe source. A clear separation of the two particle species is visible. The PSD variable is calculated as described in the section on the charge comparison method.

A clear distinction between the two particle species is visible. The FoM was calculated to be 2.77. CLYC:Ce also showed an excellent FoM when the signals were analyzed using the HF Fourier method, yielding an FoM of 10.98. The corresponding heat map is shown in Fig. 6. An analysis with the LF Fourier component does not work well, with an FoM of 0.61. The risetime method yields an FoM of 1.77.

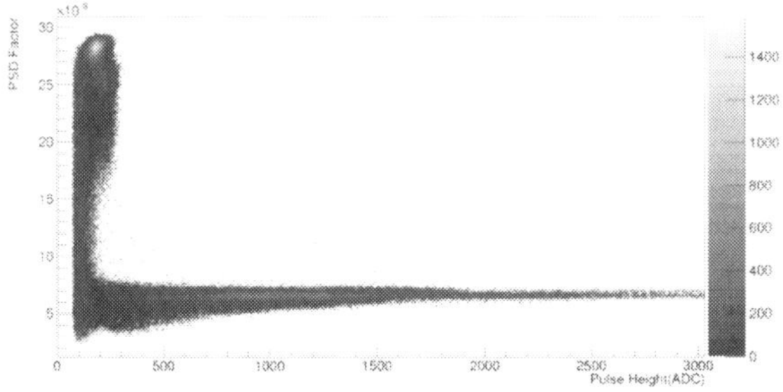

Fig. 6. Heatmap showing the distribution of the PSD variable on the y-axis versus the distribution of uncalibrated ADC channels representative of the total energy deposited in the scintillation volume on the x-axis for a CLYC:Ce crystal exposed to radiation from an AmBe source. The PSD variable is calculated for the HF Fourier method. A clear separation of the two particle species is visible.

B. CLLBC:Ce

Like CLYC:Ce, CLLBC:Ce is expected to work very well for thermal neutrons given it's predominant mode of neutron detection via capture reactions. The slightly higher density should result in a better gamma detection efficiency. CLLBC:Ce shows a significantly better discrimination between neutrons and gamma radiation compared to CLYC:Ce in the charge comparison method with a FoM of 3.69. The corresponding spectrum is shown in Fig. 7.

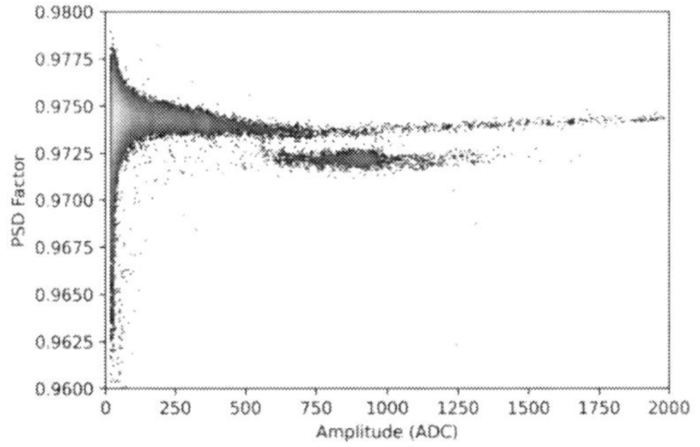

Fig. 7. Heatmap showing the distribution of the PSD variable on the y-axis versus the distribution of uncalibrated ADC channels representative of the total energy deposited in the scintillation volume on the x-axis for a CLLBC:Ce crystal exposed to radiation from an AmBe source. A clear separation of the two particle species is visible. The PSD variable is calculated as described in the section on the charge comparison method.

Due to an equipment malfunction, the HF Fourier method and the risetime method could not be evaluated. However, the LF Fourier method, depicted in Fig. 8 shows significant promise with a FoM of 4.1, exceeding the performance of the classical charge comparison method.

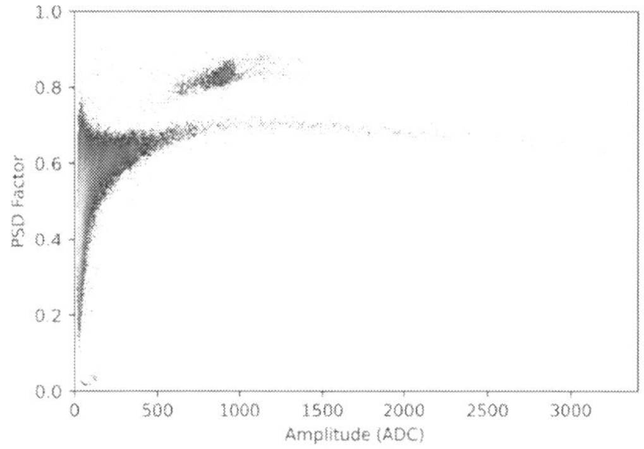

Fig. 8. Heatmap showing the distribution of the PSD variable on the y-axis versus the distribution of uncalibrated ADC channels representative of the total energy deposited in the scintillation volume on the x-axis for a CLLBC:Ce crystal exposed to radiation from an AmBe source. The PSD variable is calculated for the LF Fourier method. A clear separation of the two particle species is visible.

C. EJ-276

EJ-276 is, in contrast to CLYC:Ce and CLLBC:Ce an organic scintillator. Given the low density and low average atomic number, its performance in the detection of gamma radiation is expected to be poor. The material should also be better suited to faster neutrons and show some weakness in the detection of thermal neutrons. This is indeed observed, e.g as shown in Fig. 9 for the charge comparison method, yielding a FoM of only 0.85.

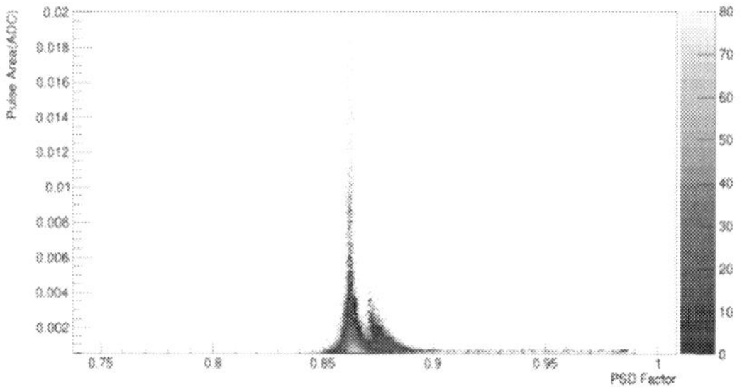

Fig. 9. Heatmap showing the distribution of the PSD variable on the x-axis versus the distribution of uncalibrated ADC channels representative of the total energy deposited in the scintillation volume on the y-axis for a EJ-276 scintillator exposed to radiation from an AmBe source. Only a poor separation of the two particle species is visible. The PSD variable is calculated as described in the section on the charge comparison method.

The LF Fourier component analysis failed to yield a measurable discrimination at all. The risetime component is also not promising, with a FoM of 0.57. An improvement, as with all investigated materials, can be seen in the frequency domain as shown in Fig. 10. The corresponding FoM is 1.19, showing a satisfactory separation between neutrons and gamma radiation.

https://doi.org/10.1051/epjconf/202328806010

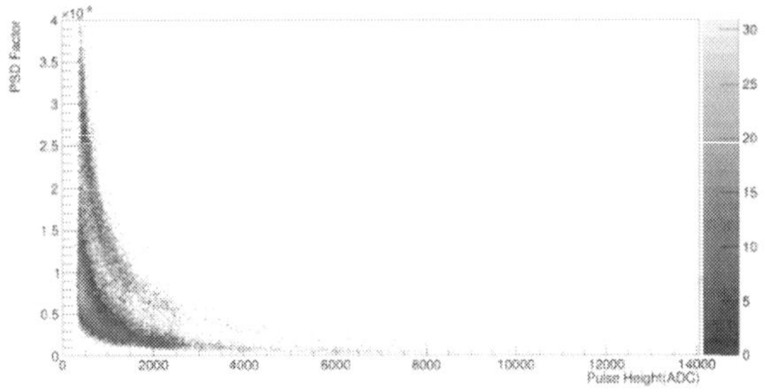

Fig. 10. Heatmap showing the distribution of the PSD variable on the y-axis versus the distribution of uncalibrated ADC channels representative of the total energy deposited in the scintillation volume on the x-axis for a EJ-276 scintillator exposed to radiation from an AmBe source. The PSD variable is calculated for the HF Fourier method. A satisfactory separation of the two particle species is visible.

V. Summary and conclusion

Three solid state scintillation crystals with pulse shaping capabilities have been tested using fast and thermal neutrons in the presence of a gamma radiation field. These scintillators have been coupled to an array of SiPM readout by a waveform digitizer, allowing a variety of offline PSA algorithms to be applied to discriminate between neutron and gamma radiation. Two inorganic crystals, CLYC and CLLBD, have been tested in comparison with an organic scintillator, EJ-274.

A figure of merit has been introduced to compare the performance of each detector material and each PSA algorithm. The results are summarized in Table II. The performance of the inorganic crystals CLLBC and CLYC using the charge comparison method when exposed to thermal neutrons yields a FoM of 3.69 and 2.77 respectively. As expected from the material properties, the response of EJ-276 to thermal neutrons is worse, yielding a FoM of 0.85. This can be significantly improved by using different analysis techniques, most notably the HF Fourier component analysis, which increases the FoM for EJ-276 to 1.19 and for CLYC to 10.98. For technical reasons, the HF Fourier component analysis for CLLBC could not yet be performed, but the LF Fourier component analysis also shows an improvement yielding a FoM of 4.1. It is illustrative to compare the performance of Fourier transform methods in the case of CLYC and EJ-276. In both cases, the HF Fourier component method provided significant improvements, while the LF Fourier component analysis provided a poor result (FoM of 0.611) for CLYC and failed to provide a separation for EJ-276, indicating that great care has to be taken to optimize each analysis algorithm to the detector material at hand.

In summary, it could be shown that coupling a scintillator exhibiting pulse shape characteristics to an array of SiPM yields a capable neutron detector with strong neutron/gamma discrimination. The performance of each detector material was found to be within expectations, i.e. EJ-276 as an organic scintillator excelled for fast neutrons, while being significantly outperformed by both in-organic scintillation crystals for thermal neutrons.

The introduction of a waveform digitizer allows the study and optimization of a variety of PSA algorithms which, with sufficient care will yield the best performance of a specific detector and read-out combination.

References

[1] D. Renker, "*Geiger-mode avalanche photodiodes, history, properties and problems*," *Nuclear Instruments and Methods in Physics Research A*, vol. 567, pp. 48–56, 2006.

[2] N. D'Olympia, P. Chowdhury, C. Lister, J. Glodo, R. Hawrami, K. Shah, and U. Shirwadkar, "Pulse-shape analysis of CLYC for thermal neutrons, fast neutrons, and gamma-rays," *Nuclear Instruments and Methods in Physics Research Section A: Accelerators, Spectrometers, Detectors and Associated Equipment*, vol. 714, pp. 121–127, jun 2013.

[3] A. Giaz, L. Pellegri, F. Camera, N. Blasi, S. Brambilla, S. Ceruti, B. Million, S. Riboldi, C. Cazzaniga, G. Gorini, M. Nocente, A. Pietropaolo, M Pillon, M. Rebai, and M. Tardocchi, "The CLYC-6 and CLYC-7 response to γ-rays, fast and thermal neutrons," *Nuclear Instruments and Methods in Physics Research, Section A: Accelerators, Spectrometers, Detectors and Associated Equipment*, vol. 810, pp. 132–139, 2016.

[4] C. Liao and H. Yang, "Pulse shape discrimination using EJ-299-33 plastic scintillator coupled with a Silicon Photomultiplier array," *Nuclear Instruments and Methods in Physics Research, Section A: Accelerators, Spectrometers, Detectors and Associated Equipment*, vol. 789, pp. 150–157, 2015.

[5] E. Van Loef, G. Markosyan, U. Shirwadkar, M. McClish, and K. Shah, "Gamma-ray spectroscopy and pulse shape discrimination with a plastic scintillator," *Nuclear Instruments and Methods in Physics Research, Section A: Accelerators, Spectrometers, Detectors and Associated Equipment*, vol. 788, pp. 71–72, 2015.

[6] M. J. Safari, F. A. Davani, H. Afarideh, S. Jamili, and E. Bayat, "Discrete Fourier Transform Method for Discrimination of Digital Scintillation Pulses in Mixed Neutron-Gamma Fields," *IEEE Transactions on Nuclear Science*, vol. 63, no. 1, pp. 325–332, 2016.

[7] G. Liu, M. J. Joyce, X. Ma, and M. D. Aspinall, "A digital method for the discrimination of neutrons and γ rays with organic scintillation detectors using frequency gradient analysis," *IEEE Transactions on Nuclear Science*, vol. 57, no. 3 PART 3, pp. 1682–1691, 2010.

[8] B. D'Mellow, M. Aspinall, R. Mackin, M. Joyce, and A. Peyton, "Digital discrimination of neutrons and γ-rays in liquid scintillators using pulse gradient analysis," *Nuclear Instruments and Methods in Physics Research Section A: Accelerators, Spectrometers, Detectors and Associated Equipment*, vol. 578, pp. 191–197..

[9] G. Ranucci., "An analytical approach to the evaluation of the pulse shape discrimination properties of scintillators.," *Nuclear Instruments and Methods in Physics Research Section A: Accelerators, Spectrometers, Detectors and Associated Equipment*, vol. 354(2-3), pp. 389–399, 1995.

.

Simulation design of online Beam Monitoring for the MULTISCAN 3D Laser-Plasma X-Ray Source

Eric Garrido[1,*], David Tisseur[2], Cedric Carasco[1] and Bertrand Perot[1]
[1] CEA, DES, IRESNE, DTN, SMTA, Nuclear Measurement Laboratory, France
[2] CEA, DES, IRESNE, DER, SPESI, Dosimetry, Sensors and Instrumentation Laboratory, France

(*) eric.garrido2@cea.fr

Abstract— **Many private and public actors in the security field use imaging systems with embedded X-ray accelerators for rail, seaports and border controls. These accelerators produce X-rays by the Bremsstrahlung effect, leading the output beam to have a very wide energy spectrum which precise form is difficult to know, especially the end-point energy. Mastering this energy distribution is nevertheless essential for certain applications like material discrimination. In addition, fluctuations over time or even from one pulse to another can disturb the analysis of the images and it is expected that the new X-ray sources based on laser-plasma accelerators, which are developed in the frame of MULTISCAN 3D H2020 project, will also increase this beam instability. In this context, we study a direct beam measurement at the exit of the target with a minimum footprint and a fast response, so that the results can be used in real time. A 50 mm × 50 mm × 2 mm scintillator is irradiated on the side by the X-ray beam (i.e. on a 50 mm × 2 mm face) and filmed from the top (i.e. on a 50 mm × 50 mm face). The light intensity of the image obtained by the camera is directly proportional to the dose deposition in this scintillator, and the depth profile is related to the X-ray energy spectrum. A study on different materials (PVT, CsI, BGO) is presented in order to determine the optimal scintillator. Several unfolding algorithms are examined to reconstruct the X-ray spectrum from monochromatic response functions of these detectors, which are calculated using MCNP code for a range of photon energies covering the expected X-ray spectrum. This study based on simulation results shows the feasibility of this technique, it appears that the Sequential Least SQuares Programming (SLSQP) minimization algorithm gives the most promising results and that the CsI scintillator provides better spectral information. It is probably due to the intermediate dose profile shape between PVT (flat falling tail) and BGO (steep rise). The binning of the depth profile (typically 100 µm or less) is also an important parameter to describe the initial steep march, and thus to keep the spectral information therein.**

Keywords **—Laser-plasma, X-ray source, monitoring, scintillator.**

I. Introduction

ALTOUGH commonly used in the security field, the energy spectrum of the X-ray beams are generally not fully mastered, both in intensity and in energy spectrum, especially the end-point energy. The current method used to evaluate their energy spectrum consists of performing several transmission measurements with increasing thicknesses of material plates, like aluminum, inserted between the accelerator and the detection system [1,2]. The transmitted signal, as a function of aluminum thickness, is unfolded to obtain the beam spectrum, see Fig. 1. However, this method requires a specific setup, is time-consuming, and therefore does not allow pulse-to-pulse online monitoring during the inspection.

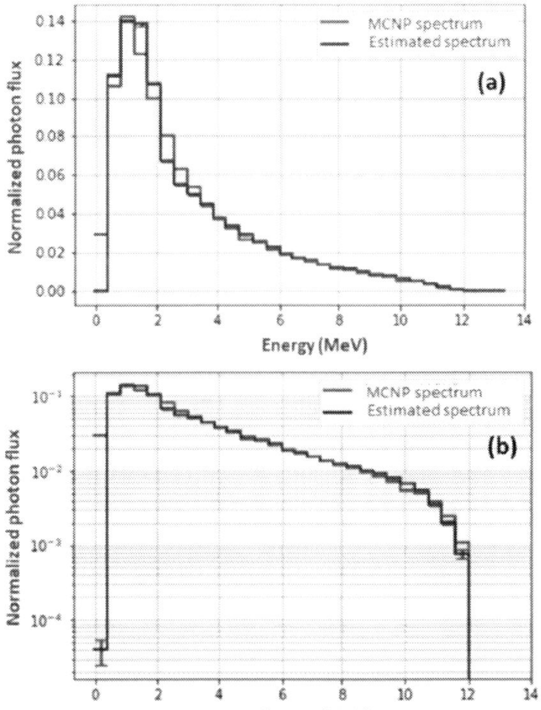

Fig. 1. Typical bremsstrahlung X-ray spectrum produced by a 12 MeV electron LINAC (linear accelerator), calculated with MCNP and estimated with the method using increasing aluminium plates [2]. (a) Spectrum in linear scale, and (b) in semi-logarithmic scale.

Having a non-invasive and instantaneous measurement would permit real-time monitoring of beam shape and intensity, allowing an adjustment or adaptation of the imaging algorithms to obtain more accurate results.

The solution studied in this work consists in measuring with

https://doi.org/10.1051/epjconf/202328806011

a camera the light emitted by a thin scintillation plate crossed by the incident beam on one of its small faces. Therefore, it makes it possible to perform a beam characterization in a single acquisition with a minimum footprint. This requires the thin crystal plate to be long enough for a sufficient dose deposition, and a scintillation material with a good light yield at the proper wavelength for the camera. The first part of this article describes the proposed setup.

The second part of this study concerns the development of the analysis code that allows, from dose deposition data, to retrieve the X-ray spectrum distribution. To this aim, we calculate a database of dose-deposition response functions for mono-energy photons covering the range of the X-ray beam spectrum, and for the different scintillation materials, using MCNP 6.1 Monte Carlo computer code [3]. Then, we study different unfolding algorithms to reconstruct the X-ray beam spectrum from a linear combination of these response functions, and we compare the results for the different scintillators.

II. Measurement principle

As mentioned in the introduction, this method utilizes the thickness of a scintillating crystal to observe the deposition within its depth, following the setup described in [1] and schematized in Fig. 2. An example of image acquisition is shown in Fig. 3 and the associated dose profile in Fig. 4. This measurement offers spectral information similar to multiple acquisitions with increasing material thicknesses between the beam and detector, but in a single acquisition, enabling real-time monitoring.

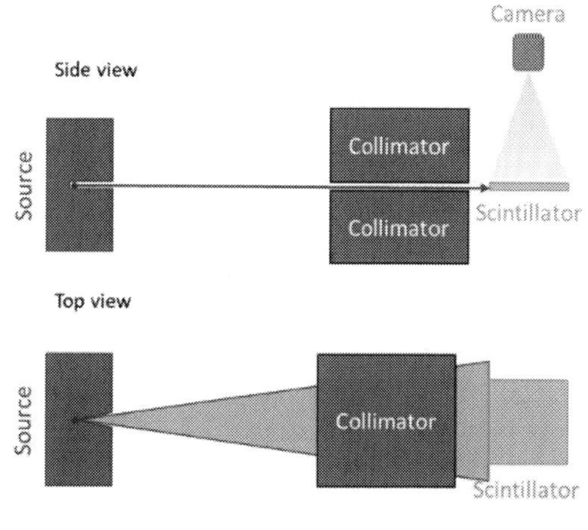

Fig 2: Set-up of the 50 mm × 50 mm × 2 mm scintillator and the camera with respect to the X-ray beam.

The poly-energetic spectrum of the X-ray beam can be discretized into a sum of mono-energetic beams, as shown in Fig. 5. Note that increasing the number of channels would improve the characterization of the X-ray spectrum, as can been seen at the low energies in Fig. 5. We consider that the dose deposition in the scintillation plate resulting from this poly-energetic spectrum is a weighted sum of the dose depositions from each of these mono-energetic beams, as formalized by equation (1):

$$D = \sum_i P_i \cdot D_i \qquad (1)$$

where D is the total dose deposition profile (Fig. 4), P_i is the weighting of the mono-energy contribution i in the poly-energetic spectrum, and D_i is the dose deposition profile of energy i (see Fig. 5, bottom panel).

Fig. 3: Example of a dose deposition image in a CsI scintillator irradiated by the bremsstrahlung X-ray beam of a 9 MeV electron LINAC simulated with MCNP 6.1. The line with the double arrow indicates the axis for the associated dose profile shown on Fig. 4.

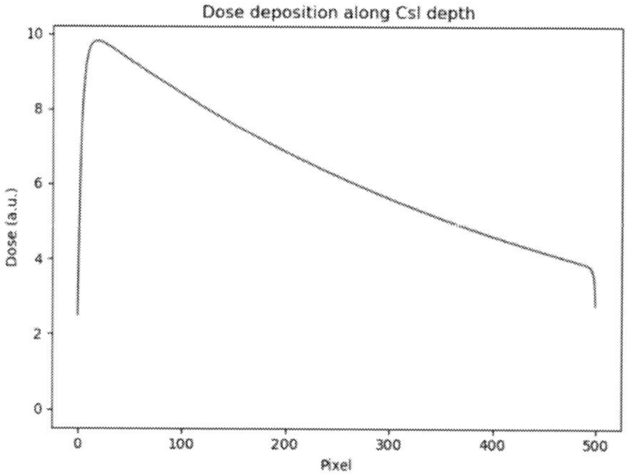

Fig 4: Dose profile along the blue arrow axis on the Fig. 3.

By knowing the dose deposition profiles of each mono-energetic contribution (database of response functions), we can determine their respective weights by minimizing the difference between the total measured dose and the dose resulting from their weighted sum. We thus obtain a spectral characterisation of the poly-energetic beam.

III. Unfolding algorithm

We first build a database of dose deposition profiles in the three 50 mm × 50 mm × 2 mm scintillation plates (CsI, BGO, PVT) for mono-energetic photons covering the expected 0-10 MeV range of the X-ray beam, using MCNP 6.1 simulations. To remain consistent with the experiment described in reference [1], the pixels of the camera and of the simulated dose profiles have a size of 100 µm. This parameter and its associated limitations will be discussed later.

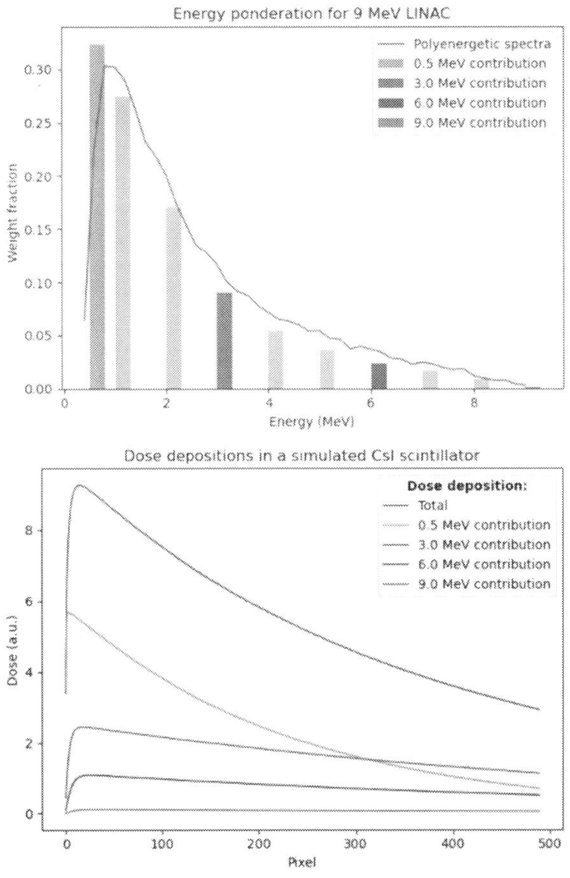

Fig. 5: Segmentation of a poly-energetic spectrum into a histogram of 10 mono-energetic contributions (top picture). Dose deposition profile simulated on a CsI scintillator for the poly-energetic spectrum, and a few examples of dose profiles for mono-energy photons (bottom picture).

An example for the simulation results is given in Fig. 6 for the CsI detector. The MCNP model reproduces an X-ray fan beam impinging a 50 mm × 50 mm × 2 mm scintillator irradiated on one of its thin faces (i.e. a 50 mm × 2 mm side) by a monochromatic photon source from 500 keV to 10 MeV.

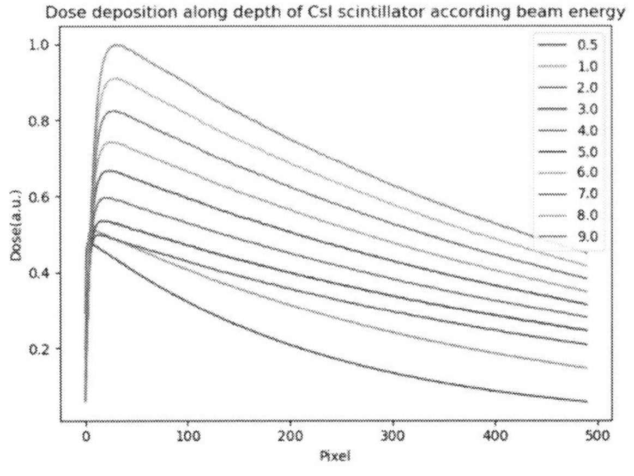

Fig. 6 : Dose deposition profiles in the CsI scintillator for single-energy photons

Fig. 7 compares the dose deposition profiles in PVT, CsI and BGO scintillators for the 1 MeV and 6 MeV beams, showing differences caused by their respective atomic numbers and densities listed in Table I. A high density and high atomic number material like BGO (7.13 g.cm^{-3}, $Bi_4Ge_3O_{12}$ – Z_{eff} = 74) exhibits a sharp, decreasing exponential profile, while the low density and low atomic number PVT (1.05 g.cm^{-3}, $[CH_2\text{-}CH\text{-}CH_3]_n$ – Z_{eff} = 5.6) shows a slow linear decay. CsI with a 4.51 g.cm^{-3} density and Z_{eff} of 54 appears to be intermediate. It is also important to note that the deformation of the dose deposition as a function of energy becomes more significant with the increase of the density and atomic number of the scintillator as the comparison of the top and bottom figures on Fig. 7 depicts that.

TABLE I
CHARACTERISTICS OF SCINTILLATOR MATERIALS

Material	Density (g.cm^{-3})	Z_{eff}	Maximum wavelength	Light-yield (photons/keV)
PVT	1.05	5.6	423 nm	10
CsI	4.51	54	550 nm	60
BGO	7.13	74	480 nm	8.5

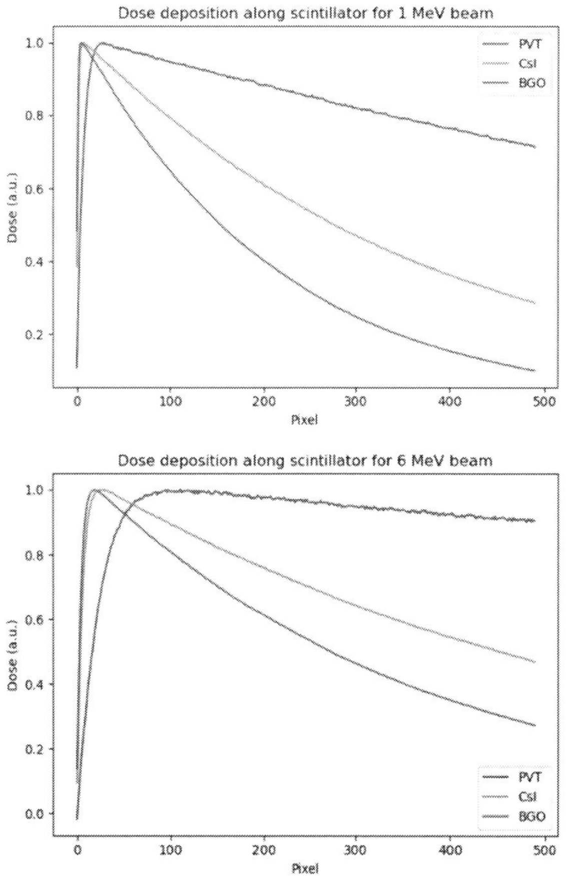

Fig. 7 : Dose depositions in different scintillators obtained from simulations at energies of 1 MeV (top picture) and 6 MeV (bottom picture).

The unfolding consists in determining the weights of each mono-energetic contribution that best approximates the dose deposition obtained with the poly-energetic beam. For each iteration where the algorithm generates a new weights vector for the mono-energetic contributions, it calculates the residual distance, or second-order norm ($norm_{L2}$), along all depth pixels j between the actual dose d and the reconstructed dose d', as described below:

$$norm_{L2} = \left[\sum_j (d_j - d'_j)^2\right]^{1/2} \quad (2)$$

We opted for a turnkey code that enabled us to obtain a quick response. Therefore, we chose the SciPy suite [3], which allows to test different minimization methods. This library, simple and quick to implement, required only a few parameters as input such as the minimization function equation (2), its tolerance and the maximum number of iterations. We have defined constraints on the output X-ray spectrum, which follows the well-known bremsstrahlung spectrum shape with a maximum intensity at low energy, and the fact that no mono-energetic contribution can be negative.

A minimization method need to be selected, we decide to chose one of those implemented by default in the SciPy library: Nelder-Mead, Powell, CG, BFGS, L-BFGS-B, TNC, COBYLA, SLSQP, and trust-constr. In order to determine the most suitable one, we compare all of these methods for the three tested scintillators, and Table II reports the obtained residual distance described in equation (2). It appears that the SLSQP method stands out from the others with lower values (below 0.1) for all three scintillators. We can add that CsI and BGO dose profiles are better fitted than that of PVT.

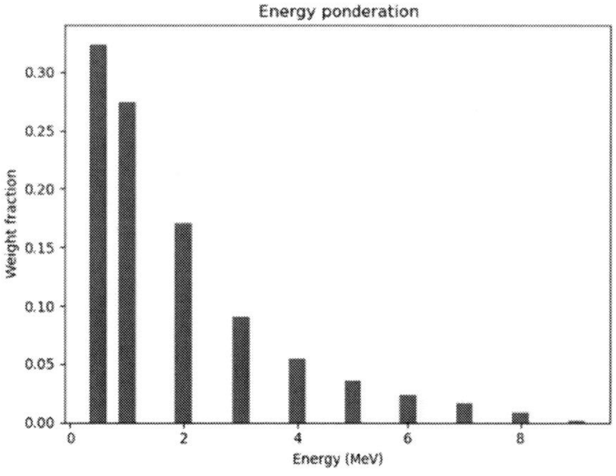

Fig 8: Resampling on 10 energy channels of the spectrum shown in Fig. 5.

Fig. 9 presents the SLSQP results (fit of the dose profile resulting from the unfolding) for each scintillator. The bottom plot focusses on the first 5 mm of the scintillator, showing that the number of pixels in the rising part, up to the dose peak, is quite limited. Therefore, thinner pixels than those of the camera (100 µm as in [1]) are necessary to describe this steep rise with a sufficient number of samples, especially for the CsI and BGO scintillators. This is confirmed by other studies (not reported here) showing that using pixels down to 10 µm significantly improves the fitting results for high-density scintillators.

TABLE III
RESIDUAL DISTANCE BETWEEN THE MEASURED AND THE FITTED DOSE DEPOSITION PROFILES

Method	PVT	CsI	BGO
Nelder-Mead	3.468	2.687	8.851
Powell	1.256	1.107	0.405
CG	4.643	1.396	0.492
BFGS	10.42	24.71	10.88
L-BFGS-B	7.723	0.864	0.421
TNC	0.689	0.282	0.203
COBYLA	1.470	7.154	2.856
SLSQP	0.044	0.007	0.016
trust-contr	4.224	0.670	0.322

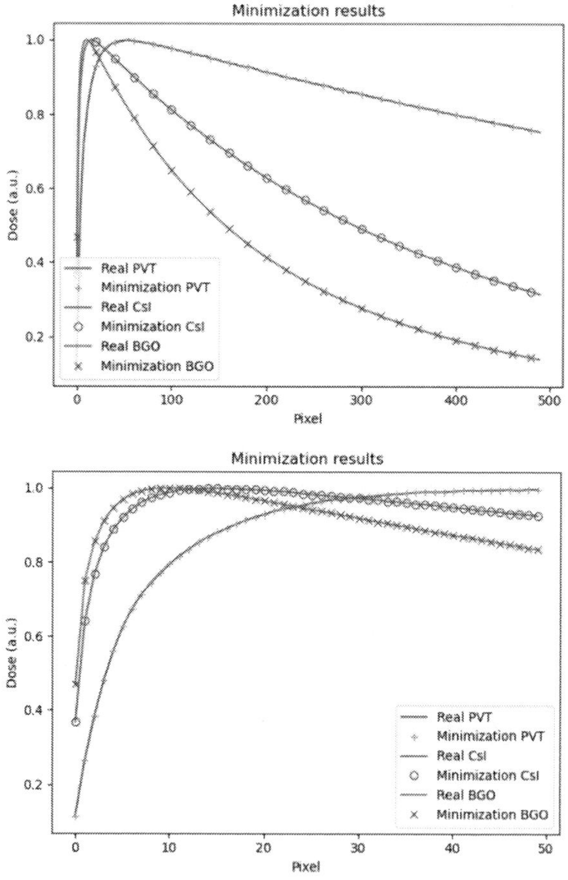

Fig. 9: Results of the minimization obtained with the SLSQP method for each of the tested scintillators (only 1 result point out of 20) on the whole depth of the scintillator (top) and focused on the 50 first pixels (up to 5 mm depth) with all the results points (bottom).

Finally, we can globally observe a satisfactory agreement in Fig. 10 between the original X-ray spectrum ("Real") and the result of the algorithm, which is also described by the residual distance in Table III.

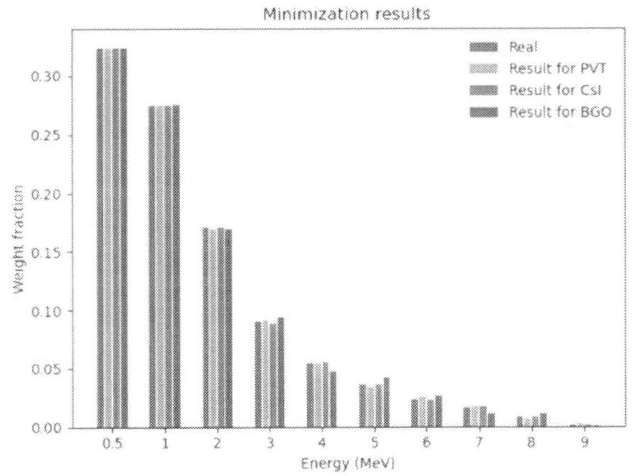

Fig 10: Comparison of the final weights calculated by the minimization code using the SLSQP method for each of the tested scintillators.

TABLE IIIII
RESIDUAL DISTANCE BETWEEN THE REAL AND THE UNFOLDED SPECTRA

Method	PVT	CsI	BGO
SLSQP	4.233e-3	2.862e-3	1.189e-2

The best agreement between the real and estimated X-ray spectra is obtained with CsI, as for the fit of the dose deposition profile. However, the spectrum reconstructed with BGO is not as good as that of PVT, probably because of insufficient sampling of the steep rise in the beginning of the dose profile. Due to its low density, low light-yield, and emission wavelength ill-suited for the camera [1], the noise on the PVT dose profile may however lead to a poorer reconstructed spectrum in a pulse-to-pulse analysis. As mentioned earlier, the overall shape of the dose deposition is directly related to the density and average atomic number of the scintillator. We will check in future work our guess that excessively steep rising shapes, for high density scintillators, or excessively flat falling tails, for low density, are not optimal for efficiently performing these types of calculations. Therefore, it is worth exploring the range of intermediate densities and atomic numbers, such as with CsI, by studying new scintillators like CaF_2 (3.18 g.cm^{-3}) or GGAG (6.60 g.cm^{-3}). We will also decrease the pixel size (< 100 µm) to achieve a better sampling in the first few mm of the dose deposition profile, especially for high densities materials showing an extremely steep rise.

IV. CONCLUSION AND PERSPECTIVES

In this study, we have demonstrated that it is possible, using a minimization algorithm and data obtained from simulations, to reconstruct the dose deposition profile of an X-ray beam poly-energetic spectrum in PVT, CsI and BGO scintillators, by unfolding it into the dose profiles due to mono-energetic photons. We have also demonstrated that using a thin scintillator plate irradiated on its side, along a large thickness, allows recording enough data points in a single pulse acquisition to perform this type of reconstruction. This study allowed us identifying a specific minimization method (SLSQP) for the dose profile fit. However, the choice of the optimal detector is still open, and we recognize that a finer pixel size (<100 µm) is necessary to achieve more promising results, especially for materials with higher densities. The presented binning of 100 µm did not allow for an accurate reproduction of the dose rise up to its maximum for the BGO. This explain why on our study the CsI with its intermediate density, gives us the best results. On the next step, we will reduce the pixel size (< 100 µm) of the dose profiles in the simulation data and we will extend our study to new types of scintillators, such as CaF_2, LYSO, GGAG and $CdWO_4$.

In order to test the unfolding robustness, we will also add noise parameters (based on statistical fluctuations of pulse-by-pulse acquisitions) and increase the number of energy channels

Another important aspect of future investigations involves experimental measurements with the set-up depicted in Fig. 1 at the end of accelerators beamlines. The dose deposition images recorded with the camera placed above PVT, CaF_2, CsI, BGO and GGAG thin scintillator plates will enable us to compare the experimental and simulated dose profiles, as well as the reconstructed X-ray beam energy spectra.

Ultimately, we will optimize and test the capability of the algorithm to reconstruct the X-ray spectra pulse by pulse.

ACKNOWLEDGMENT

This work has received funding from the European Union's Horizon 2020 research and innovation program under grant agreement No 101020100. This text reflects only the author's views and the Commission is not liable for any use that may be made of the information contained therein.

REFERENCES

[1] N. Estre et al., "Design of a very efficient detector for High Energy Tomography," 2018 IEEE Nuclear Science Symposium and Medical Imaging Conference Proceedings (NSS/MIC), Sydney, NSW, Australia, 2018, pp. 1-3
[2] M. Maulin et al., "Characterization of the X-ray spectrum of a linear accelerator," EPJ Web conf, Volume 253, 2021.
[3] Pelowitz, Denise B et al., "MCNP6 User's Manual", Los Alamos, NM, USA 2013
[4] Pauli Virtanen et al., "SciPy 1.0: Fundamental Algorithms for Scientific Computing in Python." Nature Methods, 2020, pp. 261–272.

Thermal neutron detection based on resistive gaseous devices

G. Bencivenni[1,*], I. Balossino[2], G. Cibinetto[2], R. De Oliveira[3], R. Farinelli[2], G. Felici[1], I. Garzia[2], M. Gatta[1], M. Giovannetti[1], S. Gramigna[2], L. Lavezzi[4], G. Mezzadri[2], G. Morello[1], G. Papalino[1], M. Poli Lener[1], M. Scodeggio[2]
[1] Laboratori Nazionali di Frascati – INFN, Italy
[2]INFN - Ferrara, Italy
[3]CERN, Switzerland
[4] INFN-Torino, Italy
(*) giovanni.bencivenni@lnf.infn.it

Abstract— **In the framework of the uRANIA (u-Rwell Advanced Neutron Imaging Apparatus) project, we are developing innovative thermal neutron detectors based on resistive gaseous devices such as micro-Resistive WELL (μ-RWELL) and surface Resistive Plate Counter (sRPC).**
The μ-RWELL is a single amplification stage resistive MPGD developed for HEP applications. The amplification stage, based on the same Apical® foil used for the manufacturing of the GEM, is embedded through a resistive layer in the readout board. The resistive layer is realized by sputtering the back side of the Apical® foil with Diamond-Like-Carbon (DLC). A cathode electrode, defining the gas conversion/drift gap, completes the detector mechanics. The deposition of a thin layer of ^{10}B4C on the cathode surface allows the thermal neutrons conversion into ^{7}Li and α ions, which can be easily detected in the active volume of the device. Results from tests performed with different detector layouts show that a thermal neutron (25 meV) detection efficiency up to 7% can be achieved with a single detector. A comparison between experimental data and the simulation of the detector behaviour has been performed.
In parallel, we are proposing the development of thermal neutron detectors based on a novel RPC concept. The sRPC is a revolutionary RPC based on surface resistive electrodes realized by exploiting the well-established DLC sputtering technology on thin (50μm) polyimide foils, the same used in the manufacturing of the μ-RWELL. The DLC foil is glued onto a 2 mm thick float-glass. The 2 mm gas gap between the electrodes is ensured by spacers made of Delrin®, inserted without gluing at the edges of the glass supports. By replacing DLC with ^{10}B4C sputtered electrodes, the device becomes sensitive to thermal neutrons. Different layouts of ^{10}B4C coated electrodes have been tested, allowing to achieve efficiency up to 6%. The robustness, ease of construction, and scalability of the sRPC technology should allow the construction of cost-effective large area detector units as required by applications in homeland security (such as Radiation Portal Monitor).

Keywords **— Thermal neutrons, Boron coating, MPGD, RPC**

I. Introduction

Neutrons are subatomic particles with neutral electric charge and a mass slightly greater than that of a proton. Because of their high penetration power, they are used, by means of dedicated detectors, to probe the structure and the motion of the matter in a complementary way to X-ray imaging [1]. Neutrons play an important role in the illicit traffic of radioactive material, as any abnormal neutron flux, detected with proper devices, could suggest the presence of heavy radioactive material. This is the case of Radiation Portal Monitors (RPM), installed to detect illicit traffic of radioactive material through port or airport, that require for several square meters of sensitive area.

In literature, neutron detection is categorized according to the nuclear processes, mainly neutron capture and elastic scattering (not threatened in this paper). Nuclides such as ^{3}He, ^{10}B, ^{6}Li, have a high neutron capture cross section and large probability of absorbing a neutron, as reported in Table 1.

Reaction	Q-value [MeV]	Cross section for thermal neutrons [barns]
$^{3}He + n + ^{3}H + p$	0.764	5330
$^{10}B + n \rightarrow ^{7}Li + \alpha$	2.31	3840
$^{6}Li + n \rightarrow ^{3}H + \alpha$	4.78	940

Table 1. Reactions of interest involving neutron capture, the released energy (Q-value) and their cross-section for thermal neutrons are shown, where n represents a neutron, p a proton, α an alpha particle (He^{++}).

A well-known converter used for neutron detection is the ^{3}He due to its large cross-section (5330 at 25 meV). Helium-3 proportional counters use the $^{3}He(n,p)^{3}H$ reaction for the detection of thermal neutrons, where the energy of the reaction (Q=764 keV) is carried out as kinetic energy of the daughter products, which approximately move back-to-back. In the latest years a shortage of this element increased the ^{3}He cost with a large impact on neutron detector production and their application in the fields of interest. A call for alternative solutions triggered the URANIA and URANIA-V project [2,3], where two different gaseous detector technologies (μ-RWELL and sRPC) associated with a suitable ^{10}B4C based converters have been proposed. Thanks to the neutron capture process (tab. 1), neutron detection can be activated through a thin deposition of ^{10}B on suitable detector electrodes. The reaction products (1),

^{7}Li and α particles, generated nearly back-to-back on the converter, interact with the gas creating ionization electrons that are amplified by the amplification stage of the gaseous detector. The sputtered ^{10}B is chemically stable, mechanically robust and with very good adherence on several substrates. Moreover, the deposition of ^{10}B, based on standard industrial process, can be performed with high uniformity over surface in the order of ≈ 1 m^2.

$$n + {}^{10}B \rightarrow {}^{11}B* \rightarrow \alpha + {}^{7}Li\ (6\%)$$
$$n + {}^{10}B \rightarrow {}^{11}B* \rightarrow \alpha + {}^{7}Li + 477\ keV\ \gamma\ (94\%) \quad (1)$$

II. MICRO-RWELL FOR NEUTRON DETECTION

The μ-RWELL [4], Fig. 1, is a single-amplification stage resistive MPGD composed of two elements: the cathode, a simple FR4 PCB with a thin copper layer on one side, and the μ-RWELL_PCB, the core of the detector. The μ-RWELL_PCB, a multi-layer circuit realized by means of standard photo-lithography technology, is composed of a well patterned single copper-clad polyimide foil acting as amplification element of the detector; a resistive layer, realized with a Diamond-Like-Carbon (DLC) film sputtered on the bottom side of the polyimide foil, as discharge limitation stage; a standard PCB for readout purposes, segmented as strip, pixel or pad electrodes. The neutron conversion is no longer the gas inside the detector (as in the Helium-3 proportional counter), but a thin layer (few micrometers thick) of ^{10}B4C deposited on the cathode, exploiting a well-known industrial sputtering process.

The thickness of the converter is a crucial parameter to be tuned to maximize the neutron conversion probability and at the same time to reduce the probability that the products are stopped by the same material, then not entering in the detector active volume.

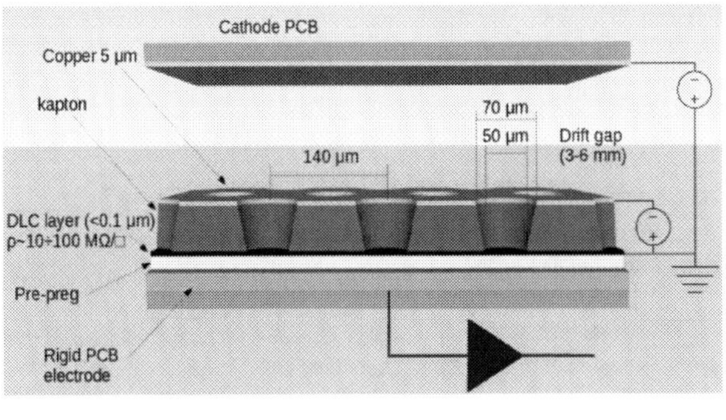

Fig.1. Sketch of μ-RWELL detector. The basic version of the neutron device is obtained with a suitable ^{10}B coating of the cathode surface. The detector is operated with Ar/CO2/CF4 (45/15/40) gas mixture.

The detector, operated with Ar/CO_2/CF_4 (45/15/40) gas mixture, is equipped with a single 10x10 cm^2 pad electrode readout with custom FEE developed by LNF electronic pool. The readout is based on the CREMAT CR-110 single channel charge sensitive amplifier (Gain = 14mV/fC - [5]), followed by the CR-200 shaping amplifier with a shaping time of 1 μs.

The simplicity and the modularity of the μ-RWELL can be exploited to adapt the technology to any desired application. Moreover, its compactness makes possible the realization of a stack of such detectors that, operating in OR-mode, sums up their efficiency.

III. NEUTRON CONVERTERS FOR μ-RWELL

The basic version of the neutron detector based on μ-RWELL technology is obtained with a suitable ^{10}B deposition on the planar cathode. The first prototypes were equipped with planar cathodes with different ^{10}B deposition thickness: 1.5, 2.5, 3.5, 4.5 μm.

The characterization of these prototypes was performed at the ENEA HOTNES facility in Frascati [6], where a ^{241}Am-B neutron source is placed in a cylindrical cavity delimited by polyethylene walls. HOTNES exploits a polyethylene shadow bar that prevents fast neutrons from directly reaching the samples. The effect of the shadow bar and the cavity walls is combined in such a way that the thermal neutron fluency is nearly uniform. The resulting fluence rate at the HOTNES reference irradiation plane is 758±16 cm^{-2}s^{-1}. Experimental measurements and simulations evaluate the neutron fraction absorbed by the mechanical structure of the detector (mainly the cathode PCB made of FR4): a fraction of 17.7% is removed from the neutron fluence rate. A sketch of the HOTNES setup is shown in Fig. 2.

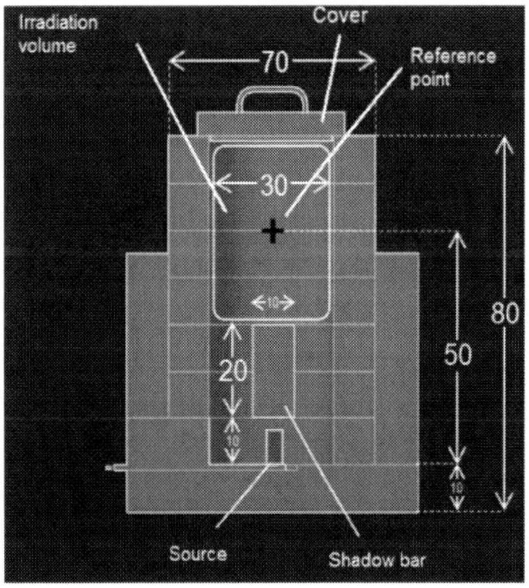

Fig. 2. Drawing of the HOTNES setup with the radioactive source and the shadow bar in the middle.

The energy spectrum of the HOTNES source is peaked at ~100 meV, with ~290 meV FWHM.

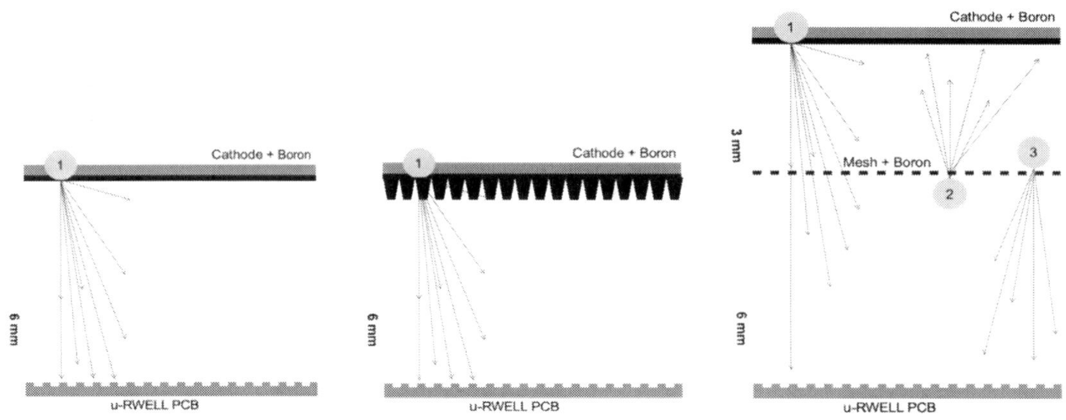

Fig.4. Sketch of the three converter layouts used for the μ-RWELL detector.

In Fig.3 the thermal neutron efficiency (as measured at HOTNES) as a function of the thickness of the ^{10}B layer deposited on the planar cathode structure is compared with the results obtained with the simulation. The detector response simulation is based on GEANT4, SRIM[7] and GARFIELD++[8] software codes used in cascade: the first one considers the physics of thermal neutron interaction with the ^{10}B layer, while the other two SW codes simulate the ionization of the heavy particles in the gas mixture and the drift of the free electrons towards the amplification channel of the detector.

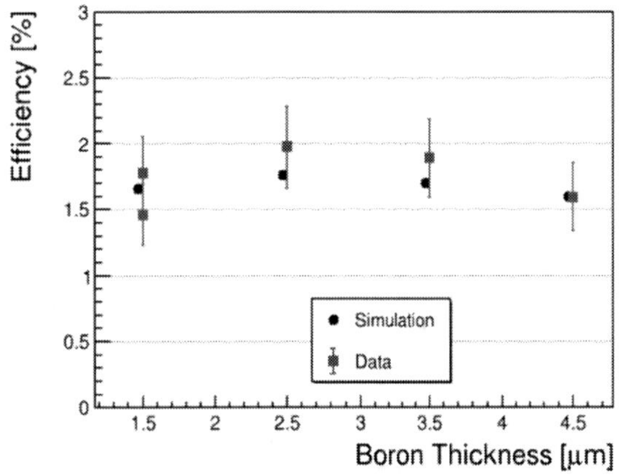

Fig. 3. Neutron detection efficiency of a μ-RWELL detector with planar boron-coated cathode as a function of the ^{10}B thickness. The measurements have been performed at the HOTNES calibrated neutron source facility.

The efficiency measured with the neutrons provided by HOTNES, about 2% for a 2.5 μm thick ^{10}B deposition, corresponds to about 4% for thermal neutrons (E_n ~25 meV).

These results clearly show that to increase the conversion efficiency it is not convenient to have a thicker boron layer, but rather increase as much as possible the converter surface. For this purpose, two different converter layouts have been considered:

- a grooved cathode
- a metallic mesh coupled with a standard planar cathode

All converter components were sputtered with a 2.5 μm thick ^{10}B layer. Fig.4 sketches the three converter layouts studied.

The summary of the results obtained with the three μ-RWELL layouts is reported in Fig.5, where the efficiency has been normalized for thermal neutron. It is worth to remind that, due to the higher neutron cross section, the efficiency for 25 meV neutron is twice the one obtained with the neutrons emitted by the HOTNES source (~100 meV). The study demonstrates that it is possible to achieve efficiency up to 8% with a single detector layer.

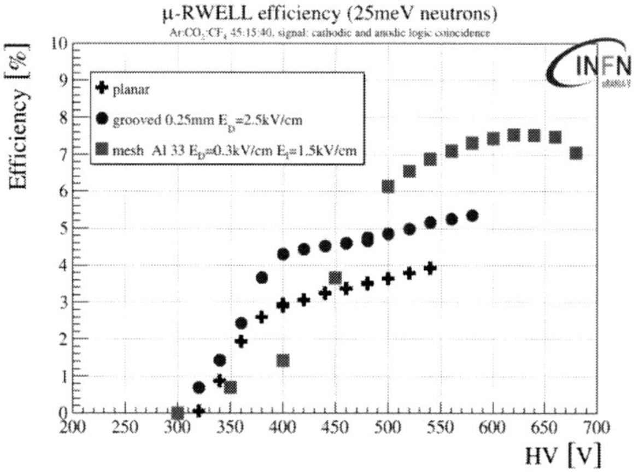

Fig.5. Efficiency obtained with the three different μ-RWELL layouts, tested at the HOTNES neutron source facility. The efficiency has been normalized to 25 meV neutrons.

IV. THE SURFACE RESISTIVE PLATE COUNTER FOR NEUTRON DETECTION

The Surface Resistive Plate Counter (sRPC- [9]) is a novel RPC based on surface resistivity electrodes, a concept completely different from traditional RPCs [10,11,12] that use electrodes characterized by volume resistivity. The electrodes of the sRPC exploit the well-established industrial DLC sputtering technology on thin (50μm) polyimide foils, which have already been introduced in the manufacturing of the μ-RWELL.

The DLC foil is then glued to a 2mm thick float-glass, which is characterized by excellent planarity. The DLC surface resistivity interesting for this detector ranges between 0.1÷1 GΩ/square. The 2 mm gas gap between the electrodes is ensured by E-shaped spacers made of Delrin®, inserted without gluing at the edges of the glass supports. The electrodes sandwich is inserted in a fiber-glass box that acts as gas volume container, Fig. 6. With this layout, operated with the usual RPC gas mixture ($C_2H_2F_4$/Iso-C_4H_{10}/SF_6 = 93.5/5/1.5), an efficiency for m.i.p. of 95-97% and a time resolution of ~1 ns have been measured.

Fig. 6. sRPC prototype: the electrodes, realized with the DLC coated Apical® foil glued on float-glass supports, are coupled by E-spacers made of Delrin® (the internal tooth of the spacer is 2 mm thick). The sRPC prototypes were readout with the same electronics developed for the μ-RWELL.

The sRPC technology, initially developed for HEP applications, can be exploited for thermal neutron detection, by replacing DLC with ^{10}B4C sputtered electrodes. Three different combinations of ^{10}B4C coated electrodes, according to the orientation of the chamber inside HOTNES, have been tested:

- Boron cathode - DLC anode
- DLC cathode - Boron anode
- Boron cathode - Boron anode

As shown in Fig. 7 with these layouts, a thermal neutron detection efficiency of 4%, 2%, and 6% has been respectively achieved.

The construction simplicity and the scalability are the main advantages of the sRPC technology, that should allow an easy implementation of high efficiency multi-stack structure exploitable for large area Radiation Portal Monitors.

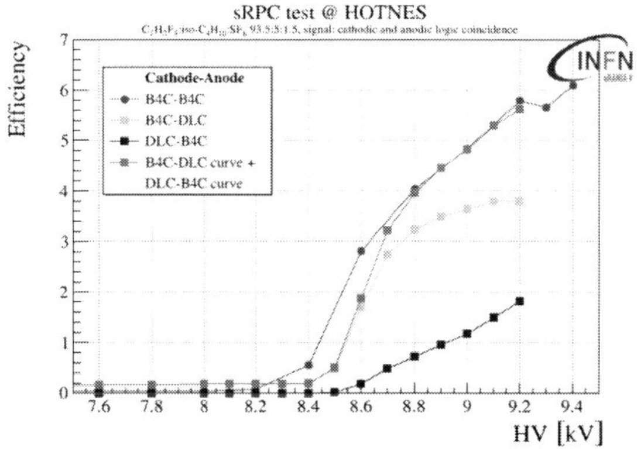

Fig.7. Efficiency obtained with the three different sRPC layouts, tested at the HOTNES neutron source facility. The efficiency has been normalized to 25 meV neutrons.

V. Conclusions

In the framework of the uRANIA project, we developed innovative thermal neutron detectors based on resistive gaseous devices such as μ-RWELL and sRPC. In such devices neutrons are converted on a ~2.5 um thick ^{10}B-coated cathode/electrode, releasing charged particles (α/7Li), which passage can be identified by the gaseous detector.

The μ-RWELL, a mature technology, with suitable converters allows to achieve a neutron detection efficiency per single layer ranging between 4÷8%.

Very promising results with ^{10}B coated sRPC have been achieved (4÷6% /single detection layer), opening the way for a cost effective, scalable thermal neutron detection technology for homeland security applications.

Neutron detection efficiency of 40÷50% seems to be achievable with both technologies by stacking order of 10 detector layers.

References

[1] J. Banhart et al., "X-ray and neutron imaging – Complementary techniques for materials science and engineering". International Journal of Materials Research, vol. 101, no. 9, 2010, pp. 1069-1079.
[2] https://phase1.attract-eu.com/showroom/project/u-rwell-advanced-neutron-imaging-apparatus-urania/
[3] I. Balossino et al., u-RANIA: a neutron detector based on μ-RWELL technology, 2020 JINST 15 C09029.
[4] G. Bencivenni, R. De Oliveira, G. Morello and M.P. Lener, The micro-Resistive WELL detector: a compact spark-protected single amplification-stage MPGD,2015 JINST 10 P02008.
[5] https://www.cremat.com/CR-110-R2.pdf
[6] A. Sperduti, et al., Results of the first user program on the Homogeneous Thermal Neutron Source HOTNES (ENEA/INFN), JINST_12_P12029 (2017).
[7] www.srim.org
[8] https://gitlab.cern.ch/garfield/garfieldpp
[9] G. Bencivenni et al., The surface Resistive Plate Counter: An RPC based on resistive MPGD technology, Nucl. Instr. & Meth., A 1046 (2023) 167728.
[10] R. Santonico, R. Cardarelli, Development of Resistive Plate Counters, Nucl. Instr. & Meth. A 377 (1981) 187.
[11] Yu. Pestov et al., A spark counter with large area, Nucl. Instr. & Meth. 93 (1971) 269.
[12] M. Anelli et al., Glass electrode spark counters, Nucl. Instr. & Meth. A 300 (1991) 572.

Flying DEMon, a newborn startup for environmental monitoring

C. Altomare*‖, L. Di Venere†‖, F. Giordano†‡‖, S. Loporchio§‖, A. Mastroserio¶‖, D. Serini†‖,

†INFN Sezione di Bari, Via E. Orabona 4 - 70125 Bari, Italy
‡Dipartimento Interateneo di Fisica dell'Università e del Politecnico di Bari, Via Amendola 173 - 70125 Bari, Italy
§Politecnico di Bari, Via Amendola 126/b - 70126 Bari, Italy
¶Università degli Studi di Foggia, Via A.Gramsci 89/91, - 71122 Foggia, Italy
‖Flying DEMon s.r.l., Via Amendola 173 - 70125 Bari, Italy
(*) corrado.altomare@ba.infn.it, info@flyingdemon.it

***Abstract*—Gamma-ray spectroscopy and dosimetry are complementary techniques used to locate and identify radioactive sources containing gamma-ray-emitting radioisotopes. Gamma-ray spectroscopy is extensively studied for various applications across multiple fields, including homeland security, environmental radioactivity monitoring, tackling illegal trade of radioisotopes, and medical sciences.**

Introducing our newly established startup, Flying DEMon s.r.l., comprised of young researchers, academic professors, and backed by university support. Our venture aims to advance project development, leveraging the grant awarded through the E-TEC2 contest initiated by ENAC. The team will showcase their comprehensive work plan, highlighting the project's competitiveness and self-sustaining potential.

The objective of our startup is to harness cutting-edge technologies in the field of gamma spectroscopy and dosimetry, adaptable for deployment via Unmanned Aircraft Systems (UAS). This innovation holds significant promise for environmental monitoring, facilitating tasks such as pinpointing widespread radioactive sources or identifying concealed and hard-to-reach nuclear waste. Additionally, this advancement holds potential for applications in military, security, and industrial oversight.

Our research focus primarily revolves around real-time and rapid gamma-ray analysis in open-field environments. Our group not only supports the core project objectives but also enables its applicability in diverse and non-traditional sectors, such as Agritech.

***Index Terms*—Gamma spectroscopy, radioactivity dosimetry, environmental monitoring, startup**

I. Introduction

The phenomenon of natural environmental radioactivity is an intricate and pervasive aspect of the Earth's composition that has captivated the attention of scientists, researchers, and environmentalists alike. From the moment our planet came into existence, various radioactive elements have been an integral part of its geological and biological makeup. These naturally occurring sources of radiation emanate from elements like uranium, thorium, and potassium, which are found within the Earth's crust, oceans, and atmosphere. While often associated with potential health risks, natural environmental radioactivity also plays a fundamental role in shaping landscapes, influencing ecosystems, and contributing to the dynamic balance of the planet's natural processes. Natural radioactivity can also be influenced by human activities, and its natural concentration can change in the vicinity of urbanized centers, industrial facilities, extractive mines, and other human-altered environments. One of the most compelling environmental issues the world is poised to confront in the twenty-first century is the disposal of nuclear waste. The remarkably prolonged half-lives and associated environmental hazards of nuclear waste have rendered its satisfactory disposal a seemingly insurmountable challenge. Radioactivity monitoring encompasses the meticulous surveillance of both man-made radioactive sources and naturally occurring environmental radioactive sources. Among the latter, a distinction is drawn between Naturally Occurring Radioactive Materials (NORM) and their evolved form, Technologically Enhanced NORM (TENORM). NORM encompasses materials harboring primordial radionuclides and natural radioactive elements like radium, uranium, thorium, and potassium, along with their radioactive decay progenies such as radium and radon. Importantly, these materials remain undisturbed by human activities, maintaining their natural state as outlined by Chaki-2008 [1]. TENORM, on the other hand, materializes as a byproduct of industrial processes that exploit natural resources, spanning fields such as coal combustion, fertilizer production, metal and oil processing, and mineral ore extraction. The International Atomic Energy Agency (IAEA), International Commission on Radiological Protection (ICRP), and Rahman-2012 [2]–[4] have collectively emphasized the significance of TENORM in this context. The crux of concern arises when the presence of unwanted substances surpasses the concentrations of the environmental radioactive background radiation levels, a situation elucidated by Chapman [5].

Italy has meticulously crafted a comprehensive legal framework to regulate and monitor radioactivity environmental levels, encompassing a range of statutes and regulations that underscore the nation's commitment to public health and environmental safety. These legal provisions, aligned with international agreements and tailored to address national priorities, establish a robust system for tracking and managing radioactive substances in the environment.

Key among these laws is Legislative Decree No. 230/95,

which transposes European Union directives on radiation protection into Italian law. This decree outlines the fundamental principles and requirements for the protection of workers, the general public, and the environment from ionizing radiation. It sets forth guidelines for the assessment of exposure levels, the establishment of dose limits, and the licensing and supervision of activities involving radioactive materials.

Legislative Decree No. 152/2006, commonly referred to as the Environmental Code, is another pivotal statute. It addresses environmental protection comprehensively and includes provisions specifically related to the monitoring of radioactivity in air, water, soil, and food. This decree establishes the legal basis for regulating radioactivity levels from various sources, including both naturally occurring and technologically enhanced radioactive materials.

The "Ispettorato Nazionale per la Sicurezza Nucleare e la Radioprotezione" (ISIN), stands as a cornerstone of Italy's commitment to ensuring the highest standards of radiological safety and protection. Operating under the purview of the Ministry of Health, ISIN plays a pivotal role in overseeing and regulating activities related to nuclear safety, radiological protection, and the management of radioactive substances.

ISIN's responsibilities encompass a wide spectrum of activities, including the regulation and supervision of nuclear facilities, radiological practices, and the transport of radioactive materials. Through rigorous inspections, audits, and assessments, ISIN ensures compliance with safety standards, thus mitigating potential hazards associated with ionizing radiation.

Moreover, ISIN collaborates closely with other regulatory bodies such as the Regional Environmental Protection Agencies (ARPA) and the National Institute for Environmental Protection and Research (ISPRA). This synergy facilitates comprehensive radiological monitoring and response across various environmental compartments.

On the other side, the "Agenzia Regionale per la Prevenzione e la Protezione dell'Ambiente" (ARPA) in upholding rigorous standards of radiological safety and environmental preservation. These agencies, distributed across different regions of Italy, are entrusted with the crucial responsibility of monitoring, assessing, and managing radioactivity levels in various environmental compartments. Through a combination of legal mandates, scientific expertise, and collaboration, ARPA agencies contribute significantly to safeguarding public health and maintaining the integrity of the environment.

Leveraging their specialized knowledge and advanced monitoring techniques, ARPA agencies systematically evaluate radioactivity levels in air, water, soil, and food, leaving no stone unturned in their pursuit of comprehensive environmental radiological control. They conduct routine surveillance, respond swiftly to emergent radiological incidents, and provide accurate, timely information to the public. By collaborating with institutions like the National Institute of Health (ISS), they contribute to Italy's collective efforts in ensuring the safety of both its citizens and the environment.

A. *Gamma spectroscpy and dosimetry*

Two primary approaches can be employed to address the challenges posed by the detection, quantification, and identification of radioactive materials in the environment. Gamma spectroscopy is a powerful analytical technique used to identify and quantify gamma-ray emitting radioactive isotopes present in various samples. Gamma rays are a type of ionizing radiation emitted during radioactive decay processes. These rays have high energy and can penetrate materials, making them useful for studying the composition of substances, ranging from environmental samples to industrial materials. Dosimetry is the measurement and assessment of radiation doses received by living organisms, objects, or the environment. It plays a crucial role in understanding and managing the potential health risks associated with exposure to ionizing radiation. There are several types of dosimetry, including personal dosimetry, environmental dosimetry, and medical dosimetry. Both gamma spectroscopy and dosimetry are integral to understanding, monitoring, and managing the potential risks and benefits associated with ionizing radiation in various contexts. They contribute significantly to ensuring the safety of individuals, the environment, and medical practices that involve exposure to radiation.

II. A Newborn Start-Up

Fig. 1. Flying DEMon logo.

Flying DEMon s.r.l. (Drone for Environmental Monitoring, see logo in Fig. 1) was established in 2023 as an innovative startup by a group of researchers from INFN (National Institute for Nuclear Physics), following the successful attainment of a project from the "Ente Nazionale per l'Aviazione Civile" known as #ETEC-Second Edition. This project aims to provide funding for startups and small medium enterprises. Subsequently, the company was registered as an innovative startup with the Chamber of Commerce.

The company's primary objective is to deliver services in the field of detection and environmental monitoring of radioactive elements. During its early months, the startup received accreditation as a spinoff of the University of Bari (Uniba). The founders' team (see Fig. 2) possesses extensive experience and expertise garnered from years of research and development activities within the field of astroparticle physics and high-energy experiments conducted in collaboration with INFN (National Institute for Nuclear Physics), the University of Bari, and the "Politecnico di Bari".

https://doi.org/10.1051/epjconf/202328806013

Fig. 2. Flying DEMon team.

III. Business Model

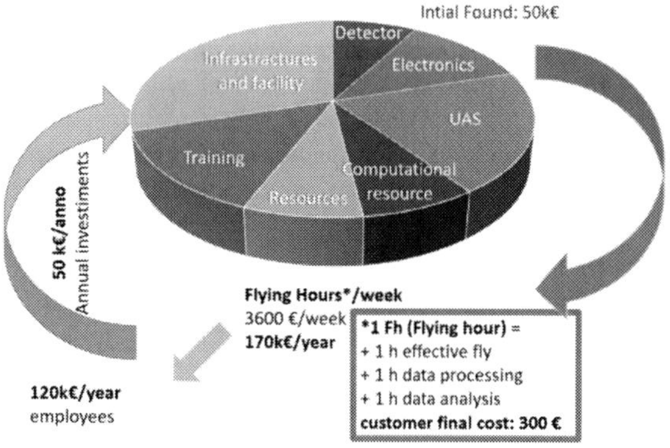

Fig. 3. Explanatory pie chart illustrating the company's expenditure, earnings, and reinvestment model.

The company's main revenues will be generated from monitoring services offered to both public entities and private enterprises. The cost of these services will be estimated based on the required flight hours and subsequent data processing, with the issuance of a measurement report interpreted by our radioactivity experts. Specifically, it is estimated that a total of three working hours for two operators will be needed for data processing and generating final results for the client for each actual flight hour. The service cost for the client will be calculated in Flying Hours, FH (1 FH = 1 actual flight hour + 2 hours of processing for two operators) (see Fig. 3). Additionally, the consultancy will offer the possibility of issuing certification by a qualified expert. A cost of 300 €/FH is anticipated in addition to fixed costs for instrument maintenance, travel expenses, etc.

A. Market

Flying DEMon s.r.l. offers its services to both public entities responsible for environmental radioactivity control (see "Introduction") and private companies that need to maintain their radioactivity levels below legally established limits, ensuring a safe working environment for the long- and short-term health of their employees. Our service is designed to provide rapid analysis of extensive or difficult-to-access areas, delivering quick results and accurate reports of the obtained data. Some examples of potentially interested public entities include the "National Inspectorate for Nuclear Safety and Radioprotection" (ISIN), which currently measures radionuclide activity deposited in the soil by collecting soil samples and analyzing them in the laboratory. This process is somewhat imprecise as it doesn't account for potential accumulation points due to landscape configurations. The utilization of drone surveys would allow for an initial analysis of large areas in these zones, providing a more reliable assessment of the executed sampling. This measurement is typically performed periodically by the entity. The map on the right in Fig. 4 illustrates the national territorial coverage of Cs-137 activity deposition measurements, highlighting limited coverage especially in central and southern Italy [9].

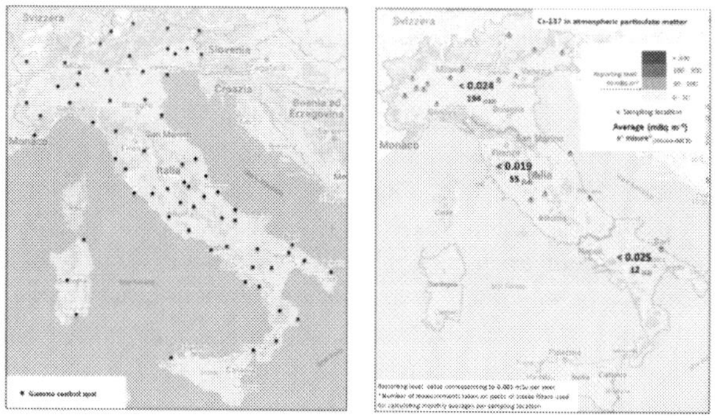

Fig. 4. Left: Gamma network central control spots of the ISIN. Right: Annual concentration of Cs-137 activity in ground deposition. [9]

The second type of clientele that the company targets are the businesses that are required by law to monitor the levels of radioactivity in their work environments. The service will be offered to consortia (for example, those in the metalworking or agri-food sectors), proposing the issuance of certifications mandated by law, provided by qualified experts. This way, customers will receive a certification with legal value. The list of activities that are legally required to monitor radioactivity is connected to the reuse and disposal of certain materials, such as processed and semi-processed metals, and is identified by a CER code ("Codice Europeo dei Rifiuti" the European Waste Code). Some of the companies currently involved in waste disposal related to these materials in Apulia have been identified: Oikos [10] s.r.l., Glob Eco [11] s.r.l., Ecologica [12] s.r.l., etc. Glob Eco s.r.l., headquartered in Molfetta (BA) in Apulia (Italy), specializes in environmental services and is focused on the collection and treatment of special waste. Oikos s.r.l., based in Bari, is a well-established service company in the environmental sector, providing consultancy, remediation, and waste disposal services. Ecologica s.r.l. offers customers a wide range of waste disposal services for various types of waste.

IV. Services offered

The company provides customers with a comprehensive range of services, starting from design and planning, all

https://doi.org/10.1051/epjconf/202328806013

the way to addressing issues related to radioactivity in specific areas. Our services encompass conducting surveys using drones to cover extensive zones "in-situ" and in "real-time". The ultimate outcome delivered by our team is a meticulous report of the analyzed area. This report includes identification of potential radioactive hotspots, quantitative assessment of radiation exposure (dosimetry), analysis of gamma spectra, and expert interpretation of the data. Additionally, if necessary, we can provide a certification issued by a qualified secondary level radioactivity expert.

A. Monte Carlo simulation and Digital twin

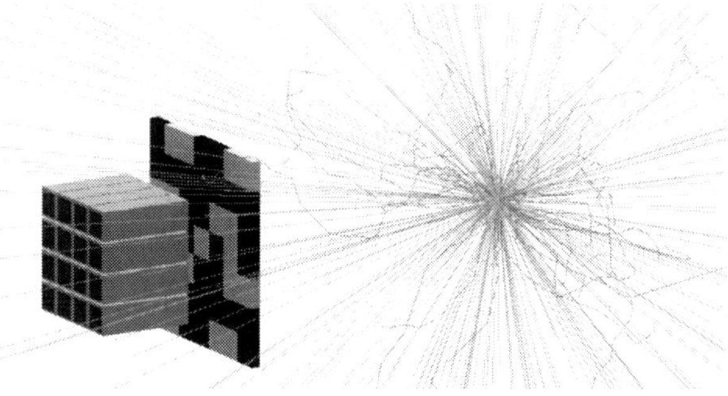

Fig. 5. Event display of a Co-60 decay example end interaction with a custom detector.

Each specific customer requirement is thoroughly examined through the utilization of a simulation software based on GEANT4 (Fig. 5), custom-developed by Flying DEMon team members [6], [7]. This enables us to professionally determine the optimal measurement plan for every customer request, considering the minimum dosimetric and spectroscopic thresholds required. This approach streamlines the verification process, ensuring efficient identification of potential radioactive sources within the analyzed area of interest.

B. Spectroscopy and dosimetry service

We employ both commercial tools and innovative instruments, designed and crafted by us [8], utilizing crystalline scintillators for conducting on-site and laboratory dosimetric analyses. Furthermore, we possess specialized equipment for radioisotope measurements using gamma spectroscopy of samples. This enables the detection of any traces of radioactivity surpassing natural radioactivity levels and facilitates the identification of present radioisotopic sources (see Fig. 6).

C. Radiometric survey "in-situ"

Our company specializes in the field of environmental monitoring services for the detection of gamma-emitting radioactive sources. We harness innovative and advanced technologies, such as drones equipped with highly sensitive detection equipment (see Fig. 7), to conduct precise and detailed surveys. Our goal is to ensure a safe and operational environment for both businesses and the public sector. We provide customized solutions to meet the specific needs of each client, delivering

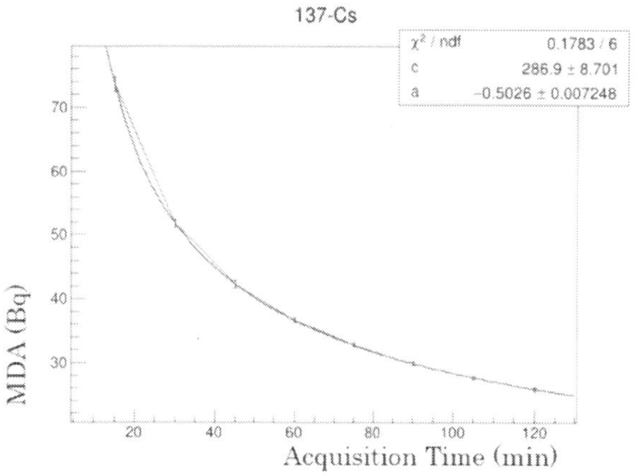

Fig. 6. Minimum detectable activity in Becquerels for a Cs-137 source measured in the laboratory as a function of acquisition time.

Fig. 7. Example of drones used and equipment for radiometric survey.

accurate and timely assessment of harmful radiation presence. Our experience and expertise allow us to effectively identify and map radioactive sources, contributing to responsible environmental management and enhanced community safety.

D. Agritech 4.0

Fig. 8. An example of "Agritech 4.0" idea.

The term "Agritech 4.0" refers to the application of Industry 4.0 technologies in the agricultural sector. Industry 4.0, also known as the fourth industrial revolution, is characterized by the widespread use of advanced digital technologies to automate and optimize production processes. Flying DEMon employs dosimetric spectroscopic techniques using UAS drones to create maps of agricultural land and study soil composition and hydration uniformity (see Fig. 8). Low-energy gamma ray measurements, generated by radionuclides like radon, can serve as secondary indicators of soil composition characteristics, thus facilitating the assessment of suitable areas for new crops. Furthermore, in conjunction with other sensors, these measurements can help address and understand the causes of unproductive or underutilized zones.

References

[1] Chaki, S. *et al.* "Technical Report on Technologically Enhanced Naturally Occurring Radioactive Materials from Uranium Mining Volume 1: Mining and Reclamation Background and Volume 2: Investigation of Potential Health, Geographic, and Environmental Issues of Abandoned Uranium Mines", Citeseer (2008)
[2] "International Atomic Energy Agency, General Environmental Impact Assessment Methodology Approach (Model Development Process) for NORM Sites Working Group; Reference approach for risk and environmental impact assessments necessary for management and remediation of NORM and legacy sites EMRAS II Norm and Legacy Working Group"
[3] "Extent of Environmental Contamination by Naturally Occurring Radioactive Material (NORM) and Technological Options for Mitigation", Technical Reports Series, 419 (2003)
[4] Abdel Rahman R. O. *et al.* "Planning and Implementation of Radioactive Waste Management System", Rehab Abdel Rahman, Radioactive Waste, IntechOpen, Rijeka (2012)
[5] Chapman P.M. *et al.* "Determining when contamination is pollution - weight of evidence determinations for sediments and effluents", Environment international, 33 4, 492—501 (2007)
[6] Altomare, C., Serini D. *et al.* "A full and customizable simulation of a scintillation tile equipped with SiPMs for Plastic Scintillator Detectors in the next generation of satellite experiments", Nucl.Instrum.Meth.A., 982, 164479 (2020)
[7] Altomare, C., Serini D. *et al.* "A complete MC optical photons tracking simulation of Plastic Scintillator Detectors for the next generation of satellite experiments", Journal of Physics: Conference Series, 012050 (2022)
[8] Altomare, C. *et al.* "A high efficiency fast-response gamma detector with mrad pointing capabilities", Nucl.Instrum.Meth.A., 1025, 166106 (2022)
[9] ISIN "La sorveglianza della radioattività ambientale in Italia", www.isinucleare.it, Report 08/2022
[10] Oikos s.r.l. "https://www.oikosambiente.it/"
[11] Glob Eco s.r.l. " https://www.globeco.info/"
[12] Ecologica s.r.l. "https://www.ecologica.it/"

3D imaging of a nuclear reactor using muography measurements with Micromegas detectors

Baptiste Lefevre[1,*], Héctor Gomez[1], Sébastien Procureur[1], David Attié[1], Laurent Gallego[2], Philippe Gonzales[3], Marion Lehuraux[1], Bertrand Lesage[4], Irakli Mandjavidze[1], Philippe Mas[1], Daniel Pomarede[1]

[1]CEA/DRF/Irfu, France
[2]CEA/DES/DDSD, France
[3]Assystem Engineering and Operation Services, France
[4]SOM-LIGERON, France
(*) baptiste.lefevre@cea.fr

Abstract—**Transmission muography is a non-invasive and non-destructive imaging method which allows to estimate the integrated density of a volume in a given direction (also referred as opacity). It relies on a reconstruction of muons tracks that crossed the studied volume compared to the corresponding open sky expectation. The experimental setup of the muography developed at CEA Irfu consists in portable muon telescopes. Each of these instruments has four Micromegas gaseous detectors, power and acquisition electronics, and an embedded computer connected to the network allowing remote control. It is then well adapted to deploy them in constrained environments. A muography measurements campaign has been carried out in the decommissioned nuclear reactors G2 and G3 at CEA Marcoule (France) which are expected to be dismantled. We were able to conduct 2D muographies and 3D analyses of inside G2's airtight concrete chamber. In this paper we describe the measurement protocol at the G2 and G3 reactors. We explain what was developed to improve the measurements between G2 and G3 campaigns. It shows how potential improvements were identified in the proof of concept and why we expect a better reconstruction for G3. Among the improvements, we describe how simulations prior to the measurements, denoising and in general automation play an important role to have a precise 3D image in reasonable time.**

Keywords—**Muography, nuclear reactor dismantling, 3D imaging.**

I. Introduction

MUOGRAPHY is a non invasive and non destructive imaging method that allows, under certain conditions, to estimate the densities inside a targeted volume. Muography is very penetrative compared to other techniques like X-rays and ultrasounds which are rapidly stopped by dense material like metal or concrete. Muography uses the natural muon flux on the contrary of X-rays or neutrography which needs radioactive sources.

2D and 3D non destructive analyses with muography are particularly interesting in constrained environments or on very opaque solids. Muography has been used in archaeology to search for hidden chambers in the Khufu's pyramid [1,2], in geophysical sciences to monitor volcanoes [3], in homeland security [3] and is currently used in many projects in the nuclear field.

Multiple techniques have been developed and can rely either on small deviations of the muons, or on their absorption or transmission [4]. Among those methods we will focus on transmission muography, which needs to place detectors underneath the studied volume. The muon flux, naturally generated in the atmosphere and mostly vertical will cross the studied volume before it reaches the detectors. Because the number of muons stopped is strongly correlated with the matter density, the volume density "shadow" will be projected on the detectors.

In each direction, this shadow represents the integrated density on a line and is called the opacity noted o as in (1), where ρ is the density integrated on the distance dl.

$$o=\int \rho \cdot dl\,, \tag{1}$$

Equation (1) can also be written in a vectorial form which is easier to implement in a program. To do so, the volume to image is modeled by voxels (3 dimensional pixels). In (2), o is the opacity of a line going though the volume, A is the vector of distances traveled in voxels by the line and $\vec{\rho}$ is the vector of densities of the voxels.

$$o=A\cdot\vec{\rho}\,, \tag{2}$$

Equations (1) and (2) are defined in [5] and used in [6].

Therefore, a muography is firstly represented as a 2D image, where each pixel counts the number of incoming muons for a given direction as in Fig. 1. This muography is then compared with a simulation of the muons expected for each direction in open-air conditions, so it is possible to compute an opacity image. On the opacity image, each pixel represents the integrated density for the studied direction as in Fig. 2.

To produce such images, multiple types of detectors exist. Emulsion films are sensitive to muon hits and are developed afterwards, scintillators work in real time and directly give pixelized images, whereas gaseous detectors work in real time and measure on copper strips the energy deposit induced by the gas ionization.

The muography team at CEA Irfu has developed muon telescopes composed of 4 Micromegas gaseous detectors [7].

These detectors use an Argon-based mixture and high voltages to detect the ionizations that charged particles create in gases. The electrons detached from gas atoms are accelerated and create a cascade that deposit signal on a readout plane. To increase the spatial resolution with less electronics, we use genetic multiplexed Micromegas [8]. The signal's amplitude is measured and analyzed in real time to trigger and store muon-candidate events. The electronic is controlled by an embedded computer in order to have an autonomous and remotely controllable system.

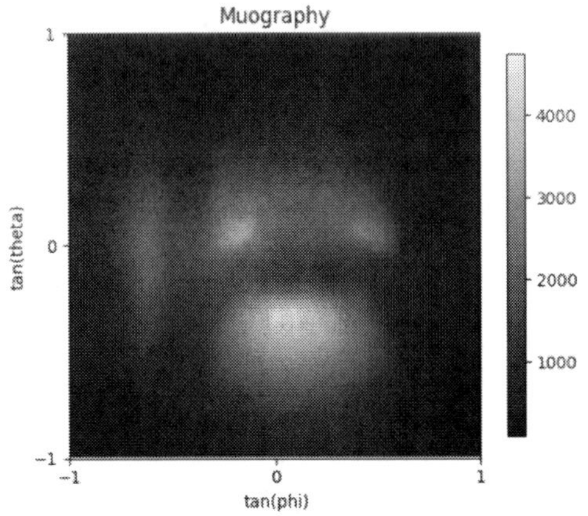

Fig. 1. Muography from below the charging block of the G3 reactor in Marcoule. Each pixel is a direction in space. The scale indicates the number of muons coming from the targeted direction. Orange to yellow pixels are the direction where the muon flux is the highest.

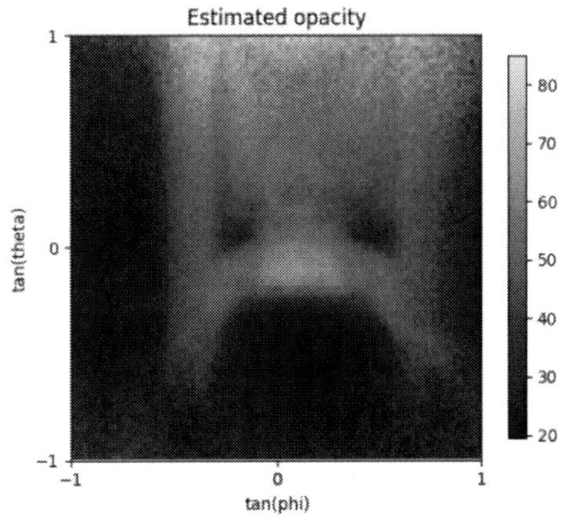

Fig. 2. Computed opacity (g/cm^2) from below the charging block of the G3 reactor in Marcoule. The opacity is the result of a comparison between the muography of Fig. 1 and an open-sky simulation. The most opaque areas (orange to yellow) show the concrete shielding of the reactor, the charging block and some graphite.

On the telescope, the embedded computer is in charge of maintaining the 4 detectors in condition of operation and of doing some of the preprocess steps. Among those steps, the detector data has to be demultiplexed in order to recover the position of the muon. This step is still subject to new developments which involve machine learning [9].

As it will be described later, the telescopes (composed of 4 Micromegas detectors) are very portable, which makes it possible to do muographies from multiple points of view of the same object, allowing to run 3D analyses by combination of muographies. Whereas the opacity measured in one muography only gives the sum of the densities in one direction, it might be possible to estimate separately all those densities.

II. G2G3 Project

The G2 and G3 reactors in Marcoule were the first french nuclear reactors connected to the grid. Both are UNGG type reactors (Natural Uranium, Graphite, Gas) and were started in 1958 and 1959 [10]. They worked respectively until 1980 and 1984, producing electricity and plutonium during more than 20 years. After their shutdown, the combustible and most of the supporting functions around the reactor were dismantled. Since 1996, G2 and G3 are monitored and maintained safe, before the full dismantling planned in the next decades.

In the context of this monitoring, the muography was foreseen to conduct a characterization of the reactors. The goals of the team in charge of the monitoring and dismantling (CEA DES/DDSD) are described in [10] : "compare the internal structures [...], characterize the civil engineering [...] and identify potential anomalies".

The project was launched in 2017 and started with simulations. Whereas X-ray and ultrasounds were not enough penetrative, the simulations demonstrated that the reactor could be imaged by muography.

It was decided to conduct a measurement campaign with telescopes under G2 as a proof of concept.

The complete telescopes weight around 40 kg, measure 60 cm × 60 cm and are 1 m high, one of them is shown on Fig. 3. It makes them very portable which allows to do multiple muographies with simple logistic.

Fig. 3. Muon telescope under the G3 reactor in Marcoule. The cubic telescope is mounted on 4 extension feet to step over a small wall.

Four telescopes were deployed at G2 during the first campaign. On site, the telescopes only need power, gas (a non inflammable Argon-based mixture) and to be connected together with a network router. The G2 G3 site operators were able to place, displace and setup the telescopes, whereas the data taking was controlled remotely by the CEA Irfu team.

This data taking procedure minimizes the intervention time on-site. This is only required to displace the telescopes and if possible to check on the gas consumption. All the data taking is already remotely handled. Ultimately, in more constrained or more radiating environments, our telescopes could be installed on a robot to avoid manual displacements.

Each of the 4 Micromegas on a telescope gives a hit position. With multiple hits we can reconstruct the original track of the muon. In the typical situation described in Fig. 4 the hits are not perfectly aligned. In the following analysis we will be interested in the track direction and the intersection between the track and the bottom Micromegas (called the X0Y0 point).

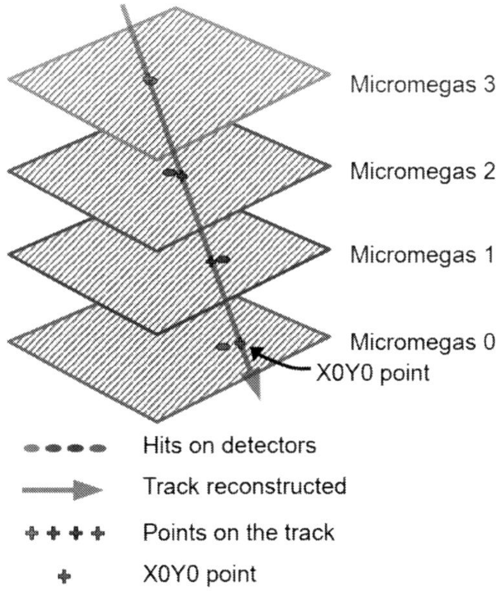

Fig. 4. Drawing of 4 Micromegas detectors with hits (colored points). A track is then reconstructed (gray arrow). The track crosses the detectors on slightly different points (colored crosses). The point on the lowest detector is called X0Y0 (blue cross).

III. Measurements

The first measurement campaign (dedicated to the G2 reactor) was a proof of concept of the methodology to manage the telescopes, take data, and do the 3D analysis.

To estimate the densities inside the reactor, it is necessary that the same regions in space are observed from different points of view. For practical reasons, it was decided to use only four telescopes like the one on Fig. 2. but to displace them to different positions. Once they are in position they can be remotely controlled over a secured network connection. In-person operation was only needed to displace the telescopes, once a month in general.

The remote controlling also allowed to have the muographies in real time during the campaign. With all the G2 muographies ready, we worked on a proof of concept of 3D reconstruction [6]. The idea is to use the SART algorithm to solve the ill-posed algebraic problem coming from the generalization of the formula (2). In (3), multiple rays in space are studied at the same time. The opacity is now a vector $\vec{o}$, A is the matrix of distances crossed in voxels by rays and $\vec{\rho}$ is still the vector of the voxel's opacities.

$$\vec{o} = A \cdot \vec{\rho}, \tag{3}$$

The experience acquired during the G2 campaign and the results of SART have been taken into account to define goals and potential improvements for the G3 campaign.

Firstly a new goal was to focus on the analysis of the graphite core because a small displacement of this part of the reactor was suspected. For this purpose we adapted the telescope's positions with simulations of a simplified graphite core.

Secondly, to adapt the telescopes to the G3 constraints we had to build new mechanical parts that facilitated the orientation of the telescopes in narrow spaces like on Fig. 5. With the compact mechanic illustrated below, we were able to place telescopes in narrow corridors at high angles to observe our targets.

Fig. 5. Muon telescope oriented at 45° in a 1 m large corridor. We recognize the telescope of Fig. 3 but mounted on rotation plates instead of its feet.

Finally we focused on the improvement of the muographies themselves. Indeed the noise present on them tends to be propagated in 3D and has a major effect during the densities estimation with the SART algorithm. Whereas the denoising was a very time consuming task for G2's reconstruction, we discuss here an automatic method and describe it's performance.

The improvement of the muographies is also possible with a better positioning of the particle hits on the detector. This step is linked to the detector demultiplexing, which made progress thanks to machine learning [9], but will not be discussed in this paper.

IV. Results

A. Preliminary simulations for the graphite core

In order to focus the analysis on G3's graphite core we did preliminary simulations to compare different scenario and optimize the reconstruction with a given amount of measurement time.

For graphite core, we were wondering if it was better to keep the telescopes longer on a small amount of positions or to do more positions with less statistics. To arbitrate this trade-off between number of positions and statistical significance we studied 8 scenarios with 4 telescopes available for 2 months. They are illustrated in Fig. 6 and described in Table I.

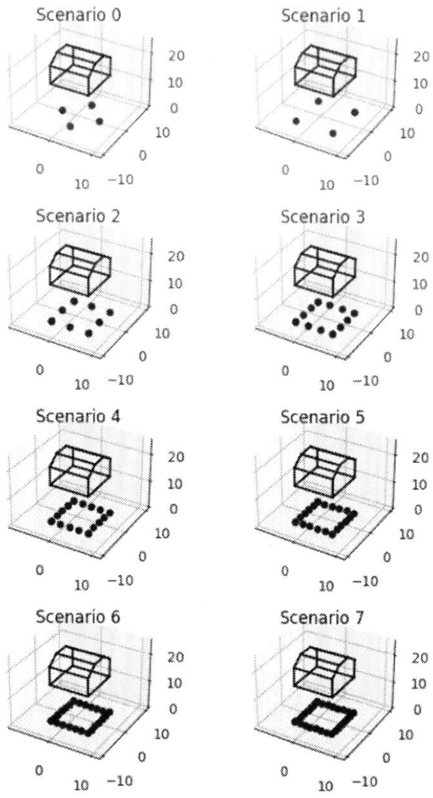

Fig. 6. Illustration of the 8 scenarios where the blue points represent the position of the telescopes under the shape of the graphite core.

TABLE I
SCENARIOS FOR A 2-MONTHS STUDY OF THE GRAPHITE CORE

Scenario	Number of positions	Duration per position (days)	Position below the graphite core
0	4	60	* Under each face's center
1	4	60	* Under each corner
2	8	30	* Corners * faces centers
3	12	15	* Corners * 3 positions per face
$N \in [4,7]$	N * 4	60 / (N+1)	* Corners * N positions per face

Every scenario is associated with a number and describe where (and how long) to put the 4 telescopes. The two first scenarios focus on 4 positions only (1 position per telescope). The other scenarios distribute the telescopes under the faces as illustrated in Fig. 6.

For each of these 8 scenarios, we simulated what muographies would be with the graphite core alone, which has a cubic shape with 2 big chamfers on the top as illustrated on Fig. 6.

To understand the trade-off between the number of positions and their duration, the effect of the measurement time increase may be seen on Fig. 7 where a position under a corner of the graphite core was simulated for 1, 5 and 25 days.

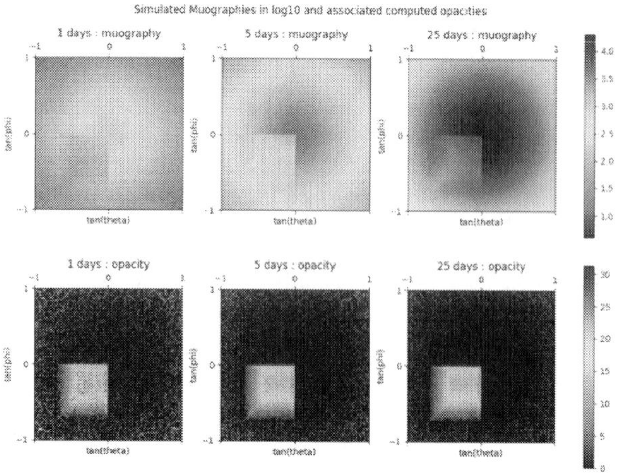

Fig. 7. Simulated muography under a corner of the graphite core for 1, 5 and 25 days. The second row shows the associated computed opacity (g/cm^2) matrices. We show that the increase of muon number in the muography induce a decrease of noise in the opacity.

On the muographies in Fig. 7, the distributions of muons stay the same when the number of muons increase linearly during time. This increase has few effect on the center of the image because muons are already statistically enough. However, on the edges of the image the opacity calculation needs longer simulations because there are less muons. Moreover the overall opacity image is less noisy with a longer muography.

To sum up, if we do less positions, the muographies will have more statistics which implies that the estimated opacity will be less noisy and significantly better on the edges.

To study the trade-off we ran the simulations mentioned in Table I and ran 3D reconstructions with the SART algorithm. We obtained the integrated densities displayed in Fig. 8. To simplify the visualization, only the pixel above 50% of the maximum of each image are displayed. The last row is the geometry used to run the simulation, ideally the projected densities should have the same shape and value.

Those simulations and reconstructions show that, even if the statistic is lower on each muography , the reconstruction is better with more positions. The improvement is visible on 3 different aspects: the overall shape of the graphite core is more correct, the density estimated in the graphite is more homogeneous and the density value is closer from the expected graphite density.

However we can see that the shape is still suffering some artifacts on the sides and below the core. Moreover the density value is underestimated.

B. Positions chosen at G3

Even if the best case simulated involves moving the telescopes every 7 days, 16 measurements of 14 days each were carried out for practical reasons. After those measurements, the telescopes were spread on 30 other positions to scan a wider range of the reactor.

The main limitation to the number of positions was that each displacement needed the intervention of the team on site.

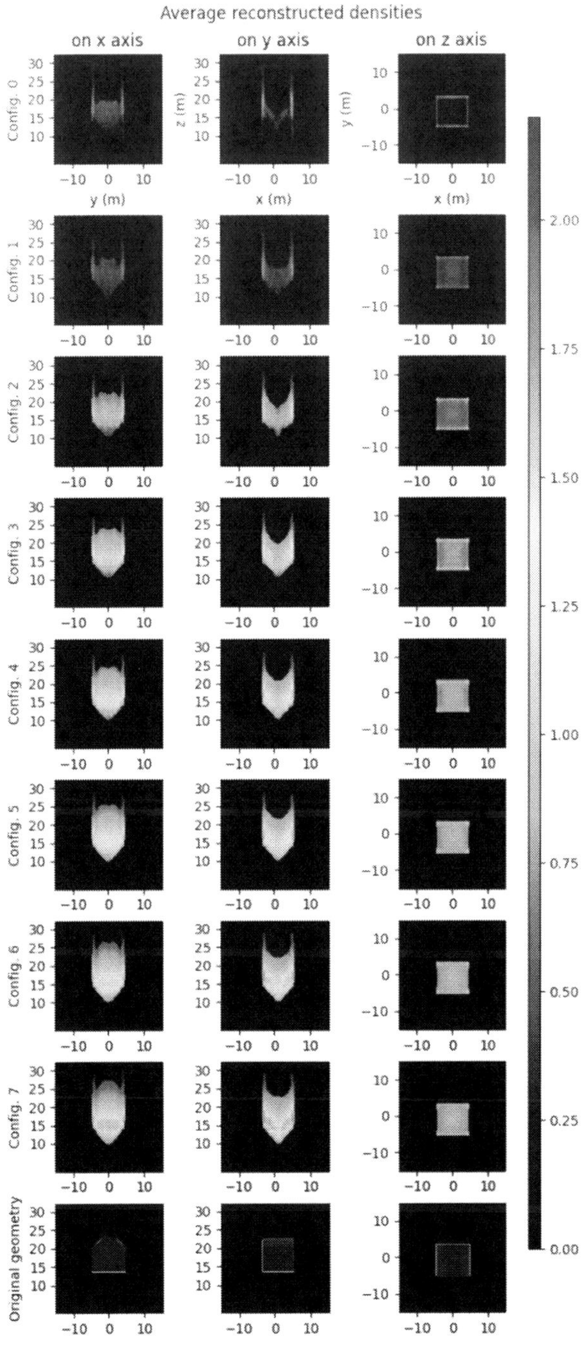

Fig. 8. Eight different scenarios results compared to the real volume simulated (last row). The columns are respectively the average density (g/m^3) of the shape in the x, y and z axis. The common color scale shows the inhomogeneities inside the reconstructed volumes.

In one year of acquisition we were able to realize 46 positions, meaning that each of the four telescopes was moved in average every month. In comparison, the G2 reconstruction in [6] uses only 27 positions from data taken in 2 years.

The 46 positions at G3 are displayed on Fig. 9 with the shape of the concrete cylinder which contains the graphite core.

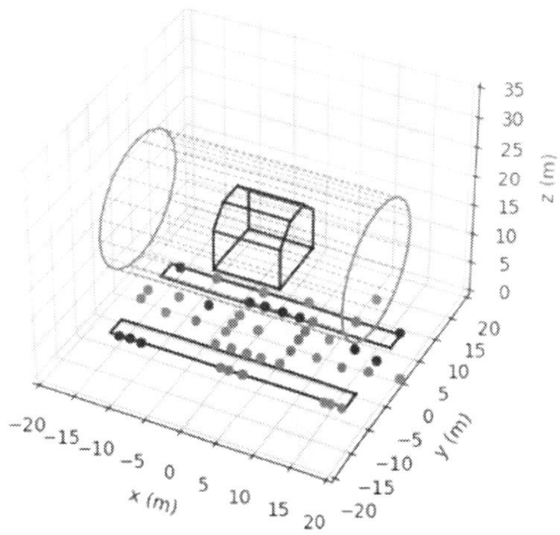

Fig. 9. Map of the 46 positions at G3. The reactor is the cylinder represented in gray. It contains the graphite core represented in black. On the ground 2 black rectangles represent the footprint of the reactor.

We recognize the 16 positions under the graphite core edges. Also most positions are directly under the concrete cylinder. Some positions are outside the cylinder footprint (represented by black rectangles on the ground) in narrow corridors along the reactor as on Fig. 5.

Given the amount of positions, the data preparation before the reconstruction plays a major role. During the G2 campaign most steps of the analysis were separated scripts using different configuration files and had to be started independently. This procedure was revealed not optimal to have results rapidly.

C. Improvement on muography processing

For G3 we standardized our tools in the form of C++ source codes wrapped into Python libraries thanks to Cython. With this method, the core of our analysis is efficient C++ using the CERN's Root library. But the usage benefits of Python's simplicity and allows to write scripts to automatically analyses the amount of data we produce.

Among the tasks that we wanted to automate, the noise detection and the consequent denoising are very important. Indeed, noise in a muography reconstructs fake muons. If there is an excess of muons in one direction, the opacity calculation will result in an under-estimated value. Indeed, more muons means that the direction is easier to cross for the muons, which means that the opacity is lower. Those under-estimated spots will then be propagated in 3 dimensions by the SART algorithm.

To denoise the muographies without signal loss, we propose a method using some properties of the reconstructed muon tracks. When the Micromegas detectors are subject to noise, signal appears on some detectors spontaneously on specific spots. If the signals on the detectors are not aligned in space, no track can be reconstructed and the event is dropped. However some of these spots may be enough aligned to be considered as muon tracks. In that case, the tracks connecting the same noisy spots will be close in space.

Tracks close in space will have the same direction and show up as points on the muography like in Fig. 10. But tracks close in space will also have close points of intersection on each planes. We will note (X0, Y0) the points of intersection between the tracks and a virtual plane on the lowest Micromegas. Those points can be used to construct what we call the X0Y0 histogram.

In previous works, ad-hoc conditions were found for each noisy position. To tackle this problem for the G3 analysis, every run of data will be subject to the same test during the track reconstruction step (the 46 positions produced more than 200 runs of data). Firstly the X0Y0 histogram of the studied run is realized with 1 mm large bins. This allows to see in which regions of the telescope the muons are estimated to cross the bottom detector. On this histogram we apply a blurring with a 5 × 5 matrix. After this blurring we consider empirically that pixels above 3.5 times the average are regions of space with noise. We can then select the events crossing this region of the bottom detector.

As an example, the Fig. 10 shows a 5-weeks muography and the associated X0Y0 histogram.

The muography is subject to noise with directions that counts more than 10k muons. When we look at the X0Y0 histogram we see that a pixel at the bottom right is concentrating 10 times more events than the average, it can be filtered to denoise the muography. During the analysis this X0Y0 histogram has 500 × 500 bins in order to select precisely the areas to filter. When applied with the parameters described above, 50 mm² (0.0002 % of the detector) was banned. It deleted around 50k events (0.005% of the events) and gave the result presented in Fig. 1.

This denoising is a necessary step in order to recover the exploitable data from the detector noise. Because this noise is easily recognizable from the exploitable data, we consider it as "visible noise". Therefore we measured qualitatively the effect of this denoising method as reported in Table II.

TABLE II
PERFORMANCE OF THE X0Y0 DENOISING ON THE NOISY POSITIONS

No effect on visible noise	Partial cleaning on visible noise	Complete cleaning of visible noise
19 %	37 %	44 %

Results of the described denoising on the visible noise.

In nearly 45% of the noisy positions this methods corrects the visible noise. In the other case it cleans partially the muography, and in less than 20% of the cases it does not detect the noise. Fortunately this method still have improvement potential like a better smoothing and a smarter choice of threshold to determine noisy bins.

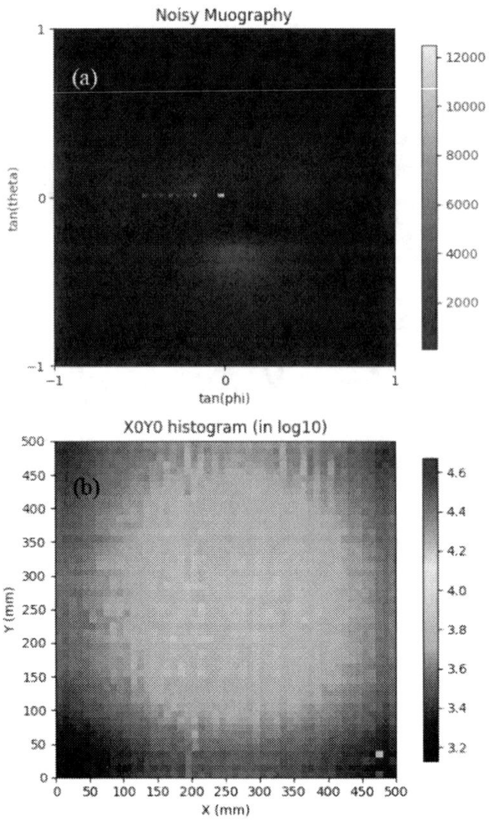

Fig. 10. Muography (a) and associated X0Y0 histogram (b). Noisy pixels appear on the muography and one pixel in the X0Y0 histogram is highly activated.

It is worth mentioning that this method does not bias directly the muography. Indeed, in the muography pixels correspond to direction in space, but the regions filtered correspond to positions on a detector. It means that even if we filter some areas on the detector, the direction in space can still be observed the same way. To study the impact of filtering areas of the detector we observed the effect of a very conservative case. In Fig. 11 half of the detector is filtered (compared to less than 1% normally). We then display the muography using only one half of the detector, and the ratio between this muography and the original muography (with the full detector). This examples demonstrates that even with a half detector it is still possible to have similar muographies.

However Fig. 11 shows a correlation between the directions (on the muography) and the hit positions (on the X0Y0 histogram). Indeed muons coming at high angles tends to hit the detector on the opposite edge. That is why when we filter half of the detector, the muography looses some events on the opposite side (red area on the ratio plot of Fig. 11).

This effect is normally negligible because during the denoising process, less than 1% of the detector surface is filtered. This filtering is then very selective and allows to remove noise with a negligible impact on the muography.

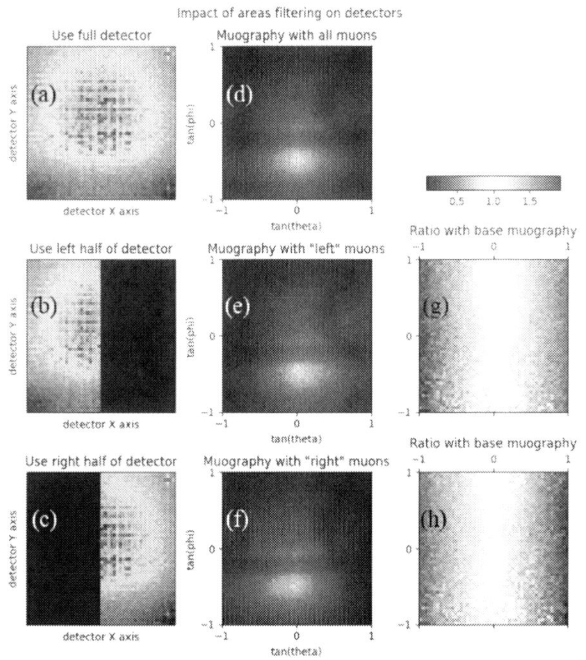

Fig. 11. Left row (a, b, c) shows X0Y0 histogram with filtered areas. The center row (e, f, g) shows the consequences on the related muography. The effect is mainly a loss of muons on one side, which is visible in the ratios in the right row (g, h).

V. Conclusion

Thanks to the experience acquired during the G2 proof of concept, we have developed methods and tools to realize more efficient measurements campaigns.

For the G3 reactor, in the scope of a PhD work, we ran simulations to help us focus on the graphite core. We were able to take data on 46 different positions. Some of those positions were accessible thanks to new mechanical developments.

To process this large amount of data, we also started to automate the different steps between the raw data (hits on the Micromegas detectors) and the opacity computations. We focused on the automation of the most time consuming tasks, like the denoising of the data.

Finally we now have the data and tools to realize the 3D reconstruction of the G3 reactor, which is expected in the next year.

References

[1] K. Morishima, M. Kuno, A. Nishio et al, "Discovery of a big void in Khufu's Pyramid by observation of cosmic-ray muons.", Nature, vol. 552, pp. 386–390, 2017, 10.1038/nature24647.

[2] S. Procureur, K. Morishima, M. Kuno et al, "Precise characterization of a corridor-shaped structure in Khufu's Pyramid by observation of cosmic-ray muons.", Nat Commun vol. 14, no. 1144, 2023, 10.1038/s41467-023-36351-0.

[3] IAEA, "Muon Imaging Present Status and Emerging Applications", [online]. Available : https://www-pub.iaea.org/MTCD/Publications/PDF/TE-2012web.pdf.

[4] R. Kaiser, "Muography:overview and future directions.", Phil. Trans. R. Soc. A., vol. 377, no. 20180049 , 10.1098/rsta.2018.0049.

[5] A.H. Andersen, A.C. Kak, "Simultaneous Algebraic Reconstruction Technique (SART): A superior implementation of the ART algorithm.", Ultrasonic Imaging, vol. 6, no. 1, pp 81-94, 1984, 10.1016/0161-7346(84)90008-7.

[6] S. Procureur et al, "3D imaging of a nuclear reactor using muography measurements", Science Advances, vol. 9, no. 5, 2023, 10.1126/sciadv.abq8431.

[7] S. Bouteille et al, "A Micromegas-based telescope for muon tomography: The WatTo experiment.", Nuclear Instruments and Methods in Physics Research Section A, vol. 834, pp. 223-228, 2016, 10.1016/j.nima.2016.08.002 .

[8] S. Procureur, R Dupre, S. Aune, "Genetic multiplexing and first results with a 50×50 cm2 Micromegas", Nuclear Instruments and Methods in Physics Research Section A, vol. 729, pp. 888-894, 2013, 10.1016/j.nima.2013.08.071.

[9] B. Lefevre, "Convolutional neural networks demultiplexing in large Micromegas detectors for muography purposes", Machine learning workshop IN2P3-IRFU, 2022, [online], available : https://webcast.in2p3.fr/video/convolutional-neural-networks-demultiplexing-in-large-micromegas-detectors-for-muography-purposes.

[10] H. Gomez Maluenda, L. Gallego, S. Procureur, "Muography is present at the heart of nuclear reactors", 2020, [online] available : https://irfu.cea.fr/en/Phocea/Vie_des_labos/Ast/ast.php?t=fait_marquant&id_ast=4888.

https://doi.org/10.1051/epjconf/202328807002

A comparative study on the influence of the type of acid used for ^{55}Fe and ^{63}Ni determination in steel samples through Liquid Scintillation Counting

Maria Letizia Cozzella[1,*], Giada Gandolfo[1], Giovanna La Notte[1], Giuseppe Augusto Marzo[1]
[1]ENEA Italian National Agency for New Technologies, Energy and Sustainable Economic Development

(*) letizia.cozzella@enea.it

Abstract **In decommissioning of nuclear facilities the presence of activated steel requires the determination, along with other radionuclides, of ^{55}Fe and ^{63}Ni. These can be considered as hard-to-measure-radionuclides because of their low energy beta particles or X radiation emitted. Furthermore, those radionuclides are usually mixed with a relevant fraction of non-radioactive Fe and Ni. This implies that, in those cases involving very low radioactive samples close to background activity level, a significant mass of steel sample must be physically and chemically treated and analyzed to achieve the Minimum Detectable activity Concentration (MDC) required by the Authority for clearance purposes. Due to colour quenching the acquisition time is particularly long. Minimizing the colour effect is therefore important for laboratories that must perform many determinations in a reasonable time, achieving the MDC in accordance with the Authority requests. For this reason, the choice of solvent can be an important information to save analyses time. The intensity of the colour quenching is, in fact, strongly dependent on the type of acid used to solubilize the sample before the mixing with the liquid scintillation cocktail. Some types of acid can even increase the phenomenon while others can drastically reduce such an effect. The aim of this work is to offer, through a correlation study between sample concentration in the LSC vial and type of solvent used in dilution, a reliable method for quantifying ^{63}Ni and ^{55}Fe in steel samples with very low activity, close to the background level.**

Keywords **—Liquid Scintillation Counting, decommissioning of nuclear installations, neutron activated steel.**

I. Introduction

Decommissioning of nuclear fission and fusion facilities leads to production, among others, of waste including activated steel. Among the most occurring radionuclides, neutron activation of steel could cause the production of gamma emitter radionuclides (e.g., ^{54}Mn, ^{58}Co, ^{60}Co) along with ^{55}Fe and ^{63}Ni, which can be considered as hard-to-measure-radionuclides because of their low energy beta particles or X radiation emitted. Both ^{63}Ni and ^{55}Fe are neutron activation products. ^{63}Ni mainly exists in steel because of the high concentration of Ni in these materials. It is a pure beta emitting radionuclide with maximum beta energy of 66.95 keV and half-life of 100.1 years. Due to its low beta energy, the measurement by gas flow proportional counter gives low counting efficiency (2.6–20%) [1].

Fe is produced by neutron activation reactions of two major stable iron isotopes: ^{54}Fe (n, γ) ^{55}Fe and ^{56}Fe (n, 2n) ^{55}Fe. Iron is the steel main component, and its concentration is relatively high in several reactor materials [2]. In the first few years after reactor operation has stopped ^{55}Fe is a main contributor to the radioactivity of nuclear waste. ^{55}Fe ($t_{1/2}$ = 2.7 years) decays via electron capture to stable ^{55}Mn with the emission of Auger electrons and low energy X-rays (5.89 keV, 16.9%). Measurement of ^{55}Fe can be carried out by low energy gamma and X-ray detector or gas flow proportional counter, but their counting efficiencies are usually very low (< 1%) [2].

For reaching higher detection efficiencies and lower MDC in ^{55}Fe and ^{63}Ni determination, Liquid Scintillation Counting (LSC) is the most suitable technique, but it requires a chemical separation which produces strongly coloured ^{63}Ni and ^{55}Fe fractions inevitably mixed with non-radioactive Ni and Fe, still present into the materials in high mass percentage. LSC technique is very sensitive to colour quenching effect, which occurs when coloured substances, present in the sample, absorb the light emitted by scintillator before it can escape the vial and reach the photomultipliers. Therefore, minimizing the colour effect is important for laboratories that must carry out many measurements, often close to background activity level, in a reasonable time for achieving the MDC in accordance with the Authority requirements. When MDC is particularly low, a large amount of steel sample must be physically and chemically treated and, due to the colour quenching and efficiency reduction, the acquisition time is consequently dramatically increased. To save analysis time, the choice of solvent can be very important. The intensity of the colour quenching is, in fact, strongly dependent on the type of acid used to solubilize the sample before mixing with the liquid scintillation cocktail. Some types of acid can even increase the phenomenon while others can drastically reduce the effect. The aim of this work is to offer, through a correlation study between sample concentration in the LSC vial and type of solvent used in dilution, a reliable method for quantifying ^{63}Ni and ^{55}Fe in steel sample with very low activity level, close to the background. To do this, many samples by simulating the amount of Fe and

Ni contained in a steel AISI 304L (Fe = 69.06% w/w, Ni = 9.25% w/w) were prepared and dissolved in 5 different acids. Coordination chemistry is the basic chemistry of metal ions which receive electron pairs from ligands around it. Several approaches have been adopted to reduce the intensive colour quench of Fe (solvent extraction [3] or treatment by reducing agents [4]), but it has not been possible to overcome this effect completely, because the reduced Fe^{2+} is not very stable. Nickel is one of transition metals and abounds in nature as nickel (II) because more stable than nickel (0), nickel(I), nickel (III), and nickel (IV). Its salts are commonly green due to the presence of hexa aquanickel(II) ion [Ni $(H_2O)_6^{2+}$]. Chemistry of iron aqueous solution is based on two oxidation states, Fe^{2+} and Fe^{3+}. Both forms are thermodynamically stable and kinetically reactive [7]. Their compounds can take a wide range of colours depending on the oxidative state.

II. Materials and Methods

Iron chloride ($FeCl_3$), nickel chloride ($NiCl_2$), hydrochloric acid (HCl), phosphoric acid (H_3PO_4), nitric acid (HNO_3), phosphonic acid (H_3PO_3), sulfuric acid (H_2SO_4) and sulfurous acid (H_2SO_3) are considered in this work, along with radioactive standard solutions of ^{63}Ni and ^{55}Fe. 24 samples, containing iron and nickel in the same proportion as AISI 304L and different concentrations, were prepared starting from iron chloride and nickel chloride and dissolved respectively in HCl, H_3PO_4, HNO_3, H_3PO_3, H_2SO_3 and H_2SO_4 (Fig.1 – phase 1). The prepared samples simulate iron and nickel fractions resulting from chromatographic separation. For this reason, the two elements were studied separately, 24 samples simulating the iron fraction and 24 samples simulating the nickel fraction were prepared.

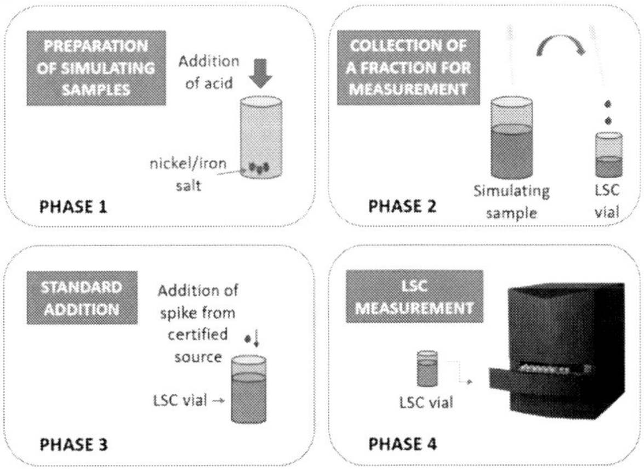

Fig. 1 Phases for sample preparation and analysis

Different fractions of each simulating sample were then collected in LSC vials varying sample concentration to evaluate the measurement performances in terms of LSC efficiency and acquisition time needed for reaching the required MDC (Fig.1 – phase 2). A LSC analysis was performed on all the vials before the internal standard calibration to evaluate the blank contribution.

To each vial, a known amount of ^{55}Fe (17 ± 1 Bq) and ^{63}Ni (34 ± 2 Bq) standard were finally added for calibration (Fig.1 – phase 3). For the radiometric measurements of ^{55}Fe and ^{63}Ni, a Hidex 300 SL automated system was used and LSC efficiency was evaluated (Fig.1 – phase 4).

Based on previous chemical purification experiences in recovering Fe and Ni, an average chemical yield of (60±6) % for each element was assumed and the needed acquisition time was calculated from Currie [5] fixing MDC to 0.3 Bq/g.

Uncertainty evaluation on iron and nickel masses and on LSC efficiencies was based on the propagation of uncertainties related to each variable involved (Fig.2-3).

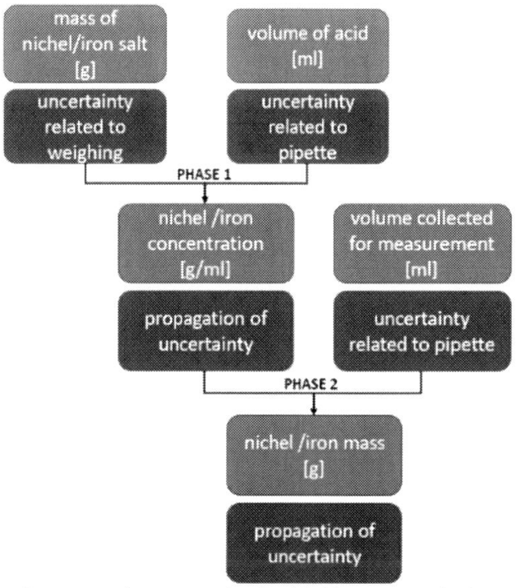

Fig. 2 Uncertainty propagation for nickel/iron mass evaluation

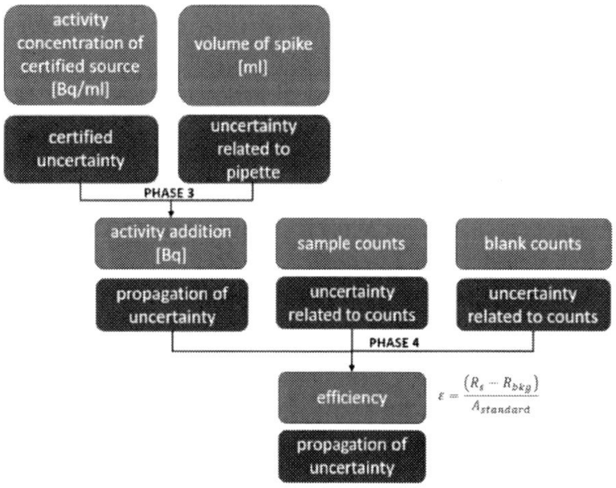

Fig. 3 Uncertainty propagation for efficiency evaluation. R_s is the counting rate related to spiked sample, R_{bkg} is the counting rate related to blank sample, $A_{standard}$ is the activity added as standard for calibration.

III. Results

The results in term of LSC efficiency and acquisition time for ^{55}Fe and ^{63}Ni determination are reported in Tables I-II. Empty cells indicate that under this condition the efficiency is too low to be evaluated: colour quenching effect is too intense, or more phases or a milky solution appeared (Fig.4).

Fig. 4 colour and immiscibility effects

TABLE I

EFFICIENCY AND ACQUISITION TIME FOR ^{55}Fe DETERMINATION

Acid	Conc. Molar	Fe mass (mg)	LSC eff. %	acq. Time (days)
HCl	2	21.5±0.2	8.2±0.3	879.1
		43.0±0.4	3.5±0.2	1194.1
		86.0±0.9	0.9±0.1	4738.0
		172.0±1.7		
H_2SO_3	0.375	21.5±0.2		
		43.0±0.4		
		86.0±0.9		
		172.0±1.7		
H_2SO_4	2	21.5±0.2	30.2±1.1	75.3
		43.0±0.4	21.5±0.8	35.1
		86.0±0.9	4.4±0.2	188.3
		172.0±1.7		
H_3PO_3	1	21.5±0.2	38.8±1.4	43.5
		43.0±0.4	36.1±1.3	12.7
		86.0±0.9	33.6±1.2	3.5
		172.0±1.7	27.4±1.0	1.3
H_3PO_4	2	21.5±0.2	37.2±1.4	48.2
		43.0±0.4	37.7±1.4	11.6
		86.0±0.9	34.0±1.3	3.5
		172.0±1.7	28.1±1.0	1.3
HNO_3	2	21.5±0.2		
		43.0±0.4		
		86.0±0.9		
		172.0±1.7		

TABLE II

EFFICIENCY AND ACQUISITION TIME FOR ^{63}NI DETERMINATION

Acid	Conc. Molar	Ni mass (mg)	LSC efficiency %	acq. time (hours)
HCl	2	6.0±0.1	67.8±4.4	161.4
		12.0±0.1	65.8±4.3	42.5
		24.0±0.2	62.6±4.1	11.1
		36.0±0.4	60.1±3.6	5.4
H_2SO_3	0.375	6.0±0.1		
		12.0±0.1		
		24.0±0.2		
		36.0±0.4		
H_2SO_4	2	6.0±0.1	68.8±4.4	157.1
		12.0±0.1	67.9±4.4	41.2
		24.0±0.2	65.4±4.3	11.0
		36.0±0.4	63.6±4.2	5.1
H_3PO_3	1	6.0±0.1	67±4.4	163.5
		12.0±0.1	65.5±4.3	42.7
		24.0±0.2	62.3±4.0	12.0
		36.0±0.4	59.1±4.0	5.7
H_3PO_4	2	6.0±0.1	69.0±4.5	155.4
		12.0±0.1	67.2±4.4	40.0
		24.0±0.2	64.0±4.2	10.8
		36.0±0.4	61.5±4.1	5.3
HNO_3	2	6.0±0.1	63.0±4.1	188.1
		12.0±0.1	60.8±3.9	49.8
		24.0±0.2	56.1±3.7	14.2
		36.0±0.4	52.7±3.6	7.4

Correlation between mass, type of solvent and efficiency is shown. The associated uncertainty has been calculated following the scheme in Fig.2-3. The aspect of vials for increasing iron or nickel mass concentration and for different types of acid is shown in Tables III-IV.

TABLE III

ASPECT OF VIALS FOR ^{55}Fe DETERMINATION

Acid	Conc. Molar	Fe mass (mg)	
HCl	2	21.5±0.2	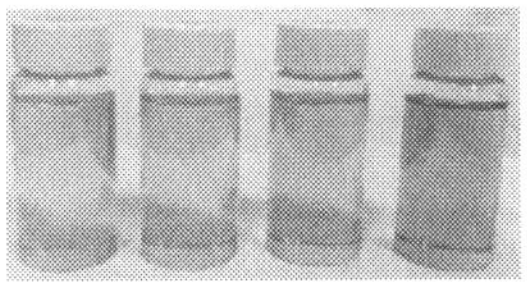
		43.0±0.4	
		86.0±0.9	
		172.0±1.7	
H_2SO_3	0.375	21.5±0.2	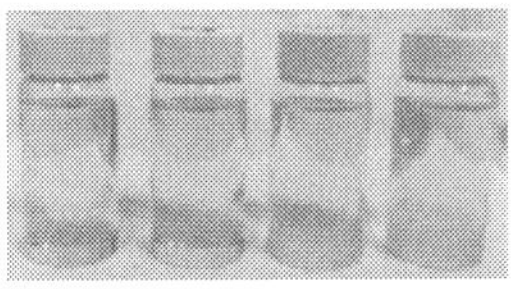
		43.0±0.4	
		86.0±0.9	
		172.0±1.7	
H_2SO_4	2	21.5±0.2	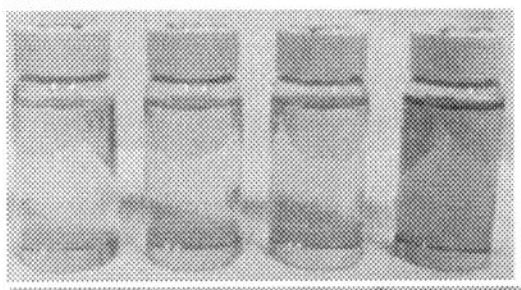
		43.0±0.4	
		86.0±0.9	
		172.0±1.7	
H_3PO_3	1	21.5±0.2	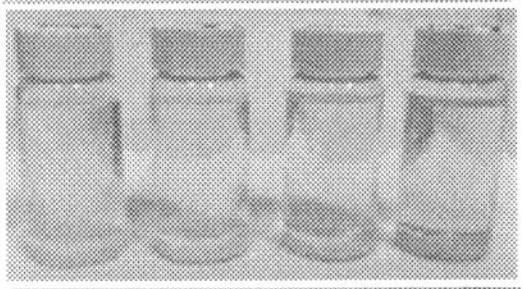
		43.0±0.4	
		86.0±0.9	
		172.0±1.7	
H_3PO_4	2	21.5±0.2	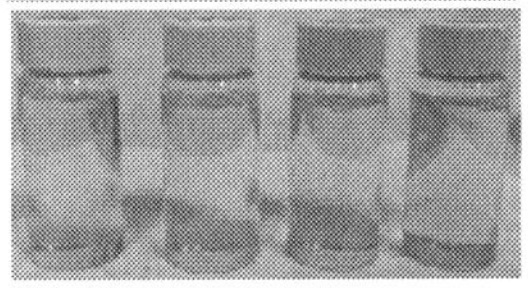
		43.0±0.4	
		86.0±0.9	
		172.0±1.7	
HNO_3	2	21.5±0.2	
		43.0±0.4	
		86.0±0.9	
		172.0±1.7	

TABLE IV

ASPECT OF VIALS FOR ^{63}NI DETERMINATION

Acid	Conc. Molar	Ni mass (mg)	
HCl	2	6.0±0.1	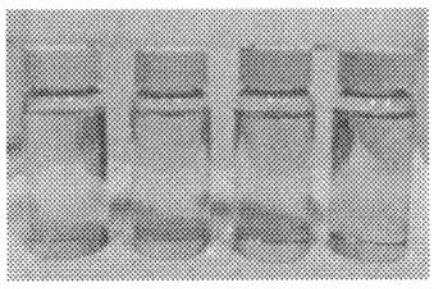
		12.0±0.1	
		24.0±0.2	
		36.0±0.4	
H_2SO_3	0.375	6.0±0.1	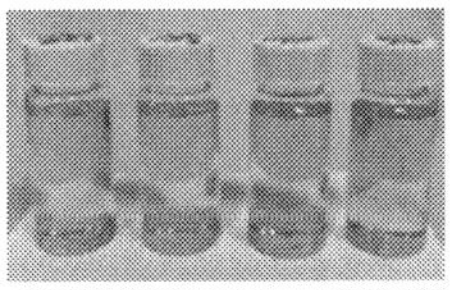
		12.0±0.1	
		24.0±0.2	
		36.0±0.4	
H_2SO_4	2	6.0±0.1	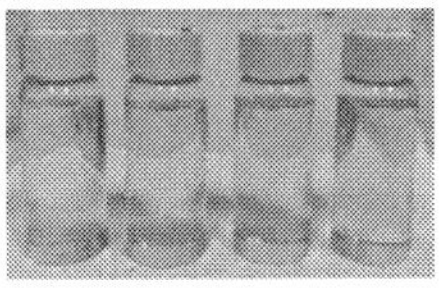
		12.0±0.1	
		24.0±0.2	
		36.0±0.4	
H_3PO_3	1	6.0±0.1	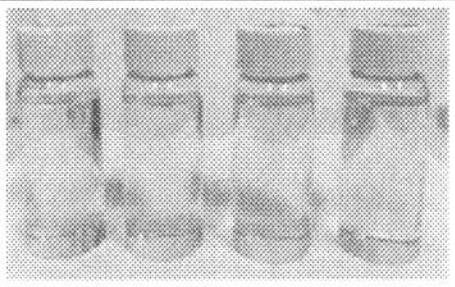
		12.0±0.1	
		24.0±0.2	
		36.0±0.4	
H_3PO_4	2	6.0±0.1	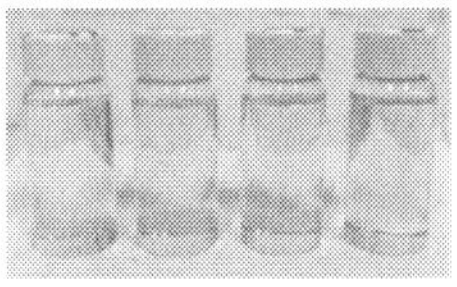
		12.0±0.1	
		24.0±0.2	
		36.0±0.4	
HNO_3	2	6.0±0.1	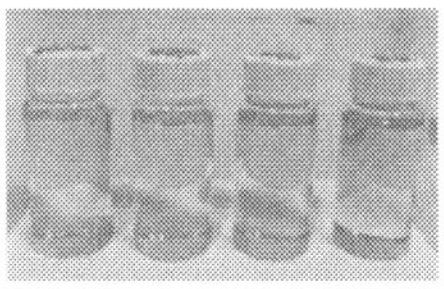
		12.0±0.1	
		24.0±0.2	
		36.0±0.4	

IV. DISCUSSION

In many cases vials containing iron fractions are coloured of different yellow levels. Several approaches have been adopted in literature to reduce the colour quench of iron, such as solvent extraction [3,6] and reduction of Fe^{3+} to Fe^{2+} by ascorbic acid [4]. It has not been possible to overcome the colour quench of iron completely, because the reduced Fe^{2+} is not very stable and it will be oxidized to Fe^{3+} very slowly. The most common acid used in LSC is HCl: it shows a high range sample uptake (only water is better) and the lowest misclassification range, but in the case study the results with HCl display unsatisfactory performances and other acids can be more suitable for a reliable LSC measurement such as H_3PO_3 and H_3PO_4. The solutions obtained with those acids are colourless for both iron and nickel solutions and provide the best performances in terms of efficiency and acquisition time. Instead, the iron solutions in HCl are coloured causing the worsening of outcoming efficiency values. Among the wide range of results, this study confirms for ^{55}Fe determination what is reported in [8] and [2] regarding the possibility to use H_3PO_4 to decolourize Fe^{3+} solutions and the behaviour of HNO_3 as a high quenching agent that can significantly reduce the counting efficiency.
In most cases, vials containing nickel fraction are weakly coloured of green. The resulting colour quenching affects the LSC efficiency in a similar way for all the type of acids except for H_2SO_3, whose corresponding solutions are strongly coloured and produce very low counting rates not discriminable from the background level. Some hours are sufficient for acquiring data for ^{63}Ni determination in accordance with Authority request respect to days needed for ^{55}Fe.

V. CONCLUSION

As a chromatographic purification is required to separate iron and nickel from the radioactive steel matrix, the purpose of this study is to offer an overview on the influence of different types of elution solvent on some of the LSC key parameters (i.e., sample concentration in the LSC vial, detection efficiency, quenching effect) to minimize the acquisition time needed to reach the MDC when the activity levels are lower than the unconditional release level and the materials are expected to be recycled or reused. HCl and HNO_3 are the most used acids in

^{55}Fe / ^{63}Ni isolation [3], but the results showed in Table I provide evidence that H_3PO_3 or in H_3PO_4 might be suitable alternatives for efficiently dissolving iron samples. Nickel fractions display a common behaviour towards a wider range of acids (except for H_2SO_3), showing only small variation in the considered parameters.

References

[1] E. Holm, B. Oregioni, D. Vas, H. Pettersson, J. Rioseco, U. Nilsson, "Nickel 63: radiochemical separation and measurement with an ion implanted detector". J. Radioanal. Nucl. Chem. Articles 138, (1990) 111.

[2] X.Hou, L.R. F. Østergaard, S.V. Nielsen "Determination of 63Ni and 55Fe in nuclear waste samples using radiochemical separation and liquid scintillation counting "*Anal. Chim.Acta* 535, (2005) 297-307

[3] F.J. Cosilito, N. Cohen, H.G. Petrow "Simultaneous Determination of Iron-55 and Stable Iron by Liquid Scintillation Counting" *Anal. Chem.* 40 (1968) 213.

[4] W. Konig, R. Schupfner, H. Schuttelkopf," A fast and very sensitive LSC procedure to determine Fe-55 in steel and concrete."*J. Radioanal. Nucl. Chem.* Articles 193 (1995) 127.

[5] Currie L.A "Limits for Qualitative Detection and Quantitative Determination, Application to Radiochemistry" Analitical Chem. Vol 40 (1968) N.3

[6] C. Yonezawa, M. Hoshi, E. Tachikawa. "Determination of specific activity of iron-55 by spectrophotometry and liquid scintillation counting with bathophenanthroline complex"*Anal. Chem.* 57 (1985) 2961.

[7] M. Sanchez,L. Sabio, N. Galvez, M.Capdevila,J. M.Dominguez-Vera"Critical Review Iron Chemistry at the Service of Life" *International Union of Biochemistry and Molecular Biology* Vol. 69, N° 6, (2017), 382–388.

[8] P.E. Warwick, I.W. Croudace, M.E.D. Bains, *Radioactiv. Radiochem.* 9 (1998) 19.

Assessing performance of Nucleco Passive/Active System (PANWAS) through a long-term verification campaign

Filippo Gagliardi[1,2,*], Alessandro Ciotoli[1], Andrea De Angelis[1], Edoardo Gorello[1], Luigi Meschini[1], Marco Pagliuca[1], Egidio Mauro[1]
[1]Nucleco spa, via Anguillarese 301, Rome, Italy
[2]La Sapienza – University of Rome, Italy
(*) gagliardi@nucleco.it

Abstract— **Nucleco SpA is responsible for the management of radioactive waste produced in Italy by industrial research, medical processes and decommissioning of nuclear installations. Nucleco SpA has an integrated gamma/neutron facility for the radiological characterization of the radioactive waste. This comprises high resolution gamma spectrometry systems using several possible measurements approaches (open geometry, segmented scanning, tomographic scanning) and a combined Passive/Active Neutron Waste Assay System (PANWAS). PANWAS system is used for quantification of Special Nuclear Materials in standard drums and cemented overpacks, thanks to its ^{3}He detectors, a deuterium-tritium (D-T) neutron generator, a ^{252}Cf neutron source and complex electronics. Starting from end of 2019, along with the scheduled and extraordinary maintenance, and in consideration of the standard operations to be performed for decommissioning activities, PANWAS system is undergoing to a performance study measuring a Plutonium standard source repetitively, to prevent subtle malfunctions (although not triggering daily quality assurance checks). In addition, thanks to the presence of a weather station installed in the measurement room, the same measurements have been used for studying the system performances as a function of the environmental conditions, namely temperature, humidity, pressure. In this study an interesting correlation was found between the temperature and the system response and the impact that the replacement of a neutron generator can have in quantifying the result.**

Keywords **— neutronic system, nuclear decommissioning, nuclear instrumentations, neutrons, assessment, plutonium quantification.**

I. Introduction

NUCLEAR decommissioning concerns places that can be very different from each other. Facilities where radioactive waste is generated range from power and reprocessing plants to healthcare and research institutions, and of course also the radiological inventory diversifies based on the type of plant. In general, we can say that we can divide contaminant radionuclides present in the waste into two categories: fuel-cycle radionuclides and non-fuel-cycle radionuclides.

All those radionuclides that derive from the nuclear fuel chain belong to the first group, like some isotopes of Actinide elements (e.g. uranium and plutonium isotopes, etc.). Non-fuel-cycle radionuclides, instead, are fission products (e.g. ^{137}Cs, ^{133}Ba, ^{90}Sr, etc.) and activation products (e.g. ^{60}Co, ^{54}Mn, ^{22}Na, etc.).

The easiest way to quantify the type and the amount of radioactivity in a waste is through a non-destructive approach that usually is gamma spectrometry if the observed radionuclides are good gamma emitters. In most cases, however, nuclear fuel-cycle radionuclides like plutonium isotopes (e.g. ^{238}Pu, ^{239}Pu, ^{240}Pu, etc.) are not good gamma emitters (i.e. the emission probability per each de-excitation gamma photon is very low) if not in large quantities. Nevertheless, nuclear materials that are accounted for in the nuclear fuel cycle emit neutrons as well as gamma rays, with a neutron emission rate that is high enough to produce an easily measurable signal. Neutron emission can happen by spontaneous fission, more likely in the even-even radionuclides (even number of protons, even number of neutrons), also called fertile radionuclides, or through a low-energy neutron-induced fission, whose cross section is very high in the radionuclides with odd neutron number, also called fissile radionuclides [1].

Nucleco S.p.A. is responsible for the management of radioactive waste produced in Italy, ensuring its radiological characterization and correct treatment and conditioning.

Regarding non-destructive radiological characterization, Nucleco has several gamma and imaging systems suitable for measuring different types of objects (e.g. drums, overpacks, boxes, etc.), as well as having a neutronic investigation system, called PANWAS (Passive Active Neutron Waste Assay System), with the aim of quantifying fissile and fertile material in standard drums and cemented overpacks deriving from plants related to the nuclear fuel cycle.

In general, a neutron system is a very complex system, in which the many variables involved, including those relating to environmental conditions, can have a significant impact on the result.

For this purpose, we have conducted a long-term verification campaign in the period 2019-2022 with the aim of monitoring the stability of system and understanding the possible impact on the system response by external variables, such as the change in atmospheric conditions or maintenance interventions.

https://doi.org/10.1051/epjconf/202328807003

II. System description

Neutrons are emitted from materials with a wide spectrum of energies and the way they interact with matter is very complex.

However, neutron detectors usually do not preserve information about the energy of neutrons detected, so neutron assay consists of counting the number of emitted neutrons in specific time gates related to the main event.

The PANWAS uses passive neutron coincidence counting to determine the quantity of spontaneous fission radionuclides in the waste and it uses the active neutron differential die-away (DDA) technique to quantify the fissile content in the waste [2, 3]. Therefore, in each assay two types of measurements are carried out, first a passive assay for the search of fertile content, and then an active assay, in which the neutron generator is used, to detect any fissile radionuclides.

Neutron detection is done by a set of cylindrical He-3 filled proportional counters. There are a total of 117 He-3 detectors located in a cadmium shielded polyethylene moderator in three of the four lateral side of the chamber; there are no detectors in the rear of the chamber (where is located neutron generator tube), nor in the chamber roof and base. In Fig. 1 the cut-away side view of the PANWAS is shown.

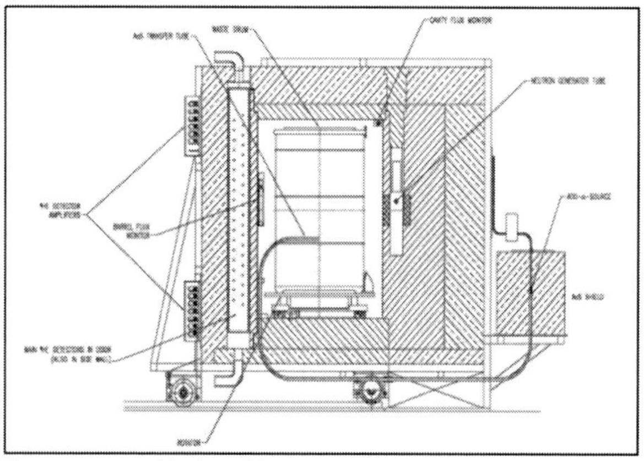

Fig. 1. Cut-away side view of the PANWAS System

An automated drum transfer mechanism is used to load a waste drum into the assay system onto a rotator mechanism built onto the inside of the chamber door. The rotator mechanism is used to turn the drum during the assay measurement to consider any inhomogeneities in the fissile material or matrix distribution. In Fig. 2 it is possible to see the photograph of PANWAS, where the conveyor system and a drum loaded on the rotating table are shown.

The Add-a-Source (AAS) technique is used to correct for the effect of the waste matrix on the passive neutron measurement.

The AAS technique makes use of a small ^{252}Cf neutron source (10^5 neutrons per second) that is driven into the chamber into a position near the waste drum during the assay measurement. The perturbation of the neutron count rate from the ^{252}Cf caused by the waste matrix (relative to that for an empty drum) is used to infer the effect of the matrix on the neutron signal from any fissile material in the drum [2, 4, 5].

The source of pulsed neutrons for the DDA measurement is

Fig. 2. Photograph of the PANWAS System

a Zetatron type neutron generator that makes use of the deuterium-tritium (D-T) reaction

$$^{3}\mathrm{He} + n \rightarrow {}^{3}\mathrm{H} + p$$

to produce 14 MeV neutrons at a rate of up to 10^8 neutrons per second in 10 μs duration pulses. The neutron generator is installed in the back wall of the chamber.

At the end of each measurement, the result is expressed in terms of equivalent masses. For passive assay a $^{240}Pu_{eff}$ mass is calculated, that coincides with the effective mass that would produce the same spontaneous fission signal (produced by ^{238}Pu, ^{240}Pu, ^{242}Pu and ^{238}U). At the same time, for active assay the result is expressed as $^{239}Pu_{eq}$ mass, that corresponds to the equivalent mass that would produce the same induced-fission signal (produced by ^{239}Pu, ^{241}Pu and ^{235}U).

III. Equipment and Materials

For this long-term verification campaign we used a 220 liter standard drum with a heterogeneous matrix inside (e.g. paper and plastic materials) with a density of 0.11 kg/l. Inside the drum there is a structure made with tubular guides, placed at fixed radial positions (0, 1/4R, 1/2R, 3/4R, R), within which it is possible to insert some certified radioactive sources (Fig. 3).

For the purpose of this work, we used a P.I.D.I.E. (Plutonium Isotopic Determination Inter Comparison Exercise, [6]) radioactive source, with a nominal mass to be observed of 26.3 mg of $^{240}Pu_{eff}$ and 85.1 mg of $^{239}Pu_{eq}$ (values are already corrected for self-shielding effect). During the measurements the source was placed at middle height position in the central tube (R = 0, indicated by yellow arrow in Fig. 3). The radioactive source and the standard drum were the same throughout the verification campaign.

In order to contextualizing the environmental conditions for each measurement, a weather station located in the room of the system was used (it should be noted that the weather station is placed on a wall of the room and not inside the measurement cavity).

IV. Results

The purpose of this work was to try to understand if and how maintenance interventions as well as changes in atmospheric conditions can impact on the response of the neutron system.

Fig. 3 Inside of the standard drum with heterogeneous matrix, with indication of the tube used for the radioactive source.

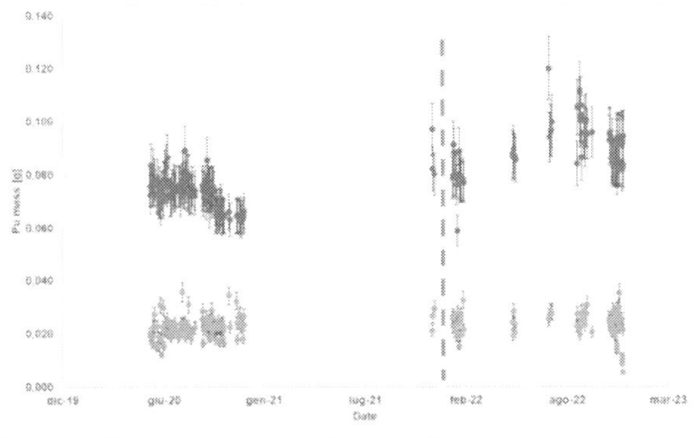

Fig. 4 Time evolution of the plutonium quantification. Blue dots indicate $^{239}Pu_{eq}$, green dots indicate 240Pu_{eff}. The dotted line represents the generator replacement date. (In 2021 data were not collected because of Covid-19 pandemic).

A. Neutron generator tube replacement

In general, PANWAS system undergoes a series of ordinary maintenance interventions, which usually occur once a year.

During these interventions, the technicians adjust all the parameters in the best possible way to make the machine work correctly. In addition to these ordinary interventions, at the beginning of 2022 the neutron generator tube was replaced. As shown in Fig. 4, the ordinary maintenance interventions didn't impact system response much, otherwise the generator tube replacement implied a relevant discontinuity in $^{239}Pu_{eq}$ quantification, both in terms of measured value, which is closest to the expected one, and in terms of slope. Indeed, evidently a near exhausted generator was leading to a slow downward drift in quantifying $^{239}Pu_{eq}$.

B. Assessment of the impact of atmospheric conditions

In this verification campaign atmospheric data was collected for each measurement using a small weather station present in the PANWAS room.

The most interesting result was that relating to the temperature. In particular, two effects can be observed looking

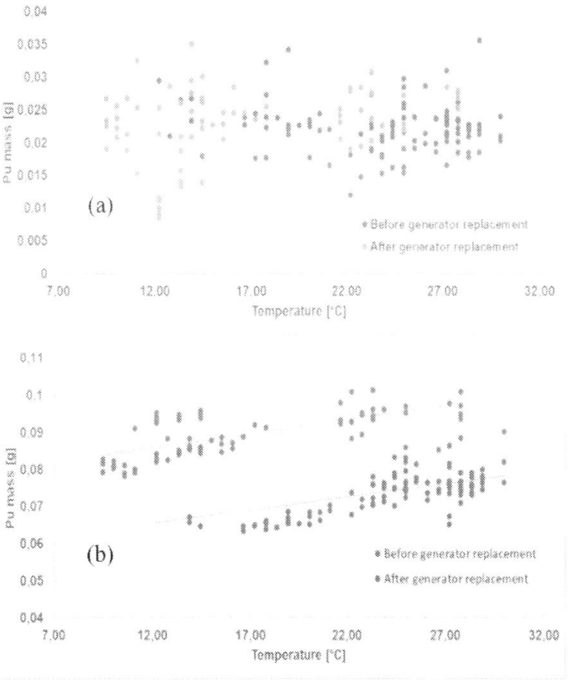

Fig. 5 Mass quantification as a function of the temperature for $^{240}Pu_{eff}$ (a) and $^{239}Pu_{eq}$ (b).

at Fig. 5b. The first one is that as the temperature increases, a linear increase in the quantification of $^{239}Pu_{eq}$ is seen. The other one is that the overall $^{239}Pu_{eq}$ mass distribution can be split into two datasets, one before the generator replacement and one after that, in which both the trendlines (1) and (2), referring respectively to before and after generator replacement, maintain almost the same slope.

$$y = 0.0007x + 0.0571 \quad (1)$$
$$y = 0.0008x + 0.0765 \quad (2)$$

Regarding $^{240}Pu_{eff}$ quantification as a function of the temperature, instead, there appears to be no correlation effect, as can be seen in Fig. 5a, therefore we can assume, in the first instance, that only the operation of the neutron generator seems to undergo a dependent effect on temperature changes.

A preliminary study was also conducted on the possible correlation between plutonium quantification and the parameters of pressure and humidity. In these cases, since the variation is less extensive than that of the temperature, we have averaged the plutonium mass results obtained for each pressure and humidity value. However, as can be seen from the distributions shown in Fig. 5a, the system response seems to be particularly stable with respect to pressure and humidity variations.

C. Radiological checks

To evaluate the dose rate near the cavity and to verify the possible presence of outgoing secondary photons (n-γ reactions), in according with radiation expert, we periodically perform a mapping of the interested area with dose rate

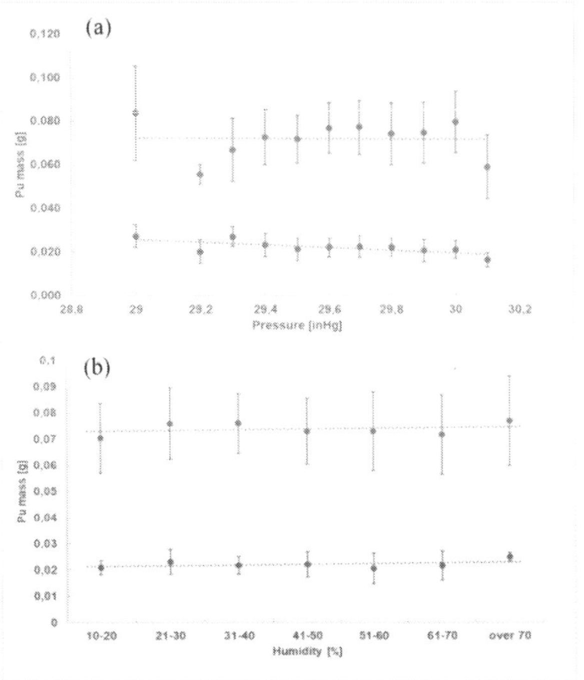

Fig. 6 Mass quantification as a function of the pressure (a) and humidity (b). Red dots are for $^{240}Pu_{eff}$, blue dots are for $^{239}Pu_{eq}$.

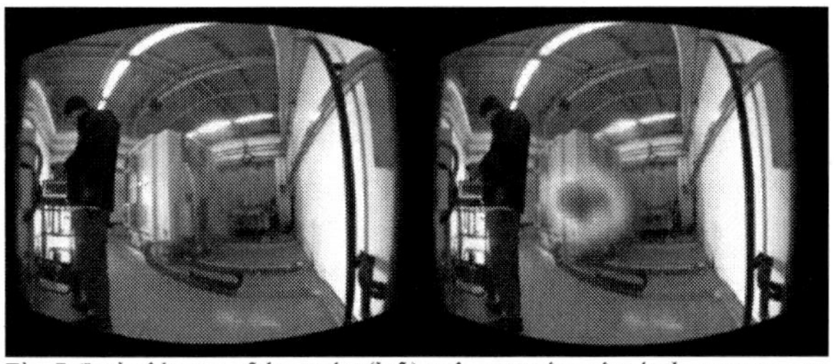

Fig. 7 Optical image of the cavity (left) and gamma imaging in the energy range 0.25÷3 MeV (right).

measurements, smear test and gamma imaging. As can be seen in Fig. 7, the signal due to secondary photons, emitted during de-excitation after neutron interactions with cadmium and hydrogen, totally comes from the cavity.

V. Conclusions

This work has made it possible to understand and quantify the impact that maintenance operations and the variation of atmospheric parameters have on neutronic system's response.

The important thing that emerged from this study is certainly the fact that it is essential to know the life cycle of the neutron generator, as an almost exhausted generator can lead to a degradation in performance and to a bad quantification of radiological fissile content.

Another interesting aspect to take into consideration concerns the correlation found between the temperature and the response of the system regarding the quantification of $^{239}Pu_{eq}$, a phenomenon which therefore seems to affect only the part of the system related to neutron generator.

As future developments, it would certainly be interesting to better investigate the dependence on atmospheric parameters, carrying out more accurate measurements with a weather station suitable for installation inside the measurement cavity.

The latter, together with another series of improvements to the system, is part of a system upgrade, which is currently under discussion with the manufacturer.

References

[1] N. Ensslin, "The Origin of Neutron Radiation," in *Passive Nondestructive Assay of Nuclear Materials*, Los Alamos National Laboratory, NM, USA, March 1991, pp 337-356. [Online]. Available: https://www.nrc.gov/docs/ Accessed on: August 01, 2023.

[2] E. Alvarez, C.G. Wilkins, S. Croft, M.F. Villani, A. Ambrifi, G. Simone, "PANWAS: A Passive/Active Neutron Waste Assay System for the Radiological Characterization of Waste Packages at the Nucleco Facility at Casaccia", in *WM'06 Conference*, Tucson, AZ, USA, 2006.

[3] J. Caldwell, R. Hastings, G. Herrera, W. Kunz, E. Shunk, "The Los Alamos Second-Generation System for Passive and Active Neutron Assays of Drum-Size Containers", Los Alamos National Laboratory, LA10774, September 1986.

[4] H.O. Menlove, L.A. Foster, J. Baca, "NBC Operation Manual including the Multiposition Add-A-Source Function", Los Alamos National Laboratory, LA-12737-M, 1994.

[5] B.H. Armitage, *et al.*, "A scan a source matrix correction technique for use with a reconfigurable passive neutron plutonium assay system", Nucl. Instr. Methods in Physics Research A, vol. 354, pp. 522-526, 1995.

[6] R.J.S. Harry, "PIDIE, Plutonium Isotopic Determination Inter-Comparison Exercise", in 31st Annual Meeting of the Inst. of Nucl. Mat. Management", Los Angeles, CA, USA, July 1990.

CLEANDEM, a Cyber physicaL Equipment for unmAnned Nuclear DEcommissioning Measurements

Maugan MICHEL[1,*], Guillaume AMOYAL[1] and Vincent SCHOEPFF[1], on behalf of the CLEANDEM consortium
[1]Paris-Saclay University, CEA, List, Palaiseau, France
(*)maugan.michel@cea.fr

***Abstract*—Human intervention is still required nowadays for most operations conducted during the Dismantling & Decommissioning (D&D) steps, which cover a wide range of radiological conditions: from the harsh initial conditions, nearly identical to when operating, to the final decommissioning steps where radioactivity has been removed.**

The goal of the three years EU-funded CLEANDEM project, led by CEA List, is to deliver a unique platform which will support the end-users' operations, from the initial radiological assessment to the final characterization of the facility, while enabling their continuous monitoring during the D&D operations.

Ten leading actors from four European countries' nuclear industry and research, have joined their expertise and efforts in the CLEANDEM consortium to develop a mobile unmanned ground platform (UGV), equipped with upgraded highly-mature detection technologies for 3D-localized radiological measurements. These will complete the facilities' available data into a 3D and fully detailed Digital Twin of the surveyed area, thus improving the planning and traceability of the D&D operations.

***Keywords* — Dismantling & Decommissioning, Euratom, Radiation Protection, Gamma imaging, Radiological mapping, Contamination monitoring, Shape-sensing, OSL, Distributed dose measurement, UGV, Robot, Autonomous navigation and Localization, SLAM, LIDAR, Digital Twin, Training, In-situ demonstration.**

I. Introduction

THE CLEANDEM consortium, led since its kick-off in March 2021 by CEA List, gathers the expertise from European nuclear industry (CAEN, ORANO DS, ANSALDO, RINA, TECNALIA, SOGIN, AiNT) and research (CEA List, INFN, ENEA), to address one of the limitations in the field of Dismantling & Decommissioning (D&D), which is the radiation exposure of human operators at the different steps of the operations.

Until the ADARA principle (As Decomissionable as Reasonably Achievable) is widely implemented during the design phase of nuclear facilities, significant improvements could be achieved during these phases regarding human exposure to ionizing radiation, the duration of operations and their cost.

Obviously, any operation that can be achieved remotely will reduce the dose uptake of operators and improve their safety. Consequently, following the advancements in robotics capabilities over recent years, a significant focus of interest for the stakeholders is the utilization of robots to be integrated in D&D operations. Also, whereas battery autonomy keeps on increasing, the power consumption of the embedded equipment also will decrease the available time for remote operations. Both subjects are actively researched and developed to achieve cheaper, faster, and safer D&D.

Reducing the dose uptake, time and cost of operations are addressed in CLEANDEM by improving the radiological assessment continuously undertaken during D&D: from the initial characterization (often synonymous of high-dose rates), to the final characterization where all radioactivity has been removed, with the update following every clean-up action undertaken in the meantime. The proposed solution aims at upgrading existing radiological measurement systems, with Technology Readiness Levels (TRL) ranging from 5 to 7, and embedding them on an Unmanned Ground Vehicle (UGV). The 3D reconstruction of the environment, required for the autonomous localization and navigation of the UGV, are sent along with the radiological measurement, by the robot, to build and/or update the Digital Twin of the facility. The latter, containing all available radiological data (legacy as well as those updated with the sensors), can then be used to track past and plan future D&D operations.

The work achieved by the consortium, will be presented infra following their division in the different Work Packages (WP), addressing:

- specifications' collection from the stakeholders to define the Concepts of Operations (WP2 "CONOPS")
- improvement of low-cost sensors for rapid neutron and gamma and distributed gamma dose rate mapping, as weel as gamma spectro-imaging (WP3).
- neutron/gamma detection and identification sensors improvements (WP4);
- air and surface contamination monitoring (WP6).
- unmanned modular mobile platform equipped with a robotic arm (WP7);
- data fusion and Digital Twin development.

Also, two additional WPs are dedicated to estimating the social and economic impact of the UGV platform on the market (WP5), and on the concept for the training programme and the system's final demonstration for CLEANDEM's end-users network (WP9).

The overview of the CLEANDEM solution is presented in Fig. 1.

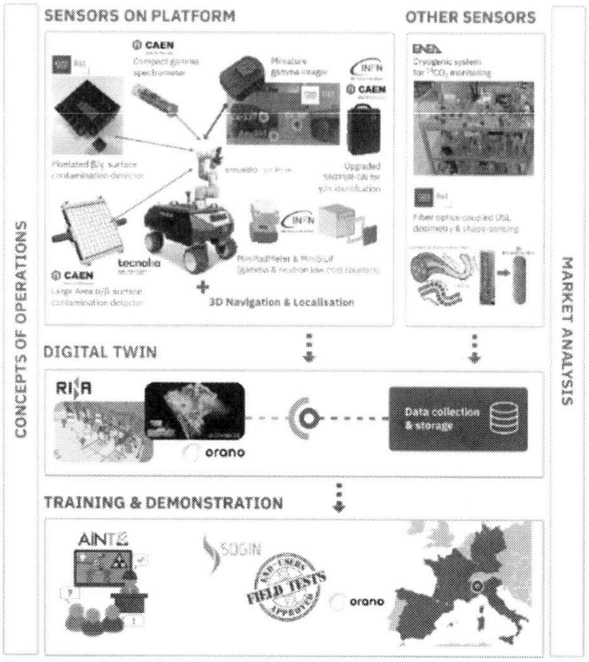

Fig. 1: overview of the CLEANDEM solution.
Sensors on platform: Pixelated β/γ surface contamination detector (CEA List), Compact gamma spectrometer (CAEN), Miniaturized gamma imager (CEA List), Upgraded SNIPER-GN for γ/n identification (CAEN and INFN), MiniRadMeter & MiniSiLiF (gamma & neutron low cost counters, INFN), Large Area α/β surface contamination detector (CAEN). Robotnik's RB-VOGUI platform with a Universal Robots' UR 5e robotic-arm (TECNALIA and ANSALDO respectively).
Other sensors: cryogenic system for $^{14}CO_2$ monitoring (ENEA) and fiber optics-coupled OSL dosimetry & shape-sensing (CEA List).
Digital Twin with PoStLAM (RINA and ORANO DS) and **Training** (AiNT) **and Demonstration** at the ENEA Casaccia facility (SOGIN).
Work Packages **Concepts of Operations** and **Market Analysis** operate in parallel to the others.

II. Concepts of Operations (WP2)

During the first months of the project, ORANO DS has been in charge of delivering the input information necessary for the technical developments to be set on their respective tracks. These information were defined following interviews with multiple stakeholders and operators of D&D to write the Technical Specifications expected from the final system developed by the consortium. With the help of SOGIN, ORANO DS was able to establish the base of the D&D scenarios to which the CLEANDEM robot shall be able to address, each considering different phases of the operations (early, day-to-day and final characterization).

A third deliverable which is a European Guideline for D&D Operation containing a list of recommendations for different case of D&D operations specific to robotic operations for radiological monitoring shall be available to the public at the end of the project to benefit from the consortium's experience gained from the project.

III. Low-cost sensors for rapid neutron, gamma and distributed dose rate mapping (WP3)

Three families of sensors are developed within the frame of WP3. The first one is a low-cost quick dose-rate monitor by INFN, combining the MiniRadMeter (miniature gamma counter/spectrometer) and MiniSiLiF (miniature neutron counter). Shown in Fig. 2, the MiniSiLiF uses a ^{6}LiF slow neutrons converter layer after they have been slowed down by the moderator (6 cm × 6 cm × 6 cm polyethylene box), taking advantage of the

$$^6Li + n_{slow} \rightarrow {}^7Li \rightarrow \alpha + {}^3H$$

reaction, where either alpha or triton particles are detected by a silicon diode. In case one were only interested in thermal neutrons, the moderator can be removed leaving the bare SiLiF detector in operation [1].

The selected scintillator for gamma detection is a CsI(Tl) coupled to a Silicon Photomultiplier (SiPM). The scintillator choice is based on its high-quality detection properties in terms of detection efficiency (i.e., high density), light yield, and energy resolution at a reasonable cost [1].

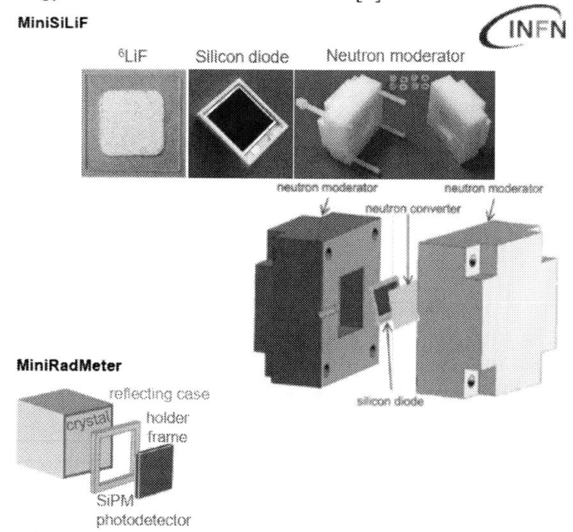

Fig. 2: parts and exploded 3D view of the INFN's MiniSiLiF neutron counter and exploded view of the MiniRadMeter (CsI(Tl) crystal).

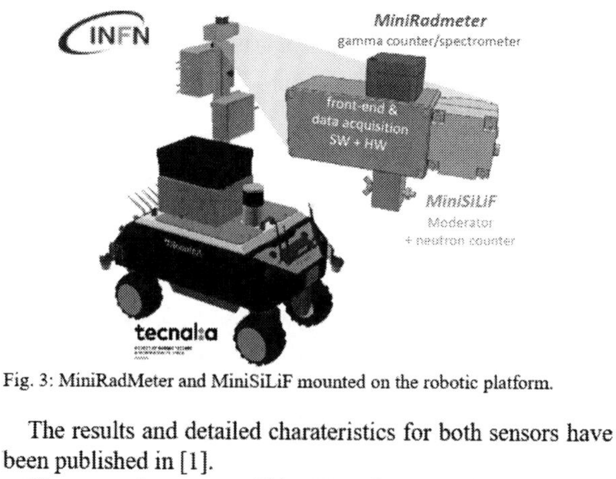

Fig. 3: MiniRadMeter and MiniSiLiF mounted on the robotic platform.

The results and detailed charateristics for both sensors have been published in [1].

The second system within WP3 is CEA List's Nanopix, currently the smallest coded-aperture gamma imager in the world (10 cm × 7 cm × 5.5 cm). Shown in Fig. 4, it's low weight (412 g) allows it to be mounted on the robotic arm and scan in any direction to localize gamma hotspots. Coupled with a miniature RGB camera, it combines the visible image with the post-processed data from a pixellated Timepix sensor to produce the superimposition of both images with spectrometric information of the source, as visible on the right in Fig. 4. It is also shown in a preliminary mounting configuration on the robotic arm in Fig. 7. A soon-to-be-addressed challenge will be the integration of Timepix3 technology to Nanopix for improved performances.

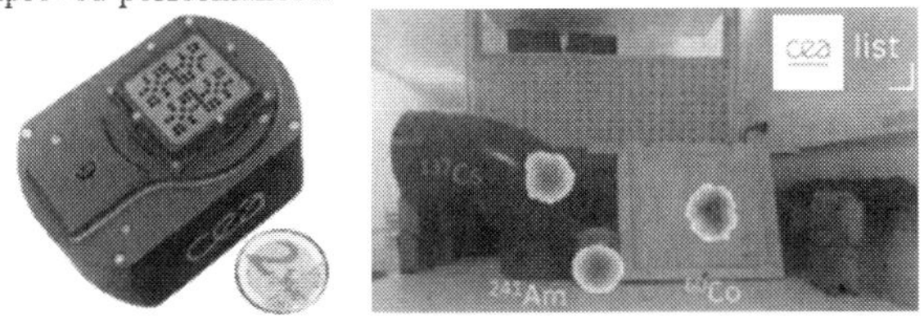

Fig. 4: Nanopix gamma spectro-imager and demonstration of the hotspots localization capabilities using different radionuclides (^{137}Cs in plastic pipe, ^{241}Am in steel drum and ^{60}Co behind 2 cm of concrete).

The last of the developments within this work package, one of the two which are not not to be mounted on the robot, is the joint use of Optically Stimulated Luminescence (OSL/FO) dosimetry and optical fiber shape sensing. The final aim of this innovative approach for D&D applications is to gather both topographic and radiological data within hard-to-access and hidden zones useful for 3D reconstruction using Monte-Carlo (MC) codes. OSL/FO dosimeters are small enough to be pushed through existing access in order to provide radiological investigation in hard-to-access zones while sparing heavy duty that would otherwise be necessary to give access to conventional dosimeters. However, the geolocalization of the miniature OSL/FO detectors is still based on original drawings that are often erroneous or even unavailable or lost. Shape sensing is a solution likely to provide missing topographic data and ensure proper 3D reconstruction to be made.

Shape sensors are already used in medical applications on short length (< 1 m) [2–3]. The CLEANDEM development aims to use shape-sensing concepts (materials, methods and algorithms) for D&D applications and to check their compliance with their specifications, particularly long-range monitoring (typically up to 10 – 13 m) and radiation hardness.

Fiber Optic Shape Sensing (FOSS) is a cutting-edge optical fiber sensing technology that employs a fiber-based flexible cable to follow the 3D shape of an object in real time without requiring eye contact. It consists in the ability to dynamically track position and shape of any point on an optical fiber cable in 3D space, relying upon bending-induced strain along multicore (MC) fibers or special cables made of several single mode fibers. Longitudinal strain profiles are recorded along each fiber core and an algorithm reconstructs the shape of the cable using strain profiles as input data. The shape sensing within CLEANDEM relies on a 7-core Multi-Core Fiber (MCF, where six fibers are evenly distributed around one). The latest results, shown in Fig. 5 (right), achieve shape reconstruction over a length of 6.2 m.

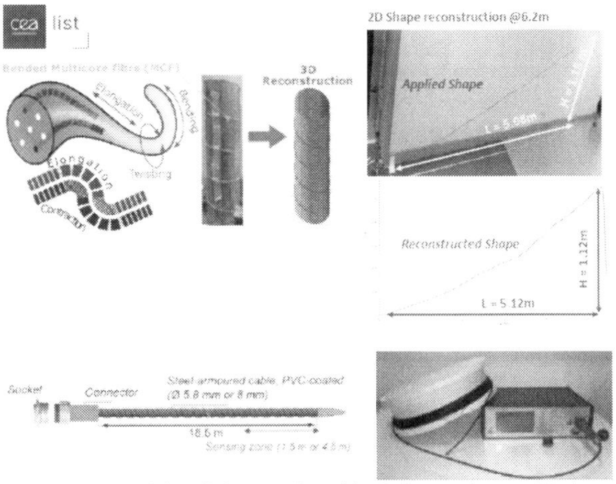

Fig. 5: (top) principles of shape-sensing with 3D shape and recent 2D shape reconstruction of a 6.2 m MCF and (bottom) probe design and INSPECT OSL/FO readout-unit designed by the CEA List within the PIA-Andra INSPECT Project.

IV. THE NEUTRON/GAMMA DETECTION AND IDENTIFICATION SENSORS (WP4)

The second sensors-dedicated work package includes the Sniper GN and Gamon systems from CAEN which will be mounted on the robot and the development of a compact sensor to be mounted on the robotic arm.

The Sniper GN for gamma and neutron measurement is composed of a 38 mm × 38 mm $CeBr_3$ crystal and a 51 mm × 51 mm stilbene scintillator. Whereas the NaI(Li) from the Gamon is able to achieve both gamma spectrometry and neutron detection using Pulse Shape Discrimination (PSD) with one device; it achieves it with a lower gamma energy resolution than the $CeBr_3$. But it's combination with the Sniper GN allows for better gamma/neutron mapping with the ability to identify radionuclides based on the neutron gamma ratio they exhibit (see Fig. 6).

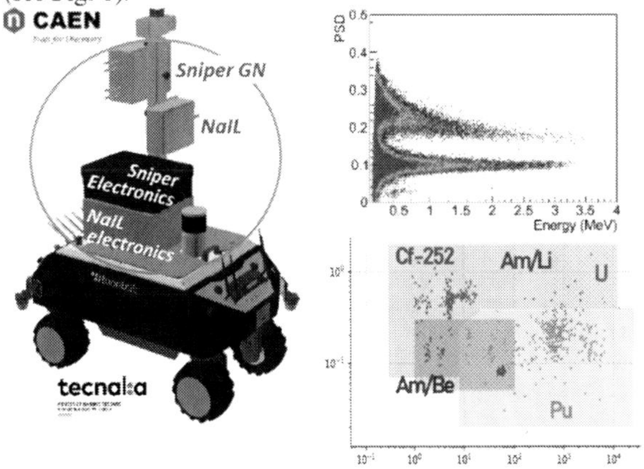

Fig. 6: CAEN's Sniper GN ($CeBr_3$ and stilbene scintillators) and Gamon (NaIL™ scintillator) mounted on the UGV (left) and example of gamma pulse-shape discrimination (top right). The latter, in combination with neutron measurement, enables to identify Nuclear Materials (bottom right) even when hidden behind gamma or neutron shielding.

Shown in Fig. 7, the last sensor developed within WP4 is a large volume hemispheric CZT detector to be mounted on the arm (probe size: L × diam. = 12 cm × 5 cm). Following hotspot localization (eg. from the Nanopix gamma spectro-imager), it will enable a deeper gamma inspection during the first step of the survey and be have to access places harder to reach for larger sensors.

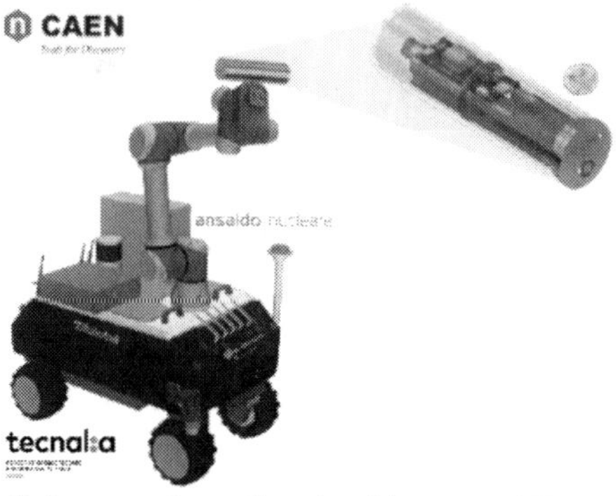

Fig. 7: preliminary mounting configuration of the compact CZT sensor (CAEN) and Nanopix (CEA List) on the UR 5e robotic arm (Universal Robots).

V. Air & surface contamination monitoring (WP6)

Two large surface contamination monitors are developed within CLEANDEM by CAEN and CEA List, respectively measuring alpha/beta and beta/gamma particles. The last system, developed by ENEA is in charge of monitoring radiocarbon dioxide ($^{14}CO_2$).

Whereas CAEN's large surface (576 cm^2) PSD phoswich contamination monitor is capable of $\alpha/\beta/\gamma$ measurement, this goes with heavy shielding (50 kg) and scintillator. Considering the β/γ capability of CEA List's system, the development was refocused on an α/β-dedicated system to benefit from this complementarity; thus allowing both systems (α/β and β/γ) to be mounted on the robotic arm. The 25 µm ZnS(Ag) coating (alpha detection) is coupled to a PVT scintillator (β detection), and both are coupled to a lucite light guide leading to the photomultiplier. Because the area of the sensor affects the light output and because working in coincidence also reduces noise level, it was chosen a geometry with two photomultiplier tubes (PMT) on opposite sides, as visible in Fig. 8.

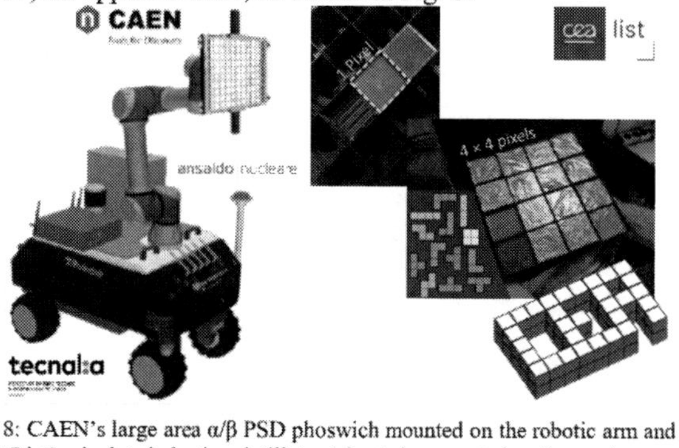

Fig. 8: CAEN's large area α/β PSD phoswich mounted on the robotic arm and CEA List's pixelated plastic scintillator (also to be mounted on the arm).

Also shown in Fig. 8 and based on PSD of phoswich scintillators, CEA List's β/γ system is made of 5 cm × 5 cm plastic scintillators inside a frame which can then be assembled as needed to fit the measured surface. Each scintillator, coupled to a SiPM and readout electronics are mounted on springs to allow them to adapt for the scintillator to be as close to the surface as possible, even if it is uneven.

VI. UGV equipped with a robotic arm (WP7)

Following the definition of Concepts of Operations and technical specifications, the first major equipments to be purchased were the the robotic arm and, above all, the unmanned robotic platform on which it has to be mounted. From the conclusions of WP2, Tecnalia and Ansaldo Nucleare respectively selected Robotnik's RB-VOGUI+, i.e. the RB-VOGUI visible in Fig. 9 equipped with Universal Robots' UR-5e robotic arm.

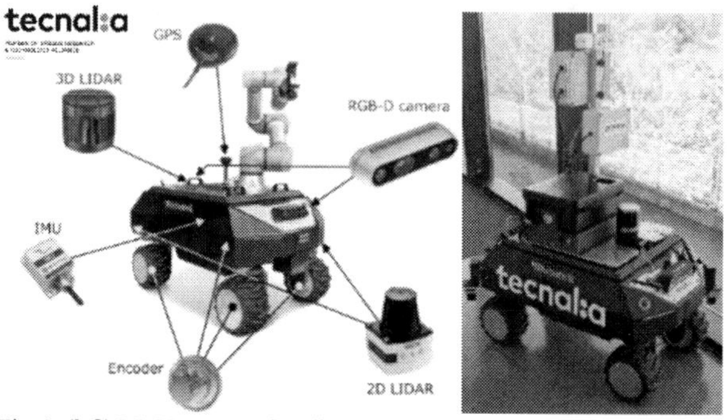

Fig. 9: (left) RB-VOGUI (Robotnik), and sensors used by Tecnalia to achieve autonomous navigation and localization of the platform. The platform's dimensions are 104 cm × 65 cm × 56 cm.
(Right) photograph of the UGV with the MiniSiLiF, MiniRadMeter (INFN), and Sniper GN and Gamon (CAEN) and their electronics, shown in Fig. 3 and Fig. 6.

In charge of the robotic arm and the interfacing of the sensors mounted on it, Ansaldo Nucleare and Tecnalia manage the data collection from the different sensors developers with the necessary connectors and communication protocols to allow their unhindered transfer to RINA, in charge of data fusion and the Digital Twin.

VII. The data fusion and Digital Twin development (WP8)

The last technological development, led and achieved by RINA, is the fusion of all data and their combination in the Qpro^2 Multiplatform to build a Digital Twin (DT). The DT design aims at providing a unique tool able to collect and display (in 3D format) all the useful data for D&D operations acquired from different data sources (sensors, materials samples, legacy database, etc.), as shown in Fig. 10. It also considers the definition of a grid for the chemical and radiological classification of waste. Still under development, its objectives are to achieve direct monitoring of the raw data acquisition to generate alarms and warnings in case of deviation, to represent acquired data as a dashboard and as a 3D representation directly in the plant model, definition and representation of suitable indicators, analysis of the legacy data. Its last and not least functionality is to support D&D decisions in defining strategy and scenarios.

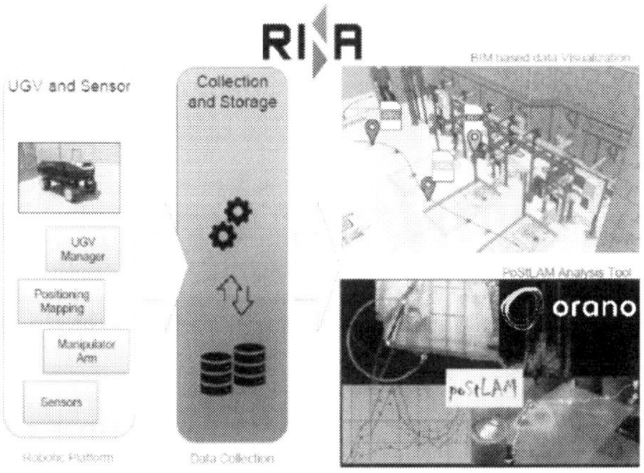

Fig. 10: Functional representation of data flow in data acquisition approach.

To achieve higher-level radiological data post-processing and better exploit 3D point clouds with positioned radiological data to prepare interventions, Qpro[2] is combined with ORANO's PoStLAM software. Based on the 3D rendering of the area built by the robot, it can display a radiological map overlay for better visualization, as well as, using avatars, to anticipate dose uptake by the operators and improve the scenario of human intervention regarding the ALARA principle. The overlay and avatar positioning in a 3D reconstructed scene are shown in Fig. 11.

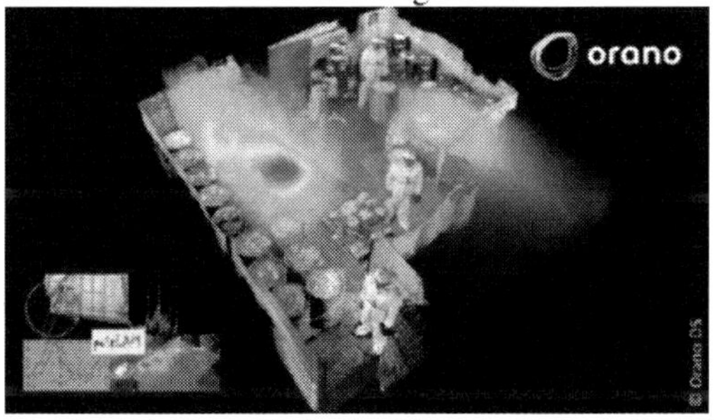

Fig. 11: overlay and avatar positioning in a 3D reconstructed scene in PoStLAM (ORANO).

VIII. Training and Demonstration (WP9)

Led by SOGIN, this work package is in charge of the training of end-users (AiNT) as well as the in-situ demonstration of the technical developments emerging from the project (SOGIN)

Aachen Institute for Nuclear Training (AiNT) is in charge of developing, test a training programme in close collaboration with the other partners of the consortium, as well as with CLEANDEM's End-Users Network (operators, stakeholders, etc.), as each user may have specific requirements and therefore the training needs to be customized accordingly. Following the CONOPS, AiNT established the training programme concept and organized two webinars for the End-Users Network, where the CLEANDEM developments were presented and feedback from the participants analyzed in order to pinpoint their needs. Once the training programme evaluated, AiNT will carry the training out to end-users.

The CONOPS having been established with different realistic scenarios, different sites were identified by SOGIN to host the in-situ demonstration of the final systems developed by WP3, WP4, and WP6–8, with the aim to demonstrate their combined capability to:

- measure in hard to access areas with remote solutions;
- give real time information on the facility status;
- perform the initial radiological characterization of an area (in presence of hot spots and very low contamination level).

In coordination with the consortium, a facility of the EUREX Saluggia plant was selected to demonstrate the capabilities of the final systems. One the areas identified to run the final demonstration is shown in Fig. 12.

Fig. 12: photograph of the host location for the final in-situ demonstration of the CLEANDEM developments.

IX. Market & Costs analysis (WP5)

Led by CAEN, this working group, including ORANO and SOGIN, is focused on the social and economic impact of the UGV platform on the market. The impact based on nowadays used technologies has to be evaluated then compared to the impact based on the UGV platform and the digital twin performances. The effectiveness of the platform will be estimated thanks to the use cases described in the CONOPS.

To do so, WP5 has developed a comparative database of the existing or commercially available and the innovative technologies identified in the CLEANDEM consortium; as well as an evaluation methodology useful for decommissioning cost estimation based on different scenario approaches (business cases).

The cost analysis of the robot on different business cases and extraction of market performance indexes shall be completed in the near future.

X. Conclusions

The developments undertaken by the CLEANDEM consortium have been presented, along with how they are implemented in order to build a Digital Twin with radiological mapping capabilities, to reduce the time and cost of D&D operations, and the dose uptake by human operators, which is a major interest of stakeholders.

Now nearly two and a half years out of the three of the

project, most technological developments are over and the consortium is preparing the final field tests for late 2023, before the final workshop at the EUREX Saluggia site, expected early 2024.

ACKNOWLEDGMENT

The still ongoing aftermath of the international sanitary crisis and the ensuing shortage in multiple areas of manufacturing could have induced major delays in the different developments of CLEANDEM. The authors are very grateful to the consortium members and end-users who made it possible to stay on par with our common goal, thanks to their proactivity, dynamism, critical thinking and support.

This project has received funding from the European Union's Horizon 2020 research and innovation programme under grant agreement n° 945335.

REFERENCES

[1] Rossi, F.; Cosentino, L.; Longhitano, F.; Minutoli, S.; Musico, P.; Osipenko, M.; Poma, G.E.; Ripani, M.; Finocchiaro, P. "The Gamma and Neutron Sensor System for Rapid Dose Rate Mapping in the CLEANDEM Project". *Sensors*, 23, 4210, 2023.

[2] N. Kosa et al., "Draw-tower gratings in multicore fiber: A paradigm shift in curvature sensing", Medical Design Briefs, April 2017.

[3] Documentation FBGS, Fiber Optic Shape Sensing, www.fbgs.com.

Gamma-ray spectroscopy for the characterization of uranium contamination in nuclear decommissioning

Francisco Salvador[1,*], Thomas Marchais[1], Bertrand Pérot[1], Pierre-Guy Allinei[1], Frédéric Morales[1], Olivier Gueton[2], Julien Venara[3], Marie Cuozzo[3], Frédéric Mayet[4]

[1]CEA, DES, IRESNE, DTN, SMTA, Nuclear Measurement Laboratory, F-13108 Saint-Paul-Lez-Durance, France
[2]CEA, DES, IRESNE, Nuclear Technology Department, 13108 Saint-Paul-lez-Durance, France
[3] CEA, DES, ISEC, DMPE, SEIP, LNPA, Univ. Montpellier, Marcoule, F-30207 Bagnols-sur-Cèze Cedex, France
[4] Laboratory of Subatomic Physics and Cosmology, CNRS/IN2P3, Grenoble Alpes University, 38026 Grenoble, France
(*) Francisco.SALVADORBARBA@cea.fr

Abstract— Decommissioning is the last step in the life cycle of a nuclear facility. After the evacuation of the facility components, the remaining structures such as concrete walls and floors must be surveyed to ensure that no residual contamination remains. It is a costly and time consuming activity, for which CEA develops fast alpha and beta detection methods allowing a full scanning of very large areas (hundreds of thousands of square meters) in legacy uranium enrichment plants. To support these developments, we present here complementary high-resolution gamma-ray spectroscopy analyses of a contaminated area at the gaseous diffusion uranium enrichment facility UDG, currently under decommissioning at Pierrelatte nuclear plant, France. Long measurements are performed with a High-Purity Germanium (HPGe) detector on the contaminated surface, and in a clean area to assess the natural gamma background of the concrete ground. The surface activity of uranium is 16.6 ± 6.0 Bq.cm^{-2}, mainly due to ^{234}U and ^{238}U, most of the uncertainty coming from the non-uniform distribution of the contamination on the ground. This measurements also allowed us estimating the uranium enrichment of the contamination, which amounts to (0.80 ± 0.13) % of ^{235}U mass fraction, consistently with the range of the Low Enrichment Plant where this measure was performed. Eventually, the background spectrum allowed us to determine the mass fractions of natural uranium, thorium and potassium in the concrete ground, which respectively amount to 3.8 ± 0.2. ppm$_U$ (i.e. 3.8 mg of uranium per kg of concrete), 7.4 ± 0.7 ppm$_{Th}$, and (2.6 ± 0.1) %$_K$ of potassium.

Index Terms— HPGe detector, gamma-ray spectroscopy, MCNP simulation, uranium contamination, Decommissioning, Dismantling, UDG

I. Introduction

THE gaseous diffusion plant (UDG) of Pierrelatte nuclear site was the first uranium enrichment facility in France. It provided enriched uranium for both pressurised water reactors and for the development of nuclear weapons. The facility was operated from 1960 until its shutdown in 1996. Shortly after, the decommissioning operations began and the enrichment equipment has since been evacuated. As of today, a complete decommissioning of the UDG has yet to be approved by the nuclear safety authorities. Before the site is given clearance for decommissioning, a verification of the radiological cleanliness of the remaining concrete structures must be carried out to ensure the absence of residual uranium contamination. A full scan of UDG soils and walls, equivalent to 700 000 m^2, has to be performed to verify that the radioactivity level stay below a clearance alpha activity threshold of 0.4 Bq.cm^{-2}, established by the French Nuclear Safety Authority (ASN).

This calls for the development of fast measurement methods, capable of reaching low detection limits in very short acquisition times to measure large areas at a high rate. This can be achieved by alpha and beta contamination monitors, which are used as 1st level detectors, but when a contamination is detected, within a few seconds, a more precise quantification of the surface activity is needed as large interpretation uncertainties may occur due to the low range of alpha and beta particles in matter, especially in concrete. Therefore, the Nuclear Measurement Laboratory of IRESNE Institute, at CEA Cadarache, studies the use of gamma spectroscopy as an upper level quantification method of uranium contamination.

Gamma rays emitted by uranium contamination are mainly due to ^{238}U, ^{235}U and ^{234}U isotopes, and their direct descendants having a short enough half-life for radioactive equilibrium to be reached. Taking into account a decay period of 40 years as reference, corresponding to the UDG service life, the radioisotopes present in UDG contamination are thus ^{238}U, ^{234}Th, ^{234m}Pa, ^{235}U, ^{231}Th and ^{234}U. Their main gamma emissions (energy and intensity with their respective uncertainties) are presented in Table I.

TABLE I
MAIN GAMMA EMISSIONS FROM ISOTOPES PRESENT IN A URANIUM CONTAMINATION AFTER A 40 YEARS DECAY PERIOD [1]. AN ASTERISK (*) INDICATES THE STUDIED GAMMA RAYS.

Decay chain	Isotope	Gamma energy (keV)	Gamma intensity (%)
^{238}U	^{238}U	49.55 (6)	0.0697 (26)
	^{234}Th	63.30 (2) *	3.75 (8)
		92.38 (1)	2.18 (19)
		92.80 (2)	2.15 (19)
		112.81 (5) *	0.215 (22)
	^{234m}Pa	766.361 (20) *	0.323 (4)
		1 001.026 (18) *	0.847 (8)
^{235}U	^{235}U	143.767 (3) *	10.94 (6)
		163.356 (3) *	5.08 (3)
		185.720 (4) *	57.0 (3)
		205.316 (4) *	5.02 (3)
	^{231}Th	25.64 (2)	13.9 (7)
		84.2140 (13)	6.70 (7)
^{234}U	^{234}U	53.20 (2) *	0.1253 (40)

These gamma emissions compose the contamination signal and are predominantly present in the low energy region of the

https://doi.org/10.1051/epjconf/202328807005

gamma spectra, between 40 keV and 210 keV (see further Fig. 6). Only the gamma rays of ^{234m}Pa are outside this interval. During the uranium gaseous enrichment process, natural uranium is filtered in porous membranes to increase the proportion of ^{235}U and ^{234}U with respect to that of ^{238}U [2]. While the mass proportion of ^{234}U is very small, from about 0.005 to 0.006 % in natural uranium to 1 % in highly enriched uranium (HEU) [3], it brings more than 50 % of the alpha activity at low enrichment and up to almost 100 % for HEU, as a results of its short half-life of 245.10^{3} years compared to 704.10^{6} years for ^{235}U and 447.10^{7} years for ^{238}U [1]. The UDG plant is made up of increasing enrichment facilities: the Low Enrichment Plant up to about 2 % (^{235}U mass fraction), the Medium Enrichment Plant (MP) up to about 8 % ^{235}U, the High Enrichment Plant (HP) up to about 25 %, and the Very High Enrichment Plant (VHP) up to more than 90 % [4]. Depending on the Enrichment Plant, the main contributors to alpha radioactivity significantly differ, as well as gamma emission rates.

Gamma detection is also subject to an important background emitted by the natural radioisotopes found in concrete, either in the natural decay chains of ^{238}U, ^{235}U, and ^{232}Th (uranium and thorium elements are present in concrete in ppm quantities, and their radioactive chains are in secular equilibrium), or by ^{40}K naturally present in potassium. Table II shows the most intense natural gamma emissions.

TABLE II
MAIN NATURAL GAMMA RAYS EMITTED IN CONCRETE, FROM ^{238}U, ^{235}U, ^{232}TH AND THEIR RADIOACTIVE DAUGHTERS, AND FROM ^{40}K [1]. A HASH (#) INDICATES THE STUDIED GAMMA RAYS.

Decay chain	Isotope	Gamma energy (keV)	Gamma intensity (%)
^{238}U	^{234}Th	63.30 (2)	3.75 (8)
		92.38 (1)	2.18 (19)
		92.80 (2)	2.15 (19)
	^{214}Pb	295.224 (2)	18.414 (36)
		351.932 (2) #	35.60 (7)
	^{214}Bi	609.312 (7) #	45.49 (19)
		1 120.287 (10) #	14.91 (3)
		1 238.111 (12)	5.831 (14)
		1 764.494 (14) #	15.31 (5)
^{232}Th	^{212}Pb	238.632 (2) #	43.6 (5)
	^{228}Ac	338.320 (5) #	11.4 (4)
		911.196 (6) #	26.2 (8)
		968.960 (9) #	15.9 (5)
	^{208}Tl	583.187 (2) #	30.6 (1)
		2 614.511 (10) #	35.84 (7)
	^{212}Bi	727.330 (9)	6.65 (4)
^{40}K	^{40}K	1 460.822 (6) #	10.55 (11)

We will show in next sections that this natural gamma background of concrete may represent a significant contribution in certain gamma rays, which must be subtracted to correctly estimate the activity of the uranium surface contamination. The measurement of this background and the detection of uranium contamination is reported in this work with a high resolution gamma spectroscopy detector, in the UDG Low Enrichment Plant.

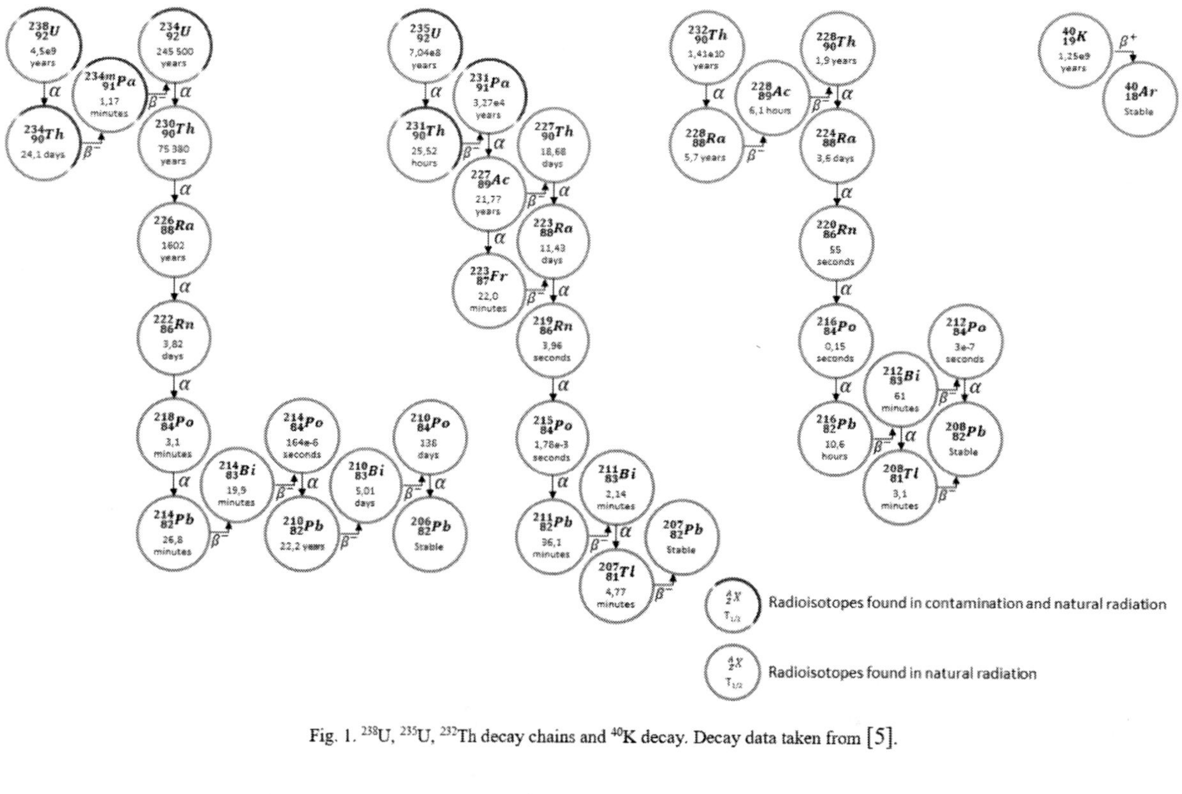

Fig. 1. ^{238}U, ^{235}U, ^{232}Th decay chains and ^{40}K decay. Decay data taken from [5].

II. High resolution gamma spectroscopy for fine contamination characterization

A. Measured areas

The measurements took place in a diffusion group of the UDG Low Enrichment Plant, especially on the reference area presented in Fig. 2, chosen because of the presence of a proven uranium contamination deposited over the concrete floor, which allows to compare different detection methods (alpha, beta, gamma).

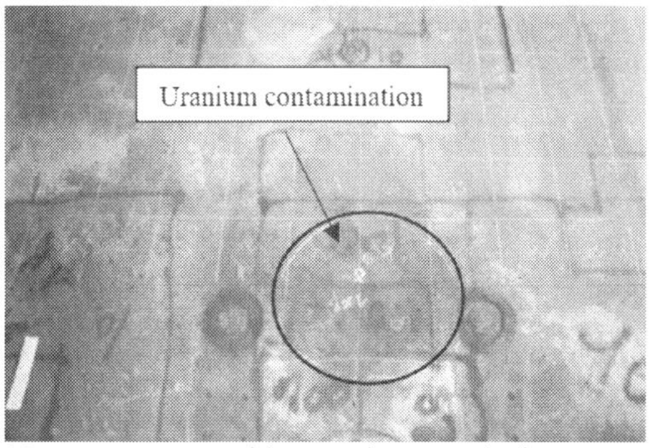

Fig. 2. Measurement area containing a uranium contamination spot.

The measurement area was separated into several measurement points (marked by the blue squares on Fig. 2) with an area of 25 cm × 25 cm each, which corresponds to the detection surface seen by our instruments. A label was attributed to each point and uranium contamination is located inside measurement points # 6, 7, 8 and 9. Other measurements were carried out on other points outside this region, which are supposed free of contamination, to estimate the concentration of natural gamma emitters in concrete (U, K, Th).

B. Experimental setup and MCNP model

High resolution gamma spectroscopy studies were conducted using a Falcon 5000 High Purity Germanium (HPGe) detector. This model includes a planar germanium crystal with a 6.5 cm diameter and 3 cm thickness, as well as an integrated multi-channel analyzer and an electrical cooling system [6]. Two measurement configurations were used, the first of which was used to measure uranium contaminated areas with the detector lifted 12.4 cm from the floor, thanks to a manual stacker (configuration # 1, see Fig. 3).

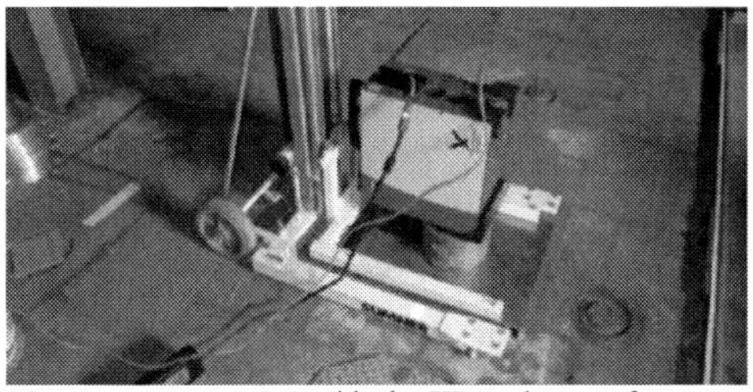
Fig. 3. Measurement setup # 1 with the HPGe detector for contamination measurement.

This setup led to a detection surface of 530 cm^2, similar to that of larger alpha and beta detectors used during the same measurement campaign, so that results could be compared. However, positioning the HPGe detector at this height above the ground reduces its detection efficiency. On the other hand, 5 cm thick lead rings were used to shield the germanium crystal against background radiation coming from neighboring soil surfaces but also from concrete walls.
In the second configuration, the HPGe detector is closer to the floor with a height of 6.7 cm between its entrance window and the ground (configuration # 2).

Fig. 4. Setup # 2 of the HPGe detector for background measurement.

This configuration # 2 allowed us to increase detection efficiency for the soil background measurements, while keeping the detector shielding. Monte-Carlo simulation models of both setups were developed with MCNP computer code [7] in order to calculate the detection efficiency of gamma rays emitted by uranium contamination, in surface, or by the bulk concrete soil for the natural background. These efficiencies are then used for activity calculations as detailed further in (1). The next figure shows a graphical representation of these MCNP models.

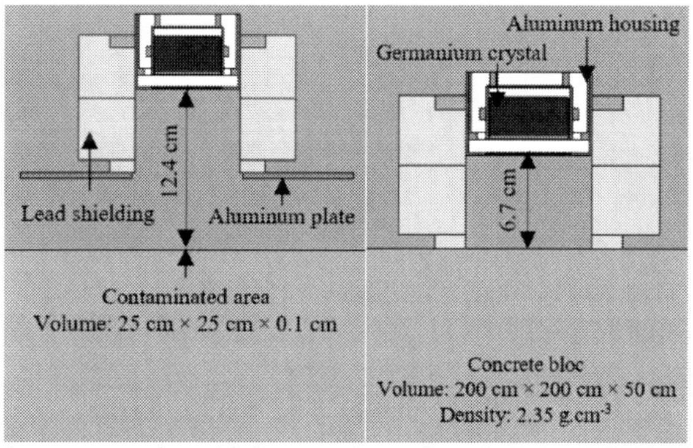

Fig. 5. Monte-Carlo models for germanium configurations # 1 and # 2 used for detection efficiency calculations.

For configuration # 1, the thin contaminated area (1 mm thickness) above the concrete ground is not visible in Fig. 5. It is also made of concrete and it is as a volume source in MCNP, while for configuration # 2, the source is the bulk concrete soil (50 cm thickness). The model of the detector and lead shielding is the same for both configurations.

C. Contamination activity estimation and enrichment percentage

As mentioned above, MCNP calculations are used to determine the gamma detection efficiencies allowing to convert the net measured count rates into activities (in Bq). Using detection setup # 1, an acquisition of 64.3 hours was performed on contaminated point # 7 located inside the contaminated area of Fig. 2. The obtained spectrum is given in Fig. 6.

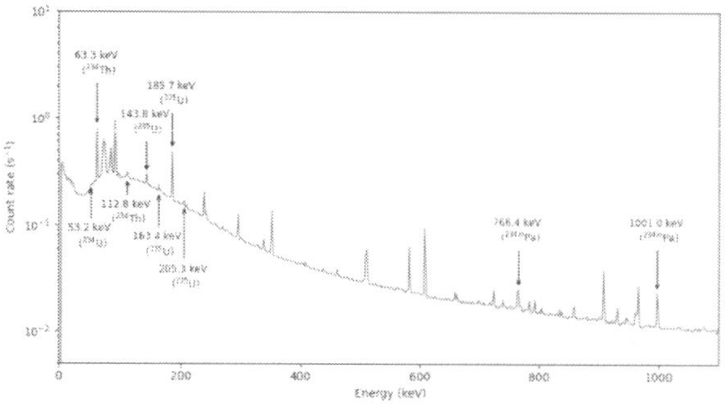

Fig. 6. Uranium contamination gamma spectrum measured in point # 7 with the HPGe detector. The main gamma peaks due to uranium contamination are indicated with arrows.

Uranium characteristic rays are visible on the spectrum, such as the 63.3 keV and 1001 keV lines emitted by ^{234}Th and ^{234m}Pa, respectively. Both isotopes are part of the ^{238}U contamination decay chain (see Table I). Peaks emitted by ^{235}U are also visible, for instance at 143.8 keV and 185.7 keV. A small 53.2 keV gamma from ^{234}U is also present (see Fig. 7).

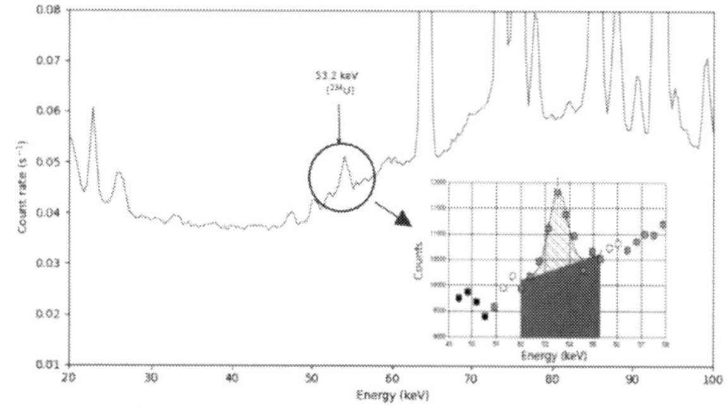

Fig. 7. Uranium contamination gamma spectrum measured in point # 7 with the HPGe detector zoomed between 20 and 100 keV to show the peak fit obtained with the Genie2000 gamma analysis software [8] for the 53.2 keV peak.

The activity of the emitting isotopes is estimated by analyzing gamma rays at different energies using the following equation:

$$A_S(E) = \frac{S_n(E)}{Eff(E)\ \times I(E)\ \times T_c\ \times S_{contamination}} \quad (1)$$

With:

- A_S the calculated surface activity (in Bq.cm^{-2}) calculated with peak of energy E (it is thus possible to calculate an average activity for the isotopes emitting several detectable gamma rays),
- $S_n(E)$ the net area under the gamma peak after subtraction of the Compton continuum under the peak, and of the net area of the natural background presented in Fig. 8 when it is significant, such as for the 185.7 keV peak of ^{235}U that is interfered by the 186.2 keV peak of ^{226}Ra in the background.
- Eff(E) the simulated detection efficiency (counts per source particle at energy E), calculated with the first model of Fig. 5. The migration depth of the uranium contamination was set to 1 mm to simulate a surface contamination. Other depth lengths from a few µm to a few mm were also considered but with a limited effect on simulated efficiency with a maximum relative difference of 4 % at low energy (53.2 keV),
- I(E) the emission intensity at energy E (number of gamma emitted per disintegration),
- T_c the active counting time (in seconds),
- $S_{contamination}$ the area of the contamination taken into account in the MCNP model, here 25 cm × 25 cm (see Fig. 5).

This led us to an estimation of the weighted average activity from the analysis of multiple gamma rays emitted by each of ^{238}U and ^{235}U isotopes. The average activities were calculated with the formula below:

$$\overline{A_x} = \frac{\sum_{i=1}^{N} \frac{A_s(E_i)}{\left(\frac{\sigma_{S_n}(E_i)}{S_n(E_i)}\right)^2}}{\sum_{i=1}^{N} \frac{1}{\left(\frac{\sigma_{S_n}(E_i)}{S_n(E_i)}\right)^2}} \quad (2)$$

With:

- $A_{s,i}$ the calculated surface activity from gamma peak at energy E_i (in Bq.cm^{-2}, refer to (1)) emitted by ^{238}U or ^{235}U,
- S_n the net area of the gamma peak at energy E_i
- σ_{Sn} the absolute statistical standard deviation for the gamma peak at energy E_i taking into account the background Compton continuum B under the peak, $\sigma_{Sn} = \sqrt{S_n + 2B}$ with S_n the net area and B the background.

From (2), we obtain the following weighted average activities: A_{238} = 8.4 ± 4.5 Bq.cm^{-2} and A_{235} = 0.44 ± 0.23 Bq.cm^{-2}. The estimation of ^{234}U activity relies on the only exploitable gamma peak at 53.2 keV, with an emission intensity of 0.13 %. Its net area fluctuates between 3711 and 5146 counts depending on fit parameters used to estimate the background under the peak (pink area in Fig. 7), with a relative standard deviation of 11 % and an average net area of 4090 counts resulting from 10 different fits. We added this standard deviation through a quadratic sum to the statistical standard deviation ($\sqrt{S_n + 2B}$), and thus estimate a total relative standard deviation of 17 % on the net area of the small 53.2 keV peak. Finally, we obtain the following activity for ^{234}U: A_{234} = 7.8 ± 4.1 Bq.cm^{-2}. We list below the different causes of uncertainty taken into account to estimate the previous confidence intervals on ^{234}U, ^{235}U, and ^{238}U activities:

- the representativeness of the MCNP model, especially the real distribution of the contamination, which was measured as non-uniform by autoradiography [9]. Therefore, we calculated efficiencies for multiple distribution hypothesis ranging from a hot spot (i.e. a point source in the middle of the collimator solid angle) to a surface of 30 cm × 30 cm exceeding the field of view of the detector inside the collimator (disk with a diameter of about 20 cm, i.e. a detection surface around 300 cm²). These calculations (not reported here) show that if the contaminated area is smaller (for instance a hot spot or a 10 cm × 10 cm surface) than the detector field of view, the efficiency can be almost twice than that used for interpretation (calculated for a contaminated area of 25 cm × 25 cm) or those calculated with surfaces larger than the field of view (we studied from 20 cm × 20 cm to 30 cm × 30 cm). From this study and due to the lack of knowledge of the real contamination distribution, we consider an arbitrary relative uncertainty of 50 % on efficiency,

- this last also includes the uncertainty on the HPGe detector MCNP model, which is however less than 10 % from our feedback [10]. Even if this uncertainty is much smaller than the previous one, we will check the HPGe detector model through MCNP vs. experiment comparisons of precise measurements with calibration point sources, on the wide energy range of interest (53.2 keV peak of ^{234}U up to 1001 keV peak of ^{234m}Pa),
- the other uncertainties on the MCNP model, such as the limited knowledge of the concrete block density and chemical composition, are smaller and also included in the abovementioned 50 % relative standard deviation,
- statistical uncertainties, including the variability associated to the net area extraction for the small 53.2 keV peak of ^{234}U, and the dispersion of the activities obtained with the different peaks of multi-gamma emitters, calculated through the weighted average (2).

Summing the individual ^{238}U, ^{235}U and ^{234}U activity values, we obtain a total uranium activity of 16.6 ± 6.0 Bq.cm^{-2}.
HPGe measurements also allowed for an estimation of the enrichment percentage of the contamination. Individual activities of ^{238}U, ^{235}U, and ^{234}U are divided by their respective mass activities of 12.4×10^3 Bq.g^{-1}, 79.9×10^3 Bq.g^{-1} and 230×10^6 Bq.g^{-1} [1], to obtain the following masses m_{238} = 0.44 ± 0.05 g, m_{235} = (3.5 ± 0.4)×10^{-3} g and m_{234} = (2.1 ± 0.3) ×10^{-5} g. An ^{235}U enrichment of 0.80 ± 0.13 % is thus deduced using (3):

$$\%_{Enrichment} = \frac{m_{235}}{m_{235} + m_{238}} \times 100 \qquad (3)$$

This 0.8 % enrichment is consistent with the range of the Low Enrichment Plant, which extends from depleted uranium up to about 2 % [4], as mentioned in Section I. Since m_{235} and m_{238} are estimated from the measurement, only statistical uncertainties are considered but not those due to the possible non-uniformity of contamination.

D. Natural radiation measurements and U, K and Th concentration estimations

A 23 hour background measurement was done in a clean area using the HPGe detector in experimental configuration # 2 (see Fig. 4 and Fig. 5). The background spectrum is shown in Fig. 8, in which the well-known natural gamma rays are identified, such as the 609 keV peak of ^{214}Bi (bottom of ^{238}U decay chain), 1460 keV of ^{40}K, and 2614 keV of ^{208}Tl (end of ^{232}Th chain).

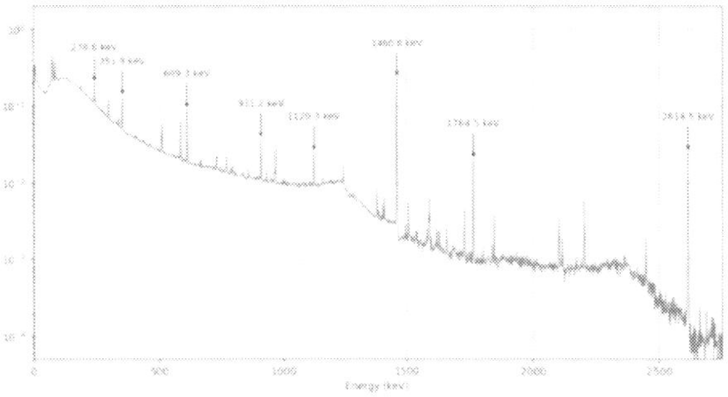

Fig. 8. Background spectrum measured with the HPGe detector with configuration # 2.

The efficiency is calculated with MCNP and the second model of Fig. 5 which considers a 200 cm × 200 cm × 50 cm concrete block of density 2.35 g.cm^{-3} as gamma source, which is a sufficient volume to have an almost infinite-equivalent geometry, i.e. no significant additional signal would come from a larger volume. We will investigate other geometries in future work, in particular to take into account new information about the real thickness of the ground at UDG Low Enrichment Plant. Note that this geometry leads to similar calculated activities for all gamma peaks (on a wide energy range) of the isotopes present in the ^{238}U and ^{232}Th natural chains, which are in secular equilibrium. The individual activities of each isotope is calculated as follows:

$$A_V = \frac{S_n(E)}{Eff(E) \times I(E) \times T_c} \qquad (4)$$

With:

- A_V the volume activity (in Bq) of ^{238}U or ^{232}Th calculated with the peak of energy E,
- $S_n(E)$ the net area under the gamma peak, after subtraction of the Compton continuum,
- Eff(E): simulated detection efficiency (counts per source particle at energy E), calculated with the second model of Fig. 5,
- I(E) and Tc as in (1).

Using (2), we obtain the following weighted average activities with the different peaks of ^{238}U and ^{232}Th chains: $A_{nat\ 238U}$ = 219 ± 14 kBq and $A_{nat\ 232Th}$ = 141 ± 14 kBq. On the other hand, the activity of ^{40}K estimated with the 1460 keV is $A_{nat\ 40K}$ = 3726 ± 192 kBq. The relative uncertainties are here ranging between about 5 and 10 %, which is much smaller than for contamination activity because natural U, Th and K elements are supposed uniformly distributed in the volume of concrete. The confidence intervals are therefore mainly due to counting statistics. The lack of knowledge of the real density and composition of the concrete ground has an effect smaller than 5 % on the energy range used to assess the above activities. From these activities, we can deduce the masses of U, Th and K elements present in the simulated concrete block, using the mass activities of the ^{238}U, ^{232}Th and ^{40}K isotopes (12.4 × 10^3, 4.07 × 10^3 and 265 × 10^3 in Bq.g^{-1} respectively, from [1]) and their natural abundance (99.3 % for ^{238}U, 99.9 % for ^{232}Th and 0.012 % for ^{40}K): m_U = 17.8 ± 1.1 g, m_{Th} = 34.6 ± 3.5 g and m_K = 120.4 ± 6.2 kg.

https://doi.org/10.1051/epjconf/202328807005

Taking into account the mass of the simulated concrete block, these masses corresponds to mass fractions of 3.8 ± 0.2 ppm_U for uranium (1 ppm is 1 mg of uranium per kg of concrete), 7.4 ± 0.7 ppm_{Th} for thorium and (2.6 ± 0.1) $\%_K$ for potassium. These mass fractions are within the range of typical U, K and Th proportions in the continental crust [11].

III. Conclusion

Experimental tests have been performed at the UDG uranium gaseous enrichment plant for the detection of residual uranium contamination, using high resolution gamma spectroscopy with an HPGe detector. This measurement of 64 hours allowed us to evaluate the surface activity of a contaminated area to 16.6 ± 6.1 $Bq.cm^{-2}$. The large uncertainty is mainly due to the limited knowledge of the real contamination distribution, which has been demonstrated to be non-homogenous thanks to autoradiography analysis [9]. We also measured the ^{235}U mass enrichment, estimated to (0.80 ± 0.13) %, and the natural background over a clean area of the facility. From the study of this background spectrum, we evaluated the natural U, Th and K mass fractions to 3.8 ± 0.2 ppm_U, 7.4 ± 0.7 ppm_{Th} and (2.6 ± 0.1) $\%_K$ inside the facility concrete, which is consistent with typical continental crust levels [11]. Future works will focus on low resolution gamma spectroscopy with an NaI(Tl) scintillator on the same contaminated area. The goal being to have a faster characterization (about 15 min) of the contaminated surfaces detected by alpha or beta contamination monitors.

References

[1] M.-M. Bé, V. Christé, and C. Dulieu, "NUCLÉIDE-LARA - Bibliothèque des émissions alpha, X et gamma," *Rapp. CEA R-6201.*

[2] W. Jeff, "Uranium conversion and enrichment," in *Nuclear Fuel Cycle Science and Engineering*, Elsevier, 2012, pp. 151–176. doi: 10.1533/9780857096388.2.151.

[3] S. Croft *et al.*, "The specific (α ,n) production rate for 234 U in UF 6," *Nucl. Instrum. Methods Phys. Res. Sect. Accel. Spectrometers Detect. Assoc. Equip.*, vol. 954, p. 161608, Feb. 2020, doi: 10.1016/j.nima.2018.11.067.

[4] S. Philippe and A. Glaser, "Nuclear Archaeology for Gaseous Diffusion Enrichment Plants," *Sci. Glob. Secur.*, vol. 22, no. 1, pp. 27–49, Jan. 2014, doi: 10.1080/08929882.2014.871881.

[5] J. Peterson, M. MacDonell, L. Haroun, and F. Monette, "Radiological and Chemical Fact Sheets to Support Health Risk Analyses for Contaminated Areas," Argonne National Laboratory, Environmental Science Division, Mar. 2007.

[6] Mirion Technologies, "Falcon 5000® Portable HPGe-Based Radionuclide Identifier." Accessed: Jul. 26, 2022. [Online]. Available: https://mirion.s3.amazonaws.com/cms4_mirion/files/pdf/spec-sheets/falcon-portable-hpge-based-identifier.pdf?1557257239

[7] J. Kulesza *et al.*, "MCNP® Code Version 6.3.0 Theory & User Manual," LA-UR-22-30006, 1889957, Sep. 2022. doi: 10.2172/1889957.

[8] Mirion Technologies, "GENIE™ 2000 Gamma Analysis Software."

[9] P. Fichet, S. Leblond, R. Laumonier, P. Sardini, and S. Billon, "MAUD Project - Characterization by Autoradiography Tech-nique of facilities under dismantlement," 2018.

[10] T. Marchais *et al.*, "Detailed MCNP Simulations of Gamma-Ray Spectroscopy Measurements With Calibration Blocks for Uranium Mining Applications," *IEEE Trans. Nucl. Sci.*, vol. 65, no. 9, pp. 2533–2538, Sep. 2018, doi: 10.1109/TNS.2018.2797312.

[11] K. S. Heier, "Radioactive Elements in the Continental Crust," *Nature*, vol. 208, no. 5009, pp. 479–480, Oct. 1965, doi: 10.1038/208479b0.

Development of a specialised calibration and infra-red emissivity correction to enable the temperature monitoring of intermediate level nuclear waste using thermal imaging

W. Bond*, M. Hayes, Dr R. Simpson, Dr S. Korniliou, Prof G. Machin, National Physical Laboratory, UK and A. Allen, Sellafield Ltd. UK
(*) wesley.bond@npl.co.uk

Abstract— **The temperature measurement of intermediate level nuclear waste containers is vital to monitor thermal stability and track temperature trends over time. The use of thermal imaging for this monitoring has numerous advantages compared to using a single spot radiation thermometer, and Sellafield wish to employ this approach for their '3 m³ box' intermediate level waste containers. In order to achieve this objective with confidence a custom temperature calibration is required for low uncertainty temperature measurement. Another key consideration for practical non-contact temperature measurement is the emissivity of the surface being measured at the relevant infra-red wavelengths, as such the emissivity of the Sellafield 3 m³ box materials was measured and then corrected for. The implementation of a calibrated instrument and emissivity correction were validated using temperature controlled, 1:10 scale model 3 m³ box containers within a climatic chamber. The surface temperature of the scale containers was determined to an uncertainty of ± 6.5 °C (k = 2.1) for the container walls, ± 3.5 °C (k = 2.1) for the container lid and ± 1.5 °C (k = 2.1) for the container vents. The measurement techniques used to determine the temperature of the 1:10 scale 3 m³ box containers were therefore successful and there is a clear development path to monitoring the containers in storage using thermal imaging.**

***Keywords* — Nuclear decommissioning waste radiation temperature thermal imaging monitoring safety stability**

I. Introduction

A significant portion of radioactive material stored at the Sellafield site in Cumbria, UK is classified as intermediate level waste. This waste was traditionally stored at the base of open pools, but it is now being moved to medium term storage containers.

The primary container to be used for this storage is the '3 m³ box'. The container system consists of a vented inner skip holding the immersed nuclear waste and an outer 3 m³ box with lid vents to allow for cooling and gas release. An exploded view diagram of this box can be seen in Fig. 1.

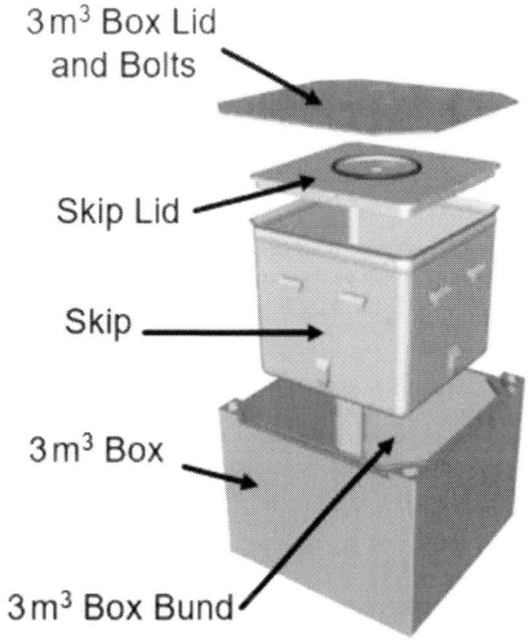

Fig. 1. Diagram showing structure of 3 m³ box internal structure with waste skip and external box [1].

The activity level of intermediate level waste mandates that a method of temperature monitoring is used to track the thermal evolution of the 3 m³ waste packages over time. The use of thermal imaging for this monitoring enables the detection of box non-uniformity, hot spots, and other visual artefacts. However, for confidence in the measured temperature the thermal imager must be calibrated to traceable national standards and in a way that accounts for the anticipated environmental temperature range within the store which will alter the device's operating temperature, impacting measurements.

Similarly, the surface emissivity of the 3 m³ box materials at the wavelength, temperature, and angle captured by the thermal imager must be measured, this enables the correction of measurements when dealing with the apparent total infra-red radiation from the box surfaces as captured by a thermal imager. This is because the reflected radiation from the measurement surface will represent the temperature of the environment, because of this the proportion of infra-red radiation from each source, emitted or reflected, must be known. A robust knowledge of the surface emissivity of major 3 m^3 box materials is therefore a critical factor in the reliable measurement of the container's surface temperature.

A custom temperature calibration, box material emissivity measurements and emissivity correction were all addressed within this activity to enable future 3 m^3 box temperature measurement and monitoring.

II. Experiments & Methods

A. Thermal imager calibration

A low cost, off the shelf, low size weight and power thermal imager was used for this research. The same model of thermal imaging core is envisaged to be used by Sellafield for the final application as their low cost makes them easily replaceable when they succumb to radiation damage. However, before any thermal imager measurements can be gathered a custom calibration needed to be performed.

The thermal imager was calibrated within a climatic chamber across an environmental temperature range of 7.5 °C to 27.5 °C, with this representing the nominal temperature range within the operational 3 m^3 box store. Humidity was not controlled as this improved the temperature stability within the chamber. The temperature calibration range was from 7.5 °C to 37.5 °C, with this range representing the expected apparent temperature of the 3 m^3 boxes. This temperature range was determined using earlier thermal imager measurements of a 3 m^3 box with an artificial internal heat source.

The thermal imager was calibrated using a high emissivity cavity blackbody positioned outside the climatic chamber and viewed through a porthole. An example calibration image can be seen below in Fig. 2, the region of interest used to gather raw 16-bit digital value data is shown in red.

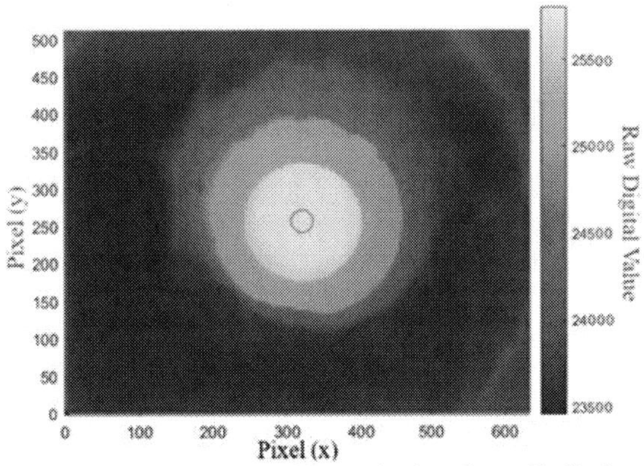

Fig. 2. Example calibration thermal imager showing reference blackbody.

B. Emissivity measurements

Samples of the vent metallic mesh, wall steel, and lid steel from 3 m^3 boxes were provided to the National Physical Laboratory (NPL) by Sellafield Ltd., these were cut to size and dispatched to Physikalisch-Technische Bundesanstalt (PTB), the National Metrological Institute of Germany to complete the spectral directional emissivity measurements of the samples using their infra-red spectrometer [2]. The dispatched vent material is shown in Fig. 3.

PTB completed emissivity measurements at 15 °C and 60 °C at a wavelength range of 5 µm to 14 µm and at seven angles of observation ranging from 10° to 70° from the normal to the sample surface. These results were delivered to NPL in full for use in the measurement correction for non-perfect emissivity.

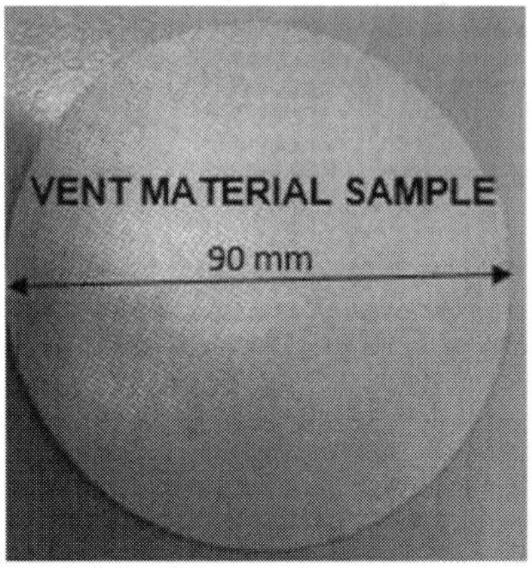

Fig. 3. 90 mm discs of ~ 1 m thickness vent material sent to PTB for surface emissivity measurements.

C. Scale box measurements

A 1:10 scale storage environment was set up inside the NPL thermal imaging team's climatic chamber. The temperature of the environment was controllable across the required temperature range, with a temperature stability of nominally ± 0.1 °C.

Plywood panels, fitted within the chamber, were given a ~1 mm coating of Secrete to mimic the concrete surfaces of the storage facility walls, ceiling and floor, this provided a more realistic background environment with estimated surface emissivity of approximately 0.95. These panels were cut to size and attached to the climatic chamber walls, ceiling, and floor, this ensured that the first material reflected in the lid, vent, and outer wall of each scale 3 m^3 box was a concrete analogue, as would be observed in the operating store. The surfaces of the 1:10 scale boxes were constructed with steels that matched the composition and finish of the 3 m^3 box wall and lid, with actual vent material used for the four vents on each lid. The scale boxes and false concrete walls of the climatic chamber are shown in Fig. 4.

Each 1:10 scale box was fitted with a 12-Watt heater which was used to increase the internal temperature to generate an external surface temperature matching that of a full-size container.

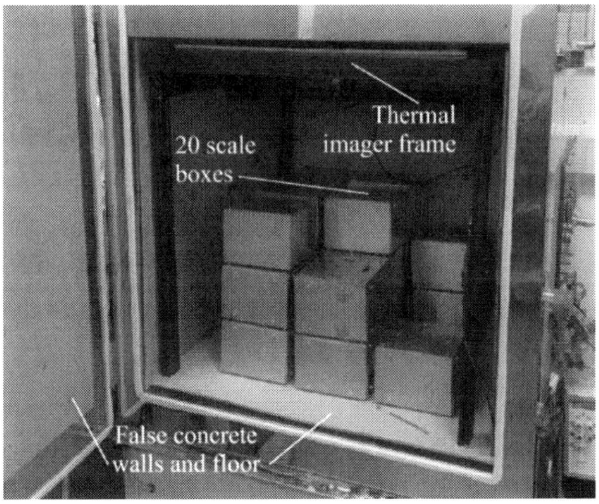

Fig. 4. External view of climatic chamber with false concrete walls and modified 1:10 scale boxes in place.

Four of the scale boxes were chosen to be measurement boxes, these were instrumented with surface mounted thermocouples to provide temperature references and their internal heaters were wired on a separate loop to be able to simulate elevated internal temperatures. These four boxes were given the arbitrary labels of "Red", "Blue", "Orange" and "Purple". The position of these boxes is shown below in Fig. 5. Thermal images were captured in 9 different positions above the four measurement boxes.

A total of 28 contact temperature sensors – measuring scale box surface temperature, chamber wall surface temperature and air temperature – were placed throughout the climatic chamber and boxes, in addition to a calibrated humidity sensor and chamber control sensor.

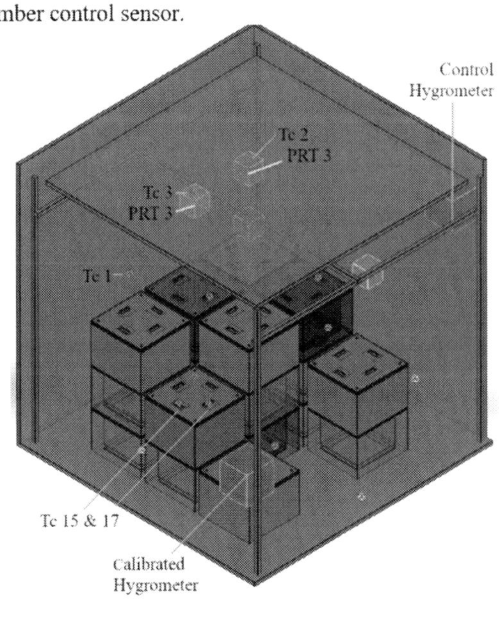

Fig. 5. 3D Diagram of climatic chamber measurement setup. The four instrumented and thermally imaged boxes are shown (Red, Orange, Blue & Purple) in addition to select thermocouple (Tc), Platinum Resistance Thermometer (PRT) and hygrometer locations.

For the measurement campaign nine different conditions were simulated using three environment temperatures (16 °C, 20 °C and 25 °C) and three box simulated temperatures. The non-measurement boxes remained at their expected normal temperature throughout. The measurement box external surfaces were thermally imaged at the expected surface temperature and two elevated surface temperatures of 2.5 °C and 5 °C above normal, with this simulating the expected change that would be seen in a full-scale container with an internal temperature 5 °C and 10 °C above normal, with the normal temperature assumed to be 45 °C

D. *Emissivity correction*

A correction for non-perfect emissivity was applied to the thermal imager measured temperature using a simplified representation of the total infra-red radiation – emitted and reflected – that the thermal imager captured, with this total producing the apparent 3 m^3 box temperature. Under this simplification the reflected component was taken to have the temperature of the chamber walls, ceiling, or floor. In practice this meant the temperature of the closest chamber surface thermocouple to where the angle of incidence met the surface. For simplicity the chamber surfaces were assumed to have a perfect emissivity (i.e.,1.0), as opposed to the actual value of around 0.95. Using the apparent temperature and the approximated reflected component the emitted component could then be calculated.

The data used for the emissivity correction is shown in Fig. 6. The emissivity of the three main box materials and responsivity of the thermal imager are all shown with respect to wavelength.

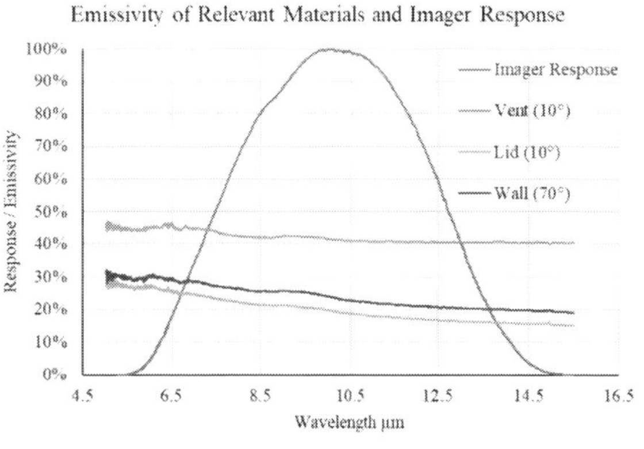

Fig. 6. Relative response of thermal imager and 3 m^3 box material emissivity results at relevant wavelengths

III. Results & Discussion

A. *Thermal imager calibration*

The thermal imager was calibrated with an uncertainty of ± 0.91 °C (k = 2.1, 95% confidence interval). This represents the measurement uncertainty of the instrument measuring apparent surface temperature. The main contributors to this uncertainty were the non-uniformity of the thermal imager array and the repeatability of measurements when viewing a stable source, the results of the calibration validation are shown in Table 1.

This validation of the calibration was performed by measuring the temperature of the reference blackbody using the thermal imager at a non-calibration blackbody temperature and environmental temperature and then calculating the temperature differential between the two.

TABLE 1
VALIDATION OF THERMAL IMAGER CALIBRATION

Thermal imager array temperature (°C)	Reference blackbody temperature (°C)	Thermal imager measured temperature (°C)	Temperature differential (°C)
15.4	7.41	7.43	-0.02
28.0	30.00	30.11	-0.11
17.8	35.00	35.07	-0.06
33.1	35.00	34.88	0.12
17.9	9.98	10.19	-0.21
33.0	9.98	10.02	-0.04
22.9	20.00	19.99	0.02

B. *Emissivity measurements*

All three sample materials were successfully measured in terms of their directional spectral emissivity at two sample temperatures. The uncertainty of these measurements was a maximum of $\pm$ 0.03 ($k = 1$), this is an excellent measurement uncertainty for material infra-red emissivity, but the material emissivity uncertainty remained one of the main contributors to the emissivity correction's overall uncertainty.

C. *Scale box measurements*

The nine environmental conditions and nine thermal imager measurement positions resulted in a data set of 81 thermal imager measurement series. Twelve measurement images were captured at each measurement position within a 30 second period. These were imported into MATLAB and the image series was averaged into a single frame. The data was captured in the same 16-bit format used for calibration and the calibration was used to convert the averaged frame digital values to temperature values.

Two examples of the thermal images captured are shown in Fig. 7, the higher apparent temperature of the four vents is clearly visible in addition to the reflection of the thermal imager itself in the box lids. An area of high emissivity tape was used on half of some vents to provide a thermal imager measured value closer to the true surface temperature.

The region of interests used to measure the vent temperature and the high emissivity surface temperature are shown as black squares. The regions used to measure the lid surface temperature are shown as red circles and these were positioned either side of the surface mounted reference thermocouples.

A visual reference for these features can be seen in Fig. 8.

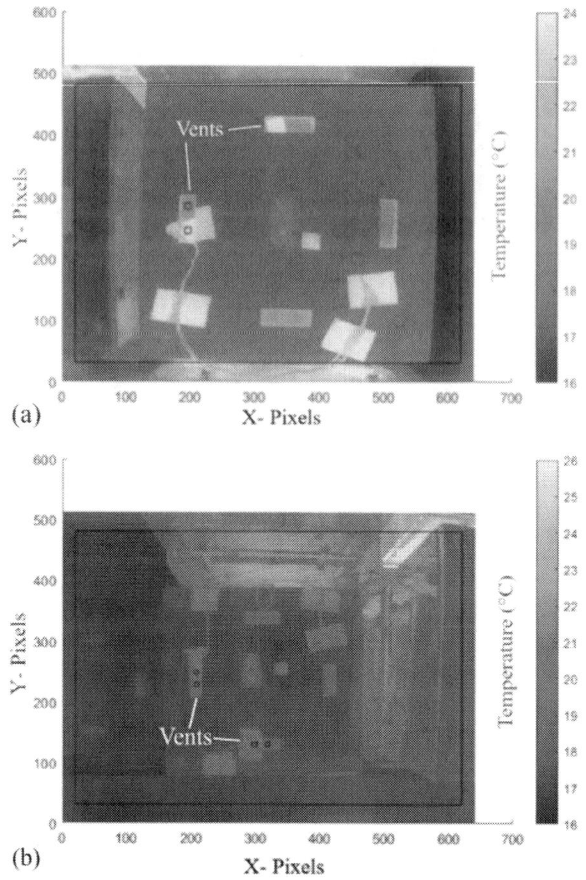

Fig. 7. (a) Thermal image of 'Red' box lid. (b) Thermal imager of 'Orange' box lid. Both images were taken from an overhead position with scale showing apparent temperature.

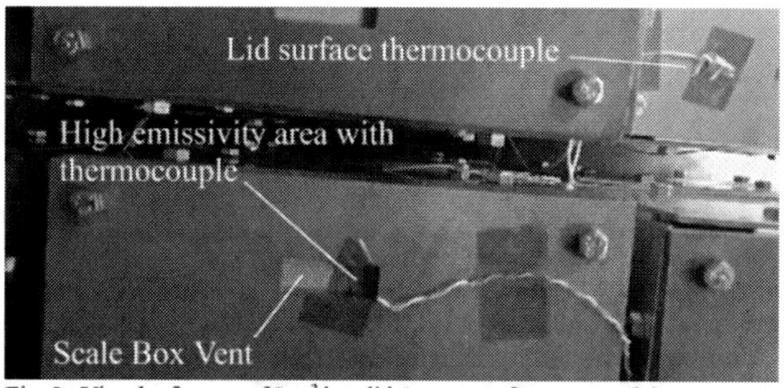

Fig. 8. Visual reference of 3 m^3 box lid & vent, surface mounted thermocouple and high emissivity area are highlighted.

The (calibration corrected, but not emissivity corrected) apparent temperature results of the measurement boxes showed that an elevated box surface temperature was detectable using a thermal imager at all environmental temperatures. The observed apparent temperature change was less than was measured using the reference thermocouples, but this was to be expected due to the non-perfect emissivity of the box materials. An example of these apparent temperature results is shown in Fig. 9.

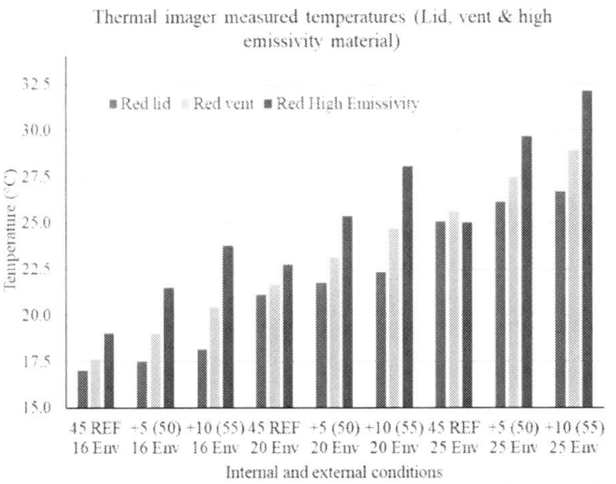

Fig. 9. Apparent temperatures of box upper surfaces as measured by thermal imager. X-axis labels show the modelled box internal conditions in top row, 45°C internal reference temperature "45 REF", 5 °C above normal "+5(50)" & 10 °C above normal "+10 (55)". The chamber environmental conditions are shown in the second row, 16 °C Environment "16 Env", 20 °C Environment "20 Env" and 25 °C Environment "25 Env".

D. *Emissivity correction*

Example results showing the impact of emissivity correction for the 'Red' box vent and 'Blue' box lid are shown in Fig. 10 & 11 respectively. It can be clearly seen that the corrected temperature value is much closer to that measured by the reference thermocouple. Changes in surface temperature are also more detectable, with a larger temperature differential measured for the elevated condition versus the normal condition.

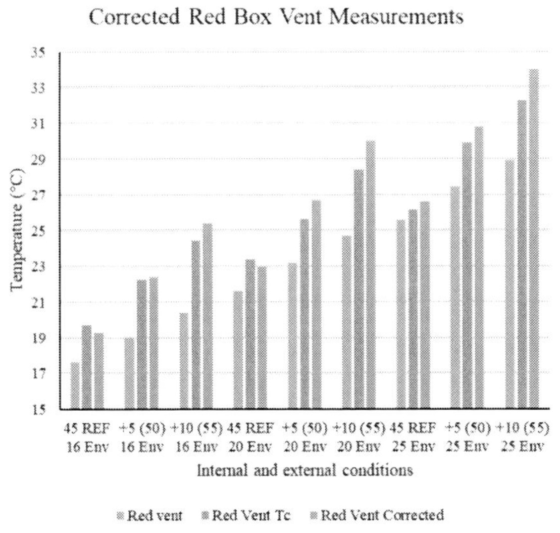

Fig. 10. Comparison of apparent surface temperature measured for vent with reference thermocouple value and emissivity corrected value

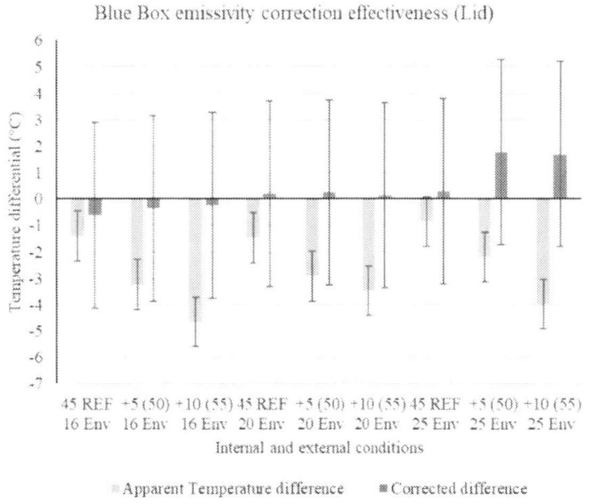

Fig. 11. Comparison of apparent temperature with surface temperature calculated using emissivity correction, the reference thermocouple temperature is taken as the true surface temperature, represented as zero on the y-axis

The average remaining measurement differential between the contact temperature measurement and the thermal imager measured temperature was 0.81 °C, down from 2.8 °C for the uncorrected, apparent box temperatures. The results across the different measurement boxes are shown in Table 2. The emissivity correction reduced the differential between the measured temperature and the reference thermocouple temperature for all measurement boxes and their surfaces

TABLE 2
RESULTS OF EMISSIVITY CORRECTION

Measurement and reference thermocouple location	Non-emissivity corrected differential (°C)	Emissivity corrected differential (°C)
Red lid	3.7	0.8
Red vent	2.6	0.9
Red outer Wall	5.0	0.7
Blue Lid	2.7	0.6
Blue Vent	3.5	0.8
Orange lid	2.4	1.3
Orange Vent	2.2	0.4
Purple outer Wall	2.7	0.9
Red box inner wall	1.7	1.1
Orange box inner wall	2.7	0.8
Blue box outer wall	1.7	0.6
Average	**2.8**	**0.8**

The uncertainty of the emissivity correction was also calculated, and this equated to a total application measurement uncertainty of ± 3.5 °C (k = 2.1) for the box lid, ± 1.55 °C (k = 2.1) for the box vents and ± 6.5 °C (k = 2.1) for the box walls. Key uncertainty contributors were the material emissivity measurement and the uncertainty in determining the background temperature that generated the reflected infra-red component.

IV. Conclusions

The temperature measurement of intermediate level nuclear waste containers is vital for condition monitoring and inspection. The use of thermal imaging, employing a bespoke calibration and surface emissivity correction approach have been successfully validated using temperature controlled, 1:10 scale model waste storage containers within a controlled mimic environment (climatic chamber). The surface temperature of the scale containers was determined to an uncertainty of ± 6.5 °C ($k = 2.1$) for the container walls, ± 3.5 °C ($k = 2.1$) for the container lid and ± 1.5 °C ($k = 2.1$) for the container vents.

The results show that thermal imaging can successfully identify scale boxes that have an elevated surface temperature under a range of environmental conditions. The use of low uncertainty emissivity and low uncertainty measurement data ensured the emissivity correction was effective at reducing the differential between the thermal imager measured temperature and the surface temperature measured by a reference thermocouple in a simulated environment. This capability will aid in 3 m^3 box intermediate level waste container monitoring, which can only be achieved by correcting for changing conditions.

Looking ahead to further development and the deployment of thermal imagers to monitor the 3 m^3 store no technical roadblocks were identified.

References

[1] M. Finnerty, "Fuel Cycle Operating Experience from a Global Perspective", Office for Nuclear Regulation presented at NRC Regulatory Information Conference, 13 March 2019

[2] C. Monte *et al*, "Radiation Thermometry and Emissivity Measurements Under Vacuum at the PTB", International Journal of Thermophysics, no. 30, pp. 203–219, June. 2008, DOI 10.1007/s10765-008-0442-9

Multi-source contamination mapping on the ground: a novel approach

Dor Raz[1,*], Ohad Fitoussi[2], Nadav Ben-David[1], Eran Vax[2], Alon Osovizky[3], Guy Zaidner[2]
[1]Israel Atomic Energy Commission, Israel
[2]Negev Nuclear Research Center , Israel
[3]Rotem Industries Ltd, Israel
(*) dra@post.bgu.ac.il

***Abstract*—Multi-source contamination mapping is a critical aspect of radiation detection and environmental monitoring. This paper introduces an innovative algorithm for accurate and efficient multi-source contamination mapping. The algorithm comprises two main components: (1) search area identification and (2) source localization using a particle filter. The search area identification involves extrapolating directional measurements to create search areas, employing clustering algorithms to refine these areas, and generating final search areas based on weighted factors. The incorporation of directional measurements significantly enhances performance by reducing parameter search space of the particle filter and thus improving practical application potential. The particle filter-based source localization employs a four-stage process, involving generation of initial guesses, comparison of expected and acquired measurements, employing genetic algorithms for particle selection, and a convergence criterion. The algorithm's efficacy is demonstrated through four simulated experimental scenarios, each with varying numbers and densities of point radioactive sources. Results illustrate successful search area identification and source localization with an average error of approximately 1 meter and source activity estimation of within 25% of the ground-truth values. Further improvements to the area clustering algorithm, improved mission planning for online applications, and extensive testing of the algorithm are planned in the future. With these improvements, the algorithm presented here could enhance the ability to respond to radioactive contamination incidents swiftly and effectively, ensure timely mapping of contaminated areas, and maintain human safety in the event of an incident.**

***Keywords*—Particle filter, clustering, mapping, contamination**

I. Introduction

Radiometric surveys are required for nuclear sites, whether they are active, decommissioned, or scheduled for decommissioning. Inspecting a nuclear site ensures personnel safety, facilitates systematizing the inspection process, allows repeatable monitoring missions, enables access to highly radioactive environments, and makes rapid and comprehensive searches at a low cost possible [1]. The current approach to contamination mapping and decommissioning relies on human operators, which is inefficient, dangerous, and time-consuming [2, 3]. In order to reduce the exposure of workers to high radiation doses during decommissioning, it is necessary to calculate where radioactive contamination may occur. As a result, it is imperative that technologies be developed and implemented that can inspect sites and detect radiation hot spots [4]. Site inspection generally involves the mapping of radiation and the mapping of space. As a result, a number of approaches to nuclear site inspection have been described in the literature, largely based on ground robots, unmanned aerial vehicles, and underwater robots [1] which are a natural fit for the task.

Decommissioning and post-operational clean-up activities may benefit greatly from high-resolution radiation mapping [5]. Several attempts have been made recently to develop algorithms and platforms for the mapping of environmental radiation, usually by creating a 3D model of an environment and overlaying a radiation field map on top. [3, 4, 6, 7, 8, 9, 10]. However, unlike mapping radiation fields, mapping contamination involves the precise estimation of each hotspot's location and activity, which in turn affects the estimation of all other hotspots. The ability to precisely map contamination is necessary for drawing a detailed radiation distribution map [9] as well as for efficient and successful decommissioning [11].

A number of contributions have been made to the topic of contamination mapping, including those in [12-17]. The methods used in these studies are maximum likelihood estimation in [14, 15, 17], and numerical adjoints used in a Bayesian formulation in [13]. In the cases described above, either the methods are capable of locating only a single source in the environment (utilizing the inverse square law) or have a limited degree of accuracy in estimating the location of sources, especially as the number of sources and the dimension of the search space increases.

One attempt has been made to map contamination using a Particle Filter (PF) [18]. Using sparse measurements from a Geiger-Müller detector, Gao et al. localize two radiation sources in a mixed multi-modal radiation field. Although the authors note this limitation of the computationally intensive PF to operate in areas with high complexity in the paper, their limited field experiment does not demonstrate the ability to map such areas. In order to successfully apply a PF to a high-dimensional system, additional information is necessary for limiting the size of the search space. To fill this gap in the literature, our goal in the present study was to develop and test a novel algorithm that generates a contamination map of multiple point sources located on the ground. To test this algorithm, a simulator of a ground mobile robot moving in 2D space equipped with a directional detector implemented and experimental scenarios are simulated. Our aim is to show that our algorithm is able to localize multiple sources in space quickly and accurately.

II. Materials and methods

This section describes the software modules of the system, which include data filtering, identification of search areas, and localization of sources using a particle filter. A search area is generated based on the filtered data, and the particle filter algorithm is then used to generate a set of particles for which there is a single source of radiation in each suspected area as well as a background source that estimates the background during the experiment.

A. Search area identification

We developed an algorithm to identify suspected contaminated areas that may contain radioactive sources in order to constrain the search space. As the mapping becomes more complex in terms of the number of sources, the size of the area and the number of visited locations, decreasing the search space greatly reduces the run time of the particle filter. There are four steps involved in the identification of an area.

Initially, the direction of the highest activity source is extrapolated up to a preset distance for each location visited by the robot. The cutoff threshold distance ensures that we do not identify false hotspots (false positives), as well as that we do not miss any sources (false negatives). Once the intersection points between the measurement lines have been calculated, they are given a weight according to the total count measurement at each location and the distance between it and the intersection point. The weight is calculated using formula (1).

$$\mathrm{wt} = \sigma_0 \frac{d_1 d_2}{d_1\sqrt{C_2}+d_2\sqrt{C_1}} \quad (1)$$

Where σ_0 is the statistical sensor noise constant independent of the measurement which does not affect the *relative* quality of the intersections, d_1 and d_2 are the distances between the intersection point and the two origins of the directional measurement, and C_1 and C_2 are the total counts measured at each origin. This calculation of the weight ensures the prioritization of measurements near high activity radioactive sources, which are highly reliable. In addition, far away readings or readings pointing to low-activity sources are filtered, so this parameter should be adjusted to suit the specific environment. The intersection points are filtered so that only those with high weights remain.

Thirdly, a search area is created around each cluster of the filtered intersection points. A KMeans clustering algorithm is used to locate clusters. This clustering depends on the accuracy of the detector direction and on other noises in the measurements, and works best when there are a large number of data points compared to the number of sources. A general rule of thumb is to use half the number of measurements as the number of clusters to look for in an environment where the number of sources is unknown.

In the final step, the multiple areas found for each cluster of intersections are combined into the final search areas that are input into the particle filter algorithm. First, area quality factors are applied to these areas, which are based on the weights of the intersections comprising them, as well as variance and distance between their points. These quality factors are then translated into changes in the size of the areas, with high quality areas being shrunk and low quality areas being enlarged. Afterwards, the filtered areas are grouped according to the overlap of the resized areas.

B. Source localization using a particle filter

The particle filter is a powerful tool for estimating the state of a system based on noisy measurements. It works by generating a set of particles, each representing a possible state of the system. These particles are then propagated through time using a model of the system dynamics and updated based on measurements taken at various times. The result is an estimate of the state of the system that is more accurate than what would be obtained by using only the measurements themselves. This approach is particularly useful in situations where direct measurement of the source is difficult, as is our case of mutual interference.

The scheme by which the particle filter operates can be seen in Fig. 1.

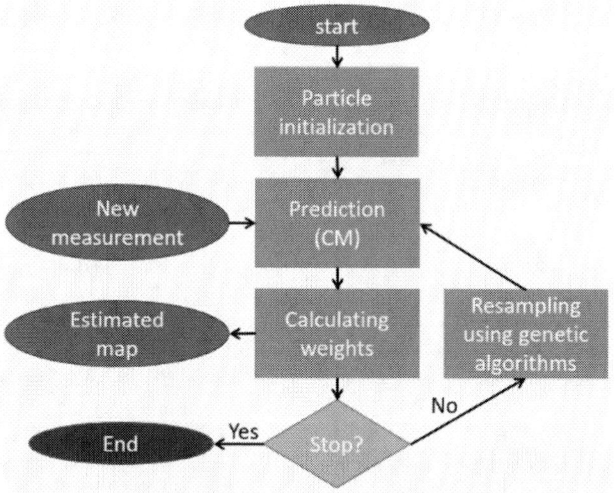

Fig. 1. Flow chart describing the particle filter algorithm

To locate the sources using the particle filter, a four-stage process is performed.

1) An initial set of guesses (particles) is generated. Each particle is composed of a set of sources (one for each search area), each of which has an activity level and a location on the map
2) For each particle, the expected readings in the locations visited by the robot are compared to the acquired readings. The expectations are calculated using the calibration matrix of the detector. The particle is given a weight based on the match to the true measurements.
3) Only the top 50% of particles move on to the next generation, using genetic algorithms.
4) A stop criterion is defined by the weight of the solutions or after a stagnation of 50 generations.

Depending on the selected percentage of particles that move onto the next generation, the algorithm will take longer converge and will be more dependent on the initial guess. Using this parameter, along with the probability and extent of mutations in the particle values, allows us to avoid local minima in exchange for run time, and to prioritize exploration over exploitation.

C. *Data acquisition*

For all simulations, a robot enters a suspected contaminated environment and begins collecting measurements in the area. The measurements include the total counts of the radiation field reading, and the direction of the radioactive sources in the area, weighted by their activities and distances to the detector. After saving the data of the location and the radiation field measurements, the robot moves to the next measurement point. The process is repeated until the area has been sufficiently explored.

D. *Experimental setup*

We simulated a total of four experimental scenarios. All locations are in meters, and the activity levels are in mCi units. Measurements had added noise levels emulating the Poisson nature of the measurement.
1) Experiment 1: two sources - In the first scenario, 16 robot measurements are simulated for an environment with two radioactive sources located 8 meters apart.
2) Experiment 2: two dense sources - In the second scenario, 16 robot measurements are simulated for an environment with two radioactive sources located 4 meters apart
3) Experiment 3: two extremely dense sources - In the third scenario, 16 robot measurements are simulated for an environment with two radioactive sources located 2 meters apart.
4) Experiment 4: three sources - In the fourth scenario, 40 robot measurements are simulated for an environment with three radioactive sources located 8 meters apart.

III. RESULTS

The purpose of this section is to describe and visualize the four experimental scenarios, together with the calculated source locations and activities as well as the ground-truth sources and activities. All simulations used 100 particles per generation, a stagnation end condition of 50 iterations without progress, and a measurement time of 10 seconds per location. The supplementary materials provide detailed information regarding the other parameters used.

In the experiment visualization plots, the purple triangles triangle indicates the direction in which the robot is oriented. At each visited location, the numbers represent the total count measurement received by the detectors. The arrows at each location indicate the direction of the weighted radiation field readings, and the weights are calculated based on the source activities and distances from the measurement site. The color of these arrows indicates the strength of the radiation field measurement in relation to the other measurements in the experiment, ranging from the lowest (white/light yellow) to the highest (black/dark red). Red stars indicate ground truth sources, and green plusses indicate sources localized by the algorithm.

Experiment 1: two sources

The visualization of Experiment 1 can be seen in Fig. 2. Although certain directional measurements in this experiment indicated uninformative orientations, the search areas were successfully identified. Both radioactive sources were located at a mean distance of 0.7 meters from the ground truth source in 61 iterations. Source activities were calculated to be 20% below ground truth activities. An explanation for this can be found in the estimated background radiation levels of 1.7 $\frac{counts}{s}$ (cps) in a simulation without background radiation.

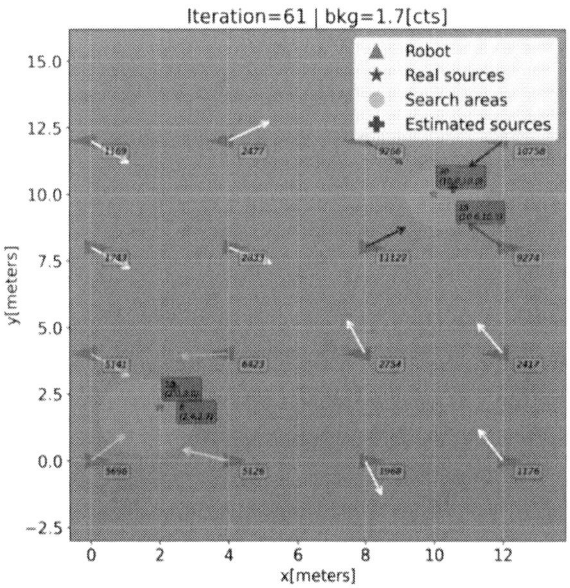

Fig. 2. Visualization of the Experiment 1 results. The ground-truth sources are $x_{1,GT}, y_{1,GT}, A_{1,GT} = (2,\ 2, 10)$ and $x_{2,GT}, y_{2,GT}, A_{2,GT} = (10, 10, 20)$. After reaching convergence at 61 iterations, the estimated sources are $x_{1,est}, y_{1,est}, A_{1,est} = (2.4, 2.7, 8)$ and $x_{2,est}, y_{2,est}, A_{2,est} = (10.6, 10.3, 16)$, with a calculated background radiation of 1.7 cps.

A. *Experiment 2: two dense sources*

The visualization of Experiment 2 can be seen in Fig. 3. Search areas are calculated correctly even when sources are four meters apart. The source with the higher activity was localized precisely, while the source with the lower activity was localized with an error of 1.6 meters. The higher level of symmetry in the data may have contributed to the longer convergence time of 103 iterations relative to Experiment 1, since the symmetry allows for a larger number of solutions to the problem and hence increases the search space. The source activities were calculated to be 10-20% lower than the ground truth activities.

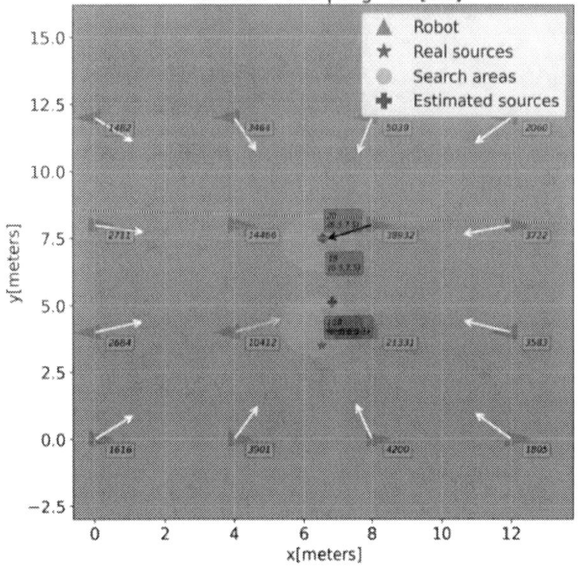

Fig. 3. Visualization of the Experiment 2 results. The ground-truth sources are $x_{1,GT}, y_{1,GT}, A_{1,GT} = (6.5, 3.5, 10)$ and $x_{2,GT}, y_{2,GT}, A_{2,GT} = (6.5, 7.5, 20)$. After reaching convergence at 103 iterations, the estimated sources are $x_{1,est}, y_{1,est}, A_{1,est} = (6.8, 5.1, 8)$ and $x_{2,est}, y_{2,est}, A_{2,est} = (6.5, 7.5, 18)$, with a calculated background radiation of 1.2 cps.

B. *Experiment 3: two extremely dense sources*

The visualization of Experiment 3 can be seen in Fig. 4. At a distance of 2 meters between the sources, the algorithm finds only one search area containing a single source, and the average error in localization is 1 meters. The source's equivalent activity level was calculated to be less than 20% below the sum of the ground truth activities in only 41 iterations, which can be attributed to the low complexity of the search space.

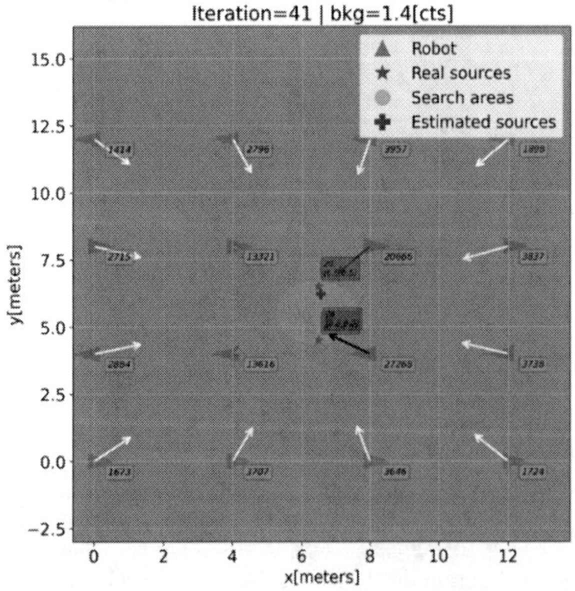

are $x_{1,GT}, y_{1,GT}, A_{1,GT} = (6.5, 4.5, 10)$ and $x_{2,GT}, y_{2,GT}, A_{2,GT} = (6.5, 6.5, 20)$. After reaching convergence at 41 iterations, the estimated source is $x_{1,est}, y_{1,est}, A_{1,est} = (6.6, 6.2, 26)$ which appears to be an average of the two ground truth sources, with a calculated background radiation of 1.4 cps.

C. *Experiment 4: three sources*

The visualization of Experiment 4 can be seen in Fig. 5. The algorithm detected all three radioactive sources and located them near their respective ground truth sources, with an average localization error of 1 meter. The source activities were calculated to be 20-30% lower than the ground truth activities, since the calculated background level was high.

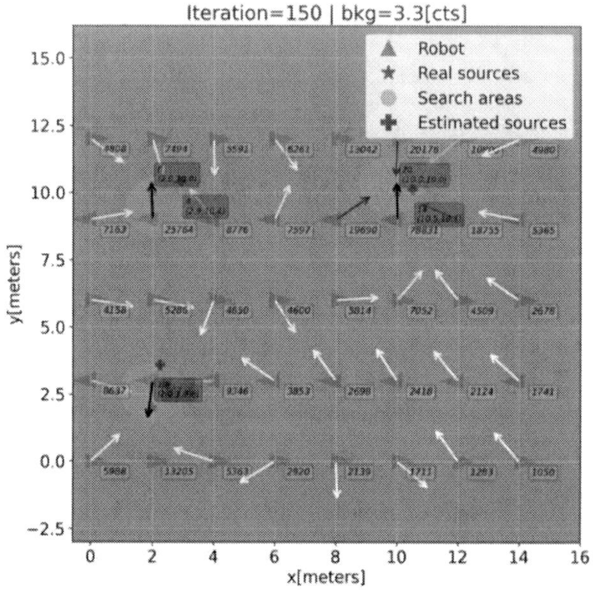

Fig. 5. Visualization of the Experiment 4 results. The ground-truth sources are $x_{1,GT}, y_{1,GT}, A_{1,GT} = (2, 2, 10)$, $x_{2,GT}, y_{2,GT}, A_{2,GT} = (10, 10, 20)$ and $x_{3,GT}, y_{3,GT}, A_{3,GT} = (2, 10, 6)$. After reaching convergence at 150 iterations, the estimated sources are $x_{1,est}, y_{1,est}, A_{1,est} = (2.3, 3.6, 6)$, $x_{2,est}, y_{2,est}, A_{2,est} = (10.5, 10.1, 14)$ and $x_{3,est}, y_{3,est}, A_{3,est} = (2.9, 10.4, 4)$, with a calculated background radiation of 3.3 cps.

The simulation results for all four scenarios are presented in Table I.

TABLE I
TRENDS IN LOCALIZATION AND SOURCE ACTIVITIES ERRORS.

Experiment name	Average Localization error (meters)	Average activity value error (% of ground truth value)	Estimated background radiation (cps)
Two sources	0.73±0.01	20±0.1	1.7±0.1
Two dense sources	0.81 ± 0.01	15±0.1	1.2±0.1
Two extremely dense sources	1.00 ± 0.01	32.5±0.1	1.4±0.1
Three sources	1.04 ± 0.01	34.4±0.1	3.3±0.1

As the experiments became more complex, the localization and activity level errors increased. We also notice a correlation between the estimated background and the error in the activity values for experiments 1, 2 and 4.

IV. Discussion

In this study, we present a novel algorithm for multi-source contamination mapping using a simulation of a mobile robot equipped with a directional detector. The research aimed to improve contamination mapping and decommissioning processes by accurately locating and estimating the activities of multiple point radioactive sources on the ground.

Simulated experiments demonstrated the effectiveness of the proposed algorithm. In Experiments 1, 2 and 4, the algorithm successfully detected the number of contamination sources on the map, demonstrating its effectiveness of the search area identification. In Experiment 3, a distance of 2 meters proved too little for the algorithm to differentiate between the sources. A possible solution for this is to improve the area clustering algorithm to allow for smarter clustering and noticing subtle subgroups of clusters.

The algorithm was successful in localizing the sources within 0.9 meters of the ground truth locations on average. Additionally, the algorithm achieved an accuracy of 25.5% for activity level, which can be attributed to the estimated background. Because there was no simulated background radiation, any calculated background level will decrease the levels of source activity to create a good fit to the data. This tradeoff is evident in the correlation between the level of background radiation and the percentage of activity level inaccuracy for experiments 1, 2 and 4. The average runtime of each iteration was 4 seconds, resulting in a mean runtime of 6 minutes for each simulation. These findings indicate the algorithm's potential for real-world applications, in which these values could be used to plan a mission for collecting data near the sources, thus improving localization and facilitating further steps.

Small improvements in localization are likely to lead to greater improvements since an error in localization for one source is often compensated by an increased error in another source's location if the resulting source set creates a radiation field which better fits the data.

The particle filter was a central component of the algorithm, enabling the estimation of hotspot locations and activities based on noisy measurements. The use of the directional readings provided crucial information about the direction of the sources, which enabled the calculation of search areas and reduced the parameter search space of the particle filter. By incorporating directional information into the algorithm, it greatly improved its performance and contributed to its practical application in real-world settings.

Despite the success of the proposed algorithm, there are some limitations and challenges that need to be addressed. A number of improvements are planned for the area clustering algorithm in the future, as well as extensive testing of the algorithm. Secondly, source differencing techniques can increase the accuracy of the resulting map. Thirdly, it is important to note that the performance of the algorithm is influenced by parameter choices, such as the threshold for intersections and the method for grouping the search areas. Finding the optimal parameter values for specific scenarios requires careful consideration.

We expect that an updated version of the directional measurements which points to the strongest point source in implementation of these suggestions, the algorithm presented in this paper, could provide a robust response to radioactive contamination incidents on-site, where rapid area contamination mapping is critical and human safety is of paramount importance.

V. Conclusion

In this study, we developed an algorithm for efficient and accurate multi-source contamination mapping using a mobile robot equipped with a directional detector. The combination of directional measurements and particle filter-based source localization allowed for successful mapping of multiple point sources located on the ground. Provided sufficient testing and the development of a physical setup, this system could be quickly deployed in response to radioactive contamination incidents to ensure quick area contamination mapping while prioritizing human safety. Future research should focus testing more scenarios with various densities and activity levels, improving the area clustering algorithm and improving the mission planning for online applications in order to provide a more accurate map and greater applicability in real-world scenarios.

References

[1] F. Mascarich, T. Wilson, C. Papachristos, and K. Alexis, "Radiation Source Localization in GPS-Denied Environments Using Aerial Robots," in 2018 IEEE International Conference on Robotics and Automation (ICRA), Brisbane, QLD: IEEE, May 2018, pp. 6537–6544. doi: 10.1109/ICRA.2018.8460760.

[2] I. Tsitsimpelis, C. J. Taylor, B. Lennox, and M. J. Joyce, "A review of ground-based robotic systems for the characterization of nuclear environments," Progress in Nuclear Energy, vol. 111, pp. 109–124, Mar. 2019, doi: 10.1016/j.pnucene.2018.10.023.

[3] B. Bird et al., "A Robot to Monitor Nuclear Facilities: Using Autonomous Radiation-Monitoring Assistance to Reduce Risk and Cost," IEEE Robot. Automat. Mag., vol. 26, no. 1, pp. 35–43, Mar. 2019, doi: 10.1109/MRA.2018.2879755.

[4] S. R. White et al., "Radiation Mapping and Laser Profiling Using a Robotic Manipulator," Front. Robot. AI, vol. 7, p. 499056, Nov. 2020, doi: 10.3389/frobt.2020.499056.

[5] P. G. Martin et al., "3D unmanned aerial vehicle radiation mapping for assessing contaminant distribution and mobility," International Journal of Applied Earth Observation and Geoinformation, vol. 52, pp. 12–19, Oct. 2016, doi: 10.1016/j.jag.2016.05.007.

[6] Martin, Peter G. & Scott, Thomas & Payton, Oliver & Fardoulis, John. (2017). High-Resolution Aerial Radiation Mapping for Nuclear Decontamination and Decommissioning.

[7] P. Martin, J. Moore, J. Fardoulis, O. Payton, and T. Scott, "Radiological Assessment on Interest Areas on the Sellafield Nuclear Site via Unmanned Aerial Vehicle," Remote Sensing, vol. 8, no. 11, p. 913, Nov. 2016, doi: 10.3390/rs8110913.

[8] Abd. H. Zakaria, Y. M. Mustafah, J. Abdullah, N. Khair, and T. Abdullah, "Development of Autonomous Radiation Mapping Robot," Procedia Computer Science, vol. 105, pp. 81–86, 2017, doi: 10.1016/j.procs.2017.01.203.

[9] Y. Sato et al., "Radiation imaging using a compact Compton camera inside the Fukushima Daiichi Nuclear Power Station building," Journal of Nuclear Science and Technology, vol. 55, no. 9, pp. 965–970, Sep. 2018, doi: 10.1080/00223131.2018.1473171.

[10] K. Vetter, R. Barnowksi, A. Haefner, T. H. Y. Joshi, R. Pavlovsky, and B. J. Quiter, "Gamma-Ray imaging for nuclear security and safety: Towards 3-D gamma-ray vision," Nuclear Instruments and Methods in Physics Research Section A: Accelerators, Spectrometers, Detectors and Associated Equipment, vol. 878, pp. 159–168, Jan. 2018, doi: 10.1016/j.nima.2017.08.040.

[11] Y. Sato, Y. Terasaka, S. Ozawa, Y. Tanifuji, and T. Torii, "A 3D radiation image display on a simple virtual reality system created using a game development platform," J. Inst., vol. 13, no. 08, pp. T08011–T08011, Aug. 2018, doi: 10.1088/1748-0221/13/08/T08011.

[12] T. Baca et al., "Gamma Radiation Source Localization for Micro Aerial Vehicles with a Miniature Single-Detector Compton Event

Systems (ICUAS), Athens, Greece: IEEE, Jun. 2021, pp. 338–346. doi: 10.1109/ICUAS51884.2021.9476766.
[13] K. D. Jarman, E. A. Miller, R. S. Wittman, and C. J. Gesh, "Bayesian Radiation Source Localization," Nuclear Technology, vol. 175, no. 1, pp. 326–334, Jul. 2011, doi: 10.13182/NT10-72.
[14] H. Wan, T. Zhang, and Y. Zhu, "Detection and localization of hidden radioactive sources with spatial statistical method," Ann Oper Res, vol. 192, no. 1, pp. 87–104, Jan. 2012, doi: 10.1007/s10479-010-0805-z.
[15] H. E. Baidoo-Williams, "Maximum Likelihood Localization of Radiation Sources with unknown Source Intensity." arXiv, Oct. 10, 2016. Accessed: Aug. 02, 2023. [Online]. Available: http://arxiv.org/abs/1608.00427
[16] H. E. Baidoo-Williams, R. Mudumbai, E. Bai, and S. Dasgupta, "Some theoretical limits on nuclear source localization and tracking," in 2015 Information Theory and Applications Workshop (ITA), San Diego, CA, USA: IEEE, Feb. 2015, pp. 270–274. doi: 10.1109/ITA.2015.7309000.
[17] G. Cordone et al., "Improved multi-resolution method for MLE-based localization of radiation sources," in 2017 20th International Conference on Information Fusion (Fusion), Xi'an, China: IEEE, Jul. 2017, pp. 1–8. doi: 10.23919/ICIF.2017.8009626.
[18] W. Gao, W. Wang, H. Zhu, G. Huang, D. Wu, and Z. Du, "Robust Radiation Sources Localization Based on the Peak Suppressed Particle Filter for Mixed Multi-Modal Environments," Sensors, vol. 18, no. 11, p. 3784, Nov. 2018, doi: 10.3390/s18113784.

Nanoreinforced concrete composites for possible nuclear waste confinement - a preliminary study

Maria Rita Mancini[1], Daniele Mirabile Gattia[1], Fabio Girardi[2], Donatella Ferri[2], Giuseppe A. Marzo[2], Domenica Masci[2 *]
[1]SSPT-PROMAS-MATPRO, ENEA, Italy
[2]FSN-FISS-CRGR, ENEA, Italy
(*) domenica.masci@enea.it

Abstract— **Cementitious materials are the most abundant building materials even nowadays. Similarly, in the nuclear industry, concrete is the most used material for the confinement of radioactive waste, it has favorable properties, both chemical and physical, making it a desirable matrix for the encapsulation of radioactive and toxic wastes. Cemented wastes are preferred for the storage of a large varieties of fission products, such as low and intermediate level radioactive waste e.g., β—emitter ^{90}Sr. Anyway, because of the intrinsic fragility of ordinary Portland Cement (OPC) a lot of efforts have been spent to enhance his mechanical resistance.**

Graphene Oxide (GO) represent an opportunity to overcome this problem. In fact, there is much evidence that GO can influence the toughness and the strength of the cement composites by regulating the morphology of the cement hydration products.

^{90}Sr, the main strontium isotope in radioactive waste, is a by-product of the fission of uranium and plutonium in nuclear reactors. In this study we investigate the effects of strontium on mechanical properties of cement (OPC) and GO-reinforced cement. Different samples have been prepared with fixed concentration of GO. It has been observed that small quantities of GO, significantly [1] increases the compression strength of cement composites. The addition of Sr (65.0, 67.6 and 70.2 ppm) resulted in an increase of the compressive strength of 6.78%, 3.02% and 0.70% for those samples respectively, in comparison to OPC.

Compression strength in samples of OPC with GO remained quite constant even in presence of Sr.

Keywords: **Radioactive waste, Graphene Oxide, Compressive strength, Strontium.**

I. Introduction

DUE to its favourable chemical and physical properties, cement is extensively utilized in the nuclear industry for confining low and intermediate-level radioactive waste, such as those containing β-emitter ^{90}Sr. However, the traditional Ordinary Portland Cement (OPC) possesses inherently brittle characteristics, and numerous efforts have been made to enhance its mechanical strength. In recent decades, fibre-reinforced and nanoreinforced concretes have been developed. However, significant progress has not been achieved in addressing the brittle performance of this material. Graphene Oxide (GO) provides an opportunity to overcome this issue by influencing the toughness and strength of cement composites through the regulation of cement hydration products morphology. This study investigates the combined effects of GO and stable strontium, the latter being used as a simulant for ^{90}Sr (a radioactive isotope), on the mechanical properties of OPC. It is noteworthy that ^{90}Sr is a by-product of uranium and plutonium fission in nuclear reactors and is considered a chemical analogue for ^{226}Ra in performance assessment studies for cements in radioactive waste repositories [2-3]. The cement specimens were characterized using various techniques, including mechanical tests, thermogravimetric analysis (TGA) and scanning electron microscopy (SEM). This study provides new insights into the use of concrete in radioactive waste storage by investigating the influence of GO and stable strontium on the mechanical properties of OPC. The synthesis of GO has been approached with the goal of developing and optimizing the synthesis process, considering economic and environmental sustainability by improving the efficiency of the different process stages.

II. Materials and Methods

The synthesis of GO has been addressed using the Hummers method [4] as a reference base, as well as modified versions [5–13], with variations to enhance the overall efficiency.

Modified Hummers method starts from a commercial graphite powder upon which is performed a chemical oxidation using sulfuric acid (H_2SO_4), potassium permanganate ($KMnO_4$) and hydrogen peroxide (H_2O_2).

The GO, once produced, was mixed to the cement in solution form, obtained dispersing a fixed quantity in distilled water and exposing the solution in an ultrasonic bath at the frequency of 37 kHz with a power of 100 W for 30 minutes.

Increasing concentrations of strontium (67.6, 70.2 and 71,6 ppm) were added to the GO solution and then mixed with cement. The concentrations used were chosen in a range close to the maximum solubility limit of the strontium in cement [14]. To realize 1 cm^3 cubic specimens was used OPC CEMII/BLL 32.5R. The water cement (w/c) ratio have been of 0.5. The GO was added to the cement in a fixed quantity equal to 0.03% w/v. In Fig. 1 is shown the flow sheet of the cubic specimen preparation.

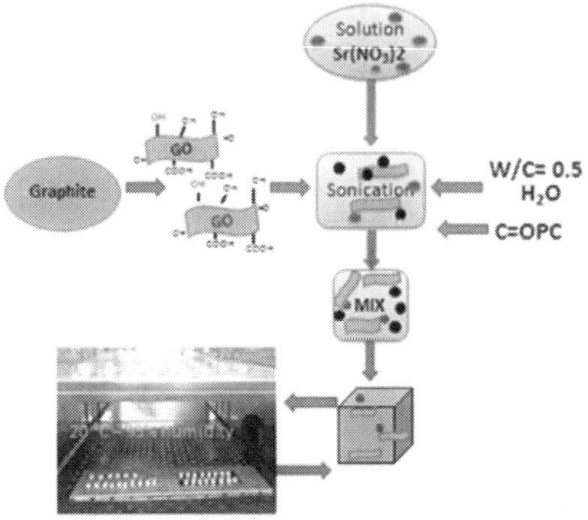

Fig. 1. Flow sheet of the cubic specimen preparation with and without GO and/or stable strontium.

After the preparation, the samples were cured in an environmental testing chamber at 20° C and 95% of relative humidity following a standard 28-day procedure Fig. 2 [1].

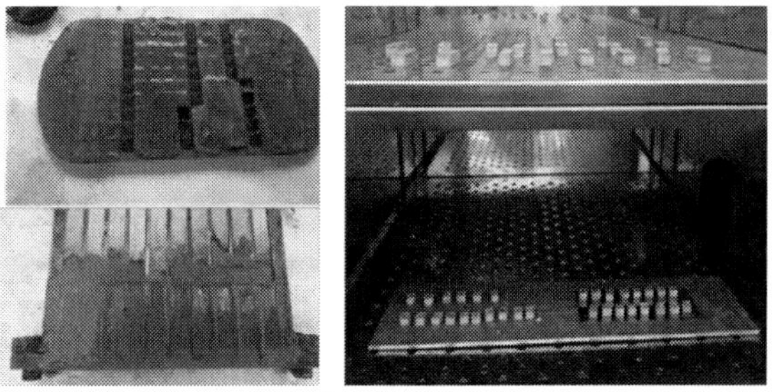

Fig. 2. Sample preparation and curing environmental test chamber.

The compressive strength test was performed using a uniaxial press (Fig. 3). Initially, several tests were conducted to optimize the compression speed based on the geometry of the prepared samples. The chosen speed was 5 mm/min. Fig. 3 shows the setup of the Tecnotest® equipment used for the compression tests, while Fig. 4 shows a sample between the plates of the press.

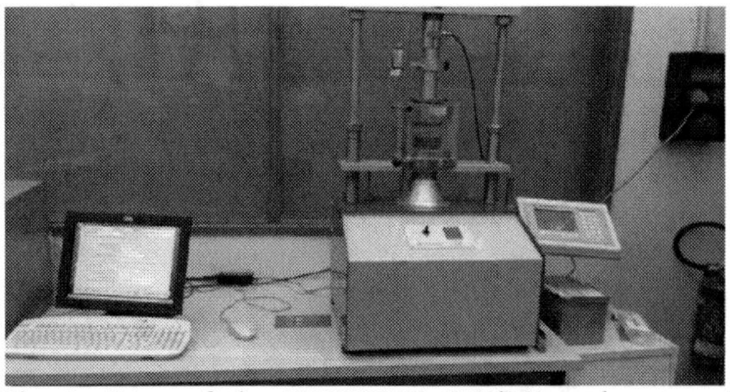

Fig. 3. Set-up of the TECNOTEST T051/B65 equipment used for compression tests.

To investigate the effects of GO and Sr on cement hydration products have been performed the thermogravimetric analysis (TGA). Thermal analyses were run on a Simultaneous TG/DTA NETZSCH STA Jupiter. at 10 °C/min from 25 °C to 1200 °C in 100 cm^3/min of air flow, in order to study the effect of Sr and 0.03% GO addition on the hydration characteristics of concrete matrix.

Fig. 4. Specimen before the compression test.

The microscopic analysis of the samples, following the mechanical tests, was carried out using a Zeiss MA 15 scanning electron microscope (SEM).

III. Results and Discussion

A. Scanning electron microscopy

The microstructure of concrete with the Sr and GO addition has a significant effect on mechanical characteristics. The SEM analysis was used to identify the relation between microstructure and the improvement of the compressive strength. The SEM image of representative sample I (OPC + 0.03% GO + 71.5 ppm Sr) in Fig. 5 reveals the hydration products. The analysis confirmed the intimate connection of the GO with the cement hydration products (Fig. 5-6).

The GO's layers are positioned as bridge to the cement hydration products endorse the hypothesis the role on the structural effects of the cement-GO composite.

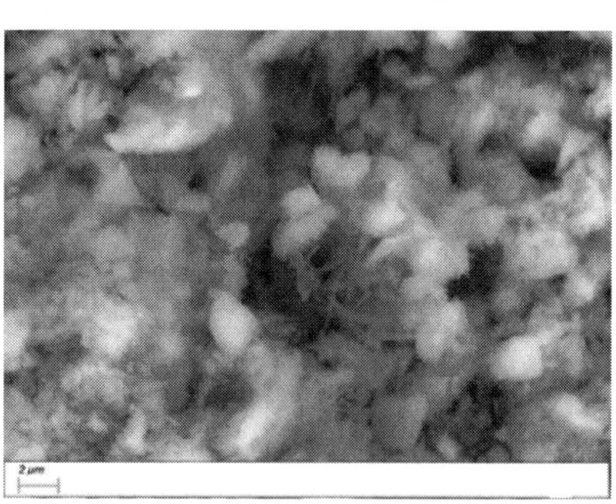

Fig. 5. SEM image of sample (I) prepared modifying OPC with 0.03% of GO and 71.6 ppm of Sr (___ 2 μm).

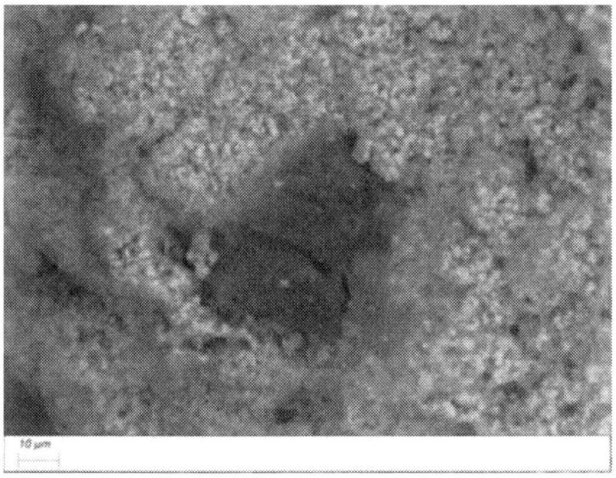

Fig. 6. Sem image of sample (F) prepared modifying OPC with 0.03% of GO (▁▁ 10 μm).

B. Thermogravimetric analysis

The thermogravimetric and derivate thermogravimetric curves of concrete (OPC) and OPC added with Sr and Sr + GO are shown in Fig. 7 and Fig. 8 respectively.

At 120 °C, a weight loss of approx. 5wt% are attributed to water desorption. Subsequent mass losses between 120-650 °C are attributed to water from the decomposition of the C-S-H (450-550 °C) and dihydroxylation of the $Ca(OH)_2$. Finally, between 650-900 °C the mass loss is attributed to the decarbonation of carbonate.

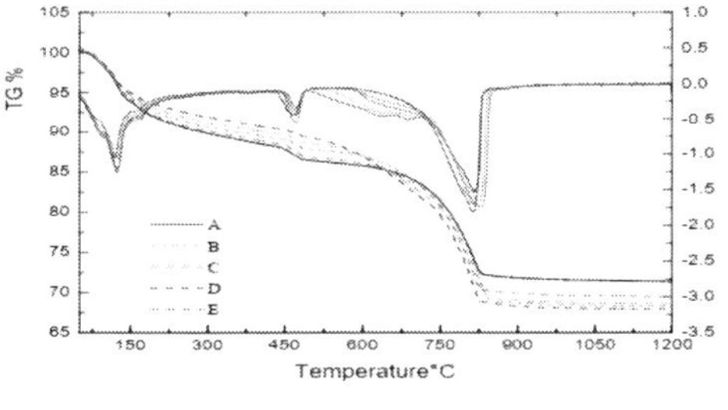

Fig. 7. TG/DTG curves of samples hydrated for 28 days at 25°C: A (OPC), B, C, D, E (OPC + %Sr) concrete specimens.

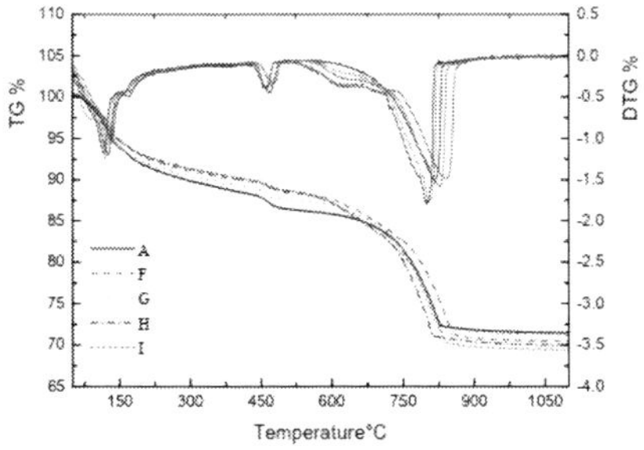

Fig.8. TG/DTG curves of samples hydrated for 28 days at 25 °C: A (OPC), F (OPC + 0.03% GO), G, H, I (OPC + 0.03% GO + %Sr) concrete specimens.

The effect of the Sr at different concentration and of the 0.03 w% GO content on the degree of hydration of the samples (α) was determined using Eq. (1) α(%) = WB/0.24; where WB is non evaporable water content of the samples [15].

The values of hydration degree of the different samples are shown in Fig. 9.

The data demonstrate that the addition of Sr (65.0, 67.6, 70.2 and 71.6 ppm) improved the hydration degree of 24.71%, 23.07%, 20.58% and 18.13% in comparison to the concrete OPC. The degree of hydration for OPC is 35.41%. While hydration degree in samples of OPC with 0.03% of GO is 14.5% and decreases anomalously at 4.5% for sample G (GO + 67,6 ppm Sr) and remained quite constant at 23% for sample H and I with GO + 70.2 ppm Sr and GO + 71.6 ppm Sr respectively. The findings demonstrate that the 0.03% of GO improved by 29% the degree of hydration of the sample E with the highest concentration of Sr 71.6 ppm.

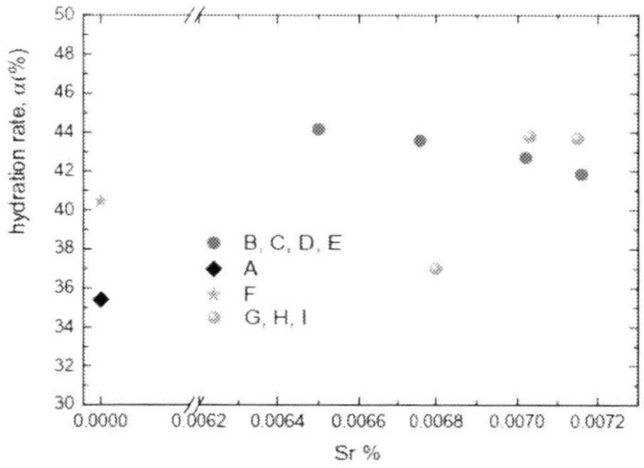

Fig. 9. Hydration rate versus %Sr of: A (OPC), F (OPC + 0.03% GO), B, C, D, E) (OPC + %Sr) and G, H, I (OPC + 0.03% GO + %Sr) concrete specimens.

C. Compressive strength test

The preliminary results of the compressive strength test showed performed by composite GO specimens showed only a partial enhancement of the mechanical properties capable to hinder the decline of mechanical performances induced by the increasing amounts of strontium. In Table I are displayed the results of compressive strength test of the specimen and the corresponding density.

TABLE I
COMPRESSIVE TEST STRENGTH AND DENSITY OF SPECIMENS

Sample symbol	Description	Compressive strength test (MPa) (μ±SD)	Density (g/cm³)
A	OPC	69.3 ± 5	1.68
B	OPC + 65.0 ppm Sr	74.1 ± 6	1.70
C	OPC + 67,6 ppm Sr	71.5 ± 4	1.66
D	OPC + 70,2 ppm Sr	69.9 ± 6	1.63
E	OPC + 71,6 ppm Sr	67.8 ± 6	1.61
F	OPC + 0.03% GO	70.5 ± 4	1.68
G	OPC + 0.03% GO + 67.6 ppm Sr	69.0 ± 3	1.68
H	OPC + 0.03% GO + 70.2 ppm Sr	70.4 ± 6	1.67
I	OPC + 0.03% GO + 71.6 ppm Sr	70.4 ± 3	1.66

OPC = Ordinary Portland Cement CEMII/B-LL32.5R Sr = $Sr(N0_3)_2$ GO = Graphene Oxide.

The specimens, once compressed as illustrated in Fig. 10,

take on a distinctive hourglass shape upon fracture.

Fig. 11 presents the graphic depicting the relationship between the applied compression on the specimen (N/mm^2) and the elapsed time (s) with specific emphasis on the breaking point.

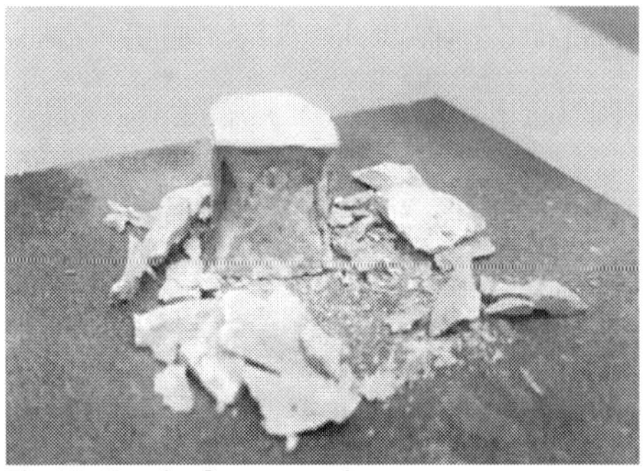

Fig. 10. Sample after compression

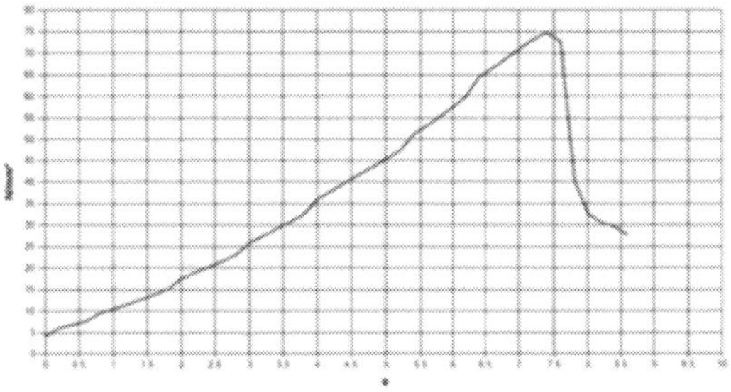

Fig. 11. Compression diagram

The bulk density decreased with the increasing amount of Sr in the mixture. for Sr = 71.6 ppm the decrease of density is by about to 7% to OPC.

The addition of 0.03% of GO decreases the density by 3% compared to OPC, while adding the same amount of GO to OPC + Sr samples induces an increase the density of about 2% (Fig. 12).

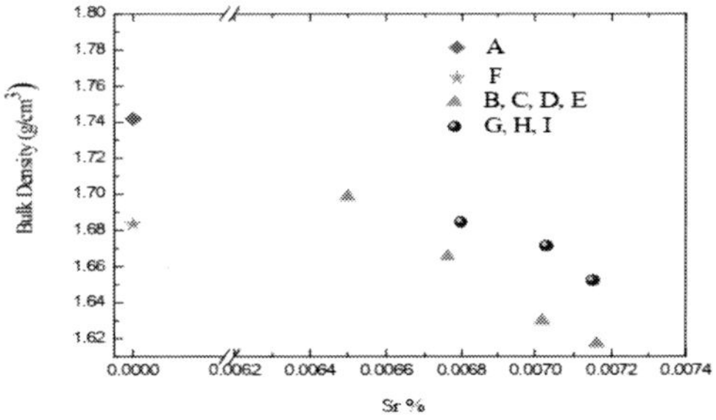

Fig. 12. Bulk density versus %Sr of: A (OPC), F (OPC + 0.03% GO), B, C, D, E (OPC + %Sr) and G, H, I (OPC + 0.03% GO + %Sr) concrete specimens.

The addition of Sr (65.0, 67.6 and 70.2 ppm) resulted in an increase of the compressive strength of 6.78%, 3.02% and 0.70% for B, C, D samples respectively, in comparison to OPC. While at higher concentrations of Sr (71.6 ppm) a reduction of 2.24%, in sample E, has been observed. Compression strength in samples of OPC with GO remained quite constant even in presence of Sr (Fig. 13).

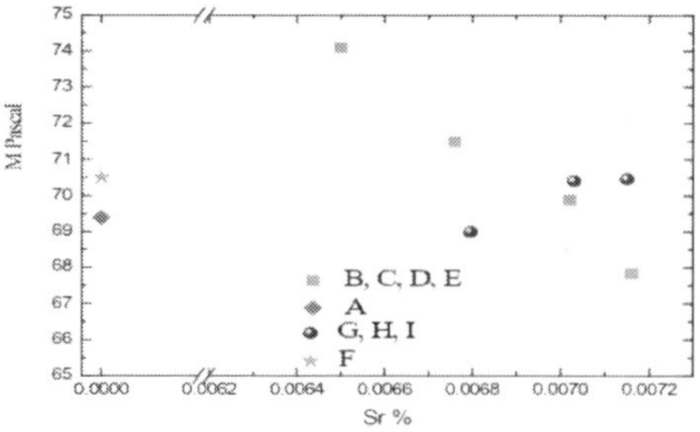

Fig. 13. Compressive strength versus %Sr of: A (OPC), F (OPC + 0.03% GO), B, C, D, E) (OPC + %Sr) and G, H, I (OPC + 0.03% GO + %Sr) concrete specimens.

IV. Conclusions

Confinement of nuclear wastes requires new research activities focused on stable and reinforced matrixes for allowing the entrapment of low and intermediate level radioactive isotopes. The addition of graphene oxide (GO) to cement CEMII/B-LL32.5R has been found to increase the compressive strength. The addition of increasing concentrations of strontium to cement CEMII/B-LL32.5R resulted in a decrease of the compressive strength. However, it appears that the addition of GO stabilizes compressive strength in cements with strontium, potentially mitigating the negative effect of strontium on such a parameter. These initial results are promising; however, further work is required to add to the obtained data and investigate the effect of different concentrations of Sr and GO on migration of Sr in cementitious materials.

References

[1] W. Yahui et al. "Effect of Graphene Oxide on Mechanical Properties of Cement Mortar and its Strengthening Mechanism," Materials, 12, 3753 (2019).

[2] J. Kittnerová, B. Drtinová, K. ˇStamberg, D. Vopàlka, N. Evans, G. Deissmann, S. Lange, "Comparative study of radium and strontium behaviour in contact with cementitious materials," Appl. Geochem., Vol. 122 no. 104713 (2020).

[3] U. Berner, "Radionuclide concentration limits in the near-field of a repository for spent fuel and vitrified high-level waste" TECHNICAL REPORT 02-10, Paul Scherrer Institut, Wettingen, Switzerland, PSI, ISSN 1015-2636.(2002).

[4] N.E. Sorokina, M.A. Khaskov, V.V. Avdeev., I.V. Nikolskaya, "Reaction of graphite with sulfuric acid in the presence of KMnO4", Russian J. of Gen. Chem. 75(2), 162 (2005).

[5] L.Y. Shahriary, A.A. Athawale, "Graphene Oxide Synthesized by using Modified Hummers Approach," Int. J. of Renewable Energy and Environmental Engineering Vol. 2 no. 1 issue 58 (2014).

[6] J. Song, X. Wang, C. Chang, "Preparation and Characterization of Graphene Oxide," J. of Nanomaterials vol. 1 (2014).

[7] T. Wu, J. Ting, "Preparation and characteristics of graphene oxide and its thin films," Surface & Coatings Technol. 231, 487 (2013).

[8] K. Krishnamoorthy, M. Veerapandian, K. Yun, S.J. Kim, "The chemical and structural analysis of graphene oxide with different degrees of oxidation," Carbon 53, 38 (2013).

[9] N.I. Zaabaa et al., "Synthesis of Graphene Oxide using Modified Hummers Method: Solvent Influence", Procedia Engineering 184, 469 (2017).

[10] B. Paulchamy, G. Arthi and B.D. Lignesh, "Simple Approach to Stepwise Synthesis of Graphene Oxide Nanomaterial", Nanomedicine & Nanotechnology 6(1), 253 (2015).

[11] F.J. Tolle, K. Gamp, R. Mulhaupt, "Scale-up and purification of graphite oxide as intermediate for functionalized graphene," Carbon 75, 432 (2014).
[12] Shao G., et al., "Graphene oxide: the mechanisms of oxidation and exfoliation", J. Mater. Sci. 47, 4400 (2012).
[13] J. Chen, B. Yao, C. Li, S. G., "An improved Hummers method for eco-friendly synthesis of graphene oxide" Carbon 64, 225 (2013).
[14] H. P. Susetyo, R. Widya, "Performance of Cement for Immobilizing Strontium Waste in Saline Environment," Materials Sci. and Appl., 4, 7-11 (2013).
[15] P. Mounanga, A. Khelidj et al, "Predicting $Ca(OH)_2$ content and chemical shrinkage of hydrating cement pastes using analytical approach" Cem. and Concr. Res. 34, 255-265 (2004).

Neutron imaging with Micromegas detectors with optical readout

A. Cools[1,*], S. Aune[1], F.M. Brunbauer[2], T. Benoit[1], A. Corsi[1], E. Ferrer-Ribas[1], F.J. Iguaz[3], C. Mazoin[4], E. Oliveri[2], T. Papaevangelou[1], E.C. Pollacco[1], M. Potignon[4], L. Ropelewski[2], A. Sari[5]

[1]DRT, IRFU, CEA, Université Paris-Saclay, F-91191 Gif-sur-Yvette, France
[2]European Organization for Nuclear Research (CERN), 1211 Geneva 23, Switzerland
[3]SOLEIL Synchrotron, L'Orme des Merisiers, Départementale 128, 91190 Saint-Aubin, France
[4]DG, SPRE, CEA, Université Paris-Saclay, F-91191 Gif-sur-Yvette, France
[5]DRT, LIST, CEA, Université Paris-Saclay, F-91191 Gif-sur-Yvette, France
(*) antoine.cools@cea.fr

Abstract. **Optical readout of Micromegas gaseous detectors has been achieved by implementing a Micromegas detector on a glass substrate with a glass anode and a CMOS camera. Efficient X-ray radiography has been demonstrated due to the integrated imaging approach inherent to optical readout. High granularity values have been reached for low-energy X-rays from radioactive sources and X-ray generators taking advantage of image sensors with several megapixel resolution. Detector characterization under X-ray radiography opens the way to different applications from beta imaging to neutron radiography. Here we will focus on one application: neutron imaging for non-destructive examination of highly gamma-ray emitting objects. This article reports the characterization of the detectors when exposed to a low activity neutron source. The response of the detector to thermal neutrons has been studied with different field configurations and gap thicknesses.**

I Introduction

The optical readout Micromegas detector is a brand new imaging device taking advantage of the CMOS sensor technology of high granularity and sensitivity. A Micromegas detector [1] of high intrinsic gain and spatial resolution has been implemented on a transparent anode. When operated with a mixture of Carbon tetrafluoride (CF_4) and Argon, visible light is produced by scintillation when ionization occurs. This detector has shown high spatial resolution under X-ray irradiation and was able to detect very low activity β-ray emitting samples [2, 3].

Coupling MicroPattern Gaseous Detectors (MPGDs) with a metallic cathode coated with $^{10}B_4C$, a neutron to charge converter, leads to a thermal neutron detection device, with efficiency of few percent and high gamma suppression capability [4]. Such a neutron converter has been used with an optical readout Micromegas detector aiming at performing real-time imaging of neutron sources with 100 μm spatial resolution.

Neutron radiography allows to characterize reactor materials like radioactive waste or nuclear fuel in a high γ-flux environment. First images of neutron events from a weak neutron source were acquired and reported in [2]. The Micromegas cathode (an aluminum plate) is coated with 1.5 μm of $^{10}B_4C$ which converts the thermal neutrons into ^7Li and α of 0.8 MeV and 1.4 MeV respectively, having a range in the gas of few mm. The present article describes the analysis of the tracks giving information about the diffusion in the gas and the neutron incident position on the cathode. Different field configurations were tested with a 10.5 mm conversion gap. A new setup with a 250 μm drift gap was tested in order to reduce the track range and to perform double stage amplification.

II Neutron radiography

The main goal of the glass Micromegas detector is to reach high spatial resolution for real-time neutron imaging. The capacity of the camera to integrate the light makes the detection of the neutron impact position feasible with very few data processing which allows to display the neutron image in real-time. However, the spatial resolution of the device is strongly affected by the diffusion and the particle range inside the gas. In the first parts of this section, the analysis of the ^7Li and α tracks from neutron events is described. In a second step, the reconstruction of the neutron impact position with two different drift gap thicknesses is explained and illustrated.

II.A Experimental setup

This test was performed at the CERN Gas Detector Development (GDD) laboratory with an Americium–Beryllium source, emitting 1–14 MeV neutrons with a detection rate of 4 Hz in average. The main characteristics of the glass Micromegas detector, of the chosen gas mixture and the CMOS camera are fully described in previous work [2]. The drift gap is of 10.5 mm in order to contain the full

*e-mail: antoine.cools@cea.fr

 https://doi.org/10.1051/epjconf/202328807009

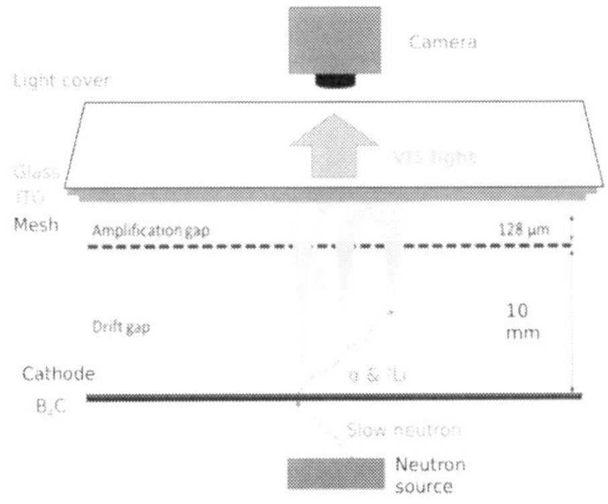

Figure 1: Sketch of the glass Micromegas detector setup for neutron radiography. Neutrons are interacting in the $^{10}B_4C$ layer producing ^7Li and α particles emitted back-to-back. The gas is ionized along the ^7Li or α tracks and the produced electrons drift towards the amplification gap. In the amplification gap, they are multiplied, producing also visible light captured by the camera.

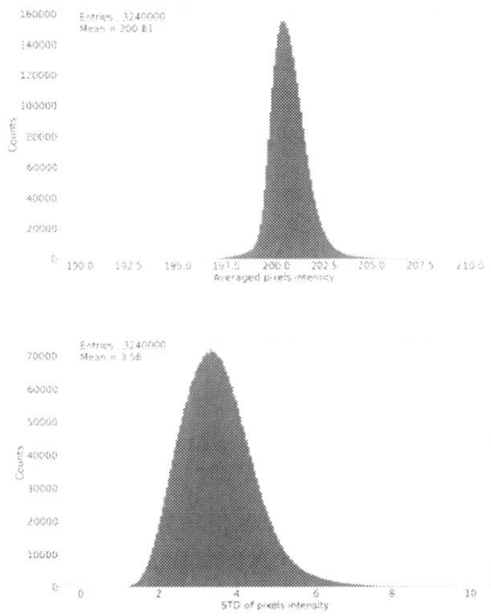

Figure 2: Mean (top) and standard deviation (bottom) histograms of pixels intensity from dark images.

tracks. The amplification gap is 128 µm and the applied voltage 570 V (Figure 1). At this voltage value, discharges occur rarely and bring almost no dead-time of the detector while the ITO layer remains undamaged. Due to the low rate of the source, 5 s exposure time images were acquired in order to collect several tracks in one frame, while avoiding overlap of tracks. Background images of the same integration time were recorded without any applied voltage or neutron source. Matrices of the average and standard deviation of every pixel of dark images are then computed for background suppression.

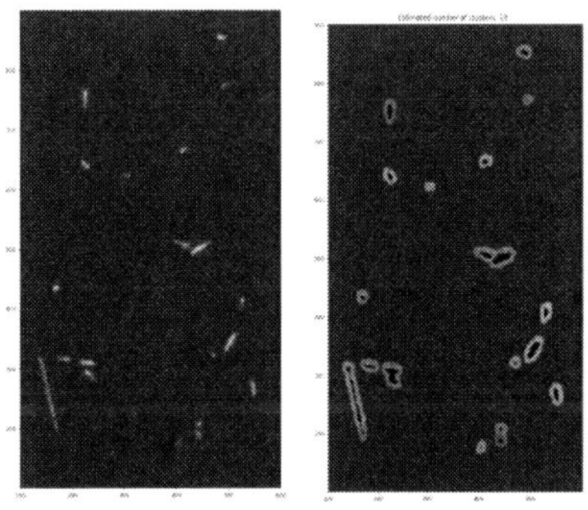

Figure 3: Neutron image before (left) and after (right) clusters identification by the DBSCAN algorithm.

II.B Tracks analysis and results

The analysis of the images is done in Python and aims to localize the tracks in a first step. The background noise coming from the dark current and the readout noise of the camera constitutes the pedestal of an image. The pixels intensity (0 to 65536) of dark images correspond to 0 photons and varies from one pixel to another as shown in Figure 2. These histograms represent the mean (μ_i) and standard deviation (σ_i) of every pixel i intensity. First, the μ_i are suppressed from the neutron images and a threshold of $3\sigma_i$ is applied on every pixel in order to reject most of the background noise. To increase the signal to noise ratio, the new neutron image is rescaled to a 1024×1024 pixels frame by applying a 2×2 binning.

The clustering algorithm employed for track identification is the so-called Density-Based Spatial Clustering of Applications with Noise (DBSCAN)[5]. This algorithm measures the distance between points and generates three categories of points :

- A core point is localized in an area of high density of points and is generally situated in the center of a cluster.
- A reachable point is at a distance ϵ of a core point and constitutes the cluster surrounding.
- A noise point is not reachable by any other points within the distance ϵ.

Only the clusters containing a minimum number of points *MinPts* are kept. Hence, this algorithm requires only the matrix of pixel intensities as an input and ϵ and *MinPts* as parameters. No optimization of the ϵ and *MinPts* parameters has been done yet. However, most of the tracks are being well identified and separated (Figure 3).

From the clustering algorithm, we access the coordinates of all the tracks. It is then possible to measure the width of the tracks by integration of the 2D track image over the track length. The Gaussian fit of the integrated intensity directly gives a σ value corresponding to the diffusion in the gas. The transverse diffusion is given by the equation $\sigma_D = \sqrt{2Dt}$ where D is the diffusion coefficient, function of the electric field (E), and t is the drift time.

According to Magboltz [6] simulations, the smallest transversal diffusion coefficient value corresponds to E $\approx$ 700 V/cm for a gas mixture with 80% of Ar and 20% of CF_4 (Figure 4). First measurements were taken with the same gas mixture at three different drift fields: 50 V/cm,

800 V/cm and 2000 V/cm. The measured diffusion was around 130 μm for a 800 V/cm drift field whereas the σ value reached higher values for the other two drift field configurations (Figure 4). The simulated diffusion evolution agrees within 2σ with the extracted points from measured track widths meaning that the broadening of the tracks is mainly due to diffusion.

However, the tendency of the measured data of having lower values than the simulated ones could be justified by the following: in the simulation, we consider that every electron is emitted at the cathode level and diffuses all along the drift gap. In reality, free electrons are generated all along the track and a majority is produced at the end of the track, at a distance from the cathode. By drifting a shorter distance, their diffusion is smaller, so the extracted values for the coefficient are systematically lower. In fact, the diffusion increases with the depth x as $\sqrt{x}$ which brings a difference of few tens of % between the diffusion at the cathode level and at the end of the track. Another consideration that may have an effect is the absence of monitoring of the temperature and pressure conditions that strongly affect the diffusion.

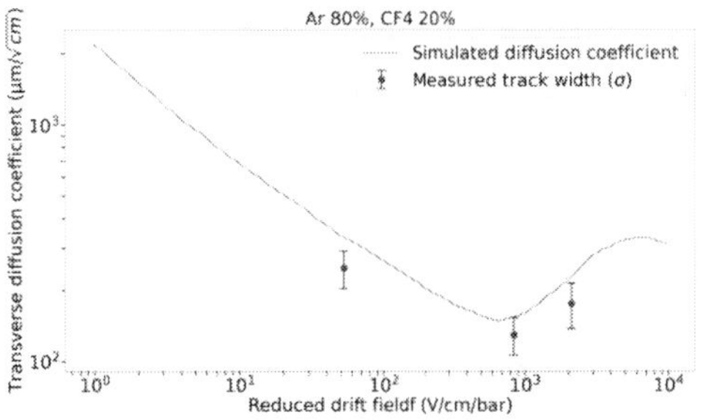

Figure 4: Simulated transverse diffusion in 80% of Ar and 20% of CF_4. Average track width (σ) from measurements.

II.C Neutron position reconstruction

It is possible to overcome the incertitude on the neutron position brought by the track length by fitting the Bragg peak of a track. In fact, the ^{7}Li and α particles deposit more energy in the gas at the end of their path. This means that the neutron position of interaction with the cathode is localized at the lowest point of the Bragg curve. 60 000 events were analyzed and associated to a position of interaction by fitting the Bragg peak with a linear distribution and represented on an histogram (Figure 5). A copper tape band was glued on the boron side of the cathode so that ^{7}Li and α particles are absorbed in one side only of the detector. The spatial resolution is then calculated by fitting the edge with a step function f described in (1) where μ is the edge position and A is the maximum amplitude. A spatial resolution of σ = 1420 (±124) μm was obtained.

$$f(x) = \frac{A(1 + erf(p))}{2} \qquad p = \frac{(x-\mu)}{\sqrt{2}\sigma} \qquad (1)$$

The image on Figure 6 represents the light intensity profile obtained by integrating the light coming from the tracks. Here the precision on the neutron position measurement is affected by the particles range and should lead to a larger spatial resolution. However, the spatial resolution obtained from the reconstruction method and from the light integration method (1198 (±263) μm) are of the same order. This means that the fit of the Bragg peak fails in some cases. Further study on the simulation of the ^{7}Li and α tracks and on the reconstruction algorithm are needed.

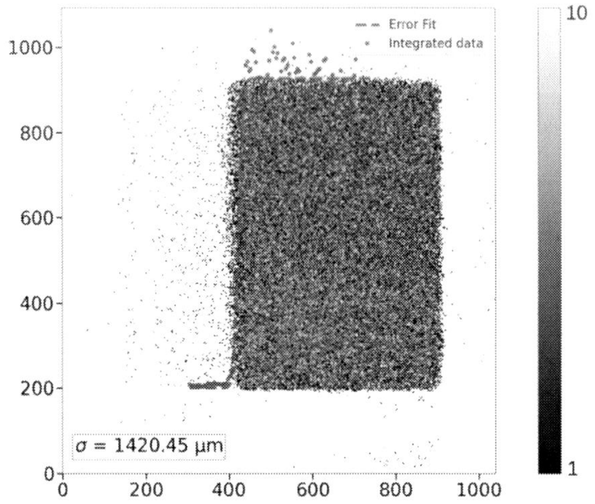

Figure 5: Neutron position histogram for a 10.5 mm drift gap.

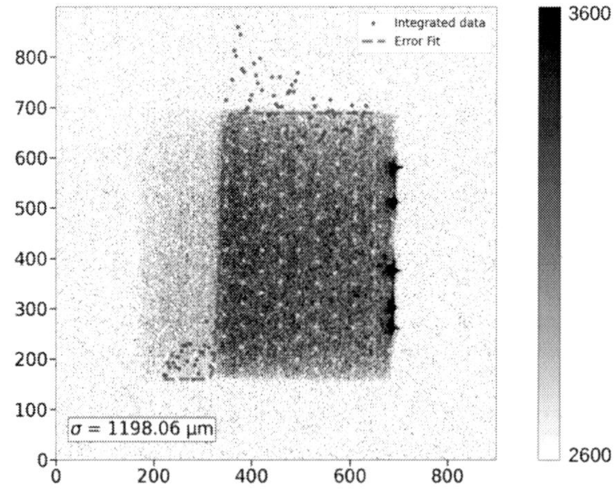

Figure 6: Neutron light intensity profile for a 10.5 mm drift gap.

A new configuration with a 250 μm drift gap was tested in order to reduce significantly the particles range in the gas. Then the error on the position computation brought by the track length is expected to decrease. As a consequence, less charges are produced in the gas, decreasing the signal intensity. By generating an electric field of 20 kV/cm in the drift gap, a double stage amplification leads to a higher gain and compensate the lack of primary electrons. Moreover, the gain increases with the avalanche length which means that primary electrons that are produced close to the cathode are amplified further. This mechanism enhances the signal at the neutron impact position. In this configuration, clusters with a Gaussian shape were recorded instead of track shapes. Hence, implementing a $2D$ Gaussian fit of the clusters allows to reconstruct the neutron position with a better precision (Figure 7).

It is not possible to use a copper tape in this setup since the copper thickness is not negligible and would affect the electric field in such a small gap. However, boron evaporation on the cathode with a mask with patterns will be realized in the future for spatial resolution measurement without affecting the electric field. A boron band was placed outside of the detector on one side with the ambition of visualizing an edge. Because of the large angle neutrons, an important number of events were detected on the side where the boron was put, making the edge image very blurred. On the right side of the histogram, the density of hits is not homogeneous along the y-axis. Mechanical constraints on the cathode brought defaults on its planarity of few tens of μm which caused a non uniform gain in the drift gap during double stage amplification.

However, the 500 μm diameter pillars are well resolved in the right side of this histogram. The assessment of the pillars position is an indication that a spatial resolution of a few hundred μm can be achieved. The low spreading of the charges in this configuration will allow us to perform real time neutron radiography by simply integrating the light intensity over time for high flux sources.

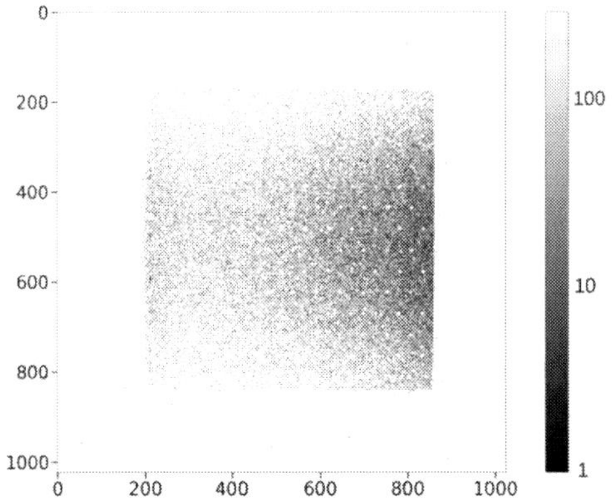

Figure 7: Neutron position histogram for a 250 μm drift gap with double stage amplification.

III Conclusion

Neutron events were detected with a glass Micromegas detector and studied under different field configurations. The diffusion from the ionization of ^{7}Li and α particles was measured and compared to computed diffusion from simulation. Two methods were explored in order to increase the spatial resolution. The first one consisted in the reconstruction of the neutron impact position on the cathode. A spatial resolution better than 1500 μm was achieved. In a second step, double stage amplification was performed with a 250 μm drift gap in order to reduce the particle range and to emphasize the neutron position. The spatial resolution was not measured in this configuration because of the lack of statistics and the non-parallelism of the neutrons. In the near future, a neutron beam test will be conducted at a neutron facility with a high flux source of parallel thermal neutrons for precise estimation of the spatial resolution with different field configurations and gap thicknesses.

References

[1] Y. Giomataris. "Development and prospects of the new gaseous detector "Micromegas"". In: *Nuclear Instruments and Methods in Physics Research Section A: Accelerators, Spectrometers, Detectors and Associated Equipment* 419.2 (1998), pp. 239–250. ISSN: 0168-9002. DOI: https : / / doi . org / 10 . 1016 / S0168 - 9002(98) 00865 - 1. URL: https : / / www . sciencedirect . com / science / article / pii / S0168900298008651.

[2] A. Cools et al. "Neutron and beta imaging with Micromegas detectors with optical readout". In: *Nuclear Instruments and Methods in Physics Research Section A:Accelerators, Spectrometers, Detectors and Associated Equipment* 1048 (2023), p. 167910. ISSN: 0168-9002. DOI: https://doi.org/10.1016/j.nima.2022.167910. URL: https://www.sciencedirect.com/science/article/pii/S0168900222012025.

[3] A. Cools et al. *X-ray imaging with Micromegas detectors with optical readout*. 2023. arXiv: 2303.17444 [physics.ins-det].

[4] T. Papaevangelou et al. "ESS nBLM: Beam Loss Monitors based on Fast Neutron Detection". In: *Proc. 61st ICFA Advanced Beam Dynamics Workshop (HB'18), Daejeon, Korea, 17-22 June 2018* (Daejeon, Korea). ICFA Advanced Beam Dynamics Workshop 61. https://doi.org/10.18429/JACoW-HB2018-THA1WE04. Geneva, Switzerland: JACoW Publishing, July 2018, pp. 404–409. ISBN: 978-3-95450-202-8. DOI: doi : 10 . 18429 / JACoW - HB2018 - THA1WE04. URL: http : / / jacow . org / hb2018/papers/tha1we04.pdf.

[5] Martin Ester et al. "A Density-Based Algorithm for Discovering Clusters in Large Spatial Databases with Noise". In: *Proceedings of the Second International Conference on Knowledge Discovery and Data Mining*. KDD'96. Portland, Oregon: AAAI Press, 1996, pp. 226–231.

[6] S. F. Biagi. "Monte Carlo simulation of electron drift and diffusion in counting gases under the influence of electric and magnetic fields". In: *Nucl. Instrum. Meth. A* 421.1-2 (1999), pp. 234–240. DOI: 10.1016/S0168-9002(98)01233-9.

Neutron-gamma survey system for decommissioning and dismantling activities

Jessica C. Delgado[1,2], Felix Pino[1], Erica Fanchini[3], Alessandro Iovane[3],
Daniela Fabris[4] and Sandra Moretto*[1,4]

[1]Department of Physics and Astronomy "Galileo Galilei", University of Padova, Italy
[2]Department of Physics and Earth Science, University of Ferrara, Italy
[3]CAEN S.p.A., Viareggio, Italy
[3]INFN- Padova Section, Padova, Italy

(*) sandra.moretto@unipd.it

***Abstract*—The nuclear plant decommissioning and dismantling (D&D) operations will amount to €200 billion in costs over decades around the world, with three-fourths coming from Europe. Decommissioning includes activities such as planning, physical and radiological characterization, facility and site decontamination, dismantling, and materials management. This work is focused on the development of a compact, light and low-power consumption neutron-gamma survey system which could be easily mounted on an remotely operated vehicle. It is made up of a 4"x4"x2" NaIL (NaI:Tl + 1% ^{6}Li [95% enriched]) neutron/gamma scintillation detector coupled to a SiPM array. Digital pulse processing techniques were implemented to acquire and process the signals, by means of a CAEN DT5780 unit. A comprehensive characterization of this system, based on experiments and Monte Carlo simulations, is reported. The system can be used as a secondary inspection tool, useful for identifying radioactive and special nuclear materials in hotspots.**

***Keywords* — NaIL detector, D&D operations, radiation monitoring, neutron/gamma discrimination, hotspots.**

I. Introduction

THE decommissioning and dismantling (D&D) operations of nuclear facilities are crucial processes to safely retire and remove nuclear facilities (nuclear power plants, research reactors, etc.) from service. These operations involve a series of steps to ensure the safe management of radioactive materials, environmental protection, and the restoration of the site to a condition suitable for other uses. The D&D operations will cost an important amount of money and effort over decades around the world, in particular, in Europe will be more significant. As an example, in France, some thirty facilities are in the process of being dismantled or are pending to be dismantled. A determining part of the decommissioning includes the planning and preparation of physical and radiological characterization, which then, can facilitate the next steps of the operation: site decontamination, dismantling, materials management, etc. For that reason, the development of innovative systems to perform radiological surveys, taking advantage of the new technological advances in radiation detection, can save time, reduce costs, and minimize human intervention while increasing safety [1]. The development of techniques to characterize complex and/or poorly accessible structures, such as pipes, vessel internals, underground tanks with liquid radioactive waste, etc., are of particular interest currently [1]. Further development and use of robots, remotely operated vehicles, and drones equipped with radiation detectors could offer new options [2], especially in high-intensity radiation areas.

Different kind of sensors, such as gas filled, scintillation and solid state detectors have been used for radiation measurements (dose rate, gamma spectroscopy, estimation of activity, etc.) in the field of D&D of nuclear facilities. They have proven to be well developed and reliable. Very complex systems such as gamma cameras are currently employed to perform imaging in radiation monitoring applications, those can be based on semiconductor [3] or scintillation [4] detectors. Recently, the development of new scintillation materials, that offer excellent performance in terms of energy resolution [5], combined with the capability to detect thermal neutrons [6, 7], and taking advantage of the SiPM technology, has let to produce light, compact, and low power consumption systems [8], suitable to be deployed in D&D operations.

This work is focused on the preliminary study about the possibility to use a NaIL scintillation detector to perform neutron-gamma survey measurements. Specifically, the detector is made up of a 4"x4"x2" NaIL crystal coupled to a SiPM array. This detector offers thermal neutron counting and gamma-spectroscopic information. This system could be used as a secondary inspection tool of the site, mounted in a fixed position on an unmanned ground vehicle. It could be useful for characterizing and identifying the radioactive sources in hotspots located in large areas. Besides, thanks to the thermal neutron detection capability it could be used for recognition of special nuclear materials (such as: HEU, Pu, etc). Digital pulse processing of the detector signals is performed. The system has been extensively tested in laboratory conditions with point-like gamma sources, and neutron/gamma sources. A Monte Carlo simulation study, using GEANT4 v10.7 toolkit [9], is also reported. Specifically, the experimental gamma and thermal neutron responses of the detector are reported, including the high gamma background scenario. Besides, the absolute full-energy peak (@662keV) efficiency as a function of the angular position of the source has been studied. Finally, the energy resolution as a function of the interaction point of the gamma-ray in the NaIL crystal is reported.

II. Experimental details

A 4"x4"x2" NaIL (NaI:Tl + 1% ^{6}Li [95% enriched]) neutron/gamma scintillation detector coupled to a SiPM array purchased from Saint-Gobain Crystals company was used (see Fig. 1). The assembly detector includes an adapter board to simplify the power supply and data collection. A micro USB connector is used to provide the required +5VDC to the read-out board, and a coaxial MCX connector for the output signal.

A set of gamma-ray point-like calibration sources (with activities ~300 kBq) was employed: ^{137}Cs, ^{60}Co, ^{241}Am and ^{22}Na. Besides, neutron/gamma sources, ^{252}Cf and AmBe, with neutron emission rate of 2x10^4 s^{-1} and 2x10^5 s^{-1} respectively were used for the neutron measurements.

A CAEN DT5780 Digital Multi Channel Analyzer (100 MS/s, 14-bit, 16k Digital MCA) was used for the acquisition and processing of the NaIL pulses. The DT5780 unit features the DPP-PHA firmware, usually employed for gamma spectroscopy applications. Furthermore, a pile-up filter is included in the firmware, in order to obtain good results in situations of high counting rates. The unit is fully-controlled by CoMPASS Software.

Monte Carlo simulations of the gamma-ray and thermal neutron detection efficiencies of the NaIL detector were performed using the GEANT4 v10.7 toolkit. The *QGSP_BERT_HP* physics list was used, which contains high precision data for the transport of fast neutrons.

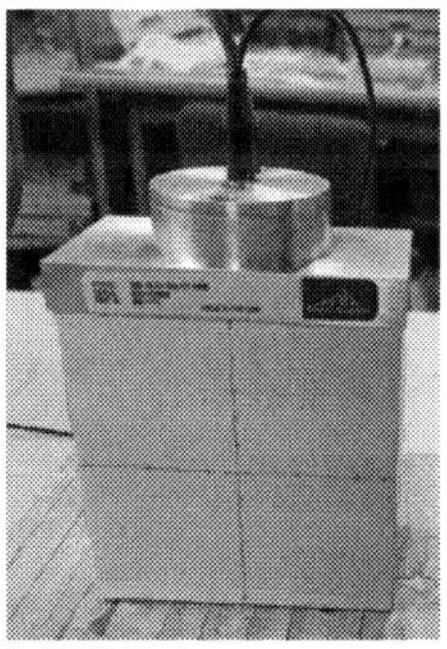

Fig. 1. Picture of the 4"x4"x2" NaIL detector

III. Results and discussion

A. Gamma-ray response

In Fig. 2a the energy spectra, corresponding to a ^{137}Cs source and a background, acquired with the NaIL detector are given. After the background subtraction, the energy resolution (% of FWHM) at 662keV obtained was 7.8%. Since the DPP-PHA firmware is based on the trapezoidal filter, the processing parameters were optimized in order to obtain the best results in terms of energy resolution. The optimal values are: rise time 3µs, flat top 1µs, pole zero 45 µs, peaking time 80%, and 4 samples of the flat top to get the amplitude of the trapezoid.

In Fig. 2b the energy calibration curve is presented. The explored energy range is 0.06-4.44 MeV. The best result was obtained by fitting the quadratic equation: $E[keV] = a \cdot Ch + b \cdot Ch^2$. As can be seen a small non linearity is exhibited by the detector. This behavior is commonly observed in large volume scintillators when a large energy range is taken into account.

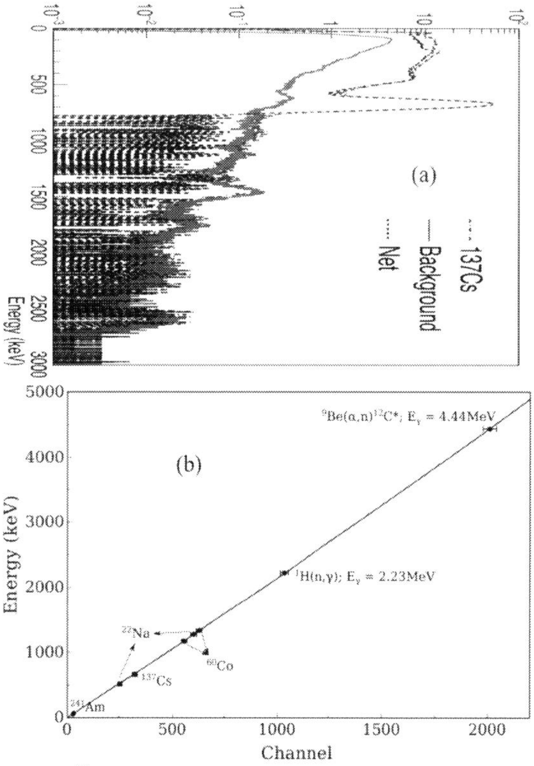

Fig. 2. (a) ^{137}Cs spectrum acquired with the NaIL detector (FWHM[%] @662keV = 7.8 %). (b) Energy calibration curve of the NaIL detector. Fitted parameters are: a = 2.068, b = 7.127 x 10^{-5}.

B. Thermal neutron response

The NaIL detector responses corresponding to AmBe and ^{252}Cf neutron sources are shown in the Fig. 3. Each source was placed at 25 cm, and centered, from one of the larger faces of the detector. The fast neutrons emitted were moderated with 6cm of polyethylene. The associated "thermal neutron peak" is observed at 3.2 MeVee. Besides, in the case of the AmBe source, the 4.44 MeV peak, and the single escape (S.E.) peak can be also observed. It is important to mention that this structure can be used to identify neutron sources based on ^{9}Be(α,n)^{12}C reaction. As output, the read-out board provides a positive going charge sensitive tail-pulse with a 50 µs decay, so, pulse shape discrimination by performing pulse shape analysis of the output signal is not possible to execute. However, since the thermal neutron detection signature is observed at high energies (3.2 MeVee full-energy peak), for most of cases it is not necessary to perform neutron/gamma discrimination to get good results in terms of thermal neutron detection capability.

A Monte Carlo simulation of the intrinsic thermal neutron detection efficiency was performed using the GEANT4 v 10.7 toolkit. Thermal neutrons with energies of 0.025 eV were directed to one of the larger faces of the NaIL detector. The origin of the thermal neutrons was 25 cm away from the detector face. All the face was illuminated by the neutrons. The efficiency was estimated to be around 60%.

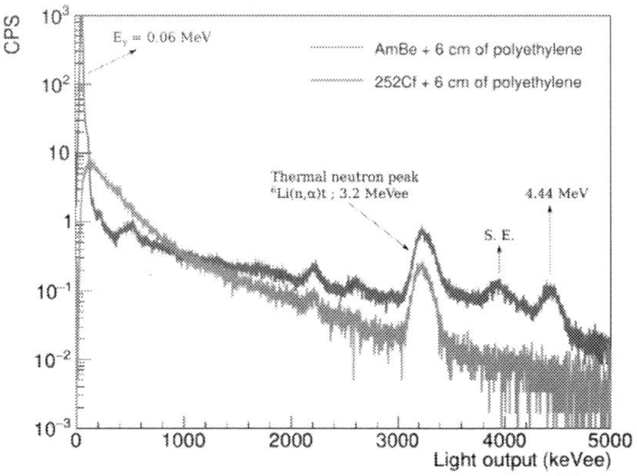

Fig. 3. NaIL detector responses associated to AmBe and ^{252}Cf sources moderated with 6 cm of polyethylene.

C. *Thermal neutron detection against the counting rate*

The main reasons that can cause a loss of thermal neutron detection efficiency are the effects produced by a high counting rate, such as: pile-up events, dead time, summing events, etc., so, it is important to assess the response of the NaIL detector as a function of the measured counting rate.

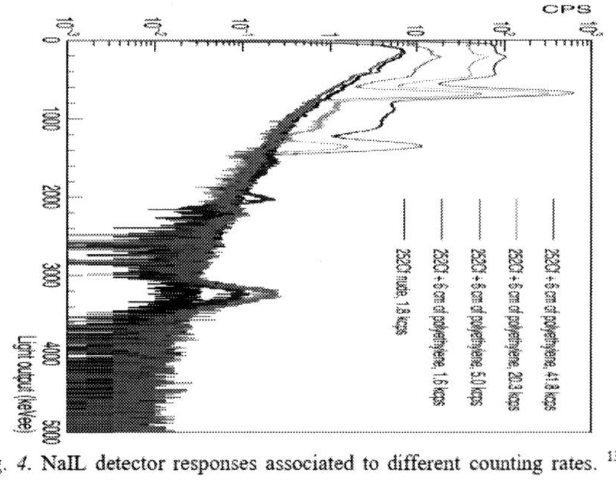

Fig. 4. NaIL detector responses associated to different counting rates. ^{137}Cs source was used to modify the counting rate.

The measurements were performed using the ^{252}Cf moderated with 6 cm of polyethylene, combined with a ^{137}Cs source. The last one was used to increase the counting rate. In Fig. 4 several detector responses at different counting rates are shown. As can be seen the thermal neutron peak at 3.2 MeVee reduced its intensity as the counting rate increase (excluding the spectrum corresponding to the non-moderated ^{252}Cf source).

Fig. 5 shows the behavior of the thermal neutron detection efficiency as a function of the counting rate. Specifically, it is reduced by 50% when the measured counting rate of the system is around 40 kcps.

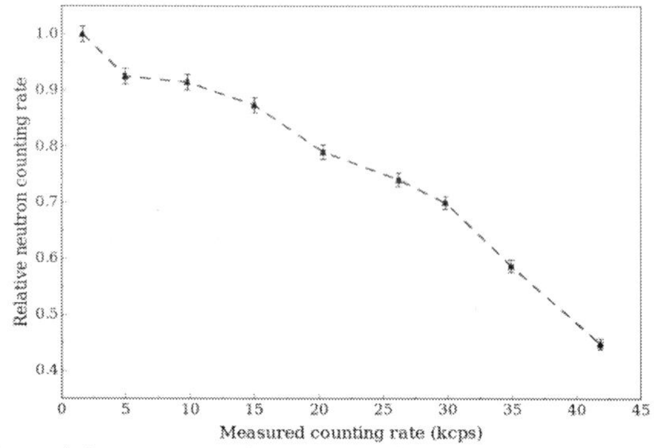

Fig. 5 Relative neutron counting rate against the measured counting rate.

D. *Full-energy peak gamma efficiency vs angular position*

In radiation monitoring applications, the response of the detector as a function of the position of the source with respect to the detector is an important scenario to be studied.

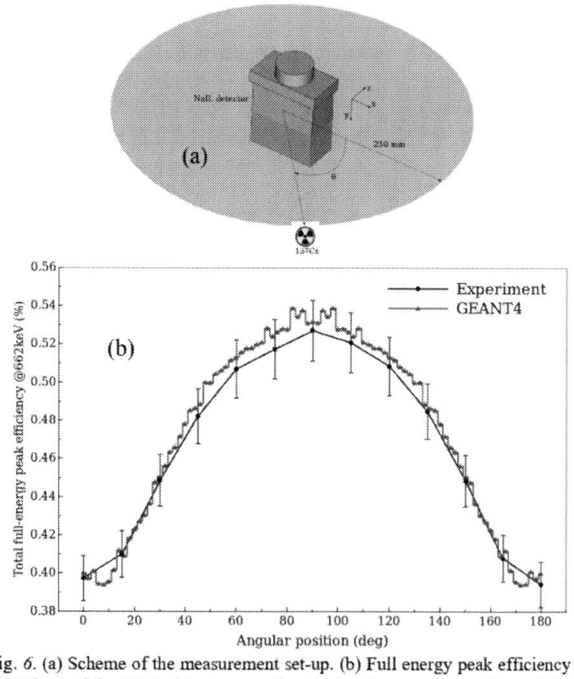

Fig. 6. (a) Scheme of the measurement set-up. (b) Full energy peak efficiency @662keV of the NaIL detector as a function of the angular position of the source.

In this case, a ^{137}Cs source was placed at 25 cm from the center of the active volume of the detector (see Fig. 6a). The angular position of the source, θ, was changed in the range 0-180°. The absolute full-energy peak gamma efficiency at

662keV as a function of the angular position of the source is given in the Fig. 6b. Excellent agreement between the experimental and simulation results was obtained. The results suggest that at 1 meter, the minimum detectable activity (MDA) [10], of a ^{137}Cs source is between 3 and 4 kBq (taking into account a measurement time = 1 min), depending on the angular position of the source.

E. Energy resolution vs interaction point

The energy resolution at 662keV as a function of the initial interaction point of the gamma-ray was also studied. Figure 7a depicts a schematic representation of the setup. This setup involves the placement of a ^{137}Cs source within a collimator made of tungsten. The collimator is carefully positioned in front of both the smaller and larger faces of the detector, each separately. This arrangement is designed to examine how the energy resolution behaves in relation to the interaction point within the detector. A worsening behavior (up to 20% with respect to the best value) is observed as the interaction point becomes closer to the SiPM array, see Fig. 7b. In fact, the poorer energy resolution is observed when the initial interaction point is located in the smaller face and close to the height of the SiPM array. Taking into account that the mean free path of the 662keV photons in NaI is around 3.8 cm, it seems that an important amount of light is produced in the corners of the scintillator, making difficult its collection by the read-out device. No dependance is seen in the cross-line direction, at least at half height of the scintillator. In this configuration the best energy resolution is obtained, meaning that most of the scintillation light yield is efficiently collected.

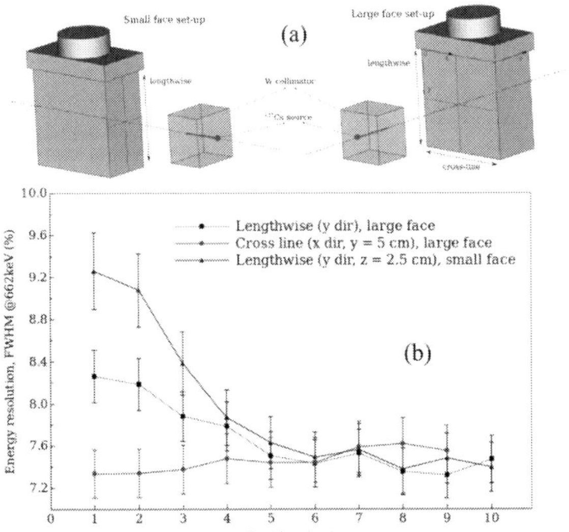

Fig. 7. (a) Scheme of the measurement set-up. (b) Energy resolution (% of FWHM @662keV) associated at different interaction points.

IV. CONCLUSIONS

A comprehensive study of a large NaIL scintillator coupled to a SiPM array has been performed. It exhibits an energy resolution of 7.8% @662keV, and shows a slightly non-linear energy calibration when considering an energy range 0.06-4.44MeV. By Monte Carlo simulation the intrinsic thermal neutron efficiency was estimated to be 60%. The thermal neutron peak is clearly observed at 3.2 MeVee, and as the gamma background increase the thermal neutron detection efficiency decrease. Specifically, it is reduced 50% when the measured counting rate is about 40 kcps.

There is a strong dependance of the gamma detection efficiency as a function of the angular position of the source, and it has been well reproduced by the Monte Carlo model (difference lower than 5%). Taking into account the maximum value of gamma efficiency, the minimum detectable activity of ^{137}Cs was estimated to be 3 kBq (1 m away from the detector).

Finally, the energy resolution of the NaIL detector as a function of the initial interaction point was studied. In particular, a worse performance is observed when the interaction point is close to the SiPM array, meaning that in this situation it is hard to achieve a good collection of light.

According to the obtained results, we can conclude that a neutron-gamma survey system based on a high-efficient NaIL scintillation detector, that coupled to a SiPM array offers a compact, light and low-power-consumption solution, can fulfill the current requirements of the systems designed to perform the radiological characterizations in D&D operations. In the near future the integration of this system with a unmanned ground vehicle to perform a field test will be realized.

REFERENCES

[1] INTERNATIONAL ATOMIC ENERGY AGENCY, Global Status of Decommissioning of Nuclear Installations, IAEA Nuclear Energy Series No. NW-T-2.16, IAEA, Vienna (2023).

[2] K.A. Pradeep Kumar, G.A. Shanmugha Sundaram, B.K. Sharma, S. Venkatesh, R. Thiruvengadathan, "Advances in gamma radiation detection systems for emergency radiation monitoring," Nuclear Engineering and Technology, Volume 52, Issue 10, 2020, pp. 2151-2161, 10.1016/j.net.2020.03.014.

[3] V. Paradiso, K. Amgarou, N. Blanc de Lanaute, F. Bonnet, O. Beltramello, E. Liénard, "3-D localization of radioactive hotspots via portable gamma cameras," Nuclear Instruments and Methods in Physics Research Section A, Vol. 910, 2018, pp. 194-203, 10.1016/j.nima.2018.09.081.

[4] D.-H. Han, S.-J. Lee, H.-J. Lee, J.-O. Kim, K.-H. Jung, D.-E. Kwon, C.-H. Baek, "Development and evaluation of a compact gamma camera for radiation monitoring," Nuclear Engineering and Technology, Vol. 55, Issue 8, 2023, pp. 2873-2878, 10.1016/j.net.2023.04.029.

[5] M. Ye, P. Gong, S. Wu, Y. Li, C. Zhou, X. Zhu, X. Tang, "Lightweight SiPM-based $CeBr_3$ gamma-ray spectrometer for radiation-monitoring systems of small unmanned aerial vehicles," Applied Radiation and Isotopes, Vol. 176, 2021, 109848, 10.1016/j.apradiso.2021.109848.

[6] M. Polo, F. Pino, J.C. Delgado et al., "Assessment of a NaIL detector performance for radiation monitoring applications," Eur. Phys. J. Spec. Top., 2023, 10.1140/epjs/s11734-023-00896-4.

[7] F. Pino, C.L. Fontana, J. Delgado, D. Fabris, G. Nebbia, et al., "Characterization of a medium-sized CLLB scintillator: Single neutron/Gamma detector for radiation monitoring," J. Inst. 16 (11), 2021, P11034, 10.1088/1748-0221/16/11/P11034.

[8] C. M. Whitney, L. Soundara-Pandian, E. B. Johnson, S. Vogel, B. Vinci, M. Squillante, J. Glodo, J. F. Christian, "Gamma–neutron imaging system utilizing pulse shape discrimination with CLYC," Nuclear Instruments and Methods in Physics Research Section A, Vol. 784, 2015, pp. 346-351, 10.1016/j.nima.2014.09.022.

[9] S. Agostinelli, et al., "Geant4—a simulation toolkit," Nuclear Instruments and Methods in Physics Research Section A, Vol. 506, Issue 3, 2003, pp. 250-303, 10.1016/S0168-9002(03)01368-8.
[10] L. A. Currie, "Limits for qualitative detection and quantitative determination," Application to radiochemistry. Anal. Chem., vol. 40, 3, pp. 586–593, 1968, 10.1021/ac60259a007.

New 3D coded aperture collimator for X/gamma-ray wide-field imaging

Yaroslav Susaiev[1,*], Vincent Schoepff[1], Olivier Limousin[2]
[1]Université Paris-Saclay, CEA, List, F-91120 Palaiseau, France
[2]Université Paris-Saclay, CEA, Irfu, F-91191 Gif-sur-Yvette, France
(*) yaroslav.susaiev@cea.fr

Abstract—**Remote localization of radioactive sources is vital to the nuclear industry and high-energy astrophysics. One widely used method to localize radioactive sources relies on indirect imaging with coded aperture cameras. It involves using a coded aperture collimator called the "coded mask" accompanied by a position-sensitive X/gamma-ray detector with a dedicated deconvolution algorithm. However, a known disadvantage of such cameras is their limited field of view, resulting in a restricted capture area when analyzing complex contaminated scenes or transient astrophysical events. Our paper introduces a new non-planar coded mask prototype designed to expand the field of view of standard coded aperture cameras. Our 3D coded mask is a tantalum hemisphere with a spiral-shaped hole. We use the Maximum Likelihood Expectation Maximization algorithm to reconstruct the radioactive source images. We test our prototype on data acquired with Timepix3 hybrid pixel detector equipped with our mask.**

Keywords **—Gamma imaging, 3D coded mask, Timepix3**

I. Introduction

Remote localization of radioactive sources is a crucial challenge in nuclear instrumentation physics, with different applications ranging from the nuclear industry [1,2] to high-energy astrophysics [3,4]. Identifying radioactive source position and intensity in complex contaminated scenes is gaining significant relevance following the Fukushima nuclear disaster in Japan and the recent Zaporizhzhia Nuclear Power Plant occupation in Ukraine.

Localization of radioactive sources by direct imaging using compact systems of lenses and mirrors is not feasible due to the fundamental nature of X/gamma-ray photons. Fortunately, indirect imaging methods offer commonly used solutions, with two primary approaches available: Compton imaging and coded aperture imaging. The first approach relies on the Compton scattering phenomenon and can be applied to imagine radioactive sources of energies above 250 keV [5]. This method benefits from a wide field of view from 2π to 4π steradians while being capable of providing an angular resolution of approximately 10°. This paper focuses on the second approach of coded aperture imaging, which allows for imaging radioactive sources across a wide energy range, from a few keV to several MeV [6]. Historically, this method originates from the classical pinhole camera, and its idea is as follows. A coded aperture mask, a multi-pinhole collimator, modulates the X/gamma-ray photon flux emitted by a radioactive source. The opaque elements of the mask absorb the incoming X/gamma rays, while the transparent elements (holes) allow them to pass through. After, the holes cast the shadow of the mask pattern onto the detector plane called the "shadowgram". Finally, we can decode the shadowgram using deconvolution algorithms to obtain the reconstructed image of the radioactive source.

There is a wide variety of different coded mask patterns, such as Uniformly Redundant Arrays (URAs) [7], Modified Uniformly Redundant Arrays (MURAs) [8], and random arrays [9,10]. These patterns are primarily used in coded masks of planar geometries, limiting corresponding imaging characteristics. In principle, for a given position of a point source of photons in space, any unambiguous pattern projecting a unique shadowgram onto the detector can be considered to create a coded mask. Of course, the pattern design depends on various factors, such as observed flux intensity, photon energy, background noise, and mathematical reconstruction methods. Usually, coded aperture cameras equipped with the planar coded masks feature a field of view up to 50° × 50° and an angular resolution above 1°. Table I provides the parameters of some existing X/gamma-ray imaging systems.

TABLE I
CHARACTERISTICS OF EXISTING X/GAMMA-RAY IMAGING SYSTEMS

Imaging system	Mask type	Field of view	Angular resolution
Integral / IBIS [11]	Random	29.1° × 29.4°	0.2° (12 arcmin)
iPIX [12]	MURA	48.8° × 48.8°	2.5° (150 arcmin)
NuVISION [2]	MURA	45.0° × 45.0°	3.5° (210 arcmin)
Nanopix [13]	MURA	48.8° × 48.8°	2.5° (150 arcmin)
SpidX [14]	Random	55.0° × 55.0°	1.0° (60 arcmin)
SVOM / Eclairs [4]	Random	90.0° × 90.0°	0.5° (30 arcmin)

Therefore, the ideal coded aperture camera would have a field of view as wide as 2π steradians and an angular resolution of less than 1°, which is practically unachievable with currently available technologies. Noteworthy endeavors to reach these parameters can be found in literature, such as radial hole coded masks [15], rotating scatter masks [16], and active coded masks [17].

Our research introduces a novel method based on a non-planar coded mask aperture system called the "3D coded mask". Its primary objective is to achieve nearly 2π steradians field of view with a single planar position sensitive detector while maintaining an angular resolution in the range of the state-of-the-art systems as in Table I. We manufactured the first

3D coded mask by the tantalum machining process. To reconstruct radioactive sources in a wide field of view, we apply a deconvolution algorithm that relies on a database of the mask pattern projections (the "projectors") explicitly generated for our geometry using an analytical approach.

The paper is structured as follows. Section II introduces the 3D coded mask, its characteristics, and expected performance. Section III describes the camera prototype based on the 3D coded aperture equipped with the Timepix3 hybrid-pixel detector. Sections IV describes the method of generating the projection dataset specific to our mask, which is later used in Sections V and VI to analyze the simulation and experimental results.

II. 3D CODED MASK

Our 3D coded mask is a tantalum hemisphere with a spiral-shaped hole. Its unique design aims to show how an unambiguous arbitrary mask pattern can provide a wide field of view of nearly 2π steradians. The mask is presented in Fig. 1.

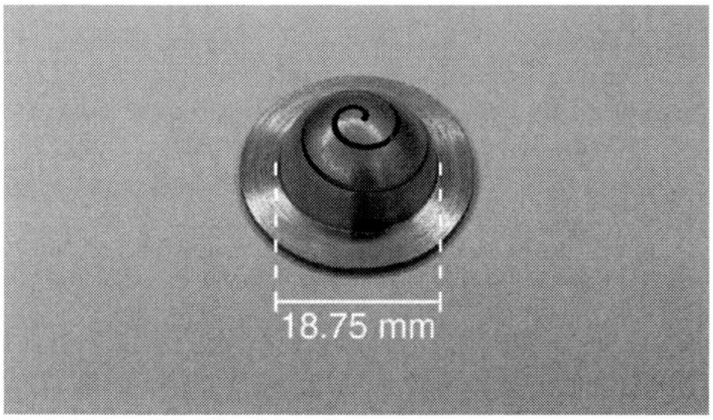

Fig. 1. Picture of the 3D coded mask.

The 3D coded mask is manufactured by tantalum machining process, which offers precise cutting and shaping of the metal. Tantalum is often used as coded mask material due to its capability of absorbing high-energy photons. With its high atomic number ($Z = 73$) and density ($D = 16.69$ g/cm³), the total probability of shielding the ^{241}Am 59.54 keV photons by the opaque elements of the mask is 99%.

The opaque elements of the mask form a hemispherical shape. Such geometry ensures the uniform mask surface illumination by X/gamma rays coming from any direction in the sky. The diameter of the hemisphere is 18.75 mm, or 28 mm in total, including the outer brim. The latter is used to fasten the aperture in front of the detector accurately. The mask has a thickness of 0.75 mm, matching the width of the spiral hole.

The transparent elements of the mask form a continuous opening that has a shape of a one-arm Archimedean spiral. The reason for choosing this unique pattern geometry was to achieve significant pattern asymmetry while retaining the manufacturing advantages of a continuous hole, thus simplifying the initial phase of our work. Mathematically, the Archimedean spiral can be defined by (1) in polar coordinates.

$$r = a + b\theta = a + \frac{1}{2\pi n_t} \cdot \theta \tag{1}$$

Here, a and b are real numbers, and n_t is the number of successive turns of the spiral. By design of our mask pattern, $a = 0$ and $n_t = 2.5$. Then the Archimedean spiral equation represented by (1) takes the form (2).

$$r = \frac{1}{5\pi} \cdot \theta \tag{2}$$

Table II summarizes the parameters of the 3D coded mask spiral pattern.

TABLE II
PARAMETERS OF THE 3D CODED MASK SPIRAL PATTERN

Parameter	3D CODED MASK
Pattern geometry	Hemispherical Archimedean spiral
Number of spiral arms	1
Number of spiral turns (n_t)	2.5
Spiral diameter	16.875 mm
Spiral hole width	0.75 mm
Open fraction	10.2%

III. TIMEPIX3 CAMERA PROTOTYPE

We developed an ultra-compact X/gamma-ray coded aperture camera prototype equipped with the 3D coded mask to conduct measurements of radioactive sources in a wide field of view. The camera is based on state of the art Timepix3 photon sensor, a part of the Timepix family of hybrid pixel detectors developed in the frame of the Medipix3 collaboration at CERN [18]. In our work, we utilize a Timepix3 interface called MiniPIX TPX3 designed by Advacam. It is a small 77 mm × 21 mm × 10 mm device with an ARM processor and USB interface for data pre-processing and readout.

Timepix3 is 14.08 mm × 14.08 mm large, with a 1 mm thick cadmium telluride crystal. The detector features a high pixel resolution matrix with a 256 × 256 pixels detection surface and 55 µm pixel pitch. Timepix3 can operate in two modes simultaneously: Time-over-Arrival, which provides the particle arrival time, and Time-over-Threshold, which records the particle deposited energy. This work focuses on the second mode when collecting data.

To assemble the X/gamma-ray camera, we position our 3D coded mask in front of the Timepix3 detection surface, ensuring alignment along the axis that intersects the mask zenith and the detector center. The mask should be placed directly on top of the detector to provide the maximum field of view, with the upper detection plane being coplanar with the mask brim. In such a setup, the camera only has the fully coded field of view when the coded aperture fully modulates the photon flux.

However, in our camera setup, this ideal configuration is not achievable due to the MiniPIX TPX3 plastic body, which creates a gap between the upper detection plane and the 3D coded mask brim. This configuration is illustrated in Fig. 2.

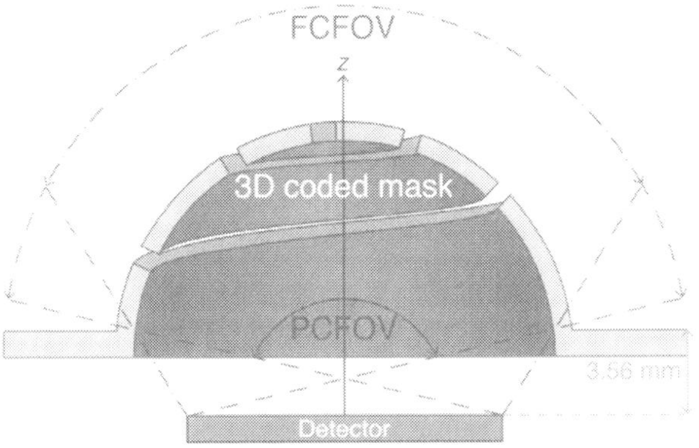

Fig. 3. Illustration of the 3D coded mask camera setup (side view). FCFOV stands for "fully coded field of view" (blue-colored range), and PCFOV corresponds to "partially coded field of view" (red-colored range).

The smallest gap between the mask and the detector is 3.56 mm, resulting in a reduced field of view. In such an imaging system, we have the partially coded field of view, including the observation range when the mask modulates the photon flux only partially.

From geometrical considerations, the maximum field of view of our Timepix3-based camera is 154.3° × 154.3°. However, the effective field of view is not constant and depends on the spiral hole orientation to the observable source. According to the mask hole width, the detector pixel pitch, and mask-to-detector distance (radius of the mask and the gap), the best angular resolution achievable by our camera is 3.5°. However, the spiral hole is continuous along in azimuth direction but segmented in polar. Thus, our X/gamma camera has two distinct angular resolutions, azimuthal and polar, which also depend on the reconstruction algorithm.

IV. Projection Dataset and Reconstruction Algorithm

Projection datasets are essential in reconstructing radioactive source shadowgrams captured by the 3D coded mask camera. In planar coded aperture imaging systems, the traditional reconstruction procedure relies on a single projection of the whole mask pattern onto the detection plane, representing the mask pattern itself. However, when the mask geometry is not planar, every position of the observable source in the sky is associated with a projection onto the detector plane that has to be carefully calculated.

To address this issue, we developed a Python code that provides spiral pattern projections for any given position of the source in the sky from an analytical description. Given the 3D model of our mask, the code generates orthogonal projections of the spiral hole onto the detector surface. The algorithm is not Monte Carlo-based, offering the flexibility of altering the 3D coded mask geometry and a fast computation time. Fig. 3 shows the projectors generated for two different positions of the observable source.

The trade-off is the somewhat simplified model, where the 3D coded mask is infinitely thin. Therefore, the absence of the vignetting effect might bias the reconstruction as the effective open fraction of the 3D coded aperture is smaller in reality than

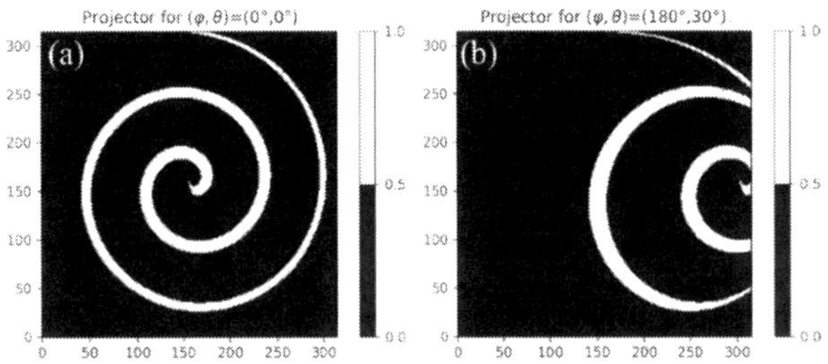

Fig. 2. Mathematical projections of the spiral pattern onto the detector plane for two different positions of a photon source in the sky: (a) observable source is in front of the camera, (b) source is shifted on the zenith angle of 30°.

in the model. The model also does not consider the attenuation or scattering effects in the projectors. The projection datasets generated by the code are later used for reconstruction and detector response simulation.

Reconstructing radioactive sources involves solving an inverse problem, where the goal is to decode the original image from the encoded image represented by a detector response. For this purpose, we use the Maximum Likelihood Expectation Maximization (MLEM) algorithm, which conveniently relies on projection datasets independent from the mask pattern geometry. MLEM is an iterative algorithm initially developed for tomographic reconstruction [19], which iteratively maximizes the probability $p(D|S)$ of observing detector response D provided the reconstructed image S. For each iteration n, the update formula for the optimal image $\hat{S}$ is given by (3).

$$\hat{S}_j^{n+1} = \hat{S}_j^n \sum_i \frac{p(i|j)D_i}{\sum_{j'} p(i|j')\hat{S}_{j'}^n} \qquad (3)$$

Here, $p(i|j)$ is the probability of detecting a photon emitted from sky pixel j in the detector pixel i. With each iteration of MLEM, the resulting image quality increases at the cost of the computational time, which may be impractical when real-time radioactive source reconstruction is required.

V. Simulations with the 3D Coded Mask

Now we can simulate the Timepix3 detector response using the previously generated projection dataset. In our model, we define the photon source position in the sky using the horizontal coordinate system (θ, φ), where θ is the zenith angle, and φ is the azimuth angle. Choosing the zenith angle over the polar allows us to define the zenith point at $(\theta, \varphi) = (0°, 0°)$. Photon counts are simulated by filling the projections with Poisson distribution $\text{Pois}(\lambda)$, where λ is a real number that corresponds to the number of counts N in a specific time interval. In the model, N also depends on the source zenith angle and follows a cosine-squared low $N = \cos^2(\vartheta)$. The number of counts is maximum when the photon source is in the camera zenith.

Fig. 4 shows the simulated Timepix3 responses alongside the corresponding MLEM reconstruction of a far-field point source located at two positions: $(\varphi_1, \theta_1) = (180°, 15°)$, and $(\varphi_2, \theta_2) = (180°, 45°)$. The number of registered counts is

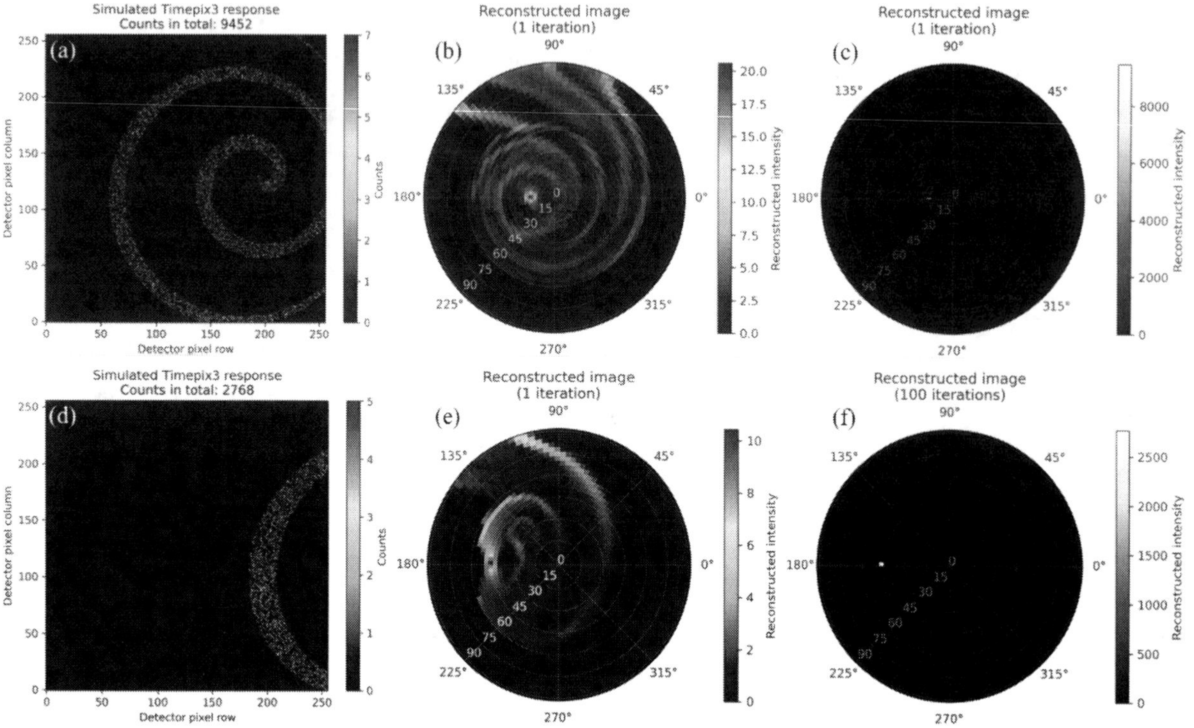

Fig. 4. Simulated Timepix3 responses (a),(d) alongside with the reconstructed images obtained by MLEM after 1 iteration (b),(e) and 100 iterations (c),(f). The top row corresponds to $(\theta_1, \varphi_1) = (180°, 15°)$, and the bottom row to $(\theta_2, \varphi_2) = (180°, 45°)$. The reconstructed images are discretized with $120° \times 30°$ positions, where each pixel corresponds to the angle of $3°$.

9452 and 2768 for the camera at $(\theta_1, \varphi_1) = (180°, 15°)$ and $(\theta_2, \varphi_2) = (180°, 45°)$, respectively. MLEM localized the observable source at its true position for both simulations.

Fig. 4e shows that the reconstructed noise becomes more substantial with the observable source moving further from the zenith along the polar direction. With no additional simulated background, this noise is intrinsic to our 3D coded aperture imaging system due to the uneven effective open fraction of the mask. Indeed, when the detector response is represented only by the arc-shaped part of the spiral pattern, as in Fig. 4d, the MLEM algorithm attributes it to more than a single projector and generates noise. If the observable source is located at the zenith angles $\theta > 30°$, we can improve the imaging quality by performing more MLEM iterations to reach an almost noise-free image (Fig. 4f) or by rotating the mask to a more favorable position featuring more explicit spiral shadow (Fig. 4a).

VI. Measurements with the 3D coded mask

We conducted a series of measurements with the Timepix3 camera equipped with the 3D coded mask to compare the acquired data with the simulations. In these measurements, we observed a single ^{241}Am source with an activity of 74 MBq, which was placed one meter away from the camera at two positions: $(\theta_1, \varphi_1) = (180°, 15°)$ and $(\theta_2, \varphi_2) = (180°, 45°)$. Such a setup is similar to the simulated examples in Fig 4. The acquisition time is 100 seconds at one frame per second. The number of registered counts is 63293 and 41162 for the observable source located in $(\theta_1, \varphi_1) = (180°, 15°)$ and $(\theta_2, \varphi_2) = (180°, 45°)$, respectively.

Fig. 5 shows the Timepix3 responses and the associated MLEM reconstruction images after 1 and 100 iterations. The algorithm localized the ^{241}Am source at its actual position for both measurements.

We observe that the reconstructed source is elongated in Fig. 5e, roughly three pixels wide in azimuth. While the setup alignment is supposed to be perfect in the measurement, a slight shift of the radioactive source affects the reconstruction quality. In addition, neglecting the vignetting effect in the projector database might also contribute to this uncertainty and will be studied further in future works. Thus, the observed inequality between the angular resolutions justifies the need to attribute two angular resolutions to our mask, azimuthal and polar. These parameters also depend on the observable source position and the reconstruction algorithm. According to these data, the angular width of the point spread function after 100 MLEM iterations is approximately 3° in the zenith angle as expected, and 9° in the azimuth angle.

VII. Conclusions

A hemispherical coded mask with a spiral-shaped hole was developed and characterized with a single Timepix3 chip. We aimed to demonstrate how the traditional limitations of standard

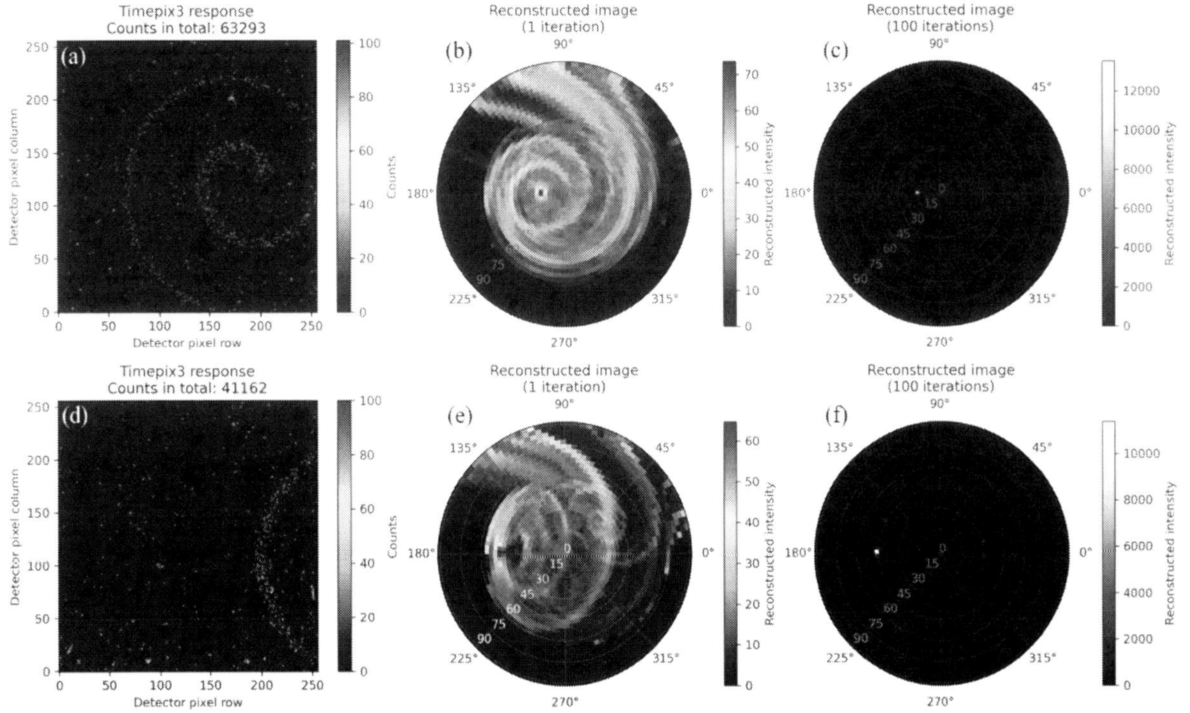

Fig. 5. Measured Timepix3 responses (a),(d) obtained after 100 seconds of data acquisition with ^{241}Am (74 MBq) alongside with the reconstructed images obtained by MLEM after 1 iteration (b),(e) and 100 iterations (c),(f). The top row corresponds to $(\theta_1, \varphi_1) = (180°, 15°)$, and the bottom row to $(\theta_2, \varphi_2) = (180°, 45°)$.The reconstructed images are discretized with 120° × 30° positions, where each pixel corresponds to the angle of 3°. Finally, we apply an upper threshold of 100 counts to (a),(d) to cover the noise.

X/gamma-ray cameras can be addressed using a non-planar coded mask with an unambiguous pattern.

Our imaging system has demonstrated accurate localization capability within the effective field of view up to 90° × 90° according to the analytical simulations with a far-field point source and measurements with ^{241}Am with the activity of 74 MBq. Such camera holds potential for use in the nuclear industry and high-energy astrophysics applications, particularly when the radioactive source location is unknown.

We have effectively implemented the MLEM algorithm to reconstruct the images of radioactive sources. The zenith angle increase results in more prominent artifacts and increased azimuthal angular resolution, thus affecting the source localization precision. A higher number of MLEM iterations or an additional measurement with the 3D coded mask rotation can partially solve this issue. The uneven effective open fraction of the one-arm spiral pattern also affects the quality of reconstruction. It signifies the need of further aperture pattern optimization.

Future work will focus on the next 3D coded mask design and development, which features an enhanced coded pattern and geometrical characteristics aiming to overcome the imperfections of its predecessor. We plan to optimize the projector generation and reconstruction codes as we move forward. Finally, we anticipate conducting simulations and measurements with several radioactive sources of different activities.

References

[1] F. Carrel *et al.*, "GAMPIX: A new gamma imaging system for radiological safety and Homeland Security Purposes," in *2011 IEEE Nuclear Science Symposium Conference Record*, Valencia, Spain: IEEE, Oct. 2011, pp. 4739–4744. doi: 10.1109/NSSMIC.2011.6154706.

[2] G. Montemont *et al.*, "NuVISION: a Portable Multimode Gamma Camera based on HiSPECT Imaging Module," in *2017 IEEE Nuclear Science Symposium and Medical Imaging Conference (NSS/MIC)*, Atlanta, GA: IEEE, Oct. 2017, pp. 1–3. doi: 10.1109/NSSMIC.2017.8532713.

[3] F. Lebrun *et al.*, "ISGRI: The INTEGRAL Soft Gamma-Ray Imager," *Astron. Astrophys.*, vol. 411, no. 1, pp. L141–L148, Nov. 2003, doi: 10.1051/0004-6361:20031367.

[4] O. Godet *et al.*, "The X-/Gamma-ray camera ECLAIRs for the Gammay-ray burst mission SVOM," 2014, doi: 10.48550/ARXIV.1406.7759.

[5] C. G. Wahl *et al.*, "The Polaris-H imaging spectrometer," *Nucl. Instrum. Methods Phys. Res. Sect. Accel. Spectrometers Detect. Assoc. Equip.*, vol. 784, pp. 377–381, Jun. 2015, doi: 10.1016/j.nima.2014.12.110.

[6] R. Accorsi, "Design of a near-field coded aperture cameras for high-resolution medical and industrial gamma-ray imaging," Thesis, Massachusetts Institute of

Technology, 2001. Accessed: Dec. 16, 2022. [Online]. Available: https://dspace.mit.edu/handle/1721.1/8684

[7] E. E. Fenimore and T. M. Cannon, "Coded aperture imaging with uniformly redundant arrays," *Appl. Opt.*, vol. 17, no. 3, p. 337, Feb. 1978, doi: 10.1364/AO.17.000337.

[8] S. R. Gottesman and E. E. Fenimore, "New family of binary arrays for coded aperture imaging," *Appl. Opt.*, vol. 28, no. 20, p. 4344, Oct. 1989, doi: 10.1364/AO.28.004344.

[9] R. H. Dicke, "Scatter-Hole Cameras for X-Rays and Gamma Rays," *Astrophys. J.*, vol. 153, p. L101, Aug. 1968, doi: 10.1086/180230.

[10] J. G. Ables, "Fourier Transform Photography: A New Method for X-Ray Astronomy," *Publ. Astron. Soc. Aust.*, vol. 1, no. 4, pp. 172–173, Dec. 1968, doi: 10.1017/S1323358000011292.

[11] P. Ubertini *et al.*, "IBIS: The Imager on-board INTEGRAL," *Astron. Astrophys.*, vol. 411, no. 1, pp. L131–L139, Nov. 2003, doi: 10.1051/0004-6361:20031224.

[12] K. Amgarou *et al.*, "A comprehensive experimental characterization of the iPIX gamma imager," *J. Instrum.*, vol. 11, no. 08, pp. P08012–P08012, Aug. 2016, doi: 10.1088/1748-0221/11/08/P08012.

[13] V. Schoepff *et al.*, "Nanopix, a smart and highly miniaturized gamma camera for robotic radiological characterization," presented at the SFEN DEM 2021 – International Conference on Decommissioning Challenges: Industrial Reality, Lessons learned and Prospects, Avignon, France, Jun. 2021.

[14] R. Le Breton *et al.*, "The Spid-X gamma camera: A miniature gamma ray integral field spectrometer for nuclear industry applications," *Nucl. Instrum. Methods Phys. Res. Sect. Accel. Spectrometers Detect. Assoc. Equip.*, vol. 1047, p. 167722, Feb. 2023, doi: 10.1016/j.nima.2022.167722.

[15] J. E. Grindlay and J. Hong, "Optimizing wide-field coded aperture imaging: radial mask holes and scanning," presented at the Optical Science and Technology, SPIE's 48th Annual Meeting, O. Citterio and S. L. O'Dell, Eds., San Diego, California, USA, Jan. 2004, p. 402. doi: 10.1117/12.506260.

[16] J. G. M. FitzGerald, "A Rotating Scatter Mask for Inexpensive Gamma-Ray Imaging in Orphan Source Search: Simulation Results," *IEEE Trans. Nucl. Sci.*, vol. 62, no. 1, pp. 340–348, Feb. 2015, doi: 10.1109/TNS.2014.2379332.

[17] D. Hellfeld, P. Barton, D. Gunter, L. Mihailescu, and K. Vetter, "A Spherical Active Coded Aperture for $4\pi $ Gamma-Ray Imaging," *IEEE Trans. Nucl. Sci.*, vol. 64, no. 11, pp. 2837–2842, Nov. 2017, doi: 10.1109/TNS.2017.2755982.

[18] T. Poikela *et al.*, "Digital column readout architectures for hybrid pixel detector readout chips," *J. Instrum.*, vol. 9, no. 01, pp. C01007–C01007, Jan. 2014, doi: 10.1088/1748-0221/9/01/C01007.

[19] L. A. Shepp and Y. Vardi, "Maximum Likelihood Reconstruction for Emission Tomography," *IEEE Trans. Med. Imaging*, vol. 1, no. 2, pp. 113–122, Oct. 1982, doi: 10.1109/TMI.1982.4307558.

Pure Beta Emitters in Water Samples from Neutron Irradiation Facilities: Activity Determination Via Vacuum Distillation and Liquid Scintillation Counting

D. Masci[1,*], D. Ferri [1], G. Gandolfo[1], L. Lepore[1], G.M. Contessa[2], A. Pietropaolo[1], M. Pillon[1], G.A. Marzo[1]
[1] ENEA, Department of Fusion and Technology for Nuclear Safety and Security, Rome, Italy
[2] Italian National Institute of Health, Rome, Italy
(*) domenica.masci@enea.it

Abstract- **In the decommissioning of nuclear facilities, it is necessary to assess the radioactivity inventory for waste classification and management. For nuclear fusion experiments or accelerator-driven neutron sources, the main contribution to the radiological inventory comes from activation products and tritium. When water is used as coolant of components, it may be activated by neutrons with the consequent production of tritium, and short-lived oxygen and nitrogen isotopes. In most cases several chemical elements may be in solution or suspension in water, because of metallic pipes corrosion when occurring, or functional reasons, e.g. chemical conditioning, addition of antifreeze agent, etc. In this case other activation products could be present and need to be characterized. In this case study, a sample of the coolant of the ENEA Frascati Neutron Generator (FNG) has been radiologically, chemically and physically characterized. In such a facility, during the operation, a beam of deuterons is accelerated up to 300 keV and impacts on a tritiated target for producing 14 MeV neutrons exploiting the fusion reaction T(d,n)α. To avoid overheating in the area around the target, a water-cooling loop is used. After several years of operation, a sample of the fluid has been extracted to undergo a qualitative and quantitative analysis of the radionuclides inventory. Gamma spectrometry has not shown any presence of gamma emitters. On the other hand, Liquid Scintillation Counting (LSC) has provided evidence of the presence of beta emitters. It is known that tritium is expected to occur in the sample because of releases from the tritiated titanium target. This work concerns the development of a method to discriminate the tritium contribution with respect to the total beta activity within the FNG coolant. Since it is known that an unspecified percentage of an unknown anticorrosive agent was added to the water coolant, vacuum distillation has been used to extract purified water (containing only tritium) from the coolant sample leaving any (activated) non-volatile compound in the undistilled fraction. After distillation, any fraction has been investigated by Raman spectroscopy, to determine: a) the chemical composition of the colored anticorrosive agent, b) the efficiency of the separation. Afterwards, both fractions have been analysed by means of LSC, and results showed that the main contribution to sample activity is due to tritium and the activity values are consistent with the theoretical H atomic concentration in both fractions.**

Keywords — **Liquid Scintillation Counting, Tritium, distillation, neutron generator**

I. Introduction

During the decommissioning of a nuclear facility, it is necessary to determine the activity of various radionuclides to assess the total radiological inventory. In nuclear fusion experiments with accelerator-driven neutron sources, the main contribution to the inventory comes from activation products and tritium [1]. When water is used as a coolant for components, it could be activated by neutrons, resulting in the production of tritium and short-lived isotopes of oxygen and nitrogen, or contaminated by gaseous tritium that has a high capacity to permeate through metals. Additionally, various chemical elements may be present in solution or suspension in water due to potential corrosion of metallic pipes or for functional reasons, such as chemical conditioning or the addition of refrigerant agents. It is crucial to find a simple and convenient way to assess the contribution of tritium with respect to any other radionuclides when advanced technologies and equipment are not available.
Among the various techniques used for tritium separation from aqueous solutions, the easiest and most inexpensive method is the vacuum distillation [2]. In this study, a sample of water from the ENEA Frascati Neutron Generator (FNG) cooling system was analyzed. Specifically, this work illustrates a method for discriminating the contribution of tritium from the total beta activity of the FNG cooling water.

II. Materials and methods

FNG target consists of a 50 mm diameter copper plate with a titanium substrate on which tritium is adsorbed. During operations, a beam of deuterons impacts on the tritiated surface to produce 14 MeV neutrons through the fusion reaction T(d,n)α. To prevent overheating of the target, a 150-liter water cooling loop is used. A simplified scheme of the circuit is shown in Fig.1. An unspecified percentage of a colored anticorrosive agent was added to the cooling water. The formulation of this additive was initially unknown. After 30 years of operation without changing or refilling the water, a sample was collected and characterized through qualitative and quantitative analyses of the chemical content and radionuclide inventory.

Before any treatment, a preliminary non-destructive analysis through low-background gamma spectrometry on the original sample was performed to quantify, if present, gamma emitter radionuclides.
The sample was then analyzed to determine the composition of the colored anticorrosive agent, considering parameters such as alkalinity, viscosity, refrigeration power, and water miscibility. These characteristics led to consider the presence of diols in the sample. To confirm this hypothesis, the sample was initially subjected to a vacuum distillation to obtain a complete separation of the sample into water and colored anticorrosive agent residue. The determination of the density of the residual sample, compared with the mixture before distillation, allowed to characterize the colored anticorrosive agent and its percentage in the aqueous mixture. Raman analysis was performed to determine the chemical composition of the anticorrosive agent.
LSC analysis on both fractions, distilled water and residue, was then carried out to assess the radioactivity distribution after distillation.

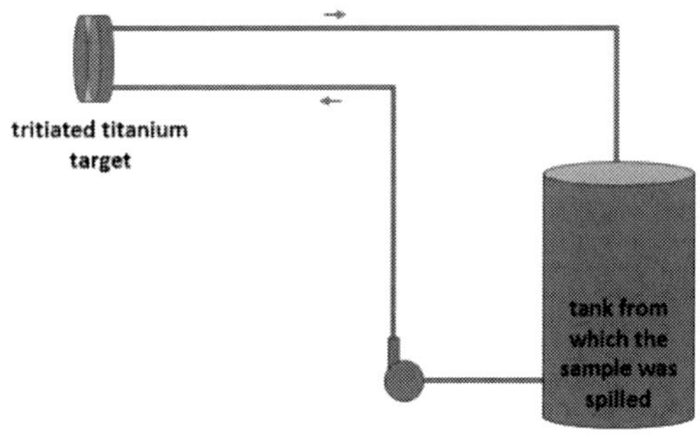

Fig. 1. *Simplified scheme of FNG cooling circuit.*

To investigate the nature of the anticorrosive agent used in the FNG cooling circuit and the quantity added to the water to obtain the coolant solution, the following steps were followed. A sample of the coolant solution was taken and weighed using an analytical balance. Additionally, the volume of the sample was measured with a calibrated pipette. Twenty measurements were performed to obtain a statistical average. The sample was then subjected to distillation using the distillation apparatus shown in Fig.2, where a refrigerant cooled the water to 4 °C through a chiller. During distillation, the system was connected to a vacuum pump.

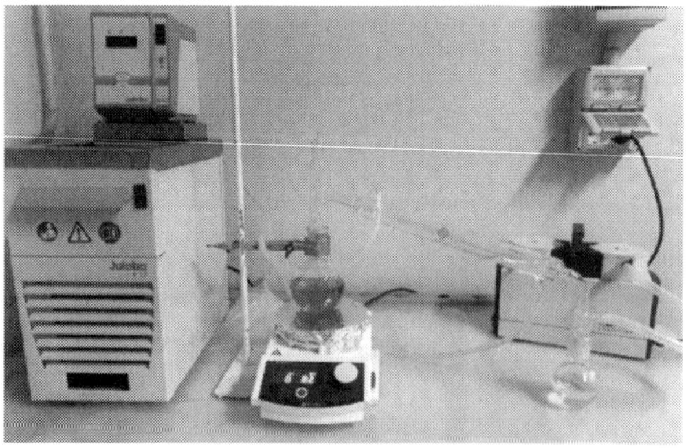

Fig. 2. Distillation apparatus.

After distillation, both the distilled sample and the residue (anticorrosive agent) were weighed using the analytical balance. The volume was measured again using calibrated pipette. Again, twenty samples were taken, and a statistical average was calculated.
To evaluate the nature of the anticorrosive agent, both the distilled fraction and the residue were analyzed using a Raman spectrometer with a laser source at 535 nm coupled with a fiber optic probe. To quantify the beta activity concentration in the sample, both before and after distillation, a Hidex 300SL LS counter was used. A volume of 1 ml of each sample was diluted in 4 ml of distilled water, mixed with Ultima Gold AB scintillation cocktail in a 20 ml glass vial, and then counted for 3600 s. To evaluate any possible interference caused by the blue coloration of the sample in LSC measurements (color quenching), a preliminary study on the original sample was performed, using hydrogen peroxide as bleaching agent [ref] and carrying out LSC analysis on bleached and unbleached samples.

III. Results and discussion

A. *Density evaluation*

In Table I results from density evaluation are reported. The density of the original sample (1.063 g/ml) is consistent with the hypothesis that ethylene glycol was the anticorrosive agent, with a concentration of 40% by weight.

TABLE I
DENSITY EVALUATION

		Density (g/ml)
Before distillation	Sample (water + anticorrosive agent)	1.063 ± 0.007
After distillation	Transparent distilled fraction (water)	1.003 ± 0.007
	Blue undistilled fraction (anticorrosive agent)	1.108 ± 0.003

B. *Raman analysis*

Raman spectra were performed on the different fractions of the distillation: the distilled fraction and the residue. We can assert that the distillation was complete, and that the anticorrosive agent was definitely ethylene glycol. In fact, from the blue line in Fig. 3 the typical Raman spectrum of ethylene glycol (residue of the distillation) is observed while the red line highlights the typical Raman spectrum of water (distilled fraction). In Fig. 4-5 the spectra of pure substances ethylene glycol and distilled water are shown as reference. Therefore, the Raman analysis combined with density data confirms that the anticorrosive agent was ethylene glycol.

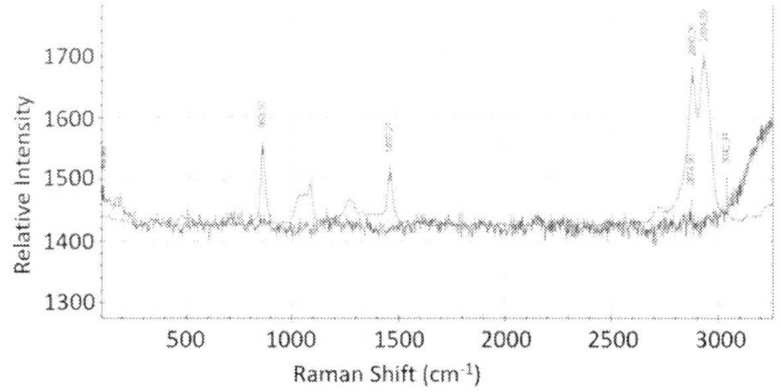

Fig. 3. Raman spectra: A) the blue line is undistilled fraction (anticorrosive agent), B) the red line is the water fraction.

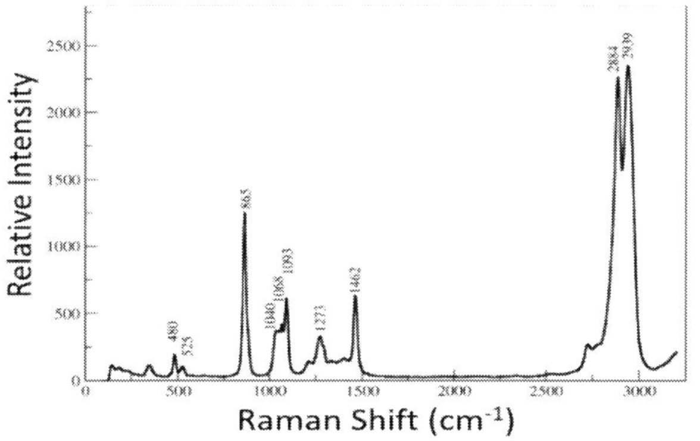

Fig. 4. Raman spectrum of pure ethylene glycol [3]

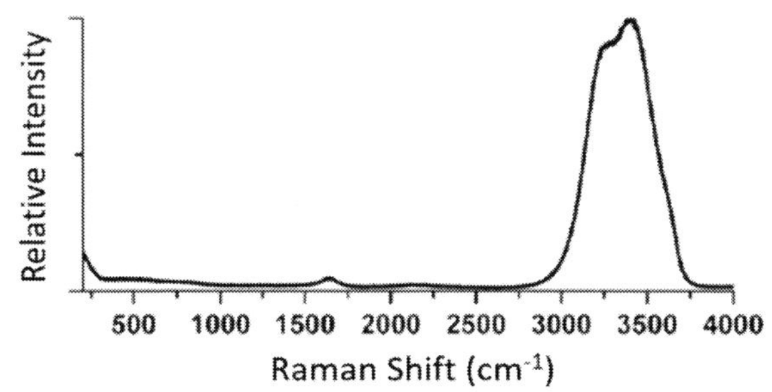

Fig. 5. Raman spectrum of distilled water [4]

C. *Activity evaluation through LSC*

Before carrying out the LSC analysis, low-background gamma spectrometry was performed, and no gamma emitters were detected above the minimum detectable activity (about 0.02 Bq/ml).

Fig.6 shows LSC spectra from HIDEX 300SL obtained in the preliminary study on the influence of the blue coloring of the sample before and after the bleaching with hydrogen peroxide. These graphs and activity evaluation are consistent within the uncertainty range, as shown in Table II, indicating that the blue coloration of the sample does not affect liquid scintillation results. In fact, the absorption of light photons by the blue solution occurs in the range from 490 nm to 800 nm, as shown by further analyses through UV-Vis spectrophotometer (Fig.7). The region of greatest interest for LSC is between 300 nm and 400 nm, and it does not overlap with the sample absorption region.

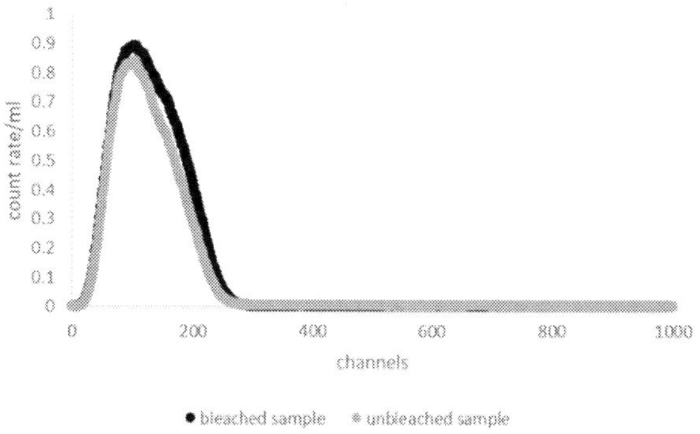

Fig. 6. LSC spectra for bleached and unbleached sample

TABLE II
LSC RESULTS FOR BLEACHED AND UNBLEACHED SAMPLE

Sample	Activity concentration (Bq/ml)
Unbleached sample	334 ± 53
Bleached sample	336 ± 51

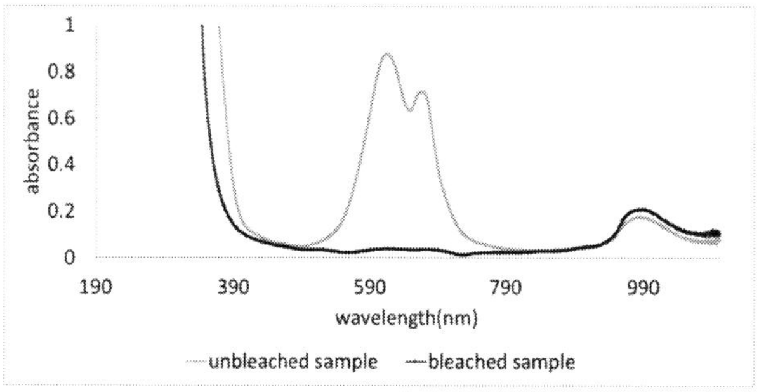

Fig. 7. UV-Vis spectra for bleached and unbleached (colored) sample

In Table III results from LSC analysis before and after distillation are reported.

TABLE III
LSC RESULTS FOR BLEACHED AND UNBLEACHED SAMPLE

Sample	Activity concentration (Bq/ml)	Theoretical H atomic concentration
Original sample	334 ± 53	64.1%*
Transparent distilled fraction (water)	330 ± 50	66.7%
Blue distilled fraction (anticorrosive agent)	315 ± 47	60.0%

*evaluated considering an ethylene glycol concentration of 40% w/w

LSC outcomes suggest that the measured activity in the original sample is mainly due to tritium. The activity concentration of the distilled fraction (purified water) containing only tritium is in fact consistent with the theoretical H atomic concentration in the undistilled fraction and in the original sample.

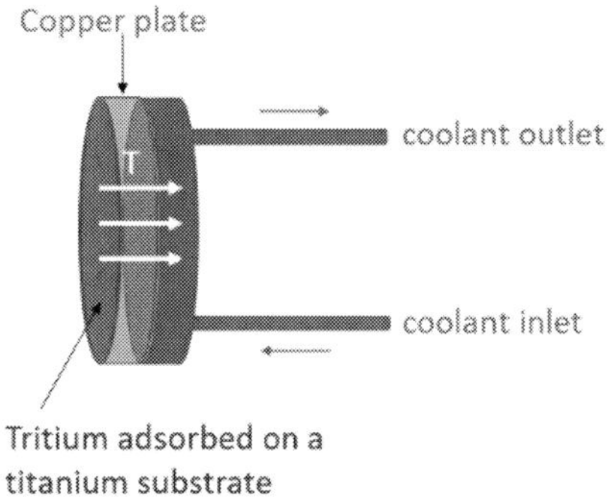

Fig. 8. Tritium permeation into FNG titanium substrate and copper plate

Fig. 8 illustrates the permeation of tritium into the FNG target. Subsequently, hydrogen-tritium exchange occurs in water and ethylene glycol molecules, providing a possible explanation for the observed phenomenon. The significant values in activity concentrations are likely derived from this diffusion process rather than solely from neutron activation. In fact, characteristic neutron fluxes, irradiation times and materials involved in the facility [5] does not justify the production of such amount of tritium through neutron activation processes. As tritium is the radioactive isotope of hydrogen its chemical properties are identical to those of hydrogen. Whether of natural or man-made origin, tritium is extremely mobile in the environment and in all biological systems. It exists in different chemical forms, the most of one are tritiated water (HTO) and gaseous tritium (HT). Moreover, tritium as hydrogen can also migrate through metals [6,7]. For these reasons we suppose that the tritium migrates from the metallic target, reaches the circulating cooling liquid and finally replaces the hydrogen in the water molecules and in the OH functional groups of ethylene glycol, creating exchangeable organically bound tritium (OBT). Tritium can in fact replace hydrogen in organic compounds, forming exchangeable OBT if the bonds involve sulphur, phosphorus, nitrogen, or oxygen, as in ethylene glycol, and non-exchangeable OBT if the bonding is with carbon. In exchange reactions, tritium bonds to oxygen, sulphur, phosphorous, or nitrogen atoms as hydroxides, thiols, phosphides, or amines, respectively [8].

IV. CONCLUSIONS

A rapid method to discriminate tritium content with respect to the total beta activity in water mixture, used as coolant of components operating in neutron fields, was established, by using a simple distillation apparatus. Several analyses were performed to assess the radionuclide content of the sample taken from the coolant of the Frascati Neutron Generator.
First result seems to confirm that the main contribution is due to tritium. Since no complete radiochemical separation has been carried out and beta energies cover a broad region, it is not possible to discriminate among contributions from different beta emitters. For these reasons the presence of traces of other beta emitting radionuclides cannot be completely excluded.

REFERENCES

[1] Sandri, S.; Contessa, G. M.; Guardati, M.; Mariano, G.; Villari, R., "The radiological source terms in a nuclear fusion experimental facility", Proceedings of the 15th International Congress of the International Radiation Protection Association, January, 2021.
[2] Hou, Xiaolin, "Radiochemical analysis of radionuclides difficult to measure for waste characterization in decommissioning of nuclear facilities" (2007), 273 (1), pp. 43 – 48, DOI: 10.1007/s10967-007-0708-x
[3] Burikov, S., Dolenko, T., Patsaeva, S., Starokurov, Y., Yuzhakov, V. "Raman and IR spectroscopy research on hydrogen bonding in water–ethanol systems." Molecular Physics. 108. 2427-2436, 2010. DOI 10.1080/00268976.2010.516277.
[4] K Krishnan, RS Krishnan - Proceedings of the indian academy of sciences, 64, 111–122 (1966) Springer Verlag
[5] M. Martone, M. Angelone, M. Pillon, The 14 MeV Frascati neutron generator, J. Nucl. Mat. 212-215, 1661 (1994); M. Angelone et al., Rev. Sci. Instr. 67, 2189 (1996)
[6] T. Tanabe et al., "Hydrogen ion driven permeation trough metals", Journal of Nuclear Materials, vol.145-147, pag.305-308, (1987)
[7] T Otsuka, T Hoshihira and T Tanabe, "Visualization of hydrogen depth profile by means of tritium imaging plate technique:determination of hydrogen diffusion coefficient in pure tungsten" Phys. Scr.vol.2009, T138 , 014052 (4pp), (2009)
[8] S.B. Kim, V. Korolevych, "Quantification of exchangeable and non-exchangeable organically bound tritium (OBT) in vegetation", Journal of Environmental Radioactivity 118, 9-14 (2013), https://doi.org/10.1016/j.jenvrad.2012.11.006

Seeing radioactivity: gamma-ray imaging technique applied to TRIGA RC-1 Research Reactor in ENEA Casaccia

Filippo Gagliardi[1,2,*], Alessandro Ciotoli[1], Edoardo Gorello[1], Luigi Meschini[1], Marco Pagliuca[1], Egidio Mauro[1], Nadia Cherubini[1,3], Luigi Lepore[3], Giada Gandolfo[3], Giuseppe A. Marzo[3], Luca Falconi[3], Davide Formenton[3], Andrea Roberti[3],
[1]Nucleco spa, via Anguillarese 301, Rome, Italy
[2]La Sapienza – University of Rome, Italy
[3]ENEA, Italian National Agency for New Technologies, Energy and Sustainable Economic Development, Casaccia Research Center, via Anguillarese 301, Rome, Italy
(*) gagliardi@nucleco.it

***Abstract*—Current radiological characterization protocols applied in nuclear installations provide a valuable set of information used to assess level and distribution of radioactivity, which is a crucial information in the plant management. Nevertheless, typical routines may present limitations that novel technologies can alleviate. In this work, a new-generation 3D pixelated CdZnTe gamma imaging and spectrometer detector has been used at TRIGA RC-1 Research Reactor in ENEA Casaccia Research Centre to assess its performances in several possible tasks and effectively understand this technology readiness. In particular, three possible tasks have been considered: survey of potentially contaminated area (with determination of known and unknown hot spots), source identification and localization (including neutron sources via their gamma emissions), and quantification of isotope-specific activities in generic components thanks to the point-kernel-based algorithm the system is equipped with. The experience has demonstrated the capability to achieve reliable results in these tasks, with consequences that could impact not only the current characterization protocols but also the way some Nuclear Safeguards and Non-Proliferation operations are performed.**

***Keywords* — Gamma imaging, TRIGA research reactor, radiological survey, neutron sources, quantitative gamma spectrometry, nuclear decommissioning, waste management.**

I. Introduction

In the perspective of proper management of any nuclear site, a comprehensive understanding of the radiological status of buildings and surrounding areas, from identification to localization of radioactive sources and contamination spots, is critical and necessary. Actions that may seem simple, such as planning and executing routine operations, maintenance, and inspection activities, can become complex without full knowledge of the radiological scenario. In line with the ALARA (As Low As Reasonably Achievable) principle [1], having a general overview of the occurring radioactivity can improve the implementation of action plans in terms of correct procedures, precautions, or restrictions for operators, as long as helping to define specific procedures for managing radioactive sources, and to ensure appropriate radiation exposure of personnel and radiation protection of the environment according to international standards.

The radiological state of a nuclear installation is typically ensured through characterization protocols involving H*(10) dose-rate measurements, total superficial contamination measurements and smear tests, on areas and components with the higher expected risk to be contaminated or activated, possibly complemented by other non-destructive assays (mainly, gamma spectrometry) and destructive analyses for determination of Difficult-To-Measure (DTM) Radionuclides [2]. Such protocols have two main limitations. Firstly, they cannot be applied to every single component or area of the installation (it would require far too much time, money, and personnel efforts), but they are just limited to the main parts of components and areas known/expected to be radioactive, i.e. some radioactivity spots may remain unidentified (e.g. an unknown leakage of radioactive liquid). Secondly, they provide data that "remain on paper" and are even prone to arise difficulties in taking them into account: a nuclear installation is a complex and spatially wide environment, where, for example, the irradiation due to a high-radioactivity hot spot may influence operations done in far area, for which that hot spot has not been taken into account. Another example is the cutting of a pipe, where eventual hot spots should be ensured to not be cut. In both cases, radiological data can be available, but their accounting is entrusted completely to the technicians planning the task to be perform, which may commit a trivial clerical error and neglect them.

These aspects are in contrast with the indication of the state-of-art in plant management [3], whose requirement is to integrate as many details as possible in a unique digital environment where all data are easy to be accessed, understood and taken into account. Such environment, for example, would permit to define the approach to be followed

https://doi.org/10.1051/epjconf/202328807013

in a specific task (e.g. cutting a pipe or move a high-activity source) in terms of dose conferred to operators and potential release of radioactivity. Finally, radioactivity distribution in waste can affect the reliability of other quantitative measurement techniques, thus its knowledge is crucial for a proper waste route definition [4].

In this perspective, nuclear industry is seeking novel characterization systems exploiting modern technologies to improve the level of details current characterization procedures can provide. System equipped with imaging capabilities, for example, can provide results in a data format naturally easy to be understood and integrated in digital models. Moreover, modern electronics and algorithms allow for measuring in a large (or even in the entire) field of view, ensuring the possibility to detect radioactivity located in unexpected areas or components.

A new-generation gamma-ray imaging and spectrometric system has been tested in the real-world case of a TRIGA research reactor to assess its performance in a variety of tasks and effectively understand this technology readiness, for potentially taking part in standard characterization tasks and routines.

II. Experimental Equipment and Methods

The measurements were conducted at the TRIGA RC-1 research reactor in the ENEA Casaccia Research Centre (Rome). It is a pool-type, natural-convection cooled, 1 MW_{th} maximum power reactor, operated for running irradiations and experiences using thermal neutron spectra and fluxes in several experimental channels available. An H3D, Inc. H420 3D-pixelated CdZnTe gamma-ray imaging and spectrometer was used for acquisition and analysis [5]. It has roughly 20 cm^3 of CdZnTe crystal with better than 1.1% full-width-at-half-maximum energy resolution at 662 keV (see Fig. 1). Images of moderate/high- and low-energy photons were generated using Compton imaging and coded aperture imaging (CAI) respectively. The H420 has an angular resolution of $\approx 5°$ FWHM in coded-aperture field of view (CAI imaging) and $\approx 20°$ FWHM for all 4π (Compton imaging). Quantitative analysis of H420 gamma-ray spectra was conducted using SourceTerm tool in Visualizer software [6] SourceTerm estimates geometry-specific measurement efficiency for required energies, via a point-kernel-based calculation and measured detector efficiency, to convert photopeak count rates to absolute activities, with reliability close to the one ensured by the golden standard of detectors based upon High Purity Germanium (HPGe) crystals [7].

The combination of the CdZnTe crystal's good energy and angular resolution, isotope-specific imaging and quantification capability, and detector compact design, allowed to realize an accurate survey and characterization campaign of the most radiologically significant zones and components within TRIGA-RC1 facility. Measurements were conducted in areas where a more relevant radioactivity is to be expected. As a first step once entering in a new area, some preliminary acquisitions are taken to collect general information about the level of environmental background and radiological condition, allowing for proper planning of following detailed measurements. Then, based on the information obtained from these preliminary investigations, the focus shifted on individual components and more confined areas, physically bringing the detector closer to the objects suspected/detected to be radioactive. This approach aimed to optimize the time of each measurement and to locate the radioactivity more precisely within the components.

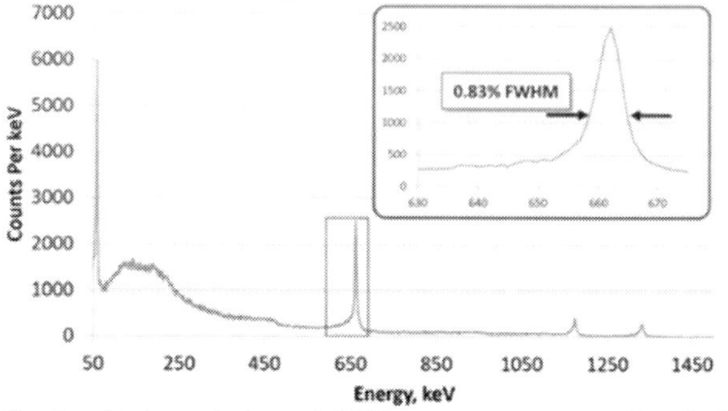

Fig. 1. Background-subtracted H420 gamma-ray spectra with mixed radionuclides. ^{137}Cs energy resolution was measured at 0.83% FWHM.

III. The TRIGA RC-1 Research Reactor

The TRIGA RC-1 at the ENEA Casaccia Research Centre is a thermal neutron spectrum, pool-type, water-cooled Research Reactor, with MARK III type containment, originally having a thermal power of 100 kW when built in 1960, increased to 1 MW based on an in-house developed ENEA design in 1967.

TRIGA RC-1 utilization has covered in the past, and covers currently, a wide range of fields:

- Neutron Activation Analysis (NAA).
- Neutron imaging.
- Forensic studies and food traceability.
- Material studies via neutron scattering and diffractometry.
- Stress-test for materials and electronic components for space applications.
- Radionuclide production, e.g., for industrial applications or precursor of radiopharmaceuticals such as of ^{99m}Tc ^{177}Lu, ^{161}Tb.
- Detector testing.
- Education and training.

Many experimental channels are available, i.e.:

- Thermal Column: used for validation of novel detectors.
- Tangential Piercing Channel: optimized for performing neutron radiography and tomography.
- Radial Channels: historically used for neutron diffraction experiments. Channels A, B and C are undergoing refurbishment before various new activities can occur.
- Lazy Susan: activities include NAA, tests for space applications, and food traceability mainly.
- Central Thimble: activities include precursor

radionuclides production for radiopharmaceuticals, NAA, tests for space applications, and food traceability.
- Rabbit Channel: used for NAA mainly.

All such activities involve the irradiation of samples, container capsules, supports and metal parts of the reactor and experimental setup both. Neutron activation is carefully evaluated case by case during the design phase of the experiment (to understand feasibility) and verified during and after the experience by direct measurements. Moreover, neutron activation of fluids (liquids, and air mainly) and dust could produce further contributions to the personnel exposure, external irradiation, and contamination either.

The rationale is that in the normal life of the TRIGA RC-1 reactor, more than half of the day is a rest-time that could be used to set gamma imaging inspections to manage events before, eventually.

IV. Results

The H3D H420 pixelated CdZnTe gamma camera was used at the ENEA TRIGA RC-1 Research Reactor for an experimental campaign from January 2023 to June 2023. More than 50 spectral acquisitions were carried out, with a duration spanning from few minutes up to almost 70 hours depending upon the level of occurring radioactivity and the desired sensitivity.

Measurements covered the 'Reactor Top' and related equipment, several views of the Reactor Hall ground floor, some views of the Pump Room in the basement, and direct measurements of a fresh fuel element (known quantity of low-enriched uranium mass) at the Fresh Fuel Storage Room.

According to the experience achieved along the whole campaign, three main goals have been demonstrated to be accomplished by means of modern gamma imaging systems:

A. Survey of potentially contaminated areas and identification of radioactive items.
B. Identification and localization of radioactive sources.
C. Quantification of gamma-emitting radionuclides in generic components.

These tasks will be addressed in detail in the following discussion.

A. Survey of potentially contaminated areas

Knowing the radiological state of a nuclear installation is typically ensured through characterization protocols involving H*(10) dose-rate measurements and smear test on areas and components with the higher expected risk to be contaminated or activated, eventually complemented by other non-destructive assays (mainly, gamma spectrometry) and destructive analyses for determination of DTMRs.

In this context, gamma imaging has been investigated as a supporting technique to help reactor operators in manage risks regarding activated materials, to check and "look for" unexpected activations or contaminations events especially once the experience is done. Gamma imaging could become a novel protocol in safety management to deepen and improve the knowledge of radiation fields, and to enhance radiation protection definitively.

Regarding this last aspect, the most interesting outcomes obtained are related to the items that were not expected to be radioactive but turned out to be, as reported for example in the case of a cabinet storing samples and materials from reactor experience over decades (see Fig. 2).

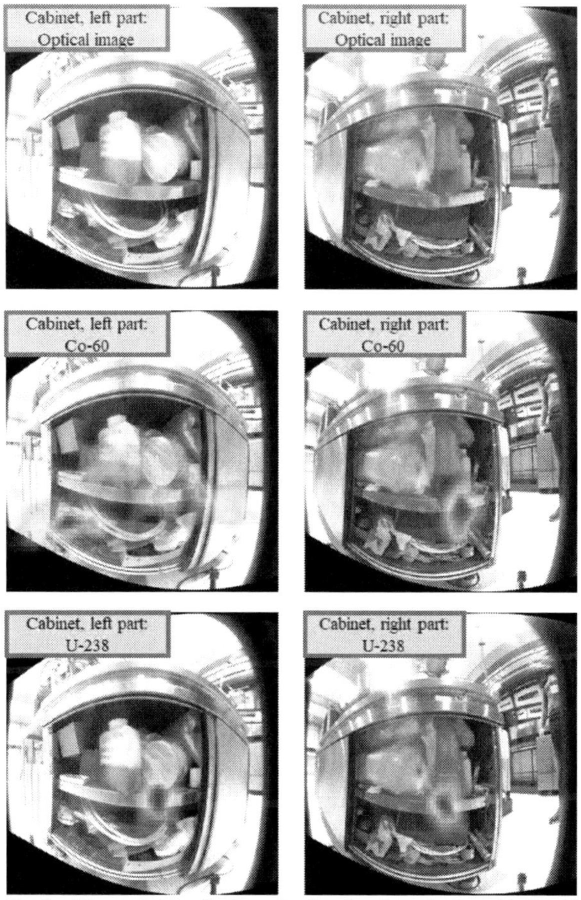

Fig. 2. Unexpected localization of radioactive items in the cabinet, with identification of ^{238}U and ^{60}Co (both reconstructed via Compton imaging).

An even more interesting result is obtained by looking carefully at the distribution of ^{60}Co in the measure of the left part of the cabinet: the tail in the distribution exceeding the physical cabinet space is, in fact, not due to a poor-statistics reconstruction, but to another contaminated component placed on the back of the detector, as clearly visible in Fig. 3: looking to the opposite direction with respect to the cabinet, a small canister containing a ^{60}Co contaminated liquid is dominating the reconstruction, although a spot corresponding to the contribution of left part of the cabinet is still present.

Fig. 3. Distribution of ^{60}Co obtained pointing the detector in the opposite direction with respect to Fig. 2. The spot in the dark region on the right is corresponding to the ^{60}Co reconstructed on the left part of the cabinet.

B. *Source identification and localization*

Within the TRIGA RC-1 Reactor Hall, radioactive materials generating radiation fields exceeding the ambient H*(10) dose rate are stored in a specific area called 'the Bunker'. This structure has concrete walls and no ceiling, allowing to move items in and out by means of the Reactor Hall crane. The Bunker contains activated materials (steel components, essentially) coming from reactor operations, and radioactive sources in their storage cans. Among those items, there are both known sources (present in the installation radiological inventory) and radioactive objects whose information is not available. Moving those items, as well as entering in the area, would imply a significant exposure for the operators by external irradiation, meaning that such operations have to be optimized as much as possible. This is possible only if the components generating the higher dose rate and the sources' locations are known.

In order to seek this information using the imaging system, several acquisitions have been taken from different sides and positions, and looking inside the Bunker from the top. Fig. 4 reports some of setups used and the main results. The presence of one ^{226}Ra and two ^{241}Am hot spots can be easily distinguished. Furthermore, a higher photon yield in the energy region from 1.7 MeV up to 3 MeV is detected (probably dominated by Compton continuum due to the 4.4 MeV delayed gammas emitted after Beryllium transmutation), as long as the 2.2 MeV photon characteristic of a (n, γ) reaction on hydrogen (deeply used for shielding of neutron sources, e.g., paraffin or polyethylene): both these ranges images are spatially matching the reconstruction of ^{226}Ra and ^{241}Am, thus strongly suggesting the fact the sources are actually RaBe and AmBe neutron sources, which is in agreement with the radiological inventory of the installation. Moreover, the reconstruction of 2.2 MeV line is actually bigger, suggesting the presence of more neutron shielding in the area (and so possibly another low-activity or highly shielded neutron source). Finally, ^{137}Cs is imaged as widely spread over components present in the Bunker, while ^{60}Co reconstruction is dominated by a single activated item (the photon yield is high enough to permit the reconstruction using the Coded Aperture Mask for such high energy).

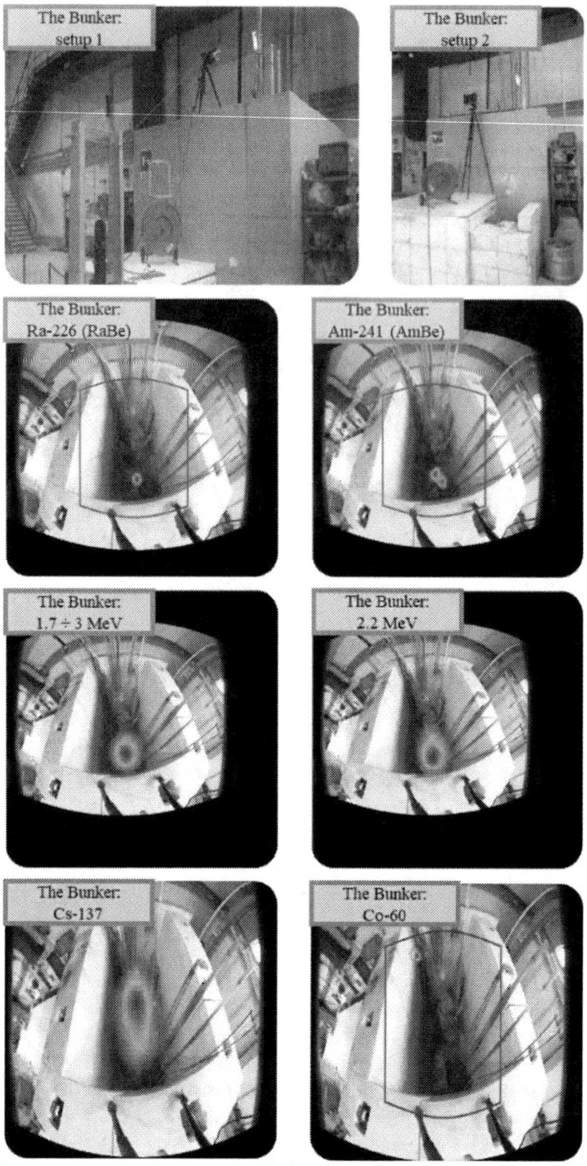

Fig. 4. Detected radionuclides inside the Bunker area.

The Bunker reserved other two "lessons learned" while pursuing the goal of surveying in its surrounding. The first was the capability to detect a small ^{60}Co signal (turned out to come from a source stored in a small shield) in a reconstruction dominated by the signal coming from activated item before (see Fig. 5).

The second was the detection of a steady signal coming from a large polyethylene shielding (measurement done in setup 2 of Fig. 4). Three clues suggest this signal is coming from a AmBe neutron source (see Fig. 6): the presence of the 2.2 MeV peak characteristic of (n, γ) reaction on hydrogen (i.e. a neutron source interacting in polyethylene) and a large count rate in the high-energy range (due to the continuum of 4.4 MeV delayed photons), both being reconstructed in the same position (corresponding to the polyethylene shield), and the absence of any other identified gamma peak (a significant ^{226}Ra activity would be easy detectable, thus the alpha emitter is expected to be ^{241}Am according to the installation inventory).

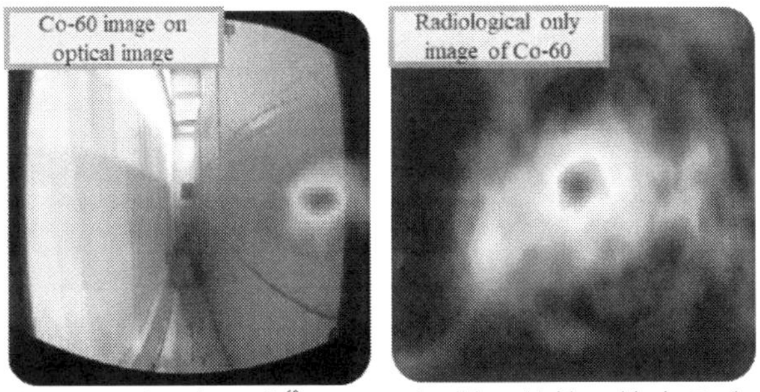

Fig. 5. Detection of a small ^{60}Co source placed in a shield outside the Bunker area as a tailing effect of the main reconstructed dominated by the activated components.

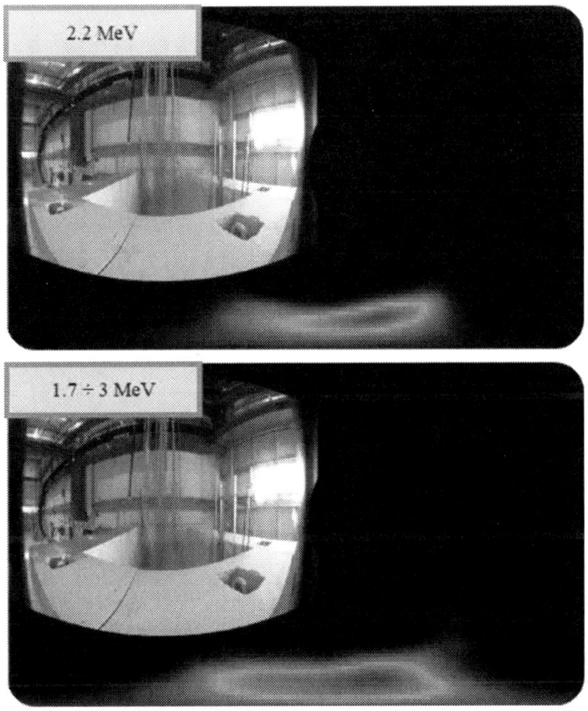

Fig. 6. Meaningful reconstructions due to a AmBe neutron source at the external side of the Bunker.

In the reactor building there is another shielded well, on the Reactor Top actually, used for storing radioactive sources and small hot samples. In particular, two spaces are filled with radioactive materials according to the installation historical data. As Fig. 7 shows, ^{60}Co and ^{137}Cs are identified: the first is clearly located in one pit, while ^{137}Cs is spread over a larger area suggesting (despite the spreading effect due to the scattering introduced by the shield) its presence in more than one pit

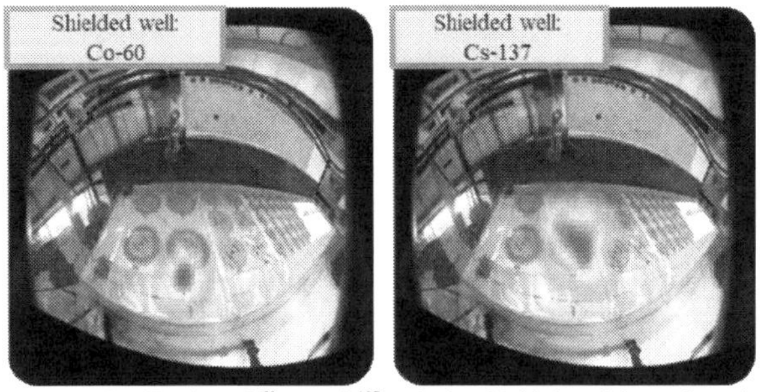

Fig. 7. Identification of ^{60}Co and ^{137}Cs in the shielded well located on the Reactor Top.

C. *Quantitative characterization of components*

As already mentioned, the gamma-ray imaging system is equipped with a post-processing tool capable of spectrum analysis and efficiency calculation by means of a point-kerned based algorithm. Its performance is close to the industrial standard available for other characterization system, e.g. ISOCS by Mirion Technologies [8]. This feature is worth to be further explored, thanks to the imaging systems native capability to determine the geometrical distribution of radioactivity, which is one of the main unknowns affecting gamma spectrometry uncertainties in assaying components of generic complexity.

Two interesting cases were studied. The first is the analysis of a tank containing ion exchange resins placed in a small, separated room and measured from the top (see Fig. 8): ^{137}Cs and ^{60}Co were identified and quantified assuming a standard resins compound uniformly filling the tank. The quantified activity 8.02 ± 0.38 MBq for ^{137}Cs and 1.29 ± 0.06 MBq for ^{60}Co will be validated when the tank will be replaced and then characterized.

Looking carefully, ^{137}Cs and ^{60}Co distributions are slightly different, with the first one showing a spreading toward the floor below the tank also. Although very small, assuming mass and radioactivity distributions are uniform in the tank, this effect is not expected to be a statistical uncertainty: in fact, ^{137}Cs has more than doubled the statistic of ^{60}Co, whose high-energy photons penetrate more in matter (i.e. generating a "more uniform" image than ^{137}Cs over the tank volume), which cause also a slightly higher intrinsic uncertainty in the imaging reconstruction. These considerations point toward the fact that (assuming the uniformity in tank) would imply the ^{60}Co reconstruction should be larger than ^{137}Cs one, whereas the result is the opposite. This effect could be explained assuming a loss of water is happened from time to time from the tank and has contaminated the floor below: the higher activity of ^{137}Cs in the resins would explain why its presence is more relevant than ^{60}Co and thus influence more the reconstruction.

Fig. 8. Imaging and quantification of radioactivity in a tank containing ion exchange resins for primary loop water purification.

The second interesting case is a fuel element from the 100-kW operation phase measured in the Fresh Fuel Storage Room. The element weights 3.4 kg and is made of HZr-U expected to contain 34 gr of ^{235}U. Other details about the elements and its assembly are known from the historical knowledge of the TRIGA reactor [9]. The reconstructions are reported in Fig. 9: ^{235}U (key line at 185.7 keV) is imaged at the center of the element in agreement with the reconstruction of low-energy emissions from ^{234}Th (a ^{238}U son in secular equilibrium) decay. The non-uniformity in the ^{235}U vertical distribution is expected to be due to a non-corrected distance-effect and low-energy gamma-ray self-absorption. Differently, the image of high-energy emissions from ^{234m}Pa (another son of ^{238}U in secular equilibrium) are instead reconstructed in the backward direction because of the presence of a large amount of natural uranium elements stored in a the immediate surrounding.

The quantitative analysis returned a mass of ^{235}U of 37.02 ± 0.49 g against the 34 g expected, whilst the quantification of ^{238}U was not possible because of the interferences and will be done once a proper background acquisition will be acquired. Analysis and imaging were performed without subtracting any background, which is known to be non-negligible and could explain the ^{235}U overestimation (although the agreement level is more than enough for many applications) and the interference in ^{238}U imaging via ^{234m}Pa (its high energy emissions are less affected by the absorption due to walls than ^{234}Th and thus are largely detected by the detector).

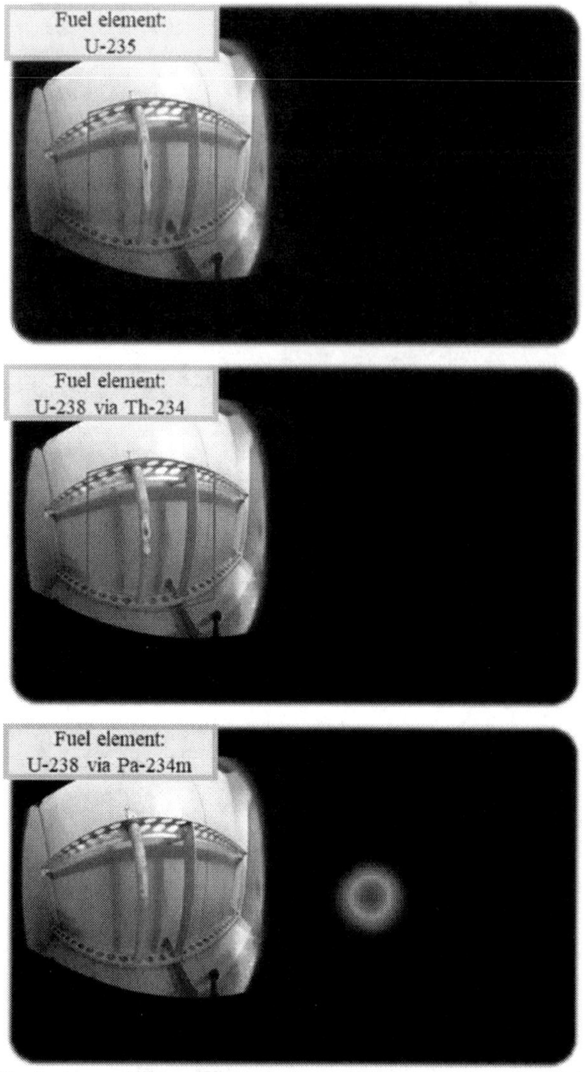

Fig. 9. Image of ^{235}U and ^{238}U (using low- and high-energy lines) for a fuel element used in the 100-kW operation phase.

V. CONCLUSIONS

A 3D pixelated CdZnTe gamma imaging and spectrometry system has been extensively tested at TRIGA RC-1 Research Reactor at the ENEA Casaccia Research Center to evaluate its capability to complement typical Radiation Protection routine. The experience has demonstrated the system is suitable for certain surveying activities (e.g. identifying hotspot and unknown radioactive item, or loss of containment). Furthermore, it has been used to retrieve many detailed information related to radioactive materials (i.e. sources and waste) all around the plant, often inside appropriate shields to reduce their outcome and dose rate. Finally, its post-processing software, thanks to the spectrum analysis and efficiency computation tools, has shown promising quantification capabilities also in complex and delicate situations like characterizing Special Nuclear Materials (namely, uranium) contained in a fuel element, foreseeing future applications also in the field of Safeguards and Non-Proliferation fields. All these tasks can be considered to be within the capabilities of the used system. Moreover, the native feature to make radioactivity "visible" will enhance the installations safety and reliability in radioactive material management.

Research is still on going to further evaluate system performance in an even bigger variety of situations and tasks, like characterizing the water of the reactor pool, or clearly identifying the radioactive items found during the survey, validating the presence of neutron source (and if the compound effectively has beryllium, because the detection of the 4.4 MeV photons is not possible with the current imaging system due to its analog readout electronics) and verifying the amount of radioactivity circulating in the primary loop water-cooling system during reactor operations.

Moreover, a new hardware and software add-on is currently under development and validation in order to fill some gaps the current 2D imaging system has: this add-on feeds data from a LIDAR and a camera to a SLAM-based algorithm capable to determine at each timeframe detector's position and pose (i.e. rotation angle respect to the three cartesian axis) in space and also to figure out the space in the format of a point-cloud. This will provide the imaging system the spatial awareness needed for projecting 2D reconstructions in 3D space, thus obtaining 3D distance-corrected radioactivity reconstruction and also useful information for efficiency modelling.

References

[1] International Commission on Radiological Protection. (1997). Recommendations of the ICRP. Pergamon press for the International Commission on Radiological Protection. Available: https://www.icrp.org/publication.asp?id=icrp%20publication%2026. Accessed on: Aug. 5, 2023.

[2] INTERNATIONAL ATOMIC ENERGY AGENCY, Determination and Use of Scaling Factors for Waste Characterization in Nuclear Power Plants, IAEA Nuclear Energy Series No. NW-T-1.18, IAEA, Vienna (2009).

[3] INTERNATIONAL ATOMIC ENERGY AGENCY, Application of Plant Information Models to Manage Design Knowledge through the Nuclear Power Plant Life Cycle, IAEA-TECDOC-1919, IAEA, Vienna (2020)

[4] INTERNATIONAL ATOMIC ENERGY AGENCY, Policies and Strategies for Radioactive Waste Management, IAEA Nuclear Energy Series No. NW-G-1.1, IAEA, Vienna (2009)

[5] C. G. Wahl, W. R. Kaye, W. Wang, F. Zhang, J. M. Jaworski, A. King, et al., "The Polaris-H imaging spectrometer," Nuclear Instruments and Methods in Physics Research A, vol. 784, no. 1, pp. 377–381, 2015.

[6] H3D, "Visualizer with SourceTerm", https://h3dgamma.com/sourceTermSpecs.pdf?, Accessed August 2023.

[7] F. Gagliardi, D. I. Goodman, E. Gorello, W. Kaye, E. Mauro, M. Pagliuca, "Reliability of 3D Pixelated CdZnTe system in the quantitative assay of radioactive waste: a demonstration," presented at the *2022 IEEE Room Temperature Semicondutor Detector Conference*, Milan, Italy, Nov. 7-11, 2022

[8] R. Venkataraman, F. Bronson, V. Abashkevich, B. M. Young, M. Field, "Validation of in situ object counting system (ISOCS) mathematical efficiency calibration software," Nuclear Instruments and Methods in Physics Research A, vol. 422, no. 1-3, pp. 450-454, 1999.

[9] L. Di Palo, et al., "Progetto Definitivo e Rapporto di Sicurezza del Reattore TRIGA RC-1 1 MW C.R. ENEA Casaccia", 1966

https://doi.org/10.1051/epjconf/202328807014

System-level testing and qualification of environmental sensors for nuclear decommissioning

A. Urena-Acuna[1*], J. Favrichon[1], E. Cantrel[1], P. A. Robin[2], A. Balier[2], T. Maraine[3], F. Saigné[3], J. Boch[3]

[1] French Alternative Energies and Atomic Energy Commission (CEA, DES, ISEC, DPME, SEIP), Univ. Montpellier, Marcoule, France
[2] ISYMAP society, France
[3] Université de Montpellier, IES-UMR UM/CNRS 5214, France
(*) alejandro.urenaacuna@cea.fr

Abstract— **In this work, we show the first steps of a system-level qualification methodology for a series of Commercial off-the Shelf (COTS) environment sensors subjected to Total Ionizing Dose (TID). We demonstrate a correlation between the degradation of electrical signals such as consumption current and functional failures in sensor measurements. The sensors were irradiated using high-energy X-rays and ^{60}Co under different biasing cases in order to determine their worst-case scenario.**

Keywords — **Consumption current, Environmental Sensors, Qualification Methodology, System-level test, Total Ionizing Dose,**

I. Introduction

Decommissioning of nuclear facilities uses electronic equipment (sensors, mobile vectors, robots, remotely operated arms, etc.) in an irradiating environment. When using Simultaneous Localization And Mapping (SLAM) systems, we require the measurement of multiple sensors to measure the environment and collect data. However, the effect of radiation is one of the main factors affecting the reliability of electronic components and systems. The degradation in the performance of the different blocks could induce electrical disturbances that affect the reliability of the entire system. The effects of the TID in elementary devices such as Metal-Oxide-Semiconductor Field Effect Transistors (MOSFET) and Bipolar Junction Transistors (BJT) are well known [1-3]. Typical effects manifest as a threshold voltage shift (ΔV_{th}), as a variation in the Subthreshold Slope (ΔSS), as an increase in leakage currents (ΔI_{off}) in the case of MOSFETs, and as a loss in the factor gain (β) and the Enhanced Low Dose Rate Sensitivity (ELDRS) in the case of BJTs [2,3]. In complex systems, this causes an impact on the operating frequency, an increase in power consumption, and functional errors in addition to partial or total failures [4]. For instance, it is essential to qualify these systems in order to determine their dose resistance and to ensure the reliability of the measurements during a mission of nuclear decommissioning.

Several standards to qualify COTS devices under TID, such as MIL-STD-883 method 1019, the European Space Component Coordinate (ESCC) standard 22900, and JEDEC Solid State Technology Association, have been established but they only focused on the component level and the dose rates are adapted to space environments. Nevertheless, system-level testing and qualification methodologies for COTS systems under TID are current challenges due to the complexity of actual systems. Their radiation response strongly depends on the type of environment, the biasing conditions, and the different blocks that compose the system (i.e. power, logic, memory, etc) in addition to factors inherent to the manufacturing processes such as lot-to-lot variability. Then, to characterize their radiation response, it is essential to design a strategy that allows us to qualify and determine the blocks prone to fail during a mission in a civilian nuclear environment.

Therefore, the goal of this work is to explore a system-level testing and qualification methodology for COTS environmental sensors. This paper is organized as follows: Section II describes the tested devices, the experimental setup, the radiation conditions as well as the qualification criteria. Section III shows the electrical and functional degradations of the irradiated sensors. In Section IV, we determine the worst-case biasing as well as the qualification of the irradiated devices. Section V explains the first steps of the proposed qualification methodology. Finally, the conclusions of this work are shown in Section VI.

II. Tested devices and Experimental conditions

A. Tested devices

The tested devices were MCP9808 digital temperature sensor and VL53L0X Time of Flight (ToF) distance sensor [5, 6]. Both sensors can communicate via I2C protocol with a host computer. Previous studies have shown the resilience to ionizing radiation of these sensors [7, 8]. In [7] a series of VL53L0X ToF distance sensors were exposed to a 60Co source. Here a progressive increase in the consumption current is shown and at 5.8 kGy, the sensors were still working. In [8], a study is presented where the MCP9808 digital temperature sensor was irradiated. The sensor presents functional failures in its measurements from 80 Gy, however, the possible correlation with the consumption current of the circuit is not shown.

B. Experimental setup

Forty sensors of two different lots (L1 and L2) were irradiated to qualify their radiation response. The tested devices were 20 MCP9808 digital temperature sensors and 20 VL53L0X Time of Flight (ToF) distance sensors. Functional and electrical parameters were monitored before, during, and after irradiation. The consumption current (I_{cons}) was monitored using a Keysight N6705C power analyzer. The devices communicate under the I²C protocol with a microcomputer (Nvidia Jetson-nano) and a script was designed to acquire two measurements per second in a CSV file for post-data processing as shown in Fig 1. The VL53L0X was positioned at 160 mm from a target to get a distance reference.

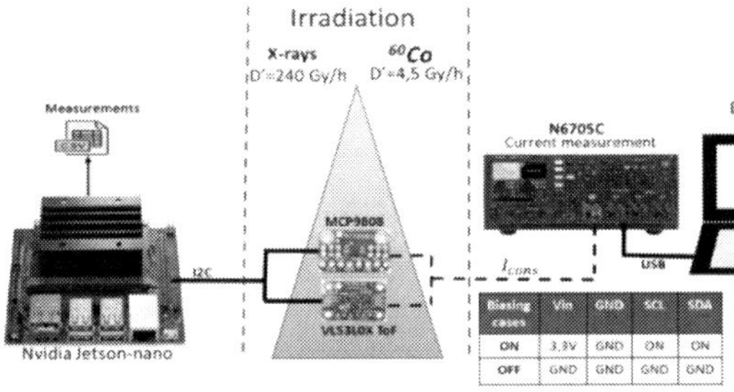

Fig. 1. Experimental setup.

C. Radiation and biasing conditions

All devices were irradiated with the PRESERVE platform at the Institut of Electronics and Systems (IES) in Montpellier, France with a 320 keV X-rays generator with a 2 mm aluminum filter to eliminate photons with low energies and under a ^{60}Co source (Fig. 3). The applied dose rate was 240 Gy/h (SiO_2) for X-rays and 4,5 Gy/h for ^{60}Co. All sensors were biased in two different modes during irradiation to determine the worst-case of biasing as depicted in Fig. 1. The OFF state, with all pins connected to the ground, and the ON state, where the sensor operates in normal biasing mode. The number of tested devices can vary in each experiment. The radiation was done in several steps to monitor the TID-induced degradation until the sensor's failure was reached.

D. Qualification criteria

We seek to demonstrate that the increase in consumption current produced by the TID beyond the datasheet current limit can induce total or partial loss of functions of the sensors. For this reason, the first parameter to monitor is the consumption current of the sensor as a function of the absorbed dose. Therefore, to qualify these sensors, we based on three criteria: a) the communication with the microcomputer, b) The consumption current (I_{cons}), where the limit is set according to the maximum current limit (I_{limit}), reported by the device's manufacturer (datasheet) and c) we propose a tolerance of 10% on the accuracy of the sensor measurements in order to keep good reliability in the functional measurements. We define dose tolerance as the dose limit at which at least one of these three criteria is not satisfied.

III. Results

A. MCP9808 digital Temperature sensor

We observe in Figure 2 the radiation response of the MCP9808 temperature sensor. The I_{limit} for this sensor is 0.4 mA according to the datasheet [5]. In Fig. 2(a) we show the I_{cons} measurements of L1 and L2 irradiated with X-rays. L1 was irradiated in step-by-step mode while L2 was in continuous mode (CM) with one measurement per minute. We noticed an increase in I_{cons} for both lots when the dose increases beyond 120 Gy. We observe functional failures manifested as errors in the accuracy of the temperature measurements when I_{cons} exceed I_{limit}, as shown in Figure 2(b). Devices from L2 were irradiated with the ^{60}Co source. The degradations of these devices are similar to those irradiated under X-rays. For this reason, only X-ray irradiated devices from L2 are shown in Figures 2 and 3.

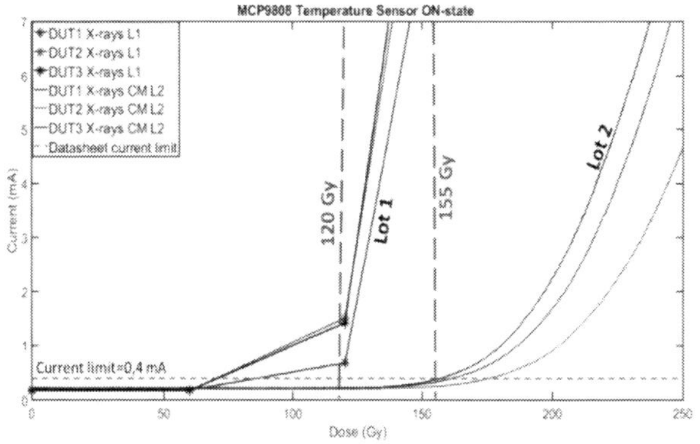

(a) Consumption current of the MCP9808 digital temperature sensor irradiated in ON state

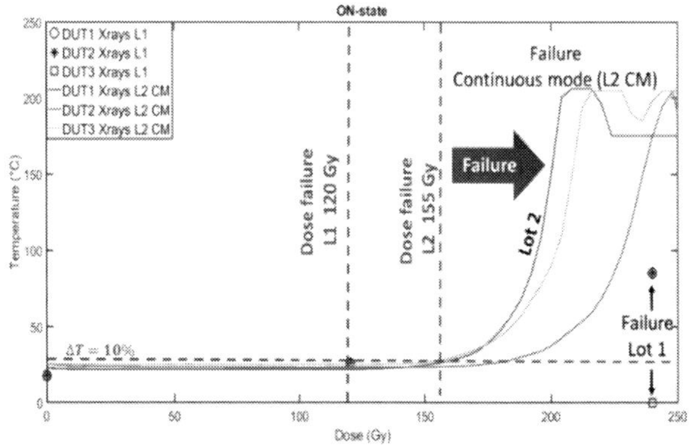

(b) Temperature measurements of the MCP9808 digital temperature sensor irradiated in ON state

Fig. 2. Consumption current and temperature measurements of MCP9808 digital temperature sensor

Figure 3 shows the MCP9808 temperature sensor irradiated in the OFF state. In Fig.3 (a) the increase in I_{cons} is much less important than the case in the ON state. This is due to the absence of an electric field during irradiation. As reported in [9], the probability of non-recombination of charges created by ionizing radiation decreases in the absence of an electric field. That is, the charges created are close to each other, stimulating their prompt recombination. Therefore, the radiation damage in the device is less important.

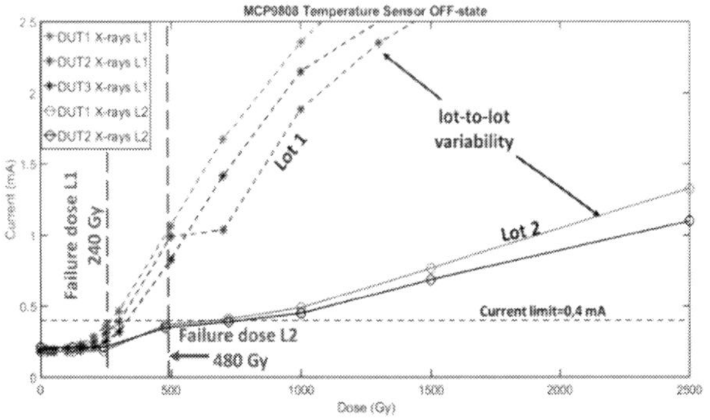

(a) Consumption current of the MCP9808 digital temperature sensor irradiated in the OFF state

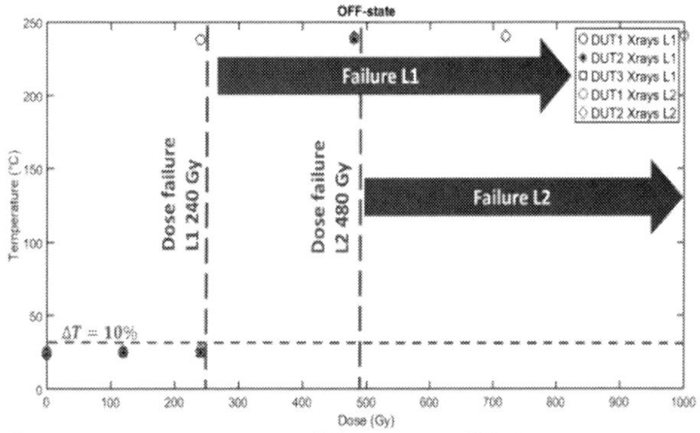

(b) Temperature measurements of the MCP9808 digital temperature sensor irradiated in the OFF state

Fig. 3. Consumption current and temperature measurements of MCP9808 digital temperature sensor irradiated in OFF state

In Figure 3(a) and (b) we noticed the lot-to-lot variability. In Figure 3 (a) we notice that L1 and L2 exceed I_{limit} at 240 Gy and 480 Gy respectively. This has an impact on the functional measurements depicted in Fig 3(b), where the temperature increase from 24 °C to 240 °C at these dose levels. The effect of biasing is clearly identified, where the dose tolerance is higher than the devices irradiated in the ON state.

B. VL53L0X ToF distance sensor

Figure 4 shows the electrical and functional measurements of the VL53L0X distance sensors irradiated in the ON state. The I_{limit} established by the manufacturer is set at 40 mA [6]. We observe in Fig 4 (a) an increase in I_{cons} beyond the 40 mA limit for devices of L1 at 480 Gy. This has an impact on the communication with the host microcomputer. As shown in Figure 4(b), a total loss of measurements is reported for L1 at 480 Gy. L2 irradiated with X-rays and ^{60}Co in both biasing modes never exceeded the current limit so, its functionality and accuracy were not affected during the irradiation. This is due to the lot-to-lot variability of the devices. The devices of L1 and L2 irradiated in the OFF state never exceed the current limit, so, their functionality was not affected at the maximum dose level reached of 3250 Gy.

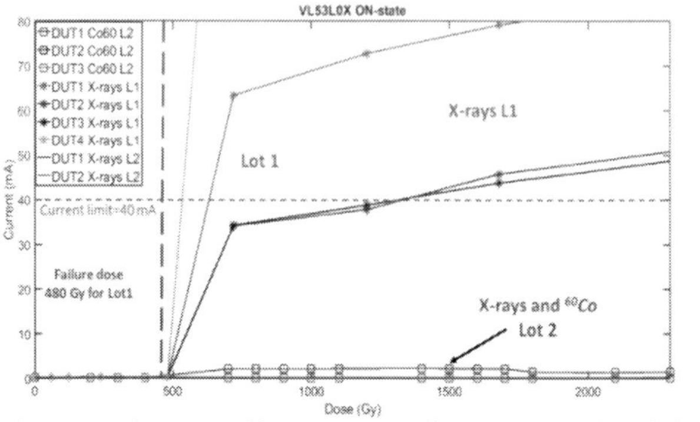

(a) Consumption current of the VL53L0X ToF distance sensor irradiated in the ON state

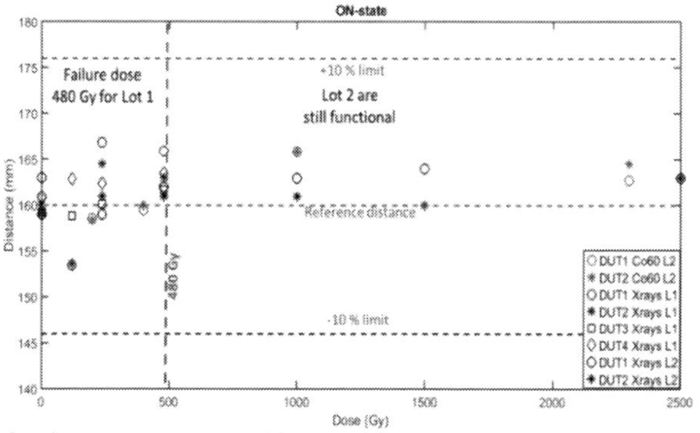

(b) Distance measurements of the VL53L0X ToF distance sensor irradiated in ON state

Fig. 4. Consumption current and temperature measurements of VL53L0X irradiated in the ON state

IV. WORST-CASE BIASING

As observed in Figures 3 and 4, sensors irradiated in the ON state showed the highest variation in consumption current and lower dose tolerance than those irradiated in the OFF state. This is well known according to the literature, where the application of an electric field during radiation prevents the prompt recombination of charges in the insulating oxides of single transistors. This factor causes a higher electrical degradation in the device's performance [9]. Based on the previously established qualification criteria, Table 1 shows the qualified dose tolerance for both sensors for the ON and the OFF states.

TABLE I
QUALIFIED DOSE TOLERANCE OF THE DIFFERENT SENSORS

Sensor	Dose Tolerance (Gy)		Worst-case Biasing
	ON	OFF	
MCP9808	120 Gy L1 155 Gy L2	240 Gy L1 480 Gy L2	ON
VL53L0X	480 Gy L1 3000 Gy L2	>3000 Gy L1 and L2	ON

V. SYSTEM-LEVEL METHODOLOGY

The study carried out in this work represents the first steps for a system-level testing and qualification methodology. This work allows us to establish some general rules:

1. - Determine the radiation environment in order to determine the dose rate.
2. - Analyze the system under test (DUT). Radiation effects differ depending on the type of technology. Therefore, certain effects such as (ELDRS) must be considered when using BJTs based devices.
3. - Determine the auxiliary signals to monitor the subsystem's performances (power, logic, etc). This allows the detection of the radiation-sensitive subsystems.
4. - Establish electrical and functional margins and conditions. The limits of the electrical parameters must be chosen according to the manufacturer's datasheet. In addition, to reduce or eliminate lot-to-lot variation, it is recommended to use single-lot systems.
5. - Irradiate the components with an X-rays generator under different biasing modes to determine the worst case.
6. - Confirm the effects with an irradiation in ^{60}Co. An inconvenience of the existing test methodologies arises from the use of ^{60}Co as an irradiation reference. In general, the use of this means of radiation implies a high cost, both in the facilities and in radioprotection. In the long term, the use of high-energy X-rays could be convenient due to their simplicity in terms of maintenance, radioprotection, and speed in obtaining results.

VI. CONCLUSIONS

In this work, we show the first steps of a system-level test methodology. We have demonstrated the correlation between an increase in the consumption current and functional failures in the sensor measurements. This fact is a good indicator to diagnose that a functional failure is approaching when the consumption reaches the established current limit. In addition, we noticed a lot-to-lot variability that influences the radiation response of the devices. A strategy to avoid this situation is the use of systems with similar characteristics (manufacturer, supplier, package material, origin, etc) and only representative samples would undergo qualification testing. In order to validate the test methodology, we confirm the results by a ^{60}Co irradiation. Nevertheless, in the long term, we seek to use only high-energy X-rays. This could represent multiple advantages in terms of cost, speed of testing, maintenance and radioprotection.

ACKNOWLEDGMENT

The authors thank the Occitanie region for funding this work within the framework of the PRRI CASIMIR Project. X-rays and ^{60}Co tests were performed using the PRESERVE platform, which was funded by the Occitanie Region and the EU via ERDF funds.

REFERENCES

[1] Oldham, T. R., & McLean, F. B. Total ionizing dose effects in MOS oxides and devices. IEEE transactions on nuclear science, 50(3), 483-499. 2003.
[2] Colder, A., Croitoru, N., et al. Study of radiation effects on bipolar transistors. Nuclear Instruments and Methods in Physics Research Section B: Beam Interactions with Materials and Atoms, 179(3), 397-402. 2001.
[3] Hjalmarson, H. P., Pease, R. L., et al. Mechanisms for radiation dose-rate sensitivity of bipolar transistors. IEEE Transactions on Nuclear Science, 50(6), 1901-1909. 2003.
[4] Hughes, H. L., & Benedetto, J. M. Radiation effects and hardening of MOS technology: Devices and circuits. IEEE Transactions on Nuclear Science, 50(3), 500-521. 2003.
[5] Microchip Technology Inc. MCP9808 Digital Temperature Sensor Datasheet. 2011.
[6] STMicroelectronics, N. V. World smallest time-of-flight ranging and gesture detection sensor, VL53L0X. Datasheet—production data. 2016.
[7] Chesnevskaya, S., Da Vià, C., Utting, B., Hughes, H. L., & Watts, S. J. Radiation testing of robotic systems–LiDAR as a case study.
[8] Tsiligiannis, G., Touboul, A., Bricas, G., Maraine, T., Boch, J., Wrobel, F.,& Girones, P. Evaluation and analysis of technologies for robotic platforms for the nuclear decommissioning. In 2020 15th Design & Technology of Integrated Systems in Nanoscale Era (DTIS) (pp. 1-6). IEEE. 2020.
[9] Schwank, J. R., Shaneyfelt, M. R., Fleetwood, D. M., Felix, J. A., Dodd, P. E., Paillet, P., & Ferlet-Cavrois, V. Radiation effects in MOS oxides. IEEE Transactions on Nuclear Science, 55(4), 1833-1853. 2008.

The MICADO Integrated Gamma Station for Radioactive Waste Packages radiological characterization

L. Lepore[1,*], G. Gandolfo[1], G. A. Marzo[1], N. Cherubini[1], D. Casagrande[1], E. Palumbo[1], N. Ratto[1], E. Fanchini[2], A. Pepperosa[3], F. Giordano[2] , M. Morichi[2]
[1]ENEA, Nuclear Material Characterization Laboratory and Nuclear Waste Management, Casaccia Research Centre, Rome, Italy
[2]CAEN spa, Viareggio, Italy
[3]ISIN, Ispettorato Nazionale per la Sicurezza Nucleare e la Radioprotezione, Rome, Italy
(*) luigi.lepore@enea.it

***Abstract* — The MICADO (Measurement and Instrumentation for Cleaning And Decommissioning Operations) Project of the H2020 Research and Innovation Programme aims to propose a cost-effective solution for non-destructive characterization of nuclear waste, implementing a digitization process that could become a referenced standard facilitating and harmonizing the methodology used for the in-field Waste Management and Dismantling & Decommissioning operations. It employs instruments based on different technologies: an active and passive neutron measurement system, a photo-fission facility, and the 'Integrated Gamma Station'.**

Within the MICADO Project, an entire work package has been dedicated to the design and realization of the 'Integrated Gamma Station' obtained by combining different gamma detection technologies supporting each other to a comprehensive and effective non-destructive gamma characterization, able to accommodate Radioactive Waste Packages of different sizes. The techniques implemented are i) dosimetry measurements, count rate inspection and raw spectroscopy with the CAEN RadHAND, ii) gamma imaging in open geometry with the CEA Nanopix gamma camera, iii) high resolution gamma spectrometry with the ENEA Tomographic Gamma Scanner (TGS). The latter able to carry out different characterization methodologies, i.e., Open Geometry, Segmented Gamma Scanning, Angular Scanning, and Tomography.

This paper describes the layout of the Integrated Gamma Station conceived, its features and detection capabilities, and part of the measurement campaign realized during the MICADO Project.

***Keywords*— Radioactive Waste, Integration and Digitization, Gamma Spectrometry Characterization.**

I. Introduction

THE main aim of MICADO (Measurement and Instrumentation for Cleaning And Decommissioning Operations, Project of the H2020 Research and Innovation Programme) has been to define an innovative methodology for the characterization and long term monitoring of Radioactive Waste Packages (RWP), with a toolbox of measurement devices, simulation, software and analysis tools providing a comprehensive solution for the global digitization in real-time of all nuclear waste relevant information dose-rate, gamma spectroscopy, gamma imaging-hot spot identification [1] [2], active and passive neutron measurements [3] [4], photofission techniques [5] [6], long term monitoring grid [7] [8], sample description, origin, scaling factor, etc.) and define a global procedure for waste characterization and uncertainty management [9].

Along with technical implementation, study and integration of RWP characterization and monitoring methodologies, the RCMS (Radiological Characterization & Monitoring System) DigiWaste Platform is the main output of the MICADO Project, a completely independent database connected to all measurement stations and monitoring systems [10]. It allows:

- to provide a data continuum and a single digital identity (ID) to each RWP. Data resulting from characterization actions collected on different sites/techniques are integrated in a single database record,
- faster execution of radiological measurements (cross-combination of technical information, e.g., weight, waste matrix content and density, expected radionuclides from other methodologies, etc.),
- optimized characterization of a RWP results with lower uncertainties, by combining outputs of all non-destructive methods and tools already used as reference,
- accurate tracking and long-term monitoring of RWP neutron and gamma emission,
- efficient digitization of the full characterization and logistical processes.

Within the MICADO Project, an entire work package has been dedicated to the realization of the 'Integrated Gamma Station' (IGS), to accommodate Radioactive Waste Packages of different sizes, obtained by combining different gamma detection technologies, supporting each other to a comprehensive non-destructive gamma characterization. The techniques implemented are i) raw spectroscopy and dosimetry, ii) gamma imaging in open geometry; iii) high resolution gamma spectrometry to different characterization methodologies, i.e., Open Geometry, Segmented Gamma

Scanning, Angular Scanning, and Tomography.

The combination of three separate and independent systems based on gamma detection, the i) dosimeter and raw spectroscopy CAEN RadHAND, ii) CEA Nanopix gamma camera, iii) ENEA Tomographic Gamma Scanner, could provide:

- increased reliability of the characterization results, by comparing and sharing information and different outcomes, improving the overall characterization quality and speeding up the full analysis;
- more automated procedures reducing human intervention, minimizing the possibility of human error in the whole procedure, and increase of conventional safety and radiation protection for the operator or user;
- reduction in the timespan to carry out the complete analysis, because of the integration of techniques and the adoption of an automatic system using "decision parameters" for selecting the most suitable gamma characterization technique to be applied to the RWP, automatically. Expert user intervention will remain an option for those cases the central automatic control system cannot manage by itself.

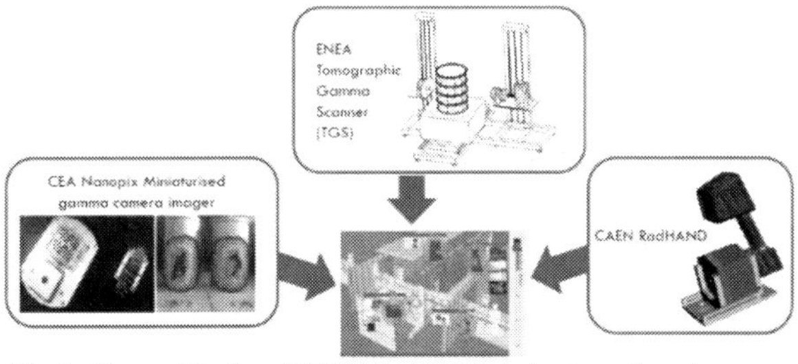

Fig. 1. The combination of CAEN RadHAND dosimetry and spectroscopy, CEA Nanopix gamma camera, ENEA Tomographic Gamma Scanner within the MICADO IGS.

The IGS will be able to provide:

- multiple RWP sizes accommodation and characterization;
- determination of the waste matrix material and density spatial distribution based on the evaluation of linear attenuation coefficients (transmission measurements);
- radioactive "hot-spot" localization and count rate distribution estimation;
- radioisotopes identification, and spatial distribution of the radioactive sources;
- decision about the best characterization technique to be applied, proper detector calibration, and activities estimation based on high resolution gamma spectrometry.

The key issue for the coordination of data is to keep devices and information connected to a central storage unit – the RCMS DigiWaste Database – from which the IGS automatic control system can retrieve information from, execute the characterization required for the RWP, and upload the obtained results.

II. The MICADO Characterization Campaign

During the MICADO Final Demonstration Event (FDE) on 2023 January 23rd–27th, all MICADO technologies available at the ENEA Casaccia Research Center site (Rome, Italy), have been tested on some mock-up RWPs just prepared.

In particular, at the ENEA Radiochemistry Laboratory, the IGS was installed (Work Package 4) along with the neutron measurement system (Work Package 5) by CEA DEN Cadarache. The latter focused only on the passive mode because of licensing limitations relative to use of the neutron generator.

The main aim of the test was to demonstrate that different techniques can cooperate to a comprehensive and complete characterization of the Radioactive Waste Package to be analysed.

Radioactive Waste packages adopted to test the full procedure are described in the following section, and they are targeted to cover effective needs of the nuclear industry, reproducing typical wastes from:

A. Reprocessing Plant/MOX Fuel production facility (U and Pu in metal and burnable materials, essentially),
B. Research Reactor/Research Laboratory (fission and activation radionuclides within burnable materials),
C. Legacy Waste (various radionuclides in different matrices).

For the sake of brevity, just the 'Legacy Waste' mock-up drum C is illustrated in this work. Its main characteristics are the following:

- Waste matrix: metal, burnable (paper, plastic, cotton, etc.) simulated with the ENEA mock-up Multi-material Reference Drum, Fig. *2*;
- Radiological content:
    - Plutonium, ~5 g di Pu, at R3 (Fig. *2*) in wood slice (~55 cm from bottom),
    - Plutonium, ~5 g di Pu and Uranium, ~7 g with 10% enrichment in U-235, at R0 (Fig. *2*) in metal slice (~33 cm from bottom),
    - Cs-137 linear source, ~10 MBq at R2 (whole height),
    - Co-60 point source, ~300 kBq at R4 (Fig. *2*) in metal slice (~55 cm from bottom),
    - Eu-152 point source, ~700 kBq (Fig. *2*) at R5 in metal slice (~55 cm from bottom).

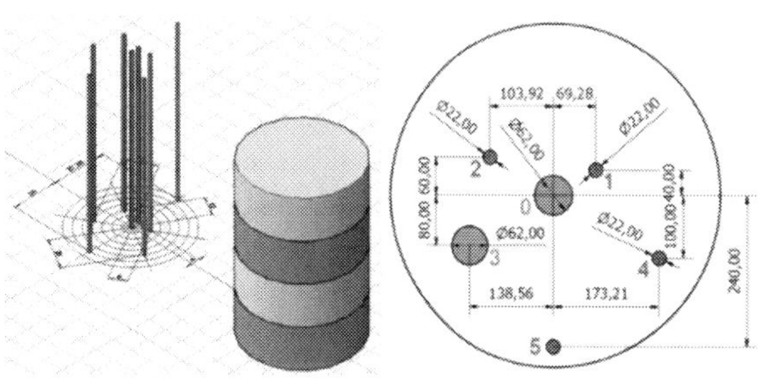

Fig. 2. The ENEA Multi-material Reference 220 liter drum. From the bottom to the top: concrete, metal, wood, neoprene. Penetration channels from the top allow to insert radioactive sources at different R0…R5 positions (dimension in mm).

The 'Legacy Waste' mock-up drum C is covering the most complex case, where Special Nuclear Materials are mixed with fission products and activation products, within a heterogeneous waste matrix. Other MICADO mock-up drums are simpler cases in terms of radioactive sources and waste matrix material contained.

The characterization procedure has been divided in three phases, initially.

1) *Screening phase*: RCMS database is filled with RWP information available, and the RFID (Radio Frequency IDentification) is applied to unique identification of the item. The integration of technologies in the station is expecting the RadHAND and the Nanopix gamma camera at the start-up phase mainly. Information about heterogeneity in spatial distribution of the radiological content (e.g., hotspots) is investigated. Dose-rate inspection and raw gamma spectrometry (3-minutes each) at fourteen different positions are carried out, and gamma camera imaging is run for a single 30-minutes acquisition. If heterogeneity in radioactive content is found, decision about the characterization technique to be applied to the RWP (Open Geometry, OG; Angular Scanning, AS; Segmented Gamma Scanning, SGS; Emission/Transmission Tomography, ECT/TCT) can be taken immediately.
2) *Preliminary phase*: information about heterogeneity in spatial distribution of waste materials is searched for with the ENEA TGS transmission measurements, and hypotheses about the characterization technique to be applied to the RWP are complemented with the previous step:
    a. the OG is suitable for homogenous (both waste matrix and distribution of radioactive content) RWPs. The collimation and the detector-to-RWP distance are selected in such a way that the RWP can be seen entirely by the detector in a single shot measurement, while the RWP is rotating;
    b. the AS is performed for a RWP with a homogeneous matrix and a heterogeneous distribution of radioactivity. By acquiring spectra at different angles and height, the variation of the count rate allows the reconstruction of the spatial distribution of the radiological content;
    c. SGS is used for a RWP with a slightly heterogeneous matrix and a heterogeneous distribution of the radioactive source. The RWP is divided into slices, and gamma spectra are acquired measured uninterruptedly while rotating.
    d. ECT/TCT is the most complex and general approach that can be used for any kind of RWP. Transmission tomography can obtain the spatial distribution of waste matrix materials. The Emission tomography allows the reconstruction of the radioactivity spatial distribution by back-projection method reconstruction.
3) *Characterization phase*: it is the execution itself, of the most suitable technique, to determine the activity of those radionuclides identified to occur in the RWP, and results are recorded on the RCMS platform, to be shared with other techniques.

Cooperation of different technologies is achieved not only with the RCMS database, but also by integration of detectors on the IGS by a common National Instrument LabView platform prepared on the PC controlling the automation of the whole IGS station.

III. An Example of Characterization Results

The section describes the results achieved in radiological characterization of MICADO 'Legacy Waste' mock-up drum C. From the initial inspection, the vertical distribution of the dose rate is given by the HPGe detector analysing vertical slices of the drum while it is in rotation, Fig. *3*.

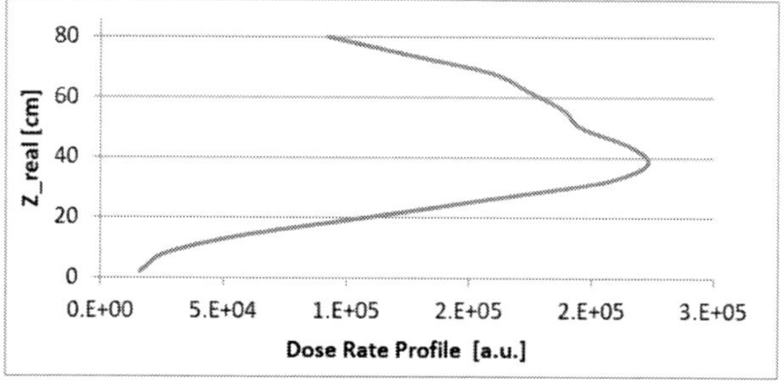

Fig. 3. Mock-up drum "C". Vertical dose profile by HPGe detector.

The vertical coordinate of the hotspot identified is ~40 cm from the bottom. At this height, a rotational inspection gave the rotational dose profile, Fig. *4*.

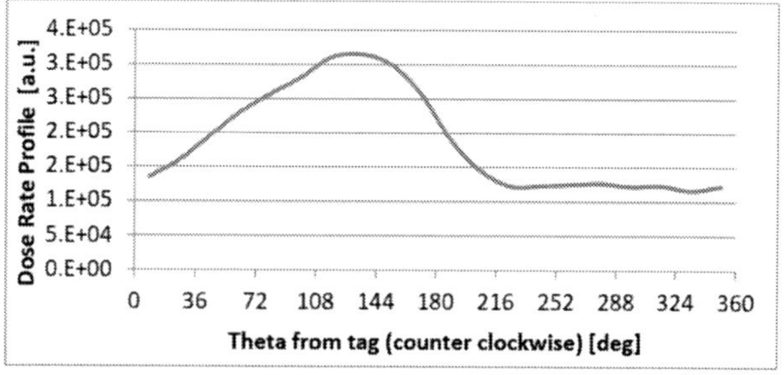

Fig. 4. Mock-up drum "C". Angular dose profile by HPGe detector at the vertical coordinate of the hotspot identified.

The angular coordinate of the hotspot is found at ~140 °. The identified maximum surface dose-rate hotspot point is placed in front of the RadHand, to determine:

- the maximum dose rate at contact;
- the maximum dose rate at 1m (once the RadHand is placed 1 m away from the drum surface).

RadHand inspection of 12 sectors followed (the drum is conceptually divided in 3 slabs and 4 angular sectors), giving for each position:

- one dose rate value, 3-min averaged;
- one gamma spectrum (acquisition of 3 min) for initial radionuclides identification and to qualitative activity distribution estimation, Fig. 5.

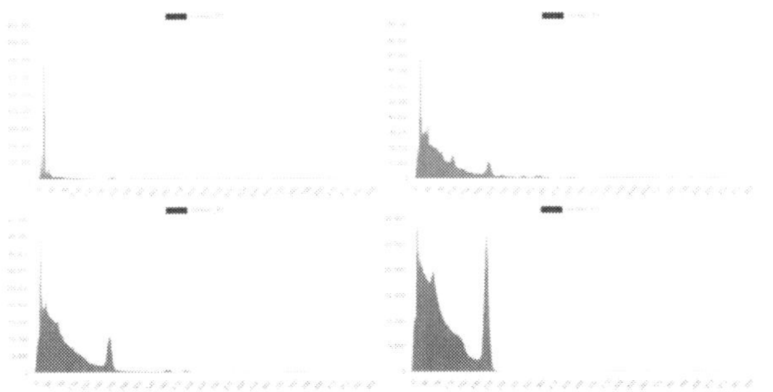

Fig. 5. Mock-up drum "C". Some of the spectra collected from CAEN RadHand. Heterogeneity in spatial distribution of radionuclides is clearly evident: the 4 plots qualitatively illustrate the spectral variation vs. some detector positions.

Then RadHand and the Nanopix Gamma Camera are then placed 1 m away from the surface of the drum (corresponding to the hotspot position identified) at the middle height of the drum and a 30 min acquisition is adopted for both detectors.

The 30 min acquisition by RadHand retrieves a quick and global identification of present radionuclides, while Nanopix gamma camera estimates the surficial extension of the radioactive source, if the gamma rate emission is strong enough. Am-241, Eu-152, U-235, Co-60 and Cs-137 are identified, and the gamma distribution by Nanopix is reported below, Fig. 6.

Fig. 6. Mock-up drum "C". CEA Nanopix imaging.

Because of the heterogeneity of the activity distribution inside the drum, it is decided to perform ECT as final radiological characterization technique. The 3-D view and some 2-D sketches of results obtained are reported in Fig. 7 and Fig. 8.

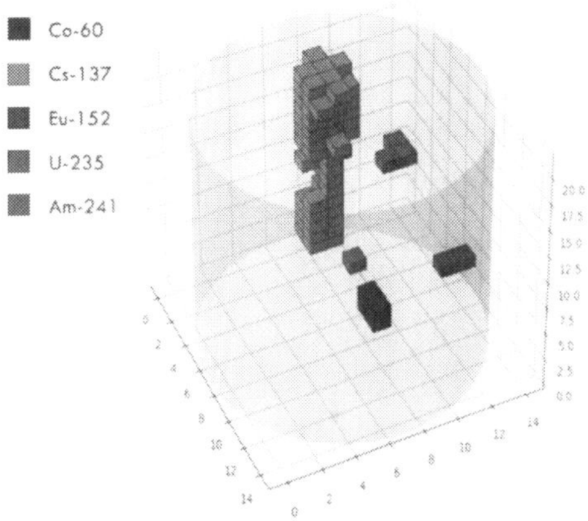

Fig. 7. 3D view of sources identified within the mock-up drum "C".

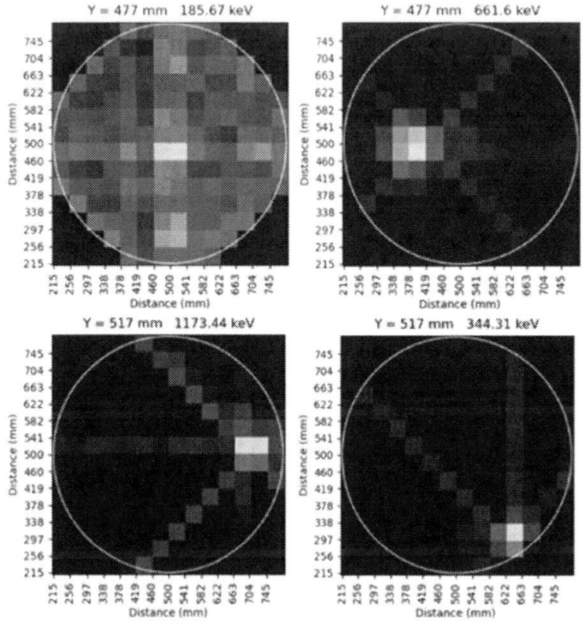

Fig. 8. Mock-up drum "C". Mock-up drum "C". 2-D plot from tomography. Four of the twenty-five hotspots identified: in the order U-235, Cs-137, Co-60, and Eu-152.

Table I reports the calculated activity values for identified radionuclides against the corresponding certified values. The largest bias identified is for the Eu-152, due to the calculation method: at the current stage of development the tomography is considering the 344 keV-line only, instead of using all statistically significant peaks of Eu-152. The other large bias is for the Am-241, mainly due to its low energy 59 keV line that is strongly suffering the waste matrix attenuation effects. Anyway, the order of magnitude of the activity is well reproduced.

Table II is reporting the Minimum Detectable Activities and Critical Level achieved for radionuclides identified in mock-up drum "C" vs. drum waste matrix material.

TABLE I
ACTIVITY VALUES DETERMINED IN MOCK-UP DRUM "C" WITH THE MICADO WP4 INTEGRATED GAMMA STATION VS. CERTIFIED VALUES. THE LAST COLUMN IS PROVIDING BIASES

RADIO NUCLIDE	CALCULATED ACTIVITY (Bq)	CERTIFIED ACTIVITY (Bq)	BIAS (%)
Co-60	$(2.68 \pm 0.69) \cdot 10^{5}$	$(2.50 \pm 0.03) \cdot 10^{5}$	+7.2
Cs-137	$(8.97 \pm 0.49) \cdot 10^{6}$	$(11.1 \pm 0.8) \cdot 10^{6}$	-19.2
Eu-152	$(7.31 \pm 0.41) \cdot 10^{5}$	$(4.46 \pm 0.05) \cdot 10^{5}$	+63.9
U-235	$(6.68 \pm 0.80) \cdot 10^{4}$	$(6.70 \pm 0.47) \cdot 10^{4}$	-0.3
Am-241	$(12.40 \pm 0.80) \cdot 10^{10}$	$(9.04 \pm 0.30) \cdot 10^{10}$	+37.2

TABLE II
MINIMUM DETECTABLE ACTIVITIES AND CRITIC LEVEL FOR RADIONUCLIDES IDENTIFIED VS. DRUM WASTE MATRIX MATERIAL

RADIO NUCLIDE	MATRIX	MDA (Bq)	CRITICAL LEVEL (Bq)
Co-60	Metal	$5.99 \cdot 10^{3}$	$2.98 \cdot 10^{3}$
Cs-137	Neoprene	$4.13 \cdot 10^{2}$	$2.06 \cdot 10^{2}$
	Wood	$3.17 \cdot 10^{2}$	$1.58 \cdot 10^{2}$
	Metal	$7.88 \cdot 10^{2}$	$3.93 \cdot 10^{2}$
	Concrete	$7.53 \cdot 10^{3}$	$3.76 \cdot 10^{3}$
Eu-152	Metal	$3.53 \cdot 10^{3}$	$1.77 \cdot 10^{3}$
U-235	Metal	$1.26 \cdot 10^{3}$	$6.29 \cdot 10^{2}$
Am-241	Wood	$8.89 \cdot 10^{6}$	$4.44 \cdot 10^{6}$
	Metal	$4.76 \cdot 10^{7}$	$2.38 \cdot 10^{7}$

IV. CONCLUSIONS

Within the MICADO Project, an entire work package has been dedicated to the realization of the Integrated Gamma Station, to accommodate Radioactive Waste Packages of different sizes, obtained by combining different gamma detection technologies, supporting each other to a comprehensive and thorough radiological characterization.

The combination of three separate and independent systems based on gamma detection at the MICADO IGS (CAEN RadHAND dosimeter and raw spectroscopy, CEA Nanopix gamma camera, ENEA Tomographic Gamma Scanner) has allowed to develop a new approach to RWP non-destructive gamma characterization. obtaining increased reliability of the results by comparing and sharing information and outcomes among different measurement techniques.

Implementation of the IGS has been driven by the following general criteria: i) automation of measurement procedures, ii) reduction of the human intervention and minimization of the human error, iii) increasing of conventional safety and radiation protection for the operator or user, iv) increased quality and reliability of the characterization results with, at the same time, reduction of the timespan for carrying out the complete analysis of the drum. Expert user intervention (e.g., manual actions and human decision) will remain an option for those cases where the automatic control system might potentially fail.

To test the IGS capabilities and performances, several characterization campaigns have been set up during the development of the MICADO Project, and part of the results obtained have been presented here.

The realization of suitable mock-up drums representing real RWPs occurring in the nuclear industry has allowed to demonstrate that the IGS is fully operative and cooperation of different techniques can speed-up the characterization phase and increase the quality of results, truly.

The MICADO IGS is installed at the ENEA Radiochemistry Laboratory at the Casaccia Research Center, Rome, Italy. It is currently used as an experimental prototype to develop and strengthen gamma-based methodologies applied to non-destructive RWPs characterization techniques.

REFERENCES

[1] G. Gandolfo, L. Lepore, G. A. Marzo, N. Ratto, D. Casagrande, N. Cherubini, E. Fanchini e A. Pepperosa, «Report D4.3: Intermediate report on the integrated gamma spectroscopic system,» 2022.

[2] E. Fanchini, A. Pepperosa, L. Lepore, G. Gandolfo e G. Marzo, «Report D4.6: The Integrated Gamma Spectroscopit and hot spot characterization system,» 2022.

[3] Q. Ducasse, C. Eleon e B. Perot, «Report D5.3: Commissioning and Calibration of the neutron measurement system in the laboratory,» 2022.

[4] Q. Ducasse, C. Eleon e B. P. e. al., «Design of MICADO advanced passive and active neutron measurement system for radioactive waste drums,» *Nuclear Inst. and Methods in Physics Research, A,* vol. 1005, n. 165398, 2021.

[5] I. Meleshenkovskii, A. Sari, A. Elayeb e R. De Stefano, «Report D6.3: Summary of the experimental photofission validation step,» 2022.

[6] I. Meleshenkovskii, A. Sari, A. Elayeb e R. De Stefano, «Report D6.4: The LINAC measurement system,» 2022.

[7] P. Finocchiaro, L. Cosentino, F. Longhitano, G. E. Poma, M. Simonetto e F. Giordano, «Report D7.2: Report on the monitoring grid system laboratory,» 2021.

[8] P. Finocchiaro, L. Cosentino, F. Longhitano, G. E. Poma e S. Amaducci, «Report D10.1: The Waste Monitoring Grid System in a nuclear waste repository,» 2022.

[9] A. Bielen, S. Boden e e. al., «Report D8.3: Quantification of the sources of uncertainty,» 2021.

[10] E. Fanchini e D. Ninci, «Report D9.4: Digital software Platform demonstrator».

An electronic dosimeter for diagnostic X-ray beams based on a differential amplifier circuit with MOSFETs

Luiz C. Gonçalves Filho[1,2,*], and Luiz A. P. Santos[1,3]
[1]CRCN-NE/CNEN, Brazil
[2]UFPE, Brazil
[3]SCIENTS, Brazil
(*) lasantos.scients@gmail.com.br

***Abstract*—It is known that the device MOSFET works as a dosimeter in radiation beams for cancer radiotherapy. Basically, the radiation beam generates defects in a way that produces in the MOSFET the variation in its threshold voltage, V_T, which is proportional to the accumulated radiation dose. Recently, MOSFETs were also tested under X-ray beams commonly used in the energy range applied to medical diagnoses, which is lower energy than radiotherapy. Indeed, it was already shown that the MOSFET drain current varies with the radiation dose for energies in the range applied to diagnostic radiology. In this paper, an innovative method based on a differential amplifier circuit with two electronically matched MOSFETs is presented. Therefore, before irradiation the differential output voltage signal, $V_{dif} = 0$. Actually, it is a differential amplifier circuit with one of the MOSFETs shielded from the ionizing radiation beam. Therefore, if such a circuit is irradiated the difference between I_{D1} and I_{D2}, from MOSFET1 and MOSFET2 respectively, will result $V_{dif} \neq 0$. To evaluate the response of this electronic dosimeter it was tested under a diagnostic radiology beam typically used in clinics to take radiography of patients. The results showed that the proposed dosimeter works and varies linearly with the radiation dose. The technology developed in this paper will allow research to be carried out for optimizing such electronic dosimeter to make it a wearable device and can be able to use it by workers exposed to ionizing radiation in nuclear power plants or even in outer space.**

***Keywords* — X-ray, Detector, MOSFET, Dosimeter.**

I. Introduction

MOSFET has been used as an X-ray detector for over 40 years, mainly as a dosimeter in radiation beams normally used for cancer radiotherapy [1-2]. Basically, the radiation beam generates defects in the silicon oxide of the device in a way that produces in the MOSFET variation in its threshold voltage, ΔV_T [3], which is proportional to the accumulated radiation dose. Generally, ΔV_T varies by tens or hundreds of mV per unit dose (Gy) delivered to the patient. Recently, MOSFET was tested under X-ray beams commonly used in the energy range applied to medical diagnoses [4-6], which is lower energy than radiotherapy. Also, unlike the method used in radiotherapy, the measurement technique is based on variation of the drain current, I_D. Indeed, it was previously shown that the drain current varies with the radiation dose for energies in the range applied to diagnostic radiology [4]. In this work, knowing that the so-called differential amplifier circuit provides an improvement in the characteristics of a typical amplifier, an innovative dosimetry technique is proposed. Actually, the motivation for developing this work comes from the fact that a simple and versatile circuit can be used in workers' uniform in nuclear power plants or even by astronauts in outer space, that is, providing a circuit optimization process it can become a wearable electronic system.

II. Material and methods

A. Differential amplifier circuit

The proposed circuit is similar to a typical operational amplifier. Actually, it is a differential amplifier circuit with one of the MOSFETs shielded from the ionizing radiation beam as shown in Fig. 1. The circuit has two electronically matched MOSFETs: T_1 and T_2. In this way, initially, the differential output voltage signal (V_{dif}) will be zero before irradiation or even can have any offset, which can be canceled by varying R_{D1} or R_{D2}, for example. As previously shown the MOSFET drain current (I_D) varies each time the device is exposed to diagnostic X-ray beam [4]. Therefore, if the differential amplifier circuit is irradiated the difference between I_{D1} and I_{D2}, from T_1 and T_2 respectively, will produce $V_{dif} \neq 0$. The signal V_{dif} passes through an instrumentation amplifier (IA) to be digitized by an ADC circuit to be registered in a memory block (MB). In that way, one can have the dose history of a patient or worker on a chip card.

B. MOSFET

A printed circuit board (PCB – Fig. 2) was made to connect the two transistors. The MOSFETs were selected from a sample of 50 MOSFETs to obtain matched devices. To analyse the new dosimeter, two PCBs were assembled to perform experiments in duplicate. The chosen MOSFET for this research was FQP2N06C, Infineon®, just based on the fact that such a device has already been used previously [4].

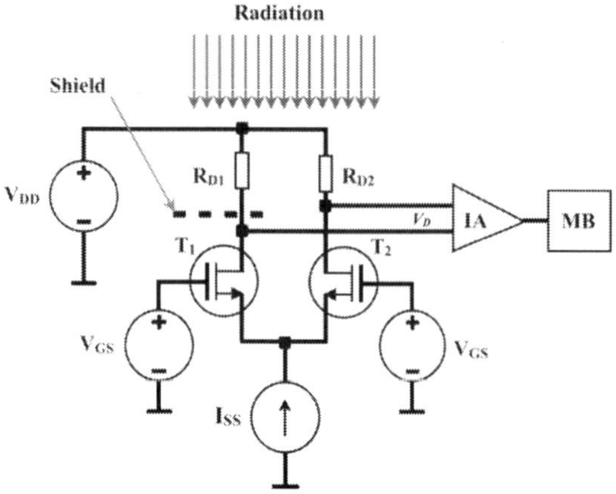

Fig. 1. Electronic dosimeter based on a differential amplifier circuit. T_1 and T_2 are electronically matched MOSFETs; V_{dif} is the differential output signal to be amplified by an instrumentation amplifier (IA); the signal is digitized by an ADC and registered in the memory block (MB).

Fig. 2. PCB assembled to connect T_1, T_2 and the IA circuit. T_2 connector, which is not shielded from radiation, is positioned outside the PCB area. The output signal, V_{dif}, can be evaluated by a typical multimeter or directly recorded in a microcontroller to be displayed on a small LCD screen.

C. X-ray equipment, radiation pulses and procedures

X-ray pulses were generated from a Polymat 30/50 Plus clinical equipment, Siemens®. The pulse duration was set to be 1600 ms to minimize the uncertainties. To assess how the dosimeter works in the energy range applied to radiodiagnosis the X-ray tube potentials were set to be the following values: 77kV, 102kV and 125kV, in order to make a comparison with other recent publications. Also, to evaluate the influence of the X-ray tube current two experiments was carried out by selecting two workloads in the control panel of the X-ray generator: 100mAs and 200mAs. Therefore, each X-ray pulse yields the radiation dose as shown in Table I. The dosimetric circuit was positioned at approximately 32.5 cm from the radiation focus. For each X-ray tube peak potential and for each selected workload value, measurements of the V_{dif} variation were made with two assembled PCBs. An EFF1705 source-meter, Scients®, was used for real-time measurements of V_{dif}. Before carrying out the experiments described above, dosimetry of each X-ray pulse was performed using a 10X5-6 ionization chamber, Radcal®, which is the reference detector and it has an uncertainty of 0.02 mGy.

TABLE I
RADIATION PULSE DOSE AS A FUNCTION OF X-RAY TUBE PARAMETERS

Peak kilovoltage (kV)	Workload (mAs)	Radiation pulse dose (mGy)
77	100	39.84
102	100	68.31
125	100	97.78
77	200	75.96
102	200	131.29
125	200	188.49

III. RESULT AND DISCUSSIONS

Firstly, Fig. 3 shows the result of the X-ray pulse dosimetry. The size of the marker corresponds to uncertainties in the *x* and *y* axes, which is less than 2% in both. As can be seen in the potential range applied to diagnostic radiology, the dose rate is practically linear because of the inherent filtration of clinical X-ray equipment. For this reason, the peak potential of ≈100kV ± 25% was chosen to carry out the V_{dif} measurement experiments for the two chosen workloads: 100 mAs and 200 mAs (Fig. 4 and Fig. 5), *i.e.*, the photonic intensity doubles in value when the X-ray tube current is doubled, since the pulse duration is held constant. It can be seen that the increase in V_{dif} is practically doubled when the dose rate doubles. This last result shows the variation in V_{dif} for only three X-ray pulses and there is no offset canceled in the circuit. A sequence of 10 X-ray pulses with a peak potential equal to 102kV for the two workloads are shown in Fig. 6 and Fig. 7, respectively, where there is offset canceled. It can be observed that in any case of dose rate corresponding to 100 mAs and 200 mAs, the increase in V_{dif} is a linear function of the dose with coefficients R^2=099892 and R^2=0.99854, respectively. Then each X-ray pulse generates a small step in V_{dif}.

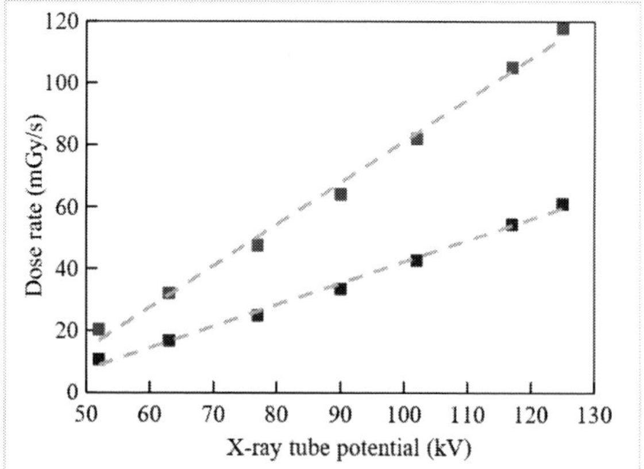

Fig. 3. Dose rate at 32.5 cm from the X-ray tube focus as a function of the X-ray tube potential and parameterized by the workload: 100 mAs (maroon) and 200 mAs (red).

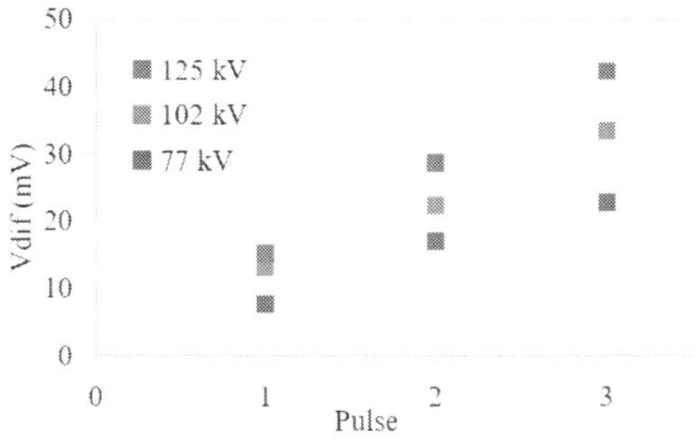

Fig. 4. Variation in V_{dif} (there is no offset canceled) for three X-ray pulses with potential corresponding to ≈100kV ± 25%, the workload is 100 mAs.

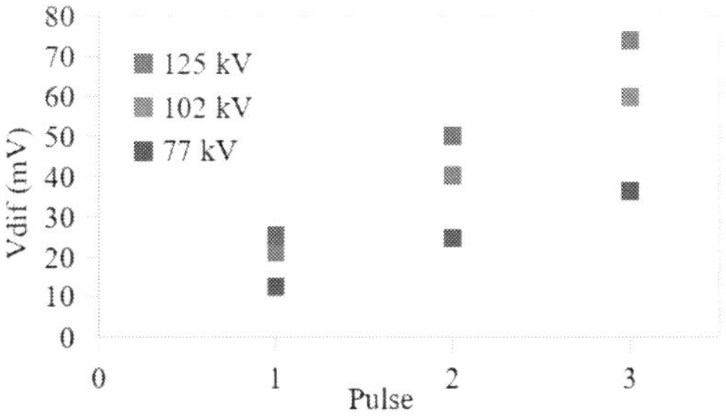

Fig. 5. Variation in V_{dif} (there is no offset canceled) for three X-ray pulses with potential corresponding to ≈100kV ± 25%, the workload is 200 mAs.

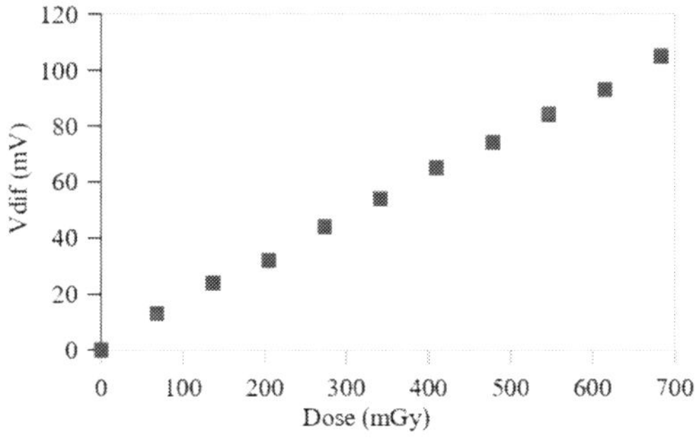

Fig. 6. Variation in V_{dif} (there is offset canceled) for a sequence of 10 X-ray pulses with a peak potential equal to 102kV@100 mAs, R^2=0.99892.

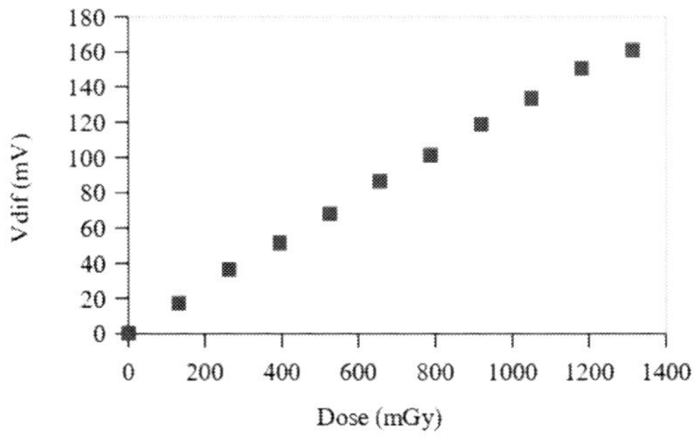

Fig. 7. Variation in V_{dif} (there is offset canceled) for a sequence of 10 X-ray pulses with a peak potential equal to 102kV@200 mAs, R^2=0.99854.

One can observe that the step of the curve in Fig. 5 practically doubles in value compared with Fig. 4, since it corresponds nominally to twice the value of the dose received by the circuit regardless of the potential applied to the X-ray tube. Note that the same happens for irradiation repeated 10 times in each workload. In other words, each X-ray pulse regularly generates steps in the dose range up to 1.4 Gy. From the point of view of lung radiography, which is the most frequent in radiography examinations and corresponds to a dose lower than 1 mGy [7], if the circuit does not present saturation, it would be possible to use the circuit more than 1,400 times in the radiography clinic. Actually, the MOSFET used in this circuit showed linearity with doses of up to 70 Gy (see Fig. 8 in [4]) and by canceling the offset or even adjusting it to a certain negative value, the circuit can have a much higher voltage compliance for V_{dif}. Therefore, the new dosimeter can be reused more than 70000 times. In addition, the IA circuit can also be adjusted to increase the voltage compliance of the differential circuit. As the circuit proposed in this article is not complex, it is suggested that an optimization study can be carried out to become a wearable circuit and then be printed on patients' clothing, or on the uniforms of nuclear power plant workers, or even astronauts. In a microcontrolled system, it is possible to easily have the dose history of the patient or worker exposed to radiation. Such a system has already been applied for a patent in 2021 and the application has already been published by the Brazilian patent office recently [8].

IV. Conclusions

Based on the results found, the innovative differential circuit with MOSFET works as a linear dosimeter in the X-ray range applied to diagnostic radiology. The useful life of such an electronic dosimeter will be a function of the dose rate to which the circuit is exposed and also depend on the voltage compliance of V_{dif}. Therefore, this will depend on the type of MOSFET, its response as a function of dose and even on the circuit adjustments, that is, applying a negative offset to gain compliance in V_{dif}. Actually, as the circuit is shielded from radiation, it is only necessary to change the irradiated device to reuse the electronic dosimeter.

Acknowledgment

The authors would like to thank the "Unidade de Diagnostico por Imagem" at University Hospital (HC-UFPE/EBSERH) for the collaboration in the irradiation procedures. We also thank the company Scients (contract 200313) and the Brazilian agency CNPq for financial supports (grant 305017/2021-7).

References

[1] R. C. Hughes, D. Huffman, J. V. Snelling, T. E. Zipperian, A. J. Ricco, and C. A. Kelsey, "Miniature radiation dosimeter for in vivo radiation measurements," Int. J. Radiat. Oncol. Biol. Phys., vol. 125, no. 2, pp. 963–967, May. 1988, DOI: 10.1016/0360-3016(88)90019-3.

[2] L. J. Asensio, M. A. Carvajal, J. A. López-Villanueva, M. Vilches, A. M. Lallena, and A. J. Palma, "evaluation of a low-cost commercial mosfet as radiation dosimeter," Sensors and Actuators A: Physical, vol. 14, no. 5, pp. 288–295, Jan. 2006, DOI: 10.1016/j.sna.2005.08.020.

[3] G. Sarrabayrouse, and S. Siskos, "Radiation dose measurent using MOSFETs." IEEE Instrumentation & Measurement Magazine, vol. 1, no. 2, pp. 26–34, Jun. 1998, DOI: 10.1109/5289.685494.
[4] L. C. Gonçalves Filho, D. S. Monte, F. R. Barros, and L. A. P Santos, "Radiation dose response of n-channel MOSFET submitted to filtered X-ray photon beam," IEEE Transactions on Nuclear Science, vol. 65, no. 9, pp. 2607–2610, Sept. 2018, DOI: 10.1109/TNS.2018.2846668.
[5] D. S. Monte, and L. A. P Santos, "A method for measuring ionizing radiation dose by analysing hybrid-π parameters of transistors," 2020 I2MTC, pp. 1-6, DOI: 10.1109/I2MTC43012.2020.9129059.
[6] F. A. Cavalcani, and L. A. P. Santos, "A MOSFET based method for measuring peak kilovoltage (kVp) in diagnostic X-ray beams, vol. 1, pp. 1–7, Aug. 2022, DOI: 10.1109/OJIM2022.3196725.
[7] IAEA. Applying Radiation Safety Standards in Diagnostic Radiology and Interventional Procedures Using X rays. IAEA; Vienna, Austria: 2006. (Safety Report Series No. 39).
[8] L. C. Gonçalves Filho, and L. A. P Santos, "Dosimetro eletronico de radiação ionizante com MOSFET na configuração diferencial," BR Patent 1020210141212, Jan. 31, 2023.

An optocoupler-based method for dosimetry in low energy X-ray beams

Jonatas V. Silva[1], Luiz C. Gonçalves Filho[2], and Luiz A. P. Santos[2,3,*]
[1]UFPE, Brazil
[2]CRCN-NE/CNEN, Brazil
[3]SCIENTS, Brazil
(*) lasantos.scients@gmail.com

***Abstract*—Semiconductor electronic devices have been used as detector in X-ray beams for decades, and the principle of operation is based on measuring the electrical current generated while such a device is under the ionizing radiation. Some devices that can be used are: photodiode, phototransistor, bipolar junction transistor, MOSFET, among others. In this work, a method based on an optocoupler device is presented for radiation dose real-time measurement in low-energy X-ray beam, which is used in medical radiology. Such an optoelectronic device has an LED and a phototransistor inside it, in general, for signal communication between two circuits that have to be electrically decoupled. Therefore, the light emitted by the LED excites the phototransistor. To develop an innovative method for radiation detection, the output signal from the phototransistor is used to feed back the current source of the LED so that the phototransistor is kept at a constant operating point by the photonic intensity of the LED, which is shielded from the ionizing radiation beam to be detected. When the optocoupler is irradiated an additional signal on the phototransistor is superimposed on the signal originated by the LED. Then, a time-integrated error signal is digitized and stored in real-time. Each step of the error signal is proportional to the accumulated radiation dose during the device exposure. This innovative method was tested under different conditions varying both peak kilovoltage (kV) and workload (mAs) in order to compare with results from other electronic devices. The experimental results showed that the optocoupler-based method for X-ray dose measurement works and consequently may become an option for X-ray dosimetry.**

***Keywords* — Optocoupler, X-ray, Dosimeter.**

I. Introduction

Semiconductor devices such as photodiode, MOSFET, bipolar junction transistor (BJT), and other devices are widely used as an ionizing radiation detector [1-5]. When the *pn* junction (photodiode) is under X-rays, pairs of electron-holes are generated and a current in the device $i_d(t)$ [1] can be measured. i_d is proportional to the intensity of the radiation beam and also the energy of the incident photons. While on the subject, regarding the effective energy of X-ray photons used in medical diagnosis, it can vary between 50keV and 100keV depending on the peak potential applied to the X-ray tube, generally between 70kV and 150kV, respectively. Furthermore, the mean energy of the X-ray beam also depends on the radiation filtration, which can change the spectrum from soft to hard X-ray [6-7]. Indeed, the probability of interaction of X-ray photons depends on the material. For low atomic numbers and energies between 50keV and 100keV pratically more than 90% of the interactions corresponds to the Compton effect and less than 10% are the other effects [8].

A typical phototransistor, which can be seen as a BJT with transparent encapsulation, has the base-collector *pn* junction operating as a photodiode and the photocurrent, i_{ph}, is actually the input bias or base current ($I_B = i_{ph}$), which is amplified by the transistor gain, β, resulting in the output collector current, $I_C = \beta \cdot I_B$. Generally, a typical phototransistor has only two terminals: collector and emitter, and it is called as floating base device (Fig. 1a), since only the photocurrent i_{ph} is needed to make the device operate as a light photon detector.

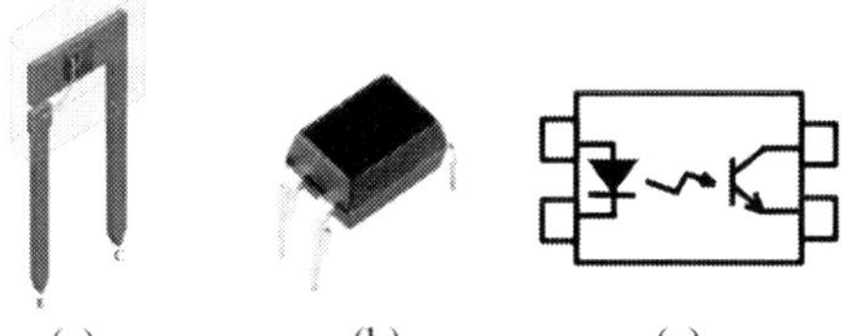

(a) (b) (c)

Fig. 1. a) Transparent floating base phototransistor [4]; b) Typical optocoupler which has an opaque package; c) An optocoupler symbol.

If the phototransistor is used as an X-ray detector, in a typical device with transparent package, there will be two components of the noisy current, i_n: 1) photocurrent produced by the ambient light; 2) current due to the electron-hole pairs produced by the effect of room temperature. A phototransistor that has an opaque package (an ambient light filter) will obviously not have the first noisy component or at least it can be about 40dB weaker than the desired output signal. In fact, it was one of the reasons why the optocoupler was chosen to develop the dosimetry method for X-ray used in diagnosis. Following Santos [4], one can assign the current due to X-ray photons as I_{BX}, which is amplified and results in I_{CX}, i.e., the current due only to the interaction of X-rays on the phototransistor chip. It is important to note that, in the case of X-rays incident on the transistor, the thickness of its package contributes to the build-up cap effect [9], and I_{CX} would be actually a current greater than that of a transistor without a package, even if it is opaque or transparent. Therefore, I_{CX}, is produced by the device as a whole (package + chip) [4].

If a BJT is used as a detector to determine some parameter of a radiation beam, such as the photon energy, collimation techniques must be used to avoid errors due to scattered photons. However, for dose measurement in a patient, it is important to measure the total dose delivered to him, with the contribution of the scattered radiation. Therefore, the X-ray beam applied to the diagnosis must not be collimated in order to can perform dosimetry in a patient, or even in a certain position within a phantom.

Another important thing is about the effect of the directional dependence. The angle of incidence of the X-ray photon beam on the transistor changes its response, *i.e.*, I_{CX} varies with the device angle. It was observed that as the photon incidence angle increases (up to ≈50°) on a BJT there is an increase in the current I_{CX} [4]. This is due to the increase in the path of the X-ray photon in relation to the face of the chip itself, *i.e.*, the increase in the thickness of the device package (build-up cap effect) because of the angle change. For angles greater than 50° the current I_{CX} start to decrease [4]. Based on the explanations given above, a dosimetry method using an optocoupler was developed, as detailed next.

II. Material and Methods

A. Basic circuit

In general, an optocoupler has the pin diagram as shown in Fig. 1c, and terminals A and C are anode and cathode, respectively, corresponds to the LED where the input current, I_{LED}, flows from a current source. The output signal is often from a floating base phototransistor, and terminals C and E are collector and emitter, respectively. Fig. 2 shows a circuit to measure the output current, I_C. For I_C measurement a series ammeter must be connected with the voltage source (source-meter instrument). The ratio between I_C and I_{LED} is referred to the current transfer ratio (CTR) of the optocoupler, which varies nominally between 50 and 500.

B. Optocouplers

Two types of optocouplers were selected to carry out the experiments: 1) SFH620A-2, Vishay®, with CTR≈130; 2) HMHA2801A, Infineon®, with CTR≈100. The last one has mini-flat SMD (Surface Mounted Device) type package and is soldered directly to its specific printed circuit board (PCB). The first one is for AC input signal, therefore has two LEDs, one for positive bias and other for negative bias. This choice allows testing the sensitivity of the phototransistor from two LEDs, providing lower relative uncertainties. In addition, it has a DIP-type package and this will allow for some comparison with the SMD-type package optocouplers.

C. PCB and shielding of the LED

Two types of PCBs were made: one for the SMD-type device and other for the DIP-type device. In the last one a DIP socket was soldered in the PCB, which was fixed in a plastic holder (Fig. 3). This technique was required for the following reasons: 1) to be able to hold the sample to be irradiated at a given angle, θ, 2) to allow the LED to be shielded from the incident X-ray photons; 3) to exchange the optocoupler device with practicality and make statistics. As can be seen in Fig. 3, the optocoupler is placed so that the LED must be at the end of the PCB to shield only it.

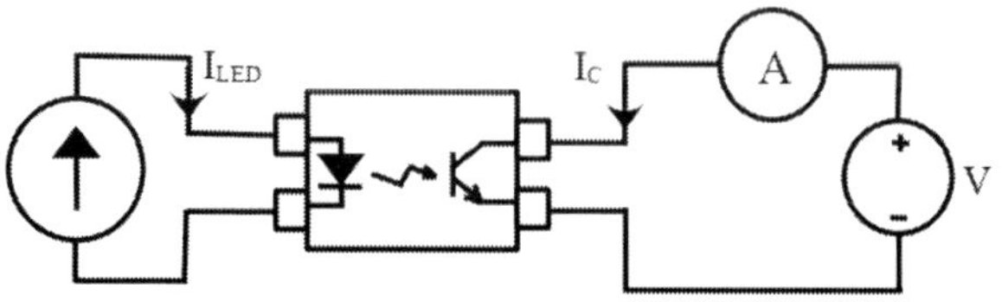

Fig. 2. Basic circuit for measuring the output current I_C of an optocoupler.

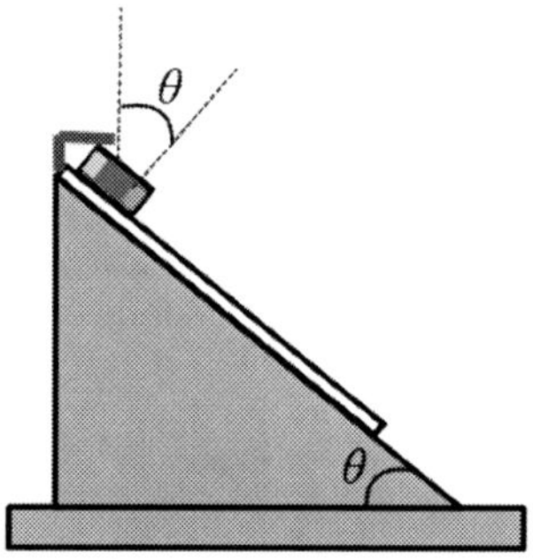

Fig. 3. Illustration of a plastic holder (gray); with PCB (white); LED shield (red); and the optocoupler with its LED (orange) and phototransistor (green).

D. Radiation generator and X-ray tube parameters

The radiation generator was a typical clinical equipment for X-ray beam applied to the diagnosis: Polymat 30/50 plus, Siemens®. The X-ray tube parameters chosen in the control panel of the clinical equipment were: 1) 1000 ms pulse time; 2) 73kV, 102kV and 125kV peak kivoltages; 3) 50 mAs, 100 mAs and 200 mAs for workloads. Each point of each graph presented in results corresponds to an average of two readings.

E. Experimental setup

Fig. 4 illustrates the experimental setup for irradiation procedures. θ is the incidence angle of the radiation beam in relation to the surface of the phototransistor chip face in the optocoupler. Therefore, the photons of light coming from the LED arrive perpendicularly at the phototransistor chip face, whereas X-ray photons reach the chip with the angle θ.

F. Instruments

The main instruments used in the experiments are: 1) 2450 source-meter, Keithley®, working as a current source, I_{LED}. 2) a 6430 source-meter, Keithley®, working as voltage source, V_{CE}, and current meter, I_C; 3) an EFF1705, 2 channel source-meter, Scients®, for real-time measurements in the clinic.

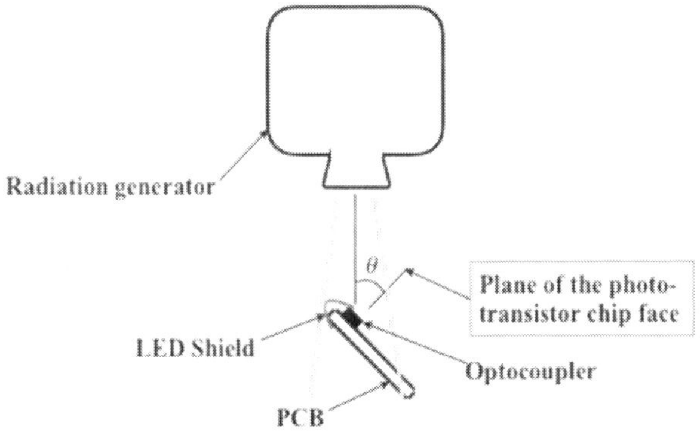

Fig. 4. Experimental setup for irradiation procedures.

G. Basic methods

First of all, an evaluation was carried out to check which angle of incidence, θ, the optocoupler should have for an optimized efficiency, *i.e.*, we are interested in obtaining the highest value of current I_{CX}. The values of angles chosen for this experiment were: 30°, 45° and 60°, following Santos [4].

Secondly, in dosimetric techniques, generally there are two important parameters to take into account: 1) the detector response as a function of the X-ray tube peak potential; and 2) the detector response as a function of dose rate, which is equivalent to say the device response as a function of X-ray tube current. For example, the X-ray photonic intensity with workload Q_1=100mAs is 50% less than the workload of Q_2=200mAs, both during a time of 1s. However, I_{CX} could not be linearly proportional. In other words, I_{CX} does not necessarily obey a linear function so that $I_{CX1}=I_{CX2}/2$. The X-ray tube peak potential chosen were: 73 kV, 102 kV, and 125 kV. These choices represent approximately 100kV±25%. The workload was set to be: 50 mAs, 100 mAs, and 200 mAs. In general, 50 mAs is lower workload normally used at the clinic, therefore this will be the worse case.

The third method is to assess how the phototransistor responds to the radiation dose step. Although knowing that the response of the detector must be analyzed during the exposure of the radiation beam, as explained above, the most important in this study is the variation of the electrical state of the phototransistor inside the optocoupler, since it is known that such a device degrades after each radiation dose step [4]. In this experiment one must deliver systematically dose steps to the device, so that be possible to analyze if there is a correlation with the variation of I_C and the accumulated radiation dose. Therefore, the characteristic curves of $I_C \times I_{LED}$ were plotted before and after irradiation to make comparisons.

H. Proposed technique

The three initial methods described above will culminate in the proposal of this work. Fig. 5 can help us clarify that. Based on the results of the phototransistor response under radiation beams, it can be seen that I_C decreases with the accumulated radiation dose [4]. It is actually a decrease in the transistor gain after each X-ray dose. Therefore, at each radiation pulse, an error signal can be registered on a microcontroller (μC), for example. The μC corrects the signal and changes the bias, I_{LED}, so that the phototransistor operating point is returned, *i.e.*, as V_{CE} is constant, the current of the LED is increased so that the I_C returns to the previous value. Practically, this is a digital feedback. Using the timer in the μC and the registered error signal, both dose rate and dose on the irradiated device can then be measured, after a calibration procedure.

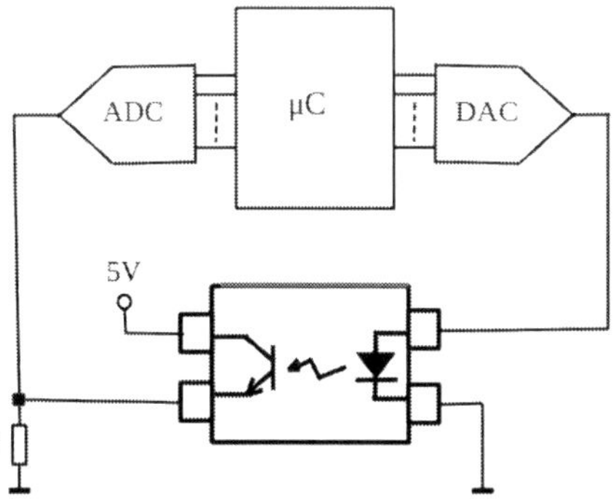

Fig. 5. Simplified circuit schematic of the optocoupler-based method for dosimetry in low energy X-ray beams.

III. Result and Discussions

First of all, it was observed that at the angle of θ=30° I_{CX} was less than 50% compared with θ=45°. At the angle θ=60° the LED shield itself covered part of the phototransistor and I_{CX} also decreased. Therefore, the chosen angle was 45°.

Fig. 6 and 7 show the real-time I_{CX} measurements for the chosen peak potentials and workloads, for the two types of optocouplers. Note that it does not depend on the type of optocoupler, i.e. thicker (DIP) or thinner (SMD) package. Also note that the behavior of the detector response is independent of the CTR, that is, it does not matter what type of optocoupler is used as the response to the X-ray beams is similar. However, a higher CTR can provide a higher output signal.

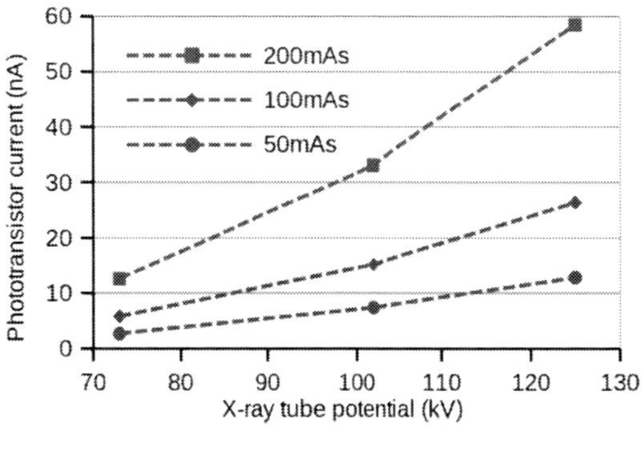

Fig. 6. I_{CX} as function of X-ray tube potential, parameterized by the workload for the HMHA2801A optocoupler (SMD type).

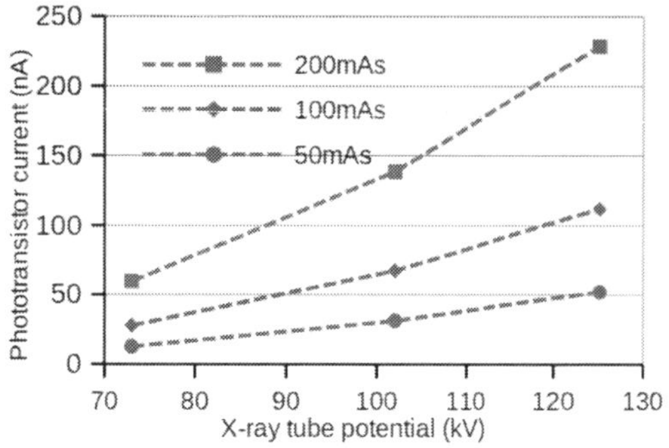

Fig. 7. I_{CX} as function of X-ray tube potential, parameterized by the workload for the SFH620A-2 optocoupler (DIP type).

Also, one can observe that the DIP type optocoupler has a higher I_C output signal than the SMD type, as this is due to the buildup cap effect since the CTR of both are similar.

The most important result of this work is shown in Fig. 8. It is I_C as a function of I_{LED}, parameterized by each 1Gy dose step. These data correspond to a SFH620A-2 optocoupler. It is worth noting that 1Gy dose is equivalent to more than one thousand (lung) radiography exams [6]. Therefore, it can be said that, in X-ray beams applied to medical diagnosis, the optocoupler lifetime could be long, i.e., it can be used at least for 3000 exposures.

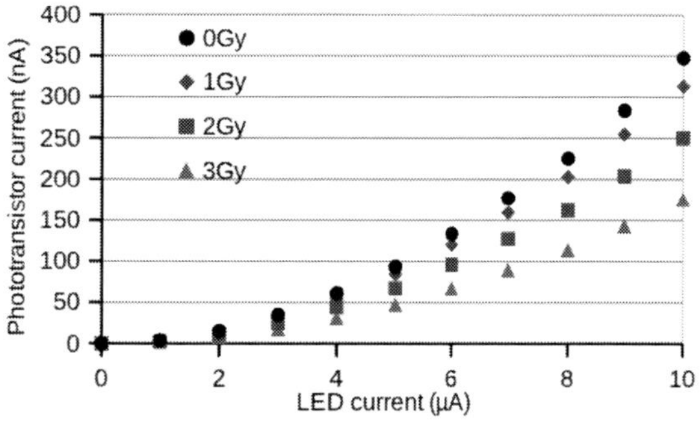

Fig. 8. I_{CX} as function of X-ray tube potential, parameterized by the workload for the SFH620A-2 optocoupler (DIP type).

IV. Conclusions

An optocoupler-based dosimetry method for low-energy X-ray beams was presented. For this purpose, the optoelectronic device must be placed at 45° angle in relation to the face of the phototransistor chip inside the optocoupler. As there is gain degradation in a transistor after irradiation, basically, the phototransistor operating point must be corrected from a digital feedback applied to the LED inside the optocoupler. Also the LED must be shielded. In this way, each X-ray pulse generates an error signal that is recorded. The results show that each dose step corresponds to the LED current step stored by the digital circuit, which is previously calibrated as a function of the dose. It can be concluded that this innovative method works and may become an option for low energy X-ray dosimetry in the near future.

Acknowledgment

The authors would like to thank the "Unidade de Diagnostico por Imagem" at University Hospital (HC-UFPE/EBSERH) for the collaboration in the irradiation procedures. We also thank the company Scients (contract 200313) and the Brazilian agency CNPq for financial supports (grant 305017/2021-7).

References

[1] G. Lutz, Semiconductor Radiation Detectors, Heidelberg, GER: Springer Berlin, 2007, pp. 79–258.
[2] G. Sarrabayrouse, and S. Siskos, "Radiation dose measurement using MOSFETs," IEEE Instrumentation & Measurement Magazine, vol. 1, no. 2, pp. 26–34, Jun. 1998, doi: 10.1109/5289.685494.
[3] M. Spahn, "Flat detetors and their clinical applications," Eur. Radiol. Vol. 15, pp. 1934–1947, Apr. 2005, doi:10.1007/s00330-995-2734-9.
[4] L. A. P. Santos, "An overview on bipolar junction transistor as a sensor for X-ray beams used in medical diagnosis," Sensors, vol. 22(5), pp. 1923, Mar. 2022, doi: 10.3390/s22051923.
[5] L. A. P. Santos, C. M. S. Magalhães, J. O. Silva, J. Antonio Filho, E. F. Silva Jr., W. M. Santos, "A feasibility study of a phototransistor for the dosimetry of computerized tomography and stereotactic radiosurgery beams," Radiation Measurements, vol. 43, pp. 904–907, June 2008, doi: 10.1016/j.radmeas.2007.11.065.
[6] IAEA, Diagnostic Radiology Physics: a handbook for teachers and students, IAEA: Vienna, Austria, 2014.
[7] D. McLean, "X-ray Spectra and beam qualities," In Proceedings of the joint ICTP-IAEA advanced school on dosimetry in diagnostic radiology, Trieste, Italy, May 2009.
[8] A. Lechner, "Particle interactions with matter," In Proceedings of the CAS-CERN Accelerator School: Beam injection, Extraction and Transfer, Erice, Italy, Mar. 2017, CERN yellow reports, Geneve, Switzerland, 2018.
[9] J. E. Turner, Atoms, Radiation, and Radiation Protection, 3rd ed., Weinheim, GER: Willey-VCH Verlag GmbH & Co. KgaA, 2007.

Prompt gamma-ray spectroscopy in conjunction with the Monte Carlo Library Least Squares approach: applications to range verification in proton therapy

Kyrre Skjerdal[1,*], Toni Kögler[2,3], William Lionheart[4], Kristian Smeland Ytre-Hauge[5], and Ilker Meric[1]
[1] Department of Computer Science, Electrical Engineering and Mathematical Sciences, Western Norway University of Applied Sciences, Norway
[2] OncoRay – National Center for Radiation Research in Oncology, Faculty of Medicine and University Hospital Carl Gustav Carus, Technische Universität Dresden, Helmholtz-Zentrum Dresden - Rossendorf, Germany
[3] Helmholtz-Zentrum Dresden - Rossendorf, Institute of Radiooncology – OncoRay, Germany
[4] Department of mathematics, The University of Manchester, UK
[5] Department of Physics and Technology, University of Bergen, Norway

(*) kyrre.skjerdal@hvl.no

Abstract— **Prompt Gamma-ray Spectroscopy (PGS) in conjunction with the Monte Carlo Library Least Squares (MCLLS) approach was investigated for the purposes of range monitoring in proton therapy through Monte Carlo simulations. Prompt gamma-rays are produced during treatment and can be correlated to the range of the proton beam in the tissue. In contrast to established approaches, MCLLS does not rely on the identification of specific photopeaks. Instead it treats each individual constituent as a library spectrum and calculates coefficients for each spectrum, and therefore takes both the photopeaks and the Compton continuum into account. It can thus be applied to organic scintillators traditionally not used for energy spectroscopy due to their low Z number and density. Preliminary results demonstrate that the proposed approach returns a strong linear correlation between the range of the primary proton beam and the calculated library coefficients, depending on the composition of libraries. This can be exploited for range monitoring.**

***Keywords* —proton therapy, spectroscopy, Monte Carlo simulations, range verification**

I. Introduction

PARTICLE therapy is a promising alternative to traditional radiotherapy with photons. A particle beam deposits most of its energy towards the end of its range, as opposed to photon beams which deposits energy all the way through the tissue. Particle therapy is therefore a more precise form of treatment, as long as one can control the range of the beam with high precision. The range of a proton beam depends on its initial energy and physical properties of the target, such as electron density and excitation potential [1]. The proton range in a patient is normally estimated based on a CT scan of the patient. Because of uncertainties in conversion from Hounsfield units to proton stopping power, heterogeneities in the tissue organ motion as well as patient motion and positioning, there will be uncertainties in the range of the proton beam [2]. As of today, there are no systems in clinical that are routinely in daily clinical use for monitoring the proton range in proton therapy. Since the protons stop inside of the patient, the range cannot be measured directly, and one must rely on the detection of secondary radiation escaping the patient. As the particle beam traverses the tissue, it will produce neutrons and gamma-rays at a rate correlated with the range and intensity of the beam [3]. In the NOVO project (Neutron and gamma-ray imaging for real-time range verification and image guidance in particle therapy) we propose to use The NOVO Compact Detector Array (NOVCoDA) to image fast neutrons and prompt gamma-rays (PG) to monitor the range of the proton beam in real-time [4]. A prototype of the NOVCoDA is under construction. The prototype will have a total of 16 detector elements, where each detector element consists of a bar shaped organic scintillator [5] and silicon photomultipliers (SiPM) for dual-ended signal readout.

Monte Carlo studies have shown that to detect changes range shifts of the order of 1 mm with the NOVCoDA, 2×10^7 protons per spot is needed [4]. In proton therapy, especially in pencil beam scanning, it is desirable to be able to detect range changes with intensities as low as 10^6 protons per spot. Therefore, we are investigating ways to improve the efficiency of the NOVCoDA. One possibility is to use the detector for energy spectroscopy, as a large proportion of detected PGs will scatter only once or twice and can therefore not be used for imaging. The use of Prompt Gamma-ray Spectroscopy (PGS) for range monitoring in particle therapy has previously been proposed by several groups [6]–[8] and clinical trials of at least one clinical prototype are ongoing [7]. These systems make use of the specific photopeaks in the acquired energy spectra as the intensities of these have been shown to correlate with the proton beam range in tissue.

Organic scintillators are usually not used for energy spectroscopy due to their low effective atomic number, Z_{eff}, and density, which means that photoelectric absorption is very unlikely in these materials, especially at the PG energies of several MeV present in proton therapy. The library least squares (LLS) method makes use of the whole detected spectrum and not only the photopeaks as in conventional spectroscopy. This will reduce the dependency on being able to identify the peaks in the spectrum for quantitative analysis. The LLS method relies on so called libraries, which are PG energy spectra from the individual elements of the sample that is irradiated. These library spectra are challenging to generate experimentally, and in this study we therefore use Monte Carlo (MC) simulations to generate the libraries needed. This is called the Monte Carlo library least squares (MCLLS) approach [9]–[13]. The MCLLS approach has previously been applied to numerous inverse radiation analyser problems [12], but never in the context of range monitoring in proton therapy. In this study we want to investigate whether it is possible to use the NOVCoDA for quantitative spectroscopy in conjunction with the MCLLS approach. If successful, the NOVCoDA could then also be used for PGS, despite the fact that it is based on the use of organic scintillators. Successful application of the MCLLS approach would allow it to be used as an alternative to conventional spectroscopy in the context of range monitoring in proton therapy.

II. Methods

A. The Monte Carlo Library Least Squares approach

As mentioned in the introduction the LLS approach makes use of the whole PG spectrum, both the specific energy peaks and the Compton continua [11]. This automatically treats unresolved spectral features and returns residuals that may reveal whether or not there are missing spectral contributions in the quantitative analysis. The main assumption is that the total PG spectrum is a linear combination of the relevant library spectra, i.e., the individual spectra of each constituent in a given sample. The counts in each channel (energy histogram bin) can then be formulated mathematically as follows [10]:

$$R_i = \sum_{j=1}^{m} \alpha_j R_{j,i} + e_i,\ i = 1,2,3,\dots,n \qquad (1)$$

Here R_i is the total counts in channel i for the unknown sample, α_j is the multiplier coefficient for component j in the unknown sample, m is the number of elements, $R_{j,i}$ is the counts in channel i for component j (the library spectrum of the jth component), e_j the statistical uncertainty in channel i, and n is the number of channels. The coefficients α_j are obtained through a least squares analysis by minimizing the reduced chi-square:

$$\chi_v^2 = \sum_{i=1}^{n} \frac{e_i^2}{(n-m)\sigma_i^2} \qquad (2)$$

Here $(n-m)$ is the number of degrees of freedom and σ_i is the standard deviation in channel i [10]. The minimization of the chi-square values is done by setting the derivatives with respect to the library multiplier coefficients equal to zero. When the coefficients are obtained, the unknown spectrum should be reconstructed by using the library coefficients in (1). The residuals in each channel can be calculated with the formula

$$Res_i = \frac{R_i - \sum_{i=1}^{m} \alpha_j R_{j,i}}{\sigma_i}. \qquad (3)$$

The linearity assumption is normally violated since the formation of composite PG spectra is an inherently non-linear process. The contribution to the library spectrum from one element will not only depend on the amount of that element, but also all other elements [12]. In the Monte Carlo library least squares approach the elemental library spectra are therefore generated with accurate forward Monte Carlo (MC) radiation transport simulations, which will take into account the non-linear matrix effects mentioned above and generate accurate library spectra [11].

The MCLLS approach proceeds in three steps [11]:

1. Generate elemental library spectra with MC simulations for a sample of an assumed composition.
2. Obtain the total PG spectrum of an unknown sample of the same elements (normally obtained experimentally).
3. Obtaining fitting coefficients by performing a least squares analysis using the libraries generated in step 1 and the gamma-ray spectrum in step 2.

B. Simulations

A Geant4 (v11.0.2) MC simulation framework for generation of the elemental library spectra [14]–[16] was applied in the present study. The pre-packaged reference physics list QGSP_BIC_HP was used in the simulations.

The simulation geometry, shown in Fig. 1, consists of a phantom and a monolithic detector.

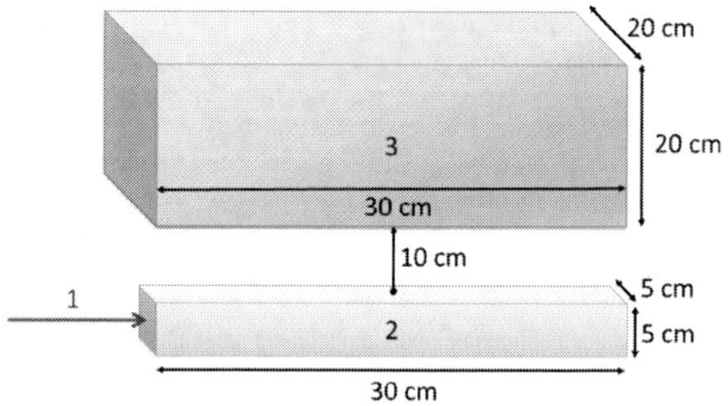

Fig. 1. The simulated geometry. The incoming proton beam (1) is impinging on the short end of the phantom (2), and the prompt gamma-rays are detected by the detector (3).

The phantom has the dimensions 5 cm × 5 cm × 30 cm and is made of polymethyl methacrylate (PMMA) with the chemical formula $C_5O_2H_8$. The detector is a 20 cm × 20 cm × 30 cm organic glass scintillator (OGS). The composition of the OGS material is implemented as 53.2 % carbon (C), 45.6 % hydrogen (H) and 1.2 % silicon (Si), with a density of 1.09

g/cm³. The detector is placed 10 cm lateral to the phantom and centred at 15 cm depth for the reference case without range shift.

The known PG spectra were generated by simulating a pencil beam of 10^9 protons with an energy of 150 MeV, impinging on the short end of the phantom, as shown in Fig. 1. To introduce range shifts PMMA was added or removed from the phantom where the beam enters. The PMMA was added or removed in steps of 2 mm, from -10 mm to 10 mm. For each geometry, 10^9 primary protons were simulated.

The deposited energies of the gamma-rays interacting in the detector was recorded and resampled from a Gaussian distribution to account for the energy resolution of the detector, using the ROOT data analysis framework (v6.18.04) [17]. The resolution defined by the FWHM was calculated with the formula [18]

$$\frac{\Delta E}{E} = \sqrt{A^2 + \frac{B^2}{E} + \frac{C^2}{E^2}}, \quad (4)$$

where E is the deposited energy, and $A = 0.0497 \pm 0.0031$, $B = 0.0000 \pm 8.4221$, and $C = 0.0349 \pm 0.0011$ are measured from an OGS bar of 1.0 cm × 1.0 cm × 10 cm [4], [19].

C. *Analysis of the generated prompt gamma-ray spectra*

The spectra generated were analysed using the code CEARLLS [13], [20], which is developed by The Center for Engineering Applications of Radioisotopes (CEAR), for MCLLS analysis. The elemental spectra were divided into two library spectra, one for the sum of the spectra coming from scatters of hydrogen, carbon, and oxygen (HCO), and one from prompt gamma-rays coming from other processes (OP), including events with pile-up of several PG's interacting in the detector simultaneously. Further, the combination of the sum of the spectra from scatters of hydrogen and carbon (HC), and the sum the spectra of scatters of oxygen and from other process (OOP) has been evaluated. Only two library spectra were used because the spectra from some of elements, especially the spectrum from scatters of hydrogen, contains significantly fewer entries than other spectra.

The elemental libraries were generated from the reference case with no range shift, and the total spectra of all detected PGs from each of the range shifted cases were used as the unknown spectra.

The analysis was done using both all 10^9 primary protons, and by choosing 10^8 primary proton events randomly. In the latter case, the process was repeated 10 times and the average fitting coefficients calculated.

The uncertainty in each channel is taken from Poisson statistics as the square root of the counts in each channel, $\sigma_i = \sqrt{R_i}$. This is used for the weighting in the calculation of the reduced chi-square value.

III. Results and Discussion

A. *Produced and detected prompt gamma-ray spectra*

The generated PG energy spectra are shown in Fig. 2a, and in Fig. 2b the detected PG spectra including detector resolution, coming from scatters from the different elements in the PMMA phantom, are shown. The PG spectra produced include the expected specific energy peaks on top of the continuum. As expected, these peaks are smeared out when the gamma-rays are detected by the organic scintillator detector.

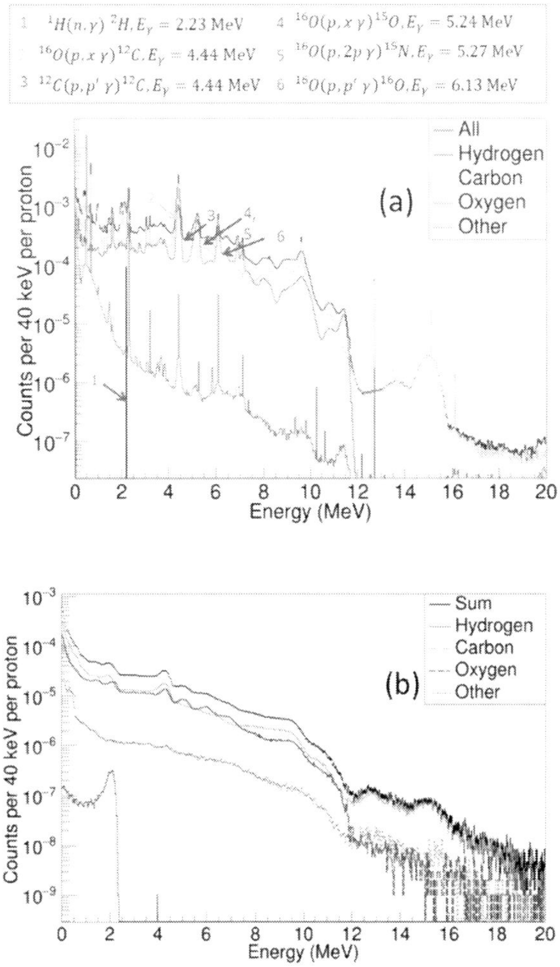

Fig. 2. The energy spectra of the generated (a) and detected (b) prompt gamma-rays.

B. *Results from 10^9 primary protons*

In Fig. 3 the original total PG spectrum, and the spectrum reconstructed using the library coefficients from the MCLLS analysis and the library spectra, are shown. The residuals, calculated with (3), are indicated in the lower subplot.

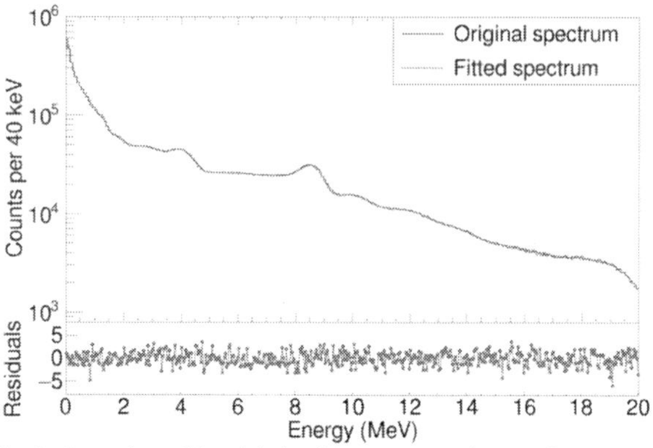

Fig. 3. Comparison of the original and the reconstructed spectra for 10 mm range shift for 10^9 primary protons. The residuals in number of standard deviations are indicated in the lower subplot.

The fitting coefficients for 10^9 primary protons for range shifts from -10 mm to 10 mm in steps of 2 mm are shown in Fig. 4. The reduced chi-square values for each range shift are indicated in the lower subplot. For each of the two libraries the fitting coefficients are linearly dependent on the range shifts, with $R^2 = 0.992$ for the HCO library spectrum and $R^2 = 0.953$ for the OP library spectrum.

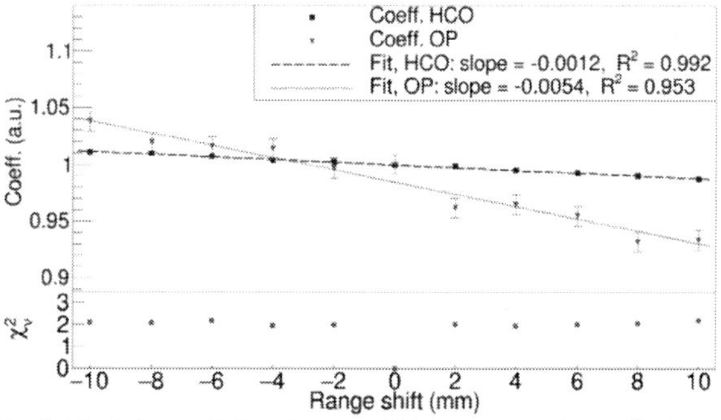

Fig. 4. The fitting coefficients for range shifted PG spectra from 10^9 primary protons. The chi-square values for each of the geometries are indicated in the lower subplot.

In the OP library spectrum, the slope of the linear fit is steeper than in the HCO library spectrum and should therefore be more sensitive to changes in the range than the HCO library spectrum. On the other hand, the OP library spectrum has larger uncertainties (about one order of magnitude) in the fitting coefficients, and they are more sensitive to small changes in the initial conditions, also changes that does not come from range shifts. This is due to the limited size of the OP library, which make up 5.3% of the detected prompt gamma-rays, compared to the HCO library.

C. *Results from 10 iterations of 10^8 primary protons*

In Fig. 5 and Fig. 6 the results from the analysis of the 10 samples of 10^8 primary protons are shown. In Fig. 5 a comparison of the original spectrum for one sample of 10 mm range shift is shown.

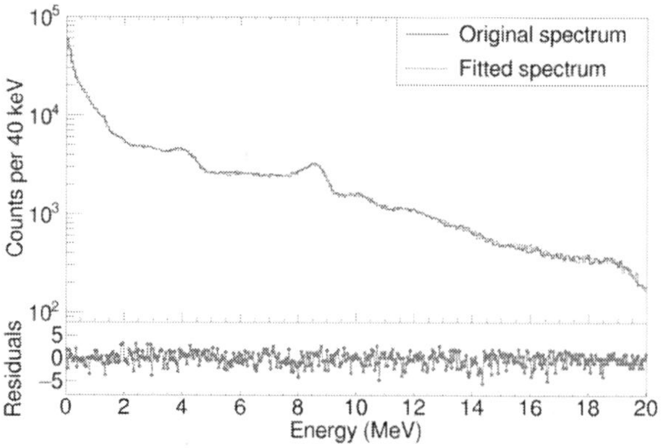

Fig. 5. Comparison of the original and reconstructed spectra for 10 mm range shift for 10^8 primary protons. The residuals in number of standard deviations are indicated in the lower subplot.

The average fitting coefficients are shown in Fig. 6, with the standard deviation indicated by the error bars. The fitting coefficients are still linearly dependent on the range shift with $R^2 = 0.983$ for the HCO library spectrum and $R^2 = 0.949$ for the OP library spectrum.

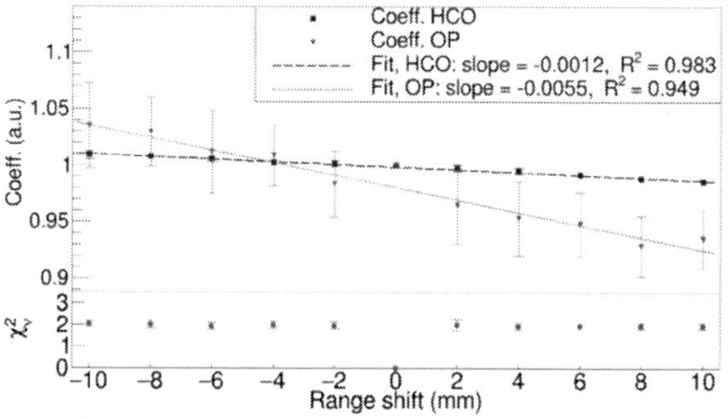

Fig. 6. The average fitting coefficients for range shifted PG spectra from 10 iterations 10^8 primary protons. The average chi-square values for each of the geometries are indicated in the lower subplot.

The same trend as in the previous section can be observed when using 10^8 primary protons. There is a linear correlation for both the HCO library spectrum and the OP library spectrum, with a steeper slope for the latter. Especially for the OP library spectrum, there is a large variation for a single range shift between the various samples, as indicated by the error bars. It was therefore decided to use an average of 10 different samples. This comes again from the fact that the OP library spectrum only makes up approximately 5% of the detected PGs and is therefore much more sensitive to small random changes.

D. *Using another library composition*

In the library used above the HCO library spectrum contains in the order of 95% of the detected PGs. To make the two libraries more equal in size, using the library spectra of the sum of the spectra from scatters of hydrogen and carbon (HC), and the sum the spectra of scatters of oxygen and from other process (OOP) has been evaluated. The HC library spectrum makes up approximately 56% of the detected PGs, while the OOP library spectrum makes up the other 44%. In Fig. 7 the comparison of the original and fitted spectra with residuals indicated underneath for the 10 mm range shifted spectra, is shown. The fitting coefficients for all the range shifted spectra are shown in Fig. 8. The fitting coefficients are still linearly dependent on the range shift, but with weaker correlation for the library of the sum of oxygen and others. The R^2 is 0.949 for the HC library, and 0.748 for the OOP library.

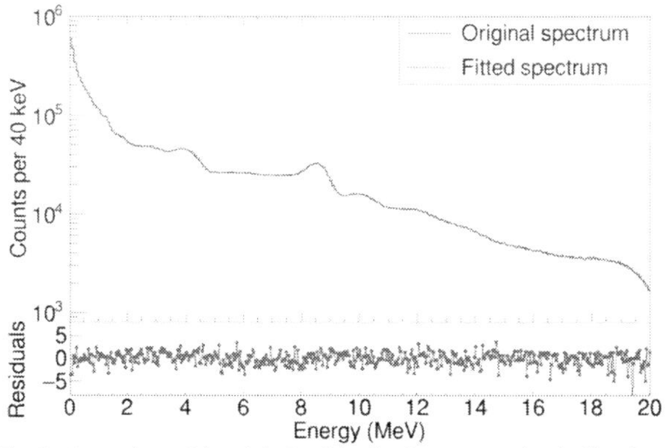

Fig. 7. Comparison of the original and the fitted spectrum using the libraries HC, and OOP. The residuals in number of standard deviations are indicated in the lower subplot.

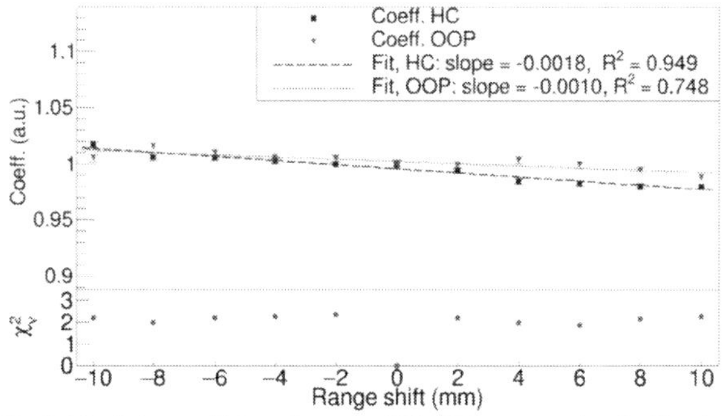

Fig. 8. The fitting coefficients for range shifted PG spectra using the libraries HC, and OOP. The chi-square values for each of the geometries are indicated in the lower subplot.

In this case, the results are more similar for the two library spectra, with uncertainties of the same order of magnitude, but with a weaker linear correlation for the OOP library spectrum. This library spectrum also has a very low slope, almost equivalent to zero, and it is therefore unclear it is feasible to apply this library spectrum for range shift detection.

IV. Conclusions and Outlook

The purpose of this work is to investigate if it is possible to use the NOVCoDA detector for PGS in conjunction with the MCLLS approach, despite it being based on organic scintillators, in the aspect of range verification in proton therapy.

A framework for simulation of a proton beam and recording the energies of the PGs generated in a phantom, has been built. Prompt gamma-ray spectra were successfully generated for 11 different scenarios, emulating a reference case and 10 different scenarios with range shifts from −10 mm to 10 mm in steps of 2 mm increments.

For the analysis of the PG spectra from 10^9 primary protons the MCLLS fitting coefficients have a linear dependence on the range shifts for both the HCO and OP library spectra, where the HCO library spectrum contains the energies detected prompt gamma-rays from scatters of hydrogen, carbon and oxygen, and the OP library spectrum contains the energies of detected PGs from other processes. The slope is steeper for the latter library, but with larger uncertainties. For the analysis using 10 iterations of 10^8 primary protons, the same trend can be observed. The linear correlation of the library coefficients with the change in range can be exploited for detecting range shifts in proton therapy, by measuring these spectra and detect deviations in the library coefficients.

When using the library spectra HC and OOP, the results and uncertainties of the library coefficients are more similar for the two library spectra. However, the linear correlation is weaker for the OOP library spectrum, and therefore it is uncertain if this library composition can be used for detecting range shifts. This needs to be investigated further with a statistical classification analysis.

As explained in the introduction the NOVCoDA will be a detector array consisting of bar shaped detector elements, and not a monolithic detector as simulated in this work. Each detector element will be read out separately, and therefore the different detector elements will have different gain drift. This will have to be accounted for in future work, both in simulation studies and when doing experiments. In addition, effects such as event pile-up might become a challenge when using experimental data. These events should be filtered out based on the pulse shape.

Acknowledgment

We would like to acknowledge the financial support by The Research Council of Norway (Grant no. 301459).

References

[1] W. D. Newhauser and R. Zhang, 'The physics of proton therapy', *Phys Med Biol*, vol. 60, no. 8, pp. R155-209, Apr. 2015, doi: 10.1088/0031-9155/60/8/R155.
[2] A. J. Lomax, 'Intensity modulated proton therapy and its sensitivity to treatment uncertainties 1: the potential effects of calculational uncertainties', *Phys. Med. Biol.*, vol. 53, no. 4, p. 1027, 2008.
[3] K. S. Ytre-Hauge, K. Skjerdal, J. Mattingly, and I. Meric, 'A Monte Carlo feasibility study for neutron based real-time range verification in proton therapy', *Sci. Rep.*, vol. 9, no. 1, p. 2011, Feb. 2019, doi: 10.1038/s41598-019-38611-w.
[4] I. Meric *et al.*, 'A hybrid multi-particle approach to range assessment-based treatment verification in particle therapy', *Sci. Rep.*, vol. 13, no. 1, Art. no. 1, Apr. 2023, doi: 10.1038/s41598-023-33777-w.
[5] N. P. Giha *et al.*, 'Organic glass scintillator bars with dual-ended readout', *Nucl. Instrum. Methods Phys. Res. Sect. Accel. Spectrometers Detect. Assoc. Equip.*, vol. 1014, p. 165676, Oct. 2021, doi: 10.1016/j.nima.2021.165676.
[6] J. M. Verburg, K. Riley, T. Bortfeld, and J. Seco, 'Energy- and time-resolved detection of prompt gamma-rays

for proton range verification', *Phys. Med. Biol.*, vol. 58, no. 20, p. L37, Sep. 2013, doi: 10.1088/0031-9155/58/20/L37.
[7] F. Hueso-González, M. Rabe, T. A. Ruggieri, T. Bortfeld, and J. M. Verburg, 'A full-scale clinical prototype for proton range verification using prompt gamma-ray spectroscopy', *Phys. Med. Biol.*, vol. 63, no. 18, p. 185019, Sep. 2018, doi: 10.1088/1361-6560/aad513.
[8] P. Magalhaes Martins, H. Freitas, T. Tessonnier, B. Ackermann, S. Brons, and J. Seco, 'Towards real-time PGS range monitoring in proton therapy of prostate cancer', *Sci. Rep.*, vol. 11, no. 1, Art. no. 1, Jul. 2021, doi: 10.1038/s41598-021-93612-y.
[9] T. He, R. P. Gardner, and K. Verghese, 'The Monte Carlo—Library Least-Squares approach for energy-dispersive x-ray fluorescence analysis', *Appl. Radiat. Isot.*, vol. 44, no. 10, pp. 1381–1388, Oct. 1993, doi: 10.1016/0969-8043(93)90089-S.
[10] R. P. Gardner and L. Xu, 'Status of the Monte Carlo library least-squares (MCLLS) approach for non-linear radiation analyzer problems', *Radiat. Phys. Chem.*, vol. 78, no. 10, pp. 843–851, Oct. 2009, doi: 10.1016/j.radphyschem.2009.04.023.
[11] I. Meric, G. A. Johansen, M. B. Holstad, J. Wang, and R. P. Gardner, 'Produced water characterization by prompt gamma-ray neutron activation analysis', *Meas. Sci. Technol.*, vol. 22, no. 12, p. 125701, Oct. 2011, doi: 10.1088/0957-0233/22/12/125701.
[12] I. Meric, G. A. Johansen, M. B. Holstad, J. Mattingly, and R. P. Gardner, 'On the treatment of ill-conditioned cases in the Monte Carlo library least-squares approach for inverse radiation analyzers', *Meas. Sci. Technol.*, vol. 23, no. 5, p. 055603, Apr. 2012, doi: 10.1088/0957-0233/23/5/055603.
[13] R. P. Gardner, A. Sood, Y. Y. Wang, L. Liu, P. Guo, and R. J. Gehrke, 'Single peak versus library least-squares analysis methods for the PGNAA analysis of vitrified waste', *Appl. Radiat. Isot.*, vol. 48, no. 10, pp. 1331–1335, Oct. 1997, doi: 10.1016/S0969-8043(97)00127-9.
[14] S. Agostinelli *et al.*, 'Geant4—a simulation toolkit', *Nucl. Instrum. Methods Phys. Res. Sect. Accel. Spectrometers Detect. Assoc. Equip.*, vol. 506, no. 3, pp. 250–303, Jul. 2003, doi: 10.1016/S0168-9002(03)01368-8.
[15] J. Allison *et al.*, 'Geant4 developments and applications', *IEEE Trans. Nucl. Sci.*, vol. 53, no. 1, pp. 270–278, Feb. 2006, doi: 10.1109/TNS.2006.869826.
[16] J. Allison *et al.*, 'Recent developments in Geant4', *Nucl. Instrum. Methods Phys. Res. Sect. Accel. Spectrometers Detect. Assoc. Equip.*, vol. 835, pp. 186–225, Nov. 2016, doi: 10.1016/j.nima.2016.06.125.
[17] R. Brun and F. Rademakers, 'ROOT — An object oriented data analysis framework', *Nucl. Instrum. Methods Phys. Res. Sect. Accel. Spectrometers Detect. Assoc. Equip.*, vol. 389, no. 1, pp. 81–86, Apr. 1997, doi: 10.1016/S0168-9002(97)00048-X.
[18] H. Klein and S. Neumann, 'Neutron and photon spectrometry with liquid scintillation detectors in mixed fields', *Nucl. Instrum. Methods Phys. Res. Sect. Accel. Spectrometers Detect. Assoc. Equip.*, vol. 476, no. 1, pp. 132–142, Jan. 2002, doi: 10.1016/S0168-9002(01)01410-3.
[19] J. Turko *et al.*, 'Characterization of organic glass scintillator bars and their potential for a hybrid neutron/gamma ray imaging system for proton radiotherapy range verification', *Submitted to. JINST.*
[20] I. Miller, T. W. Holmes, and R. P. Gardner, 'An analytical approach for treating background in spectral analysis measurements', *Radiat. Phys. Chem.*, vol. 116, pp. 87–91, Nov. 2015, doi: 10.1016/j.radphyschem.2015.01.018.

Radiological characterization of high-energy proton Tantalum targets from CERN-ISOLDE Facility

Nabil Menaa[1*], Aurore Boscher[1], Richard Catherall[1], Gerald Dumont[1], Matteo Magistris[1], Sven De Man[1], Renaud Mouret[1], Paolo Pisano[1], Sebastien othe[1], Simon Stegemann[1], Chris Theis[1] and Joachim Vollaire[1]
[1]CERN - European Organization for Nuclear Research, Switzerland
(*) Nabil.menaa@cern.ch

***Abstract*—Accelerator-based techniques are considered among the leading methods to produce radioactive nuclei. The ISOLDE facility (Isotope mass Separator On-Line DEvice) at the European Organization for Nuclear Research (CERN) is a unique source of beams of radioactive nuclides that are used in a wide range of research domains, from nuclear astrophysics to life sciences. Over 400 ISOLDE targets are currently stored at CERN and an average of 30 targets are irradiated every year. All these targets are planned to be dismantled and disposed of as radioactive waste in a dedicated repository in Switzerland. This paper provides an overview of the challenges related to the radionuclide activity predictions, the number of activation scenarios, dismantling and conditioning of the Radioactive Waste (RW) using a hot cell (HC), the high dose rate, uncertainties related to unknown geometry parameters of the RW packages, and finally industrializing a complex RW elimination process as ITEP for an accelerator complex. The performed work addresses each of these challenges and offers technical solutions based on state-of-the-art computational codes, statistical techniques, and state of art Non-Destructive Assay (NDA) techniques using high energy-resolution gamma spectrometry. The methodology followed can be of guidance for the development of similar processes at other facilities.**

***Keywords* —ISOLDE, Non-Destructive Assay, NDA, Radioactive Waste Elimination, Radioactive Characterization.**

I. Introduction

The ISOLDE facility (Isotope mass Separator On-Line DEvice) at CERN (European Organization for Nuclear Research) is a unique source of beams of radioactive nuclides that are used in a wide range of research domains, from nuclear astrophysics to life sciences [1]. The radioactive nuclides are produced via spallation, fission, and fragmentation reactions originating from a 1.4 GeV proton beam that impinges on specifically designed thick targets. For more than 50 years, the ISOLDE facility at CERN [2] has been delivering radioactive ion beams to users from all around the world, extending the knowledge of radionuclides for a large panel of applications. To extend the benefits of this knowledge to societal applications, the MEDICIS project was initiated by CERN in 2010 [3][4]. The CERN-MEDICIS experimental facility (MEDical Isotopes Collected from ISOLDE) is dedicated to the production of non-conventional radionuclides for research and development in medical imaging, diagnosis, and radiation therapy. Most of the MEDICIS operation is driven by the ISOLDE facility from which it receives protons with the same energy and intensity.

Over 300 ISOLDE targets are currently stored at CERN and an average of 30 targets are irradiated every year. All these targets are planned to be dismantled and disposed of as radioactive waste in a dedicated repository in Switzerland following the tripartite agreement between CERN, France and Switzerland (Host States) [5]. The first elimination phase consists of dismantling 12 so-called Tantalum targets and categorize them into structural (Aluminum), target core (Tantalum), and other structural materials. The resulting materials are packaged in dedicated waste drums depending on the material type and undergo a complete characterization. The corresponding radiological acceptance criteria require an accurate knowledge of the levels of induced radioactivity and contamination of the targets. The radiological characterization of the ISOLDE and MEDICIS targets is therefore a necessary step in view of their disposal.

II. Materials and Methods

The following sections describe the ISOLDE and CERN-MEDICIS targets, hot cell (HC) dismantling and drum packaging, and methodologies used to estimate and quantify the radionuclide inventory.

A. ISLODE and CERN-MEDICIS Targets

At the ISOLDE facility, radioactive ion beams (RIBs) are produced via spallation, fission, or fragmentation reactions in a thick target, irradiated with a proton beam from the Proton Synchrotron Booster (PS-Booster) at an energy of 1.4 GeV. What is usually referred to simply as "target" is a complex assembly that comprises the target core and its container, the ion source, the target base, and the auxiliary components. Since each physics experiment requires a different target and ion source configuration, different target units are manufactured and irradiated every year. Figure 1 shows a 3D drawing and a representative picture of the ISOLDE targets. More details on the target design and operation can be found in [6][7][8]. Different materials can be used for the target core (as UC2-Cx, ThC2-Cx, Pb, Ta, etc.) depending on the experimental needs. As of January 2021, the most used target material (70-80 % of all targets) is the uranium carbide (UC2-Cx) simply referred to as UC2C. A target can be expected to decommissioned after having received about 5.0E+18 protons over an irradiation period spanning from a few days to a couple of weeks depending on the beam availability and the operation schedule.

Fig. 1. 3D model rendering and picture of an ISOLDE target assembly. The so-called casserole, partially transparent in the drawing, was removed in the picture on the right to show the target core (circled in red) and other auxiliary components.

The MEDICIS experimental facility (MEDical Isotopes Collected from ISOLDE) is designed to produce radionuclides for medical applications. The MEDICIS operation is driven by the ISOLDE facility and, as for the latter, it receives protons from the Proton Synchrotron Booster with the same energy and intensity. As of February 2021, the MEDICIS irradiation station is located downstream, with respect to the beam traveling direction, of the ISOLDE High-Resolution Separator station (HRS). Thus, a MEDICIS target being placed downstream of an ISOLDE target, receives the fraction of primary protons which have not interacted with the ISOLDE target plus the secondary particles produced by the beam interaction with the latter [9].

For this paper, only Tantalum targets are considered even though the nuclide inventory calculations are performed for a wide variety of target core materials. The latter will be the subject of future publications.

B. Radionuclide Inventory and Scaling Factor Estimation Method

1) FLUKA modeling and Particle Fluence Spectra

To estimate the nuclide inventory for the multiple irradiation configurations and conditions, to which the ISOLDE and MEDICIS targets can be exposed, a multiple-step scheme is used. First, with the Monte Carlo particle transport code FLUKA [10] [11] [12], different irradiation configurations are modeled and simulated. An irradiation configuration is defined by combinations of the following parameters:

- The ISOLDE and MEDICIS target core materials,
- The type of target placed upstream of the MEDICIS target (for MEDICIS targets only),
- The use of the so-called neutron converter [8].

The aim is to calculate, using FLUKA, the proton, neutron, positive and negative pion fluence spectra produced by the interaction of the primary proton beam with the target material Tantalum (Ta) given the target area geometry for each of the possible irradiation configurations. For the characterization of the core materials, it is also necessary to score the photon fluence spectra inside the core material. These irradiation configurations are used to obtain two sets of particle fluence spectra:

- A first set representative for all the structural materials (e.g. aluminum, copper, stainless steel, etc.) calculated in a region outside the target core,
- A second set, including the photon fluence, for the Tantalum core material calculated inside the target core.

Once the particle fluence spectra are calculated for all the relevant irradiation configurations, the material activation calculations are performed with ActiWiz Creator [13] . ActiWiz Creator, hereinafter simply ActiWiz, is an analytical software whose core functionality allows calculating the nuclide inventory of a user-defined material given the following parameters:

- Neutron, proton, photon, positive and negative pion fluence spectra,
- Material elemental composition and density,
- Irradiation and cooling times,
- Beam intensity expressed in protons per second.

The ActiWiz analytical calculations are less computationally demanding compared with the Monte Carlo FLUKA simulations. It is, therefore, possible to perform a large set of calculations, covering all possible irradiation scenarios. In particular, for every irradiation configuration simulated in FLUKA, we calculate the residual radionuclide inventories for all the relevant cooling times and for any given activated material. Finally, the material-specific nuclide inventories obtained for all the core and structural materials, given all the combinations of irradiation configurations and scenarios are statistically analyzed to extract relevant information such as specific activities.

A simplified FLUKA model of the ISOLDE-MEDICIS target unit and its surroundings was produced. As can be seen from Figure 2, the model consists of an inner cylinder containing the target core material (Ta). The target core material is pure tantalum, with a density of 5.0 g/cm^3. The latter is surrounded by a second cylinder which represents the target unit structure (i.e. casserole, target flange, etc.). The distance between the inner and the outer cylinder is representative of the average distance between the target core and the other structures of the target unit. The geometry dimensions are presented in Table I which includes the distance from the

ISOLDE and MEDICIS target that is placed upstream (see Figure 3). The target unit thus modeled is placed inside a concrete box, whose walls are 200 cm thick, which models the walls, floor, and ceiling of the irradiation station (see Figure 3). The position of the target model inside the concrete box is representative of the actual position of the target unit in the irradiation station. The concrete walls and the beam dump are expected to affect the particle fluence spectra inside the target unit by means of back-scattered radiation (e.g. thermal neutrons back-scattered into the target from the concrete walls). Therefore, it was decided to include them in the model.

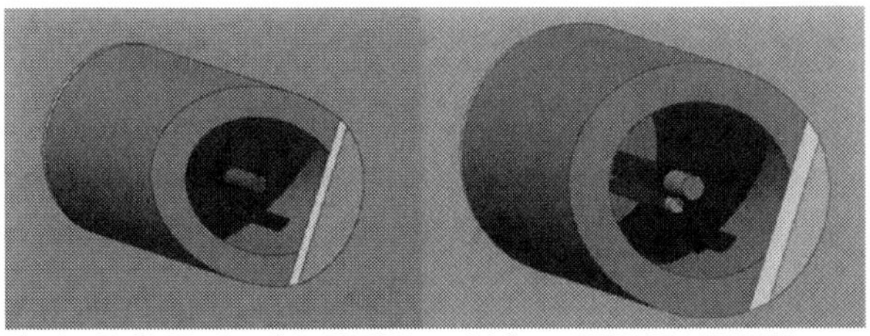

Fig. 2. (Left) 3D plot of the FLUKA model for an ISOLDE-MEDICIS target unit showing the target container, the outer cylinder, and its base plates. (Right) 3D plot of the FLUKA model for an ISOLDE target featuring a neutron converter placed below the target container.

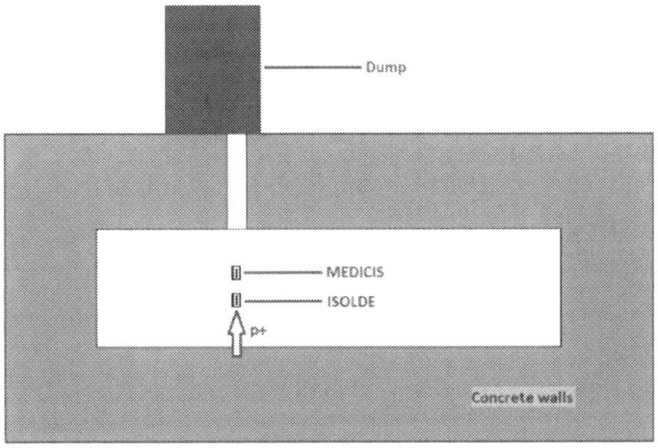

Fig. 3. Top view of the FLUKA model showing the ISOLDE and MEDICIS targets surrounded by the concrete walls with the opening to the beam dump.

TABLE I
MAIN DIMENSIONS OF THE FLUKA MODELS USED FOR THE FLUENCE SPECTRA CALCULATIONS. THE "DIST FROM TARGET CONTAINER" IS THE DISTANCE BETWEEN THE AXIS OF THE TARGET CONTAINER CYLINDER AND THE AXIS OF THE NEUTRON CONVERTER CYLINDER. THE "DIST. FROM ISOLDE TARGET" IS THE DISTANCE BETWEEN THE CENTRE OF THE ISOLDE TARGET CONTAINER AND THE MEDICIS TARGET CONTAINER.

Element	Dimension	Value [cm]
Target container	Radius	0.9
	Height	20
Outer cylinder	Inn. radius	6.45
	Out. radius	9.32
	Height	29
Out. cylinder plates	Radius	9.32
	Height	0.5
Neutron converter	Radius	0.6
	Height	12.5
	Dist. from the target container	1.75
MEDICIS target	Dist. From ISOLDE target	55.5

The base plates of the outer cylinder representing the target unit are made of aluminum as the actual target casserole. Whereas the body of the outer cylinder is made of a dummy material obtained by combining the four structural materials which dominate the total mass of the target unit. These are aluminum alloys (AW 6082 and AW 5083), stainless steel alloys (304 and 316), copper (CU ETP C11000) and pure tantalum. Hence, for the elemental compositions of the structural material, a representative composition was obtained by estimating the Main structural materials found in an ISOLDE-MEDICIS target (see Table II).

In one configuration, the proton beam energy is set to 1.4 GeV. The beam impinges on the target core after having passed through the aluminum base plate of the outer cylinder. Otherwise, the beam impacts on the neutron converter, after having passed through the aluminum base plate, for the corresponding irradiation configurations.

As stated previously, the fluence spectra for protons, neutrons, positive and negative pions are scored in FLUKA simulations. The scoring regions are the target container (i.e. inner cylinder in Figure 2 Left) and the outer cylinder representative of the unit structural materials (i.e. outer cylinder in Figure 2).

TABLE II
ELEMENTAL COMPOSITION OF THE DUMMY MATERIAL USED FOR THE FLUENCE SPECTRA CALCULATION AND REPRESENTATIVE OF THE MAIN STRUCTURAL MATERIALS FOUND IN THE ISOLDE-MEDICIS TARGET

Element	**Mass fraction**	**Element**	**Mass fraction**
Aluminum	5.95E-01	Magnesium	5.57E-03
Copper	1.93E-01	Zinc	1.24E-03
Iron	8.84E-02	Titanium	6.19E-04
Tantalum	6.39E-02	Nitrogen	1.25E-04
Chromium	2.54E-02	Phosphorous	5.64E-05
Nickel	1.25E-02	Oxygen	4.33E-05
Silicon	7.13E-03	Carbon	3.76E-05
Manganese	6.84E-03	Sulphur	3.76E-05

Structural Material Irradiation Configurations

For the radiological characterization of the ISOLDE structural materials the following irradiation configurations are considered:

- Direct irradiation of a tantalum target core, and
- Indirect irradiation of a tantalum target core through the neutron converter.

The irradiation station for MEDICIS targets is located downstream of the ISOLDE irradiation station with respect to the beam travelling direction. Therefore, the MEDICIS targets can either receive an unperturbed proton beam (no ISOLDE target located upstream) or a proton beam which has interacted with an ISOLDE target located upstream plus the produced secondary radiation (i.e. neutron, secondary protons, and pions). In the former case, the MEDICIS target irradiation configuration is equivalent to an ISOLDE target, whereas, in the latter, the MEDICIS target irradiation depends upon the type of ISOLDE target located upstream. Therefore, for the radiological characterization of MEDICIS structural materials the following irradiation configurations are considered:

- Indirect irradiation of a tantalum target core through a UC2C ISOLDE core, and
- Indirect irradiation of a tantalum target core through the ISOLDE neutron converter.

Target containers, made of tantalum, are used to accommodate the target material that interacts directly with the proton beam. The target container is located only a few millimeters away from the beam interaction point (i.e. from the target core). Dedicated FLUKA simulations and ActiWiz Creator calculations showed that the particle fluence spectra scored inside the tantalum core can be used to produce conservative and representative estimates of the container activation. Therefore, the activity of the target container is calculated with the particle fluence spectra scored inside a tantalum core for the irradiation configurations described above.

Core Material Irradiation Configurations

The activation of the ISOLDE and MEDICIS core materials is largely dominated by the nuclear reactions induced by the primary proton beam. The only exception is when the proton beam impinges on the neutron converter leading to the production of a higher neutron fluence per primary proton. Therefore, for the radiological characterization of the core materials, the following two irradiation configurations are used for both the ISOLDE and MEDICIS targets:

- Direct irradiation of the core material: this configuration is representative of the direct irradiation of any tantalum core material in the ISOLDE and MEDICIS targets,
- Indirect irradiation of the core material through the neutron converter: this irradiation is representative of the indirect irradiation of the tantalum core material in the ISOLDE and MEDICIS targets.

2) Activation Calculations With ActiWiz Creator

A necessary set of input parameters is required by the tool ActiWiz Creator to calculate the nuclide inventories for the considered core and structural materials given the particle fluence spectra obtained with FLUKA. The material compositions and densities of the structural and core materials used to perform the ActiWiz Creator calculations are provided hereafter. The ISOLDE and MEDICIS targets are irradiated for several days depending on the proton beam availability. As the cooling times are typically in the order of years, the irradiation time can be considered as instantaneous (i.e. 1 second) for activation studies. The total number of protons on target (POT) generally does not exceed 5.0E+18. The cooling times range from 3 to 30 years in one-year steps.

3) Scaling Factor Calculation

A scaling factor (SF) is the ratio of the activity of a nuclide that cannot be measured via gamma spectroscopy (e.g. pure beta emitter) and one that, if present in the waste, is systematically and readily measured (e.g. Co-60, Na-22). The former are categorized as difficult-to-measure (i.e. DTM) depending on their radiological characteristics, whereas the latter is the so-called key nuclides (i.e. KN) belonging to the list of easy-to-measure isotopes (i.e. ETM).

Therefore, by measuring the activity of the ETMs via gamma spectrometry, one can estimate the activity of the DTMs by multiplying the activity of the selected key nuclide by the corresponding scaling factor.

Scaling factors are calculated based on the expected activities in a material or waste package (i.e. mixture of waste materials) upon verification of the existence of a mathematical correlation between the activity levels of the DTMs and the chosen key nuclide

4) Key Nuclide Selection

Given the radionuclide inventories and activity levels discussed previously, the first step towards the calculation of the scaling factors consists of selecting a key nuclide for each material. Generally, a key nuclide shall satisfy the following criteria [14] :

- It can be detected by gamma spectrometry,
- It is expected to have an activity higher than its detection limit (also known as Minimum Detectable Activity, MDA),
- Its activity value is correlated with the activity of the DTM of interest,
- It has a relatively long half-life (e.g. years rather than days),
- It is exclusively produced by the activation of the considered material, if possible.

C. Hot Cell dismantling and Drum packaging

The targets are dismantled in a dedicated hot cell (HC) (see Figure 4 Left). It enables the disassembly without cutting or grinding. The choice to perform such a process in a hot cell is due to the targets high dose rate (up to 2 mSv/h at contact) and the associated contamination risk. A hot cell offers a two-way protection by:

- Shielding from radiation with a 150 mm lead shielding all around and 150 mm thick glass, and
- Confining contamination with a leak-tight envelope and a lower atmospheric pressure compared surrounding volumes.

The mechanical telemanipulators are the main instruments to action anything inside the HC.

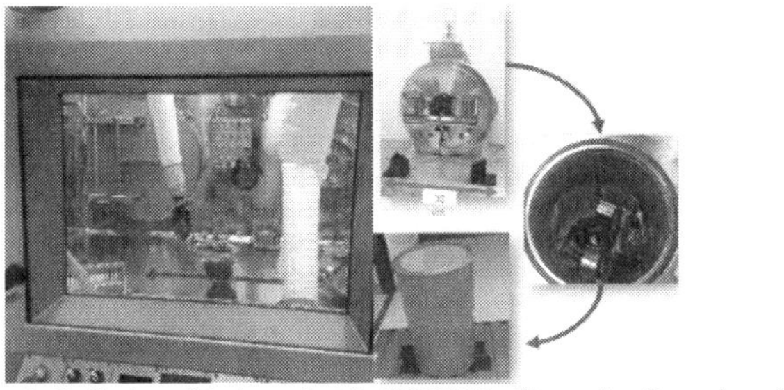

Fig. 4. (Left) view of the hot cell at the ISOLDE facility used to dismantle and package the targets into drums. (right) a process showing the packaging of the target aluminum containers into a 107 L drum.

The dismantled targets are then packaged into three waste drums belonging to the following categories as shown in the right of Figure 4:

- Structural material in 107 L drums: Aluminum,
- Target material in 21 L drums: Tantalum,
- Other structural materials in 21 L drums.

D. *Gamma Spectrometry (GS) Setup*

GS is a widely deployed non-destructive assay technique at CERN to quantify the residual activity of gamma emitters or ETM nuclides in various items, ranging from small-volume samples in a laboratory to large objects such as unitary radioactive waste packages. For this study, we use high energy resolution germanium detectors Falcon 5000 from Mirion technologies (Canberra) as shown in Figure 5. To generate the efficiency calibration curves, one needs to know several geometry parameters including the dimensions of the material, the distance between the detector and its composition (material information). ISOCS (In Situ Counting Object System) and LabSOCS (Laboratory Sourceless Calibration Software) from Mirion Technologies (Canberra) are applied in the laboratory for creating efficiency calibrations of good quality without using radioactive standards at the laboratory [16] [17] [18] [19].

Fig. 5. View of the gamma spectrometry setup.

III. Results and Discussion

A. *Radionuclide Inventory and Scaling Factor*

For the purpose of this study, it was decided to choose the third statistical quartile of the activity ratio distribution as the retained scaling factor value. This choice is considered robust as it does not require any specific distribution assumption of the SF (e.g. log-normal) and is less sensitive to data outliers. Figure 6 shows a third quartile example obtained for the SF distribution for the ratio H-3/Na-22. In this case, H-3 is the DTM while Na-22 nuclide is considered as the KN.

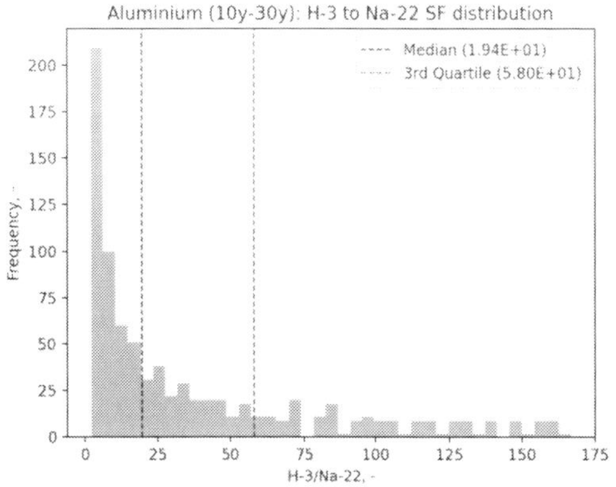

Fig. 6. Histogram of the H-3/Na-22 activity ratios for the ISOLDE aluminum alloys for the cooling time period 10y-30y.

The selected key nuclides for the materials considered in this study are listed in Table III. The materials EN AW 6082 and EN AW 5083 are grouped in "aluminum alloys", and AISI 304, AISI 304L, AISI 316, S235JR in "steel alloys".

TABLE IIII
LIST OF SELECTED KEY NUCLIDES.

Material	Key nuclide
Al2O3	Na-22
Aluminum alloys	Na-22
Brass	Co-60
CU-ETP-C11000	Co-60
Molybdenum	Nb-94
NiFe42	Co-60
Steel alloys	Co-60
Tantalum	Hf-178n
Tungsten	Co-60

The scaling factors for the tantalum are defined as the mass-weighted mean of the third quartile of the distributions obtained for the structural materials, and the third quartile of the distributions obtained for the core material. Given the material content allowed in each waste category, it is possible to produce three sets of combined scaling factors for each waste category. That is, one set per cooling time period (i.e. 3y-10y, 10y-30y,

https://doi.org/10.1051/epjconf/202328809004

and 3y-30y) per waste category. The combined scaling factors are obtained by taking the highest scaling factor for each DTM expected to be found in each waste category for a given key nuclide. For instance, the waste packages belonging to the category "Structural material: aluminum" are expected to contain the key nuclide Na-22. This key nuclide is related to the presence of both Al_2O_3 and aluminum alloys. Since the DTM nuclide H-3 is produced by the activation of both materials, two values of the H-3/Na-22 scaling factor are available – one for Al_2O_3 and one for the aluminum alloys. A conservative estimate of the H-3 activity in the waste is obtained by selecting the highest scaling factor between the aforementioned values.

Similarly, if a DTM is related to two or more key nuclides found in the waste category, the DTM total activity shall be estimated by summing the values derived from each scaling factor. For example, the waste packages belonging to the category "Structural material: aluminum" are expected to contain the key nuclides Na-22 and Co-60. The former is due to the presence of Al_2O_3 and aluminum alloys, whereas the latter is due to steel alloys. For instance, the DTM nuclide Cl-36 is produced by the activation of both the aluminum and steel alloys. Hence, the Cl-36 activity contained is a waste package shall be calculated by summing the activity estimate based on the Cl-36/Na-22 scaling factor (aluminum alloys), and the one estimated via the Cl-36/Co-60 scaling factor (steel alloys). The scaling Factor values for 10 to 30 years decay time that are above 2% are shown in Figure 7.

Nuclides	Na-22
H-3	1.44E+02
C-14	4.26E-01
Ti-44	2.32E-02
Al-26	1.05E-02

Nuclide	Nb-94
H-3	7.34E+02
Nb-93m	3.88E+02
Nb-91	1.60E+02
Mo-93	2.91E+01
Tc-99	7.78E-01
Sr-90	6.63E-01
Kr-85	1.39E-01
Kr-81	1.93E-02

Nuclide	Hf-178n
H-3	1.11E+02
Pm-145	5.92E+00
Gd-148	2.75E+00
Ba-133	2.49E+00
Tb-157	1.99E+00
Ta-179	5.96E-01
Hf-172	5.78E-01
Lu-174	5.17E-01
Lu-173	1.10E-01
Ho-163	6.23E-02
Ni-63	4.22E-02
Nb-93m	3.13E-02
Ar-42	1.66E-02
Rh-101	1.56E-02
Eu-150	1.31E-02
Ar-39	1.13E-02
Fe-55	1.09E-02
Kr-85	1.08E-02

Nuclide	Co-60
H-3	1.50E+04
Ta-179	6.72E+02
Hf-172	2.05E+02
Pm-145	1.51E+02
Lu-173	4.00E+01
Lu-174	3.46E+01
Ba-133	1.65E+01
Ho-163	1.24E+01
Fe-55	1.08E+01
Ni-63	7.05E+00
Kr-85	2.13E+00
Eu-150	7.11E-01
Ar-42	4.96E-01
Cd-113m	3.91E-01
Eu-152	2.56E-01
C-14	2.44E-01
Ti-44	2.26E-01
Rh-101	2.08E-01
Ar-39	1.83E-01
Rh-102	1.62E-01
Nb-93m	1.12E-01
Pm-146	1.07E-01
Ag-108m	8.19E-02
Sm-145	2.79E-02
Pt-193	2.04E-02
Tl-204	1.01E-02

Fig. 7. Scaling Factor values for 10 to 30 years decay time. Due to the large size of the tables, only SF values above 2% are included.

B. *ETM nuclides activity Estimation*

The GS technique presents several challenges during both the acquisition and the analysis steps. The challenges are summarized when establishing the following:

- The nuclide library,
- The γ-spectrometry counting time and the allowed data acquisition dead time limit, and
- The best geometry model for efficiency calibration needed for activity calculations.

Regarding the dead time, the performed studies show that there is no systematic activity bias beyond 10% if the GS acquisition dead time was maintained below a value of approximately 10%. A typical GS spectrum is shown in Figure 8 and a typical GS geometry model, used for efficiency calibrations, is shown in Figure 9.

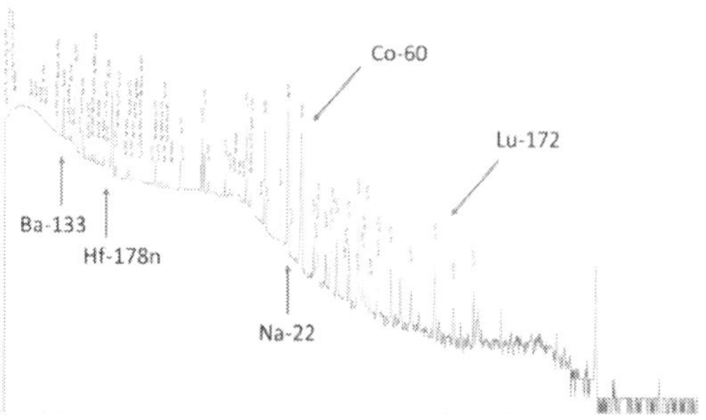

Fig. 8. A typical GS spectrum for the Ta 21 L drum.

Fig. 9. A typical GS ISOCS geometry model for the Ta 21 L drum.

To validate the ETM estimation using GS, two targets ID#239 and #308 are dismantled in the HC. During this process, targets have been separated into three drum categories for disposal (Structural material: aluminum 107 L drum, target material: tantalum 21 L drum, and other structural materials 21 L drum). The sum of activities obtained for the target ID#239 and #308 (target unit modeling) and for the three drums (Al-Drum, Other-Drum and Ta-Drum modeling) are compared. The drums and targets activity ratio illustrated in Figure 10 is comprised between 0.69 and 1.48. Therefore, for targets ID#239 and #308, the measured activity ratio is within ± 50%, which is consistent with the estimated activity ratio uncertainty at k=2 (see Figure 9).

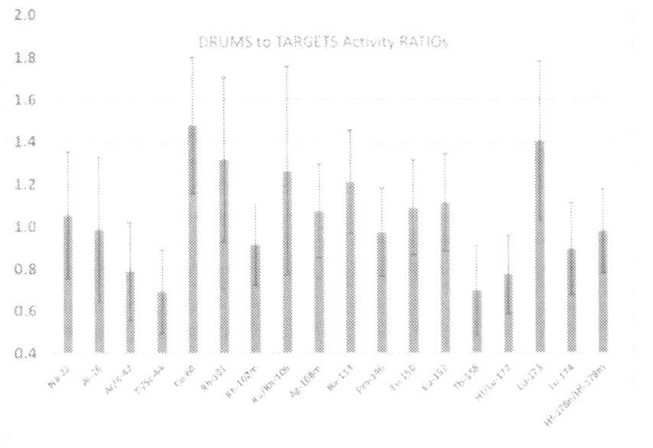

Fig. 10. Drums to targets activity ratio and error bars (the uncertainties are reported at 1 σ).

IV. Conclusions

ISOLDE and MEDICIS are two experimental facilities located at CERN where fixed targets are irradiated by protons beam to produce radioactive ion beams. Currently, over 400 targets are stored at CERN (as of 15/08/2023) and, on average, 30 targets are irradiated every year. All these targets will be dismantled and disposed of as radioactive waste in a dedicated waste repository in Switzerland.

In this study, a three-step scheme was developed to estimate the nuclide inventories resulting from the activation of the core and structural materials of which ISOLDE and MEDICIS targets are made. The steps can be summarized by:

- Predicting the radionuclide inventory and associated DTM scaling factors taking into account the targets irradiation conditions and material elemental compositions,
- Dismantling and conditioning the waste into the various drum types, and
- Performing the Gamma Spectrometry analysis to estimate the ETM radionuclides and deduce the activity values of the other DTM nuclides.

This study describes the methodology used to estimate the complete radionuclide inventory and activity levels of present and future targets based on extensive analytical calculations and the associated non-destructive assay (NDA) technique by evaluating the counting conditions and establishing the efficiency calibration curves using ISOCS (In Situ Counting Object Software) from Mirion Technologies (Canberra). The study also compares the ETM activities obtained by the NDA technique before and after packaging the targets following the segregation step into the three material types. This paper addresses each of these challenges and offers technical solutions based on computational codes, statistical techniques and state of the NDA technique using high energy resolution gamma spectrometry. We believe that the methodology here presented can be applied to activation and characterization studies that are characterized by a relatively large number of possible activation scenarios and input parameters.

Acknowledgment

The authors would like to thank the CERN radiological analysis service, the COFLU computing cluster administration team, and the operational teams of radioactive waste management at CERN for their support in performing the calculations and providing the measurement results.

References

[1] Catherall, R. et al. The ISOLDE facility. Journal of Physics G: Nuclear and Particle Physics 44, 094002 (2017).
[2] Borge, M. J. & Jonson, B. ISOLDE past, present and future. Journal of Physics G: Nuclear and Particle Physics 44, 044011 (2017).
[3] Augusto, R. M. d. S. et al. CERN-MEDICIS (medical isotopes collected from ISOLDE): a new facility. Applied Sciences 4, 265–281(2014).
[4] Duchemin, C. et al. CERN-MEDICIS: A review since commissioning in 2017. Frontiers in Medicine 8 (2021).
[5] Accord entre le Conseil Fédéral Suisse, le Gouvernement de la République Française, et l'Organisation Européenne pour la Recherche Nucléaire relatif à la Protection contre les rayonnements ionisants et à la Sûreté des Installations de l'Organisation Européenne pour la Recherche Nucléaire, conclu le 15 novembre 2010, entré en vigueur par échange de notes le 16 septembre 2011. Number RS 0.814.592.2, Recueil O-ciel n. 34 du 23 août 2011, page 3825. Décret n. 2011-1024 du 24 août 2011, JORF n. 0199 du 28 août 2011 page 14594. 2011.
[6] M. Brice, An ISOLDE target unit. Cible ISOLDE, CERN-EX-0204023.
[7] T. Stora, Radioactive Ion Sources. CAS-CERN Accelerator School: Ion Sources - Proceedings. 10.5170/CERN-2013-007.331 (2014).
[8] A. Gottberg et. al., Experimental tests of an advanced proton-to-neutron converter at ISOLDE-CERN, Nuclear Instruments and Methods in Physics Research Section B: Beam Interactions with Materials and Atoms, Volume 336, 2014, Pages 143-148, ISSN 0168-583X, https://doi.org/10.1016/j.nimb.2014.04.026.
[9] C. Duchemin et al., CERN-MEDICIS: A Unique Facility for the Production of Nonconventional Radionuclides for the Medical Research, 11th International Particle Accelerator Conference (2020).
[10] CERN,online:https://fluka.cern/,(Last accessed 15 August 2023).
[11] Ahdida, C. et al.New capabilities of the FLUKA multi-purpose code.Front.Phys.9,14 (2022).
[12] G. Battistoni et al., Overview of the FLUKA code, Annals of Nuclear Energy 82, 10-18 (2015).
[13] H. Vincke and C. Theis, Actiwiz 3 - an overview of the latest developments and their application, Journal of Physics Conference Series, no. 1046(1):012007 (2018).
[14] IAEA, Determination and use of scaling factors for waste characterisation in nuclear power plants, IAEA Nuclear Energy Series No. NW-T-1.18 (2009).
[15] B. Zaffora et al., Radionuclide inventory and scaling factors of metallic, TFA radioactive waste, EDMS 1868767 v.1.2 (2019).
[16] Patrycja Dyrcz, Thomas Frosio, Nabil Menaa, Matteo Magistris, and Chris Theis. Qualification of the activities measured by gamma spectrometry on unitary items of intermediate-level radioactive waste from particle accelerators. Appl. Radiat. Isot., 167:109431, 2021. doi:10.1016/j.apradiso.2020.109431
[17] F. Bronson. ISOCS, a Laboratory Quality Ge Gamma Spectroscopy system you can take to the Source for Immediate High Quality Results. Proceedings of the Rapid Radioactivity Measurements in Emergency and Routine Situations Conference, UK, October 1997.
[18] N. Menaa, A. Bosko, F. Bronson, R. Venkataraman, W. R. Russ, W. Mueller, V. Nizhnik, and L. Mirolo. Mathematical e-ciency calibration with uncertain source geometries using smart optimization. In 2011 2nd International Conference on Advancements in Nuclear Instrumentation, Measurement Methods and their Applications, pages 17, 2011. doi: 10.1 109/ANIMMA.2011.6172913.
[19] Thomas Frosio, Nabil Menaa, Philippe Bertreix, Maeva Rimlinger, and Chris Theis. A novel technique for the optimization and reduction of gamma spectroscopy geometry uncertainties. Applied Radiation and Isotopes, 156:108953, 2020. ISSN 0969-8043. doi: https://doi.org/ 10.1016/j.apradiso.2019.108953. URL https://www.sciencedirect.com/science/article/pii/S0969804319306852.

Systematic Radioecological Experiments in the United Arab Emirates (UAE) for Nuclear Safety and Emergency-Preparedness

Prasoon Raj[1,2,*], Nemeer Padiyath[1,3], Francois Foulon[1*], Maryam Almakrani[1]

[1]Emirates Nuclear Technology Center (ENTC), Department of Nuclear Engineering, Khalifa University, Abu Dhabi, United Arab Emirates (UAE)

[2]Commonwealth Fusion Systems (CFS), Devens, Massachusetts, United States (USA)

[3]MS Student, Mechanical Engineering Department, Stanford University, United States (USA)

(*) prasoon.raj@outlook.com, francois.foulon@ku.ac.ae

***Abstract*—Nuclear power and other radioactivity-producing industries like oil and gas, desalination, etc. are rapidly maturing in the arid countries, most importantly in the United Arab Emirates (UAE). Thus, it becomes vital to study and model the transfers and impacts of radionuclides' release into the local environments. Baseline or predictive studies on this subject are extremely scarce for the UAE-like desert-marine ecosystems. In global literature, the retention and transfer mechanisms of natural and anthropogenic radionuclides in arid soils and plants are grossly underrepresented. We debrief the recent research activities in the UAE and nearby nations on background radioactivity measurements in soils and plants, as well as those estimating the radioecological transfer factors or concentration ratios. Soil to plant transfer is a key sought after data for arid region radioecology. We further outline some systematic field and laboratory experiments executed by Khalifa University of the UAE, which includes numerous field surveys, identifying radionuclides and crop categories of importance, sampling in representative localities in the Abu Dhabi emirate, developing protocols for sample management and processing, and radiometry using gamma-spectrometry and inductively coupled plasma mass-spectrometry.**

***Keywords* —Radioecology, Soil, Uptake, Emirates, NORM.**

I. Introduction

Robust radiological environmental impact assessments (REIA) are essential to establish and enhance the safety cases at different stages of a nuclear power program [1]. It is necessary for design and operational decisions throughout a nuclear plant's lifecycle, from reactor conceptualization to site selection, radioactivity effluent releases, all the way to plant dismantling and waste management. REIA, in simple words, quantifies the radiological and physicochemical characteristics of a region's environmental matrices, primely, the bedrocks, topsoil, groundwater, surface and marine waters, plants, crops, winds, rains, etc. It measures the susceptibility of the ecosystem to accumulate and/or transfer the radionuclides which are either operationally or accidentally released/deposited in the region. Through the various food chain routes, e.g., cows consuming contaminated grass crops followed by humans consuming cows' milk, traces of the environmental radionuclides reach animals and humans. Radioecological experiments involve in-situ or controlled measurements of these transfer processes, defined using *transfer factors* (TF). As an example, a *plant-soil TF* is defined as the ratio of measured radioactivity of a selected radionuclide (e.g., ^{226}Ra) in plant to that in soil [2]. The TFs then serve as input for biosphere-scale modeling and REIA.

Radioecological experiments are of utmost importance for the United Arab Emirates (UAE), as it is the first Arab country to set forth on long-term nuclear energy with now operational Barakah Nuclear Power Plants (BNPP) [3]. Radionuclides of natural and anthropogenic origins have historically impacted some habitat zones of the Middle East and North Africa (MENA) region, and the Arabian Peninsula [4]. But scientific knowledge remains very scarce and inconsistent for the radionuclides' dispersion, retention, and transport among the region-specific air, bedrocks, soil/sand, water, plants, animals, etc. Increasing risks with the upcoming nuclear sector has made this information critical. Interlinked industries, like oil and gas, modern agriculture (e.g., via fertilizers), seawater desalination, etc., also add radioactivity transmission channels under some conditions, which are poorly studied. International agencies like the International Atomic Energy Agency (IAEA) [5] and International Union of Radioecologists (IUR) [6] have thus recommended immediate research for a rigorous 'arid region radioecological (ARRE) database'. This research will advocate the safety and emergency-readiness of reactors in the UAE, as well as prepare us to deal with similar issues in the ever-increasing warmer and drier climates worldwide.

For the UAE's nuclear power program, the Emirates Nuclear Technology Centre (ENTC) of Khalifa University (KU) [7] based in Abu Dhabi (UAE) is the major academic stakeholder from the research and knowledge management points of view. It began facility and skill development in REIA since the last decade and took part in the IAEA MODARIA II [8], as well as is engaging in the ongoing IAEA CRP K41022 [9]. The center is conducting several field sampling and controlled plant growth experiments, to generate UAE-specific TFs for natural and main fallout nuclides, while also developing a knowledge base for dispersion of BNPP-origin radionuclides in the marine, atmosphere, and terrestrial domains.

In this paper, we present a concise overview of the status-quo of environmental radiation measurements in the UAE. While negligible imprints are found of artificial isotopes like ^{137}Cs and ^{90}Sr, naturally occurring radioactive material (NORM) isotopes like ^{40}K, ^{238}U, ^{232}Th are frequently detectable in the UAE environs but below their respective global average values [10]. Such limited field-scale studies have been done on soils, rocks, farm products like dates, vegetables, and groundwaters. Baseline soil to plant TF studies on fruits and vegetables have been started in the ENTC since 2020. Laboratory studies are underway for more difficult measurements, like uptake of cesium and strontium in major crops like tomatoes under a hypothetical accidental scenario. For radiometry, gamma spectroscopy is being most employed, but techniques like gross alpha-beta, and sequential extraction of trace radionuclides are being developed. Plasma mass spectrometry is under use for isotopic analysis, albeit not for radioisotopes yet.

II. Available Data for Arid Region Radioactivity

Located between the Persian Gulf and the Gulf of Oman, sharing borders with Saudi Arabia and Oman, ~80% of the UAE's land area is desert, which includes plains, sand sheets, sand dunes, sabkhas, rocky outcrops, and mountains [11]. However, seas and coasts, oases and mangroves also make important parts of the UAE's ecosystems. The major population density is near the coastlines in Dubai, Sharjah, and Abu Dhabi. In terms of activities causing radionuclide releases, a large portion of oil and gas exploration as well as nuclear power production is held in the Abu Dhabi emirate, with many industries localized in the western region (Al Dhafra). The UAE has low annual rainfalls, high temperatures and loose and infertile soil. The UAE soils are primarily sandy Aridisols and Entisols, with median sand component of 90 to 95% [11].

More than 90% of the daily produced/used water in the UAE is desalinated seawater, while a small fraction (~1% of total consumption) is groundwater for use in farming and forest sustenance. In terms of human food sources in the UAE, about 82% is imported, but a slowly increasing share of 18% is locally produced, including fresh vegetables and fruits, some meats, poultry, and dairy products. Most farming within the Abu Dhabi emirate happens in Al Ain, Liwa and Ghayathi. A major part of crop production is done in the northern emirates as well, most importantly in Ras al Khaimah. If one focuses on radionuclide migration within national boundaries, i.e., ignoring the transfers from imported goods, then soil to plant transfer for fodder crops (which then connects to the animal kingdom), fruits, vegetables, etc. are important. So are the water to soil to plant transfers.

In Table I, data from literature is reported to highlight frequently measured radionuclides in UAE's environment, based on the national radiological environmental monitoring program (REMP) report of the year 2015 [12]. Data is reported as range of measured *activity concentrations* (*AC*, in Bq kg^{-1}) and their averages in the brackets. In a previous review, we have collated a few other academic studies in the UAE on this subject [4]. For the soil data sources in this review, we get averages of radionuclide concentrations (in Bq kg^{-1}) as 13.6 to 15.5 for ^{238}U decay chain, 3.3 to 8.3 for ^{232}Th chain, and 98.2 to 349.7 for ^{40}K. Some instances of global fallout radionuclide from nuclear weapons' testing, most commonly ^{137}Cs, are also recorded.

Table I: Literature values for radioactivity (Bq kg^{-1}) in the UAE soil and the major farm products: date fruits and cucumbers. Source of data: [12] ; BDL means below detection limit.

	Soil	Date fruit	Cucumber
^{238}U chain	7.7-23.4 (15.4)	BDL-0.8 (0.66)	BDL
^{232}Th chain	2.8-12.7 (6.5)	BDL-0.23 (0.15)	BDL
^{40}K	133-364 (231.4)	191-362 (277)	42.4
^{137}Cs	0.03-0.93 (0.38)	BDL	BDL

Date palms are a categorically unique agricultural product from the UAE and Arab nations [13]. Radiological studies on date fruits and trees, especially their tendencies to transfer radionuclides from soil and water to food, are extremely rare in global literature. The REMP report (Table 1) suggests that they have a higher average of ^{40}K than the soil, with plant-soil TF~1.2 if calculated using estimated average values. Likewise, TF~0.04 for ^{238}U and TF~0.02 for ^{232}Th chains. Among the vegetables grown in the UAE, cucumbers have been measured in literature, mostly below gamma detection limits, but a TF (^{40}K)~0.18 could be estimated based on averages in Table I. Note that the plant samples are measured at fresh/wet weights.

Some more data on plant-soil TF in arid areas are found in literature from nearby countries as reviewed in [4], and extracted in Table II. These values range from extremely low uptakes, like TF~1.7e-4 for fruits to possible bioconcentration (TF>1) regimes like TF~1.1 for non-leafy vegetables. Also, the standard deviations are very high for these averages, making us conclude that the data from these limited studies do not lead to a consensus and it is dangerous to apply literature values in REIA. This forms the fundamental motivation for the UAE-specific experiments developed at ENTC, as reported here.

Table II: Estimated plant-soil TF from literature of arid Arab countries for fruits and non-leafy vegetables (NLV). Source of data: [4]; Values are geometric means (GM) with geometric standard deviations (GSD) in brackets.

	^{238}U chain	^{232}Th chain	^{40}K	^{137}Cs
Fruits	1.7e-4 (1.5)	0.033 (2.2)	0.6 (2.8)	0.011 (1.6)
NLV	0.032 (2.5)	0.11 (3.6)	1.1 (5.4)	0.058 (3.5)

III. Experimental Methods

We present the ENTC experiments designed to measure the baseline radioactivity in soils and plants of the UAE, and to quantify the soil to plant transfer process through *concentration ratios (CR)*. We define the CR, a measure of TF, as the ratio of measured radioactivity (Bq kg^{-1} units) in plant parts (often measured for edible parts like fruits or leaves) to that in the soil.

Based on a detailed radioecological strategy formulated in 2020 [4], we decided to focus our studies on the Abu Dhabi emirate: as it constitutes 85% of the land area, 56% of the crop area, and 55% of the agricultural economic value generated in the UAE [11]. Within that, we conducted field work in Bahya (representing semi-urban farms), traditional farming and oases areas in Al Ain and Liwa, both also known for large-scale farming and exports, and Ghayathi and Sila where farming is minor activity, but they are geographically closer to BNPP. Some of the sampling areas are denoted in Fig. 1 (top). Crops which make it to the markets in larger quantities were prioritized, consisting of 46% fodder (although no experiment on fodders have been concluded to be reported yet), 24% vegetables, 30% fruits, and native trees. Finally, the focus isotopes were gamma-emitters from the NORM chains [10]. We report four experiments as discussed in the following.

We conducted in-situ CR measurements for date fruits and leaves and for three top vegetable crops in Abu Dhabi. Dates are unique in the sense that they are economically and culturally important to the region, they can grow in harsh desert environments, and the data on their soil to plant transfer of radionuclides is largely unavailable in the literature. For this study, seven sampling spots were selected in the Sas al Nakhl area of Abu Dhabi near to ENTC campuses [13]. For each tree studied, large (~5-10 kg) portions of fruits, pedicels and leaves were sampled. Each of the plant samples were then processed through a protocol developed in-house (see Fig. 1 bottom [14]), ultimately leading to ~1 kg dried powders of fruits and leaves, packed in radon-tight plastic bottles/beakers.

For vegetables, three top grown crops were selected: cucumbers, tomatoes, and bell peppers. They were sampled in >10 kg sample sizes per spot from multiple commercial farms in Bahya, Liwa and Al Ain. In study of the vegetables fresh samples were packaged by grinding each of them to < 1mm grain size. To convert our sample masses from wet to dry weights, which is part of our adopted protocol [6], small aliquots of each sample were dried to obtain average wet to dry ratios. Like plants, rhizosphere soils were sampled for both the experiments. Typically, rooting depth varies depending on plant species and the nature of soil and irrigation. But for consistent comparison across regions, agencies like IUR recommend ~20 cm depth from topsoil [6]. In each field study, soils were Auger cored to a representative root depth of 20-22 cm.

Besides fruits and vegetables, a first of its kind NORM radioactivity estimation has been done for leaves of the Ghaf tree (*prosopis cineraria*) through sampling in Abu Dhabi, Al Ain-Dubai highway and Ghayathi. Ghaf is a native tree found all over the Arab region, and for its geographical prevalence, vitality in sustainability of the desert lands, and for being a crucial link to nomad camels, ghaf is an important non-food plant to study for REIA. Leaves were sampled from various heights, composited, and processed as in Figure 1 (bottom).

For date palm, vegetables, and ghaf leaves, all plant and soil samples were at first stored after radon-tightening for about 4-5 weeks to establish secular equilibrium. Then blank samples and calibration standards were measured to obtain efficiency calibration and minimum detectable activities (MDA). Then the samples were measured using high-resolution HPGe detectors, over counting times of 8 to 24 hours to obtain Bq Kg^{-1} of progenies of ^{238}U, ^{232}Th and ^{40}K. Artificial radionuclides could not be quantified confidently in our measurements, but $^{137/137}Cs$ gamma peaks were occasionally observed.

For artificial radionuclides, a controlled potting experiment of tomatoes, cucumbers, and eggplants was conducted [15]. For this, surveys of productional farms were conducted in Abu Dhabi, and representative soil mixture, chemical use and irrigation regimes were ascertained. The experiment, involving spiking of the soil using solutions of stable Sr and Cs, ran for 3-4 months. The obtained fruit, leaf, shoot, and root samples will be studied using ICP-MS techniques, for which they have been processed through multi-acid microwave digestion. Weekly sampling of the spiked soil was also done, which are under physicochemical and radiometric studies, employing leaching and sequential extraction methods.

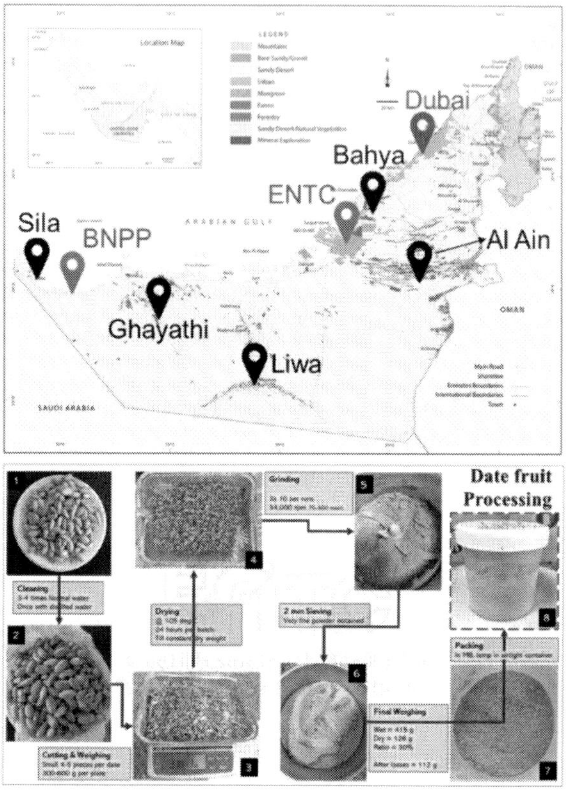

Fig. 1. (top) geographical map of the UAE with red markers showing locations of BNPP, ENTC and Dubai, and black markers showing areas of sampling. (Bottom) an example sample processing protocol developed at ENTC. Here it shows processing and packaging of date fruits in Marinelli (MB) or similar plastic bottles/beakers for gamma spectrometry.

IV. Tentative Results

The gamma spectrometry of all samples showed presence of all NORMs. Counting times were optimized for quantification of the three decay chains but many of them had large uncertainties. Most frequently measured progenies were ^{214}Pb or ^{214}Bi for ^{238}U chain, and ^{212}Pb or ^{212}Bi or ^{228}Ac for ^{232}Th chain. A typical gamma spectrum with peak analyses and nuclide identification is reported in Fig. 2 (shown sample is of ghaf leaves collected at a spot between Al Ain and Dubai highway).

Drying procedures concluded that 90-96% of vegetables are moisture/water, while that for date fruits was 24-34%, for date leaves ~38%, for ghaf leaves 51-72%, and for the soil samples it is ~10%. For the measurements done with fresh samples (vegetables), each sample's AC data has been divided by its respective drying fraction to obtain its equivalent dry mass AC.

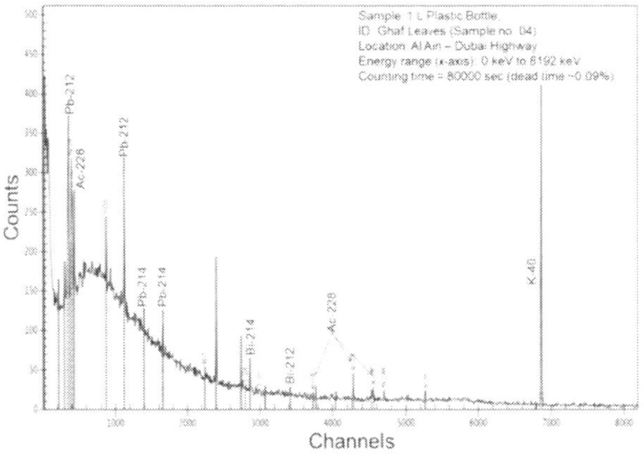

Fig. 2. Example of a measured and analyzed gamma spectrum as obtained in the reported study of ghaf leaves.

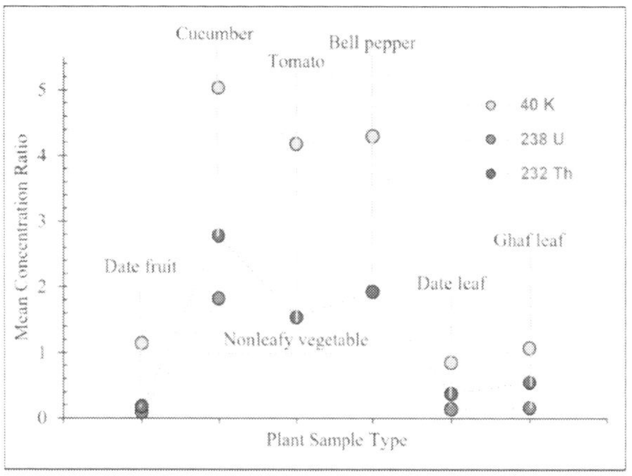

Fig. 3. Average CR values for various plant sample types and the three NORM chains, as measured or estimated through field experiments in this work and literature soil data. Dotted lines connecting the markers do not show a trend and are used for visual guidance only.

In the field sampling experiments, we collected and analyzed soil samples from more than 25 geographical spots, each being a homogenous composite of multiple sub-spots within a section of the farm or trees plantation. From date-palm studies, average ACs (in Bq kg^{-1}) of NORMs in the soil are around 15.5 for ^{238}U, 8.3 for ^{232}Th, and 278.9 for ^{40}K. Measurements from vegetable farms (tentative values, data is still under scrutiny) lead to similar average ACs (in Bq kg^{-1}): 13.0 for ^{238}U, 5.6 for ^{232}Th, and 296.3 for ^{40}K. Our soil measurements compare well with the literature values (see Table I) and follow the necessary rigid protocol of sampling for consistent comparison globally.

Detailed analyses of date fruits and leaves were reported in ref. [13]. For date fruits, indicative plant-soil CRs have been measured as 0.08 for ^{238}U, 0.17 for ^{232}Th, and 1.12 for ^{40}K. Respective CRs for palm leaves, another commodity, are 0.13, 0.36, and 0.77.

For the studies on vegetables and ghaf leaves, the data is under scrutiny and will be published elsewhere. However, tentative analyses of this data are presented here in Fig. 3 for a discussion, with the caveat that the final numerical values may alter after detailed analyses. The plotted data shows how the nuclide specific average CRs for different crops vary. In this estimation, GM of date fruits and leaves are used from ref. [13], while arithmetic average are taken for vegetables' CRs. For Ghaf leaves, the soil was not sampled. To suffice for CR estimation of the ghaf leaves, average soil ACs from the nearest regions were extracted from the national report in ref. [12].

Among the NORM chains, the CRs follow sequence ^{40}K>^{232}Th>^{238}U. Higher uptake of potassium is frequently observed in arid lands due to higher ^{40}K in soil, as well as K being a macronutrient, and large application of NPK containing fertilizers. Absolute values of U- or Th-CRs are difficult to come by due to large uncertainties, and we expect more complex radiometric techniques will be required to establish these numbers with confidence. Tree samples, namely, date fruits, leaves and ghaf leaves have CRs smaller than vegetables. Comparing date fruits with vegetables, the latter have CRs larger by 2 to 27 times of the former. Mechanisms of uptake between smaller seasonal vegetable plants and desert evergreen trees like date palms need to be compared to elucidate this data.

In-house potting experiments simulated contamination of the irrigation water, with ~1000 ppm of ^{133}Cs and 150 ppm of ^{89}Sr. Rough estimates of translocation factors have been obtained for tomatoes. Cesium is accumulated at 30% in leaves, 30% in fruit, 19% in stalk and 21% in roots. Likewise, strontium is absorbed 51% in leaf, 22% stalk, 25% in roots, and negligibly small ~1% in fruits. On tomatoes, this study with stable analogues of ^{137}Cs and ^{90}Sr have provided their rather high TFs as 12.1 and 2.1, respectively. Newer plantation experiments have been conceptualized based on this learning experience.

V. Proposed Future Experiments

For the missing gaps in ARRE and the UAE's local radioecology, several systematic experiments are necessary to be implemented. First, field experiments need to be extended to all kinds of crops, grasses, native trees etc., and to all regions of the UAE. The controlled potting experiment reported here served only for development of an experimental methodology. For measurement of CR for artificial radionuclides, further controlled planting in farm like setting needs to be conducted. Among other missing studies are the atmospheric dispersion measurements with UAE's wind patterns, and wet deposition dynamics for fogs which occur more often than rains in this region. Vertical and lateral migration speeds of radionuclides deposited on sand dunes/sheets need to be formally evaluated. There are some existing and underway studies on groundwaters and marine waters, but those need to be extended as well. Finally, completely amiss data for the UAE is on the transfer to animals. Among them, camels are an important subject of study. Such empirical data will find immense use in predicting radiation transmission pathways and impacts on biota and humans in the UAE.

VI. Conclusions

Arid region radioecological data, especially on radioactivity of soils and plants and plant-soil concentration ratios are vital for the radiological protection of the UAE's environs, animals, and humans. There is a major gap in this data worldwide, and literature values of the CRs vary over several orders of magnitude. Therefore, at ENTC Abu Dhabi, several field-based sampling and controlled lab potting experiments are conducted. Stringent protocols are developed through experimentation for sample collection, transport, storage, cleaning, drying, processing, and packaging for gamma-spectrometry. Likewise, for ICP MS, sample leaching and acid digestion protocols are developed. Soil and plant radioactivity for NORM radionuclides and their respective plant-soil CRs are measured for date fruits and leaves, top vegetables and ghaf leaves. They are tentatively concluded to range 0.08-1.82 for ^{238}U, 0.17-4.66 for ^{232}Th, and 0.77- 5.26 for ^{40}K. Rough estimates of uptake CR and translocation for Cs and Sr isotopes in tomatoes have been obtained with controlled potting experiment. Finally, some crucial next steps are outlined to continue and enhance the discussed experiments.

ACKNOWLEDGMENT

The authors would like to thank the kind support received from Dr. Natalia Semioshkina and Dr. Gabriele Voigt of the erstwhile r.e.m. GbR (Munich, Germany) for the discussions and advice in building the protocols for the experiments presented in this manuscript. Many of these experiments have been conducted using instrumentation at the Environmental Laboratories of Nawah Energy Company in Ruwais (Abu Dhabi, UAE), Quality and Conformity Council (QCC) Laboratories (Abu Dhabi, UAE), and International Center for Biosaline Agriculture (ICBA Dubai, UAE). The authors would like to acknowledge the scientists from these labs, particularly Dr. M Millinor and Dr. A Maddison of Nawah Labs, Mr. Abdulla Yousuf Almarzooqi of QCC labs and Dr. Ahmed el-Naggar of ICBA's soil analysis lab.

DISCLAIMER

This work was carried out under theme-3 (radiation safety and environment) of the Emirates Nuclear Technology Center (ENTC) of Khalifa University Abu Dhabi (UAE). The expressed opinions and thoughts reflect only the authors' views, and do not represent the views of their affiliation institutions.

REFERENCES

[1] D. Copplestone, C.-M. Larsson, P. Strand, and M. K. Sneve, "Protection of the environment in existing exposure situations," *Ann. ICRP*, vol. 45, no. 1_suppl, pp. 91–105, Jun. 2016, doi: 10.1177/0146645316630167.

[2] N. Caplin and N. Willey, "Ionizing Radiation, Higher Plants, and Radioprotection: From Acute High Doses to Chronic Low Doses," *Front. Plant Sci.*, vol. 9, p. 847, Jun. 2018, doi: 10.3389/fpls.2018.00847.

[3] ENEC, "Emirates Nuclear Energy Corporation," 2022. https://www.enec.gov.ae/ (accessed May 25, 2022).

[4] P. Raj *et al.*, "Conceptualization of arid region radioecology strategies for agricultural ecosystems of the United Arab Emirates (UAE)," *Sci. Total Environ.*, vol. 832, p. 154965, Aug. 2022, doi: 10.1016/j.scitotenv.2022.154965.

[5] IAEA TECDOC 1979, *Soil-Plant Transfer of Radionuclides in Non-Temperate Environments*. in Technical Document, no. 1979. Vienna: International Atomic Energy Agency, 2021.

[6] "IUR: International Union of Radioecology." https://www.iur-uir.org/en/

[7] "Emirates Nuclear Technology Center, Khalifa University, Abu Dhabi (UAE)." https://www.ku.ac.ae/entc

[8] J. Brown, D. Telleria, T. Yankovich, T. Cabianca, and C. Halsall, "Overview of the MODARIA programme (and comments on implications for future programmes of work) *," *J. Radiol. Prot.*, vol. 42, no. 2, p. 020505, Jun. 2022, doi: 10.1088/1361-6498/ac5bdd.

[9] IAEA CRP K41022, "New CRP: Transfer of Radionuclides in Arid and Semi-Arid Environments for Radiological Environmental Impact Assessment (K41022)," Apr. 29, 2021. https://www.iaea.org/newscenter/news/new-crp-transfer-of-radionuclides-in-arid-and-semi-arid-environments-for-radiological-environmental-impact-assessment-k41022 (accessed May 25, 2022).

[10] J.-F. Lecomte *et al.*, "ICRP Publication 142: Radiological Protection from Naturally Occurring Radioactive Material (NORM) in Industrial Processes," *Ann. ICRP*, vol. 48, no. 4, pp. 5–67, Dec. 2019, doi: 10.1177/0146645319874589.

[11] UAE MOCCAE, "Open Data | UAE Ministry of Climate Change and Environment;," 2022. https://www.moccae.gov.ae/en/open-data.aspx#page=1 (accessed May 25, 2022).

[12] R. L. Conatser *et al.*, "Radiological Environmental Monitoring in the United Arab Emirates, First Report, January 1 - December 31, 2015; Federal Authority for Nuclear Regulations (FANR);," *Available Online HttpsbitIy2DhwvVQ Accessed 25 August 2022*, [Online]. Available: https://www.fanr.gov.ae/en/Lists/Publications/Attachments/12/FANR%20REMP%20English.pdf

[13] P. Raj *et al.*, "Transfer of Natural Radionuclides from Soil to Abu Dhabi Date Palms," *Sustainability*, vol. 14, no. 18, p. 11327, Sep. 2022, doi: 10.3390/su141811327.

[14] Prasoon Raj et al, "Baseline soil-to-plant transfer of NORM radionuclides in Abu Dhabi date-palms," presented at the 9th Intl. Conf. Radiation in Various Fields of Research (RAD 2021) 14–18.06.2021 | Montenegro, 2021. doi: https://doi.org/10.21175/rad.abstr.book.2021.34.14.

[15] Nemeer Padiyath et al, "Soil-to-plant transfer factors in arid regions using stable isotopes of caesium and strontium in local vegetables," presented at the 9th Intl. Conf. Radiation in Various Fields of Research (RAD 2021) 14–18.06.2021 | Montenegro, 2021. doi: https://doi.org/10.21175/rad.abstr.book.2021.34.10.

An innovative detector for X-ray beams made of polyethylene terephthalate built in a 3D printer

Daniel F. Gomes[1], Edvane B. Silva[1], and Luiz A. P. Santos[2,3,*]
[1]UFPE, Brazil
[2]CRCN-NE/CNEN, Brazil
[3]SCIENTS, Brazil
(*) lasantos.scients@gmail.com

***Abstract*—3D printers have been increasingly used to create different objects and tools for the most diverse areas of knowledge. In this paper, we bring results of the response of an X-ray detector that was made on a 3D printer with polyethylene terephthalate glycol (PET-G) material. The manufacturing process has practically two steps, which consists of building a composite of two materials: a build-up cap made of PET-G; and a semiconductor film made of carbon-doped PET. The experimental methods performed were basically to compare the signals from the new PET-based X-ray detector with the signals from typical detectors. Clinical X-ray beams were used to test such an innovative detector. Two X-ray tube parameters were varied to analyse the response of the PET-based X-ray detector: potential (kV) and workload (mAs). The study points out the PET-G detector works and is easy to manufacture, although the signal is weaker than a typical semiconductor detector. Also, the noise signal is at least one order of magnitude smaller than silicon detectors. Furthermore, the density of PET is almost equivalent to human tissue, which is an advantage for dosimetry in phantoms. Finally, such a detector could become an option for X-ray beams in the near future.**

***Keywords* — Polyethylene terephthalate, X-ray, Detector.**

I. Introduction

SINCE the last century, some international institutions of both radiological protection and health [1-2], have been warning about some possible radiological accidents due to unnecessary radiation doses or overdoses in patients undergoing medical radiodiagnosis. In fact, these types of events have already happened and some researchers are looking to develop new detectors or dosimetry systems to be applied in radiodiagnosis. Furthermore, the modern digital radiology equipment itself operates with high dose rates [3], although in this type of equipment the patient is usually exposed to radiation in a short time when submitted to radiography exams, for example. Also, there are several researches in the dosimetry area that use detectors inserted inside human body simulator [4] (e.g. Alderson Random anthropomorphic phantom) to avoid performing measurements in humans (in vivo dosimetry) that could evidently be harmful to patients.

Regarding the methods that use numerical dosimetry with phantoms, the detectors are not always made with equivalent tissue material, such as silicon, which is widely used in semiconductor detectors. Furthermore, it is known that the manufacturing process of the silicon detector is complex and in some cases it can take long time to build it. However, 3D printers have been increasingly used to create different objects and tools for the most diverse areas of knowledge, among which we can exemplify: mechanical engineering, medicine, information technology, in addition to objects such as toys, souvenirs, etc. For the reasons explained above, we were motivated to create an innovative X-ray detector, which is built of polyethylene terephthalate Glycol (PET-G), for applications in X-ray beams for medical radiology.

II. Material and Methods

A. Building the PET detector

Firstly, PET material was chosen to build the X-ray detector due to the fact that it is the raw material for 3D printer. In this way, the PET itself becomes the detector's buildup cap. Furthermore, PET can be doped with carbon to become a semiconductor, which will provide the conduction of the current signal produced during irradiation. Fig. 1 shows an illustration of the detector manufactured on a 3D printer. Basically the detector is made in two steps: 1) PET-G chip and wires are impregnated on the edges; 2) a thin carbon-doped PET film is deposited onto the chip in the region between the two wires. The PET chip volume dimensions (L×W×H) can vary depending on the photon energy or desired output current. Therefore it can be from 0.1 cm^3 up to 1 cm^3, for while.

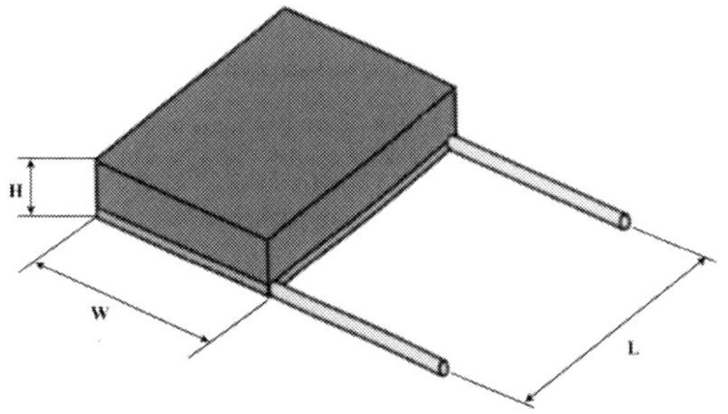

Fig. 1. Illustration of the PET X-ray detector: PET (orchid); carbon-PET semiconductor film (grey); metallic wire (silver).

B. Biasing the PET detector and instruments

The buildup cap has the purpose of increasing the electron rain due to the numerous interactions that occur there (see Fig. 15 in [5]). Actually, the voltage bias is for the carbon-PET semiconductor film and it is where the current generated by the electron rain will be conducted between the wires of the detector. The detector biasing circuit schematic is illustrated in Fig. 2. A 6430 source-meter, Keithley®, was used for biasing the detector with 20V in SVMI mode (voltage-source and current-measurement). Although the detector can be biased with voltages between 5 and 35V, it was chosen the average value to improve the signal-to-noise ratio and avoid hysteresis in the detector's I×V curve. One way to characterize the PET X-ray detector is to trace its I×V curve on a semiconductor parameter analyzer, 4200A-SCS, Keithley®. Such a technique can also show us how much the PET detector is damaged by the accumulated radiation dose.

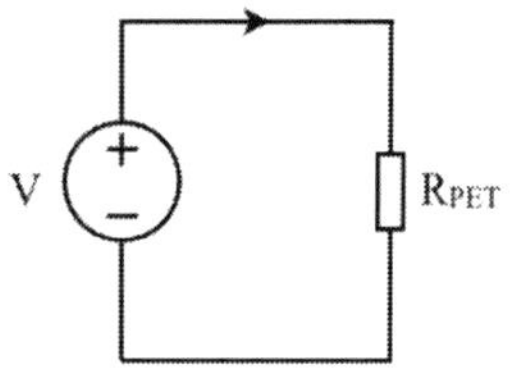

Fig. 2. Circuit schematic to bias the PET X-ray detector.

C. Testing the PET detector under X-ray beams

To perform tests with the PET-G detector, a clinical X-ray generator was used: Polymat 30/50 Plus, Siemens®. In the control panel of the clinical equipament it is possible to select the main X-ray tube parameters: 1) Exposure time (ms); 2) Potential (kV); 3) Workload (mAs). Table 1 displays the parameters and their respective selectable values. In order to compare the PET detector measurements with other types of semiconductor detectors, experiments were done varying the kilovoltage and the workload. For all experiments the X-ray pulse duration was held constant at 3200 ms. Experiments were performed using PET X-ray detectors with following dimensions: 1×1×0.3 cm^3. Each point on each graph in the results is to the average of two readings. Each experiment was repeated twice to check reproducibility. Fig. 3 illustrates the setup for carrying out the PET detector irradiation procedures.

D. Basic methods

First, the PET-G detector was analyzed under an optical coherence microscope, THORLABS®, to see if the carbon-doped PET film was uniform. Second, I×V curve of the device was analyzed to confirm that the detector behaves like a resistor, here called R_{PET}. Finally, under pulses of diagnostic X-rays, PET detector output signal measurements were made.

The first part of the irradiation procedures consisted in verifying how the PET detector responds to beams with different X-ray tube potentials. In this procedure, the workload was varied for the following values: 100 mAs and 200 mAs. In order to have comparison with silicon based devices two additional experiments were done: 1) FQP20N06C MOSFET measurements; 2) ZTX851 NPN Bipolar Junction Transistor (BJT) measurements.

TABLE I
X-RAY TUBE PARAMETERS AND SELECTABLE VALUES

X-ray tube parameter	Selectable values
Potential (kV)	52, 63, 77, 82, 90, 102, 117, 125
Workload (mAs)	50, 62.5, 80, 100, 125, 160, 200, 250, 320
Pulse time (mS)	50, 100, 200, 400, 800, 1000, 1600, 2000, 3200

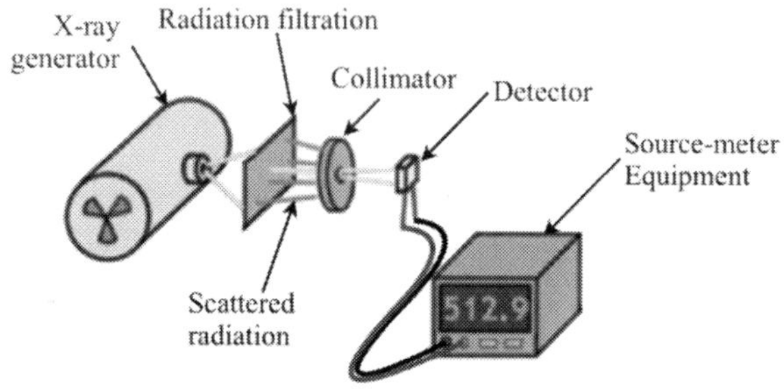

Fig. 3. Experimental setup for PET X-ray detector irradiation [5].

III. RESULT AND DISCUSSIONS

As shown in Fig. 4, the carbon-doped PET semiconductor film did not have uniformity in its deposition on the PET-G chip and, consequently, in its thickness within the region between the two wires. In addition to the deposition of the carbon-PET film not being perfect, the face of the PET-G chip itself is also not perfect, unlike the surface on the silicon crystal.

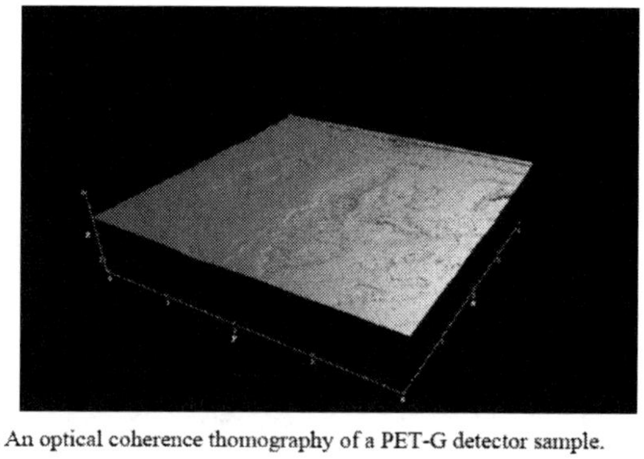

Fig. 4. An optical coherence thomography of a PET-G detector sample.

Fig. 5 shows the PET detector is like a 2.5±0.2GΩ resistor. Furthermore, it appears to have a capacitance (to be studied later), as the behavior is not exactly a straight line. Prior to irradiating the PET detector samples, dosimetry of the X-ray pulses was performed with our reference detector, 10X5-6 ion chamber, Radcal®, to measure the dose rate of each X-ray tube potential in Table 1. The pulse dosimetry results are shown in Fig. 6.

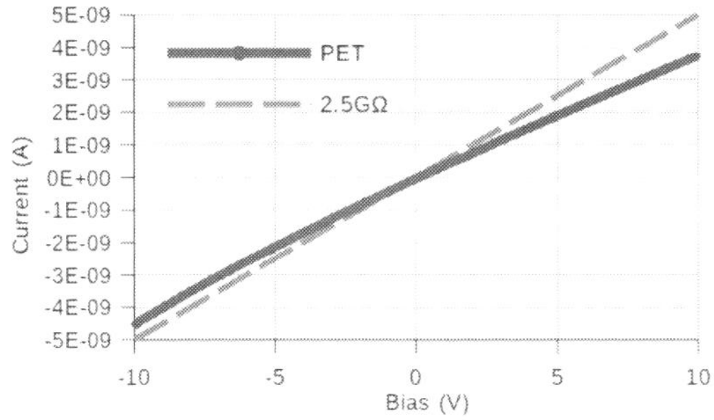

Fig. 5. I×V curve of a PET detector sample: R_{PET}=2.5±0.2 GΩ.

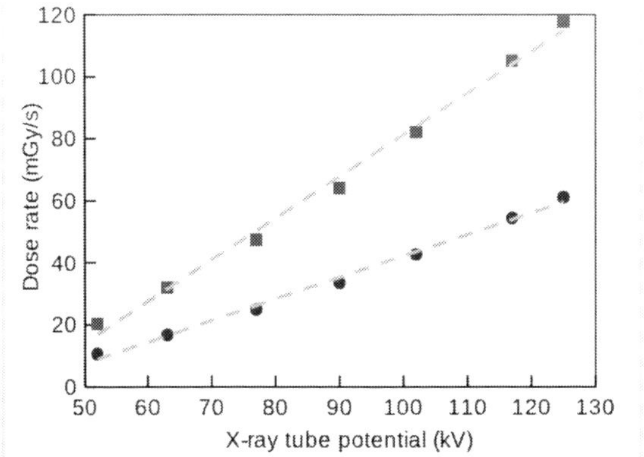

Fig. 6. Dose rate as a function of X-ray tube potential at the position where the PET detector was placed to perform measurements during irradiation: 100mAS (maroon); 200mAS (red).

The first comparison, which was made with a MOSFET, is shown in Fig. 7 and Fig 8. Note that, although the responses of each type of detector closely resemble the response of the reference detector, the signal produced by the PET detector is weaker than the signal produced by the MOSFET. Actually, if the latter is compared with other silicon semiconductor devices, a MOSFET has the weakest signal than a BJT or phototransistor, for example [6]. However, the PET detector signal is about 20-40 dB stronger than a typical ionization chamber. On the other hand, at room temperature, the noise signal produced by a PET detector is around 0.2 nA, whereas the signal from a silicon semiconductor device can vary between 2 nA and 20 nA, for example. That is, the noise in the silicon device is at least 20dB greater than the R_{PET} resistor.

Another comparison was made based on the effect of radiation on a BJT. Although the output signal of a BJT is 40-60 dB stronger than PET output signal when it is irradiated, a degradation in transistor gain systematically occurs [5]. It can be easily seen in Fig. 9 and Fig. 10 the effect of irradiation on the BJT and PET detector, respectively. One can observe that the transistor gain decreases and consequently the transistor output signal also decreases, whereas the PET detector output signal remains practically constant. Note that the PET detector received double the dose received by the BJT. PET samples were sequentially irradiated with a dose of approximately 7.86Gy (30 X-ray pulses of ≈262 mGy) and no traces of degradation were measured by the source-meter. This total dose corresponds to about 7860 exposures (typical lung radiograph ≈1 mGy) [1], i.e., the innovative detector could be inserted into a phantom to measure more than 7800 times the dose rate there.

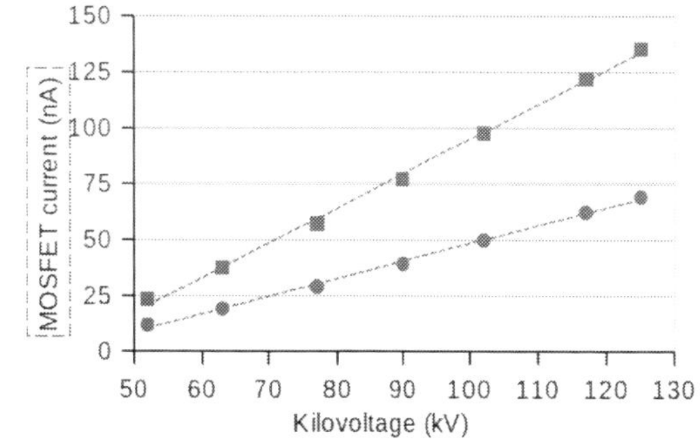

Fig. 7. MOSFET detector output signal as a function of X-ray tube potential, for workload of 100mAs (maroon), and 200mAs (red).

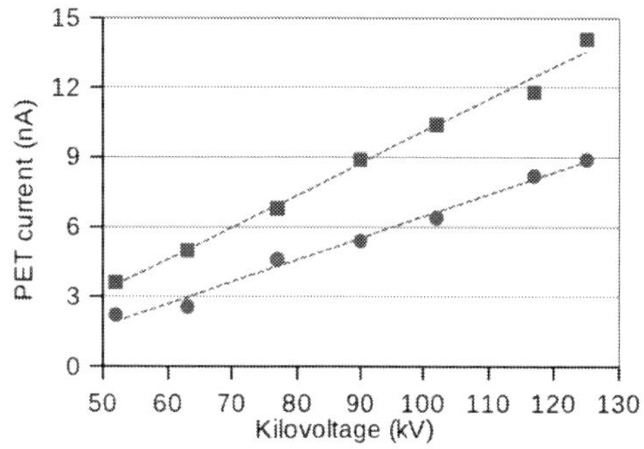

Fig. 8. PET detector output signal as a function of X-ray tube potential, for workload of 100mAs (maroon), and 200mAs (red).

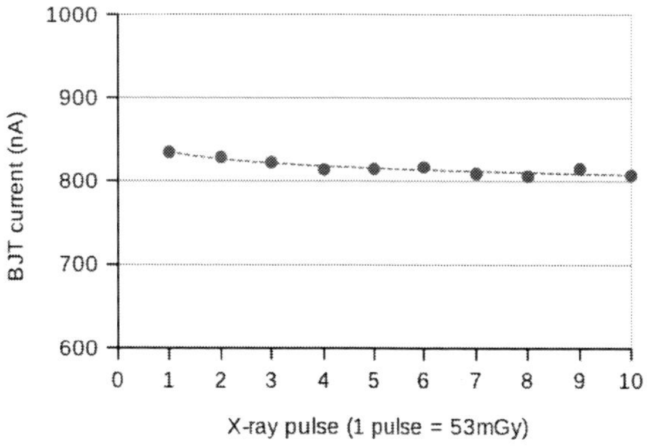

Fig. 9. BJT output signal after 10 X-ray pulses; D = 530 mGy.

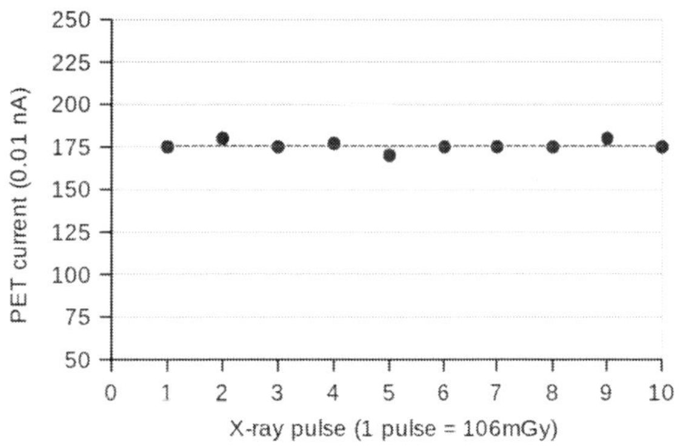

Fig. 10. PET detector output signal after 10 X-ray pulses; D = 1060 mGy.

IV. Conclusions

An innovative radiation detector for X-ray beams applied to radiodiagnosis was presented. The output signal of this new device is about 20dB weaker than a MOSFET, which is the weakest signal among silicon-based detectors. On the other hand, the PET detector has some advantages for using it as a detector in radiation beams. First of all, it is a device that can be easily manufactured by a 3D printer in a practical and fast way, unlike silicon semiconductor detectors. Second, the PET detector behaves like a typical resistor, which has a weaker noise signal than a silicon semiconductor at room temperature. One of the main advantages of PET detector is radiation hardness as it can be used repeatedly without degradation. Furthermore, the PET detector has a density that makes it equivalent to human tissue, which is an important feature for it to be usable inside phantoms. Also, it is important to point out that the device's semiconductor film between the wires is irregular, without impairing its functioning as a radiation detector. Finally, such a detector could become an option for X-ray beams in the near future.

Acknowledgment

The authors would like to thank the "Unidade de Diagnostico por Imagem" of the University Hospital (HC-UFPE/EBSERH) for the collaboration in the irradiation procedures. We also thank the company Scients (contract 200313) and the Brazilian agency CNPq for financial support (grant 315017/2021-7).

References

[1] IAEA, "Applying radiation safety standards in diagnostic radiology and interventional procedures using X rays," IAEA., Vienna, Austria, Safety Report Series, No. 39., Oct. 2006.

[2] ICRP, "The 2007 Recommendations of the International Commission on Radiological Protection," ICRP Publication 103, Vienna, Austria, Annals of the ICRP, 2007.

[3] C. J. Martin, "Management of patient dose in radiology in the UK," *Rad. Protec. Dosim.*, vol. 147, pp. 355–372, Nov. 2011, doi: 10.1093/rpd/ncr386.

[4] G. Brix, U. Lechel, R. Veit, R. Truckenbrodt, G. Stamm, E. M. Coppenrath, J. Griebel, H. -D. Nagel, "Assessment of a theoretical formalism for dose estimation in CT: an anthropomorphic phantom study," *Eur. Radiol.*, vol. 14, pp. 1275–1284, 2004, doi: 10.1007/s00330-004-2267-7.

[5] L. A. P. Santos, "An Overview on Bipolar Junction Transistor as a Sensor for X-ray Beams Used in Medical Diagnosis" Sensors, vol. 22, no. 5, 1923, Mar. 2022, doi: 10.3390/s22051923.

[6] C. P. V. Valença, L. C. Gonçalves Filho, A. N. Alves, M. A. Macedo, D. N. Souza, L. A. P. Santos, D. A. A. Santos, "The comparison of a thin-film ZnO nanodevice with silicon-based electronic devices for diagnostic X-ray beam detection," Radiation Effects and Defects in Solids, vol. 177, pp. 642-654, June 2022. doi: 10.1080/10420150.2022.2073878

Radiation monitoring with plant-based biotas and an automated micronucleus scoring approach

F. d'Errico[1,*], F. Barco[1], T. Butini[1], M.G. Cascone[1], R. Ciolini[1], M. Quattrocchi[3], E. Rosellini[1], T. Hulber[4], J.A. Torres Novaes[2], M.N. Xavier[2], S. de Souza Lalic[2]
[1]School of Engineering, University of Pisa, Italy
[2]Department of Physics, Federal University of Sergipe, Brazil
[3]Department of Radiology, St. Luca Hospital, Lucca, Italy
[4]Radosys, Budapest, Hungary

(*) francesco.derrico@unipi.it

Abstract— **In this work, we investigated the response of the stem cells from the roots of Allium Cepa (meristems) to ionizing and non-ionizing radiations of different qualities. Allium cepa (Onion) is a well-established in-vivo standard model, widely used in cytogenetic studies for different environmental pollutants. Endpoints, such as chromosomal aberrations (CAs), micronuclei (MNs), and disturbance in the mitotic cycle of root meristematic cells of onion are frequently used to determine the cytotoxic and genotoxic potential of different environmental pollutants. Traditionally, these studies have been carried out to assess chemical toxicity, while the toxicity of ionizing radiation has been studied less extensively. We also examined the repurposing of a previously described optical microscope system originally designed for automated non-fluorescent micronucleus (MN) scoring in binucleated peripheral lymphocytes. The microscope system relies on hardware and software layers in parallel in order to optimize the performance in automated MN scoring.**

Keywords —**Biota, micronucleus, ionizing radiation, non-ionizing radiation.**

I. Introduction

THE analysis of cytotoxic effects of ionizing radiation in the flora and fauna is of interest in agriculture, ecology, health care, and space exploration. Humans are continuously exposed to radiation from a variety of sources throughout their lives, such as solar radiation, natural background radiation, or medical devices. Typically, absorbed dose assessment is done using physical dosimeters, but these tools are not commonly available. This has led to the need to develop new strategies to assess the damage, especially to DNA, caused by such radiation. There was a need to identify parameters (endpoints) that would allow the effects of radiation on cells, particularly on genetic material, to be assessed and to correlate these effects with the dose received. The relationship between endpoints and absorbed dose is known as biodosimetry [1], and relies on the quantification of biomarkers such as micronuclei and chromosomal aberrations to assess the dose received by a subject.

The effects of ionizing radiation in plants [2] may provide an efficient warning system to avoid harm to human health. Plants are unlikely to exhibit the same stochastic effects of ionizing radiation that are observed in animals. However, higher plants present effects that may mimic those of other biological systems, since the critical target is the DNA, common to all organisms. Indeed, previous studies have shown that chemical genotoxic agents produce comparable effects in plant and animal systems in terms of genetic abnormalities [3]. Additionally, these studies on so-called plant-based biotas avoid the controversial use of animal models. Thus, they represent a new frontier of radioecology, an emerging field of environmental ecology that investigates the presence and the effects of radioactivity in Earth's ecosystems. The investigated effects are typically DNA damage visualized as chromosomal aberrations when cells divide, or as the presence of micronuclei in inter-phasic cells.

The present study focuses on biodosimetry of ultraviolet (UVB and UVC) and ionizing radiation (high and low linear energy transfer, LET) particularly alpha particles and X-rays, using a plant organism, Allium cepa (commonly known as "onion"), as an alternative method to animal models. Among biological species, Allium cepa is a good bioindicator for damage estimation because of its high sensitivity not only to chemicals but also to radiation [4-7]. The use of this organism is very important because it is a diploid species (2n = 16) with a low number of large chromosomes, which makes it easy to observe defects occurring in these chromosomes. In addition, it has a predominantly aqueous chemical composition, similar to that of human soft tissue.

The genotoxicity of radiation was studied using the micronucleus test [5]. The micronucleus test consists of microscopic analysis of cells during mitosis, taken from an organism exposed to a potentially mutagenic agent. Its purpose is to identify the presence of scattered fragments of DNA in the cytoplasm that are not incorporated into the nucleus during the final stages of cell division due to the mutagenic agent.

The experimental work based on the use of Allium cepa seeds started from a known method [7] but was optimized and standardized through the utilization of a device that mechanically compresses the apical meristem of Allium cepa. This technique allows obtaining the required monolayer of cells for observing DNA damage (micronuclei, aberrations, etc.), thereby enhancing the reproducibility of the entire procedure.

II. MATERIALS AND METHODS

The method for preparing the samples necessary to perform damage analysis resulting from irradiation (particularly for the identification of micronuclei and aberrations) can be divided into five main phases: germination of Allium Cepa seeds, irradiation, slide preparation, sample squashing, and microscope analysis. All operations are carried out under sterile conditions, operating within a laminar flow hood.

After sterilization, the sample preparation procedure begins as follows:

Germination: In this study, organic onion seeds of the Tropea variety were used. The seeds are placed, maintaining a distance of 1 cm from each other, inside Petri dishes in which moistened filter paper sheets are placed as a culture medium. Finally, the plates are sealed and placed in an incubator at 25 ±1 °C for 3 days to obtain sprouts approximately 3-4 mm in length.

Irradiation: For irradiating the meristems of Allium Cepa, a lamp consisting of three UV tubes, one for each wavelength (254 nm, 302 nm, 365 nm), was used. Dosimetric studies were conducted using three radiometers. In the first phase, the three detectors were placed at the same distance from the lamp to obtain the normalization point of the radiometers' response. Subsequently, the irradiance profile was evaluated, observing that the measurements follow a 1/r profile, consistent with the behavior of a linear source. For UVC irradiation, the plates were positioned at a distance of 7.3 cm from the lamp, corresponding to an irradiance of 0.818 mW/cm^2, and doses of 123 mJ/cm^2, 245 mJ/cm^2, 368 mJ/cm^2, and 491 mJ/cm^2 were administered. For UVB irradiation, the plates were positioned at a distance of 13 cm, resulting in an irradiance of 0.690 mW/cm^2, and doses of 104 mJ/cm^2, 207 mJ/cm^2, 311 mJ/cm^2, and 414 mJ/cm^2 were administered.

Alpha irradiation was conducted at the laboratories of the University of Pisa using a 241Am source with a dose rate of 7.92 mGy/min. The absorbed doses were 20, 40, 60, and 80 mGy. Five samples were analyzed for each dose point.

Regarding X-rays, the experimentation was conducted at the San Luca Hospital Center in Lucca, where a linear accelerator was used to produce 6 MV X-ray photons with a dose rate of 100 monitor units (MU)/min. MU is defined as the monitoring unit, representing the time required for the accelerator's monitor chamber to deliver a certain dose to the tissue. The system is calibrated so that 100 MU corresponds to 1 Gy. The absorbed doses were 150, 300, 450, and 600 mGy, and these values were obtained based on the assessment of relative biological effectiveness. Literature data indicate that, at the same dose, there is an approximately one-order-of-magnitude difference in the effect between high and low linear energy transfer (LET) radiation. Therefore, doses for X-rays were chosen to be approximately one order of magnitude higher than those for alpha particles.

After irradiation, the germinated seeds are returned to the incubator for 24 hours. This time is necessary to allow the cells in the apical portions of the sprouts to complete at least one cell cycle and permit the potential formation of micronuclei and chromosomal aberrations. After 24 hours, the sprouts are placed in Falcon tubes containing the fixative Carnoy (which halts the cell cycle) and left for 24 hours at a the seeds to Falcon tubes containing 70% ethanol, where they are stored at a temperature of 4 ±1 °C until they are used for slide preparation.

Slide preparation: The germinated seeds are placed inside a Falcon tube with 1M HCl and placed in a water bath at a temperature of 60°C for 5 minutes. Subsequently, the seeds are covered with distilled water for 10 minutes to hydrate them, and then soaked in drops of acetic acid for 3 minutes to facilitate the breaking of the cell wall [8]. Finally, the sprouts are stained using acetic orcein for 15 minutes.

Squashing of the apical meristem of Allium Cepa: Each seed is placed on a glass slide, and using a scalpel, the mitotically active apical region is separated from the rest of the sprout. It is then covered with a coverslip. The squashing of the meristem, traditionally done manually by pressing the slide with the thumb, was instead performed using a device designed and developed for this thesis work to standardize the procedure and obtain perfectly reproducible samples. After squashing, the slides are sealed and stored in a refrigerator at a temperature of 4 ±1 °C.

Reading of the slides: The slides, removed from the refrigerator 20-30 minutes before reading, are placed under a Nikon Eclipse TS100 optical microscope and analyzed at a total magnification of 400x. The analyzed sample is considered valid for analysis if it contains 1000 cells [6], as the micronucleus test aims to determine the number of cells with micronuclei out of 1000 cells counted. To reduce time and minimize the risk of errors during reading, a double reading approach was implemented, one on the computer where the counting of 1000 cells is performed using a photo of the region of interest, and one under the microscope, necessary to discriminate micronuclei and chromosomal aberrations.

A device for crushing sprouts was created with the aim of "engineering" and standardizing one of the most delicate phases of the procedure, which allows obtaining a monolayer of cells to be subsequently analyzed under a microscope. The traditional method involves compressing the sprouts by crushing them with the thumb, without any control over the applied pressure (which can vary from time to time).

The functioning of the device is as follows: the glass slide with the sprout is placed in a dedicated housing on a cart. This cart, guided by two linear guides, moves under a bridge to which the system for applying force is hinged. The force application system consists of a roller connected to an adjustable spring with a nut. By compressing or loosening the spring, the force that the roller will apply to the glass slide during the cart's passage can be modulated.

To evaluate the actual force applied by the roller, a scale has been designed and implemented within the structure. It consists of a load cell mounted near the cart, which contains the electronic circuitry inside. This load cell is controlled by a dedicated chip (HX711), which is connected to the "brain" of the system, namely a compatible Arduino Nano board. The board is also connected to a display for visualizing the result and a button for taring.

Besides the manual microscope mentioned above, an automated system was also examined relying on hardware and software operating in feedback-mode in order to optimize the performance in automated MN scoring [9]. A MN assay automatically identified. The image processing module of the microscope is an integral part of the system and works in synergy with the image acquisition. This module identifies the contour of the objects of interest (namely cell, nucleus and MN) stepping through the scanned fields of view of a slide. This module also provides feedback to the illumination and focusing functions. The image segmentation process of a single field of view can be divided into three subsequent steps. Cells are identified first, then nuclei and finally micronuclei. Each step of the segmentation comprises two substeps: first, the algorithm creates a pool of candidate objects that still includes a significant number of false positives. Afterward, in the so-called classification step, the artifacts are eliminated based on their geometrical, morphological and staining intensity features. The device was previously demonstrated to offer the accuracy of visual scoring aided by minimal user interaction. The possibility of fully automatic scoring with an accuracy suitable for triage purposes is also available in case a surge of cases is to be managed.

III. RESULTS

A. Alpha particle and X-Ray Irradiations

The parameters considered for evaluating the damage resulting from irradiation were the number of cells with at least one micronucleus and the number of cells with multiple micronuclei. Analyzing the results obtained on the apical meristems of Allium cepa (Figure 1), it can be observed that both in the case of alpha particles and X-rays, the number of cells with micronuclei increases with the absorbed dose. Therefore, a dependence of the number of cells with micronuclei on the absorbed radiation dose can be hypothesized.

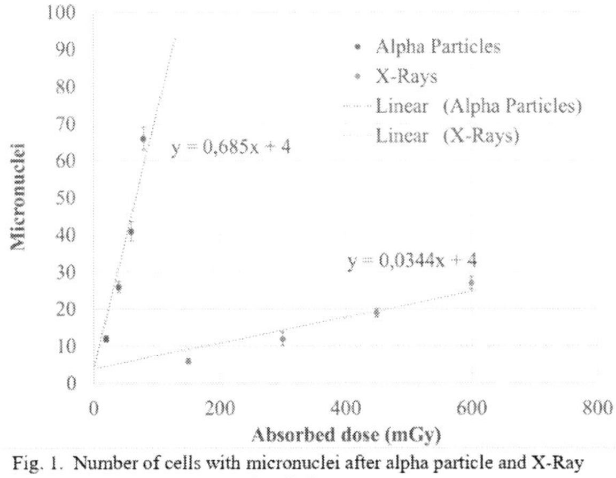

Fig. 1. Number of cells with micronuclei after alpha particle and X-Ray irradiations.

In particular, it is observed that in the case of alpha radiation, there is a much more rapid growth of the number of cells with micronuclei, concentrated in a narrower range of absorbed doses, while in the case of X-rays, the growth is much slower and occurs over a wider range of doses. This result is in perfect agreement with the fact that high linear energy transfer (LET) radiation (alpha particles), as known, is capable of causing more DNA damage compared to low LET radiation (X-rays).

This confirmation is evident from the analysis of the ratio between the two plotted lines (which do not follow the actual data but are drawn as a guide for visualization). Considering the ratio between the slopes, it is possible to define a ratio of approximately 20.

In the determination of the effective dose of radiation, the "radiation weighting factor" is introduced, assigned based on their potential harm. Low LET radiation is associated with a value of 1, while high LET radiation is assigned a weighting factor of 20. Therefore, the ratio between the factors related to high and low LET radiation is precisely 20, as observed in the ratio between the slopes of the previously described lines. This confirms the reliability of the obtained results.

Additionally, as a parameter, the number of cells with two or more micronuclei was analyzed. The presence of a higher number of DNA defects within a cell can be a symptom of greater damage to the genetic material. Once again, it can be observed that alpha radiation is capable of generating a higher number of "polymicronucleated" cells compared to X-rays. Specifically, considering all the analyzed samples, 40 cells with a double micronucleus (including one with three micronuclei) were counted out of a total of 20,000 cells counted in the case of alpha particles, while only two cells with a double micronucleus were counted out of 20,000 cells in the case of X-rays.

B. Ultraviolet B and C Irradiations

To compare the two types of UV radiation used, Figure 2 shows the variation in the number of cells with micronuclei as a function of absorbed dose. The two lines in the figure do not represent the actual interpolation of the data but are shown for visualization purposes. Analyzing the trends of these lines, it can be observed that both UVC and UVB irradiation lead to an increase in the number of cells with micronuclei as the dose increases, which is considered an indicator of radiation-induced damage. The two trends are comparable, indicating that there does not seem to be a microscopic difference between the two types of radiation.

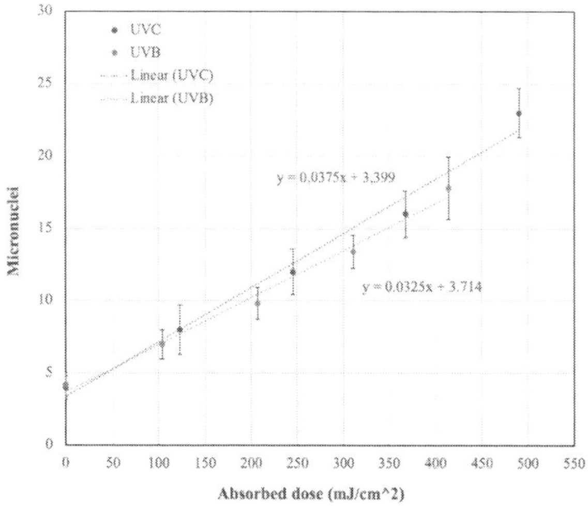

Fig. 2. Number of cells with micronuclei after UV irradiation.

It can be presumed, therefore, that the difference in damage caused by UVB and UVC radiation is not related to the deposited dose but rather to the type of action exerted by these radiations on the cells. What has been observed is a difference in terms of macroscopic damage: for UVC radiation, cells irradiated to 491 mJ/cm^2 show intact cytoplasmic membrane integrity (Fig. 3), this holds even at higher doses and the only detected damage is at the microscopic level involving DNA.

On the other hand, for UVB radiation, cells irradiated to a similar dose of 414 mJ/cm^2 show altered morphology characterized by cytoplasmic rupture and nucleus degeneration (Fig. 4).

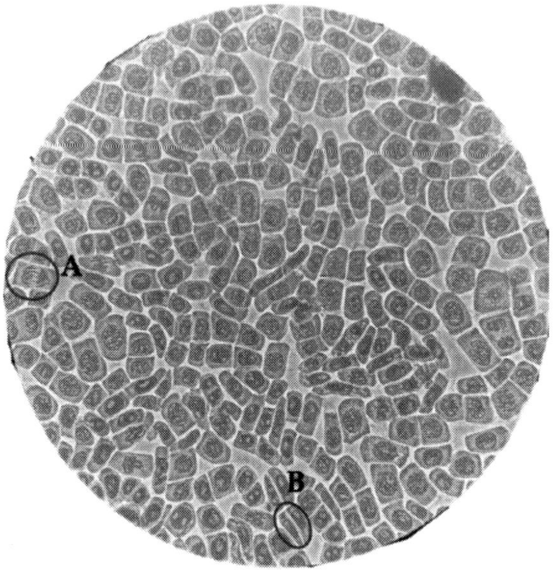

Fig. 3. Allium cepa meristems irradiated to 491 mJ/cm^{2}of UVC radiation, a bridge (A) and a micronucles (B) can be observed in the image.

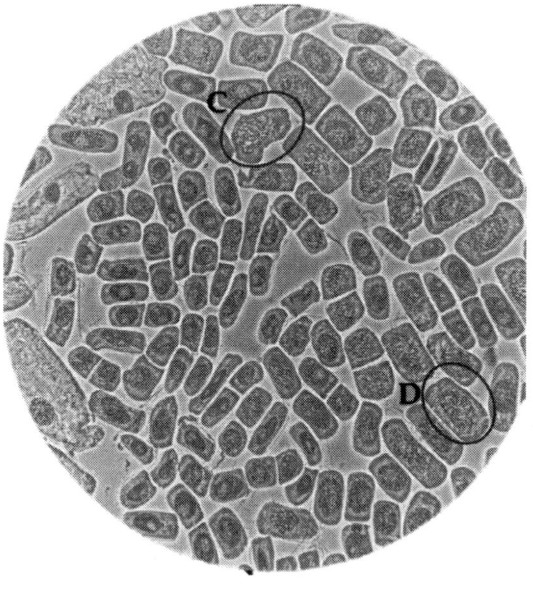

Fig. 4. Allium cepa meristems irradiated to 491 mJ/cm^{2}of UVC radiation, altered morphology (C) and nucleus degeneration (D) can be observed in the image.

By observing the cell morphology after irradiation, it can be inferred that UVC radiation primarily exerts a particle-like effect, characterized by targeted ionizing action on DNA and photochemical action. In contrast, UVB radiation predominantly exhibits a photochemical action, characteristic of this type of radiation, which leads to morphological alterations of the entire cell.

Indeed, UV radiation can damage DNA in two ways: directly and indirectly [10]. Direct DNA damage primarily involves the creation of dimerized pyrimidines, and this type of damage is likely the most significant factor in causing cancer. This dimer formation process depends on the absorption of UV light by the DNA bases, particularly in the UVB and UVC ranges causing the formation of electronic excited states. For instance, germicidal lamps emitting light at 254 nm effectively cause the dimerization of pyrimidines. However, this process is still quite effective in the UVB range (280–320 nm). The dimerization process leads to the formation of cis-syn cyclobutane pyrimidine dimers, which are the most common abnormalities in double-stranded DNA, or pyrimidine (6–4) pyrimidone photoproducts, which are the second most frequent DNA lesions in the UVB range. In addition to the formation of pyrimidine dimers, UVA and UVB indirectly contribute to the generation of oxidized DNA base damage, often triggered by light-absorbing molecules, known as chromophores, leading to photosensitization reactions.

IV. Conclusions

By comparing the results obtained from alpha radiation in this study with those documented in the literature [7], it can be concluded that Allium cepa has the potential to serve as a viable alternative model for detecting genotoxic effects from low doses of alpha radiation, instead of relying on animal models. Furthermore, this plant model has demonstrated the ability to show distinct effects caused by different types of radiation, indicating its potential to differentiate between various linear energy transfer (LET) values.

Furthermore, Allium cepa shows promise as a substitute for animal models in understanding X-ray radiation-induced biological effects. However, because this study represents a novel exploration not previously conducted by others, further investigation and data collection are needed to validate the results and hypotheses obtained.

The results of this study also indicate, for the first time, that Allium cepa has potential as a model for evaluating the biodosimetry of ultraviolet radiation. It was observed that both UVC and UVB radiation result in an increase in the number of cells with micronuclei as the dose increases. UVC radiation primarily acts on DNA at the microscopic level, whereas UVB radiation predominantly induces photochemical actions that alter cell morphology. These observed effects motivate further research, including analysis of the effects of UVA radiation, to understand the full spectrum of ultraviolet radiation, and opens the possibility of performing analysis of strictly wave radiation, such as electromagnetic radiation.

References

[1] IAEA in *Cytogenetic Dosimetry: Applications in Preparedness for and Response to Radiation Emergencies, Emergency Preparedness and Response*, Vienna, 2011.

[2] Ludovici GM, Oliveira de Souza S, Chierici A, Cascone MG, d'Errico F, Malizia A. "Adaptation to ionizing radiation of higher plants: From environmental radioactivity to chernobyl disaster." J Environ Radioact. 2020 Oct;222:106375. Epub 2020 Aug 10. PMID: 32791372.

[3] S.B Tedesco e H.D.L. IV. "Bioindicator of Genotoxicity: The Allium Cepa Test", in Enviromental Contamination, Jatin Kumar Srivastava, 2012.

[4] D. Feretti, I. Zerbini, C. Zani, E. Ceretti, M. Moretti e S. Monarca, "Allium cepa chromosome aberration and micronucleus tests applied to stu- dy genotoxicity of extracts from pesticide-treated vegetables and grapes." *Food Additives and Contaminants* , pp. 561-572, 2007.

[5] S. G. Vaijapurkar, D. Agarwal, S. K. Chaudhuri, K. R. Senwar e P. K. Bhatnagar, "Gamma-irradiated onions as a biological indicator of radiation dose." *Radiation Measurements,* pp. 833-836, 2001.

[6] M. Saghirzadeh, M. Gharaati, S. Mohammadi e M. Ghiassi-Nejad, "Evaluation of DNA damage in the root cells of Allium cepa seeds growing in soil of high background radiation areas of Ramsar – Iran." *Journal of Environmental Radioactivity,* vol. 99, pp. 1698-1702, 2008.

[7] M. Xavier, S. Pantaleao, R. Scher, R. Ciolini, F. d'Errico e S. Souza, "Allium cepa used as a dosimetry system in nuclear and radiological emergencies." *The European Physical Journal Plus,* vol. 136, 2021.

[8] H. Nefic, J. Musanovic, A. Metovic e K. Kurteshi, "Chromosomal and nuclear alterations in root tip cells of allium cepa L. Induced by alprazolam" *Med Arch* , pp. 388-392, 2013.

[9] Timea Hülber, Zsuzsa S. Kocsis, Judit Németh, Enikõ Kis, Francesco d'Errico, Géza Sáfrány & Csilla Pesznyák (2021): "Influence of sample preparation optimization on the accuracy of dose assessment of an automatic non-fluorescent MN scoring system", *International Journal of Radiation Biology.*

[10] Gerd P. Pfeifer. (2020) "Mechanisms of UV-induced mutations and skin cancer". *Genome Instability & Disease.* Vol. 1, pp. 99–113 (2020) https://doi.org/10.1007/s42764-020-00009-8

Model-Based Deep Learning Algorithm for Pulse Shape Discrimination in High Event Rates

Itai Morad[1,2,*], Max Ghelman[1,3], Dimitry Ginzburg[4], Alon Osovizky[3,4], and Nir Shlezinger[1]
[1]Ben-Gurion University of the Negev,
[2]Israel Atomic Energy Commission (IAEC),
[3]Nuclear Research Center Negev (NRCN),
[4]Radiation Detection Department, Rotem Industries Ltd
(*) itaimor@post.bgu.ac.il

***Abstract*—Pulse shape discrimination (PSD) is at the core of radioactive particles monitoring. Conventional PSD methods are geared towards low event rates, and struggle in the presence of pileups resulting from high rate. In this work we develop a PSD algorithm that combines classic approaches with deep learning techniques, that is highly suitable for coping with the dramatic challenges associated with classifying pulses in high event rates. Common PSD algorithms for high event rates limit their research to two piled-up pulses. Our algorithm is designed and tested under severe pileup conditions, where three or more pulses were piled-up. We tested the algorithm on simulated data based on $Cs_2LiYCl6:Ce$ (CLYC) based detector pulse shapes and compare its performance to both traditional PSD algorithms and data-driven deep neural network (DNN) based algorithms. In high event rates, ranging up to 10 Mcps, the algorithm demonstrates up to 8 times fewer miss-classifications than the traditional normalized cross-correlation (NCC) approach, and up to 1.7 times fewer miss-classifications than a purely data-driven DNN-aided method.**

***Keywords*—Neutron, gamma ray, pulse shape discrimination, pulse pileup, deep learning, digital signal processing, high rate.**

I. Introduction

THE process of radioactive decay may result in the emission of various radioactive particles such as neutrons, gamma rays, alphas, and betas. Some detectors have the ability to detect multiple types of radiation. Different particles interacting with these detectors produce different pulse shapes, and can therefor be distinguished using pulse shape discrimination (PSD). It is often important to be able to distinguish these different types of radiation for homeland security, nuclear safeguard, and neutron scattering applications. In some cases, these applications require accurate counting of multiple radiations in high event rates [1]–[4]. Operation within these high event rates often causes severe pileup conditions, notably distorting the shape of the pulses, in contrast to operation within low event rates [1], [3], [5]–[8]. It is therefor particularly important to design dedicated algorithms able to distinguish radioactive particles arriving at high event rates.

Various PSD techniques were proposed in the literature. Traditional digital signal processing PSD methods include charge integration [9], slice-fitting algorithms [10], histogram-difference methods [11], and discrete wavelet transforms [12], [13]. PSD methods based on traditional machine learning techniques include Gaussian mixture models [14], support vector machines [15], and k-means [16]. Finally, many authors have proposed PSD methods based on deep neural networks (DNNs) [4]–[6], [17]–[20]. Even though many approaches to PSD have been proposed over the years, the vast majority is designed to discriminate particles at low event rates. Therefor, the discrimination capabilities of these approaches degrade when discriminating highly distorted pulse shapes. PSD approaches geared towards pileup conditions often limit their scope to cases where at most two pulses are piled-up, discarding severe conditions with three or more piled-up pulses [3]–[6], [21].

As detailed above, current approaches to PSD in high event rates can be generally divided into model-based methods and data-driven methods. Model-based methods (based on signal processing techniques) are often designed for specific pileup scenarios and may struggle to discriminiate highly distorted pulse shapes. Purely data-driven methods (based on machine learning and deep learning tools) on the other hand lack knowledge about the underlying physics of the detector, missing out on important information for discrimination.

In this work, we present a PSD algorithm combining classic PSD approaches with deep learning techniques following model-based deep learning methodology [22]–[24]. By doing so we design a scheme that is both highly suitable for coping with the dramatic challenges associated with classifying pulses in high event rates while preserving the interpretability and generalization capabilities of classic PSD methods.

To derive our scheme, we first simulate the detector signal. Although the data used in this work is based on neutron and gamma ray interactions with a $Cs_2LiYCl6:Ce$ (CLYC) detector, the designed scheme can be generalized to other detectors and radiations. Next, we design our PSD algorithm intended for operation in high event rates by discriminating based on both classic features known to be useful in low rates and the raw signal, which are balanced based on features that are informative of the pileup conditions. The latter enables operation in both low rates and in high rates using the same model. Our experimental study focused on simulating the detector signal under various neutron and gamma rate conditions. We show that our proposed method successfully classifies the arriving pulses, and demonstrate its effectiveness compared to competing methods.

The rest of this article is organized as follows. section II presents the system model and problem formulation. section III details the proposed method and the training procedure. In section IV, the proposed method is experimentally evaluated. The work is concluded in section V.

II. System Model

A. Problem Formulation

We consider the PSD task realized on the digitized output of the detector. We focus our work on neutrons and gamma rays interacting with the detector. Therefor, the label space $\mathcal{S} = \{0, 1\}$ consists of two elements, '0' for neutrons and '1' for gamma rays. The raw signal produced by the detector following the arrival of the kth particle, either a neutron or a gamma ray, is defined as $\boldsymbol{x}_k$, and its label is denoted by s_k. For some classification a lgorithm $g(\cdot)$ w e d efine th e algorithm's total accuracy, neutron accuracy, and gamma accuracy tested on K arriving particles in (1), (2), and (3) respectively.

Accordingly, by letting $\mathbb{1}_{\mathcal{A}}$ denote the indicator function over event $\mathcal{A}$, our main performance measures are the total discrimination accuracy, defined as

$$\text{Total accuracy}(g) = 100 \cdot \frac{\sum_{k=1}^{K} \mathbb{1}_{s=s_k} g(\boldsymbol{x}_k)}{K}, \quad (1)$$

while the particle-specific neutron and gamma accuracies are respectively defined as

$$\text{Neutron accuracy}(g) = 100 \cdot \frac{\sum_{k=1}^{K} \mathbb{1}_{s=s_k, s_k=0} g(\boldsymbol{x}_k)}{\sum_{k=1}^{K} \mathbb{1}_{s_k=0}}, \quad (2)$$

and

$$\text{Gamma accuracy}(g) = 100 \cdot \frac{\sum_{k=1}^{K} \mathbb{1}_{s=s_k, s_k=1} g(\boldsymbol{x}_k)}{\sum_{k=1}^{K} \mathbb{1}_{s_k=1}}. \quad (3)$$

B. Data

To design our algorithm, the detector signals used in this work are based on simulation. In order to simulate the pulse shapes, we leveraged the established analysis of the pulse shape characteristics of neutron and gamma ray interactions with CLYC scintillators coupled to a photomultiplier tube (PMT) [5], [25]–[30]. While shape characteristics differ due to the usage of different size and shape scintillators, different PMTs, and varying environmental conditions, the analyzing and modeling the pulse shapes allows for the creation of simulated pulses which preserve the characteristics of the experimental pulses.

Specifically, the simulated data used in our work is based on the characterization proposed in [5], with the specific configuration presented in [5, Eqn. (1), Tables 2-3]. The detector signal was simulated in the following manner: First, the arrival times of the neutrons and gammas were first drawn from two independent Poisson point processes; Next, the detector signal was created by summing simulated neutron pulses at the neutron arrival times, simulated gamma pulses at the gamma arrival times, and additive white Gaussian noise (AWGN). Simulating the detector signal under severe pileup conditions using the proposed additive model resembles a real detector signal under severe pileup conditions because the anode current of the PMT is proportional to the magnitude of the incident radiation flux, which would be linearly added when there are multiple pulses arriving within a short time window of tens of nanoseconds [4], [31]. A simulated detector signal under severe pileup can be seen in Fig. 1.

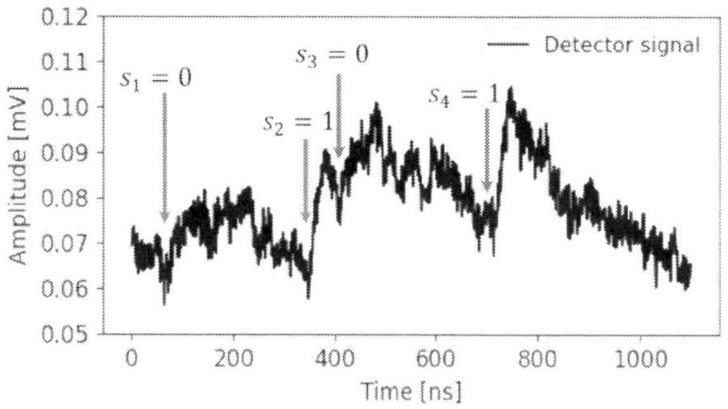

Fig. 1. A simulated detector signal under severe pileup conditions. Arriving particles and their corresponding label are marked with blue arrows. The simulated event rate is 2 Mcps, and pulse shapes were simulated based on the characterization proposed in [5], with the specific configuration presented in [5, Eqn. (1), Tables 2-3].

C. Effects of Pileup

We focus on the task of performing PSD when operating within high event rates, which is particularly challenging due to pulse pileup. Pileup occurs when multiple pulses arrive within a short period of time. The resulting overlap highly distorts the shape of the pulse, as illustrated in Fig. 1, making it challenging to discriminate. The main factors dictating the severity of pileup conditions are the pulse shapes, lengths, and arrival rate giving rise to the following challenges:

Challenge I: Mathematical modeling. PSD is usually performed based on extracted features, e.g., rise time, amplitude, and decay constants, that represent the pulses based on the mathematical description that makes them useful for discrimination. For non-piled-up pulses, the uncertainty depends solely on the signal-to-noise ratio (SNR) and the accurate representation of the underlying physics of the detector. However, extracting features of a piled-up pulse requires additional mathematical manipulations. While in cases where two pulses are piled-up, feature extraction can still be relatively accurate, in moderate and high pileup conditions where three or more pulses are piled-up onto each other, accurate feature extraction becomes extremely challenging.

Challenge II: Temporal correlation. A time window containing a non-piled-up pulse with AWGN is statistically independent of all other time windows. A time window containing a piled-up pulse on the other hand only contains fractions of the pulses which are piled-up onto the one that is being observed. The rest of these pulses lie within consecutive time windows, resulting in statistical dependencies between these time windows.

Challenge III: Interference. In cases where pulses are not piled-up onto each other, PSD capabilities highly depend on the SNR. Low energy pulses are usually harder to discriminate due to noise interference. Pileup causes additional interference caused by overlapping pulses. In severe pileup conditions, distortions caused by overlapping pulse interference are much more dominant than the distortions introduced by noise.

The aforementioned challenges motivate deriving a PSD method that both exploits the available useful features and expected temporal dependence, while leveraging the abstractness of DNNs to cope with the challenging modelling and interference, as detailed in the following section.

III. Method

In this section, we introduce our PSD algorithm that follows model-based deep learning methodology [22]–[24]. The proposed method offers a hybrid technique, that combines knowledge about the underlying physics of the detector with a data-driven DNN to benefit from the advantages of both approaches. As our design is geared towards high rates, features providing relevant information in severe pileup conditions are used.

The proposed DNN-based method utilizes the raw pulse, classical features, and weighing features for classification, based on the following rationale: The raw pulse samples and classical features, which characterize the pulse, are fed into a DNN which classifies the pulse; The weighing features characterize the pileup conditions and are used to learn a weighing vector that weighs out which parts of the raw pulse and classical features hold the most important information for classification in a specific case. We next elaborate on these features and their processing procedure.

1) Classical Features: We use the term *classical features* to refer to features extracted from the raw pulses that are conventionally employed for classification in standard low-rate settings. The classical features used in this paper are only a handful of the overall features that can be extracted to represent these pulses [10]–[12]. Although **Challenge I** makes feature extraction difficult under pileup conditions, some feature extraction methods are less prone to error than others.

In order to cope with **Challenge I**, we analyzed various features in order to understand which of them preserve their information gain under pileup conditions. The key difference between general classical features and the classical features used in our method is the preservation of information in severe pileup rates. For example, estimating the decay constants of highly piled up pulses is much more difficult than estimating their rise time. In particular, for classifying an arrival at time t, we use the vector $\boldsymbol{y}_t$ comprised of the following classical features:

- Y1 Rise time.
- Y2 Correlation coefficients: computing the correlation between the pulse with a neutron template, and the correlation between the pulse and a gamma template.
- Y3 Amplitude.
- Y4 Baseline: computing the baseline at the moment of arrival.
- Y5 Energy: calculating the integral in different sections of the pulse.

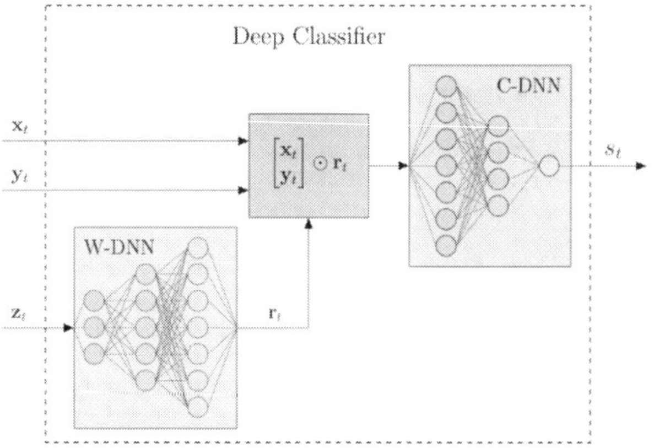

Fig. 2. Architecture illustration of the Deep Classifier.

2) Weighing Features: We use the term *weighing features* to denote features that are informative of the current pileup conditions. The rationale is to use such information to balance the reliability of the classical features, that are known to be useful in the absence of pileup.

Specifically, **Challenge II** and **Challenge III** introduce interference which is dependent both on the classifier pulse and consecutive time windows. Therefore, in order to cope with **Challenge II** and **Challenge III**, the weighing features for classifying an arrival at time t, denoted $\boldsymbol{z}_t$ are fed into a DNN which outputs a vector that is used to weight the classical features, denoted $\boldsymbol{r}_t$. Essentially, a mapping between the pileup conditions and the importance of the raw pulse and classical features under those conditions is learned. The weighing features vector $\boldsymbol{z}_t$ is comprised of:

- Z1 Distance from the pulse detected previously.
- Z2 Distance from the pulse detected next.
- Z3 The ratio between the energies of the current pulse and the pulse detected previously.
- Z4 The ratio between the energies of the current pulse and the pulse detected next.
- Z5 The ratio between the energy of the current pulse and the baseline at the time of arrival.

3) Architecture: The architecture of the Deep Classifier, illustrated in Fig. 2, consists of two DNNs, each being a multi-layered perceptron. The weighing features $\boldsymbol{z}_t$ are fed into the weighing deep neural network (W-DNN) which outputs a weighing vector $\boldsymbol{r}_t$. The resulting vector then weights the stacking of the raw pulse $\boldsymbol{x}_t$ and the classical features $\boldsymbol{y}_t$ via Hadamard product, which in turn is fed into the classifying deep neural network (C-DNN). The C-DNN has a single output neuron with sigmoid activation, such that the output of the Deep Classifier with trainable parameters $\boldsymbol{\theta}$, raw pulse $\boldsymbol{x}_t$, classical features $\boldsymbol{y}_t$, and weighing features $\boldsymbol{z}_t$, is $f_{\boldsymbol{\theta}}(\boldsymbol{x}_t, \boldsymbol{y}_t, \boldsymbol{z}_t) \in [0, 1]$. A threshold is set on the output neuron, values above it are classified as neutrons, and values under it are classified as gammas. The PSD procedure of the Deep Classifier is summarized as Algorithm 1.

The Deep Classifier is using a data set $\mathcal{D}$ comprised of tuples of $\boldsymbol{x}, \boldsymbol{y}, \boldsymbol{z}$ and their corresponding label $s \in \{0, 1\}$,

Algorithm 1: Deep Classification

Init: θ; Threshold THR
Input : Raw pulse x_t
Calculate classical features Y1-Y5 as y_t ;
Calculate weighing features Z1-Z5 as z_t ;
Apply Deep Classifier $\hat{s}_t = f_\theta(x_t, y_t, z_t)$;
if $\hat{s}_t > THR$ **then**
  Output: Neutron at t
else
  Output: Gamma at t

representing gamma and neutron, respectively. Training is done in two stages. In the first stage, the weighing vector is not computed and only the C-DNN is trained to map the raw pulse and classical features to the output. After a basic mapping between the raw pulse and classical features to the output has been learned, the second stage begins. In this second stage, the C-DNN and the W-DNN are jointly trained, boosting the performance of the Deep Classifier by weighing the data used for classification based on the pileup conditions. The binary cross-entropy (BCE) loss, which is arguably the most common loss used for binary classification tasks, is used for training, resulting in the following loss measure:

$$\mathcal{L}_\mathcal{D}(\theta) = \frac{1}{|\mathcal{D}|} \sum_{(x_l, y_l, z_l, s_l) \in \mathcal{D}} s_l \cdot \log\left(f_\theta(x_l, y_l, z_l)\right) + (1 - s_l) \cdot \log\left(1 - f_\theta(x_l, y_l, z_l)\right). \tag{4}$$

The overall two-stage training algorithm based on conventional mini-batch stochastic gradient descent is summarized as Algorithm 2.

Algorithm 2: Training the Deep Classifier

Init: Set θ randomly
  Fix learning rate $\mu > 0$ and epochs i_{mid}, $i_{\max}$
Input: Training set $\mathcal{D}$
for $i = 0, 1, \ldots, i_{mid} - 1$ **do**
  Fix $\mathbf{r}_t = \mathbf{1}$;
  Randomly divide $\mathcal{D}$ into Q batches $\{\mathcal{D}_q\}_{q=1}^{Q}$;
  for $q = 1, \ldots, Q$ **do**
    Apply model with parameters θ to $\mathcal{D}_q$;
    Compute batch loss $\mathcal{L}_{\mathcal{D}_q}(\theta)$ by (4);
    Update $\theta \leftarrow \theta - \mu \nabla_\theta \mathcal{L}_{\mathcal{D}_q}(\theta)$;
for $i = i_{mid}, i_{mid} + 1, \ldots, i_{\max} - 1$ **do**
  Randomly divide $\mathcal{D}$ into Q batches $\{\mathcal{D}_q\}_{q=1}^{Q}$;
  for $q = 1, \ldots, Q$ **do**
    Apply model with parameters θ to $\mathcal{D}_q$;
    Compute batch loss $\mathcal{L}_{\mathcal{D}_q}(\theta)$ by (4);
    Update $\theta \leftarrow \theta - \mu \nabla_\theta \mathcal{L}_{\mathcal{D}_q}(\theta)$;
return θ

4) Discussion: The usage of dedicated features to extract weighing coefficients is inspired by the emerging technique of context-aware modulation [32], which realizes a form of hypernetworks whose mapping is not fixed but is affected by additional features [33]. Accordingly, we identify that the classification mapping and how it balances the classical features y_t and the pulse data should vary between different pileup conditions. Hence the incorporation of the weighing features, that are selected due to their being informative regarding pileups. The validity and usefulness of our approach are numerically demonstrated in section IV.

IV. Results

The proposed Deep Classifier a longside f our o ther PSD algorithms were tested under various event rates. The event rate at which pileup becomes severe is dependent on the length of the pulses produced by the detector. Therefor, the event rates under which the proposed Deep Classifier w as tested were adjusted to the simulated CLYC detector data used in this work. For the CLYC detector, high event rates where PSD becomes challenging are considered to be in the range of several Mcps [1], [3], [5]–[8]. The data set $\mathcal{D}$ used for testing consisted of $50,000$ simulated neutrons and $50,000$ simulated gammas.

Firstly, the proposed Deep Classifier i s c ompared t o two classic PSD algorithms, e.g., the charge integration (CI) method and the normalized cross-correlation (NCC) method. The parameters of both methods were optimized for the simulated event rates and the algorithms were slightly altered to better fit operation in high event rates, using operations such as baseline subtraction to boost performance. The CI based algorithm was realized using a short and a long integration gate, while the NCC algorithm was realized by computing two templates from the neutron and gamma distributions respectively.

In addition, the proposed method is compared to two DNN based methods. The first method is a DNN applied to the raw pulse, which is essentially x entering the C-DNN. The second method is a DNN applied to the stacking of the raw pulse and the classical features, e.g., Y1-Y5, which is essentially x and y entering the C-DNN. Comparing the Deep Classifier to a DNN applied to the raw pulse allows us to validate the effectiveness of our method compared to a purely data-driven DNN based method which is widely used. Moreover, comparing the Deep Classifier t o b oth m ethods a llows u s to demonstrate and validate the contribution of the different parts of our method. Comparison between the DNN applied to the raw pulse and the DNN applied to the stacking of the raw pulse and the classical features demonstrates the contribution of the classical features and their validity in high event rates. Comparison between the DNN applied to the stacking of the raw pulse and the classical features to our proposed method demonstrates the contribution of the weighing features and the boost in performance achieved by the mapping between the pileup conditions and the importance of the raw pulse and classical features.

The first simulation was carried out as follows: The detector's signal was simulated as detailed in Subsection II-B. The neutrons and gammas arrived at identical count rates, making the total event rate vary between 500 kcps and 10 Mcps. The total accuracy of the methods defined in (1) was calculated for each event rate. The results of the proposed Deep Classifier and the four benchmarks in the first simulation are displayed in Fig. 3.

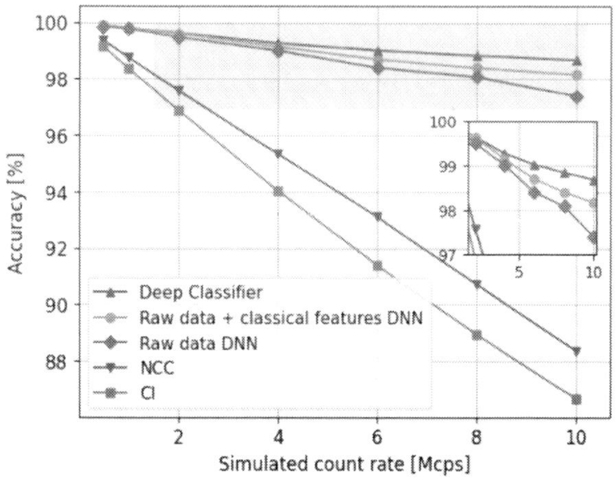

Fig. 3. The total accuracy achieved by the Deep Classifier and the four benchmark algorithms at event rates ranging between 500 kcps and 10 Mcps.

Observing Fig. 3, we note that for the lowest event rate displayed, 500 kcps, all five methods achieve a high accuracy of over 99%. The accuracies of the classic methods are slightly lower than the accuracies of the DNN based methods. As the event rate goes higher, the accuracy of all five methods becomes smaller, as expected. A clear trend can be seen where the accuracy of the classic methods falls dramatically behind the accuracy of the DNN based methods. The NCC based method achieves slightly better results than the CI based method. This result shows that data-driven DNN based methods may be more suitable than classical NCC and CI based methods for PSD in high event rates.

The zoomed-in area highlighted in light purple in Fig. 3 provides a better view of the accuracy achieved by the DNN based methods. It can be seen that all DNN based methods achieve similar accuracies at event rates of 1 Mcps and under. As the event rate goes higher, a clear trend can be seen where the lowest accuracy is achieved by the DNN applied to the raw pulse, followed by the DNN applied to the stacking of the raw pulse and the classical features, and finally, the best accuracy is achieved by the proposed Deep Classifier. These results validate the effectiveness of our method at high event rates and demonstrate the contribution of both the classical features y and the weighing features z.

For most applications, the neutron and the gamma fluxes are not the same. In particular, the gamma flux is often larger than the neutron flux. The following simulation was conducted to assess and characterize the accuracy of DNN based methods across various ratios between the gamma and neutron flux. The accuracies of the methods were evaluated under ratios ranging from 1:1 up to 25:1 between the gamma flux and the neutron flux respectively. In all cases, the total event rate was 4 Mcps. The resulting total accuracies defined in (1) are displayed in Fig. 4. The neutron accuracies defined in (2) are displayed in Fig. 5 and the gamma accuracies defined in (3) are displayed in Fig. 6.

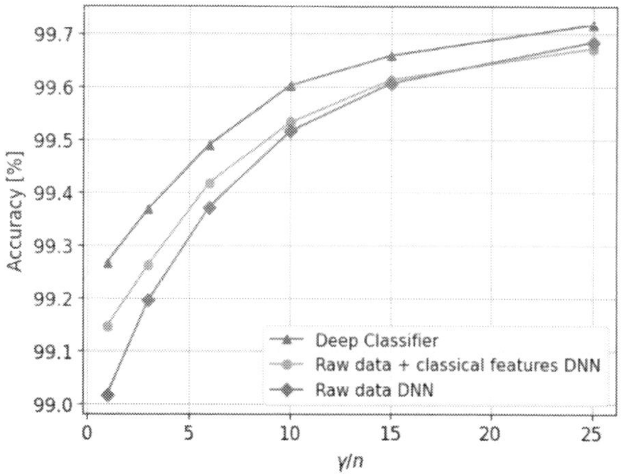

Fig. 4. The total accuracy achieved by the Deep Classifier and the two DNN based benchmark algorithms for ratios between the gamma and the neutron flux ranging from 1:1 up to 25:1 respectively. The total event rate for all ratios is 4 Mcps.

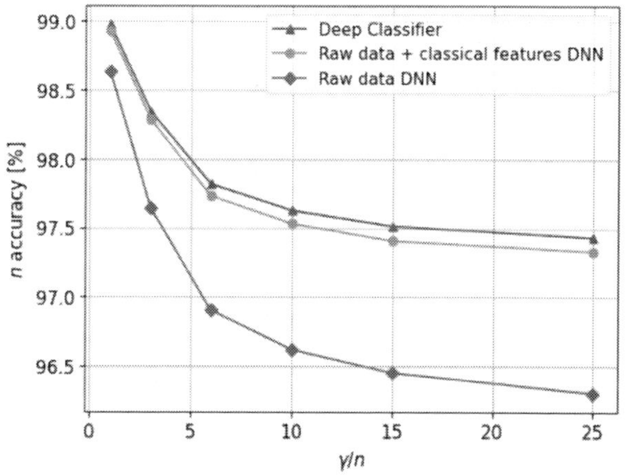

Fig. 5. The neutron accuracy achieved by the Deep Classifier and the two DNN based benchmark algorithms for ratios between the gamma and the neutron flux ranging from 1:1 up to 25:1 respectively. The total event rate for all ratios is 4 Mcps.

Firstly, it can be seen in Fig. 4 that the total accuracy of all three methods improves as the relative amount of gammas increases. This observation could suggest that the methods find it comparatively less challenging to classify gamma pulses as opposed to neutron pulses. This presumption gains further support upon examining Fig. 5 and Fig. 6, which clearly demonstrate that, across all γ/n ratios, the neutron accuracy consistently remains below that of gamma accuracy. From Fig. 4 it can be seen that as the γ/n ratio grows, the difference in total accuracy between the methods grows smaller and begins plateauing. Of particular interest is the observation in Fig. 4, where the total accuracy of the DNN applied to the

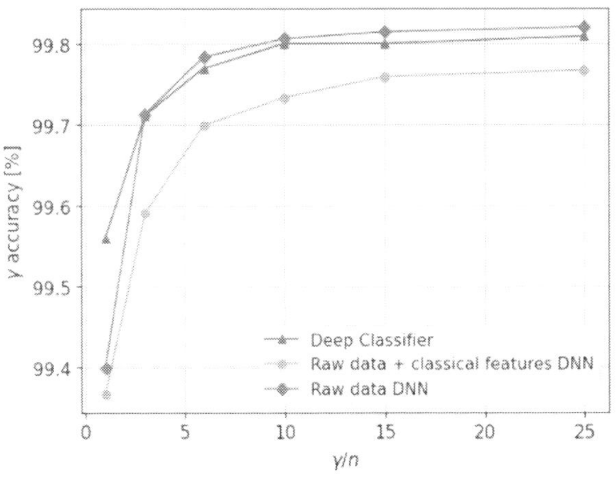

Fig. 6. The gamma accuracy achieved by the Deep Classifier and the two DNN based benchmark algorithms for ratios between the gamma and the neutron flux ranging from 1:1 up to 25:1 respectively. The total event rate for all ratios is 4 Mcps.

raw pulse eventually surpasses the total accuracy of the DNN applied to the stacking of the raw pulse and the classical features. Moreover, it can be seen from Fig. 6 that the DNN applied to the raw pulse achieves the highest gamma accuracy for almost all γ/n ratios, surpassing the gamma accuracy of the proposed Deep Classifier. Yet, this fact is accompanied by a trade-off, as evident in Fig. 5, where the DNN applied to the raw pulse shows noticeably lower neutron accuracy compared to the proposed Deep Classifier.

V. Conclusion

We developed general purpose PSD algorithm based on model-based deep learning methodology, which is designed for severe pileup conditions due to high event rates. The proposed method combines classical PSD features, relevant for high event rates, alongside a unique DNN based weighing architecture, inspired by the emerging technique of context-aware modulation. Discrimination capabilities were evaluated using simulated data from a CLYC based detector and compared to the discrimination capabilities of other methods. The proposed method provides superior discrimination capabilities compared to CI, NCC, and DNN based methods. Moreover, the effectiveness and contribution of the different blocks inside the proposed method are evaluated. While the results in this work are demonstrated on simulated data from a specific detector, the algorithm is beneficial for other detectors with long scintillation decay times operating in high event rates.

References

[1] A. Dutta, P. Chandhran, K. E. Holbert, and E. B. Johnson, "Using decay time to discriminate neutron and gamma ray pulses from a clyc detector," in *2015 IEEE Nuclear Science Symposium and Medical Imaging Conference (NSS/MIC)*. IEEE, 2015, pp. 1–7.

[2] S. Richards, G. Sykora, and M. Taggart, "High count rate pulse shape discrimination algorithms for neutron scattering facilities," *Nuclear Instruments and Methods in Physics Research Section A: Accelerators, Spectrometers, Detectors and Associated Equipment*, vol. 989, p. 164946, 2021.

[3] A. Dutta and K. E. Holbert, "Discrimination of neutron-gamma ray pulses with pileup using normalized cross correlation and principal component analysis," *IEEE Transactions on Nuclear Science*, vol. 63, no. 6, pp. 2764–2771, 2016.

[4] C. Fu, A. Di Fulvio, S. Clarke, D. Wentzloff, S. Pozzi, and H. Kim, "Artificial neural network algorithms for pulse shape discrimination and recovery of piled-up pulses in organic scintillators," *Annals of nuclear energy*, vol. 120, pp. 410–421, 2018.

[5] J. Han, J. Zhu, Z. Wang, G. Qu, X. Liu, W. Lin, Z. Xu, Y. Huang, M. Yan, X. Zhang *et al.*, "Pulse characteristics of clyc and piled-up neutron–gamma discrimination using a convolutional neural network," *Nuclear Instruments and Methods in Physics Research Section A: Accelerators, Spectrometers, Detectors and Associated Equipment*, vol. 1028, p. 166328, 2022.

[6] S. Peng, Z. Hua, Q. Wu, J. Han, S. Qian, Z. Wang, Q. Wei, L. Qin, L. Ma, M. Yan *et al.*, "Piled-up neutron-gamma discrimination system for cllb using convolutional neural network," *Journal of Instrumentation*, vol. 17, no. 08, p. T08001, 2022.

[7] E. B. Johnson, C. Whitney, S. Vogel, J. F. Christian, K. Holbert, and P. Chandhran, "High event rate, pulse shape discrimination algorithm for clyc," in *2015 IEEE International Symposium on Technologies for Homeland Security (HST)*. IEEE, 2015, pp. 1–7.

[8] X. Wen and A. Enqvist, "Pulse shape discrimination of cs2liycl6: Ce3+ detectors at high count rate based on triangular and trapezoidal filters," *Nuclear Instruments and Methods in Physics Research Section A: Accelerators, Spectrometers, Detectors and Associated Equipment*, vol. 866, pp. 129–133, 2017.

[9] Z. Zhang, Y. Zhang, C. Hu, S. Feng, J. Zhu, B. Liao, D. Liu, Q. Zhou, and H. Wang, "Detection and discrimination of α-and β-radiation with plastic scintillators based on pulse shape discrimination," *Journal of Instrumentation*, vol. 15, no. 04, p. P04008, 2020.

[10] J. Polack, M. Flaska, A. Enqvist, C. Sosa, C. Lawrence, and S. Pozzi, "An algorithm for charge-integration, pulse-shape discrimination and estimation of neutron/photon misclassification in organic scintillators," *Nuclear Instruments and Methods in Physics Research Section A: Accelerators, Spectrometers, Detectors and Associated Equipment*, vol. 795, pp. 253–267, 2015.

[11] R. M. French, M. Thevenin, M. Hamel, and E. Montbarbon, "A histogram-difference method for neutron/gamma discrimination using liquid and plastic scintillators," *IEEE Transactions on Nuclear Science*, vol. 64, no. 8, pp. 2423–2432, 2017.

[12] M. Hammad, H. Kasban, R. Fikry, M. I. Dessouky, O. Zahran, S. M. Elaraby, and F. E. A. El-Samie, "Digital pulse processing algorithm for neutron and gamma rays discrimination," *Analog Integrated Circuits and Signal Processing*, vol. 101, pp. 475–487, 2019.

[13] H. Singh and R. Mehra, "Discrete wavelet transform method for high flux n-\gamma discrimination with liquid scintillators," *IEEE Transactions on Nuclear Science*, vol. 64, no. 7, pp. 1927–1933, 2017.

[14] B. Blair, C. Chen, A. Glenn, A. Kaplan, J. Ruz, L. Simms, and R. Wurtz, "Gaussian mixture models as automated particle classifiers for fast neutron detectors," *Statistical Analysis and Data Mining: The ASA Data Science Journal*, vol. 12, no. 6, pp. 479–488, 2019.

[15] M. Gelfusa, R. Rossi, M. Lungaroni, F. Belli, L. Spolladore, I. Wyss, P. Gaudio, A. Murari, and J. Contributors, "Advanced pulse shape discrimination via machine learning for applications in thermonuclear fusion," *Nuclear Instruments and Methods in Physics Research Section A: Accelerators, Spectrometers, Detectors and Associated Equipment*, vol. 974, p. 164198, 2020.

[16] E. Doucet, T. Brown, P. Chowdhury, C. Lister, C. Morse, P. Bender, and A. Rogers, "Machine learning n/γ discrimination in clyc scintillators," *Nuclear Instruments and Methods in Physics Research Section A: Accelerators, Spectrometers, Detectors and Associated Equipment*, vol. 954, p. 161201, 2020.

[17] T. Tambouratzis, D. Chernikova, and I. Pzsit, "Pulse shape discrimination of neutrons and gamma rays using kohonen artificial neural networks," *Journal of Artificial Intelligence and Soft Computing Research*, vol. 3, no. 2, pp. 77–88, 2013.

[18] J. Griffiths, S. Kleinegesse, D. Saunders, R. Taylor, and A. Vacheret, "Pulse shape discrimination and exploration of scintillation signals using convolutional neural networks," *Machine Learning: Science and Technology*, vol. 1, no. 4, p. 045022, 2020.

[19] M. Astrain, M. Ruiz, A. V. Stephen, R. Sarwar, A. Carpeño, S. Esquembri, A. Murari, F. Belli, and M. Riva, "Real-time implementation of the neutron/gamma discrimination in an fpga-based daq mtca platform using a convolutional neural network," *IEEE Transactions on Nuclear Science*, vol. 68, no. 8, pp. 2173–2178, 2021.

[20] X. Fabian, G. Baulieu, L. Ducroux, O. Stézowski, A. Boujrad, E. Clément, S. Coudert, G. de France, N. Erduran, S. Ertürk *et al.*, "Artificial neural networks for neutron/γ discrimination in the neutron detectors of neda," *Nuclear Instruments and Methods in Physics Research Section A: Accelerators, Spectrometers, Detectors and Associated Equipment*, vol. 986, p. 164750, 2021.
[21] N. M. Michels, A. J. Jinia, S. D. Clarke, H.-S. Kim, S. A. Pozzi, and D. D. Wentzloff, "Real-time classification of radiation pulses with piled-up recovery using an fpga-based artificial neural network," *IEEE Access*, 2023.
[22] N. Shlezinger, J. Whang, Y. C. Eldar, and A. G. Dimakis, "Model-based deep learning," *Proc. IEEE*, vol. 111, no. 5, pp. 465–499, 2023.
[23] N. Shlezinger, Y. C. Eldar, and S. P. Boyd, "Model-based deep learning: On the intersection of deep learning and optimization," *IEEE Access*, vol. 10, pp. 115 384–115 398, 2022.
[24] N. Shlezinger and Y. C. Eldar, "Model-Based Deep Learning," *Foundations and Trends® in Signal Processing*, vol. 17, no. 4, pp. 291–416, 2023.
[25] B. S. Budden, A. J. Couture, L. C. Stonehill, A. V. Klimenko, J. R. Terry, and J. O. Perry, "Analysis of cs 2 liycl 6: Ce 3+(clyc) waveforms as read out by solid state photomultipliers," in *2012 IEEE Nuclear Science Symposium and Medical Imaging Conference Record (NSS/MIC)*. IEEE, 2012, pp. 347–350.
[26] B. S. Budden, L. C. Stonehill, J. R. Terry, A. V. Klimenko, and J. O. Perry, "Characterization and investigation of the thermal dependence of cs_2liycl_6 : ce^{3+} (clyc) waveforms," *IEEE Transactions on Nuclear Science*, vol. 60, no. 2, pp. 946–951, 2012.
[27] N. D'olympia, P. Chowdhury, C. Lister, J. Glodo, R. Hawrami, K. Shah, and U. Shirwadkar, "Pulse-shape analysis of clyc for thermal neutrons, fast neutrons, and gamma-rays," *Nuclear Instruments and Methods in Physics Research Section A: Accelerators, Spectrometers, Detectors and Associated Equipment*, vol. 714, pp. 121–127, 2013.
[28] N. Dinar, D. Celeste, M. Silari, V. Varoli, and A. Fazzi, "Pulse shape discrimination of clyc scintillator coupled with a large sipm array," *Nuclear Instruments and Methods in Physics Research Section A: Accelerators, Spectrometers, Detectors and Associated Equipment*, vol. 935, pp. 35–39, 2019.
[29] T. Huang and Z. Zhang, "Characterization of 1-inch clyc scintillator coupled with 8× 8 sipm array," *Nuclear Instruments and Methods in Physics Research Section A: Accelerators, Spectrometers, Detectors and Associated Equipment*, vol. 999, p. 165225, 2021.
[30] C. Wang, X. Sun, J. Yu, J. Lv, Y. Deng, J. Li, H. Jiang, L. Luo, X. Fan, F. Gu *et al.*, "n/γ discrimination of sipm array-based clyc detector," *Nuclear Instruments and Methods in Physics Research Section A: Accelerators, Spectrometers, Detectors and Associated Equipment*, vol. 1036, p. 166870, 2022.
[31] R. W. Engstrom, *Photomultiplier handbook*. RCA Corporation, 1980.
[32] N. Ding, Y. Qin, G. Yang, F. Wei, Z. Yang, Y. Su, S. Hu, Y. Chen, C.-M. Chan, W. Chen *et al.*, "Parameter-efficient fine-tuning of large-scale pre-trained language models," *Nature Machine Intelligence*, vol. 5, no. 3, pp. 220–235, 2023.
[33] V. K. Chauhan, J. Zhou, P. Lu, S. Molaei, and D. A. Clifton, "A brief review of hypernetworks in deep learning," *arXiv preprint arXiv:2306.06955*, 2023.

 https://doi.org/10.1051/epjconf/202328810002

A model of spatial resolution uncertainty for Compton camera imaging

Yanting Ma*, Joshua Rapp, Petros Boufounos, Hassan Mansour
Mitsubishi Electric Research Laboratories (MERL), USA
(*) yma@merl.com

Abstract—**Compton cameras use a pair of detectors to record the energy deposited during Compton scattering and photoabsorption of gamma ray photons. By inverting the forward model for these detection events, the 3D radioactive source distribution can be recovered computationally. However, existing methods of describing the forward model ignore the ambiguity in the Compton cone axis introduced by the detector pixel size. In this paper, we introduce approximations to the axis ambiguity that make the description of the source ambiguity computationally tractable and improve the accuracy of the detection forward model. Specifically, we model the pixel active areas as discs, so the intersection between all possible Compton cone axes and a plane parallel to the detectors is also a disc. Because of ambiguity in the Compton cone axis, the gamma source position is constrained to lie within a more complicated volume that we approximately bound by ellipses or hyperbolas at discrete depth slices. We perform simulations of single point gamma sources using the Geant4 software. Our forward model leads to improved source distribution recovery using both backprojection and iterative reconstruction methods, demonstrating that accurate localization can be performed from a smaller number of detected photons.**

Keywords—**Compton camera, pixel detectors, list-mode maximum likelihood, spatial resolution, source localization**

I. Introduction

Compton scattering of high-energy photons, such as gamma rays, can be modelled as an inelastic collision, transferring energy and momentum to a single electron [1]. This phenomenon is used in so-called "Compton cameras" to measure the location and angle of arrival of gamma photons. Typically constructed from two or more radiation-sensitive layers, these imagers detect and record the location and deposited energy of high-energy photons, which theoretically constrains the gamma source to lie on the surface of a cone. Compton cameras have largely been used for astronomy and medical imaging, although an important growing application area is localization of nuclear waste, both within power plants and externally, e.g., as a result of the Fukushima Daiichi accident.

Initial approaches to Compton imaging were influenced by astronomical applications and were predominantly focused on two-dimensional (2D) reconstructions at a fixed plane [2], [3] or projections onto a focal sphere surrounding the detector [4], [5]. Recent work has aimed to solve the more challenging tomography problem of reconstructing the 3D distribution of radioactive sources [6], [7], [8], [9], [10], [11]. A fundamental hurdle in Compton tomography is accurately constructing the measurement matrix, namely accounting for the uncertainty in the Compton cone based on ambiguity in the detector. As identified in early works, the main sources of ambiguity are the energy resolution, spatial resolution, and thickness of the detectors, and the distance of the radioactive source to the imager [12]. Research efforts have begun to tackle a number of these issues, including the photon's incident angle [13], [6], [7], the finite extent of the detector [8], and the detector's energy resolution [9]. However, the position uncertainty due to the detector spatial resolution has largely been ignored or oversimplified for 3D reconstruction.

Compton imaging in nuclear medicine requires only short-range measurements and can use high-resolution silicon strip detectors, so ambiguity due to detector spatial resolution is low and has been specifically neglected (e.g., in [9]). The Compton cone axis is thus assumed to be known exactly, leading to the intersection of the Compton cone with a plane at a fixed depth from the camera to be idealized as an ellipse. However, applications such as robotic radiation source localization (e.g., [14]) require lower-resolution scintillator-based detectors that are more rugged, lower cost, more efficient, and larger area [15]. Significant uncertainty in accurately determining the axis of the Compton cone is incurred due to the larger scintillator-based detector elements, so instead of being restricted to the surface of a cone, the gamma photon's source location lies in a volume of ambiguity. Furthermore, methods such as [6], which chooses the 3D voxel size based on the detector pixel size, naively assume the 3D position uncertainty is on the order of the detector pixel size, ignoring the increase in ambiguity as a function of distance from the imager. Several works have analyzed the effect of position resolution in 3D-position-sensitive semiconductor detectors, which have generated growing interest because their z-axis resolution enables Compton imaging with single-layer devices, e.g., [16], [17]. One approach is to obtain an empirical uncertainty model by simulating the effect of random detection locations within the detector volume [5], but the resulting model cannot be easily incorporated into a 3D reconstruction algorithm. Another approach, proposed in the context of 4π Compton imaging, describes the uncertainty region on the focal sphere as a ring centered at the projection of the cone surface. The amount of the spread is derived in either polar or Cartesian coordinates and the spread is symmetric in the inward and outward directions from the cone surface [16], [18], [19]. Although the symmetry assumption may be sufficiently accurate for 4π imaging, our work shows significant asymmetries when similar rings are projected onto a plane parallel to the detector, as is desired for 3D reconstruction.

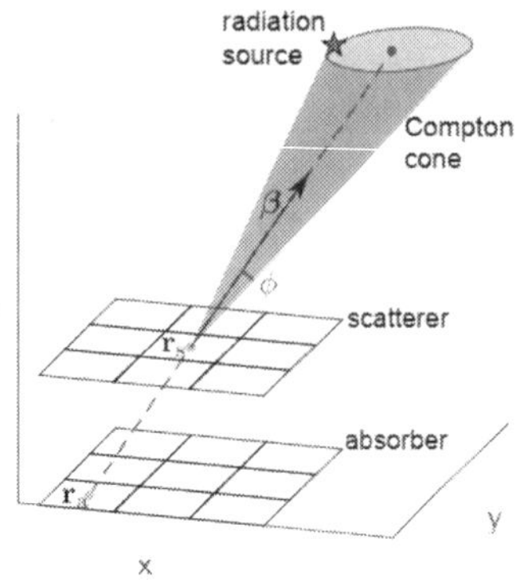

Fig. 1: Measurements of energy deposited at $\mathbf{r}_s$ on the scatterer and $\mathbf{r}_a$ on the absorber restrict the gamma photon source to lie on the Compton cone defined by axis vector $\boldsymbol{\beta}$ and angle ϕ. Coarse detector spatial resolution causes ambiguity in $\boldsymbol{\beta}$.

In this paper, we explicitly model the ambiguity in the cone axis due to the detector pixel size projected onto plane parallels to the detector. By approximating the active pixel area by a disc, we show instead that at each depth plane, plausible source locations fall in regions approximately bounded by ellipses and hyperbolas. We incorporate the spatial resolution uncertainty to construct an improved measurement matrix, which we use in both back-projection and an iterative reconstruction procedure to show improved 3D source localization from fewer photon detections.

II. Compton Imaging Model

A. Compton Cone

A classical model of Compton camera detection is depicted in Fig. 1. Ideally, an incoming gamma photon undergoes Compton scattering at the first detector array (scatterer) and is completely absorbed at the second detector array (absorber). We call this pair of interactions (scattering and absorption) an *event*. Let E_γ be the incident photon energy and let $\tilde{E}_s$ and $\tilde{E}_a$ be the energy deposited at the scatterer and absorber, respectively. Suppose that the gamma photon hits the scatterer at position $\tilde{\mathbf{r}}_s$ and absorber at position $\tilde{\mathbf{r}}_a$. Then the position of the incident gamma photon is known to be located on the surface of a cone, where the cone is determined by $(\tilde{\mathbf{r}}_s, \tilde{\mathbf{r}}_a, \tilde{E}_s, \tilde{E}_a)$. Specifically, the vertex is $\tilde{\mathbf{r}}_s$, and the axis vector $\tilde{\boldsymbol{\beta}}$ and the half-angle $\tilde{\phi}$ are defined as

$$\tilde{\boldsymbol{\beta}} = \frac{\tilde{\mathbf{r}}_s - \tilde{\mathbf{r}}_a}{\|\tilde{\mathbf{r}}_s - \tilde{\mathbf{r}}_a\|} \tag{1}$$

$$\cos\tilde{\phi} = 1 - \frac{m_e c^2 \tilde{E}_s}{(\tilde{E}_s + \tilde{E}_a)\tilde{E}_a}, \tag{2}$$

where m_e is the mass of an electron at rest ($m_e c^2 = 0.5510$ MeV).[1] We observe that the cone angle $\tilde{\phi}$ depends on the detector energies, and the effect of the energy resolution on 3D reconstruction has been investigated previously, e.g., [20]. On the other hand, the cone axis $\tilde{\boldsymbol{\beta}}$ depends on the position of the Compton scattering and photon absorption events. Typically, these positions are assigned to be the center of the pixels in which these interactions are detected, which we denote $\mathbf{r}_s$ and $\mathbf{r}_a$. The errors between the assigned and actual event positions ($\|\mathbf{r}_s - \tilde{\mathbf{r}}_s\|$, $\|\mathbf{r}_a - \tilde{\mathbf{r}}_a\|$) depend on the spatial resolution, i.e., the size of the detector pixels. Because $\boldsymbol{\beta}$ is affected by errors in both $\mathbf{r}_s$ and $\mathbf{r}_a$, the cone axis error is magnified as a function of the distance of the radiation source relative to the Compton camera.

[1] We use letters marked with a tilde to denote the idealized values assuming infinite spatial and energy resolution, whereas letters without tilde will be used to denote the measured value by a practical Compton camera.

B. Forward Model

The radiation source can be described as an intensity distribution $\boldsymbol{\lambda}$ in three spatial dimensions. The goal of the inverse problem is to reconstruct the intensity distribution given the Compton camera measurements. The most important step toward inversion is carefully defining the forward model that describes how gamma rays generate Compton camera measurements.

We first define a computational domain that specifies the extent of the volume in which to reconstruct the intensity distribution and discretizes this volume into voxels. Because of the high dimensionality of the photon coincidence, high-resolution discretization or "binning" of the measurement space leads to an extremely large measurement matrix, and the number of coincidence events is typically low, so most bins would have zero associated events. As a result, we follow the convention of using list-mode data [21], separately storing the measurement 4-tuple $(\mathbf{r}_s, \mathbf{r}_a, E_s, E_a)$ for each event. We can also represent a measurement as a 3-tuple $(\mathbf{r}_s, \mathbf{r}_a, \phi)$, where ϕ is computed from E_s and E_a according to (2).

The system matrix $\mathbf{T}$ relates the source distribution to the measurement space, with element $\mathbf{T}_{i,j}$ defined as the probability that a photon emitted from the jth voxel in the computational domain is measured as the ith event. We also define the sensitivity vector $\mathbf{s}$, where element $\mathbf{s}_j$ is the probability that a photon emitted from the jth voxel is measured by the camera.

1) General Form of the System Matrix: A general form of system matrix has been proposed in [9]. Suppose that the ith event is associated with measurement $(\mathbf{r}_s^{(i)}, \mathbf{r}_a^{(i)}, \phi^{(i)})$, then the $(i,j)^{th}$ element of the system matrix $\mathbf{T}$ is defined as

$$\begin{aligned}\mathbf{T}_{i,j} = \frac{1}{\text{vol}(V_j)} \int_{V_j} \int_{\mathbb{R}^3} \Bigg\{ & \frac{|\cos[\angle(\mathbf{x} - \tilde{\mathbf{r}}'_s)]|}{\|\mathbf{x} - \tilde{\mathbf{r}}'_s\|_2^2} p_{\mathbf{r}_s|\tilde{\mathbf{r}}_s}(\mathbf{r}_s^{(i)}|\tilde{\mathbf{r}}'_s) \\ & \int_{\mathbb{R}^3} \frac{|\cos[\angle(\tilde{\mathbf{r}}'_a - \tilde{\mathbf{r}}'_s)]|}{\|\tilde{\mathbf{r}}'_a - \tilde{\mathbf{r}}'_s\|_2^2} p_{\mathbf{r}_a|\tilde{\mathbf{r}}_a}(\mathbf{r}_a^{(i)}|\tilde{\mathbf{r}}'_a) \\ & K(\phi(\mathbf{x}; \tilde{\mathbf{r}}'_a, \tilde{\mathbf{r}}'_s)|E_\gamma) \\ & p_{\phi|\tilde{\phi}}\left(\phi^{(i)}|\phi(\mathbf{x}; \tilde{\mathbf{r}}'_a, \tilde{\mathbf{r}}'_s)\right) \Bigg\} \mathrm{d}\tilde{\mathbf{r}}'_a \, \mathrm{d}\tilde{\mathbf{r}}'_s \, \mathrm{d}\mathbf{x},\end{aligned} \tag{3}$$

where $K(\phi|E_\gamma)$ is the angular distribution given by the Klein-Nishina formula [22], $V_j \subset \mathbb{R}^3$ is the jth voxel in the computational domain, $\phi(\mathbf{x}; \tilde{\mathbf{r}}'_a, \tilde{\mathbf{r}}'_s)$ is the angle between $\mathbf{x} - \tilde{\mathbf{r}}'_s$

and $\tilde{\mathbf{r}}'_s - \tilde{\mathbf{r}}'_a$, $\angle\mathbf{a}$ is the angle between vector $\mathbf{a}$ and the z-axis, and the conditional probability distributions $p_{\mathbf{r}_s|\tilde{\mathbf{r}}_s}$, $p_{\mathbf{r}_a|\tilde{\mathbf{r}}_a}$, and $p_{\phi|\tilde{\phi}}$ model, respectively, the spatial ambiguity on the scatterer and absorber, and the half-angle ambiguity due to limited energy resolution. According to [9], the sensitivity can be approximated as

$$\mathbf{s}_j = \sum_k \frac{|\cos[\angle(\mathbf{v}_j - \tilde{\mathbf{r}}_s^{(k)})]|}{\|\mathbf{v}_j - \tilde{\mathbf{r}}_s^{(k)}\|_2^2}, \tag{4}$$

where the summation is over all scatterer pixel center positions $\tilde{\mathbf{r}}_s^{(k)}$ and $\mathbf{v}_j$ denotes the center of the voxel V_j. In [9], spatial ambiguity is ignored, i.e., $p_{\mathbf{r}_s|\tilde{\mathbf{r}}_s}(\mathbf{r}_s|\tilde{\mathbf{r}}_s) = \delta(\mathbf{r}_s - \tilde{\mathbf{r}}_s)$ and $p_{\mathbf{r}_a|\tilde{\mathbf{r}}_a}(\mathbf{r}_a|\tilde{\mathbf{r}}_a) = \delta(\mathbf{r}_a - \tilde{\mathbf{r}}_a)$, and the half-angle ambiguity $p_{\phi|\tilde{\phi}}(\phi|\tilde{\phi})$ is modeled as Gaussian with mean $\tilde{\phi}$.

2) Proposed Implementation of Cone Surface Integral: If we assumed no ambiguity in the measurements, then all $p_{\mathbf{r}_s|\tilde{\mathbf{r}}_s}$, $p_{\mathbf{r}_a|\tilde{\mathbf{r}}_a}$, and $p_{\phi|\tilde{\phi}}$ are Dirac delta functions, and the forward operator becomes the conventional cone surface model

$$\mathbf{T}_{i,j} = K(\phi^{(i)})|E_\gamma) \frac{|\cos[\angle(\mathbf{r}_a^{(i)} - \mathbf{r}_s^{(i)})]|}{\|\mathbf{r}_a^{(i)} - \mathbf{r}_s^{(i)}\|_2^2} \frac{1}{\mathrm{vol}(V_j)} \int_{V_j} \frac{|\cos[\angle(\mathbf{x} - \mathbf{r}_s^{(i)})])|}{\|\mathbf{x} - \mathbf{r}_s^{(i)}\|_2^2} \delta(\phi^{(i)} - \phi(\mathbf{x}; \mathbf{r}_a^{(i)}, \mathbf{r}_s^{(i)}))\, \mathrm{d}\mathbf{x}, \tag{5}$$

which integrates over all points on the cone surface within V_j. Numerous methods have been proposed for computing the cone surface integrals for Compton imaging. Here we propose a convenient approximation, which we use for the baseline comparison that ignores spatial resolution ambiguity.

Assuming sufficiently small voxels, we can approximate the integral in (5) by evaluating based only on the value at the voxel center $\mathbf{v}_j$ and whether the cone surface intersects V_j:

$$\mathbf{T}_{i,j}^{\mathrm{surf}} \propto K(\phi^{(i)})|E_\gamma) \frac{|\cos[\angle(\mathbf{r}_a^{(i)} - \mathbf{r}_s^{(i)})]|}{\|\mathbf{r}_a^{(i)} - \mathbf{r}_s^{(i)}\|_2^2} \frac{|\cos[\angle(\mathbf{v}_j - \mathbf{r}_s^{(i)})])|}{\|\mathbf{v}_j - \mathbf{r}_s^{(i)}\|_2^2} \mathbb{I}_{\mathrm{cone}}(V_j; \mathbf{r}_a^{(i)}, \mathbf{r}_s^{(i)}, \phi^{(i)}), \tag{6}$$

where the indicator function is defined as

$$\mathbb{I}_{\mathrm{cone}}(V_j; \mathbf{r}_a^{(i)}, \mathbf{r}_s^{(i)}, \phi^{(i)})) = \begin{cases} 1, & \text{cone defined by } (\mathbf{r}_a^{(i)}, \mathbf{r}_s^{(i)}, \phi^{(i)}) \text{ intersects } V_j \\ 0, & \text{otherwise.} \end{cases} \tag{7}$$

To determine $\mathbb{I}_{\mathrm{cone}}(V_j)$, we first describe the surface of the cone as the zero-level set of the equation

$$g(\mathbf{x}; \mathbf{r}_s, \boldsymbol{\beta}, \phi) = \left\langle \frac{\mathbf{x} - \mathbf{r}_s}{\|\mathbf{x} - \mathbf{r}_s\|}, \boldsymbol{\beta} \right\rangle - \cos(\phi), \tag{8}$$

i.e., $g(\mathbf{x}) = 0$ for any $\mathbf{x}$ on the cone surface. Points that lie inside the cone have $g(\mathbf{x}) < 0$, and points that lie outside the cone have $g(\mathbf{x}) > 0$. Because of the continuity of the cone surface, if there are some points in voxel V_j that are inside the cone and some points in V_j that lie outside the cone, then the cone surface must pass through V_j.

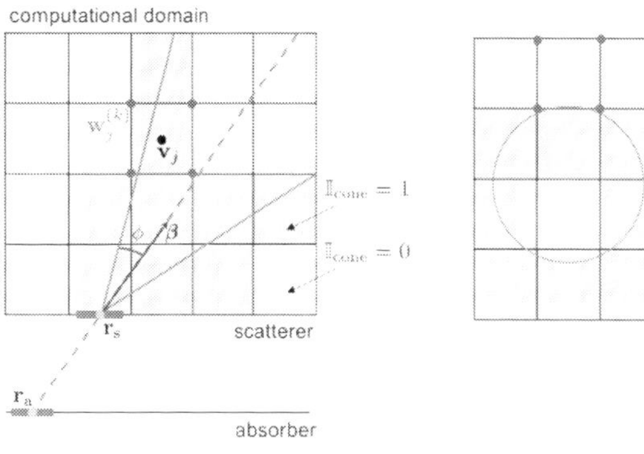

(a) side view (b) top view

Fig. 2: (a) The cone indicator function can be determined by testing whether all voxel corners lie inside or outside the cone surface. (b) The top view also shows a possible missed case that is highly unlikely to occur.

We now develop a simple test based on the voxel corners, as illustrated in two dimensions in Fig. 2. For voxel V_j with center $\mathbf{v}_j$, we define the set of 8 voxel corner points $\{\mathbf{w}_j^{(k)}\}_{k=1}^8$, where the positions are given as $\mathbf{w}_j^{(k)} = \mathbf{v}_j + (\pm d_x/2, \pm d_y/2, \pm d_z/2)$ for voxels with dimensions $d_x \times d_y \times d_z$. If there exist corner indices $m, n \in \{1, \ldots, 8\}$ such that $g(\mathbf{w}_j^{(m)}) < 0$ and $g(\mathbf{w}_j^{(n)}) > 0$, then the cone surface must pass through V_j, and $\mathbb{I}_{\mathrm{cone}}(V_j) = 1$. Otherwise, we set $\mathbb{I}_{\mathrm{cone}}(V_j) = 0$, which is true with high probability. It is possible for the cone surface to intersect the edge of a voxel despite all 8 corners lying outside the cone as shown in the top view in Fig. 2b; however, we consider this probability to be negligible, especially for small voxels and large cone radii.

3) Proposed system matrix accounting for spatial ambiguity: Due to the limitation of spatial resolution of the detector arrays, when a pair of scatterer index and absorber index is recorded, there is an ambiguity of the location where the photon hits the scatterer and absorber within those detector pixels. In other words, a measurement does not determine a cone, but a region surrounding the cone. Ignoring this ambiguity can lead to unreliable reconstruction, especially when the number of measurements is small and when the source is situated far from the detector. In our work, we assume the energy recordings are sufficiently accurate and ignore the half-angle ambiguity, thus $p_{\phi|\tilde{\phi}}(\phi|\tilde{\phi}) = \delta(\phi - \tilde{\phi})$. We ignore the thickness of the detector and model the detector pixel as the largest disc contained in that pixel whose radius is denoted by r_{pix}. That is, we define $p(\mathbf{r}_s|\tilde{\mathbf{r}}_s)$ as uniform for $\tilde{\mathbf{r}}_s$ in the disc centered at $\mathbf{r}_s$, and a similar definition is applied to $p(\mathbf{r}_a|\tilde{\mathbf{r}}_a)$.

$$\mathbf{T}_{i,j} = \frac{K(\phi^{(i)}|E_\gamma)}{\mathrm{vol}(V_j)} \int_{V_j} \int_{\mathrm{disc}(\mathbf{r}_s^{(i)}, r_{\mathrm{pix}})} \left\{ \frac{|\cos[\angle(\mathbf{x} - \tilde{\mathbf{r}}'_s)])|}{\|\mathbf{x} - \tilde{\mathbf{r}}'_s\|_2^2} \int_{\mathrm{disc}(\mathbf{r}_a^{(i)}, r_{\mathrm{pix}})} \frac{|\cos[\angle(\tilde{\mathbf{r}}'_a - \tilde{\mathbf{r}}'_s)]|}{\|\tilde{\mathbf{r}}'_a - \tilde{\mathbf{r}}'_s\|_2^2} \delta(\phi^{(i)} - \phi(\mathbf{v}_j; \tilde{\mathbf{r}}'_a, \tilde{\mathbf{r}}'_s)) \right\} \mathrm{d}\tilde{\mathbf{r}}'_a \mathrm{d}\tilde{\mathbf{r}}'_s\, \mathrm{d}\mathbf{x}, \tag{9}$$

where $\mathrm{disc}(\mathbf{c}, r)$ denotes the disc centered at $\mathbf{c}$ and with radius r. Finally, we use the center $\mathbf{v}_j$ to represent the voxel V_j and evaluate whether it lies in the ambiguity region. Therefore, our proposed system matrix accounting for spatial ambiguity is a special case of (3) and can be written as

$$\mathbf{T}_{i,j}^{\mathrm{amb}} \propto K(\phi^{(i)}|E_\gamma)\frac{|\cos[\angle(\mathbf{v}_j - \tilde{\mathbf{r}}_\mathrm{s}')]|}{\|\mathbf{v}_j - \tilde{\mathbf{r}}_\mathrm{s}'\|_2^2} \frac{|\cos[\angle(\tilde{\mathbf{r}}_\mathrm{a}' - \tilde{\mathbf{r}}_\mathrm{s}')]|}{\|\tilde{\mathbf{r}}_\mathrm{a}' - \tilde{\mathbf{r}}_\mathrm{s}'\|_2^2}\mathbb{I}_{\mathrm{amb}}(\mathbf{v}_j; \mathbf{r}_\mathrm{a}^{(i)}, \mathbf{r}_\mathrm{s}^{(i)}, \phi^{(i)}), \quad (10)$$

where

$$\mathbb{I}_{\mathrm{amb}}(\mathbf{v}_j; \mathbf{r}_\mathrm{a}^{(i)}, \mathbf{r}_\mathrm{s}^{(i)}, \phi^{(i)}) = \begin{cases} 1, & \mathbf{v}_j \text{ in ambiguity region} \\ 0, & \text{otherwise.} \end{cases} \quad (11)$$

In the following, we discuss how to approximate the indicator function $\mathbb{I}_{\mathrm{amb}}(\mathbf{v}_j)$.

III. Approximated Spatial Ambiguity Model

In this section, we introduce an approximation to the ambiguity region. Our goal is to have a model that is sufficiently accurate for reliable reconstruction and the representation is simple enough for efficiently building the forward operator. A graphic illustration of our model is shown in Figure 3.

A. Cone Axis Ambiguity

Given a measurement tuple $(\mathbf{r}_\mathrm{s}, \mathbf{r}_\mathrm{a}, \phi)$, the cone axis direction can be from any point within $\mathrm{disc}(\mathbf{r}_\mathrm{a}, r_{\mathrm{pix}})$ on the absorber to any point within $\mathrm{disc}(\mathbf{r}_\mathrm{s}, r_{\mathrm{pix}})$ on the scatterer. At any z-plane in the computational domain, the intersection of complete set of cone axes with the z-plane also forms a disc, corresponding to the cone axis ambiguity at that distance. Denote the positions of the centers of the current pixels on scatterer and absorber by $\mathbf{r}_\mathrm{s} = (x_s, y_s, z_s)$ and $\mathbf{r}_\mathrm{a} = (x_a, y_a, z_a)$, respectively. At a z-plane, the cone axis ambiguity disc has center $\mathbf{p}_0 = (x_0, y_0, z)$ and radius r_z given by

$$\begin{aligned} x_0 &= x_s + \frac{z - z_s}{z_s - z_a}(x_s - x_a) \\ y_0 &= y_s + \frac{z - z_s}{z_s - z_a}(y_s - y_a), \\ r_z &= \frac{z - (z_s + z_a)/2}{(z_s + z_a)/2 - z_a} r_{\mathrm{pix}}. \end{aligned} \quad (12)$$

The boundary of the cone axis ambiguity disc is the circle on the z-plane that can be parameterized by $\theta \in [0, 2\pi)$ as

$$x_\mathrm{c}(\theta) = x_0 + r_z \cos(\theta), \quad y_\mathrm{c}(\theta) = y_0 + r_z \sin(\theta). \quad (13)$$

For a fixed boundary point $\mathbf{p}_\mathrm{c}(\theta) = (x_\mathrm{c}(\theta), y_\mathrm{c}(\theta), z)$, we can trace back to find the corresponding cone vertex $\mathbf{v}(\theta)$ on the scatterer and the cone axis direction $\boldsymbol{\beta}(\theta)$, which along with an angle ϕ fully expresses the cone.

B. Source Ambiguity Region

We next describe how the ambiguity in the cone axis affects the possible locations of a radiation source, given an angle measurement ϕ. We focus on the possible source locations for the most extreme cone axis positions, which lie on the boundary of the disc, and thus define the boundary of the source ambiguity region. In fact, these cones define both an exterior and interior boundary, which we fit with ellipses or hyperbolas, although in some cases, only the exterior boundary will be used.

1) Boundary Points: To find the approximate interior and exterior boundaries of the ambiguity region, we define points $\mathbf{p}_\mathrm{i}(\theta) = (x_\mathrm{i}(\theta), y_\mathrm{i}(\theta), z)$ and $\mathbf{p}_\mathrm{e}(\theta) = (x_\mathrm{e}(\theta), y_\mathrm{e}(\theta), z)$ for each point $\mathbf{p}_\mathrm{c}(\theta)$ on the disc boundary; see the red cone in Fig. 3b. These points are defined by two constraints: 1) they are on the line connecting $\mathbf{p}_0$ and $\mathbf{p}_\mathrm{c}$ and 2) the angle between $\mathbf{v}(\theta) + \boldsymbol{\beta}(\theta)$ and either $\mathbf{p}_\mathrm{i}(\theta)$ or $\mathbf{p}_\mathrm{e}(\theta)$ is ϕ. To distinguish the interior and exterior points, we define $\mathbf{p}_\mathrm{i}(\theta)$ to be on the ray from $\mathbf{p}_\mathrm{c}(\theta)$ toward $\mathbf{p}_0(\theta)$, whereas $\mathbf{p}_\mathrm{e}(\theta)$ is on the ray from $\mathbf{p}_0(\theta)$ toward $\mathbf{p}_\mathrm{c}(\theta)$. Mathematically,

$$\langle \mathbf{p}_\mathrm{i}(\theta) - \mathbf{p}_\mathrm{c}(\theta), \mathbf{p}_\mathrm{c}(\theta) - \mathbf{p}_0(\theta)\rangle < 0 \quad (14)$$

$$\langle \mathbf{p}_\mathrm{e}(\theta) - \mathbf{p}_\mathrm{c}(\theta), \mathbf{p}_\mathrm{c}(\theta) - \mathbf{p}_0(\theta)\rangle > 0, \quad (15)$$

where $\langle\cdot\rangle$ denotes the inner product. For each point $\mathbf{p}_\mathrm{e}(\theta)$ and $\mathbf{p}_\mathrm{i}(\theta)$, we solve for coordinates (x, y) such that

$$y - y_0 = \frac{y_c(\theta) - y_0}{x_c(\theta) - x_0}(x - x_0) \quad (16)$$

$$\cos(\phi) = \left\langle \frac{\mathbf{v}(\theta) + \boldsymbol{\beta}(\theta)}{\|\mathbf{v}(\theta) + \boldsymbol{\beta}(\theta)\|}, \frac{(x, y, z)}{\|(x, y, z)\|} \right\rangle. \quad (17)$$

In cases where the cone axis is less tilted and/or the cone angle is sufficiently small, we obtain solutions for both $\mathbf{p}_\mathrm{e}(\theta)$ and $\mathbf{p}_\mathrm{i}(\theta)$, whereas in other cases, we only obtain a solution only for $\mathbf{p}_\mathrm{i}(\theta)$.

2) Solid Ellipse vs Ring Regions: We repeat the same calculation for $\mathbf{p}_\mathrm{e}(\theta + \pi)$ and $\mathbf{p}_\mathrm{i}(\theta + \pi)$ associated with $\mathbf{p}_\mathrm{c}(\theta + \pi)$ (the yellow cone in Fig. 3b). Suppose that we obtained two solutions for both cones as shown on the bottom right in Fig. 3b. If the two points of the yellow cone and the two points of the red cone do not interlace (e.g., for a small cone angle), then all of the points inside the exterior boundary are possible locations for the radiation source. This implies a solid ambiguity region, in this case an ellipse, so the interior boundary is discarded. If the points for the θ and $\theta + \pi$ cones do interlace (e.g., for large cone angle), then there is an open portion in the interior of the ambiguity region whose points do not lie on any plausible Compton cone surface. We then call the ambiguity region a "ring," which has both interior and exterior boundaries. As we sweep θ from 0 to π, the indication of solid ellipse or ring determined by the cones at θ and $\theta + \pi$ may be mixed, in which case, we approximate it as a solid ellipse.

3) Fitting Ellipses or Hyperbolas: We observed that the collection of points $\mathcal{X}_\mathrm{i} = \{(x_\mathrm{i}(\theta), y_\mathrm{i}(\theta), z) : \theta \in [0, 2\pi)\}$ can be approximated by an ellipse when there are two solutions for all θ, whereas when there is only one solution for θ in a continuous interval in $[0, 2\pi)$, it can be approximated as a

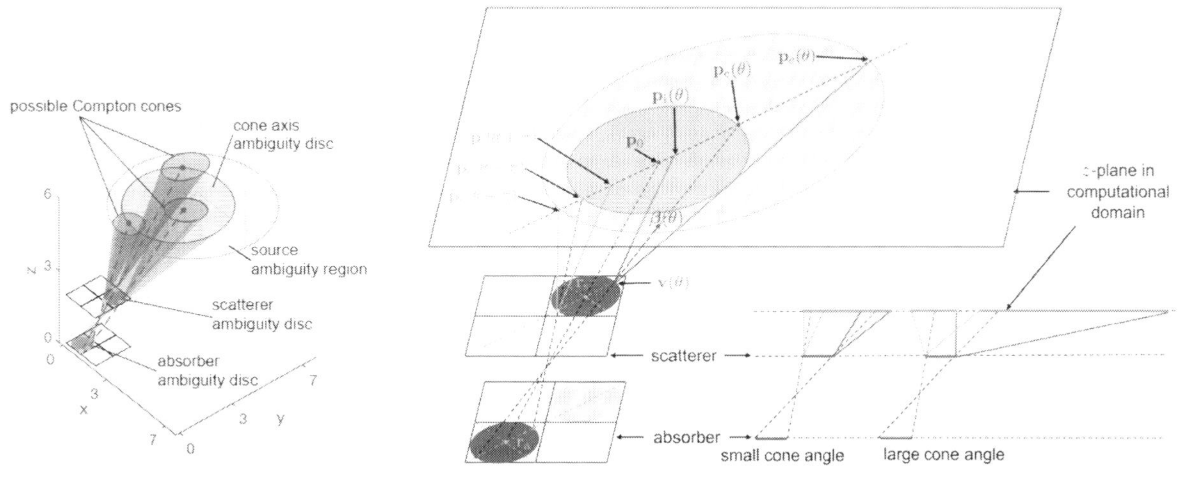

Fig. 3: Illustration of spatial ambiguity in the computational domain due to limited spatial resolution of the detector arrays. (a) Detector pixels are approximated by discs. The resulting possible Compton cones (purple) have axes whose intersection with any z-plane is a cone axis ambiguity disc (blue). Depending on the angle ϕ, the set of possible Compton cones traces out a larger region (green) corresponding to the ambiguity in the radiation source position. (b) To fit the source ambiguity region with a conic section, we solve for interior and exterior boundary points. If there are two intersections for both cones (red and yellow), the ambiguity region boundary is approximated as an ellipse; else if only one solution exists, the region boundary is approximated by a hyperbola. The interior intersections determine whether the ambiguity region is solid or an open ring.

hyperbola; we call this the interior shell of the ambiguity region. Similarly, $\mathcal{X}_e = \{(x_e(\theta), y_e(\theta), z) : \theta \in [0, 2\pi)\}$ defines the exterior shell. Therefore, we can obtain the parameters for ellipse or hyperbola for the interior (resp. exterior) shell by least squares fit to the points $\mathcal{X}_i$ (resp. $\mathcal{X}_e$) computed from a discrete set of θ angles. We used 100 equally spaced θ values in $[0, 2\pi)$ for the simulation results presented in Section V. Specifically, we first fit a conic section

$$Ax^2 + Bxy + Cy^2 + Dx + Ey + 1 = 0. \tag{18}$$

If $B^2 - 4AC < 0$, it is a shifted and rotated ellipse:

$$f_{\text{ellipse}}(x, y) = \frac{[(x - c_x)\cos(\alpha) + (y - c_y)\sin(\alpha)]^2}{a^2} + \frac{[(x - c_x)\sin(\alpha) - (y - c_y)\cos(\alpha)]^2}{b^2} = 1.$$

If $B^2 - 4AC > 0$, it is a shifted and rotated hyperbola:

$$f_{\text{hyperbola}}(x, y) = \frac{[(x - c_x)\cos(\alpha) + (y - c_y)\sin(\alpha)]^2}{a^2} - \frac{[(x - c_x)\sin(\alpha) - (y - c_y)\cos(\alpha)]^2}{b^2} = 1.$$

We can then relate the center (c_x, c_y), clock-wise rotation angle α, major axis radius a, and minor axis radius b to the estimated parameters (A, B, C, D, E). Specifically, for both ellipse and hyperbola,

$$c_x = \frac{2CD - BE}{B^2 - 4AC}, \quad c_y = \frac{2AE - BD}{B^2 - 4AC}$$
$$\alpha = \frac{1}{2}\arctan\left(\frac{B}{A - C}\right).$$

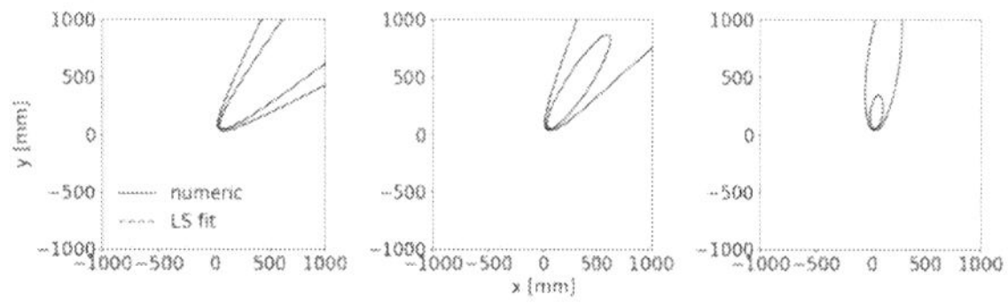

Fig. 4: Least squares fit for hyperbola and ellipse. The black solid lines represent $\mathcal{X}_i$ and $\mathcal{X}_e$, whereas the red dashed lines represent the ellipse or hyperbola obtained by least squares fit.

Let

$$F = Ac_x^2 + Bc_xc_y + Cc_y^2 - 1, \tag{19}$$

then we have

$$a^2 = \frac{F\cos(2\alpha)}{A\cos^2(\alpha) - C\sin^2(\alpha)}, \quad b^2 = \frac{\pm F\cos(2\alpha)}{C\cos^2(\alpha) - A\sin^2(\alpha)},$$

where in the expression for b^2, the '+' sign is for ellipse, whereas the '−' sign is for hyperbola. Note that a hyperbola has two branches and they lie in two different half-spaces separated by the line perpendicular to the major axis and passing through the center (c_x, c_y). The branch that we are interested in is the one in the half-space whose normal and the vector $(x_s - x_a, y_s - y_a)$ form an angle less than 90°. Fig. 4 shows an example for the least squares fit results.

Once the parameters for the ellipse or hyperbola have been determined, it is convenient to test whether an arbitrary point $\mathbf{v}_j$ in the computation domain is in the ambiguity region, and thus whether $\mathbb{I}_{\text{amb}}(\mathbf{v}_j)$ equals 0 or 1. For example, consider the $\mathbf{v}_j = (x_j, y_j, z)$'s on a z-plane, if the ambiguity region

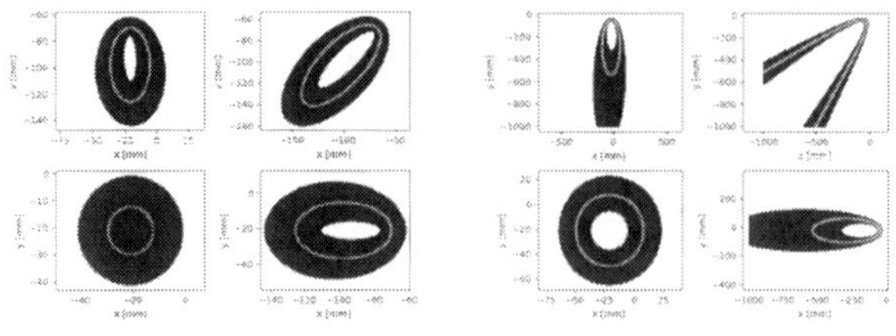

(a) Cone angle $\phi = 10°$. (b) Cone angle $\phi = 30°$.

Fig. 5: Cone surfaces (yellow) and ambiguity regions (blue) surrounding them at the plane $z = 50$ mm. Both scatterer and absorber have 8×8 detection elements and the distance between scatterer and absorber is 30 mm. The size of each element is 5×5 mm (i.e., $r_{\text{pix}} = 2.5$ mm) and there is a 1 mm gap between neighboring elements. The scatterer position is fixed at the bottom-left corner and the absorber locations vary over the four corners of an 8×8 array.

is between two hyperbola as in the first plot in Fig. 4, then $\mathbb{I}_{\text{amb}}(\mathbf{v}_j) = 1$ if the following conditions are satisfied:

$$f^{\text{i}}_{\text{hyperbola}}(x_j, y_j) \leq 1, \quad f^{\text{e}}_{\text{hyperbola}}(x_j, y_j) \geq 1,$$
$$\langle (x_j - c_x, y_j - c_y), (x_s - x_a, y_s - y_a) \rangle \geq 0,$$

where $f^{\text{i}}_{\text{hyperbola}}$ and $f^{\text{e}}_{\text{hyperbola}}$ denotes the hyperbola fitted by $\mathcal{X}_{\text{i}}$ and $\mathcal{X}_{\text{e}}$, respectively. As another example, if the ambiguity region is between an ellipse and a hyperbola as in the second plot in Fig. 4, then $\mathbb{I}_{\text{amb}}(\mathbf{v}_j) = 1$ if the following conditions are satisfied:

$$f^{\text{i}}_{\text{ellipse}}(x_j, y_j) \geq 1, \quad f^{\text{e}}_{\text{hyperbola}}(x_j, y_j) \geq 1,$$
$$\langle (x_j - c_x, y_j - c_y), (x_s - x_a, y_s - y_a) \rangle \geq 0.$$

Other cases can be defined in a similar way.

An example for the ambiguity region with various scatterer-absorber locations is shown in Fig. 5. We notice that a solid ellipse is obtained at the bottom-left plot in Fig. 5a. This is the case when the cone angle is comparatively small and the scatterer pixel is directly above the absorber pixel, thus the cone axes are not so tilted throughout the ambiguity region. The other extreme is shown at the top-right plot in Fig. 5b, where we obtained a region in between two hyperbolas. This is the case when the cone angle is comparatively large and the scatterer pixel is horizontally farther away from the absorber pixel, thus the cone axes are very tilted. Additional cases between these two extremes are shown in the rest of the plots in Fig. 5, where the ambiguity region is a ring between two ellipses.

IV. Inverse Problem

A. Measurement Selection

Although the modeling assumes a perfect event is recorded other than the spatial ambiguity, with Compton scattering and photon absorption happening at a single pixel in each of the respective detectors, actual Compton camera measurements themselves are highly ambiguous. One major challenge is that the sequence of photon hits cannot usually be tracked, so there are often multiple plausible sequences (e.g., Compton scattering occurs at the absorber, and the photon is back-scattered and absorbed by the scatterer) that result in dramatically different cones [23], [24]. Other ambiguities arise if the energy from an interaction is deposited in multiple pixels of the same detector, or if the energy is incompletely deposited, so that $E_{\text{s}} + E_{\text{a}} \neq E_\gamma$. To avoid modelling all of these additional ambiguities, we follow convention in selecting only events that meet the following criteria, based on assumptions of a ^{137}Cs radiation source (with characteristic emission of 662-keV gamma photons) and Cesium Iodide (CsI) detectors:

1) **Total energy:** the gamma photons used in our simulation have energy $E_\gamma = 0.662$ MeV, and we only use measurements with $E_s + E_a > 0.66$ MeV.
2) **Excluding fluorescence X-ray contamination:** for CsI detectors, X-rays with energy approximately 0.045 MeV escape from the absorber and are then absorbed by the scatterer [25, pp. 472]. This has nothing to do with Compton scattering and the angle of the X-ray is completely random. We observed in our simulation that the measurements with $E_s \in [0.025, 0.045]$ MeV are likely due to X-ray contamination and we exclude them for reconstruction.

B. Sequencing

For the recorded events satisfying the above criteria, we observed that with our configuration of the Compton camera, the number of back-scattering events (i.e., the photon hits the absorber first and then is back-scattered to the scatterer, thus the role of the scatterer and the absorber is interchanged) is non-negligible. Therefore, we include the following sequencing techniques in our method for more reliable reconstruction:

1) **Compton edge test:** for a given event, if the recorded energy is greater than the Compton edge energy, which is defined as

$$E_{\text{edge}} = E_\gamma \left(1 - \frac{1}{1 + \frac{2E_\gamma}{m_e c^2}} \right), \tag{20}$$

then this cannot be the first interaction, because this is the largest possible energy deposited during Compton scattering.
2) **Probabilistic combination:** when both recorded energies pass the Compton edge test, either of them can be the first interaction. It has been shown [16] that the (non-normalized) probability of (E_1, ϕ) being the first hit can be computed as

$$K(\phi; E_\gamma) \frac{2\pi m_e c^2}{(E_\gamma - E_1)^2} \frac{1}{2\pi \sin(\phi)}, \tag{21}$$

where $K(\phi; E_\gamma)$ denotes the Klein-Nishina formula. However, in our experiments, the sequence determined directly by this probability does not seem to be very accurate. We observed that a large number of events that are supposed to be forward scattering have higher probability of being back scattering. We found that an ad-hoc correction by scaling the back-scattering probability by 0.6 seems to result in more reasonable sequences. Therefore, in all our simulations, we apply this ad-hoc modification to the probability.

https://doi.org/10.1051/epjconf/202328810002

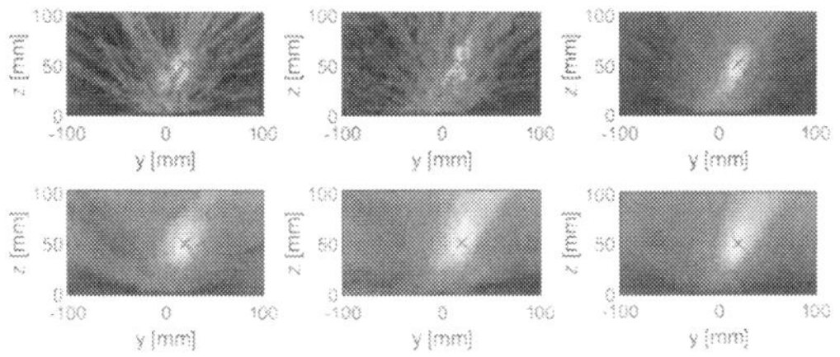

(a) Back-projection

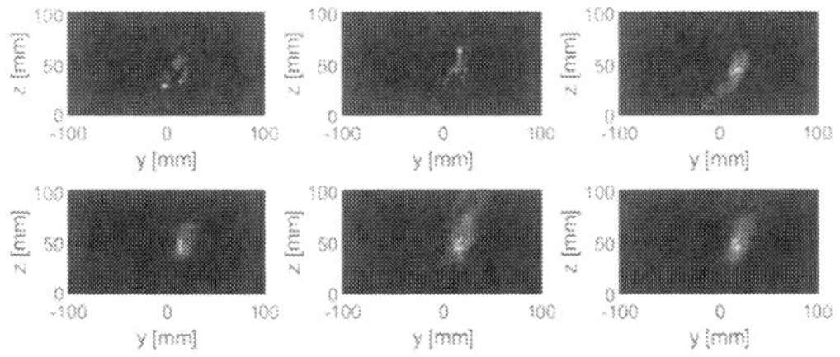

(b) LM-MLEM

Fig. 6: Reconstruction of the yz-slice at the true x position by the back-projection algorithm (panel (a)) and the LM-MLEM algorithm (panel (b)). In each panel, the first row shows the results for the forward operator constructed with the cone surface model, and the second row uses our ambiguity region model. From left to right: the number of measurements is 64, 132, 1375, respectively. The red cross shows the true source position.

C. List-Mode Maximum Likelihood Expectation Maximization

With some initial estimate of the source intensity distribution $\lambda^{(0)}$, the algorithm proceeds as

$$\boldsymbol{\lambda}^{(t+1)} = \frac{\boldsymbol{\lambda}^{(t)}}{s} \odot \left(\mathbf{T}^{\mathsf{T}} \frac{\mathbf{1}}{\mathbf{T}\boldsymbol{\lambda}^{(t)}} \right), \tag{22}$$

where $\odot$ denotes coordinate-wise multiplication and the divisions are all coordinate-wise.

V. Simulation Results

We now test our reconstruction method on Geant4 simulated data. A single point source with energy 0.662 MeV is utilized. Both the scatterer and the absorber have 8×8 pixels made of Cesium Iodide (CsI) and of size $5\times5\times5$ mm^3. There is a 1 mm gap between pixels. The scatterer is placed at depth $z = 0$ mm and the absorber is placed at depth $z = -30$ mm. In the first set of simulations, we fix the source location at $(20, 20, 50)$ mm and vary the number of measurements. We select measurements that satisfy the criteria in Section IV-A as well as $E_s < E_{\text{edge}}$. All these measurements are treated as forward scattering events. The computational domain is a $64 \times 64 \times 33$ volume and the physical size of the volume is shown in the figures. The reconstruction results obtained by a simple back-projection algorithm and the List-Mode Maximum Likelihood Expectation Maximization (LM-MLEM) algorithm are shown in Fig. 6a and Fig. 6b, respectively. In both Fig. 6a and Fig. 6b, the first row and the second row show the results when the forward operator is constructed with the cone surface model and our ambiguity region model, respectively. We can see from Fig. 6 that by using our ambiguity model instead of the cone surface model in the forward operator, we are able to obtain higher quality reconstruction with the same number of measurements. Alternatively, we can achieve good source distribution reconstructions with a smaller number of detected photons.

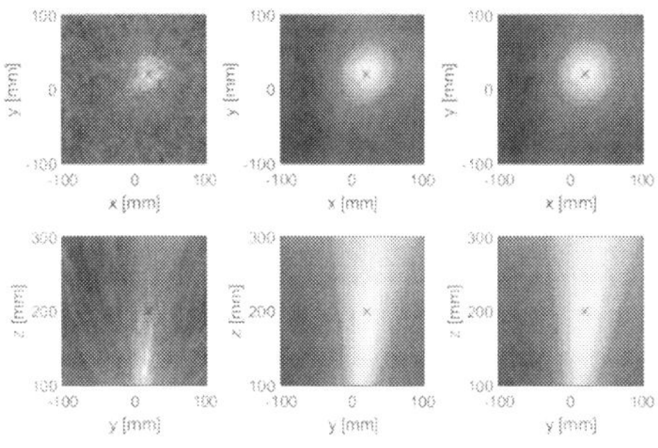

(a) Back-projection

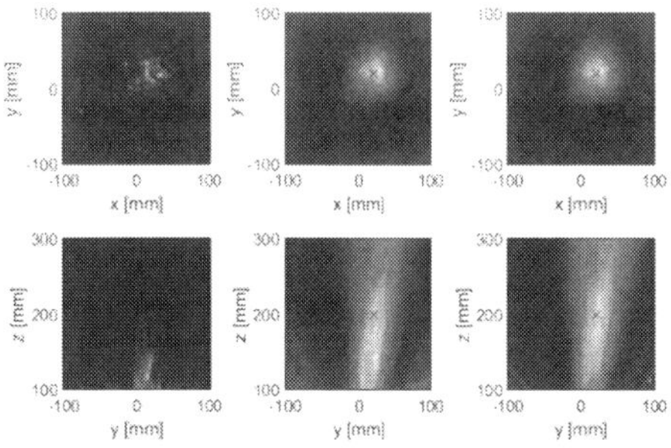

(b) LM-MLEM

Fig. 7: Reconstruction of the xy-slice at the true z position (first row in each panel) the yz-slice at the true x position (second row in each panel) by back-projection (panel (a)) and LM-MLEM (panel (b)). First and second columns: cone surface model and ambiguity region model, respectively, with all selected measurements treated as forward scattering after Compton edge test for E_s. Third column: ambiguity region model with probabilistic combination of forward- and back-scattering events.

In the second set of simulation, we move the source farther away from the detector and set it at $(20, 20, 200)$ mm. The computational domain is a $64 \times 64 \times 64$ volume and the physical size of the volume is shown in the figures. The reconstruction results obtained by the back-projection algorithm and the LM-MLEM algorithm are shown in Fig. 7a and Fig. 7b, respectively. In both Fig. 7a and Fig. 7b, the first row shows the reconstruction of the the xy-slice at the true z position and the second row shows the reconstruction of the yz-slice at the true x position. The first two columns show the results when all selected measurements are treated as forward scattering after Compton edge test. In the first column, the cone surface model is used to construct the forward operator, and in the second column, our ambiguity region model is used. The third column shows the results with our ambiguity region model and probabilistic combination of forward- and back-scattering events. As we can see from the first column that the cone surface model results in a significant bias towards the detector along the z-direction. Comparing the second and the third columns, we notice that the error due to back-scattering is non-negligible along the z-direction and a probabilistic combination of forward- and back-scattering events yields more accurate reconstruction when the source is far from the detector.

VI. Conclusion

We have demonstrated that accounting for ambiguities due to the spatial resolution of a Compton camera can improve source distribution reconstruction, enabling the use of a smaller number of detected photons. Our ambiguity model is based on an approximation of the detector pixels by a disc, and then approximating the resulting ambiguity regions as being bounded by conic sections for a given depth plane. Our method is not detector-specific, thus can potentially be applied to other pixelated detectors, such as semiconductor-based detectors, to improve reconstruction performance. However, because the pixel size is usually smaller in semiconductor-based detectors, the improvement may not be as significant. Future work may use more accurate models that also take the pixel depth into account, as well as investigating the combination of ambiguity models for both spatial and energy resolution.

References

[1] A. H. Compton, "A quantum theory of the scattering of X-rays by light elements," *Physical Review*, vol. 21, no. 5, pp. 483–502, May 1923.

[2] S. Wilderman, N. Clinthorne, J. Fessler, and W. Rogers, "List-mode maximum likelihood reconstruction of Compton scatter camera images in nuclear medicine," in *IEEE Nuclear Science Symposium and Medical Imaging Conference*, Nov. 1998, vol. 3, pp. 1716–1720 vol.3.

[3] S. Wilderman, J. Fessler, N. Clinthorne, J. LeBlanc, and W. Rogers, "Improved modeling of system response in list mode EM reconstruction of Compton scatter camera images," *IEEE Transactions on Nuclear Science*, vol. 48, no. 1, pp. 111–116, Feb. 2001.

[4] V. Schoenfelder, H. Aarts, K. Bennett, H. de Boer, J. Clear, W. Collmar, A. Connors, A. Deerenberg, R. Diehl, A. von Dordrecht, J. W. den Herder, W. Hermsen, M. Kippen, L. Kuiper, G. Lichti, J. Lockwood, J. Macri, M. McConnell, D. Morris, R. Much, J. Ryan, G. Simpson, M. Snelling, G. Stacy, H. Steinle, A. Strong, B. N. Swanenburg, B. Taylor, C. de Vries, and C. Winkler, "Instrument description and performance of the Imaging Gamma-Ray Telescope COMPTEL aboard the Compton Gamma-Ray Observatory," *The Astrophysical Journal Supplement Series*, vol. 86, pp. 657, June 1993.

[5] C. Lehner, Z. He, and F. Zhang, "4pi Compton imaging using a 3-D position-sensitive CdZnTe detector via weighted list-mode maximum likelihood," *IEEE Transactions on Nuclear Science*, vol. 51, no. 4, pp. 1618–1624, Aug. 2004.

[9] V. Maxim, X. Lojacono, E. Hilaire, J. Krimmer, E. Testa, D. Dauvergne, I. Magnin, and R. Prost, "Probabilistic models and numerical calculation of system matrix and sensitivity in list-mode MLEM 3D reconstruction of Compton camera images," *Physics in Medicine and Biology*, vol. 61, no. 1, pp. 243–264, Dec. 2016.

[6] S. R. Tornga, M. W. Rawool Sullivan, and J. P. Sullivan, "Three-dimensional Compton imaging using list-mode maximum likelihood expectation maximization," *IEEE Transactions on Nuclear Science*, vol. 56, no. 3, pp. 1372–1376, June 2009.

[7] V. Maxim, M. Frandeş, and R. Prost, "Analytical inversion of the Compton transform using the full set of available projections," *Inverse Problems*, vol. 25, no. 9, pp. 095001, Aug. 2009.

[8] V. Maxim, "Filtered backprojection reconstruction and redundancy in Compton camera imaging," *IEEE Transactions on Image Processing*, vol. 23, no. 1, pp. 332–341, Jan. 2014.

[10] F. Terzioglu, P. Kuchment, and L. Kunyansky, "Compton camera imaging and the cone transform: a brief overview," *Inverse Problems*, vol. 34, no. 5, pp. 054002, May 2018.

[11] F. Terzioglu, "Exact inversion of an integral transform arising in Compton camera imaging," *Journal of Medical Imaging*, vol. 7, no. 3, pp. 032504, Mar. 2020.

[12] C. J. Solomon and R. J. Ott, "Gamma ray imaging with silicon detectors — A Compton camera for radionuclide imaging in medicine," *Nuclear Instruments and Methods in Physics Research Section A: Accelerators, Spectrometers, Detectors and Associated Equipment*, vol. 273, no. 2-3, pp. 787–792, Dec. 1988.

[13] A. Hoover, R. Kippen, J. Sullivan, M. Rawool-Sullivan, W. Baird, and E. Sorensen, "The LANL prototype Compton gamma-ray imager: design and image reconstruction techniques," *IEEE Transactions on Nuclear Science*, vol. 52, no. 6, pp. 3047–3053, Dec. 2005.

[14] I. Tsitsimpelis, C. J. Taylor, B. Lennox, and M. J. Joyce, "A review of ground-based robotic systems for the characterization of nuclear environments," *Progress in Nuclear Energy*, vol. 111, pp. 109–124, Mar. 2019.

[15] B. D. Milbrath, A. J. Peurrung, M. Bliss, and W. J. Weber, "Radiation detector materials: An overview," *Journal of Materials Research*, vol. 23, no. 10, pp. 2561–2581, Oct. 2008.

[16] D. Xu, Z. He, C. E. Lehner, and F. Zhang, "4-pi Compton imaging with single 3D position-sensitive CdZnTe detector," in *Hard X-Ray and Gamma-Ray Detector Physics VI*. International Society for Optics and Photonics, 2004, vol. 5540, pp. 144–155.

[17] D. Turecek, J. Jakubek, E. Trojanova, and L. Sefc, "Single layer Compton camera based on Timepix3 technology," *Journal of Instrumentation*, vol. 15, no. 01, pp. C01014–C01014, Jan. 2020.

[18] D. Xu, *Gamma-ray imaging and polarization measurement using 3-D position-sensitive CdZnTe detectors*, Ph.D. thesis, University of Michigan, 2006.

[19] D. Xu and Z. He, "Gamma-ray energy-imaging integrated spectral deconvolution," *Nuclear Instruments and Methods in Physics Research Section A: Accelerators, Spectrometers, Detectors and Associated Equipment*, vol. 574, no. 1, pp. 98–109, Apr. 2007.

[20] M. Hirasawa and T. Tomitani, "An analytical image reconstruction algorithm to compensate for scattering angle broadening in Compton cameras," *Physics in Medicine and Biology*, vol. 48, no. 8, pp. 1009–1026, Apr. 2003.

[21] L. Parra and H. Barrett, "List-mode likelihood: EM algorithm and image quality estimation demonstrated on 2-D PET," *IEEE Transactions on Medical Imaging*, vol. 17, no. 2, pp. 228–235, Apr. 1998.

[22] O. Klein and Y. Nishina, "The scattering of light by free electrons according to Dirac's new relativistic dynamics," *Nature*, vol. 122, no. 3072, pp. 398–399, Sept. 1928.

[23] A. C. Zoglauer, *First Light for the Next Generation of Compton and Pair Telescopes*, Ph.D. thesis, Technische Universität München, 2005.

[24] C. Wahl, *Imaging, Detection, and Identification Algorithms for Position-Sensitive Gamma-Ray Detectors*, Ph.D. thesis, University of Michigan, 2011.

[25] D. Aitken, B. Beron, G. Yenicay, and H. Zulliger, "The fluorescent response of NaI (Tl), CsI (Tl), CsI (Na) and CaF_2 (Eu) to X-rays and low energy gamma rays," *IEEE Transactions on Nuclear Science*, vol. 14, no. 1, pp. 468–477, 1967.

A new technique for quick identification of defective region inside γ-ray detector

Biswajit Das*, R. Palit, A. Kundu, P. Dey, V. Malik, S. K. Jadav, B. S. Naidu, and A. T. Vazhappilly

Department of Nuclear and Atomic Physics
Tata Institute of Fundamental Research, Mumbai - 400005, India

(*) biswajit.das@tifr.res.in

***Abstract*—The γ-ray detection efficiency of a detector decreases over time due to factors like radiation damage or an increase in the thickness of the inactive dead layer. For large γ-ray detector facilities, it is crucial to assess the health condition and performance of the inner regions of the detector crystals over time. In this study, we have introduced a method using GEANT4 simulation to detect defective regions within thick γ-ray detectors. In the experimental phase, a scanning setup was employed, comprising a single-crystal High Purity Germanium (HPGe) detector and a position-sensitive GAGG:Ce detector for coincidence measurements, using a ^{22}Na source. The 2D images were reconstructed from the front-face and side-face scans of the single-crystal coaxial HPGe detector, employing an energy gate set at 511 keV. A position gate applied to a specific section of those 2D images allowed for the mapping of γ-ray interactions along a conical path within the HPGe detector. The methodology involved the comparison and analysis of histograms generated from various sector gates, facilitating the identification of the defective region's position. In the GEANT4 simulation, a defective region was defined within the crystal, and that was effectively represented in the corresponding scanned image, which exhibited reduced efficiency. It's important to note that this method's effectiveness is restricted by the absorption profile of the 511 keV γ-ray, limiting its applicability to a depth of approximately 4 cm from the surface of the HPGe crystal. However, this approach can offer a swift and convenient method for inspecting γ-ray detectors, making it a valuable tool for the detector industry.**

***Keywords* — defective/damaged region identification of HPGe crystal; γ-ray tracking; position-sensitive GAGG; GEANT4 simulation**

I. Introduction

Since the discovery of nuclear decay, γ-ray detection has been extensively utilized for various nuclear structure investigations. Additionally, it also plays a crucial role in the field of nuclear medicine [1–4]. In contemporary times, the utilization of γ-ray detectors in various medical imaging applications has grown to be of paramount importance [5, 6]. Positron Emission Tomography (PET) is one of the most powerful setup with large number of inorganic or organic γ-ray detectors in an array for this purpose [7–9]. From a fundamental perspective, the progress in nuclear and particle physics experimental studies has been made achievable through research and development focused on the technical aspects of efficient and high-resolution γ-ray detector arrays [10–17]. The efficiency in detecting γ-ray and collecting charge carriers in HPGe detectors, particularly at lower gamma energies, tends to diminish over time. The primary cause for this decline is radiation damage, which may result from factors like neutron exposure during in-beam experiments or prolonged periods of cooling for older HPGe detectors. As a consequence, the thickness of the inactive dead layer increases as time passes [18–22]. For the case of the scintillator detector, the possible effects of radiation damage are i) radiation-induced absorption, ii) radiation-induced phosphore-scence, and iii) damage in scintillation mechanism. Among these, the color center formations and radiation-induced absorption are dominant for the radiation damage effect of the scintillator crystal [23]. To maintain a continuous assessment of the detectors in large detector facilities, it is imperative to monitor the responses of scintillator, segmented, or single-crystal HPGe γ-ray detectors over time. To obtain the variations in efficiency across the crystals, measurements of surface scans have been conducted using collimated low and high-energy γ-rays [24]. Indeed, these measurements require considerable time and precise alignments. Scanning a γ-ray detector by projecting interaction points onto a position-sensitive detector in coincidence provides a relatively expedient approach to assess performance across the crystal. Additionally, the pulse shape analysis (PSA) method serves as a valuable means to gauge the collection efficiency of charge carriers at specific positions within the crystal [25]. The electronic response, hole mobility, and features of the grid search algorithm have been investigated by L. Lewandowski, et al [26], and B. De Canditiisa et al [27].

In this work, we have scanned a single-crystal coaxial HPGe detector by a position-sensitive cerium-doped GAGG (Gadolinium Aluminum Gallium Garnet ($Gd_3Al_2Ga_3O_{12}$:Ce)) scintillator-based scanning setup with a ^{22}Na source. By analyzing and comparing the 2D scanned images from different sector (position) gates, we have examined the efficiency, actively seeking out defective regions along the γ-ray interaction path within the crystal. The method has been demonstrated through GEANT4 simulation, showcasing its effectiveness in identifying defective regions within the detector. In practical terms, this approach offers a straightforward, rapid, and effective means to evaluate the performance and the current health status of a γ-ray detector of interest.

II. INSTRUMENTATION AND EXPERIMENTAL SETUP

A cerium-doped GAGG crystal has been coupled with a position-sensitive photomultiplier tube (Hamamatsu-8500C) to perform scans on a single-crystal coaxial reverse-electrode HPGe detector. The GAGG:Ce crystal offers several advantages, including minimal internal radioactivity, high light output (46,000–50,600 photons/1 MeV), a high density of 6.63 gm/cm^3, excellent radiation hardness, rapid timing characteristics, non-hygroscopic properties, and relative affordability. The crystal has dimensions of 5.08 × 5.08 × 0.6 cm^3. The position-sensitive photomultiplier tube consists of 64 readouts, organized in an 8 x 8 mesh of horizontal and vertical anode wires. These 64 readouts have been reduced to just 4 readouts by adding an appropriate external resistive chain network. The interaction positions of the γ-rays within the GAGG:Ce crystal were determined based on the four anode signals, which were previously discussed in our reports [28,29]. The single-crystal HPGe and GAGG:Ce detectors were biased at -3000 and -1100 V, respectively. For coincidence measurements, a ^{22}Na radioactive point source with an activity of approximately 270 kBq was employed. This source simultaneously emits two 511 keV γ-rays in opposite directions following the electron–positron annihilation of para-Positronium [30]. Both detectors were positioned on opposite sides of the source within the same horizontal plane and in the same coaxial line, maintaining suitable distances between them. The basic geometry of the setup is illustrated in Fig. 1.

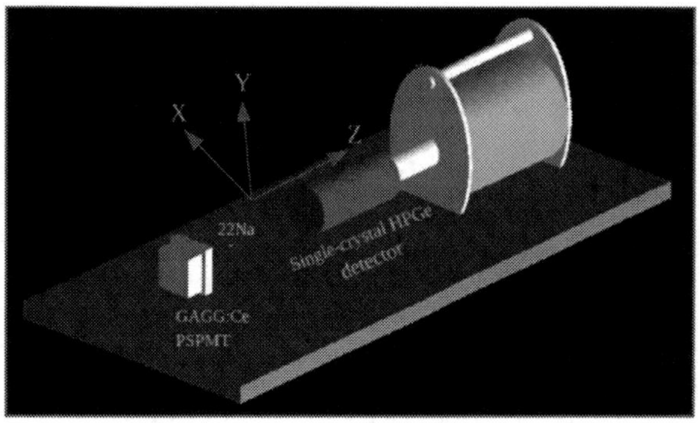

Fig. 1. Setup in GEANT4 geometry with a position-sensitive GAGG:Ce and a single-crystal HPGe detector. The ^{22}Na source is placed at 6 cm from GAGG:Ce and 12 cm from the single-crystal HPGe detector in the same line on a same horizontal plane.

The dynode and anode signals underwent amplification and shaping via spectroscopic amplifiers. A master gate was generated using the Constant Fraction Discriminator (CFD) outputs of the GAGG:Ce dynode and the single-crystal HPGe signal. The four anode signals (positions) were recorded when they fell within the specified time window of the master gate. The analog outputs from the spectroscopic amplifiers were digitized using an 8-channel, 12-bit Analog to Digital Converter (ADC) module. Experimental data were collected through the Linux Advanced Multi-Parameter System (LAMPS) [31] package. The complete dataset was analyzed using a C++-based macro within the ROOT-CERN software [32].

III. GEANT4 SIMULATION

The simulation performance was achieved using an object-oriented Monte Carlo simulation toolkit known as GEANT4, specifically version 4.10.05 [33]. Different parts of the setup were constructed using classes such as G4Box, G4Tubs, G4UnionSolid, G4SubtractionSolid, G4EllipticalTube, and G4Sphere within GEANT4. Fig. 1 provides a visualization of the setup's geometry in GEANT4. The ^{22}Na source, enclosed by a plastic disc, was defined within the simulation. In each event of a run, two simultaneous opposite 511 keV and a 1274 keV γ-rays were emitted in random directions from the source. Compton-scattered γ-rays, electrons, and positrons were expected to be generated during interactions of γ-rays with matter. In the PhysicsList, all potential physics processes for γ-rays, electrons, and positrons were considered. The primary interactions of γ-rays with matter included Compton scattering, the photoelectric effect, pair production, and Rayleigh scattering. Electrons and positrons produced during these interactions could undergo processes such as ionization, multiple scatterings, and emit Bremsstrahlung radiation. In addition, the positrons can annihilate with the surrounding electrons and emit two simultaneous 511 keV γ-rays. Within the PhysicsList, classes such as G4PhotoElectricEffect, G4Compton-Scattering, G4GammaConversion, and G4Ray-leighScattering were included for simulating γ-ray interactions. For electron and positron interactions with matter, the code considered classes such as G4eMultipleScattering, G4eIonisation, G4eBremsstrahlung, and G4eplusAnnihilation. The simulation results were stored in a ROOT NTuple on an event-by-event basis and were analyzed using a ROOT macro.

IV. ANALYSIS AND RESULT

The energy calibrations of the γ-ray spectra from both detectors were performed using a ^{152}Eu radioactive source with well-known gamma energies ranging from 100 to 1500 keV. Notably, there was no significant internal radioactivity detected in the GAGG:Ce scintillator. The (X, Y) interaction points on the GAGG:Ce crystal were calculated based on the four anode signals using two equations [28]. A Full Width at Half Maximum (FWHM) of approximately 3 mm was determined with a 2 mm collimated ^{22}Na source positioned at the center of the crystal. The FWHM exhibited a slight deterioration as one moved away from the center towards the edge. The energy resolutions at 511 keV for GAGG:Ce and HPGe were determined to be approximately 15% and 0.2%, respectively. The 2D scanned images of the front-face and side-face were reconstructed with the energy gates set of 505 keV < $E_{\gamma\, HPGe\ \&\ GAGG}$ < 516 keV. Both experimental (with a full bias of -3000 V) and simulated 2D images of the front scan are presented in Fig. 2. Notably, the reduced efficiency in the middle of the images highlights the presence of an axial cavity.

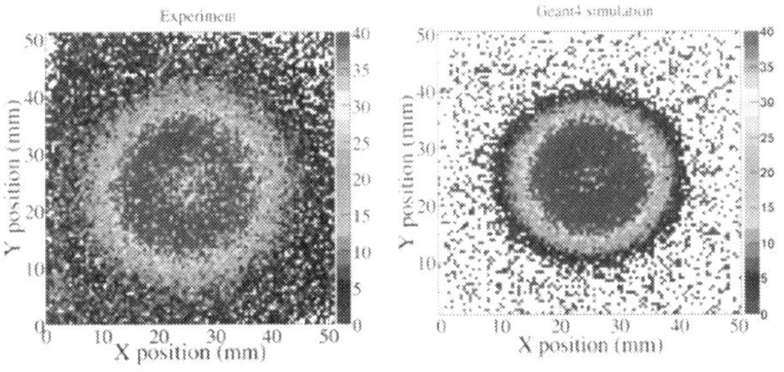

Fig. 2. Measured (left) and simulated (right) scanned images of the front face reconstructed with an energy gate of 511 keV ($E_{\gamma\, HPGe\ \&\ GAGG}$: 506-516 keV) on both the detectors. In the measured image, some background counts are there throughout the GAGG:Ce crystals because of the lower amplitude anode noise and random coincidences.

The incidence direction of the γ-ray in the HPGe crystal is well defined by a position gate on a sector (white mark on white rectangle on inset of Fig. 3) of the scanned image. Along that defined path, the 511 keV γ-ray interacts with the HPGe crystal via the photoelectric effect or Compton scattering. Irrespective of the nature of the interaction, the first interaction points will lie along this defined path. Depending upon the energy deposition condition in the HPGe detector, the effective depth of the trajectory will be different. This is simulated in GEANT4 and has been shown in Fig. 3. The image actually shows a 2D projection of the trajectory of the γ-rays having the first interaction within the HPGe crystal which are in coincidence with the GAGG:Ce detector. The interaction positions have been filled in the 2D histogram (Y-Z plane of single HPGe detector) in each step of interaction based on energy deposition in the crystal.

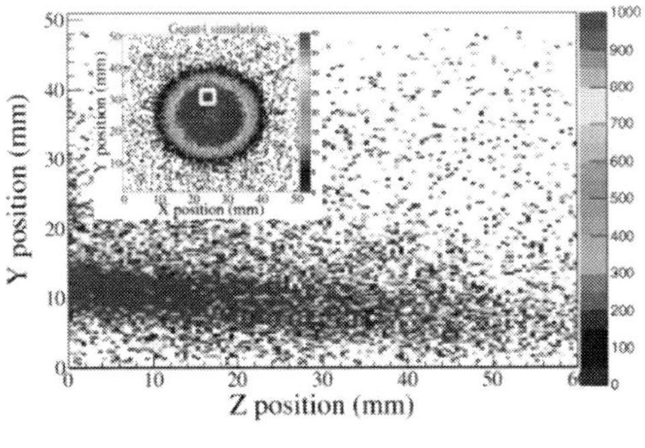

Fig. 3. The simulated interaction positions (in X-Z plane) of the γ-rays inside the single HPGe crystal obtained with a small position sector gate (shown in a white square of the inset 2D image) in the image formed by the GAGG detector.

A demonstration of the scanning results in the presence of a defective region within the crystals was illustrated through GEANT4 simulation. In this simulation, a dead region measuring 1 cm^3 was introduced into the HPGe crystal, as depicted in Fig. 4(a). The simulation was conducted for various locations (P1 to P6, marked as black dots) within the dead region.

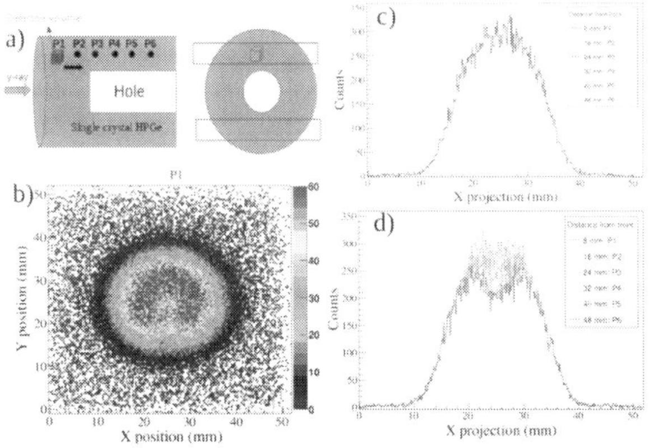

Fig. 4. a) The schematic of the single-crystal HPGe. A small (1 cm^3) dead region (in red) has been introduced on the upper side. The simulation has been run for different position of the dead region from P1 to P6, b) The 2D scanned image when the dead region is placed at P1. The projections of the slices of 2D image on c) upper portion where all histograms superposed, d) lower portion which has been shown up with lower efficiency due to presence of dead region.

In this scenario, a 2D scanned image of the front face (refer to Fig. 4(b)) was reconstructed by the GAGG:Ce detector in coincidence with the HPGe detector. The X-projections of a slice on the upper side of this 2D image, considering different locations of the defective region, are displayed in Fig. 4(c). It's noteworthy that all the projected histograms overlap, signifying the absence of a dead region on the lower portion of the HPGe crystal. However, the projected histograms of the bottom slices (Fig. 4(d)) reveal deviations in efficiency, indicating the presence of a dead region within the upper volume of the HPGe crystal. The method allows for the identification of such deviations by adjusting gates on all sides. These histograms also highlight that the described method possesses a sensitivity limit for detecting the existence of a defective region, approximately 1 cm^3 in volume, within the crystal. This limitation arises due to the absorption profile of 511 keV γ-rays within the crystal. Based on the present GEANT4 results, the sensitivity limit extends up to a depth of 40 mm from the crystal's entry surface. This limit can be adjusted depending on the density and material of the crystal. For larger crystals, scanning different surfaces may be required to surpass this limit. Notably, from the experimental 2D image, no defective regions were identified within the used single-crystal HPGe detector.

V. Conclusions

A single-crystal coaxial HPGe detector has been scanned by a position-sensitive GAGG:Ce detector. The 2D scanned images have been obtained with an energy gate of 511 keV. The simulated scanned images are consistent with the experimentally measured images. Due to some low amplitude noise of the anode signals and random coincidence, the experimental image has random background throughout the crystal. The incident direction of the γ-ray in the HPGe crystal is well defined by a position gate on a sector on the GAGG:Ce detector. Irrespective of the nature of the interaction, the first interaction points will lie along a defined path. In the GEANT4 simulation, a defective region has been introduced within the HPGe crystal. Coincidence images have been generated for varying locations of this defective region. The defective region is clearly discernible in the corresponding image as it exhibits reduced efficiency, which is further reflected in the projection of the simulated 2D spectrum. Analysis of the experimentally obtained images reveals uniform efficiency around the center, indicating the absence of any defective regions within the employed single-crystal HPGe. However, the simulation results suggest that this method has a sensitivity limitation for identifying defective regions, typically in the vicinity of 1 cm^3 in volume, within the Ge crystal. This limitation is primarily restricted to a depth of 40 mm from the surface but may vary when applied to different crystal types, making it a subject of ongoing interest.

Acknowledgment

Authors would like to thank the funding agency, Department of Atomic Energy (DAE), Government of India (Project Identification No. RTI 4002) and Department of Science and Technology (DST), Government of India (project No. IR/S2/PF-03/2003-II), respectively. B.D would also like to thank Infosys-TIFR Leading Edge Travel Grant.

References

[1] R. W. Todd, J. M. Nightingale, and D. B. Everett, "A proposed γ camera", Nature **251**, 132–134, (1974).

[2] S. Jan, G. Santin, D. Strul, S. Staelens, K. Assié, et al., "GATE: a simulation toolkit for PET and SPECT," Phys. Med. Biol., vol. **49**, no. 19, pp. 4543–4561, (2004).

[3] S. R. Cherry, T. Jones, J. S. Karp, J. Qi, W. W. Moses, and R. D. Badawi, "Total-Body PET: Maximizing Sensitivity to Create New Opportunities for Clinical Research and Patient Care", J. Nucl. Med., vol. **59**, no. 1, pp. 3–12, 2018.

[4] P. Moskal, K. Dulski, N. Chug, et al., "Positronium imaging with the novel multiphoton PET scanner", Sci. Adv. 2021; 7: eabh4394.

[5] T. Tanimori, Y. Mizumura, A. Takada, *et al.*, "Establishment of Imaging Spectroscopy of Nuclear Gamma-Rays based on Geometrical Optics", Sci. Rep. **7** 41511 (2017).

[6] E. K. Leung, M. S. Judenhofer, S. R. Cherry, and R. D. Badawi, "Performance assessment of a software-based coincidence processor for the EXPLORER total-body PET scanner", Phys. Med. Biol., vol. **63**, no. 18, p. 18NT01, 2018.

[7] Terry Jones and David Townsend, "History and future technical innovation in positron emission tomography", Journal of Medical Imaging, vol. **4**, no. 1, p. 011 013, Mar. 2017.

[8] N. A. Mullani, J. Gaeta, K. Yerian, W. H. Wong, R. K. Hartz, et al., "Dynamic Imaging with High Resolution Time-of-Flight PET Camera-TOFPET I", IEEE Transactions on Nuclear Science **31**, 1, 609-613 (1984).

[9] P. Moskal, D. Kisielewska, C. Curceanu, E. Czerwiński, K. Dulski, A. Gajos, et al., "Feasibility study of the positronium imaging with the J-PET tomograph", Phys. Med. Biol., vol. 64 no. 5, 055017 (2019).

[10] G. S. Li, C. Lizarazo, J. Gerl, I. Kojouharov, H. Schaffner, M. Górska, N. Pietralla, S. Saha, M. L. Liu, and J. G. Wang, "Simulated characteristics of the DEGAS γ-detector array", Nucl. Instrum. Methods A **890** (2018) 148-154.

[11] M. Rudigier, Zs. Podolyák, P. H. Regan, A. M. Bruce, S. Lalkovski, R. L. Canavan, E. R. Gamba, O. Roberts, I. Burrows, D. M. Cullen, L. M. Fraile, L. Gerhard, J. Gerl, M. Gorska, A. Grant, J. Jolie, V. Karayonchev, N. Kurz, W. Korten, I.H. Lazarus, C. R. Nita, V. F. E. Pucknell, J.-M. Régis, H. Schaffner, J. Simpson, P. Singh, C. M. Townsley, J. F. Smith, and J. Vesic, "FATIMA — FAst TIMing Array for DESPEC at FAIR", Nucl. Instrum. Methods A **969** (2020) 163967.

[12] S. Akkoyun, et al., "AGATA—Advanced GAmma Tracking Array", Nucl. Instrum. Meth. A **668**, 26 (2012).

[13] J. Simpson, "The Euroball Spectrometer", Z. Phys. A **358**, 139 (1997).

[14] I. Y. Lee, "The GAMMASPHERE", Nucl. Phys. A **520**, 641 (1990).

[15] R. Palit, "Recent results from digital INGA at BARC–TIFR Pelletron Linac Facility and future plans", Pramana **83**, 719-728 (2014).

[16] I. Y. Lee and J. Simpson, "AGATA and GRETA: The Future of Gamma-Ray Spectroscopy", Nucl. Phys. News Intl. **20**, 23-28 (2010).

[17] J. Eberth, et al., "MINIBALL: A Ge detector array for radioactive ion beam facilities", Prog. Part. Nucl. Phys. **46**, 389 (2001).

[18] J. Ródenas, A. Pascual, I. Zarza, V. Serradell, Josefina Ortiz, and L. Ballesteros, "Analysis of the influence of germanium dead layer on detector calibration simulation for environmental radioactive samples using the Monte Carlo method", Nucl. Instrum. Methods A **496** (2003) 390–399.

[19] A. Elanique, O. Marzocchi, D. Leone, L. Hegenbart, B. Breustedt, and L. Oufni, "Dead layer thickness characterization of an HPGe detector by measurements and Monte Carlo simulations", Appl. Radiat. Isot. **70** (2012) 538–542.

[20] E. Andreotti, M. Hult, G. Marissens, G. Lutter, A. Garfagnini, S. Hemmer, K. von Sturm, "Determination of dead-layer variation in HPGe detectors", Appl. Radiat. Isot. **87** (2014) 331–335.

[21] Truong Thi Hong Loan, Vu Ngoc Ba, Truong Huu Ngan Thy, Huynh Thi Yen Hong, Ngo Quang Huy, "Determination of the dead-layer thickness for both p- and n-type HPGe detectors using the two-line method", Journal of. Radioanal. Nucl. Chem. **315** (2018) 95–101.

[22] Mikael Hult, Stef Geelen, Mark Stals, Guillaume Lutter, Gerd Marissens, Heiko Stroh, Sonja Schreurs, Wouter Schroeyers, Michel Bruggeman, Leen Verheyen, "Determination of homogeneity of the top surface deadlayer in an old HPGe detector", Appl. Radiat. Isot. **147** (2019) 182–188.

[23] Ren-yuan Zhu, "Radiation damage in scintillating crystals", Nucl. Instrum. Methods A **413** (1998) 327-311.

[24] B. De Canditiis,G. Duchene, M. H. Sigward, M. Filliger, F. Didierjean, M. Ginsz, D. Ralet, "Full-volume characterization of an AGATA segmented HPGe gamma-ray detector using a ^{152}Eu source", Eur. Phys. J. A **57** (2021) 223.

[25] F. C. L. Crespi, F. Camera, B. Million, M. Sassi, O. Wieland, A. Bracco, "A novel technique for the characterization of a HPGe

detector response based on pulse shape comparison", Nucl. Instrum. Methods A **593** (2008) 440–447.

[26] L. Lewandowski, P. Reiter, B. Birkenbach, B. Bruyneel, E. Clement, J. Eberth, H. Hess, C. Michelagnoli, R.M. Perez-Vidal, and M. Zielinska, "Pulse-Shape Analysis and position resolution in highly segmented HPGe AGATA detectors", Eur. Phys. J. A (2019) **55**: 81.

[27] B. De Canditiisa, G. Duchêne, "Simulations using the pulse shape comparison scanning technique on an AGATA segmented HPGe gamma-ray detector", Eur. Phys. J. A **56** (2020) 276.

[28] Biswajit Das, R. Palit, R. Donthi, A. Kundu, Md. S. R. Laskar, P. Dey, D. Negi, F. S. Babra, S. Jadhav, B. S. Naidu, and A. T. Vazhappilly, "Development of a position-sensitive fast scintillator ($LaBr_3$(Ce)) detector setup for gamma-ray imaging application", EPJ Web Conf. **253** (2021) 11005.

[29] Biswajit Das, A. Kundu, R. Palit, V. Malik, P. Dey, D. Negi, S.K. Jadhav, A. T. Vazhappilly, and B. S. Naidu, "Development of a GAGG(Ce)-based compact 3D scanning setup for assessment of active volume in γ-ray detectors", Nucl. Instrum. Meth. A **1048** (2023) 167928.

[30] NNDC, 2022, http://www.nndc.bnl.gov/nudat2.

[31] TIFR homepage URL http://www.tifr.res.in/~pell/lamps.html.

[32] Root-CERN homepage URL https://root.cern/.

[33] Geant4-CERN homepage URL https://geant4.web.cern.ch/.

Active Collimator for the BGO Anti-Compton Shield of the HPGe Clover Detector

Biswajit Das[1,*], N. R. Khan Chowdhury[1], S. Saha[2], Md. R. S. Laskar[1], F. S. Babra[1], V. Malik[1], A. Kundu[1], P. Dey[1], S. K. Jadav[1], B. S. Naidu[1], A. T. Vazhappilly[1], S. Sihotra[3], and R. Palit[1,*]

[1]Department of Nuclear and Atomic Physics
Tata Institute of Fundamental Research, Mumbai - 400005, India
[2]School of Advanced Sciences, VIT University, Vellore-632014, India
[3]Department of Physics, Panjab University, Chandigarh-160014, India

(*) palit@tifr.res.in / biswajit.das@tifr.res.in

***Abstract*—An active collimator has been designed to replace the heavy metal collimator of the BGO anti-Compton shield to improve the performance of the hybrid Indian National Gamma Array (INGA) at TIFR, Mumbai. The active collimators of the 24 clover detectors, along with the BGO anti-Compton shields and 18 $LaBr_3(Ce)$ detectors, can act as a γ-ray multiplicity filter. This configuration will be beneficial for studying weak band structures decaying by high γ-ray multiplicity events. In addition, the energy deposited in all the detectors can be added to get the sum energy for certain measurements aiming to use the hybrid array for calorimetric application. We have characterized an active collimator coupled with one of the anti-Compton shields and an HPGe clover detector with various radioactive sources. The peak-to-total ratio for the active collimator configuration is similar to the standard design with the heavy metal configuration with the ^{60}Co source. Additionally, we have evaluated the Compton cross-talk between the clover HPGe detector and the anti-Compton shield/active collimator. The total multiplicity distribution for a cascade of eleven γ-rays with an energy range of 70-2000 keV has been obtained through GEANT4 simulation. The simulation results indicate the possibility of conducting a meaningful study of decay chains with a multiplicity of twelve.**

***Keywords* — γ-ray detector development, design and characterization of an active collimator; benefits for higher multiplicity decay study in INGA; GEANT4 simulation**

I. Introduction

Gamma-ray spectroscopy continues to give new insight into the nuclear structure and remains a topic of interest over the decades [1-5]. The High Purity Germanium (HPGe) detector, having very good energy resolution is a powerful instrument for discrete γ-ray spectroscopy and remains the best apparatus for nuclear structure and astrophysics studies by making of HPGe array [6-11]. A hybrid array of Compton-suppressed Clover HPGe detectors and Cerium doped Lanthanum Bromide ($LaBr_3(Ce)$) based scintillator detectors is operational in the BARC-TIFR Pelletron Linac Facility at TIFR Mumbai for nuclear structure studies [12-15]. The array has been designed to have a maximum of 24 Compton-suppressed clover detectors and 18 $LaBr_3(Ce)$ detectors. A combination of clover HPGe and $LaBr_3(Ce)$ detectors array can extend the lifetime measurement from 100 pico-second to a few nano-second by the triple-fold γ-ray coincidence [16]. A heavy metal collimator made of tungsten has been placed in front of the BGO anti-Compton shields to prevent the direct hit of the γ-rays emitted from the target positions during the experiment. This heavy metal collimator reduces the false vetoes arising from an event in a reaction with high γ-ray multiplicity, where one γ-ray strikes an HPGe crystal giving a full-energy pulse, and another hits the BGO shield detector. But, the main concern of the HPGe detector is, it suffers from low peak-to-total ratio. A mono-energetic γ-ray spectrum consists of a narrow full-energy peak and a slowly varying background arising from the Compton scattering in the HPGe crystal of its surroundings. The Compton background of all the γ-rays makes it difficult to hunt for low energy γ-rays sitting on the accumulated Compton backgrounds from high-energy peaks. The Compton background is generally reduced by placing an escape-suppression shied around the Ge detector [17,18]. Events were vetoed when scattered gammas escaped from the Ge crystal and were detected in the surrounding scintillation BGO detector providing a signal. Even with a large Ge crystal and active Compton suppression, there will be residual background from incomplete suppression plus photons scattered from adjacent materials to the Ge crystal [19]. The resolving power is the overall capacity of the array to permit the identification of weak cascades of γ-rays in a complex spectrum. In order to increase the detection inefficiencies for the study of weak band structures having high multiplicity gamma decay events, the replacement of heavy metal to active collimator is our main interest for the present work.

In this work, the heavy metal collimator of the BGO anti-Compton shield of a clover detector has been replaced by an active collimator of BGO crystal to improve the performance of the hybrid Indian National Gamma Array (INGA) at TIFR, Mumbai. We have designed and tested one of the anti-Compton shields coupled with the active collimator and an HPGe clover detector with various radioactive sources. The benefits of the use of active collimators for higher multiplicity events in INGA setup have been demonstrated through GEANT4 simulation.

II. Instrumentation and Experimental Setup

The active collimator detector consists of Bismuth Germanium Oxide (BGO) crystal and has been manufactured by SCIONIX [20]. The choice of BGO crystal as an active collimator is because of its high density (7.12 gm/cm^3), very good radiation hardness (the parameters remain stable for high radiation dose [21]), high scintillation efficiency, cost-effectiveness, non-hygroscopic, and mechanically strong. The BGO crystal produced ~9000 photons per 1 MeV gamma energy deposition. The crystal emits scintillation photons of a wavelength range of 375-650 nm with a peak at 480 nm [21]. The active collimator consists of 8 BGO crystals, which are coupled with same type of Photo Multiplier Tubes (PMT: Hamamatsu R9880-10) fabricated by SCIONIX, Holland [20]. Each BGO crystal has a cross-sectional dimension of 3.35 cm × 1.7 cm. The whole assembly of the BGO crystals and PMTs of the active collimator are encapsulated within an aluminum encasing having a thickness of 0.5 mm. A schematic diagram of the cross-sectional, side view, with Al housing and actual image, have been shown in Fig. 1.

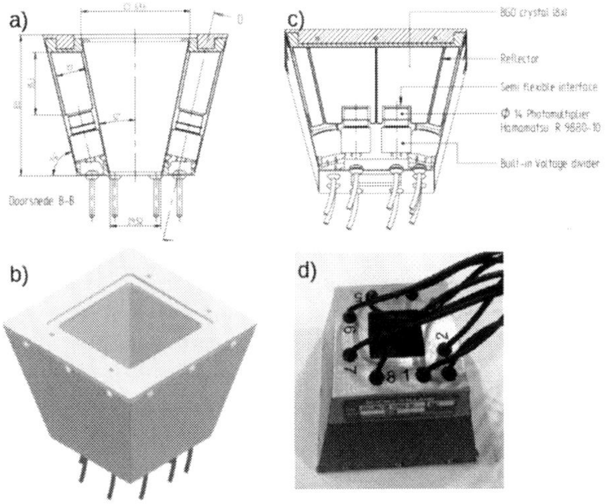

Fig. 1. Schematic of active collimator: a) cross-sectional middle view, b) one side view, c) with Al housing. d) Ensembles active collimator used in the experiment.

The detector has 8-channel outputs, which are further converted to one from OR condition. The BGO crystals have a small self-activity caused by minor contamination of the ^{207}Bi. However, few BGO crystals incidentally have higher internal radioactive background with activity up to 10 Bq/kg [22]. This background contribution is because of the ^{210}Po contamination and has a technological origin. The Clover HPGe detector has been surrounded by BGO anti-Compton shield. A collimator made of heavy metal in front of each BGO anti-Compton shield has been placed to stop the direct hit of γ-rays in the BGO crystal. In this experiment, the set-up has been made with two configurations, one is with heavy metal collimator, and other one is with a BGO active collimator detector. The geometry of the setup with HPGe clover, active collimator, and BGO anti-Compton shield detectors has been shown in Fig. 2. The PMTs of the active collimator have been biased at +900 V through a voltage divider. The gains of all the photo-multiplier tubes within the BGO annulus are matched identically by adjusting the high voltage of each individual PMT. This is accomplished by placing a mono-energetic γ-ray source (^{137}Cs) on each side of the BGO annulus. We have used the standard radioactive source like ^{241}Am, ^{57}Co, ^{137}Cs, and ^{54}Mn for energy calibration of the BGO anti-Compton shield and active collimator. A ^{60}Co source has been used for further characterization of the active collimator. We used weak sources to avoid pile-up and other count rate dependent effects. The detector noise has also been checked for the BGO element by using a ^{241}Am source which emits a primary γ-ray at 59.4 keV and the threshold is safely set at 15-20 keV. The experimental raw signals have been processed and digitized with a XIA-LLC based Digital Data Acquisition System with a 12-bit, 100 MHz, PIXIE-16 module connected in a PCI/PXI crate. The γ-ray energy and time information have been extracted by pulse processing through DAQ and were transferred to a host PC by a controller module connected in the same PCI/PXI crate. The experimental data have been acquired by a software, TIFR Digital Data Acquisition Software for Nuclear Structure Studies (TIDES) [23].

III. GEANT4 Simulation

The GEometry ANd Tracking 4 (GEANT4, version 4.10.05) simulation toolkit has been used for making the geometry and Monte-Carlo calculation [24]. G4Sphere, G4Box, G4Tubs, G4Trd, G4Polycone, G4Polyhedra, G4UnionSolid, and G4SubtractionSolid classes have been used to construct the different parts of the detector and the set-up. The GEANT4 geometry of one of the HPGe clover detectors with active collimator and BGO shield has been shown in Fig. 2. The simulation of the INGA with 24 Compton-suppressed HPGe clover detectors has been thoroughly studied by Saha, et al [14]. Later on, the simulation code has been extended by adding 18 $LaBr_3$(Ce) detectors with INGA to make it hybrid. The placement of $LaBr_3(Ce)$ and clover HPGe detectors can be found in the Ref. [15,25]. The 24 BGO-shielded HPGe clover detectors have been arranged at different angles with respect to the beamline along with the 18 $LaBr_3(Ce)$ detectors ($2'' \times 2''$) in 4π geometry. We have simulated the responses of a Compton-suppressed clover detector with an active collimator. In the simulation code, we have replaced all the heavy metals with the active collimators in order to study the combined performance of the hybrid INGA. In PhysicsList of GEANT4 simulation, we have included the classes of all possible interaction mechanisms of γ-ray, electron and positron with matter. The γ-ray photons, electrons, and positrons are expected to be produced on the path of γ-ray while interacting with the matter. The considered interaction mechanisms for the γ-photons are photoelectric effect, Compton scattering, pair production, and Rayleigh scattering. The electrons and positrons which produced during interaction can undergo multiple scatterings, ionization, and emit Bremsstrahlung radiation. In addition, the positron can undergo annihilation with the material's electrons. For γ-ray interaction with matter, the considered process classes have been used in physics list are G4PhotoElectricEffect, G4ComptonScatte-ring, G4Gamma-Conversion, and G4RayleighScattering.

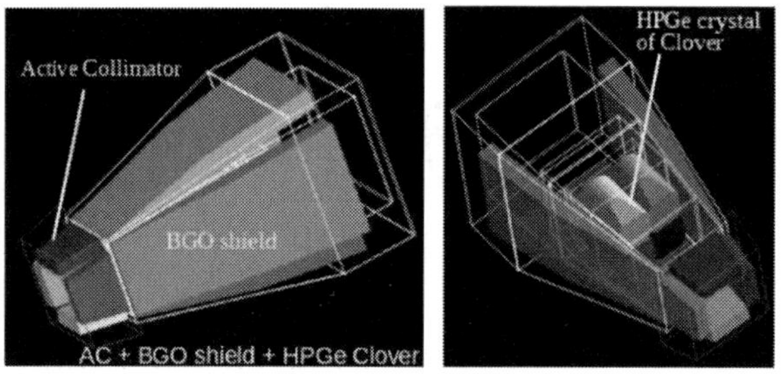

Fig. 2. GEANT4 Geometry of one of the HPGe clover detectors with an active collimator and BGO anti-Compton shield (left), and the same with one BGO slab taken out to show the internal part of the detector setup.

IV. Analysis and Result

The characterization of the BGO as an anti-Compton shield of HPGe clover detector in INGA setup has been studied in our previous work [14]. Here, we have studied the effect of γ-ray detection performance while replacing the heavy metal collimator with an active collimator. Also, how this replacement will be effective for a large array of HPGe clover detectors has been simulated. The experimental data have been sorted by a MultipARameter time-stamped based COincidence Search (MARCOS) [12,26] code, developed at TIFR, and used to generate the matrix. The energy calibration of the BGO anti-Compton shield and active collimator has been done with the single γ-ray energy emitting radioactive sources. The photo-peak energies were plotted against their corresponding channel numbers and then fitted with a quadratic function. The calibration fits are depicted and the calibration parameters are tabulated. The responses of the crystals have been found almost linear with the γ-ray energy. The energy resolution of the active collimator has been found to be ~18.8 % at 662 keV. We have tested the performance of the HPGe clover detector with three different configurations, i) HPGe+BGO Compton shield, ii) HPGe + BGO Compton shield + heavy metal, and iii) HPGe + Compton shield + active collimator. The γ-ray energy spectrum of ^{60}Co in the third configuration has been shown in Fig. 3. The Peak-to-total ratios have been obtained (Table-I) in anti-coincidence mode along with the BGO annular shield.

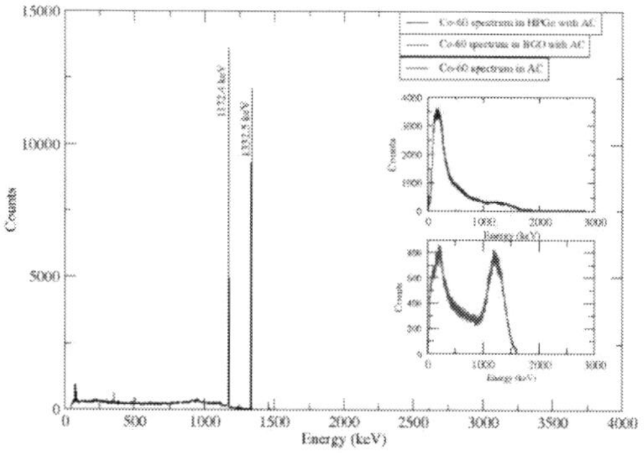

Fig. 3. Measured γ-ray spectrum of ^{60}Co in the HPGe Clover detector (shown in black). In inset, γ-ray spectrum of: BGO anti-Compton shield (blue) and active collimator (red).

Table-I: The performance of the different configurations of the setup with ^{60}Co point source has been placed at the target position.

Configuration	P/T ratio
HPGe (no veto)	0.224
HPGe + BGO (without HM)	0.343
HPGe + BGO + HM	0.429
HPGe + BGO + AC	0.439

The table demonstrates that the replacement has not had an impact on the detection capability of the HPGe clover detector in single-mode measurements. However, it does prove effective in coincidence measurements for improving detection efficiency, especially in scenarios involving higher-order multiplicity events or cascade gamma decay. This enhancement results in an increased coincidence efficiency. Figure 4 displays the coincidence spectrum of 60Co using an HPGe clover detector configuration in conjunction with one active collimator/BGO shield. The coincidence spectrum of ^{60}Co obtained between active collimator/BGO anti-Compton shield and one HPGe clover detector configuration has been shown in Fig. 4(inset). This indicates that there is an impact of active collimation in the coincidence measurements.

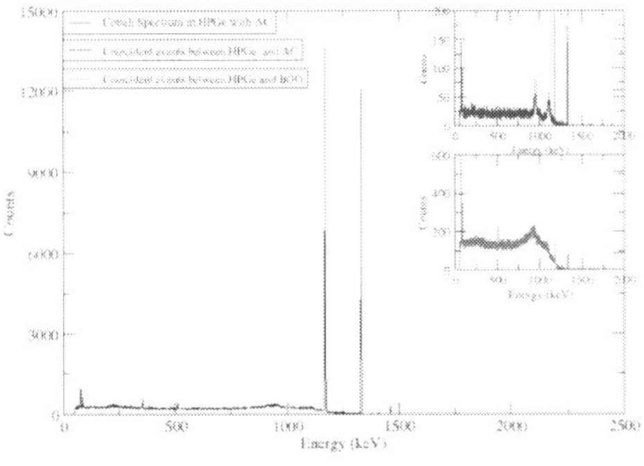

Fig. 4. The ^{60}Co γ-ray spectrum of HPGe clover detector (shown in black). In inset, the HPGe clover γ-ray spectrum in coincidence with Active Collimator (blue), in coincidence with BGO anti-Compton shield (red).

In the simulation, we have generated a cascade of 11 γ-rays in the range of 70 keV to 2 MeV. The multiplicity has been obtained from the simulation in INGA setup consisting of 24 Compton suppressed HPGe clover,18 $LaBr_3(Ce)$ detectors with 24 active collimators has been shown in Fig.5.

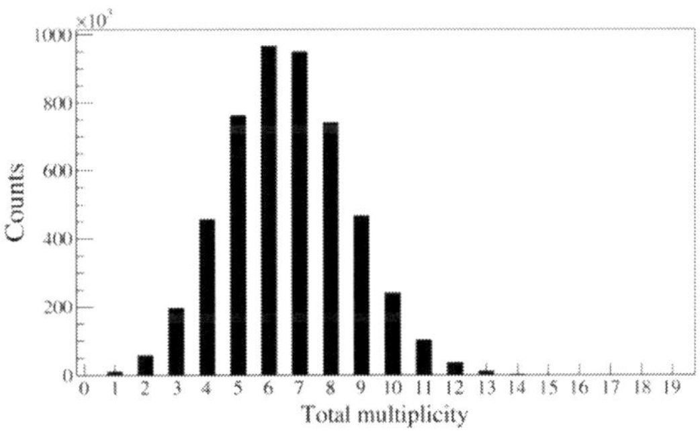

Fig. 5. Simulated multiplicity of the hybrid INGA consisting 24 HPGe clover, 24 BGO anti-Compton shield, 18 $LaBr_3(Ce)$, and 24 active collimator.

V. CONCLUSIONS

The performance of an active collimator with a HPGe clover detector surrounded by BGO Compton-shield has been designed and tested experimentally. The characterization of the active collimator has been done. The benefits of the presence of active collimators instead of heavy metal collimators has been discussed through GEANT4 simulation for a hybrid array consisting of 24 Compton-suppressed clover detectors and 18 $LaBr_3(Ce)$ detectors. With an inner layer of BGO shield detectors, we can not only use it as a veto circuit but also it allows a measure of the total γ-ray energy and the efficiency in higher order multiplicity of the γ-rays. The active collimator will be efficient for those nuclear γ-ray spectroscopy experiment where higher-order multiplicity gate is the matter of concern. We can use the inner active collimator to select events associated with long cascades of γ-rays, which are generally the cascades involving high spin states and increases the detection efficiency. The limitation of using the active collimator is the detection of γ-ray events in lower multiplicity of high energy, where there is a chance of Compton scattering of the γ-ray from active collimator and to HPGe clover crystal and hence the Compton background increases.

ACKNOWLEDGMENT

This work is supported by the Department of Atomic Energy (DAE) and Department of Science and Technology (DST), Government of India, under project No. 12-R&D-TFR-5.02-0200, and IR/S2/PF-03/2003-II, respectively. The author would also like to thank Infosys-TIFR Leading Edge Travel Grant.

REFERENCES

[1] P. Gorodetzky, J. J. Kolata, J. W. Olness, A. R. Poletti, and E. K. Warburton, "Gamma-Ray Spectroscopy in A ≈ 40 Nuclei via Heavy-Ion-Induced Reactions: High-Spin States in ^{40}K and ^{41}Ca", Phys. Rev. lett. **31**, 17 (1973).

[2] P. S. Haskins, F. E. Dunnam, R. L. Coldwell, A. C. Rester, R. B. Piercey, M. L. Muga, H. A. Van Rinsvelt, R. W. Smart, H. J. M. Aarts, J. D. Fox, L. C. Dennis, and C. B. Saw, "In-beam gamma-ray spectroscopy of ^{82}Sr", Phys. Rev. C **32**, 6 (1985).

[3] J. A. Cameron, M. A. Bentley, A. M. Bruce, R. A. Cunningham, W. Gelletly, H. G. Price, J. Simpson, D. D. Warner, and A. N. James, "High-spin gamma spectroscopy of recoil-separated ^{49}Cr, ^{49}V, and ^{46}T", Phys. Rev. C **44**, 5 (1991).

[4] A. Gade, R. V. F. Janssens, D. Bazin, P. Farris, A. M. Hill, S. M. Lenzi, J. Li, D. Little, B. Longfellow, F. Nowacki, A. Poves, D. Rhodes, J. A. Tostevin, and D. Weisshaar, "In-beam γ-ray spectroscopy of ^{62,64}Cr", Phys. Rev. C **103**, 014314 (2021).

[5] P . Dey, D. Negi, R. Palit, P. C. Srivastava, Md. S. R. Laskar, B. Das, F. S. Babra, S. Bhattacharya, Biswajit Das, K. Rojeeta Devi, R. Gala, U. Garg, S. S. Ghugre, E. Ideguchi, S. Kumar, A. Kundu, G. Mukherjee, S. Muralithar, S. Nag, S. Nandi, Neelam, M. Kumar Raja, R. Raut, R. Santra, A. Sharma, S. Sihotra, A. K. Singh, R. P. Singh, and T. Trivedi, "Experimental investigation of high-spin states in ^{90}Zr", Phys. Rev. C **105**, 044307 (2022).

[6] S. Akkoyun, et al., "AGATA—Advanced GAmma Tracking Array", Nucl. Instrum. Meth. A **668**, 26 (2012).

[7] R. Palit, "Recent results from digital INGA at BARC–TIFR Pelletron Linac Facility and future plans", Pramana **83**, 719-728 (2014).

[8] C. Louchart-Henning, G. Henning, J. Gerl, N. Pietralla, and the DESPEC collaboration, "Simulations for DEGAS detectors at FAIR", GSI Scientific Report, NUSTAR-GS-05 (2013).

[9] I. Y. Lee, "The GAMMASPHERE", Nucl. Phys. A **520**, 641 (1990).

[10] J. Simpson, "The Euroball Spectrometer", Z. Phys. A **358**, 139 (1997).

[11] J. Eberth, G. Pascovici, H. G. Thomas, N. Warr, D. Weisshaar, D. Habs, & Miniball Collaboration. "MINIBALL A Ge detector array for radioactive ion beam facilities". Progress in Particle and Nuclear Physic **46**(1), 389-398 (2001).

[12] R. Palit, S. Saha, J. Sethi, T. Trivedi, S. Sharma, B. S. Naidu, S. Jadhav, R. Donthi, P. B. Chavan, H. Tan, and W. Hennig, "A high speed digital data acquisition system for the Indian National Gamma Array at Tata Institute of Fundamental Research", Nucl. Instrum. Methods Phys. Res. A **680**, 90 (2012).

[13] R. Palit, S. Saha, J. Sethi, T. Trivedi, B.S. Naidu, P.B. Chavan, R. Donthi, S. Jadhav, "Performance of Indian National Gamma Array (INGA) Coupled with a Fast Digital Data Acquisition System for Nuclear Structure Studies", Journal of Physics: Conference Series **420** (2013) 012159.

[14] S. Saha, R. Palit, J. Sethi, S. Biswas and P. Singh, "Geant4 simulation study of Indian National Gamma Array at TIFR", JINST **11**, P03030 (2016).

[15] Biswajit Das, S. Saha, A. Kundu, Md S. R. Laskar, P. Dey, R. K. Santra, D. Negi, F. S. Babra, R. Donthi, S. Jadhav, B. S. Naidu, A. T. Vazhappilly, S. Sihotra, R. Palit, "Measured and Geant4 Simulation Performance of the Hybrid Array of Compton Suppressed HPGe Clover and $LaBr_3$(Ce) Detectors at TIFR", Jl. of the Instrum. Soc. of India, **51**, 1-4 (2021) ISSN No. 0970-9983.

[16] Md. S. R. Laskar, R. Palit, E. Ideguchi, T. Inakura, S. N. Mishra, F. S. Babra, S. Bhattacharya, D. Choudhury, Biswajit Das, B. Das, P. Dey, U. Garg, A. K. Jain, A. Kundu, D. Kumar, D. Negi, S. C. Pancholi, S. Rajbanshi, and S. Sihotra, "Enhanced B(E3) strength observed in ^{137}La", Phys. Rev. C **104**, L011301 (2021).

[17] M. P. Carpenter, T. L. Khoo, I. Ahmad, R. V. F. Janssens, T. Lauritsen, G. A. Annan, A. M. Baxter, M. E. Bleich, S. Harfenist, E. F. Moore, M. Torelli, D. W. Visser, "Test and performance of a BGO Compton-suppression shield for GAMMASPHERE", Nucl. Instrum. Methods A **353**, 2(1994).

[18] F. Camera, A. BraCCO, M. Mattiuzzi, G. LoBianCo, M. Pignanelli, J. J. Gaardhoje, I. G. Bearden, Z. Zelazny, A. Maj, "The response of a HPGe detector with a BGO shield to high energy γ-rays" Nucl. Instrum. Methods A **351**, 401-405 (1994).

[19] S. K. Mattews and L. E. Ussery, "The performance ofa single-crystal BGO Annulus as a Compton-Suppression detector", **LA-UR-91-3443 DE92**,003872.

[20] SCIONIX, Holland webpage, http://scionix.nl/.

[21] Peter Kozma and Petr Kozma Jr, "Radiation resistivity of BGO crystals due to low-energy gamma-rays", Nucl. Instrum. Methods A **501**, 449 (2003).

[22] D. N. Grigoriev, V. F. Kazanin, G. N. Kuznetcov, I. I. Novoselov, P. Schotanus, B. M. Shavinski, S. N. Shepelev, V. N. Shlegel, Ya. V. Vasiliev, "Alpha radioactive background in BGO crystals", Nucl. Instrum. Methods A **623**, 999 (2010).

[23] Md. S. R. Laskar, Abraham T Vazhappilly, R. Palit, S. Jadhav, R. Donthi, B. S. Naidu, F. S. Babra, P. Dey, Biswajit Das, B. Das, C. S. Palshetkar, L. P. Singh, and H. Tan, "TIFR Digital Data Acquisition Software for Nuclear Structure Studies (TIDES)", Proceedings of the DAE Symp. on Nucl. Phys. **63** (2018).

[24] S. Agostinelli et al. And GEANT4 collaboration, "GEANT4: a simulation toolkit", Nucl. Instrum. Meth. A **506**, 250 (2003).

[25] Biswajit Das, S. Saha, R. Palit, A. Kundu, P. Dey, V. Malik, S. K. Jadav, B. S. Naidu, A. T. Vazhappilly, and S. Sihotra, "Cross-talk estimation of the hybrid γ-ray detector array through GEANT4 simulation", Proceedings of the DAE Symp. on Nucl. Phys. **66** (2022) 1082.

[26] R. Palit and S. Saha, "Investigation of exotic modes of spinning nuclei near ^{90}Zr", Pramana **82** (2014) 649.

Analog Pulse Shape Discrimination Circuit for High Event Rates and Fast Scintillators with a Dynamic Deadtime

Itai Morad[1*], Max Ghelman[2], Alon Osovizky[2,3], Amir Ellenbogen[2], Rami Seif[2], Eran Vax[2], Amir Broide[2], and Ron Harn[2]
[1]Israel Atomic Energy Commission (IAEC),
[2]Nuclear Research Center Negev (NRCN),
[3]Radiation Detection Department, Rotem Industries Ltd
(*) itaimorad1111@gmail.com

***Abstract*—We developed an analog pulse shape discrimination (PSD) topology based on the well-established charge integration (CI) method, featuring two novel functional blocks beneficial for high event rate operation. The topology is designed for high-speed scintillators. The demonstrated analog design is potentially better suited than digital methods, when considering both processing time and power consumption aspects. The topology was tested using both experimental alpha and beta pulses from a plastic scintillator with a layer of ZnS(Ag) coupled to a PMT, and a fast digital emulator to simulate controlled high event rate scenarios. The discrimination capabilities of the topology were optimized and evaluated using a traditional figure of merit (FoM) approach. The topology achieved over 99% correct classifications when evaluated using the experimental pulses recorded. Additionally, the dedicated blocks resulted in a fourfold reduction in miss-classifications of slow pulses at an event rate of 100 kcps of fast pulses, while also providing a dynamic deadtime proportional to the pulse charge.**

***Keywords*—pulse shape discrimination, high rate, charge integration, analog circuit, plastic scintillator.**

I. Introduction

THE term PSD is a general name for techniques used to discriminate between different types of radiation based on differences between the nuclear pulse shapes. PSD can be applied to detectors of ionizing radiation with the ability to interact with multiple kinds of radiation. It is often important to distinguish different types of radiation for homeland security, nuclear safeguard, and neutron scattering applications [1]–[4]. PSD can either be performed directly on the detector signal in the analog domain, or on the sampled detector signal in the digital domain. PSD methods realized in the analog and digital domain vary in their discrimination capabilities and power consumption. Traditional digital signal processing algorithms include slice-fitting, tail-sum, cosine similarity, histogram-difference, and Fourier analysis based methods [3]–[10]. In addition, recent advancements in the field of artificial intelligence have given rise to PSD techniques based on traditional machine learning [11]–[14] and neural networks [2], [15]–[20]. Digital PSD methods require high-speed analog to digital converters (ADCs) which increases power consumption. In addition, some of these algorithms are complex and require long processing times, resulting in a limited count rate and increasing deadtime. Performing PSD in the analog domain may be more suitable for applications with limited hardware that require low power consumption. This is particularly true when working with fast decay scintillators, especially for battery-based applications.

Discrimination capabilities of PSD techniques applied in the analog domain are limited to what is achievable with integrated circuits (ICs). Pulse height discrimination (PHD) techniques discriminate the pulses based on amplitude or charge [21], while pulse width discrimination (PWD) techniques such as time over threshold (ToT) discriminate the pulses based on their duration [22]. Such techniques are often simple to realize and suffer from a considerable number of miss-classifications in low signal-to-noise ratio (SNR). In addition, operation within high event rates may cause an additional number of miss-classifications due to pileup.

In this work, a novel analog PSD topology for fast scintillators is presented. Discrimination is based on the well-established digital CI method, providing superior discrimination capabilities compared to analog methods based on PHD and PWD. The topology introduces two dedicated blocks designed to improve neutron count rate when operating within high event rates. The double$^+$ counting rejection (D^+CR) block was designed to reject gammas arriving consecutively, resulting in fewer neutron miss-classifications in high gamma fluxes. Moreover, the dynamic deadtime (DDT) block was designed to produce a deadtime proportional to the energy of the pulse, lowering the total deadtime and improving detection efficiency.

Alpha and beta pulses interacting with a plastic and ZnS(Ag) scintillator coupled to a photomultiplier tube (PMT) were recorded and analyzed in order to design the analog topology. The topology was implemented on an FR-4 printed circuit board (PCB) designed for high frequency operation. Experiments with both radioactive sources and a fast digital emulator were carried out in order to assess the discrimination capabilities of the method. Results show excellent discrimination capabilities and successfully validate the contribution of the D^+CR block and the DDT block in high event rate scenarios.

The rest of this article is organized as follows. The experimental setup, data analysis, and the challenges associated with performing analog PSD for fast scintillators under high event rates are described in section II. section III provides a detailed description of the proposed PSD method. Discrimination capabilities of the proposed method under various conditions are displayed in section IV. Finally, we provide a summary and discuss some future possibilities in section V.

II. Data

A. Experimental Setup and Data Analysis

The experimental setup used is described as follows. Alpha and beta sources were placed in front of a plastic scintillator with a ZnS(Ag) layer. The alpha particles interact with the ZnS(Ag) layer, while the beta particles interact with the plastic. The scintillator was coupled to a PMT. 20,000 alpha and 20,000 beta pulse shapes were recorded using a fast digital oscilloscope. The amplitudes and decay constants of the recorded pulses were digitally fit using non-linear optimization and are scattered in Fig. 1. The amplitudes of both the alpha and beta pulses fall within a similar range; therefore, they cannot be discriminated using PHD methods. The fitted decay constants of beta pulses extend up to around 10 ns, while the fitted decay constants of alpha pulses range between approximately 10 ns and 70 ns, establishing a distinct separation of alpha and beta pulses within the decay constant domain. The variance of the fitted decay constants higher for low energy pulses due to SNR, which results in noisy fitting. Considering these pulse characteristics, the selection of the discrimination method employed in the topology is grounded in the well-established CI method, commonly utilized in the digital domain. Discrimination using the CI method relies significantly on the decay constant parameter, rendering it a suitable choice for our data set.

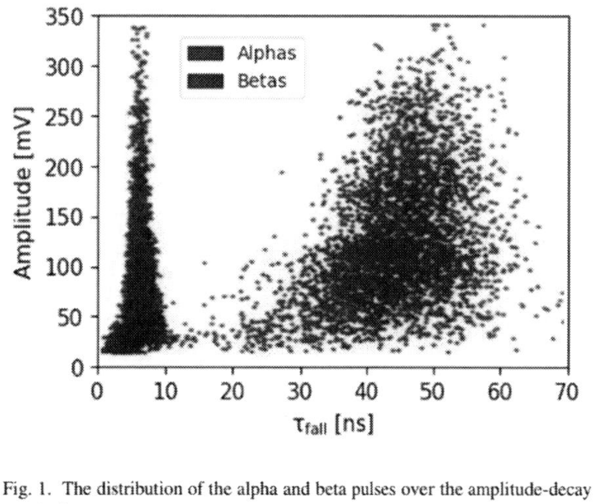

Fig. 1. The distribution of the alpha and beta pulses over the amplitude-decay constant span.

B. Challenges

Although the CI method is based on simple principles and has been shown to possess excellent PSD capabilities [23], implementing the method in an analog fashion for fast scintillators while operating in high event rates poses several challenges.

Challenge I: Fast scintillation responses. As can be seen in Fig. 1, scintillation responses of beta or gamma particles in plastic scintillators decay within several ns. Processing and discrimination of these responses demand high-speed components and careful analog design. Small timing jitters within the order of ns, and other timing uncertainties like walk or drift, highly affect the processing of these pulses.

Challenge II: Miss-classifications in high event rates. Discrimination based on the CI method is based on a ratio between two integral values referred to as the *PSD ratio*. In the case of two beta pulses arrive consecutively, the resulting PSD ratio is prone to resemble the PSD ratio of an alpha pulse, resulting in a miss-classification. This event is more likely to occur in high beta event rates. Applications such as neutron scattering demand accurate neutron counting, while the accuracy constraint for gamma counting is less strict [2]. Therefor, accurately counting neutron arrivals in high gamma fluxes is challenging.

Challenge III: deadtime in high event rates. A constant deadtime introduces a trade-off between counting accuracy and detection efficiency. For short deadtimes, long-lasting alpha scintillation responses will re-trigger the system multiple times, resulting in the classification of a single alpha pulse as multiple alpha pulses. For long deadtimes proportional to the length of the longest lasting alpha pulses, which are in the order of μs, events arriving during the deadtime will be rejected and missed [24].

III. Proposed Method

In this section, we introduce our analog PSD topology for fast scintillation responses which is based on the CI method. Discrimination relies on pulse energy, delineating three distinct regions where the most suitable method is applied for discrimination within each. Discrimination features are extracted in the block referred to as the *discrimination block*. The block where the output decision pulses are produced, based on the discrimination features, is referred to as the *decision block*.

As our design is geared towards high rates, two dedicated blocks for **Challenge II** and **Challenge III** are introduced. The D^{+}CR block is designed to improve classification accuracy in high rates by detecting and rejecting the event of two beta pulses arriving consecutively, which is prone to be miss-classified as an alpha. The DDT block is designed to improve detection efficiency by introducing a dynamic deadtime, proportional to the charge of the arriving pulses. Moreover, a varying constant deadtime is introduced in the discrimination block to further improve detection efficiency in high rates. A block diagram of the topology can be viewed in Fig. 2. The nuclear pulse arriving from the PMT is fed into the discrimination block, the D^{+}CR block, the DDT block, and the block referred to as the *control signals block*. The design of the control signals block aims to detect the incoming nuclear pulse and generate digital signals. These digital signals are instrumental in controlling the timing of various processes that take place within the other blocks, and

https://doi.org/10.1051/epjconf/202328810005

are therefore fed into the other blocks. The outputs from both the discrimination block and the D$^+$CR block are channeled into the decision block due to their direct relevance to the discrimination decision-making process. Next, we elaborate on the discrimination method and the blocks dedicated to improve operation in high rates.

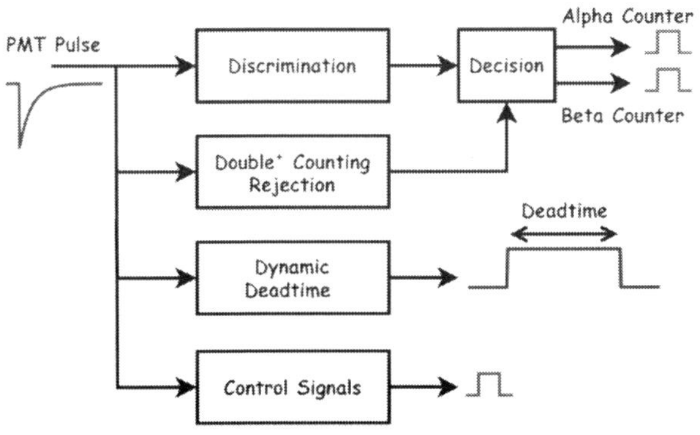

Fig. 2. Block diagram of the proposed method.

A. Discrimination Method

A scheme of the proposed PSD topology based on the CI method is shown in Fig. 3. The nuclear pulse is amplified and shaped. Then, the pulse is fed into two parallel channels referred to as the *partial channel* and the *full channel*, where the partial channel performs integration to compute $V_{partial}$ and the full channel performs integration to compute V_{full}. $V_{partial}$ and V_{full} are then fed into three comparators, which combined with logic gates discriminate the nuclear pulse as either an alpha or a beta. Two examples of signal shape at different nodes of the electronic topology for an alpha pulse and a beta pulse can be seen in Fig. 4. The digital signals produced by the control signals block which time the different components can be seen in Fig. 5.

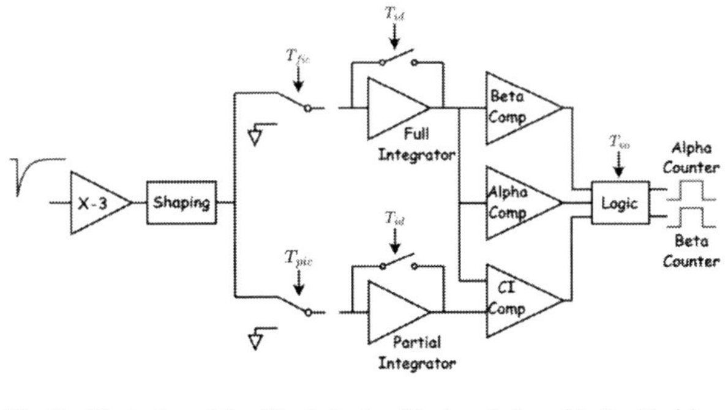

Fig. 3. Illustration of the Discrimination block and alongside the Decision block.

The nuclear pulse is fed from the detector to a high bandwidth amplifier. It is important for the pre-amplifier to have high amplification, the ability to recover from saturation, and large bandwidth. A sufficiently large bandwidth ensures the shape of the pulses is preserved and discrimination capabilities are not harmed. Therefore, the pre-amplifier should

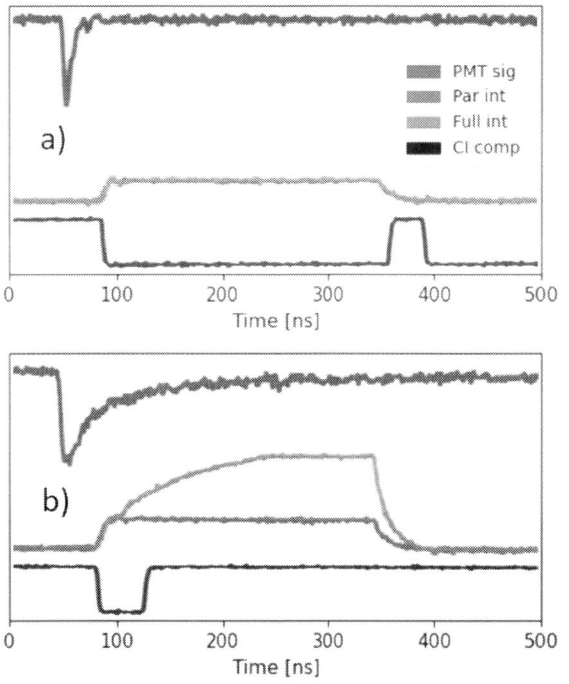

Fig. 4. Signal shape at different nodes of the discrimination block (a) for a beta pulse, (b) for an alpha pulse.

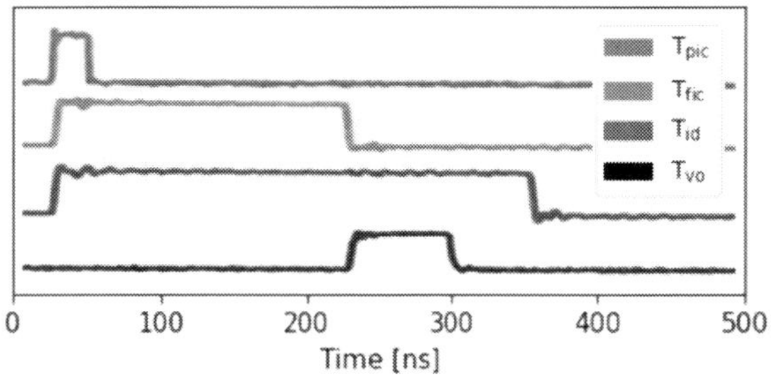

Fig. 5. Signal shape at different nodes of the control signals block. When T_{pic} is high the pulse is injected into the partial integration channel. When T_{fic} is high the pulse is injected into the full integration channel. When T_{id} is low both integrators are discharged. When T_{vo} is high the decision block is enabled to produce the output pulse.

be carefully selected, by referring to its bandwidth and noise level, compared to the detector's pulse shape and internal noise level. The amplifier output is fed into a detection channel and a classification channel.

The detection channel is responsible for both detection of the pulses and monitoring of various processes within the circuit. Pulse detection is carried out by feeding the amplifier output into a comparator level that generates a digital pulse. The threshold level is set as low as possible such that the detector noise does not cause false alarms. Monitoring of the various processes within the circuit is carried out by feeding the comparator digital signal into a chain of monostable multivibrators, which produce digital signals. Examples of these signals and their associated roles are depicted in X.Fig. 5. It can be seen that T_{pic} lasts 25 ns. Due to the lack of monostable multivibrators able to produce such short signals, a custom made topology mimicking the behavior of a monostable multivibrator utilizing a RC network, a comparator, and an AND logic gate was realized.

The classification c hannel i s r esponsible f or classification. First, the amplified p ulse i s f urther p rocessed a nd shaped. Then, it is fed into two channels, a partial integration channel and a full integration channel. Each channel consists of a single pole double throw (SPDT) analog switch responsible for feeding the pulse into the integrator during its respective integration period, and ground otherwise, an integrator, and a single pole single throw (SPST) analog switch responsible for discharging the integrator. There is a time delay introduced from the moment of detection up until the integrators begin integrating the pulse. The time delay is dictated by the propagation delay of the following components:

$$T_{\text{detection comparator}} + T_{\text{monostable multivibrator}} + T_{\text{SPDT switch}} \quad (1)$$

Due to **Challenge I**, by the time the pulse is fed into the integrator, fast scintillation responses that decay within several ns diminish and relevant information for discrimination is lost. To address this issue, the amplified pulse undergoes an additional delay and shaping process, ensuring the preservation of its information. This results in attenuation of the high frequency components, causing beta pulses to look more similar to alpha pulses, which increases the probability of miss-classification. A t rade-off b etween p reserving relevant information for discrimination and making alpha and beta pulses more similar is introduced.

The tuples of $V_{partial}$, V_{full} measured for alpha and beta pulses obtained during the experiment described in section II are scattered in Fig. 6. The PSD ratio is computed via (2) and is used for discrimination in the CI region visualized in Fig. 6 as the region where 0.3 V $< V_{full} <$ 1.7 VV.

$$PSD\ ratio = \frac{V_{full}}{V_{partial}} \quad (2)$$

Implementing analog division, which is needed in order to calcualte the PSD ratio in (2), often relies on the exponential characteristics of PN-junctions and requires several components. Fortunately, our task doesn't involve calculating the PSD ratio, but rather entails comparing it to a threshold to discriminate between alpha and beta pulses. This comparison can be effectively executed using the CI comparator. In this setup, the (-) input is supplied with $V_{partial}$, while the (+) input receives V_{full} after undergoing voltage division to yield $V_{full}\ /\ THR$.

However, not all pulses require the CI method. The cgharge of the pulse is directly proportional to V_{full}. As depicted in Fig. 6, it becomes evident that only beta pulses fall within the range of $V_{full} <$ 0.3 V (to the left of the blue dotted line), and thus, they are automatically discriminated by the Beta comparator without the necessity of the CI method. Similarly, only alpha pulses fall within the range of $V_{full} >$ 1.7 V (to the right to the red dotted line) and thus, they are automatically

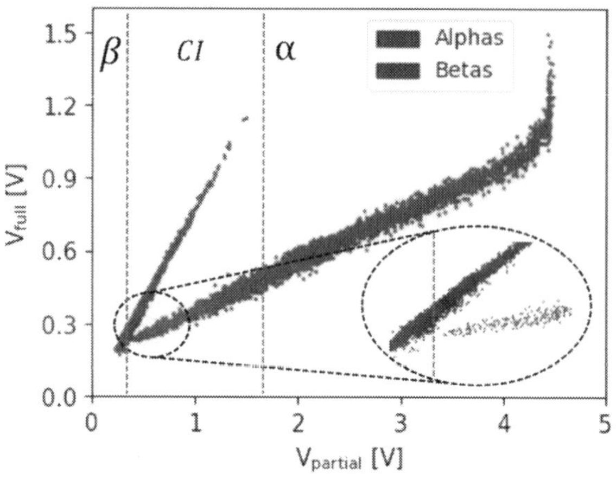

Fig. 6. Scattered V_{full} and $V_{partial}$ measured from alpha and beta pulses. The three regions where pulses are classified as either alphas, betas, or based on the CI method are illustrated.

TABLE I
FOR EVERY POSSIBLE OUTPUT COMBINATION OF THE THREE COMPARATORS "ALPHA COMP","BETA COMP", AND "CI COMP", THE DEADTIME PRODUCED BY THE CIRCUIT IS INTRODUCED ALONG WITH THE CLASSIFICATION DECISION OF THE CIRCUIT AND THE CASE THAT IS CONSIDERED.

Case	Beta Comp	Alpha Comp	CI Comp	Deadtime [ns]
Beta	0	0	0/1	300
Beta	1	0	0	300
Small Alpha	1	0	1	500
Medium Alpha	1	1	0/1	700
Not Possible	0	1	0/1	-

discriminated by the Alpha comparator. In addition, it can be seen that V_{full} of high energy pulses saturate at the power supply of the intergrator, which is about 4.5 V. If the pulses in this region would have been discriminated based on their PSD ratio, they would eventually be miss-classified, but since they are automatically classified a s a lpha p ulses, s aturation does not pose any difficulties. M oreover, m aking u se o f t he Beta comparator and the Alpha comparator extends the dynamic range of the circuit making it suitable for a wider range of incident particles in contrast to solely using the CI comparator. Discrimination of the input pulse is performed by the 3 comparators combined with logic gates and is described in Table I.

B. Double^{+} Counting Rejection

The dedicated D^{+}CR block was designed to cope with **Challenge II** and improve neutron counting within high gamma fluxes. More specifically, the D^{+}CR block is designed to detect the specific c ase o f t wo g amma p ulses arriving consecutively. This event is both prone to be classified a s a neutron and is more likely to occur the higher the gamma flux is.

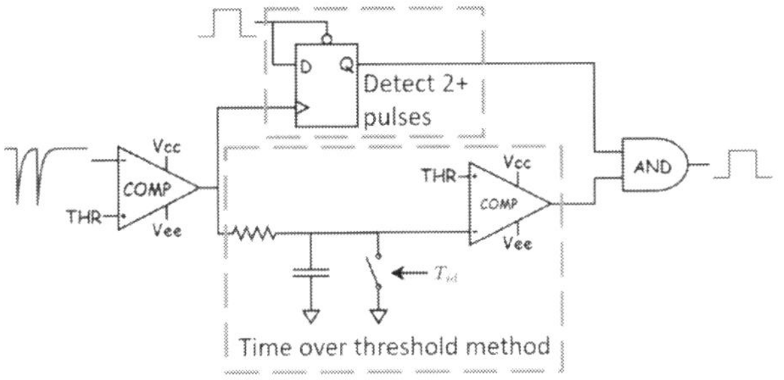

Fig. 7. Illustration of the D$^+$CR block.

An illustration of the D$^+$CR block can be seen in Fig. 7. The block is fed with the output of the comparator used for detection, which is then fed into two separate channels. The first channel is composed of a D flip-flop (D-FF) which outputs a "high" level whenever two or more pulses are detected. Note that low energy neutrons may produce a pulse that crosses the discrimination level several times, unintentionally triggering a "high" output in this channel. The second channel is composed of an RC network, a SPST switch, and a comparator, which implements a ToT method. The detection comparator output charges the (-) input of the ToT comparator via the RC network. Two consecutive low energy gamma pulses are likely to charge the RC network to a low level, while low energy neutrons are not. A "high" output in both channels causes the D$^+$CR block to reject the event. Two examples of signal shapes at different nodes of the D$^+$CR block for a low energy alpha pulse and two beta pulses arriving consecutively can be seen in Fig. 8. While the low energy alpha pulse does trigger the D-FF due to the tail re-triggering the detection comparator, the ToT is long enough so that the pulse is not rejected. The two consecutive beta pulses on the other hand both trigger the D-FF and have a low ToT, therefor the output of the D$^+$CR goes high and the event is successfully rejected instead of being classified as an alpha.

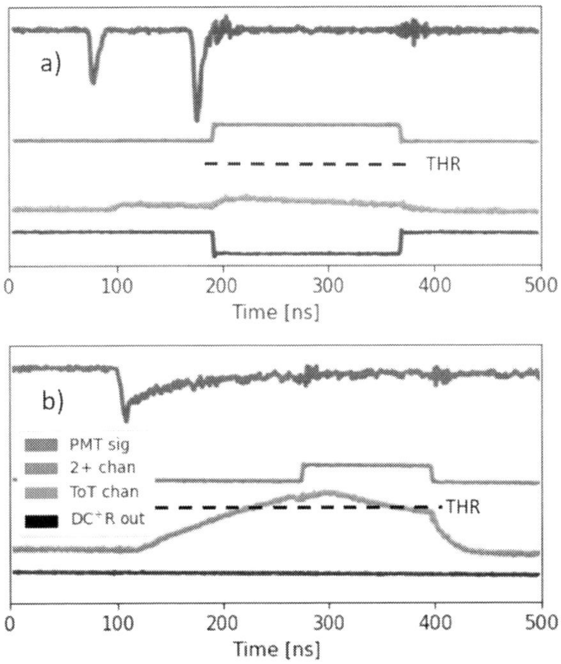

Fig. 8. Signal shape at different nodes of the D$^+$CR block (a) for two beta pulses arriving consecutively, (b) for a low energy alpha pulse.

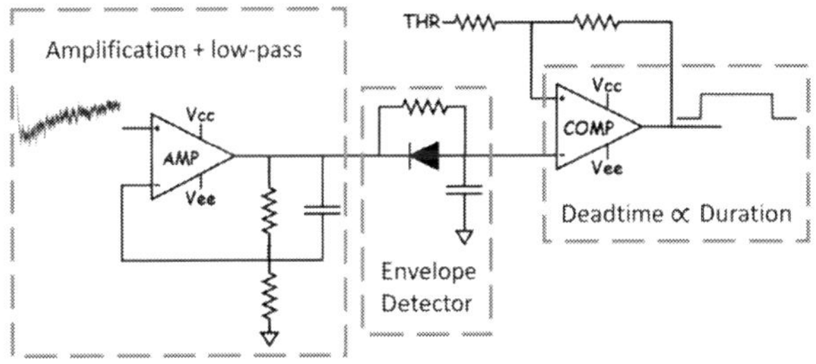

Fig. 9. Illustration of the DDT block.

C. *Dynamic deadtime*

In order to cope with **Challenge III**, a varying constant deadtime and a dynamic deadtime is employed. The varying constant deadtimes were introduced in Table I. The DDT block is designed to produce a dynamic deadtime that is proportional to the energy of the arriving pulse. Instead of deploying a simple ToT method which fails to produce a deadtime lasting during afterglow, a method based on peak-detection is deployed. An illustration of the DDT block can be seen in Fig. 9. The pulse from the detector is fed into a high speed amplifier which acts both as an amplifier and as a low-pass filter, capturing the energy deposited. Next, the amplified pulse is fed into an envelope detector. Rising edges of the pulse are immediately passed forward while falling edges aren't, and the output of the envelope detector is discharged using a slow RC network. This way, sharp peaks resulting from afterglow "re-trigger" the deadtime until the slow RC network discharges them.

The detector signal, the amplified signal, and the amplified signal after the envelope detector can be seen in Fig. 10. The detector signal produces an afterglow lasting up to 50 μs. Using the ToT method on the detector signal or the amplified signal would unsuccessfully result in a 10 μs or a 30 μs deadtime, respectively. On the other hand, using the ToT method on the amplified signal after the envelope detector successfully results in a 50 μs deadtime, which is proportional to the afterglow produced by the detector signal.

IV. Results

The topology parameters were adjusted based on the alpha and beta pulse characteristics displayed in section II so that pulses with a decay constant of under 11 ns would be classified as betas, and pulses with decay constants above 11 ns would be classified as alphas.

In order to evaluate the discrimination capabilities of the circuit solely based on the PSD ratio, the following experiment

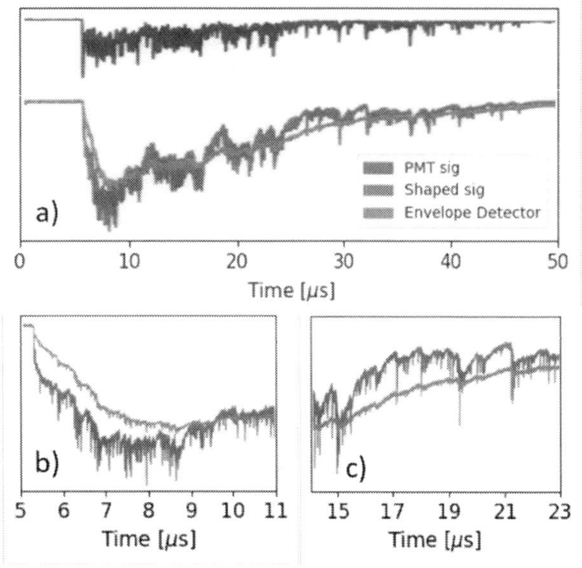

Fig. 10. (a) signal shape at different nodes of the DDT block. (b) Zoomed-in view on the rising edge of the shaped signal, visualizing the envelope detector successfully following the shaped signal. (c) Zoomed-in view on the decaying part of the shaped signal, visualizing the envelope detector signal successfully decaying based on a RC constant, instead of following the shaped signal.

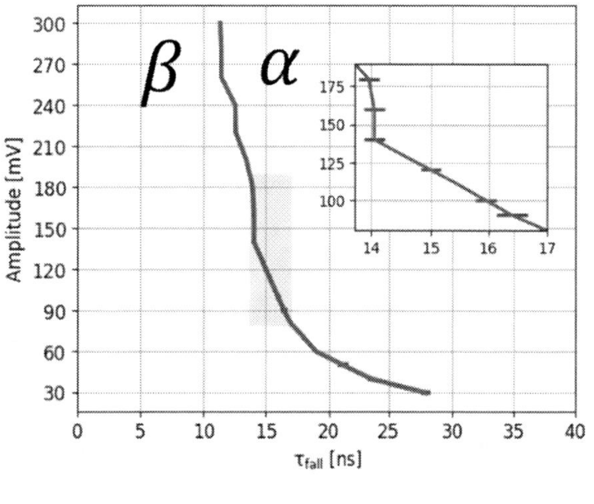

Fig. 11. The decay time programmed in the emulator at which the circuit's discrimination changes for different amplitudes. Error bars are also displayed.

was conducted. Exponential pulses from a fast digital emulator were injected into the circuit at amplitudes ranging between 30 and 300 mV, in 10 mV increments. For each amplitude, The decay constant at which the output of the CI comparator changed its value was found. At least five decay constants near the flipping point were tested in order to evaluate the uncertainty in the flipping region. The results are displayed in Fig. 11.

Firstly, it can be seen in the zoomed-in region highlighted in light purple that the uncertainty in the flipping region is relatively small for all regions, being under 1 ns for all amplitudes above 100 mV. At high amplitudes, the discrimination line is relatively stable around the decay constant of 11 ns. As the amplitude of the pulses goes lower, the discrimination capabilities of the circuit deteriorate and the discrimination line shifts towards higher decay constants. This effect is likely due to a constant charge injection being injected into the integrator capacitor, which introduces the same additive constant to both $V_{partial}$ and V_{full}. pulses with smaller amplitudes have smaller $V_{partial}$ and V_{full} values, leading to a larger effect from this constant voltage.

A well-established and widely accepted method for quantitative comparison of various PSD techniques is to evaluate the FoM [25]. The method is to fit a double-Gaussian function, in which, each one of the Gaussian functions has a corresponding centroid and a full width at height maximum (FWHM). The FoM is defined in (3).

$$FoM = \frac{|\mu_n - \mu_g|}{FWHM_n + FWHM_g} \quad (3)$$

In order to evaluate the discrimination capabilities of the circuit on the detector introduced in section II, the output of the partial integrator and the full integrator were recorded for 20,000 alpha pulses and 20,000 beta pulses from the experiment detailed in section II. The PSD ratio was then calculated using sampled $V_{partial}$ and V_{full} values. Fig. 12 displays a histogram of the PSD ratios obtained, with a calculated FoM of 2.35.

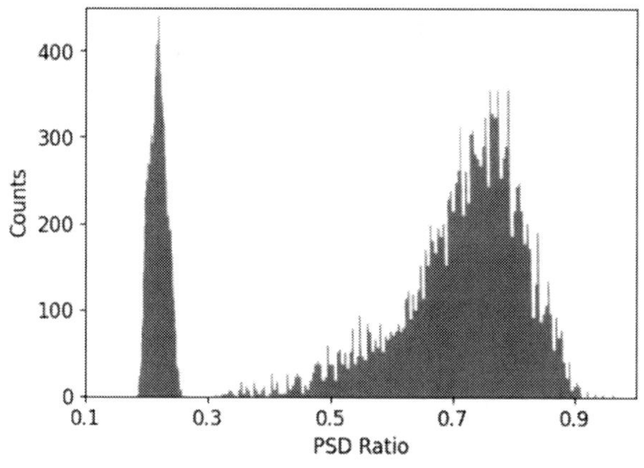

Fig. 12. Histogram of the PSD ratio measured for the 20,000 alpha pulses and 20,000 beta pulses discriminated by the circuit.

In order to evaluate the discrimination capabilities at different amplitudes the FoM was calculated for several amplitude bins. The FoM calculated at 10 equally spaced points between 20 mV and 200 mV are displayed in Fig. 13. At each of these points A_i, the FoM was calculated using pulses with amplitudes lying in the interval $A_i \pm 10$ mV. As expected, it can be seen that the FoM improves as the amplitude of the pulses, which is proportional to the SNR, increases.

The goal of the D^+CR block is to reject cases where two gamma pulses arrive consecutively in order to reduce neutron miss-classifications. The effectiveness of the D^+CR was tested

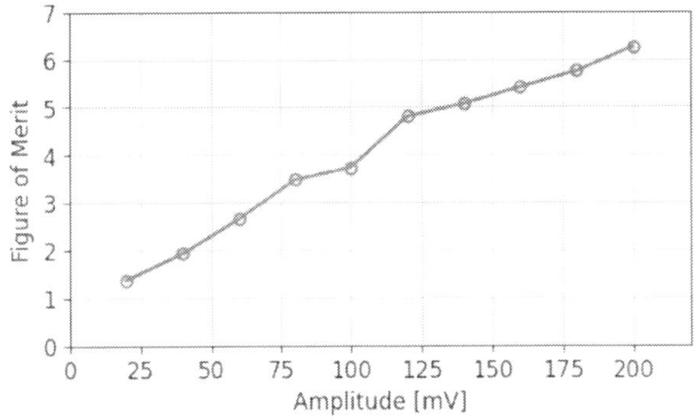

Fig. 13. The FoM calculated from the PSD ratios displayed in Fig. 12 at several amplitude ranges. The amplitude range corresponding to each amplitude point A_i displayed is $A_i \pm$ 10 mV.

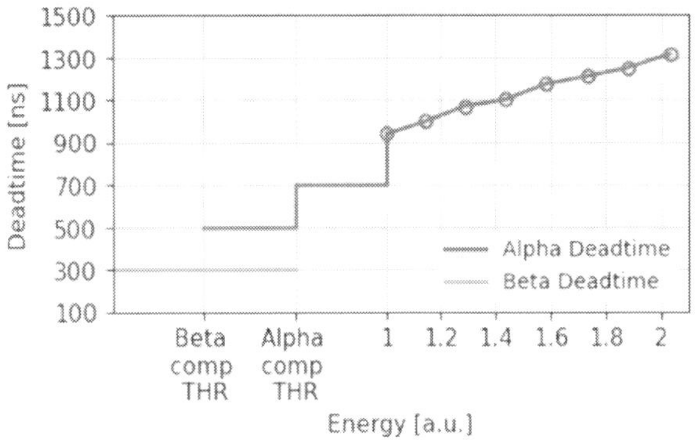

Fig. 15. The measured deadtime produced by the circuit as a function of both the charge of the arriving nuclear pulse and the discrimination made by the circuit.

using the fast digital emulator. The emulator was set to produce beta signals with the arrival process resembling a Poisson point process, and the number of pulses classified by the circuit as alphas were counted for beta fluxes ranging between 1 kcps and 100 kcps. Ideally, the desired output would be no alpha classifications for all event rates, but due to **Challenge II** as the beta flux increases so does the number of false alpha detections. The results obtained both when the D^+CR block was connected and disconnected can be seen in Fig. 14.

It can be seen that for all beta fluxes the D^+CR block successfully lowers the number of false alpha detections by a factor of up to 4. The remaining miss-classifications result from either the occurrence of three or more beta pulses within a single integration window, or the presence of two closely spaced beta pulses that cannot be rejected by the D^+CR block.

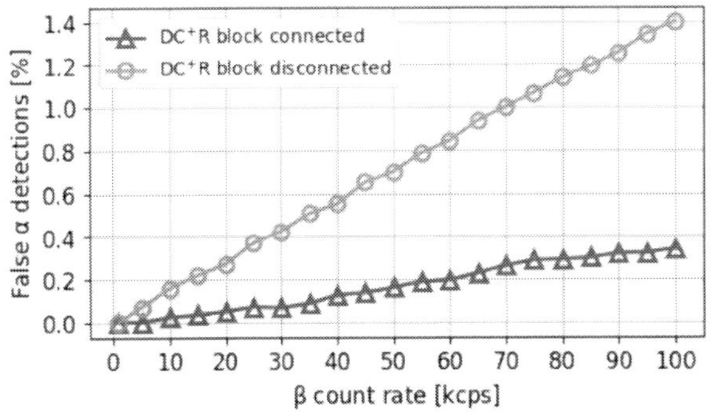

Fig. 14. The percentage of false α detections for different beta fluxes with the D^+CR block connected and disconnected.

The effectiveness of the DDT block can be seen in Fig. 10, where the produced signal successfully creates a deadtime proportional to the length of the afterglow produced by the initial pulse. The fast digital emulator was used in order to validate the expected behavior of the block, where higher energy pulses are required to produce longer deadtimes. Pulses with varying total charge were injected into the circuit and the resulting deadtime was measured. The results can be seen in Fig. 15. Pulses with total charge which is not high enough to trigger the DDT block yield varying constant deadtimes which are detailed in Table I. For higher energy alpha pulses, the DDT block successfully produces a deadtime that is proportional to the energy of the alpha pulse. By successfully producing a deadtime proportional to the energy of the alpha pulses, the circuit is ensured to have a smaller overall deadtime, which yields more pulse detections in total.

V. Conclusion

A novel analog PSD topology based on the well-known CI method designed for fast scintillators has been developed and implemented on an FR-4 PCB. A FoM of 2.35 was calculated for alpha and beta pulses from a plastic scintillator with a ZnS(Ag) layer coupled to a PMT. The topology has a dedicated block designed to reject fast pulses arriving consecutively, resulting in up to four times less slow pulse miss-classifications in fast pulse event rates ranging up to 100 kcps. Moreover, the dedicated block that yields a dynamic deadtime proportional to the energy of the pulse is proved to be beneficial for lowering the total deadtime and improving detection efficiency. The topology does not require fast ADCs, therefor it may be suitable for low power applications. The topology is versatile and can easily be adapted for different detectors with different pulse shapes by adjusting several passive electronic components.

References

[1] A. Dutta and K. E. Holbert, "Discrimination of neutron-gamma ray pulses with pileup using normalized cross correlation and principal component analysis," *IEEE Transactions on Nuclear Science*, vol. 63, no. 6, pp. 2764–2771, 2016.

[2] C. Fu, A. Di Fulvio, S. Clarke, D. Wentzloff, S. Pozzi, and H. Kim, "Artificial neural network algorithms for pulse shape discrimination and recovery of piled-up pulses in organic scintillators," *Annals of nuclear energy*, vol. 120, pp. 410–421, 2018.

[3] A. Dutta, P. Chandhran, K. E. Holbert, and E. B. Johnson, "Using decay time to discriminate neutron and gamma ray pulses from a clyc detector," in *2015 IEEE Nuclear Science Symposium and Medical Imaging Conference (NSS/MIC)*. IEEE, 2015, pp. 1–7.

[4] S. Richards, G. Sykora, and M. Taggart, "High count rate pulse shape discrimination algorithms for neutron scattering facilities," *Nuclear Instruments and Methods in Physics Research Section A: Accelerators, Spectrometers, Detectors and Associated Equipment*, vol. 989, p. 164946, 2021.

[5] J. Polack, M. Flaska, A. Enqvist, C. Sosa, C. Lawrence, and S. Pozzi, "An algorithm for charge-integration, pulse-shape discrimination and estimation of neutron/photon misclassification in organic scintillators," *Nuclear Instruments and Methods in Physics Research Section A: Accelerators, Spectrometers, Detectors and Associated Equipment*, vol. 795, pp. 253–267, 2015.
[6] M. Nakhostin, "A general-purpose digital pulse shape discrimination algorithm," *IEEE Transactions on Nuclear Science*, vol. 66, no. 5, pp. 838–845, 2019.
[7] R. M. French, M. Thevenin, M. Hamel, and E. Montbarbon, "A histogram-difference method for neutron/gamma discrimination using liquid and plastic scintillators," *IEEE Transactions on Nuclear Science*, vol. 64, no. 8, pp. 2423–2432, 2017.
[8] M. Safari, F. A. Davani, H. Afarideh, S. Jamili, and E. Bayat, "Discrete fourier transform method for discrimination of digital scintillation pulses in mixed neutron-gamma fields," *IEEE transactions on nuclear science*, vol. 63, no. 1, pp. 325–332, 2016.
[9] H. Singh and R. Mehra, "Discrete wavelet transform method for high flux $n-\gamma$ discrimination with liquid scintillators," *IEEE Transactions on Nuclear Science*, vol. 64, no. 7, pp. 1927–1933, 2017.
[10] M. Hammad, H. Kasban, R. Fikry, M. I. Dessouky, O. Zahran, S. M. Elaraby, and F. E. A. El-Samie, "Digital pulse processing algorithm for neutron and gamma rays discrimination," *Analog Integrated Circuits and Signal Processing*, vol. 101, pp. 475–487, 2019.
[11] B. Blair, C. Chen, A. Glenn, A. Kaplan, J. Ruz, L. Simms, and R. Wurtz, "Gaussian mixture models as automated particle classifiers for fast neutron detectors," *Statistical Analysis and Data Mining: The ASA Data Science Journal*, vol. 12, no. 6, pp. 479–488, 2019.
[12] A. Glenn, Q. Cheng, A. D. Kaplan, and R. Wurtz, "Pulse pileup rejection methods using a two-component gaussian mixture model for fast neutron detection with pulse shape discriminating scintillator," *Nuclear Instruments and Methods in Physics Research Section A: Accelerators, Spectrometers, Detectors and Associated Equipment*, vol. 988, p. 164905, 2021.
[13] M. Gelfusa, R. Rossi, M. Lungaroni, F. Belli, L. Spolladore, I. Wyss, P. Gaudio, A. Murari, and J. Contributors, "Advanced pulse shape discrimination via machine learning for applications in thermonuclear fusion," *Nuclear Instruments and Methods in Physics Research Section A: Accelerators, Spectrometers, Detectors and Associated Equipment*, vol. 974, p. 164198, 2020.
[14] E. Doucet, T. Brown, P. Chowdhury, C. Lister, C. Morse, P. Bender, and A. Rogers, "Machine learning n/γ discrimination in clyc scintillators," *Nuclear Instruments and Methods in Physics Research Section A: Accelerators, Spectrometers, Detectors and Associated Equipment*, vol. 954, p. 161201, 2020.
[15] T. Tambouratzis, D. Chernikova, and I. Pzsit, "Pulse shape discrimination of neutrons and gamma rays using kohonen artificial neural networks," *Journal of Artificial Intelligence and Soft Computing Research*, vol. 3, no. 2, pp. 77–88, 2013.
[16] J. Griffiths, S. Kleinegesse, D. Saunders, R. Taylor, and A. Vacheret, "Pulse shape discrimination and exploration of scintillation signals using convolutional neural networks," *Machine Learning: Science and Technology*, vol. 1, no. 4, p. 045022, 2020.
[17] M. Astrain, M. Ruiz, A. V. Stephen, R. Sarwar, A. Carpeño, S. Esquembri, A. Murari, F. Belli, and M. Riva, "Real-time implementation of the neutron/gamma discrimination in an fpga-based daq mtca platform using a convolutional neural network," *IEEE Transactions on Nuclear Science*, vol. 68, no. 8, pp. 2173–2178, 2021.
[18] X. Fabian, G. Baulieu, L. Ducroux, O. Stézowski, A. Boujrad, E. Clément, S. Coudert, G. de France, N. Erduran, S. Ertürk *et al.*, "Artificial neural networks for neutron/γ discrimination in the neutron detectors of neda," *Nuclear Instruments and Methods in Physics Research Section A: Accelerators, Spectrometers, Detectors and Associated Equipment*, vol. 986, p. 164750, 2021.
[19] J. Han, J. Zhu, Z. Wang, G. Qu, X. Liu, W. Lin, Z. Xu, Y. Huang, M. Yan, X. Zhang *et al.*, "Pulse characteristics of clyc and piled-up neutron–gamma discrimination using a convolutional neural network," *Nuclear Instruments and Methods in Physics Research Section A: Accelerators, Spectrometers, Detectors and Associated Equipment*, vol. 1028, p. 166328, 2022.
[20] S. Peng, Z. Hua, Q. Wu, J. Han, S. Qian, Z. Wang, Q. Wei, L. Qin, L. Ma, M. Yan *et al.*, "Piled-up neutron-gamma discrimination system for cllb using convolutional neural network," *Journal of Instrumentation*, vol. 17, no. 08, p. T08001, 2022.
[21] M. Flaska and S. A. Pozzi, "Identification of shielded neutron sources with the liquid scintillator bc-501a using a digital pulse shape discrimination method," *Nuclear Instruments and Methods in Physics Research Section A: Accelerators, Spectrometers, Detectors and Associated Equipment*, vol. 577, no. 3, pp. 654–663, 2007.
[22] M. Nakhostin and P. Walker, "Application of digital zero-crossing technique for neutron–gamma discrimination in liquid organic scintillation detectors," *Nuclear Instruments and Methods in Physics Research Section A: Accelerators, Spectrometers, Detectors and Associated Equipment*, vol. 621, no. 1-3, pp. 498–501, 2010.
[23] D. Cester, M. Lunardon, G. Nebbia, L. Stevanato, G. Viesti, S. Petrucci, and C. Tintori, "Pulse shape discrimination with fast digitizers," *Nuclear Instruments and Methods in Physics Research Section A: Accelerators, Spectrometers, Detectors and Associated Equipment*, vol. 748, pp. 33–38, 2014.
[24] G. J. Sykora, E. M. Schooneveld, and N. J. Rhodes, "Zno: Zn/6lif scintillator—a low afterglow alternative to zns: Ag/6lif for thermal neutron detection," *Nuclear Instruments and Methods in Physics Research Section A: Accelerators, Spectrometers, Detectors and Associated Equipment*, vol. 883, pp. 75–82, 2018.
[25] M. J. Balmer, K. A. Gamage, and G. C. Taylor, "Comparative analysis of pulse shape discrimination methods in a 6li loaded plastic scintillator," *Nuclear Instruments and Methods in Physics Research Section A: Accelerators, Spectrometers, Detectors and Associated Equipment*, vol. 788, pp. 146–153, 2015.

Characterization and use of Stilbene scintillator for neutron metrology and spectrometry from 100 keV to 22 MeV

Diane QUEVAUVILLERS[1], Laurent OTTAVIANI[2], Michael PETIT[1,*], Christelle REYNARD-CARETTE[2]
[1]Institut de Radioprotection et de Sûreté Nucléaire (IRSN), PSE-SANTE/SDOS/LMDN
13115 Saint Paul Lez Durance, France
[2]Aix Marseille Univ, Université de Toulon, CNRS, IM2NP, Marseille, France
(*) michael.petit@irsn.fr

***Abstract*—Stilbene scintillators have now the capability to extend the neutron energy measurement from 100 keV to 1 MeV. To be used as spectrometers, scintillators have to be characterized in entire energy range covered. For the photon characterization, usual calibration source determined the calibration and the resolution function. To do the photon matrix, MCNP PTRAC simulations were done between 59 keV up to 7 MeV. Using MCNP PTRAC allows the calculation of the photon response matrix from the tracking detail of the electron energy deposition in the crystal, including the effect of the Aluminum protection around. Resolution and sensibility obtained from gamma sources are applied later to the calculated photon matrix.**

For neutron characterization, Time-Of-Flight measurements on white neutron spectra were realized. However, the fluence reference on these spectra begins at 1 MeV. So, the shape of the response function was measured using stilbene scintillators on 100 keV- 1 MeV decade. After, the monoenergetic neutron reference of AMANDE facility was used to give neutron fluence normalization below 1 MeV. This method could characterize these scintillators for neutron energy from 100 keV up to 22 MeV.

***Keywords* —Scintillator characterization, digital acquisition, neutron, Time-Of-Flight, stilben**

I. INTRODUCTION

THE laboratory for micro-irradiation, neutron metrology and dosimetry (LMDN) is in charge of establishing metrological references in France for neutron fluence energy distributions and associated dosimetric quantities, on behalf of the LNE which is the French National Metrological Institute (NMI). One of the laboratory purposes is to maintain and improve metrological references and calibrations.

The laboratory has two facilities for establishing metrological references. The first, named AMANDE [1], has an ion accelerator which can produce mono-energetic neutron fields between 2 keV and 20 MeV. Metrological references by spectroscopy are difficult to cover below 1 MeV due to the lack of efficiency of usual scintillators [2]. For the same reason, Time-Of-Flight references are usually not covered below 1 MeV.

Stilbene with higher purity demonstrated their capability to detect neutrons energy until less than 100 keV [3]. So, the use of stilbene scintillator could allow the extension of the reference measurements, with or without Time-Of-Flight, to the 100 keV-1 MeV energy range.

The second facility, named CEZANE, has two sources of reference (^{241}AmBe and ^{252}Cf) used as calibration standard [ISO 8529 standard]. As ^{252}Cf source is isotopically impure [20], regular source calibration [4], not possible on site, is required. To overcome this difficulty while maintaining metrological references, the use of reference scintillators for source decay monitoring is planned.

To cover these needs, two stilbene scintillators, coupled to a digital acquisition system, were purchased. However, neutrons characterization of these stilbene scintillators is needed. As all neutron fields are accompanied with a photon field, the stilbene scintillator characterization for photons is interesting. Ultimately, this work will result in a high-resolution neutron spectrometer with a large range of energy [3] and a low-resolution photon spectrometer [5].

II. STILBENE SCINTILLATORS AND DIGITAL ACQUISITION

A. Stilbene scintillators

Stilbene scintillators are 2"x2" cylindric organic scintillators composed of Stilbene crystal ($C_{14}H_{12}$). Photomultipliers were adapted for each scintillator energy range target. The new stilbene growth method allows a more homogeneous crystal [6-7] with a size compatible with our needs. This purer crystal can detect neutron energy down to 100 keV and discriminate with photon [3, 8-9].

The LMDN stilbene scintillators (named SX and SX2) were purchased at Scionix [10] in 2020. SX has a fast photomultiplier response time [11] and is dedicated to a large range of energy (500 keV-22 MeV). SX2 has a large light collection photomultiplier with relatively low time response and is dedicated to low neutron energy range (100 keV-3 MeV).

B. Digital acquisition

The digital acquisition is a DT5730 [12] distributed by CAEN [13]. The measure with scintillators at higher energy range is limited by the DT5730 digital range amplitude (0-2 Volt). In order to increase the operating range of the SX, the signal at the scintillator anode is split in two signals both with halved amplitude. One signal amplitude is reduced with a factor of four by a passive attenuator before being digitized. The second signal is digitized on the same digitizer directly after the splitter.

This method makes a channel for the "high energy" neutron and the second channel for "low energy" neutron with a full energy range recovery. The high tension of scintillator and the passive attenuator are optimized and chosen to have a low channel energy limit upper than 7 MeV and a high channel energy limit lower than 7 MeV [14]. The entire energy range of SX can be measured in one go.

The discrimination between neutron and photon is realized with the Pulse Shape Discrimination (PSD) using the Charge Comparison Methods (CCM) as described in [3]. The PSD is directly obtained by CoMPASS software [15]. The long integration gate was adjusted to limit statistical drift at low energy due to an incorrect estimate of the baseline. The fast gate was adjusted to optimize the discrimination threshold. The Constant Fraction Discrimination (CFD) of both scintillators were optimized by a dedicated study on a ^{22}Na photon source [14].

Scintillator main characteristics are presented in Table I.

TABLE I
SCINTILLATOR CHARACTERISTICS

Characteristics	SX (2 channel)	SX2 (1 channel)
Used neutron energy range	500 keV – 22 MeV	100 keV – 3 MeV
Temporal response	1 ns	2 – 4 ns

III. THE PHOTON CHARACTERIZATION

The electron energy deposition will be calculated by a simulation. Many experimental photon spectra will determine sensitivity and resolution of the measurement device. Photon response matrix will be obtained after parametrization of calculation with experimental sensitivity and resolution.

A. The electron energy deposition by simulation

The electric signal at the scintillator anode is named "Light Output (LO)".

The simulation is realized with Monte Carlo N-Particle code (MCNP) [21]. In the simulation, the scintillator is modelled with the size, composition and density (1.15 g.cm^{-3}) of IRSN stilbene and a 0.4 mm thickness of Al covering the crystal. The air around the detector is modelled, and the composition and density are those presented in [16]. The isotropic photon source, assumed to be point-like, is positioned at 20 cm from the crystal on the cylinder edge. The photon energy varies between 50 keV up to 7 MeV with a step of 1 keV.

MCNP PTRAC is used to determine the electron energy deposited in the crystal. Indeed, even if MCNP F8 tally is often used to simulate the photon response matrix [3], it is a pseudo-kerma which does not calculate the exact electron energy deposit.

A root program reads the output files and determines the electron energy deposited in the crystal, considering all contributions, including the Al cover contribution. Different cases are possible and must be considered.

The first case is the simplest one: the photon interacts with an electron of the crystal, which deposes all its energy in it. In the second case, the photon interacts with an electron of the crystal, which gets out it and so without depositing all its energy. In the third case, the photon interacts with two or more electrons in the crystal. In this case, the electron energy deposits in the crystal are summed. In the fourth case, the photon does not interact with electron but creates electron pair. The electron/positron pair is considered for energy deposition as previous case. An additional contribution from the two 511 keV photon annihilation from positron is also considered as they can lead to electron energy deposition in the crystal. The contribution of Al is also considered, with photon pair creation or electron coming in crystal.

Cases are resumed on Fig. 1. The aluminum protection is in grey, and the crystal is in blue. The black arrows correspond to the photons emitted by the source. These photons collide in dark blue with electrons in red, or pair off in orange with a positron emitted in green. Several cases can occur during the same interaction.

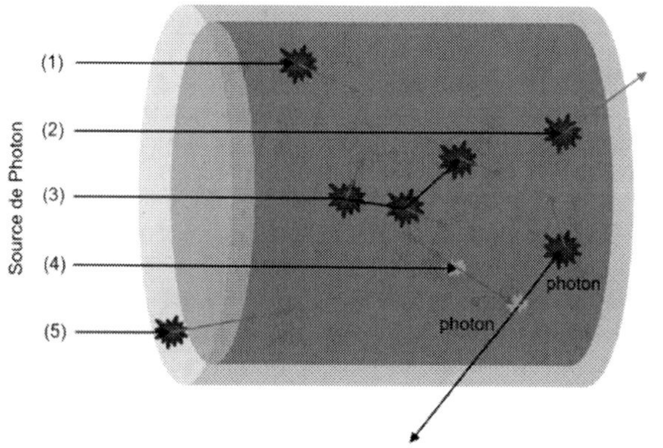

Fig. 1: Different cases in the crystal and aluminum

A spectrum for each energy between 50 keV and 7 MeV with a step of 1 keV, will be obtain. These spectra will be used to construct a photon response matrix without sensitivity and resolution.

The experimental spectrum from a ^{137}Cs source positioned at 20 cm is shown on Fig. 2. It is presented with the result of the equivalent simulation at 662 keV from the MCNP PTRAC and the equivalent MCNP tally 8. All simulation is presented without considering the stilben resolution and ith absolute normalization considering the well know source activity (379.59 kBq).

The Compton front on the MCNP PTRAC analysis and MCNP tally 8 can be identify. The MCNP tally 8 is clearly too low up to the Compton edge, especially at low electron energy. Whereas the MCNP PTRAC analysis enables to reproduce accurately the Compton shape.

https://doi.org/10.1051/epjconf/202328810006

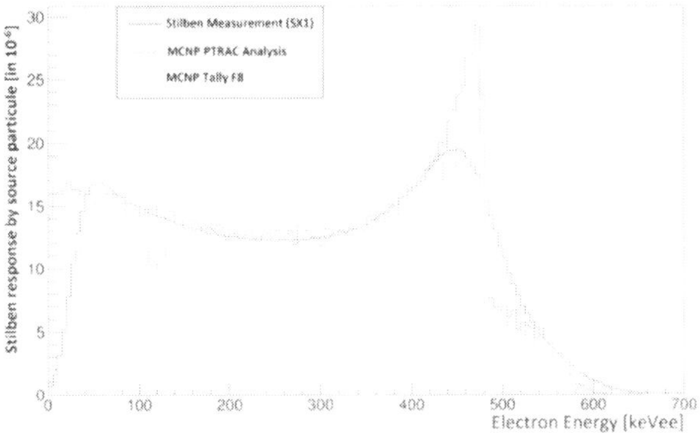

Fig. 2: Electron energy response of stilben for 137Cs source

B. The resolution and the sensitivity of scintillators

In order to complete the photon response matrix, the calculated photon matrix is convolved with experimental parameter. The sensitivity and the resolution function of scintillator are deduced by monoenergetic photon spectra. The list of photon sources and rays is shown in Table II.

For all sources except ^{241}Am-Be with polyethylene, the sources are at 20 cm from scintillators in the same configuration as the MCNP-Simulation. The ^{241}Am-Be with polyethylene source size is a cube of around 20 cm edge and the edge is only 1 cm from scintillators.

TABLE II
LIST OF PHOTON SOURCES

Photon Sources	Photon rays used [keV]
^{241}Am	59
^{133}Ba	80, 302 et 356
^{137}Cs	661
^{22}Na	511 et 1274
^{207}Bi	569, 1063 et 1770
^{60}Co	1173 et 1332
^{241}Am-Be	4438
^{241}Am-Be + Poly	2223
^{152}Eu	121, 244, 344, 778, 964, 1085, 1112 et 1408

The sources, except the ^{152}Eu, are used to determine the sensitivity and the resolution of scintillators at different energies. Due to the photon energy range of SX, the 59 keV peak of the ^{241}Am and the 80 keV peak of the ^{133}Ba are used only for SX2. Due to the photon energy range of SX2, only energy below 511 keV can be used. When the photon response matrix is completed, the spectrum will be used to check the deconvolution method and the matrix. The ^{152}Eu will be used as a test spectrum for the deconvolution [3].

The MCNP PTRAC simulation result is convoluted with gaussian function which width at half height function is adjusted to fit the experimental Compton edges or photoelectric peak. Using the same methods as described in [5], the sensitivity and the resolution of the scintillators will be determined.

IV. THE NEUTRON CHARACTERIZATION

The neutron characterization will be experimentally made using a Time-of-flight (TOF) method from white spectra. TOF is the metrological method to determine the neutron energy. The neutron speed, so its energy, is determined by the time difference between its emission and its detection, knowing the distance between both places. White spectra are issued from Neutron For Science (NFS at Caen in France) and Physikalisch-Technische Bundesanstalt (PTB at Braunschweig in Germany). These white spectra are obtained with a pulsed beam interacting with a thick target [17-18]. White spectra at PTB and NFS are shown on Fig. 3.

A. Time-Of-Flight

The scintillators will be positioned between 10 and 30 meters away at NFS and 14 meters at PTB. The distance at NFS and the beam pulsation are adjusted to allow the low neutron detection limit before the photons of the next pulse may arrive. Indeed, photons are produced at the same time as neutrons by particle beam interaction in the target. The neutron energy, E is calculated according to (1) [19]; with m is the neutron mass, c the velocity of light and L is a distance between the detector and the target.

$$E = mc^2\left(\frac{1}{\sqrt{1-\beta^2}} - 1\right) \ with \ \beta = \frac{v}{c} = \frac{L}{cT} , \qquad (1)$$

T is given by (2) but is calculated with (4). Indeed, the photon peak time gives the time reference production, knowing the distance between the target and the detector (3). In this equation, t_0 is the arrival time of the pulsed beam in the target. It is assumed that both neutrons and photons are emitted from the target at t_0. t_{photon} and $t_{neutron}$ are respectively the photon and the neutron time detection in the scintillator, respect to the beam pulsation.

$$T = t_{neutron} - t_0 , \qquad (2)$$

$$t_{photon} - t_0 = L/c , \qquad (3)$$

$$T = t_{neutron} - t_{photon} + \frac{L}{c} , \qquad (4)$$

B. The white Spectra

In order to determine the neutron response of the scintillator at a given energy, time windows are made in neutron time spectrum. The width of the time windows gives the energy resolution of the response matrix. To be valid, the number of events must be sufficient. Also, a compromise between energy resolution and statistics must be found. All the neutron-induced events in the time window are grouped together to obtain the LO profile assumed at the average neutron energy windows. All the windows are used to obtain, after correction, the response matrix over the whole energy range.

To limit the pile up effect and thus having to make a large and difficult correction, it is necessary to limit the number of events per pulse. In practice, an event rate lower than 1 per 100 pulses seems to be a good compromise.

After a detection, the time to return to the operating point depends on the energy deposit and the type of particle. The greater the energy deposit is, the longer the return to the operating point will be. In this case, a secondary trigger is possible, so the acquisition blocking time must be increased.

Fluence references are established on the white spectrum of the PTB by its reference detector. This reference will therefore be used to determine all the reference fluences, including beyond its high energy limit using both spectra results, extrapolation, and simulation.

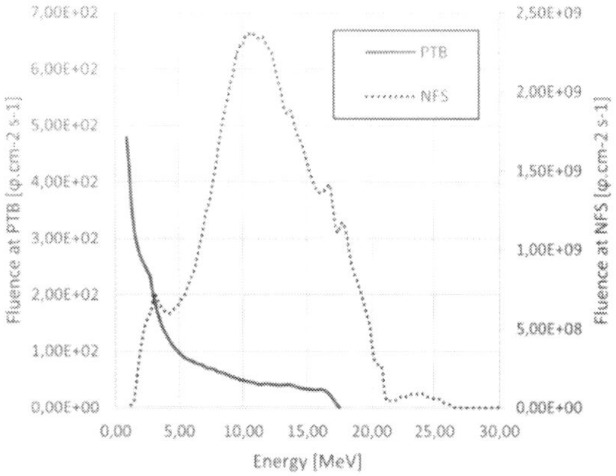

Fig. 3: White Spectra at NFS and PTB

For neutron characterization between 100 keV and around 1 MeV, there is no fluence reference at PTB or NFS as neutron reference detector have no efficiency. However, neutron at this energy obviously exists both at PTB and NFS. So, the LO shape of the neutron detectors responses will be measured. The fluence determination for LO shapes will be done using monoenergetic neutron fields from at AMANDE facility.

V. Conclusions

In neutron metrology, scintillators are often used as secondary standards due to their specific qualities - very fast response, high neutron efficiency and response, neutron-photon discrimination, etc. However, their "high" detection threshold around 1 MeV induces that white (or large energy range) neutron fields with a component below this limit relatively unknow. Recent progress in crystallography on stilbene crystals has led to the development of scintillators with threshold detection around 100 keV, so one decade lower. Neutron characterization of such stilbene scintillators will provide a directional neutron spectrometer over an extremely wide energy range (100 keV-22 MeV), so including 100 keV-1 MeV decade. As this energy range is particularly wide, various experimental techniques are required. Data analysis will also require a specific study to establish neutron response matrices and carry out deconvolution. Photon characterization of the same scintillators will enable the evaluate the contribution of photons in a neutron field. All of this work will complement the time-of-flight references on AMANDE facility and improve the characterization of the metrological references of LMDN.

Acknowledgements

The authors would like to thank the LNE, the French National Metrology Institute (NMI), for its financial and technical support to this study.

References

[1] V. Gressier *et al.*, 'AMANDE: a new facility for monoenergetic neutron fields production between 2 keV and 20 MeV', Radiation Protection Dosimetry, vol. 110, no. 1–4, Art. no. 1–4, Aug. 2004, doi: 10.1093/rpd/nch185.
[2] L. Dioni, 'Development of a Multi-purpose Fast Neutron Spectrometric Capability in the MASURCA Facility', AMU ED352, 2017.
[3] A. Di Chicco, 'Characterization of a stilbene organic scintillator for use as a broadband neutron spectrometer in mixed radiations fields', AMU ED 352, 2022
[4] N. Roberts *et al.*, Recent developments in neutron metrology at the National Physical Laboratory, Neutron Metrolgy Group, NPL, March 2009
[5] A. Di Chicco *et al.*, 'Gamma-response characterization of a solution-grown stilbene-based detector assembly in the 59 keV–4.44 MeV energy range; an alternative low-resolution gamma spectrometer', Nuclear Inst. and Methods in Physics Research, vol. 1034, July 2022, 166740
[6] L. Carman *et al.*, 'The effect of material purity on the optical and scintillation properties of solution-grown trans-stilbene crystals', Journal of Crystal Growth, vol. 368, pp. 56–61, Apr. 2013, doi: 10.1016/j.jcrysgro.2013.01.019.
[7] N. Zaitseva *et al.*, 'Scintillation properties of solution-grown trans-stilbene single crystals', Nuclear Instruments and Methods in Physics Research Section A: Accelerators, Spectrometers, Detectors and Associated Equipment, vol. 789, pp. 8–15, Jul. 2015, doi: 10.1016/j.nima.2015.03.090.
[8] M. Sénoville, F. Delaunay, M. Pârlog, N. L. Achouri, and N. A. Orr, 'Neutron-γ discrimination with organic scintillators: Intrinsic pulse shape and light yield contributions', Nuclear Instruments and Methods in Physics Research Section A: Accelerators, Spectrometers, Detectors and Associated Equipment, vol. 971, p. 164080, Aug. 2020, doi: 10.1016/j.nima.2020.164080.
[9] R. L. Craun and D. L. Smith, 'Analysis of response data for several organic scintillators', Nuclear Instruments and Methods, vol. 80, no. 2, Art. no. 2, Apr. 1970, doi: 10.1016/0029-554X(70)90768-8.
[10] https://scionix.nl/
[11] W. M. Steinberger, M. L. Ruch, A. Di-Fulvio, S. D. Clarke, and S. A. Pozzi, 'Timing performance of organic scintillators coupled to silicon photomultipliers', Nuclear Instruments and Methods in Physics Research Section A: Accelerators, Spectrometers, Detectors and Associated Equipment, vol. 922, pp. 185–192, Apr. 2019, doi: 10.1016/j.nima.2018.11.099.
[12] 'User Manual UM3148 DT5730/DT5725'. May 04, 2021.
[13] https://www.caen.it/
[14] A. Alves, Etude de scintillateurs stilbène pour la métrologie des neutrons, IRSN/LMDN, 2021
[15] 'User Manual UM5960 CoMPASS'. CAEN, Feb. 17, 2021.
[16] 'Inrad Optics Newsletter Signup Homepage'. https://form.jotform.com/210256397469162
[17] W. Dep. 6, 'Cyclotron', Apr. 27, 2016. { https://www.ptb.de/cms/en/ptb/fachabteilungen/abt6/fb-64/641-ion-accelerators/cyclotron.html }
[18] D. Ackermann *et al.*, 'Whitebook Science requirements, New Ganil injector, June 2021
[19] M.-A. Cognet and V. Gressier, "Development of a measurement reference standard for neutron energies between 1 MeV and 20 MeV using time of flight method at the AMANDE facility", Metrologia vol. 47, June 2010, doi:10.1088/0026-1394/47/4/004
[20] N. J. Roberts* and L. N. Jones, "The content of ^{250}Cf and ^{248}Cm in ^{252}Cf neutron sources and the effect on the neutron emission rate", Radiation Protection Dosimetry (2007), Vol. 126, No. 1–4, pp. 83–88
[21] https://mcnp.lanl.gov/

Characterization of 3D micro-structured TIMEPIX detectors for neutron imaging

Matteo Polo[1,2,*], Roberto Mendicino[3], Alberto Quaranta[1,2], Gian-Franco Dalla Betta[1,2]

1- Dipartimento di Ingegneria Industriale, Università di Trento, via Sommarive 9, 38123, Trento, Italy
2- INFN , TIFPA, Trento, via Sommarive 14, 38123, Trento, Italy
3- Center for Sensing Solutions, Eurac Research, Via A. Volta 13A, 39100 Bolzano, Italy
(*) polo.matteo1@gmail.com

Abstract - **Neutron imaging provides additional information to X-ray imaging and can be used in many applications, for example, nuclear engineering, non-industrial diagnostics and homeland security. This paper presents the latest development results of a new 3D structured pixel detector for thermal neutron detection and imaging. The device is based on Medipix/Timepix read-out chip family. The detector, without converter materials, was initially tested in the laboratory with a ^{241}Am source, to simulate events that are induced by reaction products of active material in the silicon bulk. Thanks to the analysis of the size of event clusters and deposited energy, α particles and noise signals were discriminated. The detector was also tested with β and γ sources, to study the effect of other radiation on the device. These results can be useful as a starting point for future experiments, where a neutron source will be used, after filling the cavities with active materials, such as ^{6}Li or ^{10}B. In this paper, the results of the characterization and a comparison between experimental data and Monte Carlo simulations are shown.**

Keywors - **Neutron imaging, 3D Radiation detector, Solid-state neutron detectors**

1 Introduction

Non-destructive techniques can be helpful in revealing the interior of many materials and engineering components. The X-ray imaging is the most common solution, but in recent years neutron imaging has become a complementary method since it can provide additional information, as compared to X-rays [1]. In fact, X-rays interact primarily with high Z material, while neutrons have the opposite behavior, so they can use for studying low Z elements, which are the main components of organic materials [2]. In the past, the most common solution for detecting neutrons was the ^{3}He tube, but due to the lack of this gas other detectors had to be developed [3], as an example based on inorganic scintillators, where an active conversion material is used [4, 5]. Semiconductor detectors can also be used for thermal neutron detection: the first solution is based on neutron-sensitive semiconductor materials (LiInSe, LiSe, BN), where most of the charge from the neutron reaction is available for transduction within the sensor itself [6]. Another approach uses semiconductor sensors coated with thin films of neutron converter materials, as described in [7]. The best performance in terms of detection efficiency are offered by 3D detectors, in which the silicon bulk is drilled [8, 9, 10, 11, 12, 13]. The cavities are filled with an active material, where the neutron interacts producing charged particles, which can be detected by the sensitive volume around the cavities. In this paper, we report the latest results of the characterization of the detector proposed in the INFN HYDE project (HYbrid DEtectors of neutrons) [8]. A schematic overview of the device is reported in Fig. 1. The detector is a hybrid 3D micro-structured pixel detector module made of 256×256 pixels of 55×55 μm^2 size. The nominal resistivity is in the range from 10 to 30 kΩ·cm. On the back side, a p$^+$ layer and an Al layer are deposited. After that, the surface is drilled using the Deep reactive-ion etching (DRIE) technique. For the passivation of the cavities, that are not doped, a thin layer of Al_2O_3 (∼ 10 nm) is deposited [8]. Due to heat dissipation problems during the DRIE process, it was not possible to drill holes deeper than 25-30 μm. The characterization is divided into two parts: in the first part some diode test structures have been used, and in the second part, the complete devices have been tested.

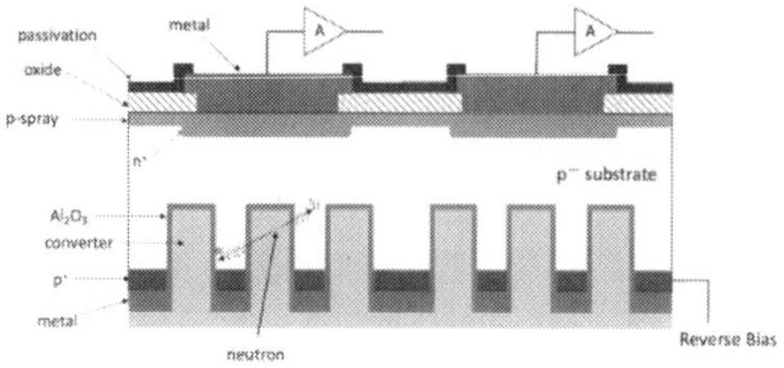

Figure 1: Graphical representation of the HYDE2 device, not to scale [14].

In these preliminary tests, the cavities have not yet been filled with active material. Therefore the measurements were carried out with alpha sources, in order to simulate the effects of the conversion of neutrons into charged particles. The main results and a comparison between experimental data and simulations will be reported in the following sections.

2 Monte Carlo simulation

Monte Carlo simulations are performed to determine the optimal geometry and active material to fill the hole. Geant4 v.10.0.3 has been used [15]. The two main parameters that must be optimized are the lateral sizes of the cavities and of the sensitive regions around them, aimed at maximizing the neutron capture and charged particle detection. Two different active materials were tested, i.e., ^{10}B and ^{6}Li. These two materials are able to capture thermal neutrons and convert them into charged particles. In particular the two reactions are respectively $^{10}B(n,\alpha)^{7}Li$ with a Q_{value} = 2.8 MeV and a cross-section of $3.6 \cdot 10^3$ b and $^{6}Li(n,\alpha)^{3}H$ with Q_{value} = 4.8 MeV and a cross-section of $8.7 \cdot 10^2$ b. In one case there is a greater cross-section, so a higher interaction probability, while in the other a higher Q_{value} which allows easier detection of charged particles due to the larger number of electron-hole pairs generated. In figures 2 and 3 the simulation results are shown.

As expected with ^{10}B it is necessary to use narrow cavities because the α particles have less energy, and consequentially a smaller range. In silicon, an α particle with 1.01 MeV (from ^{10}B reaction) can travel for 4μm, compared to 12 μm of the α particle of 2.7 MeV from the ^{6}Li reaction [1]. If the cavities are too large the α particle generated in the center can not reach the silicon region around, so they can not be detected. In the case of ^{10}B with small cavities the efficiency can reach $\sim$30%. This is only a theoretical value because, due to their high aspect ratio, the cavities are not so easy to fill completely, as reported in [14].

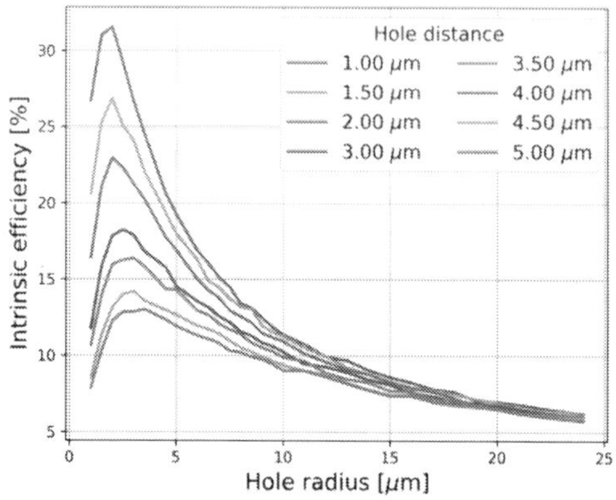

Figure 2: Geant4 simulation of detection efficiency with ^{10}B as active material. The maximum efficiency is reached using cavities with a radius of 3μm.

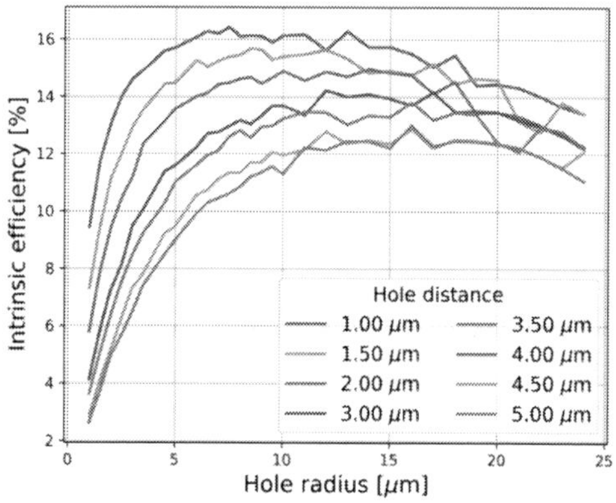

Figure 3: Geant4 simulation of detection efficiency with ^{6}Li as active material.

3 Measurements and results

To study the performance of the sensor two different devices are used:

- Diode structures, that are large surface diodes with guard rings, that are suitable

[1] The range of α particles have been determined with Lise++ v. 16.3.11

for preliminary tests. Their dimensions are $\sim 7\times 7$ mm^2.

- Pixel devices, which are pixel detectors, their overall dimension are $\sim 14\times 14$ mm^2. The pixel pitch is 55μm to be compatible with the Medipix/Timepix electronics.

Some of the devices are drilled on the backside, as reported in Fig. 1, with different geometries as studied during the simulation process. Not all the devices are drilled so that the effects of the cavities in the devices can be studied.

3.1 Diode test

The first test performed is the current-voltage (I-V) characterization of the diode test structures, in order to determine the breakdown voltage and consequently the bias voltage (V_{bias}) that can be applied during the following tests. To perform this measurement a Keithley 4200A-SCS Parameter Analyzer was used, with a current compliance set to (c.c.) to 10μA. The I-V curves of two diodes, with and without drilled backside, are shown in Fig. 4.

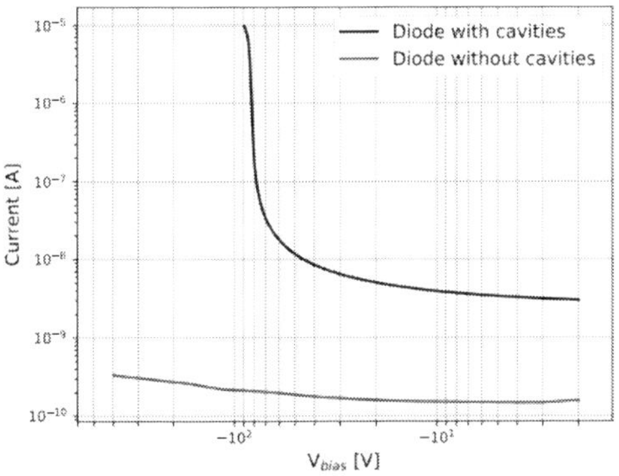

Figure 4: I-V curves of two diodes. The blue line is the I-V curve of the diode with cavities on the back side, it is possible to notice the current increase at -80 V. The red line is the I-V curve of the diode without cavities, the current change from 0.15 nA to 0.35 nA and there is no breakdown up to -400 V.

As it can be seen, while a rapid increase of the current occurs in the diode with the drilled backside at V_{bias}=-80 V, the current remains almost constant in the diode without cavities, going from 0.15nA (V_{bias}=-5 V) to 0.35nA at V_{bias}=-400 V (the maximum value allowed by the instrument). Hence, the current increase at -80 V in the case of drilled diode is likely not a breakdown (that in these devices normally occurs at the surface), but can be attributed to an insufficient passivation of the cavities, where a significant generation of electron-hole pairs leads to an increase of the leakage current when the depletion region begins to spread in the area where the cavities are present. Due to this problem with passivation, it is not possible to further increase the bias beyond -80 V in the diodes with cavities. The α detection capability of the diode was then studied. For this measurement, the diode, which was previously placed on a support electronic board, was connected to a Cremat CR-150-R5 evaluation board, where a charge-sensitive preamplifier Cremat CR-110-R2 is mounted. The CAEN DT5780 Dual Digital Multi-Channel Analyzer manages the data acquisition and all the power supplies. In the DT5780 is installed the DPP-PHA firmware, which is able to perform the trapezoidal filtering [16]. For this experiment an ^{241}Am radioactive source was used, that emits α particle at 5.5 MeV [17]. To reduce the loss of the α particle energy in the air, the diode and the evaluation board were placed inside a vacuum chamber. With this apparatus, it was possible to acquire the energy spectrum (Fig. 5)and the signal (Fig. 6). The measurement was carried out by moving the source first to one side (front) and then to the other (back, which is the drilled side). It is possible to notice a difference in the shape of the spectrum and of the signal between the two cases. In particular, the spectrum is broader for back-side exposure, with a peak at lower energy. The explanation for this effect is two-fold: on one hand, when alpha particles impinge from the back side, they can enter the sensitive volume of the devices in different regions (i.e., closer to the surface or closer to the bottom of the cavities). Since it is not possible to bias the diode at the optimum voltage, because of the leakage current increase, the generated charges must move by diffusion before reaching the depleted region, leading to slower signals (e.g., as shown in Fig. 6), and to increased recombination, eventually to a dispersion in the spectrum.

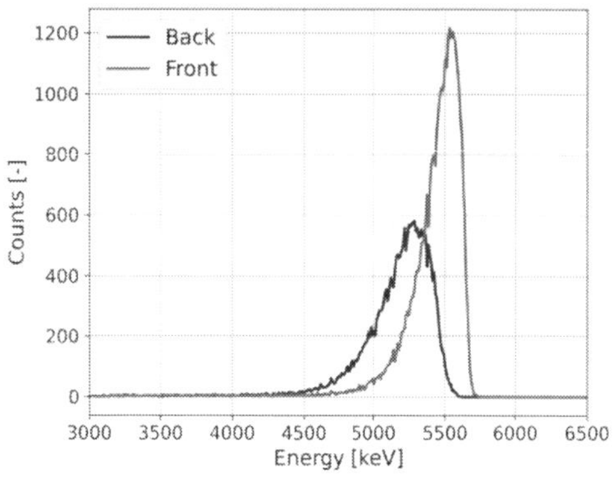

Figure 5: ^{241}Am energy spectrum acquired for front-side and back-side source exposure with a drilled diode polarize with V_{bias}=-80 V. The signal energy is determined from signal charge integration. The same calibration parameters were used for both sides.

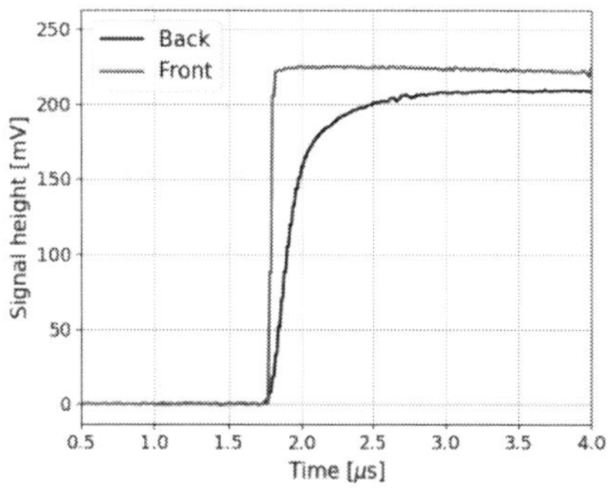

Figure 6: Details of voltage signals at the output of the preamplifier in response to an α particle for front-side and back-side source exposure. As in the previous case the diode is polarized with V_{bias}=-80 V.

3.2 Pixel measurement

After performing the diode characterization, pixel detectors were tested. A photo of the final assembly is reported in Fig. 7.

Figure 7: Final detector assembly. It can be connected directly to a computer using a USB cable.

The data acquisition is managed by Pixelman software [18]. With this software, it is possible to set all the parameters for the data acquisition and also see in real-time the acquired data. As reported in the previous section, the cavities are not yet filled with the converter material, so, as for the diodes, only the ^{241}Am α source was used. The detector is bump bonded to a Timepix chip, where the particle energy is not determined by integrating the total charge but it is based on the Time Over Threshold (ToT) [19]. The use of this approach allows to reduce the readout costs in the case of multi-channel readouts [20]. Also in this measurement, there is a high leakage current, so it is not possible to bias the device at the optimum voltage. The high leakage current introduces a lot of noise, so it is necessary to use a very high threshold; by doing so, the ToT of an α event is quite small, of the order of 10 clock cycles, as shown in the energy spectrum of Fig. 8.

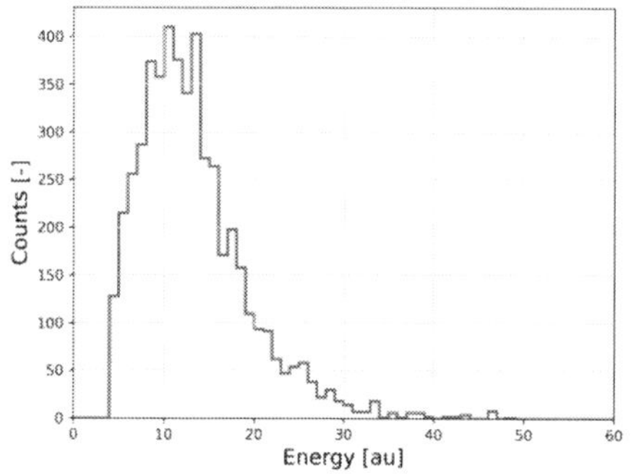

Figure 8: ToT spectrum of ^{241}Am source acquire with the final detector assembly. The detector is polarize with -80 V. If there is charge sharing between adjacent pixels, the total energy of the event is determined by the sum of the ToT of the involved pixels.

Despite the high level of noisy signals, spatial resolution was studied. Figure 9 shows that the device is able to reconstruct the spatial distribution of the events, this is a basic requirement for an imaging detector.

As shown in Fig. 9 the signal from α particle does not release all the energy in a single pixel but in some cases, there is a creation of a cluster, due to charge sharing events, so a single interaction can be detected from several adjacent pixels. In Fig. 10 the different probabilities to create a non-single pixel event are shown.

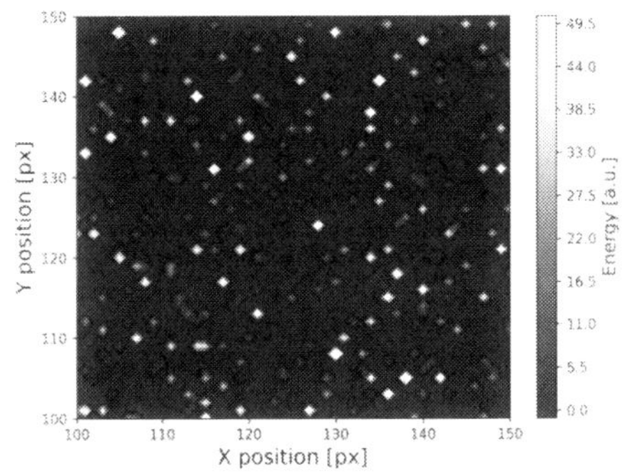

Figure 9: Map showing where alpha particles interacted with the sensor. In this case, a central area of the sensor has been reported.

Starting from Fig. 10, it can be assumed that from the cluster dimension is possible to discriminate different particles, based on the different dE/dx in silicon of different particles. For this reason, different radioactive sources as ^{90}Sr for β particles and ^{133}Ba for γ photons were used, but, due to the high noise level, it was not possible to discriminate these events from the noise.

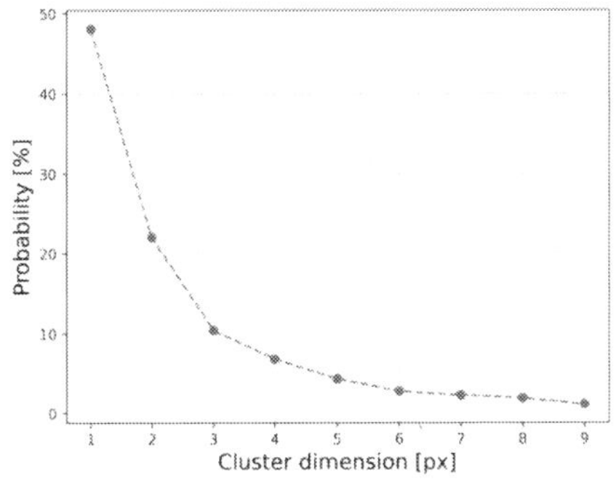

Figure 10: Probability distribution of the size of the clusters generated by the interaction of the α particles with the sensor.

4 Conclusion

Neutron imaging can be used as a new technique for non-destructive analysis of material and components. In this paper, a new device for neutron imaging is proposed, where a silicon detector with cavities filled with active materials are used. It is used for neutron conversion in charge particles. The detailed description of the device is reported in Sec. 1.

In Sec. 2 are reported the results of the Monte Carlo simulation done with GEANT4. In the simulation, the geometry and the active material have been optimised.

After the simulation in Sec. 3 are reported the preliminary measurement results. From the measurement of the I-V curve it is observed that due to the passivation problem of the cavities is not possible to polarize correctly the devices, so the maximum V_{bias} that can be used is -80 V, which is much lower respect the case of a planar detector of the same size, where the breakdown is better than -400 V. After the I-V characterization, some diode tests with the ^{241}Am source are performed. In this test, it is possible to notice that there are some differences by moving the source first to one side (front) and then to the other (back). This problem is caused by the low bias voltage being applied to the sensor. This means that in one case the charges move primarily by drift (source in front) and by diffusion (back). Finally, a test with the complete pixel devices is performed, in order to evaluate the spatial resolution capabilities of the sensor. For α particles, the sensor functions accurately despite significant noise levels. However, due to the substantial energy released by α particles, addressing this issue is possible by applying a high threshold. On the contrary, in tests performed with other sources, ^{90}Sr (β) and ^{133}Ba (γ) particle-noise discrimination was not possible.

Despite the previously discussed problems, the device seems to be a good candidate to perform neutron imaging measurements. The next step is to fill the cavities with active material, such as ^{10}B or ^{6}Li depending on the geometry of the cavities. After that, it is planned to carry out neutron detection measurements using for example a ^{252}Cf source or through the use of an accelerator that exploits the reaction, such as for example the CN accelerator of the Legnaro National Laboratories.

References

[1] N. Kardjilov, I. Manke, A. Hilger, M. Strobl, and J. Banhart, "Neutron imaging in materials science," *Materials Today*, vol. 14, no. 6, pp. 248–256, 2011.

[2] K. S. Krane, *Introductory nuclear physics*. John Wiley & Sons, 1991.

[3] M. Henske, M. Klein, M. Köhli, P. Lennert, G. Modzel, C. Schmidt, and U. Schmidt, "The 10B based jalousie neutron detector–an alternative for 3he filled position sensitive counter tubes," *Nuclear Instruments and Methods in Physics Research Section A: Accelerators, Spectrometers, Detectors and Associated Equipment*, vol. 686, pp. 151–155, 2012.

[4] F. Pino, M. Polo, J. C. Delgado, G. Mantovani, S. M. Carturan, D. Fabris, D. Brunelli, L. Pancheri, A. Quaranta, and S. Moretto, "Evidence of fast neutron detection capability of the CLLB scintillation detector," *Radiation Physics and Chemistry*, vol. 202, p. 110494, 2023.

[5] M. Polo, F. Pino, J. C. A. Delgado, D. Fabris, and S. Moretto, "Assessment of a nail detector performance for radiation monitoring applications," *European Physical Journal: Special Topics*, 2023.

[6] K. Ahmed, R. Dahal, A. Weltz, J. J.-Q. Lu, Y. Danon, and I. B. Bhat, "Solid-state neutron detectors based on thickness scalable hexagonal boron nitride," *Applied Physics Letters*, vol. 110, no. 2, p. 023503, 2017.

[7] D. S. McGregor, M. Hammig, Y.-H. Yang, H. Gersch, and R. Klann, "Design considerations for thin film coated semiconductor thermal neutron detectors—i: basics regarding alpha particle emitting neutron reactive films," *Nuclear Instruments and Methods in Physics Research Section A: Accelerators, Spectrometers, Detectors and Associated Equipment*, vol. 500, no. 1-3, pp. 272–308, 2003.

[8] R. Mendicino and G.-F. Dalla Betta, "Three-dimensional detectors for neutron imaging," *Nuclear Instruments and Methods in Physics Research Section A: Accelerators, Spectrometers, Detectors and Associated Equipment*, vol. 878, pp. 129–140, 2018. Radiation Imaging Techniques and Applications.

[9] J. Shultis and D. McGregor, "Design and performance considerations for perforated semiconductor thermal-neutron detectors," *Nuclear Instruments and Methods in Physics Research Section A: Accelerators, Spectrometers, Detectors and Associated Equipment*, vol. 606, no. 3, pp. 608–636, 2009.

[10] G.-F. Dalla Betta and M. Povoli, "Progress in 3d silicon radiation detectors," *Frontiers in Physics*, vol. 10, p. 927690, 2022.

[11] S. L. Bellinger, R. Fronk, W. McNeil, T. Sobering, and D. McGregor, "Improved high efficiency stacked microstructured neutron detectors backfilled with nanoparticle ^{6} lif," *IEEE Transactions on Nuclear Science*, vol. 59, no. 1, pp. 167–173, 2011.

[12] Q. Shao, L. Voss, A. Conway, R. Nikolic, M. Dar, and C. Cheung, "High aspect ratio composite structures with 48.5% thermal neutron detection efficiency," *Applied Physics Letters*, vol. 102, no. 6, 2013.

[13] K.-C. Huang, R. Dahal, J. J.-Q. Lu, Y. Danon, and I. B. Bhat, "High detection efficiency micro-structured solid-state neutron detector with extremely low leakage current fabricated with continuous pn junction," *Applied Physics Letters*, vol. 102, no. 15, 2013.

[14] R. Mendicino and G.-F. Dalla Betta, "A CMOS-MEMS pixel sensor for thermal neutron imaging," *Micromachines*, vol. 14, no. 5, 2023.

[15] S. Agostinelli, J. Allison, K. Amako, *et al.*, "Geant4—a simulation toolkit," *Nuclear Instruments and Methods in Physics Research Section A: Accelerators, Spectrometers, Detectors and Associated Equipment*, vol. 506, no. 3, pp. 250–303, 2003.

[16] CAEN, "Caen tools for discovery - official site," *https://www.caen.it/*, Access Feb 2023.

[17] M. Basunia, "Nuclear data sheets 107, 3323," *National Nuclear Data Center, Upton*, 2006.

[18] D. Turecek, T. Holy, J. Jakubek, S. Pospisil, and Z. Vykydal, "Pixelman: a multi-platform data acquisition and processing software package for Medipix2, Timepix and Medipix3 detectors," *Journal of Instrumentation*, vol. 6, no. 01, p. C01046, 2011.

[19] R. Ballabriga, M. Campbell, and X. Llopart, "ASIC developments for radiation imaging applications: The Medipix and Timepix family," *Nuclear Instruments and Methods in Physics Research Section A: Accelerators, Spectrometers, Detectors and Associated Equipment*, vol. 878, pp. 10–23, 2018. Radiation Imaging Techniques and Applications.

[20] S. Sharma, J. Chhokar, C. Curceanu, E. Czerwiński, M. Dadgar, K. Dulski, J. Gajewski, A. Gajos, M. Gorgol, N. Gupta-Sharma, *et al.*, "Estimating relationship between the time over threshold and energy loss by photons in plastic scintillators used in the j-pet scanner," *EJNMMI physics*, vol. 7, no. 1, pp. 1–15, 2020.

Performance comparison of different SiPM arrays coupled to neutron-γ scintillators

Jessica Delgado[1,2,*], Felix Pino[1], Sandra Moretto[1,3], Daniele Corti[3], Alessandro Griggio[3], Lucio Pancheri[4,5], Daniela Fabris[3]

[1]Department of Physics and Astronomy "Galileo Galilei", University of Padova, Italy
[2]Department of Physics and Earth Science, University of Ferrara, Italy
[3]INFN- Padova Section, Italy
[4]Department of Industrial Engineering, University of Trento, Italy
[5]INFN-TIFPA Trento Section, Italy
(*) jessicadelgadoalvarez@gmail.com

Abstract— **The market for SiPMs has grown rapidly in recent years, driven by their increasing use in applications such as medical imaging, nuclear and particle physics. Due to the growing demand, many manufacturers are now offering SiPMs, each of them with unique features and performance characteristics. As a result, it is essential to conduct comparative studies to evaluate and compare the performance of different SiPM arrays. Therefore, this work aims to compare the performance of assembled detectors based on two different SiPM arrays (AdvanSiD-NUV hybrid array ASD-NUV4S-P-4x4TD 17 mm x17 mm and the MPPC Hamamatsu S14161-6050HS-04 25 mm x 25 mm) coupled to two organic scintillators, a liquid scintillator EJ-309 (50 mm diameter x 50 mm thickness) and a plastic scintillator EJ-276G (25 mm diameter x 25 mm thickness). The assessment of the performance was made in terms of their energy resolution, time resolution, and capability to discriminate between γ-rays and fast neutrons. The outcomes show that the Hamamatsu SiPM array performs better for all of the characteristics studied, with the best configuration being when it was coupled to the EJ-309 liquid scintillator, obtaining the lowest energy and time resolutions, (0.108 ±0.004) and (0.599 ± 0.004) ns respectively, and the highest Figure of Merit (2.04 ± 0.01). A good Figure of Merit value that ensures an effective discrimination between fast neutrons and γ-rays.**

***Keywords*—SiPM-arrays, n/γ discrimination, neutron detection, scintillators.**

I. Introduction

The rapid advancement of silicon photomultiplier (SiPM) technology has led to an opportunity to investigate the replacement of commonly used photomultipliers (PMTs) by SiPMs as light converters in various applications, particularly in radiation detectors to be coupled to scintillators. Silicon Photomultipliers are high-sensitivity, high-efficiency light sensors that detect light ranging from near ultraviolet to near-infrared. They are made up of a matrix of sensitive micro-cells that are all connected in parallel. Each micro-cell is a Geiger-Mode avalanche photo-diode that operates beyond the breakdown voltage and incorporates a resistor for passive quenching [1].

The wide variety of applications on SiPM-based detectors goes from high-energy physics, particle physics, medical imaging, astronomy, to hazard and threat detection, among others, making SiPMs an active research target. Some of the most interesting characteristics of SiPMs are high photon detection efficiency, high gain at low operating voltages, compactness, and insensitivity to magnetic fields [2]. It should also be noted that SiPMs' prices are declining even as their technological advancements are accelerating. However, the SiPMs have some properties that make precise electronic readout difficult. These include a longer single photon response duration and relatively higher noise contributions from dark pulses, and after-pulsing [3]. Where a dark pulse is the result of a thermally generated electron that initiates an avalanche in the high field region, while the after-pulsing is produced by the release of carries that are trapped in the silicon crystal, and a second avalanche is created in the same microcell, the occurrence of this effect depends on the pixel recovery time [4]. Nevertheless, despite the noise that comes with SiPMs, they have demonstrated amazing results in terms of sensitivity, efficiency, and particle discrimination.

Among the most common SiPM arrays on the market are: 1. Hamamatsu SiPM arrays, which are manufactured by one of the leading SiPM array manufacturers. Their arrays have a high photon detection efficiency, a fast-timing resolution, and a low dark count rate [5]. 2. SensL, another manufacturer that offers SiPM arrays with high photon detection efficiency and low cross-talk [4]. 3. KETEK SiPM arrays, which have a low dark count rate and a low probability of after-pulsing [6]. 4. AdvanSiD, a company that provides SiPM arrays with high gain and a low after-pulsing probability [7]. Given that each SiPM array manufacturer has distinct characteristics and performance parameters, the choice of SiPM array depends on the specific application and experimental conditions.

Different investigations have been going on in order to compare these distinct characteristics, for example in a work done by M. Grodzicka-Kobylka [8], a SensL and a Hamamatsu SiPM arrays were compared in terms of γ spectrometry, using three inorganic scintillators (LSO, BGO, and CsI:Tl). In that study, the energy resolution, linearity, and stability of the two arrays were studied in function of the temperature and bias voltage applied. The results indicate that both SiPM arrays demonstrate satisfactory energy resolution and linearity, but the SensL array performed better at low temperatures.

Another important parameter to be studied is the discrimination performance when a SiPM array is coupled to a scintillator with n/γ-ray discrimination capability. It represents an important feature in different fields, such as radiation monitor applications [9] and homeland security for identifying SNM (Special Nuclear Materials) [10]. The PSD (Pulse Shape Discrimination) technique using SiPM arrays is a challenging task, mainly because it is based on the reconstruction of the time profile of the scintillation decay, and, it must be taken into account that the conversion of the light into charge in a SiPM does not occur in the same way as in a PMT. Nevertheless, from a previous work conducted by our group, we demonstrated the possibility of doing n/γ-ray discrimination using large-sized scintillators coupled to large area SiPM arrays [7]. In that case, three commercial scintillators were used (EJ-276, EJ-309, and EJ-301, with diameters ranging from 20 to 50 mm) in conjunction with two large areas of SiPM arrays from AdvanSiD (NUV and RGB). Good n/γ-ray discrimination was obtained for all combinations, with results comparable to the ones obtained using a PMT as a read-out. Taking these advancements into consideration, we continue our research, but this time using and compare different types of SiPM arrays.

In the current paper, we compare the performance of assembled detectors based on two organic scintillators with n/γ-ray discrimination capabilities, a plastic scintillator EJ-276G (25 mm dia. x 25 mm thick.) and a liquid scintillator EJ-309 (50 mm dia. x 50 mm thick.), both with cylindrical shape and from Eljen Technology–Texas USA, coupled to two different SiPM arrays, the AdvanSiD-NUV hybrid array ASD-NUV4S- P-4x4TD 17 mm x17 mm and the MPPC Hamamatsu S14161-6050HS-04 25 mm x25 mm. The performance was assessed in terms of the energy resolution, time resolution, and n/γ discrimination. The purpose of this work is to provide the performance characteristics of the four possible configurations in order to aid future studies toward a better selection of a SiPM array for a specific experiment or application.

II. Methods and Experimental Set-up

A. SiPM arrays

The two types of SiPM arrays tested are the following: AdvanSiD Hybrid TD Array NUV-SiPM (16 channels, 4 mm x 4 mm active area per channel) and Hamamatsu S14161-6050HS-04 (16 channels, 6 mm x6 mm active area per channel), a picture of them can be observed in Fig. 1, and their main characteristics can be seen in Table I, respectively. The Photon Detection Efficiency (PDE) of the SiPMs with respect to the wavelength and the emission curves of the plastic and liquid scintillators are illustrated in Fig. 2. The values of the integrated PDE of each SiPM multiplied by the normalized emission spectrum, which represent the probability that a scintillation photon that impinges the active area of the SiPM is being converted into a measurable current by the SiPM, can be seen in Table II. It is visible that the PDE of the two SiPM arrays covers the wavelength emission range of both organic scintillators, and the best match is seen with the Hamamatsu SiPM and the emission peak of the liquid scintillator.

Fig. 1. Pictures of the AdvanSiD-NUV and Hamamatsu SiPM arrays.

TABLE I
MAIN CHARACTERISTICS OF THE SIPM ARRAYS TESTED

	ADVANSID-NUV	Hamamatsu
Array size	16.8 mm x 17.4 mm	25 mm x 25 mm
Single SiPM size	4 mm x 4 mm	6 mm x 6 mm
Channel number	16 channels	16 channels
No. of cells/channel	9340	14331
Cell size	40 μm x 40 μm	50 μm x 50 μm
Terminal capacitance Ct	90 fF	2000 pF
Peak sensitivity wavelength	420 nm (43%)	450 nm
Breakdown voltage, typ	26 V	38 V
Dark count rate @4V OV	< 100kHz/mm2	7.5 μA

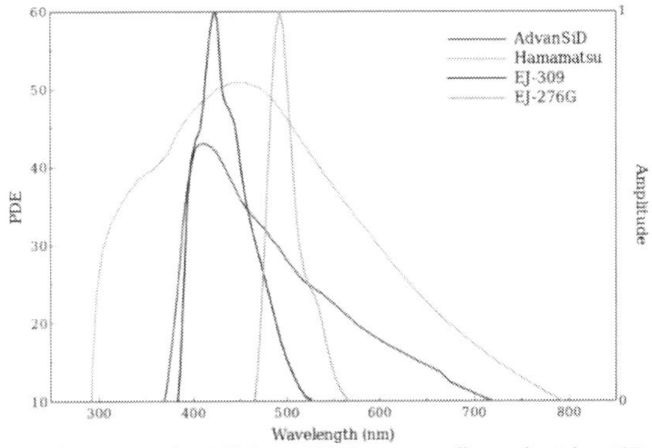

Fig. 2. Photon Detection Efficiency (PDE) corresponding to the AdvanSiD and Hamamatsu SiPM arrays (left axis) compared with the emission curves of the plastic and liquid scintillators (right axis). Information taken from the manufactures [5, 6, 11].

TABLE II
INTEGRATED PHOTON DETECTION EFFICIENCY MULTIPLIED BY THE NORMALIZED EMISSION SPECTRUM FOR EACH SIPM ARRAY/SCINTILLATOR COMBINATION

Detector + SiPM array	$\sum_{\lambda=250\,nm}^{900\,nm} \mathrm{PDE}(\lambda)\cdot \mathrm{I}(\lambda)\cdot \Delta\lambda$
EJ-276G + AdvanSiD	26.7
EJ-309 + AdvanSiD	36.6
EJ-276G + Hamamatsu	43.4
EJ-309 + Hamamatsu	48.4

https://doi.org/10.1051/epjconf/202328810008

B. *Organic scintillators*

In order to test the SiPM arrays, two organic scintillators with n/γ-ray discrimination capabilities were used, a plastic scintillator EJ-276G (25 mm dia. x 25 mm thick.) and a liquid scintillator EJ-309 (50 mm dia. x 50 mm thick.), both with cylindrical shape and from Eljen Technology–Texas USA. A small portion of optical grease was used for coupling the detectors to the SiPM arrays (see Fig. 3).

Fig. 3. Picture of the EJ-276G and the EJ-309 scintillators coupled to the AdvanSiD-NUV and Hamamatsu SiPM arrays, respectively.

C. *Readout board of the SiPM arrays*

The readout of the SiPM array output, and the bias voltage supply, was performed using a dedicated board, which follows the sequence, of four sets, where each set is made up of four single SiPMs connected in parallel. The output signals of each set are amplified using four trans-impedance preamplifiers based on an ultra-low noise, high-speed OpAmp (LMH6629). Then, the four amplified signals are summed giving the total output. For a better comprehension of the circuit a simple diagram of the readout board is shown in Fig. 4. Voltages of +/- 2.9 V are applied to each OpAmp. The applied bias voltage is between 30 and 40 V (depending on the SiPM array used). The same bias voltage is applied to each single SiPM.

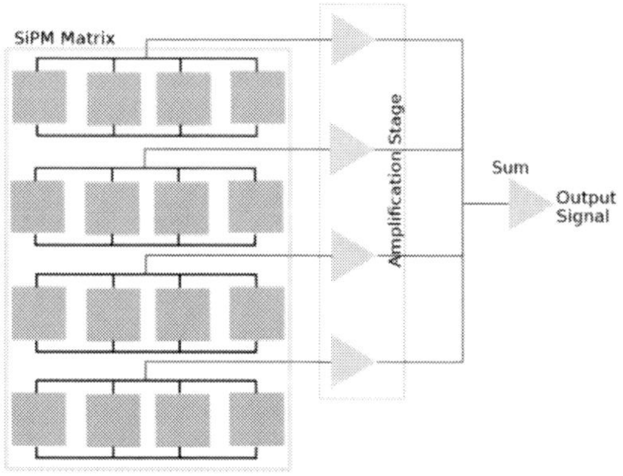

Fig. 4. Circuit diagram of the SiPM arrays' readout board.

D. *Experimental details*

The readout boards of the SiPM arrays were powered with an MT180T power supply. The overvoltage (OV) applied to each single SiPM was determined by optimizing the energy resolution corresponding to each assembled detector. The bias voltage (+800 V) applied to the $LaBr_3$:Ce detector, that is used for coincidence purposes, was delivered by a V6533 CAEN power supply unit controlled through a USB Controller (CAEN V1718). The output signals were digitized by a V1730 CAEN digitizer, which has a sampling rate of 500 MSamples/s, and an ADC resolution of 14-bit. Modern digitizers are based on a programmable FPGA (Field Programmable Gate Array), where the digitized waveform can be pre-processed online, the algorithm implemented in the digitizer provided for each triggered event, among others, the timestamp, and the total (Q_{total}) and partial (Q_{short}) integrals of the waveforms. Also, it is possible to acquire the digitized waveforms and analyze the data offline. Optical fiber connection was used to control the data acquisition. The digitizer was connected to the PC using a CAEN A4818 USB 3.0 to CONET2 Adapter. The ABCD (Acquisition and Broadcast of Collected Data) software [12, 13], which is provided as an open-source project (https://github.com/ec-jrc/abcd), was used to handle all aspects of the electronics and data acquisition. The following sources were used for the measurements: two γ-ray sources ^{22}Na and ^{137}Cs (~ 300 kBq), and one n/γ-ray source Am-Be (neutron emission rate ~ 2×105 s^{-1}).

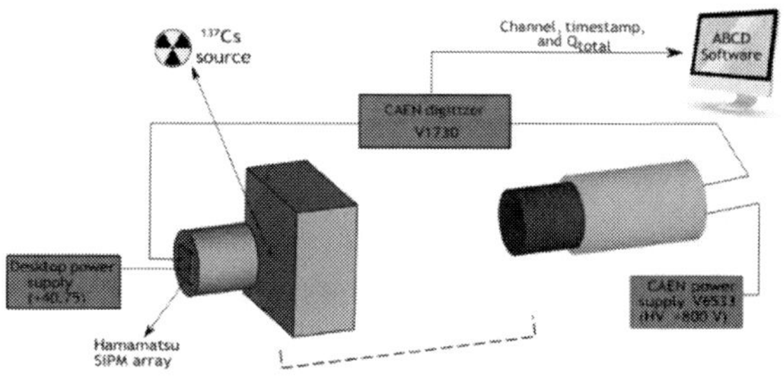

Fig. 5. Schematic view of the experimental setup used for the determination of the energy and time resolution, using the Hamamatsu SiPM array coupled to the EJ-309 detector in coincidence with the LaBr3:Ce detector, and placing the source and the collimator between both detectors. The electronic components used for data acquisition and high voltage supply are also displayed.

E. *Determination of the energy and time resolution*

The determination of the energy resolution of each assembled detector was conducted using a Compton coincidence technique, the details can be found in [14]. The method consists in setting up the organic detector under study in coincidence with a reference detector. In this case, the reference detector is a $LaBr_3$:Ce detector (50 mm diam. x 50 mm thick., from Saint-Goban). It was placed in front of the organic scintillator (face-to-face) at ~ 20 cm. A lead collimator with a cuboid form (10x10x5 cm^3) was located between both detectors, with a diameter of aperture of ~ 1 cm. A ^{137}Cs source (with activity of ~ 300 kBq) was attached to the collimator's hole (on the side of the organic detector). A schematic view of the experimental setup can be seen in Fig. 5. The method takes advantage of the fact that a γ-ray coming from the source (661.7 keV) can undergo a scattering of 180° on the active volume of the organic detector, depositing approx. 477.4 keV, and then, the back-scattered γ-ray goes through the collimator's hole, reaches the reference detector, and deposits all its energy (184.3 keV). Finally, the main idea is to build the energy spectrum of the events, recorded by the organic detector under study, which are in coincidence with the full-energy events corresponding to the back-scattered γ-rays detected by the LaBr detector. In this case, the organic detector spectrum mainly consists of the Compton edge events. Then, a gaussian fit is applied, and the energy resolution of the detector is obtained by $R_E = FWHM/E_o$, where E_o=477.4 keV.

For the time resolution determination, the same experimental setup was employed, but in this case a ^{22}Na source (with activity of ~ 300 kBq) was used. An off-line analysis (coincidence filter) was performed on the digitized waveforms coming from each detector, with the purpose to optimize the parameters (fraction and delay) of the Digital Constant Fraction Discrimination (DCFD), the parameters were optimized to obtain the best time resolution. The delay parameter was varied from 10 ns to 100 ns, and for the fraction parameter, three values were considered 25%, 50%, and 75%. Then, the final time spectrum is built by considering the events in the full-energy peak (@511 keV) of the LaBr detector, and the events with energies greater that 150 keV of the organic detector.

F. Determination of the Pulse Shape Discrimination (PSD) capabilities

The n/γ-ray discrimination capability of each assembled detector under study was determined. An Am-Be neutron source was used for this purpose. The double gate integration method was employed to perform the pulse shape discrimination. It consists in calculating the pulse shape parameter (PSP), defined as PSP=$(Q_{long} - Q_{short})/Q_{long}$, where Q_{short} is the integral over a short time window including the fast rise time region and part of the fastest decay component of the pulse, and the Q_{long} is the integral over a long gate, encompassing the majority of the pulse. By analyzing the PSP distribution, it is possible to compute the Figure of Merit (FoM), defined as FoM $=\Delta/(\delta_n + \delta_\gamma)$, where Δ represents the difference between the mean values of the neutron and γ-ray distributions, while δ_n and δ_γ represent the Full Width at Half Maximum (FWHM) of the neutron and γ-ray distributions, respectively. The short and long integration gates needed to be optimized, until finding the maximum value of the FoM, which represents the best results in terms of n/γ-ray discrimination. The FoM values given in this work were determined using only the events located in a light output range corresponding approximately to Compton edge of the gamma line emitted by the ^{137}Cs source.

III. Results and analysis

A. Signals comparison of both SiPM arrays

As the first step, a comparison of the output signals coming from both SiPM arrays was performed, for this purpose a ^{137}Cs source was used, and both SiPM arrays were coupled to the EJ-276G detector. In Fig. 6 it can be seen the typical waveforms for both cases. The signals correspond to an event with an energy equal to the Compton edge of the γ-ray emitted by the ^{137}Cs source. From the shape of the waveforms, in both cases the rising time is quite similar, while the decay time is so much faster in the case of the AdvanSiD-NUV array. Behavior that depends on the recharge time constant of the SiPM micro cells, which is proportional to the capacitance of the micro cell, ~RC (see Table I), and this parameter describes how quickly the micro cells can be recovered after detecting a scintillation photon.

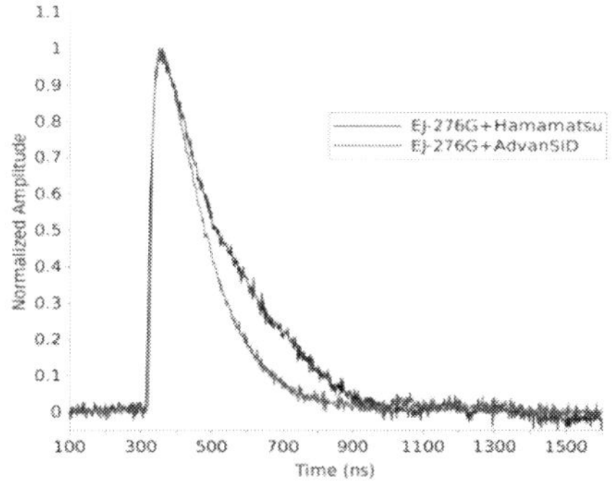

Fig. 6. Typical waveforms of the EJ-276G scintillator coupled to the Hamamatsu and AdvanSiD- NUV SiPM arrays. The signals correspond to an event of 477.4 keV, so the Compton edge of a 661.7 keV γ-ray.

B. Energy and Time Performance

The overvoltage applied to each scintillator/SiPM array configuration was optimized in function of the energy resolution, where the best performance was found for the voltages shown in Table III. The energy resolution determination was conducted using a Compton coincidence technique, as it was explained in Subsection 2.5. An illustration of the results obtained for one of the configurations can be seen in Fig. 7. Fig. 7a shows the energy spectrum of the LaBr detector in coincidence, wherein a red line series is seen as the energy selection corresponding to the back-scattered γ-ray full- energy peak at approx. 184.3 keV. Then, in Fig. 7b the energy spectrum of the EJ-309/Hamamatsu SiPM array assembly is shown, besides, the energy spectrum of the events that are in coincidence with the full-energy events of the back-scattered γ- rays are delineated with a green line. To get the energy resolution at 477.4 keV, the last-mentioned spectrum was fitted with a Gaussian function. Finally, using the set-up described in Subsection 2.5, the time spectrum obtained with the coincidences between the two γ-rays of 511 keV emitted by the ^{22}Na source is given in Fig. 7c.

TABLE III
BIAS VOLTAGE APPLIED TO EACH SCINTILLATOR/SIPM ARRAY CONFIGURATION

Scintillator/SiPM array	OV (V)
EJ-276G/AdvanSiD	4
EJ-276G/Hamamatsu	2.75
EJ-309/AdvanSiD	3.5
EJ-309/Hamamatsu	2.75

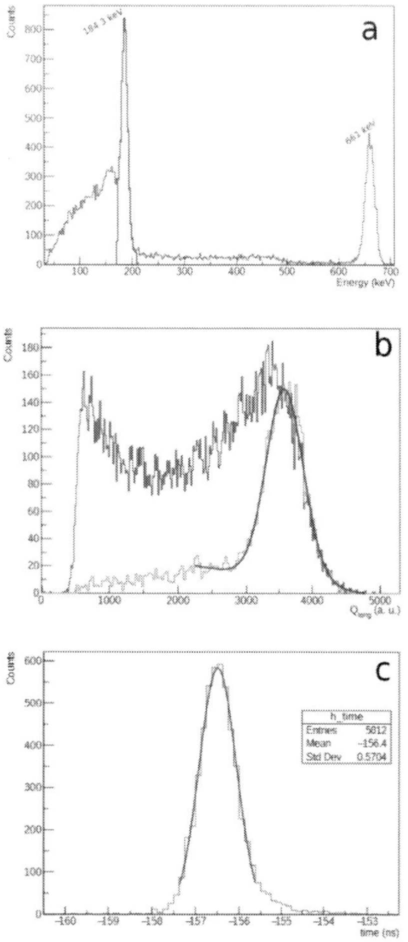

Fig. 7 (a) Energy spectrum of the LaBr detector in coincidence using the 137Cs source, wherein a red line it is shown the energy selection corresponding to the back-scattered γ-ray full-energy peak at approx. 184.3 keV. (b) Energy spectrum of the EJ-309/Hamamatsu SiPM array assembly in coincidence using the 137Cs, in green line the events that are in coincidence with the full-energy events of the back-scattered γ- rays. (c) Time spectrum of the coincidence experiment using the same detectors and the 22Na source.

On the other hand, for the time resolution determination, an offline analysis of the waveforms was carried out, where the DCFD's parameters were optimized in order to get the best time resolution. The optimal results were found using 120 ns of delay and 50% of fraction for the organic detectors/SiPM arrays assemblies, and 20 ns of delay and 20% of fraction for the LaBr detector. Then, another filter was implemented, by selecting the energy ranges of interest in each detector's spectrum, which corresponds to the 511 keV peak for the LaBr and all events with energies greater than 150 keV for the organic detector. A Gaussian fit is performed to the time spectrum, where the total time resolution (FWHM) of the system is obtained. The total time resolution corresponds to the following expression: $FWHM^2_{total} = FWHM^2_{LaBr} + FWHM^2_{Scintillator/SiPM}$, where the time resolution of the LaBr detector is 0.366±0.002, and in this way, the time resolution of the organic detector/SiPM array can be easily obtained. The same procedure for determining time and energy resolution was used for each organic scintillator/SiPM array assembly.

The results obtained can be seen in Table IV and Fig. 9. Despite the pulse's decay time of the AdvaSiD is faster than in the case of the Hamamatsu SiPM array (mentioned in Subsection 3.1), the Hamamatsu SiPM performances better in terms of energy and time resolution, for both organic scintillators. In particular, with the plastic scintillator, where the energy resolution of the Hamamatsu improved by 35% with respect to the AdvanSiD, and the time resolution improved by 40%.

TABLE IV
VALUES OF THE ENERGY AND TIME RESOLUTION OBTAINED FOR EACH SCINTILLATOR/SIPM ARRAY CONFIGURATION

Scintillator/ SiPM array	ENERGY RESOLUTION	TIME RESOLUTION (NS)
EJ-276G/ AdvanSiD-NUV	0.290 ± 0.002	1.84 ± 0.01
EJ-276G/ Hamamatsu	0.188±0.004	1.09 ± 0.01
EJ-309/ AdvanSiD-NUV	0.20 ± 0.02	1.30 ± 0.01
EJ-309/ Hamamatsu	0.108 ± 0.004	0.599± 0.004

C. *Neutron/γ-ray discrimination capability*

The discrimination capability of each combination scintillator/SiPM array was explored, using an Am-Be source and the double integration method. First, for each assembly, an optimization of the short and long integration gates was conducted, until finding the maximum value of the FoM. For each configuration, a 2D-PSP and PSP plots were obtained. The 2D-PSP plot corresponds to 2D histogram of the PSP values against the Q_{long} (it is proportional to the total light output) values. The PSP distribution plot is built with all the events that are in a light output range corresponding to the Compton edge associated to the γ-ray of 661.7 keV (^{137}Cs emission). The FoM value is calculated by analyzing the PSP distribution, as it was explained in Subsection 2.6. Fig. 8 shows the 2D-PSP and the PSP plots of the EJ-276G coupled to the Hamamatsu SiPM array, using an Am-Be source. In the 2D-PSP plot is highlighted in red the range of the events used for the construction of the PSP distribution. The FoM and the optimized values of the w_{long}, and the w_{short}, for each configuration can be seen in Table V. The FoM values of the Hamamatsu configuration outperform the AdvanSiD-NUV results by 43% with the liquid scintillator and 56% with the plastic scintillator.

TABLE V
OPTIMIZED INTEGRATION WINDOWS (W_{LONG} AND W_{SHORT}), AND FOM VALUES OBTAINED FOR EACH SCINTILLATOR/SIPM ARRAY CONFIGURATION

Scintillator/SiPM array	w_{long}	w_{short}	FoM
EJ-276G/AdvanSiD-NUV	904	240	0.76 ± 0.02
EJ-276G/Hamamatsu	590	220	1.19 ± 0.02
EJ-309/AdvanSiD-NUV	586	122	1.15 ± 0.01
EJ-309/Hamamatsu	730	180	2.04 ± 0.01

In Fig. 9 it can be seen an illustration of all the results obtained in this work, where it is notable the higher performance of the Hamamatsu SiPM array, in terms of energy resolution, time resolution, and discrimination capabilities. In particular, the best results were obtained with the EJ-309/Hamamatsu SiPM array assembly. The superior performance of the Hamamatsu SiPM array can be explained by several factors: The array size is bigger with respect to the AdvanSiD's area (see Table I), thus more light of each event can be collected. The Hamamatsu array covers the whole face of the plastic scintillator and ~28% of the liquid's face area, while the AdvanSiD covers ~60% and ~15% respectively. Higher PDE (Photon Detection Efficiency), then the higher signal-to-noise ratio. And good matching between the PDE of the Hamamatsu SiPM and the emission spectrum of the detectors (see Table I and Fig. 2), increasing in that way the probability of detection.

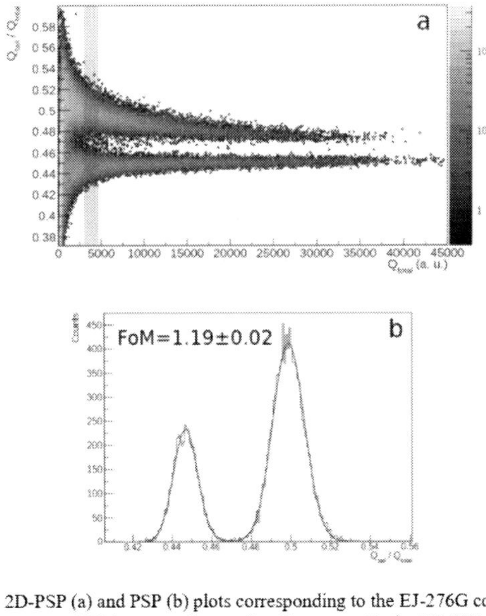

Fig. 8. 2D-PSP (a) and PSP (b) plots corresponding to the EJ-276G coupled to the Hamamatsu SiPM array, using an Am-Be source. In the 2D-PSP plot it is highlighted in red the range of the events used for the construction of the PSP distribution.

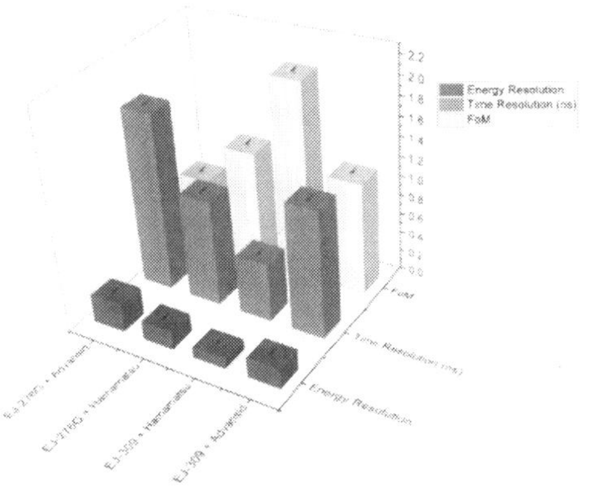

Fig. 9. Energy resolution, time resolution and FoM values of the EJ-309 and EJ-276G scintillators coupled to the AdvanSiD-NUV and Hamamatsu SiPM arrays.

IV. CONCLUSIONS

In this paper, two types of SiPM array sensors were compared in terms of energy resolution, time resolution, and discrimination capabilities between neutrons and γ-rays. Specifically, the SiPM arrays investigated are the AdvanSiD hybrid array ASD- NUV4S-P-4x4TD 17 mm x17 mm and the MPPC Hamamatsu S14161-6050HS-04 25 mm x 25 mm. To conduct the tests, we coupled the SiPM arrays to two organic scintillators that possess n/γ-rays discrimination capabilities, a plastic scintillator EJ-276G (25 mm dia. x 25 mm thick) and a liquid scintillator EJ-309 (50 mm dia. x 50 mm thick), both from Eljen Technology, Texas USA. Our objective is to provide a detailed performance analysis of these four configurations, considering that by studying the performance characteristics of SiPMs, researchers can gain a better understanding of their capabilities and limitations. Furthermore, as SiPMs are becoming increasingly popular in research and industry, it is important to have a thorough understanding of their properties in order to make informed decisions regarding their selection and use. To determine the energy and time resolution, a coincidence experiment was conducted, using a LaBr detector. For the energy resolution determination, a Compton coincidence technique was applied, where the energy resolution of each configuration was obtained at 477.4 keV, which represents the energy deposited of a γ-ray of 661.7 keV (coming from a 137Cs source) after suffering a back-scattering interaction in the active volume of the detector. From all the configurations studied, the EJ-309/Hamamatsu SiPM array gave the best performance, with an energy resolution of ~10%, and with a time resolution of (0.599±0.004) ns, overcoming the results of the AdvanSiD-NUV array (using the same scintillator), where the values of energy resolution and the time resolution increased by 46% and 53%, respectively. Also, the bias voltage applied to the arrays was optimized in function of the energy resolution. For the determination of the time resolution, a 22Na source was used, where the annihilation 511 keV γ-rays were measured in both detectors (the organic and the LaBr).

On the other hand, the study of the discrimination capability consisted in using a mixed n/γ-ray source (AmBe source), where the performance of the SiPM arrays coupled to the organic detectors is compared in function of the FoM value, which is a quantitative measure used to evaluate the discrimination capability of a system. Once more, the Hamamatsu SiPM array with the EJ-309 liquid scintillator showed a superior performance, giving a FoM value of 2.04, 43% greater than the AdvanSiD-NUV array. From the results obtained in this work, it has been concluded that the Hamamatsu SiPM array performs better than the AdvanSiD-NUV due to its larger array size, greater Photon Detection Efficiency, higher signal-to-noise ratio, and good matching with the emission spectrum of the scintillators.

References

[1] "Introduction to silicon photomultipliers (SiPMs)," First Sensor, Tech. Rep., 2019.
[2] E. Garutti, "Silicon photomultipliers for high energy physics detectors," Journal of Instrumentation, vol. 6, no. 10, pp. C10 003–C10 003, Oct. 2011.
[3] R. M. Preston, J. E. Eberhardt, and J. R. Tickner, "Neutron-Gamma Pulse Shape Discrimination Using Organic Scintillators with Silicon Photomultiplier Readout," IEEE Transactions on Nuclear Science, vol. 61, no. 4, pp. 2410–2418, Aug. 2014.
[4] "Introduction to SiPM. technical note," SensL, Tech. Rep., 2017.
[5] "MPPCs (SiPMs) / SPADs | Hamamatsu Photonics." [Online]. Available: https://www.hamamatsu.com/eu/en/product/optical-sensors/mppc.html
[6] "Products." [Online]. Available: https://www.broadcom.com/products. Accessed on: April 15, 2023.
[7] F. Pino, J. Delgado, G. Mantovani, L. Pancheri, M. Bello, D. Fabris, C. L. Fontana, M. Polo, V. Ruiz, D. Brunelli, A. Quaranta, and S. Moretto, "Novel Detector Assembly for Neutron/Gamma-Ray Discrimination Applications Based on Large- Sized Scintillators Coupled to Large Area SiPM Arrays," IEEE Transactions on Nuclear Science, vol. 69, no. 4, pp. 668–676, Apr. 2022.
[8] M. Grodzicka-Kobylka, T. Szczesniak, and M. Moszyn´ski, "Comparison of SensL and Hamamatsu 4×4 channel SiPM arrays in gamma spectrometry with scintillators," Nuclear Instruments and Methods in Physics Research Section A: Accelerators, Spectrometers, Detectors and Associated Equipment, vol. 856, pp. 53–64, Jun. 2017. [Online]. Available: https://linkinghub.elsevier.com/retrieve/pii/S0168900217303388. Accessed on: April 15, 2023.
[9] F. Pino, C. Fontana, J. Delgado, D. Fabris, G. Nebbia, M. Turcato, D. Brunelli, L. Pancheri, A. Quaranta, and S. Moretto, "Characterization of a medium-sized CLLB scintillator: single neutron/gamma detector for radiation monitoring," Journal of Instrumentation, vol. 16, no. 11, p. P11034, Nov. 2021. [Online]. Available: https://iopscience.iop.org/article/10.1088/1748-0221/16/11/P11034. Accessed on: April 15, 2023.
[10] D. Cester, G. Nebbia, L. Stevanato, F. Pino, L. Sajo-Bohus, and G. Viesti, "A compact neutron–gamma spectrometer," Nuclear Instruments and Methods in Physics Research Section A: Accelerators, Spectrometers, Detectors and Associated Equipment, vol. 719, pp. 81–84, Aug. 2013. [Online]. Available: https://linkinghub.elsevier.com/retrieve/pii/S0168900213004373. Accessed on: April 15, 2023.
[11] "Products - Eljen Technology." [Online]. Available: https://eljentechnology.com/products. Accessed on: April 15, 2023.
[12] C. L. Fontana, M. Lunardon, F. E. Pino, L. Stevanato, A. Carnera, C. Sada, F. Soramel, and S. Moretto, "A distributed data acquisition system for signal digitizers with on-line analysis capabilities," in 2017 IEEE Nuclear Science Sym- posium and Medical Imaging Conference (NSS/MIC). Atlanta, GA: IEEE, Oct. 2017, pp. 1–5. [Online]. Available: https://ieeexplore.ieee.org/document/8533063/. Accessed on: April 15, 2023.
[13] C. L. Fontana, A. Carnera, M. Lunardon, F. E. Pino, C. Sada, F. So- ramel, L. Stevanato, and S. Moretto, "A distributed data acquisition system for nuclear detectors," International Journal of Modern Physics: Conference Series, vol. 48, p. 1860118, Jan. 2018. [Online]. Available: https://www.worldscientific.com/doi/abs/10.1142/S2010194518601187. Accessed on: April 15, 2023.
[14] L. Swiderski, M. Moszynski, W. Czarnacki, J. Iwanowska, A. Syntfeld Kazuch, T. Szczesniak, G. Pausch, C. Plettner, and K. Roemer, "Measurement of Compton edge position in low-Z scintillators," Radiation Measurements, vol. 45, no. 3-6, pp. 605–607, Mar. 2010. [Online]. Available: https://linkinghub.elsevier.com/retrieve/pii/S1350448709001917. Accessed on: April 15, 2023.

https://doi.org/10.1051/epjconf/202328810009

Comparison of CLYC vs. CsI in a hand-held detector

Monika Risse*, Hermann Friedrich, Jeannette Glabian, Theo Köble, Olaf Schumann, Thorsten Teuteberg

Fraunhofer-Institut für Naturwissenschaftlich-Technische Trendanalysen, Germany

(*) monika.risse@int.fraunhofer.de

Abstract— **The detection of radioactive and nuclear (RN) material is gaining importance due to the current political situation. Fast and reliable results from on-site measurements require devices to be transportable, quickly ready-for-use, and easily operable. This has driven the development of measurement devices based on materials which enable simultaneous detection of gammas and neutrons. A detector material with these properties is the scintillator Cs_2LiYCl_6:Ce (CLYC). It combines the characteristics of a medium resolution gamma ray detector like Sodium iodide (NaI) or Cesium iodide (CsI) and a ^{3}He proportional counter tube for the detection of neutrons. In this paper the performance of a CLYC based device was compared to one using a traditional scintillator. The exemplary study was carried out on the basis of two RadEYE devices from Thermo: the RadEye SPRD GN with a CLYC crystal and the RadEYE SPRD with a CsI crystal were compared regarding gamma detection. Moreover, the neutron detection capability of the CLYC device was investigated. Dynamic tests were performed using our qualification test system (QuTeSt). This allows repeatable tests with radioactive sources passing the device under test at controlled conditions. Applicable standards and test procedures for Radiation Isotope Identifiers (RIID) from the International Electrotechnical Commission (IEC), from the American National Standards Institute (ANSI), and those released during the ITRAP+10 program were applied. Several tests were performed using radioactive sources covering a wide energy range.**

Keywords —**Dynamic tests, CLYC, CsI, hand-held detector.**

I. Introduction

A wide variety of handheld systems for detection purpose exists. They can be assigned to different classes like Radiation Isotope Identifiers (RIID), Personal Radiation Detectors (PRD), or Spectroscopic version of PRDs (SPRD) which differ in their external appearance such as weight, size or shape, the measurement task, the measurement result output, which type of radioactivity they can detect, and also in the detector material used. In the present paper a comparison between different detector materials and their influence on the output is investigated. This is done by using handheld devices which should leave constant as many factors as possible. Two devices of the same kind but with different detector materials were chosen. The question to be answered in this paper is in particular: Do the two detector materials behave differently in identification and dynamic tests?

To answer this question, results have been obtained in tests according to International standards like International Electrotechnical Commission (IEC), American National Standards Institute (ANSI) or procedures established during the "Illicit Trafficking Radiation Assessment Program (ITRAP+10)" of the European Union. The Fraunhofer INT facility QuTeSt which had been developed in the framework of ITRAP+10 was used for these tests.

II. QuTeSt

Mainly dynamic tests were performed in the framework of the current paper which were done using the dynamic test system part of the QuTeSt system [1].

In Fig. 1 an example of a screenshot of the video observation system during dynamic testing is given. The source is attached to an extension bar to get closer to the devices. The display gives the time prior and after passing the zero position: the point of closest approach at which the source is in front of the detector. The velocity of the source is given as well.

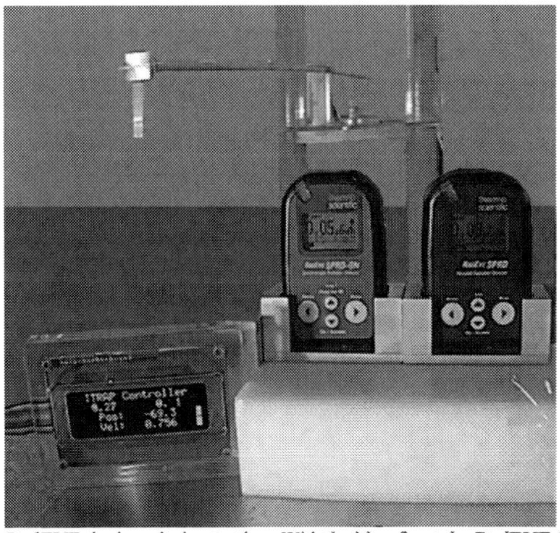

Fig. 1. RadEYE devices during testing. With the blue front the RadEYE SPRD GN with a CLYC crystal and with the black front the RadEYE SPRD equipped with a CsI crystal. The source at the extension bar is moving to the left. Below the source the control display is placed.

 https://doi.org/10.1051/epjconf/202328810009

III. Dynamic Tests for Gammas

Dynamic testing is foreseen in the time to alarm tests for gammas and neutrons. The specifications for the gamma testing are listed in Table I. The ITRAP+10 test method for PRDs has a separate dynamic testing part. With a dose rate equivalent of 0.5 μSv/h a source shall be passing by with a speed of 0.6 m/s. The sources ^{241}Am, ^{137}Cs, and ^{60}Co shall be used. 30 trials are foreseen and a device will pass the test if an alarm is generated in at least 29 out of 30 trials. The time to alarm is not specified.

TABLE I
Time to Alarm Specifications for Gammas

Device class	ANSI			IEC			ITRAP +10		
	[s]	condi-tions	trials	[s]	condi-tions	trials	[s]	condi-tions	trials
RIID	1	0.5 m/s 0.8-1.2 m	9/10	2	0.5 m/s 1 m	9/10	3	step change	29/30
PRD	2	1.2 m/s 1.5 m	19/20	5	1.2 m/s 1.5 m	8/10	2	step change	29/30
SPRD	2	1.2 m/s 1.5 m	19/20	20	step change	8/10	-	-	

All test to be performed with ^{241}Am, ^{60}Co, and ^{137}Cs with 0.5 μSv/h, except for the ANSI RIID testing with 0.1 μSv/h. In dynamic testing, the source passes the device. The first columns indicate the time within the alarm is to be generated. The conditions are v_0 and d_0 for dynamic tests or states a step change test. The trials column gives the criterion for passing as well as the number of trials to be performed.
ANSI values taken from [5], [6] and [7].
IEC values taken from [2], [3] and [4].
ITRAP+10 values taken from [8] and [9].

Only the dynamic tests within the ITRAP+10 do not foresee a specified distance d_0 for the source. In the other cases the distance for the given dose rate varies from 1 m to up to 1.5 m. To fulfill the criteria for tests according to different standards a greater variety of different source activities is necessary.

To achieve the foreseen dose rate equivalent at the point of closest approach with a limited number of sources, the distance d of the radiation source is adjusted. The dose rates are measured in a static measurement with a calibrated device. If the source used does not fulfill the foreseen criterion, the conditions can be changed. The change in the distance has to be compensated in the dynamic tests by changing the velocity v_0 stated in the standard (see Table I) to v using (1).

$$v = v_0 \times d/d_0 \tag{1}$$

IV. Test Devices RadEYE SPRD and RadEYE SPRD GN

The qualification tests were carried out using two hand-held devices of the same kind but with different detector materials. These are the spectroscopic personal radiation detector RadEYE devices from Thermo [10], [11]. The RadEYE SPRD is equipped with a CsI crystal and the RadEYE SPRD GN with a CLYC crystal. Due to the incorporated nuclear reaction ^{6}Li(n,t)α the latter enables the simultaneous measurement of gammas and neutrons. Table II gives an overview of the crystal related specifications. The Cesium iodide crystal is a bit larger and has a larger surface than the CLYC detector. Fig. 2 shows the relative energy response for the two devices as given in the manufacturer manuals.

TABLE II
Specifications of RadEYE SPRD and RadEYE SPRD GN

		RadEYE SPRD	RadEYE SPRD GN

	material	CsI	CLYC
crystal	volume [cm^3]	6.8	4.8
	front face	∅: 1.6 cm	1.25 cm x 1.25 cm
	can detect	gamma	gamma und neutron

Information taken from the manuals [10], [11].

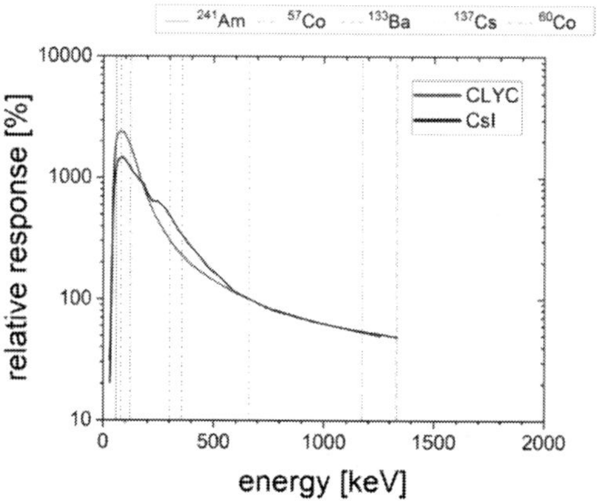

Fig. 2. Energy response for CLYC in red and CsI in blue taken from the instruction manuals of the devices given by Thermo [10], [11]. The response is given relative to the response for ^{137}Cs which is set to 100 % at the corresponding energy line of 661.6 keV. In addition, the positions of the prominent energy lines of the relevant nuclides are added as indicated in the figure.

V. Static Identification Measurements – Simultaneous Nuclides

In order to evaluate the performance concerning identification, static measurements with ^{137}Cs, ^{60}Co, ^{133}Ba, and ^{152}Eu were carried out. These nuclides were chosen to cover the whole energy region in which the crystals are sensitive. The sources were measured one by one and in combination. In a first set of measurements the sources were placed in a distance in which each lead to a dose rate of 0.5 μSv/h like it is foreseen for the dynamic testing. Unfortunately, except for ^{133}Ba, the dose rates were too low to obtain an identification result. Therefore, the sources were moved closer and showed in the single measurements 1.5 μSv/h ^{60}Co, 0.9 μSv/h ^{137}Cs, 0.7 μSv/h ^{152}Eu, and 0.4 μSv/h ^{133}Ba. Fig. 3 a) and b) show the gamma spectra for the nuclides measured in combination. For comparison purpose a gamma spectrum obtained with a Ge detector is given in a). The one obtained with the CLYC containing RadEYE SPRD GN is given in red in b) and the

RadEYE SPRD with the CsI in blue in b) too. There is a difference in the count height due to the different crystal sizes. The shape of the ^{60}Co peaks is different and worse for the CLYC compared to CsI. The lines for ^{137}Cs are similar. In addition, the lines from ^{133}Ba with lower energies and for ^{152}Eu in the whole energy region are marked.

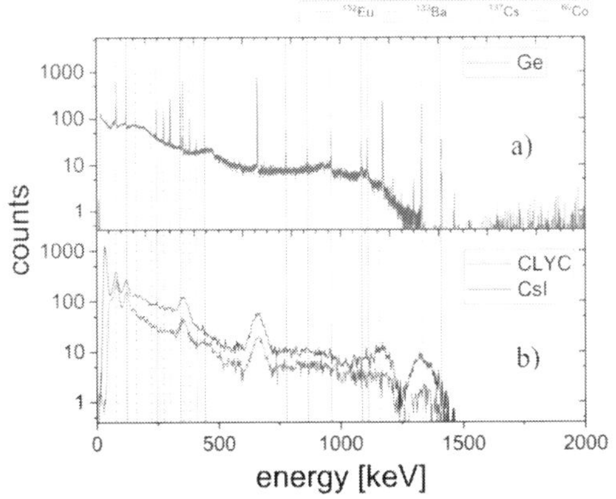

Fig. 3. Simultaneous measurement of 1.5 µSv/h ^{60}Co, 0.9 µSv/h ^{137}Cs, 0.7 µSv/h ^{152}Eu, and 0.4 µSv/h ^{133}Ba. The corresponding energy lines are marked like indicated. a) Spectra obtained with a germanium detector, b) spectra obtained with RadEYE SPRD with CsI in blue and with RadEYE SPRD GN with CLYC in red.

Both devices were used in the ratemeter mode. The identification results from the single source measurements for the RadEYE SPRD with CsI yielded always the correct result. The RadEYE SPRD GN was correct for the ^{133}Ba and the ^{152}Eu. The ^{137}Cs measurements indicated always the correct nuclide, but the result was partly uncertain. On the other hand, the measurement with the ^{60}Co instead never indicated ^{60}Co which reflects the poor quality of the peaks in the spectrum. The combined task was not fulfilled by both devices. Surprisingly, however, the combined measurements with the RadEYE SPRD GN also identified ^{60}Co. All results obtained in 3 respectively 4 measurements are listed in Table III.

TABLE III
COMBINED IDENTIFICATION MEASUREMENTS

nuclide	RadEYE SPRD GN CLYC			RadEYE SPRD CsI			
	70 s	95 s	100 s	50 s	50 s	50 s	50 s
^{60}Co	x	x	x	x	x	x	x
^{137}Cs	x	-	-	x	x	x	x
^{133}Ba	-	-	-	-	x	x	-
^{152}Eu	-	x	x	-	-	-	-
i. a.	^{57}Co	^{57}Co	^{57}Co	u. n.	u. n.	u. n.	u. n.

Measurement results in 3 measurements with RadEYE SPRD GN with CLYC und 4 measurements with RadEYE SPRD with CsI with 4 sources: 1.5 µSv/h ^{60}Co, 0.9 µSv/h ^{137}Cs, 0.7 µSv/h ^{152}Eu, 0.4 µSv/h ^{133}Ba. The noted times are rounded measurement times until the result is displayed in the ratemeter mode. When indicated in the result it is marked with an x if not with -. In the line i.a. results are indicated which are given in addition to the given nuclides. u.n. stands for unknown nuclides.

VI. Dynamic tests

A. Pass-by alarm tests

The devices' performance regarding alarm generation upon a moving probe was investigated in a series of pass-by tests. These tests were carried out according to RIID ANSI and IEC as well as ITRAP+10 standards. These were carried out for the nuclides ^{241}Am, ^{60}Co, ^{137}Cs, ^{133}Ba, ^{226}Ra, ^{232}Th, and ^{57}Co (only for ITRAP+10) using both RadEYE devices. For all tests the dose rate at the detector position at the position of closest approach was 0.5 µSv/h as prescribed by the standards. Comprehensive results for the dynamic tests are shown in Table IV, where the number of triggered alarms A compared to the number of trials B is listed as A/B.

TABLE IV
DYNAMIC TESTS

nuclide	RIID ANSI and IEC			ITRAP+10	
	velocity [m/s]	RadEYE SPRD GN CLYC	RadEYE SPRD CsI	RadEYE SPRD GN CLYC	RadEYE SPRD CsI
^{241}Am	0.05	10/10	10/10	30/30	30/30
^{60}Co	0.15	6/10	10/10	2/30	30/30
^{137}Cs	0.15	10/10	10/10	8/30	30/30
^{133}Ba	0.16	10/10	10/10	12/30	30/30
^{226}Ra	0.18	10/10	10/10	4/30	30/30
^{232}Th	0.26	2/10	10/10	3/30	30/30
^{57}Co	-	-	-	30/30	30/30

Dose rate at detector position 0.5 µSv/h. The velocity for the RIID ANSI and IEC tests is modified according to the compensation rule for changing the distance to reach 0.5 µSv/h. The tests according to ITRAP+10 were performed with 0.6 m/s. A/B means: an alarm was triggered A times in a total of B trials.

Tests according to ITRAP+10 were carried out at a velocity of 0.6 m/s. It is clear that the CsI detector device detected all tested nuclides reliably, i.e. the alarm set off in every trial. In contrast, the CLYC detector could achieve this only for ^{241}Am and ^{57}Co. For all other nuclides less than 50% of the trials resulted in the alarm setting off.

In the RIID ANSI and IEC case velocities were determined according to the compensation of distance changes to achieve the dose rate of 0.5 µSv/h at the detector position like given in (1). The results are qualitatively similar to the ITRAP+10 results. Again, the CsI detector raised an alarm in all trials. On these conditions with the lower velocity the CLYC crystal performed notably better than before, setting off the alarm on every trial for four out of six tested nuclides. For ^{60}Co and ^{232}Th, however, only few trials (6 out of 10 and 2 out of 10, respectively) set off the alarm.

B. Maximum reading ^{226}Ra

For ^{226}Ra the maximum dose rates displayed by the RadEYE SPRD GN device versus the trial number are presented in Fig. 4. Additionally, the background of 0.06 ± 0.02 µSv/h is shown. The device was used with factory settings, i.e. with an alarm threshold of 0.5 µSv/h. It can be seen, that only a single dose readout within the 30 trials surpassed the threshold, but in total 5 trials resulted in an alarm. This stems from the fact that the device alerts upon detection of artificial radiation regardless of the determined dose rate. In addition, the display does no longer show the dose rate value after alarming, instead a text message is displayed. Therefore, the dose rate values in the alarm cases could be higher, but could not be read from the display. However, it remains unclear why only in a small number of sub-threshold trials an alarm from detection of artificial radiation is set off.

 https://doi.org/10.1051/epjconf/202328810009

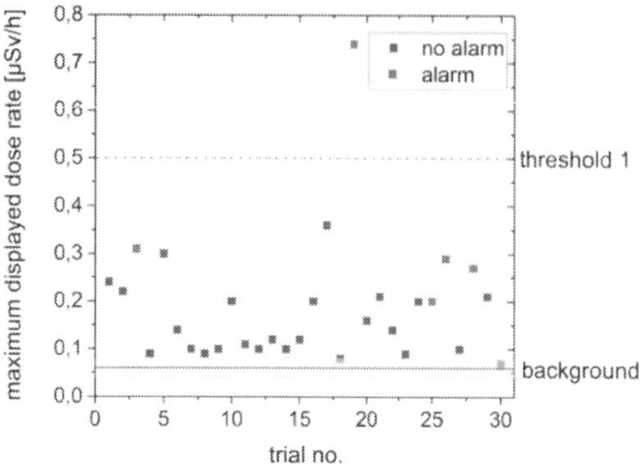

Fig. 4. Maximum displayed dose rate value which could be read at the display of the RadEYE SPRD GN device. ^{226}Ra, 0.5 µSv/h, 0.6 m/s, 30 trials, background 0.06 ± 0.02 µSv/h, device settings according to manufacturer's specification/factory setting, alarm threshold 1: 0.5 µSv/h (dashed line). Alarm is also triggered when artificial radiation is detected.

C. ^{137}Cs – readings at display for ANSI and IEC RIID testing

For ^{137}Cs the measured dose rates of both RadEYE devices were compared during the progress of 10 trials within the ANSI and IEC setting. Dose rates read from the device (SPRD GN with CLYC in red, SPRD with CsI in blue) with respect to measuring time are plotted in Fig. 5. Additionally, the time when the alarm is set off is indicated there with the open symbols. It is clear, that in all cases the CsI device raised the alarm earlier with a time difference in the order of seconds. Moreover, in all cases the CsI device already raised the alarm before the source had passed the detector whereas the CLYC device did not. This can also be seen in Fig. 6 where a sample trial is shown. It can be seen that the CsI device alerts already seconds before the source passes the detector position, whereas the CLYC device alerts slightly after the source has passed. Nevertheless, both devices successfully alerted upon each trial.

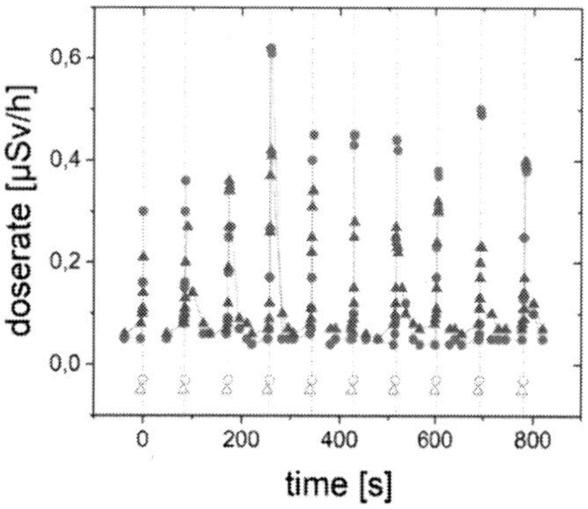

Fig. 5. Maximum displayed dose rate values read from the detector display at different timepoints during a series of trials. ^{137}Cs, 0.5 µSv/h, 0.15 m/s, RIID test ANSI and IEC. Dotted vertical lines mark the timepoints when the source passes in front of the detector. RadEYE SPRD GN with CLYC in red, RadEYE SPRD with CsI in blue. Open symbols indicate the moment when the alarm is triggered.

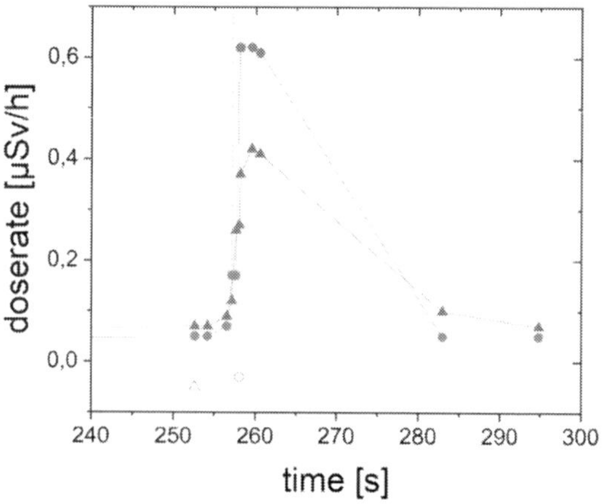

Fig. 6. One trial from the diagram of the previous Fig. 5 enlarged.

VII. Energy response

To analyze the energy response of both detector materials the gamma spectra of ^{60}Co were recorded both with CsI and CLYC crystals. Resulting spectra showing number of counts versus energy are shown in Fig. 7. Since the CsI crystal employed is larger than the CLYC counterpart, a larger total number of counts was recorded. Also, the gamma spectrum obtained with the CsI crystal is clearly of better quality than the other one recorded with the CLYC crystal. The peaks at 1173.2 keV and 1332.5 keV in the CsI spectrum are much more pronounced and distinguishable from the background than those in the CLYC spectrum. This might also explain the superior performance of CsI compared to CLYC in the dynamic test for ^{60}Co. Similarly, the relatively high number of counts in the low energy region corresponds to a remarkably better performance for ^{241}Am.

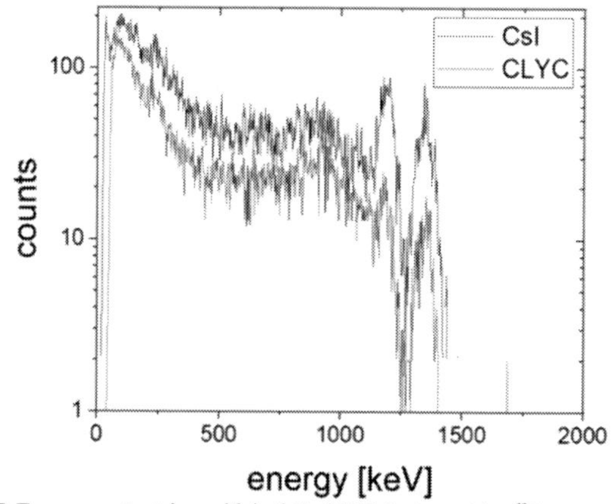

Fig. 7. Energy spectra taken with both RadEYE devices with a ^{60}Co source converted to the same measurement times.

VIII. Neutron detection – ^{252}Cf

In order to test the detection capability for neutrons, a ^{252}Cf source for both RadEYE devices was used. The SPRD GN with CLYC crystal enables the simultaneous detection of gammas and neutrons. It was mounted on a Polymethylmethacrylate (PMMA) phantom according to the standards in order to simulate a person wearing this device. The SPRD with CsI does not contain any neutron detection capability, but it can detect the accompanying gammas from the ^{252}Cf source and was therefore employed for comparison.

At a probe velocity of 0.6 m/s without any moderator both devices generated only a few alarms (5/30 and 3/30 in the CLYC and CsI cases, respectively). The CLYC detector correctly indicated these as neutron alerts, the CsI detector raised gamma alerts. At a reduced velocity of 0.42 m/s similar results were obtained. The CsI performance was essentially the same, the CLYC detector raised a higher number of alerts with respect to the number of trials, but still only in 4 out of 10 cases.

Significant improvement was achieved when 5 cm of polyethylene were placed between neutron source and detector. In this case the CLYC detector set off neutron alerts in 19 out of 20 trials. Since the CLYC detector material is sensitive for thermal neutrons, a moderator material like polyethylene is essential for successful detection of neutrons.

IX. Conclusions

The comparative behavior of RadEYE detector devices using either CLYC or CsI detector materials was investigated with respect to identification and dynamic detection.

In the static identification of nuclides both materials performed quite differently. Simultaneously exposed to four radiation sources (^{60}Co, ^{137}Cs, ^{133}Ba, and ^{152}Eu), both detector devices identified ^{60}Co. Which was not the case in the single measurements for the CLYC device. However, the CsI device could also find ^{137}Cs and ^{133}Ba, whereas the CLYC device detected the ^{152}Eu and even indicated the presence of ^{57}Co. None of the detector materials could reliably identify all nuclides if four of them were presented simultaneously.

Substantial differences between both detector materials could also be observed for the sensitivity with respect to the radiation energy. On ITRAP+10 conditions the CsI detector could reliably detect gammas for all tested nuclides, whereas the CLYC device could only achieve this for low energy emitters. Similarly, in the RIID ANSI and IEC tests CsI performed equally flawless, whereas CLYC performed significantly better for several nuclides, only struggling with the high energy emitters ^{60}Co and ^{232}Th. The presence of more pronounced peaks in the ^{60}Co gamma spectrum for CsI compared to CLYC supports this idea. Hence, further tests covering a wide energy region are required.

Finally, the CLYC detector allows for the simultaneous detection of gammas and neutrons. Although the detection rate is poor for unmoderated neutrons, with a moderator present neutron radiation from ^{252}Cf is detected and assigned reliably.

Overall, the RadEYE SPRD GN device with CLYC crystal seems to perform less sensitive in gamma detection and slower in alarm generation than the RadEYE SPRD device with CsI detector crystal, but it includes the possibility to also detect neutrons (moderated). And the answer to the question whether the two detector materials behave differently in identification and dynamic tests or not is yes, they do.

References

[1] M. Risse, P. Clemens, J. Glabian, O. Schumann, T. Koeble, H. Friedrich, W. Berky, M. C. Bornhoeft, S. Chmel, "Qualification Test System for Radiation Detection Devices QuTeSt", in ANIMMA 2021 - Advancements in Nuclear Instrumentation Measurement Methods and their Applications, Red Hook, NY, 2021, pp. 321-325.

[2] Radiation Protection Instrumentation - Hand-held Instruments for the Detection and Identification of Radionuclides and for the Estimation of Ambient Dose Equivalent Rate from Photon Radiation, IEC Standard 62327, 2017

[3] Radiation Protection Instrumentation - Alarming Personal Radiation Devices (PRDs) for the Detection of Illicit Trafficking of Radioactive Material, IEC Standard 62401, 2017

[4] Radiation Protection Instrumentation - Spectroscopy-based Alarming Personal Radiation Detectors (SPRD) for the Detection of Illicit Trafficking of Radioactive Material, IEC Standard 62618, 2013

[5] American National Standard Performance Criteria for Handheld Instruments for the Detection and Identification of Radionuclides, ANSI Standard N42.34-2015

[6] American National Standard Performance Criteria for Alarming Personal Radiation Detectors for Homeland Security, ANSI Standard N42.32-2016

[7] American National Standard Performance Requirements for Spectroscopic Personal Radiation Detectors (SPRDs) for Homeland Security, ANSI Standard N42.48-2018

[8] C. Carrapiço, G. Eklund, M. Marin-Ferrer, P. Peerani, F. Rosas, A. Tomanin and H. Tagziria, "Illicit Trafficking Radiation Detection Assessment Program ITRAP+10 Phase II - Round Robin: Radiation Isotope Identifiers Device (RIID) - Test Method", JRC Technical Reports, 2017

[9] C. Carrapiço, G. Eklund, M. Marin-Ferrer, P. Peerani, F. Rosas, A. Tomanin and H. Tagziria, "Illicit Trafficking Radiation Detection Assessment Program ITRAP+10 Phase II - Round Robin: Personal Radiation Detector (PRD) - Test Method", JRC Technical Reports, 2017

[10] RadEye SPRD Spektroskopischer Strahlungsdetektor, Thermo Fisher Scientific Messtechnik GmbH, Erlangen, Germany, 2018.

[11] RadEye SPRD-GN Gamma/Neutron Personal Radiation Detector, Thermo Fisher Scientific Messtechnik GmbH, Erlangen, Germany, 2018.

https://doi.org/10.1051/epjconf/202328810010

Design of a reliable numerical model of a plastic scintillation detector for beta spectrometry

L. Fleres[1*], J. Venara[1], M. Cuozzo[1], C. Mahe[1], S. Fargier[1], E. El Haber[1], F. Carrel[2]
[1]CEA, DES, ISEC, DPME, SEIP, LNPA, Univ. Montpellier, Marcoule, F-30207 Bagnols-sur-Cèze Cedex, France
[2]Université Paris-Saclay, CEA, List, F-91120, Palaiseau, France
(*) lorenzo.fleres@cea.fr

Abstract— **The accurate measurement of beta emitters is of utmost importance in various fields, including environmental monitoring, radiation protection and dismantling and decommissioning (D&D) operations. Non-destructive measurement techniques are crucial as they enable assessments of radioactivity without causing further harm or releasing additional hazardous materials. Due to the limited range of beta particles in matter, their characterization requires measurements in contact with the source term. In this paper, we present a study on the performance evaluation of a plastic scintillation detector designed for *in-situ* beta spectrometry measurements. The MCNP6.2 numerical model of the detector is detailed in the measurement configuration with beta, gamma and internal conversion electrons emitting sources. The calibration process and Gaussian Energy Broadening (GEB) parameters evaluation are discussed. Comparisons between the MCNP6.2 simulations and experimental measurements are exposed, both for gamma and beta spectra. While the gamma-ray detector response shows good agreement, greater discrepancies are observed in the beta ray response. To shed light on these discrepancies, a sensitivity analysis considering various sources of uncertainties is planned for future work.**

Keywords — **Beta spectrometry, Gaussian energy broadening, Monte Carlo, Plastic scintillator.**

I. INTRODUCTION

THE characterization of pure beta emitting radionuclides is difficult due to the very short path of electrons in matter. Their detection is usually performed in laboratory conditions by destructive analysis. Liquid scintillation counting, is still the reference method for characterizing pure beta emitters but requires numerous upstream operations [1]. These operations are often time-consuming and entail specific constraints in terms of safety and security. Additionally to destructive techniques, *in situ* measurement methods can be used to obtain initial knowledge of the radiological state of the materials to be characterized [2]–[5]. Plastic scintillators are currently the best candidates for *in situ* beta particle detection as they have the lowest density and effective atomic number of all solid scintillators. These properties make them less sensitive to gamma radiation thus enhancing the ratio of beta signal to gamma background, which is always present during *in situ* measurements.

In previous works, CEA has developed a methodology based on a portable beta plastic scintillator spectrometer for *in situ* radiological characterization of ^{90}Sr in contaminated matrices [5]. We plan to use this methodology for the *in situ* characterization of beta emitting radionuclide mixtures using deconvolution techniques. This requires performing numerical simulations to obtain an accurate response function of the detector for different radionuclides and measurement configurations.

In this study, beta and gamma responses of the plastic scintillator were evaluated using MCNP6.2 [6] simulations and compared with laboratory measurements. In section II, we describe the measurement methodology, the numerical model of the detector and the method used for the energy calibration and energy resolution computation. In section III, the results and the comparison between measurements and MCNP6.2 simulations for two standard sources of ^{137}Cs and ^{207}Bi are provided. In section IV, we summarize the key findings of the study and outline potential areas for further research and improvements.

II. MATERIALS AND METHODS

A Measurement methodology

The plastic scintillation detector for beta spectrometry was designed in previous work [5]. The sensitive part of the detector consists of a 4 mm thick EJ200 scintillator with an 18 micrometer Mylar window used to avoid ambient light sensitivity. A 30 mm thick Poly(methyl methacrylate) (PMMA) light guide couples the scintillator to the photocathode of a photomultiplier. For data acquisition and analysis, the probe is connected via a LYNX multichannel analyzer (MIRION Technologies, CA, USA) to a computer equipped with the Genie 2000 software [7]. Fig. 1 shows the experimental set-up. The measurement methodology requires to perform two measurements [4], [5], [8]: a first one with the bare detector and a second one with a 4 mm thick aluminum cover placed over the detector. This cover was designed to stop energetic beta particles up to 2.3 MeV and to ensure that the detector only measures gamma rays if exposed to $^{90}Sr/^{90}Y$ contaminated environment. The subtraction of the two acquired spectra provides the beta component as shown in Fig. 2, which illustrates the spectra acquired for a ^{137}Cs source. The blue line represents the response of the detector to the photons emitted

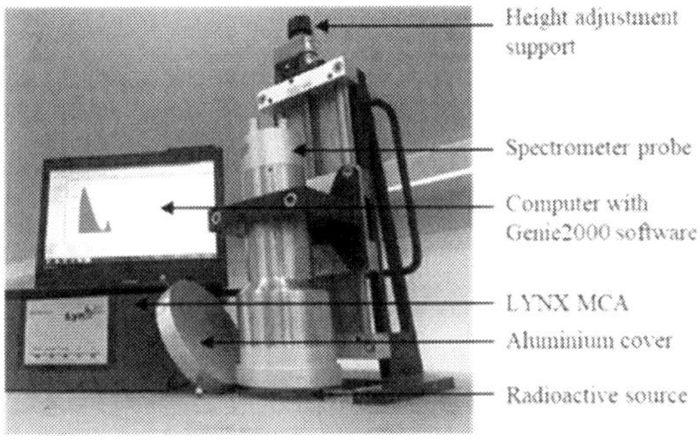

Fig. 1. Beta spectrometry probe and instrumentation devices.

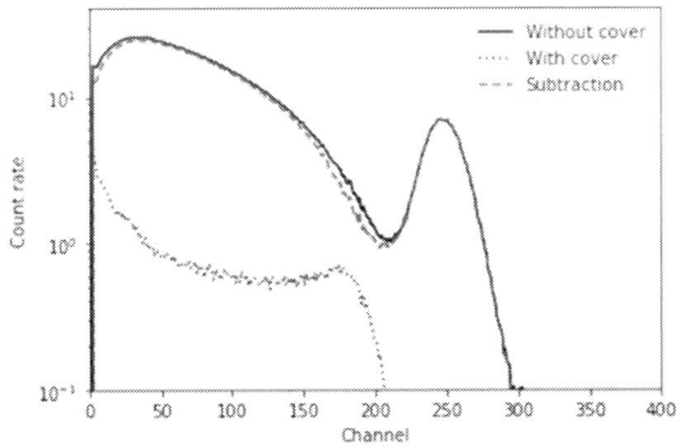

Fig. 2. Measured spectra for a ^{137}Cs source placed at a distance of 1 cm from the detector.

by the ^{137}Cs source. The photons emitted at 661.66 keV mainly interact by Compton scattering and the resulting spectrum has no full absorption peak. The red line, obtained by subtraction of the two measurements, is a convolution of the continuous β- component and the peak arising from the internal conversion electrons emitted at energies of 624.22 and 655.67 keV. The poor energy resolution of the detector does not allow to discriminate the two peaks.

B MCNP6.2 numerical model

In previous studies [5], the MCNP6.1 code was used to create a numerical model of the beta spectrometry detector. We updated the model for MCNP version 6.2 and changed the design of the calibration source since we have used a different set-up. We procured new radioactive sources from the Electron Sources Series of Eckert & Ziegler [9]. These sources are prepared by evaporation of the radionuclide on very thin surface foil, and protected by a 100-200 µg.cm^{-2} acrylic cover. The source modelling was based on the geometric characteristics available in the product catalog. The source term was assumed to be homogeneously distributed in a cylinder of air of a very small thickness (0.1 µm) placed between the substrate and the acrylic window. Fig. 3 shows the numerical model with the calibration source placed at a distance of 10 mm from the detector.

In this work we used a ^{137}Cs, a ^{207}Bi and a ^{241}Am gamma ray source. The characteristics of the sources are summarized in Table I in which square brackets indicate the emission intensities.

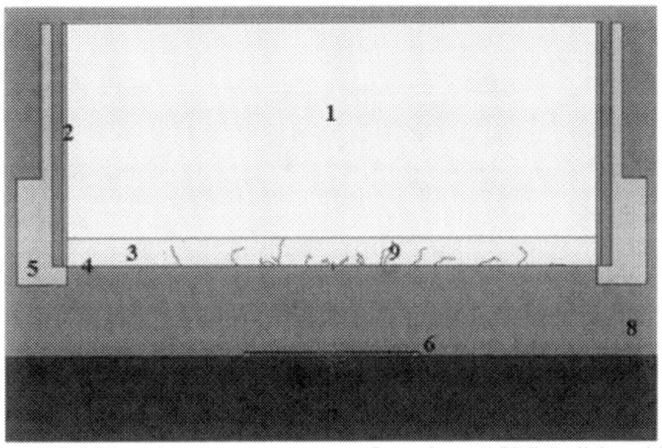

Fig. 3. MCNP6.2 model of measurement configuration. (1) PMMA light guide, (2) Polytetrafluoroethylene (PTFE) reflector, (3) EJ200 scintillator, (4) Mylar window, (5) aluminium case, (6) calibration source, (7) concrete floor, (8) ambient air, (9) 1 MeV electron collision point.

Emission data for different radionuclides were taken from the Joint Evaluated Fission and Fusion (JEFF) nuclear data library version 3.3 [10] except for β- emissions. For the latter, we considered data from the BetaShape program [11] in order to take into account experimental shape factors of beta emitters.

The numerical model allows to assess the performance of the detector and to compare with experimental results. To obtain the energy distribution of pulses in the sensitive part of the detector we applied a "F8" pulse height tally to the cell representing the EJ200 material. The MCNP code assumes that the detector is ideal and does not consider all the physical processes that determine its energy resolution. To take account of the statistical dispersion in MCNP simulations, the GEB (Gaussian Energy Broadening) option must be activated in MCNP for the F8 tally. The function relating the detector's energy resolution ($FWHM$ – full width at half maximum) to a given energy (E) is expressed as follows:

$$FWHM(E) = a + b\sqrt{E + c \times E^2}, \quad (1)$$

The procedure for evaluating parameters a, b and c will be presented in section II.C.

We used the default parameters for photon and electron transport (PE mode) in all simulations.

C Energy calibration and energy resolution of the detector

Following a different approach compared to the previous study [5], we carried out the energy calibration process and energy resolution evaluation according to the response of the detector to gamma sources. The method used is the one proposed by Kim *et al.* [12]. Through an iterative process of Monte Carlo simulations, energy calibration and energy resolution update, convergence is achieved. At the first iteration the calibration is carried out according to the positions of the measured Compton maxima and the energy resolution is evaluated with a Gaussian adjustment on the Compton maxima. From the second iteration onward, the Gaussian adjustment is performed on the position of the Compton edges as shown in Fig. 4 for a ^{137}Cs source. In the case of the ^{241}Am source, Gaussian adjustment was performed on the full absorption peak resulting from the 59.5 keV gamma ray emission. Although for energies greater than 30 keV the Compton effect is prevalent

TABLE I
CHARACTERISTICS OF CALIBRATION SOURCES

Radionuclide	^{137}Cs	^{207}Bi	^{241}Am
Activity (kBq)	9.07 ± 2.5%	9.37 ± 2.5%	151.0 ± 5.0%
Eckert & Ziegler type	Mono-Energetic Series	Mono-Energetic Series	Alpha Wide Area
Gamma emission (keV)	661.66 [85.05%]	569.70 [97.76%] 1063.66 [74.58%]	59.54 [35.92%]
Discrete electron emission (keV)	624.22 [7.78%] 655.67 [1.70%]	481.69 [1.55%] 555.25 [0.43%] 975.66 [7.11%] 1049.20 [1.84%] 1060.49 [0.44%]	-
β- transitions (keV)	Forbidden 1st unique: 513.97 [94.61%] Forbidden 2nd non-unique: 1175.63 [5.39%]	-	-

Activities are provided at the measurement date.
Internal conversion electrons emitted by ^{241}Am are omitted as we are interested only in the gamma response.
EC, electron capture.

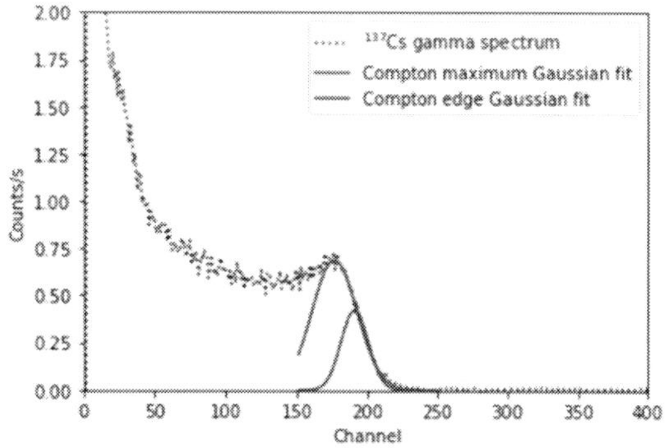

Fig. 4. Gaussian adjustments on the ^{137}Cs gamma spectrum.

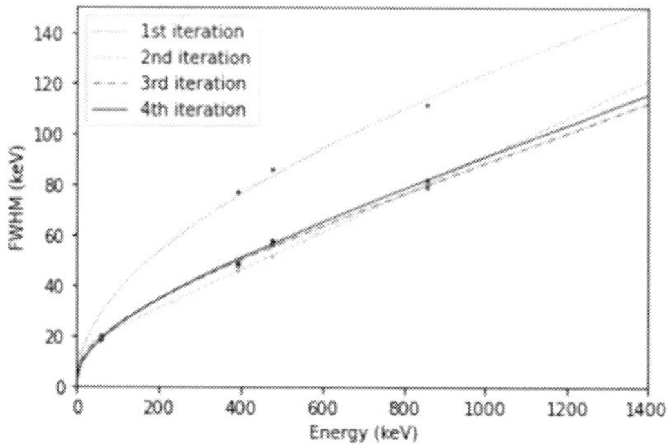

Fig. 5. FWHM curves as a function of the iteration number.

over the photoelectric effect in the EJ200, a full absorption peak is distinguishable in the measured ^{241}Am spectrum (results are not shown here).

We carried out all the adjustment using the Curve Fitter Toolbox [13] available in MATLAB and the Linear Least Squares method.

III. RESULTS AND DISCUSSION

A Energy calibration and GEB parameters computation

Following the procedure described in section II.C we achieved convergence just after 4 iterations. The resulting energy calibration is:

$$Energy\ (MeV) = 0.014 + 0.00245 \times Channel, \qquad (2)$$

Fig. 5 shows the evolution of the energy resolution curves obtained at each iteration. The dotted points represent the FWHM values evaluated by Gaussian fitting at the energy values of 59.54, 477.34, 393.31 and 857.65 keV. The first value corresponds to the energy of the full absorption peak of ^{241}Am, the last three values are the theoretical Compton edge [14] energies for photons emitted by ^{137}Cs and ^{207}Bi listed in Table I. The values obtained for the *a*, *b* and *c* parameters of the GEB function at each iteration are summarized in Table II.

We compared the obtained results with the evaluation of the energy resolution from the electron response of the detector. We focused on the region between 525 and 725 keV of the ^{137}Cs beta spectrum, where the signal produced by the internal

TABLE II
GEB PARAMETERS AS A FUNCTION OF THE ITERATION NUMBER

Iteration	a (MeV)	b (MeV^{-1})	c (MeV^{-1})
1	0.0003	0.1171	0.1135
2	0.012	0.0289	6.69
3	0.0033	0.065	0.7131
4	0.0011	0.0702	0.6385

conversion electrons is dominant. In this range, the total response is given by the convolution of three components. The first is a continuous component associated with the β- emission. The second and third are the two components associated with the two internal conversion electrons. We then defined a custom function to perform a deconvolution as the sum of a linear component and two Gaussians of relative amplitudes equal to the emission ratio of the internal conversion electrons. The result of deconvolution is shown in Fig. 6. The FWHM values obtained are 66.63 keV and 69.60 keV for the two Gaussians centered at 620.12 and 652.36 keV, respectively. Calculating the FWHM values for the energies of 620.12 and 652.36 keV based on equation (1) with the obtained a, b and c parameters given in Table II results in 66.64 keV and 68.56 keV. The relative error of less than 1.5% supports our energy calibration and energy resolution evaluation process based on gamma ray sources.

An evaluation of the energy resolution can be obtained using the internal conversion peaks around 975 keV of the ^{207}Bi source but the presence of three internal conversion electrons

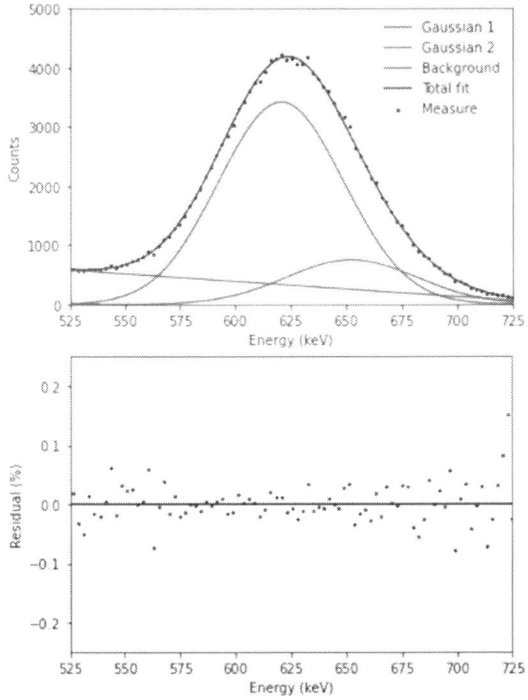

Fig. 6. (a) The analysis of ^{137}Cs beta spectrum in the region of the internal conversion peak. The peak is fitted with two Gaussian functions (red) and a linear background (green). (b) Residuals expressed in percentage between measurement and adjustment.

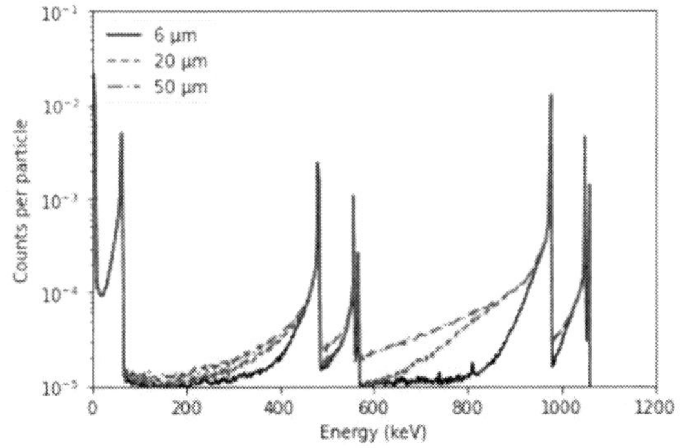

Fig. 7. Electron energy flux coming from the ^{207}Bi source as a function of platinum substrate thickness. Results obtained using an F1-type tally in MCNP6.2.

makes deconvolution more complex. Moreover, compared to the ^{137}Cs source where the substrate is a thin Mylar foil, the ^{207}Bi source substrate is made of platinum. Such a high-density and high atomic number material strongly impacts the backscattering and self-absorption of electrons in the source, as well as the production of Bremsstrahlung photons. The impact of the platinum thickness on the electron flux leaving the source is shown in Fig 7. To obtain a correct estimate of the resolution based on the detector response to a ^{207}Bi source, a high-quality source must be used in order to limit the emission spectrum degradation.

B Comparison of MCNP6.2 simulations and measurements

The comparison between experimental measurements and Monte Carlo simulations is the last step in the qualification phase of a detector numerical model. We used the activity value of the various sources under 4π at the measurement date as the multiplication factor to compare simulated and measured responses. The previously obtained energy calibration and GEB function were applied to F8-tally for all the simulations. The spectra for ^{137}Cs and ^{207}Bi sources are shown in Fig. 8: gamma ray spectra represent the measurement with the cover and beta spectra are obtained as the subtraction of the two measurements (without and with the cover).

A qualitative analysis shows good agreement between simulated and measured gamma spectra. In the MCNP numerical model, the aluminium cover was added on the detector to take into account the attenuation and build-up of photons in the cover itself. Regarding the beta spectra, the shape of the ^{137}Cs spectrum appears to be accurately rebuilt in the region of the internal conversion emission with a larger gap at low energy (50 – 500 keV). On the other hand, looking at the spectrum of ^{207}Bi, the gap between simulation and measurement is important, again in the lower energy region but especially in the region of the internal conversion peak around 975 keV. The largest deviation for the ^{207}Bi simulation is associated with the uncertainty related to the platinum substrate on which the radionuclide is deposited.

The quantitative analysis was carried out by comparing the integrated counts over different energy intervals, focusing on the global spectra and the internal conversion electrons emission regions. The relative deviations between measured and simulated counts for ^{137}Cs and ^{207}Bi sources are listed in Table III. For the total spectrum electron responses, we considered a low energy threshold at 50 keV for ^{137}Cs and 100 keV for ^{207}Bi. Below these energies, the attenuation of X-rays emitted by the sources during the two measurements must be taken into account. The relative deviations of gamma responses are lower than 2% which confirms the reliability of the detector to the measurement of gamma-emitting radionuclides. Larger differences are encountered for the electron responses with a 22% maximum deviation for ^{207}Bi.

The discrepancy between Monte Carlo simulations and β-spectra measurements may be due to several factors. Among the most important of which are: the algorithm and physical electron transport parameters, the modelling and quality of the source, the nuclear radionuclide emission data and the cross-

TABLE III
COMPARISON OF THE MEASURED AND SIMULATED SPECTRA OF THE CALIBRATION SOURCES.

Radionuclide	Response	Energy range (keV)	Relative deviation {S/M – 1}
^{137}Cs	γ [(a)]	50 – 700	+0.0
	e^- [(a)]	50 – 1200	+11.0
	e^- [(b)]	525 – 725	+8.0
^{207}Bi	γ [(a)]	40 – 1100	+1.8
	e^- [(a)]	100 - 1300	+2.0
	e^- [(b)]	400 – 600	-6.7
	e^- [(c)]	850 – 1200	+21.7

[(a)] Total spectrum. [(b)] Internal conversion 1st peak. [(c)] Internal conversion 2nd peak.

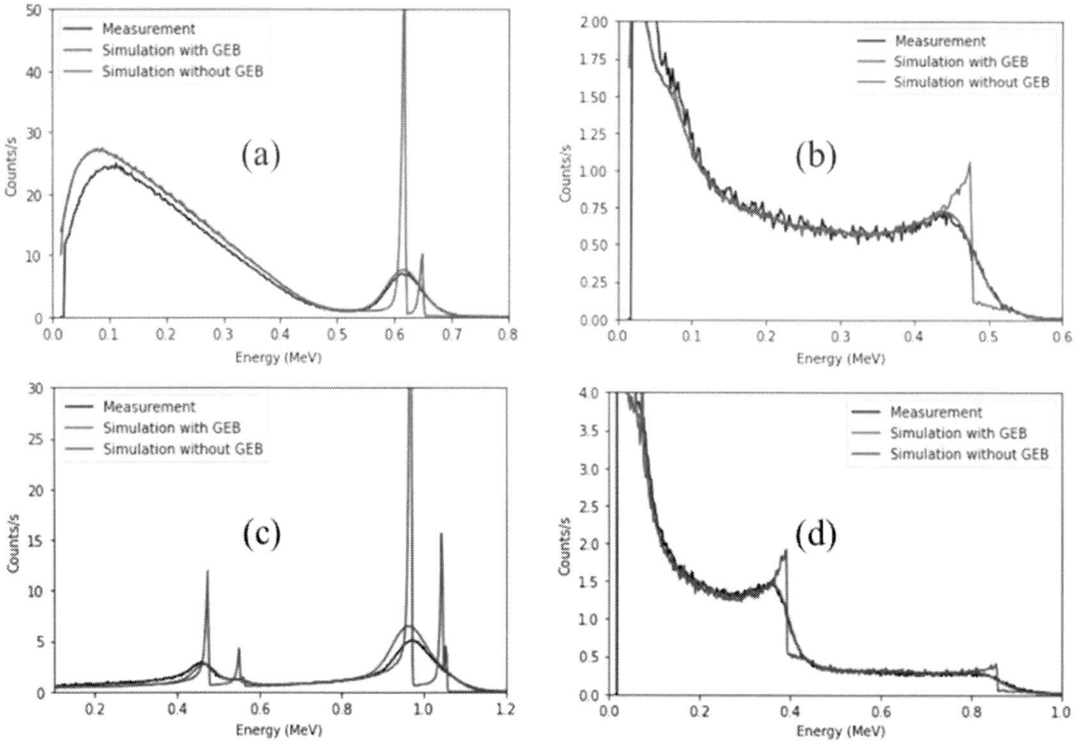

Fig. 8. Comparison of the measured (black) and simulated spectra with Gaussian energy broadening (red) and without (green) : (a) ^{137}Cs beta contribution, (b) ^{137}Cs gamma contribution, (c) ^{207}Bi beta contribution, (d) ^{207}Bi gamma contribution.

sections used in our simulations. We plan to carry out in future works a detailed sensitivity analysis to assess the impact of different sources of uncertainties on our numerical detector model.

IV. Conclusions

In this study, we assessed the performance of a plastic scintillation detector to measure gamma radiation and beta particles. Comparison between Monte Carlo simulations using MCNP6.2 and laboratory measurements showed good agreement for the gamma-ray detector response. Regarding the detector response to electrons, we still encountered significant discrepancies between measurements and simulations. These deviations will be investigated in the future by means of an in-depth parametric analysis and by comparison of Monte Carlo numerical models using independent Monte Carlo codes. The conclusions of this study contribute to the understanding of the plastic scintillation detector's capabilities and limitations and pave the way for further improvements in the numerical modeling of this type of detector.

References

[1] K. A. Arslanov, T. V. Tertychnaya, and S. B. Chernov, 'Problems and Methods of Dating Low-Activity Samples by Liquid Scintillation Counting', *Radiocarbon*, vol. 35, no. 3, pp. 393–398, Jan. 1993, doi: 10.1017/S0033822200060409.

[2] A. Schilk, K. H. Abel, D. Brown, R. Thompson, M. Knopf, and C. Hubbard, 'Selective, high-energy beta scintillation sensor for real-time, in situ characterization of uranium-238 and strontium-90', 1995, doi: 10.1007/BF02041923.

[3] B.-K. Seo, C.-H. Park, K.-W. Lee, D.-G. Lee, and C.-H. Jung, 'Development of the In-situ Monitoring System for Pipe Internal Contamination Measurement in the Decommissioning Site', *Journal of Nuclear Science and Technology*, vol. 45, pp. 500–502, Aug. 2014, doi: 10.1080/00223131.2008.10875900.

[4] V. Potapov, A. G. Volkovich, O. Ivanov, V. Stepanov, S. Smirnov, and V. G. Volkov, 'Development of Portable Beta Spectrometer for Sr-90 Activity Measurements in Field Conditions and Its Application in Rehabilitation Activities at RRC Kurchatov Institute', *undefined*, 2006, Accessed: Jan. 27, 2022. [Online]. Available: https://www.semanticscholar.org/paper/Development-of-Portable-Beta-Spectrometer-for-Sr-90-Potapov-Volkovich/57324a32fe9a3543ba592526ce7de536f0bbd4fd

[5] J. Venara *et al.*, 'Design and development of a portable β-spectrometer for 90Sr activity measurements in contaminated matrices', *Nuclear Instruments and Methods in Physics Research Section A: Accelerators, Spectrometers, Detectors and Associated Equipment*, vol. 953, p. 163081, Feb. 2020, doi: 10.1016/j.nima.2019.163081.

[6] C. J. Werner *et al.*, 'MCNP Version 6.2 Release Notes', LA-UR--18-20808, 1419730, Feb. 2018. doi: 10.2172/1419730.
[7] Z. Laili, 'Genie TM 2000 spectroscopy software operations, Canberra 52 (1) (2016) 1–361.', 2016.
[8] D. E. Martz, B. L. Rich, and L. O. Johnson, 'A Portable Beta Spectrometer for Tissue Dose Measurement', *Radiation Protection Dosimetry*, vol. 14, no. 2, pp. 183–186, Feb. 1986, doi: 10.1093/oxfordjournals.rpd.a079644.
[9] 'Eckert & Ziegler Strahlen- und Medizintechnik AG', *Eckert & Ziegler Strahlen- und Medizintechnik AG*, Jun. 17, 2022. https://www.ezag.com/home/ (accessed Jun. 24, 2022).
[10] A. J. M. Plompen *et al.*, 'The joint evaluated fission and fusion nuclear data library, JEFF-3.3', *The European Physical Journal A*, vol. 56, no. 7, p. 181, Jul. 2020, doi: 10.1140/epja/s10050-020-00141-9.
[11] X. Mougeot, 'BetaShape: A new code for improved analytical calculations of beta spectra', *EPJ Web Conf.*, vol. 146, p. 12015, 2017, doi: 10.1051/epjconf/201714612015.
[12] C. Kim, Y. Kim, M. Moon, and G. Cho, 'Iterative Monte Carlo simulation with the Compton kinematics-based GEB in a plastic scintillation detector', *Nuclear Instruments and Methods in Physics Research Section A: Accelerators, Spectrometers, Detectors and Associated Equipment*, vol. 795, pp. 298–304, Sep. 2015, doi: 10.1016/j.nima.2015.06.007.
[13] 'Curve Fitting Toolbox'. https://www.mathworks.com/products/curvefitting.html (accessed Aug. 07, 2023).
[14] C. L. Fontana, 'Compton edge calculator', *Cristiano Lino Fontana*, Mar. 16, 2018. https://userswww.pd.infn.it/~moretto/fontana/project/software/2018/03/16/fontana/project/software/2018/03/16/compton-edge.html (accessed Jul. 31, 2022).

Evaluation of neutron-induced activation gamma-rays creation and transport in a $LaBr_3$ inorganic scintillator with PHITS

C. Besnard-Vauterin[1,*], R. De Stefano[2], G. Amoyal[2], V. Schoepff[2], J-C. Angélique[3]
[1] Université Paris-Saclay, CEA, LIST, Laboratoire National Henri Becquerel, F-91129, Palaiseau, France
[2] Université Paris-Saclay, CEA, LIST, F-91129, Palaiseau, France
[3] Normandie Univ, ENSICAEN, UNICAEN, CNRS/IN2P3, LPC Caen, 14000 Caen, France
(*) clement.besnardvauterin@cea.fr

Abstract—**In the framework of its R&D activities to support the metal industry, the Laboratory for Integration of Systems and Technology of CEA Paris-Saclay started studies with the aim of characterizing scrap metal by means of neutron activation analysis. This involves irradiating samples of scrap metal with a pulsed-neutron source in order to determine the copper composition (mainly Cu-65 and Cu-63). In this scope, the use of a $LaBr_3$ detector with an energy resolution of 14.6 keV at 661 keV is hereby investigated to carry out acquisitions during and between the irradiation pulses. As pointed out in literature, the $LaBr_3$ inorganic scintillator might suffer some damage in a neutron-rich environment. Understanding the degradation of the energy resolution of such detector due to a high dose environment is essential to analyse the recorded signal. In this context, the work described in the present paper explores the coherent creation of neutron induced activation products as well as their respective delayed gamma rays in the quite well established LaBr3 inorganic scintillator by the mean of two steps Monte Carlo simulations performed first with the DCHAIN code, and secondly with the PHITS general purpose Monte Carlo particle transport code. Coherence was shown between neutron-induced activation gamma rays and the isotopes created in the crystal, and between the isotopes created and their decay through time. Thus, this study validate the two-step calculation scheme with PHITS and DCHAIN for both the time and energy aspects.**

Keywords — **Neutron activation analysis, PHITS, DCHAIN.**

I. INTRODUCTION

Neutron activation analysis (NAA) is a technique widely employed for the non-destructive determination of elemental composition in various materials, including several types of radioactive waste packages but also metals. This technique involves subjecting the samples of interest to neutron irradiation, leading to the production of radioisotopes, and subsequently measuring the gamma radiation emitted during their radioactive decay. By analysing these gamma rays, valuable information regarding the elemental composition of the samples can be obtained. Neutron activation analysis finds applications in various fields due to its ability to provide accurate and precise elemental analysis. Some of the fields where NAA is predominantly used include archaeology, environmental and earth sciences, forensics, food and agriculture, nuclear science and reactor analysis, homeland security and finally material science [1].

Within the context of supporting the metal industry, the Laboratory for Integration of Systems and Technology at CEA Paris-Saclay has initiated research and development activities focused on characterizing scrap metal using neutron activation analysis [2]. This innovative approach involves subjecting samples of scrap metal to neutron irradiation using a pulsed-neutron source. The primary objective is to determine the composition of copper within the samples, with a particular focus on identifying the isotopes Cu-65 and Cu-63. By leveraging neutron activation analysis, valuable insights can be gained regarding the copper content in the scrap metal, enabling enhanced recycling processes and efficient utilization of resources.

In this context, gamma spectrometry is an extensively used tool for the identification and quantification of the radioisotopes present in the samples. $LaBr_3$ inorganic scintillators are well-established detectors for gamma spectrometry, thanks to their high light output and favourable energy resolution of 2 to 3% at 662 keV, although lower than High Purity Germanium (HPGe) detectors, which are considered the gold standard for gamma spectrometry and offer resolutions around 0.3%. $LaBr_3$ scintillators provide a compelling alternative to HPGe detectors, as they do not require cooling, are cost-effective, easily deployable, and portable. However, it has been reported in [3] that $LaBr_3$ scintillators may suffer some damage in a neutron-rich environment, which can lead to degradation of their energy resolution. Therefore, understanding the impact of neutron flux on the performance of $LaBr_3$ detectors is essential for accurate characterization of waste or metals using neutron activation analysis.

In this article, we investigate the coherent creation of neutron induced activation products as well as their respective delayed gamma rays if the $LaBr_3$ scintillators using two steps Monte Carlo simulations performed with the DCHAIN code (version 3.25) and the PHITS (version 3.27) general-purpose Monte Carlo particle transport code. DCHAIN [4, 5] is a software package designed to calculate the radioactive decay and transmutation chains of a wide range of isotopes, including fission products and actinides, produced by neutron irradiation of materials. It can also calculate the gamma-ray spectrum resulting from the decay of these isotopes, and the neutron spectrum resulting from neutron-induced reactions. PHITS [6] is a general-purpose Monte Carlo particle transport code that simulates the interactions of various particles with matter, including neutrons, photons, electrons, and heavy ions. It can also simulate the transport of radiation through complex geometries, such as those encountered in nuclear facilities or medical applications.

Simulated results were compared with data from the Nuclear Data Sheets for bromine and lanthanum to validate the coherent calculation of neutron induced activation gamma rays [7-8-9]. There half-lives were also calculated and compared to databases [10-11].

II. METHODS AND NUMERICAL APPROACH

A detailed model of a $LaBr_3$ inorganic scintillator with dimensions of 7.62 × 7.62 cm² was implemented for this study. The scintillator crystal in the model corresponds to a cylindrical volume doped with 5% cerium, resulting in a $LaBr_3$:Ce composition. To capture the emitted light from the scintillator, a cylindrical aluminum photomultiplier (PMT) was coupled to the crystal. The PMT was depicted by a volume of 7 cm height, 3 cm radius , and 0.64 mm thickness. A one millimetre thick PVC seal joins the last two parts together. Figure 1 depicts the three-dimensional geometry of the detector, providing a visual representation of its structure.

For the material composition of the detector, the natural isotopes present in the respective elements were considered. The bromine component consisted of Br-79, which accounts for 50.69% of the natural abundance, and Br-81, which makes up the remaining 49.31%. In the case of lanthanum, the material description included La-138, which has a natural abundance of 0.09%, and La-139, which comprises 99.91% of the isotopes. Lastly, cerium was represented by the isotope Ce-140, which is present at a 100% abundance as a first approximation. By incorporating these specific isotopes into the material description, the simulation accurately reflects the natural composition of the elements used in the $LaBr_3$:Ce detector. This level of detail enhances the realism of the simulation and ensures that the generated results align as closely as possible with the characteristics and behaviour of the actual detector system.

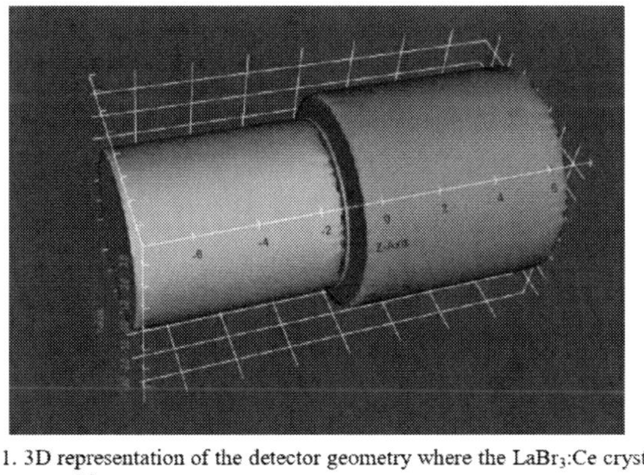

Fig. 1. 3D representation of the detector geometry where the $LaBr_3$:Ce crystal is represented in green.

In this section, we focus on analysing the activation response of a $LaBr_3$:Ce detector to neutron flux. To accomplish this, we employed a two-step approach using Monte Carlo simulations with the DCHAIN and PHITS codes. These simulations enabled us to investigate both the creation and transport of neutron-induced activation gamma-rays within the detector, as well as the subsequent decay of activation products.

Due to a limitation in the functionality of the DCHAIN code, the calculation was divided into two steps. While DCHAIN can simulate the gamma spectra resulting from the decay of activation products generated by neutrons, it provides only a description of these spectra based on energy group structures through the IGGRP (option of gamma-ray spectrum group) card output option. This limitation makes it challenging to perform a more detailed analysis of each individual gamma-ray generated in the process.

To overcome this limitation and obtain a more comprehensive understanding of the neutron-induced activation response, we incorporated the PHITS code into the workflow. The PHITS code allowed us to capture the detailed characteristics of the activation gamma-rays and their subsequent transport within the detector. By employing this two-step process, we could bridge the gap between the initial creation of activation products at different output times and their subsequent decay, enabling a more in-depth analysis of the generated gamma-ray spectra.

The flowchart presented in Figure 2 illustrates the sequential nature of this two-step process. It highlights the simulation workflow, starting with the initial creation of activation products through neutron interactions and their subsequent transport within the detector, followed by their decay and the resulting gamma-ray spectra.

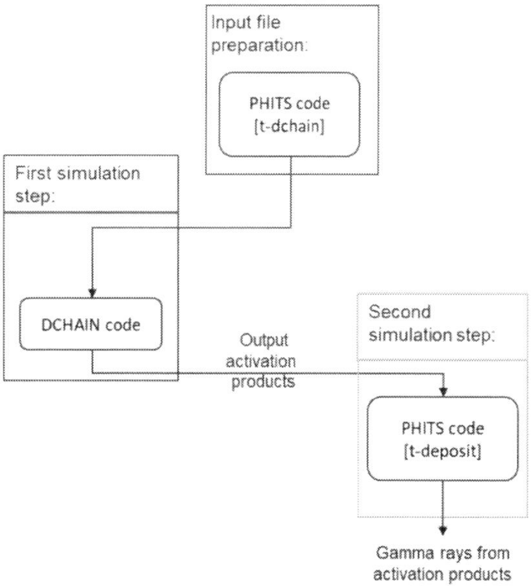

Fig. 2. Flowchart of the two-steps process for the investigation of the activation products and their decay induced by neutrons in a $LaBr_3$:Ce detector.

During the initial step, the PHITS tally [T-Dchain] was used to facilitate the generation of the required input files for the

DCHAIN code. The PHITS tally served as an automated tool, streamlining the process of file generation and ensuring the necessary files were appropriately prepared for subsequent analysis using DCHAIN.

To initiate the simulation process, a preliminary step involved subjecting the system to a 2-hour irradiation using a Cf-252 source. The source was positioned at a distance of 10 cm in front of the detector. During the irradiation, the Cf-252 source emitted neutrons with an average energy of 2.3 MeV, following a Watt spectrum. The neutron flux emitted by the source had an intensity of 2.4×10^4 neutrons per second within a 4π sr sphere.

Following the 2-hour irradiation, the subsequent step involved extracting descriptions of the created isotopes at different times over a 72-hours time period. The extracted descriptions of the isotopes were registered in a systematic manner. To provide a visual representation of the extraction process, the chronogram in Figure 3 illustrates the time intervals at which the descriptions were captured throughout the entire 72-hours period.

By conducting the simulation in this manner, the study aimed to closely monitor and analyse the behaviour of the isotopes during the phase following the initial irradiation.

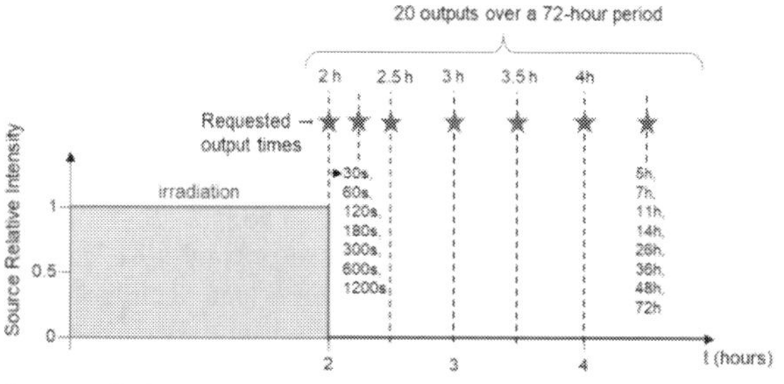

Fig. 3. Irradiation and radioactive decay monitoring chronogram.

Following the extraction of isotopic descriptions, a second calculation step was performed to consider the activated $LaBr_3$:Ce detector as a source. To achieve this, the PHITS RI-source function was employed, enabling the generation of photon sources with energy spectra corresponding to radioisotope (RI) decay. By specifying the activity and name of the radioisotopes, PHITS utilized the nuclear decay database DECDC, which is equivalent to ICRP107 [10], to accurately model the decay process.

To ensure a reliable and accurate assessment, a passive approach was adopted by simulating a second detector placed nearby the primary activated $LaBr_3$:Ce. This approach was chosen in order to limit potential effects arising from the simulation of a region used as both a source and a detector. To implement this configuration, the geometry reported in Figure 4 was employed.

In this setup, the activated primary $LaBr_3$:Ce crystal, located on the left side, emitted gamma-rays, which were then detected by the secondary $LaBr_3$:Ce crystal positioned on the right side.

The emitted gamma-rays from the activated $LaBr_3$:Ce crystal exhibited a consistent energy distribution. This indicates that there was no inherent bias introduced by considering the region as both the source and detector. The simulation accurately computed the energy distribution of the emitted photons, ensuring a reliable representation of the system's behaviour.

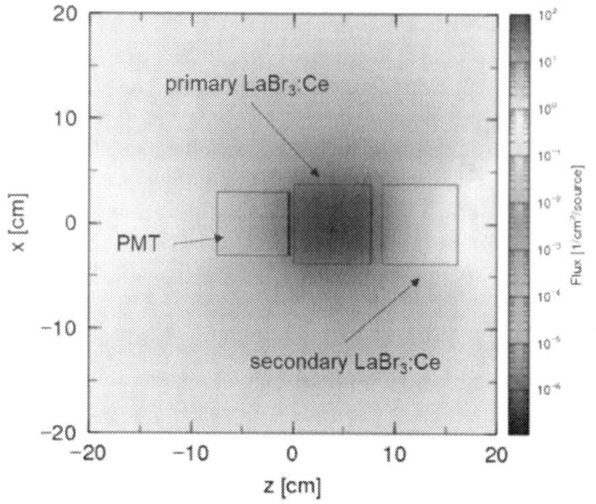

Fig. 4. Simulation of the photon mesh using PHITS [t-track] tally and showing gamma-rays emitted by the primary LaBr3:Ce after induced activation products calculated with DCHAIN where used as the source description for the second calculation step.

III. RESULTS AND DISCUSSIONS

A. Analysis of the activation gamma rays energy spectra

To illustrate the simulated gamma-ray spectrum, a specific time point of 15 seconds after the cooling phase was chosen. At this stage, the activated crystal had undergone a partial decay, resulting in a distinct gamma-ray signature. Figure 5 represents this gamma spectrum and was computed with the PHITS tally [T-Deposit] used to estimate the deposited energies in certain regions. Here it was applied to the activated crystal itself.

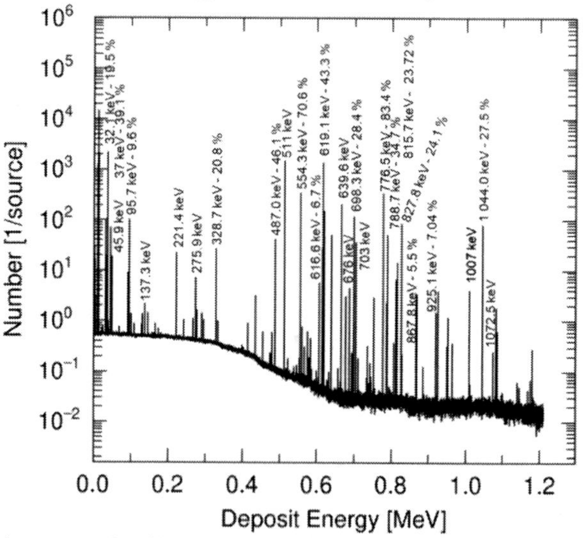

Fig. 5. Energy deposition of photons in the activated $LaBr_3$:Ce crystal without energy resolution.

Figure 6 showcases the simulated gamma spectrum after

applying the energy resolution of the $LaBr_3$ detector, which is known to be around 2.2% at 662 keV. This resolution value accounts for the inherent limitations of the detector and provides a realistic representation of the measured energy spectrum.

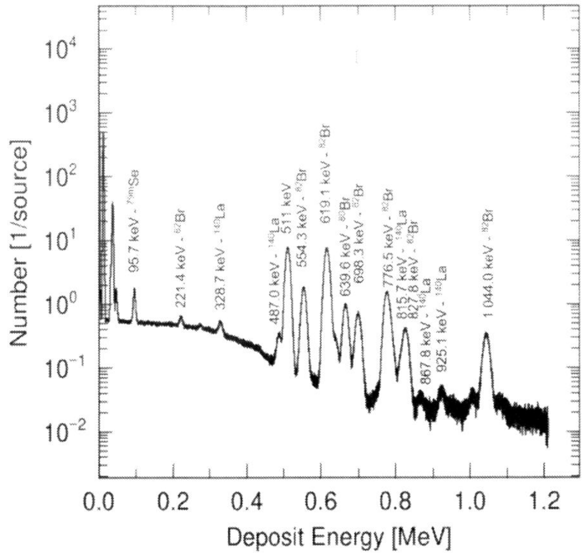

Fig. 6. Energy deposition of photons in the activated $LaBr_3$:Ce crystal taking into account its 2.2% resolution at 662 keV.

The observed spectrum exhibits several peaks corresponding to gamma-rays emitted during the decay of specific activated isotopes. For instance, among others, we can evidence a significant presence of Br-82, resulting from the (n, γ) capture of Br-81 in the spectrum. Additionally, the presence of La-140, produced through the (n, γ) capture of La-139, and Br-80, produced via the same nuclear reaction from Br-79, can be identified. Another noteworthy activation product is Se-79m, which is likely generated through the electron capture of Br-79. Those results are conform to what was reported in the publication of Y. Lu et al. [3].

B. Analysis of the temporal behaviour of the activated $LaBr_3$:Ce

By analysing the gamma-rays at different time intervals during the cooling calculation, a verification process was conducted to confirm that the decay of the activation products follows the expected Poisson law for radioactive decay (Eq. 1).

$$N(t) = N_0 . \exp(-\lambda . t) \quad (1)$$

Where $N(t)$ represents the number of radioactive nuclei at time t, N_0 is the initial number of nuclei, λ is the decay constant, and t represents the elapsed time. This fundamental law describes the exponential decrease in the activity or number of radioactive nuclei over time. By comparing the experimental data with the expected decay behaviour following Eq. 1, the consistency and validity of the simulation results can be assessed. The decay constant is a fundamental parameter used to describe the rate at which radioactive isotopes decay. It is closely related to the concept of the half-life of decaying products. The relationship between the decay constant (λ) and the half-life ($T_{1/2}$) can be defined using the following equation:

$$\lambda = \frac{ln\,2}{T_{1/2}} \quad (2)$$

The fitting procedure involved utilizing a negative exponential function of the form $a.\exp(-b.t)$, where ' a ' and ' b ' are parameters to be optimized, and ' t ' represents the time variable. By fitting this function to the data, the goal was to capture the underlying trends and check if the dataset was following the Poisson's law.

To quantify the uncertainties associated with the optimized parameters, the covariance matrix of the optimized parameters was used. The covariance matrix provides valuable information about the interdependencies between the parameters and their respective variances. Specifically, the diagonal elements, which represent the variances of the optimized parameters. By extracting the variances from the diagonal elements of the covariance matrix, the uncertainties associated with each parameter were calculated. In particular, the standard deviations of the parameters were determined by taking the square roots of their respective variances. It is also to mention that the statistic uncertainty for the Monte-Carlo calculations are less than 1% in most of the cases.

The estimation of parameter uncertainties is crucial in evaluating the reliability and robustness of the fitted function. These uncertainties provide insights into the sensitivity of the fitted function to changes in the input data and aid in understanding the stability and precision of the fitting procedure.

Starting with the nucleus exhibiting the shortest half-life, Figure 7 depicts the temporal evolution of gamma ray counts at 96 keV originating from Se-79m, alongside its corresponding fitted curves. The calculated half-life for Se-79m is 3 minutes and 58.47 seconds with a precision of ±1.37 seconds, while the CEA-LNHB LARA library for gamma and alpha spectrometry (Ref. [11]) lists its half-life as 3 minutes and 55.2 seconds.

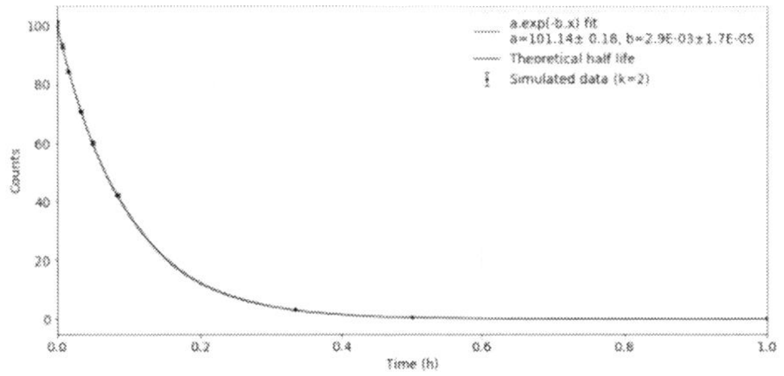

Fig. 7. Temporal variation of 96 keV gamma ray counts from Se-79m and fitted curves.

Figure 8 reports the temporal evolution of gamma ray counts at 46 keV emitted from Br-82m, complemented by fitted curves. Br-82m's calculated half-life is 6 minutes and 18.10 seconds (±3.57 seconds), while the LARA library records its half-life as 6 minutes and 7.8 seconds.

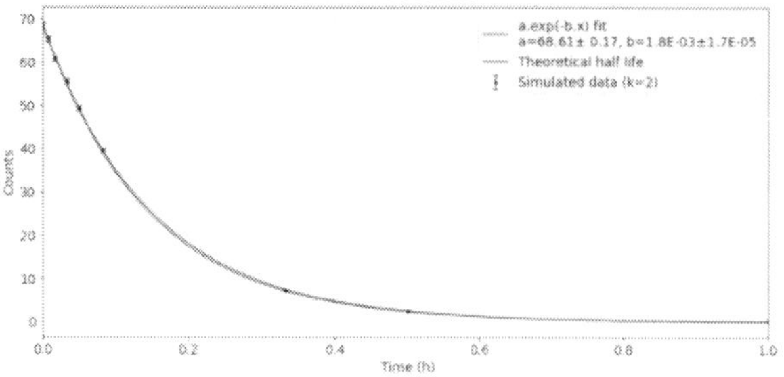

Fig. 8. Temporal behaviour of 46 keV gamma ray counts from Br-82m and fitted curves.

Figure 9 displays a distinct scenario involving the temporal behaviour of gamma ray counts at 639 keV originating from two isotopes, namely Br-80 and Br-80m. This case holds particular significance due to the contrasting half-lives of the isotopes and their decay processes. Despite Br-80m showing a longer half-life than Br-80, it assumes a crucial role in the decay chain, effectively being a precursor to the decay of Br-80. To comprehensively understand this dependence, a double exponential fit was employed, offering an analytical approach that takes into account the contributions from both Br-80 and Br-80m to the observed gamma ray counts. The derived half-life measurements for Br-80 and Br-80m were found to be 17 minutes and 41.14 seconds (±0.75 seconds) and 4 hours, 24 minutes, and 31 second (±1 minute, 15 seconds), respectively. Notably, the LARA library values for Br-80 and Br-80m were found as 17 minutes, 40.8 seconds and 4 hours, 25 minutes, 10 seconds respectively.

This particular case exemplifies the importance of carefully considering the interplay between different isotopes and their decay processes, especially when one isotope influences the decay of another. The observed temporal behaviour, coupled with distinct half-life measurements, unveils the intricacies of nuclear decay and provides valuable insights for further studies in nuclear physics and related disciplines.

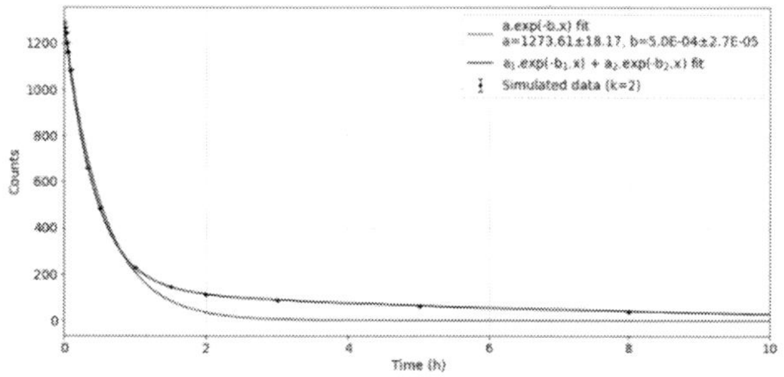

Fig. 9. Temporal behaviour of 639 keV gamma ray counts from br-80 and br-80m with fitted curves. The red curve represents a single exponential fit, inadequate to account for contributions from both isotopes.

Transitioning to isotopes with longer half-lives, Figure 10 illustrates the temporal evolution of the 221 keV gamma ray emitted by the Br-82 isotope. The calculated half-life is determined to be 36 hours, 56 minutes, and 32 seconds, accompanied by an uncertainty of 1 hour, 10 minutes, and 48 seconds. Conversely, the LARA library attributes a half-life of 35 hours, 18 minutes, and 0 seconds to Br-82.

The presence of a 4.65% relative difference between the calculated and library-based half-life values can be rationalized by statistical fluctuations occurring in the simulation data points during the initial hour after irradiation. Nevertheless, it is noteworthy that the simulation results stay in good agreement with theoretical expectations.

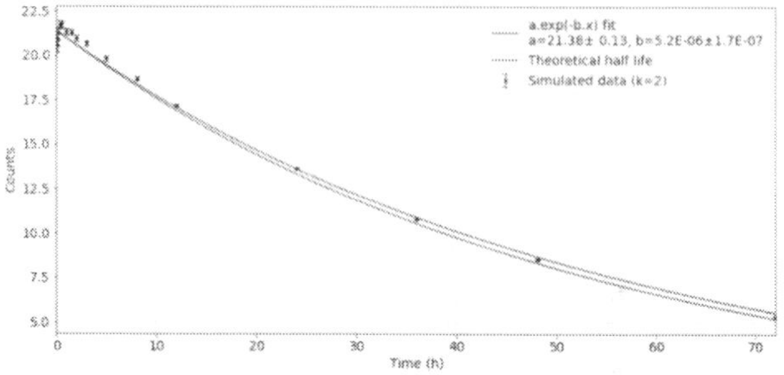

Fig. 10. Temporal behaviour of 221 keV gamma ray counts from Br-82 and fitted curves.

The La-140 isotope, so far neglected, is presented in Figure 11. The fitting procedure enables the determination of its half-life as 40 hours, 1 minute, and 43 seconds, accompanied by an uncertainty of 14 minutes and 3.97 seconds. This computed value demonstrates concordance with the LARA-based half-life of 40 hours, 17 minutes, and 9 seconds.

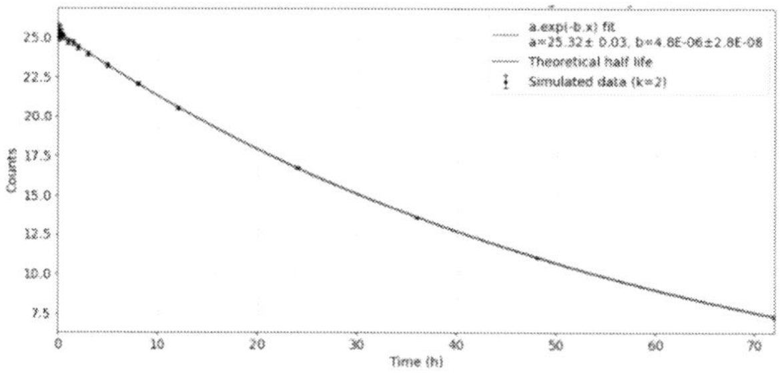

Fig. 11. Temporal behaviour of 329 keV gamma ray counts from La-140 and fitted curves.

Table I summarizes, the results of calculated half-lives based on simulations with the theoretical half-lives from the CEA-LNHB LARA library. Remarkably, a high degree of concordance is observed between the calculated and library-based half-life values, which significantly increase our confidence in the accuracy and reliability of the numerical model employed in this study.

TABLE I
COMPARISON OF CALCULATED AND THEORETICAL HALF-LIVES FOR ISOTOPES CREATED IN THE ACTIVATED $LaBr_3$

Isotopes created	Calculated half-lives	Uncertainty of the calculated half-lives	Theoretical half-life from Ref. [7]	Relative differences
Se-79m	3m 58.47s	1.37s	3m 55.20s	1.39 %
Br-82m	6m 18.10s	3.57s	6m 7.80s	2.80 %
Br-80	17m 41.14s	0.75s	17m 40.80s	0.03 %
Br-80m	4h 24m 31s	1m 15s	4h 25m 10s	0.25 %

Br-82	36h 56m 32s	1h 10m 48s	35h 18m 0s	4.65 %
La-140	24h 37m 30s	14m 3.97s	40h 17m 9s	0.64 %

We would like to emphasise that the important presence of Br-82, with a half-life of 23 hours, 2 minutes, and 22 seconds, and La-140, with a half-life of 24 hours, 37 minutes, and 30 seconds, should be acknowledged due to their relatively long half-lives. These longer-lived isotopes have the potential to contribute to the internal gamma background noise of the detector after activation especially because of the high intensity on some gamma-rays like at 1596 keV (95.4%) and 487 keV (46.1%) for LA-140 and at 776 keV (83.4%), 554 keV (70.6%) and 619 keV (43.3%) for Br-82 [11]. When the detector has been activated and these isotopes are present, the internal activity levels can remain elevated for a considerable period of time. It could take several days for the activity levels to gradually decrease and return to normal background levels. During this period, the presence of Br-82 and La-140 will contribute to the gamma radiation emitted by the detector, resulting in an increased internal gamma background noise.

It is essential to consider the impact of these longer-lived isotopes when analysing the gamma spectra obtained from the activated detector. Their presence can influence the overall detection sensitivity and may require additional data processing techniques or corrections to accurately distinguish and quantify other gamma rays of interest.

Our analysis sheds light on the effects of neutron-induced activation on the $LaBr_3$:Ce detector and validates the calculation scheme with PHITS and DCHAIN for both the energy spectra aspect and the temporal behaviour of activation gamma-rays. In addition, we discuss the implications of our findings for the accurate characterization of materials using neutron activation analysis.

IV. Conclusions

In conclusion, our study has provided valuable insights into the behaviour of $LaBr_3$ scintillators in gamma spectrometry for neutron activation analysis, particularly under neutron flux conditions. Through comprehensive simulations using the DCHAIN and PHITS codes in a two-steps process, we have demonstrated the coherence between neutron-induced activation gamma-rays, the resulting isotopes in the $LaBr_3$:Ce crystal, and their decay characteristics over time. This validation of the numerical approach consisting in the coupled use of PHITS along with DCHAIN strengthens the reliability and accuracy of calculations in this field. Moreover, our study showcases the versatility and adaptability of these codes in addressing complex research questions in the field of neutron activation analysis. It is a compelling demonstration of the correct use of the DCHAIN and PHITS codes in two-step computations.

As perspectives, it is crucial to further study and explore the combined effects of activation and (n, n'γ) reactions in the simulation and to compare those results with experimental measurements of gamma spectra in a neutron field. This will led to a better comprehension of the degradation of the energy resolution of the $LaBr_3$ scintillators when exposed to a neutron flux.

Acknowledgment

The authors gratefully acknowledge the fruitful collaboration with the PHITS team, and in particular, express their sincere appreciation to Tatsuhiko-san for his valuable support, availability, and prompt responses, which significantly contributed to this work.

References

[1] E. Witkowska, K. Szczepaniak, M. Biziuk, "Some applications of neutron activation analysis: A review", *Journal of Radioanalytical and Nuclear Chemistry*, vol. 265, no. 1, pp. 141-150 (2005).
[2] R. De Stefano, A. Elayeb, A. Sari et al., "Evaluation of linac-based delayed gamma neutron activation technique for copper characterization in scrap metal by means of Monte Carlo modeling", *Nucl. Inst. and Meth. in Phys. Res. Sect. A*, vol. 1045, no. 167441 (2023).
[3] Y. Lu, Z. Song, G. Li, et al., "Study of neutron radiation effect in LaBr3 scintillator", *Chinese Phys. C*, vol. 39, no. 106003 (2015).
[4] K. Tasaka, "DCHAIN 2: A Computer Code for Calculation of Transmutation of Nuclides: Tech. Rep. JAERI-M 8727", Japan Atomic Energy Research Inst. (1980).
[5] N. Hunter et al., "Modernization of the DCHAIN-PHITS activation code with new features and updated data libraries", *Nucl. Inst. And Meth. in Phys. Res. B*, vol. 484, pp. 29-41 (2020).
[6] T. Sato, Y. Iwamoto, S. Hashimoto, et al., "Features of particle and heavy ion transport code system (PHITS)", *J. Nucl. Sci. Technol.*, vol. 55, no. 6, pp. 684-690 (2018).
[7] B. Singh, Nuclear Data Sheets for A = 79, *Nuclear Data Sheets*, vol. 96, np.1, pp. 1-176 (2002).
[8] Baglin, Coral M., Nuclear Data Sheets for A = 81. *Nuclear Data Sheets*, vol. 109, no. 10, pp. 2257-2437 (2008).
[9] Joshi, Paresh K., et al., Nuclear Data Sheets for A = 139. *Nuclear Data Sheets*, pp. 1-292 (2016).
[10] A. Endo, Y. Yamaguchi and K.F. Eckerman, "Nuclear decay data for dosimetry calculation - Revised data of ICRP Publication 38", JAERI 1347 (2005).
[11] Nucléide-LARA, 2018. http://www.nucleide.org/Laraweb/index.php. (Accessed 3 May 2023).

 https://doi.org/10.1051/epjconf/202328810012

Fast timing detectors for the muon system of a muon collider experiment

C. Aimè[1,2], S. Calzaferri[2], D. Fiorina[2], C. Riccardi[1,2], P. Salvini[2], I. Vai[1,2,*] and P. Vitulo[1,2] on behalf of the International Muon Collider Collaboration
[1]Dipartimento di Fisica, Università di Pavia, Pavia, Italy
[2]Istituto Nazionale di Fisica Nucleare sez. di Pavia, Pavia, Italy
(*) ilaria.vai@unipv.it

Abstract—**The muon collider offers a great discover potential for the future of high energy physics. Indeed, it combines the advantages of a lepton collider, with clean signatures and maximum available center of mass energy, with those of a hadron collider, such as low synchrotron radiation. However, it poses interesting challenges, manly related to the fact that muons decay and their product interact with the material of the machine producing the so-called Beam-induced-Background. For this reason, a careful design of the experiments running at a muon collider must be performed, starting from simulation, and moving to the R&D on dedicated technologies. This contribution will focus on the muon system of a muon collider experiment: the current design limitations will be presented, together with the alternative solutions that are being considered. In particular, we are proposing to use the Picosec Micromegas technology for a dedicated timing layer in the muon system, which will improve the muon reconstruction performance in that region. The R&D on this technology will be discussed and preliminary results will be presented.**

Keywords —**Muon collider, time resolution, Picosec-Micromegas.**

I. Introduction

THE future of high-energy physics lies in the ability to make collisions at a multi-TeV center-of-mass energy and to measure the products of the collisions with high precision. This will be crucial both for exploring new physics and for deepening the understanding of possible discoveries made by the LHC at high luminosity. Muon colliders thus offer great potential for discovery in the multi-TeV energy range: first, the large mass of muons, compared with that of electrons, implies enormous suppression of synchrotron radiation compared with electron beams of the same energy. In addition, the leptonic nature of muons poses a cleaner analysis environment, free of the accumulation problems characteristic of hadron colliders. The development of a muon collider obviously poses some major technological challenges, arising from the short muon lifetime, 2.2 μs, and from the difficulty of producing large numbers of muons in groups with small emittance. In addition, the Beam-induced Background (BIB) due to muon decay affects the design of the machine and detector. Simulations performed with FLUKA show that the BIB is mainly composed of neutrons, photons, and low-energy electrons/positrons. They deposit energy diffusely throughout the detector volume, with a significant dispersion of their arrival time with respect to crossing the bunch, due to their different velocities [1]. All these characteristics must be considered for proper detector design.

II. Simulation Of The Experiment

The existing simulation framework is based on the iLCSoft framework, previously adopted by the CLIC Collaboration [2], and updated for Muon Collider developments. Currently, the simulation includes a tracking system based on multiple layers of silicon detectors, followed by electromagnetic and hadronic calorimeters. These three components are contained within a solenoidal magnet, which provides a field of 3.57 T. Outside the solenoid extends the muon system, based on multiple layers of gaseous detectors in both the barrel and endcap regions, as shown in Fig.1.

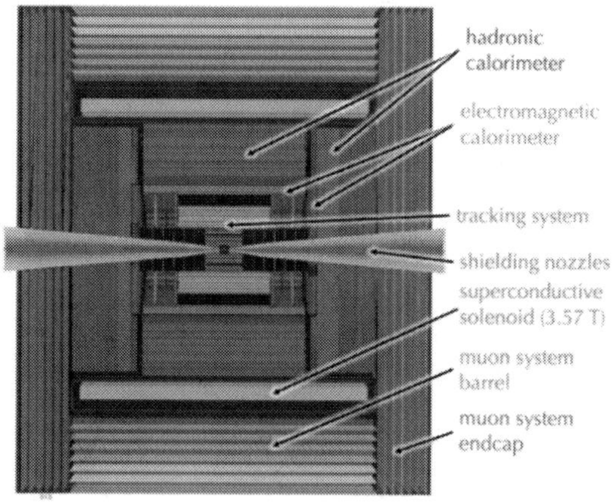

Fig. 1. Baseline detector design for a muon collider experiment [2].

The muon system is currently instrumented in the simulation with glass-RPC technology, however performed standalone simulations show that the hit rate expected in the endcap regions closest to the beamline are at the limit of the current performance reachable by such a detector (Fig. 2) [3].

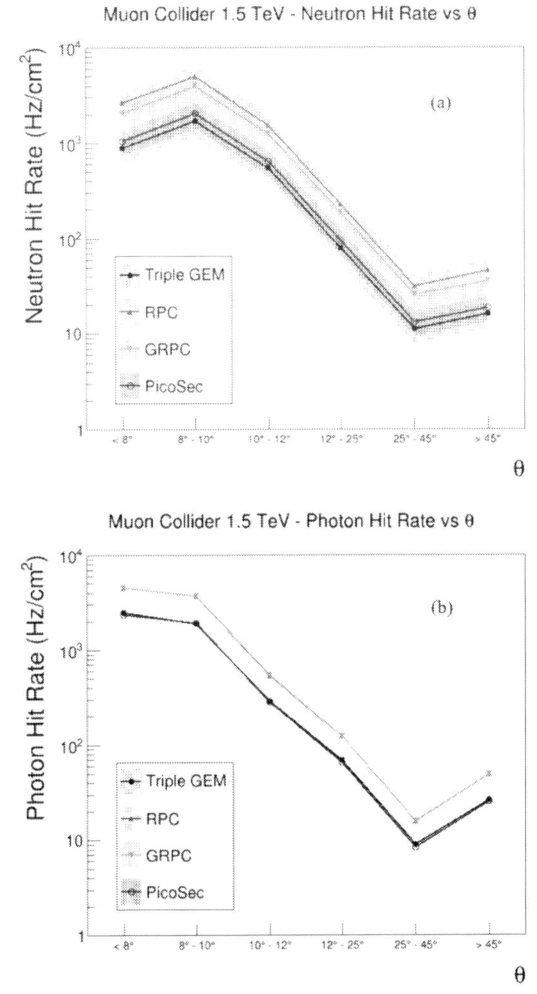

Fig. 2. Neutron (a) and photon (b) expected hit rate in the different regions of the muon system endcap [3].

For this reason, we are proposing an alternative design of the muon system fully based on Micropattern Gaseous Detectors (MPGD).

III. Alternative Muon System Design

To reduce the combinatorial background due to BIB, a proposal is to implement an out-to-in muon reconstruction approach. Muons are initially identified as segment in the muon system and then matched with hits in the tracker system for the reconstruction of the full muon track. Fig. 3 outlines this process: it requires a muon system characterized by an excellent space and time resolution. The segment reconstruction could be performed with a multilayer tracking system based on MPGDs, such as Triple-GEM, which can assure a ~100 μm space resolution. The timing instead must be comparable with the one in the inner tracking system, thus reaching ~100 ps. A dedicated timing layer is thus needed.

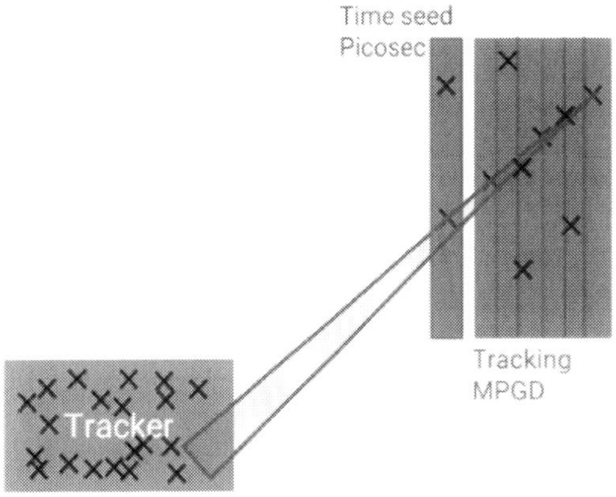

Fig. 3. Out-to-in muon reconstruction approach.

Classical MPGDs, such as Triple-GEM, cannot reach such a high time resolution due to fluctuations on the position of the first ionization cluster in the drift gap. The drift gap of a classical MPGD is indeed usually 3-5 mm-thick: the large uncertainty on the position of the first ionization cluster translate in a large fluctuation of the time of arrival of the electrons to the amplification layer and therefore in a large fluctuation of the time of arrival of the signal. Due to this effect, the time resolution of classical MPGDs is usually limited to 4-5 ns. To address this issue, the Picosec Micromegas collaboration [4] developed the detector shown in Fig. 4. The Picosec detector [5] is composed by a Cherenkov radiator (MgF_2), followed by a proper photocathode and a micromegas with reduced-thickness drift gap (~100 μm instead of few mm). When the particle passes through the radiator, Cherenkov photons are emitted and then converted into electrons by the photocathode.

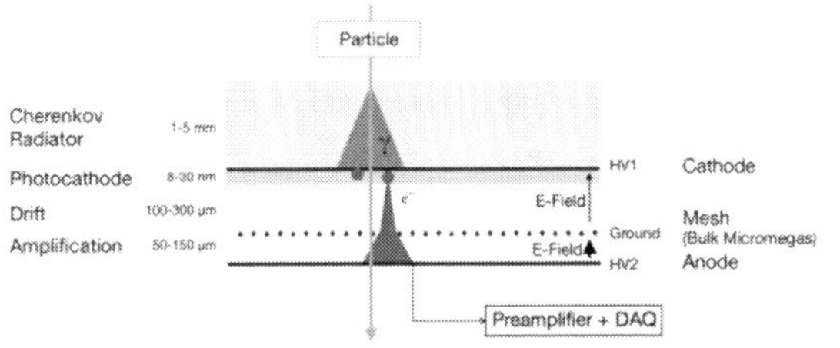

Fig. 4. Picosec micromegas detector [5].

The electrons then enter the micromegas and are amplified, generating the visible signal. The advantage is that all the electrons are produced in the same position, at the level of the photocathode, thus zeroing the fluctuations. The time resolution reachable with this configuration is of the order of ~20 ps [5].

IV. Dedicated R&D On Picosec Detector

In order to use such a technology in the context of a muon collider experiment, few dedicated R&Ds are needed, which are summarized in Fig. 5. The current baseline for the Cherenkov radiator is MgF_2, which has a high UV transparency, but is fragile, costly and can be produced with a maximum size of the order of ~100 cm^2. The main alternatives considered are quartz and sapphire, which up to now could not return the desired performance. Indeed, for example quartz used in combination with CsI as photocathode has a measured number of photoelectrons/MIP ~ 2, seven times lower than MgF_2 with CsI, which turns out in a measured time resolution of a least a factor four lower.

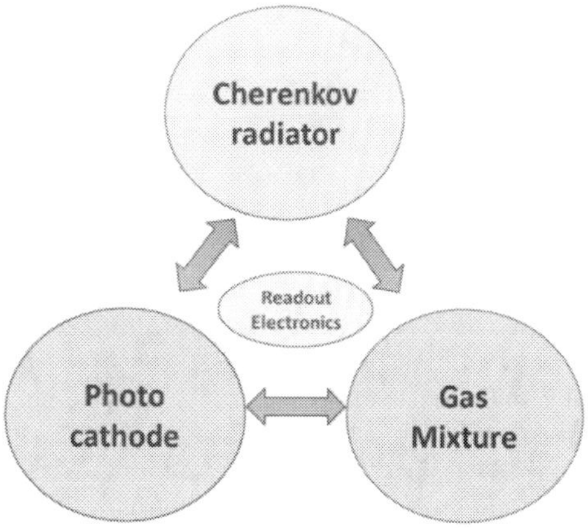

Fig. 5. Outline of the R&D activities on Picosec micromegas detector.

Moving to the photocathode, the current baseline is CsI, which has a high quantum efficiency to UV (~10 p.e./MIP) and is easy to coat by Chemical Vapour Deposition. However, it's hygroscopic, which forces a sealed operation with dry gas, and can be easily damaged by ion bombardment. Alternatives considered includes metallic (Al, Cr, Au) and carbon-based photocathodes (DLC, B4C). This last category is being currently investigated, but it seems to be still limited by the number of p.e./MIP produced, which is at least a factor 3 lower than for the baseline configuration.

The most interesting results so far have been obtained by the search of alternative gas mixtures. The baseline, $Ne/C_2H_6/CF_4$ (80/10/10), is costly, flammable, and most importantly has a very high Global Warming Potential (see Table I).

TABLE I
DETAILS OF CONSIDERED GAS MIXTURES

Gas Mixture	100-year Global Warming Potential (normalized to CO2)
$Ne/C_2H_6/CF_4$ (80/10/10)	740
Ne/iC_4H_{10} (94/6)	0.2
Ar/CO_2 (93/7)	0.07
$Ar/CO_2/iC_4H_{10}$ (93/5/2)	0.11

As alternative, Ar-based and Ne-based gas mixtures are being tested, as listed in Table I. Ar-based gas mixtures turned out to have a too small operation range: it is therefore extremely difficult to find and operational configuration in which the detector is stable and the signal is visible. On the other hand, interesting results were obtained with Ne/iC_4H_{10}: Fig. 6 compares the time resolution measured with the baseline mixture (a) and with Ne/iC_4H_{10} (b). The RMS of the two plots show that the time resolution is of the same order of magnitude between the two and it's around the value expected in a configuration with 6 nm B4C photocathode (which has only 3 p.e./MIP and therefore gives a lower time resolution than CsI).

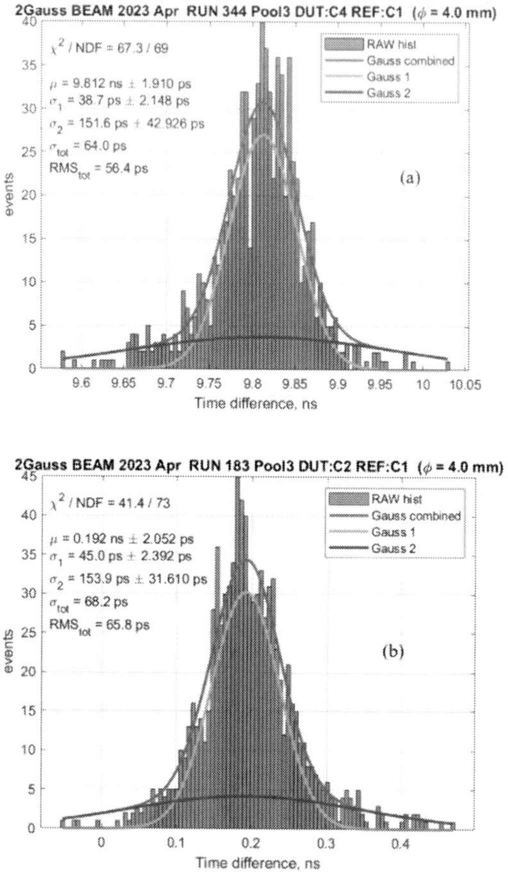

Fig. 6. Time resolution with (a) $Ne/C_2H_6/CF_4$, (b) Ne/iC_4H_{10}

This preliminary result is only the first of a systematic study that is being performed on this item and that will continue in the next few years.

Finally, it is worth to mention that other two elements must but developed to prepare Picosec for the installation on a large-scale experiment at the muon collider. On one side, dedicated integrated electronics able to sustain such a high time resolution must be developed. On the other side, a careful design of the mechanics of the detector must be carried out: indeed, so far, the largest prototypes reach a surface of 10x10 cm^2. Increase the dimension of a single Picosec detector could turn out in a reduced time resolution, due to the difficulty in maintaining a flat geometry in the micromegas and in the production of a proper radiator. A possibility could be to couple several 10x10 cm^2 detectors to cover large areas: a dedicated R&D is needed and will start in the next few months.

V. Conclusions

The muon collider is an interesting opportunity for the future of high energy physics. It poses however challenging operational conditions, mainly related to the presence of BIB. The design of a proper experiment is ongoing: MPGDs are being considered as candidate detectors for the muon system. Picosec is a candidate technology for a timing layer in the muon system: a dedicated R&D is ongoing, mainly focused on optimization of the Cherenkov radiator, photocathode and gas mixture. Future R&D activities will have to focus on the readout electronics, as well as on the mechanics needed for the instrumentation of large surfaces.

Acknowledgment

We warmly thank the RD51 Picosec-Micromegas Collaboration for the precious support and help.

References

[1] F. Collamati et al, "Advanced assessment of beam-induced-background at a muon collider," Journal of Instrumentation, Volume 16, November 2021, 10.1088/1748-0221/16/11/P11009.
[2] L. Linssen et al, "Physics and detectors at CLIC: CLIC conceptual design report," arXiv:1202.5940[physics.ins-det].
[3] C. Accettura et al, "Towards a muon collider", arXiv:2303.08533 [physics.ins-det].
[4] Picosec Micromegas Collaboration, https://gdd.web.cern.ch/activities-picosec
[5] J. Bortfeldt et al, "PICOSEC: Charged particle timing at sub-25 picosecond precision with a Micromegas based detector," Nuc. Instr. And Meth. A, Volume 903, 21 September 2018, Pages 317-325, 10.1016/j.nima.2018.04.033.

Influence of base material thickness on spectrometry of semiconductor detectors based on semi-insulating GaAs

Andrea Šagátová[1,*], Nikola Kurucová[1], Soňa Kotorová[1], Eva Kováčová[2] and Bohumír Zaťko[2]

[1]Institute of Nuclear and Physical Engineering, Faculty of Electrical Engineering and Information Technology, Slovak University of Technology, SK-812 19 Bratislava, Slovak Republic

[2]Institute of Electrical Engineering, Slovak Academy of Sciences, Dúbravská cesta 9, SK-814 04 Bratislava, Slovak Republic

(*) andrea.sagatova@stuba.sk

Abstract—**The bulk semi-insulating GaAs material was used for preparation of pad radiation detectors with circular contacts of 1 mm diameter. The spectrometric properties of a semiconductor detector depend on the quality of the base material and on the deposited metallization. Another factor affecting the detector spectrometry is the applied bias controlling the electric collection field. With increasing bias, the charge collection efficiency of particular detector grows. However, this spectrometric property should be changing with detector thickness, which affects the intensity of electric collection field at constant bias applied through the detector sandwich structure. In this paper we have studied the electrical and spectrometric properties of semi-insulating GaAs detectors as a function of their thickness. The measured saturation reverse current was in the range of 3 – 30 nA, increasing with decreasing detector thickness at a substrate resistivity of about of 10^7 Ωcm. The maximal obtainable charge collection efficiency evaluated from ^{241}Am gamma spectra grew with decreasing detector thickness from 50% for a 450 μm thick detector to 80% for a 230 μm thick detector.**

Keywords —**GaAs detector, bulk semi-insulating, detector thickness, charge collection efficiency.**

I. Introduction

Semiconductor detectors of ionizing radiation are a rapidly developing field with expanding applications. The most common semiconductor material is silicon, however, there are other alternatives used for detector preparation like semi-insulating (SI) GaAs [1-9]. The SI GaAs is a material with a high resistivity of about 10^8 Ωcm and a high charge carrier mobility of up to 8000 and 400 $cm^2V^{-1}s^{-1}$ for electrons and holes, respectively, at room temperature (RT) leading to a high detector reaction rate. Its wide bandgap of 1.42 eV enables detector operation at RT. In comparison to silicon, the GaAs exhibits much better radiation hardness [10-16] and has higher detection efficiency for photons thanks to its higher electron density. However, in comparison to silicon detectors, its spectrometry quality is a limiting factor. The spectrometry properties are affected by the base material quality and also by the detector structure. In the case of single pad SI GaAs detectors with a circular electrode the collecting field of the charge generated by the radiation is affected by detector thickness and applied bias influencing the spectrometric properties of the detector. Typical thickness of SI GaAs wafer for detector preparation ranges from 500 μm down to 230 μm, obtainable by wafer polishing.

In this paper we have prepared the SI GaAs detectors of various thicknesses 230, 350 and 450 μm and evaluated their important electrical and spectrometric properties with respect to the detector thickness. The charge collection efficiency (CCE) during the spectrometry of alpha particles and gamma rays from ^{241}Am is presented as well as current-voltage characteristics measured for prepared detectors.

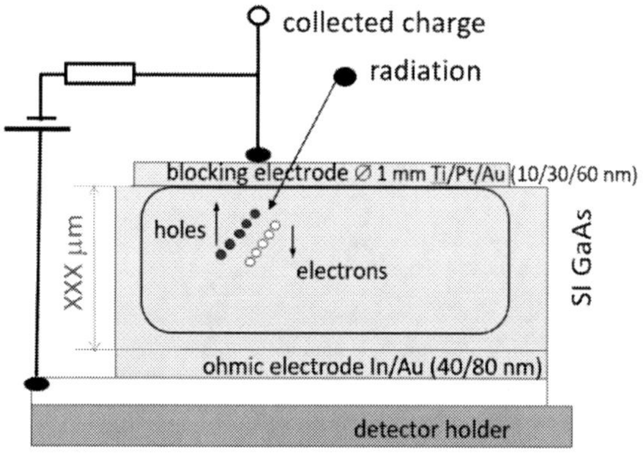

Fig. 1. The cross-section view of prepared SI GaAs radiation detector. The different thicknesses are marked as "XXX".

II. Detector structure

The detectors were fabricated at the Institute of Electrical Engineering of SAS in Bratislava from bulk undoped SI GaAs substrates double-side polished to three various thicknesses: 230, 350 and 450 μm. Top blocking electrode was realized as circular Schottky contact of a 1 mm diameter made of Ti/Pt/Au (10/30/60 nm) multilayer evaporated through a metal contact mask. The bottom ohmic electrode was covering whole substrate area and was prepared as In/Au (40/80 nm) double layer by evaporating in a high vacuum chamber. The cross-section view of the detector structure is depicted in Fig. 1.

III. Experiment and discussion

A. Current-voltage measurements

Foremost, the current-voltage characteristics of the fabricated SI GaAs detectors were measured at RT to determine the correct operation of the detectors. The reverse current depicted in Fig. 2 affects the value of the noise signal during radiation detection. Its saturation value was in the range of 3 to 30 nA depending on the detector thickness. The lowest saturation current was measured for the thickest (450 μm) detector and as the detector thickness decreases, the saturation reverse current increases, which is understandable as the resistivity of all compared SI GaAs substrates was similar of 10^7 Ωcm.

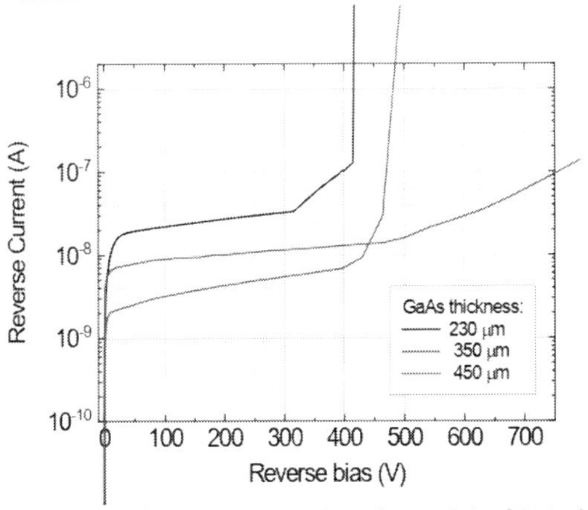

Fig. 2. The typical reverse current–voltage characteristic of GaAs detectors for various detector thicknesses.

B. Alpha spectrometry

The detectors were fabricated for spectrometry of alpha particles and gamma rays and examined by combined ^{241}Am alpha and gamma source. Measured alpha spectra of ^{241}Am are shown in Figs. 3. – 5. for various biases applied. The shift of the peak to higher channels with increasing bias can be observed for all three detector thicknesses. However, this phenomenon is the most intensive with the thinnest detector, which reaches the highest channels corresponding to the highest CCE and reaching also the best peak height. As the thickness of detector grows, the peak becomes smaller and wider indicating inhomogeneities in charge collection.

The CCE as a function of reverse bias for all three types of detectors is depicted in Fig. 6. The increase of CCE with applied bias was reached due to higher charge drift velocity obtained by increasing intensity of electric collection field. The CCE during alpha detection is affected dominantly by electron collection, as alpha particles interact close to the Schottky electrode. The 5.485 MeV alpha particles from ^{241}Am had to penetrate 6 mm of air and the metallization of top Schottky contact before reaching the active detector volume in our experiment. According to results of simulation in SRIM [17], the particles penetration depth is 17 μm in GaAs [16]. The electrons as charged carriers in GaAs exhibit very high mobility of 8000 $cm^2V^{-1}s^{-1}$ and the effect of detector thickness on CCE is minimal. On the other hand, the 350 and 450 μm thick detector substrates have measured electron Hall mobility @ 300K: 6200 cm^2/Vs and the 230 μm thick ones: 7119 cm^2/Vs, which led to lower CCE of thicker detectors.

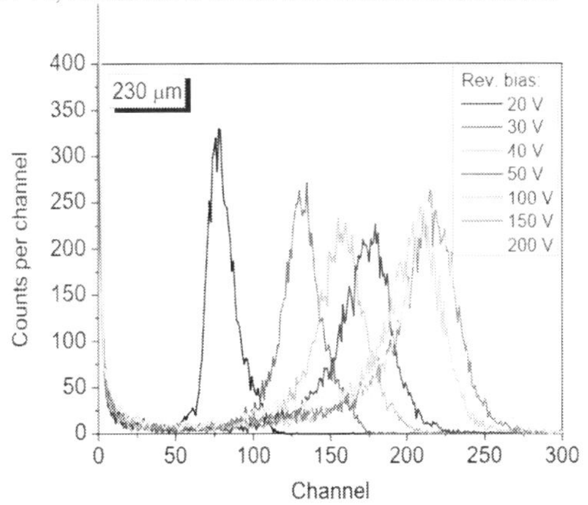

Fig. 3. The alpha spectra of 5.5 MeV alpha particles from ^{241}Am measured by a 230 μm thick SI GaAs detector at various reverse biases.

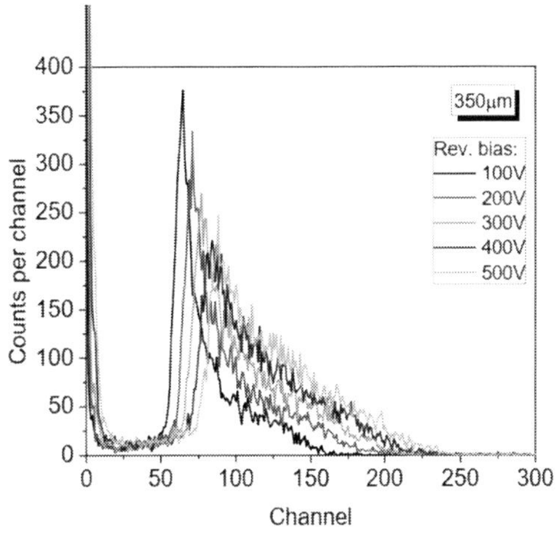

Fig. 4. The alpha spectra of 5.5 MeV alpha particles from ^{241}Am measured by a 350 μm thick SI GaAs detector at various reverse biases.

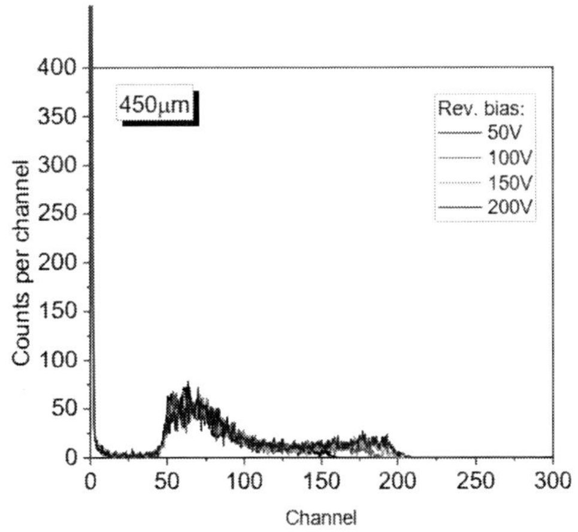

Fig. 5. The alpha spectra of 5.5 MeV alpha particles from ^{241}Am measured by a 450 μm thick SI GaAs detector at various reverse biases.

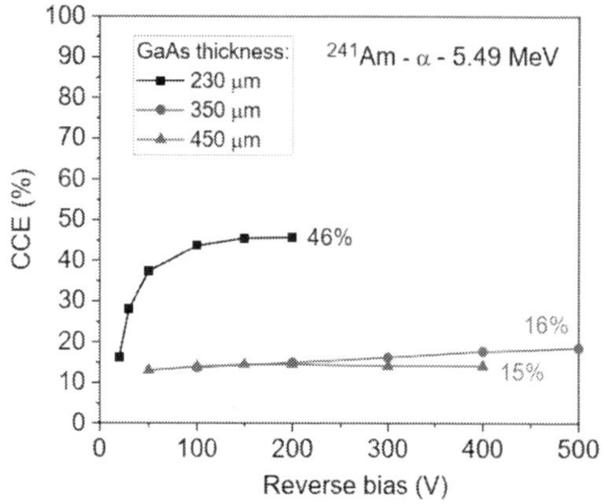

Fig. 6. The CCE for alpha spectrometry as a function of reverse bias for three different detector thicknesses.

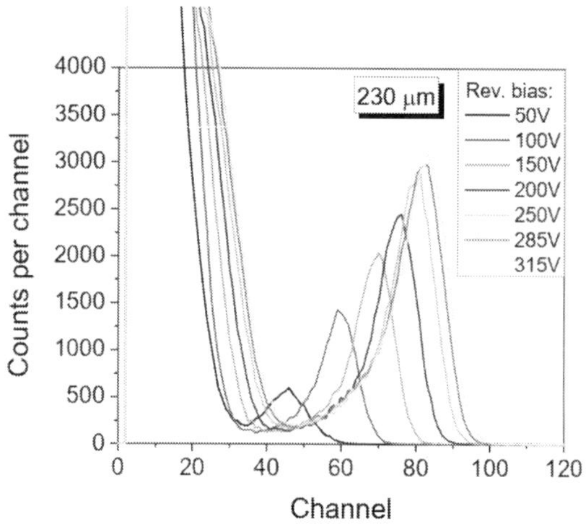

Fig. 7. The gamma spectra of 60 keV gamma rays from ^{241}Am measured by a 230 μm thick SI GaAs detector at various reverse biases.

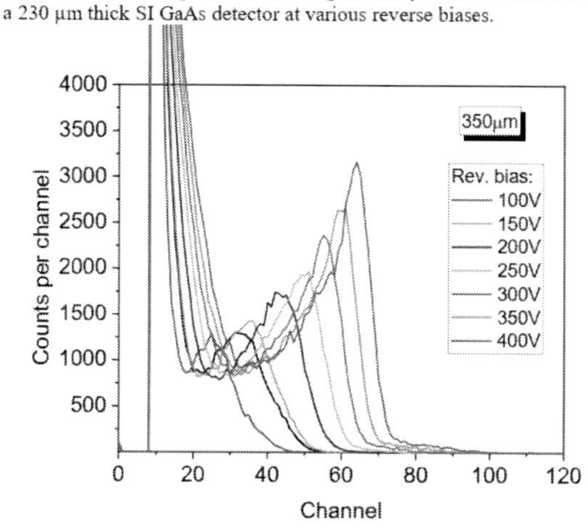

Fig. 8. The gamma spectra of 60 keV gamma rays from ^{241}Am measured by a 350 μm thick SI GaAs detector at various reverse biases.

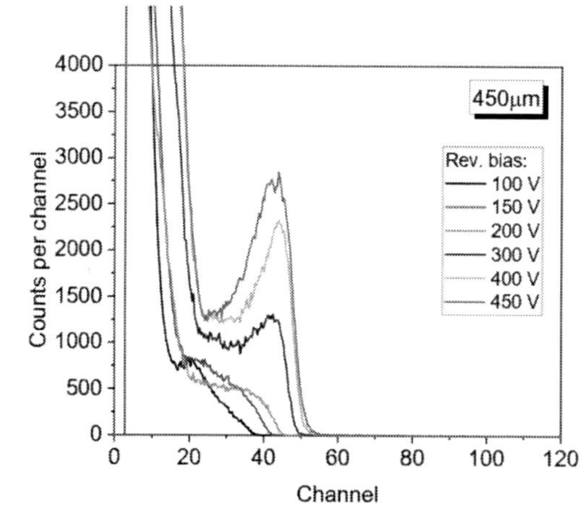

Fig. 9. The gamma spectra of 60 keV gamma rays from ^{241}Am measured by a 450 μm thick SI GaAs detector at various reverse biases.

C. Gamma spectrometry

The gamma spectra of ^{241}Am were measured with thin aluminum foil to stop the alpha particles and with higher preamplifier gain to detect the 60 keV gamma rays. The gamma spectra measured at various reverse biases by detectors with different thicknesses are shown in Figs. 7.-9. Similarly to alpha spectra, higher bias leads to higher CCE shifting the photopeak to higher channels. The thinnest detector reaches the peaks in the highest channels. On the contrary to the alpha spectrometry, the gamma rays penetrate along whole detector thickness and are absorbed through whole detector volume. Thus, both types of charged carriers are participating at charge collection, also the holes with lower mobility than electrons. This might affect the maximal obtainable CCE during gamma spectrometry, which decreases with increasing detector thickness. Limited charge carriers' lifetime will prevent holes from transporting too large detector thickness reducing reachable CCE.

In Fig. 10, we can observe the CCE as a function of the applied reverse bias for all three types of detectors with different thickness. The thinnest (230 μm) detector has reached saturated CCE of 80.4% at 300 V, whereas the middle (350 μm) detector CCE goes up only to 60.4% and the thickest (450 μm) detector did not exceed 51% CCE even at much higher bias applied (450 V). A similar tendency was observed with Cr-doped n-type GaAs detectors in [18], where 60 keV ^{241}Am peak floated to higher channels at 500 V with 715 μm thick detector than peak measured by 830 μm thick detector at 600 V and this one has had slightly higher CCE than 1200 μm thick detector at 800 V. Mentioned CCEs were in the range from 70% up to 80%.

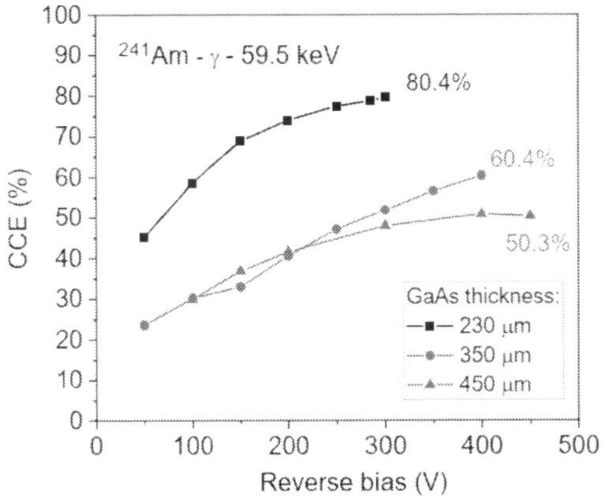

Fig. 10. The CCE for gamma spectrometry as a function of reverse bias for three different detector thicknesses.

IV. Conclusions

We have fabricated SI GaAs pad detectors for spectrometry of alpha particles and gamma rays of various thicknesses but with the same electrode-size of 1 mm in diameter. Obtained saturation reverse current was in the range of 3 – 30 nA, increasing with decreasing detector thickness from 450 down to 230 µm.

The spectrometric properties of fabricated SI GaAs detectors were evaluated from the gamma and alpha spectra of ^{241}Am with respect to detector thickness. An improvement in the charge collection efficiency with increasing reverse bias was observed in both experiments, in alpha and also gamma spectrometry. However, during alpha spectrometry the CCE as a function of reverse bias depended dominantly on electron Hall mobility, which was significantly higher for 230 µm thick detectors leading to lower CCE of thicker detectors. In the case of gamma spectrometry, where charge collection does not depend dominantly on electron transport in GaAs, the hole lifetime affected the results and thicker detectors suffered lower maximal obtainable CCE.

Acknowledgment

This work was partially supported by grants of the Slovak Research and Development Agency No. APVV-SK-CZ-21-0116, APVV-22-0382, APVV-18-0273 and of the Scientific Grant Agency of the Ministry of Education of the Slovak Republic and the Slovak Academy of Sciences No. 2/0084/20.

References

[1] D. S. McGregor *et al.*, "Thin-film-coated GaAs detectors for thermal and fast neutron measurements," Nucl. Instrum. & Meth. in Phys. Res. A, vol. 466, pp. 126-141, June 2001.
[2] A. J. Kordyasz *et al.*, "GaAs detectors with an ultra-thin Schottky contact for spectrometry of charged particles, " Nucl. Instrum. & Meth. in Phys. Res. A, vol. 545, issue 3, pp. 716-720, June 2005.
[3] B. Zatko, F. Dubecky, "Performance of a Schottky surface barrier radiation detector based on bulk undoped semi-insulating GaAs at reduced temperature," IEEE Trens. Nucl. Sci., vol. 53, issue 2, pp. 625-629, April 2006.
[4] A. Sagatova-Perdochova *et al.*, "GaAs detectors with LiF layer for detection of thermal neutrons," Nucl. Instrum. & Meth. in Phys. Res. A, vol. 591, pp. 98-100, June 2008.
[5] B. Zatko *et al.*, "Development and evaluation of semi-insulating GaAs detectors in hot plasmas diagnostics, " Nucl. Instrum. & Meth. in Phys. Res. A, vol. 633, pp. S131S133, May 2011.
[6] B. Zat'ko *et al.*, "Detection of fast neutrons using detectors based on semi-insulating GaAs.," Journal of Instrumentation, vol. 6, Dec. 2011, Art no C12047.
[7] A. Sagatova *et al.*, "Semi-insulating GaAs based detector of fast neutrons produced by D-T nuclear reaction," Journal of Instrumentation, vol. 11, Dec. 2016, Art no C12002.
[8] S. V. Chernykh *et al.*, "GaAs detectors with an ultra-thin Schottky contact for spectrometry of charged particles, " Nucl. Instrum. & Meth. in Phys. Res. A, vol. 845, pp. 52-55, February 2017.
[9] A. Sagatova *et al.*, "From single GaAs detectors to sensor for radiation imaging camera," Applied Surface Science, vol. 461, pp. 3-9, December 2018.
[10] K. Afanaciev *et al.*, "Investigation of the radiation hardness of GaAs sensors in an electron beam," Journal of Instrumentation, vol. 7, Nov. 2012, Art no P11022.
[11] A.G. Torres *et al.*, "Analysis of radiation effects on some properties of GaAs:Cr and Si sensors exposed to 22 MeV electron beam, " Nucleus, vol. N64, pp. 4-9, Dec. 2018.
[12] A. Sagatova *et al.*, "Radiation hardness limits in gamma spectrometry of semi-insulating GaAs detector irradiated by 5 MeV electrons," Journal of Instrumentation, vol. 15, Jan. 2020, Art no C01024.
[13] U. Kurchonak *et al.*, "Radiation hardness of GaAs: Cr and Si sensors irradiated by electron beam," Nucl. Instrum. & Meth. in Phys. Res. A, vol. 975, Sep. 2020, Art no. 164204.
[14] A. Sagatova *et al.*, "Gamma spectrometry of different energies by radiation-degraded SI GaAs detectors," AIP Conf. Proc., vol. 2411, Nov. 2021, Art no. 080013.
[15] A. Sagatova *et al.*, "Semi-insulating GaAs detectors degraded by 8 MeV electrons up to 1500 kGy," Journal of Instrumentation, vol. 16, Dec. 2021, Art no. C12032.
[16] A. Sagatova *et al.*, "Alpha spectrometry by radiation-degraded semi-insulating GaAs detectors," Materials Today: Proceedings, vol. 53, pp. 293-298, March. 2022.
[17] J.F. Ziegler, M.D. Ziegler, J.P. Biersack, "SRIM - The Stopping and Range of Ions in Matter," Nucl. Instrum. & Meth. in Phys. Res. B, vol. 268, pp. 1818–1823, June 2010.
[18] A. V. Tyazev *et al.*, "GaAs radiation imaging detectors with an active layer thickness up to 1 mm," Nucl. Instrum. & Meth. in Phys. Res. A, vol. 509, pp. 34–39, August 2003.

Inorganic scintillator surface enhancements with 2-D photonic crystals to improve light collection

Stuti Surani[1,*], Faruk Logoglu[1], Patrick Albert[2], Dr. Douglas Wolfe[2], and Dr. Marek Flaska[1]
[1]Ken and Mary Alice Lindquist Department of Nuclear Engineering
[2]Department of Materials Science and Engineering
Pennsylvania State University, United States of America
(*) srs6376@psu.edu

Abstract— **Inorganic scintillators are widely used in various applications of gamma spectroscopy such as nuclear nonproliferation and safeguards, medical applications, space applications, and astronomy. This is due to good energy resolution, stable performance, somewhat low cost, and relatively high detection efficiency. However, many inorganic scintillators have high refractive indices and suffer significant light losses due to total internal reflection (TIR). This project proposes using optimized periodic nanostructures called photonic crystals to recover some of the light originally lost due to TIR. Photonic crystals provide an optical bridge (constructive interference) between the scintillator and the photosensor for the trapped light photons. Improving the light extraction can improve the energy and time resolutions of the scintillator, allowing for a wider range of research and industry applications. Photonic crystals can be optimized in terms of their dimensions, shapes, and materials to maximize the light extraction. Preliminary optimization tests were performed using a LYSO scintillator coupled with Si_3N_4 photonic crystals. First, a realistic light input source is obtained by simulating the scintillation process in Monte Carlo code Geant4. The simulated scintillation photons are collected at the LYSO-PMT boundary to obtain their energy and angular distributions. In the next step, a deterministic code OptiFDTD is used to simulate light interactions with different nanostructures. Currently, the simulations are limited to 2-D block nanostructures. The optimization tests vary the height, width, and spacing of the photonic crystals. Preliminary optimization tests show an improvement in the light transmission by more than 60%. The optimized geometry will be manufactured in the lab using various manufacturing techniques such as ion milling, electron beam lithography, or 3D printing. Various gamma sources will be used to experimentally characterize the LYSO scintillators with and without photonic crystals. These experiments will also be used to validate the simulations and demonstrate the effectiveness of the photonic crystals in improving the energy resolution. Once validated, the simulations will be used to determine optimized photonic crystals for other inorganic scintillators, such as bismuth germanate, sodium iodide, and lanthanum bromide.**

Keywords — **Photonic crystals, nanostructures, light output, inorganic scintillators**

I. Introduction

Inorganic scintillators are essential for various applications, including nuclear security, medical applications such as PET, high energy physics, and industrial control [1]. In recent years, scintillator development has been focused on improvements in energy resolution for homeland security applications and time resolution for medical applications. Inorganic scintillators are more versatile than semiconductors but typically lack the high energy resolution of semiconductors. A scintillator's energy and time resolution can be improved by improving the light output [2]. While most inorganic scintillators have a high light yield, the light output is lower due to various light loss mechanisms at play, such as absorption within the crystal, crystal defects, refraction through the remaining sides of the crystal, and most importantly, total internal reflection (TIR) at the scintillator photosensor interface.

Photonic crystals (PHCs) are nanostructures theorized to improve light output by reducing light loss due to TIR. PHCs have been used for various light applications, such as LEDs, optical fibers, and solar cells [3,4]. These nanostructures of high refractive index material act as periodic alternating dielectric media at the crystal surface. PHCs, thus, provide an optical bridge for photons of a specific wavelength through interference and diffraction [2,3]. The nanostructure pattern for the PHCs must be optimized for the light emission spectrum of the given scintillator. To do this, we must simulate both the scintillation process within the scintillator and the light modes produced by the PHCs at the scintillator-photosensor interface.

Manufacturing such nanostructures can be expensive and time-consuming depending on the required spatial resolution of the resulting nanostructures. We have chosen a LYSO scintillator and a BGO scintillator for our initial tests. LYSO was chosen for its prevalence in literature for PHC development [5,6]. The scintillators are coupled with a pillar array geometry, using silicon nitride (Si_3N_4) as the PHC material due to its inherently high refractive index. Once manufactured, the scintillators will be characterized with a gamma source to note relative improvements in light output and energy resolution. This paper will discuss the simulation and optimization of PHCs, the manufacturing techniques, and preliminary efforts to characterize the scintillators with a gamma source.

II. Methodology

The project is broadly divided into four parts: macroscale light transport simulations in Geant4 [7], PHC optimization simulations in OptiFDTD [8], manufacturing the optimized PHC geometry using electron beam lithography, and characterization of scintillators by radiation measurements with a gamma source, as shown in the flowchart in Fig. 1.

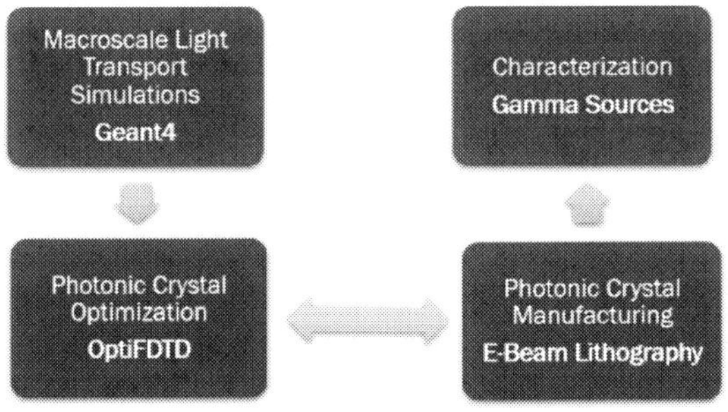

Fig. 1. Flowchart for the project.

The simulations consist of macroscale light transport and nanoscale light transport, a division required to fully simulate the light transport through the scintillator and the effects at the nanoscale level with the PHCs. The next step is manufacturing the optimized PHC geometry and characterizing it using a gamma source.

The radiation measurements provide us with validation for the simulation model, along with helping identify issues with the fabrication process. All the steps are interlinked as the simulations provide the manufacturing template, and the radiation measurements and manufacturing provide data to make the simulations more realistic and account for the manufacturing constraints.

We use a LYSO and a BGO scintillator for our initial tests coupled to a block structure Si_3N_4 PHC. The material data for LYSO and BGO is given below in Table. I.

TABLE I
MATERIAL DATA FOR SCINTILLATORS

Properties	LYSO	BGO
Dimensions	10 mm × 10 mm × 3 mm	10 mm × 10 mm × 2.5 mm
Refractive Index	1.81	2.15
Light Output (Photons/keV)	33	9
Decay Time (ns)	36	300

The two scintillators have similar dimensions and emission spectra. Thus, we expect similar energy and angular distribution of light at the scintillator photosensor boundary. BGO has a higher refractive index than LYSO, so it has a higher percentage of internally reflected photons. We thus expect to observe a more significant improvement in BGO transmission than LYSO transmission for the optimized PHC geometry.

A. Simulations

PHCs provide an optical bridge for light photons of a specific wavelength to escape the scintillator and reach the photosensor. However, the PHC geometry must be optimized for the scintillator material and geometry. Simulating light transport through the PHCs allows us to optimize the PHC geometry for a given material.

We first simulated the macroscale light transport within the scintillator using a Monte Carlo code called Geant4. Geant4 simulates the scintillation process and light transport in the scintillator, as shown in Fig. 2. A point gamma source simulates the Cs-137 source used for the radiation measurement process. Each gamma interaction within the scintillator yields an isotropic production of optical photons within the scintillator consistent with the emission spectrum of the scintillators. These photons are then transported as particles through the scintillator material, and they undergo absorption, reflection, refraction, and scattering in the material. The photon's energy and angular distribution is then tallied at the scintillator-photosensor interface. These energy-angle maps produce the light input source for the nanoscale simulations [9]. The material data for these simulations is obtained from UV-vis measurements.

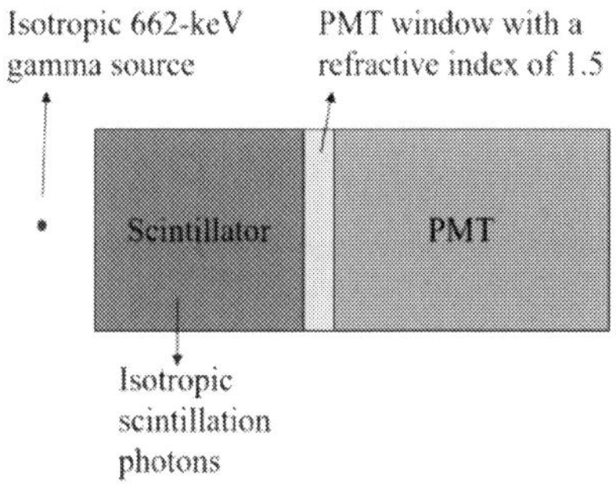

Fig. 2. Geant4 simulation geometry for macroscale light transport within the scintillator.

The nanoscale simulations are done in OptiFDTD, a deterministic code that models light as electromagnetic waves, thus allowing for a more complex treatment of light transport through the PHCs. The input source for OptiFDTD is obtained from the Geant4 simulations. At this point, the source only consists of the first pass of light through the scintillator-photosensor interface. The light reflected from the sides of the scintillator is not accounted for. Future simulations will include multiple reflections, making the simulations more realistic and allowing for more light transmission.

The optimization simulations are done as shown in Fig. 3. For these initial tests, a 2D block structure geometry is selected for the PHCs. The input source obtained from Geant4 is modeled as a planewave within the scintillator. It then interacts with the PHCs and is detected within the PMT, modeled as a dielectric material with a refractive index of 1.5. The material data for these simulations are obtained from ellipsometry measurements. There are three variables of interest:

https://doi.org/10.1051/epjconf/202328810014

- The height of the PHC (PHC z)
- The thickness of the PHC (PHC x)
- The spacing between the PHC (Gap)

The simulations vary these variables from 0.1 um to 1 um for ten iterations each. The results are then tallied as a function of the percentage of light transmission to the PMT. The spacing between the PHC can be modeled as a vacuum to simulate a dry coupling or as optical grease to simulate a grease coupling. For our initial tests, we only model dry coupling.

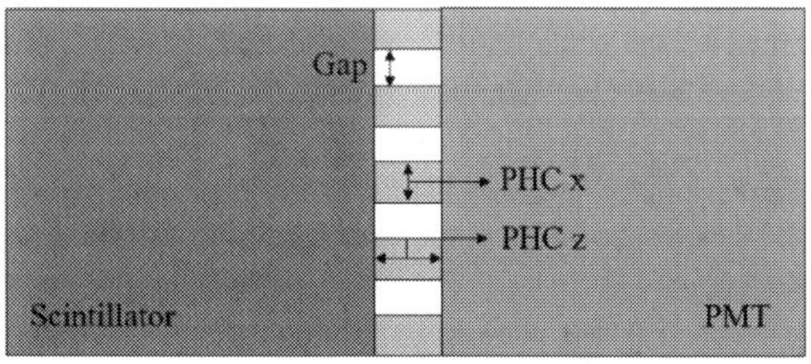

Fig. 3. OptiFDTD simulation geometry for nanoscale light transport and PHC optimization.

B. *Manufacturing*

For this effort, the manufacturing of PHCs is carried out using various techniques broadly divided into either top-down or bottom-up approaches, as shown in Fig. 4. Bottom-up processes include self-assembly of polystyrene nanospheres via spin casting. This method is both scalable and inexpensive. Top-down approaches explored include ion milling, laser write methodologies, and electron beam lithography. Each approach has its associated advantages and disadvantages. The top-down processes can have the advantage of extremely high pattern resolution, which allows for greater control over manufacturing the pattern of interest. Electron beam lithography, while marginally more expensive than the bottom-up processes and the other aforementioned top-down approaches, offers the ability to pattern nanoscale features with high spatial resolution.

Furthermore, it is an industry-available technology that can be scaled up for manufacturing if required. Our initial tests focus on electron beam lithography and dry etching to replicate the modeled geometry and verify the simulations. However, future work will focus on 3D simulations of nanospheres to explore the economic feasibility for different applications.

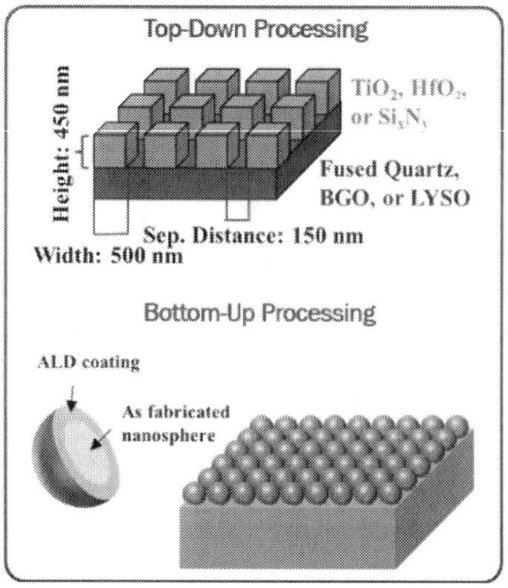

Fig. 4. Manufacturing approaches for PHCs.

PHCs can be made from any dielectric material with a high refractive index as compared to the refractive index of the base scintillator—some favorable materials for this type of processing are TiO_2, HfO_2, and Si_3N_4. For LYSO, we chose Si_3N_4 for its refractive index of approximately 2.01 at 415 nm, relative to the refractive index of LYSO (1.83). The complete PHC manufacturing methodology is explained in the flowchart in Fig. 5. First, we identify our substrate, which is the LYSO scintillator material. We then perform radiation baseline measurements to provide a reference light yield and energy resolution. After cleaning the crystals, we deposit a thin (300 nm) Si_3N_4 film onto the scintillator using radio frequency (RF) reactive sputtering at room temperature. The PHC geometry, or pattern, is then transferred onto this thin film through electron beam lithography. The first step is to deposit a resist layer, or polymer-based film, using spin casting onto the surface of the silicon nitride. Next, a thin gold layer is deposited to aid in conductivity during the electron beam writing process to help mitigate charge build-up. Finally, we can use an electron beam lithography tool to write the pattern, or PHC geometry, into the photoresist to cure. The sample is then transferred into a reactive ion etch chamber to mill or etch away the silicon nitride, leaving behind the pillar geometry. The gold and photoresist layers are then removed to reveal the pattern on the substrate of our finished PHC.

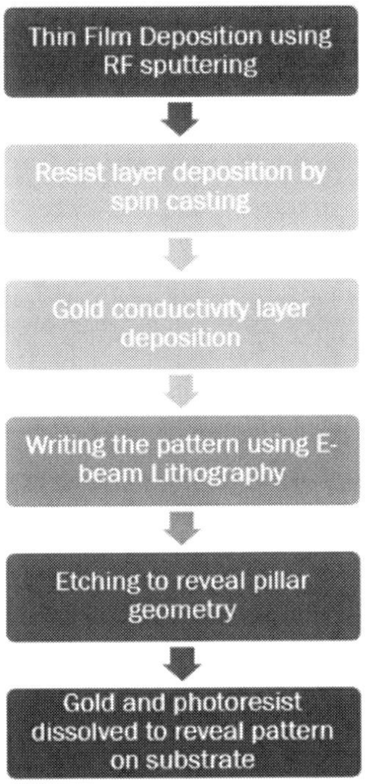

Fig. 5. Flowchart for manufacturing PHCs using dry etching and electron beam lithography.

C. Radiation Measurements

The radiation measurements for the scintillators are done using a Cs-137 gamma source. The experimental geometry is shown in Fig. 6. The scintillator is placed on a Hamamatsu H10580 PMT and kept stable using a 3D-printed PMT holder. The PMT is connected to a CAEN DT5730 digitizer which processes the analog signals from the PMT and converts them to digital signals processed by a computer software called CoMPASS, also provided by CAEN. The PMT is also connected to a high-voltage power supply set to -1200 volts.

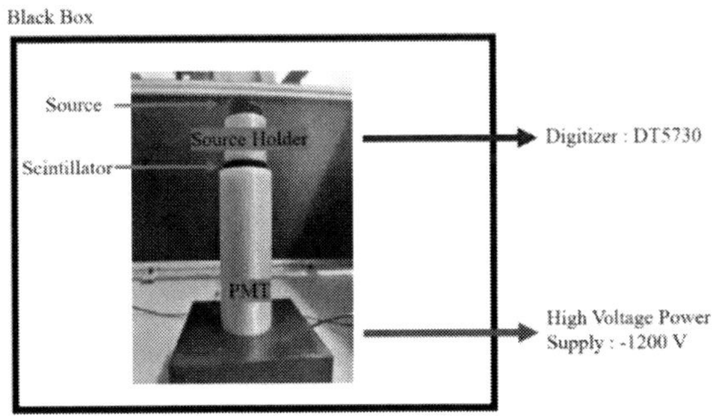

Fig. 6. Radiation measurements setup.

The source is placed on the scintillator in a 3D-printed source holder. The source consists of four 5uCi Cs-137 sources stacked on each other and taped together, as given in Fig. 7. This creates a stronger source to reduce the signal-to-noise ratio and improve the statistics. The scintillator responses are measured with and without optical grease. Measurement is done for peak counts greater than 100000 to reduce the uncertainty to less than 0.1%.

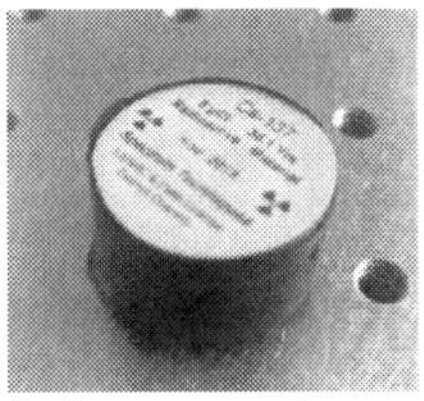

Fig. 7. Radiation source for the measurements.

Baselines measurements for bare LYSO crystals have been done with an average energy resolution of 13.3% for crystals without optical grease, and 10 % for crystals with optical grease. These measurements were done without using any reflectors to improve energy resolution. Future radiation measurements will be compared to these data to note improvements in light output and energy resolution.

Furthermore, radiation baseline measurements were done after some of the manufacturing steps to ensure that the crystal's optical properties are consistent with expected optical degradation, and to aid in influencing the optimal manufacturing flow path.

For example, one of the first manufacturing steps for PHCs is to deposit a Si_3N_4 thin film onto the scintillator (LYSO). Before formalizing the manufacturing in *section B*, the thin films were deposited via direct current (DC) reactive magnetron sputtering at 200°C with an applied substrate bias. At this stage, we performed radiation measurements to compare the crystal's optical properties before and after thin film deposition. The radiation measurements showed a significant decline in the light output of the scintillator. While a decrease in performance was expected, some optical effects were also observed on visual inspection, as shown in Fig. 8. These optical effects can result from coating overspray leading to thickness gradients along the edges, or the temperature and substrate bias could damage the crystal.

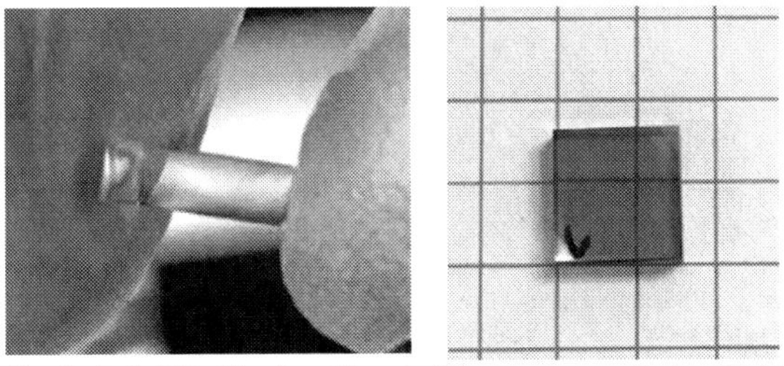

Fig. 8. Left: Thin film deposition via DC magnetron sputtering. Right: Thin film deposition via RF sputtering.

To resolve this, we investigated the optical effects when using RF sputtering for thin films. RF Sputtering, like DC sputtering, is done under a high vacuum, creating high-density, low-defect optical films. However, RF sputtering tends to have a slower deposition rate, allowing us to mitigate further defects, and we were able to establish equivalent optical properties without needing to deposit at 200°C and substrate bias. We tested various baseline deposition conditions to study the sputtering process as a function of power, nitrogen flow, and time as shown in Fig. 9.

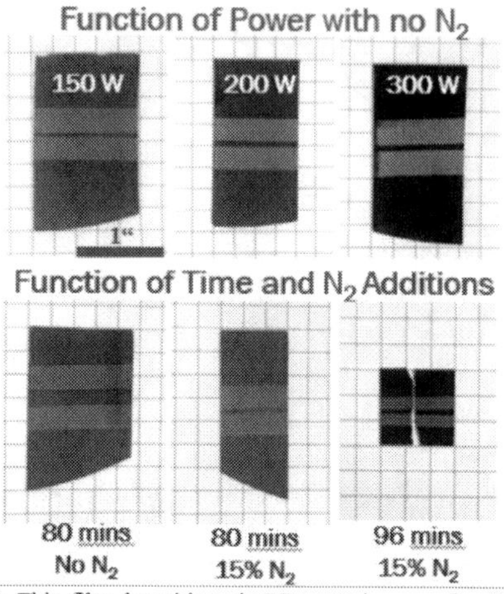

Fig. 9. Left: Thin film deposition via RF sputtering as a function of power, time, and nitrogen flow to investigate optimal processing conditions.

We found the optimal conditions to be 300 W power, 15% nitrogen flow, and a 3.125 nm/min deposition rate for our Si_3N_4 thin film. Scanning electron microscopy and ellipsometry were used to look for microstructure changes and to observe the coating thickness and refractive index. By performing radiation measurements after the thin film deposition, we identified issues with our initial process and corrected them before those effects were categorized in the final assessment of the PHC structure.

III. Results

A. Optimization Simulations

The Geant4 simulations were used to produce energy-angle maps for LYSO and BGO. These data were then used to create the input source for the OptiFDTD simulations [9].

The optimization simulation results are given in Fig. 10 for a BGO scintillator and in Fig. 11 for a LYSO scintillator. The results are also summarized in Table. II.

As discussed in the methodology section, LYSO and BGO have comparable geometry and emission spectra. Thus, they both have similar optimized PHC geometry. The BGO has a higher refractive index than LYSO, resulting in a more significant improvement in light transmission with the PHCs. For LYSO, the optimized PHC geometry yields an improvement in light transmission of 46.8 %, while for BGO, the improvement is 91.9 %. For LYSO, further simulations were done with finer optimization steps to obtain a light transmission of 59%, which is an improvement of 62.8%. Improvement is calculated as relative change between the initial and final light transmission in percentage.

The simulation results are consistent with our initial expectations and show that light output improvement due to PHCs is better for higher refractive index materials.

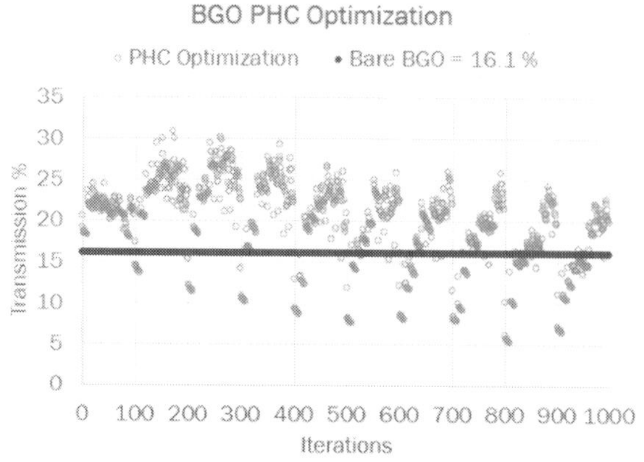

Fig. 10. Optimization simulation results for a BGO scintillator.

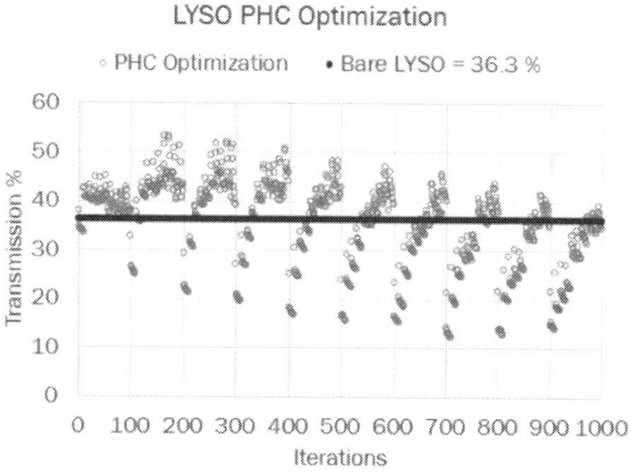

Fig. 11. Optimization simulation results for a LYSO scintillator.

Table II
Optimization Results for LYSO and BGO

Material	Initial Transmission	Gap (μm)	PHC x (μm)	PHC z (μm)	Final Transmission	Improvement
LYSO	36.3 %	0.2	0.7	0.2	53.2 %	46.8 %
BGO	16.1 %	0.2	0.8	0.2	30.9 %	91.9 %

B. Manufactured PHC structures

The simulations provide data for manufacturing PHCs. Five PHC geometries were selected for the initial tests, given in Table. III. The first four are from the optimization simulations, and the last is inspired by literature [10]. These specific geometries were selected to help validate the modeling efforts and experimentally explore the impact of the Gap between the PHCs on the light transmission. The spacing between the features determines the required manufacturing resolution.

TABLE III
PHC GEOMETRICAL PARAMETERS FOR FABRICATION

	Gap (μm)	PHC x (μm)	PHC z (μm)	Transmission	Improvement
1	0.22	1	0.3	59.0 %	62.8 %
2	0.5	1	0.3	48.3 %	33.2 %
3	0.7	1	0.3	45.6 %	25.9 %
4	0.8	1	0.3	43.6 %	20.2 %
5	1	1	0.3	Literature Inspired [10]	

Fig. 12 illustrates the efforts to baseline and optimize the electron beam writing behavior on our prepared LYSO substrates. Fig. 12 (A) is a digital image showing 60 different permutations of varied processing parameters patterned onto the surface of one crystal to assess the optimal write conditions. The SEM micrographs are increasingly magnified in Fig. 12 (B-D) to look at the various patterns deposited and subsequently etched with plasma etching.

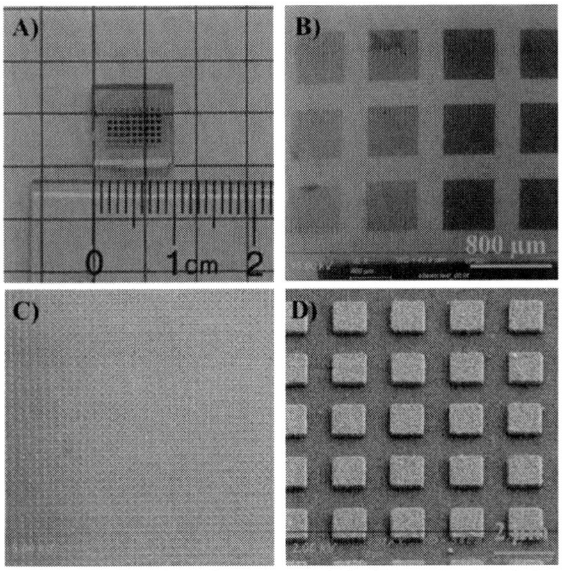

Fig. 12. E-Beam Lithography process optimization on A) a 10x10x3 mm^3 LYSO substrate, and increasing magnification of SEM micrographs B, C, & D) to observe the pattering.

The key exercise was to establish our optimal exposure energy and dose with respect to our pattern's density, resist material, resist thickness, and conductivity behavior of our resist-coated gold on silicon nitride. These variables must be optimized to achieve the pattern fidelity and resolution of interest. The optimal electron beam lithography processing window is highlighted in Fig. 13. The pillar arrays were etched and developed with good resolution. However, initial observations revealed that the lateral accuracy was off by ~15%. This will be corrected in the subsequent manufacturing. Furthermore, our target PHC z- thickness was 300 nm, but our etching rate will also need to be adjusted as our total milled depth was only ~130 nm, as observed in Fig. 13(Top right).

Fig. 13. Various SEM micrographs of optimal milled structures from the preliminary electron beam lithography trials.

These minor adjustments will be implemented for the final manufacturing process to prepare the five samples listed in Table. III.

IV. CONCLUSIONS

This project is a collaborative effort to simulate, manufacture and test the enhancement of light output from inorganic scintillators using PHCs. The simulations are done for a LYSO and a BGO scintillator to optimize the 2D block structure PHC geometry for Si_3N_4. Optimized PHC geometries show an improvement in light transmission of 46.8% for LYSO and 91.9% for BGO. For LYSO, further simulations were done to improve light transmission by 62.8%. These optimized geometries are then manufactured in the lab.

The manufacturing process has been discussed in detail. We did 60 permutations of varied processing parameters to understand the optimal write conditions. We obtained a high-resolution, low-defect PHC geometry using electron beam lithography. Radiation measurements are also done between manufacturing steps to account for the optical defects produced by each step. This process allowed us to identify that DC magnetron sputtering damaged our crystal's optical properties. This was resolved using RF sputtering to deposit Si_3N_4 thin films, a primary manufacturing step.

The future work will characterize these PHC geometries with radiation measurements as outlined in the paper. We will also stimulate and validate other PHC shapes and structures for different scintillator materials.

ACKNOWLEDGMENT

This research is sponsored by the Defense Threat Reduction Agency (DTRA) as part of the Interaction of Ionizing Radiation with Matter University Research Alliance (IIRM-URA) under contract number HDTRA1-20-2-0002

References

1. C. Dujardin *et al.*, "Needs, Trends, and Advances in Inorganic Scintillators," in *IEEE Transactions on Nuclear Science*, vol. 65, no. 8, pp. 1977-1997, Aug. 2018, doi: 10.1109/TNS.2018.2840160.
2. M. Salomoni, R. Pots, E. Auffray, and P. Lecoq, "Enhancing light extraction of inorganic scintillators using photonic crystals," *Crystals*, 8(2), 78, 2018, *https://doi.org/10.3390/cryst8020078*
3. J.D. Joannopoulos, S.G. Johnson, J.N. Winn, and R.D. Meade, *Photonic Crystals: Modeling the flow of light* 2nd ed. Princeton University Press, 2008, pp 2, 156-161.
4. C. Wiesmann, K. Bergenek, N. Linder, and U.T. Schwarz. (2009). Photonic crystal LEDs – designing light extraction. Laser & Photon. Rev., 3: 262-286. https://doi.org/10.1002/lpor.200810053
5. M. Kronberger, E. Auffray and P. R. Lecoq, "Improving light extraction from heavy inorganic scintillators by photonic crystals," in *IEEE Transactions on Nuclear Science*, vol. 57, no. 5, pp. 2475-2482, Oct. 2010, doi: 10.1109/TNS.2010.2063438.
6. A. Knapitsch, E. Auffray, C.W. Fabjan, J. Leclercq, P. Lecoq, X. Letartre, C. Seassal, "Photonic crystals: a novel approach to enhance the light output of scintillation based detectors," *Nuclear Instruments and Methods in Physics Research Section A: Accelerators, Spectrometers, Detectors and Associated Equipment*, Volume 628, Issue 1,2011, Pages 385-388, ISSN 0168-9002, https://doi.org/10.1016/j.nima.2010.07.007
7. J. Allison *et al.*, "*Recent developments in Geant4*," Nuclear Instruments & Methods in Physics Research. Section A, Accelerators, Spectrometers, Detectors and Associated Equipment, vol. 835, (C), pp. 186-225, 2016.
8. Optifdtd overview (Jun 2020). https://optiwave.com/optifdtd-overview/
9. S. R. Surani, P. E. Albert, F. Logoglu, D.E. Wolfe, and M. Flaska, "Light-extraction chracterization and optimization for 2-D photonic crystals," in *Proc. INMM & ESARDA joint annual meeting*, Vienna, Austria, 2023
10. A. Knapitsch *et al.*, "Results of Photonic Crystal Enhanced Light Extraction on Heavy Inorganic Scintillators," in *IEEE Transactions on Nuclear Science*, vol. 59, no. 5, pp. 2334-2339, Oct. 2012, doi: 10.1109/TNS.2012.2184556.

Investigating the Thickness-Dependent Scintillator-PMT Interface Reflection Coefficients with GAGG:Ce^{3+} Crystals Using the Dual-PMT Setup

Faruk Logoglu*, Stuti Surani and Marek Flaska
Ken and Mary Alice Lindquist Department of Nuclear Engineering
The Pennsylvania State University
(*) ful51@psu.edu

Abstract - **Recently, it has been shown through both Monte Carlo simulations and experiments that scintillator-PMT interface reflection coefficients could depend on crystal thickness. It has been argued that the thickness-dependency on the interface reflection coefficient is a result of bulk attenuation and surface reflections. So far, only LYSO:Ce^{3+} scintillators have been tested to investigate thickness-dependent reflection coefficients. In this work, the simulations and experiments are extended to GAGG:Ce^{3+} crystals. Moreover, a new experimental technique (the dual-PMT setup) has been tested to measure the interface reflection coefficients, and it has been shown through the dual-PMT setup that GAGG:Ce^{3+}-PMT interface reflection coefficients are thickness-dependent with more than 1σ confidence. These results are also supported with extensive Monte Carlo simulations.**

Keywords -- **Scintillators, Geant4 simulations, reflection coefficients, light yield**

I. Introduction

LIGHT reflection coefficient explicitly depends on the refractive indices of two media in addition to the photons' angle of incidence [1]. In scintillators, photons arrive at the scintillator-PMT boundary at various angles due to isotropic scintillation emission. It has recently been argued and shown through Monte Carlo simulations that photons' angular distribution at the scintillator-PMT boundary changes as a function of crystal thickness, which is a result of different absorption and lateral transmission probabilities as a function of photon angle. It has further been argued that the distribution dependency on crystal thickness leads to a change in reflection coefficients according to the Fresnel reflection equations. Although the theory and simulations suggested reflection coefficient dependency on crystal thickness, experiments could not show this dependency with high confidence due to the low energy resolution of the experimental setup. Moreover, the simulations and experiments were only performed for LYSO:Ce^{3+} scintillators. In this work, GAGG:Ce^{3+} crystals are tested using a similar approach outlined in [2, 3].

A new experimental method called the dual-PMT setup has been tested to show the reflection coefficient dependency on crystal thickness. The dual-PMT setup involves measuring the scintillation light with two identical PMTs situated opposite to each other. The dual-PMT setup is also claimed to yield larger interface reflection coefficients compared to the single-PMT setup due to a modified angular distribution at the scintillator-PMT boundaries. The reason the interface reflection coefficients change in different experimental setups is identified to have similar roots as to why they change as a function of crystal thickness. We argue that as the opposite PMTs share the available light generated in the scintillator, the angular distribution of photons on each scintillator-PMT boundary skews toward larger angle photons with respect to the interface normal, which increases the interface reflection coefficients according to the Fresnel reflection equations.

II. Method

Although the extended 2D-model is useful to describe the light transport in scintillators and is suitable for investigating thickness-dependent reflection coefficients, it is rather complicated due to both vertical and horizontal position dependency as well as the angular dependency. Furthermore, the floor functions in its integral formulation make its numerical evaluation computationally expensive and time consuming. All the scintillation photons can be assumed to be generated along the central y-axis, and it can be shown that this assumption introduces negligible errors (1 reflection on average) on the results. The extended 2D-model with this assumption is given by

 https://doi.org/10.1051/epjconf/202328810015

$$LO = \frac{LY(1-R_{\mathrm{i}})}{\pi H}\left[\int_0^{\frac{\pi}{2}}\int_0^H \frac{R^{\frac{y|tan\theta|}{W}}e^{-\alpha y|sec\theta|}}{1-R_{\mathrm{i}}R^{\frac{2H|tan\theta|}{W}+1}e^{-2\alpha H|sec\theta|}}dyd\theta + \int_{\frac{\pi}{2}}^{\pi}\int_0^H \frac{R^{\frac{(2H-y)|tan\theta|}{W}+1}e^{-\alpha(2H-y)|sec\theta|}}{1-R_{\mathrm{i}}R^{\frac{2H|tan\theta|}{W}+1}e^{-2\alpha H|sec\theta|}}dyd\theta\right] \quad (1)$$

where LO and LY are light output and light yield defined as the number of light photons transmitting through the scintillator-PMT interface and the number of light photons produced in the scintillator per unit energy deposited. The reflection coefficient R is the reflection coefficient of lateral surfaces, whereas R_i is the reflection coefficient of the interface. W, H and α are the crystal width, height and optical absorption coefficient, respectively.

Although (1) works for the single-PMT setup, it needs to be modified for the dual-PMT setup. Since the top surface has the same (ideally) interface reflection coefficient as the bottom surface, one of Rs in the integral formulation needs to be replaced by R_i. Upon this substitution, the modified 2D-model for the dual-PMT setup is given by

$$LO = \frac{LY(1-R_{\mathrm{i}})}{\pi H}\left[\int_0^{\frac{\pi}{2}}\int_0^H \frac{R^{\frac{y|tan\theta|}{W}}e^{-\alpha y|sec\theta|}}{1-R_{\mathrm{i}}^2R^{\frac{2H|tan\theta|}{W}}e^{-2\alpha H|sec\theta|}}dyd\theta + \int_{\frac{\pi}{2}}^{\pi}\int_0^H \frac{R^{\frac{(2H-y)|tan\theta|}{W}}R_{\mathrm{i}}e^{-\alpha(2H-y)|sec\theta|}}{1-R_{\mathrm{i}}^2R^{\frac{2H|tan\theta|}{W}}e^{-2\alpha H|sec\theta|}}dyd\theta\right] \quad (2)$$

where the top surface reflections are replaced by R_i. The top and bottom surfaces are defined according to the single-PMT setup where bottom refers to the scintillator-PMT boundary. In the dual-PMT setup, both the bottom and the top surfaces are scintillator-PMT boundaries.

The method of estimating interface reflection coefficients involves measuring light outputs of scintillators as a function of their thickness. However, since two PMTs face the same crystal and placed horizontally to the ground, only the grease coupling configuration yields consistent measurement values. Therefore, the reference interface reflection coefficient must be estimated through Monte Carlo simulations. Then, the interface reflection coefficients of the other thicknesses are estimated through (2) by comparing the relative light outputs between the crystals with different thicknesses and using the reference interface reflection coefficients estimated through the Monte Carlo simulations. In order to improve the reliability of the Monte Carlo simulations, UV-Vis transmission/reflection spectroscopy has been performed to estimate the optical absorption lengths, optical scattering lengths and the refractive indices of the samples. Multiple reflection method has been used to calculate the optical constants to obtain more accurate results [3, 4, 5].

III. MONTE CARLO SIMULATIONS

Monte Carlo simulations using Geant4 simulation toolkit have been performed to tally the angular and energy distributions of photons on the scintillator-PMT boundaries. These simulations have been carried out for four different thicknesses (2.5 mm, 5 mm, 7 mm, and 10 mm) and a single cross-sectional area of 10 mm x 10 mm. These are the dimensions of the crystals experimentally tested in this work. In all the simulations, experimentally-determined absorption coefficients, scattering coefficients, and refractive indices are used. These parameters have been obtained through UV-Vis spectroscopy, and Geant4 input has been prepared with wavelength-dependent optical constants to improve the accuracy of the simulations. The UV-Vis experiments have been carried out with a 0.5 mm x 10 mm x 10 mm GAGG:Ce^{3+} crystal to minimize multiple photon scatterings. This crystal has not been tested in radiation detection experiments due to its small thickness and impracticality in handling.

A point 662 keV gamma-ray source is simulated in all simulations to mimic the ^{137}Cs radioactive source used in the experiments. A single rectangular parallelopiped volume representing scintillating crystals is placed next to the silicone volume, which represents silicone grease with a refractive index of 1.46 [6]. Then, the PMT window volume is placed next to the silicone grease volume and is assumed to have a constant refractive index of 1.5. The angular and energy distributions of photons on the PMT window are shown in Fig. 1 for the 2.5-mm and 10-mm-thick GAGG:Ce^{3+} crystals.

As expected from the theory and the previous results [3], the angular distribution shifts towards smaller values (cosine of the angle increases) as the crystal thickness increases. This phenomenon is explained through the larger absorption and lateral surface transmission probabilities of the larger angle photons. However, compared to the single-PMT setup, there are differences in the angular-energy distributions of optical photons on the scintillator-PMT interface. Since there are two PMTs situated opposite to each other and facing the same crystal, the photons emitted in the direction away from the PMT of interest (the PMT for which the contour plots are generated) are more likely to transmit through the opposite PMT. As the transmission probability is larger for small angle photons (with respect to the surface normal), those photons are removed from the distribution before they arrive at the PMT of interest, which is the reason of the abrupt cutoff in the distribution along the critical angle line. As a result, the remaining photons arriving at the PMT boundary are more likely to have larger angles compared to the single-PMT setup, which increases the interface reflection coefficients. It should be noted that the critical angle line is not straight but inclined due to wavelength-dependent refractive indices of GAGG:Ce^{3+} crystals. Also, in GAGG:Ce^{3+} crystals, the absorption tail fades off at ~530 nm based on our UV-Vis results; therefore, absorption is not assumed to take place above this wavelength. These results are also supported in literature [7].

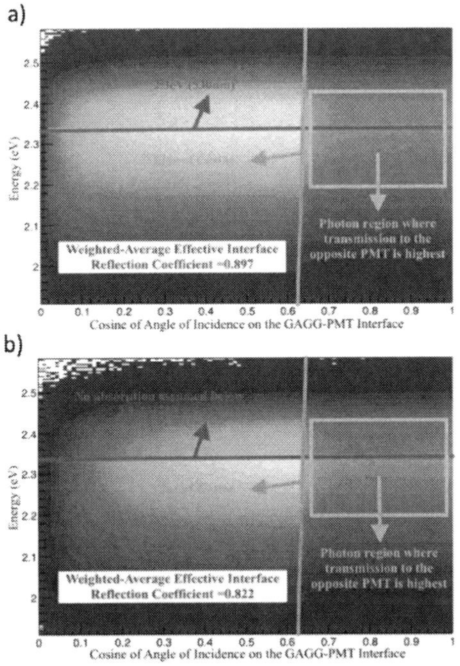

Figure 1. Energy-cosine of angle of incidence distribution of light photons on the PMT window for a) 2.5mm and b) 10-mm-thick GAGG:Ce^{3+} crystals in the dual-PMT setup.

IV. Experiments

GAGG:Ce^{3+} crystals with five different thicknesses and the same cross-sectional area (square base 100 mm^2) have been purchased from EPIC Crystals (China). These thicknesses are 0.5 mm, 2.5 mm, 5 mm, 7 mm, and 10-mm. The 0.5-mm-thick crystal has been used for UV-Vis measurements to estimate the optical properties of the crystals. The other four crystals have been tested under irradiation with a ^{137}Cs gamma-ray source. Experimental schematic with two PMTs is shown in Fig. 2.

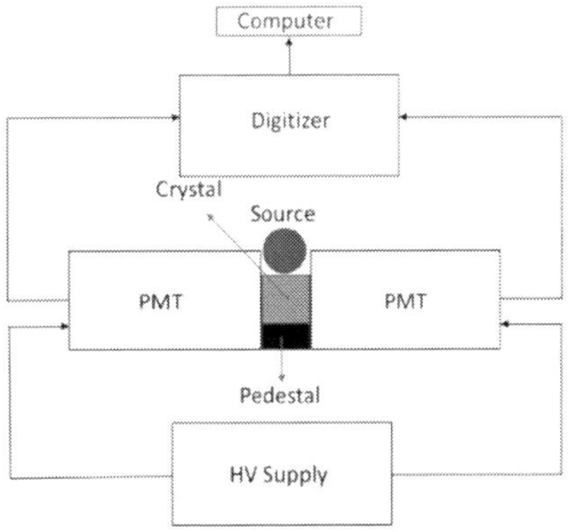

Figure 2. A schematic for the dual-PMT setup. Two PMTs situated facing the same crystal. Pedestals with different heights are used to position the crystals facing the centers of the PMTs.

Pedestals with different heights have been 3D-printed so that all crystals face the PMTs' centers. Identical PMTs have been used in the experiments, and they are the model H10580 PMTs from Hamamatsu. These PMT assemblies have good pulse linearity because of their linear-focused dynode structure and active voltage divider circuit. The PMTs are calibrated using the 10-mm-thick crystal, and the 662 keV photopeaks have been used to adjust the applied high voltages. One of the PMTs (PMT-1) is operated at -1200 V and the other PMT (PMT-2) is operated at -1183 V.

The experimental spectra obtained with the two PMTs are shown in Fig. 3. As expected from the light transport theory in scintillators, the light output decreases as the thickness increases due to the higher probability of absorption and lateral surface transmission. As a result of calibration, two PMTs yield similar light outputs within the resolution of the system. The x-axis is normalized to the 10-mm-thick crystal measured with PMT-1 since it is the configuration that has yielded the lowest light output. Taking $R_i = 0.822$ (see Fig. 1), the interface reflection coefficients of the other configurations are calculated using Eq. 2. The calculation procedure involves estimating the reference light output using $R_i = 0.822$ for $H = 10$ mm, then using the relative light outputs to estimate the interface reflection coefficients of the other configurations. In order to analyze the spectra in a consistent manner, Gaussian functions are fitted to photopeaks.

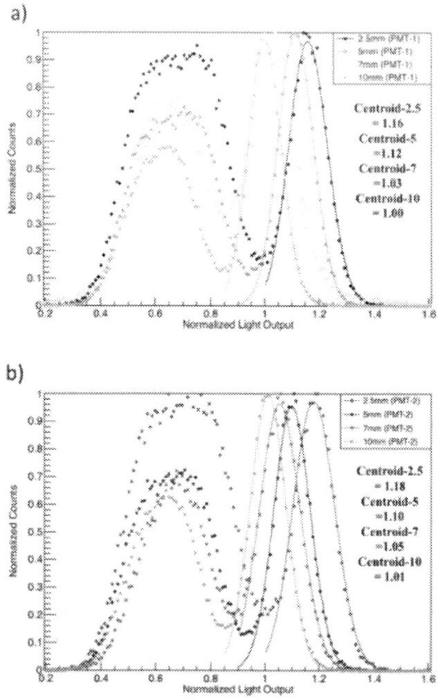

Figure 3. Normalized light output spectra with two PMTs situated at the opposite sides of the crystals. a) PMT-1 b) PMT-2. The two PMTs yield similar light outputs within the resolution of the experimental system.

V. Results and Discussion

Scintillator-photosensor interface reflection coefficients estimated through Eq. 2 using the relative normalized light outputs are presented in Table 1. The reference interface reflection coefficient was taken to be 0.822 for the 10-mm-thick crystal measured with PMT-1. Error margins are 1σ and they are the averages of the upper and lower bounds of errors. The lower and upper error bounds are different due to the asymmetric shape of the model functions. As a result, the errors are smaller when the scintillator-PMT interface reflection coefficient is larger.

It can be seen from Table 1 that the reflection coefficients estimated with the two PMTs are similar and agree within 1σ error margins. This is expected since two identical PMTs were used in the experiments and they were calibrated before the experimental data were collected.

Table 1. Experimentally-determined interface reflection coefficients of GAGG:Ce^{3+} crystals with the dual-PMT setup.

R_i / Thickness (mm)	PMT-1 Reflection Coefficients	PMT-2 Reflection Coefficients
2.5	0.932 ± 0.013	0.929 ± 0.014
5	0.879 ± 0.021	0.884 ± 0.021
7	0.864 ± 0.022	0.858 ± 0.024
10	**0.822**	0.819 ± 0.028

It is also seen that the interface reflection coefficients decrease with increasing crystal thickness as predicted from the light transport theory in scintillators. Although the reference interface coefficient could not be obtained as in [3] due to the horizontal placement of PMTs, reflection coefficients estimated for other thicknesses closely match with the simulation results. As an example, the reflection coefficient estimated through the Monte Carlo simulations is 0.897 for 2.5-mm-thick crystal, whereas this number is 0.929-0.932 in the experiments. It is observed that the experiments consistently predict larger reflection coefficients than the simulations, which can be explained through perfect geometries assumed in the simulations. In experiments, the observed reflection coefficient can be influenced by imperfections. Moreover, the observed reflection coefficients are the effective reflection coefficients of scintillator-grease-window-photocathode systems. Therefore, it is expected that the experimental interface reflection coefficients are larger.

VI. Conclusions

In this paper, we showed the relative light outputs obtained with GAGG:Ce^{3+} crystals using the dual-PMT setup. We demonstrated that the scintillator-photosensor interface reflection coefficients are thickness-dependent with more than 1σ confidence. We explained this phenomenon through the Monte Carlo simulations and demonstrate that the scintillators have thickness-dependent interface reflection coefficients through experiments by comparing the relative light outputs of crystals with different thicknesses. In order to relate the experimental observations to reflection coefficients, we used the modified 2D-model, and showed that the Monte Carlo results and experimental results match within 1σ. In future work, we are going to extend this methodology to different crystals.

[1] E. Hecht, *"Optics,"* (5th ed.) Boston: Pearson Education, Inc, 2017.

[2] F. Logoglu *et al.*, *"Experimental and Computational Evaluation of Thickness-Dependent Reflection Coefficients of Scintillator-Photosensor Interfaces,"* IEEE Nuclear Science Symposium and Medical Imaging Conference (NSS/MIC), 2022.

[3] F. Logoglu *et al.*, "*Introduction of a thickness-dependent scintillator-PMT interface reflection coefficient to improve absolute light yield calculations for inorganic scintillators,*" Optical Materials, vol. 138, 2023.

[4] C. Look and J. H. Leach, "*On the accurate determination of absorption coefficient from reflectance and transmittance measurements: Application to Fe-doped GaN,*" Journal of Vacuum Science and Technology. B, Nanotechnology & Microelectronics, vol. 34, (4), pp. 4, 2016.

[5] F. Logoglu *et al.*, "*Experimental and Simulation Investigation of Micro- and Nano-Structured Neutron Detectors,*" EPJ Web of Conferences, vol. 253, pp. 11010, 2021.

[6] G. Romanchek *et al.*, "*Performance of Optical Coupling Materials in Scintillation Detectors Post Temperature Exposure,*" Sensors (Basel, Switzerland), vol. 20, (21), pp. 6092, 2020.

[7] Uchida *et al.*, "*Attenuation characteristics of a Ce:Gd3Al2Ga3O12 scintillator,*" Nuclear Instruments & Methods in Physics Research. Section A, Accelerators, Spectrometers, Detectors and Associated Equipment, vol. 986, pp. 164725, 2021.

https://doi.org/10.1051/epjconf/202328810016

Methodology for an experimental drift velocity determination using a µTPC. Example in a 700mbar 95%He+5%CO_2 mixture

Thibaut Vinchon[1,*], Michaël Petit[1]

[1] Institut de Radioprotection et de Sûreté Nucléaire (IRSN), PSE-SANTE/SDOS/LMDN, Cadarache 13115 Saint-Paul-Lez-Durance, France
(*) thibaut.vinchon@irsn.fr

Abstract: The micro-irradiation, neutron metrology and dosimetry laboratory of the Institute for Radiological Protection and Nuclear Safety (IRSN) is a National Metrological Institute associated laboratory, is in charge of the neutron fluence distribution references and produces neutron reference fields.
The laboratory is developing a primary neutron-detector based on the use of Time Projection Chamber technology. This detector, called µTPC, can use different types of gas over a range of precise pressures. The analysis of µTPC data requires the determination of both the energy and the track direction of the recoil particles. The drift velocity in the gas is a key data point for analysis. Based on several experimental measurements performed using the µTPC with monoenergetic neutron fields, a direct methodology to determine the drift velocity can be proposed. These results are compared with those found using theoretical drift velocity from the Magboltz simulation.

Keywords -- Neutron, Fluence, Time Projection Chamber (TPC), drift velocity, Magboltz

I. INTRODUCTION

The French Institute for Radiological Protection and Nuclear Safety (IRSN), as a designated institute by the French National Metrological Institute (LNE) [1], develops primary measurement standards using a Time projection detector (µTPC). Data collected using this detector leads to nuclear recoil track reconstruction. For this purpose, the analysis requires a precise value for drift velocity. The MAGBOLTZ simulation calculates an initial value of 9.55 µm/ns. However, after the analysis of µTPC data, this value appears to be incorrect. Taking into account the known physical properties of neutron scattering, an experimental method is proposed to determine the drift velocity in the detector.

II. THE LNE-IRSN/MIMAC MICROTPC DETECTOR

A. Description

In the neutron detection field, there exists at least two models of µTPC. The LNE-IRSN-MIMAC detector (also called µTPC) is a time projection chamber that uses a pixelated Micromegas anode, coupled to a fast-self-triggered electronic system. Figure 1 is a schematic illustration of the µTPC working principle with the charge collection and the ion track. Each event has a charge collection profile measured by the grid (bottom left in figure 1) and a series of $Nb_{Slices}(i_{event})$ obtained on the pixelated anode (bottom right in figure (1)).

The detector active zone is a rectangular parallelepiped volume measuring 10.8 cm x10.8 cm x17.7 cm. It is filled with an adjustable gas mixture. A homogeneous electric field of adjustable intensity is applied through a wired field cage. This field draws the electrons toward the grid.

The total charge is collected by the grid. In the amplification zone, which is located between the grid and the anode, there is a higher electric field which produces a Townsend avalanche, which amplifies the signal. The anode is composed of 256 strips in each direction with a pitch of 424 µm. The Micromegas pixelized anode is read at a frequency of 50 MHz using a self-triggered electronic system developed at the Subatomic Physic and Cosmology Laboratory in Grenoble (C.N.R.S/L.P.S.C, France) [4].

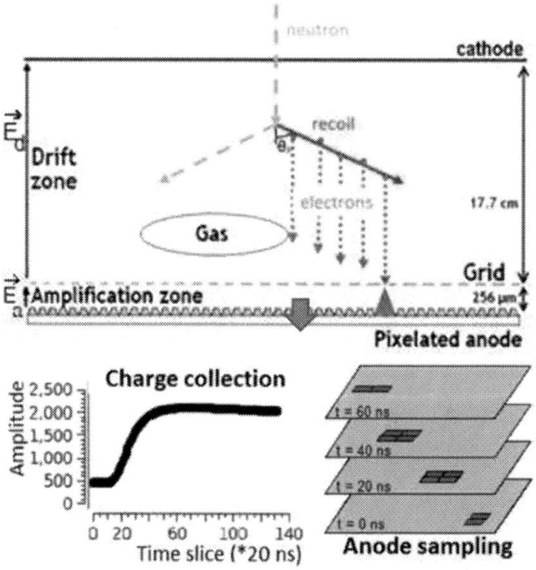

Fig 1 :Schematic view of the microTPC working principle

The µTPC can use different types of gas at a precise pressure within a limited range. This limitation is due to specific detector constraints. The gas mixture used here was 95% ^{4}He and 5 % CO_2 at an absolute pressure of

700 mbar. The mixture is pre-made by the supplier which introduces a certain level of uncertainty, but the same shipment is used for all our experiments to guarantee reproducibility.

The analysis presented below is done with data from a measurement campaign on 1.2 MeV and 5 MeV monoenergetic neutrons fields. The nuclear recoils of interest are those of the alpha, which are the result of (n, α) elastic scattering reactions.

B. *Use of the µTPC in neutron metrology*

In neutron metrology, the µTPC detector can also be used for the determination of the incident neutron energy based on knowledge of the kinematic parameters of the scattering reaction. These parameters are used in equation (1) for the neutron energy (E_n) computation, where (m_n) and (m_r) are respectively the neutron and the recoil nucleus mass, (E_r) is the kinetic energy of the nuclear recoil and (Θ_{nr}) is the angle between the incident neutron direction and that of the nuclear recoil.

$$E_n = \frac{(m_n + m_r)^2}{4.m_n.m_r} \cdot \frac{E_r}{\mathrm{Cos}^2(\theta_{nr})} \qquad (1)$$

Each of these parameters can be deduced from the acquired raw data. The initial direction of a neutron is imposed by the experimental setup. But it can also be deduced using a statistical approach, as this reaction conforms to a specific angular distribution as a function of $\cos(\Theta_{nr})$. Obviously, the neutron energy calculation should not be dependent on the scattering angle. In the following, to simplify, we note:

$$P_z = \cos(\theta_{nr}) \qquad (2)$$

We can assume an approximate P_z by the ion direction component relative to the detector main symmetry axis (positive from cathode to anode). It can be justified by the fact that, in the neutron fields analyzed in the present work, the angle between the detector z axis and the direction of an emitted neutron is statically limited to +/- 1°. Moreover, this distribution is symmetrical with respect to 0°. This simplification has a negligible impact on the method presented below. The full metrological fluence measurement capability is described in reference [5].

C. *Available observable quantities and associated corrections*

From the raw data acquired by the detector, the recoil ionization energy is computed, event by event, from the associated charge collection profile. In order to estimate the recoil kinetic energy E_r, the following sequence is applied: first, the ionization energy $E_{ionization}$ is obtained from the charge collection profile by applying a linear calibration, then E_r is deduced from $E_{ionization}$ using the Ionization Quenching Factor (*IQF*) which is ionization energy dependent [6].

A three-dimensional reconstruction of the charged particle track is also performed event by event [7][8][9]. The pixelated anode provides X and Y direction coordinates. The third direction coordinate Z is deduced from the anode read period ΔT, the number of slices $Nb_{Slices}(i_{event})$ of the associated event and the drift velocity V_{Drift} using the relation given by (3):

$$Z = Nb_{Slices}(i_{event}).\Delta T.V_{Drift} \qquad (3)$$

The produced electron drift velocity is a key data for the computation of the recoil ion track angle and thus, the neutron energy E_n. Moreover, the correlation between track length and energy deposit is the basis for identifying and selecting alpha recoil events [7][8][9].

D. *Drift velocity simulation in the gas mixture*

The drift velocity can be simulated by MAGBOLTZ [10] which computes drift gas properties in a given electric field by numerically integrating the Boltzmann transport equation (i.e., simulating an electron bouncing around, inside a gas). The MAGBOLTZ simulation computes the drift velocity by tracking virtual electron propagation step by step. The MAGBOLTZ simulation was performed for the µTPC assuming a uniform electric field of ~156.5 V/cm in the defined gas mixture. The MAGBOLTZ calculated drift velocity was evaluated at 9.55 µm/ns.

III. METHODOLOGY: DETERMINING DRIFT VELOCITY

A. *Exploitation of the recoil nuclei energy/direction correlation*

The proposed methodology is based on the use of the energy/direction correlation for a group of recoil nuclei induced by a monoenergetic neutron field. In the following, we calculate the recoil nucleus energy using the IQF law computed using the TRIM option of the SRIM software [11]. The ionization energy is deduced from the charge collection profile using a calibration of the µTPC carried out using the alpha from the $^{10}B(n_{th},{}^{7}Li)^{4}He$ reaction [12].

The reconstructed average neutron energy, normally, should not be dependent on the direction of the recoil. But this is not the case if we use the drift velocity calculated by the Magboltz simulation. So, we decided to perform a parametric analysis as a function of V_{Drift} to accurately determine the optimal drift velocity V_{Drift}^{opt}.

We performed several analyses with different drift velocities as input. In the following figure 2, we compare the bidimensionnal histograms linking the computed P_z (on the ordinate) and the computed neutron energy (on the abscissa) for two V_{Drift} value. We observed a loss in resolution when decreasing P_z. This is related to recoils with a smaller track length and therefore a greater difficulty to estimate the initial nuclear recoil direction.

Indeed, electron diffusion becomes an important phenomenon making the estimation of the initial recoil direction more and more difficult. So angles larger than 60° (where $P_z < 0.5$) are not considered for the proposed methodology. The blue line on each graph shows the average neutron energy calculated as a function of P_z.

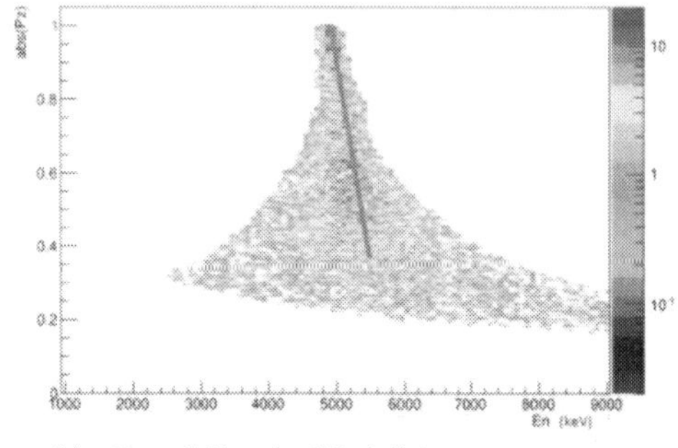

(a) V_{Drift} =9.55 µm/ns (Magboltz)

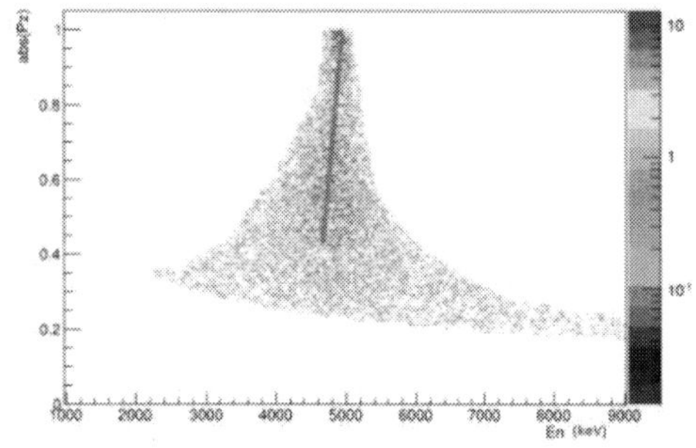

(b) V_{Drift} =10.2 µm/ns

Fig. 2. Bidimensionnal histograms of the (E_n(keV), $|P_z|$) distribution for different input drift velocities V_{Drift} (in blue a visual estimation of average) obtained with the IQF correction issued from SRIM/TRIM [11]

Visually, we can see on chart (a) associated with the Magboltz drift velocity estimation (i.e. 9.55 µm/ns), that the calculated average energy increases as P_z decreases. On the other hand, when drift velocity is increased to 10 µm/ns, the calculated energy decreases as P_z decreases (see chart (b)). So, the drift velocity is too high. In order to get accurate quantitative results, a sampling method by band of P_z is proposed. The full methodology used to estimate the optimal drift velocity is presented in the next section.

B. Methodology: Quantitative V_{Drift} determination

The following method aims at a systematic and effective determination of an optimal drift velocity. First, for a set of drift velocity values, the calculation of the average neutron energy is performed as a function of the recoil direction P_z. Then, for each drift velocity a straight-line slope is deduced from the curve relating calculated mean energy as a function of recoil direction P_z. So, for each V_{Drift}, a slope can be found through a linear fit on the curve relating the mean neutron energy E_{mean} and the mean direction cosine P_z^{mean}. This approach enables us to build a graph relating slopes with their associated uncertainties as a function of V_{Drift}. Finally, the optimal drift velocity is deduced via the search for a zero slope (which ensured the independence of the mean neutron energy calculation as a function of recoil direction).

The following sampling sequence was applied:

- For a set of V_{Drift} values:
  - A mobile selection zone of $|P_z|$ was defined as $[P_z^{min}, P_z^{max}]$ ensuring a constant statistic for the number of events ($Nb_{EventInPzSlice}$). The mean value P_z^{mean} is determined for each selection zone $[P_z^{min}, P_z^{max}]$.
  - The neutron energy deduced from E_{recoil} and angular scattering is computed for each event. By $|P_z|$ zone, a gaussian fit on the projected histogram is performed to get the E_{mean} value. Uncertainties on E_{mean} is estimated by the sigma of the gaussian fit done with ROOT software [13].
  - A linear fit of $E_{mean} = f(P_z^{mean})$ for a given drift velocity value is determined leading to $Slope(V_{Drift})$ and its associated uncertainty $\Delta Slope(V_{Drift})$ deduced from the linear fit for this drift velocity.
- Resulting slopes are then plotted on a graph as a function of V_{Drift}. A linear fit is performed on that graph which enables the deduction of the optimal V_{Drift} value.

This methodology was applied to two measurements performed at the National Physical Laboratory (N.P.L, Teddington, U.K.) in monoenergetic neutron fields at 5 MeV and 1.2 MeV and for different V_{Drift} values.

We applied the methodology to the 565 keV monoenergetic neutron field data analysis. The results were more difficult to obtain and at a much higher uncertainty level. These difficulties have various possible causes:

- First, at that neutron energy, the differential cross section of the (n, α) reaction peaks steeply at 0° i.e. at P_z=1, leading to a very limited zone for the linear fit.
- Secondly, we suspect a high relative bias induced by the choice of the applied *IQF* curve. This point will be discussed briefly in section 4 and more precisely in a future publication.
- Third, the reduced track lengths result in poor angular resolution.

We realized a parametric analysis as a function of the target statistic in the selection zone. The final impact of this target statistic on the average slope associated to a given drift velocity appears to be negligible.

C. Results: 5 MeV monoenergetic neutrons field

Figures (3) and (4) show results of the methodology with the 5 MeV neutron field. Figure (3) presents E_{mean}

variation as a function of P_z^{mean} for different V_{Drift} values. For each V_{Drift}, a $Slope(V_{Drift})$ can be derived with its associated uncertainty $\Delta Slope(V_{Drift})$ from the fit realized on the series of points (P_z^{mean}, E_{mean}) computed for that V_{Drift} taking into account the ΔE_{mean}.

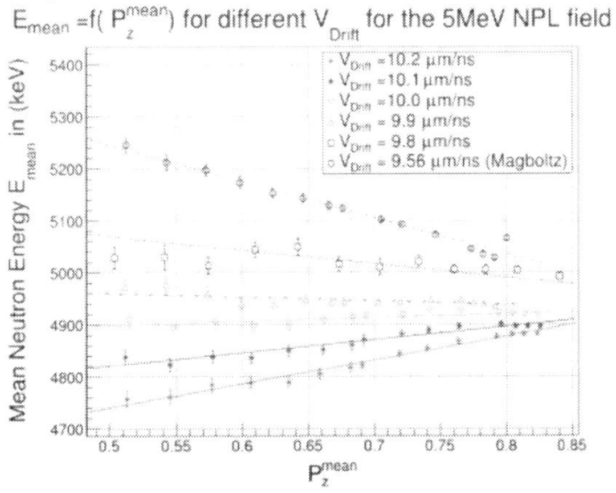

Fig.3: Multigraph showing (P_z^{mean}, E_{mean}) points for different V_{Drift} values and the associated linear fits using measurements from the 5MeV NPL reference field

The results of the fit for each value a V_{Drift} are listed in the table 1:

TABLE 1

FIT RESULTS FOR EACH DRIFT VELOCITY

V Drift (µm/ns)	Slope(+/-ΔSlope) (keV)	Offset(+/-ΔOffset) (keV)
10.2	459.6(+/-24.0)	4510.5 (+/-18.3)
10.1	253.4(+/-22.3	4694.3 (+/-17.1)
10.0	76.2(+/-23.7)	4860.6 (+/-18.0)
9.9	-68.0(+/-28.0).	4994.0 (+/-20.5)
9.8	-260.8(+/-23.6)	5199.8 (+/-19.7)
9.56	-695.7(+/-27.0)	5592.8 (+/-19.5)

As shown in figure 4, the set of (V_{Drift}, $Slope(V_{Drift})$) and their associated uncertainties $\Delta Slope(V_{Drift})$ are plotted on a second graph. A second linear fit is then applied to this set of points taking uncertainties into account. The optimal value of V_{Drift} is defined as a slope value equal to zero. From this fit, we estimate an accurate V_{Drift} value of 9.95µm/ns

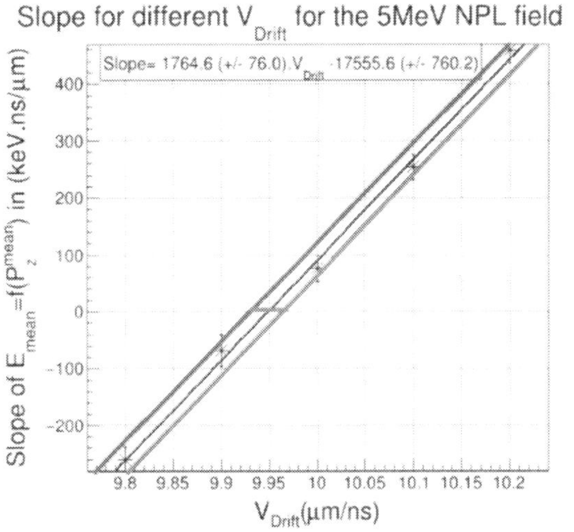

Fig.4: Graph showing the set (V_{Drift}, $Slope(V_{Drift})$), their associated uncertainties $\Delta Slope(V_{Drift})$ and the associated linear fits (in red: fit with Slope+ΔSlope, in green : fit with Slope-ΔSlope, in black: the initial fit) for the 5MeV NPL reference field.

To quantify the uncertainty associated with this estimation, we perform two additional fits:

- One for the set of points given by (V_{Drift}, $Slope(V_{Drift}) + \Delta Slope(V_{Drift})$),
- Another for the set of points given by (V_{Drift}, $Slope(V_{Drift}) - \Delta Slope(V_{Drift})$),.

With this approach, the optimal V_{Drift} value is estimated to be in the range [9.93-9.96] µm/ns

D. Results: 1.2 MeV neutrons

The same methodology was then repeated on the measurements performed at 1.2 MeV reference field. From this analysis, we can estimate an optimal V_{Drift} value close to 10.07µm/ns. With the uncertainty approach described above, the optimal V_{Drift} value is estimated to be in the range [10.06-10.09] µm/ns.

IV. DISCUSSION

The drift velocity in a µTPC is a key parameter in the proper reconstruction of charged particles tracks. It has been shown that the drift velocity value computed by the Magboltz simulation leads to a result incompatible with the physics which requires the mean average energy to be constant as a function of P_z.

The drift velocity in a gas mixture (95.0%He + 5.0% CO_2) was determined using measurements in two monoenergetic neutron fields at 5 MeV and 1.2 MeV. Results were compared with the Magboltz simulation.

Some differences in V_{Drift} estimations from both fields were observed:

- V_{Drift_opt}(5 MeV) is between 9.93 and 9.96 µm/ns, 4.1% *higher* relative to the Magboltz estimation (9.55 µm/ns),

- V_{Drift_opt}(1.2 MeV) is between 10.06 and 10.09 µm/ns, 5.3% *higher* relative to the Magboltz estimation.

A V_{Drift} sensitivity study performed with Magboltz near the nominal operating point (95%He/ HV_{Drift}= 2770V/ 700 mbar/20°C) shows that V_{Drift} sensitivity relative to:

A) stoichiometry is +/- 1.2% for +/-1% of ^{4}He,
B) high voltage is 0.046% per Volt,
C) pressure is 0.24% per mbar,
D) temperature is 0.28% per °C.

Combining these sensitivities with the associated uncertainties leads to a cumulative relative error of 0.31% on V_{Drift}. Therefore, we conclude that it is not possible to attribute the gap of 1.2% between the two estimations, to variations in the experimental conditions, alone, between the two measurements.

One possible explanation for this gap could stem from the assumption, in the Magboltz simulation, of a uniform electric field. Indeed, simulations previously carried out in reference work [14] using a different gas mixture have demonstrated that there was a slight deformation of the electric field near the grid. As the tracks in the 5 MeV field are longer, one can postulate a slight difference in the influence of this deformation on the tracks in the drift cage.

Another explanation could come from the use of an Ionization Quenching Factor (IQF) curve computed using the TRIM option of SRIM [11]. Indeed, some measurements [9] done at the AMANDE facility lead us to believe that the TRIM software, as it is designed, underestimates the IQF at low recoil energy. We tested the application of a preliminary IQF curve to the V_{Drift} determination process and this led to an additional increase in the drift velocity.

Even though some additional measurements are needed to provide a consolidated estimation, increasing the drift velocity compared to the Magboltz estimation is a necessity.

V. Conclusion and Perspectives

A drift velocity measurement with a Time Projection Chamber has been performed in this study. The measured value is slightly higher, around +5%, above the value computed by Magboltz. The sensitivity study cannot explain the discrepency, in state. In order to conclude, the proposed method could be completed and improved by other studies:

- Additional measurement of the drift velocity with the cathode signal as in reference [15],
- Further measurements using the AMANDE monoenergetic neutrons fields with enhanced statistics
- Using the charge collection profile on the anode grid of the µTPC as in reference [16].

A better estimation of IQF through experimental campaigns is required to obtain a consolidated estimation of the drift velocity.

Acknowledgment

This project has received funding from the LNE.

References

[1] V. GRESSIER et al., "New IRSN facilities for neutron production", NIM A, 505 370-373, 2003.
[2] D. SANTOS et al., "MIMAC: MIcro-tpc MAtrix of Chambers for dark matter directional detection", Proceedings of the 4th International Conference on Directional Dark Matter Detection CYGNUS2013, held in Toyoma (Japan), 2013.
[3] I. GIOMATARIS et al., "Micromegas in a bulk", NIM A, vol. 560, pp. 405-408, 2006.
[4] O. BOURRION et al., "Data acquisition electronics and reconstruction software for real time 3D track reconstruction within the MIMAC project", JINST 6 C11003, 2011.
[5] M. PETIT et al ; "A Time projection chamber (TPC) : a primary detector concept for neutron metrology between 24 keV and 19 MeV", to be published
[6] O. GUILLAUDIN et al., "Quenching factor measurement in low pressure gas detector for directional dark matter search", EAS Publ.Ser. 53 (2012) 119-127.
[7] D. MAIRE et al., "First Measurement of a 127 keV Neutron Field with a -TPC Spectrometer", IEEE TNS (2014), Volume: 61 , Issue: 4 , Part: 2, pp. 2090-2096.
[8] D. MAIRE et al., Rad. Prot. Dos., Vol. 161, Nos 1–4, pp. 245-248 (2014).
[9] B. TAMPON, PhD Thesis, (2017), " Qualification expérimentale de la µTPC LNE-IRSN-MIMAC comme instrument de référence pour les mesures en énergie et en fluence de champs neutronique entre 27keV et 6,5 MeV "
[10] S.F. BIAGI, "Monte Carlo simulation of electron drift and diffusion in counting gases under the influence of electric and magnetic fields", Nuclear Instruments and Methods in Physics Research A 421 (1999) 234—240,
[11] J. F. ZIEGLER, J. P. Biersack and U. Littmark, Pergamon Press New York (1985), www.srim.org.
[12] D. SANTOS et al. ,"Neutron spectroscopy from 1 to 15 MeV with Mimac-FastN, a mobile and directional fast neutron spectrometer and an active phantom for BNCT and PFBT", EPJ Web of Conferences 231, 05003 (2020)
[13] « ROOT a Data analysis Framework | ROOT a Data analysis Framework ». https://root.cern.ch/
[14] G. PUILL, " Le Développement de Micromégas, un nouveau détecteur gazeux de position à microgrille", PhD Thesis 2000 directed by B. TAMAIN
[15] C. COUTURIER et al., "Cathode signal in a TPC directional detector: implementation and validation measuring the drift velocity", Journal of instrumentation, 2017, 12 (11), pp.P11020
[16] J. BILLARD et al.," Measurement of the electron drift velocity in CF4 and CHF3 gas mixtures in the context of upcoming directional Dark Matter detectors", Journal of instrumentation, May 2013 9(01)

Particles discrimination and tracking through a matrix of detection coupled with a voltage-controlled oscillator

Karine Coulié[1,*], Wenceslas Rahajandraibe[1], and Laurent Ottaviani[1]
[1] Aix Marseille Université, CNRS, Université de Toulon, IM2NP UMR 7334, Marseille, France
(*) karine.coulie@im2np.fr

***Abstract*— A particle detection chain based on a CMOS-SOI VCO circuit associated to a 3x5 matrix of detection is presented. The solution is optimized for the recognition and tracking of low energy particles. Two ions are considered: an alpha and a lithium. These two ions were chosen because there are the secondary particles generated by the neutron-boron 10 interaction.**

***Keywords* — Sensor, detection, VCO, radiation.**

I. Introduction

RAPID development of digitalization and automation in the electronics industry, along with the use of memory-based elements in technologically advanced products, is continuously challenging the systems reliability. Increasing usage of semiconductor devices in the automotive fields and in other systems where reliability is mandatory is a key factor driving semiconductors development. Semiconductor memories are an inherent part of modern System-on-Chip (SoC) designs. The increase in computing capacity induces a proportional increase in the amount of data to be processed in SoCs, thus creating larger memory requirements. The latest technologies make these memories denser and thus defects due to the manufacturing process are more prone to occur. Among different types of memories, Static Random-Access Memories (SRAMs) are still dominant in SoCs. This is because such devices can provide high access performance and allow a fast integration in a design due to the use of CMOS technology. However, the dense structure of SRAMs makes them to be more prone to physical defects.

Semiconductor radiation detectors, also called solid-state detectors, can be classified into several categories, depending on the application field and the particles to be detected. These detectors are based on charge collection and amplification using a semiconductor volume such as a p-n junction [1-2]. They exhibit several advantages in comparison with other types of detectors such as gas-filled counters, as faster charge-collection time which provide the ability to process higher counting rates. Another advantage is their compactness which allows the measurement of intensity variations over small distances improving the measurement resolution. A pixel organization including several sensors could be used when a high detector sensitivity is required [2-8]. Thus, the goal of this work is to study the mechanisms involved in particle interaction with SRAM through a matrix of detection based on a typical SRAM junction.

Particles counting and/or recognition require the conditioning of the signal generated by the sensor. Most of the existing conditioning chains are based on a direct reading of the detected currents created by the crossing of an ionizing particle through the matter. When an embedded reading system is necessary, this solution could not be suitable for the detection of low energy particles where the current signature could be a narrow pulse with a duration of few nanoseconds. Indeed, the readout chain must ensure sufficiently high resolution for an appropriate post-processing. Moreover, a reading amplifier exhibiting a fast response time is required. This induces a higher current consumption and complex circuit implementation. Designing high precision, low-power wideband amplifier is a critical issue. So, most practical detectors are based on direct reading of the detected currents using Charge Sensitive Amplifiers (CSA) [9]. In this case, the circuit does not deliver information about the electrical signature of the particles. For applications such as gas prospection or medical therapy, the characteristics of the electrical signature of the particles must be known. Moreover, the knowledge of the generated current shape at the output of the detector allows easier post processing of the signal.

In this context, an approach based on the use of a ring oscillator was developed with an innovative conditioning method [10-17]. This approach is based on the reading of the information related to indirect output parameters of the detection chain signal (*i.e.* voltage variation), instead of directly measuring the current from the sensor. This solution avoids most of the design problems described upper. Then, the information is extracted by correlating the initial oscillating signal of the system with the oscillating signal after the particle has passed the detector. The only requirement to allow particle recognition is then to link the output information (*i.e.* voltage variation of the oscillating signal) to the input information (current stimuli).

The full detection system (VCO + matrix) is described in section II. Section III is dedicated to three case studies: an alpha particle, a lithium particle and the neutron-boron 10 interaction. The last section concludes the paper.

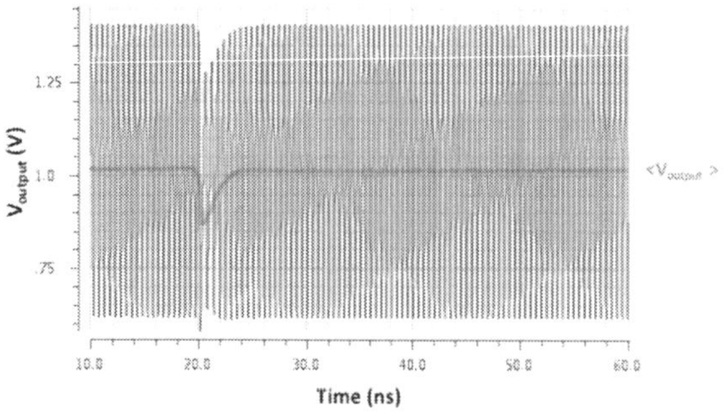

Fig.1. Variation of the output voltage versus time for the VCO based detection chain.

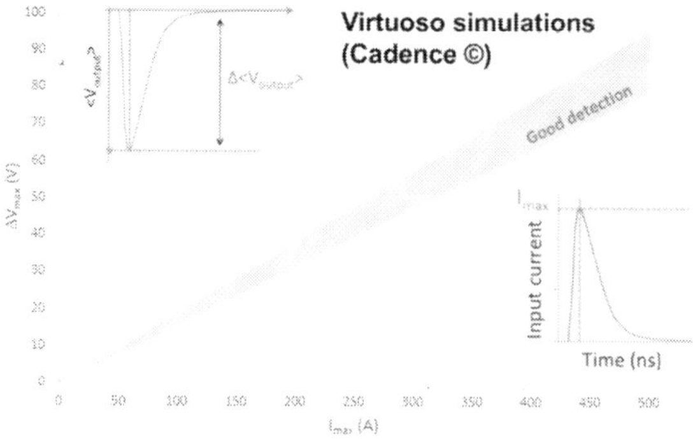

Fig.2. Variation of the average output signal (ΔV_{max}) versus the maximum of the input current (I_{max}).

II. Characterization of the System

The proposed detection system is composed of a detection matrix together with an innovative readout circuit based on a high frequency voltage-controlled oscillator. The operating is synthesized in this section.

A. The VCO detection chain

The detection chain has been designed and implemented on 130nm CMOS SOI technology, then simulated at circuit level using "Spectre" simulator (SPICE-based) under Cadence *Virtuoso* © CAO tool.

The concept itself was presented in [10-13]. The VCO chain is composed of three parts: a CMOS based pixel detector, a shaping circuit based on a Voltage Controlled ring Oscillator (VCO) and a system for frequency and amplitude detection (Fig. 1). This system allows the evaluation of the circuit sensitivity to radiation by measuring the oscillator responses. In the very first versions of the system, the operating frequency of 0.35µm Bulk Silicon substrate CMOS VCO was 1MHz. Thanks to a new solution based on another CMOS process (130nm SOI), it reaches now 4.8GHz. This increase of the operating frequency makes possible the development of new methods for signal recognition. As actual operating frequencies are reaching 4.8GHz, the shape of the signal is directly reproduced at the output of the VCO. Then, the output parameter is now the average voltage variation (Fig.1). The advantage of this solution is a direct recognition of the signal shape as an amplitude modulation of the VCO output frequency.

B. Calibration of the VCO chain

The key point is how the output parameters of the VCO chain can give information related to the input current. This could be done through the analysis of various characteristics extracted from the average voltage variation (Fig. 1). In this paper, we study the variation of the average output signal (ΔV_{max}) versus the maximum of the input current (I_{max}) for different particle charges. In [15], a linearity curve linking input to output parameters was determined for a 2GHz VCO by using standard currents. A new linearity curve has been obtained for this new 4.8GHz VCO (Fig. 2). In [13], it has been demonstrated that two different currents with the same maximum value get the same corresponding ΔV_{max}. Then ΔV_{max} does not depend on the shape of the current but only on the current peak. The measurement of ΔV_{max} appeared then to be a reliable parameter to recognize the current peak of the input signal. Through calibration curves, the output parameters can be linked to the input currents, which could allow the incoming particle identification.

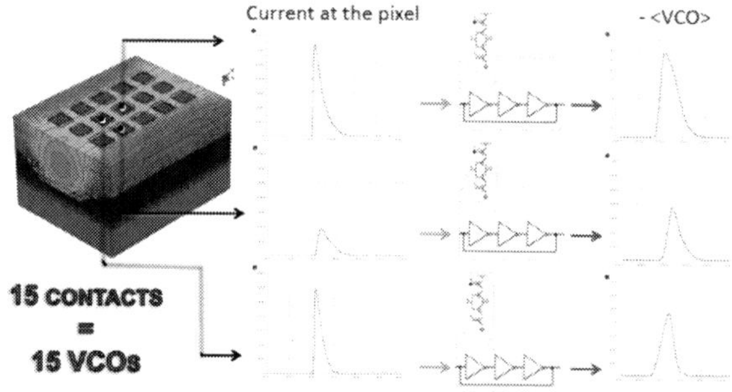

Fig.3. Left: Simulation of the 3x5 matrix; middle: currents generated by the particle at three pixels; right: corresponding signal at the output of the VCO

To get reliable results, the resolution of the system has to be checked. Thus, the results are valuable if the points are included in the "good detection" zone. This zone corresponds to a linear response of the system in the range of ±10 %. When the results are situated out of this zone, the resolution is clearly degraded [15] and the pulse characteristics are also not suitable with the resolution of our system. In this condition the system can only be used as a particle counter.

C. The detection matrix

In order to improve the efficiency of the detection and the tracking ability, a 3x5 matrix is modeled and analyzed (Fig.3). The input current is no longer the simplified current used for the VCO characterization. Now, the current source comes from the realistic simulation of the matrix using TCAD device simulation tools (Synopsis ©). The effect of the ion strike is simulated using the *Heavy Ion* module of Synopsis [18], considering an electron-hole pair column centered on the ion track axis. The Linear Energy Transfer is defined as the energy lost by the particle, by unit of length and varies along the track depending on the initial energy of the particles [19]. An actual variation of the LET was integrated in our simulations, based on the value given by SRIM tables [20]. For many years, neutron – boron 10 interaction is a concern for microelectronics

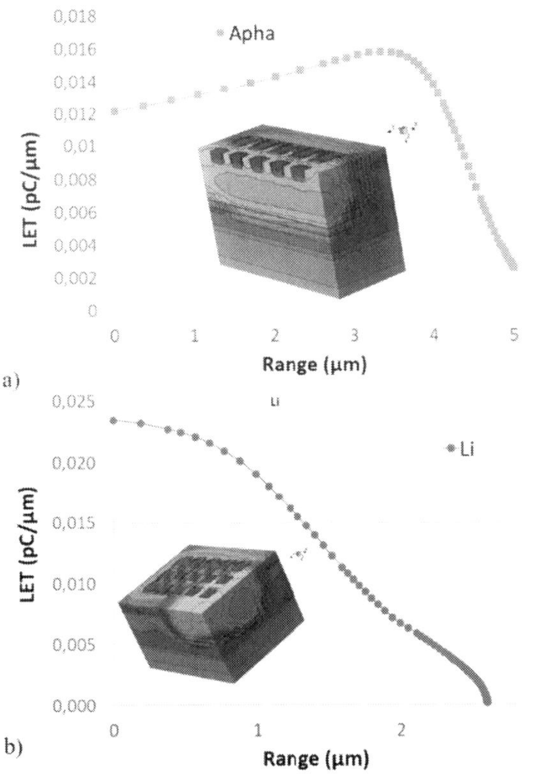

Fig.4. a) LET variation versus range and charge density in the matrix for the 1.5 MeV Alpha particle. b) LET variation versus range and charge density in the matrix for the 0.9 MeV f Lithium particle

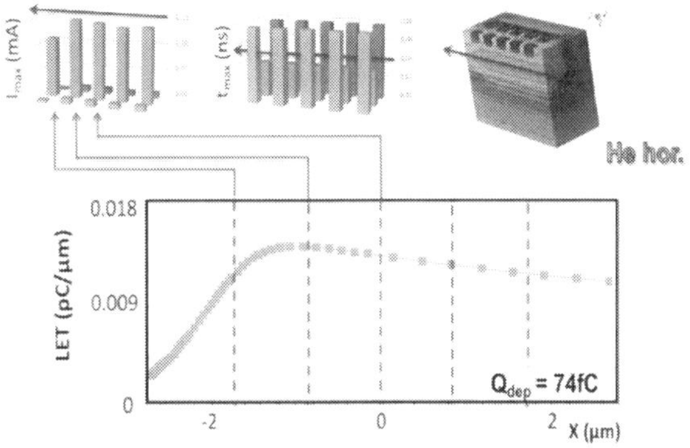

Fig.5. Alpha particle 1,5MeV, bottom: electron density versus LET; current peak versus pixels (left), current delay versus pixels (middle).

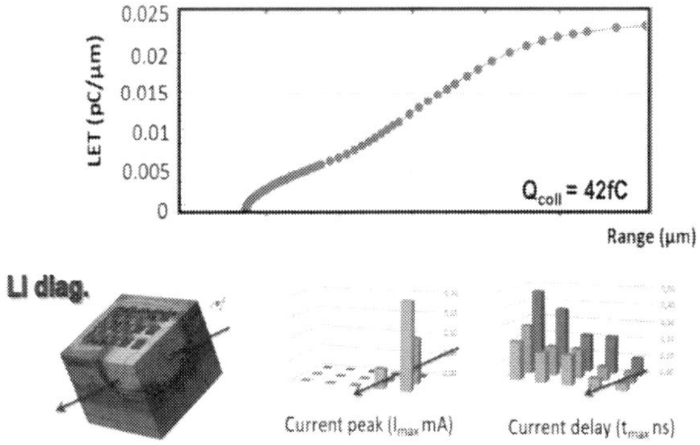

Fig.6. Lithium particle 0,9MeV, top: electron density versus LET; current peak versus pixels (middle), current delay versus pixels (right).

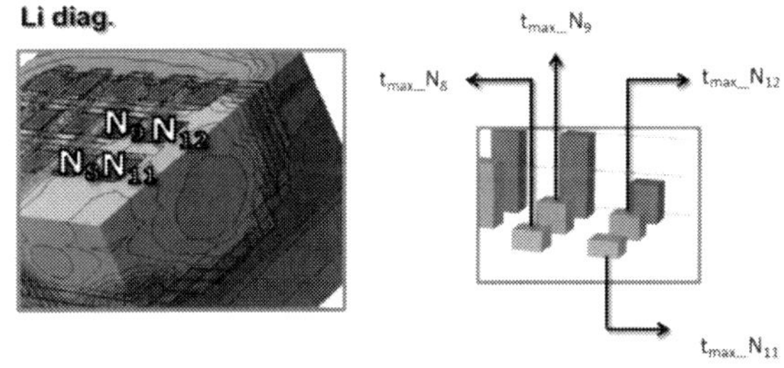

Fig.7. Lithium particle 0,9MeV, zoom on the current delay versus pixels.

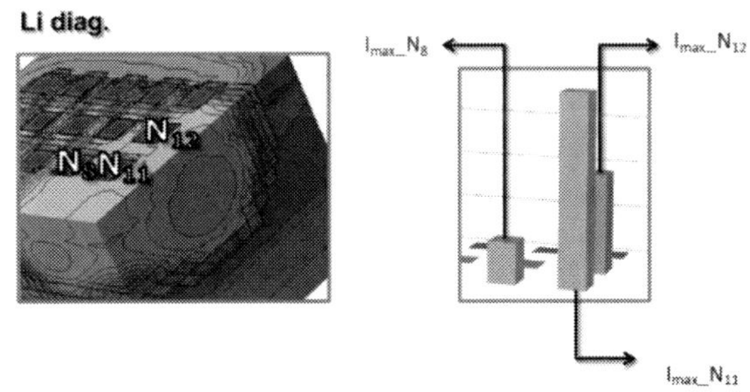

Fig.8. Lithium particle 0,9MeV, zoom on the current peak versus pixels.

reliability. Boron is used at different step of the process in microelectronics devices [23]. As thermal neutrons can be found at ground level, their interaction with terrestrial electronics systems can considerably affect reliability. Then we decided to study the particles generated by the main interaction of thermal neutron with boron-10, i.e. one Li with an initial energy of 0.9 MeV and one Alpha with an initial energy of 1.5 MeV. The LET variations of these particles along the track are shown in Fig. 4 a) and b). The particle is generated at 50 ps. The total deposited charge for the alpha and lithium are 74fC and 42fC, respectively.

An example of the generation of a particle in the whole structure is shown in Fig. 3. This figure presents the currents generated by the particle at three pixels and the corresponding signal at the output of the VCO. If a 3x5 matrix is simulated, then fifteen currents will be collected and injected at the input of the VCO. So fifteen voltage variations will be analyzed at the VCO output.

III. Case study

A. Tracking of an alpha particle

First case-study aims to test the matrix response for an alpha particle crossing the 3x5 matrix as shown in Fig.5, 0.8µm under the contacts. The configuration is presented at the right top of the figure. At the bottom, the electron density corresponding to the crossing of the alpha particle is visible. The current peak I_{max} generated at the 15 pixels (left) and the current delay t_{max}, corresponding to the maximum time at current peak (middle) are also plotted.

In a previous study [16], it was shown that the highest current at a pixel corresponds to the highest electron density and the lowest current corresponds to the lowest charge density. Thus, the current variation can provide an information about the charge density in the detection cell. It was also observed that the current shape can give information about the distance between the charge generation and the contact. Indeed, the contact located just above the ion corresponds to the shortest current delay.

By comparing I_{max} and t_{max} variations depending on the pixels, the symmetry of the track is visible. More, the current peaks of the middle contacts depict directly the charge density as they are located just above the ion track. As expected, the highest current correspond to the highest electron density and the lowest current corresponds to the lowest charge density.

The t_{max} shape versus pixel position also gives information about the distance between the charge generation and the contact. Indeed, the t_{max} corresponding to middle contacts, appears earlier than the one corresponding to the contacts at the edge of the matrix.

B. Tracking of a lithium ion

The second case study deals with the crossing of a lithium particle in the 3x5 pixels structure, 0.8µm under the contacts (Fig.6). The configuration is presented at the left bottom of the figure. At the top, the electron density corresponding to the crossing of the lithium particle is visible. The current peak generated at each pixel is plotted on the middle and on the left is the current delay, corresponding to the current peak.

The behavior of the current delay is similar to that of the electron density, which means that the VCO detects all the variations of the delay. Fig.7 presents a zoom on the delay in the pixels of interest in correlation with the corresponding configuration. The location of the various pixels regarding to the contact location is clearly visible. The highest delay corresponds to the farthest contact and the shortest delay corresponds to the closest contact. So, we got an information about the location and direction of the particle.

The behavior of the current peak is not so obvious. Indeed, a highest current peak can be due to a highest current density or to a closest contact. Then the analysis of the currents versus time at contacts N8, N11 and N12 is necessary (Fig.8). By comparing N8 and N11, we observe that the current peak is higher for N11 whereas the t_{max} are similar. Indeed, we can assume that the electron density at contact N11 is higher than the one at contact N8. Now, by comparing N11 and N12, we observe that $t_{max}\_N_{11}$ is lower than $t_{max}\_N_{12}$, then we could assume that the charge is generated next to contact N11. As $I_{max}\_N_{11}$ is higher than $I_{max}\_N_{12}$, this highest current peak could be due to the location of contact N11 or to a highest charge density. Finally, no clear information about the charge density variation is available from this analysis.

In Fig.10, the collected charge versus time at each significant

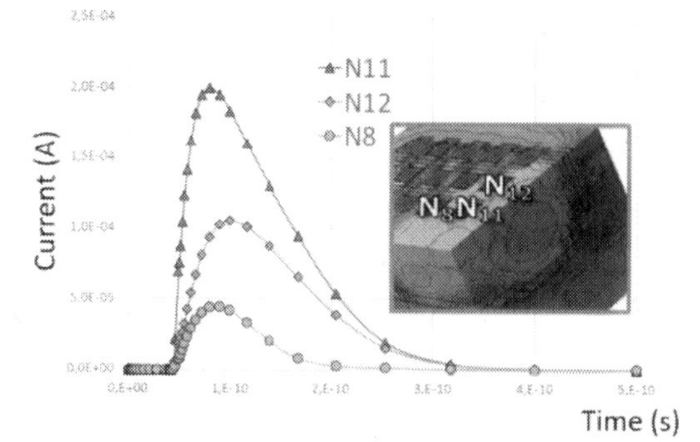

Fig.9. Lithium particle 0,9MeV, current versus time at contacts N8, N11 and N12.

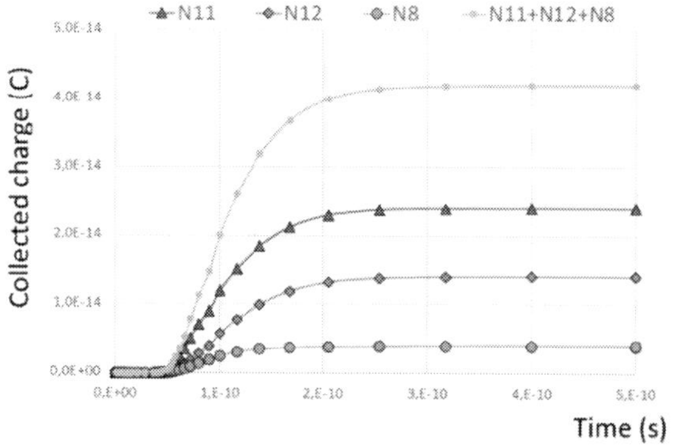

Fig.10. Lithium particle 0,9MeV, collected charge versus time at contacts N8, N11 and N12.

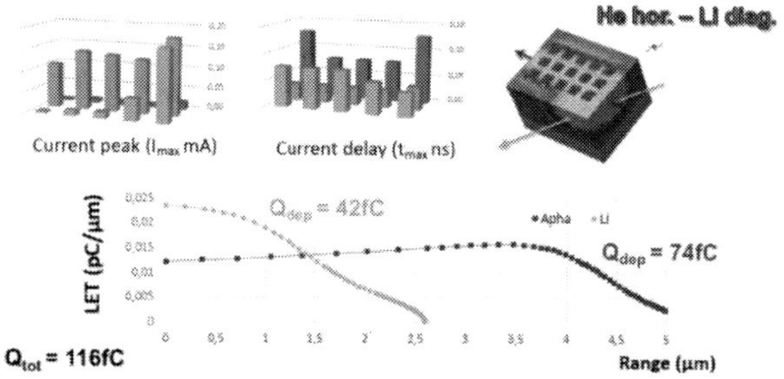

Fig.11. Li and He from neutron-boron 10 interaction, bottom: electron density versus LET; current peak versus pixels (left), current delay versus pixels (middle).

electrode is plotted. The addition of the three collected charges corresponds to 42fC (yellow curve), which is the expected charge for the studied lithium. Then the whole charge is collected at these three electrodes. For this configuration, we do not need information from the other pixels.

C. Tracking of a neutron-boron 10 interaction

The last case study deals with the crossing of an alpha ion and a lithium particle in the 3x5 matrix, 0.8µm under the contacts (Fig.11). The configuration is presented at the right top of the figure. As visible, the alpha (red arrow) and the lithium (orange arrow) enter the matrix at the same point. At the bottom, the electron density corresponding to the crossing of the alpha particle is visible. The current peak (left) and the

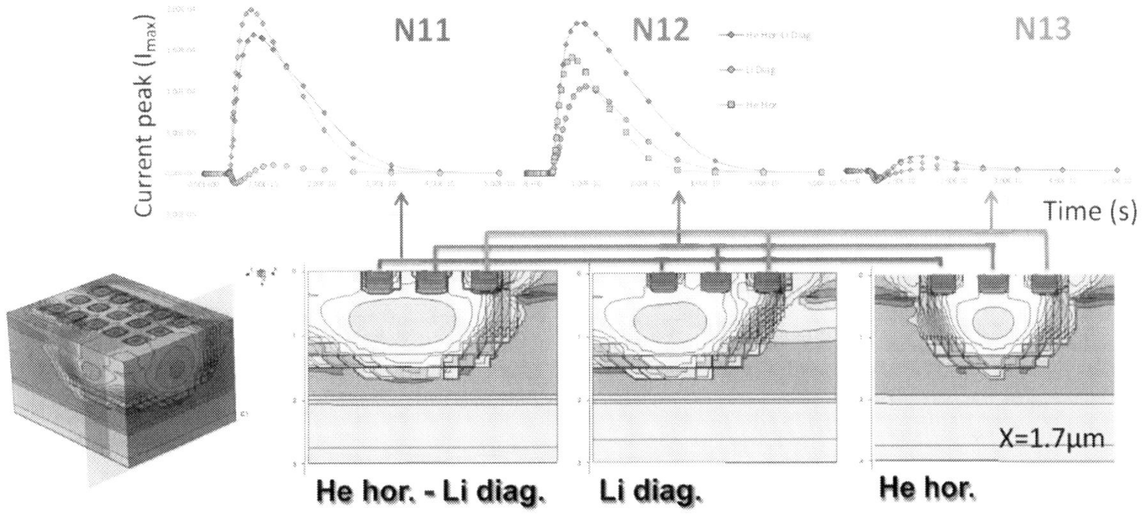

Fig.12. Current versus time at contacts N11, N12, N13 (top), electron density for the three case-studies (bottom), for the maximum time generation (50 ps).

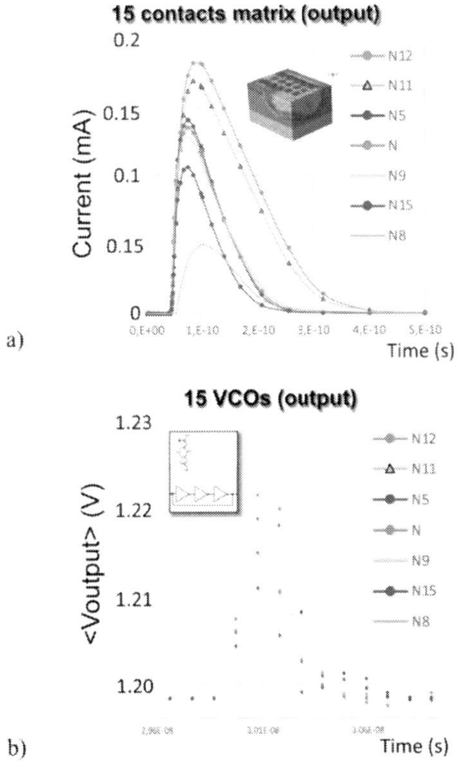

Fig.13. a) Current versus time at the output of the 3x5 matrix, b) Average voltage versus time at the output of the VCOs.

current delay (middle) generated at the 15 pixels are also plotted. The variations induced by simultaneous generation of the alpha and the lithium are visible on both, current peak and current delay. Then the 3x5 matrix is able to detect the crossing of secondary particles generated by the neutron-boron interaction.

Fig. 12 presents a comprehensive analysis of the current profiles within the three specified case studies. Figures displayed at the bottom correspond to the vertical-cut shown on the left. As previously shown in Fig. 11, the alpha and the lithium are introduced in the matrix at the same point. The electron density for the three case-studies is presented at the bottom, for the maximum time generation (50 ps). The current versus time at contacts N11, N12, N13 are shown on the top. N12 is the middle contact. As expected, given the configuration of particles, the highest current observed at N12 is attributed to the n-^{10}B interaction, while the lowest corresponds to the lithium one. A similar behavior is observed at the N13 contact. The currents are lower because the generations are far from this contact. On the contrary, at contact N11, the highest current originates from lithium interaction, not the n-^{10}B interaction as anticipated. The charge sharing, visible on the electron density at 50ps, can explain that phenomenon. The charge generation involved in the n-^{10}B interaction is locally denser because of the simultaneous creation of Li and He, leading to an influence on the charge transport towards the contacts. Comparing to the lithium alone, as it is visible by observing the corresponding currents, the charge reaches more slowly the N11 contact.

D. Detection based on the VCOs

Both helium (He) and lithium (Li) particles have been introduced into the VCO system. The resulting currents observed at the primary output terminals of the matrix are depicted in Figure 13.a. Additionally, the associated average output voltages from the VCOs are shown in Figure 13.b. Notably, the observed variations exhibit consistent behavior across the different contacts examined. This correlation underscores a clear and direct relationship between the two parameters, namely, the current and the voltage.

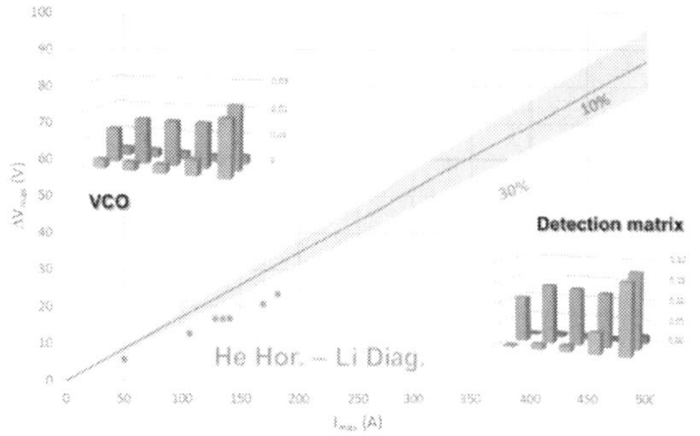

Fig.14. Linearity curve and detected currents concerning the alpha and lithium crossing in a fifteen contacts matrix.

The metrics corresponding to these currents and voltages have been reported in Fig.14. All the currents are correctly reproduced with an accuracy of 30%. Thus, the 3x5 matrix is suitable for the tracking of neutron-boron 10 interaction.

IV. Conclusion

The detection chain presented in this work relies on an indirect particle detection through a voltage-controlled oscillator. The work in [16] demonstrated the feasibility of identifying input signals from specific particles by analyzing the average output voltage of the VCO chain.

This method offers significant advantages, notably the comprehensive acquisition of input signal characteristics, encompassing current waveform, magnitude, and duration, all achieved with a high sensitivity in detectionIn this paper, we have explored the possibility to detect two particles simultaneously and then to track the neutron-boron 10 interaction. Furthermore, we establish that the chosen operating frequency of 4.8 GHz is well-suited for particle tracking purposes. The analysis of the current delay gave us information about the distance between the contact and the particle and the current peak study gave information about the charge density variation. The cumulative collected charge provided information about the deposited energy.

Now, our aim is to optimize the discrimination method in order to improve the tracking and to simulate different interactions to allow the method improvement. Another aspect of our work is to investigate the feasibility of utilizing a greater number of pixels in terms of power consumption and addressing associated technological challenges. This pursuit encompasses a comprehensive approach to elevate the capabilities and robustness of our detection and tracking methodology.

References

[1] C. Fiorini, A. Pullia, E. Gatti, A. Longoni, and W. Buttler, "A monolitic implementation of a switch-current 'wheel' amplifier for multi-channel silicon drift detectors, *IEEE Trans.* Nucl. Sci., vol. 46 n°3 p. 161, 1999.

[2] D. Hussona,, A. Bozierb, S. Higueretc, T.D. Lec, A. Nourreddinea, "AlphaRad, a new integrated CMOS System-on-Chip for high efficiency alpha particles counting" , *Nuclear Instruments and Methods in Physics Research A 569*., pp. 845-852 161, 2006.

[3] I. Perić, C. Kreidl, P. Fischer, "Particle pixel detectors in high-voltage CMOS technology–New achievements", Nuclear Instruments and Methods in Physics Research A, Vol. 650, Issue 1, p. 158-162, 2011.

[4] M. Trocmé, S. Higueret, D. Husson, A. Nourreddine, T.D. Lê, "A new compact device for efficient neutron counting using a CMOS active pixel sensor", Radiation Measurements, pp. 1100-1103, 2008.

[5] D. Passeri, P. Placidi, L. Verducci, P. Ciampolini, G. Matrella, A. Marras, G/M. Bilei, "High-resolution CMOS particle detectors: design and test issues", 331-334 Vol.1, *2003* IEEE Nuclear Science *Symposium Conference Record*.

[6] M. Campbell, V. Havranek, E. HEIjne, T. Holy, J. Idarraga, J. Jakubek, C. Lebel, C. Leray, X. Llopart, J. Novotny, S. Pospisil, L. Tlustos, Z. Vykydal, "Charge collection from proton and alpha particle tracks in silicon pixel detector devices", 2007 IEEE Nuclear Science *Symposium Conference Record*.

[7] M. J. Baudot ; G. Bertolone ; A. Brogna ; G. Claus ; C. Colledani et al., "First test results Of MIMOSA-26, a fast CMOS sensor with integrated zero suppression and digitized output", 2009 IEEE Nuclear Science *Symposium Conference Record*, 2009.

[8] Tomas Baca ; Martin Jilek ; Petr Manek ; Petr Stibinger ; Vladimir Linhart ; Jan Jakubek ; Martin Saska, "Timepix Radiation Detector for Autonomous Radiation Localization and Mapping by Micro Unmanned Vehicles", 2019 IEEE/RSJ International Conference on Intelligent Robots and Systems (IROS), 3-8 Nov. 2019.

[9] Ying Zhang, Christine Hu-Guo, Daniel Husson, The-Duc Le , Stéphane Higueret, Yann Hu, "A High-Sensitivity Low-Power CMOS Sensor for a Future Neutron Personal Dosimeter", *in IEEE Trans. On Nucl. Sc.*, vol.59, no.4, pp. 1465-1471, August 2012.

[10] Rahajandraibe W., Aziza H., Coulié-Castellani K., Micolau, "Device and Method for Detecting Radiation Particles" G. Patent AMU/CNRS/UAPV – Septembre 17, 2015 WO2015136220, https://patentscope.wipo.int/search/en/detail.jsf?docId=WO2015136220

[11] K. Castellani-Coulié, H. Aziza, W. Rahajandraibe, G. Micolau, J.-M. Portal, "Development of a CMOS Oscillator Concept for Particle Detection and Tracking", *IEEE Trans Nucl. Sci.*, Vol. 60 , Issue 4 , pp. 2450 - 2455 , Aug. 2013. DOI: 10.1109/TNS.2013.2254723

[12] K. Coulié-Castellani, W. Rahajandraibe, G. Micolau, H. Aziza, J.-M. Portal, "Improvement of a detection chain based on a VCO concept for microelectronic reliability under natural radiative environment", *IEEE LATW Conference*, 2015.

[13] S. Ben Krit, K. Coulié-Castellani, W. Rahajandraibe, L. Ottaviani, G. Micolau, H. Aziza, J.-M. Portal, "Comparison of a Readout Chain Dedicated to the Signal Conditioning of a Particle Detector and an Innovative Chain Based on a VCO Concept", *RADECS 2015*, Moscou, Russia, September 2015. DOI: 10.1109/RADECS.2015.7365592

[14] K. Coulié, W. Rahajandraibe, H. Aziza, G. Micolau, R. Vauché, "Detection limit of a VCO based detection chain dedicated to particles recognition and tracking", *2005 8th European Conference on Radiation and Its Effects on Components and Systems*, 2005. DOI: 10.1109/RADECS.2005.4365629.

[15] K. Coulié, W. Rahajandraibe, L. Ottaviani, "Detection limit of a VCO based detection chain dedicated to particles recognition and tracking", *2005 8th European Conference on Radiation and Its Effects on Components and Systems*, 2005. DOI: 10.1109/RADECS.2005.4365629.

[16] K. Coulié, W. Rahajandraibe and L. Ottaviani, "On the Improvement of VCO Based Matrix Particle Detector for High Resolution Particles Recognition and Tracking," 2020 20th European Conference on Radiation and Its Effects on Components and Systems (RADECS), Toulouse, France, 2020, pp. 1-6, doi: 10.1109/RADECS50773.2020.9857723.

[17] K. Coulié, W. Rahajandraibe and L. Ottaviani, "Particle Identification and Tracking by the Use of a Pixel-Based Semiconductor Radiation Detector Coupled with Voltage Controlled Oscillators", EPJ Web Conf., 253 (2021) 09007, DOI: https://doi.org/10.1051/epjconf/202125309007

[18] TCAD Sentaurus user's manual (Synopsys).

[19] A. Akkerman, J. Barak, D. Emfietzoglou, "Ion and electron track-structure and its effects in silicon: model and calculation", *Nuclear Instruments and Methods in Physics Research*, pp. 319-336, 2005.

[20] www.srim.org

[21] M. Gedion, F. Wrobel, F. Saigné, R. D. Schrimpf and J. Mekki, "Monte Carlo Simulations to Evaluate the Contribution of Si Bulk, Interconnects, and Packaging to Alpha-Soft Error Rates in Advanced Technologies," in IEEE Transactions on Nuclear Science, vol. 57, no. 6, pp. 3121-3126, Dec. 2010, doi: 10.1109/TNS.2010.2085446.

Radioactive Direction of Arrival Estimation Using Neural Networks Approach

Yossi Salomon[1,2*], Eran Vax[2], Alon Osowizky[2], Yakir Knafo[2], Nadav Ben David[3], and Dan Vilenchick[1]

[1]Ben Gurion University of the Negev, Israel
[2]Electronics and Control Laboratories, Nuclear Research Center-Negev (NRCN), Israel
[3]Israel Atomic Energy Commission (IAEC), Israel
(*) Yosefsal@post.bgu.ac.il

Abstract—**In this paper, we present a comprehensive investigation into improving Direction of Arrival (DOA) estimation for gamma-emitting isotopes using deep neural networks. The direction of arrival estimation is most valuable for Home Land Security (HLS) applications or increased safety in Decontamination and Decommissioning (D&D). Traditional methods, such as beamforming (BF), have limitations in accuracy and sensitivity to noise and background variations. In recent years, data-driven approaches utilizing deep neural networks, including Convolutional Neural Network (CNN) and Gated Recurrent Unit (GRU) models, have shown promise in enhancing DOA estimation. By considering the full energy spectrum and augmenting recorded data, our neural network models outperform traditional BF methods and exhibit greater resilience in diverse background scenarios. The 2-layer CNN model, in particular, achieves up to 40% improvement in estimation accuracy. Our research provides a reliable and data-driven approach for precise DOA estimation with potential applications in nuclear security and safety in D&D.**

Keywords—**Radiation detection, Direction of Arrival (DOA) estimation, convolution neural network (CNN), Reinforcement neural network (RNN), localization of radiation sources.**

I. Introduction

Direction of Arrival (DOA) estimation plays a critical role in radiation detection across various applications, such as preventing homeland security (HLS), medical imaging, and personnel safety in Decontamination and Decommissioning (D&D). The localization of radiation sources is crucial to identifying potential threats and mitigating their impact. In the context of HLS, DOA estimation anticipates actions against concealed radioactive sources and monitors Special Nuclear Materials (SNM) trafficking, e nhancing s ecurity t hrough improved camera direction for prompt threat responses. In post-event scenarios, like malfunctions or radioactive incidents, DOA estimation becomes a means to minimize personnel radiation exposure by expediting hotspot identification, which is crucial for time-sensitive recovery efforts. For instance, in public areas where radioactive materials are detected, knowing the source direction enables authorities to quickly evacuate people and contain the situation. Furthermore, accurate detailed operational information such as DOA estimation, reduces radiation exposure by allowing individuals to move away from the source or take other protective measures.

DOA estimation using sensors array is an important problem that has been extensively researched in fields such as wireless communications, radar, and sonar [1]–[5]. Traditional DOA estimation techniques rely on signal processing algorithms that extract features from the received signals at multiple sensors and then use these features to estimate the direction of the source. Methods such as multiple signal classification (MUSIC) [6], generalized cross-correlation (GCC) [7], beamforming (BF) [8], [9], and Maximum Likelihood (ML) [10] methods have been commonly used for DOA estimation tasks. However, these techniques have limitations in terms of accuracy, especially in noisy or complex environments. One key challenge is the reliance on a calibration matrix that is either calculated analytically, by simulation, or measured. These methods are sensitive to noise and environmental background, as the calibration matrix needs to precisely match the specific environment in which the estimation is performed. In scenarios with varying environmental conditions, the performance of these traditional techniques may deteriorate, affecting accuracy.

Over the last decade, there has been a substantial shift toward using data-driven methods for DOA estimation, particularly in signal and speech recognition [11]. Recent advances in machine learning, especially neural networks, have yielded promising results in DOA estimation. Neural networks can learn to extract features from raw signals and estimate the DOA more accurately and with a higher Signal-to-Noise Ratio (SNR) than traditional techniques. Various neural network architectures have been explored for DOA estimation, including Convolutional Neural Networks (CNNs) [12]–[14], Feed-forward Neural Networks (FNNs) [15], and Recurrent Neural Networks (RNNs) [16].

In this paper, we present a comprehensive investigation into enhancing the accuracy of DOA estimation for radiation isotopes using deep neural networks. Our primary focus is to leverage the power of CNN and Gated Recurrent Unit (GRU) models in capturing spatial features from the full energy spectrum, thereby achieving superior performance compared to traditional BF methods. We train and evaluate our CNN and GRU models to handle various challenging scenarios and background conditions by utilizing recorded data from directional detectors and augmenting it to introduce diversity. This research aims to provide a reliable and data-driven approach for precise DOA estimation under various conditions, in order to enable robust applications in nuclear security, D&D, and beyond. Through a rigorous evaluation of our models and comparisons against existing methods, we aim to demonstrate the efficiency of deep learning techniques in advancing the field of DOA estimation for radiation isotopes.

Our study showcases substantial improvements by utilizing neural networks, with accuracy enhancements ranging from 15% to 20%, and particularly notable improvements of up to 40% observed for ^{60}Co isotopes. This robust enhancement underscores the effectiveness of our proposed approach in overcoming the limitations of traditional algorithms. By exploiting the capabilities of deep learning, we not only confirm the superiority of neural networks over conventional methods but also highlight the potential for precision and accuracy improvements in DOA estimation tasks. These results affirm the promising avenue that data-driven techniques offer in achieving highly accurate and efficient radiation source localization, setting a strong foundation for applications in HLS and D&D.

II. Methodology

In this study, we adopt a data-driven approach, employing deep neural networks to address the challenge of DOA estimation for radiation isotopes. Traditional methods like BF and ML rely on handcrafted features and are sensitive to environmental variations, posing limitations in accuracy and adaptability. To overcome these issues, we harness the power of neural networks, particularly CNN, to automatically learn intricate spatial features from the input spectrum data. Additionally, our directional detector's calibration matrix, obtained from pre-recorded data, provides essential insights into its setup and sensitivity to known signal directions. By synergizing this information with our CNN-based DOA estimation models, we strive to achieve precise and real-time detection and localization of radiation isotopes, unlocking potential applications in nuclear security and personnel safety in D&D.

A. Data-set

The data set used in this study comprises recordings of four distinct radiation isotopes: ^{60}Co, ^{57}Co, ^{137}Cs, and ^{241}Am. Each radiation isotope was recorded for 60 seconds using a directional detector consisting of seven cylindrical detectors made of NaI(Tl) scintillators [17]. The detector arrangement included one central detector positioned higher than the surrounding six detectors, arranged in a non-symmetrical configuration. This design takes advantage of the mutual-shielding effect and offers distinct vantage points, improving the DOA estimation. The NaI(Tl) scintillators used in the detector have $\sim$7% FWHM energy resolution which is well beyond the required for our applications. The high density of the NaI(Tl) (3.67 g/cm^3) gives excellent sensitivity to gamma radiation even for the higher energies, ensuring relatively high count rates for all the radiation isotopes.

During the data acquisition, the directional detector was mounted on a step motor, enabling automated precise rotation of the detector. Each radiation isotope was recorded while being rotated through a complete rotation with a 1° step interval, resulting in 360 labeled samples for each isotope. The data acquisition system utilizes seven Time Over Threshold Multi-Channel Analyzers (MCAs), capturing 50 bins of energy histogram for each of the seven scintillators, these samples of 7 by 50 channels provide insights into the signals' spatial and energy distribution.

As part of our data augmentation strategy, we employed techniques to enhance the variability and quality of the recorded data. Samples with different Signal-to-Noise Ratios (SNRs) and Signal-to-Background Ratios (SNBs) were intentionally captured for each radiation isotope, simulating real-world scenarios and increasing dataset robustness. It is important to note that our testing samples, used for model evaluation, were kept separate from this augmentation process and were recorded with varying SNBs to ensure a comprehensive evaluation of the models' performance. Data augmentation techniques were employed to introduce further variability to each labeled sample by randomly selecting and integrating 20-second segments from the available 60-second spectrum. This approach takes advantage of the stochastic and independent nature of the non-overlapping energy spectrum measurements taken every second. By selecting random segments, we ensure that the neural network models are exposed to diverse combinations of energy patterns within the recorded samples. This augmentation process effectively increases the diversity and coverage of the training dataset, allowing the models to learn a wide range of spatial and temporal characteristics associated with the DOA of different radiation isotopes. Consequently, this data augmentation process significantly expanded the dataset from 500 to 70,000 samples per isotope (as shown in Table I), fostering a rich and diverse training dataset to boost the neural network models.

TABLE I:
Effect of Data Augmentation on Database Size

Isotope:	Recorded Samples:	After Augmentation Training:	Validation:	Testing:
^{60}Co	743	57,000	14,250	470
^{57}Co	473	53,550	13,400	520
^{137}Cs	464	53,450	13,370	460
^{241}Am	489	56,330	14,090	480

Each input sample in our data set is labeled with the true DOA and the corresponding isotope. The input sample is a 20-second spectrum, comprising 7 measurement channels (from each detector) and 50 energy bins. To prepare the input for the neural network, we aggregate the counts for each energy bin across all channels and reorder the matrix into a 7x50 input matrix. This reprocessing step enables the network to effectively learn spatial and energy features across the detectors and energy bins. As for the beam-forming benchmark, we utilize the total count for each detector over the 20-second duration and all 50 energy bins, resulting in a vector of size 7. This vector captures the total count for each detector, facilitating a performance comparison between the neural network models and the traditional BF approach.

B. DOA Estimation by BF Method

BF is a widely used technique in array signal processing for estimating a signal source's DOA. Its primary objective is to enhance the signal coming from a specific direction while suppressing the noise and interference from other directions. In the context of DOA estimation, BF leverages an array of sensors or antennas to spatially process the incoming signals and estimate the angle of arrival of a signal source. By adjusting the weights and summing the weighted signals received at each sensor, the BF output signal power is largest toward the direction of the incoming signal. This process enables to search over the entire DOA space by steering the signal amplification using the weights effectively localizing the source's direction. One can look at the weights as a matched filter with the signals from the correct direction.

In this study, we employ conventional BF as a benchmark method to evaluate the performance of our DOA estimation models based on Deep Neural Networks architectures for DOA. Our Conventional BF relies on a calibration matrix derived from pre-recorded data to determine the fixed weights for the directional detector array. This calibration matrix is crucial in characterizing the directional detector's setup, providing insights into its sensitivity to known signal directions and the resulting patterns. By utilizing this calibration matrix, conventional BF can effectively estimate the DOA of radiation isotopes based on the signals received by the array of sensors. The calibration matrix serves as a reference for the BF method, enabling it to precisely map incoming radiation signals to their corresponding DOA angles.

Pseudo Code Beam-Forming

1. Obtain Calibration Matrix A
2. Measure Total Counts Vector
3. Calculate the sample covariance Matrix:
   $R_x = \sum_{t_i} V_{t_i} \cdot V_{t_i}^T$
4. **for** each $\theta \in [0, 360]$ **calc** the BF output:
   $likelihoodPerAngle(\theta_i) = \frac{\overline{a}(\theta_i)^T \cdot R_x \cdot \overline{a}(\theta_i)}{\overline{a}(\theta_i)^T \cdot \overline{a}(\theta_i)}$
5. $estimateDOA = \text{argmax}_{\theta_i}(likelihoodPerAngle)$

C. CNN-based architecture

DOA estimation can be formulated as a classification task [18], aiming to classify incoming data into one of the 360 possible degrees representing the angle of arrival. An effective solution to this classification task is leveraging CNNs to achieve 1° accuracy, making it suitable for leveraging CNNs as an effective solution. CNNs have demonstrated remarkable success in image classification tasks, primarily due to their ability to capture spatial features and learn hierarchical representations from input data. Although our data is not in the form of traditional images, the correlation features that need to be extracted bear similarities to image processing tasks. By utilizing CNNs, we can exploit their capacity to analyze spatial patterns and dependencies within the input signals. This enables the network to learn and extract the essential features necessary for accurate classification, even in non-image data scenarios.

Our study employed a CNN architecture to estimate the DOA of radiation isotopes consisting of two layers. Our network specifically designed to effectively extract spatial features from the input spectrum data. This architecture is tailored to capture the underlying patterns and correlations present in the data, enabling it to learn and discern the intricate spatial information related to the DOA estimation problem. By leveraging the ability of CNN to analyze and extract meaningful spatial representations, we aim to improve the accuracy and performance of DOA estimation for radiation isotopes. Through our investigation of this CNN architecture, we seek to demonstrate the efficacy of utilizing deep learning approaches in solving the complex and challenging task of DOA estimation.

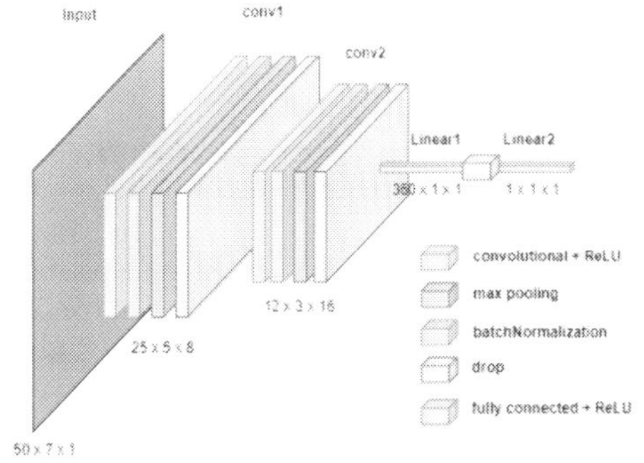

Fig. 1: CNN 2 Layers model Architecture, with 93,000 trainable parameters

For our two-layer CNN architecture (Figure 1), each layer includes a convolution layer with the Rectified Linear Unit (ReLU) activation function, followed by a max pooling layer for downsampling. To promote convergence and mitigate overfitting, batch normalization is applied after each convolution layer. Additionally, a dropout layer [19] with a rate of 0.5 is incorporated to further enhance regularization. These architectural choices facilitate the extraction of informative spatial features from the input spectrum data. Following the convolution layers, two fully connected linear layers with the ReLU activation function process the extracted features. These layers play a crucial role in mapping the learned features to the DOA and isotope classes. During training, the networks utilize an adjusted Mean Squared Error (MSE) loss function, which will be described in detail later. The Adam optimizer [20], a popular gradient-based optimization algorithm, updates the network parameters and facilitates efficient convergence.

D. RNN-based architecture

In addition to the examined CNN architecture, we evaluated the effectiveness of a Gated Recurrent Unit (GRU) network as an independent neural network model for DOA estimation. The GRU network, designed to capture temporal patterns, operates on a distinct input sequence representing the energy bins of 7 detectors over 20 seconds. Comprising a single layer, the GRU network leverages the GRU architecture to effectively model sequential information. It includes two fully connected linear layers, augmented by a dropout layer for regularization. By processing the input data, the GRU network captures temporal dependencies and maps them to the estimated DOA. The inclusion of the GRU network as an alternative model allows for comparison against the CNN architecture, providing insights into the relative performance and advantages of different neural network approaches for DOA estimation.

E. Loss Function

The accurate estimation of DOA is a critical task in various applications. However, this task poses a unique challenge due to the symmetrical and circular nature of the DOA space. Traditional loss functions, such as MSE, are not well-suited to handle this circularity, which can lead to false results and inaccurate estimation metrics.

In a symmetrical scenario, the error between an estimated angle of 10° and the true ground angle of 350° degrees should not be calculated as a 340° error, as MSE would suggest. Instead, the error measurement should reflect the circular nature of the problem, resulting in a 20° error. Failing to account for this circularity can lead to misleading results and an inaccurate assessment of the DOA estimation performance.

To address this limitation, we introduce the Symmetrical Mean Squared Error (S-MSE) loss function and ensure that the error metric provides a more precise and reliable measure of the angular difference between the estimated and true ground angles. The S-MSE takes into consideration the absolute circular difference between the angles, resulting in improved estimation metrics that align with the circular nature of the problem.

The modified MSE is given by:

$$\text{S-MSE} = \frac{1}{N}\sum_{i=1}^{N}(((180^\circ + \theta_{\text{est}_i} - \theta_{\text{true}_i}) \mod 360^\circ) - 180^\circ)^2 \quad (1)$$

Here, est_i represents the estimated DOA angle for the ith sample, and $true_i$ represents the true ground angle for the ith sample. By utilizing the S-MSE loss function during the training phase, we ensure that the error metric accurately captures the circular nature of the DOA space, leading to improved accuracy in DOA estimation.

F. Evaluation Metrics

To evaluate the performance of our DOA estimation models, we conduct two sets of testing that encompassed different scenarios and challenges. These testing sets allowed us to comprehensively assess the capabilities and robustness of our models in various conditions.

The first testing set is comprised of recordings obtained from different laboratory environments, for each of the four radiation isotopes. This set provides us with a comprehensive evaluation of the model's ability to handle variations in energy spectra and accurately estimate DOA angles across different isotopes. We calculated the Mean Angular Error (MAE) for each model on this testing set, which measures the average absolute angular difference between the estimated and true DOA angles. Furthermore, we compared the performance of our models against the benchmark BF method on this set, providing a baseline for performance comparison.

It is important to note that traditional loss functions, such as MSE, are not suitable for our task due to their lack of tolerance to the symmetrical and circular properties of the DOA angles. Therefore, we employ the MAE and use it for two informative metrics: the loss metric and the accuracy metric.

The loss metric is given by:

$$180^\circ - \frac{1}{N}\sum_{i=1}^{N}|((\theta_{\text{est}_i} - \theta_{\text{true}_i}) \mod 360^\circ) - 180^\circ| \quad (2)$$

while the Accuracy metric is given by:

$$\frac{1}{N}\sum_{i=1}^{N}\frac{|((\theta_{\text{est}_i} - \theta_{\text{true}_i}) \mod 360^\circ) - 180^\circ|}{180} \quad (3)$$

This metric represents the relative angular error normalized to the range of 0 to 1, with 1 indicating perfect accuracy and 0 indicating the wrong and opposite estimation. The normalization of the error to the range of 0 to 1 accounts for the symmetry of the problem, where the maximum possible error is 180 degrees.

Through these testing, we can thoroughly evaluate the performance of our DOA estimation models in diverse scenarios. The MAE evaluation metric allows us to quantify the estimation errors and assess the accuracy of the model's predictions, providing a reliable measure of their performance. The results obtained from these evaluations allow us to compare the performance of our models against each other and the benchmark BF method. This comprehensive evaluation provides valuable insights into the strengths and limitations of our models, guiding their practical application in radiation isotope detection and localization.

III. Results

In this section, we present the results of our experiments evaluating the performance of the DOA estimation models based on deep neural networks compared to the traditional DOA estimation method, conventional BF. As described earlier, the evaluation metrics used are the MAE and the Accuracy based one MAE output.

Figure 2 depicts the comparison between different methods for DOA estimation across various isotopes using the MAE as the evaluation metric. We can observe that our convolution neural network model (2-layer CNN) consistently outperformed the BF method and the GRU network for all isotopes, indicating a significant reduction in the estimation error. Our results showcase a remarkable 30-40% improvement in accuracy for ^{60}Co and ^{241}Am isotopes, contributing to an average improvement of 20-25% across all tested isotopes.

Among the isotopes, the best improvement was achieved with ^{60}Co and ^{241}Am using the 2-layer CNN model, demonstrating their effectiveness in capturing complex spatial features and accurately estimating the DOA angles for both high-energy and low-energy spectrum isotopes.

Furthermore, when comparing the performance of the deep neural network models with the GRU network, we found that the CNN architectures were better suited for the DOA estimation task. The CNN model's ability to capture spatial dependencies and hierarchical features from the input spectrum data contributed to their improved accuracy over the GRU model.

To quantify the improvement achieved by our deep neural network models, we examined the percentage improvement in MAE compared to the BF and GRU methods. Figure 2b displays the percentage improvement for each isotope. Notably, we achieved a substantial 40% improvement in the estimation accuracy for both ^{60}Co and ^{241}Am isotopes. These findings demonstrate that our deep neural network models significantly enhance the DOA estimation accuracy for both high-energy and low-energy spectrum isotopes, making them highly effective for radiation detection and localization tasks across a wide range of isotopic sources.

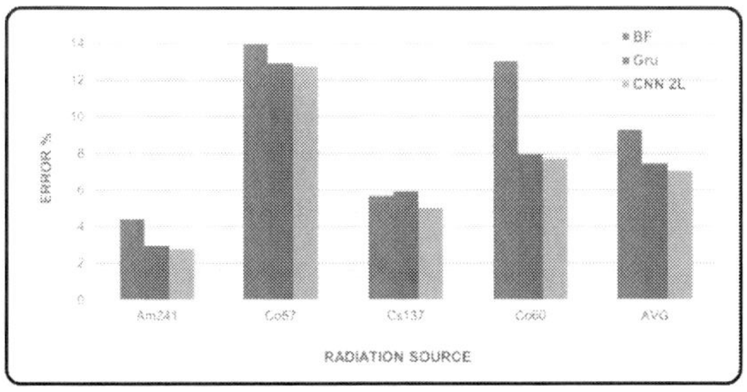

(a)

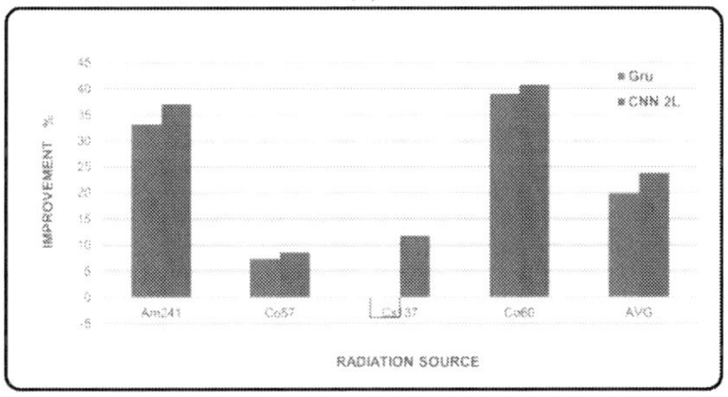

(b)

Fig. 2: Comparison of Different Radiation Isotopes (a) Error: DNN models outperform BF method for all isotopes. (b) Improvement: DNN models achieve significant accuracy enhancement.

In this results section, we evaluated the performance of our DOA estimation models based on deep neural networks, comparing them to the conventional BF method. The 2-layer CNN model consistently outperformed both BF and the GRU model across different isotopes. These results highlight the effectiveness of deep learning approaches for precise DOA estimation in radiation isotopes and pave the way for further discussions in the subsequent section.

IV. Conclusion

In conclusion, our study demonstrates the effectiveness of deep neural networks, particularly CNN, in enhancing the accuracy of DOA estimation for radiation isotopes. By considering the full energy spectrum in our neural network models, we achieved improved accuracy compared to traditional BF methods, which rely solely on total counts. Our investigation showcases substantial improvements by utilizing neural networks, with accuracy enhancements ranging from 15% to 20%, and particularly notable improvements of up to 40% observed for ^{60}Co isotopes. This robust enhancement underscores the effectiveness of our proposed approach in overcoming the limitations of traditional algorithms.

The augmentation of our recorded data significantly increased the dataset size and diversity, providing the models with exposure to a broader range of scenarios. This, in turn, enabled our CNN models to better generalize and handle various challenging conditions and variations, resulting in stable and consistent improvements in accuracy compared to the BF and GRU models across different SBR ratios.

To further improve the performance and generalization of our deep neural network models, we propose several avenues for future work. Further enhancement of data augmentation techniques, incorporating more diverse and realistic simulation data to provide the models with exposure to rare or extreme scenarios that may occur infrequently in real-life measurements. Thereby enhancing the accuracy and robustness of the DoA estimation models by increasing the similarity between the recorded and simulated data sets.

Furthermore, we propose extending our evaluation to encompass a broader range of Signal-to-Background (SNB) scenarios. Different background conditions can significantly influence the performance of DoA estimation methods, and testing NN models across a spectrum of SNB ratios would provide valuable insights into their robustness and reliability.

Additionally, exploring advanced and deeper neural network architectures along with model configurations could further boost performance. The continuous advancement in deep learning offers new possibilities for model innovation and optimization. By leveraging state-of-the-art architectures and training techniques, we can potentially achieve even higher accuracy and better performance, bolstering the practical applicability of our deep neural network-based DOA estimation approach.

Overall, our findings underscore the potential of neural networks in enabling precise and real-time detection and localization of radiation isotopes, with applications in nuclear security and environmental monitoring. As the field of deep learning continues to progress, we anticipate continuous improvements in accuracy and performance, unlocking new possibilities for DOA estimation and beyond.

References

[1] H. Krim and M. Viberg, "Two decades of array signal processing research: the parametric approach," *IEEE signal processing magazine*, vol. 13, no. 4, pp. 67–94, 1996.

[2] H. L. Van Trees, *Optimum array processing: Part IV of detection, estimation, and modulation theory*. John Wiley & Sons, 2002.

[3] A. Nehorai and E. Paldi, "Vector-sensor array processing for electromagnetic source localization," *IEEE transactions on signal processing*, vol. 42, no. 2, pp. 376–398, 1994.

[4] W. Cash, "Parameter estimation in astronomy through application of the likelihood ratio," *Astrophysical Journal, Part 1, vol. 228, Mar. 15, 1979, p. 939-947.*, vol. 228, pp. 939–947, 1979.

[5] J. Shi, F. Wen, and T. Liu, "Nested mimo radar: coarrays, tensor modeling, and angle estimation," *IEEE Transactions on Aerospace and Electronic Systems*, vol. 57, no. 1, pp. 573–585, 2020.

[6] R. Schmidt, "Multiple emitter location and signal parameter estimation," *IEEE transactions on antennas and propagation*, vol. 34, no. 3, pp. 276–280, 1986.

[7] C. Knapp and G. Carter, "The generalized correlation method for estimation of time delay," *IEEE transactions on acoustics, speech, and signal processing*, vol. 24, no. 4, pp. 320–327, 1976.

[8] B. D. Van Veen and K. M. Buckley, "Beamforming: A versatile approach to spatial filtering," *IEEE assp magazine*, vol. 5, no. 2, pp. 4–24, 1988.

[9] J. Litva and T. K. Lo, *Digital beamforming in wireless communications*. Artech House, Inc., 1996.

[10] P. Stoica and K. C. Sharman, "Maximum likelihood methods for direction-of-arrival estimation," *IEEE Transactions on Acoustics, Speech, and Signal Processing*, vol. 38, no. 7, pp. 1132–1143, 1990.

[11] A. Barthelme and W. Utschick, "A machine learning approach to doa estimation and model order selection for antenna arrays with subarray sampling," *IEEE Transactions on Signal Processing*, vol. 69, pp. 3075–3087, 2021.

[12] S. Chakrabarty and E. A. Habets, "Broadband doa estimation using convolutional neural networks trained with noise signals," in *2017 IEEE Workshop on Applications of Signal Processing to Audio and Acoustics (WASPAA)*. IEEE, 2017, pp. 136–140.

[13] ——, "Multi-speaker doa estimation using deep convolutional networks trained with noise signals," *IEEE Journal of Selected Topics in Signal Processing*, vol. 13, no. 1, pp. 8–21, 2019.

[14] A. M. Elbir, "Deepmusic: Multiple signal classification via deep learning," *IEEE Sensors Letters*, vol. 4, no. 4, pp. 1–4, 2020.

[15] E. Ozanich, P. Gerstoft, and H. Niu, "A feedforward neural network for direction-of-arrival estimation," *The journal of the acoustical society of America*, vol. 147, no. 3, pp. 2035–2048, 2020.

[16] J. Chung, C. Gulcehre, K. Cho, and Y. Bengio, "Empirical evaluation of gated recurrent neural networks on sequence modeling," *arXiv preprint arXiv:1412.3555*, 2014.

[17] Y. Ifergan, E. Vax, Y. Yehuda-zada, B. Sarussi, D. Smadja, S. Dadon, Y. Levin, T. Mazor, Y. Kadmon, Y. Knafo, *et al.*, "Improving the radiation detection level and operational data of pre and post event monitoring systems," in *International Conference on Nuclear Security: Commitments and Actions. Summary of an International Conference. Companion CD-ROM*, 2017.

[18] T. Hastie, R. Tibshirani, J. H. Friedman, and J. H. Friedman, *The elements of statistical learning: data mining, inference, and prediction*. Springer, 2009, vol. 2.

[19] N. Srivastava, G. Hinton, A. Krizhevsky, I. Sutskever, and R. Salakhutdinov, "Dropout: a simple way to prevent neural networks from overfitting," *The journal of machine learning research*, vol. 15, no. 1, pp. 1929–1958, 2014.

[20] D. P. Kingma and J. Ba, "Adam: A method for stochastic optimization," *arXiv preprint arXiv:1412.6980*, 2014.

Resource Constrained Electronics and Signal Processing for UAV Radiation Sensors

Andrea Chierici[1,*], Riccardo Ciolini[1], Andrea Malizia[2], Francesco d'Errico[1]
[1]Department of Industrial and Civil Engineering, University of Pisa, Italy
[2]Department of Biomedicine and Prevention, Faculty of Medicine and Surgery, University of Rome Tor Vergata, Italy
*andrea.chierici@ing.unipi.it

***Abstract*—This paper details the development of an inexpensive, power-efficient, and lightweight radiation detection system specifically designed for deployment on Unmanned Aerial Vehicles (UAVs). The system addresses the need for swift, remote radiation detection capabilities during radiological and nuclear emergencies, a demand emphasized by events such as the Fukushima accident in 2011. The design is based on readily accessible components, which lessens reliance on costly, specialized hardware like Field-Programmable Gate Arrays (FPGAs) and Application-Specific Integrated Circuits (ASICs). The proposed system comprises two types of detectors: inorganic scintillators, specifically CsI:tl and GAGG:ce, paired with Silicon Photomultipliers (SiPMs) for gamma-ray detection, and solid-state detectors provided with a ^{6}Li converter for thermal neutron detection. These detectors operate independently, increasing redundancy and system reliability. The system's performance, evaluated through extensive testing, has shown significant improvements in linearity and noise characteristics over previous iterations. Beyond its primary function in emergency response, the system could find applicability in diverse fields such as environmental radiation monitoring, geological surveys, industrial inspections, and scientific research.**

***Keywords* —UAVs, Gamma and Neutron Detectors, Mobile Radiation Detection System, Emergency Response.**

I. Introduction

Following high-profile events such as the Fukushima Daiichi nuclear power plant accident in 2011 [1, 2, 3], radiation monitoring has grown in importance as a component of worldwide risk mitigation efforts; additionally, such incidents have highlighted the importance of fast, accurate, and accessible detection systems to quickly identify and locate radioactive contaminants during and after emergencies. As such, easily deployable radiation monitoring systems may result essential not only for immediate disaster response, but they may also play an important role during standard operations or over the course of recovery and decommissioning phases following the shutdown of a nuclear power plant [4, 5]. Indeed, radiation contamination data is an essential information for decision-making, especially regarding the safe reoccupation of interested areas and the extent of environmental remediation which could be required; moreover, in the context of ongoing radioactive monitoring, these systems may provide valuable data that help to understand long-term impacts on the environment and form potential future prevention strategies [6].

In response to these needs, robotics advancements and the downsizing of sensor technologies have increased the employment of Unmanned Aerial Vehicles (UAVs) as a critical instrument in dealing with Chemical, Biological, Radiological, and Nuclear (CBRN) threats [7, 8, 9]; in fact, these units provide an efficient way of investigating and assessing potentially hazardous environments, while at the same time greatly reducing the risk of personnel exposure to harmful agents than traditional methods.

By focusing on radiological and nuclear (RN) threats, the role of UAVs in countering emergencies has become increasingly significant [10, 11]; indeed, they easily allow to limit the exposure of first responders and military personnel to dangerous areas which may otherwise be impossible to reach. Moreover, the capability to swiftly approach the locations of interest enhances their detection efficiency while minimizing potential signal attenuation caused by shielding materials. Thanks to great flexibility and adaptability, UAVs equipped with specific detection systems are currently employed in several sectors, such as homeland security, nuclear safety and security, environmental recovery, and military operations [12, 13].

However, the development of UAV-based remote sensing systems faces challenges related to cost, size, power consumption, and system complexity [14]. In addition to these difficulties, many existing radiation detection systems are based on expensive hardware such as Field-Programmable Gate Arrays (FPGA) and Application-Specific Integrated Circuits (ASICs), posing barriers to widespread diffusion [15].

In response to these challenges, we propose an inexpensive and lightweight radiation detection system that can be mounted on UAVs; the architecture of the unit is based on commonly available components, an aspect which reduces the requirements of expensive and very specialized hardware and thus making the system easier to deploy at large scale. In addition, in the event of contamination, the system can be easily discarded thanks to its inexpensive nature, further reducing the risks associated with decontamination processes. Finally, our system aims to not only support first responders during emergencies but also contribute to the advancements in easily accessible radiation monitoring technologies.

II. Analog Front-End and Multi-Channel Analyzer

Designing an efficient and mobile radiation detection system when faced with limited computational and memory resources requires a meticulous optimization of several aspects. These include a strategic selection and integration of hardware components, the application of efficient signal acquisition and processing techniques, and fine-tuning of the overall design methodology. This section details our methodology for tackling these challenges, ultimately leading to a radiation detection system designed for integration with low take-off weight category UAVs.

Our design includes two types of detectors, specifically inorganic scintillators attached to Silicon Photomultipliers (SiPM) for gamma-ray detection, and off-the-shelf and solid-state detectors for thermal neutron detection; those units are acquired by means of two completely independent channels to improve the reliability and robustness of our system.

Signals are digitized by a Multi-Channel Analyzer (MCA) board built around a Microcontroller Unit (MCU), specifically, the AVR128DB48 model from Microchip. While this MCU comes with inherent computational limitations, such as an 8-bit architecture and a 32 MHz clock speed, it offers a range of integrated peripherals that make up for these aspects by facilitating robust and effective data acquisition and processing.

Once collected, data is transmitted to a ground station through a 2.4 GHz radio module, ensuring reliable medium to long-range communication. Additionally, the inclusion of a GPS unit in the system facilitates the geo-localization of the detectors.

A. Radiation Detectors

The inorganic scintillators utilized in our system, namely CsI:tl and GAGG.ce, were procured from Epic Crystal company (Fig. 1). Their selection was determined by the advantageous properties they offered for our specific application requirements. CsI:tl is recognized for its high light yield of 60,000 photons per MeV, peaking at 550 nm, and it offers a good energy resolution, ranging from 5.5% to 7% at 662 keV. Its high density of 4.5 g/cm³ and effective atomic number (Z_{eff}) approximating 50 contribute to robust stopping power in a small volume. Although slightly hygroscopic, CsI:tl is well-suited for UAV use due to its compact size, low weight, and ruggedness [16]. The CsI:tl crystal was wrapped in BC-642 PTFE Reflector Tape from Saint-Gobain for optimal light reflection.

Similarly, GAGG:ce presents itself as another promising candidate for UAV applications. It features a high light yield of 50,000 photons per MeV, peaking at 520 nm, with an average energy resolution of 6% to 8.5% at 662 keV. GAGG:ce boasts an even higher density of 6.6 g/cm³, leading to very high stopping power with a Z_{eff} close to 50. Benefiting from a relatively rapid decay time of 88 ns, it offers an advantage over CsI:tl, which exhibits a slower decay time of approximately 1 µs [17]. Although GAGG:ce comes at a higher cost, its superior stopping power and faster decay time make it very valuable in high counting rate situations. For optimal light reflection, the GAGG:ce crystal was coated with 0.2 mm $BaSO_4$ and wrapped in 0.03 mm aluminum foil.

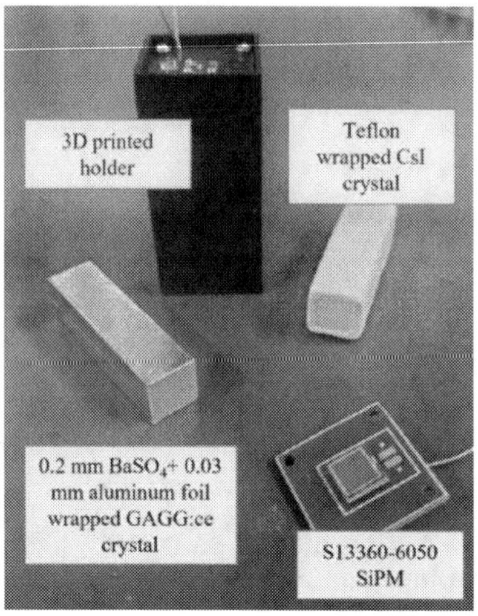

Fig. 1. Inorganic scintillators, SiPM, and 3D-printed holders used to assemble the scintillation detectors.

Both CsI:tl and GAGG:ce scintillators were individually optically coupled to their own 6 mm × 6 mm S13360-6050 SiPM from Hamamatsu, using BC-630 silicon optical grease. This SiPM exhibits good photon detection efficiency at both emission wavelengths, making it an excellent match for the scintillators. The crystals' size was chosen based on a trade-off between cost, weight, and stopping power, resulting in dimensions of 40 mm × 40 mm × 10 mm.

While gamma-ray detection is crucial for identifying many radioactive and nuclear materials—especially considering that gamma-ray sources are widespread—the ability to detect thermal neutrons can be particularly useful in homeland security applications and in certain emergencies involving nuclear accidents or uncontrolled nuclear chain reactions [18].

Thus, incorporating thermal neutron detection capabilities in our system allows for a more comprehensive and nuanced understanding of a radiation environment, enhancing its utility in various emergency response scenarios. For this purpose, we selected thermal neutron detectors sourced by Radiation Detection Technologies company, which employ a Microstructured Semiconductor Neutron Detector (MSND) technology with a ^{6}Li converter and 2 cm × 2 cm active area. In terms of performance, they offer up to 30% thermal neutron detection efficiency and can respond to thermal neutrons with energies up to 1 eV. Being radiation-hardened, these detectors can withstand high radiation levels without performance degradation. Importantly, they are gamma insensitive, allowing for specific neutron detection while reducing background noise from gamma rays. Thus, the incorporation of these detectors greatly enhances our system's ability to detect nuclear threats in complex and challenging radiation environments.

B. Analog Front-end for SiPM

The design and realization of the analog front-end (Fig. 2) for the SiPM were carried out in our laboratory. This electronics architecture incorporates a charge preamplifier to process the

SiPM current, tailored with an approximate decay constant of 100 μs. Following the preamplifier, a single inverting stage serves as a CR-RC shaper provided with pole-zero cancellation and with a 5 μs time constant. Such a time constant assures the full integration of slower scintillating pulses from the CsI:tl detector, optimizing energy resolution. However, the system may accommodate a lower time constant when employing the faster GAGG:ce detector in high counting rate scenarios.

To avoid distortion in the signal amplitude and high counting rate complications, we used DC coupling both at the input stage between the SiPM and the preamplifier and at the output of the shaper to the MCA.

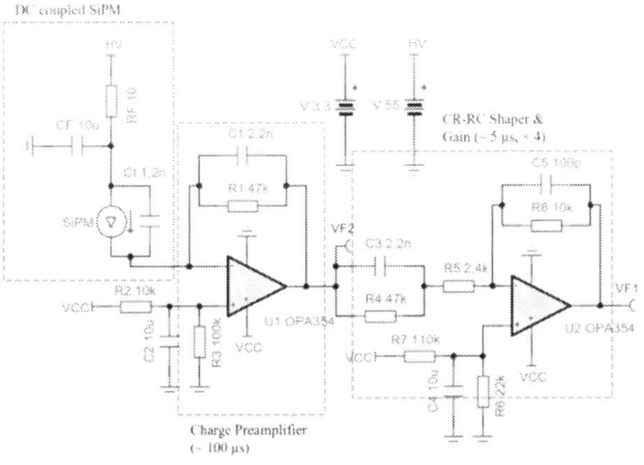

Fig. 2. Architecture of the analog front-end for SiPMs.

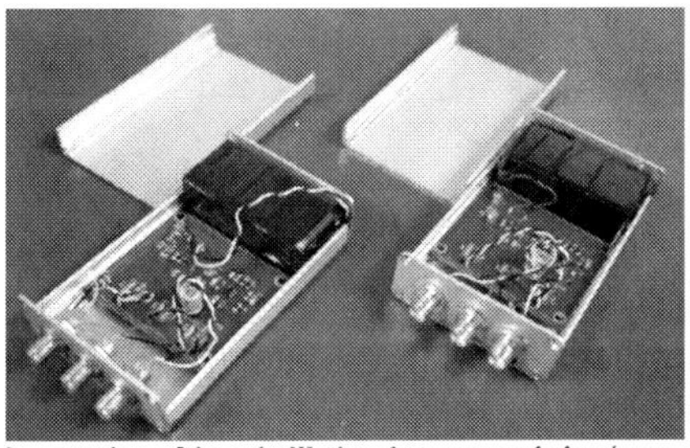

Fig. 3. Implementation of the scintillation detectors and aluminum enclosures.

Moreover, the gain across both stages was distributed to limit an excessive reduction in bandwidth and slew rate, with the total gain empirically determined based on prior testing. This strategy balances the need for signal amplification without compromising system stability or performance.

As operational amplifiers (op-amps), we selected the OPA354 from Texas Instruments. These single-supply op amps provide very high speed, with a gain bandwidth product of 250 MHz and a slew rate of 150 V/μs, while delivering an output current of up to 100 mA, all operating on a single 3.3 V power supply; therefore, they are well suited for the role of preamplifier, shaper, buffer, and other functions. Once provided with the scintillation detectors, the circuits were housed in aluminum enclosures (Fig. 3) to further mitigate the effects of environmental light and electromagnetic interference.

C. *Multi-channel analyzer board*

At the heart of our design is a custom-made MCA board. This board houses a variety of components, including two Adjustable Gain Amplifiers (AGAs), two Peak Detectors (PKDs), an MCU evaluation board, a temperature sensor, a temperature-compensated bias generator for the SiPMs, a radio communication module, a GPS module, and an SD card (Fig. 4).

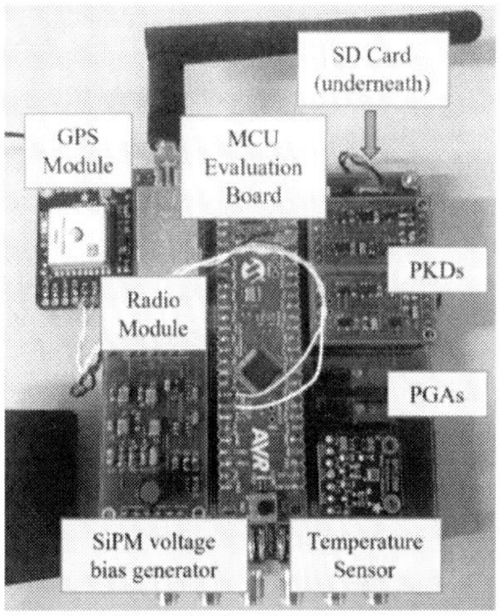

Fig. 4. Implementation of the MCU-based MCA board.

The AGAs are op-amp-based, using the same op-amp type as in the front end, with their gain adjustable using multi-turn trimmers. Following the amplification stage, signals are directed to the PKDs. These PKDs, also op amp-based, utilize the same high slew rate and current output to drive the load of the hold capacitor without inducing instability issues (Fig. 5).

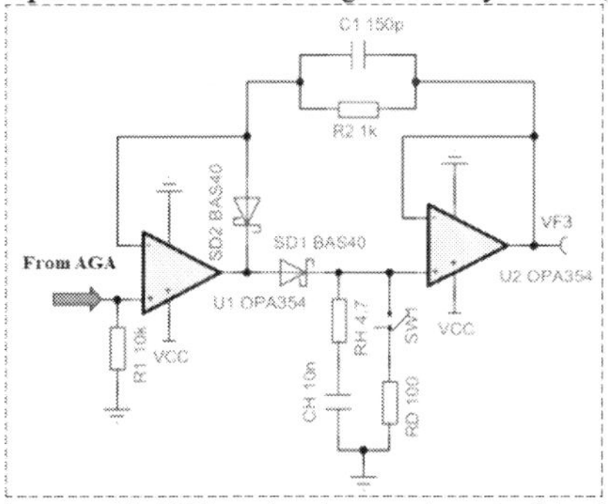

Fig. 5 Architecture of the PKD employed to store the peak of the signal pulses during digital conversion.

As will be demonstrated later, PKDs respond linearly and without distortion to input signals ranging from 60 mV (baseline) to 2.5 V (the full-scale range of the MCU internal 12-bit ADC), even with rise times down to approximately 1 μs.

Each PKD is equipped with two Single-Pole Single-Throw (SPST) switches, controlled by the MCU: when an input signal surpasses a predetermined threshold, one of the MCU internal analog comparators initiates a process to store and convert the peak value. During this conversion, a switch at the PKD input is opened to prevent subsequent pulses from interfering with the conversion process, ensuring a non-paralyzable behavior. Upon completion of the conversion, the hold capacitor is reset. Each conversion takes about 25 μs (the same duration as an incoming pulse) allowing the system to easily manage up to several thousand pulses per second. Alternatively, the MCU internal 16-bit, 32 MHz timer/counters peripherals can measure the pulse width of an incoming signal. This allows us to create a histogram of pulse widths, which can be linked to the signal's amplitude using a non-linear relationship typical of Time over Threshold (ToT) conversion.

The MCU's internal Real-Time Counter (RTC) schedules several critical functions: it orchestrates the transmission of the 1024-channel spectra through the radio module (ML01DP5, Ebyte), activates sampling of the temperature sensor (BME280, Bosch) and associated bias compensation process, and it updates data from the GPS module (MTK3339, MediaTek).

Additionally, our system incorporates a bias generator, which is designed around a MAX5026 boost converter (Analog Devices), followed by a voltage doubling circuit. The output voltage, ranging from 20 to 70 V, can be controlled by the MCU through an MCP4725 Digital-to-Analog Converter (DAC) (Microchip Technology), in conjunction with a resistor network.

At the ground station, users can monitor and download data in real time through a dedicated user interface. This interface provides access to all relevant information and allows modification of various parameters, including the operating voltage of the SiPMs.

Furthermore, once the GPS module achieves a satellite fix, the system records radiation and geolocation data in GeoJSON format, a geospatial data format based on JavaScript Object Notation (JSON), which encodes various geographic data structures, including coordinates, shapes, and other features. GeoJSON is largely compatible with different GIS (Geographical Information System) software like QGIS, ArcGIS, or simple web-based platforms such as Google Maps, Mapbox, and OpenStreetMap, and it easily allows users to visualize and analyze the geospatial radiation data.

Finally, the MCA board is compact and lightweight with dimensions of 10 cm × 10 cm × 5 cm and a weight of less than 200 grams, and it is optimally suited for portable applications. Moreover, the average current draw remains below 50 mA, an aspect that further emphasizes the efficiency of our system and its suitability for long-term, battery-powered operations.

III. Results and Discussion

In the following subsections, the results obtained from the tests conducted on our radiation detection system are described. The initial evaluation focused on the electronic subsystems to get insights into the system's noise performance and linearity.

Subsequently, we exposed the scintillation detectors to gamma-ray sources to assess the quality of the detectors' response and calculate the energy calibration factors. These evaluations, comprising both electronic characterization and gamma radioactive source exposure, took place at the Laboratory of Nuclear Measurements, University of Pisa, Italy. Further evaluation was conducted at the Polytechnic University of Milan, Italy, where the response of thermal neutron detectors was investigated. Finally, an operational field test was carried out at the French Alternative Energies and Atomic Energy Commission (CEA) in Saclay, France. For this, the detection system was installed onto a UAV and tested under real-world conditions.

A. Noise and Linearity Analysis

To thoroughly assess the linearity of the response along with the noise levels imparted by the acquisition chain, a trapezoidal waveform characterized by a 10 ms period, a 20 ns rise time, and a 5 ms fall time was generated using a LeCroy Wavesurfer 3104z oscilloscope. This waveform served to inject a variable charge - ranging from 50 pC to 1.85 nC - into the preamplifier's input via a 1 nF coupling capacitor (Fig. 6).

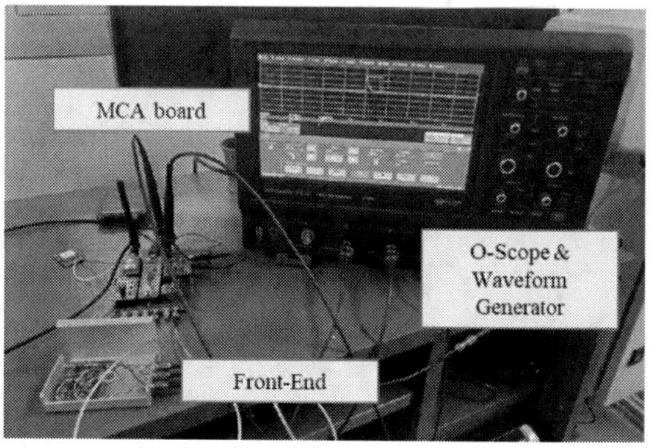

Fig. 6. Setup used to investigate the potential non-linearity and noise level introduced by the electronic acquisition.

The output pulses originating from the front end were subsequently digitized using the PKD and the ADC integrated within the MCU (as depicted in Fig.7(a)). AGAs were set to unitary gain.

In examining the system's response, we found a notable linearity relative to the injected charge, with a percentage non-linearity measuring less than 0.6% (refer to Figs. 7(c), (d)). Additionally, the noise attributed to the entire acquisition chain - expressed as charge resolution- was established to be less than 1% and approached 0.1% across the full dynamic range (see Fig. 7(b)). Given the typical energy resolution observed in scintillator materials, the noise levels reported here are notably negligible.

In comparison to our prior system iterations [19, 20], the current setup showcases an improvement of an order of magnitude in both linearity and noise performance. This substantial advancement is the result of several adjustments to the electronic subsystem, including meticulous component selection and thoughtful circuit re-design strategies.

B. Scintillation Detectors response

Upon verifying the linearity of the analog-to-digital conversion process, our investigation then focused on assessing the linear response of the system to various gamma-ray energies. Simultaneously, we evaluated the spectral features of the gamma spectra, alongside parameters such as energy resolution and non-linearity, to thoroughly evaluate the system's performance.

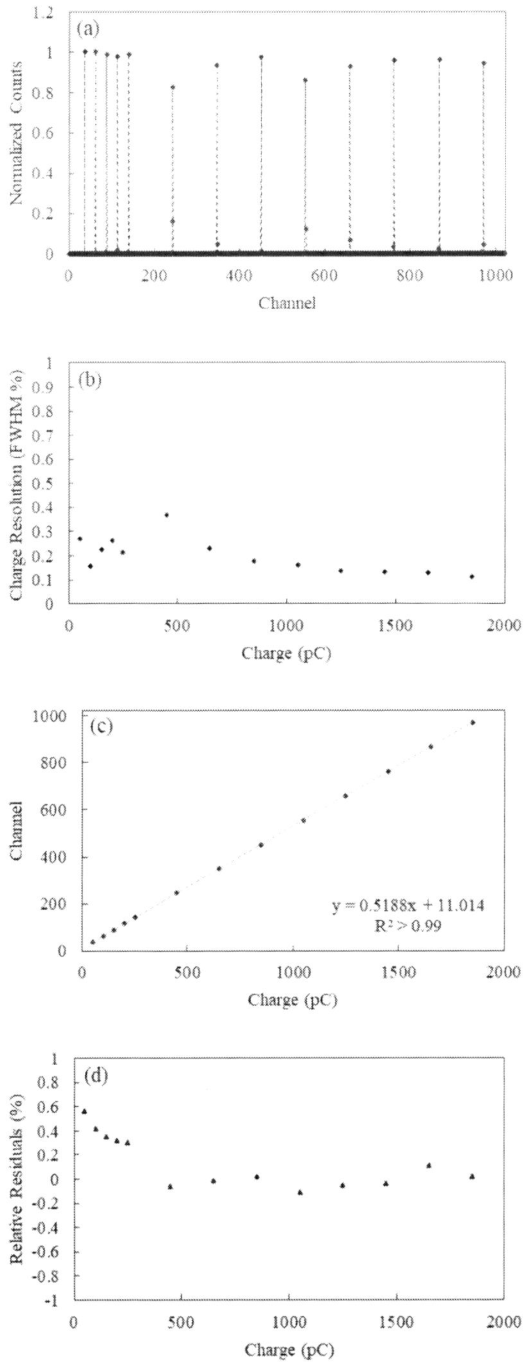

Fig. 7. Results of the electronic characterization of the analog front end and the analog-to-digital conversion process. Histogram of the amplitudes of the pulses generated by the injection of a variable charge in the preamplifier input (a). Noise level expressed as charge resolution (b). Linear fit of the collected data (c). Relative residuals of the linear fit (d).

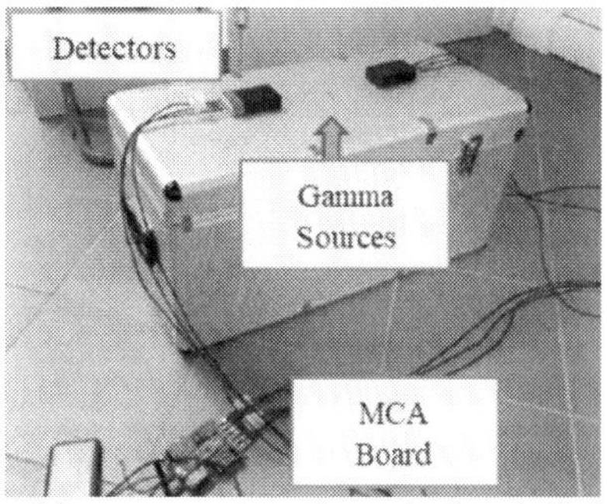

Fig. 8. Setup of the measurements to characterize the scintillation detectors' response.

The obtained data (Table I) returned a clear representation of the detectors' response and performance, as reflected in the gamma spectra (Fig. 9) and the linear response to different gamma-ray energies (Fig. 10). A noteworthy observation from these tests was that the ratio of gamma photopeak to Compton scattering was higher for GAGG due to its superior stopping power.

The measurements indicated that the detectors' responses' energy resolution and non-linearity values fall within a suitable range considering the system's intended use. Specifically, this system was engineered for deployment on UAVs and is designed to cater to emergency response scenarios. Instead, the primary requirement is that it can operate effectively under real-world conditions, which is a benchmark it successfully meets according to our testing outcomes.

TABLE I
SCINTILLATION DETECTORS RESPONSES

Energy (keV)	Measured Channel		Energy Resolution (%)		Non-Linearity (%)	
	CsI	GAGG	CsI	GAGG	CsI	GAGG
59	66	57	16.0	16.0	-1.3	0.0
662	466	426	7.0	8.7	-0.2	-0.1
1117	803	740	5.0	5.4	-0.2	-0.3
1332	904	832	4.7	4.9	0.2	0.2

Following the initial characterization of the scintillation detectors, we proceeded to estimate the minimum distance required to detect a gamma-ray radioactive source in a potential real-world scenario. This experiment was performed at the NBC Defence School in Rieti, Italy, which provided a concealed 500 MBq ^{137}Cs source situated inside a hangar. The system, equipped with two CsI:tl crystals, was mounted on a DJI Matrice 200 UAV.

From earlier testing, the detectors were known to exhibit a sensitivity of approximately 7000 cpm per μSv/h for a ^{137}Cs source. For our experiment, the detection limit was calculated as the average value of the background count rate plus three times its standard deviation. Using a counting time of 1 second, the count rate yielded a detection limit of roughly 0.28 μSv/h. This sensitivity allowed the system to detect the presence of the source from approximately 11 meters.

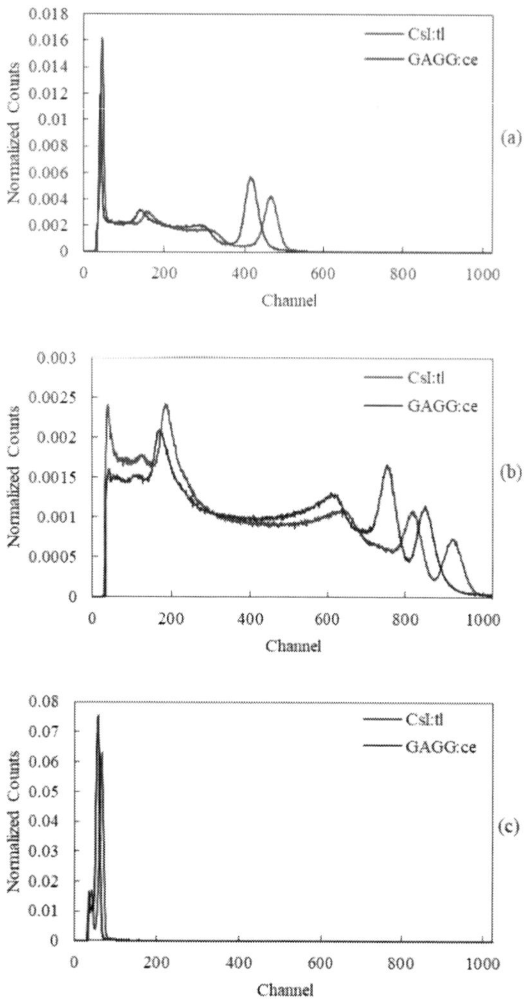

Fig. 9. Gamma spectra obtained from CsI:tl and GAGG:ce crystals when exposed to ^{137}Cs (a), ^{60}Co (b), and ^{241}Am (c).

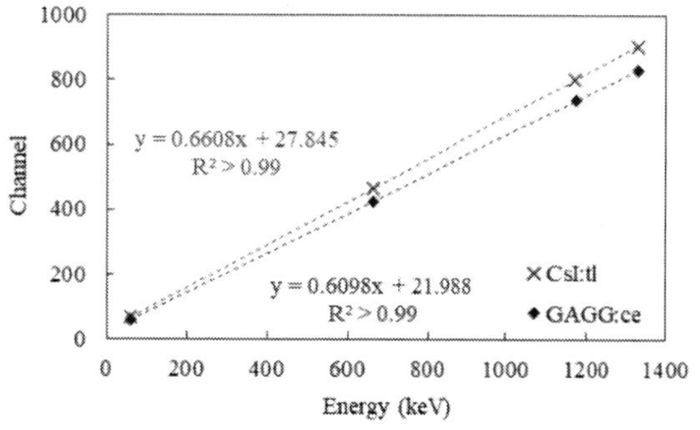

Fig. 10. Linear fit of the ADC channel versus energy for CsI:tl and GAGG:ce scintillators.

At first glance, a radioactive source with an activity of 500 MBq may seem substantial. However, the International Atomic Energy Agency (IAEA) and the U.S. Nuclear Regulatory Commission have recommended that the activity level for many radionuclides considered to be hazardous materials is within the range 37 - 370 GBq. While these activity levels may appear significant, it's important to note that radionuclides such as ^{137}Cs and ^{60}Co have specific activities on the order of TBq per gram. This means that only milligrams of these materials are needed to pose a threat [21].

C. Thermal Neutron Detectors response

To further evaluate our system's capabilities, we conducted a series of tests focusing on the response of our detectors to thermal neutrons at the Polytechnic University of Milan. For this purpose, the detectors were exposed to a polyethylene moderated Am-Be neutron source (2.14×10^6 s^{-1}) yielding a thermal neutron fluence rate (with energies up to 5 eV) equal to 437 ± 14 cm^{-2} s^{-1} (Fig. 11).

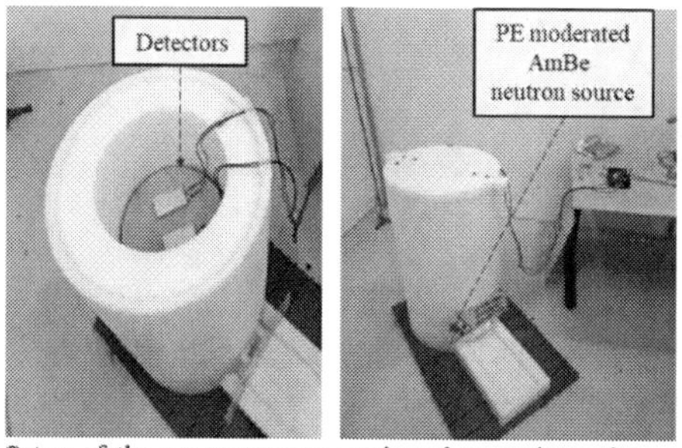

Fig. 11. Setup of the measurements conducted to evaluate the response of thermal neutron detectors.

The average detection efficiencies for the two detectors were measured to be 0.132 ± 0.012 and 0.131 ± 0.011 respectively. These results align closely with the data provided by the manufacturer, thus further confirming our detectors' performance and suitability as UAV radiation sensors.

D. Field tests

The culmination of our tests occurred during an in-field evaluation, which was conducted as part of a simulated radiological and nuclear scenario under the aegis of the Horizon 2020 INCLUDING Project at CEA, Saclay, France. This exercise aimed to simulate the identification, recovery, and decontamination of orphan sources of ^{137}Cs and ^{99m}Tc. Within this scenario, the ^{137}Cs was provided as a sealed source with very low activity, whereas ^{99m}Tc was presented in a liquid form (Fig. 12).

Despite the challenging nature of this real-world test due to the low activities of the employed sources, our UAV-mounted detection system successfully identified the presence of both radioactive sources in a matter of minutes. After only a three-minute hovering phase at approximately 4 meters from the sources, the user interface displayed a distinct photopeak for each radioactive source, demonstrating the system's ability to identify different types of radioactive materials quickly and accurately in a real-world operational environment (Fig. 13).

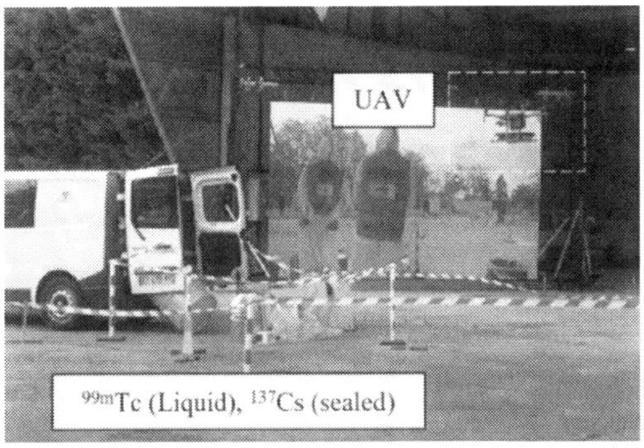

Figure 12. Use of the detection system mounted on a UAV during a simulation of an identification, recovery, and decontamination RN scenario held at CEA. Saclay, France.

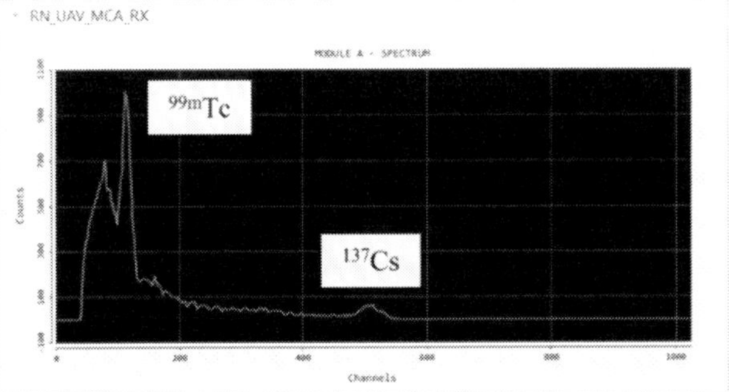

Fig. 13. User interface showing the photopeaks associated with ^{137}Cs and ^{99m}Tc after 3 minutes of hovering.

IV. Conclusions

This work focused on the design and implementation of a lightweight and resource-constrained radiation detection system tailored specifically for small-category UAVs. The development and deployment of effective radiation detection systems for UAVs are becoming increasingly important in managing radiological and nuclear emergencies; additionally, such systems are essential for immediate response to catastrophic events and play a crucial role in subsequent recovery and long-term monitoring. The proposed system is different from conventional solutions, utilizing off-the-shelf components and avoiding costly, specialized hardware like FPGAs or ASICs. This design choice enhances accessibility, facilitates deployment on a large scale, and allows for easy replacement if contamination occurs.

A dual-detector configuration has been integrated into the design to extend the radiation detection capabilities of the unit. It employs inorganic scintillators (CsI:tl and GAGG:ce) with SiPMs for gamma-ray detection, and solid-state detectors equipped with a ^{6}Li converter for thermal neutron detection. The separate acquisition channels introduce redundancy, enhancing system robustness and reliability. The core of the acquisition system is an AVR128DB48 MCU, which, while limited in terms of computational power, also offers multiple integrated peripherals which are well-suited for data acquisition and basic processing. Finally, the collected data are transmitted automatically to a ground station via a 2.4 GHz radio module, while a GPS unit tracks detectors' geolocation allowing the reconstruction of radiation contamination data.

Through extensive testing, we have confirmed the system's robustness and efficiency: performance parameters, such as linearity, noise, and detection of gamma and thermal neutrons, meet the acceptable ranges for their intended purpose as an easily deployable or disposable tool for first responders. Field tests have further validated the system's capability to detect sources from considerable distances, demonstrating its readiness for various radiation detection applications.

Beyond emergency response, the system could find use in environmental radiation monitoring, geological surveys, industrial inspections, and scientific research. By advancing drone-mounted radiation detection, we can enhance safety, improve response times to radiological incidents, and provide a practical, cost-effective solution that will influence the future of radiation detection technology.

References

[1] M. I. Ahmad, M. H. Ab. Rahim, R. Nordin, F. Mohamed, A. Abu-Samah, and N. F. Abdullah, "Ionizing Radiation Monitoring Technology at the Verge of Internet of Things," *Sensors*, vol. 21, no. 22, Art. no. 7629, 2021. [Online]. Available: https://doi.org/10.3390/s21227629.

[2] K. A. Pradeep Kumar, G. A. Shanmugha Sundaram, B. K. Sharma, S. Venkatesh, and R. Thiruvengadathan, "Advances in gamma radiation detection systems for emergency radiation monitoring," *Nuclear Engineering and Technology*, vol. 52, no. 10, pp. 2151-2161, 2020. [Online]. Available: https://doi.org/10.1016/j.net.2020.03.014.

[3] Y. Ishigaki, Y. Matsumoto, R. Ichimiya, and K. Tanaka, "Development of Mobile Radiation Monitoring System Utilizing Smartphone and Its Field Tests in Fukushima," in IEEE Sensors Journal, vol. 13, no. 10, pp. 3520-3526, Oct. 2013. [Online]. Available: https://doi.org/10.1109/JSEN.2013.2272734

[4] R. Gomaa, I. Adly, K. Sharshar, A. Safwat, and H. Ragai, "ZigBee wireless sensor network for radiation monitoring at nuclear facilities," in *6th Joint IFIP Wireless and Mobile Networking Conference (WMNC)*, Dubai, 2013, pp. 1-4. [Online]. Available: https://doi.org/10.1109/WMNC.2013.6548954.

[5] J. H. Elrefaei, H. Kunber, M.K. Shaat, A.H. Madian, M.H. Saad, "Energy-efficient wireless sensor network for nuclear radiation detection," *Journal of Radiation Research and Applied Sciences*, vol. 12, no. 1, pp. 1-9, 2019. [Online]. Available: https://doi.org/10.1080/16878507.2019.1594080.

[6] D. Sun, H. M. Wainwright, C. A. Oroza, A. Seki, S. Mikami, H. Takemiya, K. Saito, "Optimizing long-term monitoring of radiation air-dose rates after the Fukushima Daiichi Nuclear Power Plant," *Journal of Environmental Radioactivity*, vol. 220–221, Article 106281, 2020. [Online]. Available: https://doi.org/10.1016/j.jenvrad.2020.106281.

[7] F. Fumian, A. Chierici, M. Bianchelli, et al., "Development and performance testing of a miniaturized multi-sensor system combining MOX and PID for potential UAV application in TIC, VOC and CWA dispersion scenarios," *Eur. Phys. J. Plus*, vol. 136, Article 913, 2021. [Online]. Available: https://doi.org/10.1140/epjp/s13360-021-01858-2.

[8] P. Janik, M. Zawistowski, R. Fellner, and G. Zawistowski, "Unmanned Aircraft Systems Risk Assessment Based on SORA for First Responders and Disaster Management," Appl. Sci., vol. 11, no. 5364, 2021. [Online]. Available: https://doi.org/10.3390/app11125364.

[9] D. Di Giovanni, F. Fumian, A. Chierici, M. Bianchelli, L. Martellucci, G. Carminati, A. Malizia, F. d'Errico, and P. Gaudio, "Design of miniaturized sensors for a mission-oriented UAV application: A new pathway for early warning," *International Journal of Safety and Security Engineering*, vol. 11, no. 4, pp. 435-444, 2021. [Online]. Available: https://doi.org/10.18280/ijsse.110417.

[10] S. M. Syed Mohd Daud, M. Y. P. Mohd Yusof, C. C. Heo, L. S. Khoo, M. K. C. Singh, M. S. Mahmood, and H. Nawawi, "Applications of drone in disaster management: A scoping review," *Science & Justice*, vol. 62, no. 1, pp. 30-42, 2022. [Online]. Available: https://doi.org/10.1016/j.scijus.2021.11.002.

[11] P. G. Martin, J. Moore, J. S. Fardoulis, O. D. Payton, and T. B. Scott, "Radiological Assessment on Interest Areas on the Sellafield Nuclear Site

via Unmanned Aerial Vehicle," *Remote Sens.*, vol. 8, no. 11, p. 913, 2016. [Online]. Available: https://doi.org/10.3390/rs8110913.

[12] D. Popescu, F. Stoican, G. Stamatescu, O. Chenaru, and L. Ichim, "A Survey of Collaborative UAV–WSN Systems for Efficient Monitoring," *Sensors*, vol. 19, no. 21, p. 4690, 2019. [Online]. Available: https://doi.org/10.3390/s19214690.

[13] H. Wang, H. Cheng, and H. Hao, "The Use of Unmanned Aerial Vehicle in Military Operations," in *Man-Machine-Environment System Engineering: MMESE 2020*, S. Long and B. S. Dhillon, Eds., vol. 645, Singapore: Springer, 2020. [Online]. Available: https://doi.org/10.1007/978-981-15-6978-4_108.

[14] L. Marques, A. Vale, and P. Vaz, "State-of-the-Art Mobile Radiation Detection Systems for Different Scenarios," *Sensors*, vol. 21, no. 4, 1051, 2021. [Online]. Available: https://doi.org/10.3390/s21041051.

[15] M. Bouhali, F. Shamani, Z. E. Dahmane, A. Belaidi, and J. Nurmi, "FPGA Applications in Unmanned Aerial Vehicles - A Review," in *Applied Reconfigurable Computing: ARC 2017*, S. Wong, A. Beck, K. Bertels, L. Carro, Eds., vol. 10216, Cham: Springer, 2017. [Online]. Available: https://doi.org/10.1007/978-3-319-56258-2_19.

[16] M. Lowdon, P. G. Martin, M. W. J. Hubbard, M. P. Taggart, D. T. Connor, Y. Verbelen, P. J. Sellin, and T. B. Scott, "Evaluation of Scintillator Detection Materials for Application within Airborne Environmental Radiation Monitoring," *Sensors*, vol. 19, no. 18, Art. no. 3828, 2019. [Online]. Available: https://doi.org/10.3390/s19183828.

[17] Y. Zhu, S. Qian, Z. Wang, H. Guo, L. Ma, Z. Wang, and Q. Wu, "Scintillation properties of GAGG:Ce ceramic and single crystal," Optical Materials, vol. 105, Art. no. 109964, 2020. [Online]. Available: https://doi.org/10.1016/j.optmat.2020.109964.

[18] S. Mukhopadhyay, R. Maurer, P. Guss, and C. Kruschwitz, "Review of current neutron detection systems for emergency response," *Proc. SPIE 9213, Hard X-Ray, Gamma-Ray, and Neutron Detector Physics XVI*, Art. no. 92130T, Sep. 5, 2014. [Online]. Available: https://doi.org/10.1117/12.2058165.

[19] A. Chierici, A. Malizia, D. di Giovanni et al., "A low-cost radiation detection system to monitor radioactive environments by unmanned vehicles," *Eur. Phys. J. Plus*, vol. 136, art. no. 314, 2021. [Online]. Available: https://doi.org/10.1140/epjp/s13360-021-01276-4.

[20] A. Chierici, A. Malizia, D. Di Giovanni, R. Ciolini, and F. d'Errico, "A High-Performance Gamma Spectrometer for Unmanned Systems Based on Off-the-Shelf Components," *Sensors*, vol. 22, no. 3, art. no. 1078, 2022. [Online]. Available: https://doi.org/10.3390/s22031078.

[21] International Atomic Energy Agency, "Categorization of Radioactive Sources," IAEA-TECDOC-1344, Vienna, Austria, 2003. [Online]. Available: https://www.pub.iaea.org/MTCD/Publications/PDF/te_1344_web.pdf

Semiconductor-based diodes for tritium detection

Paul Eyméoud[1*], Stéphane Biondo[2], Vanessa Vervisch[1], Nadia Grillet[3], Laurent Ottaviani[3], and Wilfried Vervisch[1,3]

[1]SiClade Technologies, 2013 Chemin des Ratonneaux, 13680 Lançon-Provence, France
[2]Magdala, 1641 RN8, 13400 Aubagne, France
[3]Aix-Marseille Univ., Université de Toulon, CNRS, IM2NP, Marseille, France
(*) paul.eymeoud@siclade-technologies.com

Abstract—**In order to plan an experimental betavoltaic detection process of tritium using 4H-SiC diodes, we have performed a preliminary numerical Monte-Carlo investigation. In a first part, by evaluating the transparency of several materials to the electrons produced by tritium decay, we have shed light on: (i) the necessity to place the detection diode in close neighborhood of the tritiated sample (less than 1mm distance) or to work in vacuum, (ii) the importance to use very thin coating layers (less than 0.1μm), containing low density materials, (iii) the strong screening effect of 4H-SiC (0.4μm thickness of 4H-SiC divides the intensity flux by 4). In a second part, we have built a deposition energy cartography in PIN and Schottky diodes, confirming that the upstream surface part of the diode (less than 0.5μm depth layers) will constitute the detection region.**

Keywords — **Silicon carbide, tritium detection.**

I. Introduction

Tritium-deuterium fusion reaction constitutes the key point of ITER (world fusion nuclear program) functioning. However, tritium constitutes an important nuclear safety issue, because of its ability to substitute protium in water and biological chain [1]. It is thus important to be able to easily detect tritium in ambient air surrounding ITER installation, and in constituting tokamak materials, such as beryllium coatings and inox and copper shields, particularly during maintenance or dismantling. The techniques developed for tritium detection in ITER will be also useful to evaluate tritium rejections by more conventional nuclear plants, such as CANDU reactors [2].
Among the several experimental approaches employed to detect tritium (such as calorimetry, pyrolysis, scintillation, etc [3]), semiconductor-based betavoltaic detection is a simple, real-time and non-destructive technique. It consists in using a semiconductor diode to collect the electrons produced from tritium decay, before converting them into electrical signal:

$$^3_1\mathrm{H} \rightarrow {}^3_2\mathrm{He}^+ + {}^{\;0}_{-1}e^- + \bar{\nu}_e \qquad (1)$$

The main difficulty of this approach relies on the very low energy of the electron produced in (1) (namely, average energy of 5.7keV, and maximal energy of 18.5 keV), implying the necessity to get a strong sensitivity for the sensor.

One of the first semiconductor diode detection processes was realized for the first time in 1994 by Wampler et al., using a Si-based PIN diode with a thin oxide layer, coupled to an amplification assembly [4]. Similar detection process has been then experimentally reproduced using Si diode [5,6], and GaAs diode [7]. In addition, several numerical investigations (using Monte-Carlo and/or finite elements approaches) have been driven on semiconductor-based betavoltaic detection with GaAs [7], GaN [8,9] and InGaP [9]. Especially, Maximenko et al. [10] drew a comparative study of betavoltaic conversion efficiencies of several semiconductors, leading to good results for wide band gap semiconductors such as 4H-SiC, diamond and c-BN.
In this context, the CASCADE project proposes to implement the betavoltaic detection process using the cutting-edge 4H-SiC diodes developed by SiClade Technologies [11], that have proven their performance for various particle detection approaches [12,13].
The challenging task will be to conceive 4H-SiC-based diodes able to precisely detect and quantify the tritium contained in a given sample. In this scope, we have realized a preliminary numerical study, based on Monte-Carlo calculations, in order to answer to the following practical questions:

(i) If detection is realized in ambient atmosphere, what is the sample-diode distance optimal range to manage tritium detection?
(ii) What will be the screening effect of diode materials (4H-SiC target, and oxide or metallic coatings) on electrons resulting from tritium decay?
(iii) What will be the typical energy deposition profile in a 4H-SiC-based PIN diode, or Schottky diode?

To answer to the first two questions, we have performed a transparency study of several materials (including air) to the electrons resulting from tritium decay (section 2). The question (iii) is then answered is section 3, by an energy deposition simulation.

II. Transparency of several materials to the electrons resulting from tritium decay

We have evaluated the transparency of several materials to electrons resulting from tritium decay, namely: air (corresponding to the ambient medium), 4H-SiC (the detection target material), and coating materials SiO_2 (oxide coating, in the case of a PIN diode), or Al, Ti, Ni, Au (commonly employed in the case of metal-SiC Schottky diodes [14]).

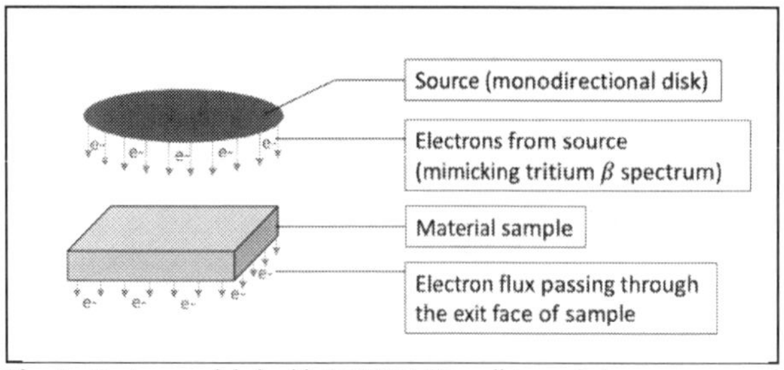

Fig. 1. System modeled with MCNPX ($1\mu m$ distance between source and target).

For that, we have employed MCNPX code [15] in order to model the devices presented on Figure 1: a monodirectional disk source emitting electrons whose spectrum reproduces tritium decay spectrum, bombarding a material sample (located at $1\mu m$ from the source). For several materials constituting the sample (air, 4H-SiC, SiO_2, Al, Ti, Ni, Au), and several thicknesses of this sample, we have collected the surface flux spectrum of electrons, and its average value, flowing through the exit face of sample. Calculations have been performed in p, e mode, using $4 \cdot 10^6$ source particles histories, and the electron source defined by Litz [8] in order to mimic tritium decay.

The results obtained for ambient air are presented on Figure 2, with the comparative case of vacuum. One can see that 1mm of air reduces the electron signal resulting from tritium decay by a factor 5.7, and the screening effect becomes quasi-total after 3mm of air, contrary to the vacuum case wherein the average surface flux remains unchanged. These first results point out an important practical conclusion for the experimental betavoltaic detection of tritium: the necessity to place the diode in near neighboring of the tritiated sample (less than 1mm distance in air), or to work in vacuum.

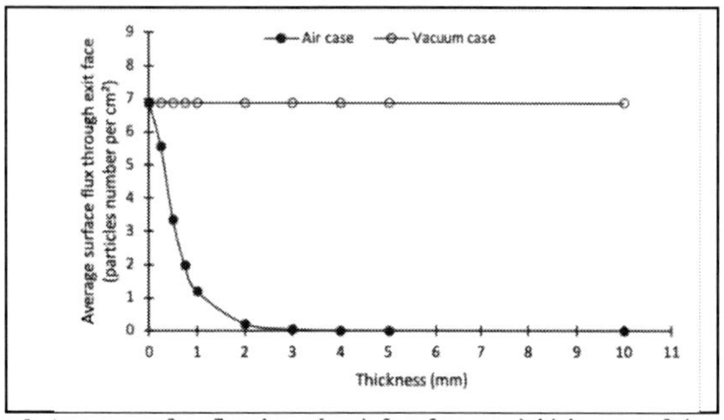

Fig. 2. Average surface flux through exit face for several thicknesses of air, and comparison with the vacuum case.

The results obtained for solid materials, candidates for the diode conception, are presented on Figure 3, top.

For the 4H-SiC (d_{bulk} = 3.21g·cm^{-3}) constituting the detection target, we can see that 400nm approximately divides the average surface flux by a factor 4 (associated attenuation for complete spectrum is presented on Figure 3, bottom). This suggests the necessity to implement a detection procedure using the upstream surface part of 4H-SiC target.

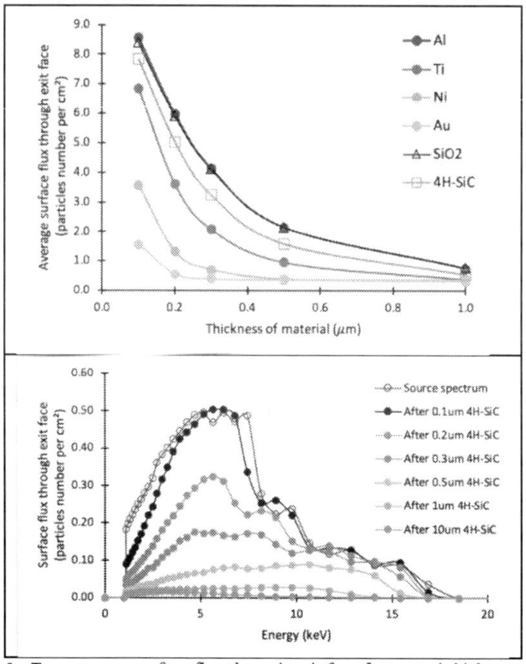

Fig. 3. Top: average surface flux through exit face for several thicknesses of target material (4H-SiC) and coating materials (SiO2, Al, Ti, Ni, Au). Bottom: detailed spectral profile for the 4H-SiC case presented on the top.

Concerning the metallic coatings, the screening effect is growing with atomic number and density: it is lower for Al (Z=13, d_{bulk} = 2.7g·cm^{-3}) than for Ti (Z=22, d_{bulk} = 4.51g·cm^{-3}), than for Ni (Z=28, d_{bulk} = 8.9g·cm^{-3}), than for Au (Z=79, d_{bulk} = 19.3g·cm^{-3}). Aluminum, which presents the lowest screening power, can be chosen as good candidate to conceive a Schottky diode for tritium betavoltaic detection. Oppositely, denser materials with higher Z, such as Ni or Au, present a too important screening effect (unless one uses very thin films, lower than 100nm thick).

For SiO_2 coating, whose bulk density (d_{bulk} = 2.65g·cm^{-3}) is close to the one of aluminum (d_{bulk} = 2.7g·cm^{-3}), the screening effect is quasi-equivalent to the one observed for this metal.

Moreover, if one compares Figures 2 and 3, one can notice that the average surface flux is greater after 0.1 μm of 4H-SiC, SiO_2, or Al, than after 1 μm of air. This is due to the numerous electron creations happening in the first layers of solid materials, mainly due to primary knock-on and Auger effect.

The main conclusion concerning coatings is the necessity to use very thin layers (if possible, no more than 100nm), constituted by low-density materials.

III. PIN and Schottky diodes: evaluation of energy deposition

We have simulated with MCNPX code the irradiation of several SiC-based diodes by a particle source emitting electrons whose spectrum reproduces tritium decay spectrum (Fig. 4, top), and evaluated the resulting energy deposition within the diode. We have built:

- A PIN diode, constituted of SiC target with $0.1\mu m$ of SiO_2 coating (in order to limit leakage currents).
- Two Schottky diodes (metal-semiconductor junction): a SiC target with $0.1\mu m$ of Al coating, and a SiC target with $0.1\mu m$ of Ti coating.

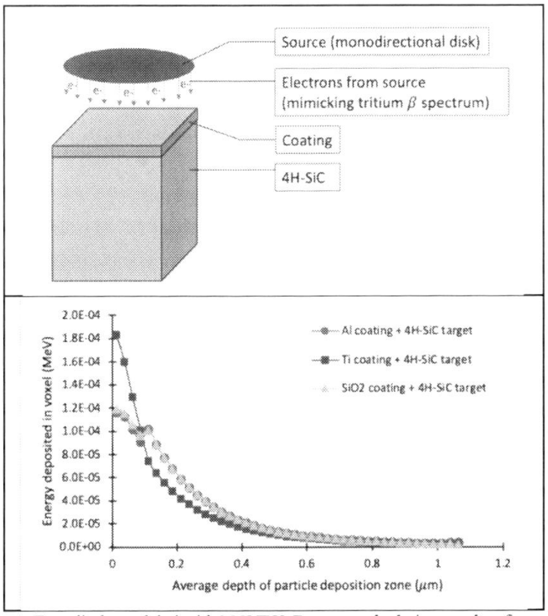

Fig. 4. Top: diode modeled with MCNPX. Bottom: calculation results of energy deposition in voxels, as a function of voxel depth.

To estimate the energy deposition, each diode has been divided into $100\mu m \times 100\mu m \times 25nm$ voxels, and the stared tally of pulse height distribution has been registered in each of these voxels. Calculations have been performed in p, e mode using $4 \cdot 10^6$ source particles histories.

The numerical values of the obtained energy depositions have been plotted on Fig. 4, bottom, and represented in a schematical way (color scale for energy deposition in voxels) on Fig. 5.

For the PIN diode SiO_2+SiC, one can see on Fig. 4, bottom and Fig. 5, top that more than 80% of the energy is deposited in the "100nm SiO_2 coating + first 500nm SiC" zone. The discontinuity that can be observed on the plot of Fig. 4, bottom is a signature of the SiO_2/SiC interface.

If one considers the energy deposition profile for Schottky diode Al+SiC, one can see that this one can be quite superimposed with the previous energy deposition profile of the PIN diode SiO_2+SiC. This is caused by the similar density of SiO_2 (d_{bulk} = 2.65g·cm^{-3}) and Al (d_{bulk} = 2.7g·cm^{-3}) bulk materials.

By comparing both Schottky diodes Al+SiC and Ti+SiC deposition profiles, one can observe a stronger energy deposition in Ti coating than in Al coating, correlated to a lower energy deposition in SiC target of Ti-coated Schottky diode, that in SiC target of Al-coating Schottky diode. Such observation, due to the greater atomic number and density of Ti with respect to Al, suggests that Al will be more suitable to conceive a Schottky diode for tritium betavoltaic detection.

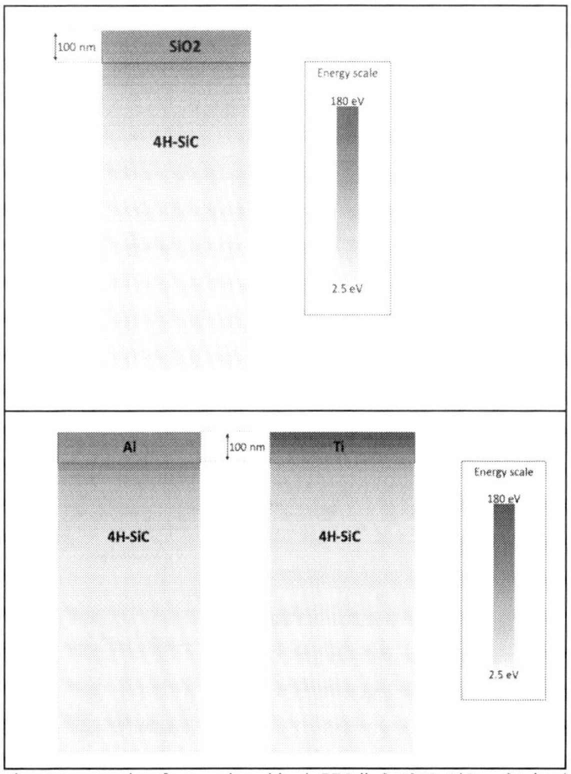

Fig. 5. Cartography of energy deposition in PIN diode SiO2+SiC, and Schottky diodes Al+SiC and Ti+SiC.

Such energy deposition in the first 600nm of the diode:

- is fully coherent with previous numerical studies from literature, realized with other kinds of semiconductors (GaAs [7], GaN [8,9] and InGaP [9] PIN diodes).
- confirms the necessity to use the upstream surface part of the diode for the detection, as conjectured in section 2.

IV. Conclusions

Our Monte-Carlo investigation on materials transparency to electrons, and energy deposition in the diode, leads to the following conclusions:

(i) To analyze the tritiated sample, one needs to place the diode in near neighboring of sample (less than 1mm distance), or to work in vacuum.

(ii) It is necessary to use very thin coating layers (less than 100nm thick) constituted of low-density materials, and to use the first 500nm of 4H-SiC (upstream surface part of the diode) to detect the electrons produced from tritium decay.

In order to prepare the future experimental conception of the diode, the next numerical task is to design the optimal thickness and concentration of the several SiC doped layers, by a finite elements approach.

ACKNOWLEDGMENT

This work received support from the French government under the France 2030 investment plan, as part of the Initiative d'Excellence d'Aix-Marseille Université A*MIDEX AMX-20-TRA-044. The authors acknowledge SiClade Technologies and Magdala companies for financial support (CASCADE project).

REFERENCES

[1] French National Institute for Radioprotection and Nuclear Safety, ``Actualisation des connaissances sur les effets biologiques du tritium'', IRSN Report 2021-00206 (2021). https://www.irsn.fr/FR/Actualites_presse/Actualites/Documents/20210506_IRSN-rapport-2021-00206-TRITIUM.pdf

[2] Canadian Nuclear Safety Commission, ``Étude sur le devenir environnemental du tritium dans l'atmosphère'', CCSN Report n°0792 (2009). https://publications.gc.ca/collections/collection_2010/ccsn-cnsc/CC172-51-2009-fra.pdf

[3] P. Fichet, A. Bultel, S. Markelj, C. Moreno, ``Review of the different techniques to analyze tritium'', CEA Report (2018). https://transat-h2020.eu/wp-content/uploads/2020/04/TRANSAT_D2_1.pdf

[4] W. R. Wampler, B. L. Doyle, ``Low-energy beta spectroscopy using pin diodes, to monitor tritium surface contamination'', Nuclear Instruments and Methods in Physics Research A **349** (1994), 473. https://doi.org/10.1016/0168-9002(94)91213-0

[5] M. Culcer, M. Iliescu, M. Curuia, A. Enache, I. Stefanescu, C. Ducu, V. Malinovschi, ``Tritium contaminated surface monitoring with a solid-state device'', International Conference Nuclear Energy for New Europe (2004). https://www.energ-en.ro/pages/article/359

[6] B. A. VanDevender, L. I. Bodine, A. W. Myers, J. F. Amsbaugh, M. A. Howe, M. L. Leber, R. G. H. Robertson, K. Tolich, T. D. Van Wechel, B. L. Wall, ``Performance of a TiN-coated monolithic silicon pin-diode array under mechanical stress'', Nuclear Instruments and Methods in Physics Research A **673** (2012), 46. https://doi.org/10.1016/j.nima.2012.01.033

[7] A. M. Barnett, J. E. Lees, D. J. Bassford, ``Direct detection of tritium and carbon-14 beta particles with GaAs photodiodes'', JINST **7** (2012), 09012. https://iopscience.iop.org/article/10.1088/1748-0221/7/09/P09012

[8] M. Litz, ``Monte Carlo evaluation of Tritium Beta Spectrum Energy Deposition in Gallium Nitride (GaN) Direct Energy Conversion Devices'', Army Research Laboratory Report n°ARL-TR-70 (2014). https://apps.dtic.mil/sti/pdfs/ADA609481.pdf

[9] M. Khan, ``Demonstration and comparison of GaN/Sapphire and InGaP p-i-n devices for Tritium Betavoltaic Power Source'', Report from Naval Surface Warfare Center Indian Head Division. https://apps.dtic.mil/sti/citations/AD1150231

[10] S. I. Maximenko, J. E. Moore, C. A. Affouda, P. P. Jenkins, ``Optical semiconductors for ^{3}H and ^{63}Ni betavoltaics'', Scientific Reports **9** (2019), 10892. https://doi.org/10.1038/s41598-019-47371-6

[11] Siclade Technologies company. https://www.societe.com/societe/siclade-technologies-893564153.html

[12] A. Tchoualak Tchamako, L. Ottaviani, W. Rahajandraibe, W. Vervisch, V. Vervisch, J.-P. Walder, ``Front end electronics for radiation detectors based on SiC: application to high dose per pulse charged particle beam current measurement'', IEEE Sensors Journal **22** (2022), 2326. https://doi.org/10.1109/JSEN.2021.3117864

[13] O. Obraztsova, L. Ottaviani, B. Geslot, G. de Izarra, O. Palais, A. Lyoussi, W. Vervisch, ``Comparison between silicon carbide and diamond for thermal neutron detection at room temperature'', IEEE Transactions on Nuclear Science **67** (2020), 863. https://doi.org/10.1109/TNS.2020.2981059

[14] J. H. Zhao, K. Sheng, R. C. Lebron-Velilla, ``Silicon carbide Schottky barrier diode'', International journal of high speed electronics and systems, **15** (2005), 821. https://ntrs.nasa.gov/api/citations/20050080762/downloads/20050080762.pdf

[15] L. S. Waters, G. W. McKinney, J. W. Durkee, M. L. Fensin, J. S. Hendricks, M. R. James, R C. Johns, D. B. Pelowitz, ``The MCNPX Monte Carlo Radiation Transport Code'', AIP Conference Proceedings **896** (2007), 81. https://inis.iaea.org/search/search.aspx?orig_q=RN:39072006

The High Efficiency fast-Response GAmma (HERGA) detector based on SiPM readout

C. Altomare[1], D. Cerasole[1,2], L. Di Venere[1]*, E. Fanchini[3], F. Gargano[1], F. Giordano[1,2], F. Loparco[1,2], S. Loporchio[1,2], A. Mastroserio[1,4], M. Morichi[3], F. R. Pantaleo[1,2], D. Serini[1], P. Spinelli[1,2], L. Swiderski[5]

[1]Istituto Nazionale di Fisica Nucleare - Sezione di Bari, Italy
[2]Dipartimento Interateneo di Fisica dell'Università e del Politecnico di Bari, Italy
[3]CAEN s.p.a. - Viareggio, Italy
[4]Università di Foggia
[5]National Centre for Nuclear Research (NCBJ), 05-400 Otwock-Swierk, Poland
(*) leonardo.divenere@ba.infn.it

***Abstract*—Gamma-ray spectroscopy and gamma-ray imaging are two complementary techniques used for the localization and the identification of radioactive sources containing gamma-ray emitting radioisotopes. The radioactivity monitoring is focused on the detection of both artificial and environmental radioactive sources like Naturally Occurring Radioactive Materials (NORM). This kind of contamination becomes dangerous when the detection of the unwanted substances exhibits a concentration significantly above the environmental radioactive background radiation levels. For this purpose, we have developed, tested and shown a High Efficiency fast-Response GAmma (HERGA) detector useful for the identification of radionuclides and for gamma-ray imaging. A first version of the gamma detector prototype was composed of 16 CsI(Tl) scintillating crystals of 3x3x10 cm3 size, arranged in 4x4 matrix coupled with standard Photomultiplier tubes (PMTs). An image reconstruction of a radioactive gamma emitter source is possible using the coded mask technique, in which a 7x7 mask, made of Plastic and Tungsten tiles, is placed in front of the detector and a pattern recognition algorithm based on classical statistical methods (Kolmogorov Smirnov) is used to reconstruct the source position. The measurements carried out showed a point spread function (PSF) of a few mrad for pointlike sources. The Minimum Detectable Activity (MDA) was also determined in the case of pointlike radioactive sources. In this contribution we will present an update of the HERGA detector prototype in which Silicon Photomultipliers (SiPMs) are used in place of the PMTs. SiPMs provide similar or even better performance compared to the standard PMT sand provide benefits in terms of lower power consumption and reduced cost and compactness. The advantages of the SiPM technology are also characterized by the robustness of the photosensor that makes the new prototype compact, portable, ideal for in-situ and real-time. We will show a comparison between the results obtained with the newest SiPM read-out technology with respect to those obtained with the PMT one, in terms of energy and spatial resolution. The imaging performance is also in phase of testing in order to localize extended radioactive sources such as for example NORM samples or to detect inaccessible or hidden nuclear waste.**

***Keywords* —Gamma rays, environmental monitoring, NORM, scintillator crystals, Silicon Photomultipliers.**

I. Introduction

THE detection of gamma rays of ambient origin is of very wide interest in several fields, such as the homeland security, environmental monitoring, detection of illegal radioisotopes trades, in medical science. The radioactivity monitoring is focused on the detection of both artificial and environmental radioactive sources. The latter include Naturally Occurring Radioactive Materials (NORM), which are materials containing any of the primordial radionuclides or radioactive elements as they occur in nature, such as radium, uranium, thorium, potassium and their radioactive decay products, such as radon [1]. This kind of contamination is dangerous when the presence of unwanted substances exhibits a concentration significantly above the environmental radioactive background radiation levels [2].

Gamma-ray spectroscopy and imaging are widely used and combined to identify and localize sources containing gamma-ray emitting radioisotopes, respectively. Gamma-ray spectroscopy is based on the measurement of the gamma-ray energy, allowing the identification of the radionuclide through its characteristic emission lines. In addition, imaging techniques can be used to measure the radioactive source position. Different techniques can be employed to reconstruct the gamma-ray directions. Among these, the coded mask technique is widely used especially in astrophysics and medical to precisely localize X-rays and gamma rays with energy from tens of keV to few MeV [3].

The scintillator technology is nowadays widely being adopted for environmental monitoring. Thanks to their high Z number, inorganic scintillators guarantee a high efficiency in the gamma-ray detection. They are usually equipped with traditional photomultiplier tubes (PMTs) or modern photon detectors like Silicon Photomultipliers (SiPMs) to detect the scintillation light and measure the gamma ray energy. An extensive review is collected in [4].

II. The HERGA prototype

In this work, we present a gamma-ray detector prototype with high efficiency and source localization capabilities based on the coded mask technique, named High Efficiency fast-Response GAmma detector (HERGA). A full description and characterization of this detector is presented in [5].

The HERGA prototype is composed of an array of 4 × 4 CsI(Tl) scintillator crystals (3 × 3 × 10 cm^3), each coupled to a PMT (ET Enterprises, 9124B, 26% quantum efficiency at 400 nm) with a Cockcroft–Walton voltage multiplier as high voltage supplier (ET Enterprises, electron tubes, HV3020CN). The total weight of the detector is ~ 8 kg. The readout system is based on a CAEN V1725 16-channel digitizer (14-bit ADC and 250 MS/s sampling frequency), which is used to generate the trigger and digitize the PMT signals. The 16 signals from the 16 PMTs are recorded and integrated with an on-line custom analysis software providing count spectra in real time. A coded mask, consisting of an array of 7 × 7 PVC and tungsten tiles, which are respectively transparent and opaque to gamma rays, is placed in front of the detector. The mask is generated from a 4 × 4 Modified Uniform Redundant Array (MURA). Its throughput, defined as the fraction of transparent pixels, is equal to 38.8%. Each tile has a size of 3 × 3 × 1 cm^3, while the whole mask has a size of 21 × 21 × 1 cm^3 and a weight of ~ 6 kg. FIGURE 1 shows a picture of the detector prototype.

The image reconstruction performances of the detector were studied with Monte Carlo simulations based on Geant4 and tested experimentally in laboratory with radioactive sources.

Fig. 1. Picture of the HERGA prototype.

A. Data acquisition and Energy calibration

Signals from the 16 CsI(Tl) scintillator crystals each one coupled to one PMT were acquired with the V1725 digitizer. Several radioactive sources were used to measure the PMT signals as a function of energy. The gamma detector was operated in self-triggering mode. A trigger was generated whenever the pulse height from any PMT exceeded a preset threshold value, which could be adjusted for individual channels. When a trigger occurred, the pulses from all PMTs were integrated and the charge values were recorded by the digitizer. The gains of the 16 PMTs and the thresholds were adjusted to equalize the individual trigger rates. The signal integration time was optimized to maximize the energy resolution and fixed to 2 μs.

The energy calibration was performed using eight radioactive sources placed at 15 cm from the detector. The gamma-ray energy spectra were fitted with Gaussian functions to describe the gamma-ray peaks and a second order polynomial function describing the background. The peak positions allow to determine the calibration functions, while the peak widths provide a measurement of the energy resolution.

Fig. 2 shows the calibration curve of one of the crystals, which exhibits a good linearity up to 1500 keV, and its energy resolution $\sigma(E)/E$. The energy resolution curve in FIGURE was well described with the function $\sigma(E)/E \approx 4.1\%/\sqrt{E(MeV)}$. A similar behavior is observed for all scintillator bars.

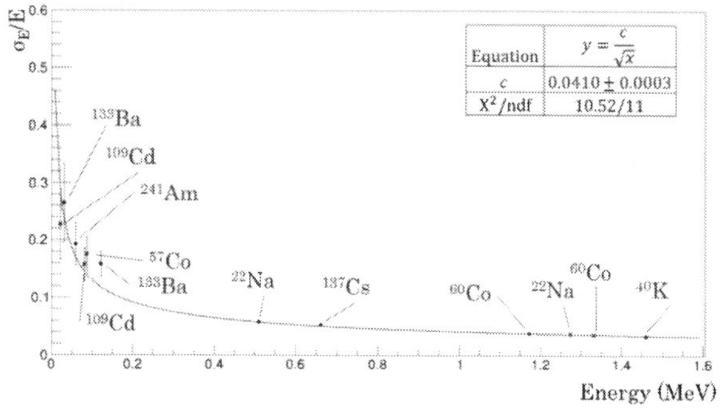

Fig. 2. Energy resolution curve for one the scintillator crystals.

B. Position reconstruction and angular resolution.

The HERGA prototype has capabilities to reconstruct the position of the gamma ray sources exploiting the coded mask technique. A Monte Carlo simulation based on GEANT4 (REF) was built to study the performance of the detector and to validate the reconstruction algorithm. Gamma rays were generated from a point-like source with isotropic emission characterized by an energy value extracted from the radioisotope energy peak distribution. The source was placed at a fixed distance z = 40 cm from the detector and its position was changed on a plane parallel to the detector surface. The mask was placed between the detector and the source, at z = 10 cm from the detector and at 30 cm from the source. In each simulation the 16 count spectra corresponding to each scintillator crystal were recorded. Three radioactive sources were used to test the reconstruction capabilities as a function of the gamma-ray energy, namely ^{60}Co, ^{137}Cs and ^{241}Am.

The reconstruction method was based on a two dimensional Kolmogorov-Smirnov test. At first, simulations were performed varying the radioactive source position. For each simulation, the image produced in the scintillator array by the radioactive source was recorded and a database of these images was built. The reconstruction of the position of an unknown source was then performed defining a similarity probability of the unknown image with the database. The weighted average position provides the optimal reconstructed position.

The theoretical resolution of the detector is estimated performing a great number of simulation runs. Fig. 3(a) shows the 2D distribution of the reconstructed positions when a source of ^{60}Co is placed in the central position. The angular separation between the reconstructed and simulated positions was then calculated for each simulation. Its distribution is shown in Fig. 3(b) and the 68% and 95% quantiles were calculated and provide an estimate of the Point Spread Function (PSF) of the instrument. Fig. 4 shows the 68% and 95% quantiles of the PSF evaluated simulating a ^{60}Co, a ^{137}Cs and an ^{241}Am source at different positions along the diagonal of the detector plane. We note that the angular resolution deteriorates as the gamma-ray energy increases, since the absorption efficiency of the tungsten pixels in the mask decreases with increasing photon energy.

The angular resolution was measured experimentally using the same radioactive sources. A database of images was built by placing the radioactive sources on a grid with a 1-cm step. The reconstruction algorithm was then tested with a source placed at the center of the detector and at a distance of 3 cm along the x- and y- axis from the center. A PSF of 4±2.5 mrad was measured for a 60Co source placed in the central position, in agreement with the Monte Carlo simulation predictions.

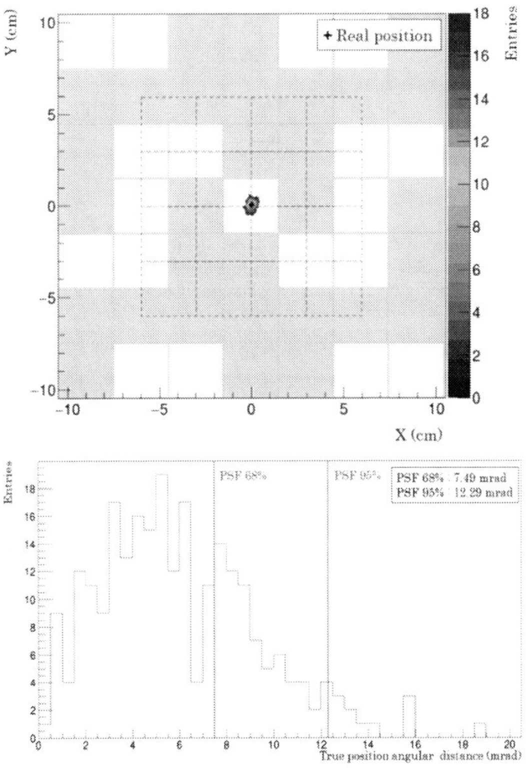

Fig. 3. (a) Two dimensional reconstructed positions for a simulated 60Co source placed at the center of the detector. (b) Angular separation between the simulated and reconstructed positions. The red lines show the 68% and 95% quantiles of the distribution.

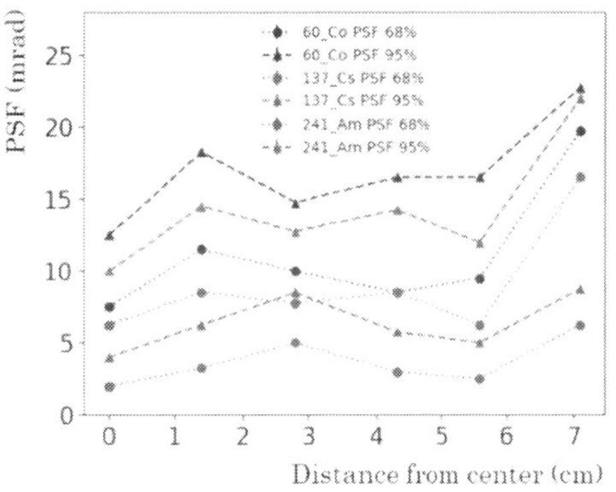

Fig. 4. 68% and 95% PSFs for three radioactive sources placed in different positions along the diagonal of the detector plane.

C. *Minimum Detectable Activity*

Finally, the minimum detectable activity (MDA) of the detector was measured. The gamma camera was placed in a 5 cm thick lead shield to reduce the environmental background. The radioactive sources were placed on the front face of the camera. The coded mask was removed to maximize the radiation throughput. Six different radioactive sources were placed at a distance of 15 cm from the detector and their spectra were measured by varying the acquisition time in the range from 15 to 120 min with 15- min steps.

To evaluate the MDA, the efficiency calibration is required, which was measured with sources with known activities: ^{241}Am, ^{57}Co, ^{137}Cs, ^{60}Co, ^{152}Eu and ^{22}Na. The spectra were analyzed with the Quantus - Quantitative Spectrometry Software by CAEN and the efficiency ϵ was calculated as the ratio between the number of detected photons and the number of emitted photons. The efficiency curve $\epsilon(E)$ was fitted with the following function:

$$\ln\big(\epsilon(E)\big) = \sum_k a_k \left(\frac{E}{E_0}\right)^k$$

where E_0 = 1 keV is a fixed scale energy. Fig. 5 shows the measured points and the best fit curve.

Finally, the MDA was measured as a function of the acquisition time using the following equation:

$$MDA(Bq) = \frac{L_D}{t_m(s) \times \epsilon \times b}$$

where t_m is the source acquisition time, b is the relative line intensity for the specific gamma-ray emission peak, which corresponds to the probability of a gamma-ray emission with the line energy in a radioactive decay event, ϵ is the calculated efficiency for the specific peak, L_D is the critical level of counts above the background to detect a source, that is given by [6]:

$$L_D = 4.65 \times \sqrt{R_b(s^{-1}) \times t_m(s)} + 3$$

where R_b is the background count rate.

Fig. 6 shows the MDA for ^{137}Cs source for data acquisition times from 15 to 120 min. The values of the MDA have been fitted with a power-law function $MDA = c\, t^a$, where t is the acquisition time. We see that the MDA is roughly proportional to the inverse of the square root of the acquisition time, as expected.

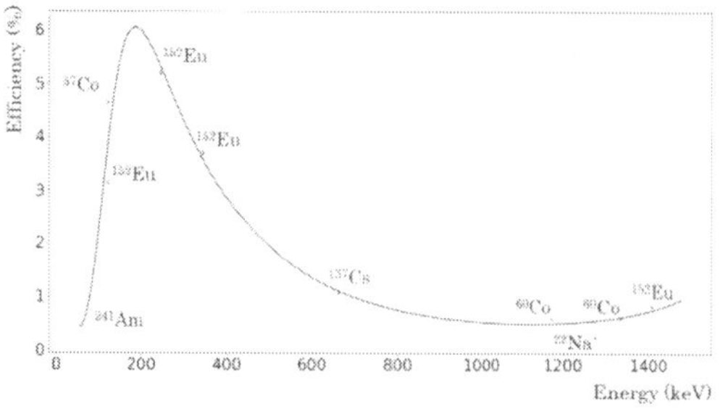

Fig. 5. Efficiency curve for the HERGA prototype measured for the MDA calculation.

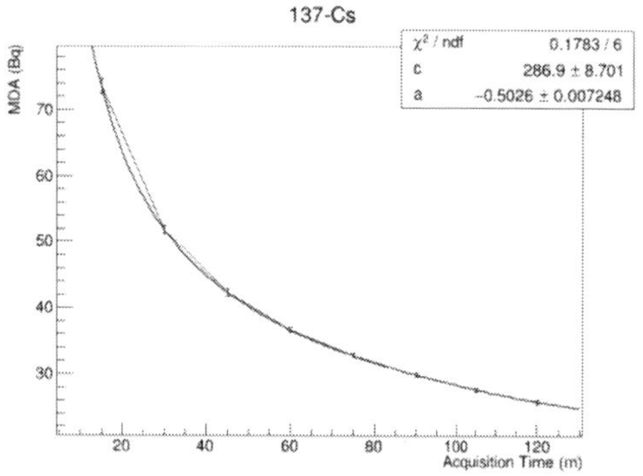

Fig. 6. MDA as a function of the acquisition time for a ^{137}Cs source

III. THE HERGA UPGRADE

The HERGA prototype proved to have excellent performance in terms of angular resolution and MDA. An upgrade of the detector is foreseen to make the instrument more compact and portable. The first step is the employment of SiPMs in place of the PMTs. This reduces the overall volume and weight of the instrument, still keeping its performances. A prototype crystal was built employing 6x6 mm^2 SiPMs (Hamamatsu S14160-6050HS) mounted on custom PCBs. Two SiPMs were connected in parallel on the small face of a 3x3x10 cm^3 CsI(Tl) crystal in order to detect enough scintillation light and increase the energy resolution. A picture of the prototype is shown in Fig. 7. The energy resolution of this prototype was measured with ^{241}Am, ^{137}Cs and ^{60}Co radioactive sources. Results are reported in Table I and are comparable with the ones obtained with the PMT prototype and shown in previous section.

TABLE I
ENERGY RESOLUTION FOR SIPM PROTOTYPE

Radioisotope	Energy (keV)	Resolution
^{241}Am	59.5	29%
^{137}Cs	661.7	10%
^{60}Co	1173	6%
^{60}Co	1332	7%

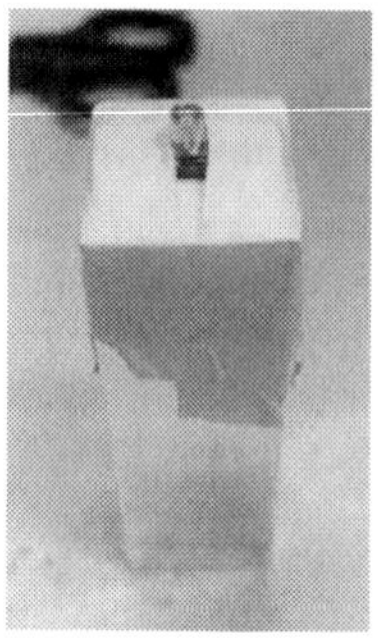

Fig. 7. CsI(Tl) crystal coupled to Hamamatsu S14160-6050HS SiPMs

IV. FUTURE PERSPECTIVES

In this work a gamma-ray detector with imaging capability was developed, based on a CsI(Tl) crystal array coupled to PMTs. A coded mask reconstruction technique was employed to reconstruct the gamma-ray source position, reaching a PSF at the level of few mrad. In addition, the large active volume of the prototype also resulted in a very low MDA. An upgrade of this prototype is foreseen based on SiPMs and a compact and low power electronics, in order to make the instrument compact and portable. A preliminary prototype was built based on 4 CsI(Tl) crystals and used for the first in-situ measurements on the Nisyros Island in Greece. Data analysis is ongoing and will be presented in a future work.

REFERENCES

[1] S. Chaki, E. Foutes, S. Ghose, B. Littleton, J. Mackinney, D. Schultheisz, M. Schuknecht, L. Setlow, B. Shroff, T. Peake, Technical Report on Technologically Enhanced Naturally Occurring Radioactive Materials from Uranium Mining Volume 1: Mining and Reclamation Background, Citeseer, 2008, URL http://citeseerx.ist.psu.edu/viewdoc/summary?doi=10.1.1.348.3026 .W.-K. Chen, Linear Networks and Systems. Belmont, CA, USA: Wadsworth, 1993, pp. 123–135.

[2] P.M. Chapman, Determining when contamination is pollution—weight of evidence determinations for sediments and effluents, Environ. Int. 33 (4) (2007) 492–501, URL: https://www.sciencedirect.com/science/article/pii/S016041200600136X.

[3] H. Anger, A new instrument for mapping gamma-ray emitters, Biol. Med. Quarterly Rep. UCRL 3653 (1957) 38.

[4] H. Al Hamrashdi, D. Cheneler, S.D. Monk, A fast and portable imager for neutron and gamma emitting radionuclides, Nucl. Instrum. Methods Phys. Res. A 953 (2020) 163253, URL https://www.sciencedirect.com/science/article/pii/S0168900219315256.

[5] C. Altomare, L. Di Venere, F.R. Pantaleo, E. Fanchini, F. Giordano, F. Loparco, M. Morichi, P. Spinelli, L. Swiderski, A high efficiency fast-response gamma detector with mrad pointing capabilities, Nuclear Inst. and Methods in Physics Research, A 1025 (2022) 166106

[6] L.A. Currie, Limits for qualitative detection and quantitative determination. application to radiochemistry, Anal. Chem. 40 (3) (1968) 586–593, URL https://pubs.acs.org/doi/pdf/10.1021/ac60259a007

Optoelectronics for neutron detectors based on superheated emulsions

Andrea Chierici[1,*], Francesco d'Errico[1,2]
[1]School of Engineering, University of Pisa, Italy
[2]Italian National Institute of Nuclear Physics, Pisa, Italy

*andrea.chierici@ing.unipi.it

Abstract—**This paper presents the design, assembly, and performance investigation of an optoelectronic readout system specifically designed to enhance the real-time readout capabilities of Superheated Drop Detectors (SDDs). SDDs as fast neutron detectors have been widely used in different areas such as homeland security, radiation protection, and personal dosimetry. In homeland security, SDDs showed the potential of detecting prompt fission neutrons from special nuclear materials in radiation portal monitor systems, possibly positioning them as a valuable alternative to He-3 detectors. In radiation protection and personal dosimetry, downsized SDDs may be employed to assess the level of exposure of workers in nuclear energy production facilities or as a tool to perform photoneutron contamination studies of medical and research accelerators. The proposed readout system utilizes light-emitting diodes to illuminate the detectors, while photodiodes are employed to record the integral value of scattered light resulting from bubbles' formation, a consequence of neutron interaction. To validate the behavior of the system, comprehensive field testing was conducted after assembling and integrating it with custom-made 3D printed structures tailored for different SDD geometry, and the reader was utilized to examine the optical and detection characteristics of different SDD compositions under various ionizing radiation conditions. The potential to actively read out SDDs in real-time represents a significant advancement for this radiation detection technology, particularly in fields such as radiation protection and industrial applications, and it provides valuable insights for future enhancements in the field of neutron radiation detection.**

Keywords —**Superheated Drop Detector, Optical Readout, Active Interrogation, Neutron Dosimetry, Homeland Security.**

I. Introduction

The increasing need for new neutron detection technologies and their diverse applications have resulted in a growing research effort across several domains, such as homeland security, radiation safety, fusion and fission research, and more [1, 2]. For instance, in the field of national security, neutron detectors can play a critical role in mitigating threats posed by the illicit transport of Special Nuclear Materials (SNM) like ^{235}U, ^{237}Np, and ^{239}Pu [3, 4, 5].

Superheated Droplet Detectors (SDDs) offer several unique advantages compared to other neutron detection technologies, such as size adaptability, passive operation, and gamma-ray insensitivity, making them invaluable in these applications [6, 7, 8]. In active interrogation techniques employed for SNM detection, SDDs' adjustable energy thresholds and gamma radiation insensitivity enables them to operate as prompt fission neutron detectors without interference from the high energy and flux of the interrogation source.

In the domains of radiation safety and medical contexts, SDDs are interesting tools for personnel dosimetry and photo-neutron contamination studies [9, 10]. The challenge of accurate neutron dose measurement, given the complexities of neutron detection physics and the broad energy range to be covered, is met well by SDDs, whose response curve aligns closely with the International Commission on Radiological Protection (ICRP) equivalent curve across the entire energy range [11]. Indeed, a 1989 research study supported this, demonstrating that the response curve of SDDs followed the ideal ICRP dose equivalent response curve within a minor deviation [12].

Leveraging the inherent property of the superheated liquid in SDDs, neutron spectrometry can also be fine-tuned by controlling the operative thermodynamic parameters of the emulsions, such as temperature and pressure. Additionally, in scenarios involving patient exposure to unwanted neutron doses, SDDs, due to their photon insensitivity, passive operation, tissue-equivalent composition, isotropic response, compact size, and cost-effectiveness, can provide reliable estimations of the spatial and energetic distribution of contamination neutrons.

This paper focuses on the design, realization, testing, and calibration of a fast neutron detection system based on SDDs equipped with dedicated optical readout systems. Different strategies have been pursued for the readout of bubble formation in SDDs, such as acoustic counting or image processing [13, 14]. However, each of these methods comes with its own set of challenges. For instance, acoustic counting, which involves the use of piezoelectric sensors or microphones, is significantly sensitive to environmental noise and vibrations. This can interfere with accurate bubble detection and counting, thus compromising the reliability of the system. On the other hand, image processing, involving the use of digital cameras such as CCD (Charge-Coupled Device) or CMOS (Complementary Metal-Oxide-Semiconductor) sensors, requires a more complex and expensive hardware setup. This complexity not only lies in the level of detection, where high-resolution cameras with precise focus control are required but also extends to the data processing stage. Large volumes of image data need to be processed and analyzed in real-time to detect and count the bubbles accurately, which demands high computational resources.

Considering these challenges, our proposed system offers a promising solution that capitalizes on optical readout systems for SDDs. The wide-ranging applicability of this proposed system spans homeland security, neutron spectrometry, personnel dosimetry, and radiation safety. This underscores its potential as a tool in the detection of prompt fission neutrons in active interrogation techniques, significantly aiding in the prevention of unauthorized SNM transport across national borders.

II. Superheated Liquids and Droplet Detectors

SDDs are engineered utilizing a liquid that exists in a metastable or *superheated* state. Superheating is a process where a liquid remains in a state of unstable equilibrium, not vaporizing even when its temperature exceeds its boiling point but is still below the critical point. This liquid can be dispersed as minuscule droplets embedded within an insoluble host gel matrix or in other words as an *emulsion*, which in turn allows to maintain the liquid's metastable state [7]. These droplets, with diameters varying between 50 μm and 200 μm, can transform into gas bubbles, typically ranging from 1 mm to 2 mm in size (Fig. 1). This vaporization process is caused by the deposition of energy following the interaction of the droplets with charged particles traveling through the emulsion [15, 8].

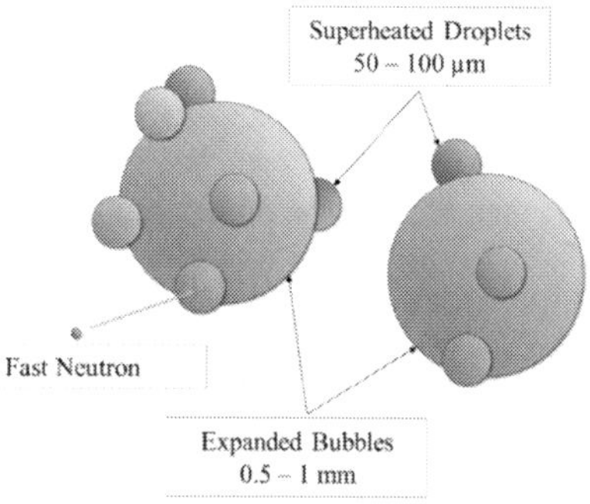

Fig. 1. Superheated Droplet Detectors.

SDDs are predominantly composed of elements such as Hydrogen (H), Carbon (C), Oxygen (O), Fluorine (F), and Chlorine (Cl). The interaction of these elements with neutron scattering leads to the production of energetic recoiled charged particles, which prompts the droplet to vaporize, forming a gas bubble. The unique characteristic of SDDs that renders them insensitive to gamma rays depends on the low linear energy transfer of gamma rays, which prevents them from depositing the requisite energy for vaporization (Fig. 2) [16].

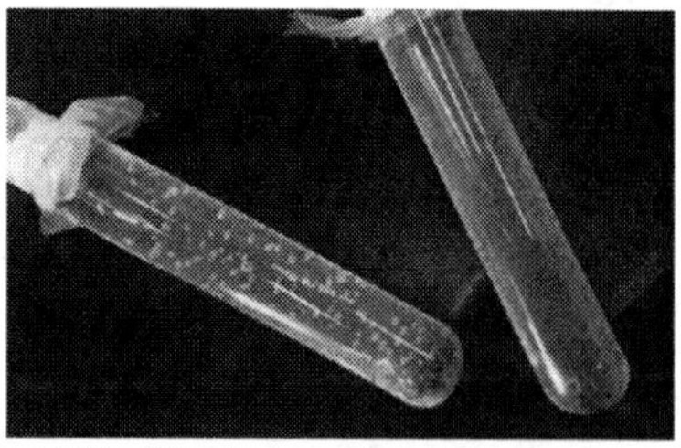

Fig. 2. Superheated Liquid in glass containers in the presence and absence of bubbles following exposure to fast neutrons.

Notably, the threshold energy for SDDs that can be detected can be adjusted not only by modifying the chemical composition of the detector but also by modifying the operating temperature and pressure of the emulsion. When the temperature rises, both the surface tension and the enthalpy of evaporation decrease, which lowers the minimum energy required for a droplet to vaporize. Similarly, by reducing the operating pressure, the energy necessary for vaporization can be decreased, as reaching the critical radius demands less work under reduced pressure conditions.

III. Optical Readout for Superheated Emulsions

Our approach leverages an optical readout scattering principle for the detection and quantification of bubbles in neutron detectors. This method involves the use of photodiode arrays and trans-impedance op-amp circuits (Fig. 3) together with LEDs for stimulation.

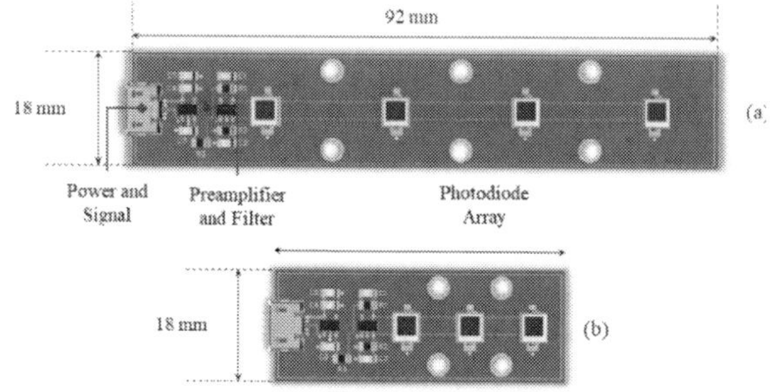

Fig. 3. Transimpedance amplifiers used to read out the lateral optical scattering in large vessels (a) and small vials (b).

Table I
Halocarbon composition and properties

Name	Chemical Formula	Boiling Point (°C)	Critical Temperature (°C)
Octafluorocyclobutane (C318)	C_4F_8	-7.0	115.2
Hexafluoropropylene (HFP)	C_3F_6	-29.4	85.2
Decafluorobutane (R610)	C_4F_{10}	-1.7	113.3

Due to the large size of the scattering particle, the scattering regime is closely tied to the optical geometry [17]. Although most of the scattering transpires in the forward direction, this light overlaps with transmitted light and does not exhibit a wavelength shift, rendering it impractical for detection purposes. Consequently, we concentrate on the side scattered light, ensuring its proper collimation to exclude the significant, but unwanted, forward component of light. Throughout the project's development, we employed two distinct container geometries: small glass vials of a few milliliters in volume, and larger tempered glass vessels capable of holding several liters, provided by Büchi Labortechnik AG. Given their compact size, the glass vials are ideal for use as personal dosimeters or as distributed detectors in radiation contamination studies. On the other hand, large vessel detectors are constructed from tempered glass that can withstand pressures up to 5 MPa. This strength allows for the adjustment of their operational pressure, enabling selective response to neutrons of various energies and/or resetting the formed bubbles back to a liquid phase. Consequently, as SDDs can be rendered gamma radiation insensitive, they are well-suited as neutron detectors for active interrogation techniques and neutron spectrometry applications. For this study, halocarbons with a moderate degree of superheat (Table I) were utilized as the superheated liquid, owing to their nucleation exclusively by energetic heavy ions, such as those released by fast neutron interactions [8]. Concerning the optical readout, blue light LEDs (HPWN-MB00, Lumileds Lightning) peaking at 450 nm were chosen because of the superior light output stability versus temperature changes and for the low absorption of blue light in aqueous media [18]. The operation of the LEDs was managed by a self-made Pulse Width Modulation (PWM) driver, which was centered around a TLV431 reference. This reference was arranged to serve as a constant current sink during the high periods of the PWM signal. The nominal current during these "on" phases was adjustable, offering a range between 1.24 mA and 12.4 mA. Furthermore, the PWM frequency could be selected between 2 kHz and 10 kHz. The duty cycle was user-configurable based on the needed average light intensity, offering a selection range from 1% to 10%. These user-configurable parameters, therefore, enabled a vast span of average LED currents, from a minimum of approximately 12.4 μA to a maximum of about 1.24 mA, thus ensuring flexibility in adjusting the LED intensity to match the requirements of each experiment.

To convert and amplify the current signals generated from incident light on blue response-enhanced photodiode arrays (BPW34B, Vishay), trans-impedance amplifiers were implemented using TLV341 operational amplifiers (Texas Instruments). The trans-impedance amplifiers consisted of two cascaded operational amplifiers. The first operational amplifier was configured to reject electrostatic coupling by utilizing the Common Mode Rejection Ratio (CMRR) of the op-amp, and it had a gain of 200 kohms to amplify the current signals effectively. The second operational amplifier acted as a buffer for the passive low pass filter set at a cutoff frequency of 15.9 Hz. The output signals from the trans-impedance amplifiers underwent subsequent filtering and digitization. This process was performed by a microcontroller-based data acquisition module. The module provided a resolution of 16 bits and a conversion speed of 1 ksps.

The data acquisition module and the controller were directly connected to a personal computer, facilitating control through a dedicated graphical user interface. This enabled convenient interaction and control over the system's functionality.

As discussed, we utilized two different container geometries for the SDDs in this study - glass vials and tempered glass vessel containers. Consequently, we designed and assembled two distinct types of 3D-printed polylactic acid holders to accommodate these different geometries. These holders serve a dual purpose; not only do they provide mechanical support for the optoelectronics printed circuit boards, but they also function as light collimators. collimators. To reduce the sensitivity to long-term drifts in the light source and the optical characteristics of the emulsion, a real-time edge detection algorithm was implemented. This was accomplished by processing the detected scattered light signals with a band-pass moving average filter. Each time an uptick in the light signal is registered, the corresponding edge is subsequently reconstituted.

A. Readout System for Vial Detectors

In Fig. 4 the design and assembly of the 3D-printed holders used for the glass vial containers are shown. Since vials are relatively small (Ø17 mm × 30 mm active volume), they are well suited to be used both as pocket personal dosimeters and as distributed detectors for radiation contamination studies.

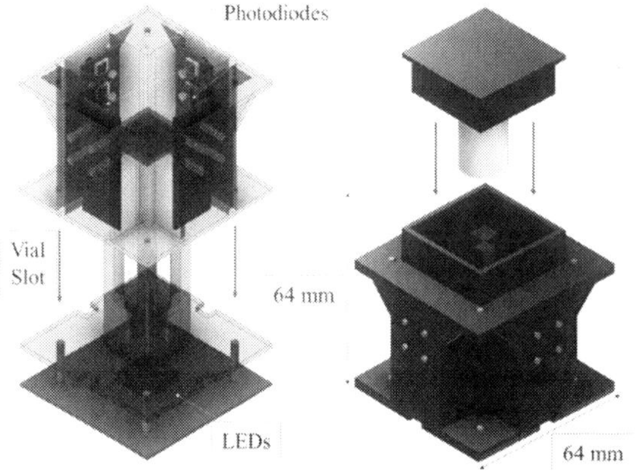

Fig. 4. Architecture of the optoelectronics readout system for small vial detectors.

Blue light LEDs expose vials from the bottom and to collect the side scattered fraction of light, 4 photodiode array boards are mechanically fixed to the top part of the system. After being collimated by holders, side-scattered light is detected radially at 0, 90, 180, and 270 deg to minimize the angular dependence of the response of the readout system and, consequently, to also increase the repeatability of the measurements.

A top cap allows to easily place and remove the vial from the reader, making the user able to process numerous detectors in a short time window; this last feature may be handy in photo-neutron contamination studies, where several vial SDDs may be displaced in the proximity of a LINAC or inside the inserts of a dosimetry phantom, and sequentially read before and after exposure.

Summarizing, the goals of the previous design were the following:

- High sensitivity, accuracy, and precision of the scattered light measurements.
- Low angular dependence of the response.
- Compactness.
- High reading rate of detectors for radiation contamination studies.

Since the readout system is also relatively compact (64 mm × 64 mm × 64 mm), it is suitable to be used as a personal alarm dosimeter (as a passive detector equipped with an active readout).

B. Readout System for Vessel Detectors

Fig. 5 illustrates the design and assembly of 3D printed holders employed for the larger vessel containers. These containers, constructed of tempered glass, can withstand pressures of up to 5 MPa and are produced by Büchi Labortechnik AG.

The active detection volume for each module is Ø43 mm × 95 mm. However, when multiple modules operate in tandem, this active volume can be significantly expanded. For bubble detection during our experiments, we used only two photodiode boards and two LED boards per module. The main objective of this system was not to achieve high counting accuracy or minimal angular response dependence, but rather to deliver a rapid binary response - indicating the presence or absence of prompt fission neutrons, as could be the case during an active interrogation of a cargo container housing SNMs.

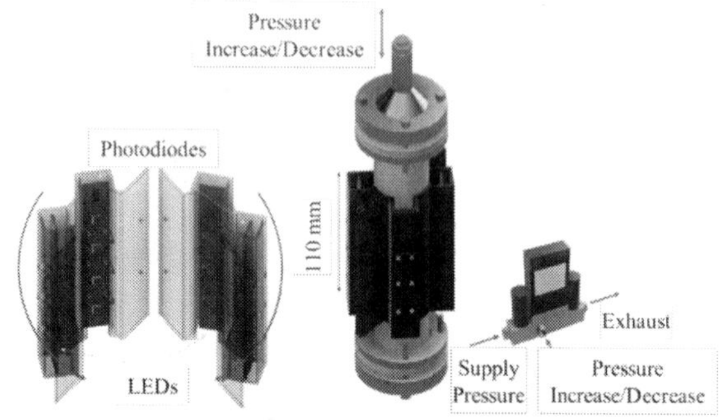

Fig. 5. Architecture of the optoelectronics readout system for large vessel detectors.

Each vessel container is equipped with a digital proportional-integral-derivative (PID) pressure controller from the PCD series by Alicat Scientific. This device serves to adjust the detector's energy threshold response to fast neutrons and, if necessary, repressurize the formed bubbles back into their liquid phase.

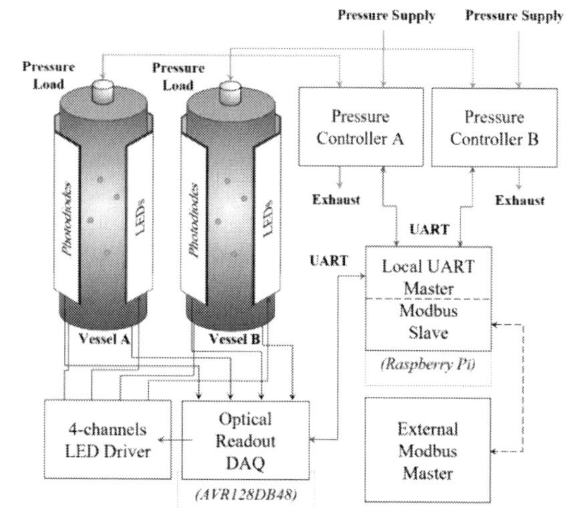

Fig. 6. Architecture of the optoelectronics readout system for large vessel detectors featuring a 100% duty cycle measuring time.

Such a configuration cannot inherently maintain a 100% duty cycle measurement time. As the recompression of bubbles demands several hours, the system is obligated to alternate between a neutron-sensitive low-pressure state and a neutron-insensitive high-pressure state. Furthermore, the base level of scattered light signals depends on both the operative pressure and temperature, as well as the optical properties of the used chemical composition, rendering the system susceptible to long-term drift.

To remedy this and ensure a continuous measurement time, we redesigned the system architecture by utilizing two separate SDDs, each equipped with a dedicated pressure controller (Fig. 6 and Fig. 7). With this setup, one pressure controller maintains a low operative pressure within the first vessel SDD while the other ensures a high operative pressure within the second vessel, and vice versa.

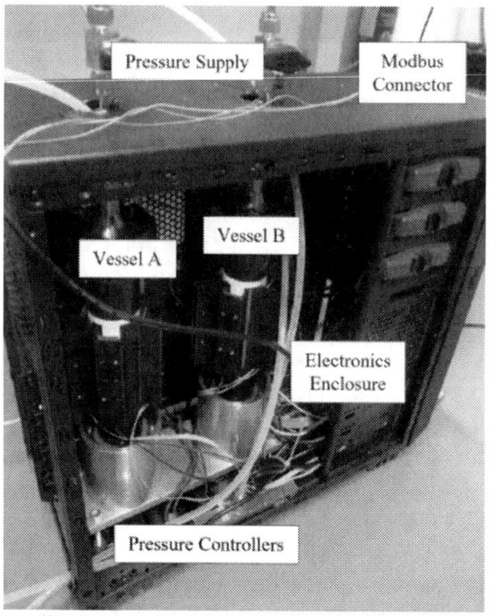

Fig. 7. Realization and assembly of optoelectronics readout system for large vessel detectors featuring a 100% duty cycle measuring time.

The responsibilities of the pressure controllers interchange after a set period, thereby achieving a 100% duty cycle measurement time. Moreover, we incorporated a server for a Modbus RTU communication protocol in the firmware to ensure interface compatibility with external industrial networks. For this task, we employed a commercial single-board computer (Raspberry Pi) and a multi-thread approach for effective control and communication between the devices. The primary threads included a Modbus slave thread for parsing requests from the Modbus master, a data polling thread for querying new data entries from pressure controllers, optoelectronic devices, and a temperature sensor every second, a serial event handler thread for managing received serial data asynchronously and storing them in dedicated data registers, and a main thread for system initialization and operation.

IV. Results and discussion

In this section, we present the findings from our experiments testing the optoelectronic readout for vial and vessel SDDs. The experiments were conducted at the Laboratory of Nuclear Measurements, Department of Industrial and Civil Engineering, University of Pisa, Italy. Four vial detectors from the same production batch (C318 halocarbon droplets), labeled #1 to #4, were exposed to a ^{241}Am-Be fast neutron source. This was done to evaluate several key characteristics:

- The signal shape generated by the readout system when bubbles are formed.
- The correlation between the recorded signal and the number of formed bubbles, which can range up to several hundred.
- The SDDs' sensitivity when paired with the optoelectronic readout system.

- The angular sensitivity of the readout system along with the consistency of its measurements.

Fig. 8 illustrates the arrangement used for these measurements.

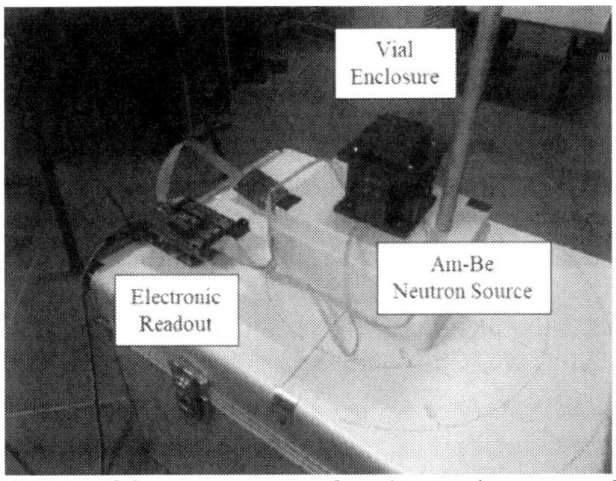

Fig. 8. Setup of the measurements performed to test the response of the readout system for vial SDDs.

Starting from vial #1, once housed within the reader, it was exposed to a ^{241}Am-Be fast neutron source (67'000 n/s) situated at a 50 mm inter-axis distance for a duration of 4 hours. Before and following each exposure, the light scattering signal was systematically measured. After the signal was successfully recorded, the vial was removed and repositioned 10 times, allowing us to account for any geometric variation that might have been introduced during the repositioning process. This process was executed with a duration of 10 seconds for each respective position. To evaluate the repeatability of our system behavior, the experiment was performed 4 times over, each time re-pressurizing the system to transition the formed bubbles back to the liquid phase. This iterative procedure allowed for a rigorous assessment of the system's consistent performance under identical conditions. All tests were performed in a controlled environment with a consistent temperature of 23.5 ± 0.5 °C.

The reconstructed scattered light signals for the four tests are shown in Fig. 9.

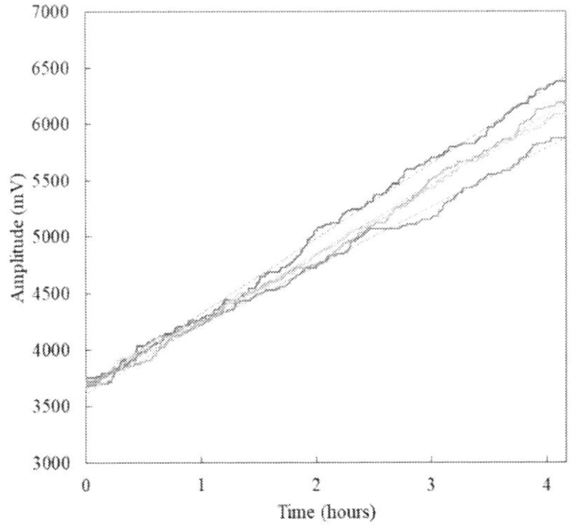

Fig. 9. Response of vial SDD N° 1 during 4-hour exposures to a ^{241}Am-Be neutron source.

During the 4-hour exposure periods, the combined signals from the photodiode arrays saw an increase of over 2000 mV. Overall, the signal amplitudes demonstrated a notably linear progression over time, and, presuming a consistent vaporization rate, over the number of formed bubbles.

Interestingly, data revealed a significant variance in the increase of signals post-exposure each time, fluctuating roughly between 2100 mV and 2700 mV. This variance corresponded with a significantly different bubble count on each occasion, ranging from 148 to 178 bubbles on average (Table II). These bubble counts were manually determined through offline processing of images captured after the experiment (average value of three pictures taken at 120 deg from each other), using ImageJ image analysis software and its multi-point tool.

TABLE III
RESPONSE OF VIAL #1 DURING 4-HOURS EXPOSURES TO A ^{241}AM-BE NEUTRON SOURCE.

Test No	R^2	ΔV (mV)	ΔN (bubbles)	Sensitivity (mV/bubble)
1	> 0.99	2650.9 ± 41.7	178 ± 7	14.89 ± 0.63
2	> 0.99	2203.5 ± 93.0	148 ± 5	14.88 ± 0.81
3	> 0.99	2452.0 ± 37.4	168 ± 6	14.59 ± 0.71
4	> 0.99	2371.9 ± 87.5	162 ± 4	14.64 ± 0.65

According to an ANOVA statistical test, it was found that the sensitivity (mV/bubble) values across the four tests did not significantly differ from each other ($p > 0.05$). As a result, it seemed reasonable to assume a single average sensitivity value for vial #1. The results indicate that the variance in signals observed across the four repetitions can be primarily attributed to differences in the number of bubbles formed post-exposure, rather than inaccuracies in the optoelectronic readout system's readings.

To assess the number of bubbles formed based on the increase of voltage due to the increase of scattered light, the system's response was modeled with an error in variables linear regression model and fitted using the method from York et al. [19] (1):

$$\begin{gathered} \Delta N = \Delta N^* + \varepsilon_N = \ k \times \Delta V^* + \varepsilon_N \\ \Delta V = \ \Delta V^* + \eta_V \end{gathered} \quad (1)$$

where ΔN and ΔV are the measured values of bubbles and voltage respectively, ΔN^* and ΔV^* are the true values of bubbles and voltage respectively, ε_N and η_V are the measurement errors, and k is a calibration factor.

An identical sequence of measurements and analytical procedures were subsequently conducted on vials #2, #3, and #4 (identical chemical composition of C318, comparable droplet sizes, and gel composition). The sensitivity determined by fitting the data for each vial revealed no significant discrepancy among the repeated tests, with ANOVA testing always returning a $p > 0.05$. This supports the assumption that it is reasonable to apply a single calibration factor for each vial.

However, a noticeable difference was detected in the average total number of bubbles formed between the different vials, suggesting that the fast neutron detection efficiency might considerably vary between different vials of the same batch since a Poisson statistic cannot entirely account for this variability.

Furthermore, a significant difference in sensitivity was also discerned ($p < 0.05$), which points towards a potential variation in the optical properties across different detectors (Table III).

TABLE IIIII
UNITS FOR SOME MAGNETIC PROPERTIES

Vial ID	Vial Sensitivity (mV/bubble)	k (bubbles/mV)	N_0 (bubbles)
#1	14.89 ± 0.37	0.0678 ± 0.0017	0.0 ± 1.3
#2	13.71 ± 0.44	0.0729 ± 0.0024	0.0 ± 1.5
#3	14.17 ± 0.44	0.0702 ± 0.0022	0.0 ± 1.5
#4	15.49 ± 0.46	0.0646 ± 0.0019	0.0 ± 1.8

N_0 expresses the number and associated error of estimated bubbles without a voltage increase.

Consequently, it would be advantageous to compute a different calibration factor for each vial, when feasible, to minimize the maximum error when estimating the number of formed bubbles using a post-exposure readout value. Both the accuracy and precision of the measurement are crucial when deducing the number of bubbles from the post-exposure voltage increase. This aspect is particularly pertinent in photo-neutron contamination studies, where a single optoelectronic device could be employed to sequentially read a substantial number of detectors positioned around the area of interest (for instance, inside a dosimetry phantom or near a LINAC).

The system's response's angular dependence was also evaluated. At the end of each exposure, the vials were manually rotated around their longitudinal axis from 0 to 360 degrees in 45-degree increments. The signal was recorded for 10 seconds and subsequently averaged over time. Although the difference between readings taken at various angles is statistically significant when compared to the baseline noise level, the variation is minimal (approximately 2%). This angular dependence of the response needs to be considered, but it's comparable to the error incurred when merely removing and re-positioning the vials. As a result, this aspect doesn't degrade the accuracy or precision of the readout system characterization as described earlier.

V. CONCLUSIONS

This study demonstrated the potential of optically stimulated SDDs as effective neutron detectors, given their versatile attributes including passive operation, flexible size, and gamma insensitivity. These characteristics, coupled with the capability of tailoring their response to different neutron energies by modifying chemical compositions and operative thermodynamic parameters, position SDDs as promising tools for multiple applications such as homeland security, neutron spectrometry, personnel dosimetry, and radiation safety.

The core of our investigation centered on an optoelectronic readout system for SDDs based on light scattering principles. This system relied on LED light beams interacting with the SDD emulsion, causing measurable attenuation and scattering, subsequently detected by photodiodes. The quantity of scattered light served as a marker for the dose delivered to the detectors.

Our project utilized SDDs in various glass containers and tailored optoelectronic readout systems to meet different design goals. For vial SDDs, we aimed for high sensitivity, accuracy, precision, linearity, and compactness. These attributes make them ideal for use as an active pocket personal dosimeter and a portable SDDs reader in photo-neutron contamination studies. Conversely, the vessel SDDs system focused on detecting bubble formation within seconds in large volumes and utilized digital pressure controllers to manipulate the operative pressure.

The vial SDDs readout system showed a very linear increase in the amount of scattered light (and hence bubble formation) over time, with minimal differences in sensitivity in repeated tests. Nevertheless, a noticeable difference was observed in the optical readout sensitivity and neutron detection efficiency among the tested vials. These findings highlight the importance of individual calibration for each vial but also suggest the potential for a global calibration factor with a higher readout uncertainty.

The developed system may emerge as a cost-effective candidate for neutron spectrometry after processing the acquired data with deconvolution algorithms.

Overall, our study underscores the potential of SDDs and optoelectronic readouts in various domains and sets the stage for further refinements and applications of this technology.

REFERENCES

[1] R. T. Kouzes, J. H. Ely, L. E. Erikson, W. J. Kernan, A. T. Lintereur, E. R. Siciliano, D. L. Stephens, D. C. Stromswold, R. M. Van Ginhoven, and M. L. Woodring, "Neutron detection alternatives to 3He for national security applications," *Nuclear Instruments and Methods in Physics Research Section A: Accelerators, Spectrometers, Detectors and Associated Equipment*, vol. 623, no. 3, pp. 1035-1045, 2010. [Online]. Available: https://doi.org/10.1016/j.nima.2010.08.021.

[2] C. W. E. van Eijk, "Neutron detection and neutron dosimetry," *Radiation Protection Dosimetry*, vol. 110, no. 1-4, pp. 5–13, Aug. 2004. [Online]. Available: https://doi.org/10.1093/rpd/nch155.

[3] R. C. Runkle, A. Bernstein, and P. E. Vanier, "Securing special nuclear material: Recent advances in neutron detection and their role in nonproliferation," *Journal of Applied Physics*, vol. 108, no. 11, p. 111101, Dec. 2010. [Online]. Available: https://doi.org/10.1063/1.3503495.

[4] T. Gozani, "The role of neutron-based inspection techniques in the post 9/11/01 era," *Nuclear Instruments and Methods in Physics Research Section B: Beam Interactions with Materials and Atoms*, vol. 213, pp. 460-463, 2004. [Online]. Available: https://doi.org/10.1016/S0168-583X(03)01590-8.

[5] T. Roy, Y. Kashyap, M. Shukla, P. Singh, and R. Baribaddala, "Fast neutron interrogation of special nuclear material using differential die-away technique," *Applied Radiation and Isotopes*, vol. 176, p. 109896, 2021. [Online]. Available: https://doi.org/10.1016/j.apradiso.2021.109896.

[6] R. Sarkar, B. K. Chatterjee, B. Roy, and S. C. Roy, "Radiation detection by using superheated droplets," *Radiation Physics and Chemistry*, vol. 75, no. 12, pp. 2186-2194, 2006. [Online]. Available: https://doi.org/10.1016/j.radphyschem.2005.10.007.

[7] F. d'Errico, "Status of radiation detection with superheated emulsions," *Radiation Protection Dosimetry*, vol. 120, no. 1-4, pp. 475-479, Sep. 2006. [Online]. Available: https://doi.org/10.1093/rpd/ncj018.

[8] F. d'Errico, "Radiation dosimetry and spectrometry with superheated emulsions," *Nuclear Instruments and Methods in Physics Research Section B: Beam Interactions with Materials and Atoms*, vol. 184, no. 1-2, pp. 229-254, 2001. [Online]. Available: https://doi.org/10.1016/S0168-583X(01)00730-3.

[9] D. T. Bartlett, R. Tanner, and D. J. Thomas, "Active neutron personal dosemeters - A review of current status," *Radiation Protection Dosimetry*, vol. 86, no. 2, pp. 107-122, 1999. [Online]. Available: https://doi.org/10.1093/oxfordjournals.rpd.a032930.

[10] F. d'Errico and A. J. Bos, "Passive detectors for neutron personal dosimetry: state of the art," *Radiation Protection Dosimetry*, vol. 110, no. 1-4, pp. 195-200, 2004. [Online]. Available: https://doi.org/10.1093/rpd/nch129.

[11] S. C. Roy, "Superheated liquid and its place in radiation physics," *Radiation Physics and Chemistry*, vol. 61, no. 3-6, pp. 271-281, 2001.

[Online]. Available: https://doi.org/10.1016/S0969-806X(01)00250-X.

[12] R. E. Apfel and Y. C. Lo, "Practical neutron dosimetry with superheated drops," *Health Physics*, vol. 56, no. 1, pp. 79-83, 1989. [Online]. Available: https://doi.org/10.1097/00004032-198901000-00007.

[13] S. Gao, G. Zhang, B. Ni, C. Zhao, H. Zhang, Y. Guan, Z. Chen, C. Xiao, C. Liu, and C. Liu, "Acoustic response of superheated droplet detectors to neutrons," *Nuclear Instruments and Methods in Physics Research Section A: Accelerators, Spectrometers, Detectors and Associated Equipment*, vol. 668, pp. 21-25, 2012. [Online]. Available: https://doi.org/10.1016/j.nima.2011.11.006.

[14] F. d'Errico, R. Nath, S. K. Holland, M. Lamba, S. Patz, and M. J. Rivard, "A position-sensitive neutron spectrometer/dosimeter based on pressurized superheated drop (bubble) detectors," *Nuclear Instruments and Methods in Physics Research Section A: Accelerators, Spectrometers, Detectors and Associated Equipment*, vol. 476, no. 1-2, pp. 113-118, 2002. [Online]. Available: https://doi.org/10.1016/S0168-9002(01)01403-6.

[15] R. E. Apfel, "The superheated drop detector," Nuclear Instruments and Methods, vol. 162, no. 1-3, pp. 603-608, 1979. [Online]. Available: https://doi.org/10.1016/0029-554X(79)90735-3.

[16] R. E. Apfel, S. C. Roy, and Y.-C. Lo, "Prediction of the minimum neutron energy to nucleate vapor bubbles in superheated liquids," *Physical Review A*, vol. 31, no. 5, pp. 3194-3198, 1985. [Online]. Available: https://doi.org/10.1103/PhysRevA.31.3194.

[17] T. Wriedt, "Mie theory: a review," in *The Mie Theory: Basics and Applications*, 2012, pp. 53-71.

[18] L. Zhao, et al., "Temperature-Dependent Efficiency Droop in GaN-Based Blue LEDs," IEEE Electron Device Letters, vol. 39, no. 4, pp. 528-531, April 2018, doi: 10.1109/LED.2018.2805192.

[19] D. York, N. M. Evensen, M. López Martínez, and J. De Basabe Delgado, "Unified equations for the slope, intercept, and standard errors of the best straight line," American Journal of Physics, vol. 72, no. 3, pp. 367–375, March 2004. [Online]. Available: https://doi.org/10.1119/1.1632486.

Methodical Optimization of Mutual-Shielding Directional Gamma Detector

Vax Eran[1,2,*], Ben David Nadav[3], Salomon Yossi[1], Knafo Yakir[1], Dadon Shay[1], Stern Adrian[2], Behar Ehud[4], Osovizky Alon[1,5]
[1]Electronics & Control Laboratories, Nuclear Research Center - Negev, Israel
[2] Ben Gurion University of the Negev
[3] Israel Atomic Energy Commission (IAEC)
[4] Technion - Israel Institute of Technology
[5] Radiation Detection Department, Rotem Industries Ltd, Israel
(*) eranv@nrcn.gov.il

Abstract— **Mutual-Shielding Directional Gamma Detector excels at sensitivity and can achieve good angular accuracy and a large field of view. In this paper, we present, in detail, a method to optimize a mutual shielding directional detector array regarding the scintillators positioning for a given number and size of detection units. The optimization goals are improvement of accuracy and sensitivity.**

The optimization uses two types of fast, non-Monte-Carlo simulations since the simulation needs to generate measurements from all directions for each array formation tested. Both these simulations are explained and demonstrated.

The presented method uses the ambiguity function to assess the accuracy and identify directions that may generate errors due to similar responses from other directions. Then, addressing these problematic directions can be fixed by changing the scintillators' positions or adding a partial shield to change the response at one of the similar directions.

Keywords **—Directional Gamma Detector, Detectors Array, Mutual Shielding.**

I. Introduction

Typical gamma radiation detector measures the photon flux passing through its volume, estimating the gamma field intensity only and indicating whether a source is present via an alarm. The next step in radiation detection is to find the Direction Of Arrival (DOA) of the Gamma photons. This new operational data can help in many scenarios, such as locating concealed radioactive sources in homeland security, searching hotspots in a post-radiological event, or during Decontamination and Decommissioning (D&D).

We can count the coded mask aperture, Compton scatter, and collimation among the various principles used for Gamma directionality. The significant disadvantages of these methods are limited Field OF View (FOV), medium to low efficiency, and limited energy response. Alternatively, a compact array structure of detectors uses mutual shielding to generate significant difference measurements from different directions. This array enables high sensitivity, which can be increased since there is no limit to the sensor size. Although mutual shielding is highly effective for low-energy Gamma, it can be easily adjusted to higher energies by selecting larger detectors or adding more detector layers to increase the total cross-section length from every direction. Another significant advantage is that there is no inherent limit to the FOV, except the need to construct an intelligent structure with a unique response from every DOA required.

A. *Previous works based on mutual-shielding*

Several previous works suggested detectors based on the mutual-shielding principle. The initial studies in [1] - [4] used 3-7 scintillators in a close-packed structure to measure the different responses of a single source on a 2-D plane and estimate its angular direction.

In 2017 [5], our group introduced a concept of raising the central detector resembling a sun-dial (Fig. 1b). This structure adds a variation of the readings in different elevations due to the occlusion from the central scintillator. Using Monte-Carlo (MC) simulations of five or six cylindrical scintillators surrounding a raised central one, we achieved a mean angular error of 7°-9° or 4°-7° respectively for all elevations. In comparison, for a flat structure, with the central scintillator at the same height as the others, the worst mean angular errors at some elevations were 23° and 19°, respectively.

The work of McCann and Chen et al. [6] [7] presented a new 2x4 close-packed structure Fig. 1c) using rectangular-cuboid CsI(Tl) scintillators to measure directionality from the whole 4π directions. The direction estimation method was simple: for each axis (x, y, and z), they calculated the total difference in count rates between the front four scintillators and the rear four scintillators and estimated the azimuth and elevation from this 3-dimensional vector. The count rates were measured for a specific energy photopeak. When simulating a one mCi ^{137}Cs point source at a 10 m distance, they calculated the angular Root Mean Square Error (RMSE). The results showed a 10°-30° error for a short measurement time (<20s), 6°-18° error for 30s measurements, and up to 8° error for 150s measurements, where the results variations were due to the angle of incident.

Rahin et al. [8] suggested putting a large number of small cubic scintillators to generate a large measurement vector, rich with DOA information. This detector aimed to identify and localize deep space Gamma Ray Bursts (GRBs), with a spectrum dominant in the low Gamma energies. Their prototype Fig. 1c) used a random configuration of 90 cubic scintillators $(9\text{mm})^3$ coupled to silicon photomultipliers (SiPMs) occupying 0.25 l volume. They also suggested a planned detector with 350 cubic scintillators occupying 1 l volume.

In continued work [9], our group improved the instrument's accuracy and sensitivity compared to the random configuration. The optimization used a fast approximate simulation of slowly modified configurations, limited only in their volume and not by the number or position of the scintillators. Using a gradient descent of a loss function combining the sensitivity and accuracy, we improved the configuration from angular MSE of 3.73° to 2.84° and ~6% in the sensitivity.

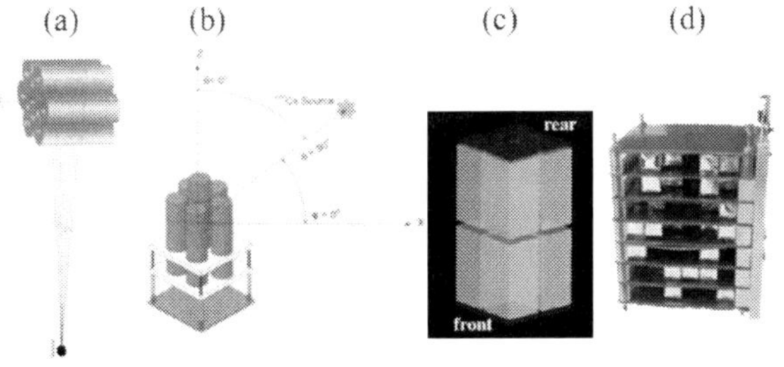

Fig. 1. Previous works of mutual-shielding directional detectors: (a) with azimuth only [1]; (b) added elevation capability for half a sphere (2π) using the 'sun-dial' array [5]; with whole 4π directionality using 2x4 cuboid scintillators array coupled to SiPMs [6]; (c) prototype of 90 small cubic scintillators randomly placed [8] for high accuracy.

In this work, we introduce an analytical methodology to improve the spatial configuration of a mutual shielding array. The optimization is aimed toward both the accuracy and the sensitivity of the mutual shielding detector array under the constraint of keeping the same number and volume of scintillators. The accuracy is improved for the expected mean angular error as well as for the best and worst angle errors. To demonstrate the optimization method, we optimize our suggested seven scintillators 'sun-dial' directional detector as a test case throughout this paper. The aim is to develop a detector to locate ground contaminations (hotspots) for post-radiological events. The detector will be located on a robot at a 1.5 m height; hence, it implies that the detector's directions of interest are the bottom half sphere (2π).

II. Optimization

We started by selecting the scintillator type and size, which will be constant throughout the optimization phase. We decided on six identical cylindrical NaI(Tl) scintillators, each with a 25 mm diameter and 50 mm height, while the central scintillator has the same diameter but double the height (100 mm), making the total scintillation volume of 0.2 l, which is comparable to a 2"x2" and 3"x3" cylindrical scintillators with volumes of 0.1 l and 0.35 l respectively.

A. simulation methods

To optimize the 7-scintillators array directional detector, we used simulations. The simulations must generate measurements from all DOAs for every array structure under evaluation. This requirement implies that the simulations cannot be slow to compute to try different structures, which means Monte Carlo (MC) physics simulations are unsuitable for array optimization.

The first type of simulation we used is called 'zero energy' simulation. Here, we simulate very low-energy gamma photons, almost entirely absorbed in the dense scintillator, similar to light photons. Given that, the simulation starts by drawing the 3-D structure of the dense materials in the detector, such as the scintillators, iron, or lead, and neglecting any light materials like plastics or aluminum. Each sensor is colored differently, and the detector is viewed from a specific viewpoint, representing the source position. By using perspective geometry, built in most plotting tools, it is easy to measure the area exposed to the source in a direct line of sight. This area is linearly linked to how many zero-energy photons will reach each sensor. Fig. 2a. illustrates the simulation. When the seven scintillators are rotated to be viewed from the side with no elevation and 50° azimuth relative to detector #1 (dark blue), the two rear scintillators, #3 (yellow) and #4 (purple), are almost entirely occluded except for a small gap between the front scintillators; hence their reading will be close to zero, while about a quarter of detector #2 (orange) is visible. The bottom numbers at each detector in Fig. 2b represent the area measured in arbitrary units, while the percentage shows the expected readings at each detector relative to the total counts of the array.

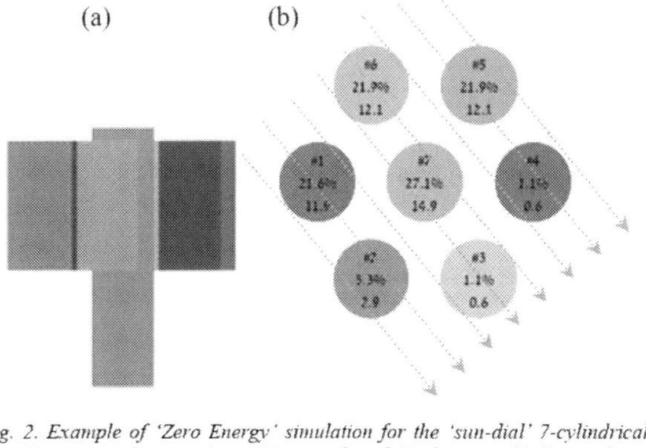

Fig. 2. Example of 'Zero Energy' simulation for the 'sun-dial' 7-cylindrical scintillators. (a) the view from 50° azimuth and 0° elevation (b) the results of the area measured (bottom number, arbitrary units) and as a percentage of the total area (middle number).

This type of simulation is accurate for low energies and is very fast to compute. We simulated every one-degree over 360° azimuth with 15° elevation steps. On the other hand, the simulation accuracy degrades as the incident photon energies are higher, depending on the scintillator's size and density.

To get an accurate spatial calibration table for higher energies, we use another simulation method we address as an 'attenuation-only' simulation. This simulation uses ray-tracing of multiple direct rays from a point source toward the array. For each ray, it calculates the cross-section length at each scintillator and the order in which the ray hits them. In an iterative loop from the first detector hit to the last, we calculate the absorption in each scintillator according to the Linear Energy Absorption Coefficient (LEAC) and the transfer percentage towards the next scintillator using the Linear Attenuation Coefficient (LAC).

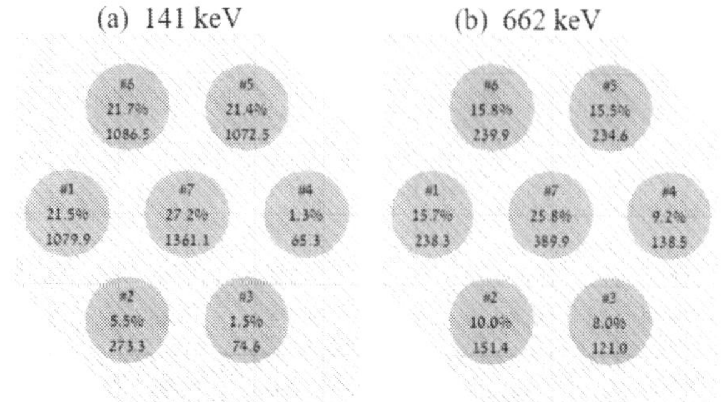

Fig. 3. Example of attenuation-only simulation results for DOA of 50° azimuth and 0° elevation through 7-scintillators array. The results show the number of rays time, their probability of interaction (bottom number, arbitrary units), and the percentage of expected events per scintillator relative to the total array measurements (middle number).

It is very clear from the percentage out of the total counts and even more so when examining the absolute expected counts (bottom number in arbitrary units), where the front detectors (1,5 and 6) show a reduction in the expected counts due to the lower probability of event in the detectors cross-section length. In contrast, the rear detectors (3 and 4) show an increase in the absolute measurement because fewer photons interacted with the front detectors, and the photon flux through the rear detectors is significantly higher.

B. *Optimization methodology*

The optimization process analyses the angular response using the ambiguity function. The ambiguity function (ρ) is a two-dimensional function that evaluates the difference between two vectors, $\bar{a}(\theta_1)$ and $\bar{a}(\theta_2)$, measured from two directions θ_1 and θ_2. In our case, each vector is the seven photo-peak counts from the seven detectors.

$$\rho = \sum_{n=1}^{N} \left| \frac{\bar{a}(\theta_1)}{\sum \bar{a}(\theta_1)} - \frac{\bar{a}(\theta_2)}{\sum \bar{a}(\theta_2)} \right| \qquad (1)$$

An example of the ambiguity function generated from the zero-energy simulation is shown in Fig. 4. Here, the x and y axes span from 0°-360° for the azimuth only, while the elevation was kept level at 0° as seen on the right side of the figure. The minimum value of the ambiguity is zero and occurs at the diagonal since the difference between a vector and itself is zero. If we treat the ambiguity value as the height over θ_1, θ_2 plane, we get a valley along the diagonal, where its steepness corresponds to the accuracy attainable from the directional detector configuration. Hence, we can select a contour (a constant value of ambiguity) and measure the distance of the contour line from the diagonal, which will give a performance assessment. The ambiguity value corresponds to testing the angular error of a specific Signal-to-noise ratio (SNR), which translates to the number of counts measured. For example, in Fig. 4a of the ambiguity for the 'sun-dial' array, by selecting an ambiguity value of 0.1, we get an expected angular error of 14.1° at azimuth 60° and a minimum error of 9.5° in other azimuths.

The advantage of the method is that running the simulation once and without adding random noise in multiple runs produces a qualitative assessment of the expected error without linking it to a specific SNR value.

III. RESULTS

A. *First optimization attempt: increasing scintillator distance*

We started the array optimization from the proposed seven scintillators packed together in a 'sun-dial' shape with a minimal 5 mm distance between the scintillators caused by the aluminum case thickness of the cylindrical detectors. As described earlier, we selected an arbitrary ambiguity value of 0.1 and measured the distance from the diagonal as a qualitative test. When we increased the spacing between the detectors from 5 mm to 10 and 15 mm, the ambiguity function (Fig. 4) showed that for most DOAs, there is a reduction in the expected azimuth error from 9.5° to 7° and 5.6° respectively. However, every 60°, there is an increase in the expected error to 18.4° and 21.2°.

The improvement caused by distancing the detectors is probably due to a sharper occlusion change for a smaller DOA shift.

To understand the decrease in accuracy every 60°, we examine the response plot (Fig. 5). It shows that every 60°, when there are three detectors fully exposed to the source and three rear detectors, a symmetry causes the same response vector from two angles, e.g., 50° response is the same as 70° response, demonstrated in the views showed at Fig. 5a, and Fig. 5. This symmetry happens again where 110° is the same as 130° and so on, as marked in the response plot on Fig. 5c.

When inspecting, the response of any single detector in a close-packed 'sun-dial' array shows there are four states for the detector: (1) fully exposed over a 120° section, (2) fully occluded over a 120° section, and (3+4) the transitions between the states each taking about 60°. When the elevation is not zero, the detector is not fully occluded except for a specific angle behind the central extended detector, as seen in Fig. 5c.

The conclusion from the analysis of our first attempt is that in order to get better accuracy and fewer potential errors, we need to (a) break the symmetry seen from every side of a front detector, (b) generate more changes to the responses of each detector to differentiate the response vector from different DOAs.

B. *Second optimization attempt*

To break the symmetry, we moved the three odd-numbered detectors at a different distance and height than the three even-numbered detectors. The resulting array is labeled 'triangles' and can be seen in Fig. 6.

The increased distance of detectors 1,3, and 5 generates a large gap that creates a peaking gap for a rear detector (resembling collimation), such as the visibility of detector 6 (light blue) through the gap between the blue and grey detectors in Fig. 6b.

These uneven gaps generate more changes in the azimuthal response, highlighted in Fig. 6c, which enhance the DOA data and should enable a tighter DOA estimation.

The ambiguity plot of the 'triangles' array clearly shows the much steeper valley along the diagonal, implying better accuracy. However, a different phenomenon has manifested.

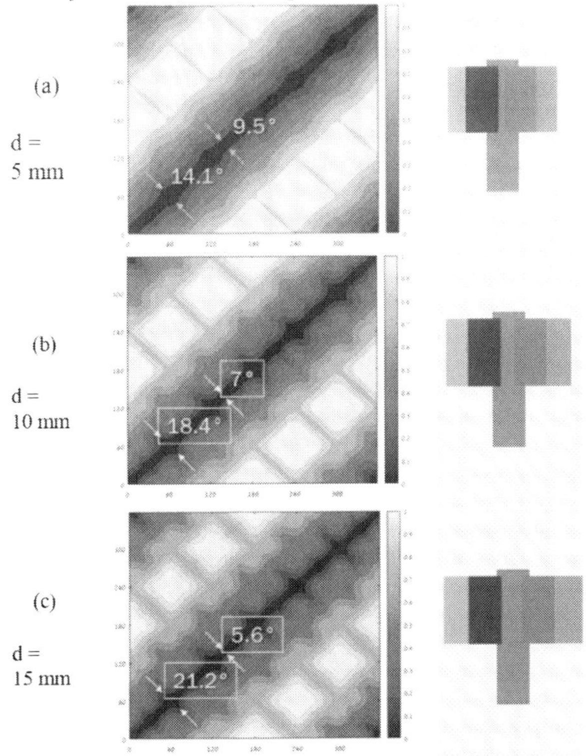

Fig. 4. Ambiguity function of azimuth response for 'sun dial' array with 5 mm (a), 10 mm (b), and 15 mm (c) distance between scintillators.

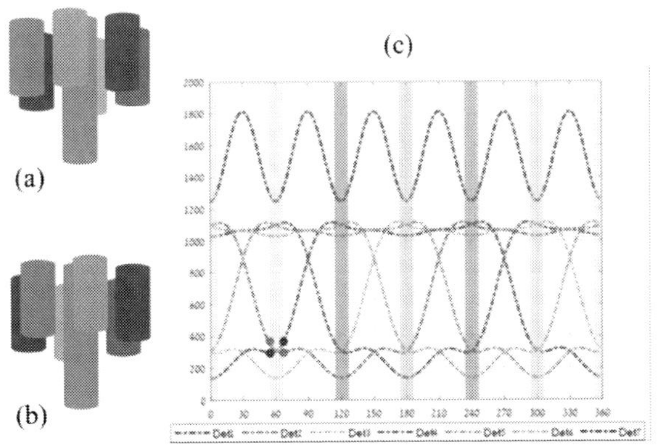

Fig. 5. Illustration of same response from different DAO using zero energy response of 15 mm spaced 'sun-dial' array: (a) view from (50°, -15°) (azimuth, elevation). (b) view from (70°, -15°). (c) response graph showing the symmetrical response on both sides of the 60°,120°..., and 360° azimuths.

Three dips exist apart from the diagonal (the plot shows six; however, the upper triangle mirrors the lower one). These dips imply that the area (in zero-energy simulation) or the rays' absorption (in attenuation-only simulation) at each detector is the same for two different views of the array. These similarities happen at specific azimuths: 28° vs. 332° (demonstrated Fig. 7b), 148° vs. 92° and 268° vs. 212°. The triangle shape and distances were found after several arrays were tried and showed promising results except for these dips.

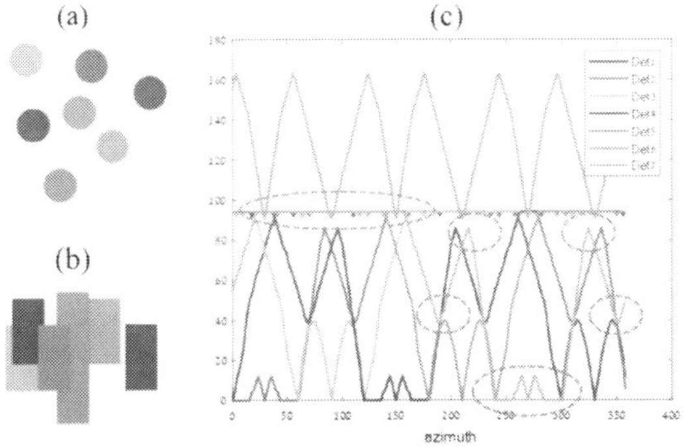

Fig. 6. 'Triangles' array (a) top view (b) side view (c) response plot at zero elevation, circling the six peaks and valleys of det #6 response.

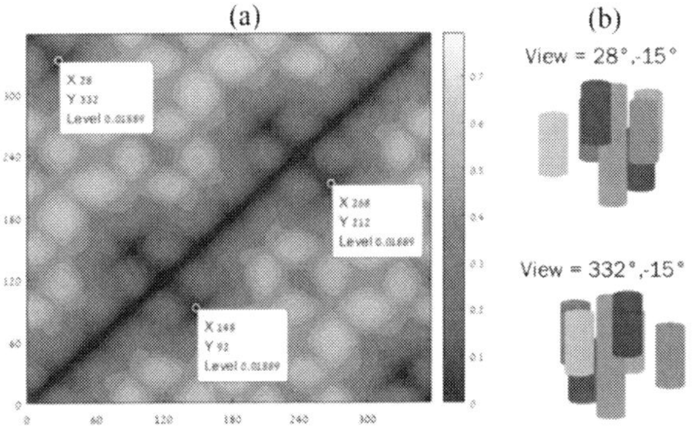

Fig. 7. 'Triangles' array: (a) ambiguity function, generated using attenuation simulation of 141 keV at 15° elevation. The three dips marked suggest the DOA estimation is ambivalent between two azimuths.
A demonstration of the views of such two DOAs is shown in (b), where the area of each detector is the same in both views, although the views are different.

C. Third optimization attempt

We needed to address the resemblances generated at the three dips to benefit from the improved accuracy of the 'triangles' array. To do so, we added three lead slabs to generate differentiation between every two similar azimuths. We carefully placed each slab in front of the front detectors so that every unique angle found at the triangle's dips shields a different detector to modify the response vector. The slabs and their shielding are demonstrated for a specific 2 DOA in Fig. 8b. The difference created by shielding even only one detector for each azimuth mitigates the problem of the dips, as shown in the ambiguity plot in Fig. 8a.

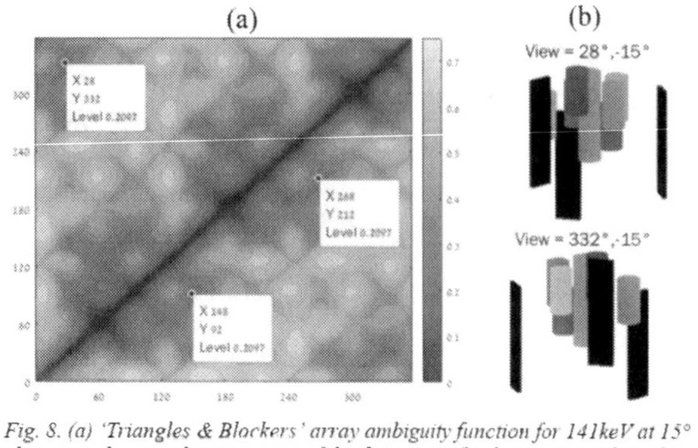

Fig. 8. (a) 'Triangles & Blockers' array ambiguity function for 141keV at 15° elevation, showing the mitigation of the dips issue. (b) demonstrates how the slab shields the yellow detector at 28° azimuth while shielding the purple detector at 332° azimuth.

D. Arrays accuracy comparison

As explained earlier, selecting a specific ambiguity threshold value and measuring the distance to the diagonal is equivalent to measuring the mean angular error for a specific SNR since the diagonal of the ambiguity plot represents a true DOA estimation. Although we cannot determine the SNR value precisely, it is a valid qualitative estimation of the expected error.

In Fig. 9, we plot a qualitative comparison based on the measured distance from true DOA at a constant ambiguity value. It is easily shown that the 'triangles' array (red) performs mostly well except for an area that can give a false DOA with huge errors. The 'sun-dial' array (blue) performance is consistent except every 60°, while the 'triangles & blockers' formation has the lowest mean error for almost all azimuths, with some larger errors every 120°, but still better than the 'sun-dial' expected errors. Since this array showed the best accuracy performance, we must examine the effect of the blockers on the array sensitivity.

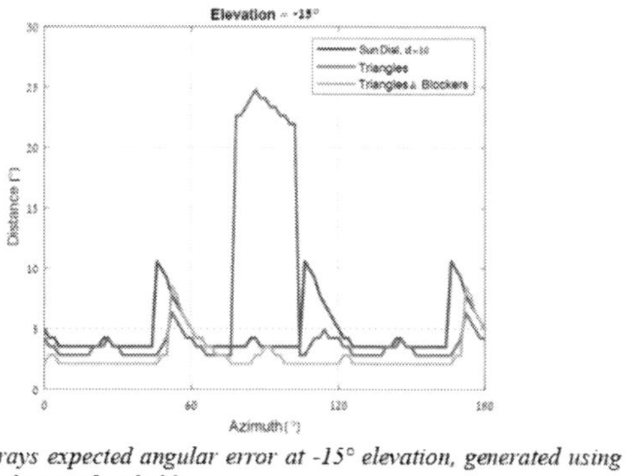

Fig. 9. The arrays expected angular error at -15° elevation, generated using the constant ambiguity threshold.

E. Sensitivity analysis

Any array sensitivity depends on the energy photopeak measured; hence, we use the attenuation-only simulation to evaluate the sensitivity. The simulation sums the probability of each ray interacting with each scintillator and outputs the scintillator absorption as an arbitrary number, which depends on the density of rays generated by the simulation. Keeping the same angular density of rays makes it possible to compare the total absorption of all scintillators in each array as a measure of the sensitivity.

When comparing the total absorption in

, it shows that as we increase the distance between scintillators (for 'sun-dial' arrays and then the 'triangles'), the sensitivity improves, both for the minimal and the average sensitivity. The absolute sensitivity is higher for the 45° elevation due to the larger cross-section area. However, due to the two heights at the 'triangles' array, the sensitivity is slightly higher at zero elevation, which is also reflected in the more significant improvement from 'sun-dial' to 'triangles' at zero elevation.

As expected, adding the lead blockers (marked "T. & B." in

.) decreases the sensitivity relative to the plain 'triangles' array. The mean and minimal sensitivity are higher than the 'sun-dial' arrays. An exception is the case of 141 keV at 45° elevation, where the mean sensitivity is on par with the 'sun-dial' array with a 15 mm distance.

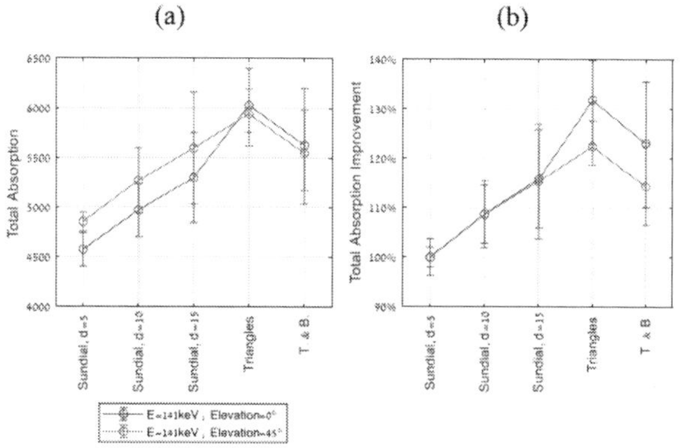

Fig. 10. Sensitivity comparison of the five arrays: (a) Total absorption in arbitrary units. (b) Total absorption relative to the close-packed 'sun-dial' array. The error bar limits mark the maximum and minimum values calculated for all DOAs.

IV. DISCUSSION

This work presents our methodical approach to improving the accuracy and sensitivity of a fixed-size directional detector array. The work was mainly done using two types of simulations: a quick 'zero-energy' simulation based on the built-in 3D image capabilities of the plotting tool, and the second we called 'attenuation only' simulation, based on ray tracing combined with LAC and LEAC parameters. Both of these simulations are much faster than a standard MC simulation, which is vital when examining many forms of detector arrays.

For the selected 25 mm diameter scintillators, it can be seen in Fig. 3. that for the ^{99m}Tc with 141 keV photopeak, the results of the 'attenuation only' simulation are very close to the zero-energy simulation, while for higher energy, such as the ^{137}Cs 662 keV photopeak, the discrepancy is noticeable. This discrepancy indicates that different spatial calibration tables

https://doi.org/10.1051/epjconf/202328810023

will be needed for different energies.

The optimization method used the ambiguity function as a primary tool to estimate the expected angular error and to identify the problematic DOAs. These angles are identified by looking at local minima in the ambiguity map. We hypothesized that increasing the distance between the scintillators will increase the accuracy since a smaller change at the incident angle will have a more significant change in the occlusion for a larger distance. This hypothesis was proved by examining the ambiguity function as sowed in Fig. 9, where the distance to the diagonal represents the noise margin for correct DOA estimation for a constant ambiguity value.

We can use the normalized or un-normalized distance for the ambiguity function. However, we need to normalize the results in actual algorithms since we compare them to the spatial calibration table, measured with a specific count rate for a specific source activity. When normalizing the measurement vector, a basic matched filter algorithm does not consider the absolute number of events; thus, the Poisson distribution effect of having better accuracy at a higher count rate is not taken into account. As a further work, we will adjust and compare several algorithms to find the best one for a single source DOA with background estimation.

We are constructing a directional detector prototype based on the suggested optimized formation – the 'triangles & blockers' array. The development will include an embedded DAQ system to measure seven low-resolution energy spectrums since we do not need to identify isotopes, only measure several energy windows. These energy windows boundaries are determined so they would have approximately constant attenuation, for example, a division of five energy bands: under 80 keV, 80-200 keV, 200-400 keV, 400-700 keV, and above 700 keV. As part of the DAQ system design, we will analyze the absorption and attenuations in the NaI scintillators and decide on the required energy bands for the measurements and that will introduce minimal errors.

This paper presented a method to optimize a mutual shielding directional detector array regarding the positioning of the scintillators using the ambiguity function and the array simulated response. The method was demonstrated on a seven-scintillators array but can be applied to any other mutual shielding array for low and high energies.

References

[1] A. Pernick, I. Orion, D. Ilzycer and H. Zafrir, "Applications of a Self-Collimating BGO Detector System to Radiological Emergency Response," in *Health physics*, 1997.

[2] Y. Shirakawa, T. Yamano and Y. Kobayashi, "Remote Sensing of Nuclear Radiation Leakage by a Direction Finding Detector," in *IEEE International Symposium on Industrial Electronics (ISIE)*, Montreal, Quebec, Canada, 2006.

[3] N. Schemm, S. Balkir, M. W. Hoffman and M. Bauer, "A Directional Gamma Ray Detector Using a Single Chip Computational Sensor," in *IEEE SENSORS Proceedings*, 2011.

[4] M. J. Willis, S. E. Skutnik and H. L. Hall, "Detection and positioning of radioactive sources using a four-detector response algorithm," *Nuclear Instruments and Methods in Physics Research Section A*, vol. 767, pp. 445-452, 2014.

[5] Y. Ifergan, E. Vax, Y. Yehuda-Zada, B. Sarussi, D. Smadja, S. Dadon, Y. Levin, Y. Knafo, T. Mazor, Y. Kadmon and A. Osovizky, "Improving the radiation detection level and operational data of pre and post event monitoring systems," in *International Conference on Nuclear Security, IAEA*, Viena, 2017.

[6] A. McCann, L. E. Sinclair, P. Saull, C. V. Ouellet, R. Fortin, C. Chen, M. Coyle, R. Mantifel, A. MacLeod, R. V. V. Brabant, J. Buckle, P.-L. Drouin, J. Hovgaard, B. Krupskyy, B. Beckman and B. Fairbrother, "Which is better, a SCoTSS gamma imager, or an ARDUO UAV-borne directional detector?," in *IEEE Nuclear Science Symposium (NSS)*, 2017.

[7] C. Chen, L. E. Sinclair, R. Fortin, M. Coyle and C. C. M. Samson, "In-flight performance of the Advanced Radiation Detector for UAV Operations (ARDUO)," *Nuclear Instruments and Methods in Physics Research Section A: Accelerators, Spectrometers, Detectors and Associated Equipmen*, 2020.

[8] R. Rahin, L. Moleri, A. Vdovin, A. Feigenboim, S. Margolin, S. Tarem, E. Behar, M. Ghelman and A. Osovizky, "GALI, a Gamma-ray burst Localizing Instrument," in *SPIE Astronomical Telescopes + Instrumentation*, 2020.

[9] N. Ben David, E. Vax, M. Ghelman, R. Rahin, S. Tarem, A. Osovizky, E. Behar and A. Zeltzer, "Spatial Optimization for Directional Detectors of γ-Rays Using Approximate Simulations," in *Nuclear Science Symposium*, Manchester, UK, 2022.

[10] S. Haykin and K. J. R. Liu, Handbook on array processing and sensor networks, Wiley-IEEE Press, 2009, pp. 91-95.

[11] V. Batiy, O. Stoyanov, D. Fedorchenko, S. Prohoretz and M. Khazhmuradov, "Mathematical modeling to support gamma radiation angular distribution measurements," 2007.

[12] N. B. David, E. Vax, M. Ghelman, R. R, E. Behar, S. Tarem and A. Osovizky, "Spatial Optimization for Directional Detectors of γ-Rays Using Approximate Simulations," in *IEEE NPSS*, Manchester, 2022.

[13] F. Carrel, R. A. Khalil, S. Colas, D. d. Toro, G. Ferrand, E. Gaillard-Lecanu, M. Gmar, D. Hameau, S. Jahan, F. Lainé, A. Lalleman, F. Lemasle, C. Mahé, J.-E. Maurer, N. Menaa, S. Normand, H. Onillon, N. Saurel, V. Schoepff and H. Toubon, "GAMPIX: A new gamma imaging system for radiological safety and Homeland Security Purposes," in *IEEE NSS*, 2011.

[14] V. Schonfelder, "The Imaging Compton Telescope Comptel on the Gamma Ray Observatory," in *IEEE NSS*, 1984.

[15] G. W.Phillips, "Gamma-ray imaging with Compton cameras," in *Gamma-ray imaging with Compton cameras." Nuclear Instruments & Methods*, 1995.

[16] "h3dgamma products," h3dgamma, 2022. [Online]. Available: https://h3dgamma.com/h100.php.

[17] D. Maier, C. Blondel, C. Delisle, O. Limousin, J. Martignac, A. Meuris, G. Daniel, P.-A. Bausson, O. Gevin, G. Amoyal, F. Carrel, V. Schoepff, C. Mahé, F. Sou, F. Soufflet and M.-C. Vassal, "Second generation of portable gamma camera based on Caliste CdTe," in *Nuclear Inst. and Methods in Physics Research, A*, 2017.

[18] *Polaris-H Measurements and Performance,* H3D Inc, 2015.

[19] C. L. Larsson and T. Jones, "Testing of a new directional gamma survey meter," Defense R&D Canada (DRDC), Ottawa, 2012.

[20] *RadCompass - Directional radiation survey meter,* Ontario, Canada: Industries, BTI - Bubble Technology, 2017.

[21] R. Accorsi, F. Gasparini and R. Lanza, "A coded aperture for high-resolution nuclear medicine planar imaging with a conventional Anger camera: experimental results," *IEEE Transactions on Nuclear Science,* vol. 48, pp. 2411-2417, 2001.

[22] M. Gmar, O. P. Ivanov, F. Laine, F. Lamadi, C. LeGoallerc, C. Mahé, E. Manach and V. E. Stepanov, "Development of a portable gamma camera with coded aperture," in *Nuclear Instruments & Methods in Physics (NIM*, 2006.

[23] J. M. Nilsson, R. R. Finck and C. L. Raaf, "A rotating-slit-collimator-based gamma radiation mapper," *Journal of Environmental Radioactivity,* no. 177, pp. 225-232, 2017.

[24] Y. Shirakawa, T. Yamano and Y. Kobayashi, "Remote sensing of nuclear accidents using a direction finding detector," in *35th Annual Conference of IEEE Industrial Electronics (IECON)*, Porto, Portugal, 2009.

[25] R. Parajuli, M. Sakai, R. Parajuli and M. Tashiro, "Development and Applications of Compton Camera—A Review." Sensors (Basel, Switzerland) 22 (2022): n. pag.," MDPI, Basel, Switzerland, 2022.

[26] M. Ghelman, N. Kopeika, S. Rotman, T. Edvabsky, E. Vax and A. Osovizky, "Design of 4π High-Efficiency Directional Radiation Detector Based on Compton Scattering," *IEEE Transactions on Nuclear Science,* no. 69, pp. 832-839, 1 April 2022.

[27] K. Kojima, T. Nakamori, D. Nemoto, S. Gunji, H. Sato, S. Ito, S.Kato, M. Yoshino, Y. Usuki and J. Kataoka, "Development of Gamma-Ray Detector Sensitive to Source Directions using GAGG(Ce) Scintillators and MPPCs," in *IEEE- NSS*, Strasbourg, France, October 2016.

 https://doi.org/10.1051/epjconf/202328811001

Education, Training and Research at Khalifa University in Support of the United Arab Emirates Nuclear Power Program

Francois Foulon*, Yacine Addad, Saeed Al Ameri, Prasoon Raj, Ahmed Al Kaabi, Akram Al Fantazi
Emirates Nuclear Technology Center, Khalifa University, Abu Dhabi, United Arab Emirates
(*) francois.foulon@ku.ac.ae

Abstract— **In 2009 United Arab Emirates (UAE) has engaged in an ambitious nuclear power program with the construction of four 1400 MWe pressurized water reactor units at the Barakah site. Fourteen years later, three units are in commercial operation, and the fourth one is in the final steps of commissioning. The four units will produce up to 5600 megawatts of electricity, supplying approximately 25% of the UAE's electricity requirements.**

To ensure in such a short period of time the building of the nuclear capabilities and support to the program, Khalifa University has been, since 2010, a key contributor to education and training activities as well as the development of R&D capabilities in the field of nuclear technology.

This paper provides an overview of Khalifa University's activities implemented within the Nuclear Engineering Department, the Gulf Nuclear Energy Infrastructure Institute and the Emirates Nuclear Technology Center (ENTC). This joint Center, which embarks the Emirates Nuclear Energy Corporation and the Federal Authority for Nuclear Regulation, is a hub that addresses the research requirements of the nuclear stakeholders for the delivery of safe and efficient nuclear energy as well as the deployment of advanced nuclear technologies in the UAE. The paper briefly describes the current research projects implemented at the ENTC and gives reference related recent publications.

Keywords — **United Arab Emirates nuclear power program, Capacity building, Academic education, Vocational training, Research program.**

I. Introduction

THIS paper provides an overview of the education, training and research activities that have been implemented, since 2010, at Khalifa University in Abu Dhabi, to support the deployment of the nuclear power program (NPP) in the United Arab Emirates (UAE). Section II briefly describes the UAE nuclear power program with its objectives and status. Section III and IV are dedicated, successively, to the academic program in Nuclear Engineering and to the vocational training in nuclear infrastructure that were implemented for capacity building in support of the deployment of the NPP program. Section V provides an overview of the research activities developed at Khalifa University and at the recently established Emirates Nuclear Technology Center (ENTC) that involves not only Khalifa University but also the Emirates Nuclear Energy Corporation (ENEC) and the Federal Authority for Nuclear Regulation (FANR). Concluding remarks and future perspectives and objectives are addressed in Section VI.

II. UAE Nuclear Power Program

In 2008, UAE has taken the decision to go nuclear and embark in an ambitious nuclear power program after releasing its policy on "The evaluation and potential development of peaceful nuclear energy" [1]. Having no previous involvement in a nuclear program, the infrastructure and capacity had to be built from scratches to achieve the goal of producing 25% of electricity with nuclear power by 2030 and contributing to the UAE Net Zero objective for 2050.

In 2009 UAE issued its nuclear energy law and contracted KEPCO for the construction of four APR1400 units, 1400 MWe pressurized water reactors. In July 2012 started the construction of the Barakah Nuclear Power Plant (BNPP) located on the sea side, 250 km to the West of the Capital Abu Dhabi.

Nine years later, in April 2021 the first unit came into commercial operation, soon followed, in March 2022 and February 2023 by the successive commercial operation of units 2 and 3, respectively. As of August 2023, the project reaches a completion rate of 98 % while the hot functional tests of unit 4 have been completed in July 2022 and commercial operation is expected in the coming months. As planned the four units are to produce a total of 5600 MWe representing 25% of the UAE electricity and contributing to decarbonization.

As a new comer in nuclear power with no previous nuclear infrastructure, UAE has demonstrated that it is possible in a period of 15 years to fully implement a nuclear power program from the implementation of the infrastructure to the commercial operation of the plant.

This very ambitious objective was achieved through an efficient establishment of the regulation, building of the capabilities as well as management of the industrial project. To ensure an optimized implementation of the project. It is also important to highlight that UAE has been following the milestone approach [2] of the International Atomic Energy Agency (IAEA) and has benefited from its assistance in the various domains of the nuclear infrastructure.

In the field of capacity building Khalifa University has been and is playing a major role supporting the nuclear power program with three pillars described in the following sections: the academic program in nuclear engineering, vocational training and research activities adjusted and focused according to the specific needs of the UAE program.

III. Nuclear Engineering Academic Program

Following the decision to go nuclear in 2008-2009, the nuclear engineering academic program was launched at Khalifa University with its Nuclear Engineering Department, in September 2010, with an MSc degree in Nuclear Engineering. It was followed in 2012 with the introduction of Nuclear Engineering Minor into the BSc degrees in other engineering disciplines such as mechanical, chemical and electrical engineering. A PhD program with a concentration in nuclear engineering was also launched in 2012. It is important to note that these programs until 2021 have been open only to UAE nationals expressing the priority given to develop the national capabilities.

The nuclear engineering program has been established and further developed in coordination with the key nuclear stakeholders, i.e. ENEC, FANR and lately Nawah, the operating organization of the BNPP. It has been supporting the development of the human resources as needed along the implementation of the nuclear power program. It is important to mention that, until 2020, a majority of the MSc students were part-time students employed by one of the key stakeholders going through nuclearization but potentially having a work experience in other engineering and energy fields.

According to the needs, areas addressed in the nuclear engineering programs includes:

- Reactor Physics,
- Neutronics, Thermal Hydraulics,
- Nuclear Fuel Cycle,
- Nuclear Safety, Security, Safeguards,
- Nuclear Materials and Chemistry,
- Radiation metrology, Radiological protection & impact,
- Nuclear waste.

To take into account the diversity of profiles identified by the stakeholders, the MSc program integrates, which combines core and elective courses, exhibits five tracks, namely: Nuclear systems and probabilistic safety assessment, Nuclear reactor design, Nuclear safeguards, security and the fuel cycle, Nuclear material and radiation damage, and Nuclear and radiation Safety.

It is also to mention that the educational program is underpinned by research activities since students are heavily involved in research projects meeting the needs of the stakeholders and implemented within the ENTC (see section V).

IV. Vocational Training In The Nuclear Field

In 2011 the Gulf Nuclear Energy Infrastructure Institute (GNEII), a Gulf and Middle East regional human resource development capability, has been launched to be hosted at Khalifa University. The institute was established as the result of a strategic partnership between UAE stakeholders, namely Khalifa University, ENEC and FANR, and USA partners, Sandia National Laboratories and the Nuclear Security Science and Policy Institute at Texas A&M University.

GNEII was initially implemented as a continuing education and research entity with a focus on the so called 3 S concept, which stands for Safety/Security/Safeguard, and with the objective of developing the nuclear energy infrastructure knowledge and expertise, a 13-week program being regularly run every year.

To meet the current needs of the UAE stakeholders, i.e. at the advanced stage of the nuclear power program, the scope and objectives of GNEII are currently been revised. From 2024 a broad range of one- to two-week training modules will be offered, covering basic nuclear knowledge to specialized nuclear topics, modules being organized under six sub-areas:

- Basic knowledge (entry modules),
- Nuclear reactors,
- Nuclear fuel cycle,
- Nuclear safety and licensing
- Nuclear security and safeguards,
- Nuclear project management.

Additionally, since 2016 Khalifa University is a designated IAEA Collaborating Center. As such and in collaboration with UAE nuclear stakeholders, Khalifa University assist the IAEA in implementing its programmatic activities, hosting workshops, training courses, fellowship trainings and scientific visits. The Center also provides the opportunity for UAE to share resources, knowledge and expertise with the IAEA Member States. This is an important role of the Center now that UAE can share its newcomer experience after having successfully managed, in a short period of time to development of its nuclear infrastructure and implementation of its nuclear power program.

Furthermore, in 2022, Khalifa University has signed a Practical Arrangement with the IAEA for the development and the implementation of training courses and a post graduate program and certificate in nuclear law. This training opportunities will address both technical aspects and the legal framework related to nuclear reactors. Such a holistic approach is necessary to facilitate the deployment of nuclear power applications including Small Modular Reactors and micro-reactors.

V. Research At The Emirates Nuclear Technology Center

Research activities in the nuclear domain have been initially performed at Khalifa University under the Nuclear Engineering Department. With the progression of the nuclear power program and the expansion of the needs in research activities the key nuclear stakeholders, i.e. ENEC, FANR and Khalifa University took the decision to launch in January 2020 the Emirates Nuclear Technology Center (ENTC). This joint Center operated and funded by the three entities is embedded in Khalifa University, which represents the legal entity. The Center located within Khalifa University, is a hub to address the research requirements to support the UAE's Nuclear Power

Program. Research activities and projects are implemented under four thematic areas whose objectives are descried here after.

Theme 1, Nuclear systems & safety, is dedicated to neutronics & thermal hydraulic studies in relation to the safe and optimal operation of NPPs with a focus on BNPP. This theme aims to assess the thermal hydraulic safety analysis codes by validating their predictions against experimental data obtained from integral effect test facilities. This is intended to identify and validate dominant thermal hydraulic phenomena occurring in NPPs under postulated accident scenarios, including rare events. The theme is also focusing in applying computational reactor physics codes for safety analysis and operation of the advanced pressurized water reactor APR1400 as well as innovative reactor designs. The main drive for selecting these topics is that thermal hydraulic and neutronics studies on, beyond design-based accident for the APR1400, is one of the immediate necessities that the UAE nuclear program needs to develop to guarantee a safe operation of these power plants.

Theme 2, Nuclear materials & chemistry, is dedicated to the optimization of the NPP operation and availability in relation to the structures, systems and components integrity keeping in mind the 60 + year operation. Most materials in NPPs undergo various degradation phenomena under operation and by ageing. The type and degree of the degradation depends on the operational conditions including environmental, mechanical, and materials conditions. Since such material degradation under operation and by ageing can affect either the maintenance program, i.e. the frequency of testing, maintenance or replacement of components, as well as the structural integrity of components, material degradation needs to be understood in a comprehensive way and managed in the most effective way.

Theme 3, Radiation safety & the environment, is dedicated to radiation measurement and the modeling of the impact of radiation on the environment and the public. The construction and operation of NPPs, as well as the development and implementation of nuclear methods and techniques, are associated with the risk of release of radioactive materials than can affect the sensitive regions and population. Experimental data and numerical tool are used to map the ecosystems during normal and abnormal situations with a particular focus on BNPP and the Gulf region. This theme also addresses studies related to the influence of the operation of the BNPP on the environment and vice versa, with the objective of reducing the impact and optimizing the NPP performances.

Theme 4, Advanced nuclear technology, has been introduced at the beginning of 2023. While the three first themes were initially developed with a focus on supporting the BNPP safe and effective operation, Theme 4 aims at broadening the research activities to a large variety of applications of nuclear technology and advanced technologies that can improve nuclear technologies. Foreseen activities encompass small modular reactors, nuclear cogeneration, nuclear security, emergency preparedness, as well as nuclear applications in the health sector, water management, agriculture and food sector, forensic, to which can be added the utilization of advanced manufacturing, Artificial Intelligence and Machine Learning and other advanced technologies that can benefit nuclear technology.

It is important to highlight that research activities implemented under the ENTC are primarily focused on the UAE-specific technology, environment and related needs. For example, there is a need to study the impact of the UAE harsh conditions, due to high temperatures, humidity, salinity, dust and sand, that impact the APR1400 reactors structures, components and operation. Also, radiation mapping, both natural and anthropogenic, and the modeling of radionuclides dispersion including environmental migration and bioaccumulation, are UAE-specific and needs to be extensively studied taking into consideration that the transfer of radionuclides in arid environments is not yet comprehensively documented.

For the first three-year cycle of the ENTC (2020-2022) goals and achievements can be summarized as follow:

- Research activities focusing on end-user and industry driven projects for ENEC, Nawah and FANR with a focus on UAE specifics;
- 15 research projects implemented and agreements finalized to start 6 additional projects at the beginning of 2023;
- Total amount of fund engaged over the period, $ 20 M, including matching funds from international partners;
- Partnership with UAE entities, such as Technology Innovation Institute, Borouge Ltd, Union Pipe Industry, as well as international partners such as Korea Atomic Energy Research Institute, Korean Advanced Institute of Science and Technology, Massachusetts Institute of Technology, University of Birmingham, University of Manchester, Institut de Radioprotection et de Sûreté Nucléaire, Institut Mines Telecom Atlantique;
- Effective capacity building with the involvement of about 30 faculty members, researchers and postdocs, as well as 30 PhD and MSc students, the latter being UAE national in a large majority;
- Development of experimental capabilities, the ENTC exhibiting 4 Laboratories, namely the Computation and simulator Lab, Heat transfer Lab, Nuclear material and chemistry Lab, Radiation metrology and Environmental Lab. Note that the ENTC also benefit from Khalifa University core laboratories and capabilities of other Centers for cross-cutting activities.

Here after is a synthetic overview of the research areas of the projects currently implemented under the four thematic areas of the ENTC. Insight into the recent development and achievements of research projects implemented under the ENTC can be found in [3 – 23]

Theme 1 - Nuclear Systems and Safety

- Evaluation and selection methodology of Accident Tolerant Fuel for the APR1400 reactor
- Code development and validation for the prediction of rare events in the APR1400
- Flexible and efficient APR1400 reactor operations through integration of energy storage
- Validation of safety analysis code and assessment of thermal-hydraulic behaviors of APR1400

Theme 2 - Nuclear Material and Chemistry

- Enhancement of corrosion and wear resistance of cooling water components for the APR1400
- Prediction of degradation in concrete structures by sulphate attack and development of remedial measures
- Development of a multi-physics computational modelling framework and experimental procedure of butt fusion welding of large HDPE pipes

Theme 3 - Radiation Safety and the Environment

- Marine environment monitoring for the BNPP
- Anti-fouling remediation strategies for water intakes for nuclear reactor cooling systems
- Biological treatment of ethanolamine-rich wastewater from NPPs
- Radionuclide dispersion modeling in the UAE environments and inverse modeling (origin/magnitude)
- Transport of radionuclides in the UAE environment (emergency preparedness, sitting and design of new nuclear facilities)

Theme 4 – Advanced nuclear technology

- User driven approach and technology development for an effective deployment of small modular reactors
- Quantum-inspired algorithms for nuclear reactor physics and computational fluid dynamics

VI. Concluding remarks and future perspectives

UAE rapidly went from the statute of new comer in nuclear power to a country operating its first nuclear power plant that will provide very soon, when the 4th unit will enter into commercial operation, 25 % of the electricity.

Starting from very limited nuclear capabilities UAE has proven that a nuclear program can be successful implemented in a period of 15 years from the development of the nuclear infrastructure to the commercial operation of the plant.

This success relied upon extensive capacity building activities in which Khalifa University played a major role with its academic program and vocational training in nuclear engineering, as well as with the deployment of research activities in support of the UAE nuclear power program.

An important step to reinforcing capacity building and to provide a dedicated support to the nuclear program was taken in 2020 with the creation of the ENTC that embarked the nuclear energy corporation - ENEC - and the regulatory body – FANR. This step forward was decisive in amplifying the research and development efforts in support of these stakeholders.

To meet the current and future UAE needs perspectives and objectives incudes the adjustment and expansion of the academic and professional training to meet the needs of the nuclear power program and the further development of nuclear applications in the UAE. For research activities, key objectives are: (1) to further develop the national research capabilities, both human and experimental, with a focus on the UAE and Gulf region specifics; (2) to expand the scope of the research activities to advanced nuclear technologies, as defined in Theme 4; (3) to further develop international collaboration to leverage capacity building and international expertise as well as to facilitate the access to experimental facilities and; (4) to become a nuclear actor sharing UAE experience and expertise in the deployment and operation of NPPs that can benefit newcomers in deploying or developing their nuclear power program.

References

[1] "Policy of the United Arab Emirates on the Evaluation and Potential Development of Peaceful Nuclear Energy", uae-peaceful-nuclear-energy-policy-5722278a2952f.pdf (enec.gov.ae).

[2] IAEA Nuclear Energy Series No. NG-G-3.1 (Rev. 1), "Milestones in the Development of a National Infrastructure for Nuclear Power".

[3] P. Raj, N. Padiyath, N. Semioshkina, Y. Addad, F. Foulon, D. Francis, and G. Voigt, "Conceptualization of arid region radioecology strategies for agricultural ecosystems of the United Arab Emirates", Science of The Total Environment, Volume 832, Aug. 2022, Art no. 154965.

[4] P. Raj, N. Padiyath, N. Semioshkina, F. Foulon, A.K. Alkaabi, G. Voigt, and Y. Addad, "Transfer of Natural Radionuclides from Soil to Abu Dhabi Date Palms", Sustainability, Vol. 14, Sept. 2022, Art no. 11327.

[5] M.A. Amidu, S.A. Olatubosun, A. Ayodeji, and Y. Addad, "Severe accident in high-power light water reactors: Mitigating strategies, assessment methods and research opportunities", Progress in Nuclear Energy, Vol. 143, Jan. 2022, Art no. 104062.

[6] I. Khurshid, I.Afgan, A.M. Alade, and Y. Addad, "Influence of corium temperature, concrete composition and water injection time on concrete ablation during MCCI: New insights", Progress in Nuclear Energy, Vol. 144, Feb. 2022, Art no. 104102

[7] M.K. Riahi, M. Ali, Y. Addad, and E. Abu-Nada, "Combined Newton–Raphson and Streamlines-Upwind Petrov–Galerkin iterations for nanoparticles transport in buoyancy-driven flow, Journal of Engineering Mathematics, Vol. 132, no. 1, pp. 22, Jan. 2022.

[8] M. Weston, D. Francis, N. Nelli, R. Fonseca, M. Temimi, and Y. Addad, "The First Characterization of Fog Microphysics in the United Arab Emirates, an Arid Region on the Arabian Peninsula", Earth and Space Science, Vol. 9, n. 2, Feb. 2022, Art no. 2021EA002032.

[9] M.A. Amidu, and Y. Addad, "The influence of the water ingression and melt eruption model on the MELCOR code prediction of molten corium-concrete interaction in the APR-1400 reactor cavity", Nuclear Engineering and Technology, Vol. 54, no. 4, pp. 1508–1515, Oct. 2021.

[10] D.Y. Kim, H.C. No, B.G. Jeon, S.G. Lim, and H.J. Yoon, "Multi-tube effect of external condensation with non-condensable gas in a tube bundle", Annals of Nuclear Energy, Vol.168, April 2022, Art no. 108894.

[11] M. Ali, A.K. Alkaabi, and Y. Addad, "Numerical investigation of a vertical triplex-tube latent heat storage/exchanger to achieve flexible operation of nuclear power plants", International Journal of Energy Research, vol. 46, no. 3, pp. 2970-2987, Oct. 2021.

[12] A. Alraisi, Y. Yi, S. Lee, S.A. Alameri, M. Qasem, C.Y. Paik, and C. Jang, "Effects of ATF cladding properties on PWR responses to an SBO accident: a sensitivity analysis", Annals of Nuclear Energy, Vol. 165, Jan. 2022, Art no. 108784.

[13] M. Alrwashdeh, and S.A. Alameri, "Preliminary neutronic analysis of alternative cladding materials for APR-1400 fuel assembly", Nuclear Engineering and Design, vol. 384, Dec. 2021, Art no. 111486.

[14] S.A. Alameri, and M. Alrwashdeh, "Preliminary three-dimensional neutronic analysis of IFBA coated TRISO fuel particles in prismatic-core advanced high temperature reactor", Annals of Nuclear Energy, vol. 163, 108551, 2021.

[15] M. Baghdad, A. Ouadha, and Y. Addad, "Effects of kinetic energy and conductive solid walls on the flow and energy separation within a vortex tube", International Journal of Ambient Energy, vol. 42, no. 3, pp. 297–313, Dec. 2021.

[16] A.Razali, A. Ouadha, M. Baghdad, and Y. Addad, "Effect of corrugating the cold orifice of a vortex tube on its flow and energy separation features", International Journal of Fluid Mechanics Research, Vol. 48, no. 4, pp. 45–64, 2021.

[17] M.A. Amidu, Y. Addad, J.I. Lee, D.H. Kam, and Y.H. Jeong, "Investigation of the pressure vessel lower head potential failure under IVR-ERVC condition during a severe accident scenario in APR1400

reactors", Nuclear Engineering and Design, Vol. 376, May 2021, Art no. 111107.

[18] O. Nesterov, M. Temimi, R. Fonseca, Y. Addad, E. Bosc, and R. Abida, "Validation and statistical analysis of the Group for High Resolution Sea Surface Temperature data in the Arabian Gulf", Oceanologia, vol. 63, no. 4, pp. 497–515, Oct. 2021.

[19] M.A. Amidu, Y. Addad, M.K. Riahi, and E. Abu-Nada, "Numerical investigation of nanoparticles slip mechanisms impact on the natural convection heat transfer characteristics of nanofluids in an enclosure", Scientific Reports, vol. 11, no. 1, Aug. 2021, Art no. 15678.

[20] J. K. Alsuwaidi, H.G. Kang, and H.J. Yoon, "Studies on the optimal mitigation strategy of extended SBO for the isolated multi-unit NPP site Annals of Nuclear Energy", vol. 164, Dec. 2021, Art no. 108615.

[21] O.S. Al-Yahia, S. Lee, H.J. Ryu, and H.J. Yoon, "The Feasibility Study on Development of Porous Ceramic Composites for Ion-Exchanger", Nuclear Engineering and Design, vol. 381, Sept. 2021, Art no. 111329.

[22] M. Ali, A.K. Alkaabi, S.A. Alameri, and Y. Addad, "Overall efficiency analysis of an innovative load-following nuclear power plant – thermal energy storage coupled cycle", International Journal of Energy, vol. 36, no. 1, pp. 98-122, Jan. 2021.

[23] K. Ko, H. Choi, Y. Yi, W. Kim, J. Kim, J. Hwang, E. Ko, and G. Cho, "Discrete Convolution-Based Energy Spectrum Configuring Method for the Analysis of the Intrinsic Radiation of 176Lu Sensors", vol. 21, Oct. 2021, Art no. 7040.

AUTHOR INDEX

Acosta, Luis ... 73
Addad, Yacine ... 561
Adrian, Stern ... 554
Aimè, C. ... 494
Al, J. Brinster And ... 126
Albert, Patrick ... 502
Allinei, Pierre-Guy ... 231, 333
Almakrani, Maryam ... 412
Alon, Osovizky ... 554
Altana, Carmen ... 73
Altomare, C. ... 306, 543
Amaducci, Simone ... 269
Ameri, Saeed Al ... 561
Amoyal, G. ... 488
Amoyal, Guillaume ... 327
Ampollini, Alessandro ... 115
Angélique, J.-C. ... 488
Angelis, Andrea De ... 323
Arbor, Nicolas ... 205
Astorino, Maria Denise ... 115
Attié, David ... 311
Aune, S. ... 356
Babra, F. S. ... 446
Bakkari, B. El ... 78
Balier, A. ... 382
Ball, R. D. ... 98
Balossino, I. ... 302
Barbagallo, Massimo ... 205
Barbot, Loic ... 168
Barbot, Loïc ... 173
Barco, F. ... 421
Bardorfer, Aleš ... 168
Batista, A. J. N ... 62
Baudry, F. ... 219
Bazzano, Giulia ... 115
Belverge, D. ... 151
Bencivenni, G. ... 302
Ben-David, Nadav ... 345
Bennett, David ... 292
Benoit, T. ... 356
Bensedik, Youcef ... 231
Besnard-Vauterin, C. ... 255, 488
Betta, Gian-Franco Dalla ... 463
Beyer, R. ... 15
Bianchi, Barbara ... 115
Biondo, Stéphane ... 539
Bisiach, Danilo ... 168
Blideanu, V. ... 255
Boch, J. ... 382
Boffy, Romain ... 143
Bond, W. ... 339
Bonzon, P. V. ... 219
Borella, Alessandro ... 280
Boscher, Aurore ... 405
Boufounos, Petros ... 433
Boulaich, Y. ... 78
Braun, Laurent ... 180
Broide, Amir ... 451
Brunbauer, F. M. ... 356
Burgio, Nunzio ... 115
Butini, T. ... 421
Byun, Soo Hyun ... 30
Calzaferri, S. ... 494
Cano, Daniel ... 138
Cantone, C. ... 37
Cantrel, E. ... 382
Carasco, Cedric ... 297
Carasco, Cédric ... 7
Cardella, Giuseppe ... 73
Carette, M. ... 89, 111, 121
Cargnelutti, Manuel ... 168
Carlesi, Tommaso ... 223
Carminati, Federico ... 205
Carminati, Giuliana Galli ... 205
Caroli, Emmanuel ... 7
Carpenter, D. ... 89
Carrel, F. ... 482
Carvalho, B. ... 62
Casagrande, D. ... 386
Cascone, M. G. ... 421
Castoldi, Andrea ... 73
Catherall, Richard ... 405
Çelik, Yurdunaz ... 280
Cerasole, D. ... 543
Ceravolo, S. ... 37
Cesaroni, Mateo ... 115
Cheng, Xingzhi ... 30
Cherubini, N. ... 386
Cherubini, Nadia ... 68, 375
Chevalier, Vincent ... 132
Cheymol, G. ... 126
Chierici, Andrea ... 531, 547
Chowdhury, N. R. Khan ... 446
Chuilon, Pierre ... 7
Cibinetto, G. ... 302
Ciolini, R. ... 421

Ciolini, Riccardo ... 531
Ciotoli, Alessandro ... 323, 375
Cirillo, Roberta ... 138
Clarke, S. D. ... 147
Clough, Matthew ... 53
Coito, Rui ... 240
Colao, F. ... 37
Collin, Jonathan ... 205
Collot, Johann ... 7
Combette, P. ... 219
Contessa, G. M. ... 371
Cools, A. ... 356
Corsi, A. ... 356
Cortese, Pietro ... 19
Corti, Daniele ... 470
Cosentino, Luigi ... 269, 274
Costa, Michele ... 73
Costa, Tiago ... 240
Cotto, Giorgio ... 19
Coulié, Karine ... 519
Cozzella, Maria Letizia ... 318
Croia, Michele ... 115
Cuozzo, M. ... 482
Cuozzo, Marie ... 333
Cvachovec, Frantisek ... 197
Cybowska, Justyna ... 185
Czakoj, Tomas ... 197
D'Errico, F. ... 421
D'Errico, Francesco ... 531, 547
Darby, F. B. ... 147
Das, Biswajit ... 441, 446
David, Nadav Ben ... 525
Delgado, Jessica C. ... 360
Delgado, Jessica ... 470
Destouches, C. ... 126, 219
Destouches, Christophe ... 42, 168
Dey, P. ... 441, 446
Dietz, M. ... 15
Diociaiuti, E. ... 37
Doan, Mai-Linh ... 7
Downey, C. ... 98
Dubus, Léo ... 42
Dumercq, Bernard ... 269
Dumont, Gerald ... 405
Dyer, Benjamin ... 30
Ehud, Behar ... 554
Ellenbogen, Amir ... 451
Eran, Vax ... 554
Eyméoud, Paul ... 539
Fabris, Daniela ... 360, 470
Fabrizio, V. ... 94
Falconi, L. ... 94
Falconi, Luca ... 115
Fanchini, E. ... 386, 543
Fanchini, Erica ... 360
Fantazi, Akram Al ... 561
Fargier, S. ... 482
Farinelli, R. ... 302
Favrichon, J. ... 382
Felice, Pierino De ... 263
Felici, G. ... 302
Fernandes, H. ... 62
Fernandes, Sofia ... 240
Ferrandis, J. Y. ... 219
Ferrer-Ribas, E. ... 356
Ferri, D. ... 371
Ferri, Donatella ... 351
Filho, Luiz C. Gonçalves ... 391, 395
Filippo, Enrico De ... 73
Finocchiaro, Paolo ... 269, 274
Finotello, Riccardo ... 22
Fiore, Salvatore ... 115
Fiorina, D. ... 494
Fitoussi, Ohad ... 345
Flaska, Marek ... 502, 509
Fleres, L. ... 482
Fondement, Valentin ... 7
Formenton, D. ... 94
Foulon, Francois ... 412, 561
Fourmentel, D. ... 219
Frajtag, Pavel ... 180
Fridrikhsen, D. ... 47
Friedrich, Hermann ... 477
Fürbringer, Jean-Marie ... 104
Gagliardi, Filippo ... 323, 375
Gaillot, S. ... 126
Gajewski, Jacek ... 138
Gallego, Laurent ... 311
Gandolfo, G. ... 371, 386
Gandolfo, Giada ... 68, 318, 375
Gargano, F. ... 543
Garrido, Eric ... 297
Garzia, I. ... 302
Gatta, M. ... 302
Gattia, Daniele Mirabile ... 351
Gaudio, Pasqualino ... 288
Gay, Georges ... 143
Geraci, Elena ... 73
Geslot, B. ... 151
Geslot, Benoit ... 180
Geulin, Eleonore ... 53
Ghelman, Max ... 426, 451
Ghninou, H. ... 78
Gianotti, P. ... 37

Ginzburg, Dimitry ... 426
Giordano, F. ... 306, 386, 543
Giovannetti, M. ... 302
Girardi, Fabio ... 351
Glabian, Jeannette ... 477
Gnoffo, Brunilde ... 73
Godat, Daniel ... 180, 189
Gomes, Daniel F. ... 417
Gomez, Héctor ... 311
Gonzales, Philippe ... 311
Gorello, Edoardo ... 323, 375
Gramigna, S. ... 302
Griggio, Alessandro ... 470
Grillet, Nadia ... 539
Gruel, A. ... 78
Gruel, Adrien ... 143
Guazzoni, Chiara ... 73
Gueton, Olivier ... 333
Haber, E. El ... 482
Hanu, Andrei ... 30
Harn, Ron ... 451
Harris, B. ... 98
Hauptman, S. ... 89, 121
Havranek, A. ... 62
Hayes, M. ... 339
Helbert, Christopher ... 269
Hmissi, Zied ... 249
Hocquet, Sebastien ... 231
Holtz, M. R. ... 98
Horodynski, Jean-Michel ... 205
Houry, Michael ... 42
Hua, M. Y. ... 147
Hulber, T. ... 421
Hursin, Mathieu ... 104
Iglesias, Daniel ... 53
Iguaz, F. J. ... 356
Iltis, Alain ... 249
Iorio, M. G. ... 94
Iovane, Alessandro ... 360
Izarra, Grégoire De ... 143, 173
Izarra, Gregoire De ... 168
Jadav, S. K. ... 441, 446
Jammes, Christian ... 143
Jiang, Yifeng ... 180
Junghans, A. ... 15
Juricek, Vlastimil ... 197
Kaabi, Ahmed Al ... 561
Kashchuk, Yu. ... 47
Kessedjian, G. ... 151
Knafo, Yakir ... 525
Köble, Theo ... 477
Kögler, Toni ... 399
Kohout, Zdenek ... 58
Kohse, G. ... 89, 121
Kormilitsyn, T. ... 47
Korniliou, S. ... 339
Kostal, Michal ... 197
Kotnik, Domen ... 84
Kotorová, Sona ... 498
Kovácová, Eva ... 58, 498
Krasilnikov, A. ... 47
Kundu, A. ... 441, 446
Kurucová, Nikola ... 498
L'Hermite, Daniel ... 22
Laboureur, Delphine ... 223
Lalic, S. De Souza ... 421
Lamirand, V. ... 147
Lamirand, Vincent ... 104, 163, 180, 189, 211
Laskar, Md. R. S. ... 446
Laureau, Axel ... 163
Le, Nhan ... 249
Lebonvallet, Guillaume ... 249
Leconte, P. ... 151
Leconte, Pierre ... 180
Ledoux, X. ... 151
Lefevre, Baptiste ... 311
Legros, Bruno ... 231
Lehuraux, Marion ... 311
Lengar, Igor ... 84
Lepore, L. ... 94, 371, 386
Lepore, Luigi ... 68, 115, 375
Lesage, Bertrand ... 311
Liedl, A. ... 37
Ligonnet, Thomas ... 104, 163
Limousin, Olivier ... 365
Lionheart, William ... 399
Llido, Olivier ... 42
Logoglu, Faruk ... 502, 509
Lombardo, Luca ... 19
Longhitano, Fabio ... 269, 274
Loparco, F. ... 543
Loporchio, S. ... 306, 543
Lucchesi, D. ... 37
Ludovici, Gian Marco ... 288
Lunardon, Marcello ... 263
Lutz, B. ... 151
Lyoussi, A. ... 78, 89, 111, 121
Lyoussi, Abdallah ... 42
Ma, Yanting ... 433
Machin, G. ... 339
Mager, Tom ... 104
Magistris, Matteo ... 405
Mahe, C. ... 482
Maiolino, Cettina ... 73

Makovac, Aurora 168
Malik, V. 441, 446
Malizia, Andrea 288, 531
Man, Sven De 405
Mancini, Maria Rita 351
Mandjavidze, Irakli 311
Manenti, Guglielmo 288
Manessi, Giacomo P. 245
Mansour, Hassan 433
Maraine, T. 382
Marchais, Thomas 7, 231, 333
Marco, Nora De 19
Marques, Luís 240
Marzo, G. A. 371, 386
Marzo, Giuseppe A. 245, 351, 375
Marzo, Giuseppe Augusto 68, 318
Mas, Philippe 311
Masci, D. 371
Masci, Domenica 351
Mastroserio, A. 306, 543
Matej, Zdenek 197
Mathieu, L. 151
Mauro, Egidio 323, 375
Mayet, Frédéric 333
Mazoin, C. 356
Meco, E. Di 37
Mégret, Patrice 223
Menaa, Nabil 405
Mendicino, Roberto 463
Meplan, O. 151
Meric, Ilker 399
Meschini, Luigi 323, 375
Michel, Maugan 327
Mieszkalski, Romain 231
Morad, Itai 426, 451
Morales, Frédéric 333
Moretto, Sandra 360, 470
Morichi, M. 386, 543
Mouret, Renaud 405
Mravec, Filip 197
Mutti, P. 151
Nadav, Ben David 554
Naidu, B. S. 441, 446
Nichols, D. 98
Nieuwenhove, Rudi Van 157
Nolte, R. 15
Notte, Giovanna La 318
Nourreddine, Abdel-Mjid 205
Novaes, J. A. Torres 421
Novák, Andrej 58
Novák, Evžen 138
Obudovsky, S. 47
Oliveira, J. 62
Oliveira, R. De 302
Oliveri, E. 356
Osovizky, Alon 345, 426, 451
Osowizky, Alon 525
Othe, Sebastien 405
Ottaviani, Laurent 42, 459, 519, 539
Pacio, Julio 223
Padiyath, Nemeer 412
Paesani, D. 37
Pagliuca, Marco 323, 375
Paglovec, Peter 168
Pakari, O. 147
Pakari, Oskari 180
Palit, R. 441
Palumbo, E. 386
Pancheri, Lucio 470
Pankratenko, A. 47
Pantera, Laurent 180
Papaevangelou, T. 356
Pastrone, N. 37
Paumel, K. 219
Pautz, A. 147
Pautz, Andreas 104, 163, 180, 189
Peltan, Tomas 197
Pepperosa, A. 386
Peric, Julijan 173
Perot, Bertrand 297
Pérot, Bertrand 7, 231, 333
Perrin, Océane 1
Petit, Michael 459
Petit, Michaël 514
Piccinelli, Ermanno 68
Pietropaolo, A. 371
Pillon, M. 371
Pino, Felix 360, 470
Pirovano, E. 15, 151
Pisano, Paolo 405
Planquart, Philippe 223
Pohlner, Georg 138
Polansky, Štepán 58
Polo, Matteo 463
Poma, Gaetano Elio 269, 274
Pomarede, Daniel 311
Potiron, Quentin 42
Pozzi, S. A. 147
Procureur, Sébastien 311
Puggioni, Carlo 19
Pungercic, Anže 138
Puyvelde, Lisanne Van 138
Quaranta, Alberto 463
Quaranta, Riccardo 288

Quattrocchi, M. 421
Quéré, Celine 22
Quevauvillers, Diane 459
Radulovic, Vladimir 173
Rahajandraibe, Wenceslas 519
Raj, Prasoon 412, 561
Rapp, B. 255
Rapp, Joshua 433
Ratto, N. 386
Raz, Dor 345
Rebaud, J. 111
Reichenberger, M. A. 98
Remigis, Paolo De 19
Reynard-Carette, C. 78, 89, 111, 121
Reynard-Carette, Christelle 42, 459
Riccardi, C. 494
Risse, Monika 477
Roberti, A. 94
Robin, P. A. 382
Rodiac, F. 151
Rodionov, R. 47
Rollins, H. 98
Romanets, Y. 15
Rosellini, E. 421
Rosenkrantz, E. 219
Rossa, Riccardo 280
Rouge, Benjamin 22
Šagátová, Andrea 58, 498
Saha, S. 446
Saigné, F. 382
Salomon, Yossi 525
Salvador, Francisco 333
Salvini, P. 494
Salvo, Jacques Di 132
Santos, Luiz A. P. 391, 395, 417
Sari, Adrien 235
Sartori, Paolo 263
Savov, Dragomir 231
Schoepff, V. 488
Schoepff, Vincent 327, 365
Schubert, Matthew 231
Schumann, Olaf 477
Seif, Rami 451
Seitz, Bjoern 292
Serini, D. 306
Shay, Dadon 554
Shin, Won Dong 104
Shlezinger, Nir 426
Silva, Edvane B. 417
Silva, Jonatas V. 395
Simon, Jan 197
Simpson, R. 339
Sirven, Jean-Baptiste 22
Sitta, Mario 19
Škabar, Matjaž 168
Skjerdal, Kyrre 399
Snoj, Luka 84
Snoussi, Hichem 249
Sousa, J. 62
Sperandio, L. 94
Starflinger, Jörg 138
Stefano, R. De 488
Stegemann, Simon 405
Stevanato, Luca 263
Surani, Stuti 502, 509
Susaiev, Yaroslav 365
Sýkora, Rudolf 58
Syromukov, S. 47
Szentmiklósi, Laszlo 138
Tabti, Nouhaila 235
Tagliapietra, Luca J. 205
Tamaazousti, Mohamed 22
Teuteberg, Thorsten 477
Thomson, Frank 292
Tichelen, Katrien Van 223
Tisseur, David 297
Tomkow, Lukasz 53
Tondut, Lionel 269
Torres, A. 62
Toubon, Hervé 231
Trnovec, Jure 168
Tromson, Dominique 235
Urena-Acuna, A. 382
Vai, I. 494
Vale, Alberto 240
Valentin, Jean-Baptiste 104
Varignier, Geoffrey 7
Vax, Eran 345, 451, 525
Vayakis, George 53
Vaz, P. 15
Vaz, Pedro 240
Vazhappilly, A. T. 441
Venara, J. 482
Venara, Julien 333
Venere, L. Di 306, 543
Vero, Luca F. Ferrante 245
Vervisch, Vanessa 539
Vervisch, Wilfried 539
Vilenchick, Dan 525
Vilimova, Eva 197
Vinchon, Thibaut 514
Vitullo, Fanny 189
Vitulo, P. 494
Volte, A. 89, 111, 121

Walkiewicz, Joanna 138
Weiss, Nicolas 163
Wolfe, Douglas 502
Xavier, M. N. 421
Yakir, Knafo 554
Yossi, Salomon 554
Younoussi, C. El 78
Ytre-Hauge, Kristian Smeland 399
Yurkov, D. 47
Zaidner, Guy 345
Zatko, Bohumír 58, 498
Zeufack, Ghislain 249
Zorloni, Gabriele 245
Zorzut, Sebastjan 168
Zugravel, Stefan Cristi 19